To Prof Sharon Zane,
my long-time friend at UM
Physics,
Enjoy this book – it has some
useful parts – even for undergrad
classes! With compliments,

Carolyne M. Van Vliet, Author.

2/1/2010.
To Prof Sharon T. Zane
With compliments.
Carolyne M. Van Vliet,
author.

Equilibrium and Non-equilibrium Statistical Mechanics

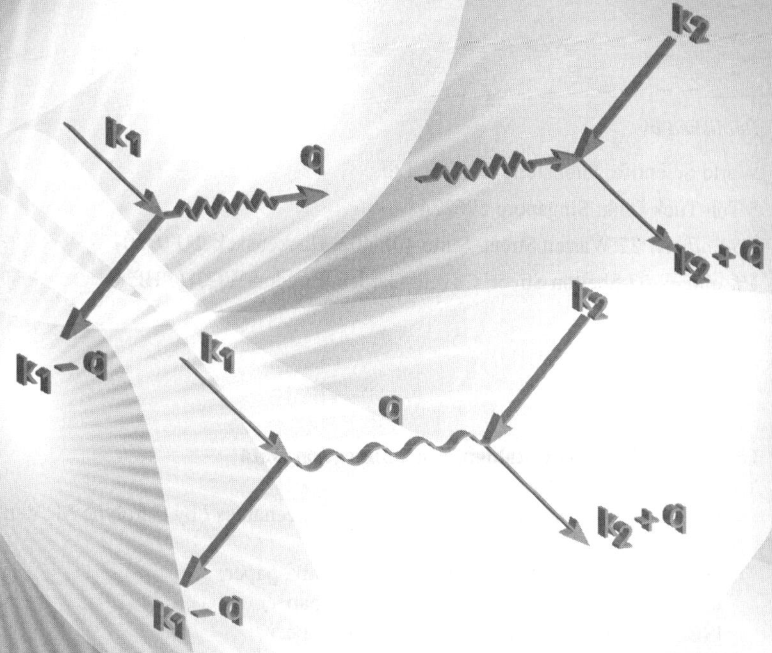

Equilibrium and Non-equilibrium Statistical Mechanics

Carolyne M. Van Vliet

Professor Emerita ofTheoretical Physics, Université de Montréal, Canada
and Adjunct Professor of Physics, University of Miami, USA

World Scientific

NEW JERSEY • LONDON • SINGAPORE • BEIJING • SHANGHAI • HONG KONG • TAIPEI • CHENNAI

Published by

World Scientific Publishing Co. Pte. Ltd.

5 Toh Tuck Link, Singapore 596224

USA office: 27 Warren Street, Suite 401-402, Hackensack, NJ 07601

UK office: 57 Shelton Street, Covent Garden, London WC2H 9HE

Library of Congress Cataloging-in-Publication Data

Van Vliet, Carolyne M.

Equilibrium and non-equilibrium statistical mechanics / by Carolyne M. Van Vliet.

p. cm.

ISBN-13: 978-981-270-477-1 (hardcover : alk. paper)

ISBN-10: 981-270-477-9 (hardcover : alk. paper)

ISBN-13: 978-981-270-478-8 (pbk. : alk. paper)

ISBN-10: 981-270-478-7 (pbk. : alk. paper)

1. Equilibrium--Statistical methods. 2. Nonequilibrium statistical mechanics. 3. Statistical mechanics. 4. Quantum statistics. I. Title.

QC174.86.E68V55 2008

530.13--dc22

2008002713

British Library Cataloguing-in-Publication Data

A catalogue record for this book is available from the British Library.

Printed in Singapore by B & JO Enterprise

To: Elsa Marianne
Mark Edward
Cynthia Joyce
Renata Annette Carolina

Preface

With statistical mechanics being one of the oldest branches of theoretical physics, going well back to the end of the nineteenth century, when the two great pioneers, Ludwig Boltzmann (1844-1906) in Graz, Munich and Vienna and J. Willard Gibbs (1839-1903) at Yale laid the foundations for the molecular approach to equilibrium and the statistical approach to many particle systems via ensemble theory, respectively, it seems a precarious task to endeavour to compose a textbook for a modern course in statistical physics, that will render homage to the past, summarise the exponential growth in the second half of the twentieth century – in particular with respect to critical phenomena in equilibrium statistical mechanics and the quantum foundations of transport processes and response theory in non-equilibrium statistical mechanics – to the student body of the twenty-first century, which typically has a two-semester slot in the graduate curriculum reserved for this topic. All this in contrast to electrodynamics and quantum mechanics, which is usually being taught in two rounds, each having two semesters available. While in these areas there is a rather common consensus about what should be taught, backed up by 'standard texts' in these fields – my own favourites being "Jackson" and "Messiah" – no such consensus exists with respect to statistical mechanics, usually named statistical physics these days, a pity, since the mechanical basis (classical or quantal) of the subject, as noticeable not only from the oeuvre of the pioneers mentioned above, but also stressed in little 'jewels' like Paul and Tatiana Ehrenfest's "The Conceptual Foundations of the Statistical Approach in Mechanics" (Leipzig, 1912, Cornell,1959) is often blatantly ignored. Instead, a great many modern texts commence with a résumé of macroscopic thermodynamics, after which a connection with the phase-space quantity $\ln\Omega$ is postulated and away we go. Needless to say that statistical mechanics is meant to provide the microscopic basis of all many-body systems, gases, liquids and condensed matter, of which the *ensuing* thermodynamic properties are established and form a basic part.

Yet, a text on statistical mechanics should not be just a text on many-body or condensed-matter physics, for which many excellent books are available; we mention e.g. Mahan's book "Many Particle Physics" of which an up-to-date third edition recently saw the light. While such books *use* statistical mechanics, part of which could be included in a statistical mechanics book as 'illustrations' of its principles, it is these principles which should be the expounded in a true text on the subject. The challenge to the writer is then how many illustrations or applications should be included. Obviously, the list is endless and a quick glance through half a dozen recent books reveals instantly the 'hobbies' of the authors.

This, then, brings me to my own views on what should be in a textbook on statistical mechanics, aimed at a two semester course. Given the fact that there is no *a priori* consensus the book should have sufficient material, so that a choice can be made as to what will be taught in a given year at a given place. Thus the book can have 900 pages, although only 600 or so can be taught. Even so, the foundations of the subject should not be undercut and more or less standard topics in equilibrium statistical mechanics, such as mean field theory, cluster and virial expansions, phase transitions, critical exponents, renormalization, quantum liquids, etc., to mention just a few, should be presented. In non-equilibrium material there is even less consensus, but most will agree that there should be an account of Boltzmann's *H*-theorem, classical transport theory and the hydrodynamic equations, the basis of quantum transport as contained in Pauli's and Van Hove's master equations, linear and nonlinear response theory, Brownian motion and other stochastic processes. For the rest the field is wide open and I have not hesitated to include my own preferences. Where I have been inspired by several textbooks, which I have from time to time recommended as reading material for my classes, I have made ample reference to these texts; however, I have always gone back to the original sources.

So let me present in this "à propos" a synopsis of my own evolution in this area. My course-packet at the Free University of Amsterdam included a full year in statistical mechanics, given by the late C.C. Jonker, who held 'the' chair in theoretical physics, as it was customary in those days; he certainly was overburdened and therefore took recourse to simply presenting the underground lectures in this area given in 1944 by his tutor, H.A. Kramers of the University of Leiden during World War II, when all university classes were suspended. They were excellent lectures! To our surprise these lectures were largely published in 1954 by D. ter Haar, who had also attended the underground lectures in 1944, in book-form: "Elements of Statistical Mechanics", (Rinehart & Co.) . Rightly he writes in his foreword, that "it is far from a platitude to say that it (the book) would never have been written but for Professor Kramers." My own pre-1954 lecture notes testify to this truth! In the mean time my research, which was concerned with fluctuation processes in semiconductors and photoconductors, turned more and more theoretical and I benefited for the theoretical part of my thesis from the scrutiny of one of Holland's most prominent theorists, the late Prof. H.B.G. Casimir; my interest in statistical mechanics was born.

In 1967 I had the pleasure to teach my first course in the field at the Physics Department of the University of Minnesota since my colleague Lewis Nosanov was on a sabbatical. He also introduced me to Van Hove's profound papers on the Pauli master equation and on the subject of irreversibility. The next spring I had the opportunity to discuss with Van Hove his so-called "diagonal singularity" in the perturbation expansion of the von Neumann equation, to which he attributed the irreversible character of the master equation, during a stay at the Theoretical Physics Institute at the University of Utrecht. These events shaped my thinking and I decided to forego my experimental endeavours. A brilliant opportunity opened up when I was offered a post in Physics at the Université de Montréal and as senior researcher at the Centre de Recherches Mathématiques (CRM). For many years I taught there courses in equilibrium and non-equilibrium statistical mechanics, and it is there that most ideas and presentations set forth in this book were developed. However, the usual conflict between teaching and writing research papers and proposals prevented me to condense my notes – several versions of which were made available for U. de M. students – to final book-form. That had to wait until after my second 'retirement', when I resumed among other topics graduate lectures in this field at the University of Miami. Needless to say, that many subjects had to be thoroughly updated or added. Nevertheless, the reader will find much of the flavour of my original lectures in this text and I tend to believe that this text still reflects more than the usual book that my origins in this field go back to Kramers and through him to the founders. So the reader will find here Gibbs' original considerations about entropy in the microcanonical ensemble, adapted, however, to quantum systems, not found in any present-day text. For the H-theorem, extended to quantum systems, I have freely borrowed from Tolman and passages of several older books, like Fowler and Fowler-Guggenheim, where still relevant, have been included. In the early days of my studies I profited much from Sommerfeld's book on the topic (Vol. 5 of his series on Theoretical Physics) and from Schrödinger's little but revealing book on "Statistical Thermodynamics", which used and explained in detail the Darwin-Fowler procedure to obtain the Boltzmann, Bose-Einstein and Fermi-Dirac distributions, all based on the microcanonical ensemble, which is the only ensemble that connects directly to classical mechanics via Poincaré invariants and Liouville's equation of 1838. Yet the emphasis in this book is on quantum statistics, and a replacement of the phase space by the appropriate Hilbert space is found on the earliest pages. Later on the Fock space and the occupation-number formalism, commonly called the second quantization procedure, is introduced, given that many-body Hamiltonians of strongly interacting particles can only properly be described with creation and annihilation operators. Since the typical student who takes a course in statistical physics has not yet had an exposure to advanced quantum courses or quantum-field theory, these developments are set forth in detail. For its introduction I still find Dirac's book, fourth edition (1958) one of the best sources.

This brings me to the organization of this text. Both the division on equilibrium statistical mechanics and on non-equilibrium statistical mechanics have three main parts each, A-C and D-F, respectively. Part A deals with the general principles of many-particle systems. There are five chapters, the first one of which is of an introductory nature, dealing with the purpose of statistical mechanics, my philosophy on the subject and a bit of thermodynamics, whereby we emphasize the fact that Gibbs' point of departure basically (may) conflict(s) with the standard views cherished in macroscopic thermodynamics, as set forth for example in Callen's book on the subject. We introduce classical and quantum ensembles, the *a*-space for 'mesoscopic' variables as found with van Kampen and in the classical text on non-equilibrium thermodynamics by de Groot and Mazur and other places, with emphasis on the microcanonical ensemble. Also, elementary and not-so-elementary topics in probability theory are discussed: transforms, generating functions, cumulants, etc. In Chapter II we thoroughly discuss the statistics of closed systems and in chapter III we deal with thermodynamics *a posteriori*, i.e., as a consequence of Gibbs' approach to the microcanonical ensemble, leading to the basic result $S = k \ln \Delta\Gamma$, where S is the entropy and $\Delta\Gamma$ is the accessible number of quantum states or the microcanonical partition function. The rest of needed thermodynamics then follows now as a desert, rather than as usual in the introductory part of a text. We then connect with Boltzmann's ideas and the famous relationship $S = k \log W$, as well as Einstein's inversion thereof. Next in Chapter IV we introduce the more common and useful ensembles, the canonical and grand-canonical ensemble. In order to avoid that all this formalism becomes sterile we liven up the story at this point with realistic applications of these ensembles and we embark on a variety of simple as well as quite complex topics, including the 1D Ising model, the 1D hard-core model of Tonk and Takahashi, dense classical gases involving cluster-integral diagrams, the cumulant expansion and the virial expansion. To do justice to my heritage I have also included the essence of Ornstein's Ph.D. thesis on the derivation of van der Waals' equation and van Kampen's later elaboration. The usual mean-field theories are discussed, including the Weiss molecular field and the Debye-Hückel theory for ionized gases. In the next chapter we make a little excursion into generalized canonical ensembles and transformation theory, as e.g. found in Münster's book; this concludes part A.

In the next part, B, we mainly consider perfect gases and their properties. The level of difficulty here, initially, is a great step backwards. So be it! It is like the mid-January thaw that usually appears to cheer up the long winters in Montreal. However, apart from the elementary (non-existing) Boltzmann gas, we soon go over to quantum gases and it is here that second quantization is set forth and elaborated. We discuss Bose-Einstein condensation and the elementary excitations in solids, phonons and electron-phonon interaction. Part C then contains the most pertinent theory of modern statistical mechanics, involving quantum systems with strong interactions, for which a quantum-field formulation is indispensable. We discuss critical phenomena and phase transitions, renormalization, the 2D Ising model, quantum liquids, the basis of superconductivity, etc. A part with the diagrammatic approach to many-body theory has been added for the more advanced student. Clearly, to do justice to all the developments of the second half of the twentieth century, for which the names of Wilson, Fisher, Widom, Kadanoff, Stanley, and many others, stand out, a book of several thousand pages would not suffice. So we have made a choice – our choice – constantly mindful of Goethe's words: *"in der Beschränkung zeigt sich der Meister."*

Now some words regarding the second division. We begin non-equilibrium statistical mechanics with what we call "Boltzmann theory", part D, with the developments including, however, quantum gases viewed from a semi-classical point of view. We prove the H-theorem for Boltzmann, B-E and F-D gases. We then discuss the hydrodynamic equations, near-equilibrium classical transport involving electrical and thermal conductivity problems, and (a novum perhaps) transport far from equilibrium, based on the papers of Yamashita and Watanabe, culminating in the Davydov and Druyvesteyn distributions for hot plasmas and hot electron gases. In the next part, E, we return to serious quantum developments and we establish the basis for quantum non-equilibrium irreversible kinetic equations, following the work of Van Hove, Zwanzig, Fano , Kubo, Mori and others. Also, we discuss linear and nonlinear response theory, the Wigner formalism and many applications. This material is based in part on articles by the author and I gratefully acknowledge the help and papers of my former graduate students and present colleagues P. Vasilopoulos, M. Charbonneau and A. Barrios. A special section has been devoted to the meagre foundations of linear response theory, discussed by van Kampen in his well-known Physica Norvegica paper, and cleared up, or at least elucidated I hope, in my response in a 1988 paper in the J. of Statistical Physics.

Lastly, part F deals with stochastic processes. I have started with the older ideas on Brownian motion by Einstein, Smoluwkowski, Kramers, Moyal, and of the Dutch School with the standard papers of Ornstein and Burger, Uhlenbeck and Ornstein, Chandrasekhar and others. A short sub-chapter is devoted to spectral analysis and besides the Wiener–Khintchine theorem, the methods of the 'short-time average' (MacDonald, Milatz), of 'elementary events' (Campbell, Carson) and the Allen-variance theorem, suitable for many forms of 'pathological noise', do the round. A single section is devoted to fluctuation processes in solids, to which I have actively contributed for many years, but, mostly, referral to the literature must suffice. A further chapter is devoted to continuous stochastic processes and branching processes and finally there is brief chapter on fluctuations in radiation fields and photons, based on work by Glauber, Sudarshan, Louisell and others with experimental verification by Zijlstra's group on counting statistics in Utrecht. We understand that the usual curriculum leaves no time for stochastic processes but for the barest principles of diffusion and Brownian motion, but we feel that stochastic processes, when microscopically founded as in this text, are an essential part of non-equilibrium statistical mechanics. Perhaps much of it can be relegated to a special seminar.

Several important topics have not been treated. We mention disordered systems and percolation theory in equilibrium statistical mechanics and out-of-equilibrium phase transitions in non-equilibrium theory; we must refer the reader to other texts.

A note on units, quite inconsequential in statistical mechanics, is in order. With rare exceptions we employ rationalized Gaussian or Heaviside–Lorentz units. The integrity of electromagnetic theory is then maintained, but the equations 'look like' as in the *Système International*; the *S-I* proponent only has to stick in ε_0 and μ_0 at appropriate places and remove some 'c's.

Problems have been added at the end of each chapter. They are generally of two kinds. First, there are exercises to verify some not explicitly derived results as well as straightforward applications. Secondly there are extensions of the theory for which space lacked in the chapter proper; the problems are an integral part of the text.

Finally, I have of course a list of acknowledgements. First of all, I thank the many graduate students who took my courses. If in earlier times I have not given enough information on certain topics, my hope remains that I have at least been able to awaken their critical faculties and their interest in this field. I also thank all who have pointed out errors and who have made known their wishes for improvements.

Since by nature I am more stimulated by auditory than visual inputs, I have greatly benefited from lectures, colloquia, and presentations at symposia and by the many visitors received yearly at the CRM. In particular stand out the lectures by the titulaires of the Chaire André Aisenstadt at the Université de Montréal, notably by the late Professors Sybren de Groot and Marc Kac. Also, by the visitors Profs. L. de Sobrino of UBC, S. Fujita of SUNY at Buffalo, Ch. G. van Weert and A.J. Kox of the Institute for Theoretical Physics at the University of Amsterdam and many others. I vividly remember and acknowledge the presentation by (and subsequent discussions with) Prof. E.G.D. Cohen of Rockefeller University in honour of Prof. van Kampen's sixtieth birthday at the University of Utrecht. I thank Prof. van Kampen for many discussions over the years regarding linear response theory and the assumptions underlying the master equation. And last but not least, I have learned much at the bi-annual conferences organized by Prof. Joel Lebowitz at Yeshiva University and later at Rutgers University.

A final word of thanks goes to Professor George Alexandrakis, former Chairman of the Physics Department of the University of Miami, who welcomed me as an Adjunct Professor in the department in 2001. I also thank many colleagues in Miami for discussions, seminars, and their impact; in particular Profs. M. Huerta, J. Nearing, O. Alvarez, H. Gordon, J. Ashkenazi and F. Zuo. Special thanks are due to Dr. A. Barrios, Mr. P. Sajnani and to Prof. Olga Korotkova for their critical reading of various chapters of the Non-equilibrium Part of this text.

Being myself illiterate in computers in the eighties and early nineties, I must acknowledge the earlier versions of parts of this text, which were ably composed in Tex or AmTex by Mme Louise Letendre of the CRM. The present version of this text being sent to the publisher was composed by myself in MS WORD 2003 and Math-type 5.2c. I am greatly indebted to the Production Manager Ms. Yolande Koh of WSPC and our Computer Scientist Mr. Marco Monti for their valuable advice.

Miami, Fall 2007

Contents

NON-EQUILIBRIUM STATISTISTICAL MECHANICS

PART D. CLASSICAL TRANSPORT THEORY

1. THE MESOSCOPIC MASTER EQUATION AND THE MOMENT EQUATIONS

2. BROWNIAN MOTION PROPER. VELOCITY FLUCTUATIONS AND DIFFUSION

3. SPECTRAL ANALYSIS

EQUILIBRIUM STATISTICAL MECHANICS

Nil fit ad nihilum. Ex nihilo nil fit. Parmenides, 500 B.C. [*Περί Φυσέ′ως*]

PART A

GENERAL PRINCIPLES OF MANY-PARTICLE SYSTEMS

Chapter I

Introduction to the State of Large Systems and Some Probability Concepts

1.1 Purpose of Statistical Mechanics

In statistical mechanics we are concerned with the physical properties of large systems. We assume the existence of the *thermodynamic limit*:

$$N \to \infty, \quad V \to \infty, \quad N/V = n = 1/\hat{v} \textit{ is finite.} \tag{1.1-1}$$

Here N is the number of microscopic constituents, V is the volume of the system, n is the average density and $\hat{v}$ is the molecular (or specific) volume. The peculiarity, which requires that the mechanics of such a system is "statistical", stems from the fact that such a system is as a rule *incompletely defined*. By this we mean that the equations of motion for such a system cannot be uniquely solved. Were this true in Gibbs' time already for the simple reason of mathematical complexity, the real problem is not computational, as is clear from interesting computer simulations nowadays available. Basically, the need for statistical methods stems from the lack of detailed information on the system.

We will illustrate this for the simple system of one harmonic oscillator. The Hamiltonian is

$$\mathcal{H}(p,q) = \frac{p^2}{2m} + \frac{1}{2} m\omega^2 q^2 \tag{1.1-2}$$

and Hamilton's equations are

$$\begin{aligned} \dot{p} &= -\partial\mathcal{H}/\partial q = -m\omega^2 q, \\ \dot{q} &= \partial\mathcal{H}/\partial p = p/m. \end{aligned} \tag{1.1-3}$$

We will solve these equations by a somewhat unfamiliar procedure. If we divide the first equation by the second one we obtain

$$\frac{dp}{dq} = -\frac{m^2\omega^2 q}{p}, \quad \text{or} \quad \frac{pdp}{m} + m\omega^2 q dq = 0,$$

which integrates to

$$\mathcal{H}(p,q) \equiv \frac{p^2}{2m} + \frac{1}{2} m\omega^2 q^2 = \mathcal{E} = c_1, \tag{1.1-4}$$

where $\mathcal{E}$ is a constant of motion. If we now substitute for p from (1.1-4) into the equation for $\dot{q}$ we easily obtain

$$dt = \frac{mdq}{\sqrt{2m\mathcal{E} - m^2\omega^2 q^2}},$$

or upon integrating

$$F(p,q) \equiv \arcsin \frac{m\omega q}{\sqrt{p^2 + m^2\omega^2 q^2}} = \omega t + \theta = \omega t + c_2. \tag{1.1-5}$$

Clearly there are two phase functions that are integrals of motion, $F(p,q)$ and $\mathcal{H}(p,q)$, the latter one being an integral of motion *proper*. Equations (1.1-4) and (1.1-5) directly yield the standard results,

$$q = \frac{1}{\omega}\sqrt{\frac{2\mathcal{E}}{m}} \sin(\omega t + \theta), \qquad p = \sqrt{2m\mathcal{E}} \cos(\omega t + \theta). \tag{1.1-6}$$

The constants $\mathcal{E}$ and θ, or c_1 and c_2, can be linked to the initial values q_0 and p_0.

The same procedure can be carried over to a system of N particles, each having r degrees of freedom, as was shown by the Ehrenfests.[1] We now have $2rN$ Hamilton's equations to be satisfied,

$$\dot{q}_i = \partial\mathcal{H}/\partial p_i, \quad \dot{p}_i = -\partial\mathcal{H}/\partial q_i, \quad i = 1 \ldots rN. \tag{1.1-7}$$

The integrals of Eqs. (1.1-7) give rise to $2rN$ constants of motion, all but the last one being proper, as is seen as follows. If we divide the first $2rN-1$ equations by the last one, then all the resulting equations are independent of time. These equations yield $2rN-1$ phase functions

$$\begin{aligned} &\varphi_1(p,q) \equiv \mathcal{H}(p,q) = \mathcal{E} = c_1, \\ &\varphi_2(p,q) = c_2, \\ &\quad\cdot \\ &\quad\cdot \\ &\varphi_{2rN-1}(p,q) = c_{2rN-1}. \end{aligned} \tag{1.1-8}$$

Upon substituting these results into the last equation we obtain by quadrature

[1] Paul and Tatiana Ehrenfest, "The Conceptual Foundations of the Statistical Approach in Mechanics", Cornell University Press, Ithaca, N.Y. 1959, pp. 17-19.

$$\varphi_{2rN}(p,q) = c_{2rN} + t. \tag{1.1-9}$$

However, the problem is only solved if all constants $c_1 \ldots c_{2rN}$ are known, or equivalently, if all initial coordinates and momenta q_{i0}, p_{i0} are known. This is of course preposterous for a system of, say, 10^{23} particles! Moreover, only certain averages have physical significance. *The purpose of statistical mechanics is to devise methods to handle incompletely known systems.*

So what can be specified for a many-particle system? First of all, for a closed system $\mathcal{E}$, V and N are fixed. Further, we assume that a representative Hamiltonian, involving the basic inter-particle interactions, can be constructed; in addition, there may be what Gibbs calls '*external parameters*', $A_1 \ldots A_p$, such as an electrical or magnetic field. For example, a system of particles with charges $\mathcal{q}_i$ in an external magnetic field of vector potential **A** with Coulomb interactions has the Hamiltonian[2]

$$\mathcal{H}(p,q;\mathbf{A}) = \sum_{i=1}^{N} \frac{1}{2m_i}\left(\mathbf{p}_i - \frac{\mathcal{q}_i}{c}\mathbf{A}\right)^2 + \frac{1}{4\pi}\sum_{i>j}^{N} \frac{\mathcal{q}_i \mathcal{q}_j}{|\mathbf{q}_i - \mathbf{q}_j|}. \tag{1.1-10}$$

Here we have ignored surface energies which go as $V^{2/3}$ and disappear in the thermodynamic limit; the above system has $3N$ degrees of freedom and V acts as a (non-holonomic) constraint on the position coordinates, while the components of the vector potential are external parameters.[3]

We now consider many systems for which the other constants of motion are non-specified but which all have the same Hamiltonian $\mathcal{H}(p,q;\{A_i\})$. A large collection of such systems is called a *microcanonical ensemble.* Such systems have the same statistical properties, as we will establish later, although their actual coordinates and momenta $q_i(t)$ and $p_i(t)$ trace out widely different paths. Later we shall introduce other *open* ensembles, which better satisfy the actual thermodynamic specification by allowing for variable energy, variable volume, or variable particle number. Although we shall, at times, introduce a Hamiltonian for such situations – like the "grand-Hamiltonian" – these concepts are artefacts of the formalism. Strictly speaking, only the microcanonical ensemble has a mechanical basis vested in Hamilton's equations.

We must now extend the discussion to quantum systems, for which a simultaneous specification of coordinates and momenta is not possible in view of Heisenberg's uncertainty relations. The system is now characterized by a dynamical state $|\psi(t)\rangle$ or by its projections on a complete set of many-body states. While later we shall employ suitable occupation-number states (chapter VII), for the present we will use the (theoretical) eigenstates $\{|\eta\rangle\}$ of the many-body Hamiltonian as a basis.

---

[2] We employ rationalized Gaussian (Heaviside–Lorentz) units throughout this text.

[3] More completely the Hamiltonian is $\mathcal{H}(p,q;V,N,\mathbf{A}) \equiv \mathcal{H}(p,q;\{A_i\})$. Sometimes we include V and N with the external parameters $\{A_i\} = A_1 \ldots A_p$.

We have the Schrödinger equation

$$\mathcal{H}\,|\psi(q_1...q_{rN},t)\rangle = \hbar i\frac{d}{dt}|\psi(q_1...q_{rN},t)\rangle. \qquad (1.1\text{-}11)$$

Now write

$$|\psi\rangle = \sum\nolimits_\eta\{|\eta\rangle\langle\eta|\}|\psi\rangle = \sum\nolimits_\eta|\eta\rangle c_\eta(t),$$

where $c_\eta(t) \equiv \langle\eta|\psi(q_1...q_{rN},t)\rangle$ is the projection on the 'axis' $|\eta\rangle$ of the Hilbert space and where we used the *closure property* $\sum_\eta|\eta\rangle\langle\eta| = 1$. Then from (1.1-11), starting with the rhs,

$$\hbar i\frac{d}{dt}\sum\nolimits_\eta|\eta\rangle\langle\eta|\psi(q_1...q_{rN},t)\rangle = \mathcal{H}\sum\nolimits_\eta|\eta\rangle c_\eta(t)\,.$$

Taking the scalar product with $|\eta'\rangle$ and using $\langle\eta'|\eta\rangle = \delta_{\eta\eta'}$,we find the *transformed Schrödinger equation,*

$$\hbar i\frac{d}{dt}c_{\eta'}(t) = \sum_\eta\langle\eta'|\mathcal{H}|\eta\rangle c_\eta(t)\,. \qquad (1.1\text{-}12)$$

The equations (1.1-12) form an infinite set of first order differential equations for the projections of the dynamical state $|\psi(t)\rangle$ on the coordinates of the Hilbert space; these take the place of Hamilton's equations of motion. Since $|\psi(0)\rangle$ is unknown, the $c_\eta(0)$ are unknown. Again, we need an ensemble to represent the system. While the c's for a particular system have no physical relevance, in the next section we will see that the average $\overline{c_\eta(t)c_\eta^*(t)}$ for all systems in the ensemble is of paramount statistical significance.

1.2 Gamma-Space and its Quantum Equivalent

J.Willard Gibbs (1839-1903) can be considered to be the father of statistical mechanics of large systems. He developed a statistical representation for such systems which, in principle, is applicable regardless of the complexity and the strength of the interactions. His 1902 book "*Elementary Principles in Statistical Mechanics*" (subtitle "*Developed with Special Reference to the Rational Foundation of Thermodynamics*") still merits reading today.[4] His gamma-space (Γ-space) or phase-space is a multi-dimensional space whose axes are the rN coordinates and rN momenta of the constituents of the system. The state of the motion in the system is at any moment represented by the set $\{q_i(t), p_i(t)\}$; in the course of time a system traces

[4] J. Willard Gibbs, Yale University Press, New Haven 1902; reprinted by Dover Publications, N.Y., 1960.

out a complex trajectory in the phase-space. Generally the Hamiltonian $\mathcal{H}[p(t),q(t),t]$ will not be separable into a sum of one-particle Hamiltonians because of inter-particle energies. In a system of a microcanonical ensemble the energy is fixed; hence the trajectory lies on the energy surface $\mathcal{H}(p,q)=\mathcal{E}$, where for conservative systems we exclude explicit time dependence in $\mathcal{H}$. The trajectory of the representative point can trace out a very complex path, but the curve never crosses itself; see Fig. 1-1. This follows from Hamilton's equations. If the curve crossed itself, there would be two or more tangents $\dot{p}_i(t),\dot{q}_i(t)$; however, for well-behaved Hamiltonians the derivatives $\partial\mathcal{H}/\partial q_i$, $\partial\mathcal{H}/\partial p_i$ are single-valued. It was felt by Gibbs, as well as by his near-contemporaries Maxwell and Boltzmann, that the system trajectory would eventually fill out the entire energy surface; this is part of the *ergodic problem* to which we come back in Section 2.2.

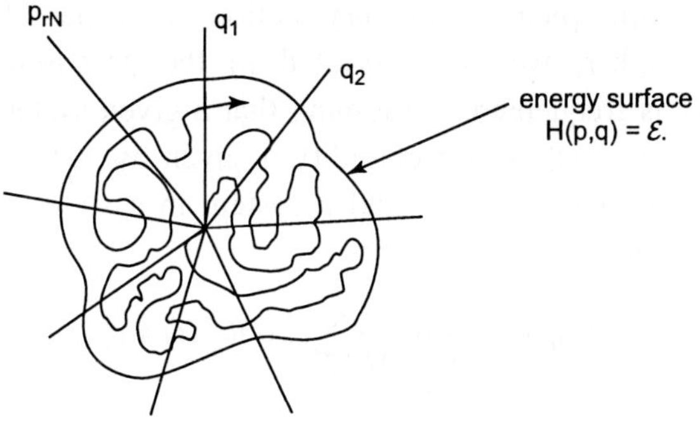

Fig. 1-1. Gamma-space with the system trajectory on the energy surface.

The concept of ensemble was introduced to permit statistics without knowing the complex trajectory traced out in the course of the time. Thus, consider at time t the representative points of a large number M of systems forming the ensemble. There is then a density of points in phase space $D(p,q,t)$. The normalized *density function* is introduced as

$$\rho(p,q,t)=\lim_{M\to\infty}\frac{1}{M}D(p,q,t). \tag{1.2-1}$$

Clearly then,

$$\int\rho(p,q,t)\prod_{i=1}^{rN}dp_i dq_i=\int\rho(p,q,t)\,d\Omega=1. \tag{1.2-2}$$

Averages are now introduced according to the usual statistical rules. Let $F(p,q)$ be a phase function; then

$$\langle F(t)\rangle=\int F(p,q)\rho(p,q,t)\,d\Omega. \tag{1.2-3}$$

For a conservative system ρ does not depend on t explicitly, $\partial\rho/\partial t = 0$; we speak of *statistical equilibrium.* The stationary ensemble average is then

$$\langle F\rangle = \int F(p,q)\rho(p,q)\,d\Omega\,. \tag{1.2-4}$$

It is clear that these averages have physical meaning only if they are the same as for the time average of a single system,

$$\tilde{F} = \lim_{T\to\infty}\frac{1}{T}\int_{-T/2}^{T/2} F[p(t),q(t)]\,dt\,. \tag{1.2-5}$$

In Section 2.1, after the discussion of Liouville's equation, we will fix $\rho(p,q)$ for a microcanonical ensemble.

In the quantum picture the very notion of a particle trajectory has to be abandoned. Rather, we are interested in the probability that the many-body eigenstate $|\eta\rangle$ is frequented. Assuming that a given system of the ensemble has a dynamical state $|\psi^i(t)\rangle$, this probability is given by $c_\eta^i(t)c_\eta^{i*}(t)$ – see any quantum text. Hence, the probability to find a system in a state $|\eta\rangle$ in the ensemble is the *ensemble probability*,

$$p(\eta,t) = \lim_{M\to\infty}\frac{1}{M}\sum_i c_\eta^i(t)c_\eta^{i*}(t) = \overline{c_\eta^i(t)c_\eta^{i*}(t)}^{M}\,. \tag{1.2-6}$$

For any operator F we now have as average in the ensemble

$$\langle F(t)\rangle = \sum\nolimits_\eta p(\eta,t)\langle\eta|F|\eta\rangle\,. \tag{1.2-7}$$

This is a "double average": first we need the quantum mechanical expectation value in the state $|\eta\rangle$, next we need to average over the probability that the state $|\eta\rangle$ occurs. The question now presses whether the above can be simplified in using a single averaging process. This is indeed the case, providing we introduce the *density operator*,

$$\rho(t) = \sum\nolimits_\eta |\eta\rangle p(\eta,t)\langle\eta|\,. \tag{1.2-8}$$

Then (1.2-7) can be replaced by

$$\langle F(t)\rangle = \mathrm{Tr}[F\rho(t)]\,. \tag{1.2-9}$$

For the proof consider the corollary

$$\mathrm{Tr}\,|\varphi\rangle\langle\psi| = \sum\nolimits_\eta \langle\eta|\varphi\rangle\langle\psi|\eta\rangle = \langle\psi|\{\sum\nolimits_\eta |\eta\rangle\langle\eta|\}|\varphi\rangle = \langle\psi|\varphi\rangle\,. \tag{1.2-10}$$

Setting $|\varphi\rangle = |\eta\rangle$ and $|\psi\rangle = F^\dagger|\eta\rangle$, the proof follows. We further note that the relationship between $\rho(t)$ and $p(\eta,t)$ can be inverted: p is the *diagonal part* of ρ:

$$\langle \eta | \rho(t) | \eta' \rangle = p(\eta, t)\delta_{\eta\eta'} . \tag{1.2-11}$$

In the above the density operator depends on *t*, but we emphasize that $\rho(t)$ is a Schrödinger operator and *not* a Heisenberg operator. For an ensemble in statistical equilibrium neither ρ nor p depends on *t*. We then have for the stationary ensemble average,

$$\langle F \rangle = \mathrm{Tr}\,(\rho F) = \mathrm{Tr}\,(F\rho) . \tag{1.2-12}$$

We shall also consider these results in a representation in which ρ is not diagonal. Let $\{|\gamma\rangle\}$ be a basis that spans the Hilbert space associated with a many-body operator that possibly does not commute with $\mathcal{H}$. Let again $|\psi^i(t)\rangle$ be the dynamical state of a given system in the ensemble. In this state we have the expectation value

$$\langle F(t) \rangle_{\psi^i} = \langle \psi^i(t) | F | \psi^i(t) \rangle. \tag{1.2-13}$$

Writing $|\psi^i(t)\rangle = \sum_\gamma c^i_\gamma(t) |\gamma\rangle$ we have,

$$\langle F(t) \rangle_{\psi^i} = \sum_{\gamma\gamma'} c^i_\gamma(t) c^{i*}_{\gamma'}(t) \langle \gamma' | F | \gamma \rangle. \tag{1.2-14}$$

Next we average over all systems of the ensemble; this requires the introduction of the *density matrix*

$$\rho_{\gamma\gamma'}(t) \equiv \langle \gamma | \rho(t) | \gamma' \rangle = \overline{c^i_\gamma(t) c^{i*}_{\gamma'}(t)}^{M} . \tag{1.2-15}$$

From (1.2-14) we find

$$\langle F(t) \rangle = \overline{\langle F(t) \rangle_{\psi^i}}^{M} = \sum_{\gamma\gamma'} \rho_{\gamma\gamma'}(t) F_{\gamma'\gamma} = \mathrm{Tr}[\rho(t) F] , \tag{1.2-16}$$

in accordance with (1.2-9), noting that the trace is always independent of the chosen representation. Many authors only speak of the density matrix as given here; on the contrary, we prefer to deal principally with the density operator, which befits better the operator algebra employed throughout this book.

1.3 The Thermodynamic State

1.3.1 *Macroscopic Thermodynamics; Callen's Entropy*

A system is distinguished from its surroundings by walls, being either introduced purposely as for the container of a fluid, or being present naturally as for the surfaces of a solid. The mechanics of the interaction with the walls will be ignored. The system proper can then be characterized by *phenomenological* or *thermodynamic* state variables. Most important among these are the extensive variables, which are quantities that double if the volume within the walls is doubled. Examples are the volume V itself, the number of molecules N^α of a given species, the electrical charge Q, the magnetization M and the polarization P. Also, in a homogeneous system, internal energy U is extensive, regardless of the coordination numbers and the details of the interactions in the system.[5] The extensive thermodynamic variables will be denoted by $X_0 ... X_s$, where we make the convention that $X_0 = U$, $X_1 = V$, $X_2 = N$ (or $X_2^\alpha = N^\alpha$ when there are various species).

Historically the concepts of internal energy, temperature and heat were developed without a microscopic basis through a combination of superb abstract analysis, laboratory and thought experiments by Clausius, Maxwell, Carnot, Joule and others. Perhaps the most readable book on 'standard' macroscopic thermodynamics is the treatment by ter Haar and Wergeland.[6] On a more axiomatic basis these concepts are introduced in Callen's exposé.[7] From the outset it is assumed that internal energy is measurable by macroscopic means and is a *potential* or *state function*, i.e., in a cyclic process $A \to B \to A$ we have $\oint dU = 0$; it satisfies the first law of thermodynamics

$$đQ = dU + đ\tilde{W} \quad , \tag{1.3-1}$$

in which $đQ$ is a small quantity of heat added to the system and $đ\tilde{W}$ is a small amount of 'external work'; it is counted positive when done *by the system*, and negative when performed *on the system* [the tilde is used to distinguish from the W used for enthalpy]. It is noted that Q and $\tilde{W}$ are not state functions, so that $đQ$ and $đ\tilde{W}$ are not mathematical total (or exact) differentials.

The walls of the system may prevent energy and particle flow to and from the environment. In that case they provide constraints for the internal energy U and for the particle numbers N^α. In the usual case that the walls are rigid, there is also a constraint for the volume V. Quite generally any extensive variable is either *constrained* or *freely varying* (*open*). The 'gadgets' for allowing volume variation and particle variation are moveable pistons and permeable membranes, respectively. If the walls have adiabatic shielding, energy can be constrained; we refer to more popular College Physics textbooks for details. In Fig. 1-2 we have sketched some possible set-ups.

If the system is open with respect to an extensive variable, there must be interaction with a reservoir; common examples are a heat reservoir (also called 'heat bath' or 'thermostat') and a particle reservoir. When an extensive variable is freely varying experience tells us that other thermodynamic variables of the system, called intensive variables or 'thermodynamic forces', balance the corresponding intensive variables of the reservoirs. For example, a gas in a cylinder with a moveable piston will assume a volume such that the pressures on both sides of the piston balance each

---

[5] This will be borne out by later detailed computations of the energy for classical and quantum systems.

[6] D. ter Haar and H. Wergeland, "Elements of Thermodynamics", Addison-Wesley, London 1966.

[7] H.B. Callen, "Thermodynamics", John Wiley & Sons, NY 1960.

other; likewise, for a system in equilibrium with a heat bath the temperature adjusts itself to that of the reservoir.[8] Below we define these intensive variables more appropriately. Here we just note that the state of any system is fully defined by the extensive variables that are constrained and by the intensive variables of the reservoirs corresponding to the variables that are open. Also, in this connection, we define *thermal equilibrium.* A closed system is said to be in equilibrium if no variable X_i depends on time; for an open system thermal equilibrium entails that there is no net transport to or from the system. The equivalence of both definitions is evident: combine two connected systems, open with respect to a variable X_i, to a composite system that is closed. For the latter we thus have,

$$\dot{X}_{i,c} = \dot{X}_{i,I} + \dot{X}_{i,II} = 0, \quad \text{or} \quad \dot{X}_{i,I} = -\dot{X}_{i,II}, \tag{1.3-2}$$

indicating that there is no net flow.

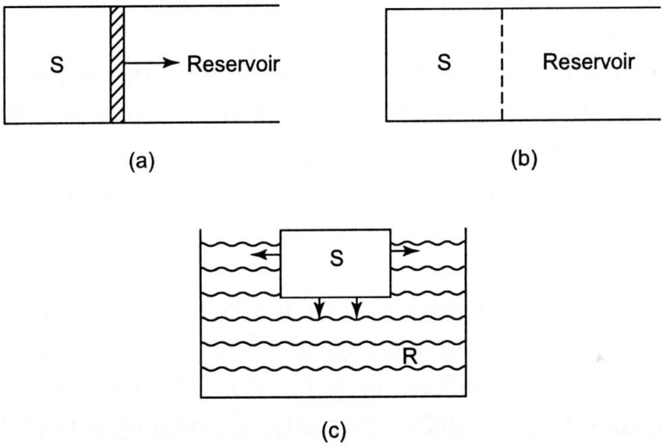

Fig. 1-2. Examples of open systems. (a) Moveable piston: S open with respect to *V*; (b) Permeable membrane: S open with respect to *N*; (c) Energy exchange with a heat bath: S open with respect to $\mathcal{E}$.

Finally, we come to the main point of this subsection. The principal problem of classical thermodynamics was to reformulate the law of energy-heat conservation (1.3-1) for a system undergoing a change in variables from $\{X_i\}_A$ to $\{X_i\}_B$, by means of a state-function differential ΔS which is independent of the path. For the most efficient heat converting cycle, the Carnot cycle, consisting of two adiabats and two isotherms, a simple computation showed that

$$\oint đQ/T = [\Delta Q_{\text{high}}/T_{\text{high}} - \Delta Q_{\text{low}}/T_{\text{low}}] = 0. \tag{1.3-3}$$

As to an arbitrary reversible (or quasi-static) cyclic process, it can always be resolved in a sum of Carnot cycles, so that the above contour integral should be likewise zero.

[8] Later we shall see that the system temperature actually is subject to rms fluctuations proportional to $V^{-1/2}$, with $\overline{T} = T_{\text{reservoir}}$.

Accordingly, entropy defined as the integral of the reduced heat $S = \int^T đQ/T$, has a perfect differential and is a state function. For quasi-static processes we thus have in addition to (1.3-1) the 'equality-form' of the second law of thermodynamics:

$$dS = \sum_i Q_i dX_i = (1/T)dU + (P/T)dV - (\varsigma/T)dN - ... \qquad (1.3\text{-}4)$$

where the $Q_i = \partial S / \partial X_i$ are the intensive variables or thermodynamic forces; *T, P* and ς refer to the usual variables: temperature, pressure and chemical potential.

If irreversible (or spontaneous) processes are also involved, the result must be modified. From Kelvin's principle it can now be deduced that the contour integral of the reduced heat will be negative, i.e., $\oint đQ/T < 0$. Thus, let in a cyclic contour the process $A \to B$ be irreversible and the process $B \to A$ be reversible. We then find

$$\int_A^B đQ_{\text{irrev}}/T + S(A) - S(B) < 0, \quad \text{or,}$$

$$S(B) > S(A) + \int_A^B đQ_{\text{irrev}}/T \,. \qquad (1\text{-}3\text{-}5)$$

In particular, this result shows that for an *adiabatic spontaneous process*, like the free expansion of a gas, *the final entropy is larger than the initial one.* We note that this result was obtained without ever introducing a nonequilibrium entropy function; we only need to go from a constrained equilibrium state to a true equilibrium state via a spontaneous process, such as the release of a piston held in place with a spring (for details, see Section 3.8, Fig 3-1).

We will yet obtain another form, known as Clausius' inequality, for a system in contact with a heat reservoir (or thermostat) of temperature T^r, volume reservoir with pressure P^r and possibly others. From (1.3-5) it follows that $\Delta S \geq \Delta Q/T^r$ where the equal sign applies if heat is added reversibly. Combining with (1.3-1), we have

$$\Delta S \geq \frac{1}{T^r}\Delta U + \frac{P^r}{T^r}\Delta V + ..., \quad \text{or} \quad \Delta S \geq \int_A^B ... \int \sum_i Q_i^r \, dX_i \,. \qquad (1.3\text{-}6)$$

In statistical mechanics it is customary to refer to Eq.(1.3-5) or (1.3-6) as "the" *Second Law of Thermodynamics*. On the contrary, Eq.(1.3-4), after multiplication by *T*, is usually referred to as the *Gibbs' Relation*. We have the explicit form[9]

$$TdS = dU + PdV - \varsigma dN - \Phi dQ - HdM - EdP - ... \qquad (1.3\text{-}7)$$

We assumed here that besides chemical work, we performed potential work to charge the system, magnetic work to magnetize the system, electric work to polarize the system, etc. The reader has undoubtedly noted that all terms have the minus sign,

[9] For simplicity we have omitted the scalar dot products ($\mathbf{H} \cdot d\mathbf{M}$, etc.) and we left out $\int ...dV$ in the magnetizing and polarizing work. The former is of no concern, since as a rule the magnetization is along the direction of the field; as to the latter, we will be mostly concerned with the entropy per unit volume, in order to effectuate the thermodynamic limit.

except the work done for volume expansion; the reason is that all other work as defined is performed *on* the system, while the pressure has historically been chosen as the force per unit area exerted on its surroundings (no need to fix this now...); thus, PdV represents work done *by* the system. Also, a note is in order with respect to the term $-\Phi\, dQ$. We consider the case that the charge Q is attached to the particles under consideration, so that $Q = \mathscr{q}N$, where $\mathscr{q}$ includes the sign of the charge. Then, upon combining this term with $-\varsigma dN$ we have the work $-\mu dN$, where μ is the electrochemical potential:

$$\mu = \varsigma + \mathscr{q}\Phi. \tag{1.3-8}$$

Finally we indicate some customary Legendre transforms of the entropy. Generally, these generalized Massieu functions are defined as

$$S_m(X_0 ... X_m, Q_{m+1} ... Q_s) = S - \sum_{i=m+1}^{s} X_i Q_i. \tag{1.3-9}$$

This yields for the total differential

$$dS_m = \sum_{i=0}^{s} Q_i dX_i - \sum_{i=m+1}^{s} X_i dQ_i - \sum_{i=m+1}^{s} Q_i dX_i = \sum_{i=0}^{m} Q_i dX_i - \sum_{i=m+1}^{s} X_i dQ_i. \tag{1.3-10}$$

Particular cases are the original Massieu function $S_M(T, X_1, X_2 ...) = S - U/T$ and the Planck function $S_{Pl}(T, P, X_2 ...) = S - U/T - PV/T$.

1.3.2 *Statistical Mechanical State Functions; Gibbs' Entropy*

So far we have followed as point of departure Callen's view in which the entropy is a function of extensive variables only. Yet this view does often not befit the framework of statistical mechanics. For a given system we must find a Hamiltonian in accord with the thermodynamic specification. The question arises as to which energies are to be included, since some energies are externally turned on. This point goes back to Gibbs who notes[10]

"*It will be observed, that although we call* $\mathscr{E}_q$ *the potential energy of the system which we are considering, it is really so defined as to include that energy which might be described as mutual to that system and external bodies* (*reservoirs*)."

As a consequence, the entropy of a system derived from Gibbs' statistical mechanics contains both extensive variables (V, N, etc.) and intensive variables such as external fields (H, E, g, etc). Clearly, it is not useful to speak of 'internal energy' – being a non-realizable *fictitious* concept of macroscopic thermodynamics; thus, U will be replaced by $\mathscr{E}$, the energy of the envisioned Hamiltonian. The ensuing Gibbs entropy

[10] J. Willard Gibbs, Op. Cit., p. 4.

has the form,

$$S_{Gibbs} = S[\mathcal{E}(A_i, V, N), V, N, A_i] \to S[\mathcal{E}(A_i), \{A_i\}]. \tag{1.3-11}$$

In the latter form we have included V and N (or N^{α}) in the external parameters $\{A^i\}$. Thus, Gibbs' entropy appears as a Legendre transform of Callen's entropy with respect to those variables, which are specified by their conjugate intensive reservoir variables. As an example, we consider a spin lattice placed in an external magnetic field $H = H_z$. The Heisenberg Hamiltonian reads:

$$\mathcal{H} = -\sum_{i,j}^{N} J_{i,j} \mathbf{S}_i \cdot \mathbf{S}_j - g(\rho_B / \hbar) H \sum_i S_i^z , \tag{1.3-12}$$

where $J_{i,j}$ is the exchange integral, $\mathbf{S}_i$ are the spin angular momenta, g is the Landé factor and ρ_B is the Bohrmagneton. This yields an entropy S_{Gibbs}=$S[\mathcal{E}(N,H),N,H]$, which for quasi-static changes satisfies the Gibbs' relation,

$$TdS_{Gibbs} = d\mathcal{E} - \varsigma dN + \overline{M} dH, \tag{1.3-13}$$

where $\overline{M}$ is an ensemble average. The appearance of '*MdH*' rather than '*HdM* ' has led some investigators[11] to consider the enthalpy, being the Legendre transform W_{magn}= $\mathcal{E} - MH$, as the appropriate state function, interpreting (1.3-13) as

$$\begin{aligned} TdS_{Callen} &= d(\mathcal{E} - MH) - \varsigma dN + MdH, \\ TdS_{Callen} &= d\mathcal{E} - \varsigma dN - HdM. \end{aligned} \tag{1.3-13'}$$

However, It should be borne in mind that the magnetic energy, mutual with the reservoir, *cannot always be separated off.* For the treatment of diamagnetism or of magneto-phonon effects the magnetic interactions appear in the kinetic energy $[\mathbf{p} - (q / c)\mathbf{A}]^2 / 2m$, cf. (1.1-10) and the magnetic enthalpy is of no avail. Therefore, our point of view is that the Gibbs entropy is not to be seen as a Legendre transform, *but as the proper entropy emerging from a microscopic statistical treatment.* The subscript 'Gibbs' will subsequently be dropped. The general Gibbs' relation is now

$$TdS = d\mathcal{E} + \sum\nolimits_i \overline{F}_i dA_i , \tag{1.3-14}$$

where the $\overline{F}_i$ are *generalized conjugate variables*; volume and number effects are included in this sum, the terms being $\overline{P}dV$ and $-\overline{\varsigma}dN$ [12]. Although from now on we have no strict need for the previously introduced X_i and Q_i, we just mention that the connections are: $\overline{F}_i = TQ_i$ for those A's that are extensive variables and $\overline{F}_i = -X_i$ for those A's which are intensive (external) variables.

---

[11] See e.g., M. Plischke and B. Bergersen "Equilibrium Statistical Physics,"3rd Ed., World Scientific, 2006, for their mean-field treatment of the Ising model, p. 65.

[12] Strictly speaking, in a statistical mechanics treatment, P and ς will appear as averages in a microcanonical ensemble; however, the bars will usually be omitted to ease the notation.

1.4 Various Ensembles and their Main State Functions

Although we started this introduction with thermodynamic considerations involving the entropy, clearly the energy $\mathscr{E}$, associated with the Hamiltonian of the systems in the ensemble, is a more appropriate state function for statistical physics purposes. We write $\mathscr{E} = \mathscr{E}[S(A_i),\{A_i\}]$, where again V and N are included in the $\{A_i\}$; specifically we set $A_1 = V,\ A_2 = N$.

For a closed system, with $\mathscr{E}$ fixed, the description involves a microcanonical ensemble, as stated previously. For a quasi-static change in the parameters of the ensemble we only need to slightly rewrite the Gibbs' relation (1.3-14):

$$d\mathscr{E} = TdS - \sum_i \overline{F}_i dA_i = TdS - PdV + \varsigma dN - \sum_{i=3}^{p} \overline{F}_i dA_i\,. \tag{1.4-1}$$

The microcanonical ensemble is the only *basic* ensemble, which directly relates to classical or quantum mechanics, without necessarily invoking the thermodynamic limit. However, the main constraints, $\mathscr{E}$, V and N fixed, make computations usually quite cumbersome. The constraints can be circumvented with the Darwin-Fowler procedure (or saddle-point method) to be discussed later, but the procedure basically implies an asymptotic series in ($1/N$) which in practice is near-equivalent to taking the thermodynamic limit. Therefore, in most cases the computations are done in other ensembles, whose foundations are less rigorous, but for which one or more constraints are *a priori* removed. These are denoted as generalized canonical ensembles.

If energy can be freely exchanged with the surroundings (or an infinite reservoir) the system is part of a *canonical ensemble.* The basic state function is the Helmholtz free energy

$$F(T,V,N,\{A_i\}) = \mathscr{E} - TS\,. \tag{1.4-2}$$

The differential for quasi-static changes of its parameters satisfies

$$dF = -SdT - PdV + \varsigma dN - \sum_{i=3}^{p} \overline{F}_i dA_i\,. \tag{1.4-3}$$

If energy and volume can be freely exchanged we need the *pressure ensemble.* The basic state function is the Gibbs free energy

$$G(T,P,N,\{A_i\}) = \mathscr{E} - TS + PV, \tag{1.4-4}$$

with differential

$$dG = -SdT + VdP + \varsigma dN - \sum_{i=3}^{p} \overline{F}_i dA_i. \tag{1.4-5}$$

Finally, if energy and particle number are variable, the description requires the *grand-canonical ensemble.* The appropriate state function is the Gibbs' function, also called the "grand potential"

$$\Omega(T,V,\varsigma,\{A_i\}) = \mathcal{E} - TS - \varsigma N, \tag{1.4-6}$$

with differential

$$d\Omega = -SdT - PdV - Nd\varsigma - \sum_{i=3}^{p} \overline{F}_i dA_i. \tag{1.4-7}$$

Of course, this does not exhaust all possibilities. In the *open ensemble* $\mathcal{E}$, V and N are variable. Further, we can have reservoirs for other extensive variables A_i, like the magnetization or the charge (large capacitor acting as charge reservoir). These can, in principle, be described by *generalized canonical ensembles*; they are found in Guggenheim's book[13] and in Chapter V of this text. Their main importance lies in the computation of fluctuations; also, 'transformation theory" will shed a new light on connecting the various ensembles of statistical physics.

1.5 Fluctuating State Variables; the *a*-Space

One of the goals of statistical mechanics is to obtain thermodynamic relationships. In Section 1.2 we considered ensemble averages for a phase function $F(p,q)$. Not every phase function, however, is an observable. For instance, the average $\langle q_i^2 \rangle$ for one particle coordinate does not relate to macroscopic measurement. In fact, any precise function of the p's and q's is as microscopic (i.e. 'fine-grained') as the coordinates and momenta which enter into the Hamiltonian. So we will introduce 'coarse-grained' variables, which are averaged in the phase space over a small volume $\Delta\Omega$, or for the quantum case cover a range of states $|\eta\rangle \in \Delta\mathcal{E}^J$ in an energy cell. We shall refer to these as "*fluctuating state variables.*"[14] Since they are not the specified, fixed thermodynamic variables of the system, they are subject to random fluctuations that can be measured with sensitive instruments, e.g., a fast-Fourier transform noise analyser (FFT). These variables, in our view, have the following properties:

i. They are coarse-grained in phase space or cover a range of quantum states;
ii. They are generally observables of subsystems of the given system, either in phase space or in the Hilbert space;
iii. The observables are symmetric in the particle attributes comprising the sub-

[13] E.A. Guggenheim, "Statistical Thermodynamics", Cambridge University Press, 1953, p. 253ff.

[14] In former times, before 'nano-science' appropriated the term, these variables were called '*mesoscopic'* from the Greek μεσοσ = in between; we shall employ this denotation esp. in the Non-Equilibrium part.

system;

iv. These observables can be either of an extensive or an intensive nature.[15,16,17]

Extensive variables are easily written as phase functions $a_i(p,q)$ or as quantum operators. For intensive variables, see below. Besides the examples for fluctuating state variables in footnote[15] we mention the energy of a subsystem with $\mathcal{H}_{sub} = \mathcal{E}'$, the kinetic energy of the (full) system, being in the momentum subspace of the phase space, $\mathcal{T} = \Sigma_{i=1}^{N} p_i^2/2m$, the orientational polarization of a subsystem, $P_z'(\{\theta_i\}) = \Sigma_{i=1}^{N'} p\cos\theta_i = pN'\overline{\cos\theta}^{N'}$. For a quantum example, consider the number of electrons in the conduction band of a multi-band solid.

At the end of this section we shall indicate a procedure, due to van Kampen, to ensure that all a_i commute. It is therefore possible to introduce a many-variate probability density function (pdf) $W(\{a_i\}) \equiv W(\mathbf{a})$. For the equilibrium state this pdf can be obtained from the density function $\rho(p,q)$ or from the density operator ρ, as will be outlined in Section 2.4. For a nonequilibrium steady-state stochastic methods can be employed. Averages in a-space are obtained by the usual rule:

$$\langle F(\mathbf{a})\rangle = \int ... \int \prod_i da_i F(\mathbf{a}) W(\mathbf{a}) . \qquad (1.5\text{-}1)$$

Although the averages so obtained must be the same as via an ensemble average in a microscopic treatment, we emphasize that the motion in a-space is not governed by precise differential equations as considered in section 1.2. Rather, the motion in a-space is of a diffusive nature, as is borne out by such stochastic equations as the master equation and the Langevin equation, discussed in Chapter XVIII.

With the a-variables we can associate an *entropy function*. This function is the entropy that would be obtained if at an instant t all variables were 'frozen' to their instantaneous values. The entropy function will be denoted as $S(\mathbf{a})$. We can now define conjugate intensive a-variables by the rule $(a_i)_{int} = \partial S / dx_i$, where $x_i(p,q)$ are extensive variables. E.g., for the temperature of a subsystem we have $T^{-1} = \partial S / d\mathcal{E}'_{sub}$. In section 3.7 we shall prove Boltzmann's famous result, engraved on his epitaph:

$$S(\mathbf{a}) = k_B \log W(\mathbf{a}) + const . \qquad (1.5\text{-}2)$$

---

[15] There is no unanimity of opinion with respect to (iv). Van Kampen (Ref.16, p.183) takes the attitude that the a_i can represent any observable whose relaxations are noticeable on a macroscopic scale, such as temperature fluctuations in a bolometer, pressure fluctuations of a membrane, etc. De Groot and Mazur (Ref. 17, p. 85) on the contrary, assume that only extensive variables, expressive as $a(p,q)$ or corresponding operators, should be considered. As an example they mention charge fluctuations.

[16] N.G. van Kampen in "Fundamental Processes in Statistical Mechanics," Vol. I (E.G.D. Cohen, Editor), North Holland, Amsterdam, 1962, p. 173.

[17] S.R. de Groot and P. Mazur, "Non-Equilibrium Thermodynamics", North Holland, Amsterdam, 1962.

Finally, we describe the coarse-graining for quantum operators after van Kampen.[16] By a decomposition of unity (closure property) we can write for any of the operators a_i,

$$a_i = \sum_{\eta\eta'} |\eta\rangle\langle\eta| a_i |\eta'\rangle\langle\eta'|, \tag{1.5-3}$$

where the $|\eta\rangle$ are energy eigenstates. We now lump these states into energy cells, $|\eta\rangle \in \Delta\mathscr{E}^J$. Next we erase matrix elements belonging to different cells; the result is,

$$a_i = \sum_J \sum_{\eta,\eta'\in\Delta\mathscr{E}^J} |\eta\rangle\langle\eta| a_i |\eta'\rangle\langle\eta'|. \tag{1.5-4}$$

Now we make a unitary transformation in each cell, $|\bar\eta\rangle_{(i)} = \sum_{\eta\in\Delta\mathscr{E}^J} c_{\bar\eta\eta} |\eta\rangle$, such that $\langle\bar\eta| a_i |\bar\eta'\rangle = \langle\bar\eta| a_i |\bar\eta\rangle \delta_{\bar\eta\bar\eta'}$ (is diagonal). Then,

$$a_i = \sum_J \sum_{\bar\eta\in\Delta\mathscr{E}^J} \{|\bar\eta\rangle\langle\bar\eta|\}_{(i)} \langle\bar\eta| a_i |\bar\eta\rangle. \tag{1.5-5}$$

These are the sought for coarse-grained operators. They commute with each other and with the Hamiltonian, which has the coarse-grained form,

$$\mathscr{H} = \sum_J \sum_{\bar\eta\in\Delta\mathscr{E}^J} |\bar\eta\rangle\langle\bar\eta| \mathscr{E}^J. \tag{1.5-6}$$

In the last two equations we can omit the bar over the states, providing the eigenstates have been chosen and arranged as indicated.

1.6 Some Mathematical Distribution Functions

In this section we shall review some standard probability distributions and relate them to some elementary problems of statistical mechanics.[18]

1.6.1 *The Binomial Distribution and Random Walk*

Suppose that in an experiment p will be the *a priori* chance that the event A will occur and $1-p$ be the chance that the only mutually exclusive event B will occur and that N trials will be performed. The probability that in these trials n trials yield event A, while $N-n$ trials yield B, depends on the number of ways we can divide N into groups n and $N-n$, viz.,

$$\binom{N}{n} = \frac{N!}{n!(N-n)!}. \tag{1.6-1}$$

[18] Sections 1.6 and 1.7 are an interlude dealing with probability concepts and generating functions. The text proper on statistical mechanics of closed systems continues in Chapter II.

Since the trials are independent, the probability sought for is also proportional to p^n and to $(1-p)^{N-n}$. Hence,

$$W(n) = \frac{N!}{n!(N-n)!} p^n (1-p)^{N-n}. \tag{1-6-2}$$

Note that this probability distribution is normalized, for by the binomial theorem

$$\sum_{n=0}^{N} \binom{N}{n} p^n (1-p)^{N-n} = (p+1-p)^N = 1. \tag{1.6-3}$$

The mean and the variance can be found by the same elementary algebra:

$$\begin{aligned}\langle n\rangle &= \sum_{n=1}^{N} nW(n) = \sum_{n=1}^{N} \binom{N-1}{n-1} (pN) p^{n-1} (1-p)^{N-1-(n-1)} \\ &= pN(p+1-p)^{N-1} = pN,\end{aligned} \tag{1.6-4}$$

$$\langle n(n-1)\rangle = \sum_{n=2}^{N} n(n-1)W(n) = p^2 N(N-1). \tag{1.6-5}$$

This yields,

$$\langle n^2\rangle = \langle n(n-1)\rangle + \langle n\rangle = p(1-p)N + p^2N^2. \tag{1-6-6}$$

We now define the variance:

$$\mathrm{var}\, n \equiv \langle \Delta n^2\rangle = \langle [n-\langle n\rangle]^2\rangle = \langle n^2\rangle - \langle n\rangle^2 .$$

For the binomial distribution this gives

$$\langle \Delta n^2\rangle = p(1-p)N. \tag{1.6-7}$$

As an example, consider the case of N spins of magnetic moment μ, oriented at random, with up or down probability ½. Then,

$$W(n_+, n_-) = \frac{N!}{n_+!n_-!}\left(\frac{1}{2}\right)^N. \tag{1.6-8}$$

Clearly, $\langle n_+\rangle = \langle n_-\rangle = \frac{1}{2}N$ and the mean magnetization $\langle M\rangle \propto N[\langle n_+\rangle - \langle n_-\rangle] = 0$, which is a meaningless result; it is only correct in the absence of a magnetic field.

In the presence of a magnetic field we must consider the energy constraint, or put the system in contact with a heat bath, which we take to consist of a large expanse containing an ideal mono-atomic gas. Let the total energy of the spin system and the gas be given by

$$\mathcal{E} = n_+\varepsilon_+ + n_-\varepsilon_- + \mathcal{T}^r = const., \tag{1-6-9}$$

where $\mathcal{T}^r$ is the kinetic energy of the mono-atomic gas, which randomizes the orientation of the spins. Note that the temperature of the heat bath $T^r \propto \mathcal{T}^r$; further $\varepsilon_\pm = \mp \rho_B H$, in which ρ_B is the

Bohrmagneton, the spin was taken to be ± ½ and g was taken to be 2. The distribution can now be written down as

$$W(n_+,n_-;\mathscr{E}) = C\int d\Omega_p^r \frac{N!}{n_+!n_-!}\left(\frac{1}{2}\right)^N \delta\left(\mathscr{E}-\mathscr{T}^r - n_+\varepsilon_+ - n_-\varepsilon_-\right), \tag{1.6-10}$$

where C is a normalization constant and where we integrate over the momentum space of the reservoir with $\mathscr{T}^r = \Sigma_i p_i^2/2m,\ i = 1...3M$. To facilitate this we take the Laplace transform (generating function) with respect to $\mathscr{E}$:

$$\Psi(s) = \int_0^\infty e^{-s\mathscr{E}} W(n_+,n_-;\mathscr{E})d\mathscr{E} = C\int d\Omega_p^r \frac{N!}{n_+!n_-!} 2^{-N} e^{-s(n_+\varepsilon_+ + n_-\varepsilon_-)} e^{-s\mathscr{T}^r(p)}. \tag{1.6-11}$$

The integral over the p's is trivial:

$$\int_{-\infty}^{\infty}...\int dp_1...dp_{3M} \prod_{i=1}^{3M} e^{-sp_i^2/2m} = \left(\frac{2\pi m}{s}\right)^{3M/2}. \tag{1.6-12}$$

The inverse transform is easily found[19]

$$W(n_+,n_-;\mathscr{E}) = C2^{-N}(2\pi m)^{3M/2} \frac{N!}{n_+!n_-!} \frac{(\mathscr{E} - n_+\varepsilon_+ - n_-\varepsilon_-)^{(3M/2)-1}}{\Gamma(3M/2)}. \tag{1.6-13}$$

In the thermodynamic limit, $M \to \infty$, $\mathscr{T}^r \to \infty$. Hence we write, noting also $\mathscr{E} \sim \mathscr{T}^r$

$$\begin{aligned}(\mathscr{E} - n_+\varepsilon_+ - n_-\varepsilon_-)^{(3M/2)-1} &\approx \mathscr{E}^{3M/2} \exp\left[\frac{3M}{2}\ln\left(1 - \frac{n_+\varepsilon_+ + n_-\varepsilon_-}{\mathscr{E}}\right)\right] \\ &\sim \mathscr{E}^{3M/2} \exp\left[-\frac{3M}{2}\frac{n_+\varepsilon_+ + n_-\varepsilon_-}{\langle\mathscr{T}^r\rangle}\right].\end{aligned} \tag{1.6-14}$$

Now from equipartition theory $\langle\mathscr{T}^r\rangle = (3M/2)k_BT^r$. We drop further the super 'r' on the reservoir temperature and we define a new constant

$$C' = C(2\pi m)^{3M/2} 2^{-N} \mathscr{E}^{3M/2}/\Gamma(3M/2).$$

The distribution (1.6-13) becomes,

$$W(n_+,n_-;T) = C'\frac{N!}{n_+!n_-!}\left(e^{-\varepsilon_+/k_BT}\right)^{n_+}\left(e^{-\varepsilon_-/k_BT}\right)^{n_-}. \tag{1.6-15}$$

Normalization for this result yields

$$1/C' = \left(e^{-\varepsilon_+/k_BT} + e^{-\varepsilon_-/k_BT}\right)^{(n_+ + n_-)} \tag{1.6-16}$$

(where $N_+ + N_- = N$.), which is the canonical partition function for this problem. Equation (1.6-15) together with (1.6-16) is again a binomial distribution. For the average spin numbers one finds,

$$\langle n_\pm\rangle = \frac{e^{-\varepsilon_\pm/k_BT}}{e^{-\varepsilon_+/k_BT} + e^{-\varepsilon_-/k_BT}} N, \tag{1.6-17}$$

---

[19] If $\mathscr{L}^{-1}[f(s)] = F(\mathscr{E})$, $\mathscr{L}^{-1}[e^{-bs}f(s)] = F(\mathscr{E}-b)$; further, $\mathscr{L}^{-1}(1/s)^k = \mathscr{E}^{k-1}/\Gamma(k)$.

while the mean magnetization is given by

$$\langle M\rangle = \rho_B \left(\langle n_+\rangle - \langle n_-\rangle\right) = N\rho_B \tanh\left(\rho_B H / k_B T\right). \tag{1.6-18}$$

This is the complete result, which is, of course, more directly obtained from molecular statistics, to be treated in Chapter VI.

This example shows that the application of elementary probability theory usually does not work in statistical mechanics, because of constraints! Therefore, while some knowledge of probability theory is necessary, statistical mechanics uses its own methods, which we will develop from Chapter II onwards.

Another, less involved example is that of random walk. Let

$$x = \sum_{i=1}^{N} s_i \tag{1.6-19}$$

be the path covered by a disoriented person after N steps. The probability for each step is

$$w(s_i) = \begin{cases} \frac{1}{2} & \text{if } s_i = \Delta, \\ \frac{1}{2} & \text{if } s_i = -\Delta. \end{cases} \tag{1.6-20}$$

Since the steps are independent, the probability distribution for x is

$$W(x) = \prod_{i}^{N} w(s_i)\,\delta_{x,\Sigma s_i}. \tag{1.6-21}$$

We get rid of the constraint by using the Fourier transform (characteristic function),

$$\begin{aligned} \Phi(u) &= \int_{-\infty}^{\infty} e^{iux} W(x)dx = \left\langle e^{iux}\right\rangle \\ &= \left\langle e^{iu\Sigma s_i}\right\rangle = \prod_{i=1}^{N}\left\langle e^{ius_i}\right\rangle = \left\langle e^{ius}\right\rangle^N. \end{aligned} \tag{1.6-22}$$

Now,

$$\left\langle e^{ius}\right\rangle = \tfrac{1}{2}e^{iu\Delta} + \tfrac{1}{2}e^{-iu\Delta} = \cos u\Delta. \tag{1-6-23}$$

Hence,

$$\Phi(u) = (\cos u\Delta)^N = 2^{-N}\sum_n \binom{N}{n} e^{iun\Delta} e^{-ium\Delta}, \tag{1.6-24}$$

where $m = N - n$. The inverse transform is found by using the well-known integral,

$$\frac{1}{2\pi}\int_{-\infty}^{\infty} e^{iu(x-x')}du = \delta(x-x'). \tag{1.6.25}$$

We thus obtain,

$$W(x) = 2^{-N} \sum_n \binom{N}{n} \delta[x - (n-m)\Delta]. \qquad (1.6\text{-}26)$$

This indicates that x has discrete values $(n-m)\Delta$, resulting from m steps to the left and n steps to the right. The discrete distribution is clearly

$$W(n,m) = \frac{N!}{n!m!} 2^{-N}, \qquad (1.6\text{-}27)$$

in accord with the binomial distribution law. We can also write down the probability for $K = n - m$ steps to the right,

$$W(K) = \frac{2^{-N} N!}{[(N+K)/2]!\,[(N-K)/2]!}. \qquad (1.6\text{-}28)$$

It is left to the reader to verify that the factorials in this expression always involve integers.

1.6.2 *The Multinomial Distribution Function*

We consider K mutually exclusive events, with *a priori* probabilities $p_1 ... p_K$, with $\Sigma p_i = 1$. Out of N trials, the joint probability for finding N_1 events 1, N_2 events 2, ... N_K events K is then by a similar reasoning as before

$$W(N_1, N_2, ..., N_K) = \frac{N!}{\prod N_j!} \prod_i p_i^{N_i}. \qquad (1.6\text{-}29)$$

The multinomial coefficient $N!/\Pi_j N_j!$ is easily derived from the binomial coefficient. For we can choose N_1 events out of N, then N_2 events out of $N - N_1$, etc. Thus the number of ways is

$$\binom{N}{N_1 N_2 ..} = \frac{N!}{N_1!(N-N_1)!} \frac{(N-N_1)!}{N_2!(N-N_1-N_2)!} .... \frac{(N-N_1-...N_{K-1})!}{N_K!(N-\Sigma N_j)!} = \frac{N!}{N_1!..N_K!}.$$

(1-6-30)

Employing the multinomial expansion theorem,

$$(a_1 + a_2 + ... a_K)^N = \sum_{all\ sets\{N_i\}} \frac{N!}{N_1!...N_K!} a_1^{N_1} a_2^{N_2} ... a_K^{N_K} \qquad (1.6\text{-}31)$$

one easily finds,

$$\langle N_i \rangle = p_i N, \quad \langle \Delta N_i^2 \rangle = p_i(1-p_i)N, \quad \langle \Delta N_i \Delta N_j \rangle = -p_i p_j N. \qquad (1.6\text{-}32)$$

The covariance for the cross-correlation is negative, as expected, since $\Sigma_i N_i = N$.

We may apply this distribution for finding the density correlations of particles, which have an equal *a priori* probability to reside anywhere in a given domain *V*. Dividing this domain in *K* equal cells Δv_i we have $p_i = \Delta v_i / V$. This gives for the variances and covariances,

$$\begin{aligned} \langle \Delta N_i^2 \rangle &= N(\Delta v_i / V)[1-(\Delta v_i / V)], \\ \langle \Delta N_i \Delta N_j \rangle &= -N \Delta v_i \Delta v_j / V^2. \end{aligned} \tag{1.6-33}$$

We define the density by $N_i = n(\mathbf{r}_i)\Delta v_i$. Then for the density fluctuations we have,

$$\begin{aligned} \langle \Delta n(\mathbf{r}_i)^2 \rangle (\Delta v_i)^2 &= N(\Delta v_i / V)[1-(\Delta v_i / V)], \\ \langle \Delta n(\mathbf{r}_i) \Delta n(\mathbf{r}_j) \rangle \Delta v_i \Delta v_j &= -N \Delta v_i \Delta v_j / V^2. \end{aligned} \tag{1.6-34}$$

We let $K \to \infty$ and $\Delta v_i \to 0$; both equations are then summarised by the form for the density-density covariance,

$$\langle \Delta n(\mathbf{r}) \Delta n(\mathbf{r}') \rangle = \langle n(\mathbf{r}) \rangle [\delta(\mathbf{r}-\mathbf{r}') - \frac{1}{V}]. \tag{1.6-35}$$

The last term stems from the constraint that $N = \langle n(\mathbf{r}) \rangle V$ is fixed. In case all particles have the same energy ε the constraint $\mathscr{E} = \Sigma N_i \varepsilon_i$ reduces to $N = \Sigma N_i$; Eq. (1.6-35) is not only good statistics, but also good statistical mechanics since the temperature clearly does not enter in. In a similar fashion one may establish for the triple correlations,

$$\begin{aligned} \langle \Delta n(\mathbf{r}_1) \Delta n(\mathbf{r}_2) \Delta n(\mathbf{r}_3) \rangle &= \langle n(\mathbf{r}) \rangle \\ &\times \left[\delta(\mathbf{r}_1 - \mathbf{r}_2)\, \delta(\mathbf{r}_1 - \mathbf{r}_3) - \frac{1}{V} \sum_{i,j\ i \neq j}^{3} \delta(\mathbf{r}_i - \mathbf{r}_j) + \frac{2}{V^2} \right]. \end{aligned} \tag{1.6-36}$$

1.6.3 *The Poisson Distribution, the Gauss Distribution and Normal Distribution*

If in the binomial distribution $N \to \infty$, $p \to 0$ and $pN = \langle n \rangle$ remains finite, we arrive at the Poisson distribution,

$$W(n) = e^{-\langle n \rangle} \langle n \rangle^n / n! \tag{1.6-37}$$

The road is easy; we have

$$(1-p)^N = \left(1 - \frac{pN}{N}\right)^N = \left(1 - \frac{\langle n \rangle}{N}\right)^N \sim e^{-\langle n \rangle}. \tag{1.6-38}$$

Further,

$$\frac{N!\,p^n (1-p)^{-n}}{(N-n)!} \to \frac{N!\,p^n}{(N-n)!} = p^n\, N(N-1)...(N-n+1) \sim p^n N^n = \langle n \rangle^n. \tag{1-6-39}$$

These two steps establish the result (1.6-37). Clearly $\Sigma W(n)=1$, so the distribution is normalized. The second factorial moment is

$$\langle n(n-1)\rangle = e^{-\langle n\rangle}\sum_{n=0}^{\infty}\frac{\langle n\rangle^n n(n-1)}{n!}$$

$$= e^{-\langle n\rangle}\sum_{n=2}^{\infty}\frac{\langle n\rangle^n}{(n-2)!} = e^{-\langle n\rangle}\langle n\rangle^2\sum_{m=0}^{\infty}\frac{\langle n\rangle^m}{m!} = \langle n\rangle^2. \tag{1.6-40}$$

For the variance we obtain,

$$\langle \Delta n^2\rangle = \langle n(n-1)\rangle - \langle n\rangle[\langle n\rangle - 1] = \langle n\rangle. \tag{1.6-41}$$

In many problems n is also large and the distribution can be further approximated by using Stirling's formula,

$$\begin{aligned} n! &\sim \sqrt{2\pi}\, e^{-n} n^{(n+\frac{1}{2})}\left[1+\frac{1}{12n}+\frac{1}{288n^2}+\ldots\right], \\ \ln n! &\sim (n+\tfrac{1}{2})\ln n - n + \tfrac{1}{2}\ln 2\pi + \ldots \sim n\ln n - n. \end{aligned} \tag{1.6-42}$$

The approximations are most easily carried out in ln $W(n)$; setting $D \equiv n - \langle n\rangle$,

$$\begin{aligned} \ln W(n) &\sim -\langle n\rangle + n\ln\langle n\rangle - (n+\tfrac{1}{2})\ln n + n - \tfrac{1}{2}\ln 2\pi \\ &= -[\langle n\rangle + D + \tfrac{1}{2}]\ln[1+(D/n)] + D - \tfrac{1}{2}\ln(2\pi\langle n\rangle) \\ &= -[\langle n\rangle + D + \tfrac{1}{2}][(D/\langle n\rangle) - (D^2/2\langle n\rangle^2) + ..] + D - \tfrac{1}{2}\ln(2\pi\langle n\rangle) \\ &= -\frac{D^2}{2\langle n\rangle} + \frac{D^3}{2\langle n\rangle^2} - \frac{D}{2\langle n\rangle} + \frac{D^2}{4\langle n\rangle^2} + \ldots - \tfrac{1}{2}\ln(2\pi\langle n\rangle). \end{aligned} \tag{1.6-43}$$

We assume that $|D| >> 1$, in addition to $|D/\langle n\rangle| << 1$; so we retain only the first and the last terms, the result being,

$$W(n) = \frac{1}{\sqrt{2\pi\langle n\rangle}} e^{-(n-\langle n\rangle)^2/2\langle n\rangle}. \tag{1.6-44}$$

Since $n - \langle n\rangle$ is a large number, we may replace it by the continuous variable x [interval $(-\infty, \infty)$]; we then arrive at the Gauss distribution (Gaussian probability density function):

$$W(x) = \frac{1}{\sqrt{2\pi\langle x\rangle}} e^{-(x-\langle x\rangle)^2/2\langle x\rangle}. \tag{1.6-45}$$

For the variance we have clearly,

$$\langle \Delta x^2\rangle = \langle x\rangle. \tag{1.6-46}$$

For various problems the distribution is bell-shaped, but its width is varying, i.e., the shape is as in (1.6-45), but the variance $\langle \Delta x^2 \rangle \equiv \sigma_x^2$ is not given by (1.6-46). These distributions are called "normal" and have the general form[20]

$$W(x) = \frac{1}{\sqrt{2\pi\sigma_x{}^2}} e^{-(x-\langle x \rangle)^2 / 2\sigma_x{}^2}. \tag{1.6-47}$$

One easily sees that $W(x)$ is normalized, while the results for $\langle x \rangle$ and $\langle \Delta x^2 \rangle$ are consistent.

To find higher order moments, we use the following tricks. First consider

$$\int_{-\infty}^{\infty} e^{-y^2} dy = \sqrt{\pi}. \tag{1-6.48}$$

Substituting $y = \alpha x^2$, we have

$$\int_{-\infty}^{\infty} e^{-\alpha x^2} dx = \sqrt{\frac{\pi}{\alpha}} \tag{1.6-49}$$

Differentiating repeatedly to α one obtains

$$\int_{-\infty}^{\infty} x^2 e^{-\alpha x^2} dx = \frac{1}{2}\sqrt{\frac{\pi}{\alpha^3}}, \tag{1.6-50}$$

$$\int_{-\infty}^{\infty} x^{2k} e^{-\alpha x^2} dx = \frac{1.3.5...(2k-1)}{2^k}\sqrt{\frac{\pi}{\alpha^{2k+1}}} = (2k-1)!!\ 2^{-k} \pi^{1/2} \alpha^{-k-1/2}. \tag{1.6-51}$$

The integrals can also be expressed in the gamma-function. If the interval is changed to (0,∞) the results are to be multiplied by (1/2).

For odd integrals we start from

$$\int_0^{\infty} e^{-y} dy = 1. \tag{1.6-52}$$

Substituting $y = \alpha x^2$, we have

$$\int_0^{\infty} x e^{-\alpha x^2} dx = 1/2\alpha. \tag{1.6-53}$$

Repeated differentiation to α yields,

$$\int_0^{\infty} x^{2k+1} e^{-\alpha x^2} dx = k!/2\alpha^{k+1}. \tag{1.6-54}$$

Needless to say that these integrals will be encountered over and over.

---

[20] The names Gaussian distribution and normal distribution are often interchanged in the physical literature.

1.6.4 *Multivariate Distributions, the Maxwell-Boltzmann Distribution and the Virial Theorem*

When we have a set of variables $x_1 ... x_K$, that are independent, their joint distribution is simply the product distribution $\Pi\, W(x_i)$. For the case that the variables are not independent, the general multivariate normal distribution has in the exponent a polynomial of the second degree:

$$W(x_1,...x_K) = Ae^{-\sum_{ij}\alpha_{ij}x_i x_j/2} = Ae^{-\tilde{\mathbf{x}}\alpha\mathbf{x}/2}, \tag{1.6-55}$$

where A is a normalization constant and α is a symmetrical matrix; in the second form the x's have been arranged in a column matrix, with the tilde denoting the transpose (row matrix). The moments can be found by diagonalizing the matrix α, or faster by a method going back to Onsager.[21] To that purpose we consider 'conjugate' variables, defined as:

$$z_i = -\partial \ln W/\partial x_i = \sum_j \alpha_{ij} x_j. \tag{1.6-56}$$

The inversion is,

$$x_j = \sum_k (\alpha^{-1})_{jk} z_k. \tag{1.6-57}$$

Hence we have,

$$\langle x_i z_k \rangle = -\int x_i \frac{\partial \ln W}{\partial x_k} W d\mathbf{x} = -\int x_i \frac{\partial W}{\partial x_k} d\mathbf{x}. \tag{1.6-58}$$

The integration over dx_k is singled out and we integrate by parts to obtain,

$$\langle x_i z_k \rangle = -\int ... \int dx_1 ... dx_{k-1} dx_{k+1} ... dx_K \int \frac{\partial W}{\partial x_k} x_i dx_k = \delta_{ik} \int W d\mathbf{x} = \delta_{ik}. \tag{1.6-59}$$

Using now (1.6-57), we arrive at the simple result,

$$\langle x_i x_j \rangle = \sum_k (\alpha^{-1})_{jk} \langle x_i z_k \rangle = (\alpha^{-1})_{ji} = (\alpha^{-1})_{ij}. \tag{1.6-60}$$

Thus, to find the covariances it suffices to invert the matrix α.

Two more results about the conjugate variables z_i are in order. First, we consider $\langle z_i z_j \rangle$. Using (1.6-56) and (1.6-59) we have

$$\langle z_i z_j \rangle = \sum_k \alpha_{jk} \langle z_i x_k \rangle = \alpha_{ij}. \tag{1.6-61}$$

For the distribution $W(z_1 ... z_K)$ we now find,

[21] L.S. Onsager, Phys. Rev. **37**, 405 (1937); ibid. **38**, 265 (1938).

$$W(z_1 \ldots z_K) = W(x_1 \ldots x_K) \left| \frac{\partial(x_1 \ldots x_K)}{\partial(z_1 \ldots z_K)} \right| . \tag{1.6-62}$$

The Jacobian is just the determinant of $(\alpha)^{-1}$. Employing (1.6-57) to rewrite the polynomial in the exponent, we directly obtain,

$$W(z_1 \ldots z_K) = A|\alpha|^{-1} e^{-\sum_{ij} (\alpha)^{-1}_{ij} z_i z_j / 2} . \tag{1.6-63}$$

The reason that a normal distribution so often occurs lies in the *central limit theorem*, which states:

For a random variable x(t) composed of N random contributions, the distribution function W(x) approaches the normal distribution if N→∞, providing none of the random contributions shall be compatible with the resultant of all others.

We will give an example of the central limit theorem. Consider a current through a conductor, connected to a capacitor, see Fig. 1-3. The current density is (V_0 being the volume of the conductor),

$$J(t) = (q/V_0) \sum_{i=1}^{N} v_i(t); \tag{1.6-64}$$

the carrier velocities are random variables, fluctuating due to electron-phonon or other collisions.

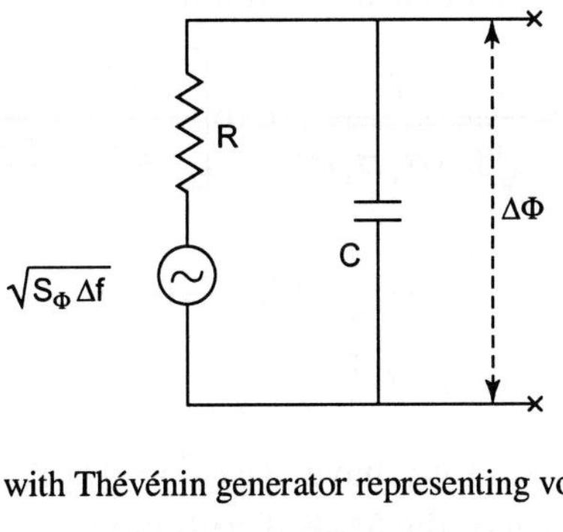

Fig. 1-3. Passive *RC* circuit with Thévénin generator representing voltage fluctuations across *R*.

Since *J* is made up of many events, the pdf for *J* will be a normal distribution, as is the distribution for the capacitor charge, *W(Q)*. In Chapter III we shall prove the Boltzmann-Einstein result

$$W(Q) = c\,e^{(S-S_0)/k_B} = c\,e^{\frac{1}{2} S''(Q) Q^2 / k_B} , \tag{1.6-65}$$

where *S(Q)* is the entropy function and S_0 is the equilibrium entropy; $S''(Q) < 0$ is the second derivative about the equilibrium state. The central limit theorem assures us that no higher derivatives need to be retained in the limit $N\to\infty$, since (1.6-65) is a normal distribution. With $TdS = -\Phi dQ$ we have

$$d^2S/dQ^2 = -d(\Phi/T)/dQ = -1/CT\,. \tag{1.6-66}$$

Comparing with (1.6-47), we find

$$\sigma_Q^2 = \langle \Delta Q^2\rangle = k_B TC \text{ and } \sigma_\Phi^2 = \langle \Delta\Phi^2\rangle = k_B T/C\,. \tag{1.6-67}$$

The voltage noise can be represented by a voltage generator per unit bandwidth, S_Φ, in series with the resistor R. For frequencies well below the inverse collision time S_Φ is independent of frequency. Therefore,

$$\langle\Delta\Phi^2\rangle = \frac{S_\Phi}{R^2}\int_0^\infty \left(\frac{d\omega}{2\pi}\right)\frac{R^2}{1+\omega^2C^2R^2} = \frac{S_\Phi}{4CR}\,. \tag{1.6-68}$$

Comparing (1.6-67) and (1.6-68) we established[22]

$$S_\Phi = 4k_B TR\,, \tag{1.6-69}$$

which is Nyquist's famous formula for thermal noise (without quantum-correction factor).

As an example of a multivariate normal distribution we mention the Maxwell-Boltzmann distribution. Using Cartesian velocity components, and using the standard practice whereby molecular distributions are normalized to N, we can write

$$n(v_x,v_y,v_z) = \frac{N}{\sqrt{8\pi^3\sigma_x^2\sigma_y^2\sigma_z^2}}\exp\left[-\frac{v_x^2}{2\sigma_x^2}-\frac{v_y^2}{2\sigma_y^2}-\frac{v_z^2}{2\sigma_z^2}\right]. \tag{1.6-70}$$

Clearly,

$$\begin{gathered}\langle v_x\rangle = \langle v_y\rangle = \langle v_z\rangle = 0\,,\\ \langle v_x^2\rangle = \langle v_y^2\rangle = \langle v_z^2\rangle = \tfrac{1}{3}\langle v^2\rangle\,.\end{gathered} \tag{1.6-71}$$

From the equipartition theorem we have $\frac{1}{2}m\langle v^2\rangle = \frac{3}{2}k_B T$, so that $\sigma_x^2 = k_B T/m$, etc. We thus obtain the usual form of the M–B distribution,

$$n(v_x,v_y,v_z) = N[m/2\pi k_B T]^{3/2}e^{-mv^2/2k_BT}\,. \tag{1.6-72}$$

Going to polar coordinates in velocity space, we have

$$n(v,\vartheta,\varphi) = n(v_x,v_y,v_z)\frac{\partial(v_x,v_y,v_z)}{\partial(v,\vartheta,\varphi)} = n(v_x,v_y,v_z)v^2\sin\vartheta. \tag{1.6-73}$$

---

[22] H. Nyquist, Phys. Rev. **32**, 110 (1928).

Integration over the polar angles yields the Maxwell distribution for the velocities:

$$n(v) = 4\pi N[m/2\pi k_B T]^{3/2} v^2 e^{-mv^2/2k_B T}. \tag{1.6-74}$$

The mean velocity is $\langle v \rangle = \sqrt{8k_B T/\pi m}$; the rms velocity is $v_{rms} = \sqrt{3k_B T/m}$. For the fourth moment we find – using the previously stated Gaussian integrals – $\langle v^4 \rangle = 15(k_B T/m)^2$. To find the energy fluctuations we note that the molecules fluctuate independently, so that their variances add. Therefore,

$$\langle \Delta\mathcal{E}^2 \rangle = N\langle \Delta\varepsilon^2 \rangle = \tfrac{1}{4} N m^2 [\langle v^4 \rangle - \langle v^2 \rangle^2]. \tag{1.6-75}$$

From the above values for the second and fourth moments one easily arrives at

$$\langle \Delta\mathcal{E}^2 \rangle = \tfrac{3}{2} k_B N\, k_B T^2 = C_V k_B T^2, \tag{1.6-76}$$

where C_V is the molar specific heat when N (= N_A) and V are fixed. These are all very elementary results, which require only the application of the central limit theorem and the energy equipartition rule.

Virial theorem

We consider the quantum mechanical Hermitean operator,[23]

$$F = \tfrac{1}{2} \sum_{i=1}^{3N} (p_i q_i + q_i p_i) \tag{1.6-77}$$

where the sum is over all translational coordinates. The commutator with the Hamiltonian (if $r > 3$, there are internal degrees of freedom)

$$\mathcal{H} = \sum_{i=1}^{3N} (p_i^2/2m_i) + \mathcal{V}(q_1 \ldots q_{rN}), \tag{1.6-78}$$

is easily worked out; one obtains

$$[F, \mathcal{H}] = \hbar i \sum_{i=1}^{3N} \left[(p_i^2/m_i) - q_i \partial\mathcal{V}(q_1 \ldots q_{rN})/\partial q_i \right]. \tag{1.6-79}$$

With the Heisenberg equation of motion this gives,

$$\frac{dF}{dt} = \sum_{i=1}^{3N} \left[(p_i^2/m_i) - q_i \partial\mathcal{V}/\partial q_i \right]. \tag{1.6-80}$$

We now take a time average over a sufficiently long time. Then, if F is bounded $d\tilde{F}/dt = 0$. Taking in addition a quantum mechanical average, we have

[23] G.H. Wannier, "Statistical Mechanics", Wiley, NY, 1966, p. 43ff.

$$\left\langle \sum_{i=1}^{3N} p_i^2 / m_i \right\rangle = \left\langle \sum_{i=1}^{3N} q_i \partial \mathcal{V} / \partial q_i \right\rangle. \tag{1.6-81}$$

Assuming all molecules to be identical, this amounts to the following identity for the molecular average of the lhs, denoted by an overhead bar:

$$(3N/m)\overline{p_i^2} = \left\langle \sum_i \mathbf{r_i} \cdot \nabla_{\mathbf{r_i}} \mathcal{V} \right\rangle \equiv -W. \tag{1.6-82}$$

Clausius called the rhs the *virial*; it is the summed scalar product of the position and the total force on each particle. The virial can be split into an external and internal virial, since

$$\mathcal{V} = \sum_i \mathcal{V}_{ext}(\mathbf{r}_i) + \mathcal{V}_{int}(q_1 \ldots q_{rN}). \tag{1.6-83}$$

accordingly we have $W = W_{ext} + W_{int}$. The external virial is easily computed. We rely on Newton's third law and consider the force exerted by all molecules on the walls of the container, rather than the forces on the particles. Using Gauss' theorem,

$$W_{ext} = -P \oiint \mathbf{r} \cdot d\mathbf{S} = -P \int (\nabla \cdot \mathbf{r}) d^3 r = -3PV. \tag{1.6-84}$$

Let at first there be no internal virial. Then we have from (1.6-82) and (1.6-84), using once more equipartition, $\overline{p_i^2}/2m = \frac{1}{2} k_B T$,

$$PV = Nk_B T, \tag{1.6-85}$$

which is the perfect gas law. More generally, if $W_{int} \neq 0$, we have

$$PV = Nk_B T + \tfrac{1}{3} W_{int}. \tag{1.6-86}$$

This is the extension of the perfect gas law. In principle the internal virial allows a development of the form

$$PV = Nk_B T \left(1 + B\frac{N}{V} + C\frac{N^2}{V^2} + \ldots \right), \tag{1.6-87}$$

where B, C,... are functions of T, which are called virial coefficients. Dividing by V and employing the thermodynamic limit with $V/N \to \hat{v}$ (the specific volume or the molecular volume) we also write the virial expansion as,

$$P = \frac{k_B T}{\hat{v}} \left(1 + \frac{B}{\hat{v}} + \frac{C}{\hat{v}^2} + \ldots \right). \tag{1.6-88}$$

This form will be encountered later in connection with condensation and van der Waals' law.

1.7 Transforms of Probability Functions

1.7.1 *Characteristic Functions*

The characteristic function is defined as follows:

$$\Phi(u) \equiv \left\langle e^{iux} \right\rangle = \int e^{iux} dP(x), \quad (-\infty < x < \infty) \tag{1.7-1}$$

where the latter expression is a Lebesgue-Stieltjes integral and $P(x)$ is the cumulative distribution function (cdf), being the probability that $x \le x_0$:

$$P(x_0) = prob(x \le x_0) . \tag{1.7-2}$$

For a continuous variable we have the connection with the pdf

$$P(x_0) = \int_{-\infty}^{x_0} W(x)dx; \tag{1.7-3}$$

Then $dP(x) = W(x)\,dx$ and the definition $\Phi(u) = \langle \exp iux \rangle$ is corroborated. For a discrete variable, for which $x = \{x_k\}$ – where these discrete values are not necessarily integers – we have

$$P(x_0) = \sum_{x_k \le x_0} W(x_k) = \sum_{x_k} W(x_k)\Theta(x_0 - x_k + 0) , \tag{1.7-4}$$

where $\Theta(x)$ is the Heaviside function, $\Theta(x<0) = 0$, $\Theta(x>0) = 1$. If we resort to the use of the delta function, then $dP(x) = \Sigma_k W(x_k)\delta(x - x_k)dx$ and

$$\begin{aligned} \int e^{iux} dP(x) &= \sum_k \int_{-\infty}^{\infty} e^{iux} W(x_k)\delta(x - x_k)dx \\ &= \sum_k e^{iux_k} W(x_k) = \left\langle e^{iux} \right\rangle, \end{aligned} \tag{1.7-5}$$

as required by the definition. The Lebesgue-Stieltjes integral therefore provides a general expression, valid for both continuous and discrete distributions.

Suppose now that we expand the exponential of the middle member of (1.7-1) into a MacLaurin series,

$$\Phi(u) = 1 + \frac{i\langle x \rangle u}{1!} + \frac{i^2 \langle x^2 \rangle u^2}{2!} + ... \tag{1.7-6}$$

This is to be compared with the standard form,

$$\Phi(u) = \Phi(0) + \frac{\Phi'(0)u}{1!} + \frac{\Phi''(0)u^2}{2!} + ... \frac{\Phi^{(n)}u^n}{n!} \tag{1.7-7}$$

Clearly, we have the moment rule,

$$m_n \equiv \langle x^n \rangle = (-i)^n \left. \frac{d^n \Phi(u)}{du^n} \right|_{u=0} . \tag{1.7-8}$$

So, once $\Phi(u)$ is known, all moments are known. Conversely, if all moments are known, $\Phi(u)$ is known and by Fourier inversion $W(x)$ is known.

We give an example. Consider a normal distribution with zero mean, then

$$\Phi(u) = \frac{1}{\sqrt{2\pi\sigma^2}} \int_{-\infty}^{\infty} e^{iux}\, e^{-x^2/2\sigma^2} dx . \tag{1.7-9}$$

The integral is computed by completing the square,

$$iux - x^2/2\sigma^2 = -\underbrace{[x/\sqrt{2\sigma^2} - iu\sqrt{\sigma^2/2}]}_{call\ z}{}^2 - u^2\sigma^2/2 . \tag{1.7-10}$$

The square bracket results in

$$\frac{1}{\sqrt{\pi}} \int_{-\infty - iu\sqrt{\sigma^2/2}}^{\infty - iu\sqrt{\sigma^2/2}} e^{-z^2} dz = \frac{1}{\sqrt{\pi}} \int_{-\infty}^{\infty} e^{-z^2} dz = 1 , \tag{1.7-11}$$

as is found by Cauchy's contour-integral theorem, see Fig. 1-4.

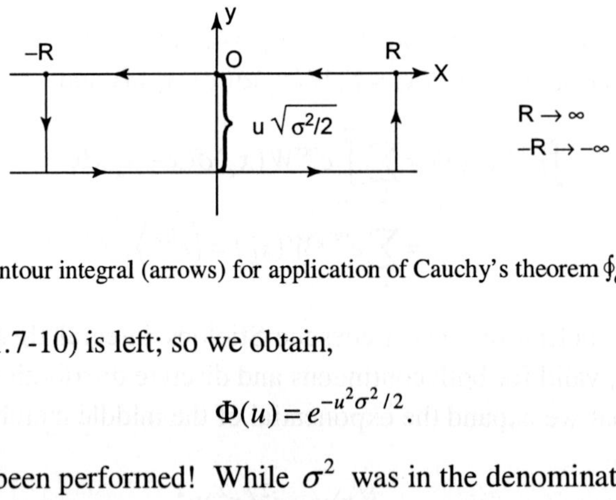

Fig. 1-4. Contour integral (arrows) for application of Cauchy's theorem $\oint_C f(z) dz = 0$.

The last part in (1.7-10) is left; so we obtain,

$$\Phi(u) = e^{-u^2\sigma^2/2} . \tag{1.7-12}$$

The miracle has been performed! While σ^2 was in the denominator of the exponent in (1.7-9), it is in the numerator in (1.7-12). Thus $\Phi(u)$ is again a normal distribution, but with the reciprocal width. If we define $(\Delta x)_{rms} = \sqrt{\overline{\Delta x^2}}$, $(\Delta u)_{rms} = \sqrt{\overline{\Delta u^2}}$, then

$$(\Delta x)_{rms} (\Delta u)_{rms} = 1. \tag{1.7-13}$$

This is of course well-known in quantum mechanics from the Gaussian wave packets $W(x)$ in coordinate-space and $\Phi(p)$ in momentum-space; it leads to Heisenberg's uncertainty principle.

Finally, we compute the moments of the normal distribution from the rule (1.7-8). We easily find $\langle x^2\rangle = \sigma^2$ as expected and

$$\begin{aligned}\langle x^{2n+1}\rangle &= 0,\\ \langle x^{2n}\rangle &= (2n-1)!!\,\sigma^{2n}.\end{aligned} \tag{1.7-14}$$

We also consider multivariate characteristic functions. The definition is, analogous to (1.7-1)

$$\Phi(u_1...u_K) = \left\langle e^{i\sum_{k=1}^K u_k x_k}\right\rangle = \int ... \int e^{i\sum_{k=1}^K u_k x_k} d^K P(x_1...x_K). \tag{1.7-15}$$

For continuous variables the rhs reduces to $W(x_1...)dx_1..dx_K$ and (1.7-15) is an ordinary Fourier integral, with inversion

$$W(x_1...x_K) = \left(\frac{1}{2\pi}\right)^K \int ... \int e^{-i\sum_{k=1}^K u_k x_k}\Phi(u_1...u_K)\prod_k du_k. \tag{1.7-16}$$

The moment rule follows from series expansion of $\Phi(u_1...u_K)$:[24]

$$\Phi(\{u_k\}) = \sum_{\{n_k\}} (i)^{\Sigma_k n_k} \prod_{k=1}^K \frac{u_k^{n_k}}{n_k!}\left\langle x_1^{n_1} x_2^{n_2} ... x_K^{n_K}\right\rangle \tag{1.7-17}$$

(where $\sum_{\{n_k\}}$ means $\sum_{n_1}...\sum_{n_K}$, subject to $\Sigma_i n_i = N$ for the term u^N), which is to be compared with

$$\Phi(\{u_k\}) = \sum_{\{n_k\}} \left.\frac{\partial^{\Sigma n_k}\Phi}{\partial u_1^{n_1}...\partial u_K^{n_K}}\right|_{\{u_k\}=0} \prod_{k=1}^K \frac{u_k^{n_k}}{n_k!}. \tag{1.7-20}$$

We thus find,

$$\left\langle x_1^{n_1} x_2^{n_2} ... x_K^{n_K}\right\rangle = (-i)^{\Sigma n_k} \left.\frac{\partial^{\Sigma n_k}\Phi}{\partial u_1^{n_1}...\partial u_K^{n_K}}\right|_{\{u_k\}=0}. \tag{1.7-21}$$

For a multivariate normal distribution the characteristic function is most easily found by diagonalizing the matrix α. One easily finds,

---

[24] Note that if (1.7-15) had been written as $\Phi = \Pi_k \langle e^{iu_k x_k}\rangle$, then the series had been

$$\Phi(\{u_k\}) = \prod_k \sum_{n_k} \frac{(iu_k)^{n_k}}{n_k!}\langle x_1^{n_1} ... x_K^{n_K}\rangle. \tag{1.7-18}$$

These results show the validity of the often used 'series-product interchangeability rule' :

$$\sum_{\{n_k\}} \prod_k ... = \prod_k \sum_{n_k} ... \tag{1.7-19}$$

Note that the left-hand-side involves a set of sums, whereas on the right-hand-side we have a single sum.

$$\Phi(u_1...u_N) = \exp[-\sum_{ij}(\alpha^{-1})_{ij}u_iu_j/2]. \tag{1.7-22}$$

Comparison with (1.6-63) indicates that this is the same distribution as that for the conjugate variables $z_1...z_N$. The variances and covariances are therefore as indicated in Section 1.6.4.

Lastly, we shall use the characteristic function to prove the central limit theorem. Let again $x = \sum_{j=1}^{N} s_i$ where the s_i are random contributions. We consider the new variable

$$y = \frac{x - \langle x \rangle}{N}, \quad \sigma^2 = \text{var } y. \tag{1.7-23}$$

Since the s_i are uncorrelated we have

$$\sigma^2 = \frac{1}{N^2}\sum_i \text{var } s_i = \frac{1}{N}\sigma_0^2. \tag{1.7-24}$$

We have hereby assumed, in accord with the enunciated statement of the theorem, that all contributions are of the same order, so that an average variance σ_0^2 exists and does not grow with N, but has an upper bound. From these results,

$$\frac{y}{\sigma} = \frac{x - \langle x \rangle}{\sigma N} = \frac{1}{\sigma_0\sqrt{N}}\sum_{i=1}^{N}(s_i - \langle s_i \rangle). \tag{1.7-25}$$

Let $\Phi(u)$ be the characteristic function of y and let $\varphi_i(u)$ be the characteristic function of $s_i - \langle s_i \rangle$. Then,

$$\Phi(\frac{u}{\sigma}) = \left\langle \exp(i\frac{uy}{\sigma}) \right\rangle = \prod_i \left\langle \exp[i\frac{u(s_i - \langle s_i \rangle)}{\sigma_0\sqrt{N}} \right\rangle = \prod_i \varphi_i\left(\frac{u}{\sigma_0\sqrt{N}}\right). \tag{1.7-26}$$

Since N is large, the argument of φ is small. Expanding to second order,

$$\varphi_i(u') = 1 - \tfrac{1}{2}\sigma_i^2 u'^2 + \mathcal{O}(u'^3), \tag{1.7-27}$$

the rhs of (1.7-26) becomes

$$\prod_i \varphi_i\left(\frac{u}{\sigma_0\sqrt{N}}\right) = \prod_{i=1}^{N}\left(1 - \frac{\sigma_i^2 u^2}{2\sigma_0^2 N}\right) \sim e^{-u^2/2}, \tag{1.7-28}$$

by Weierstrasz' product theorem. Whence for y, setting $v = u/\sigma$,

$$W(y) = \frac{1}{2\pi}\int_{-\infty}^{\infty} e^{-iyv} e^{-\sigma^2 v^2/2} dv = \frac{1}{\sqrt{2\pi}\sigma} e^{-y^2/2\sigma^2}. \tag{1.7-29}$$

Finally, going back to x we have,

$$W(x) = \frac{1}{N\sqrt{2\pi\sigma}} e^{-(x-\langle x\rangle)^2/2N^2\sigma^2}. \tag{1.7-30}$$

This is a normal distribution with variance $\operatorname{var} x = N^2\sigma^2 = N\sigma_0^2 = \Sigma_i \operatorname{var} s_i$, which completes the proof.

1.7.2 *The Generating Function According to Laplace*

Several types of generating functions are in use for statistics of variables on the interval $(0,\infty)$. First of all, the Laplace transform is employed, see already our example of the spin system in section 1.6.1. Thus we write

$$\Psi(s) = \langle e^{-sx} \rangle = \int_0^\infty e^{-sx} dP(x). \tag{1.7-31}$$

For a continuous variable $dP(x) = W(x)dx$ and the consistency of the middle member and the right-side member in (1.7-31) is clear. The inversion then is,

$$W(x) = \frac{1}{2\pi i} \int_{\gamma-i\infty}^{\gamma+i\infty} e^{sx}\, \Psi(s) ds\,, \tag{1.7-32}$$

where γ is a number such that $\Psi(s)$ is analytic for $\operatorname{Re} s \geq \gamma$. We may also choose a contour encircling all poles counter-clockwise (ccw), providing $\Psi(s) = O(s^{-k})$, $k > 1$. Some authors set $s = iw$, where $w = \omega + i\mu$ is a complex frequency. Then

$$W(x) = \frac{1}{2\pi} \oint_C e^{iwx} \Psi(iw) dw. \tag{1.7-33}$$

Both contours are indicated in Fig. 1-5.

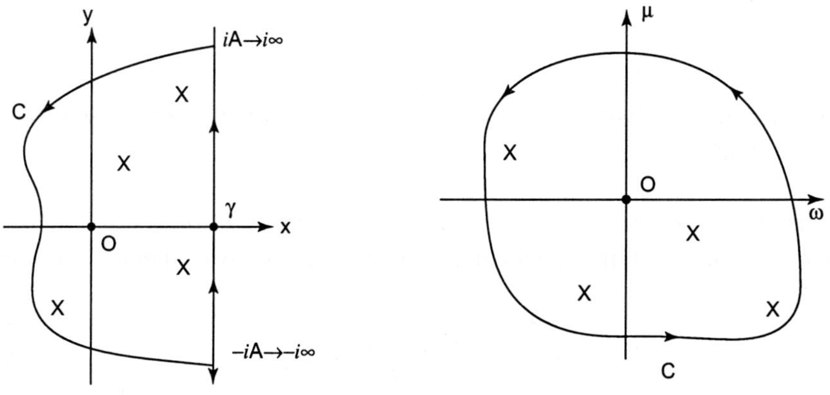

Fig.1-5. (a) Contour for Eq. (1.7-32) with poles X; (b) Contour for Eq. (1.7-33) with poles X.

For a discrete variable $dP(x)=\Sigma_k W(x_k)\delta(x-x_k)$ and the inversion rules read

$$\Sigma_k W(x-x_k)\delta(x-x_k)=\begin{cases}\frac{1}{2\pi i}\int_{\gamma-i\omega}^{\gamma+i\infty} e^{sx}\,\Psi(s)ds,\\ \frac{1}{2\pi}\oint_C e^{iwx}\Psi(iw)dw.\end{cases} \tag{1.7-34}$$

From the 'amplitudes' the discrete distribution $W(x_k)$ can be read off. However, usually the Fowler transform is better suited for discrete variables, see next subsection.

The moments follow from series expansion. From the definition (1.7-31)

$$\Psi(s)=\left\langle e^{-xs}\right\rangle=1-s\langle x\rangle+\frac{s^2}{2!}\langle x^2\rangle+\ldots \tag{1.7-35}$$

so that

$$\langle x^m\rangle=(-1)^m\left.\frac{d^m\Psi(s)}{ds^m}\right|_{s=0}. \tag{1.7-35'}$$

Next we consider multivariate transforms

$$\Psi(s_1\ldots s_K)=\left\langle e^{-\Sigma s_k x_k}\right\rangle=\int\ldots\int e^{-\Sigma s_k x_k}\,dP(x_1\ldots x_K). \tag{1.7-36}$$

For continuous variables we have $dP(x_1\ldots x_K)=W(x_1\ldots x_K)dx_1\ldots dx_K$ and the inversion is direct,

$$W(x_1\ldots x_K)=(\tfrac{1}{2\pi i})^K\oint\ldots\oint e^{\Sigma_k s_k x_k}\Psi(s_1\ldots s_K)\prod_i ds_i, \tag{1.7-37}$$

where the contours encircle all poles. The moment rule is similar to that of the characteristic function,

$$\left\langle x_1^{n_1}x_2^{n_2}\ldots x_K^{n_K}\right\rangle=(-1)^{\Sigma n_k}\left.\frac{\partial^{\Sigma n_k}\Psi}{\partial s_1^{n_1}\ldots\partial s_K^{n_K}}\right|_{\{s_k\}=0}. \tag{1.7-38}$$

1.7.3 *The Factorial Moment Generating Function, Fowler Transform and Cumulants*

For discrete distributions another type of generating function is often used, already known to Laplace, but in statistical mechanics its use was mainly introduced by Fowler[25]; we shall denote it by $\chi(z)$:

$$\chi(z)=\langle z^x\rangle=\sum_k z^{x_k}W(x_k), \tag{1.7-39}$$

[25] R.H. Fowler, "Statistical Mechanics", second Ed., Cambridge University Press, 1936, p.33ff.

see below.[26] We will call it the "generating function according to Fowler", or simply the Fowler transform. The factorial moments are generated by differentiation:

$$\langle [x_m] \rangle \equiv \langle x(x-1)...(x-m+1) \rangle = \left. \frac{\partial^m \chi(z)}{dz^m} \right|_{z=1} . \tag{1.7-41}$$

The inversion of $\chi(z)$ is a bit tricky, except when the$\{x_k\}$are integers. Then, for $x_k = n$ $n = 1,2,3...$, Eq. (1.7-39) is a power series; hence,

$$W(n) = \frac{1}{n!} \left. \frac{d^n \chi(z)}{dz^n} \right|_{z=0} = \frac{1}{2\pi i} \oint_C \frac{\chi(\zeta)}{\zeta^{m+1}} d\zeta \tag{1.7-42}$$

where C encircles the origin ccw . When the x_k are not integers, we choose a unit x_0, which makes x_k/x_0 dimensionless (otherwise z to the power x_k makes no sense); the unit x_0 should be small enough so that the x_k/x_0 values are close to integers, but large enough so that the x_k/x_0 values have no divisor in common. We will then regard (1.7-39) as an ordinary MacLaurin series, with inversion,

$$W(x_k) = \frac{1}{x_k!} \left. \frac{d^{x_k} \chi(z)}{dz^{x_k}} \right|_{z=0} = \frac{1}{2\pi i} \oint_C \frac{\chi(\zeta)}{\zeta^{x_k+1}} d\zeta . \tag{1.7-43}$$

The extension to multiple Fowler transforms – unfortunately also occurring is statistical mechanics – is straightforward. Thus let

$$\chi(z_1...z_K) = \langle z_1^{x_1} ... z_K^{x_K} \rangle = \sum_{\{x_{k_i}\}} \prod_k z_k^{x_{k_i}} W(x_{k_1}...x_{k_K}) . \tag{1.7-44}$$

The factorial moments are generated by differentiation,

$$\langle [x_{1,m_1}][x_{2,m_2}]...[x_{K,m_K}] \rangle = \left. \frac{\partial^{\Sigma m_i} \chi(z_1...z_K)}{\partial z_1^{m_1} \partial z_2^{m_2} ... \partial z_K^{m_K}} \right|_{\{z_k\}=1} . \tag{1.7-45}$$

The inversion is similar to the previous case,

$$W(x_{k_1}...x_{k_K}) = \left(\frac{1}{2\pi i} \right)^K \oint_{C_1} ... \oint_{C_k} \frac{\chi(\zeta_1...\zeta_K)}{\zeta_1^{x_{k_1}+1} ... \zeta_K^{x_{k_K}+1}} d\zeta_1 ... d\zeta_K . \tag{1.7-46}$$

As a simple application we consider once more the multinomial distribution. Then with $\{x_k\} \to \{N_k\}$ we find,

---

[26] Some authors use

$$\chi(\lambda) = \langle (1-\lambda)^x \rangle = \Sigma_k (1-\lambda)^{x_k} W(x_k) . \tag{1.7-40}$$

$$\chi(z_1 ... z_K) = \sum_{\{N_k\}} \frac{N!}{N_1! ... N_K!} \prod_k (p_k z_k)^{N_k} = (p_1 z_1 + ... + p_K z_K)^N, \tag{1.7-47}$$

where we used the multinomial expansion theorem (1.6-31). From the factorial moment rule (1.7-45) we now easily find $\langle N_i \rangle = p_i N$ and

$$\langle N_i(N_i - 1) \rangle = \frac{\partial^2 \chi}{\partial z_i^2}\bigg|_{\{z_k\}=1} = p_i^2 N(N-1), \tag{1.7-48a}$$

$$\langle N_i N_j \rangle = \frac{\partial^2 \chi}{\partial z_i \partial z_j}\bigg|_{\{z_k\}=1} = p_i p_j N(N-1), \; i \neq j. \tag{1.7-48b}$$

For the variances and covariances this yields,

$$\begin{aligned} \langle \Delta N_i^2 \rangle &= \langle N_i(N_i - 1) \rangle - \langle N_i \rangle [\langle N_i \rangle - 1] = N p_i (1 - p_i), \\ \langle \Delta N_i \Delta N_j \rangle &= \langle N_i N_j \rangle - \langle N_i \rangle \langle N_j \rangle = -N p_i p_j, \end{aligned} \tag{1.7-49}$$

in accordance with (1.6-32).

Thiele semi-invariants or cumulants

In the last three subsections we have introduced three types of transforms, whose series expansions were useful in order to obtain the moments $m_n \equiv \langle x^n \rangle$, or the factorial moments $[m_n] \equiv \langle x(x-1)...(x-n+1) \rangle$. In order for the transforms to be meaningful, we must retain these moments to all orders. Omitting e.g. all moments of order higher than $n = k$, renders the Fourier inversion extremely difficult, unless one resorts to delta function derivatives of order (k–1). However, no such catastrophes occur if we consider the logarithms of these transforms. The expansion coefficients then involve cumulants, which often can be ignored after a certain number of terms, as we now show. For a mathematical discussion see Stratonovich[27] and for a statistical physics context, see Linda Reichl's book.[28] Basically, cumulants are most often defined starting from the characteristic function $\Phi(u)$, but the results apply equally well to the other transforms with the substitutions

$$iu \to -s \to \ln z. \tag{1.7-50}$$

We now introduce the cumulant expansion,

$$\ln \Phi(u) = \sum_{n=1}^{\infty} \frac{(iu)^n}{n!} c_n, \tag{1.7-51}$$

---

[27] R.L. Stratonovich, " Topics in the Theory of Random Noise", vol. I, Gordon and Breach ,NY, 1963.

[28] L.E. Reichl, "A Modern Course in Statistical Physics", Univ. of Texas Press, Austin, TX, 1980, p. 143-144; also 2nd Ed., John Wiley &Sons, N.Y. 1998 and Wiley-VCH-Verlag, Weinheim 2004, p. 188.

or,

$$\Phi(u) = \exp\left[\sum_{n=1}^{\infty} \frac{(iu)^n}{n!} c_n\right]. \tag{1.7-52}$$

Clearly, the cumulants follow from the rule,

$$c_n = (-i)^n \left.\frac{d^n \ln \Phi(u)}{du^n}\right|_{u=0}. \tag{1.7-53}$$

For the other transforms the cumulants involve the expressions,

$$\begin{aligned} c_n &= (-1)^n \left.\frac{d^n \ln \Psi(s)}{ds^n}\right|_{s=0}, \\ c_n &= \left.\frac{d^n \ln \chi(z)}{d(\ln z)^n}\right|_{z=1}. \end{aligned} \tag{1.7-54}$$

Employing the customary definition (1.7-51), we obtain the cumulants by expanding (1.7-52) in powers of u. Comparing with (1.7-6) and equating equal powers of u, the first four cumulants are readily obtained:

$$\begin{aligned} c_1 &= m_1 = \langle x \rangle, \\ c_2 &= m_2 - m_1^2 = \langle \Delta x^2 \rangle, \\ c_3 &= m_3 - 3m_2 m_1 + 2m_1^3 = \langle \Delta x^3 \rangle, \\ c_4 &= m_4 - 3m_2^2 - 4m_3 m_1 + 12 m_2 m_1^2 - 6m_1^4. \end{aligned} \tag{1.7-55}$$

Generally, c_j has the form

$$c_j = j! \sum_{\{n_i\}} (-1)^{\Sigma_i n_i - 1} (\Sigma_i n_i - 1)! \prod_i \frac{(m_i / i!)^{n_i}}{n_i!}, \tag{1.7-56}$$

where the summation is over all sets of integers (including zero) that satisfy $\Sigma_i i n_i = j$. At the far right-hand-side in (1.7-55) we have added the "fluctuation moments", i.e., the moments of the shifted variable $\Delta x = x - \langle x \rangle$. The identities follow from simple algebra; e.g.,

$$\begin{aligned} \langle [x - \langle x \rangle]^3 \rangle &= \langle [x^3 - 3x^2 \langle x \rangle + 3x \langle x \rangle^2 - \langle x \rangle^3] \rangle \\ &= m_3 - 3m_2 m_1 + 2m_1^3. \end{aligned} \tag{1.7-57}$$

However, by a similar computation,

$$\langle [x - \langle x \rangle]^4 \rangle = m_4 - 4m_3 m_1 + 6m_2 m_1^2 - 3m_1^4. \tag{1.7-58}$$

Clearly, $c_4 \neq \langle \Delta x^4 \rangle$. The same holds for higher orders. Yet, it is to be noted that the cumulants can always be expressed in suitable combinations of fluctuation moments; e.g., $c_4 = \langle \Delta x^4 \rangle - 3\langle \Delta x^2 \rangle^2$. Since the fluctuation moments of higher order, or powers of them yielding the same order [like $\langle \Delta x^2 \rangle^2 = \mathcal{O}(\langle \Delta x^4 \rangle)$] tend to be small, it is often convenient to set the higher cumulants equal to zero. Thus the cumulant expansion is likely to have excellent convergence properties and is far more useful than the moments, or factorial moments expansions discussed before.

In statistical mechanics it will turn out that thermodynamic state functions are logarithms of partition functions, which are transforms of other, similar functions in the basic microcanonical ensemble; see in particular sections 4.9 and 5.2. We shall see time and again that the fluctuation moments, in particular the variances and covariances, are obtained as second order derivatives of the logarithm of the relevant partition functions, in agreement with the cumulant rules (1.7-53) through (1.7-55).

1.7.4 *The Mellin Transform*

Let x be a continuous or quasi-continuous variable on $(0, \infty)$. The Mellin transform is defined as

$$\Omega(p) = \left\langle x^{p-1} \right\rangle = \int_0^\infty x^{p-1} dP(x) . \tag{1.7-59}$$

For simplicity we consider only the continuous case, $dP(x) = W(x)dx$. Now let for

$$x \to 0: \ W(x) = \mathcal{O}(x^{-\sigma_0}), \qquad x \to \infty: \ W(x) = \mathcal{O}(x^{-\tau_0}). \tag{1.7-60}$$

For the integral to converge in some domain of p, both at the lower and upper limit, we must have

$$\sigma_0 < \mathrm{Re}\, p < \tau_0. \tag{1.7-61}$$

If such a region exists, (1.7-59) can be inverted[29,30]

$$W(x) = \frac{1}{2\pi i} \int_{-i\infty+\beta}^{i\infty+\beta} \frac{\Omega(p)}{x^p} dp , \tag{1.7-62}$$

where β is a point on the real axis such that $\sigma_0 < \beta < \tau_0$. Often we can close by a suitable contour. Thus if the averages $\langle x^{p-1} \rangle$ can be found along the imaginary axis $\mathrm{Re}\, p = \beta$, or along a suitable complex contour, the distribution function can be directly reconstituted from these values. Often $\sigma_0 \geq \tau_0$; in that case partial Mellin transforms must be used and the problem is more complex. We refer to the literature.

---

[29] F.M. Morse and H. Feshbach, "Methods of Theoretical Physics" Vol. I, McGraw Hill, 1953, p.469ff.
[30] R. Courant and D. Hilbert, "Methods of Mathematical Physics" Vol. I, Intersc. NY, 1953, p. 103ff.

1.8 Problems to Chapter I

1.1 Obtain the connections $S_M = -F/T$, $S_{Pl} = -G/T$, where S_M and S_{Pl} are Massieu functions [see (1.3-10)], F is the Helmholtz free energy and G is the Gibbs free energy.

1.2 For some system the entropy has the form $S(U,V,N) = (kN/\alpha)\ln(UV^{\alpha}/N^{\alpha+1})$.
(a) Find the *equations of state* $Q_i(U,V,N)$ and show that the perfect gas law, $PV = nRT$, where $n = N/N_A$ with N_A being Avogadro's number, is satisfied.
(b) Show the validity of the Gibbs-Duhem relation for this case: $-Nd\varsigma = SdT - VdP$.
(c) Obtain the specific heats C_P and C_V and show that their ratio γ equals $\alpha + 1$.

1.3 Give a detailed account of the results leading to the virial theorem, Eq. (1.6-88); in particular, establish the result for the commutator $[F, \mathcal{H}]$. Note. Do *not* use the wave mechanical operators $q_i = q_i \times$, $p_i = (\hbar/i)\partial/\partial q_i$ but employ general relationships, like $[A, BC] = [A, B]C + B[A, C]$, $[AB, C] = A[B, C] + [A, C]B$ or express the commutator into the Poisson bracket: $[F, G] = \hbar i\{F, G\}$ (this is more work).

1.4 A Bernoulli trial is defined by its possible outcomes, success or failure. We have $w(succ) = p$, $w(fail) = 1 - p$. Let $n = \sum_{i=1}^{N} x_i$, where each x_i is a Bernoulli trial. Find the Fowler generating function $\Phi(z)$ and its inversion $W(n)$. Show that $W(n)$ is the binomial distribution.

1.5 Obtain the Fowler generating function for the Poisson distribution. By differentiation find the first two factorial moments and obtain the variance.

1.6 Consider a multivariate normal distribution as in (1.6-55). Use new variables $y_i = \Sigma_j \tilde{S}_{ij} x_j$, (the tilde denotes the transpose) where S is an orthogonal matrix that diagonalizes the symmetrical matrix α. Obtain the characteristic function for $W(\{y_i\})$ and then obtain the characteristic function of $W(\{x_i\})$. By differentiation find the covariance $\langle x_i x_j \rangle$ and show that the result agrees with that obtained by Onsager's method.

1.7 Given is the bivariate normal distribution $W(x, y) = C\exp[-(x^2 + 2xy + 2y^2)/2]$. Find $\langle x^2 \rangle$, $\langle xy \rangle$ and $\langle y^2 \rangle$.

1.8 A train wagon has 22 seats, of which 15 are forward facing and 7 are backward facing. There are 18 passengers, two of which get dizzy if they ride backwards. In how many ways can you seat these passengers?

1.9 Jill has once a week "fun". She knows that the chance for catching a disease (STD) is 2%. She frequents her favourite clubs on average once a week. Conventional wisdom has it that the risk remains 'only' 2%, since all events are independent!
Prove that conventional wisdom is wrong and
(a) Compute the probability that she catches nothing in a year;
(b) She has contracted a STD. Treat each event as a Bernoulli trial.
[Answers: (a) 35%; (b) 65%.]

1.10 Employing the generating function according to Fowler, find the third and fourth order (factorial) moments for the multinomial distribution.

1.11 In a random walk, the probability for a step Δ to the right is ¼, the probability for a step Δ to the left is ¼ and the probability for a step with no progress in either direction is ½. Using the method of the generating function, find the probability that a distance $x = K\Delta$ is covered in N steps. Explain your result in terms of a trinomial distribution. *Hint*: Use the following identity:

$$e^{\alpha} + e^{-\alpha} + 2 = (e^{\alpha/2} + e^{-\alpha/2})^2. \tag{1}$$

1.12 Consider a tetravariate normal distribution, $W(x_1,...,x_4) = C\exp[-\Sigma_{ij}\alpha_{ij}x_i x_j/2]$. Show that:

$$\langle x_1x_2x_3x_4\rangle = \langle x_1x_2\rangle\langle x_3x_4\rangle + \langle x_1x_3\rangle\langle x_2x_4\rangle + \langle x_1x_4\rangle\langle x_2x_3\rangle. \tag{2}$$

NB. This *looks like* Wick's theorem! – cf. Chapter XII.

1.13 In a stochastic process the probability of taking a step $x \to x \pm dx$ is given by the *a priori* probability $\pi^{-1}[a/(x^2 + a^2)]dx$, $a > 0$. Find the probability for a displacement $W(X)$ after N steps. Does $W(X)$ satisfy the central limit theorem? Also find the first four cumulants for this distribution.

1.14 Consider the following combinatorial problem. In a set of N elements N_1 are red and $N - N_1$ are black. A sample of n elements is chosen at random.
(a) Determine the probability $W(k;n,N_1,N)$ that in this sample there are exactly k red elements. Find $\langle k\rangle$ and $\mathrm{var}\,k$. Hint: you will find the hypergeometric distribution, for which the generating function is related to the hypergeometric function.
(b) Apply this to the problem of finding k spins up out of n spins in an electron gas of N electrons, for which the number of up and down spins is rigorously equal. What happens to $W(k;n)$ if $N \to \infty$? (Your answer will explain why these statistics are not necessary in physics!)

1.15 (a) Employ formula (1.7-56) to verify the result for the fourth cumulant. Hint: you should obtain all sets $\{n_i\}$ that satisfy $1n_1 + 2n_2 + 3n_3 + 4n_4 = 4$.
(b) Obtain the form for c_5.[31, 32]

[31] See C. Rose and M.D. Smith, "Mathematical Statistics with Mathematica", Springer Verlag, NY, 2002.

[32] S.L. Lauritzen, "Thiele, Pioneer in Statistics", Oxford Univ. Press, Oxford, 2002.

Chapter II

Statistics of Closed Systems

2.1 Liouville's Theorem and the Microcanonical Density Function

We start again with classical systems, whose motion can be pictured in the gamma-space or phase space, described in section 1.2. The principal theorem regarding the change of the density function $\rho(\{p_i(t)\},\{q_i(t)\},t)$ – or for short $\rho(p,q,t)$ – is due to Liouville[1]. The basic statement of the theorem is found in standard texts on classical mechanics, like Goldstein's book.[2] Let us consider the motion of a number of points dN in the phase space from a time t_1 to a later time t_2. If $d\Omega$ is the associated volume in the phase space, then we clearly have

$$dN = \rho(p_1,q_1,t_1)d\Omega_1 = \rho(p_2,q_2,t_2)d\Omega_2. \tag{2.1-1}$$

According to classical mechanics $d\Omega$ is a Poincaré invariant, since the Jacobian of the canonical transformation $\{p_{i,1}\},\{q_{i,1}\}\rightarrow\{p_{i,2}\},\{q_{i,2}\}$ is unity (Goldstein, Op.Cit. p. 270.)[3] ; hence, $d\Omega_1 = d\Omega_2$. Equation (2.1-1) therefore entails that ρ remains unchanged when we follow along the trajectories of the system points, i.e.,

$$d\rho / dt = 0. \tag{2.1-2}$$

We note that d/dt stands for the total derivative, which in this connection is usually referred to as the *convected or hydrodynamic derivative.* Equation (2.1-2) indicates that the flow of system points in phase space is like that of an 'incompressible fluid'.

We obtain another form of Liouville's theorem when we apply the chain rule for $d\rho / dt$. We then have

$$\frac{d\rho(p,q,t)}{dt} = \frac{\partial \rho}{\partial t} + \sum_{i=1}^{rN}\left[\left(\frac{\partial \rho}{\partial p_i}\right)\dot{p}_i + \left(\frac{\partial \rho}{\partial q_i}\right)\dot{q}_i\right] = 0. \tag{2.1-3}$$

Upon applying Hamilton's equations, see (1.1-7), we find the result,

---

[1] J. Liouville "Note sur la théorie de variation de constantes arbitraires", J. Math. Pures Appl. (9) **3**, 342 (1838).

[2] H. Goldstein, "Classical Mechanics", 2nd Ed., Wiley, NY, 1980.

[3] This follows from the *symplectic condition* for canonical transformations, $\mathsf{M}\mathsf{J}\tilde{\mathsf{M}} = \mathsf{J}$, where M is the Jacobian matrix and $\mathsf{J} = \begin{pmatrix} 1 & 0 \\ 0 & -1 \end{pmatrix}$ is the symplectic matrix; clearly, $|\,\mathrm{Det}\,\mathsf{M} = 1\,|$.

$$\frac{\partial \rho}{\partial t} = -\{\rho, \mathcal{H}\}, \tag{2.1-4}$$

where the curly bracket denotes the Poisson bracket with its usual definition,

$$\{A, B\} = \sum_{i=1}^{rN} \left[\frac{\partial A}{\partial q_i} \frac{\partial B}{\partial p_i} - \frac{\partial A}{\partial p_i} \frac{\partial B}{\partial q_i} \right]. \tag{2.1-5}$$

We now proceed to find a form for ρ in the microcanonical ensemble. In equilibrium statistical mechanics 'statistical equilibrium' implies that the ensemble picture is the same at all times, i.e., the local derivative of the density function, $\partial \rho / \partial t$ is zero. Therefore also,

$$\{\rho, \mathcal{H}\} = 0, \tag{2.1-6}$$

indicating that ρ *is a proper constant of motion.* Therefore ρ must have the following form,

$$\rho(p,q) = f[\mathcal{H}(p,q), \varphi_2(p,q), \varphi_3(p,q), \ldots, \varphi_{2rN-1}(p,q)], \tag{2.1-7}$$

where the φ's are the constants of motion introduced in section 1.1. As we pointed out there, however, these constants of motion are wholly unknown; only $\varphi_1(p,q) \equiv \mathcal{H}(p,q)$ is defined in the ensemble. Therefore, the other φ's must be 'talked away!'. In classical physics this is accomplished with *ergodic theory*, which has a long history, from the founders of kinetic theory, Maxwell and Boltzmann, to more recent very mathematical considerations by Oxtoby and Ulam; we shall briefly dwell on these in the next section.

Taking it for granted that ρ is only a functional of $\mathcal{H}$, we proceed to fix its form. Given that for a closed system $\mathcal{H}$ is conserved, it follows that all systems have a trajectory that lies on the energy surface $\mathcal{H}(p,q) = \mathcal{E}$. Therefore, ρ must have the form,

$$\rho(p,q) = C\delta[\mathcal{H}(p,q) - \mathcal{E}], \tag{2.1-8}$$

where C is a constant, determined by normalization, as follows. For the volume element in phase space we write

$$d\Omega = d\mathcal{E}\, dS_{\mathcal{E}} / |\, grad_{\Omega} \mathcal{H}| \equiv d\mathcal{E}\, dM_{\mathcal{E}}, \tag{2.1-9}$$

where $dM_{\mathcal{E}}$ is called the differential surface measure.

A pictorial explanation is given in Fig. 2-1. One easily notes the relations, employing the definition of generalized gradient in curvilinear coordinates,

$$\begin{aligned} d\Omega &= dS_{\mathcal{E}}\, dn_{\perp}, \\ d\mathcal{E} &= |\, grad_{\Omega} \mathcal{H}\,|\, dn_{\perp}, \end{aligned} \tag{2.1-10}$$

which result in (2.1-9).

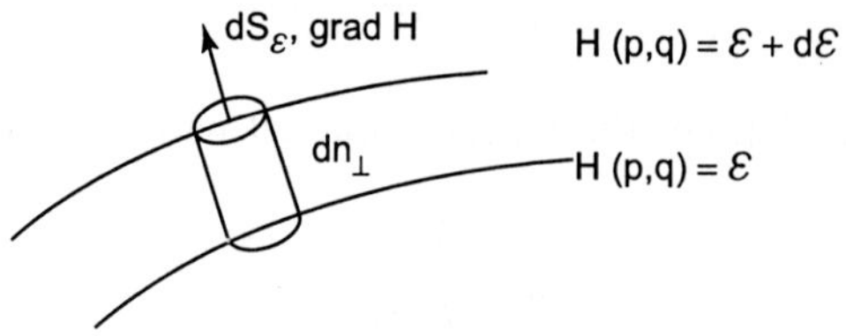

Fig. 2-1. Volume $\Delta\Omega$ between two energy surfaces.

We thus have from (2.1-8) and (2.1-9)

$$\int \rho d\Omega = C\int \delta[\mathcal{H}(p,q)-\mathcal{E}]d\Omega$$
$$= C\int d\mathcal{E}'\oint dM_{\mathcal{E}'}\,\delta(\mathcal{E}'-\mathcal{E}) = C\,M_{\mathcal{E}} = 1. \qquad (2.1\text{-}11)$$

It follows that $1/C$ is equal to the surface measure $M_{\mathcal{E}}$; hence, ρ is given by†

$$\rho(p,q) = \frac{1}{M_{\mathcal{E}}}\delta[\mathcal{H}(p,q)-\mathcal{E}]. \qquad \mathbf{(2.1\text{-}12)}$$

This is the exact form, indicating that the density is confined to the energy surface, where it is a constant (no 'loading').

Yet, for many practical computations, the energy is given a 'leeway' $\delta\mathcal{E}$. Denoting the volume between the neighbouring energy surfaces $\mathcal{E}$ and $\mathcal{E}+\delta\mathcal{E}$ by $\Delta_{\mathcal{E}}\Omega$, we then write

$$\rho(p,q) = \begin{cases} 1/\Delta_{\mathcal{E}}\Omega \quad , & \mathcal{E} \le \mathcal{H}(p,q) \le \mathcal{E}+\delta\mathcal{E}, \\ 0 \;\; elsewhere. & \end{cases} \qquad \mathbf{(2.1\text{-}13)}$$

From normalization we now have

$$\Delta_{\mathcal{E}}\Omega = M_{\mathcal{E}}\delta\mathcal{E}. \qquad (2.1\text{-}14)$$

2.2 The Ergodic Hypothesis

When one looks through Maxwell's and Boltzmann's collected papers, it becomes clear just how much these authors were concerned with the justification of the use of ensemble averages for kinetic computations. Many modern books, in which phase-space averages are 'justified' by the *ad hoc* postulate of equal *a priori* probabilities, should do well to refer to the original work of the pioneers. Boltzmann states the

† Rather than boxing equations, we have resorted to using **bold numbers** for very basic results.

ergodic hypothesis[4] as follows:[5] "The great irregularity of the thermal motion and the variety of extrinsic forces acting on bodies make it probable that, in virtue of the motion which we call heat, the atoms of bodies take on all positions and velocities compatible with the equation of energy." Maxwell states the same idea in more detail:[6] "The only assumption which is necessary for the direct proof [of Boltzmann's law] is that the system, if left to itself in its actual state of motion, will, sooner or later, pass through every phase which is consistent with the equation of energy." Clearly he came close to the enunciation of the ergodic hypothesis as it is nowadays phrased: *For a conservative system the path of the system in phase space passes through all points of the energy surface and in such a manner that this surface is covered uniformly.*

Whereas the first part of the statement seems plausible for a curve that never crosses itself (see Section 1.2) the latter part is more sticky. To clarify this point, Boltzmann derives a formula for the time interval which the phase point during its motion spends on a given element of the energy surface. First he computes the time dt that it takes for the system to move through an element of the independent variables, i.e., those p's and q's which remain time-dependent after all others are fixed by the integrals of motion, cf. Eq. (1.1-8); this calculation is straightforward, using "Jacobi's method of the last multiplier". From here he proceeds in an obscure way to obtain the time spent by a representative point of the system on an element dS of its energy surface. The asserted result is,

$$\lim_{T\to\infty} dt/T \propto dS/|\mathrm{grad}_{\mathcal{E}}\mathcal{H}| = dM_{\mathcal{E}}\,. \tag{2.2-1}$$

If this were true, the desired result is obtained, viz., that the time average over the trajectory of one system equals the ensemble average for all systems:

$$\begin{aligned}\widetilde{F(p,q)} &= \lim_{T\to\infty}\int_0^T F[p(t),q(t)]dt/T = C\oint F(p,q)\,dM_{\mathcal{E}} \\ &= C\int d\mathcal{E}'\oint F(p,q)\,dM_{\mathcal{E}'}\,\delta(\mathcal{E}-\mathcal{E}') = \langle F(p,q,)\rangle\,.\end{aligned} \tag{2.2-2}$$

However, it is certain that Boltzmann's proof was fallacious. Critique on the ergodic hypothesis was voiced by many authors in the early twentieth century, among which Rosenthal and Plancherel.[7] From a point of view of measure theory the impossibility of the hypothesis is clear: No trajectory which cannot have multiple points can fill out a multi-dimensional surface.[8] Slightly better fared the quasi-ergodic hypothesis, which states that *the trajectory will come arbitrarily close to any point on the energy*

[4] The name stems from the Greek εργοσ = work.

[5] L. Boltzmann, Wiener Berichte **63**^{2}, 679, 1871, or his collected papers No. 19.

[6] J.C. Maxwell, Proc. Cambridge Phil. Soc. **12**, 547 (1879); also collected papers **2**, p. 713.

[7] A. Rosenthal, Ann. der Physik, **42**, 796 (1913); M. Plancherel, Ann. der Physik **42**, 1061 (1913).

[8] A.I. Khintchine, "Mathematical Foundations of Statistical Mechanics", Dover, New York, 1949, p. 53.

surface. However, although Fermi showed that this is true for a large class of systems, the statement (2.2-1) remained elusive.

A more rewarding avenue was taken by Birkhoff and in a somewhat weaker form by von Neumann, both in 1931. They showed the equivalence of (2.2-2) for the condition that the energy surface is 'metrically indecomposable'. The latter property entails that $S_{\mathcal{E}}$ cannot be decomposed into two parts S_1 and S_2, both of positive measure, and both invariant under the transformation group $p_i(t), q_i(t) \to p'_i(t), q'_i(t)$. The essential argument has now been moved to another question, viz., which systems have a metrically transitive transformation group? In 1941 Oxtoby and Ulam showed that 'almost every' continuous transformation satisfies this criterion.[9]

For those who are mathematically inclined it is gratifying, that with the help of modern measure theory the alleged result of the founders has finally been shown to be correct for 'almost all' systems. However, classical systems have little place in present-day many-body quantum physics. As we will see in the next section, in most quantum systems the Hamiltonian $\mathcal{H}$ is the only constant of motion, whose states span the entire relevant Hilbert space. Therefore ρ can only be a functional of $\mathcal{H}$.

2.3 Von Neumann's Theorem and the Microcanonical Density Operator

The road for quantum statistics is beset with less trouble. Following the developments of Section 1.2, we have for the density operator in an arbitrary basis of many-body states $\{|\gamma\rangle\}$

$$\rho_{\gamma\gamma'}(t) = \lim_{M\to\infty} \frac{1}{M} \sum_{i=1}^{M} c_\gamma^i(t) c_{\gamma'}^{i*}(t) . \tag{2.3-1}$$

With $|\Psi^i(t)\rangle$ being the dynamical state of system 'i' in the ensemble, and with $c_\gamma^i(t) = \langle\gamma|\Psi^i(t)\rangle$ being its projection on $|\gamma\rangle$ we have from $\hbar i\partial|\Psi^i\rangle/\partial t = \mathcal{H}|\Psi^i\rangle$ by insertion of the closure property $\Sigma_{\gamma''}|\gamma''\rangle\langle\gamma''| = 1$,

$$\hbar i \sum_{\gamma''} (\partial/\partial t)|\gamma''\rangle\langle\gamma''|\Psi^i(t)\rangle = \mathcal{H}\sum_{\gamma''}|\gamma''\rangle\langle\gamma''|\Psi^i(t)\rangle , \quad \text{or,}$$

$$\hbar i \sum_{\gamma''} (\partial/\partial t)|\gamma''\rangle c_{\gamma''}^i(t) = \mathcal{H}\sum_{\gamma''}|\gamma''\rangle c_{\gamma''}^i(t) . \tag{2.3-2}$$

Taking the scalar product with $|\gamma\rangle$ and noting $\langle\gamma|\gamma''\rangle = \delta_{\gamma\gamma''}$, this gives the transformed Schrödinger equation in the basis $\{|\gamma\rangle\}$,

$$\hbar i(\partial/\partial t) c_\gamma^i(t) = \sum_{\gamma''} \mathcal{H}_{\gamma\gamma''} c_{\gamma''}^i(t) . \tag{2.3-3}$$

---

[9] I. Oxtoby and S. Ulam, Ann. of Math. **42**, 874 (1941).

Differentiating (2.3-1), this yields,

$$\frac{\partial \rho_{\gamma\gamma'}}{\partial t} = \lim_{M\to\infty} \frac{1}{M} \sum_i \left[\frac{\partial c_\gamma^i}{\partial t} c_{\gamma'}^{i*} + c_\gamma^i(t) \frac{\partial c_{\gamma'}^{i*}}{\partial t} \right]$$

$$= \frac{1}{\hbar i} \sum_{\gamma''} \left[\mathcal{H}_{\gamma\gamma''} \rho_{\gamma''\gamma'} - \rho_{\gamma\gamma''} \mathcal{H}_{\gamma''\gamma'} \right] = \frac{1}{\hbar i} [\mathcal{H}\rho - \rho\mathcal{H}]_{\gamma\gamma'} . \tag{2.3-4}$$

Since this holds for the matrix elements, it holds for the operators themselves. We thus obtained von Neumann's theorem,

$$\frac{\partial \rho}{\partial t} = -\frac{1}{\hbar i} [\rho, \mathcal{H}] , \tag{2.3-5}$$

where $[\rho, \mathcal{H}]$ is the commutator. Notice the minus sign compared to the well-known Heisenberg equation of motion.[10] The result is similar to the Liouville equation (2.1-4), derived earlier. We clearly have the correspondence $\{\rho, H\} \to (1/\hbar i)[\rho, H]$; this was to be expected from the general quantum condition $\{A, B\} = (1/\hbar i)[A, B]$, see Dirac.[11]

The formal solution of (2.3-5) is easily given. If $\mathcal{H}$ does not explicitly depend on time we have,

$$\begin{aligned} \rho(t) &= e^{-i\mathcal{H}t/\hbar} \rho(0) e^{i\mathcal{H}t/\hbar} = e^{-i\mathcal{L}t} \rho(0) \\ &= U(t,0)\rho(0)U^\dagger(t,0) , \end{aligned} \tag{2.3-6}$$

where U is the evolution operator and where $\mathcal{L}$ is the Liouville (super)-operator. We note that, if $\rho(0)$ commutes with $\mathcal{H}$, so does $\rho(t)$.

For an ensemble in statistical equilibrium $\partial\rho/\partial t = 0$, so that $[\rho, H] = 0$. It follows that ρ is a constant of motion; we write

$$\rho = f(\mathcal{H}) . \tag{2.3-7}$$

There is now no need to include other constants of motion, since, except for pathological cases, they only span a subspace of the Hilbert space for $\mathcal{H}$. We now use the *spectral resolution*[12] (decomposition theorem, see e.g., Messiah[13]):

$$\mathcal{H} = \sum_\eta |\eta\rangle\langle\eta| \mathcal{E}_\eta \quad \text{and} \quad f(\mathcal{H}) = \sum_\eta |\eta\rangle\langle\eta| f(\mathcal{E}_\eta) . \tag{2.3-8}$$

---

[10] The density operator is not a Heisenberg operator, but a Schrödinger operator. Also note that we used the symbol $\partial/\partial t$ for the appropriate derivative.

[11] P.A.M. Dirac, "The Principles of Quantum Mechanics", 4th Ed., Oxford University Press, 1958, p. 87.

[12] F. Riesz and B.Sz. Nagy, "Leçons d'analyse fonctionnelle ", 4th Ed., Gauthiers–Villars, Paris, 1965, Chapter 11.

[13] A. Messiah, "Quantum Mechanics, Vol. I, North Holland, Amsterdam, 1961, pp.260-272; translated from A. Messiah, "Mécanique Quantique", Dunod, Paris, 1958.

Now ρ can be represented by its diagonal part, see (1.2-8),

$$\rho = \sum_{\eta} |\eta\rangle p(\eta) \langle \eta| . \tag{2.3-9}$$

From (2.3-7) – (2.3-9) we find,

$$p(\eta) = f(\mathcal{E}_\eta), \quad |\eta\rangle \in \mathcal{E}. \tag{2.3-10}$$

The restrictions on the accessible states $|\eta\rangle$ entail that $f(\mathcal{E}_\eta)$ must be taken to be a delta-function

$$p(\eta) = A\delta(\mathcal{E}_\eta - \mathcal{E}) . \tag{2.3-11}$$

Let us write

$$\sum_{\eta} \ldots \to \int d\mathcal{E}_\eta \mathcal{X}(\mathcal{E}_\eta) \ldots , \tag{2.3-12}$$

where $\mathcal{X}(\mathcal{E}_\eta)$ is the density of states; the approximation is near-exact, since for a macroscopic system the states are very dense. From (2.3-11), (2.3-12) and normalization we have

$$\sum_\eta p(\eta) = A \int \mathcal{X}(\mathcal{E}_\eta) \delta(\mathcal{E}_\eta - \mathcal{E}) d\mathcal{E}_\eta = A\mathcal{X}(\mathcal{E}) = 1. \tag{2.3-13}$$

This fixes both $p(\eta)$ and ρ:

$$\begin{aligned} p(\eta) &= \frac{1}{\mathcal{X}(\mathcal{E})} \delta(\mathcal{E}_\eta - \mathcal{E}) , \\ \rho &= \sum_{\eta} |\eta\rangle \frac{1}{\mathcal{X}(\mathcal{E})} \delta(\mathcal{E}_\eta - \mathcal{E}) \langle \eta| . \end{aligned} \tag{2.3-14}$$

The latter expression can again be summed with the decomposition theorem (2.3-8); we then obtain the final form,

$$\rho = \frac{1}{\mathcal{X}(\mathcal{E})} \delta[\mathcal{H} - \mathsf{I}\mathcal{E}] , \tag{\textbf{2.3-15}}$$

where I is the identity operator. Obviously, Equation (2.3-15), containing a delta function of an operator, is a symbolic expression, its true meaning being just (2.3-14). Clearly, the expressions for the classical density function $\rho(p,q)$ and the quantum density operator ρ are very similar, with the surface measure $M_{\mathcal{E}}$ being replaced by the density of states $\chi(\mathcal{E})$.

Again, we can also allow for a certain tolerance $\delta\mathcal{E}$. Then, let $\Delta\Gamma(\mathcal{E}) = \chi(\mathcal{E})\delta\mathcal{E}$ be the number of states in this interval, i.e., the number of states $|\eta\rangle$ with energies $\mathcal{E} \le \mathcal{E}_\eta \le \mathcal{E} + \delta\mathcal{E}$. We refer to $\Delta\Gamma(\mathcal{E})$ as the *microcanonical partition function.* Then (2.3-15) takes the alternate form,

$$\rho = \begin{cases} \sum_\eta |\eta\rangle \dfrac{1}{\Delta\Gamma(\mathscr{E})} \langle\eta|, & |\eta\rangle \in (\mathscr{E} \le \mathscr{E}_\eta \le \mathscr{E} + \delta\mathscr{E}) \\ 0, & \textit{all other}\,|\eta\rangle. \end{cases} \tag{2.3-16}$$

To be noted is that we did not appeal to ergodic theory for the quantum case. This indicates the superiority of the quantum treatment. A device like spectral resolution is non-existing in classical mechanics. The very complex arguments of Oxtoby and Ulam, involving metrical decomposition of the energy surface, come close, however, to the considerations of spectral resolution theory of quantum operators, which, for its mathematical validity relies on very extensive properties of linear operator theory, especially when such operators are not necessarily bounded in the Hilbert space; see in particular Ref.12, Chapter VIII.

2.4 Macro-Probability in Classical and Quantum Statistics

In the previous sections we obtained the density function, or operator, for a microcanonical ensemble. The next step is to find the macro-probability for the specification of the a-variables in a system, see Section 1.5 . Suppose we have the state variables $\{a_i(t)\}$, $i = 1 \ldots P$. Let these variables lay in the domain

$$\mathscr{D}(\mathbf{a}) = \begin{cases} a_1 \le a_1(p,q) \le a_1 + da_1 \\ \vdots \\ a_P \le a_P(p,q) \le a_P + da_P \end{cases} \tag{2.4-1}$$

This specification delineates a volume $\Delta\Omega(\mathbf{a})$ in phase space, called the *accessible volume in phase space*,

$$\Delta\Omega(\mathbf{a}) = \int_{\mathscr{D}(\mathbf{a})} d\Omega\,. \tag{2.4-2}$$

The situation is pictured in Fig. 2-2.

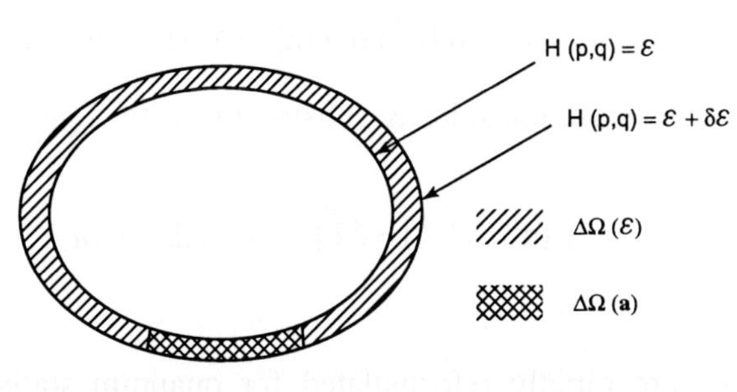

Fig. 2-2. Accessible volume for a specification (2.4-1).

We introduce a density function $G(\{a_i)\}$ by

$$\Delta\Omega(\mathbf{a}) = \int_{\mathcal{D}(\mathbf{a})} d\Omega = G(\{a_i\})da_1 da_2 ... da_P \,. \tag{2.4-3}$$

For the probability of the specification (2.4-1) we clearly have

$$W(\{a_i)\}da_1 da_2 ... da_P = \Delta\Omega(\mathbf{a}) / \Delta_{\mathcal{E}}\Omega \,. \tag{2.4-4}$$

With (2.4-3) we obtain for the probability density function (pdf):

$$W(\{a_i)\} = G(\{a_i\}) / \Delta_{\mathcal{E}}\Omega \,. \tag{2.4-4'}$$

This is the basic result for macro-probability in classical statistical mechanics.

However, if the a's are extensive variables, they are most likely fixed in the ensemble. We may then consider subsystems in which the a's fluctuate, as discussed in Section 1.4. Accordingly we write, dividing the composite system into two subsystems, noting $da_i = da_i^I = -da_i^{II}$,

$$\begin{aligned} \Delta\Omega(\mathbf{a}) &= \int_{\mathcal{D}(\mathbf{a})} d\Omega^I d\Omega^{II} \sim G(\{a_i^I\})G(\{a_i^{II}\})\prod_k da_k \\ &= G(\{a_i^I\})G(\{a_i^c - a_i^I\})\prod_k da_k \,. \end{aligned} \tag{2.4-5}$$

The reader notices that (2.4-5) is a dimensional failure, since we have P factors da too few on the rhs. This is not as grave as it looks, however. Taking the example of a hypersphere, let $\Delta\Omega(a) = Ca^{\lambda N}$, then $(d/da)\Delta\Omega(a) = G(a) = C\lambda N a^{\lambda N-1}$; or

$$\Delta\Omega(a) / G(a) = a / \lambda N \approx \delta a \,. \tag{2.4-6}$$

This indicates that $\ln(\Pi da_i) = \mathcal{O}(\ln N) \equiv \hat{c}$. Henceforth, constants that vanish logarithmically will always be denoted by $\hat{c}$. The result (2.4-5) therefore reads properly,

$$\Delta\Omega(\mathbf{a}) = \hat{c}\, G(\{a_i^I\})G(\{a_i^c - a_i^I\})\textstyle\prod_k da_k \tag{2.4-7}$$

leading to the pdf,

$$W(\{a_i\}) = \hat{c}\, G(\{a_i^I\})G(\{a_i^c - a_i^I\}) / \Delta\Omega(\mathcal{E}, \{a_i^c\}) \,. \tag{2.4-8}$$

As a direct extension, consider K subsystems with $\Sigma_\alpha a_i^\alpha = a_i^c$. Then, similar to (2.4-8),

$$W(\mathbf{a}^1 ... \mathbf{a}^{K-1}) = \hat{c} \prod_{\alpha=1}^{K} G(\mathbf{a}^\alpha) / \Delta\Omega(\mathcal{E}, \mathbf{a}^c) \,. \tag{2.4-9}$$

These results are rapidly reformulated for quantum statistics. Let $a_1 ... a_P$ be coarse grained (commuting) macroscopic fluctuating variables. Since [see (1.5-5)]

$$a_i = \sum_J a_i^{(J)} \sum_{\eta \in \Delta\mathcal{E}^J} \{|\eta\rangle\langle\eta|\}_i \tag{2.4-10}$$

we only need to count the energy cells J corresponding to the specification for the a_i to be in $(a_i, a_i + da_i)$ and the number of states in each cell. The *accessible number of quantum states* will be denoted by $\Delta\Gamma(\mathbf{a})$. This leads to a density of states in *a*-space $\Delta\Gamma(\mathbf{a}) = \chi(\mathbf{a})\Pi_i da_i$. The macro-probability density function is then given by

$$W(\{a_i\}) = \chi(\{a_i\}) / \Delta\Gamma(\mathcal{E}). \tag{2.4-11}$$

More likely, the a's will be pertaining to subsystems. The analogues of (2.4-8) and (2.4-9) are:

$$W(\{a_i\}) = \hat{c}\chi(\{a_i^I\})\chi(\{a_i^c - a_i^I\}) / \Delta\Gamma(\mathcal{E}, \{a_i^c\}) \tag{2.4-12}$$

$$W(\mathbf{a}^1, \ldots \mathbf{a}^{K-1}) = \hat{c}\prod_{\alpha=1}^{K} \chi(\mathbf{a}^\alpha) / \Delta\Gamma(\mathcal{E}, \mathbf{a}^c). \tag{2.4-13}$$

Some authors[14] absorb the constant $1/\Delta\Gamma(\mathcal{E})$, together with $\hat{c}$, in a 'big' constant C

$$C = \hat{c}\left[1/\Delta\Gamma(\mathcal{E})\right]. \tag{2-4-14}$$

This constant does not, however, vanish logarithmically! We shall refer to the quantity $W/C \equiv \tilde{W}$ as the 'thermodynamic probability'. Equations (2.4-12) and (2.4-13) then read:

$$\tilde{W}(\{a_i\}) \simeq \chi(\{a_i^I\})\chi(\{a_i^c - a_i^I\}), \tag{2.4-15}$$

$$\tilde{W}(\mathbf{a}^1, \ldots \mathbf{a}^{K-1}) \simeq \prod_{\alpha=1}^{K} \chi(\mathbf{a}^\alpha). \tag{2.4-16}$$

We note that $\tilde{W}$ is a huge number and that the thermodynamic probability *is not normalized.* Yet it is useful for some considerations, as we will see later.

Note. Quite often a semiclassical computation of $\Delta\Gamma(\mathcal{E})$ and $\Delta\Gamma(\mathbf{a})$ can be made. Let $\Delta\Omega$ be the relevant volume in phase space. According to Heisenberg's uncertainty principle p and q cannot be specified to within ~ h, i.e., $\prod_i^{rN} \Delta p_i \Delta q_i$ ~ h^{rN}. In addition the particles can be permuted, since in quantum mechanics they are indistinguishable. Therefore, the smallest volume in phase space, within which all points represent the same motion, is a *microcell* $\not{\Delta}\Omega = h^{3N} N!$ We thus surmise the following connection between accessible volume in phase space and accessible number of quantum states,

$$\Delta\Gamma = \Delta\Omega / \not{\Delta}\Omega = \Delta\Omega / h^{rN} N!. \tag{2.4-17}$$

---

[14] See e.g. L.D. Landau and E.M. Lifshitz, "Statistical Physics", Addison Wesley 1958, p.21, Eqs. (6.5) and (6.6).

Ter Haar[15], however, has pointed out that the uncertainty-principle argument leaves much to be desired, since actually $\Delta p_i \Delta q_i \geq h/4\pi$. Instead, the result follows from the connection between the Schrödinger q- and Schrödinger p-representations. From Parseval's theorem the amplitudes $\phi(q)$ and $A(k)$, $[p = \hbar k]$ are related by

$$\begin{aligned} \int \phi^*(q)\phi(q)d\Omega_q &= \frac{1}{(2\pi)^{rN}} \int A^*(k)A(k)d^{rN}k \\ &= \frac{1}{(2\pi\hbar)^{rN}} \int A^*(p)A(p)d\Omega_p = 1. \end{aligned} \tag{2.4-18}$$

According to ter Haar, the total volume in phase space corresponding to one stationary state is obtained by multiplying the probability in q-space with that in p-space and integrating over all phase space, while in addition the particles must be allowed to be permuted. We then obtain, using both equalities in (2.4-18):

$$\Delta\Omega = N! \iint \phi^*(q)\phi(q) A^*(p)A(p)d\Omega_q d\Omega_p = N!h^{rN}, \tag{2.4-19}$$

which is the desired result.

2.5 Examples of Extension in Phase Space and of Accessible Number of Quantum States

We have already remarked that the microcanonical ensemble is a poor one for actual calculations. On the other hand, it is difficult to grasp the results of the previous section without a few examples. We shall consider the simple problems of finding the energy probability density function for a subsystem of a perfect gas and for an assembly of harmonic oscillators.

2.5.1 *Ideal Gas*

Let us consider a mono-atomic ideal gas of N particles with energy $\mathcal{E}$ and volume V. The accessible volume in phase space is

$$\Delta\Omega(\mathcal{E},V,N) = \iint d\Omega_q d\Omega_p = V^N \int_{\mathcal{D}} dp_1 \ldots dp_{3N}, \tag{2.5-1}$$

where $\mathcal{D}$ is the domain bounded by

$$\mathcal{E} \leq \sum_{i=1}^{3N} p_i^2/2m \leq \mathcal{E} + \delta\mathcal{E}. \tag{2.5-2}$$

We will also write

[15] D. ter Haar, "Elements of Statistical Mechanics", Rinehart, NY, 1954, p. 60.

$$\Delta\Omega(\mathcal{E},V,N) = V^N \delta\mathcal{E} \frac{d}{d\mathcal{E}} \int_{\mathcal{D}'} dp_1 ... dp_{3N}, \tag{2.5-2'}$$

where $\mathcal{D}'$ is the domain bounded by $\sum_{i=1}^{3N} p_i^2 \leq 2m\mathcal{E}$. Clearly, $\int_{\mathcal{D}'} \Pi_i dp_i$ is the volume of a hypersphere.

To obtain this we consider the integral

$$I = \int_{-\infty}^{\infty} ... \int e^{-(x_1^2 + x_2^2 + ... x_\nu^2)} dx_1 ... dx_\nu = \left[\int_{-\infty}^{\infty} e^{-x^2} dx \right]^\nu = \pi^{\nu/2}. \tag{2.5-3}$$

We also compute this integral using 'polar coordinates'. Let $r^{\nu-1} A_\nu$ be the area of a hypersphere, where A_ν is a geometry factor (such as 4π if $\nu = 3$); then we have,

$$I = \int_0^\infty e^{-r^2} r^{\nu-1} A_\nu dr = \tfrac{1}{2} A_\nu \Gamma(\nu/2), \tag{2.5-3'}$$

where $\Gamma(y)$ is the generalized factorial or gamma function. Comparing (2.4-21) and (2.4-22) we find

$$A_\nu = 2\pi^{\frac{1}{2}\nu} / \Gamma(\tfrac{1}{2}\nu). \tag{2.5-4}$$

So, for the volume of a hypersphere with radius R we get,

$$V_\nu(R) = \int_0^R r^{\nu-1} A_\nu dr = R^\nu \pi^{\frac{1}{2}\nu} / \Gamma(\tfrac{1}{2}\nu + 1). \tag{2.5-5}$$

Setting $\nu = 3N$, $R = \sqrt{2m\mathcal{E}}$, we find for (2.5-2')

$$\Delta\Omega(\mathcal{E},V,N) = V^N \delta\mathcal{E} \frac{d}{d\mathcal{E}} \frac{(2\pi m\mathcal{E})^{3N/2}}{\Gamma(\frac{3}{2}N+1)} = V^N \delta\mathcal{E} \frac{2\pi m (2\pi m\mathcal{E})^{\frac{3}{2}N-1}}{\Gamma(\frac{3}{2}N)}. \tag{2.5-6}$$

For the number of accessible quantum states we then obtain,

$$\Delta\Gamma(\mathcal{E},V,N) = V^N \delta\mathcal{E} \frac{2\pi m (2\pi m\mathcal{E})^{\frac{3}{2}N-1}}{\Gamma(\frac{3}{2}N)\Gamma(N+1)h^{3N}}. \tag{2.5-7}$$

Since $\Delta\Gamma = \chi\delta\mathcal{E}$ this yields for the density of states,

$$\chi(\mathcal{E},V,N) = \frac{V^N 2\pi m (2\pi m\mathcal{E})^{\frac{3}{2}N-1}}{\Gamma(\frac{3}{2}N)\Gamma(N+1)h^{3N}}. \tag{2.5-8}$$

We now consider a small subsystem with attributes $\mathcal{E}'$, V' and n. Only energy exchange is considered, i.e., V' and n are fixed. From Equation (2.4-12) we find for the energy pdf of the subsystem,

$$W(\mathscr{E}') = \hat{c}\frac{V'^n(V-V')^{N-n}(2\pi m)(2\pi m\mathscr{E}')^{\frac{3}{2}n-1}[(2\pi m(\mathscr{E}-\mathscr{E}')]^{\frac{3}{2}(N-n)-1}}{V^N\delta\mathscr{E}(2\pi m\mathscr{E})^{\frac{3}{2}N-1}h^{3n}}$$
$$\times\frac{\Gamma(\frac{3}{2}N)\Gamma(N+1)}{\Gamma(\frac{3}{2}n)\Gamma(n+1)\Gamma[\frac{3}{2}(N-n)]\Gamma(N-n+1)}. \quad (2.5\text{-}9)$$

We shall only keep the essential terms associated with the subsystem. Thus we write

$$W(\mathscr{E}') = C\frac{V'^n(2\pi m)(2\pi m\mathscr{E}')^{\frac{3}{2}n-1}[2\pi m(\mathscr{E}-\mathscr{E}')]^{\frac{3}{2}N-1}}{\Gamma(\frac{3}{2}n)\Gamma(n+1)(2\pi m\mathscr{E})^{\frac{3}{2}N-1}h^{3n}}. \quad (2.5\text{-}10)$$

For the terms in $\mathscr{E}$ we have logarithmically

$$(\tfrac{3}{2}N-1)\ln\left(1-\frac{\mathscr{E}'}{\mathscr{E}}\right) \sim -(\tfrac{3}{2}N)\frac{\mathscr{E}'}{\mathscr{E}}. \quad (2.5\text{-}11)$$

Employing equipartition, $\mathscr{E} = \frac{3}{2}Nk_BT$, we arrive at

$$W(\mathscr{E}') = C\frac{V'^n(2\pi m)(2\pi m\mathscr{E}')^{\frac{3}{2}n-1}}{\Gamma(\frac{3}{2}n)\Gamma(n+1)h^{3n}}e^{-\mathscr{E}'/kT} = C\chi(\mathscr{E}')e^{-\mathscr{E}'/k_BT}. \quad (2.5\text{-}12)$$

2.5.2 *An Assembly of Oscillators*

(a) *Semiclassical*

We consider an assembly of N independent harmonic oscillators The energy is given by

$$\mathscr{E} = \sum_{i=1}^{N}(n_i+\tfrac{1}{2})\hbar\omega = (K+N/2)\hbar\omega, \quad (2.5\text{-}13)$$

where K is a sum of integers. We assume that the energy is quasi-continuous, with spacing for oscillator energy $g(\Delta\varepsilon) = \Delta\varepsilon/\hbar\omega$. For the total number of states we find,

$$\Delta\Gamma(\mathscr{E},N) = \delta\mathscr{E}\frac{d}{d\mathscr{E}}\left(\frac{1}{\hbar\omega}\right)^N\int_{\mathcal{O}} d\varepsilon_1 d\varepsilon_2 \ldots d\varepsilon_N, \quad (2.5\text{-}14)$$

where $\mathcal{O}$ is the domain

$$\varepsilon_1+\varepsilon_2+\ldots\varepsilon_N \le \mathscr{E}. \quad (2.5\text{-}15)$$

The result is easily obtained with the Dirichlet integral[16]:

[16] R.C. Tolman, "The Principles of Statistical Mechanics", Oxford University Press, 1938, p. 656.

$$\Delta\Gamma(\mathcal{E},N)=\left(\frac{1}{\hbar\omega}\right)^N \frac{1}{\Gamma(N)}\mathcal{E}^{N-1}\delta\mathcal{E}. \tag{2.5-16}$$

The Dirichlet integral is as follows. Let $\mathscr{C}$ be the domain bounded by

$$\sum_{j=1}^{\nu}(x_j/a_j)^{r_j} \le \mathscr{C}, \quad all\ x_j, a_j, r_j > 0. \tag{2.5-17}$$

Then

$$\int_{\mathscr{C}}\!\!...\!\int x_1^{i_1-1}x_2^{i_2-1}...\ x_\nu^{i_\nu-1}dx_1dx_2...dx_\nu = \frac{\left(\prod_{j=1}^{\nu}a_j^{\ i_j}\right)\left(\prod_{j=1}^{\nu}\Gamma(i_j/r_j)\right)(\mathscr{C})^{\sum_{j=1}^{\nu}(i_j/r_j)}}{\left(\prod_{j=1}^{\nu}r_j\right)\left(\Gamma\left[\sum_{j=1}^{\nu}(i_j/r_j)+1\right]\right)}. \tag{2.5-18}$$

In our example $i_1...i_\nu=1$, $a_1...a_\nu=1$, $r_1...r_\nu=1$, $\mathscr{C}=\mathcal{E}$. The result (2.5-16) then easily follows.

Again we can consider a subsystem and apply (2.4-12). We leave it up to the reader to obtain once more the form of the far rhs of (2.5-12).

We now indicate another method to evaluate integrals like (2.5-14). So consider

$$I(\mathcal{E},N)=\int_{\mathcal{O}}\!\!...\!\int d\varepsilon_1 d\varepsilon_2...d\varepsilon_N. \tag{2.5-19}$$

To rid ourselves of the energy constraint we take the Laplace transform (with $N\to n$):

$$\begin{aligned}\Phi(s)&=\int_0^\infty d\mathcal{E}\,e^{-s\mathcal{E}}\int_{\mathcal{O}}\!\!...\!\int d\varepsilon_1 d\varepsilon_2...d\varepsilon_n \\ &=\int_0^\infty d\mathcal{E}\,e^{-s\mathcal{E}}\int_0^{\mathcal{E}}d\varepsilon_n\int_0^{\mathcal{E}-\varepsilon_n}d\varepsilon_{n-1}...\int_0^{\mathcal{E}-\varepsilon_n-...\varepsilon_2}d\varepsilon_1.\end{aligned} \tag{2.5-20}$$

Intermezzo

To handle the above Laplace transform – of a type frequently occurring in statistical mechanics – we discuss the convolution theorem. Let $A_i(s)=\mathscr{L}[a_i(t)]$ denote the Laplace transform; the inverse will be denoted by the symbol $\mathscr{L}^{-1}$. The inverse for two or three factors is given by the convolution theorems,

$$\begin{aligned}\mathscr{L}^{-1}[A_1(s)A_0(s)]&=\int_0^t d\tau\, a_1(\tau)a_0(t-\tau), \\ \mathscr{L}^{-1}[A_2(s)A_1(s)A_0(s)]&=\int_0^t d\tau_2\int_0^{t-\tau_2}d\tau_1\, a_2(\tau_2)a_1(\tau_1)a_0(t-\tau_2-\tau_1).\end{aligned} \tag{2.5-21}$$

For a product of $(n+1)$ factors one may prove by induction,

$$\mathcal{L}^{-1}[A_n(s)\ldots A_0(s)] = \int_0^t d\tau_n \int_0^{t-\tau_n} d\tau_{n-1}\ldots \int_0^{t-\tau_n-\ldots\tau_2} d\tau_1 \times a_n(\tau_n)a_{n-1}(\tau_{n-1})\ldots a_1(\tau_1)a_0(t-\Sigma_i\tau_i)\,. \tag{2.5-22}$$

Needless to say, that we can also give the inverse statement,

$$\mathcal{L}[\int_0^t d\tau_n \int_0^{t-\tau_n} d\tau_{n-1}\ldots \int_0^{t-\tau_n-\ldots\tau_2} d\tau_1\; a_n(\tau_n)a_{n-1}(\tau_{n-1})\ldots a_1(\tau_1)a_0(t-\Sigma_i\tau_i)] = A_n(s)\ldots A_0(s)\,. \tag{2.5-23}$$

An important modification appears when we make the substitution of variables

$$\tau_n = t-\theta_n, \quad \tau_{n-1} = \theta_n - \theta_{n-1}, \quad \ldots \quad \tau_1 = \theta_2 - \theta_1\,. \tag{2.5-24}$$

Equation (2.5-23) then goes over into

$$\mathcal{L}[\int_0^t d\theta_n \int_0^{\theta_n} d\theta_{n-1} \int_0^{\theta_{n-1}} d\theta_{n-2}\ldots \int_0^{\theta_2} d\theta_1 \times a_n(t-\theta_n)a_{n-1}(\theta_n-\theta_{n-1})\ldots a_1(\theta_2-\theta_1)a_0(\theta_1)] = A_n(s)\ldots A_0(s)\,. \tag{2.5-24}$$

The repeated integral on the left-hand-side is called a *time-ordered integral,* since $t > \theta_n > \theta_{n-1}\ldots > \theta_1$. These results will often be needed.

We now return to Equation (2.5-20). For this case $t \to \mathcal{E}$, $\varepsilon_i \to \tau_i$; all $(N+1)$ factors a_i are 1. Hence, $A(s) = 1/s$. The result is therefore

$$\Phi(s) = 1/s^{N+1}\,, \tag{2.5-25}$$

with inversion

$$I(\mathcal{E},N) = \mathcal{E}^N/\Gamma(N+1)\,. \tag{2.5-26}$$

Whence,

$$\chi(\mathcal{E},N) = \left(\frac{1}{\hbar\omega}\right)^N \frac{d}{d\mathcal{E}} I(\mathcal{E},N) = \left(\frac{1}{\hbar\omega}\right)^N \frac{\mathcal{E}^{N-1}}{\Gamma(N)}\,, \tag{2.5-27}$$

in full accord with (2.5-16).

(b) *Quantum Mechanical*

An exact result is also readily obtained. With the energy quantization given by (2.5-13) we have

$$n_1 + n_2 + \ldots + n_N = K\,, \tag{2.5-28}$$

where $K = \mathcal{E}/\hbar\omega - N/2$. The question is now: what is the possible number of sets $\{n_i\}$ concordant with (2.5-28)? This problem will be met later in Bose-Einstein statistics. We can also formulate the question as follows: In how many ways can we distribute K entities over N boxes, with no limit to the number of entities n_i per box , except that n_i cannot exceed K. The usual way to solve this is to put the K entities on a row, after which we put in $(N-1)$ partitions to form the 'boxes'; see Fig. 2-3.

Fig. 2-3 : the boxes problem.

In this example there are five boxes with 2,3,0,4 and 2 particles in them. Clearly, altogether we have K+N–1 objects on a line, divided into groups of K and N–1. The number of states we have realised is

$$\Delta\Gamma(\mathcal{E},N) = \frac{(K+N-1)!}{K!(N-1)!} = \frac{[(\mathcal{E}/\hbar\omega)+\frac{1}{2}N-1]!}{[(\mathcal{E}/\hbar\omega)-\frac{1}{2}N]!(N-1)!}. \tag{2.5-29}$$

Dividing by $\delta\mathcal{E} = \hbar\omega$ we find the density of states,

$$\chi(\mathcal{E},N) = \left(\frac{1}{\hbar\omega}\right)\frac{(K+N-1)!}{K!(N-1)!} = \left(\frac{1}{\hbar\omega}\right)\frac{[(\mathcal{E}/\hbar\omega)+\frac{1}{2}N-1]!}{[(\mathcal{E}/\hbar\omega)-\frac{1}{2}N]!(N-1)!}. \tag{2.5-30}$$

This is the exact answer. To re-obtain the semiclassical result, assume that $\mathcal{E}$ is very large. Then we have,

$$\frac{[(\mathcal{E}/\hbar\omega)+\frac{1}{2}N-1]!}{[(\mathcal{E}/\hbar\omega)-\frac{1}{2}N]!}$$
$$\sim [(\mathcal{E}/\hbar\omega)+\tfrac{1}{2}N-1]\times\ldots\times[(\mathcal{E}/\hbar\omega)-\tfrac{1}{2}N+1] \sim (\mathcal{E}/\hbar\omega)^{N-1}. \tag{2.5-31}$$

When this is substituted in (2.5-30) the semiclassical result (2.5-27) follows, noting that $\Gamma(N) = (N-1)!$.

2.5.3 *A General Form for* $\Delta\Gamma(\mathcal{E},V,N)$ *in the Microcanonical Ensemble*

We suppose that the Hamiltonian can be expressed in the generalized coordinates and momenta:

$$\mathcal{H} = \sum_{i=1}^{rN} p_i^2/2b_i + \mathcal{V}(q_1 \ldots q_{rN}), \tag{2.5-32}$$

where the kinetic energy includes translational and internal degrees of freedom. Then,

$$\Omega(\mathcal{E},V,N)=\int_{\Sigma_i(p_i^2/2b_i)+\mathcal{V}\le\mathcal{E}} d\Omega_p d\Omega_q\,. \tag{2.5-33}$$

We set

$$p_i^2/2b_i=\varepsilon_i,\quad dp_i=d\varepsilon_i\sqrt{b_i/2\varepsilon_i}\,. \tag{2.5-34}$$

Thus,[17]

$$\Omega(\mathcal{E},V,N)=\int d\Omega_q\int\ldots\int_{\Sigma_i\varepsilon_i\le\mathcal{E}-\mathcal{V}(q)} d\varepsilon_1\ldots d\varepsilon_{rN}\prod_{j=1}^{rN}\left(\frac{b_j}{2\varepsilon_j}\right)^{\frac{1}{2}}2^{rN}. \tag{2.5-35}$$

We take the Laplace transform, noting $\mathcal{E}\ge\mathcal{V}(q)$:

$$\begin{aligned}\Phi(s,V,N)&\\ &=\int d\Omega_q\int_{\mathcal{V}(q)}^{\infty}e^{-s\mathcal{E}}d\mathcal{E}\int_0^{\mathcal{E}-\mathcal{V}(q)}d\varepsilon_{rN}\int_0^{\mathcal{E}-\mathcal{V}(q)-\varepsilon_{rN}}d\varepsilon_{rN-1}\cdots\int_0^{\mathcal{E}-\mathcal{V}(q)-\varepsilon_{rN}\ldots-\varepsilon_2}d\varepsilon_1\,2^{rN}\prod_{j=1}^{rN}\left(\frac{b_j}{2\varepsilon_j}\right)^{\frac{1}{2}}\\ &=2^{rN}\int d\Omega_q e^{-s\mathcal{V}(q)}\int_0^{\infty}e^{-s\tilde{\mathcal{E}}}d\tilde{\mathcal{E}}\int_0^{\tilde{\mathcal{E}}}\left(\frac{b_{rN}}{2\varepsilon_{rN}}\right)^{\frac{1}{2}}d\varepsilon_{rN}\int_0^{\tilde{\mathcal{E}}-\varepsilon_{rN}}\left(\frac{b_{rN-1}}{2\varepsilon_{rN-1}}\right)^{\frac{1}{2}}d\varepsilon_{rN-1}\cdots\\ &\times\int_0^{\tilde{\mathcal{E}}-\varepsilon_{rN}\ldots-\varepsilon_2}\left(\frac{b_1}{2\varepsilon_1}\right)^{\frac{1}{2}}d\varepsilon_1\,.\end{aligned} \tag{2.5-36}$$

[We set $\tilde{\mathcal{E}}=\mathcal{E}-\mathcal{V}(q)$.] Leaving the integral $\int d\Omega_q\exp[-s\mathcal{V}(q)]$ aside, the rest is amenable to the convolution theorem (2.5-23), with rN factors $(b_i/2\varepsilon_i)^{\frac{1}{2}}$ and one factor $1\to 1/s$. For the former we have, going back to a Gaussian integral,

$$\int_0^{\infty}e^{-s\varepsilon_i}\left(\frac{b_i}{2\varepsilon_i}\right)^{\frac{1}{2}}d\varepsilon_i=\int_0^{\infty}e^{-sp_i^2/2b_i}dp_i=\tfrac{1}{2}\sqrt{\frac{2\pi b_i}{s}}\,. \tag{2.5-37}$$

Accordingly $\Phi(s,V,N)$ is given by,

$$\Phi(s,V,N)=\int d\Omega_q e^{-s\mathcal{V}(q)}\sqrt{\frac{2\pi b_{rN}}{s}}\sqrt{\frac{2\pi b_{rN-1}}{s}}\cdots\sqrt{\frac{2\pi b_1}{s}}\frac{1}{s}. \tag{2.5-38}$$

The inverse is obtained as – see footnote 19 of Chapter I ,

$$\Omega(\mathcal{E},V,N)=(2\pi)^{\frac{1}{2}rN}\prod_{j=1}^{rN}\sqrt{b_j}\int_{\Omega_q=0}^{\mathcal{V}=\mathcal{E}}d\Omega_q[\mathcal{E}-\mathcal{V}(q)]^{\frac{1}{2}rN}/\Gamma(\tfrac{1}{2}rN+1)\,. \tag{2.5-39}$$

---

[17] The resulting integral must be multiplied by 2^{rN} to allow for both positive and negative values of the p_i's.

From this we have for the microcanonical partition function,

$$\Delta\Gamma(\mathcal{E},V,N)=\frac{(2\pi)^{\frac{1}{2}rN}\,\delta\mathcal{E}}{h^{rN}\Gamma(N+1)\Gamma(\frac{1}{2}rN+1)}\prod_{j=1}^{rN}\sqrt{b_j}$$
$$\times\frac{d}{d\mathcal{E}}\int_{\Omega_q=0}^{\mathcal{V}=\mathcal{E}} d\Omega_q\,[\mathcal{E}-\mathcal{V}(q)]^{\frac{1}{2}rN}. \qquad (2.5\text{-}40)$$

This is a very general result, from which we can obtain the entropy (see next chapter). Of course, the final integral is far from trivial! It was first approximated for real gases by Streiter and Mayer.[18]

2.6 Problems to Chapter II

2.1 Consider a perfect classical gas of charged particles (charge q) in a constant magnetic field $\mathbf{B}=B\hat{k}$, for which the Landau vector potential $\mathbf{A}=(0,Bx,0)$ applies. Show that the Hamiltonian is given by

$$\mathcal{H}=\sum_{i=1}^{N}[p_{i,x}^2+(p_{i,y}\pm m\omega_c q_{i,x})^2+p_{i,z}^2]/2m\,, \qquad (1)$$

where $\omega_c=|q|B/mc$ is the cyclotron frequency. The system is part of a microcanonical ensemble. Make a change of variables

$$Q_i=q_i,\quad P_{i,x}=p_{i,x},\quad P_{i,y}=p_{i,y}+m\omega_c q_{i,x},\quad P_{i,z}=p_{i,z}. \qquad (2)$$

Find the Jacobian and the accessible volume in phase space $\Delta_{\mathcal{E}}\Omega(\mathcal{E},V,N)$. Show that it does not depend on the magnetic field (nor do any thermodynamic properties depend on **B**; this failure for a classical gas to yield diamagnetism is known as the Bohr-van Leeuwen theorem, see Chapter VI).

2.2 Consider a system of (semi)-classical harmonic oscillators,

$$\mathcal{H}=\sum_i^N\left(p_i^2/2m+\tfrac{1}{2}m\omega^2q_i^2\right). \qquad (3)$$

(a) Compute the volume occupied in phase space, $\Delta_{\mathcal{E}}\Omega$, by making a change of variables, such that the volume is that between two hyperspheres [do *not* use the Dirichlet integral].
(b) Next find the associated number of accessible quantum states, $\Delta_{\mathcal{E}}\Gamma$, noting that harmonic oscillators are *distinguishable* objects, so that the factor $N!$ in the microcell volume must be omitted. Compare your result with Eq. (2.5-27).

[18] S.F. Streiter and J.E. Mayer, J. Chem. Physics **7**, 1025 (1939).

(c) Obtain the associated entropy $S = k_B \ln \Delta_{\mathscr{E}}\Gamma$. Find the temperature T and obtain the classical equipartition result $\mathscr{E} = Nk_BT$.

2.3 Consider a perfect classical gas in an adiabatic enclosure, which is divided into two subsystems, v and $V - v >> v$. We assume that the two subsystems are separated by an adiabatically lined impermeable moveable piston, so that there is free volume exchange but no energy or particle exchange. Show that the probability density function (pdf) is a gamma distribution,

$$W(v) = Cv^{n-1}e^{-Nv/V}, \tag{4}$$

where n and $N - n \gg n$ are particle numbers in the two subsystems. Find the moments $\langle v^k \rangle$ and the variance $\operatorname{var} v$.

2.4 In the same problem as above, assume that v and $V - v$ are of comparable size. Show that $W(v)$ is a beta distribution and express the moments $\langle v^k \rangle$ and the variancc $\operatorname{var} v$ as beta functions, making no approximations. Compare the answers with those of problem 2.3.

2.5 At *extremely* low temperatures the relevant eigenvalue spectrum of any system becomes discrete. Using Kronecker deltas rather than Dirac delta functions, restate the expressions for $p(\eta)$ and ρ for a microcanonical ensemble

2.6 The von Neumann equation $\partial\rho/\partial t + (1/\hbar i)[\rho, H] = 0$ is often written as

$$\partial\rho/\partial t + i\mathscr{L}\rho = 0, \tag{2.3-5'}$$

where $\mathscr{L}$ is the *superoperator* (i.e., an operator acting on an operator) defined by $\mathscr{L}K = (1/\hbar)\mathscr{L}K$. The solutions of the ordinary commutator form and of the superoperator form are indicated in (2.3-6):

$$\rho(t) = e^{-iH(t-t_0)/\hbar}\rho(t_0)e^{iH(t-t_0)/\hbar}, \text{ or } \rho(t) = e^{-i\mathscr{L}(t-t_0)}\rho(t_0). \tag{2.3-6'}$$

Show by series expansion that the two solutions are identical, given the definition of the superoperator $\mathscr{L}$ stated above.[19]

---

[19] Louisell states the following general theorem: Let A and B be two generally non-commuting operators of the same Hilbert space and let ξ be a parameter, then:

$$e^{\xi A}Be^{-\xi A} = B + \xi[A,B] + \frac{\xi^2}{2!}[A,[A,B]] + \frac{\xi^3}{3!}[A,[A,[A,B]]] + \ldots, \tag{5}$$

cf. W. Louisell "Radiation and Noise in Quantum Electronics", Mc Graw-Hill, 1964, Section 3.2, theorem 3.

Chapter III

Thermodynamics in the Microcanonical Ensemble, Classical and Quantal

3.1 Gibbs' Form of the Ergodic Density; Entropy for Classical Systems

In this chapter we shall arrive at the connection between entropy and volume in phase space $\Delta\Omega(\mathscr{E})$ (classical), and accessible number of quantum states $\Delta\Gamma(\mathscr{E})$ (quantal). In many modern texts the connection between these attributes and entropy is postulated; however, this most certainly involves entire disregard for Gibbs' oeuvre, whose purpose it was to *provide* a microscopic meaning for macroscopic entropy, as it was known since the days of Clausius and Maxwell. We shall therefore reproduce the essence of his derivation for classical systems (this section) and 'update' it for quantum systems in the next section.

The classical microcanonical density function was derived at and given in Eq. (2.1-12). Since Gibbs needed the derivative of the density function associated with a *shift* in the ensemble parameters, including the normalization 'constant' – a trick he often employed to great advantage! – he used a Gaussian density function to ease the algebra.[1] Therefore, instead of (2.1-12) we consider the density function

$$\rho(p,q) = \exp\Big(c(\omega,\{A_i\}) - [\mathscr{H}(p,q;\{A_i\}) - \mathscr{E}(\{A_i\})]^2/\omega^2\Big), \tag{3.1-1}$$

where $c(\omega,\{A_i\})$ serves to normalize the distribution. The A_i are the parameters that define the ensemble, involving volume V and particle number N, as well as other extensive and possibly intensive parameters, see our discussion in subsection 1.3.2; collectively the A_i are referred to as 'external parameters'. For $\omega \to 0$, (3.1-1) becomes a delta function. The normalization is direct; we have, writing $\{A_i\} \to A$,

$$\begin{aligned} e^{-c(\omega,A)} &= \int_{\Gamma-space} e^{-[\mathscr{H}(p,q;A)-\mathscr{E}(A)]^2/\omega^2} d\Omega \\ &= \int_{\mathscr{E}_m(A)}^{\infty} d\mathscr{E}' \oint dM(\mathscr{E}',A)\, e^{-[\mathscr{E}'-\mathscr{E}(A)]^2/\omega^2}, \end{aligned} \tag{3.1-2}$$

---

[1] J.W. Gibbs, Op. Cit. Chapter X. Dirac's delta function and its properties were of course unknown in 1902 so, for historic interest (and convenience!) we shall here follow Gibbs' treatment; however, in section 3.3 we will find it advantageous to employ the delta function.

where we used (2.1-9) to convert to an energy-shell integral. From the last expression, knowing that the Gaussian has a small width, we easily find

$$\begin{aligned}-c(\omega,A) &\approx \ln[(\tfrac{1}{2})\sqrt{\pi}\,\omega M(\mathscr{E},A)] \\ &\approx \ln[(\sqrt{\pi}\,\omega/2\delta\mathscr{E})\,\Delta\Omega(\mathscr{E},A)] \approx \ln\Delta\Omega(\mathscr{E},A).\end{aligned} \tag{3.1-3}$$

Some further notes are in order. Equation (3.1-3) gives the connection between the present normalization constant c and the surface measure M used before, with the connection (2.1-14) stated previously; we assumed that ω and $\delta\mathscr{E}$ are of the same order of smallness. The energy $\mathscr{E}_m$ is the minimum energy compatible with the external specification. As a rule the limitation, if any, is on the coordinates q and not on the momenta p; thus, $\mathscr{E}_m$ is a surface in the coordinates subspace, having no extension in the full phase space, so that $M(\mathscr{E}_m)$ can be taken to be zero. This point is extensively belaboured in Gibbs' book; we shall dispense with the discussion.

We now apply a quasi-static change to the ensemble, which involves shifts dA_i and consequently a shift in normalization dc. We differentiate separately both expressions of (3.1-2) to a particular A_i. The first one yields,

$$\begin{aligned}-\frac{\partial c}{\partial A_i} &= \int_{\Gamma-space} d\Omega\,\frac{2(\mathscr{H}-\mathscr{E})}{\omega^2}\left[-\frac{\partial\mathscr{H}}{\partial A_i}+\frac{\partial\mathscr{E}}{\partial A_i}\right]e^{c-[\mathscr{H}-\mathscr{E}(A)]^2/\omega^2} \\ &= \int_{\mathscr{E}_m}^{\infty} d\mathscr{E}'M(\mathscr{E}',A)\frac{2(\mathscr{E}'-\mathscr{E})}{\omega^2}\left[\overline{F_i}^{\mathscr{E}'}+\frac{\partial\mathscr{E}}{\partial A_i}\right]e^{c-[\mathscr{E}'-\mathscr{E}(A)]^2/\omega^2};\end{aligned} \tag{3.1-4}$$

here the 'generalized forces' (more appropriately, conjugate variables) are defined as previously discussed,

$$\overline{F_i}^{\mathscr{E}'} = -\overline{\frac{\partial\mathscr{H}(p,q;A)}{\partial A_i}}^{\mathscr{E}'}, \tag{3.1-5}$$

where the bar denotes an average for all systems with energy $\mathscr{E}'$. We now integrate by parts and find,

$$\begin{aligned}-\frac{\partial c}{\partial A_i} &= -\left\{M(\mathscr{E}',A)\left[\overline{F_i}^{\mathscr{E}'}+\frac{\partial\mathscr{E}}{\partial A_i}\right]e^{c-[\mathscr{E}'-\mathscr{E}(A)]^2/\omega^2}\right\}_{\mathscr{E}_m}^{\infty} \\ &+\int_{\mathscr{E}_m}^{\infty} d\mathscr{E}'\left[\frac{\partial\overline{F_i}^{\mathscr{E}'}}{\partial\mathscr{E}'}M(\mathscr{E}',A)+\left(\overline{F_i}^{\mathscr{E}'}+\frac{\partial\mathscr{E}}{\partial A_i}\right)\frac{\partial M(\mathscr{E}',A)}{d\mathscr{E}'}\right]e^{c-[\mathscr{E}'-\mathscr{E}(A)]^2/\omega^2}.\end{aligned} \tag{3.1-6}$$

Note that the expression in curly brackets vanishes at both limits, as argued above. We now differentiate the second expression of (3.1-2). This yields,

$$-\frac{\partial c}{\partial A_i}=\int_{\mathcal{E}_m}^{\infty} d\mathcal{E}'\left[\frac{2(\mathcal{E}'-\mathcal{E})}{\omega^2}\frac{\partial \mathcal{E}}{\partial A_i}M(\mathcal{E}',A)+\frac{\partial M(\mathcal{E}',A)}{\partial A_i}\right]e^{c-(\mathcal{E}'-\mathcal{E})^2/\omega^2}$$
$$-\left\{M(\mathcal{E}',A)\frac{\partial \mathcal{E}'}{\partial A_i}e^{c-(\mathcal{E}'-\mathcal{E})^2/\omega^2}\right\}_{\mathcal{E}'=\mathcal{E}_m}, \tag{3.1-7}$$

where the last term stems from the differentiation to the lower limit; it vanishes since $M(\mathcal{E}_m)$ was assumed to be zero. Using once more integration by parts, we obtain,

$$-\frac{\partial c}{\partial A_i}=-\left\{\frac{\partial \mathcal{E}}{\partial A_i}M(\mathcal{E}',A)e^{c-(\mathcal{E}'-\mathcal{E})^2/\omega^2}\right\}_{\mathcal{E}_m}^{\infty}$$
$$+\int_{\mathcal{E}_m}^{\infty} d\mathcal{E}'\left[\frac{\partial \mathcal{E}}{\partial A_i}\frac{\partial M(\mathcal{E}',A)}{\partial \mathcal{E}'}+\frac{\partial M(\mathcal{E}',A)}{\partial A_i}\right]e^{c-(\mathcal{E}'-\mathcal{E})^2/\omega^2}, \tag{3.1-8}$$

in which the curly bracket gives zero. Finally, we compare (3.1-6) and (3.1-8) and we let $\omega \to 0$, so that we can equate the integrands, evaluated at the peak energy $\mathcal{E}$, to obtain:

$$M(\mathcal{E},A)\frac{\partial \overline{F}_i}{\partial \mathcal{E}}+\overline{F}_i\frac{\partial M}{\partial \mathcal{E}}-\frac{\partial M}{\partial A_i}=0. \tag{3.1-9}$$

We combine the first two terms and write this in terms of Ω $[M=\partial\Omega/\partial\mathcal{E}]$:

$$\frac{\partial}{\partial \mathcal{E}}\left[\overline{F}_i\frac{\partial\Omega}{\partial \mathcal{E}}\right]-\frac{\partial^2\Omega}{\partial \mathcal{E}\partial A_i}=0. \tag{3.1-10}$$

Integrating this from $\mathcal{E}$ to $\mathcal{E}+\delta\mathcal{E}$ we find,

$$\overline{F}_i\frac{\partial\Delta\Omega(\mathcal{E},A)}{\partial \mathcal{E}}=\frac{\partial\Delta\Omega(\mathcal{E},A)}{\partial A_i}, \tag{3.1-11}$$

or,

$$\overline{F}_i=\frac{[\partial\Delta\Omega(\mathcal{E},A)/\partial A_i]_{\mathcal{E}}}{[\partial\Delta\Omega(\mathcal{E},A)/\partial\mathcal{E}]_{\{A_i\}}}. \tag{3.1-12}$$

This is our final result. Before we apply it, let us see its meaning more properly by using the "-1 theorem":

$$\left(\frac{\partial\Delta\Omega}{\partial A_i}\right)_{\mathcal{E}}\left(\frac{\partial\mathcal{E}}{\partial\Delta\Omega}\right)_{A_i}\left(\frac{\partial A_i}{\partial\mathcal{E}}\right)_{\Delta\Omega}=-1, \tag{3.1-13}$$

so that (3.1-12) also reads,

$$\overline{F}_i=-(\partial\mathcal{E}/\partial A_i)_{\Delta\Omega}. \tag{3.1-14}$$

Comparing with (3.1-5) we see that the average of the derivative of $\mathcal{H}$(p,q;A) to A_i equals the derivative of the average $\overline{\mathcal{H}} = \mathcal{E}$ to A_i, providing a new *state function* $\Delta\Omega$ is held constant. We shall presently show that $\Delta\Omega$ is linked to the entropy. However, on joining two systems, $\Delta\Omega$ is multiplicative, rather than additive, as we saw before. Thus, consider the total differential for $\ln \Delta\,\Omega$; we have by the chain rule,

$$d\ln\Delta\Omega(\mathcal{E},A) = \left(\frac{\partial \ln \Delta\Omega}{\partial \mathcal{E}}\right)_{\{A_i\}} d\mathcal{E} + \sum_{i=1}^{p}\left(\frac{\partial \ln \Delta\Omega}{\partial A_i}\right)_{\mathcal{E}} dA_i\,. \tag{3.1-15}$$

From (3.1-12),

$$\left(\frac{\partial \ln \Delta\Omega}{\partial A_i}\right)_{\mathcal{E}} = \overline{F}_i\left(\frac{\partial \ln \Delta\Omega}{\partial \mathcal{E}}\right)_{\{A_i\}}. \tag{3.1-16}$$

Substituting into (3.1-15) this yields,

$$d\ln\Delta\Omega(\mathcal{E},A) = \left(\frac{\partial \ln \Delta\Omega}{\partial \mathcal{E}}\right)_{\{A_i\}}\left[d\mathcal{E} + \sum_{i=1}^{p}\overline{F}_i\, dA_i\right]. \tag{3.1-17}$$

This is just the Gibbs relation (1.3-14), governing the change of entropy in quasi-static processes, providing we set

$$S = k_B \ln \Delta\Omega(\mathcal{E},\{A_i\}) + const\,, \tag{3.1-18}$$

where k_B is Boltzmann's constant, chosen to give S the usual dimension. In addition we found,

$$\frac{1}{T} = k_B\left(\frac{\partial \ln \Delta\Omega(\mathcal{E},\{A_i\})}{\partial \mathcal{E}}\right)_{\{A_i\}}. \tag{3.1-19}$$

We pause to reflect on the procedure just presented. In the ensemble description of statistical mechanics the most important state function is the energy, being a function of external parameters, either of an extensive or an intensive nature. The mean generalized forces (or conjugate variables) appear as energy derivatives. The entropy comes in 'via the backdoor' as the logarithm of a volume in phase space. This truly amazing feat is not *a priori* evident, but Eq. (3.1-19) is a better starting point. Temperature is a consequence of random kinetic energy $\mathcal{T}$; while in a closed system $\mathcal{E}$ is a constant, $\mathcal{T}$ is not. Once we recognize that $\overline{\mathcal{T}}$ is defined, the concept of temperature in a mechanical system is immanent. Clearly, then there is also the entropy as conjugate variable, *now defined on a mechanical basis.* This will once more be shown in section 3.1, where we do a straightforward computation of $\overline{\mathcal{T}}$, showing that it is proportional to the temperature defined here. Gibbs' derivation as presented shows the simplicity *and ingenuity* of his approach: all that was done is the manipulation of the variation of the normalization constant under quasi-static changes of the ensemble. We shall meet various other examples of his technique later, in particular in Chapter V.

3.2 Entropy for Quantum Systems

The analogue of Gibbs' derivation for quantum systems will be pursued presently.[2] We start with the density operator,

$$\rho = \exp\left(c(\omega, A) - [\mathcal{H}(A) - I\mathcal{E}(A)]^2 / \omega^2\right). \tag{3.2-1}$$

Here I is the identity operator. The normalization is expressed by

$$e^{-c(\omega,A)} = \mathrm{Tr}\, e^{-[\mathcal{H}(A) - I\mathcal{E}(A)]^2 / \omega^2}. \tag{3.2-2}$$

The trace will be evaluated in the representation $\{|\eta\rangle\}$, being the eigenstates of $\mathcal{H}$. Since the states are dense, we will also write $\Sigma_\eta \to \int d\mathcal{E}\chi(\mathcal{E})$, where $\chi(\mathcal{E})$ is the density of states. We thus obtain,

$$\begin{aligned} e^{-c(\omega,A)} &= \sum_\eta \langle \eta | e^{-[\mathcal{H}(A) - I\mathcal{E}(A)]^2 / \omega^2} | \eta \rangle \\ &= \sum_\eta e^{-[\mathcal{E}_\eta(A) - \mathcal{E}(A)]^2 / \omega^2} = \int_{\mathcal{E}_m}^{\infty} d\mathcal{E}' \chi(\mathcal{E}', A) e^{-[\mathcal{E}' - \mathcal{E}(A)]^2 / \omega^2}; \end{aligned} \tag{3.2-3}$$

note that $\mathcal{E}'$ is a dummy variable. From this expression we easily find that

$$-c(\omega, A) = \ln[(\sqrt{\pi}\omega / 2\delta\mathcal{E})\Delta\Gamma(\mathcal{E}, A)] \approx \ln[\Delta\Gamma(\mathcal{E}, A)], \tag{3.2-4}$$

which is the analogue of (3.1-3). We introduce the operators[3]

$$F_i = -\partial\mathcal{H} / \partial A_i, \tag{3.2-5}$$

in which the A's are c-numbers. Let now the ensemble undergo a shift dA_i. From (3.2-2) we then have,

$$\begin{aligned} -\frac{\partial c}{\partial A_i} &= \mathrm{Tr}\left\{ \frac{2(\mathcal{H} - \mathcal{E})}{\omega^2} \left(-\frac{\partial \mathcal{H}}{\partial A_i} + I \frac{\partial \mathcal{E}}{\partial A_i} \right) e^{c(\omega,A) - [\mathcal{H}(A) - I\mathcal{E}(A)]^2 / \omega^2} \right\} \\ &= \sum_\eta \left\{ \frac{2(\mathcal{E}_\eta - \mathcal{E})}{\omega^2} \left(\langle \eta | F_i | \eta \rangle + \frac{\partial \mathcal{E}}{\partial A_i} \right) e^{c(\omega,A) - [\mathcal{E}_\eta - \mathcal{E}(A)]^2 / \omega^2} \right\} \end{aligned}$$

---

[2] C.M. Van Vliet, *Lecture notes,* Centre de Recherches Mathématiques, U. de Mtl. (1984); unpublished.

[3] When carrying out the differentiation, noting that exponential operators are defined by series expansion, we must Hermitize the terms $(\partial / \partial A_i)[f(H)]^n$; however, since Tr(AB)=Tr(BA), this has no effect on the result.

$$= \int_{\mathcal{E}_m}^{\infty} d\mathcal{E}'\chi(\mathcal{E}',A)\frac{2(\mathcal{E}'-\mathcal{E})}{\omega^2}\left(\overline{F}_i^{\mathcal{E}'} + \frac{\partial \mathcal{E}}{\partial A_i}\right)e^{c(\omega,A)-[\mathcal{E}'-\mathcal{E}(A)]^2/\omega^2}. \tag{3.2-6}$$

This is integrated by parts,

$$-\frac{\partial c}{\partial A_i} = \int_{\mathcal{E}_m}^{\infty} d\mathcal{E}'\left\{\frac{\partial \overline{F}_i^{\mathcal{E}'}}{\partial \mathcal{E}'}\chi(\mathcal{E}',A) + \left(\overline{F}_i^{\mathcal{E}'} + \frac{\partial \mathcal{E}}{\partial A_i}\right)\frac{\partial \chi}{\partial \mathcal{E}'}\right\}e^{c(\omega,A)-[\mathcal{E}'-\mathcal{E}(A)]^2/\omega^2}. \tag{3.2-7}$$

Alternatively, we do the differentiation from (3.2-3), to obtain,

$$-\frac{\partial c}{\partial A_i} = \int_{\mathcal{E}_m}^{\infty} d\mathcal{E}'\left\{\chi(\mathcal{E}',A)\frac{2(\mathcal{E}'-\mathcal{E})}{\omega^2}\frac{\partial \mathcal{E}}{\partial A_i} + \frac{\partial \chi}{\partial A_i}\right\}e^{c(\omega,A)-[\mathcal{E}'-\mathcal{E}(A)]^2/\omega^2}. \tag{3.2-8}$$

Integration by parts gives

$$-\frac{\partial c}{\partial A_i} = \int_{\mathcal{E}_m}^{\infty} d\mathcal{E}'\left(\frac{\partial \mathcal{E}}{\partial A_i}\frac{\partial \chi}{\partial \mathcal{E}'} + \frac{\partial \chi}{\partial A_i}\right)e^{c(\omega,A)-[\mathcal{E}'-\mathcal{E}(A)]^2/\omega^2}. \tag{3.2-9}$$

We now compare (3.2-7) and (3.2-9) and we let $\omega \to 0$. We can then equate the integrands at the peak value $\mathcal{E}$. This results in

$$\frac{\partial \overline{F}_i}{\partial \mathcal{E}}\chi(\mathcal{E},A) + \overline{F}_i\frac{\partial \chi}{\partial \mathcal{E}} - \frac{\partial \chi}{\partial A_i} = 0, \quad \text{or}$$

$$\frac{\partial}{\partial \mathcal{E}}\left(\overline{F}_i\frac{\partial \Gamma}{\partial \mathcal{E}}\right) - \frac{\partial^2 \Gamma}{\partial \mathcal{E}\partial A_i} = 0, \tag{3.2-10}$$

where Γ counts all states up to $\mathcal{E}$, i.e., $\chi(\mathcal{E},A) = \partial\Gamma/\partial\mathcal{E}$. The last result is integrated from $\mathcal{E}$ to $\mathcal{E}+\delta\mathcal{E}$ and we note $\Gamma(\mathcal{E}+\delta\mathcal{E},A)-\Gamma(\mathcal{E},A) = \Delta\Gamma(\mathcal{E},A)$. The final result is

$$\overline{F}_i\left(\frac{\partial \Delta\Gamma(\mathcal{E},A)}{\partial \mathcal{E}}\right)_{\{A_i\}} = \left(\frac{\partial \Delta\Gamma(\mathcal{E},A)}{\partial A_i}\right)_{\mathcal{E}}. \tag{3.2-11}$$

We are interested in the properties of $\ln \Delta\Gamma(\mathcal{E},A)$. For the total differential we obtain, using (3.2-11):

$$d\ln \Delta\Gamma(\mathcal{E},A) = \left(\frac{\partial \ln \Delta\Gamma(\mathcal{E},A)}{\partial \mathcal{E}}\right)_{\{A_i\}}\left(d\mathcal{E} + \sum_{i=1}^{p}\overline{F}_i dA_i\right), \tag{3.2-12}$$

which is the Gibbs relation $dS = (1/T)[d\mathcal{E} + \Sigma_i \overline{F}_i dA_i]$, providing we set,

$$S = k_B \ln \Delta\Gamma(\mathcal{E},A), \tag{3.2-13}$$

$$T^{-1} = k_B\left(\partial \ln \Delta\Gamma(\mathcal{E},A)/\partial\mathcal{E}\right)_{\{A_i\}}. \tag{3.2-14}$$

The possible constant – see (3.1-18) – we omitted; the *absolute* entropy so obtained will satisfy Nernst's law, as we discuss later. Equation (3.2-13) is the main result: *the equilibrium entropy is k_B times the logarithm of the microcanonical partition function.* Similar relationships hold for other ensembles, as will be shown in the next two chapters.

3.3 The Equipartition Law

We shall proceed to give a proof for the general classical equipartition law in the microcanonical ensemble. We seek to evaluate

$$\left\langle x_i \frac{\partial \mathcal{H}}{\partial x_j} \right\rangle, \qquad x_i = p_i \text{ or } q_i\,. \tag{3.3-1}$$

We employ (2.1-12) and write

$$\begin{aligned} \left\langle x_i \frac{\partial \mathcal{H}}{\partial x_j} \right\rangle &= \frac{1}{M(\mathcal{E})} \int_{\Gamma-space} d\Omega\, \delta[\mathcal{H}(p,q)-\mathcal{E}] x_i \frac{\partial \mathcal{H}}{\partial x_j} \\ &= \frac{1}{M(\mathcal{E})} \frac{d}{d\mathcal{E}} \int_{\Gamma-space} d\Omega\; \Theta[\mathcal{E}-\mathcal{H}(p,q)] x_i \frac{\partial \mathcal{H}}{\partial x_j}\,, \end{aligned} \tag{3.3-2}$$

where $\Theta(z)$ is the Heaviside function which we met before. Therefore,

$$\left\langle x_i \frac{\partial \mathcal{H}}{\partial x_j} \right\rangle = \frac{1}{M(\mathcal{E})} \frac{d}{d\mathcal{E}} \int_{\mathcal{H}(p,q)\le\mathcal{E}} d\Omega\, x_i \frac{\partial}{\partial x_j} [\mathcal{H}(p,q)-\mathcal{E}]\,, \tag{3.3-3}$$

where we note that $\mathcal{E}$ is a *value*, not a function of p,q, so $\partial\mathcal{E}/\partial x_j = 0$. Now,

$$\begin{aligned} \int_{\mathcal{H}(p,q)\le\mathcal{E}} d\Omega\, x_i \frac{\partial}{\partial x_j} [\mathcal{H}(p,q)-\mathcal{E}] &= \int_{\mathcal{H}(p,q)\le\mathcal{E}} d\Omega \frac{\partial}{\partial x_j} \{x_i[\mathcal{H}(p,q)-\mathcal{E}]\} \\ &\quad - \delta_{ij} \int_{\mathcal{H}(p,q)\le\mathcal{E}} d\Omega [\mathcal{H}(p,q)-\mathcal{E}]\,. \end{aligned} \tag{3.3-4}$$

The first integral on the right can be written as a vanishing surface integral[4]:

$$\int_{\mathcal{H}(p,q)\le\mathcal{E}} d\Omega \frac{\partial}{\partial x_j} \{x_i[\mathcal{H}(p,q)-\mathcal{E}]\} = \oiint_{\mathcal{H}(p,q)=\mathcal{E}} dS_{\mathcal{E},j} \{x_i[\mathcal{H}(p,q)-\mathcal{E}]\} = 0\,. \tag{3.3-6}$$

---

[4] Equation (3.3-6) is the multidimensional analogue of one of Gauss' 3D theorems, viz.,

$$\int \text{dv Grad}\, \mathbf{F} = \oiint d\mathbf{S}\, \mathbf{F}, \text{ or in component form, } \int dv (\partial/\partial x_i) F_j = \oiint dS_i F_j\,. \tag{3.3-5}$$

The second part remains; we need the integral

$$\begin{aligned} \mathcal{I} &= \frac{d}{d\mathcal{E}} \int_{\mathcal{H}(p,q)\le\mathcal{E}} d\Omega[\mathcal{H}(p,q)-\mathcal{E}] \\ &\approx \frac{1}{\delta\mathcal{E}} \int_{\mathcal{H}(p,q)\in\delta\mathcal{E}} d\Omega\underbrace{[\mathcal{H}(p,q)-\mathcal{E}]}_{-\delta\mathcal{E}} = -\Delta\Omega(\mathcal{E}) . \end{aligned} \tag{3.3-7}$$

The first line indicates that the integrand, and consequently the integral, is negative; in the second line the shell $\delta\mathcal{E}$ is as always taken to be a positive quantity. The last term of (3.3-4) thus yields,

$$\frac{d}{d\mathcal{E}} \int_{\mathcal{H}(p,q)\le\mathcal{E}} d\Omega\, x_i \frac{\partial}{\partial x_j}[\mathcal{H}(p,q)-\mathcal{E}] = \delta_{ij}\Delta\Omega(\mathcal{E}) . \tag{3.3-8}$$

This is substituted into (3.3-3). We then obtain,

$$\left\langle x_i \frac{\partial \mathcal{H}}{\partial x_j} \right\rangle = \delta_{ij} \frac{1}{M(\mathcal{E})} \Delta\Omega(\mathcal{E}) = \delta_{ij} \frac{\Delta\Omega(\mathcal{E})}{\partial\Delta\Omega / \partial\mathcal{E}} = \frac{\partial\mathcal{E}}{\partial \ln \Delta\Omega} . \tag{3.3-9}$$

Employing (3.1-19) we arrive at

$$\left\langle x_i \frac{\partial \mathcal{H}}{\partial x_j} \right\rangle = \delta_{ij} k_B T . \tag{3.3-10}$$

Consider now an arbitrary system with

$$\mathcal{H} = \mathcal{T} + \mathcal{V}, \quad \mathcal{T} = \sum\nolimits_{i=1}^{rN} p_i^2 / 2b_i . \tag{3.3-11}$$

Since the p's denote generalized momenta, the quadratic components may include translational energy, (classical) rotational energy and (classical) vibrational energy. We note that $p_i \partial\mathcal{H} / \partial p_i = p_i^2 / b_i$. Hence,

$$\overline{\mathcal{T}} = \frac{1}{2} \sum_{i=1}^{rN} \left\langle p_i \frac{\partial \mathcal{H}}{\partial p_i} \right\rangle = (rN) \tfrac{1}{2} k_B T . \tag{3.3-12}$$

Clearly $\overline{\mathcal{T}} \propto T$ and the mean energy is $\frac{1}{2}k_B T$ for each degree of freedom. For the specific heat C_V of a perfect mono-atomic gas we find the molar value $(d/dT)\frac{3}{2}N_A k_B T = \frac{3}{2}R$, where N_A is Avogadro's number and R is the gas constant. For the classical model of a solid – valid at high temperatures – we may assume that the lattice gas exists of $3N$ harmonic oscillators, each having a Hamiltonian $h = (1/2m)(p^2 + m^2\omega^2 q^2)$; thus, there being $6N$ quadratic terms, the molar specific heat is found to be $3R$, in accordance with the law of Dulong and Petit.

3.4 Various Forms for the Entropy in Closed and in Open Systems

In this section we shall discuss various forms for the entropy, which are equivalent in an equilibrium situation, but which show different features, when extended to nonequilibrium states. We shall add the superscript '0' to denote equilibrium. First of all, several expressions are known as 'Gibbs entropy'. The basic result, established in Section 3.2, is

$$S^0_{Gibbs} = k_B \ln \Delta\Gamma(\mathscr{E}, \{A_i\}) . \tag{3.4-1}$$

Now consider the quantity $-\langle \ln \rho \rangle$. Using (3.2-1) we have,

$$\begin{aligned} -\langle \ln \rho \rangle &= \mathrm{Tr}\{\rho[-c(\omega, A) + (\mathscr{H} - \mathscr{E})^2 / \omega^2]\} \\ &= -c(\omega, A) + \mathrm{var}\,\mathscr{E} / \omega^2 \simeq -c(\omega, A). \end{aligned} \tag{3.4-2}$$

Employing also (3.2-4), we see that $-k_B\, c(\omega, A)$ is just the entropy. Thus we also have,

$$S^0_{Gibbs} = -k_B \mathrm{Tr}(\rho \ln \rho) . \tag{3.4-3}$$

This expression has wide applicability, since it remains valid for open systems, as well as for nonequilibrium systems, with $\rho = \rho(t)$.

Next, we look at systems with fluctuating state variables $\{a_i\}$. In section 2.4 we introduced the thermodynamic probability $\tilde{W}(a)$. We consider $\Sigma \tilde{W}(a) = (\Sigma W)/C = 1/C$, the constant C being given by (2.4-14). Since $\ln C = -\ln \Delta\Gamma$, neglecting terms of order $\ln N$, we have a third expression,

$$S^0_{Gibbs} = k_B \ln \Sigma_a \tilde{W}(\{a_i\}) . \tag{3.4-4}$$

Note that this is an equilibrium entropy, the sum depending on $\mathscr{E}$ and A_i, but not on the a_i.

Next we turn to expressions, which have an extension to the nonequilibrium entropy function to be discussed later, and which are usually endowed with the subscript 'B' for Boltzmann. Since $\Delta\Gamma = \chi \delta\mathscr{E}$, and since $\delta\mathscr{E}$ is of order $\ln N$, we have[5]

$$S^0_B = k_B \ln \chi(\mathscr{E}, \{A_i\}) . \tag{3.4-5}$$

Previously we saw that the density of states is related to the macro-probability for a given state, cf. (2.4-14). Thus, if there are macroscopic fluctuating variables with mean values $\bar{a}_i$, then the Boltzmann entropy is related to the thermodynamic probability by

[5] This expression is of course a dimensional disaster; however, such 'minor' flaws are often overlooked in statistical mechanics; see also our discussion of the Fowler transform in subsection 1.7.3.

$$S_B^0 = k_B \ln \tilde{W}(\{\bar{a}_i\}). \tag{3.4-6}$$

Comparing with the Gibbs expression (3.4-4), we suggest a reconciliation in setting $\Sigma\tilde{W} = \tilde{W}(\{\bar{a}_i\})\Pi\delta a_i$, where the 'widths' δa_i are logarithmically insignificant. Finally, we leave it up to the reader to justify the following entropy expression, due to Fierz,

$$S_{Fiertz}^0 = k_B \langle \ln \tilde{W}(\{a_i\}) \rangle. \tag{3.4-7}$$

The abundance of possible expressions for the entropy led Lorentz to speak of "l'insensibilité des fonctions thermodynamiques." They are given here in order to compare with other texts while, in addition, these expressions can be seen as limit-results of nonequilibrium entropy functions when the fluctuating variables are 'frozen' to an equilibrium state, see Section 3.7. Yet we believe that only Gibbs' results, embodied in Eqs. (3.4-1) and (3.4-3), are truly significant.

3.5 Properties of Entropy

3.5.1 *Elementary Entropies; Sackur–Tetrode Formula*

For an ideal gas we computed already $\Delta\Gamma(\mathscr{E}, V, N)$. From (2.5-7), using Stirling's formula and setting $\frac{3}{2}N - 1 \sim \frac{3}{2}N$, $\ln \delta\mathscr{E} = \mathcal{O}(\ln N)$, we easily obtain,

$$S = k_B N \ln\left\{\frac{V}{N}\left(\frac{2\pi m k_B T}{h^2}\right)^{3/2} e^{5/2}\right\}, \tag{3.5-1}$$

where we employed the equipartition result $2m\mathscr{E} = 3mNk_BT$. Since $V/N = \hat{v}$ (is given specific volume), the entropy is additive when two parts of a gas are put together, as it should be. Equation (3.5-1) is the Sackur–Tetrode formula. It cannot be used for $T \to 0$, since at low temperatures all gases are quantum gases.

When two different gases are added, there is a *mixing entropy*. Let N_1 molecules in V_1 of gas A be joined to N_2 molecules in V_2 of gas B at constant temperature. Then,

$$\begin{aligned}\Delta S/k_B &= (1/k_B)\{S_A(V_1+V_2,N_1) + S_B(V_1+V_2,N_2) - S_A(V_1,N_1) - S_B(V_2,N_2)\} \\ &= N_1 \ln\left(\frac{V_1+V_2}{V_1}\right) + N_2 \ln\left(\frac{V_1+V_2}{V_2}\right) > 0,\end{aligned} \tag{3.5-2}$$

i.e., the entropy has increased. Suppose now that we had used the classical formula $k_B \ln \Delta\Omega$ for the entropy. Instead of (3.5-1) we would have found,

$$S_{class} = k_B N \ln\left\{V\,(2\pi m k_B T)^{3/2} e^{3/2}\right\}. \tag{3.5-3}$$

The mixing entropy is the same as before, but it would *also* occur upon mixing two parts of the same gas! Per mole the mixing entropy would be

$$\Delta S_{class} = 2R\ln 2\,, \tag{3.5-4}$$

where we set $N_{1,2} = N_A$, $k_B N_A = R$. This was formerly known as 'the Gibbs Paradox'. However, Gibbs rescued the situation by postulating the necessity of dividing $\Delta\Omega$ by the factor $N!$, long before the advent of quantum mechanics!

We shall derive a result for the vapour pressure of a gas in equilibrium with a condensate, which enabled Tetrode to verify (3.5-1). The entropies of the two phases are related by

$$S - S_c = \mathcal{n} Q_s / T, \tag{3.5-5}$$

where $\mathcal{n} = N/N_A$ and where Q_s is the condensation heat per mole. Let P be the vapour pressure of the gas in equilibrium with the condensed state; it is given by the well-known Clausius-Clapeyron equation (to be derived in Section 3.6),[6]

$$\frac{dP}{dT} = \frac{(S/N) - (S_c/N_c)}{(V/N) - (V_c/N_c)} = \frac{\hat{s} - \hat{s}_c}{\hat{v} - \hat{v}_c} \approx \frac{\hat{s} - \hat{s}_c}{\hat{v}}. \tag{3.5-6}$$

The Sackur–Tetrode formula provides an integral of this equation. With the perfect gas law $(V/N) = k_B T/P$, (3.5-1) gives for S,

$$S = k_B N \ln\left\{ \left(\frac{2\pi m k_B T}{h^2}\right)^{5/2} e^{5/2} \right\} - k_B N \ln P - k_B N \ln\left(\frac{2\pi m}{h^2}\right). \tag{3.5-7}$$

Or,

$$\begin{aligned} \ln P(T) &= \ln\left\{ \left(\frac{2\pi m k_B T}{h^2}\right)^{5/2} e^{5/2} \right\} - \frac{Q_s(T)}{RT} - \frac{S_c(T)}{k_B N} - \ln\left(\frac{2\pi m}{h^2}\right) \\ &= \frac{5}{2}\ln T - \frac{Q_s(T)}{RT} - \frac{S_c(T)}{k_B N} + \ln\left\{\frac{2\pi m k_B e}{h^2}\right\}^{5/2}. \end{aligned} \tag{3.5-8}$$

Furthermore,

$$S_c(T) = \int_0^T \frac{dQ}{T'} + S_c(0) = \int_0^T \frac{\mathcal{n} C_c}{T'} dT', \tag{3.5-9}$$

where C_c is the molar heat capacity of the condensed state and where the entropy at zero temperature was set equal to zero (subsection 3.5.4). The last two equations yield,

---

[6]Throughout this text lower case entities with a caret (^) denote 'specific quantities' i.e., quantities per molecule; multiplying these quantities by N_A, we obtain the *molar* values. [Also, lower case entities without a caret denote the values per unit volume.]

$$\ln P(T) = \frac{5}{2}\ln T - \frac{Q_s(T)}{RT} - \int_0^T \frac{C_c}{RT'}dT' + \ln\left\{\frac{2\pi m k_B e}{h^2}\right\}^{3/2}. \tag{3.5-10}$$

This result can be experimentally verified from the measured temperature dependences for $P(T)$, $Q_s(T)$ and $C_c(T)$. Tetrode confirmed the validity of (3.5-10) and verified that the constant (last term) needed to fit the results gave Planck's constant within ~ 4%; this, in turn, is a strong indication that the absolute magnitude of the entropy, as given by (3.2-13), was correctly chosen.

We now compute the entropy for an assembly of oscillators, employing the quantum result (2.5-29). Using Stirling's formula we have,

$$\begin{aligned}\ln \Delta\Gamma(\mathcal{E}, N) &= \left(\frac{\mathcal{E}}{\hbar\omega} + \frac{N}{2}\right)\ln\left(\frac{\mathcal{E}}{\hbar\omega} + \frac{N}{2}\right) \\ &\quad - \left(\frac{\mathcal{E}}{\hbar\omega} - \frac{N}{2}\right)\ln\left(\frac{\mathcal{E}}{\hbar\omega} - \frac{N}{2}\right) - N\ln N.\end{aligned} \tag{3.5-11}$$

By differentiation to $\mathcal{E}$:

$$\frac{1}{k_B T} = \frac{\partial \ln \Delta\Gamma}{\partial \mathcal{E}} = \frac{1}{\hbar\omega}\ln\left(\frac{(\mathcal{E}/\hbar\omega) + \frac{1}{2}N}{(\mathcal{E}/\hbar\omega) - \frac{1}{2}N}\right) = \frac{1}{\hbar\omega}\ln\left(1 + \frac{N}{(\mathcal{E}/\hbar\omega) - \frac{1}{2}N}\right), \tag{3.5-12}$$

from which we find,

$$\mathcal{E} = \frac{N\hbar\omega}{e^{\hbar\omega/k_B T} - 1} + \tfrac{1}{2}N\hbar\omega. \tag{3.5-13}$$

This is the well-known result for the energy of a system of N harmonic oscillators. Alternatively, this result may be obtained – with a more profound rendering of quantum principles – from the Bose-Einstein statistics of the elementary excitations (photons or phonons) associated with the system of oscillators, as discussed later.

From (3.5-13) we also have,

$$\frac{\mathcal{E}}{\hbar\omega} - \frac{N}{2} = \frac{N}{e^{\hbar\omega/k_B T} - 1}, \quad \frac{\mathcal{E}}{\hbar\omega} + \frac{N}{2} = \frac{N e^{\hbar\omega/k_B T}}{e^{\hbar\omega/k_B T} - 1}. \tag{3.5-14}$$

When this is substituted into (3.5-11) we obtain,

$$S(T, N) = \frac{N e^{\alpha}}{e^{\alpha} - 1}\ln\frac{N e^{\alpha}}{e^{\alpha} - 1} - \frac{N}{e^{\alpha} - 1}\ln\frac{N}{e^{\alpha} - 1} - N\ln N, \tag{3.5-15}$$

where $\alpha = \hbar\omega / kT_B$. For $T \to 0$ or $\alpha \to \infty$, we note that $S \to 0$, in accord with most views of Nernst's law, see subsection 3.5.4.

3.5.2 *Homogeneous Entropy Form and the Gibbs-Duhem Relation*

We recall that the entropy is a homogeneous function of its extensive variables. In this subsection we shall assume that intensive external parameters are absent (or, if present, are held constant). Then if λ is any number multiplying the volume of the system,

$$S(\lambda X_0, \lambda X_1, \ldots \lambda X_s) = \lambda S(X_0, X_1, \ldots X_s). \tag{3.5-16}$$

This is differentiated with respect to λ and afterwards we set $\lambda = 1$. We thus obtain Euler's relation for first-order homogeneous functions:

$$\sum_i X_i \frac{\partial S}{\partial X_i} = S(X_0 \ldots X_s). \tag{3.5-17}$$

Substituting the thermodynamic forces Q_i, we obtain the *homogeneous form* for the entropy:

$$S(X_0 \ldots X_s) = \sum_{i=1}^{s} Q_i X_i\,. \tag{3.5-18}$$

Taking the total differential dS and comparing with Gibbs' relation $dS = \Sigma_i Q_i dX_i$, we are led to the *generalized Gibbs-Duhem relation*,

$$\sum_{1=1}^{s} X_i dQ_i = 0. \tag{3.5-19}$$

This indicates that the thermodynamic forces cannot be varied independently! For example, if $\mathcal{E}$, V and N are the pertinent variables, we have,

$$\mathcal{E}\, d(1/T) + V\, d(P/T) - N\, d(\varsigma/T) = 0. \tag{3.5-20}$$

Using the energy as the main state variable, we obtain a similar result. Considering S, V, N, and other extensive external parameters,[7] we have the homogeneous form,

$$\mathcal{E} = TS - PV + \varsigma N - \Sigma_i \overline{F}_i A_{e,i}\,. \tag{3.5-21}$$

Comparing with (1.4-1), another form of the Gibbs-Duhem relation ensues,

$$SdT - VdP + Nd\varsigma - \Sigma_i A_{e,i} d\overline{F}_i = 0. \tag{3.5-22}$$

If the only variables are S, V and N, (3.5-22) becomes the *ordinary* Gibbs-Duhem relation,

---

[7] Here and elsewhere [Sections 3.6 and 5.3–5.6] we shall denote these external parameters by $\{A_{e,i}\}$, it being implied that intensive external parameters, if present, are not varied.

$$SdT - VdP + Nd\varsigma = 0\,. \tag{3.5-23}$$

The thermodynamic intensive variables are homogeneous functions of zero-order; in the energy language,

$$\overline{F}_i = \overline{F}_i(S,V,N,\{A_{e,i}\}) = \overline{F}_i(\hat{s},\hat{v},\{\hat{a}_{e,i}\})\,. \tag{3.5-24}$$

These relations, expressing the *intensive variables as functions of the extensive variables*, are called the 'equations of state'.[8] If T, P and ς are the only relevant intensive variables, two equations of state suffice: $T = T(\hat{s},\hat{v},)$ and $P = P(\hat{s},\hat{v},)$; the third one can be found from integration of the ordinary Gibbs-Duhem relation.

3.5.3 *Maxwell Relations*

No discussion on thermodynamic state functions would be complete, without mentioning the Maxwell relations, which, moreover, sometimes show up in statistical mechanics. These relations follow from the total differentials of the principal state functions, the energy $\mathscr{E}$, the enthalpy W, the Helmholtz free energy F and the Gibbs free energy G. Considering only the standard extensive variables, we had, see Section 1.4,

$$\begin{aligned} d\mathscr{E} &= TdS - PdV + \varsigma dN, \\ dW &= TdS + VdP + \varsigma dN, \\ dF &= -SdT - PdV + \varsigma dN, \\ dG &= -SdT + VdP + \varsigma dN. \end{aligned} \tag{3.5-25}$$

From the first differential we find

$$T = \left(\frac{\partial\mathscr{E}}{\partial S}\right)_{V,N}, \quad P = -\left(\frac{\partial\mathscr{E}}{\partial V}\right)_{S,N}, \quad \varsigma = \left(\frac{\partial\mathscr{E}}{\partial N}\right)_{S,V}. \tag{3.5-26}$$

Now, assuming that $\mathscr{E}$ is a well-behaved function, we have

$$\begin{aligned} &\frac{\partial}{\partial V}\left(\frac{\partial\mathscr{E}}{\partial S}\right) = \frac{\partial}{\partial S}\left(\frac{\partial\mathscr{E}}{\partial V}\right), \quad \frac{\partial}{\partial V}\left(\frac{\partial\mathscr{E}}{\partial N}\right) = \frac{\partial}{\partial N}\left(\frac{\partial\mathscr{E}}{\partial V}\right), \\ &\frac{\partial}{\partial S}\left(\frac{\partial\mathscr{E}}{\partial N}\right) = \frac{\partial}{\partial N}\left(\frac{\partial\mathscr{E}}{\partial S}\right), \end{aligned} \tag{3.5-27}$$

It thus follows that

---

[8] We followed here the standard thermodynamic definition. We note, however, that many authors consider any set of relations between intensive and extensive variables (e.g., the perfect gas law or van der Waals' law) as an equation of state.

$$\left(\frac{\partial T}{\partial V}\right)_{S,N} = -\left(\frac{\partial P}{\partial S}\right)_{V,N}, \quad \left(\frac{\partial \varsigma}{\partial V}\right)_{S,N} = -\left(\frac{\partial P}{\partial N}\right)_{S,V}, \quad \left(\frac{\partial \varsigma}{\partial S}\right)_{V,N} = \left(\frac{\partial T}{\partial N}\right)_{S,V}. \tag{3.5-28}$$

Likewise, from the total differential dF we have,

$$S = -\left(\frac{\partial F}{\partial T}\right)_{V,N}, \quad P = -\left(\frac{\partial F}{\partial V}\right)_{T,N}, \quad \varsigma = \left(\frac{\partial F}{\partial N}\right)_{T,V}, \tag{3.5-29}$$

from which one can deduce,

$$\left(\frac{\partial S}{\partial V}\right)_{T,N} = \left(\frac{\partial P}{\partial T}\right)_{V,N}, \quad \left(\frac{\partial S}{\partial N}\right)_{V,T} = -\left(\frac{\partial \varsigma}{\partial T}\right)_{V,N}, \quad \left(\frac{\partial P}{\partial N}\right)_{V,T} = -\left(\frac{\partial \varsigma}{\partial V}\right)_{T,N}. \tag{3.5-30}$$

Six more Maxwell relations can be given, based on dW and dG. It is up to the reader to derive these results. We still note that, if N is not varied, there are just four (original) Maxwell relations. They are,

$$\begin{aligned} &\left(\frac{\partial S}{\partial P}\right)_V = -\left(\frac{\partial V}{\partial T}\right)_S, \quad \left(\frac{\partial S}{\partial V}\right)_P = \left(\frac{\partial P}{\partial T}\right)_S, \\ &\left(\frac{\partial S}{\partial V}\right)_T = \left(\frac{\partial P}{\partial T}\right)_V, \quad \left(\frac{\partial S}{\partial P}\right)_T = -\left(\frac{\partial V}{\partial T}\right)_P. \end{aligned} \tag{3.5-31}$$

The reader will recognise the first one as the reciprocal of the first result in (3.5-28); the inversion is made since derivatives to S are not easily realisable.

Finally, when terms HdM or EdP occur [or MdH and PdE], a host of 'magnetic' and 'electric' Maxwell relations can be deduced. All Maxwell relations are of importance for *response functions* [see also the Non-equilibrium Part of this text], such as the specific heats C_V and C_P, the adiabatic and isothermal compressibilities K_S and K_T, the conductivity σ and the susceptibilities χ_e and χ_m.

3.5.4 *Nernst's Law and Statistical Mechanics*

Nernst's law states: *At absolute zero temperature all changes of state occur without a change in entropy* (Nernst, 1906).[9] This means among other things: The zero-point entropy is independent of the thermodynamic variables of the system. Generally we express Nernst' law as

$$\lim_{T\to 0} \delta S(T,\{A_i\}) = 0. \tag{3.5-32}$$

For example, in the case of the gas-liquid transition (3.5-5) we must have

$$S - S_c = nQ_s / T \to 0 \text{ for } T \to 0, \tag{3.5-33}$$

[9] W. Nernst, Göttinger Nachrichte (1906) p.1.

indicating that Q_s must go to zero faster than T.

The statement (3.5-32) is equivalent with the unattainability of absolute zero temperature. To prove this, we use Clausius' definition of S as reduced heat for two situations differing by an externally invoked change in the thermodynamic variables. It then follows that,

$$S_1(T_1) - S_2(T_2) = \underbrace{S_1(0) - S_2(0)}_{\delta S(0)=0} + \int_0^{T_1} \frac{c_1}{T} dT - \int_0^{T_2} \frac{c_2}{T} dT, \qquad (3.5\text{-}34)$$

where c_1 and c_2 are heat capacities. Let the operation to lower T be carried out adiabatically and reversibly; then the lhs of (3.5-34) is zero. The integrals converge, since c goes to zero at least as fast as T for all known quantum processes. *If* T_2 could be made zero, then T_1 also must be zero, given that the c's are positive definite. It is therefore impossible to lower the temperature adiabatically to zero. For non-adiabatic processes the situation is always less favourable, since in a change 1→2 the cntropy would increase such that

$$\int_0^{T_2} \frac{c_2}{T} dT > \int_0^{T_1} \frac{c_1}{T} dT. \qquad (3.5\text{-}35)$$

The inverse statement, which makes Nernst's law a consequence of the unattainability of absolute zero, is also easily proven.

Since the entropy at $T = 0$ is a constant, it is tempting to fix the zero point entropy such that

$$\lim_{T\to 0} S(T,\{A_i\}) = 0. \qquad (3.5\text{-}36)$$

This is a stronger form of Nernst's law. As we saw for the case of an assembly of quantum oscillators, the choice $S = k_B \ln \Delta\Gamma$, without an additive constant, satisfies the statement (3.5-36). In our opinion, the stronger form of Nernst law is foreseeable for many-body quantum systems, since for *very low* temperatures the energy-spacing of quantum states should become discrete.[10] Then $\Delta\Gamma$ should be replaced by the degeneracy $g(\mathscr{E})$ of a single quantum state, i.e.,

$$S = k_B \ln g(\mathscr{E}). \qquad (3.5\text{-}37)$$

At low temperatures many systems are condensed into a cooperative state of highly reduced symmetry. At $T = 0$ the system will be in the true ground state. It is widely believed that the degeneracy of the ground state is unity when *all* interactions that lift the degeneracy are taken into account; then, $S(0) = 0$. Our view is not shared by Casimir (Op.Cit), who gives a different argument to obtain the stronger form of Nernst's law.

---

[10] As shown by Casimir, for a system of 10^{23} particles T must be lower than 10^{-20} Kelvin; H.B.G. Casimir, Zeitschr. für Physik **171**, 276 (1963).

3.6 Equilibrium and Local Stability Requirements

The conditions for a system to be in thermal equilibrium can be obtained in two ways. First, they follow easily from application of the homogeneous form for the entropy or the energy. Secondly, the extremum conditions of the state functions can be employed. We start with the former method.

We shall consider a system that is partitioned into two subsystems 'I' and 'II', the composite system being closed. For the energy in each subsystem we employ the homogeneous form (3.5-21). A small flaw must be corrected, however. In this expression the temperature T, the pressure P and the chemical potential ς appear as constants, but in reality their values fluctuate. Thus T is a mean temperature, as is clear for instance from its proportionality to the mean kinetic energy, see (3.3-12). The same applies to the pressure and the chemical potential. We shall thus add an overhead bar and, to simplify the homogeneous form, we will call $S=A_{e,0}$, $V=A_{e,1}$ and $N=A_{e,2}$, similar to our nomenclature in Section 1.4. Also, let $\overline{F}_0=-\overline{T}$, $\overline{F}_1=\overline{P}$, $\overline{F}_2=-\overline{\varsigma}$. We further assume that the subsystems are open with respect to q of the $p+1$ extensive variables, $q \le p+1$. The energy $\mathscr{E}^q$ associated with these variables has the homogeneous form $\mathscr{E}^q=-\Sigma_i^q\overline{F}_iA_{e,i}$. We then have for the reversible variations due to flow of the $A_{e,i}$ across the boundary,

$$
\begin{aligned}
&-\Sigma_i^q\overline{F}_i^{\mathrm{I}}A_{e,i}^{\mathrm{I}}-\Sigma_i^q\overline{F}_i^{\mathrm{II}}A_{e,i}^{\mathrm{II}} \\
&=-\Sigma_i^q[\overline{F}_i^{\mathrm{I}}+d\overline{F}_i^{\mathrm{I}}][A_{e,i}^{\mathrm{I}}+dA_{e,i}^{\mathrm{I}}]-\Sigma_i^q[\overline{F}_i^{\mathrm{II}}+d\overline{F}_i^{\mathrm{II}}][A_{e,i}^{\mathrm{II}}+dA_{e,i}^{\mathrm{II}}].
\end{aligned}
\tag{3.6-1}
$$

Noting that $dA_{e,i}^{\mathrm{I}}=-dA_{e,i}^{\mathrm{II}}$ and neglecting second order quantities we have,

$$\sum\nolimits_i\left(\overline{F}_i^{\mathrm{II}}-\overline{F}_i^{\mathrm{I}}\right)dA_{e,i}^{\mathrm{I}}-\sum\nolimits_iA_{e,i}^{\mathrm{I}}d\overline{F}_i^{\mathrm{I}}-\sum\nolimits_iA_{e,i}^{\mathrm{II}}d\overline{F}_i^{\mathrm{II}}=0. \tag{3.6-2}$$

The last two sums are zero in view of the Gibbs-Duhem relation for each subsystem, being in equilibrium with the other one. In the first sum the variations $dA_{e,i}$ are independent. Consequently we obtain,

$$\overline{F}_i^{\mathrm{I}}=\overline{F}_i^{\mathrm{II}}, \tag{3.6-3}$$

i.e., the intensive variables, on average, balance each other. We also look at the situation where system 'II' is very large, so that it acts as a reservoir. For an infinite reservoir the fluctuations in the intensive variables become negligibly small (shown in detail in subsection 3.8.4) and $\overline{F}_i^{\mathrm{II}}\to F_i^r$, the stable reservoir value. The equilibrium conditions for a system having exchange for q of its variables $A_{e,i}$ then become,

$$\overline{F}_i=F_i^r,\quad i\in\{q\}. \tag{3.6-4}$$

In particular, for a system open with respect to heat exchange, volume exchange, or particle exchange, the equilibrium conditions are

$$\bar{T} = T^r, \quad \bar{P} = P^r, \quad \bar{\varsigma} = \varsigma^r. \tag{3.6-5}$$

We also note that for charged particle systems we have balancing of the electrochemical potentials, which in solids are the Fermi levels. Hence,

$$\bar{\mu} \equiv \varepsilon_F = \mu^r \equiv \varepsilon_F^r. \tag{3.6-6}$$

As we indicated previously, the superscript '*r*' of the reservoir is usually omitted.

We now turn to the second type of argument.[11] To describe a system with volume and heat exchange we turn to the Gibbs free energy *G*(*T*,*P*), which must be a minimum in an equilibrium state. Unscripted quantities will refer to the system and super '*r*' to the reservoir. We consider a variation δG due to variations δS and δV associated with the exchange. Then, since

$$G(T^r, P^r, N, A_3 \ldots) = \mathcal{E} - T^r S(\mathcal{E}, V) + P^r V, \tag{3.6-7}$$

we have,

$$\begin{aligned} \delta G &= \left(\frac{\partial \mathcal{E}}{\partial S} - T^r\right)\delta S + \left(\frac{\partial \mathcal{E}}{\partial V} + P^r\right)\delta V \\ &+ \frac{1}{2}\left[\left(\frac{\partial^2 \mathcal{E}}{\partial S^2}\right)(\delta S)^2 + 2\left(\frac{\partial^2 \mathcal{E}}{\partial S \partial V}\right)(\delta S \delta V + \left(\frac{\partial^2 \mathcal{E}}{\partial V^2}\right)(\delta V)^2\right], \end{aligned} \tag{3.6-8}$$

where we expanded up to second order. For *G* to be stationary, the first-order variations must be zero; so,

$$T = \left(\partial \mathcal{E} / \partial S\right)_V = T^r, \quad P = -\left(\partial \mathcal{E} / \partial V\right)_S = P^r. \tag{3.6-9}$$

For an equilibrium state *G* must be a minimum [since $S(\mathcal{E}, V)$ is a maximum], so that

$$\left(\frac{\partial^2 \mathcal{E}}{\partial S^2}\right)(\delta S)^2 + 2\left(\frac{\partial^2 \mathcal{E}}{\partial S \partial V}\right)(\delta S \delta V + \left(\frac{\partial^2 \mathcal{E}}{\partial V^2}\right)(\delta V)^2 > 0. \tag{3.6-10}$$

We note that this is a quadratic expression of the form

$$f(u,v) = au^2 + 2buv + cv^2, \tag{3.6-11}$$

which is positive definite iff[12] $a, c > 0$ and the discriminant $D \equiv 4(b^2 - ac) < 0$. Therefore, the *stability requirements* are

$$\left(\frac{\partial^2 \mathcal{E}}{\partial S^2}\right)_V = \left(\frac{\partial T}{\partial S}\right)_V = \frac{T}{C_V} > 0, \quad \left(\frac{\partial^2 \mathcal{E}}{\partial V^2}\right)_S = -\left(\frac{\partial P}{\partial V}\right)_S = \frac{1}{VK_s} > 0; \tag{3.6-12}$$

---

[11] Cf. M. Plischke and B. Bergersen, 3rd Ed., Op. Cit. pp. 16–18. Also, L.E. Reichl, 2nd Ed., Op.Cit., pp. 57–62.

[12] 'iff' means: if and only if; in French, 'ssi': si et seulement si.

since both the specific heat and the adiabatic compressibility are positive quantities, these conditions are always satisfied. The last condition that D be negative requires

$$\frac{1}{T}VK_sC_V > \left(\frac{\partial V}{\partial T}\right)_S^2, \tag{3.6-13}$$

as is left to the reader to confirm. This inequality is a consequence of Le Châtelier's principle which states that, for a system in equilibrium, any spontaneous changes in its variables invoke processes that restore the equilibrium state. In the case considered, the Gibbs' free energy will be restored to its minimum under the specified isothermal and isobaric conditions. We also note that G is a concave function, both of P and of T. Employing results analogous to (3.6-12):

$$\left(\frac{\partial^2 G}{\partial P^2}\right)_{T,N,\{A_i\}} = \left(\frac{\partial V}{\partial P}\right)_{T,N,\{A_i\}} = -VK_T < 0, \quad \left(\frac{\partial^2 G}{\partial T^2}\right)_{T,N,\{A_i\}} = -\left(\frac{\partial S}{\partial T}\right)_{P,N,\{A_i\}} = -C_P/T < 0. \tag{3.6-14}$$

Similar statements for the Helmholtz free energy are left to the problems.

Finally, we shall apply the obtained equilibrium conditions to prove the Clausius-Clapeyron equation, which we needed before. We consider once more the equilibrium between a vapour and its condensate. Given that the chemical potentials of gas and liquid, $\overline{\varsigma}_g$ and $\overline{\varsigma}_\ell$ balance, we find from the ordinary Gibbs-Duhem relation (3.5-23) upon dividing both sides by the particle number,

$$-\hat{s}_g dT + \hat{v}_g dP = -\hat{s}_\ell dT + \hat{v}_\ell dP. \tag{3.6-15}$$

This then yields,

$$\frac{dP}{dT} = \frac{\hat{s}_g - \hat{s}_\ell}{\hat{v}_g - \hat{v}_\ell}. \tag{3.6-16}$$

3.7 Entropy and Probability

In this section and the next one we will deal with nonequilibrium states. The fluctuating variables will be denoted by $a_1 ... a_p$. As discussed in Section 1.5, these variables can, in principle, be extensive or intensive; however, here we will assume that they are extensive, so that they are variables of subsystems, given that the total system variables are fixed in the microcanonical ensemble.

3.7.1 *Boltzmann's Principle*

We introduce the nonequilibrium Boltzmann *entropy function*; by definition, the entropy function shall be the same function of the instantaneous variables as the entropy is of the equilibrium variables. Thus, upon freezing the a_i to the equilibrium variables $a_i^0 \equiv \langle a_i \rangle$, we have

$$S_B(\{a_i\}) \to S^0(\{a_i^0\}). \tag{3.7-1}$$

Guided by the relation (3.4-5), the Boltzmann entropy function is now given as

$$S_B = k_B \ln\left[\chi(\{a_i\})\chi(\{a_i^c - a_i\})\right] + k_B \ln \hat{c}, \tag{3.7-2}$$

where we assumed that the composite system contains two subsystems. Comparing with (2.4-12), we arrive at Boltzmann's famous relation[13] between entropy and W ("Wahrscheinlichkeit", German for probability), carved on his tombstone in Vienna:

$$S_B(a_1 ... a_p) = k_B \ln W(a_1 ... a_p) + const. \tag{3.7-3}$$

Note that the *constant* is fixed in the ensemble, but does not vanish logarithmically. Or, comparing with (2.4-15) we also have

$$S_B(a_1 ... a_p) = k_B \ln \tilde{W}(a_1 ... a_p). \tag{3.7-4}$$

The 'thermodynamic probability' here is not really a probability, but a measure for the number of ways to realise a given macrostate, i.e., a measure for the *disorder*.

Note. Boltzmann's entropy $'k_B \log W'$ is a thermodynamic concept *fully compatible* with Clausius' definition of entropy as the integral over the reduced heat, $\int^T đQ/T$. But, the agent that randomizes the large number of microstates associated with a specified macrostate *must always be a measure* for the Kelvin scale of temperature, as e.g. in a system of randomly oriented dipoles in a polar liquid or spin lattice; the popular concept of the entropy of tossing coins or throwing dice is, at best, only 'pseudo-science', despite its mention in numerous college physics textbooks! [13a]

3.7.2 *The Boltzmann–Einstein Principle*

A more useful quantitative relationship is obtained, if the microcanonical partition function, occurring in (2.4-12) is not shovelled under the table. We maintain (3.7-2), but obtain from (2.4-12) and (3.4-1):

$$k_B \ln W(\{a_i\}) = [S_B - S^0] + k_B \ln \hat{c}. \tag{3.7-5}$$

The term $k_B \ln \hat{c}$ is negligible in this logarithmic form, but must be retained for the inversion; we then obtain Einstein's form,

$$W(a_1 ... a_p) = \hat{c} \exp[(S_B(a_1 ... a_p) - S^0)/k_B]. \tag{3.7-6}$$

---

[13] Boltzmann never introduced "Boltzmann's constant" k_B ; this was left to Planck! See the papers mentioned in Chapter VIII, Ref. 1.

[13a] Cf. Sears and Zemansky, "University Physics", 11th Ed., Section *20.8.

Upon Taylor expansion to second order we have,

$$S(a_1...a_p) = S^0 + \frac{1}{2}\sum_{i,j} \frac{\partial^2 S}{\partial a_i \partial a_j}\bigg|_{\{a_i\}=\{a_i^0\}} \Delta a_i \Delta a_j = S^0 + \frac{1}{2}\sum_{i,j} s_{ij} \Delta a_i \Delta a_j ; \quad (3.7\text{-}7)$$

this yields,

$$W(a_1...a_p) = \hat{c}\exp[\tfrac{1}{2}\Sigma_{ij} s_{ij} \Delta a_i \Delta a_j / k_B]. \quad (3.7\text{-}8)$$

This is a multivariate normal distribution; we note that $\Delta a_i = a_i - a_i^0 = a_i - \langle a_i \rangle$ and that $[s_{ij}]$ is *negative definite.* We met an example of this procedure in the charge fluctuations of a capacitor, see Section 1.6.

Some caution is still in order. Strictly, the Boltzmann–Einstein principle, as well as Boltzmann's principle, only holds for a closed system; see e.g. Landau and Lifshitz, Op.Cit. p. 22, §7. Now let us suppose that the second subsystem with variables $a_i^c - a_i$ becomes an infinite reservoir. Then, as the example of the ideal gas, treated in subsection 2.5.1, has shown

$$\hat{c}\chi(a)\chi(a^c - a) / \Delta\Gamma(\mathcal{E}, a) \to C\chi(a)\, e^{-\mathcal{E}(a)/k_B T}. \quad (3.7\text{-}9)$$

With W being just the lhs, one finds

$$k_B \ln W(a) = S_B(a) - \mathcal{E}(a)/T + k_B \ln C = -[F(a)/T] + k_B \ln C. \quad (3.7\text{-}10)$$

Now the inverse normalization constant C^{-1} is the canonical partition function $\mathcal{Z}$, to be introduced in the next chapter; its logarithm is $-F^0/k_B T$. It thus follows that for an open system,

$$k_B T \ln[W(a)(\delta a)] = -[F(a) - F^0], \quad (3.7\text{-}11)$$

where we added the (δa) for dimensional consistency, W being a probability density function. Upon exponentiation,

$$W(a) = \hat{c}\exp[-(F(a) - F^0)/k_B T]. \quad (3.7\text{-}12)$$

Expanding F to second order, we obtain a result akin to (3.7-8); instead of adiabatic derivatives, we now need isothermal derivatives. In the case of charge fluctuations discussed before, the same result is obtained since the electrical capacitance (generally) is a purely geometrical quantity.

3.8 The Gibbs Entropy Function

3.8.1 *Gibbs' Entropy Function for a Nonequilibrium State*

Finally, we proceed to extend the Gibbs entropy to a nonequilibrium state. To this end we need the nonequilibrium density operator, $\rho(t)$. Analogous to (3.4-3) we define,

$$S_G = -k_B \mathrm{Tr}\rho(t) \ln \rho(t). \quad (3.8\text{-}1)$$

In terms of the eigenstates of $\mathcal{H}$ this also reads,

$$S_G = -k_B \sum_\eta p(\eta,t) \ln p(\eta,t) . \quad (3.8\text{-}2)$$

Again, let there be extensive variables $\mathbf{a}^\alpha = \{a_i^\alpha\}$ of K subsystems which make up the closed composite system. Let $p(\mathbf{a},t) \equiv p(\{\mathbf{a}^\alpha\},t)$ be the time-dependent probability density for all the a's; we then have,

$$p(\mathbf{a},t) \approx p(\eta,t) \prod_{\alpha=1}^{K} \chi(\mathbf{a}^\alpha) , \quad (3.8\text{-}3)$$

where on the rhs $\mathbf{a} = \langle \eta | \mathbf{a} | \eta \rangle$, cf. (1.5-5). Substituting for $p(\eta,t)$ in (3.8-2) we obtain the Gibbs entropy in terms of the $\mathbf{a}$ variables; in addition we write $\Sigma_\eta \to \int d\mathbf{a} \prod_\alpha \chi^\alpha(\mathbf{a}^\alpha)$. Hence,

$$S_G[p(\mathbf{a},t)] = -k_B \int d\mathbf{a}\, p(\mathbf{a},t) \ln \frac{p(\mathbf{a},t)}{\prod_\alpha \chi(\mathbf{a}^\alpha)} . \quad (3.8\text{-}4)$$

This coarse-grained Gibbs entropy is a *functional* of the probability $p(\mathbf{a},t)$, in contrast to the Boltzmann entropy, which varies directly with the $\{\mathbf{a}^\alpha\}$. We now take the logarithm of (2.4-13), observing (3.4-1) with $\{\Lambda_i\} \to \{\mathbf{a}_i^0\}$; this gives,

$$S^0 = -k_B \ln W(\mathbf{a}) + k_B \ln \prod_\alpha \chi(\mathbf{a}^\alpha) . \quad (3.8\text{-}5)$$

Eliminating $k_B \ln \prod \chi(\mathbf{a}^\alpha)$ from (3.8-5), the Gibbs entropy function is found to satisfy the relation:

$$S_G - S^0 = -k_B \int d\mathbf{a}\, p(\mathbf{a},t) \ln \frac{p(\mathbf{a},t)}{W(\mathbf{a})} . \quad \mathbf{(3.8\text{-}6)}$$

Note that in an approach to equilibrium $p(\mathbf{a},t) \to W(\mathbf{a})$, so that $S_G \to S^0$.

3.8.2 *Failure for the Second Law of Thermodynamics in a Precisely Defined System State*

The expression for the Gibbs entropy function S_G in a microcanonical ensemble with density operator $\rho(t)$ has as time derivative

$$\dot{S}_G = -k_B (\partial / \partial t) \mathrm{Tr}[\rho(t) \ln \rho(t)] . \quad (3.8\text{-}7)$$

Employing von Neumann's equation (2.3-5) we find,

$$\begin{aligned} \dot{S}_G &= -k_B \mathrm{Tr}\{\dot{\rho}(\ln \rho + 1)\} = (k_B / \hbar i) \mathrm{Tr}\{[\rho, H](\ln \rho + 1)\} \\ &= (k_B / \hbar i) \mathrm{Tr}\{(\rho H - H\rho) \ln \rho\} \\ &= (k_B / \hbar i) \mathrm{Tr}\{(\ln \rho)\rho H - H\rho(\ln \rho)\} \\ &= (k_B / \hbar i) \mathrm{Tr}\{[\rho(\ln \rho)]H - H[\rho(\ln \rho)]\} = 0 . \end{aligned} \quad (3.8\text{-}8)$$

We used here the property of the trace being invariant against cyclic permutation.[14] The result just obtained is a well-known catastrophe: the entropy does not change! We could also have foreseen that from the formal solution (2.3-6), which indicates that, if $\rho(0)$ commutes with $\mathcal{H}$, then $\rho(t)$ commutes with $\mathcal{H}$ for all times. The problem is that ensembles with precisely known eigenstates $\{|\eta\rangle\}$ do not exist.[15] Thus in the von Neumann equation *the Hamiltonian must be replaced with* $\mathcal{H} \to \mathcal{H}^0 + \lambda\mathcal{V}$, *where* $\mathcal{H}^0$ *is the largest Hamiltonian, representing the physical system, that can be diagonalized and* $\lambda\mathcal{V}$ *is a perturbation.* In the second part of this text, dealing with nonequilibrium situations, we shall discuss the elaborate procedures that are necessary to solve the modified von Neumann equation with projection operators and perturbation techniques. It will there be shown that, indeed, $\dot{S}_G \geq 0$, with the equal sign being applicable to the equilibrium state.

In the context of equilibrium statistical mechanics, the second law of thermodynamics can only be derived upon coarse-graining of the states, *i.e.*, we use the coarse-grained entropy function (3.8-6). This device, due to Gibbs, will be shown in the next subsection.

3.8.3 *Coarse-Graining and the Second Law*

We rewrite Equation (3.8-6) as follows:

$$S_G - S^0 = -k\int d\mathbf{a}[p(\mathbf{a},t)\ln\left(p(\mathbf{a},t)/W(\mathbf{a})\right) - p(\mathbf{a},t) + W(\mathbf{a})], \tag{3.8-9}$$

where all we did is adding two normalization integrals. We now put

$$p(\mathbf{a},t) = W(\mathbf{a})e^{x(\mathbf{a},t)}, \quad x \neq 0 \textit{ outside equil}. \tag{3.8-10}$$

Substituting into (3.8-9) we have,

$$S_G - S^0 = -k\int d\mathbf{a} W(\mathbf{a})\left[x(\mathbf{a},t)e^{x(\mathbf{a},t)} - e^{x(\mathbf{a},t)} + 1\right]. \tag{3.8-11}$$

The expression $f(x)$ in square brackets has as derivative $f_x' = xe^x$. Thus, $f(x)$ increases with x for positive x and decreases with x for negative x; it is positive definite, being zero only for $x = 0$. Hence,

$$\left[x(\mathbf{a},t)e^{x(\mathbf{a},t)} - e^{x(\mathbf{a},t)} + 1\right] \geq 0,$$

so that

$$S_G \leq S^0, \tag{3.8-12}$$

---

[14] Tr (AB)= Tr (BA); Tr $(A_1A_2\ldots A_n)$ = Tr $(A_nA_1A_2\ldots)$.

[15] An exception is formed by a system of N coupled harmonic oscillators, for which the eigenstates can be found; however, again, in practice, the coupling will involve some anharmonicity involving cubic terms that give rise to perturbations $\lambda\mathcal{V}$.

indicating that the equilibrium entropy S^0 is the state of maximum entropy. Any reader, who finds this proof 'contrived', has our sympathy....; yet, the mathematical argument is beyond reproach!

In this version of the second law we have explicitly used nonequilibrium functions. However, the law can also be given in the framework of equilibrium thermodynamics. This is accomplished by comparing the entropy in a 'constrained equilibrium' with that of a 'free equilibrium'. In Fig. 3-1 we picture a situation in which there is a spring-loaded piston, dividing the volume in two parts V_1 and V_2, held in place by removable notches. Upon pulling out the notches, a spontaneous process takes place, until the spring reaches its normal uncompressed position. In accord with Section 1.3 the second law can be stated as: *Two equilibrium situations, connected by a spontaneous (or irreversible) process, during which no heat is added or taken from the system, have entropies such that*

$$S^0(final) > S^0(initial). \tag{3.8-13}$$

This result can of course also be put in a differential form. In the above case the displacement of the piston involves work $\int PdV$, so , in accordance with (1.3-6),

$$\Delta S = \Delta S_{qs} + \Delta S_{spont} \geq \int_a^b \{[\partial U/\partial V)_T + P^r]/T^r\}dV. \tag{3.8-14}$$

For a perfect gas in equilibrium with the environment there is no Joule effect, i.e., $(\partial U/\partial V)_T = 0$; the work performed yields the molar change $\Delta S = R\ln(V_b/V_a)$. If heat is produced (because of friction), then (3.8-14) is still valid, but $\Delta S_{spont} > \int đQ/T$. Clausius called the difference 'uncompensated heat'.

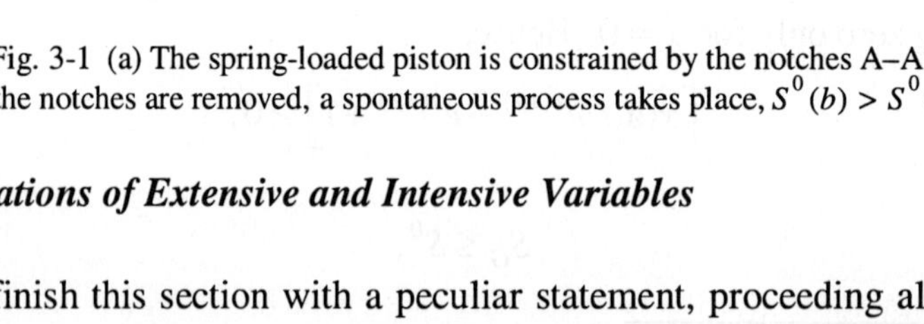

Fig. 3-1 (a) The spring-loaded piston is constrained by the notches A–A';
(b) the notches are removed, a spontaneous process takes place, $S^0(b) > S^0(a)$.

3.8.4 *Fluctuations of Extensive and Intensive Variables*

We shall finish this section with a peculiar statement, proceeding along the same lines as the proof of (3.8-12). We consider two subsystems in contact; we have,

$$S_G^c - S_G^{\mathrm{I}} - S_G^{\mathrm{II}} = 0. \tag{3.8-15}$$

We now define,

$$h^c = \ln p(\eta^c,t), \quad h^{\mathrm{I}} = \ln p(\eta^{\mathrm{I}},t), \quad h^{\mathrm{II}} = \ln p(\eta^{\mathrm{II}},t). \tag{3.8-16}$$

Following Gibbs we will call h the 'index of probability'. Clearly, $S_G = -k_B\langle h\rangle$ and (3.8-15) translates into

$$\langle h^c\rangle - \langle h^{\mathrm{I}}\rangle - \langle h^{\mathrm{II}}\rangle = 0. \tag{3.8-17}$$

We consider the quantity,

$$Y = \sum_{\eta^{\mathrm{I}},\eta^{\mathrm{II}}} \left[(h^c - h^{\mathrm{I}} - h^{\mathrm{II}} - 1)e^{h^c - h^{\mathrm{I}} - h^{\mathrm{II}}} + 1 \right] e^{h^{\mathrm{I}} + h^{\mathrm{II}}}. \tag{3.8-18}$$

By simple algebra,

$$Y = \sum_{\eta^{\mathrm{I}}\eta^{\mathrm{II}}} (h^c - h^{\mathrm{I}} - h^{\mathrm{II}})e^{h^c} - \sum_{\eta^{\mathrm{I}}\eta^{\mathrm{II}}} e^{h^c} + \sum_{\eta^{\mathrm{I}}} e^{h^{\mathrm{I}}} \sum_{\eta^{\mathrm{II}}} e^{h^{\mathrm{II}}}, \tag{3.8-19}$$

from which we deduce

$$\begin{aligned} Y &= \sum_{\eta^{\mathrm{I}}\eta^{\mathrm{II}}} (h^c - h^{\mathrm{I}} - h^{\mathrm{II}})p(\eta^c,t) - \sum_{\eta^{\mathrm{I}}\eta^{\mathrm{II}}} p(\eta^c,t) + \sum_{\eta^{\mathrm{I}}} p(\eta^{\mathrm{I}},t) \sum_{\eta^{\mathrm{II}}} p(\eta^{\mathrm{II}},t) \\ &= \left\langle h^c - h^{\mathrm{I}} - h^{\mathrm{II}} \right\rangle - 1 + 1 = 0, \end{aligned} \tag{3.8-20}$$

by virtue of (3.17). However, setting $x = h^c - h^{\mathrm{I}} - h^{\mathrm{II}}$, we see from the definition (3.8-18) that Y is semi-definite,

$$Y = \sum_{\eta^{\mathrm{I}}\eta^{\mathrm{II}}} (xe^x - e^x + 1)e^{h^{\mathrm{I}} + h^{\mathrm{II}}} \geq 0, \tag{3.8-21}$$

being equal to zero only for $x = 0$. Therefore (3.8-20) requires that $h^c = h^{\mathrm{I}} + h^{\mathrm{II}}$, or

$$p(\eta^c,t) = p(\eta^{\mathrm{I}},t)p(\eta^{\mathrm{II}},t), \text{ or also } \rho^c(t) = \rho^{\mathrm{I}}(t)\rho^{\mathrm{II}}(t). \tag{3.8-22}$$

We thus proved the somewhat paradoxical result (an adaptation from Gibbs): *Two macroscopic systems, which are in contact with each other, are still (almost) statistically independent.*[16] The explanation is not difficult. All entropy expressions in this chapter are based on the fact that surface interactions, being of order $V^{2/3}$ disappear in the thermodynamic limit. Further we note that the result (3.8-22) holds, a fortiori, for the equilibrium state.

We shall apply the above result to obtain some rules for fluctuations of the variables a_i in the ensemble. Let there be K subsystems of equal size. Then for the variance of an extensive variable,

$$\langle \Delta a^2 \rangle = \sum_{\alpha=1}^{K} \langle (\Delta a^\alpha)^2 \rangle + \sum_{\alpha\beta}^{K} \langle \Delta a^\alpha \Delta a^\beta \rangle. \tag{3.8-23}$$

---

[16] J.W. Gibbs, Op. Cit., Chapter XI, theorem VII. Gibbs' treatment, of course, deals with the classical phase-space distributions.

The last summand is zero because of statistical independence:

$$\begin{aligned}\langle \Delta a^{\alpha} \Delta a^{\beta} \rangle &= \mathrm{Tr}\{\rho^{c} \Delta a^{\alpha} \Delta a^{\beta}\} \\ &= \mathrm{Tr}(\rho^{\alpha} \Delta a^{\alpha}) \mathrm{Tr}(\rho^{\beta} \Delta a^{\beta}) = \langle \Delta a^{\alpha} \rangle \langle \Delta a^{\beta} \rangle = 0 .\end{aligned} \tag{3.8-24}$$

Whence,

$$\langle \Delta a^{2} \rangle = K \langle (\Delta a^{\alpha})^{2} \rangle . \tag{3.8-25}$$

For the relative rms fluctuations we find,

$$\sqrt{\langle \Delta a^{2} \rangle} / \langle a \rangle = (1/\sqrt{K}) \sqrt{\langle (\Delta a^{\alpha})^{2} \rangle} / \langle a^{\alpha} \rangle . \tag{3.8-26}$$

Let the size of the subsystems now be fixed, but let the total system increase in size and molecular content. Since $K \propto N$, we find that the relative rms fluctuations go to zero in the thermodynamic limit.

Fluctuations in intensive variables associated with the a's may be found from their equations of state; thus,

$$\Delta f_{i} = \textstyle\sum_{j} (\partial f_{i} / \partial a_{j}) \Delta a_{j} . \tag{3.8-27}$$

Since the a's are independent variables we have,

$$\langle \Delta f_{i}^{2} \rangle = \sum_{j} \left(\frac{\partial f_{i}}{\partial a_{j}} \right)^{2} \langle \Delta a_{j}^{2} \rangle = \sum_{j} \left(\frac{\partial f_{i}^{0}}{\partial a_{j}^{0}} \right)^{2} \langle \Delta a_{j}^{2} \rangle . \tag{3.8-28}$$

[The latter equality follows from the fact that f is the same function of the a's as the equilibrium value is of the a^{0}'s.] Assuming again that there are K subsystems of same macroscopic size, we have with $a_{j}^{0} = K a_{j}^{\alpha 0}$ from (3.8-25),

$$\langle \Delta f_{i}^{2} \rangle = \frac{1}{K} \sum_{j} \left(\frac{\partial f_{i}^{0}}{\partial a_{j}^{\alpha 0}} \right)^{2} \langle (\Delta a_{j}^{\alpha})^{2} \rangle . \tag{3.8-29}$$

Again, let the system increase in size with the volume of the subsystem being fixed, so that $\mathrm{var}\, a_{j}^{\alpha}$ is fixed. Then,

$$\mathrm{var}\, f_{i} = \langle \Delta f_{i}^{2} \rangle \to 0 . \tag{3.8-30}$$

This, then, shows that the intensive variables of infinite reservoirs are rigorously constant, a fact that we have tacitly assumed hitherto.

3.9 Problems to Chapter III

3.1 Starting from the entropy of a classical gas (Sackur–Tetrode formula) and Boltzmann's formula $S = k_B \ln W + const$, obtain the same result as in problem 2.4.

3.2 Consider a system with volume *V*, particle number *N* and energy $\mathcal{E}$. There is free energy exchange between the system and a thermal bath. Employ the Boltzmann–Einstein formula in the form (3.7-12), expanding about the equilibrium state to second order, to obtain the normal approximation to $W(\mathcal{E})$. Find $\operatorname{var}\mathcal{E}$ and show that the result is as in (1.6-76).

3.3 Consider a system of *N* particles with given temperature and fixed volume; it is characterized by the Helmholtz free energy $F(T^r, V, N, \{A_i\})$. Apply an isothermal isochoric variation δF. Obtain the equilibrium requirement(s) and the condition for local stability, expressed in response function(s). Also show that *F* is concave with respect to *T* and convex with respect to *V*, *i.e.*,

$$\partial^2 F / \partial T^2 < 0, \quad \partial^2 F / \partial V^2 > 0. \tag{1}$$

3.4 In the same system the particle constraint is removed. Apply a variation $\delta\Omega$, where Ω is the grand potential, the process being isothermal and with constant chemical potential. Find once more the equilibrium requirements and the conditions for local stability.

3.5 Instead of the correct Sackur–Tetrode formula, obtain the incorrect classical entropy (3.5-3), starting from $S = k_B \ln \Delta_{\mathcal{E}}\Omega + const$. [Do *not* include the factor N!].
(a) Mix two gases *A* and *B* with attributes (N_1, V_1) and (N_2, V_2), respectively. Obtain the mixing entropy.
(b) Let the gases be identical and mix equal volumes. Show that the mixing entropy per mole is $2R\ln 2$ ("Gibbs paradox").

3.6 At sufficiently high temperatures ($> 5000K$) a diatomic hydrogen molecule can be considered as a *dumbbell* with classical translational, rotational and vibrational motion. What is the high temperature limit of the molar specific heat? With descending temperature, first vibrational modes become quantized and later on the rotational modes disappear. Make a rough plot of molar specific heat *vs.* *T*.

3.7 Consider a classical microcanonical ensemble. Using methods similar to those of section 3.3, obtain the relative variance of the kinetic energy $\langle[\mathcal{T}-\langle\mathcal{T}\rangle]^2\rangle/\langle\mathcal{T}\rangle$. Hint: First evaluate $\langle p_i(\partial\mathcal{H}/\partial p_i)p_j(\partial\mathcal{H}/\partial p_j)\rangle$. Assume that the number of particles N is very large. Compare your result with (1.6-76).

3.8 The energy eigenstates for an N-particle system are M-fold degenerate, *i.e.*,

$$|\eta\rangle \to |\mathscr{E}_\eta, k\rangle,\ \ k = 1 \ldots M\,. \tag{2}$$

(a) Write down an exact form for the density operator in the microcanonical ensemble;

(b) Using $S = -k_B \mathrm{Tr}\rho \ln \rho$, obtain an expression for the entropy and the temperature.

Chapter IV

Ensembles in the Presence of Reservoirs. The Canonical and Grand-Canonical Ensemble

1. General Formalism and Some Quantum Applications

4.1 The Canonical Ensemble for Systems in Contact with a Heat Bath

In this chapter we shall obtain the basic density operators for ensembles pertaining to open systems. First, we consider the canonical ensemble, which describes systems for which the energy $\mathscr{E}$ is variable, while $\overline{T}$ is fixed and equal to the reservoir temperature T^r (the superscript 'r' will later be omitted, as in previous developments). Since all other extensive variables, as well as the intensive Gibbsian parameters, see subsection 1.3.2, are held constant, no external work $đW$ can be performed, so that the energy exchange is accomplished only via the exchange of heat, as indicated in the heading of this section. While the new ensembles are far more suitable for calculations, they have less physical rigour to back up their existence than the microcanonical ensemble; only the latter has a *direct* connection with the mechanical or quantum mechanical equations of motion of a conservative system. We also mention that Gibbs never fully derived these ensembles; for the canonical ensemble he just postulated that the 'index of probability', $\ln p(\eta)$, should depend linearly on the energy $\mathscr{E}_\eta$. As far as thermodynamics is concerned, the usual state variables are either fixed or averages, i.e., first order moments, in the new ensembles. All thermodynamic relationships (Gibbs' relation, Maxwell relations, etc.) remain valid in the new ensembles. Extremum principles must be adapted, as we discussed already in connection with the Boltzmann–Einstein principle, Section 3.7. In the canonical ensemble the Helmholtz free energy is a minimum in the equilibrium state. The second order moments (or cumulants), *in casu* the variances and covariances of fluctuating state variables, are different, however, from ensemble to ensemble; they are explicitly dependent on the constraints that are imposed, a fact most poignantly shown in various, rather unknown sections of Fowler's classic book[1]; we will discuss examples later on.

We now proceed to our main task, following the route that we have used already

---

[1] R.H. Fowler, "Statistical Mechanics", Cambridge Univ. Press 1936; reprinted in paperback 1966.

in various problems of the previous chapters. Let ρ^c be the density operator of a composite closed system, ρ the density operator of a subsystem open to energy exchange, and ρ^{res} the density operator of the remainder, which is labelled the reservoir. Neglecting surface interactions, the energy eigenstates can be factored, $|\eta^c\rangle = |\eta^r\rangle|\eta\rangle$. For the composite system we have, in accord with (2.3-14),

$$\rho^c = \frac{1}{\chi(\mathcal{E}_c)} \sum_{\eta^r} |\eta^r\rangle\langle\eta^r| \sum_{\eta} |\eta\rangle\langle\eta| \delta(\mathcal{E}^c - \mathcal{E}_{\eta^r} - \mathcal{E}_\eta) . \tag{4.1-1}$$

To obtain ρ for the smaller system we take the trace over the reservoir states,

$$\begin{aligned} \rho = \mathrm{Tr}_{res}\rho^c &= \sum_{\bar{\eta}^r} \langle\bar{\eta}^r|\rho^c|\bar{\eta}^r\rangle \\ &= \frac{1}{\chi(\mathcal{E}_c)} \sum_{\bar{\eta}^r,\eta^r} \langle\bar{\eta}^r|\eta^r\rangle\langle\eta^r|\bar{\eta}^r\rangle \sum_{\eta} |\eta\rangle\langle\eta| \delta(\mathcal{E}^c - \mathcal{E}_{\eta^r} - \mathcal{E}_\eta) \\ &= \frac{1}{\chi(\mathcal{E}_c)} \sum_{\bar{\eta}^r} \sum_{\eta} |\eta\rangle\langle\eta| \delta(\mathcal{E}^c - \mathcal{E}_{\bar{\eta}^r} - \mathcal{E}_\eta) . \end{aligned} \tag{4.1-2}$$

Now, as usual, we write $\sum_{\bar{\eta}^r} \to \int d\mathcal{E}^r \chi(\mathcal{E}^r)$, so that,

$$\rho = \sum_{\eta} |\eta\rangle \frac{\chi(\mathcal{E}^c - \mathcal{E}_\eta)}{\chi(\mathcal{E}^c)} \langle\eta| . \tag{4.1-3}$$

We write further,

$$\frac{\chi(\mathcal{E}^c - \mathcal{E}_\eta)}{\chi(\mathcal{E}^c)} = \exp\left[\ln \chi(\mathcal{E}^c - \mathcal{E}_\eta) - \ln \chi(\mathcal{E}^c)\right] . \tag{4.1-4}$$

We now need a Taylor expansion, noting that for a large reservoir $\mathcal{E}^c - \mathcal{E}_\eta \to \mathcal{E}^c$. *There is, however, a pitfall that occurs in a great many texts.* The argument then is,

$$\begin{aligned} \ln \chi(\mathcal{E}^c) &= \ln \chi(\mathcal{E}^c - \mathcal{E}_\eta) + \frac{\partial \ln \chi(\mathcal{E}^c - \mathcal{E}_\eta)}{\partial(\mathcal{E}^c - \mathcal{E}_\eta)} \mathcal{E}_\eta + \mathcal{O}(\mathcal{E}_\eta^2) \\ &= \ln \chi(\mathcal{E}^c - \mathcal{E}_\eta) + \frac{\partial \ln \chi(\mathcal{E}^r)}{\partial(\mathcal{E}^r)} \mathcal{E}_\eta + \mathcal{O}(\mathcal{E}_\eta^2) . \end{aligned} \tag{4.1-5}$$

The derivative in the second term of the final rhs is just $1/k_BT^r = 1/k_BT$; this will yield for (4.1-3) the *erroneous result* $\rho = \exp[(-\mathcal{H}/k_BT)]$. The problem is that the normalization constant in front of the exponential is missing; yet, we started from a normalized density operator ρ^c! In many texts this is 'remedied' by just using the '$\propto$' sign and then inserting a constant C, whose meaning is later on recovered. Others, e.g., Münster[2], believe that the terms $\mathcal{O}(\mathcal{E}_\eta^2)$ give rise to a term ln C. In our

[2] A Münster, "Statistische Thermodynamik", Springer Verlag, Heidelberg, 1956.

view the reason for the discrepancy is to be sought in the fact that one overlooks the dependence of the density of states on the other constrained variables. For instance, if the other attributes are V and N, then $\ln\chi = S/k_B$ is *not* differentiable to these variables, since generally $P \neq P^r, \varsigma \neq \varsigma^r$. Thus, more appropriately, we set

$$\ln\chi(\mathscr{E}^c, V^c, N^c, ...) = \ln\chi(\mathscr{E}^c, V^r, N^r, ...) - \Psi/k_B T^r, \tag{4.1-6}$$

where Ψ at present is unknown. The quantity $\ln\chi(\mathscr{E}^c, V^r, N^r, ...)$ is now expanded in a Taylor series to first order in the variable $\mathscr{E}$ only. We then obtain,

$$\ln\chi(\mathscr{E}^c, V^c, N^c ...) - \ln\chi(\mathscr{E}^c - \mathscr{E}_\eta, V^r, N^r, ...) = \frac{\mathscr{E}_\eta - \Psi}{k_B T}. \tag{4.1-7}$$

Eqs. (4.1-3), (4.1-4) and (4.1-7) result in:

$$\rho = \sum_\eta |\eta\rangle e^{(\Psi - \mathscr{E}_\eta)/kT} \langle\eta| = e^{(\Psi - \mathscr{H})/k_B T}, \tag{4.1-8}$$

which is the density operator of the canonical distribution. The quantity Ψ will now be elucidated. From its definition,

$$\begin{aligned} \Psi &= TS(\mathscr{E}^c, V^r, N^r, ...) - TS(\mathscr{E}^c, V^c, N^c, ...) \\ &= -T[\Delta S]_{\mathscr{E}\,fixed} = F(T, V, N, ...), \end{aligned} \tag{4.1-9}$$

where F is the Helmholtz free energy of the system in equilibrium with the reservoir. Setting $\exp(\Psi/k_BT) \equiv 1/\mathscr{Z}$, where $\mathscr{Z}$ is the canonical partition function[3], we have since $\mathrm{Tr}\,\rho = 1$,

$$\rho = \frac{1}{\mathscr{Z}} e^{-\mathscr{H}/k_B T}, \qquad \mathscr{Z} = \mathrm{Tr}\, e^{-\mathscr{H}/k_B T}, \tag{4.1-10}$$

$$F(T, V, N, ...) = -k_B T \ln \mathscr{Z}. \tag{4.1-11}$$

The partition function can either be taken as a sum over states or as an energy integral:

$$\mathscr{Z} = \sum_\eta e^{-\mathscr{E}_\eta/k_B T} \approx \int \chi(\mathscr{E}) e^{-\mathscr{E}/k_B T} d\mathscr{E}. \tag{4.1-12}$$

This result, given here for the quantum distribution, can easily be translated to the classical case. Then instead of (4.1-10), we write

$$\rho(p, q) = \frac{1}{\mathscr{Z}} e^{-\mathscr{H}(p,q)/k_B T}, \qquad \mathscr{Z} = \int d\Omega\, e^{-\mathscr{H}(p,q)/k_B T} / h^{rN} N!, \tag{4.1-13}$$

---

[3] The symbol script capital Z stems from the German *Zustandssomme*.

where $\mathcal{H}(p,q)$ is the classical Hamiltonian. We divided by the volume of a microcell, to obtain continuity between the classical and quantum case.

From $\mathcal{Z}$ all other thermodynamic functions can be obtained. For $S = -\partial F / \partial T$ one easily finds,

$$S = k_B \partial (T \ln \mathcal{Z}) / \partial T, \tag{4.1-14}$$

while for the energy we have,

$$\langle \mathcal{E} \rangle = F + TS = k_B T^2 \frac{\partial \ln \mathcal{Z}}{\partial T} = -\frac{\partial \ln \mathcal{Z}}{\partial (1/k_B T)}. \tag{4.1-15}$$

Noting further the operator relationship $\ln \rho = (F - \mathcal{H}) / k_B T = -S_{(oper)} / k_B$, we find by averaging over ρ,

$$S = -k_B \,\mathrm{Tr}(\rho \ln \rho), \tag{4.1-16}$$

which is the same expression as in the microcanonical ensemble. Finally, for the second moment of the energy we have

$$\langle \mathcal{E}^2 \rangle = \frac{\mathrm{Tr}\, \mathcal{H}^2 e^{-\mathcal{H}/k_B T}}{\mathrm{Tr}\, e^{-\mathcal{H}/k_B T}} = \frac{1}{\mathcal{Z}} \frac{\partial^2 \mathcal{Z}}{\partial (1/k_B T)^2}. \tag{4.1-17}$$

To obtain the variance we use the identity

$$\frac{\partial^2 \ln \mathcal{Z}}{\partial (1/k_B T)^2} = \frac{\partial}{\partial (1/k_B T)} \left[\frac{1}{\mathcal{Z}} \frac{\partial \mathcal{Z}}{\partial (1/k_B T)} \right] = \frac{1}{\mathcal{Z}} \frac{\partial^2 \mathcal{Z}}{\partial (1/k_B T)^2} - \frac{1}{\mathcal{Z}^2} \left(\frac{\partial \mathcal{Z}}{\partial (1/k_B T)} \right)^2, \tag{4.1-18}$$

which, in view of (4.1-18) and (4.1-15) yields the second order cumulant,

$$\langle \Delta \mathcal{E}^2 \rangle = \langle \mathcal{E}^2 \rangle - \langle \mathcal{E} \rangle^2 = \frac{\partial^2 \ln \mathcal{Z}}{\partial (1/k_B T)^2} = -\frac{\partial \langle \mathcal{E} \rangle}{\partial (1/k_B T)} = k_B T^2 \mathcal{n} C_{V,N}, \tag{4.1-19}$$

where $C_{V,N}$ is the molar specific heat for constant volume and number of molecules, and $\mathcal{n} = N / N_A$ is the number of moles in the system.

4.2 The Grand-Canonical Ensemble for Systems with Energy and Particle Exchange

We follow the same method as before, i.e., we consider a composite closed system, that is partitioned in 'the system' and another system, which acts as an energy and particle reservoir. There remains a constraint for V, which will be heeded *ab initio*.

The microcanonical partition function for the composite system is as always,

$$\Delta\Gamma(\mathscr{E}^c, V^c, N^c, \{A_i^c\}) = \chi(\mathscr{E}^c, V^c, N^c, \{A_i^c\})\delta\mathscr{E}, \tag{4.2-1}$$

where χ is the density of states. However, the counting of states must be done judiciously, since the system, as well as the reservoir, can contain any number of particles; if in the system $N = 0,1,...N^c$, then in the reservoir we have $N^r = N^c, N^c - 1,...0$. Let $\mathfrak{H}_N$ be the Hilbert space for a system with a N-particle Hamiltonian; then the applicable full Hilbert space is

$$\mathfrak{H} = \mathfrak{H}_1 \otimes \mathfrak{H}_2 \otimes ... = \prod_{N=0}^{\infty} \mathfrak{H}_N \otimes, \tag{4.2-2}$$

where the upper limit '∞' will be realized in the thermodynamic limit. The quantum states pertaining to a N-particle system will be denoted as $|\eta_N, N\rangle$ and 'Tr' in $\mathfrak{H}$ will mean $\Sigma_N \Sigma_{\eta_N}$ .. The density operator of departure for the composite system is,

$$\begin{aligned}\rho^c = & \frac{1}{\chi(\mathscr{E}^c, N^c; V^c, \{A_i^c\})} \sum_{N,N^r} \sum_{\eta_N, \eta_{N^r}^r} |\eta_N, N\rangle |\eta_{N^r}^r, N^r\rangle \\ & \times \delta_{N^c, N^r+N} \delta[\mathscr{E}^c - \mathscr{E}_{\eta^r, N^r}^r - \mathscr{E}_{\eta, N}] \langle \eta_N, N| \langle \eta_{N^r}^r, N^r|.\end{aligned} \tag{4.2-3}$$

Taking now the trace over the reservoir states for *fixed* number N [meaning also that N^r is held constant] i.e., $\mathrm{Tr}_{res,N} .. = \sum_{\bar{\eta}_{N^r}^r} \langle \bar{\eta}_{N^r}^r, N^r | .. | \bar{\eta}_{N^r}^r, N^r \rangle$, we easily obtain,

$$\rho = \frac{1}{\chi(\mathscr{E}^c, N^c; V^c, \{A_i^c\})} \sum_N \sum_{\eta_N, \bar{\eta}_{N^r}^r} |\eta_N, N\rangle \delta[\mathscr{E}^c - \mathscr{E}_{\bar{\eta}^r, N^r}^r - \mathscr{E}_{\eta, N}] \langle \eta_N, N|. \tag{4.2-4}$$

Now we write $\sum_{\bar{\eta}_{N^r}^r} \to \int d\mathscr{E}^r \chi(\mathscr{E}^r, N^r; V^r, \{A_i^r\})$ [this χ is the density of states for a given N]. We then arrive at,

$$\begin{aligned}\rho = & \sum_N \sum_{\eta_N} |\eta_N, N\rangle \exp\Big\{\ln \chi(\mathscr{E}^c - \mathscr{E}_{\eta,N}, N^c - N; V^r, \{A_i^r\}) \\ & - \ln \chi(\mathscr{E}^c, N^c, V^c, \{A_i^c\})\Big\} \langle \eta_N, N|.\end{aligned} \tag{4.2-5}$$

The entropy $S = k_B \ln \chi$ is differentiable to $\mathscr{E}$ and N, since the temperature and chemical potential are the same at both sides of the partition; however, S is not differentiable to V, since the left-side and right-side derivatives (pressures) may jump. Therefore we write,

$$\ln \chi(\mathscr{E}^c, N^c, V^c, \{A_i^c\}) = \ln \chi(\mathscr{E}^c, N^c; V^r, \{A_i^r\}) - \Phi / k_B T, \tag{4.2-6}$$

where Φ still is to be obtained. With first order Taylor expansion in $\mathscr{E}$ and N we find,

$$\ln\chi(\mathcal{E}^c, N^c, V^c, \{A_i^c\}) - \ln\chi(\mathcal{E}^c - \mathcal{E}_{\eta,N}, N^c - N; V^r, \{A_i^r\})$$
$$= \frac{\partial \ln\chi(\mathcal{E}^c - \mathcal{E}_{\eta,N}, N^c - N, ..)}{\partial(\mathcal{E}^c - \mathcal{E}_{\eta,N})}\mathcal{E}_{\eta,N} + \frac{\partial \ln\chi(\mathcal{E}^c - \mathcal{E}_{\eta,N}, N^c - N, ..)}{\partial(N^c - N)}N - \frac{\Phi}{k_B T}$$
$$= \frac{-\Phi + \mathcal{E}_{\eta,N} - \varsigma N}{k_B T}, \tag{4.2-7}$$

where T and ς are the intensive parameters of the reservoir. The result is

$$\rho = \sum_N \sum_{\eta_N} |\eta_N, N\rangle e^{[\Phi - \mathcal{E}_{\eta,N} + \varsigma N]/k_B T} \langle \eta_N, N| = e^{[\Phi - \mathcal{H} + \varsigma N_{op}]/k_B T}, \tag{4.2-8}$$

where N_{op} is the number-operator, which commutes with the Hamiltonian. Equation (4.2-8) is referred to as the grand-canonical distribution.

The meaning of Φ is easily found. From its definition in (4.2-6) and the Gibbs function (or grand potential) Ω [(Eq. (1.4-6) gives $TS = \mathcal{E} - \varsigma N - \Omega$], we have[4]

$$\Phi = -TS(\mathcal{E}^c, N^c, V^c, ...) + TS(\mathcal{E}^c, N^c, V^r, ...) = -T\Delta S|_{\mathcal{E},N} = \Omega_{system}. \tag{4.2-9}$$

We shall also write $\exp[\Phi / k_B T = 1/\mathcal{Z}$, where $\mathcal{Z}$ is the grand-canonical partition function.[5] Hence, noting $\mathrm{Tr}\rho = 1$, we obtain

$$\rho = \frac{1}{\mathcal{Z}} e^{-(\mathcal{H} - \varsigma N_{op})/k_B T}, \quad \mathcal{Z} = \mathrm{Tr}\{e^{-(\mathcal{H} - \varsigma N_{op})/k_B T}\}. \tag{4.2-10}$$

The thermodynamic connection is from (4.2-9):

$$\Omega(T, \varsigma, V, \{A_i\}) = -k_B T \ln \mathcal{Z}, \tag{4.2-11}$$

where we explicitly indicated the variables in Ω. Resorting to the density of states, we also express the partition function as,

$$\mathcal{Z} = \sum_N e^{\varsigma N/k_B T} \int d\mathcal{E} X(\mathcal{E}, N, V, ..) e^{-\mathcal{E}/k_B T}$$
$$= \sum_N e^{\varsigma N/k_B T} \sum_{\mathcal{E}} \Delta\Gamma(\mathcal{E}, N, V, ..) e^{-\mathcal{E}/k_B T}, \tag{4.2-12}$$

where the sum over $\mathcal{E}$ denotes summing over energy cells $(\mathcal{E}, \mathcal{E} + \delta\mathcal{E})$. With the *fugacities* $y_\varsigma = \exp(\varsigma / k_B T)$, $y_T = \exp(-1/k_B T)$, we also have,

$$\mathcal{Z} = \sum_N e^{\varsigma N/k_B T} \mathcal{Q}_N = \sum_N y_\varsigma^N \mathcal{Q}_N, \tag{4.2-13}$$

---

[4] We have taken the designations 'Gibbs function' and 'grand potential' to be synonymous; others (e.g. Kramers and ter Haar (D. ter Haar, Op. Cit. p. 135ff) refer to $\Phi = -(k_B T)\Omega$ as the grand potential.

[5] We use $\mathcal{Z}$ (European script capital Z) for the grand-canonical partition function.

$$\mathcal{Z} = \sum_N y_\varsigma^N \sum_{\mathscr{E}} y_T^{\mathscr{E}} \Delta\Gamma(\mathscr{E}, N, V, ..) . \qquad (4.2\text{-}14)$$

We thus note that $\mathcal{Z}$ is a Fowler transform of the canonical partition function and a double Fowler transform of the microcanonical partition function; more will be said later about this in a section on 'transformation theory' (next chapter).

We still mention the classical equivalent for the grand-canonical density function. Gibbs introduced the "specific grand-distribution function" and the "generic grand-distribution function", depending on whether interchange of identical particles is considered to be a different phase or the same phase, respectively.[6] We are only interested in the semi-classical distribution, which corresponds to Gibbs' generic distribution, except that we incorporate the phase cell h^{rN}. So,

$$\rho_{grand-can}(p,q) = \frac{1}{\mathcal{Z}} e^{-[\mathcal{H}(p,q;N) - \varsigma N]/k_B T} , \qquad (4.2\text{-}15)$$

which is to be normalized over

$$\sum_N d\Gamma_N = \sum_N \frac{d\Omega_N}{h^{rN} N!} = \sum_N \prod_i \frac{dp_i dq_i}{h^{rN} N!} . \qquad (4.2\text{-}16)$$

Consequently,

$$\mathcal{Z} = \sum_{N=0}^{\infty} \frac{1}{N!} e^{\varsigma N / k_B T} \int e^{-\mathcal{H}(p,q;N)/k_B T} \frac{dp_1 ... dq_{rN}}{h^{rN}} . \qquad (4.2\text{-}17)$$

Apart from the factor h^{rN} this is Gibbs' *generic phase integral.* Again, $\mathcal{Z} = \Sigma_N \exp(\varsigma N / k_B T) \mathcal{Q}_N$, where $\mathcal{Q}_N$ is the canonical phase integral. If various species are present, this must be generalized to

$$\mathcal{Z} = \sum_{\{N^k\}}^{\infty} \frac{1}{N^1!..N^K!} e^{\Sigma_k \varsigma^k N^k / k_B T} \int e^{-\mathcal{H}(p,q;\{N^k\})/k_B T} \frac{dp_1^1 ... dq_{r^K N^K}^K}{h^{\Sigma_k r_k N_k}} . \qquad (4.2\text{-}18)$$

For further details, see the older texts on this topic.

We return to the quantum forms. From Ω or $\ln \mathcal{Z}$ we find the other thermodynamic functions, as well as the fluctuations in the ensemble. First, the number average is given by

$$\langle N \rangle = \mathrm{Tr}(N_{op}\rho) = \frac{1}{\mathcal{Z}} \mathrm{Tr}\left\{ N_{op} e^{-(\mathcal{H} - \varsigma N_{op})/k_B T} \right\} . \qquad (4.2\text{-}19)$$

One easily verifies,

---

[6] J.W. Gibbs, Op. Cit., Chapter XV.

$$\langle N\rangle = \frac{1}{\mathcal{Z}}\frac{\partial \mathcal{Z}}{\partial(\varsigma/k_BT)} = \left(\frac{\partial \ln \mathcal{Z}}{\partial \ln y_\varsigma}\right)_{y_T}, \tag{4.2-20}$$

where we used the fugacities defined above. Also,

$$\langle N^2\rangle = \frac{1}{\mathcal{Z}}\frac{\partial^2 \mathcal{Z}}{\partial(\varsigma/k_BT)^2} = \frac{1}{\mathcal{Z}}\left(\frac{\partial^2 \mathcal{Z}}{\partial(\ln y_\varsigma)^2}\right)_{y_T}. \tag{4.2-21}$$

Now note,

$$\frac{\partial^2 \ln \mathcal{Z}}{\partial x^2} = \frac{\partial}{\partial x}\left(\frac{1}{\mathcal{Z}}\frac{\partial \mathcal{Z}}{\partial x}\right) = \frac{1}{\mathcal{Z}}\frac{\partial^2 \mathcal{Z}}{\partial x^2} - \frac{1}{\mathcal{Z}^2}\left(\frac{\partial \mathcal{Z}}{\partial x}\right)^2. \tag{4.2-22}$$

This holds, whatever the derivative to x means. Presently, with $x = \ln y_\varsigma$ we find that the variance is given by the cumulant,

$$\begin{aligned}\langle \Delta N^2\rangle &= \langle N^2\rangle - \langle N\rangle^2 \\ &= \left(\frac{\partial^2 \ln \mathcal{Z}}{\partial(\ln y_\varsigma)^2}\right)_{y_T} = \left(\frac{\partial\langle N\rangle}{\partial \ln y_\varsigma}\right)_{y_T} = k_BT\left(\frac{\partial\langle N\rangle}{\partial\varsigma}\right)_T.\end{aligned} \tag{4.2-23}$$

From the ordinary Gibbs-Duhem relation,

$$d\varsigma = \hat{v}dP - \hat{s}dT, \tag{4.2-24}$$

we have

$$(\partial\varsigma/\partial\hat{v})_T = \hat{v}(\partial P/\partial\hat{v})_T, \tag{4.2-25}$$

where we wrote $V/N = \hat{v} = 1/n$, $S/N = \hat{s}$; also, we omitted averaging bars, necessary since (4.2-24) is a macroscopic relationship, cf. footnote 12 of Chapter I. Now $\hat{v}$ can be changed by either varying N or V.[7] So,

$$\left(\frac{\partial}{\partial\hat{v}}\right)_{V,T} = \left(\frac{\partial}{\partial N}\right)_{V,T}\left(\frac{\partial N}{\partial\hat{v}}\right)_{V,T} = -\frac{N}{\hat{v}}\left(\frac{\partial}{\partial N}\right)_{V,T}, \tag{4.2-26}$$

$$\left(\frac{\partial}{\partial\hat{v}}\right)_{N,T} = \left(\frac{\partial}{\partial V}\right)_{N,T}\left(\frac{\partial V}{\partial\hat{v}}\right)_{N,T} = \left(\frac{\partial}{\partial\hat{v}}\right)_{N,T}. \tag{4.2-27}$$

Since both changes yield the same result, we obtain, also using (4.2-25),

$$-\frac{N}{\hat{v}}\left(\frac{\partial\varsigma}{\partial N}\right)_{V,T} = \hat{v}\left(\frac{\partial P}{\partial\hat{v}}\right)_{N,T} = V\left(\frac{\partial P}{\partial V}\right)_{N,T}. \tag{4.2-28}$$

With the definition of the isothermal compressibility

---

[7] Cf. Plischke and Bergersen, 3rd Ed., Op. Cit., p.43.

$$K_T = -\frac{1}{V}\left(\frac{\partial V}{\partial P}\right)_{N,T} \tag{4.2-29}$$

this yields $(\partial N/\partial \varsigma)_{V,T} = (N/\hat{v})K_T$. Since V is fixed in the grand-canonical ensemble, we find for (4.2-23)

$$\langle \Delta N^2 \rangle = k_B T \langle N \rangle K_T / \hat{v}\,. \tag{4.2-30}$$

This result shows that the fluctuations are always associated with a response function – a property we shall meet many times in linear response theory, treated in part *E* of this text. It is instructive to apply the result for a perfect gas. For the compressibility the perfect gas law $PV = Nk_BT$ gives $K_T = 1/P$, so that $\langle \Delta N^2 \rangle = \langle N \rangle$, as for a Poisson process.

Next, we consider the energy fluctuations. The most basic method goes similar as before. We have,

$$\langle \mathcal{E} \rangle = \frac{1}{\mathcal{Z}} \mathrm{Tr}\left\{\mathcal{H} e^{-(\mathcal{H}-\varsigma N_{op})/k_BT}\right\} \tag{4.2-31}$$

or,

$$\langle \mathcal{E} \rangle = -\left(\frac{\partial \ln \mathcal{Z}}{\partial(1/k_BT)}\right)_{\varsigma/T} = \left(\frac{\partial \ln \mathcal{Z}}{\partial(\ln y_T)}\right)_{y_\varsigma}. \tag{4.2-32}$$

The parallel with (4.2-20) is clear: one must differentiate to the logarithm of the appropriate fugacity, keeping the other one constant. Employing (4.2-22) one easily verifies for the variance,

$$\langle \Delta \mathcal{E}^2 \rangle = \left(\frac{\partial^2 \ln \mathcal{Z}}{\partial(\ln y_T)^2}\right)_{y_\varsigma} = \left(\frac{\partial \langle \mathcal{E} \rangle}{\partial \ln y_T}\right)_{y_\varsigma}. \tag{4.2-33}$$

Note that $[\partial/\partial(\ln y_T)]_{\varsigma/T} = k_BT^2(\partial/\partial T)_{\varsigma/T}$; clearly, the rhs of (4.2-33) is k_BT^2 times a specific heat. Formally we can denote the result as $k_BT^2 n C_{V,\varsigma/T}$. In the canonical ensemble the energy fluctuations were $k_BT^2 n C_{V,N}$. Clearly, $C_{V,\varsigma/T} > C_{V,N}$, since both heat exchange and number exchange play a role.

We shall again illustrate this for a perfect gas, for which the grand potential is easily directly obtained. Previously we showed that $\Phi = -k_BT \ln \mathcal{Z}$, while for a classical grand-canonical ensemble [see (4.2.16) and (4.2-17)]:

$$\begin{aligned}\mathcal{Z} &= \sum_{N=0}^{\infty} e^{\varsigma N/k_BT} \int e^{-\mathcal{H}(p,q)/k_BT} d\Gamma_N \\ &= \sum_{N=0}^{\infty} \frac{1}{N!}\left[e^{\varsigma/k_BT} \int e^{-\mathrm{p}^2/2mk_BT} \frac{d\Omega_1}{h^3}\right]^N = \exp\left[y_\varsigma \int e^{-\mathrm{p}^2/2mk_BT} \frac{d\Omega_1}{h^3}\right].\end{aligned} \tag{4.2-34}$$

The integral, pertaining to the phase space for one particle, is trivial; we thus find,[8]

$$-\frac{\Phi}{k_B T} = \ln \mathcal{Z} = y_\varsigma V \left(\frac{2\pi m k_B T}{h^2} \right)^{3/2} . \tag{4.2-35}$$

The first derivative (4.2-32), keeping y_ς constant, yields the familiar result,

$$\langle \mathcal{E} \rangle = \frac{3}{2} k_B T y_\varsigma V \left(\frac{2\pi m k_B T}{h^2} \right)^{3/2} = \frac{3}{2} k_B T \ln \mathcal{Z} = \frac{3}{2} N k_B T . \tag{4.2-36}$$

[The last equality identifies the chemical potential of a (mono-atomic) perfect gas, see Section 6.5.] For the fluctuations we obtain from (4.2-33),

$$\langle \Delta \mathcal{E}^2 \rangle = \frac{15}{4} (k_B T)^2 N = \frac{5}{2} k_B T^2 n C_{V,N} = k_B T^2 n C_{V,\varsigma/T} . \tag{4.2-37}$$

The energy variance, and so the molar heat capacity, is 2.5 times larger than for the canonical ensemble!

Although we made here a detailed direct computation, it is left to the problems to establish a thermodynamic relationship between $C_{V,\varsigma/T}$ and $C_{V,N}$ for any system.

4.3 Quantum Illustrations of the Canonical and Grand-Canonical Distribution

4.3.1 *The Fermi-Dirac and Bose-Einstein Results*

The Bose-Einstein and Fermi-Dirac distributions for non-interacting quantum gases are most easily obtained from the grand-canonical ensemble. The derivation is found in a great many texts, see *e.g.* ter Haar, Op. Cit., Chapter VII, § 6 and 7. We shall commence with the Bose gas.

Let the one-particle states be denoted by $|\alpha)$, their energies by ε_α and their occupancies by n_α.[9] The canonical partition function would be,

$$\mathcal{Z}_N = \sum_{\{n_\alpha\}}{}' e^{-\Sigma_\alpha n_\alpha \varepsilon_\alpha / k_B T} , \tag{4.3-1}$$

where Σ' means a restricted sum subject to $\Sigma_\alpha n_\alpha = N$. This type of sum is not easily directly evaluated (unless we use the Darwin–Fowler procedure to be discussed later), so we consider the grand-canonical partition function; it is given by

---

[8] See *e.g.* A. Isihara, "Statistical Physics", Acad. Press, NY, 1971, p. 63-64.

[9] In this text (chapter VII excepted) we use rounded kets and bras for one-particle states, $|\,)$ and $(\,|$, and we use ε for one-particle energies; for many-body states, on the contrary, we use angular kets and bras, $|\,\rangle$ and $\langle\,|$, and $\mathcal{E}$ for the system energy. For states of other entities, *e.g.* spin, we generally use angular kets and bras.

$$\mathcal{Z} = \sum_N e^{\varsigma N/k_BT} \sum_{\{n_\alpha\}}{}' e^{-\Sigma_\alpha n_\alpha \varepsilon_\alpha / k_BT}. \tag{4.3-2}$$

Clearly, the double summation $\Sigma_N \Sigma'_{\{n_\alpha\}}$ makes for an unrestricted sum over all n_α. Thus, instead of (4.3-2) we have,

$$\begin{aligned}\mathcal{Z} &= \sum_{\{n_\alpha\}} e^{-\Sigma_\alpha n_\alpha(\varepsilon_\alpha - \varsigma)/k_BT} \\ &= \sum_{\{n_\alpha\}} \prod_\alpha e^{-n_\alpha(\varepsilon_\alpha-\varsigma)/k_BT} = \prod_\alpha \sum_{n_\alpha} e^{-n_\alpha(\varepsilon_\alpha-\varsigma)/k_BT}.\end{aligned} \tag{4.3-3}$$

[Note that the latter interchange corresponds to the product of individual series.[10]] Now, summing over $n_\alpha = 0,1,...\infty$ we have for the sum of the geometrical series,

$$\sum_{n_\alpha} e^{-n_\alpha(\varepsilon_\alpha-\varsigma)/k_BT} = \frac{1}{1-e^{-(\varepsilon_\alpha-\varsigma)/k_BT}}. \tag{4.3-4}$$

Hence,

$$\mathcal{Z} = \prod_\alpha \frac{1}{1-e^{-(\varepsilon_\alpha-\varsigma)/k_BT}}, \tag{4.3-5}$$

$$\ln \mathcal{Z} = -\sum_\alpha \ln\left[1-e^{-(\varepsilon_\alpha-\varsigma)/k_BT}\right]. \tag{4.3-6}$$

The grand potential $\Omega(T,V,\varsigma) = -k_BT \ln \mathcal{Z}$ is now known, and from it all other thermodynamic functions.

For the mean number of particles in a state $|\beta)$ we have,

$$\langle n_\beta \rangle = \mathrm{Tr}(\mathbf{n}_\beta \rho) = \frac{1}{\mathcal{Z}} \sum_{\{n_\alpha\}} n_\beta e^{-\Sigma_\alpha n_\alpha(\varepsilon_\alpha-\varsigma)/k_BT}; \tag{4.3-7}$$

we use bold print for the number operators, $\{\mathbf{n}_\alpha\}$, noting that they commute with the Hamiltonian and with ρ. From the first line of (4.3-3) it is readily noted that

$$\langle n_\beta \rangle = -k_BT \frac{1}{\mathcal{Z}} \frac{\partial \mathcal{Z}}{\partial \varepsilon_\beta} = -k_BT \frac{\partial \ln \mathcal{Z}}{\partial \varepsilon_\beta}. \tag{4.3-8}$$

The differentiation to ε_β involves a *virtual* variation $\delta\varepsilon_\beta$ of the energy level; this is a purely mathematical device! From (4.3-8) and (4.3-6) we obtain the Bose-Einstein distribution for the mean occupancy of a state:

$$\langle n_\beta \rangle = \frac{e^{-(\varepsilon_\beta-\varsigma)/k_BT}}{1-e^{-(\varepsilon_\beta-\varsigma)/k_BT}} = \frac{1}{e^{(\varepsilon_\beta-\varsigma)/k_BT}-1}. \tag{4.3-9}$$

---

[10] See Eq. (1.7-19) in footnote 24 of chapter I.

For the second moment we have likewise,

$$\langle n_\alpha n_\beta \rangle = (k_B T)^2 \frac{1}{\mathcal{Z}} \frac{\partial^2 \mathcal{Z}}{\partial \varepsilon_\alpha \partial \varepsilon_\beta}. \tag{4.3-10}$$

Deviations from the mean distribution are manifested by the variances $\langle \Delta n_\alpha^2 \rangle$ and covariances $\langle \Delta n_\alpha \Delta n_\beta \rangle$. Employing once more the rule (4.2-22) we find,

$$\langle \Delta n_\alpha \Delta n_\beta \rangle = (k_B T)^2 \frac{\partial^2 \ln \mathcal{Z}}{\partial \varepsilon_\alpha \partial \varepsilon_\beta} = -k_B T \frac{\partial \langle n_\alpha \rangle}{\partial \varepsilon_\beta}. \tag{4.3-11}$$

The result, first noted by Einstein, is

$$\langle \Delta n_\alpha \Delta n_\beta \rangle = \langle n_\alpha \rangle [1 + \langle n_\alpha \rangle] \delta_{\alpha\beta}. \tag{4.3-12}$$

This result, showing 'super-Poissonian' fluctuations for bosons, indicates that bosons tend to 'bunch'; we also note that the non-correlation, expressed by the Kronecker delta, only occurs in the grand-canonical ensemble. Later we shall obtain similar results for the canonical ensemble and microcanonical ensemble. The form for the mean occupancy, Eq. (4.3-9), remains the same; however, the fluctuations are reduced due to the particle constraint $\Sigma_\alpha n_\alpha = N$ and energy constraint $\Sigma_\alpha n_\alpha \varepsilon_\alpha = \mathcal{E}$.

We still mention that Planck's law for photons is based on a special form of the Bose-Einstein distribution. Since photons can be produced or annihilated *ad libidum*, the chemical potential of a photon gas is often assumed to be zero (although some scientists disagree). Anyway, *omitting* ζ and multiplying (4.3-9) for that case with the number of modes in an interval $(\omega, \omega + d\omega)$, Planck's law for black-body radiation is obtained, see Chapter Section 8.1.

We now turn to the Fermi gas. The Fermi-Dirac distribution is established by a fully analogous reasoning. Equations (4.3-1) through (4.3-3) remain applicable. But, in summing n_α we must account for the Pauli exclusion principle, *i.e.*, $n_\alpha = 0,1$ only. Instead of (4.3-4) to (4.3-6) we now obtain,

$$\sum_{n_\alpha} e^{-n_\alpha(\varepsilon_\alpha - \varsigma)/k_B T} = 1 + e^{-(\varepsilon_\alpha - \varsigma)/k_B T}. \tag{4.3-13}$$

Consequently,

$$\mathcal{Z} = \prod_\alpha \left[1 + e^{-(\varepsilon_\alpha - \varsigma)/k_B T} \right], \tag{4.3-14}$$

$$\ln \mathcal{Z} = \sum_\alpha \ln \left[1 + e^{-(\varepsilon_\alpha - \varsigma)/k_B T} \right]. \tag{4.3-15}$$

Hereby the grand potential $\Omega(T,V,\varsigma) = -k_B T \ln \mathcal{Z}$ is known and from it all other information can be obtained. The mean distribution is once more found from the derivative of $\ln \mathcal{Z}$ to ε_α and similarly for the fluctuations. We easily establish,

$$\langle n_\alpha \rangle = \frac{1}{e^{(\varepsilon_\alpha - \varsigma)/k_B T} + 1}, \tag{4.3-16}$$

which is the well-known Fermi-Dirac distribution. Likewise, for the fluctuations we have now the 'sub-Poissonian' result – the variance being depressed by the exclusion principle,

$$\langle \Delta n_\alpha \Delta n_\beta \rangle = \langle n_\alpha \rangle [1 - \langle n_\alpha \rangle] \delta_{\alpha\beta}. \tag{4.3-17}$$

For an electron gas, or for any system of charged particles, the chemical potential ς should be replaced by the electrochemical potential μ.

Finally, we shall give the result for the mean energy for bosons as well as fermions. Combining the results for $\ln \mathcal{Z}$ we have,

$$\ln \mathcal{Z} = \mp \sum_\alpha \ln\left[1 \mp e^{-(\varepsilon_\alpha - \varsigma)/k_B T}\right]. \tag{4.3-18}$$

From this one obtains,

$$-\left(\frac{\partial \ln \mathcal{Z}}{\partial(1/k_B T)}\right)_{V,\varsigma/T} = \mp \sum_\alpha \varepsilon_\alpha \left[1 \mp e^{(\varepsilon_\alpha - \varsigma)/k_B T}\right]^{-1} = \sum_\alpha \langle n_\alpha \rangle \varepsilon_\alpha = \langle \mathcal{E} \rangle, \tag{4.3-19}$$

in accord with the general result (4.2-32) in the grand-canonical ensemble.

4.3.2 *The One-dimensional Ising Model*

We will find the partition function for a one-dimensional Ising Hamiltonian and show that no phase transition occurs, i.e., a one-dimensional ferromagnet cannot exist. This subsection will prepare the way for the two-dimensional Ising model, which is one of the rare examples involving a phase transition that can be solved exactly.

The full spin Hamiltonian , also called Heisenberg Hamiltonian, was already given in Section 1.3, Eq.(1.3-12); with a little change in notation we write, l denoting a lattice site,

$$\mathcal{H} = -\frac{1}{\hbar^2} \sum_{l,l'}^{N} J_{l,l'} \mathbf{S}_l \cdot \mathbf{S}_{l'} - g \frac{\rho_B}{\hbar} H \sum_l^N S_l^z . \tag{4.3-20}$$

The magnetic field H is, as usual, taken in the z-direction; ρ_B is the Bohrmagneton, in rationalized Gaussian units given as $\rho_B = e\hbar/2m_e c$ and g is the Landé factor, equal to 2 for all practical purposes. In the Ising model the Hamiltonian is averaged over the x- and y-directions, whereby it is assumed that the off-diagonal contributions average to zero. The Hamiltonian then depends on the spin operators S_l^z only. Further we write $\mathbf{S}_l = \frac{1}{2}\boldsymbol{\sigma}_l \hbar$, where the $\boldsymbol{\sigma}$'s are the relevant spin operators, which can be represented by the Pauli spin matrices, if desired. In the present form of the Ising model we will only need σ_ℓ^z, whose eigenstates are $|\pm 1\rangle$, having eigenvalues ± 1. Lastly, it is assumed that only nearest neighbour interactions are important. The Ising Hamiltonian for a one-dimensional chain of spins then takes the form,

$$\mathcal{H} = -J_0 \sum_\ell \sigma_\ell \sigma_{\ell+1} - \rho_B H \sum_\ell \sigma_\ell \tag{4.3-21}$$

(we omitted the superscript '*z*' on σ; further, $J_0 = \frac{1}{4}J$). Next, we apply periodic boundary conditions, *i.e.*, $\sigma_{N+\ell} = \sigma_\ell$; this is of course no restriction whatsoever, in view of the thermodynamic limit which will be taken later. The canonical partition function is then as follows,

$$\begin{aligned} \mathcal{Z} &= \sum_{\{\sigma_\ell\}} \exp\left\{ \beta \sum_{\ell=1}^N \left[J_0 \sigma_\ell \sigma_{\ell+1} + \tfrac{1}{2} \rho_B H (\sigma_\ell + \sigma_{\ell+1}) \right] \right\} \\ &= \sum_{\{\sigma_\ell\}} \prod_{\ell-1}^N \exp\left\{ \beta \left[J_0 \sigma_\ell \sigma_{\ell+1} + \tfrac{1}{2} \rho_B H (\sigma_\ell + \sigma_{\ell+1}) \right] \right\}, \end{aligned} \tag{4.3-22}$$

where we set $\beta = 1/k_B T$. This sum will be evaluated by a matrix procedure. We define a matrix of second rank P, such that,

$$P_{\sigma\sigma'} \equiv \langle \sigma | P | \sigma' \rangle = e^{\beta [J_0 \sigma\sigma' + \frac{1}{2} \rho_B H (\sigma + \sigma')]} . \tag{4.3-22'}$$

Explicitly we have,

$$\begin{aligned} &\langle 1 | P | 1 \rangle = e^{\beta (J_0 + \rho_B H)}, \quad \langle -1 | P | -1 \rangle = e^{\beta (J_0 - \rho_B H)}, \\ &\langle 1 | P | -1 \rangle = \langle -1 | P | 1 \rangle = e^{-\beta J_0} . \end{aligned} \tag{4.3-23}$$

The partition function is seen to be simply

$$\mathcal{Z} = \sum_{\sigma_1} \dots \sum_{\sigma_N} \langle \sigma_1 | P | \sigma_2 \rangle \langle \sigma_2 | P | \sigma_3 \rangle \dots \langle \sigma_N | P | \sigma_1 \rangle . \tag{4.3-24}$$

The intermediate sums can be associated with matrix multiplication; so we have,

$$\mathcal{Z} = \sum_{\sigma_1} \langle \sigma_1 | P^N | \sigma_1 \rangle = \mathrm{Tr}\, P^N = \lambda_+^N + \lambda_-^N , \tag{4.3-25}$$

where $\lambda_\pm$ are the eigenvalues of P. Hence,

$$\begin{vmatrix} e^{\beta (J_0 + \rho_B H)} - \lambda & e^{-\beta J_0} \\ e^{-\beta J_0} & e^{\beta (J_0 - \rho_B H)} - \lambda \end{vmatrix} = 0 . \tag{4.3-26}$$

The eigenvalues are found to be,

$$\lambda_{\pm} = e^{\beta J_0} \cosh(\beta \rho_B H) \pm \left[e^{2\beta J_0} \sinh^2(\beta \rho_B H) + e^{-2\beta J_0} \right]^{1/2}. \qquad (4.3\text{-}27)$$

[other expressions are found in various books[11]]. Now, noting that $\lambda_-^N \ll \lambda_+^N$, we only need to retain the + root in the thermodynamic limit. Consequently, we find for the Helmholtz free energy from $\ln \mathcal{Z}$,

$$F = -k_B TN \ln\left\{ e^{\beta J_0} \cosh(\beta \rho_B H) + \left[e^{2\beta J_0} \sinh^2(\beta \rho_B H) + e^{-2\beta J_0} \right]^{1/2} \right\}. \qquad (4.3\text{-}28)$$

In the Gibbsian sense (cf. Sect. 1.3) this is $F(T, N, H)$. Therefore, the magnetization is given by

$$M = -(\partial f / \partial H)_{n,T}, \qquad (4.3\text{-}29)$$

where f is the free energy per unit volume F/V and $n = N/V$. After some algebra the differentiation yields,[12]

$$M = \frac{N\rho_B}{V} \frac{\sinh(\beta \rho_B H)}{\left[\sinh^2(\beta \rho_B H) + e^{-4\beta J_0} \right]^{1/2}}. \qquad (4.3\text{-}30)$$

This is the final result. It is pictured in Fig. 4-1. For $H = 0$, $M = 0$, so there is no spontaneous magnetization at any finite temperature. Also note $M(-H) = -M(H)$. For saturation the spins are all lined up with the field. At very low temperatures the last term in the denominator of (4.3-30) is negligible; then M is saturated for any H, however small. We can therefore say that the phase transition, allowing for zero-field magnetization, occurs at zero temperature, with $M_{spont} = N\rho_B / V$.

Although the result of this simple model, *exact* within the assumptions made, yields no phase transition for finite temperatures, a more encouraging result is obtained from mean-field theory models, which will be discussed later.

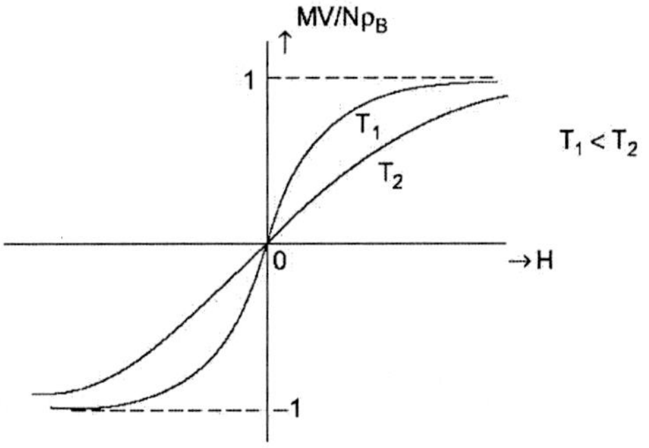

Fig. 4-1. Magnetization vs. H for two temperatures > 0.

---

[11]See e.g. K. Huang, "Statistical Mechanics", Wiley NY, 1963, Section 16.5 or 2nd Ed. Wiley NY, 1987, Section 14.6.; also, M. Plischke and B. Bergersen, 3rd Ed., Op. Cit., Section 3.6.

[12] In SI units we need $\mu_0 \rho_B = 9.27 \times 10^{-24} A\, m^2$. So M has the dimension A/m, being the same as for H.

2. Dense Classical Gases and Further Applications

4.4 Second Virial Coefficient for a Classical Gas and van der Waals' Law

As a second illustration we consider a regular classical gas with two-particle interactions given by the potential $\varphi(|\mathbf{r}_i - \mathbf{r}_j|)$. This interaction potential will look like in Fig.4-2, given below. The potential is zero for $r \equiv r_{12} = 2r_1$, where r_1 is roughly equal to the radius of a particle. For $r_{12} < 2r_1$, there is strong repulsion and $\varphi(r)$ rises steeply, while for $r_{12} > 2r_1$ the potential is attractive, having a minimum φ_0 at the equilibrium separation r_0. For later purposes we have also plotted the Mayer function $f(r)$, defined by

$$f_{ij}(r) = e^{-\beta\varphi_{ij}(r)} - 1. \tag{4.4-1}$$

This function is better behaved at the origin. In the problems we will consider various explicit forms for φ, like the Lennard-Jones potential, the Morse potential and others .

We now turn to the Hamiltonian and the canonical partition function. Using a classical description, we have

$$\mathcal{H} = \sum_i (\mathbf{p}_i^2/2m) + \mathcal{V}(\mathbf{q}_1 ... \mathbf{q}_N) = \sum_i (\mathbf{p}_i^2/2m) + \sum_{ij,i<j} \varphi_{ij}(|\mathbf{q}_i - \mathbf{q}_j|), \tag{4.4-2a}$$

$$\mathcal{Z} = \iint \frac{d\Omega_p d\Omega_q}{N! h^{3N}} \exp\{-\beta[\sum_i \mathbf{p}_i^2/2m + \sum_{i<j} \varphi_{ij}(|\mathbf{q}_i - \mathbf{q}_j|)]\}, \tag{4.4-2b}$$

where the index 'i' labels particles, *not* coordinates. The integral over the momenta is immediate; thus we define the momentum-space partition function,

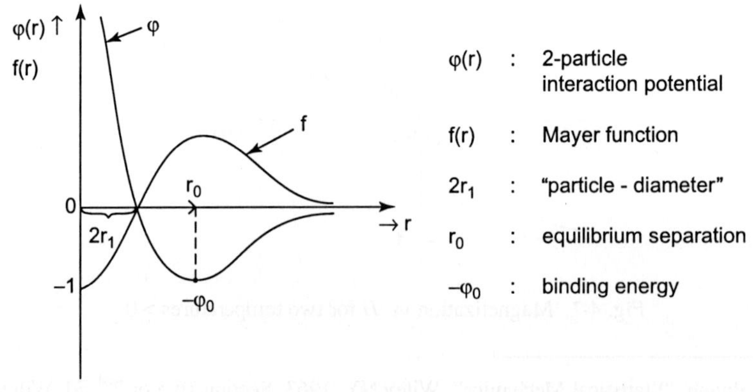

Fig. 4-2. Interaction potential $\varphi(r)$ and Mayer function $f(r)$ *vs.* r.

$$\mathcal{Z}_p = \int \frac{d\Omega_p}{h^{3N}} e^{-\beta \Sigma_i \mathbf{p}_i^2/2m} = \left(\frac{2\pi m k_B T}{h^2} \right)^{3N/2} \tag{4.4-3}$$

Setting $\mathcal{Z} = \mathcal{Z}_p \mathcal{Z}_c$, we have for the configurational partition function.

$$\mathcal{Z}_c = \frac{1}{N!} \int d\Omega_q \exp[-\beta \sum_{i,j=1,\, i<j} \varphi(|\mathbf{q}_i - \mathbf{q}_j|)]. \tag{4.4-4}$$

The quest before us is to evaluate $\mathcal{Z}_c$ and possibly obtain van der Waals' law, without knowing more than generalities about the interaction potential φ. Moreover, we must take the thermodynamic limit. The problem was first considered in detail by Ornstein in 1908, while stability criteria and an explanation for the so-called Maxwell construction of the isotherms were later re-examined by van Kampen in the sixties.[13,14] A more general method involves cluster expansions, first proposed by Urcell in 1927 and summarized in various texts.[15] We will first discuss Ornstein's method, following van Kampen and obtain van der Waals' law. Next, we describe the two-particle cluster series to obtain the simplest result for the second virial coefficient, following largely Isihara.

4.4.1 *Ornstein's Method as Elaborated by van Kampen*

Ornstein's approach involves dividing up the volume of the gas into cells $\{\alpha\}$ of size Δ, which is sufficiently large to contain many particles, but small enough, so that the two-body potential between particles in a given cell can be assumed to be constant. It is assumed that cell α contains N_α particles and is located at $\mathbf{r}_\alpha$; the interaction potential between the particles of two different cells caused by the long-range attractive tail is denoted as $-W_{\alpha\alpha'} = -W(|\mathbf{r}_\alpha - \mathbf{r}_{\alpha'}|)$. The phase-space volume occupied by one cell is

$$\omega(N_\alpha) = (\Delta - N_\alpha \delta)^{N_\alpha}, \tag{4.4-5}$$

where δ is the volume of the particle core. The partition function can now be rewritten as

$$\mathcal{Z} = \mathcal{Z}_p \frac{1}{N!} \sum_{\{N_\alpha\}}{}' \frac{N!}{\Pi_\alpha N_\alpha!} \prod_\alpha \omega(N_\alpha) \exp\left(\tfrac{1}{2}\beta \sum_{\alpha\alpha'} N_\alpha N_{\alpha'} W_{\alpha\alpha'} \right) \sim \mathcal{Z}_p \sum_{\{N_\alpha\}}{}' \exp[\Theta(\{N_\alpha\})]. \tag{4.4-6}$$

Here Σ' indicates the restriction $\Sigma_\alpha N_\alpha = N$, while further we note the occurrence of

---

[13] L.S. Ornstein, "Toepassing der Statistische Mechanica van Gibbs op Moleculair-Theoretische Vraagstukken", Ph.D. Thesis, Univ. of Leiden, (1908).

[14] N.G. van Kampen, Phys. Rev. **135**, A362 (1964); ibid. "Lectures on the Theory of Phase Transitions", Univ. of Minnesota, June 1968, unpublished.

[15] H.D. Urcell, Proc. Cambridge Phil. Soc. **23**, 685 (1927); A. Isihara, Op. Cit., Section 5.1.

the multinomial coefficient, stemming from the possible permutations of the $\{N_\alpha\}$. Also Θ, being a functional of all N_α, is defined in the thermodynamic limit; it is found to be,

$$\Theta(\{N_\alpha\}) = \sum_\alpha \left[N_\alpha \ln\left(\frac{\Delta - N_\alpha \delta}{N_\alpha}\right) + N_\alpha + \tfrac{1}{2}\beta \sum_{\alpha'} N_\alpha N_{\alpha'} W_{\alpha\alpha'} \right], \tag{4.4-7}$$

where we used Stirling's formula for the factorials. For the free energy we now have,

$$\lim_{N\to\infty} F(T,V,N)/N = -k_B T\left(\frac{3}{2}\ln\left(\frac{2\pi m k_B T}{h^2}\right) + \lim_{N\to\infty}\frac{1}{N}\ln \sum_{\{N_\alpha\}}{}' \exp[\Theta(\{N_\alpha\})]\right). \tag{4.4-8}$$

The restricted sum cannot be directly evaluated, but since we have already coarse-grained the partition function, it will suffice to pick out the maximum value of the summand[16]. Let λ be a Lagrangian multiplier for the constraint $\Sigma_\alpha N_\alpha = N$. We then find the set $\{N_\alpha\}$ from

$$\frac{\partial}{\partial N_\alpha}[\Theta(\{N_\alpha\}) - \lambda(\Sigma_\alpha N_\alpha - N)] = 0. \tag{4.4-9}$$

This requires that

$$\ln\frac{\Delta - N_\alpha \delta}{N_\alpha} - \frac{N_\alpha \delta}{\Delta - N_\alpha \delta} + \beta\sum_{\alpha'} N_{\alpha'} W_{\alpha\alpha'} = \lambda. \tag{4.4-10}$$

This condition determines N_α and by summation fixes $\lambda = \lambda(N)$.

First, however, we must assure that the set $\{N_\alpha\}$ constitutes a maximum. This means that the determinant of the second-order derivatives matrix must be negative definite; or

$$-\left|\frac{\partial^2\Theta(\{N_\alpha\})}{\partial N_\alpha \partial N_{\alpha'}}\right| = \left|\frac{\Delta^2 \delta_{\alpha\alpha'}}{N_\alpha(\Delta - N_\alpha\delta)^2} - \beta W_{\alpha\alpha'}\right| > 0. \tag{4.4-11}$$

As shown by Lee and Yang[17], a symmetric matrix of the form $a_i\delta_{ij} - b_{ij}$ with $b_{ij} > 0$ is positive definite if $a_i > \sum_j b_{ij}$ for all i. For the present case this requires,

$$\frac{\Delta^2}{N_\alpha(\Delta - N_\alpha\delta)^2} > \beta\sum_{\alpha'} W_{\alpha\alpha'} \equiv \beta W_0 \frac{\delta}{\Delta}, \quad \text{all } N_\alpha, \tag{4.4-12}$$

where W_0 is the binding energy. Whenever this is satisfied $\Theta(\{N_\alpha\})$ is concave, having a single maximum for the set $\{N_\alpha\}$ that satisfies (4.4-10). The left-hand side of (4.4-12) has a minimum for $N_\alpha = \Delta/3\delta$, its value being

---

[16] Thus, without the summation sign, the summand represents the negative of the nonequilibrium free energy function, which must be minimized.

[17] T.D. Lee and C.N. Yang, Phys Rev **87**, 410 (1952).

$$\left(\frac{\Delta^2}{N_\alpha(\Delta - N_\alpha\delta)^2}\right)_{min} = \frac{27\delta}{4\Delta}. \tag{4.4-13}$$

The critical temperature, at which the system becomes metastable (no curvature at the maximum) occurs for

$$\beta W_0 = \tfrac{27}{4}, \text{ or } kT_c = \tfrac{4}{27} W_0 . \tag{4.4-14}$$

Clearly, the lower the binding energy W_0, the lower T_c.

Returning to (4.4-10), we guess the solution. If we set,

$$N_\alpha = (N/V)\Delta = n\Delta, \quad (\text{for all } \alpha) \tag{4.4-15}$$

corresponding to a uniform distribution in each cell, then we satisfy (4.4-10) if we choose

$$\lambda(n) = \ln\frac{1-n\delta}{n} - \frac{n\delta}{1-n\delta} + \beta W_0 \delta n = const . \tag{4.4-16}$$

For $\Theta_{\max}$ we find from (4.4-7),

$$\Theta_{\max} = \sum_\alpha \left\{ n\Delta \ln\frac{1-n\delta}{n} + n\Delta + \tfrac{1}{2} n^2 \Delta\delta\beta W_0 \right\}$$

$$= N \ln\frac{1-n\delta}{n} + N + \tfrac{1}{2} N n\delta\beta W_0 , \tag{4.4-17}$$

where we noted $\Sigma_\alpha \Delta \to V$. For the free energy per molecule this finally gives

$$\hat{f}(T,n) = -k_B T \left[\tfrac{3}{2}\ln(\frac{2\pi m k_B T}{h^2}) + \ln\frac{1-n\delta}{n} + 1 + \tfrac{1}{2} n\delta\beta W_0 \right]. \tag{4.4-18}$$

The pressure is obtained as,

$$P = -\left(\frac{\partial \hat{f}}{\partial \hat{v}}\right)_T = n^2 \left(\frac{\partial \hat{f}}{\partial n}\right)_T . \tag{4.4-19}$$

The result is

$$P = k_B T \left(\frac{n}{1-n\delta} - \tfrac{1}{2} n^2 \delta\beta W_0 \right) \quad \text{or ,}$$

$$(P + \tfrac{1}{2}\delta n^2 W_0)(V - N\delta) = N k_B T, \tag{4.4-20}$$

which is van der Waals' law. Pictures for the isotherms are given in Fig. 4-3 (a) + (b). His equation, which is still one of the best approximations to the isotherms of the vapour-liquid phase transition, was proposed in his Ph.D. thesis of 1873.[18]

---

[18] J. D. van der Waals, "Over de Continuiteit van den Gas- en Vloeistoftoestand ", Ph. D. Thesis, Univ. of Leiden (1873). For a review with many current views cf. "Van der Waals Centennial Conference on Statistical Mechanics", (C. Prins, Ed.) Amsterdam 27-31 August 1973, North Holland 1974.

4.4.2 *Van der Waals' Equation; Fluctuations and Pair Correlation Function*

Van der Waals' equation is generally written as

$$\left(P+n^2\frac{a}{V^2}\right)(V-nb)=nRT. \tag{4.4-21}$$

Here a and b are parameters, specific for a given gas, and $n=N/N_A$ is the number of gram-molecules (moles) of the gas. Taking $n=1$ and denoting the molar volume by $\tilde{V}$, the isotherm $[\tilde{V}(P)]_T$ satisfies a cubic equation,

$$\tilde{V}^3-\left(b+\frac{RT}{P}\right)\tilde{V}^2+\frac{a}{P}\tilde{V}-\frac{ab}{P}=0. \tag{4.4-22}$$

For low enough temperatures the equation has three roots. In that range vapour and liquid are coexistent. If T rises to the critical temperature the roots will coalesce; we then have an inflection point, as seen in Fig. 4-3(a). From the conditions

$$\left(\frac{\partial P}{\partial \tilde{V}}\right)_{T=T_c}=0, \qquad \left(\frac{\partial^2 P}{\partial \tilde{V}^2}\right)_{T=T_c}=0, \tag{4.4-23}$$

one easily finds the values at the critical point,

$$P_c=\frac{a}{27b^2}, \qquad \tilde{V}_c=3b, \qquad T_c=\frac{8a}{27bR}. \tag{4.4-24}$$

Using new variables, $P'=P/P_c$, $V'=\tilde{V}/\tilde{V}_c$, $T'=T/T_c$, van der Waals' equation may be rewritten as the *universal equation*

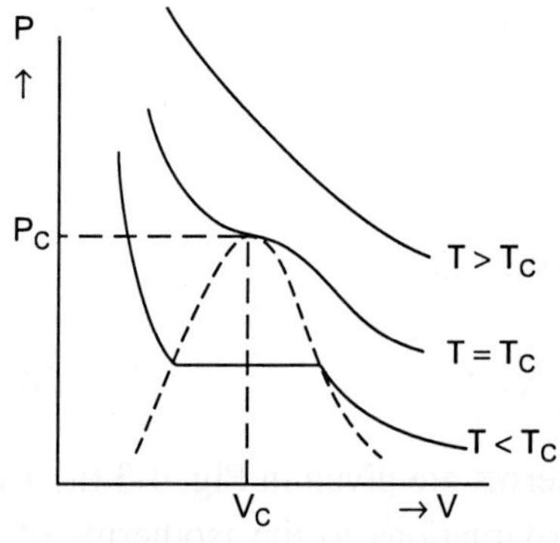

Fig.4-3(a). Isotherms for different T. (After van Kampen, 1968[14])

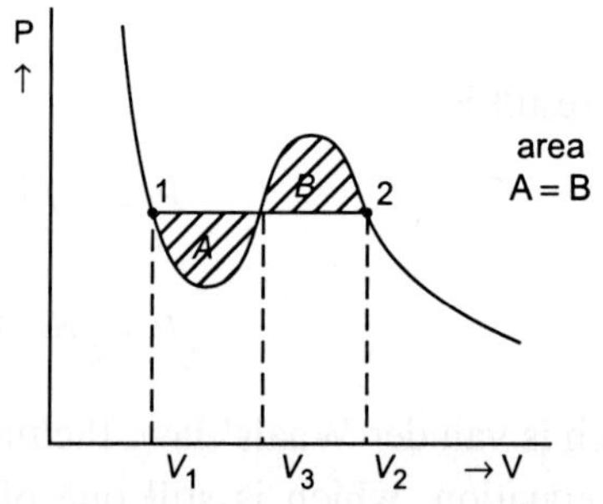

Fig.4-3(b). The Maxwell construction. The shaded areas are equal.

$$\left(P'+\frac{3}{V'^2}\right)\left(V'-\frac{1}{3}\right)=\frac{8}{3}T'. \tag{4.4-25}$$

This expression is referred to as the *law of corresponding states*. This remarkable feature is well satisfied for most gases, supporting the correctness of van der Waals' equation.

The vapour –liquid phase transition is a *first order transition*, with P and T being fixed in the condensation region (see subsection 4.9.2). As to the van der Waals isotherms, one peculiarity remains: below the critical point the $P(\tilde{V})$ behaviour is not monotonic, but shows unphysical parts with a positive slope. More clearly, the problem is pictured in the $\tilde{V}(P)$ isotherm shown in Fig.4-4(a). For given P and T the three roots for $\tilde{V}$ are at B, D and F. Stability criteria in any branch of physics or engineering imply that point D is metastable and will shift, due to any fluctuation, to either B or F. Thus the oscillatory behaviour should be replaced by the isotherm $ABFG$, with the pressure being constant in the condensation region for given T below T_c. Where to place the constant pressure segment is resolved by the Maxwell construction: the points B and F are to be placed such that the areas to the left and to the right of BF are equal. A thermodynamic proof is given in most books dealing with the topic. For a situation with P and T specified, we need the Gibbs free energy $G(T,P,N...)$. From (1.4-5) we find, for T, N and the other A_i fixed, for the molar differential: $d\tilde{G}=\tilde{V}(P;T,N_A,\{A_i\})dP$. Thus, for a change in $\tilde{G}$ along the isotherm,

$$\tilde{G}_2-\tilde{G}_1=\int_{P_1}^{P_2}\tilde{V}(P)dP, \tag{4.4-26}$$

which is the area under the curve from P_1 to P_2. A plot of the molar Gibbs free energy as a function of P is easily constructed from Fig.4-4(a) and is sketched in Fig.4-4(b). The Gibbs free energy is concave between A and C and between E and G as expected. The part CE is unstable, since the equilibrium Gibbs free energy should be a minimum. Therefore, the true curve follows the line $A(B,F)G$, from which we see that $G_B = G_F$. Employing (4.4-26) for the points B and F we have

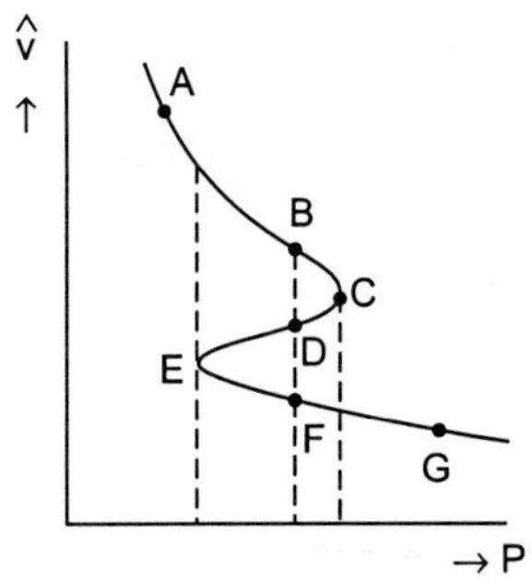

Fig. 4-4(a): van der Waals isotherm $\tilde{V}(P)$.

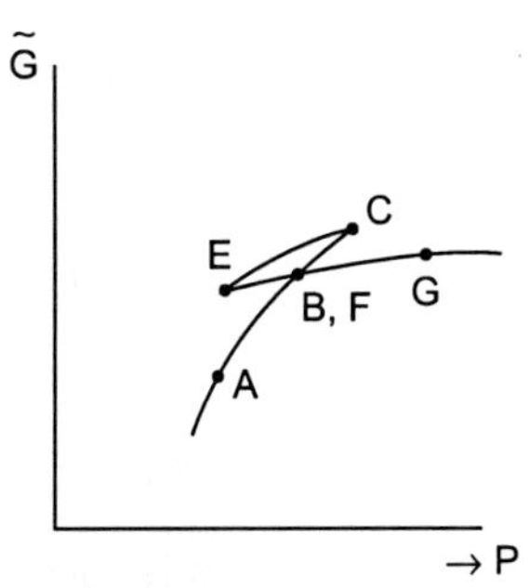

Fig. 4-4(b): Molar Gibbs free energy *vs*. P.

$$\int_{P_B}^{P_F} \tilde{V}(P)dP = 0 = \int_{P_B}^{P_C} \tilde{V}(P)dP + \int_{P_C}^{P_D} \tilde{V}(P)dP + \int_{P_D}^{P_E} \tilde{V}(P)dP + \int_{P_E}^{P_F} \tilde{V}(P)dP, \quad (4.4\text{-}27)$$

or,

$$\int_{P_B}^{P_C} \tilde{V}(P)dP - \int_{P_D}^{P_C} \tilde{V}(P)dP = \int_{P_E}^{P_D} \tilde{V}(P)dP - \int_{P_E}^{P_F} \tilde{V}(P)dP. \quad (4.4\text{-}28)$$

This indicates that the quasi-triangular areas to the right and to the left of the condensation pressure are indeed equal, so that the true isotherms of Ornstein's model are indeed the modified isotherms *ABFG*.

Fluctuations and pair correlation function

The function $\Theta(\{N_\alpha\})$ computed before, together with its second derivatives, allow us to also compute the density fluctuations. Thus from (4.4-11) we have the Gaussian approximation,

$$W(\{N_\alpha\}) = \sum_{\{N_\alpha\}}{}' \exp[\Theta(\{N_\alpha\})]$$
$$= \exp\left[\Theta_{max}(\{N_\alpha\}) - \tfrac{1}{2}\sum_{\alpha\alpha'}\left(\frac{\delta_{\alpha\alpha'}}{\langle n\rangle\Delta(1-\langle n\rangle\delta)^2} - \beta W_{\alpha\alpha'}\right)\Delta N_\alpha \Delta N_{\alpha'}\right], \quad (4.4\text{-}29)$$

where we set $N_{\alpha,max} = \langle n\rangle\Delta$. In this Gaussian form for the distribution function we write

$$\delta_{\alpha\alpha'}/\Delta \to \delta(\mathbf{r}-\mathbf{r}'), \quad W_{\alpha\alpha'} \to W(\mathbf{r}-\mathbf{r}'). \quad (4.4\text{-}30)$$

while $1/\langle n\rangle(1-\langle n\rangle\delta)^2 \equiv f(n_0)$. We now have a multivariate normal distribution of the form

$$W[\{n(\mathbf{r})\}] = C\exp\left[-\tfrac{1}{2}\iint(\mathbf{r}\,|\,Q\,|\,\mathbf{r}')\Delta n(\mathbf{r})\Delta n(\mathbf{r}')d^3r d^3r'\right],$$
$$(\mathbf{r}\,|\,Q\,|\,\mathbf{r}') = f(n_0)\delta(\mathbf{r}-\mathbf{r}') - \beta W(\mathbf{r}-\mathbf{r}'). \quad (4.4\text{-}31)$$

The dimension of the distribution, being originally (V/Δ), has now become infinite, but the result (1.6-60) for the covariance still applies. Thus formally we have,

$$\langle\Delta n(\mathbf{r})\Delta n(\mathbf{r}')\rangle = (\mathbf{r}\,|\,Q^{-1}\,|\,\mathbf{r}'). \quad (4.4\text{-}32)$$

To find the inverse kernel we consider the linear integral equation

$$\int(\mathbf{r}\,|\,Q\,|\,\mathbf{r}')X(\mathbf{r}')d^3r' = Y(\mathbf{r}), \quad (4.4\text{-}33)$$

with formal solution

$$X(\mathbf{r}) = \int(\mathbf{r}\,|\,Q^{-1}\,|\,\mathbf{r}')Y(\mathbf{r}')d^3r'. \quad (4.4\text{-}34)$$

We take the Fourier transform of (4.4-33), *i.e.* we have for $X(\mathbf{r}), Y(\mathbf{r}), \delta(\mathbf{R})$ and $W(\mathbf{R})$

$$X(\mathbf{r}) = (1/8\pi^3)\int d^3k\, X(\mathbf{k})e^{i\mathbf{k}\cdot\mathbf{r}},$$
$$Y(\mathbf{r}) = (1/8\pi^3)\int d^3k\, Y(\mathbf{k})e^{i\mathbf{k}\cdot\mathbf{r}},$$
$$\delta(\mathbf{r}-\mathbf{r}') = (1/8\pi^3)\int d^3k\, e^{i\mathbf{k}\cdot(\mathbf{r}-\mathbf{r}')},$$
$$W(\mathbf{r}-\mathbf{r}') = (1/8\pi^3)\int d^3k\, W(\mathbf{k})e^{i\mathbf{k}\cdot(\mathbf{r}-\mathbf{r}')}. \quad (4.4\text{-}35)$$

Employing also $(1/8\pi^3)\int d^3r \exp[i\mathbf{r}\cdot(\mathbf{k}'-\mathbf{k}) = \delta(\mathbf{k}-\mathbf{k}')$, we easily obtain,

$$\int d^3k[f(n_0) - \beta W(\mathbf{k})]X(\mathbf{k})e^{i\mathbf{k}\cdot\mathbf{r}} = \int d^3k\, Y(\mathbf{k})\,e^{i\mathbf{k}\cdot\mathbf{r}}. \tag{4.4-36}$$

Multiplying with $\exp(-i\mathbf{k}'\cdot\mathbf{r})$ and integrating gives $X(\mathbf{k}) = Y(\mathbf{k})/[f(n_0) - \beta W(\mathbf{k})]$. So, finally,

$$\begin{aligned} X(\mathbf{r}) &= (1/8\pi^3)\int d^3k\, e^{i\mathbf{k}\cdot\mathbf{r}} Y(\mathbf{k})/[f(n_0) - \beta W(\mathbf{k})] \\ &= (1/8\pi^3)\int d^3r' Y(\mathbf{r}')\int d^3k\, e^{i\mathbf{k}\cdot(\mathbf{r}-\mathbf{r}')}\left(1/[f(n_0) - \beta W(\mathbf{k})\right). \end{aligned} \tag{4.4-37}$$

Comparing with (4.4-34) this yields

$$(\mathbf{r}\,|\,Q^{-1}\,|\,\mathbf{r}') = (1/8\pi^3)\int d^3k\, e^{i\mathbf{k}\cdot(\mathbf{r}-\mathbf{r}')}\left(1/[f(n_0) - \beta W(\mathbf{k})\right). \tag{4.4-38}$$

This gives the covariance, see (4.4-32). With minor algebra we find, splitting off a singularity,

$$\begin{aligned} \langle \Delta n(\mathbf{r})\Delta n(\mathbf{r}')\rangle &= \frac{1}{8\pi^3}\int \frac{e^{i\mathbf{k}\cdot(\mathbf{r}-\mathbf{r}')}}{f(n_0)}d^3k + \frac{1}{8\pi^3}\frac{\beta}{f(n_0)}\int \frac{W(\mathbf{k})e^{i\mathbf{k}\cdot(\mathbf{r}-\mathbf{r}')}}{f(n_0) - \beta W(\mathbf{k})}d^3k \\ &\simeq \langle n\rangle\delta(\mathbf{r}-\mathbf{r}') + \langle n\rangle\frac{\beta}{8\pi^3}\int \frac{W(\mathbf{k})e^{i\mathbf{k}\cdot(\mathbf{r}-\mathbf{r}')}}{f(n_0) - \beta W(\mathbf{k})}d^3k, \end{aligned} \tag{4.4-39}$$

where in the last line we approximated in the magnitude $f(n_0) \approx 1/\langle n\rangle$, meaning $\langle n\rangle\delta \ll 1$. The delta function part – omitted by van Kampen – is a necessary "dimensional singularity", such as we encountered in Eq. (1.6-35).

In Section 4.10 we shall introduce the (modified) pair distribution function $w_2(\mathbf{r}_1,\mathbf{r}_2)$ as the 'probability' of finding simultaneously a particle at position $\mathbf{r}_1$ and another one at $\mathbf{r}_2$, the normalization being to V^2. Generally, $w_2(\mathbf{r}_1,\mathbf{r}_2) = w_2(|\mathbf{r}_1 - \mathbf{r}_2|)$. As shown there [(Eqs. (4.11-17), (4.11-12) and (4.11-26)] the connection with the two-point covariance function is given by:

$$\Gamma_2(\mathbf{r},\mathbf{r}') \equiv \langle \Delta n(\mathbf{r})\Delta n(\mathbf{r}')\rangle = \langle n\rangle^2[w_2(\mathbf{r},\mathbf{r}') - 1] + \langle n\rangle\delta(\mathbf{r}-\mathbf{r}'). \tag{4.4-40}$$

The quantity $p(r) = w_2(r) - 1$ is called the *pair correlation function* and its Fourier transform, $S(\mathbf{k})$, is known as the *static structure factor*, defined as $S(\mathbf{k}) = \langle n\rangle\int p(r)\exp(i\mathbf{k}\cdot\mathbf{r})d^3r$; these quantities reveal the most vital characterization of the adhesiveness in a gas or liquid.[18a] From the last two equations we thus find

$$\langle n\rangle p(|\mathbf{r}_1 - \mathbf{r}_2|) = \frac{\beta}{8\pi^3}\int \frac{W(\mathbf{k})e^{i\mathbf{k}\cdot(\mathbf{r}-\mathbf{r}')}}{f(n_0) - \beta W(\mathbf{k})}d^3k. \tag{4.4-41}$$

Whereas it has been instructive to carry Ornstein's model through to the point that we can relate the pair correlation function to the Fourier coefficient *W*(**k**) of the interaction potential, we should not expect realistic results, even if the integral were done numerically, the reason being that the $W(\mathbf{r}-\mathbf{r}')$ retained in the model only represents the long-range attractive tail, with the hard core being represented strictly only in the limitation of the volume. Thus, the rhs of (4.4-41) will not exhibit the general oscillatory behaviour, characteristic for the cohesiveness of liquids and dense gases. As shown by van Kampen, the expression is only useful for large separations, comparable to the range of the potential. Then, for small **k**-values, an expansion of *W*(**k**) is permissible which allows for a simple analytical evaluation. Therefore, let us write

$$W(\mathbf{k}) = \int W(\mathbf{r})[1 + i\mathbf{k}\cdot\mathbf{r} - \tfrac{1}{2}(\mathbf{k}\cdot\mathbf{r})^2 + \ldots]d^3r = W_0 - \tfrac{1}{6}k^2\int W(\mathbf{r})r^2d^3r = W_0 - \tfrac{1}{2}W_2k^2. \tag{4.4-42}$$

[18a] Sometimes it is convenient to define the static structure factor by $S(\mathbf{k})-1=\langle n\rangle\int d^3r\,p(r)\exp(i\mathbf{k}\cdot\mathbf{r})$.

Noting also that above and near the critical point the denominator of (4.4-41) is positive definite, we define $K^2 \equiv 2[f(n_0) - \beta W_0]/\beta W_2$. This, then, yields

$$\langle n \rangle p(|\mathbf{r}_1 - \mathbf{r}_2|) = \frac{W_0}{4\pi^3 W_2} \int \frac{e^{i\mathbf{k}\cdot(\mathbf{r}-\mathbf{r}')}}{k^2 + K^2} d^3k \,. \tag{4.4-43}$$

This integral is easily done by using $\mathbf{r} - \mathbf{r}'$ as polar axis in **k**-space. Thus with $d^3k = k^2 dk d(-\cos\vartheta) d\varphi$,

$$\text{integral} = 2\pi \int_0^\infty \frac{2\sin(k\,|\mathbf{r}-\mathbf{r}'|)}{(|\mathbf{r}-\mathbf{r}'|)(k^2+K^2)} kdk = -\frac{2\pi i}{(|\mathbf{r}-\mathbf{r}'|)} \int_{-\infty}^{\infty} \frac{e^{ik|\mathbf{r}-\mathbf{r}'|}}{(k^2+K^2)} kdk \,. \tag{4.4-44}$$

We change the integral in a contour integral, replacing the real variable k by the complex variable κ and adding a half-circle in the upper half of the complex plane of radius k_0. We then have on the arc $|\exp(i\kappa R)| = |\exp(iRk_0 e^{i\phi})| = \exp(-Rk_0 \sin\phi) \to 0$ for $0 < \phi < \pi$, so that there is no contribution from the arc. For the residue at $\kappa = iK$ one finds $res_{iK} = \frac{1}{2}\exp(-K|\mathbf{r}-\mathbf{r}'|)$. So with $R = |\mathbf{r}-\mathbf{r}'|$ we find

$$\text{integral} = 2\pi^2 e^{-KR}/R \,. \tag{4.4-45}$$

For the pair-correlation function this yields with $\langle n \rangle^{-1} = \hat{v}$,

$$p(r) = \hat{v}\,(W_0/W_2)\,[e^{-KR}/2\pi R] \,. \tag{4.4-46}$$

While similar results will be found later from Landau–Ginzburg theory (Chapter IX), the correlation length $1/K$ does not have the correct theoretical and experimentally observed temperature dependence.

Below we have given a sketch of a realistic pair distribution function $w_2(R)$, showing the typical overshoot for small R, with asymptotic value unity for large separations, see Fig. 4.5. We note that the integral of w_2 over d^3R by normalization must equal the volume V; thus also $4\pi\int p(R)R^2 dR = 0$, a condition obviously not true for the rhs in (4.46) due to the approximations made. In Fig. 4-6 we plot the structure factor for ^{36}Ar obtained from neutron scattering data by Yarnell et al. [18b] and from molecular dynamics.

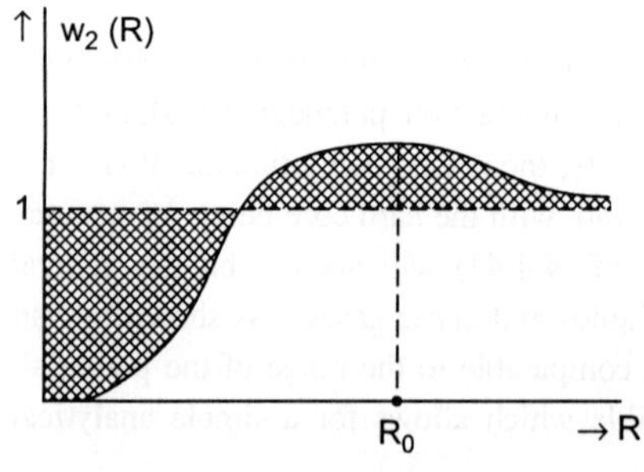

Fig. 4-5. Sketch of the expected pair distribution function for a dense gas above T_c.

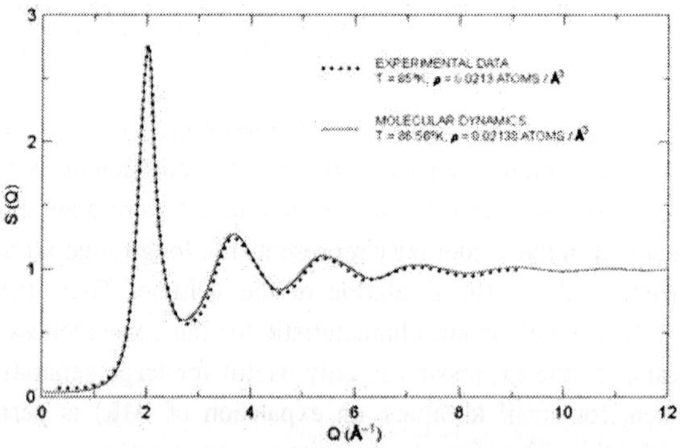

Fig. 4-6. The structure factor $S(\mathbf{k})$ vs. $|\mathbf{k}|$ for ^{36}Ar after Yarnell *et al.*[18b]. [With permission.]

[18b] J.L. Yarnell, M. Katz, R.G. Wenzel and S.H. Koenig, Phys. Rev. **A7**, 2130 (1973).

4.4.3 *Cluster-Integral Method*

We mentioned in the beginning that later considerations of the basic problem to compute the canonical partition function are based on cluster expansions of the function $\mathcal{Z}$. The basis of the method can already be found in Landau and Lifshitz, Op. Cit., §71. Although this section makes for worthwhile reading, the main problem is that they factor the partition function as $\mathcal{Z}=\mathcal{Z}_{pg}\mathcal{Z}_q$ where $\mathcal{Z}_{pg}$ is the perfect gas contribution, while also they consider only the first term in a cluster series of two molecules, so that the thermodynamic limit cannot be taken; in fact they state that "not only the density, but the quantity of the gas should be small." Their results are therefore fortuitous.

Proceeding differently, we shall as before factor the partition function as $\mathcal{Z}_p\mathcal{Z}_c$, where $\mathcal{Z}_c$ is the configurational part. Rather than working directly with φ_{ij} we use the quantities f_{ij}, defined already in (4.4-1). Letting $\mathcal{V}=\Sigma_{i<j}\varphi_{ij}(r_{ij})$, $r_{ij}=|\mathbf{r}_i-\mathbf{r}_j|$, we have

$$e^{-\beta\mathcal{V}}=\prod_{i<j}(1+f_{ij})=1+\sum_{i<j}f_{ij}+\sum_{i<j,\ k<l}f_{ij}f_{kl}+\ldots \tag{4.4-47}$$

where we expanded the product form. To obtain $\mathcal{Z}_c$ we must multiply the above by (1/*N*!) and integrate $\int\ldots\int d^3r_1\ldots d^3r_N$. The first term yields of course $(1/N!)V^N$. In the second integral we change the domain to $\int d^3r_i\int d^3r_{ij}$, where $r_{ij}=r$ is the position of the second molecule relative to the centre of the first one. Hence,

$$\int\ldots\int f_{ij}d^3r_1\ldots d^3r_N=V^{N-2}\int\int f_{ij}d^3r_id^3r_j=V^{N-1}\int_0^\infty f(r)4\pi r^2dr. \tag{4.4-48}$$

Now the number of pairs is $\binom{N}{2}\approx\frac{1}{2}N^2$, where $\binom{N}{2}$ is the binomial coefficient. Let

$$b_2=\tfrac{1}{2}\int_0^\infty f(r)4\pi r^2dr. \tag{4.4-49}$$

The contribution of the second term in (4.4-47) becomes

$$2^{nd}\text{ term contr}=(V^N/N!)\,N(Nb_2/V). \tag{4.4-49'}$$

Likewise, for the third term we have

$$\int\ldots\int f_{ij}f_{kl}\,d^3r_1\ldots d^3r_N$$
$$=V^{N-4}\int\int f_{ij}\,d^3r_id^3r_j\int\int f_{kl}\,d^3r_kd^3r_l=V^{N-2}(2b_2)^2. \tag{4.4-50}$$

The result for the third term contribution thus becomes

$$3^{rd}\text{ term contr}=\frac{V^N}{N!}\frac{1}{2!}\binom{N}{2}^2\left(\frac{2b_2}{V}\right)^2\approx\frac{V^N}{N!}\binom{N}{2}\left(\frac{N}{V}b_2\right)^2. \tag{4.4-50'}$$

We note that the factor (1/2!) on the rhs stems from the fact that products should be counted only once, i.e. in a product of two f's we divide by the factor 2! Likewise in a product of three f's we must divide by 3!, etc. All terms are evaluated similarly, making as only approximation $k!\binom{N}{k} \approx N^k$, which is good for $k << N$. *Ignoring* the errors in the higher-order terms with $k \sim N$ we then obtain a closed-form result

$$\mathcal{Q}_c \approx \frac{V^N}{N!}\left[1+\binom{N}{1}\frac{Nb_2}{V}+\binom{N}{2}\left(\frac{Nb_2}{V}\right)^2+\binom{N}{3}\left(\frac{Nb_2}{V}\right)^3+\ldots\right]=\frac{V^N}{N!}\left(1+\frac{N}{V}b_2\right)^N. \quad (4.4\text{-}51)$$

For the logarithm we find,

$$\ln \mathcal{Q}_c = N\left(\ln\frac{V}{N}+1\right)+N\ln\left(1+\frac{N}{V}b_2\right)\approx N\left(\ln\frac{V}{N}+1+\frac{N}{V}b_2\right). \quad (4.4\text{-}52)$$

Setting $V/N = \hat{v}$ we find for the free energy per particle, implying the thermodynamic limit,

$$\hat{f} = \hat{f}_p - (k_B T \ln \mathcal{Q}_c)/N = -k_B T\left(\tfrac{3}{2}\ln(2\pi m k_B T/h^2)+\ln\hat{v}+1+\frac{1}{\hat{v}}b_2\right). \quad (4.4\text{-}53)$$

The pressure is as before given by $P = -(\partial\hat{f}/\partial\hat{v})_T$, which yields

$$P = \frac{k_B T}{\hat{v}}\left(1-\frac{b_2}{\hat{v}}\right). \quad (4.4\text{-}54)$$

Comparing with (1.6-88) we see that $-b_2 = B$ is the second virial coefficient. Explicitly we have

$$B(T) = -\tfrac{1}{2}\int_0^\infty (e^{-\varphi(r)/k_B T}-1)\,4\pi r^2 dr. \quad (4.4\text{-}55)$$

To obtain an idea of what this function looks like, the integral will be split into two parts. So we write,

$$B(T) = -2\pi\int_0^{2r_1}(e^{-\varphi(r)/k_B T}-1)\,r^2 dr - 2\pi\int_{2r_1}^\infty (e^{-\varphi(r)/k_B T}-1)\,r^2 dr. \quad (4.4\text{-}56)$$

In the first integrand $\varphi(r)$ is large, so we neglect the exponential; the result is,

$$\bar{b} = 16\pi r_1^3/3, \quad \bar{b} > 0. \quad (4.4\text{-}57)$$

In the second integrand $\varphi(r)$ is small for most of the range, so we expand the exponential up to first order; this results in

$$-\bar{a}/k_B T = (2\pi/k_B T)\int_{2r_1}^\infty \varphi(r) r^2 dr, \quad \bar{a} > 0. \quad (4.4\text{-}58)$$

(note that $\varphi(r) < 0$ in this range). Finally,

$$B(T) \approx \bar{b} - \bar{a}/k_B T. \quad (4.4\text{-}59)$$

The equation of state for *P* now becomes,

$$\left(P+\frac{\bar{a}}{\hat{v}^2}\right)\approx\frac{k_BT}{\hat{v}}\left(1+\frac{\bar{b}}{\hat{v}}\right). \tag{4.4-60}$$

However, this result would not place a minimum limit on the volume when the gas is compressed. In concordance with the approximations made already in (4.4-52) we write therefore,

$$\left(1+\frac{\bar{b}}{\hat{v}}\right)\approx\left(1-\frac{\bar{b}}{\hat{v}}\right)^{-1}. \tag{4.4-61}$$

We thus find van der Waals' equation,

$$\left(P+\frac{\bar{a}}{\hat{v}^2}\right)\left(1-\frac{\bar{b}}{\hat{v}}\right)\approx\frac{k_BT}{\hat{v}}, \tag{4.4-62}$$

or, for one mole with $N = N_A$:

$$\left(P+\frac{N_A^2\bar{a}}{V^2}\right)\left(V-N_A\bar{b}\right)\approx RT, \tag{4.4-63}$$

in agreement with (4.4-21).

We still mention that the Joule effect (temperature change due to adiabatic change in volume) is associated with the interaction energy of the molecules. The cooling effect is given by $(\partial T/\partial V)_{\mathcal{E}}$. Or, using the –1 theorem and the Gibbs' relation,

$$\left(\frac{\partial T}{\partial V}\right)_{\mathcal{E}}=-\frac{(\partial\mathcal{E}/\partial V)_T}{(\partial\mathcal{E}/\partial T)_V}=\frac{P-T(\partial S/\partial V)_T}{C_V}. \tag{4.4-64}$$

With the Maxwell relation $(\partial S/\partial V)_T=(\partial P/\partial T)_V$ and (4.4-54) we obtain,

$$\left(\frac{\partial T}{\partial V}\right)_{\mathcal{E}}=-\frac{k_BT^2}{C_V\hat{v}^2}\frac{dB}{dT}. \tag{4.4-65}$$

The integral for dB/dT follows from (4.4-55). Thus measurement of the Joule effect gives information on the form for $\varphi(r)$. Likewise, the Joule-Kelvin effect $(\partial T/\partial P)_W$, where W is the enthalpy, is connected with $B(T)$. One easily confirms,

$$\left(\frac{\partial T}{\partial P}\right)_W=-\frac{k_BT^2V}{C_P\hat{v}}\frac{d(B/T)}{d(k_BT)}. \tag{4.4-66}$$

In the problems $B(T)$ will be related to several models for the two-body potential $\varphi(r)$.

4.5 Tonk's Hard-Core Gas and Takahashi's Nearest Neighbour Gas

Because of the complexity of three-dimensional gases, much attention has been given to the problem of obtaining partition functions for one-dimensional models with various interaction potentials. Most models are somewhat disappointing, since, as a rule, no phase transition occurs in one dimension. There are even several proofs that such a transition *cannot* occur. However, none of these proofs are sufficiently general and usually there is some escape clause. Van Hove[19] showed quite rigorously, using the theory of Fredholm integral equations, that no transition can occur in models in which the potential acts over a finite number of neighbours. Nevertheless, this says nothing about models in which the potential is singular (containing e.g. a delta function), such as considered by Baur and Nosanov,[20] or in which the thermodynamic limit is taken in a special way, as in the papers by Kac and Baker which we will consider at the end of this section.[21]. Here we shall only discuss a few well-known 1D models. Several of the original papers may be found in the interesting book by Lieb and Mattis, which gives a number of fully solvable models in onc dimension.[22]

The oldest paper in which hard-core interaction is considered is by Zernike and Prins.[23] Renewed attention to the hard-core model (in one, two and three dimensions) was given by Tonks and one now usually refers to the model as Tonk's hard-core gas.[24] A more general model, consisting of a hard-core and a finite attractive tail, was examined by Takahashi.[25] Presently, we will consider their 1D model

Thus, let the pair potential be given as:

$$\varphi(x) = \begin{cases} \infty \text{ for } |x| < a \text{ (hard core)}, \\ \phi(x-a) \text{ for } a < |x| < 2a, \\ 0 \text{ for } |x| > 2a. \end{cases} \tag{4.5-1}$$

For $\phi \equiv 0$ it reduces to Tonk's model. Since the particle has a diameter a, the distance between nearest neighbours (centre to centre) is $\geq a$; thus the third line assures that all contributions for farther neighbours vanish. The potential is pictured in Fig.4-7. Splitting as previously the partition function in a momentum-space part and a configurational part, we have for the latter

---

[19] L. van Hove, Physica **16**, 137 (1950).

[20] M.E. Baur and L.H. Nosanov, J. Chem. Physics **37**, 153 (1962).

[21] M. Kac, Phys. Fluids **2**, 8 (1959); also, M. Kac, G. Uhlenbeck and P. Hemmer, J. Math Phys. **4**, 216 (1964); G. Baker, Phys. Rev. **122**, 1477 (1961); ibid **126**, 2071 (1962).

[22] E. Lieb and D.C. Mattis, "Mathematical Physics in One Dimension", Acad. Press, NY. 1966.

[23] F. Zernike and J. Prins, Z. für Physik **21**, 184 (1927).

[24] L. Tonks, Phys. Rev. **50**, 955 (1936).

[25] M. Takahashi, Proc. Phys. Math. Soc. Japan, **24**, 60 (1942).

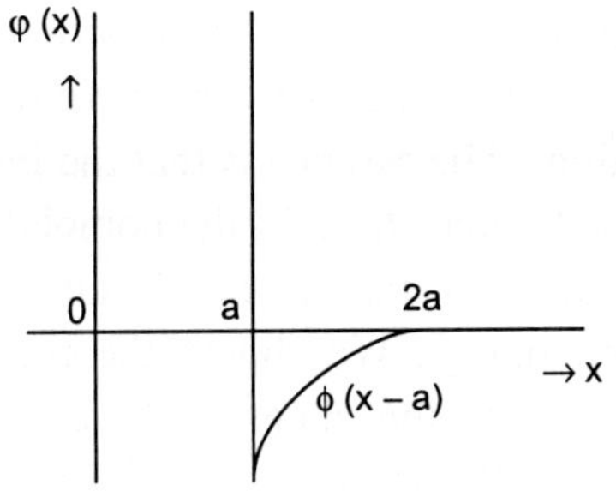

Fig. 4-7 Takahashi's interaction potential.

$$\mathscr{Z}_c = \frac{1}{N!}\int_0^L \ldots \int_0^L e^{-\beta\Sigma_i \varphi(x_i - x_{i-1})} dx_1 \ldots dx_N \, . \tag{4.5-2}$$

The integral is just $N!$ times the integral over the sub-region $0 \le x_1 \le x_2 \ldots \le x_N \le L$ (such a sub-region only exists in one dimension). We make a change to 'local variables':

$$y_j = x_j - (j-1)a, \quad \ell = L - (N-1)a \, . \tag{4.5-3}$$

The result is then,

$$\mathscr{Z}_c = \int_0^\ell dy_N \int_0^{y_N} dy_{N-1} \, e^{-\beta\phi(y_N - y_{N-1})} \ldots \int_0^{y_2} dy_1 \, e^{-\beta\phi(y_2 - y_1)} \, . \tag{4.5-4}$$

This may still be put in a slightly more familiar form by setting,

$$\lambda_N = \ell - y_N, \quad \lambda_i = y_{i+1} - y_i \quad (i = 1 \ldots N-1) \; ; \tag{4.5-5}$$

we then find for $\mathscr{Z}_c$

$$\mathscr{Z}_c = \int_0^\ell d\lambda_N \int_0^{\ell-\lambda_N} d\lambda_{N-1} \, e^{-\beta\phi(\lambda_{N-1})} \ldots \int_0^{\ell - \Sigma_{i=2}^N \lambda_i} d\lambda_1 \, e^{-\beta\phi(\lambda_1)} . \tag{4.5-6}$$

This is a convolution integral of $N-1$ exponential functions and two 'functions' 1. We therefore consider the Laplace transform with respect to ℓ. Employing (2.5-23) and denoting the Laplace transform of $\mathscr{Z}_c$ by $Z(s)$ and that of $\exp[-\beta\phi(\lambda)]$ by $K(s)$, we obtain,

$$Z(s) = s^{-2}[K(s)]^{N-1}, \tag{4.5-7}$$

with inversion

$$\mathscr{Z}_c = \frac{1}{2\pi i}\oint_C e^{\zeta\ell} [K(\zeta)]^{N-1} \zeta^{-2} d\zeta \, . \tag{4.5-8}$$

For convergence, noting that $\exp(\zeta\ell)$ increases with ζ, it is necessary that $K(\zeta)$ is a decreasing function of ζ.

Integrals of this nature are most easily evaluated with the 'saddle-point method' or 'method of steepest descent'.[26] This method, which we will need at many occasions, is discussed in the next section. The essence is that the integrand, which we write as $\exp[g(\zeta)]$, has a saddle point somewhere in the complex plane. In the above case there is a minimum along the real axis at $\zeta = \zeta_0$ and a corresponding maximum along the imaginary axis through ζ_0. We choose the contour C along that axis and make a Gaussian approximation to the integrand by expanding $g(\zeta)$ up to second order, with $g'(\zeta)$ being zero at the saddle point. The result for $\mathcal{Q}_c$ is then obtained as

$$\mathcal{Q}_c = \frac{e^{g(\zeta_0)}}{\sqrt{2\pi g''(\zeta_0)}} . \tag{4.5-9}$$

From the integrand in (4.5-8) we have

$$g(\zeta) = \ln\left\{e^{\zeta\ell}[K(\zeta)]^{N-1}\zeta^{-2}\right\} = \zeta\ell + (N-1)\ln K(\zeta) - 2\ln\zeta . \tag{4.5-10}$$

The saddle point condition is

$$g'(\zeta)_{\zeta=\zeta_0} \equiv g'(\zeta_0) = \ell + (N-1)[K'(\zeta_0)/K(\zeta_0)] - (2/\zeta_0) = 0. \tag{4.5-11}$$

For the second order derivative one finds

$$g''(\zeta_0) \approx (N-1)\left\{\frac{K''(\zeta_0)K(\zeta_0) - [K'(\zeta_0)]^2}{[K(\zeta_0)]^2}\right\}, \tag{4.5-12}$$

which indicates that $\ln g''(\zeta_0) = \mathcal{O}(\ln N)$ is of no interest in the thermodynamic limit. Thus from (4.5-9) and (4.5-10) we obtain

$$\begin{aligned} \ln \mathcal{Q}_c &= \zeta_0\ell + (N-1)\ln K(\zeta_0) - 2\ln\zeta_0 \\ &= \zeta_0 L + (N-1)\left[-\zeta_0 a + \ln K(\zeta_0)\right] - 2\ln\zeta_0 , \end{aligned} \tag{4.5-13}$$

while in terms of L the saddle-point condition becomes

$$L + (N-1)\left[-a + K'(\zeta_0)/K(\zeta_0)\right] - (2/\zeta_0) = 0 . \tag{4.5-14}$$

For the configurational free energy per unit 'volume' $(F - F_p)/L$ we now find, introducing the density $n = N/L$,

$$f_c = \lim_{L\to\infty}\left\{-\frac{k_B T}{L}\left[\zeta_0 L - (N-1)\zeta_0 a + (N-1)\ln K(\zeta_0) - 2\ln\zeta_0\right]\right\}$$

---

[26] P. M. Morse and H. Feshbach, "Methods of Theoretical Physics", Vol. I and II, McGraw-Hill, 1953, §4.6.

$$= -k_B T\left[(1-na)\right]\zeta_0 + n\ln K(\zeta_0)\,. \tag{4.5 -15}$$

The saddle-point condition in the thermodynamic limit becomes exact, *viz.*,

$$1 - na + nK'(\zeta_0)/K(\zeta_0) = 0\,. \tag{4.5-16}$$

The pressure is obtained as $P = n^2(\partial/\partial n)(f_c/n)$. This yields

$$P = k_B T\left\{\zeta_0 - n\left((1-na) + n\frac{K'(\zeta_0)}{K(\zeta_0)}\right)\frac{\partial \zeta_0}{\partial n}\right\} = k_B T\zeta_0\,, \tag{4.5-17}$$

where we used (4.5-16). This may be substituted back into (4.5-15). We then find *f*, or the Gibbs free energy per unit volume, $g = f + P$, for any interaction potential ϕ as a function of T, P and n.

We return to the hard-core gas. Then $\phi = 0$ and $K(\zeta) = 1/\zeta$. From (4.5-16) we then have for the saddle point

$$\zeta_0 = n/(1-na)\,. \tag{4.5-18}$$

For the free energy and the pressure in the hard-core model we have as final results,

$$f_c = -k_B T[n - n\ln n + n\ln(1-na)]\,, \tag{4.5-19}$$

$$P = k_B T n/(1-na)\,. \tag{4.5 -20}$$

Writing $n/(1-na)$ as a geometrical series, we have a virial expansion; however, P is a monotonic function of n and nowhere is $\partial P/\partial n$ equal to zero. There is therefore no condensation.

A tricky model, which does not suffer from this failure, was proposed by Kac and amended by Baker. They assume an infinite range potential with a hard core:

$$\varphi(x) = \begin{cases} \infty & \text{for } |x| < a\,, \\ -\alpha_0\gamma\exp(-\gamma|x|) & \text{for } |x| > a\,. \end{cases} \tag{4.5-21}$$

The partition function can be computed exactly by borrowing mathematical procedures from the theory of random walk.[27] A model in which condensation takes place results if the limit $\gamma \to 0$ is taken *after* the thermodynamic limit. Lebowitz and Penrose found, however, that an heuristic simple procedure gives the same result.[28] When γ is small, each particle interacts with all neighbours more or less uniformly and the exponential part merely adds

---

[27] Mark Kac, "Quelques problèmes mathématiques en physique statistique", Les Presses de l'Université de Montréal, 1974.

[28] J. Lebowitz and O. Penrose, J. Math. Phys. **7**, 98 (1966).

$$-\frac{N(N-1)}{2L^2}\int_{-\infty}^{\infty}\alpha_0\gamma e^{-\gamma|x|}dx=-n^2\alpha_0 \tag{4.5 -22}$$

to f_c over and above the hard-core contribution. We thus have,

$$f_c=-k_BT[n-n\ln n+n\ln(1-na)]-n^2\alpha_0\,. \tag{4.5-23}$$

This answer agrees with Kac's result. For the pressure everything works out alright also:

$$P=k_BT[n/(1-na)]-n^2\alpha_0\,, \tag{4.5-24}$$

$(n=1/\hat{v})$, which happens to be van der Waals' equation.

4.6 The Method of Steepest Descent

In statistical mechanics – and in other branches of physics – constraints are often tackled through recourse at the Laplace or Fowler generating function. The inverse then requires the evaluation of a contour integral in the complex plane, see Section 1.7. Often the integrand is sharply peaked when proceeding along the line of steepest ascent and descent. If then there is a maximum when proceeding along that line, the Cauchy-Riemann equations tell us that there is a minimum along a path perpendicular to the first line. Therefore, the extremum in question is a *saddle point.*

To present some explicit results we shall, without loss of generality, assume that the axes are chosen such that the maximum lies on a line segment that is parallel to the *y*-axis, while the minimum lies on a line segment that is parallel to the *x*-axis. Let the integrand be written as $\exp[g(\zeta)]$, where $\zeta=x+iy$. The Cauchy-Riemann equations then read,

$$\frac{\partial^2 g}{\partial x^2}+\frac{\partial^2 g}{\partial y^2}=0\,. \tag{4.6-1}$$

Thus, with $\partial^2 g/\partial y^2<0$ we have necessarily $\partial^2 g/\partial x^2>0$. The saddle point 'elevation contours' are pictured in Fig. 4-8(a). The situation is like a mountain pass: while the mountains rise up sharply from the sides of the pass which form a minimum at the point of passage, the explorer climbs to a maximum and then follows the path of descent, see Fig. 4-8(b).

For the contour integral to be meaningful and unique, it must encircle all the poles of the integrand; we can therefore shift the contour so that it includes the path parallel to the y-axis through the saddle point $\zeta_0=x_0+iy_0$. We now expand $g(\zeta)$ about the saddle point ζ_0; hence,

$$g(\zeta)=g(\zeta_0)+\tfrac{1}{2}g''(\zeta_0)(\zeta-\zeta_0)^2+\ldots \tag{4.6-2}$$

where ζ_0 is determined by the requirement,

$$g'(\zeta)|_{\zeta=\zeta_0} \equiv g'(\zeta_0) = 0. \tag{4.6-3}$$

The integral to be considered is now written as

$$I = \frac{1}{2\pi i}\int_C e^{g(\zeta)} d\zeta = \frac{1}{2\pi i}\int_{x_0-i\infty}^{x_0+i\infty} e^{g(\zeta_0)+\frac{1}{2}g''(\zeta_0)(\zeta-\zeta_0)^2+\dots} d\zeta. \tag{4.6-4}$$

Clearly along the path chosen, $d\zeta = i dy$ and $(\zeta-\zeta_0)^2 = -y^2$. If we neglect higher order terms in the expansion, then we just have a simple Gaussian integral and the result is,

$$I = \frac{1}{2\pi} e^{g(\zeta_0)} \int_{-\infty}^{\infty} e^{-\frac{1}{2}g''(\zeta_0)y^2} dy = \frac{e^{g(\zeta_0)}}{\sqrt{2\pi g''(\zeta_0)}}. \tag{4.6-5}$$

This result is general and will be used many times. Moreover, often we are only interested in $\ln I$. *If* g is of order N and $\ln g''$ of order $\ln N$, we have as first order asymptotic approximation $\ln I = g(\zeta_0)$. The main task is therefore to find the saddle point and to interpret its physical meaning.

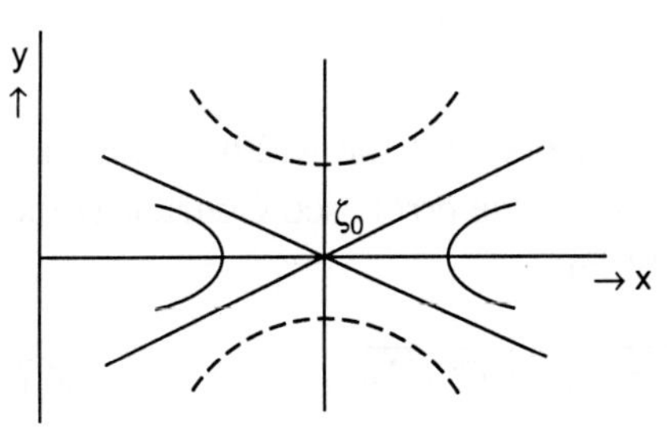

Fig. 4-8(a). Saddle point 'elevation contours'.

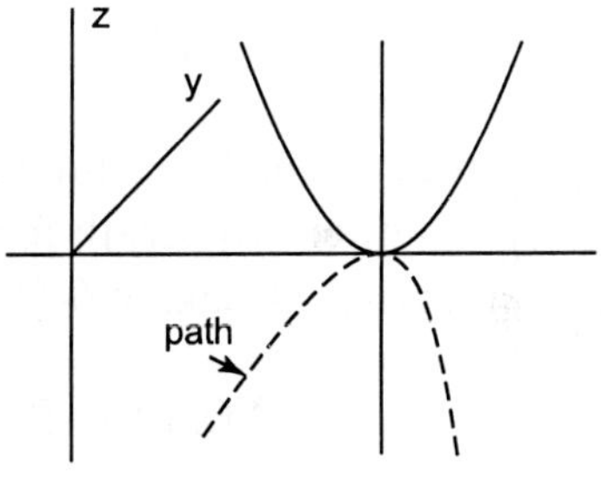

Fig. 4-8(b). Relief of saddle point.

Asymptotic expansions

The method of steepest descent can equally well be employed in order to obtain asymptotic series for functions that are defined by an integral along the real axis, having a strong maximum on that axis. We refer to Morse and Feshbach, Op. Cit. § 4.6. We shall illustrate the procedure for the gamma function, which is defined by

$$\Gamma(z+1) = \int_0^\infty e^{-t} t^z dt \overset{t\to z\tau}{=} z^{z+1}\int_0^\infty \tau^z e^{-z\tau} d\tau = z^{z+1}e^{-z}\int_0^\infty e^{-z(\tau-1-\ln\tau)} d\tau. \tag{4.6-6}$$

We now call

$$f(\tau)=\tau-1-\ln\tau. \tag{4.6-7}$$

Differentiating we have,

$$f'(\tau)=1-1/\tau, \qquad f''(\tau)=1/\tau^2. \tag{4.6-8}$$

This gives for the extremum on the real axis: $\tau_0=1,\ f''(\tau_0)=1$. So for $z>0$, this yields a maximum of the integrand of (4.6-6). There is a minimum along the imaginary axis through $\tau=1$, but that is of no concern since there are no poles for positive z (the poles of the gamma function are the negative integers $-1,-2,\ldots$) and the 'contour' will not have to be closed. Let now

$$I=\int_0^\infty e^{-zf(\tau)}d\tau \quad (z>0,\ \tau\ real). \tag{4.6-9}$$

We use the logarithmic expansion of the integrand (method of steepest descent). Hence,

$$f(\tau)=f(\tau_0)+\tfrac{1}{2}f''(\tau_0)(\tau-\tau_0)^2+\ldots=\tfrac{1}{2}(\tau-1)^2+\ldots \tag{4.6-10}$$

where we noticed $f(\tau_0)=0$. Thus, setting $\tau-1=p$, we have

$$I=\int_{-1}^\infty e^{-zp^2/2}dp\sim\int_{-\infty}^\infty e^{-zp^2/2}dp=\sqrt{2\pi/z}. \tag{4.6-11}$$

The replacement of the lower limit by $-\infty$ is acceptable if the width of the Gaussian $\sigma_p=1/\sqrt{z}$ is sufficiently small; for a discussion of the error, see Courant Hilbert, p. 522 ff.[29] From (4.6-11) and (4.6-6) we thus obtain,

$$\Gamma(z+1)\sim z^{z+1}e^{-z}\sqrt{2\pi/z}=z^{z+\frac{1}{2}}e^{-z}\sqrt{2\pi}, \tag{4.6 -12}$$

which is the leading term in Stirling's formula.

To obtain a full asymptotic expansion, i.e., an ascending series in $1/z$, we must retain higher order terms in the expansion for $f(\tau)$. This must be done judiciously, however. We set $f(\tau)=f(\tau_0)+\omega^2$. For the integral we then have,

$$I(z)\sim e^{-zf(\tau_0)}\int_{-\infty}^\infty e^{-z\omega^2}\frac{d\tau}{d\omega}d\omega. \tag{4.6-13}$$

Let now

$$\frac{d\tau}{d\omega}=\sum_{n=0}^\infty a_n\omega^n=\sum_{k=0}^\infty a_{2k}\omega^{2k}. \tag{4.6-14}$$

Each term is a Gaussian-type integral [see Eq. (1.6-51)], so that

---

[29] R. Courant and D. Hilbert, "Methods of Mathematical Physics", Vol. I and II, Interscience Publ., NY. 1953.

$$I(z) \sim e^{-zf(\tau_0)} \sqrt{\frac{\pi a_0^2}{z}} \left\{ 1 + \sum_{k=1}^{\infty} \left(\frac{a_{2k}}{a_0} \right) \frac{(2k-1)!!}{2^k} \left(\frac{1}{z} \right)^k \right\}. \tag{4.6-15}$$

The problem is now solved if we obtain the coefficients a_{2k}. This is not so simple; we refer to Morse and Feshbach, loc. cit., p. 442. For the case of the gamma function we find:

$$a_0 = \sqrt{2}, \quad (a_2 / a_0) = \tfrac{15}{8}\tfrac{1}{9}(\tfrac{1}{2})^{-3} - \tfrac{3}{2}\tfrac{1}{4}(\tfrac{1}{2})^{-2} = \tfrac{1}{6}, \quad etc. \tag{4.6-16}$$

This then yields the asymptotic series,

$$\Gamma(z+1) \sim z^{z+\frac{1}{2}} e^{-z} \sqrt{2\pi} \left\{ 1 + \tfrac{1}{12}\tfrac{1}{z} + \tfrac{1}{288}(\tfrac{1}{z})^2 + \ldots \right\}, \tag{4.6-17}$$

a result already announced in Eq(1.6-42) for integer z. See also the problems at the end of this Chapter, Section 4.12.

4.7 Dense Gases and the Virial Coefficients via the Grand-Canonical Ensemble

4.7.1 *Cluster Expansion*

Employing the grand-canonical distribution, we shall obtain a more complete result for the pressure of a dense gas than in section 4.4. In fact, we shall obtain a full result for the virial coefficients in the virial (or density) expansion of the gas. We shall not discuss the original elaborate Urcell method, which can be found in Reichl's book (first Ed. pp 355-384). Instead, we will discuss the cluster expansion in terms of reducible and irreducible diagrams, as found in Isihara and papers cited therein.[30, 31]

The semi-classical grand-canonical partition function, as defined in Eq. (4.2-17) can be written as

$$\mathcal{Z} = \sum_{N=0}^{\infty} \prod_{i=1}^{N} \int e^{-\beta \mathbf{p}_i^2/2m} \frac{d^3 p_i}{h^{3N}} e^{\beta \varsigma N} \frac{1}{N!} \int e^{-\beta \Sigma_{i<j} \varphi_{ij}(|\mathbf{q}_i - \mathbf{q}_j|)} d^3 q_1 \ldots d^3 q_N$$

$$= \sum_{N=0}^{\infty} \lambda_{th}^{-3N} y_\varsigma^N \mathcal{Q}_{N,c}\,, \tag{4.7-1}$$

where $\lambda_{th} = 1/\sqrt{2\pi m k_B T / h^2}$ is the width of the wave packet due to thermal motion as

---

[30] A. Isihara, Op. Cit., Sections 5.2–5.4; K. Husimi, J. Chemical Physics **18**, 682 (1950).

[31] G.E. Uhlenbeck and G.W. Ford, "Lectures in Statistical Mechanics", Series on Lectures in Applied Mathematics, Vol. I, The American Mathematical Society, Rhode Island 1963; ibid. in "Studies in Statistical Mechanics", Vol. I (J. de Boer and G.E. Uhlenbeck, Eds.) North Holland, Amsterdam 1962, p.119.

it results from the integrals over the momentum space, y_ς is the fugacity introduced previously and $\mathcal{Q}_{N,c}$ is the configurational canonical partition function for a gas of N particles. Explicitly we have, employing Mayer's f-function, cf. (4.4-1),

$$\mathcal{Q}_{N,c} = \frac{1}{N!}\int \prod_{i<j}(1+f_{ij})d^3q_1...d^3q_N \,. \tag{4.7-2}$$

In contrast to our previous treatment, we shall not employ the extreme rhs of (4.4-47), but we shall examine the structure of these products. Consider the following:

$$\begin{aligned}
&1\ particle: && 1,\\
&2\ particles: && (1+f_{12}),\\
&3\ particles: && (1+f_{12})(1+f_{13})(1+f_{23})\\
& && =1+[f_{12}+f_{13}+f_{23}]\\
& && +[f_{12}f_{13}+f_{12}f_{23}+f_{13}f_{23}+f_{12}f_{13}f_{23}],\\
&4\ particles: && etc.
\end{aligned} \tag{4.7-3}$$

It thus behoves us to introduce the following *cluster integrals*:

$$\begin{aligned}
b_1 &= \frac{1}{V}\int d^3q_1 = 1,\\
b_2 &= \frac{1}{2!V}\iint f_{12}d^3q_1d^3q_2\,,\\
b_3 &= \frac{1}{3!V}\iiint (f_{12}f_{13}+f_{12}f_{23}+f_{13}f_{23}+f_{12}f_{13}f_{23})d^3q_1d^3q_2d^3q_3\,,\\
&\vdots\\
b_l &= \frac{1}{l!V}\int ... \int \sum \prod f_{ij}\, d^{3l}q\,.
\end{aligned} \tag{4.7-4}$$

Note that for a 'system' with one particle, we only need b_1, for two particles we need b_1 and b_2, while for three particles we need b_1, b_2 and b_3. The cluster integrals can be represented by diagrams. The diagrams for b_2 and b_3 are given in Fig. 4-9.

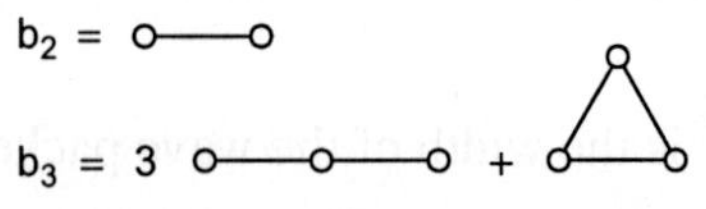

Fig. 4-9. Diagrams for the cluster integrals b_2 and b_3.

The above *connected graphs* or *clusters* show the contributions involved in each integral; the molecules form the vertices of a cluster and the f-bonds the connecting lines. The diagram for b_l would contain all possible connected graphs for l molecules. We further note that the individual connected graphs, being subdiagrams for a given b_l, contribute independently since they are not connected to each other. Now let there be m_l identical clusters of l molecules. The number of ways to put N unlabelled molecules in this arrangement is a multinomial coefficient,

$$\varpi = \frac{N!}{\Pi_l m_l!(l!)^{m_l}} \quad \text{subject to } \sum_l l\, m_l = N. \tag{4.7-5}$$

Each arrangement contributes to $\mathcal{Q}_{N,c}$ by

$$\mathcal{Q}_{m_l,c} = \frac{1}{N!}\prod_l (l!Vb_l)^{m_l}. \tag{4.7-6}$$

We thus obtain for $\mathcal{Q}_{N,c}$ the result,

$$\mathcal{Q}_{N,c} = \sum_{\{m_l\}} \mathcal{Q}_{m_l,c}\varpi = \sum_{\{m_l\}}\prod_l \frac{(Vb_l)^{m_l}}{m_l!} = \prod_l \sum_{m_l} \frac{(Vb_l)^{m_l}}{m_l!}. \tag{4.7-7}$$

The reader is encouraged to verify this result e.g., for N = 3 or 4. We note that the sum in (4.7-7) is restricted, see the condition in (4.7-5).

When we substitute this result into (4.5-1), the sum over m_l becomes unrestricted, since we sum additionally over N. Thus we now have,

$$\begin{aligned} \mathcal{Z} &= \sum_{N=0}^{\infty} \lambda_{th}^{-3N} y_\varsigma^N \prod_l \sum_{m_l=0}^{N/l} \frac{(Vb_l)^{m_l}}{m_l!} \\ &= \prod_l \sum_{m_l}^{\infty} (\lambda_{th}^{-3l})^{m_l} (y_\varsigma^l)^{m_l} \frac{(Vb_l)^{m_l}}{m_l!} = \prod_l \exp\left(Vb_l y_\varsigma^l \lambda_{th}^{-3l}\right) = \exp\left(V\sum_{l=1}^{\infty} b_l y_\varsigma^l \lambda_{th}^{-3l}\right). \end{aligned} \tag{4.7-8}$$

Further we need the grand potential $\Omega = -k_B T \ln \mathcal{Z}$. From the homogeneous form for $\Omega = \mathcal{E} - TS - \varsigma N = -PV$ we now have

$$P = k_B T \sum_{l=1}^{\infty} b_l z^l, \tag{4.7-9}$$

where we introduced the 'activity' $z = y_\varsigma \lambda_{th}^{-3}$. Finally we have for N from $\Omega(T,V,\varsigma)$: $N = (\partial \ln \mathcal{Z} / \partial \ln z)_{T,V}$. This yields,

$$n = \sum_{l=1}^{\infty} l b_l z^l, \tag{4.7-10}$$

where we applied the thermodynamic limit $N,V \to \infty$, $N/V = n$. The virial expansion can now be obtained by eliminating z from equation (4.7-10) and substituting into (4.7-9) by *successive iteration*. This straightforward procedure is carried out in a number of texts, e.g., in Landau and Lifshitz, Op. Cit., § 72.

We must first, however, dwell on the physical meaning of the results obtained. Equations (4.7-9) and (4.7-10) are called *cluster expansions of the equations of state.* At high temperatures and low densities gases are close to ideal. If the temperature drops and the density increases two-molecule clusters are formed, followed by three- and four-molecule clusters, etc. Finally, upon condensation, all molecules are linked in a large macroscopic cluster that forms the liquid. In an equilibrium state there is a distribution of cluster sizes, which can be considered to be pseudo-molecules. These do not interact with each other, since different clusters are not connected, as we saw before. In terms of these pseudo-molecules we have an ideal gas. According to Dalton's law the partial pressures add, thus we have $PV = k_B T \Sigma_l \langle m_l \rangle$. Comparing with (4.7-9) we see that the probability for a cluster of size l is proportional to $Vb_l z^l$. This interpretation is only tentative, since not all b_l are positive quantities for all gases. Roughly speaking though, we have $\langle m_l \rangle = V \,|\, b_l \,|\, z^l$.

Irreducible cluster integrals

Rather than using successive iteration, we will indicate that a *closed form solution* for the equation of state $P = P(T, 1/\hat{v})$ can be obtained. To that end we need some contour integrals. With z we shall associate a complex variable ξ and with n a complex variable w, such that (4.7-10) still holds, i.e.,

$$w = \sum_{l=1}^{\infty} l b_l \xi^l. \tag{4.7-11}$$

Also, we introduce new integrals β_k by the relationship

$$\xi = w \exp\left(-\sum_{k=1}^{\infty} \beta_k w^k \right). \tag{4.7-12}$$

A fortiori, on the real axis we have $z = n \exp\left(-\sum_{k=1}^{\infty} \beta_k n^k \right)$, or

$$\ln z = \ln n - \sum_{k=1}^{\infty} \beta_k n^k. \tag{4.7-13}$$

Since $l^2 b_l$ is the coefficient of ξ^l in the power-series expansion of $\xi dw / d\xi$, we have from Cauchy's formula,

$$l^2 b_l = \frac{1}{2\pi i} \oint \frac{d\xi}{\xi^l} \frac{dw}{d\xi} = \frac{1}{2\pi i} \oint \frac{\exp(l \Sigma_k \beta_k w^k)}{w^l} dw. \tag{4.7-14}$$

From the last result we see that $l^2 b_l$ is also the coefficient of w^{l-1} in the expansion of $\exp(l \Sigma_k \beta_k w^k)$ in powers of w. It thus follows that

$$l^2 b_l = \sum_{\{m_k\}}{}' \prod_k \frac{(l\beta_k)^{m_k}}{m_k!}, \tag{4.7-15}$$

where the prime on the sum implies the restriction

$$\sum_k k m_k = l - 1. \tag{4.7-16}$$

[*Note*: the sets $\{m_k\}$ satisfy the same condition as that used for obtaining the cumulants in (1.7-56).] From (4.7-15) and (4.7-16) one easily constructs the relations between the b_l and the irreducible integrals β_k. The first few are:

$$\begin{aligned} &b_1 = 1, \quad b_2 = \tfrac{1}{2}\beta_1, \quad b_3 = \tfrac{1}{2}\beta_1^2 + \tfrac{1}{3}\beta_2, \\ &b_4 = \tfrac{2}{3}\beta_1^3 + \beta_1\beta_2 + \tfrac{1}{4}\beta_3, \quad \ldots \end{aligned} \tag{4.7-17}$$

Employing the previously given forms for the b_l we obtain:

$$\begin{aligned} \beta_1 &= \frac{1}{1!V}\int f_{12}\, d^3q_1 d^3q_2 = \int f(r)\, d^3r, \\ \beta_2 &= \frac{1}{2!V}\int f_{12} f_{13} f_{23}\, d^3q_1 d^3q_2 d^3q_3, \\ \beta_3 &= \frac{1}{3!V}\int (3 f_{12} f_{14} f_{23} f_{34} + 6 f_{12} f_{13} f_{14} f_{23} f_{34} + f_{12} f_{13} f_{14} f_{23} f_{24} f_{34})\, d^3q_1 \ldots d^3q_4. \end{aligned} \tag{4.7-18}$$

The irreducible integrals are represented by irreducible diagrams of the form given in Fig. 4-10 below.

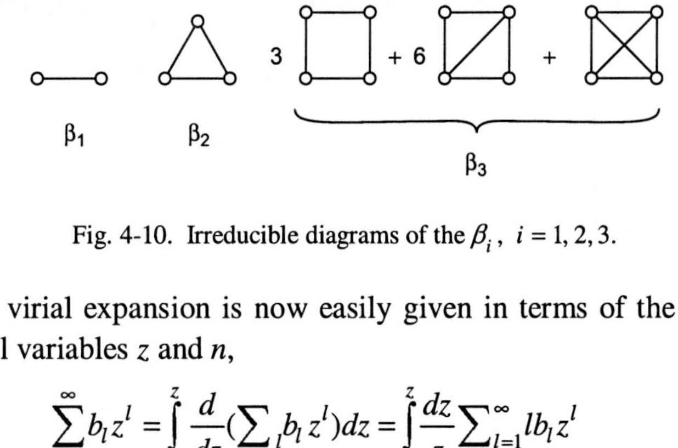

Fig. 4-10. Irreducible diagrams of the β_i, $i = 1, 2, 3$.

The density or virial expansion is now easily given in terms of the β_k. We write, reverting to real variables z and n,

$$\sum_{l=1}^{\infty} b_l z^l = \int_0^z \frac{d}{dz}\left(\sum_l b_l z^l\right) dz = \int_0^z \frac{dz}{z} \underbrace{\sum_{l=1}^{\infty} l b_l z^l}_{n}$$

$$\overset{(4.7\text{-}13)}{=} \int_0^n \frac{dn}{n}\left(1 - \sum_k k\beta_k n^k\right) n = n - \sum_k \frac{k}{k+1}\beta_k n^{k+1}. \tag{4.7-19}$$

In terms of $\hat{v} = 1/n$ this gives,

$$P = \frac{k_B T}{\hat{v}}\left(1 - \sum_k \frac{k}{k+1}\frac{\beta_k}{\hat{v}^k}\right) = \frac{k_B T}{\hat{v}}\left(1 + \frac{B}{\hat{v}} + \frac{C}{\hat{v}^2} + \ldots\right), \tag{4.7-20}$$

which is the virial expansion, cf. (1.6-88). Thus, we obtain the results,

$$B(T) = -\tfrac{1}{2}\beta_1 = -b_2 \text{ (as found before)},$$
$$C(T) = -\tfrac{2}{3}\beta_2 = 4b_2^2 - 2b_3, \quad etc. \tag{4.7-21}$$

The irreducible integrals were here introduced by the definition (4.7-12). They can also be obtained in a more direct context, see the cited paper by Husimi. Other researchers call the irreducible diagrams "star-graphs"[32]. Such a graph has the property that it can not be separated into other parts by a cut through a single connecting line. Although we gave in (4.7-21) the more conventional results in terms of the b's, what is remarkable is that *the virial coefficients involve irreducible diagrams* β_i *only*, see (4.7-20); all others apparently cancel out! In particular we note that the second virial coefficient depends only on the simplest f-bond diagram (first from the left in Fig. 4-10), while the third virial coefficient is solely given by the triangular diagram (second from the left in Fig. 4-10).

As to actual results, both of a simulated nature and experimental nature, we refer to other texts. For the hard sphere potential the virial coefficients up to seventh order have been computed, while for the Lennard-Jones ('six-twelve') potential results up to the fifth order have been reported. For the hard sphere potential we give the results in Fig. 4-11, and compare them with Monte Carlo results of Barker and Henderson[33] It is noted that the agreement is satisfactory, except for very high densities. Experimental results are available in many studies for noble gases. A comparison for the computed and experimentally extracted third virial coefficient for argon in the range of 50 K to 100 K, after Ref. 33 and Michels et al.[34] is given in Fig.4-12.

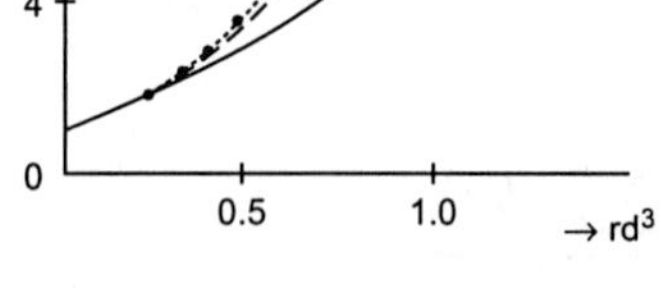

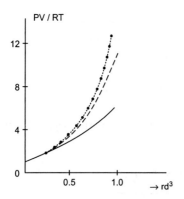

Fig. 4-11. Equation of state for hard spheres of diameter d. Solid curve: two virial coefficients; dashed curve: four coefficients; dotted curve: six coefficients. Solid circles: Monte Carlo results.[33] [With permission.]

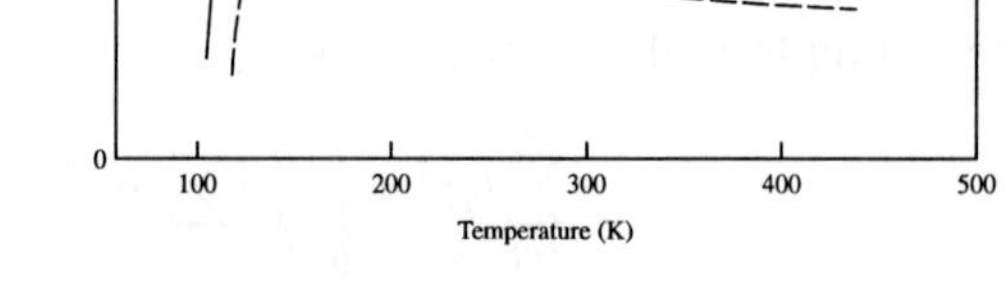

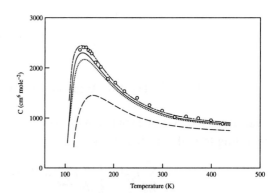

Fig. 4-12. $C(T)$ for argon. Circles: Experimental data of Michels et al.[34] Dashed curve: Lennard–Jones potential. Dotted curve: includes Axilrod–Teller interactions. Solid curve: includes dipole-quadrupole interactions. [With permission.]

[32] See M. Plischke and B. Bergersen, 3rd Ed., Op. Cit., Section 5.1.
[33] J.A. Barker and D. Henderson, Molecular Physics **21**, 187 (1971); Rev. Mod. Phys. **48**, 587 (1976).
[34] A. Michels, J.M.H. Levelt and W. de Graaff, Physica **24**, 659 (1958).

4.7.2 *Cumulant Expansion*

Some additional insight is gained from the cumulant expansion of $\ln\mathcal{Z}$. As indicated in subsection 1.7.3 the cumulants c_k are the coefficients in the expansion of the logarithm of the generating function. They are connected to the moments m_k of the distribution function by the rules (1.7-55) and (1.7-56). At this moment let us consider the Laplace generating functional for a function $\gamma(r)$; i.e., we have,

$$Z = \langle e^{-s\gamma(r)}\rangle = \int e^{-s\gamma(r)} f(r_1...r_N)\, d^3r_1...d^3r_N\,, \tag{4.7-22}$$

where $f(r)$ is normalized, $\int f(r)d^{3N}r = 1$. Expansion in a power series yields

$$Z = \sum_{k=0}^{\infty} \frac{m_k(-s)^k}{k!}; \quad \ln Z = \sum_{k=1}^{\infty} \frac{c_k(-s)^k}{k!}, \tag{4.7-23}$$

where $m_k = \langle \gamma^k \rangle$ is the *k*-th moment and c_k are the cumulants, being composed of the fluctuation moments.[35] Let us look for instance at the second order cumulant. We have

$$c_2 = \langle (\Delta\gamma)^2\rangle = \langle \gamma^2\rangle - \langle\gamma\rangle^2 \equiv \langle \gamma^2\rangle_c. \tag{4.7-24}$$

The subscript '*c*', for correlated part, at the extreme rhs indicates that *the cumulants are free from uncorrelated products of lower order moments.* So we obtain for $\ln Z$ the alternate result,

$$\ln Z = \sum_{k=1}^{\infty} \frac{(m_k)_c(-s)^k}{k!} = \langle e^{-s\gamma(r)} - 1\rangle_c. \tag{4.7-25}$$

We now return to the problem of evaluating the partition function for a dense classical gas. Let the weighting function $f(r_1...r_N)$ just be $1/V^N$. Clearly this is normalized. Hence,

$$\langle ...\rangle_N = \left(1/V^N\right)\int ... d^3r_1...d^3r_N. \tag{4.7-26}$$

This gives for the grand-canonical partition function,

$$\mathcal{Z} = \sum_{N=0}^{\infty} e^{\beta\varsigma N} \mathcal{Z}_p \mathcal{Z}_{N,c}\,, \text{ with } \mathcal{Z}_{N,c} = \left(V^N/N!\right)\langle e^{-\beta\mathcal{V}}\rangle_N\,, \tag{4.7-27}$$

where $\mathcal{V} = \Sigma_{i<j}\varphi_{ij}(r_{ij})$ is the potential energy. The sum cannot be performed exactly, so we assume that it suffices to pick the maximum value of the summand, with $N_{(\max)} \simeq \bar{N}$. We then have, employing Stirling's formula,

$$\begin{aligned} \ln\mathcal{Z} &= \ln\mathcal{Z}_p + \beta\varsigma\bar{N} + \bar{N}[\ln\hat{v} + 1 + \psi], \\ \bar{N}\psi &\equiv \ln\langle e^{-\beta\mathcal{V}}\rangle_{\bar{N}} = \langle e^{-\beta\mathcal{V}} - 1\rangle_c. \end{aligned} \tag{4.7-28}$$

---

[35] We ignore at this moment the **r**-dependence, which makes the moments functions of $r_1 \ldots r_k$.

For the grand potential per particle and for the pressure we obtain

$$\hat{k} \equiv \Omega(T,V,\varsigma)/\bar{N} = -k_B T(\ln \mathcal{Z})/\bar{N} = -k_B T[\ln \lambda_{th}^{-3} + \beta\varsigma + (\ln \hat{v} + 1 + \psi),$$

$$P = -\left(\frac{\partial \hat{k}}{\partial \hat{v}}\right)_{T,\varsigma} = k_B T\left[\frac{1}{\hat{v}} + \left(\frac{\partial \psi}{\partial \hat{v}}\right)_{T,\varsigma}\right]. \tag{4.7-29}$$

We consider the second term in an expansion of ψ, being the second order cumulant and note,

$$\langle \Delta \mathcal{V}^2 \rangle = \sum\sum [\langle \varphi_{ij}\varphi_{kl} \rangle - \langle \varphi_{ij} \rangle\langle \varphi_{kl} \rangle]. \tag{4.7-30}$$

There is no contribution, unless we have $(i,j) = (k,l)$. For the other combinations we get zero. E.g., with i,j,k,l being different indices,

$$\begin{aligned} \langle \varphi_{ij}\varphi_{kl} \rangle &= \langle \varphi_{ij} \rangle\langle \varphi_{kl} \rangle, \\ \langle \varphi_{ij}\varphi_{jl} \rangle &= \langle \varphi_{ij} \rangle\langle \varphi_{jl} \rangle. \end{aligned} \tag{4.7-31}$$

The first case represents two non-connected graphs, so obviously there is no correlated part. The second case is the connected three-particle diagram o—o—o, but it is reducible and does likewise not contribute.[36] From this observation, we arrive at the following

Cumulant theorem: Only irreducible diagrams contribute to $\langle e^{-\beta\mathcal{V}} - 1 \rangle_c$.

We will proceed to constructing the function ψ. With $\mathcal{V} = \Sigma_{i<j}\varphi_{ij}(r_{ij})$ the first term in the power series is $\langle -\beta\varphi_{12} \rangle$, multiplied by the number of pairs, $\binom{N}{2} \simeq N^2/2$. The second term stems from three particles, but only terms like $\langle \varphi_{12}\varphi_{23}\varphi_{13} \rangle$ (triangle) contribute. Setting $\binom{N}{k} \simeq N^k/k!$[37] we have for the two-particle and three-particle terms,

$$\begin{aligned} \binom{N}{2}\langle \varphi_{12} \rangle &\simeq \frac{\bar{N}n}{2!}\int \varphi_{12} d^3r, \\ \binom{N}{3}\langle \varphi_{12}\varphi_{23}\varphi_{13} \rangle &\simeq \frac{\bar{N}n^2}{3!}\int \varphi_{12}\varphi_{23}\varphi_{13} d^3r_2 d^3r_3, \end{aligned} \tag{4.7-32}$$

Doing the other terms likewise, we obtain the following result:

$$\psi(T,n) = \sum_k \frac{n^k}{(k+1)!}\sum_{\{\nu_{ij}\}}\int \prod_{1\le i<j}^{k+1} \frac{(-\beta\varphi_{ij})^{\nu_{ij}}}{\nu_{ij}!} d^3r_2 \dots d^3r_{k+1}, \tag{4.7-33}$$

---

[36] We have extended here the concept of diagrams to a products of φ's, rather than f's, with the topology of the interactions being the same in both cases.

[37] The approximations for $\binom{N}{k}$ become exact in the thermodynamic limit, $\bar{N}, V \to \infty$, $V/\bar{N} = \hat{v}$, since k is a small finite number.

where the second summation is over all irreducible diagrams with $(k+1)$ vertices. Interchanging the second sum and the product, we find for $\nu_{ij} = \nu$,

$$\prod_{i<j} \sum_{\nu=1}^{\infty} \frac{(-\beta\varphi_{ij})^{\nu}}{\nu!} = \prod_{i<j} (e^{-\beta\varphi_{ij}} - 1) = \prod_{i<j} f_{ij}, \tag{4.7 -34}$$

where f_{ij} is again the Mayer function and the product involves irreducible *f*-bond graphs only. Summing additionally over all such graphs with $(k+1)$ vertices and noticing (4.7-18), the result is,

$$\psi(T,n) = \sum_k \frac{\beta_k}{k+1} n^k = \sum_k \frac{\beta_k}{k+1} \frac{1}{\hat{v}^k}. \tag{4.7-35}$$

For *P* we find, see (4.7-29),

$$P = \frac{kT}{\hat{v}} \left(1 - \sum_k \frac{k}{k+1} \frac{\beta_k}{\hat{v}^k} \right), \tag{4.7-36}$$

in accord with our previous result for the virial expansion (4.7-20).

4.8 Mean Field Theories

4.8.1 *The Ising Hamiltonian and the Weiss Molecular Field*

We consider again the Ising Hamiltonian for interacting spins on a regular 3D lattice. For simplicity we assume that the lattice is simple cubic, so that the overlap integrals are the same in each direction. A lattice site is denoted by $\boldsymbol{l}$ and we consider nearest neighbour interaction only. From (4.3-20) and analogous to (4.3-21) we have,

$$\mathcal{H} = -J_0 \sum_{\boldsymbol{l},\hat{1}} \sigma_{\boldsymbol{l}} \sigma_{\boldsymbol{l}+\hat{1}} - \rho_B H \sum_{\boldsymbol{l}} \sigma_{\boldsymbol{l}}. \tag{4.8-1}$$

Here $\hat{1}$ is a unit vector on the lattice in the principal directions, $\sigma = \pm 1$ reflects the spin angular momentum in the *z*-direction ($/\!/ H$) and $J_0 > 0$ for a ferromagnetic system. Although no phase transition was found in the exact 1D model, simple reasoning tells us that such a transition should occur. Take at first very high temperatures; the spins are then randomly oriented and there is no net magnetization. At very low temperatures we expect, however, that the spins should line up with the field, so that the free energy is minimized. So we expect that there is a critical temperature T_c at which 'all spins follow the leader' and long-range order sets in. Whereas in sophisticated theories we will be able to identify the order parameter from first principles, here we shall only assume that such an order parameter *L* exists and that it is proportional to the magnetization along *H*, denoted by *M*, with $L = \gamma M$.

Let us now consider the contribution to $\mathcal{H}$ associated with a given spin σ_k. We thus set,

$$\mathcal{H}(\sigma_k) \equiv -\sigma_k\left(J_0 \sum_{\hat{1}} \sigma_{k+\hat{1}} + \rho_B H \right)$$
$$= -\sigma_k \rho_B \left(p\frac{m}{\rho_B^2} J_0 + H \right) - J_0 \sigma_k \sum_{\hat{1}} (\sigma_{k+\hat{1}} - \frac{m}{\rho_B}), \qquad (4.8\text{-}2)$$

where p is the coordination number for a site (in the cubic lattice $p = 6$). If we omit the second term in the last expression of (4.8-2), all spins will be independent and see an effective magnetic field indicated by the large round bracket term in (4.8-2). The term additional to H is called the Weiss molecular field.[38] We have,

$$H_m = pJ_0 m / \rho_B^2. \qquad (4.8\text{-}3)$$

The above Hamiltonian is then a *mean-field* Hamiltonian, in which the fluctuating exchange interactions are replaced by a 'smeared out field', H_m, seen by each spin and stemming from its environment. A subsidiary condition to this treatment is that the average magnetic moment per spin m, related to the magnetization M by $M = (N/V)m$, stemming from the term we omitted in (4.8-2), is now given self-consistently as,

$$m = \rho_B \langle \sigma_{k+\hat{1}} \rangle = \rho_B \langle \sigma \rangle. \qquad (4.8\text{-}4)$$

We can now proceed with the computation. Clearly we obtain,

$$m = \rho_B \langle \sigma \rangle = \rho_B \frac{\mathrm{Tr}\{\sigma \exp[-\beta\mathcal{H}(\sigma)]\}}{\mathrm{Tr}\{\exp[-\beta\mathcal{H}(\sigma)]\}} = \rho_B \tanh[\beta\rho_B (H_m + H)]. \qquad (4.8\text{-}5)$$

To see whether a phase transition can occur we take $H = 0$. We then have

$$m_{sp} = \rho_B \tanh(\beta\rho_B H_m). \qquad (4.8\text{-}6)$$

where the subscript 'sp' denotes the spontaneous magnetization. We substitute (4.8-3) into (4.8-6); this yields the transcendental equation,

$$m_{sp} = \rho_B \tanh(\beta p J_0 m_{sp} / \rho_B), \qquad (4.8\text{-}7)$$

which is equivalent to the two equations,

$$m_{sp} / \rho_B = \tanh x,$$
$$m_{sp} / \rho_B = x k_B T / p J_0 . \qquad (4.8\text{-}8)$$

It is now expedient to plot the two expressions *vs.* x. Note that the slope of the

[38] P. Weiss, Phys. Rev. **6**, 667 (1907).

straight line for the second equation decreases with decreasing T. There is therefore an intersection of the two curves only for sufficiently low T, see Fig. 4-13. The critical temperature occurs if the straight line becomes tangent to the curve for $\tanh x$. Hence,

$$kT_c = pJ_0. \tag{4.8-9}$$

When we expand the tanh up to third order we easily find

$$M = (N/V)m_{sp} = n\rho_B\sqrt{3}(T/T_c)^{\frac{3}{2}}\left(\frac{T_c - T}{T}\right)^{\frac{1}{2}}. \tag{4.8-10}$$

The order parameter L approaches zero in a singular way when $T \to T_c$ from below, vanishing asymptotically with the exponent $\beta = 1/2$. In actual ferromagnets a more complicated behaviour is observed. We leave it to the student to establish the Curie-Weiss law for the susceptibility, approached from the paramagnetic state above T_c:

$$\chi_m\big|_{H\to 0} = \frac{C}{T - T_c}, \quad T > T_c\,. \tag{4.8-11}$$

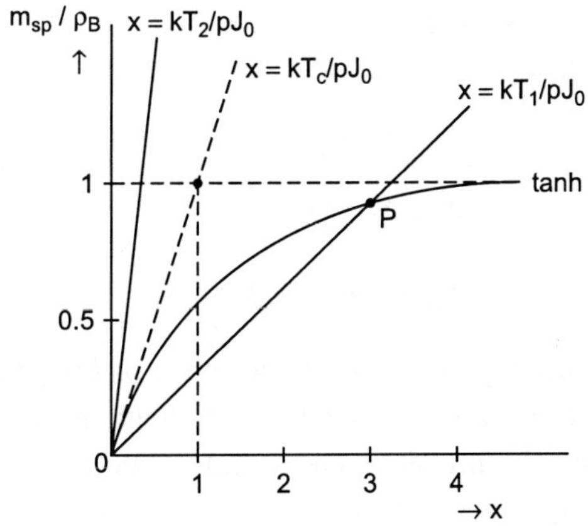

Fig. 4-13. Graphical solution for the spontaneous magnetization. $T_1 < T_c$: the intersection is the solution for m_{sp} ; $T_2 > T_c$: the paramagnetic phase; dashed line tangent: $T = T_c$.

4.8.2 *The Bragg–Williams Method. Order-Disorder Transitions*

The previous considerations can be considerably sharpened by employing the Bragg–Williams method, which, moreover, can also be adopted to other configurational problems with long-range order, such as the 'lattice gas' and a binary alloy with order-disorder transition.

First we look once more at the Ising energy. Let N_a be the number of spins with $\sigma = 1$ and N_b be the number with $\sigma = -1$. Further, let N_{aa} be the number of adjacent pairs with $\sigma = 1$, N_{bb} be the number of adjacent pairs with $\sigma = -1$ and N_{ab} be the number of pairs with opposite spins. Thus pictorially,

$$\begin{aligned} &aa \text{ is a link } \uparrow\uparrow && (\sigma_1 = 1\ ,\ \sigma_2 = 1\ ) \\ &bb \text{ is a link } \downarrow\downarrow && (\sigma_1 = -1, \sigma_2 = -1) \\ &ab \text{ is a link } \uparrow\downarrow \text{ or } \downarrow\uparrow && (\sigma_1 = \pm 1, \sigma_2 = \mp 1) \end{aligned} \tag{4.8-12}$$

The Ising energy is now expressed in the total number of links as follows,

$$\mathcal{E} = J(N_{ab} - N_{aa} - N_{bb}) - \rho_B H(N_a - N_b)\,, \tag{4.8-13}$$

In order to obtain the canonical partition function as a sum over energies we must find the degeneracy $G(N_a, N_b, N_{aa}, N_{bb}, N_{ab})$ and then perform the summation. Let p be the coordination number of a site. Then the following relations exist between these numbers,

$$\begin{aligned} N_a + N_b &= N, \\ pN_a &= 2N_{aa} + N_{ab}, \\ pN_b &= 2N_{bb} + N_{ab}. \end{aligned} \tag{4.8-14}$$

To preserve symmetry we use as new independent variables N, the long-range order parameter L and the short-range parameter ℓ, the latter being defined by

$$\left.\begin{aligned} N_a / N &= \tfrac{1}{2}(1+L), \\ N_b / N &= \tfrac{1}{2}(1-L), \end{aligned}\right\} (-1 \le L \le 1)$$

$$N_{ab} / \tfrac{1}{2} pN = \ell(1-L^2)\,. \tag{4.8-15}$$

The meaning of L is intuitively clear. In the absence of long-range order, $L = 0$ and we have that half the spins are up and half of them are down. Also note that $L = (N_a - N_b)/N$. The parameter ℓ takes into account that the immediate environment of a given spin may be different of what one expects from the overall distribution of a and b sites. All quantities can now be expressed in N, L and ℓ and one easily finds,

$$\mathcal{E}(N, L, \ell) = \tfrac{1}{2} pNJ[2\ell(1-L^2) - 1] - \rho_B HNL\,. \tag{4.8-16}$$

The Bragg–Williams method now consists in that we ignore the short-range order[39]; in other words: *each site sees around it a configuration as expected from the long-range distribution in the lattice*. This gives

$$N_{ab} \approx \langle N_{ab} \rangle_L = (N_a / N)_L (N_b / N)_L pN = \tfrac{1}{4}(1-L^2) pN\,, \tag{4.8-17}$$

[39] W.L. Bragg and E.J. Williams, Proc. Royal Soc. **A 145**, 699 (1934).

with similar expressions for N_{aa} and N_{bb}. Comparing (4.8-17) with (4.8-15), we infer that the B-W approximation amounts to setting $\ell = ½$. Consequently,

$$\mathcal{E}(N,L) = -\tfrac{1}{2} pNJL^2 - \rho_B HNL. \tag{4.8-18}$$

In the B–W approximation the degeneracy $\Delta\Gamma(L)$ is clearly $N!/N_a!N_b!$; we have for the density of states,

$$\chi(L)\Delta L = \frac{N!}{N_a!N_b!} = \frac{N!}{[\frac{1}{2}N(1+L)]![\frac{1}{2}N(1-L)]!}. \tag{4.8-19}$$

The canonical partition function becomes

$$\mathcal{Q}_N = \int_{-1}^{1} dL\, \chi(L) \exp[\beta(\tfrac{1}{2} pNJL^2 + \rho_B HNL)]. \tag{4.8-20}$$

We shall use a Gaussian approximation for the exponent in the integrand, denoted by $\exp[\beta\Theta(N,L)]$. Its maximum is obtained from

$$\frac{d}{dL}\left(\ln\chi + \beta[\tfrac{1}{2} pNJL^2 + \rho_B HNL]\right)\Big|_{L=L_0} = 0. \tag{4.8-21}$$

Employing Stirling's formula for the factorials we obtain,

$$\frac{1}{2\beta}\ln\frac{1+L_0}{1-L_0} = pJL_0 + \rho_B H. \tag{4.8-22}$$

This result is equivalent with

$$L_0 = \tanh[\beta(pJL_0 + \rho_B H)], \tag{4.8-23}$$

which is the same transcendental equation we met in (4.8-7). Identifying the molecular field by

$$H_m = pJL_0/\rho_B, \tag{4.8-24}$$

we only must connect L_0 with M. This is rapidly accomplished by evaluating

$$\mathcal{Q}_N \approx \int_{-\infty}^{\infty} \exp\left\{\beta\Theta(N,L_0) - \frac{1}{2}\beta\left|\frac{d^2\Theta}{dL^2}\right|\left[\frac{\Delta N_a - \Delta N_b}{N}\right]^2\right\} d(N_a - N_b)/N$$
$$= \exp[\beta\Theta(N,L_0)]\sqrt{2\pi k_B T/|\Theta''(L_0)|}. \tag{4.8-25}$$

Since $\Theta \propto N$ and $\ln|\Theta''| = \mathcal{O}(\ln N)$ we simply have

$$F(N,L_0) = -k_B T \ln \mathcal{Q}_N \sim -\Theta(N,L_0), \tag{4.8-26}$$

where the rhs is the negative of the integrand we maximized. We readily obtain,

$$F(N,L_0) = -k_BT\ln\chi(L_0) - \tfrac{1}{2}pNJL_0^2 - \rho_B HNL_0$$
$$= \left[\tfrac{1}{2}Nk_BT\ln\tfrac{1}{4}(1-L_0^2) + \tfrac{1}{2}Nk_BTL_0\ln\frac{1+L_0}{1-L_0}\right] - \tfrac{1}{2}pNJL_0^2$$
$$-\rho_B HNL_0 \overset{(4.8\text{-}22)}{=} \tfrac{1}{2}Nk_BT\ln\tfrac{1}{4}(1-L_0^2) + \tfrac{1}{2}pNJL_0^2 . \quad (4.8\text{-}27)$$

Although the explicit dependence on H disappeared, the field still appears in L_0. For the magnetization we have,

$$M = -\left(\frac{\partial\{F[L_0(H),H]/V\}}{\partial H}\right)_T = -\left(\frac{\partial(F/V)}{\partial L_0}\right)_T\left(\frac{\partial L_0}{\partial H}\right)_T - \left(\frac{\partial(F/V)}{\partial H}\right)_{L_0}, \quad (4.8\text{-}28)$$

where the last term is zero. Remains,

$$\frac{\partial(F/V)}{\partial L_0} = -nL_0\left[\frac{k_BT}{1-L_0^2} - pJ\right], \quad (4.8\text{-}29)$$

whereas by implicit differentiation of (4.8-22) we find,

$$\left(\frac{\partial L_0}{\partial H}\right) = \rho_B\left[\frac{k_BT}{1-L_0^2} - pJ\right]^{-1}. \quad (4.8\text{-}30)$$

This yields the expected result,

$$M = n\rho_B L_0 . \quad (4.8\text{-}31)$$

Comparison with the previous subsection shows that for the Ising model L_0 is just the mean magnetic moment per spin in Bohrmagnetons. We still note that $M = -n\rho_B L_0$ is also possible, since the free energy depends on L_0^2, see (4.8-27). The solution for the spontaneous magnetization, based on (4.8-23) for $H = 0$, is exactly the same as before, see Fig. 4-13. Other thermodynamic properties will now be discussed.

The energy $\mathscr{E}$ follows from $k_BT^2(\partial/\partial T)(-F/k_BT)$. Once more we need implicit differentiation of (4.8-22), this time to T. We find,

$$\mathscr{E} = \left(\frac{Nk_BT^2L_0}{1-L_0^2} - pNTJL_0\right)\left(\frac{\partial L_0}{\partial T}\right) + \tfrac{1}{2}pNJL_0^2, \quad (4.8\text{-}32)$$

while from (4.8-22),

$$\left(pJ - \frac{k_BT}{1-L_0^2}\right)\left(\frac{\partial L_0}{\partial T}\right) = \frac{pJL_0}{T} + \frac{\rho_B H}{T}. \quad (4.8\text{-}33)$$

Whence,

$$\mathscr{E} = -\tfrac{1}{2}pNJL_0^2 - \rho_B NHL_0, \text{ or } \mathscr{E}/V = -\tfrac{1}{2}pnJL_0^2 - MH . \quad (4.8\text{-}34)$$

We note that the energy *mutual with the reservoir*, MH, is included in $\mathcal{E}$, as is natural in Gibbs' approach, see subsection 1.3.2. The thermodynamical 'internal energy' density u is therefore,

$$u = -\tfrac{1}{2} pnJL_0^2 = \begin{cases} 0, & T \geq T_c, \\ -\tfrac{1}{2}\gamma M^2, & T < T_c. \end{cases} \qquad (4.8\text{-}35)$$

where we set $H_m = \gamma M$, $\gamma = pJ / n\rho_B^2$. The factor ½ appears here since this is a magnetic self-energy. The specific heat (per unit volume) $c = du / dT$ becomes

$$\begin{aligned} c &= 0, & T &> T_c, \\ c &= -\gamma M (dM / dT), & T &< T_c. \end{aligned} \qquad (4.8\text{-}36)$$

At the Curie temperature there is a jump. From (4.8-9), (4.8-10) and the above given constant γ we find this jump to be,

$$\Delta c \Big|_{T_c+0}^{T_c-0} = \tfrac{3}{2} k_B n. \qquad (4.8\text{-}37)$$

A plot is given in Fig. 4-14 below.

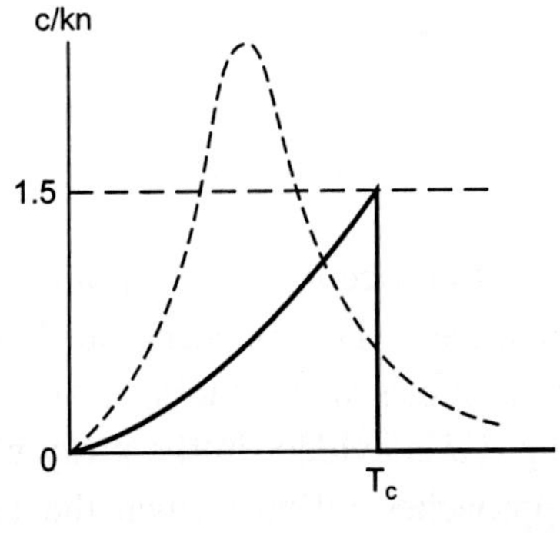

Fig. 4-14. Specific heat near the critical temperature according to the Bragg–Williams approximation (solid line). Also drawn in (near) exact result (dashed line).

An improvement over the Bragg–Williams approximation is the Bethe–Peierls approximation, in which some local order apart from long-range order is maintained.[40]

Binary alloys

There is a great parallel between a binary alloy, where we can have two types of neighbours and the Ising model, as we now show. Particularly we have in mind an

---

[40] H. Bethe, Proc. Royal Soc. **A 150**, 552 (1935); R.E. Peierls, Proc. Cambridge Phil. Soc. **32**, 477 (1936).

alloy such as β'-brass, consisting of zinc atoms and copper atoms, which occupy lattice positions in a body-centred cubic arrangement. For equal numbers of Zn and Cu, at absolute zero there is complete ordering, with all atoms of one kind occupying corner sites and atoms of the other kind occupying body-sites. The structure is pictured in Fig. 4-15, which shows that one can also look upon the structure as two interpenetrating simple cubic sublattices.

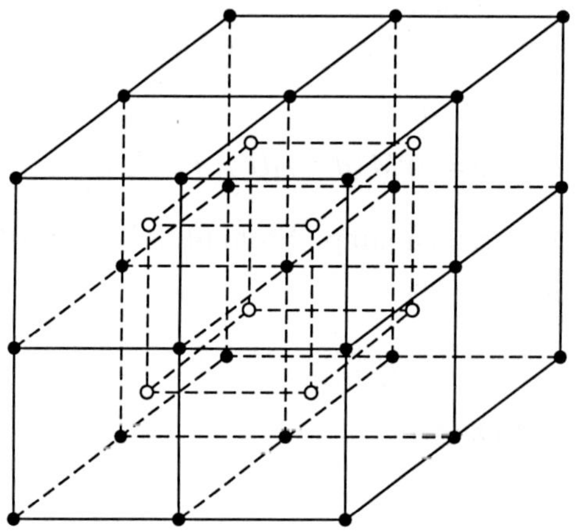

Fig. 4-15. Ordered structure for β'-brass.
•: Zn (corner points), ∘: Cu (body points).

Above 742 K a phase transition occurs. The Zn and Cu atoms are than equally distributed over the two sublattices. This is borne out by x-ray diffraction data. At low temperatures two different planes with spacing d are observed, while above 742 K a single plane with spacing $\frac{1}{2}d$ is evident in the x-ray pattern. It is also observed that the specific heat C_p approaches infinity when the temperature approaches T_c from below or above. We shall proceed to construct a statistical description. We assume that the interaction is given by the potential

$$\varphi(r) = \begin{cases} \infty \text{ for } r = 0, \\ \varepsilon_i \text{ for } 0 < r \le r_i, \quad (\varepsilon_i < 0) \ , \end{cases} \tag{4.8-38}$$

where 'i' denotes the different interactions, Zn-Zn, Cu-Cu, Zn-Cu. With similar notation as before the energy is written as

$$\mathscr{E} = N_{aa}\varepsilon_1 + N_{bb}\varepsilon_2 + N_{ab}\varepsilon_{12}. \tag{4.8-39}$$

Let N be the total number of lattice sites and N_a the number of Zn atoms and N_b the number of Cu atoms. We also need the numbers on the two sublattices, denoted by N_{a1}, N_{a2}, N_{b1} and N_{b2}, where '1' and '2' refer to each sublattice. We have the following relationships:

$$\begin{aligned} N_{a1}+N_{a2} &= N_a = c_a N, \\ N_{b1}+N_{b2} &= N_b = c_b N, \\ N_{a1}+N_{b1} &= \tfrac{1}{2}N, \\ N_{a2}+N_{b2} &= \tfrac{1}{2}N. \end{aligned} \tag{4-8-40}$$

For the sake of argument we assume $N_a \le N_b$ and we define the long-range order parameter:

$$L = (N_{a1} - N_{a2})/N_a, \quad -1 \le L \le 1. \tag{4.8-41}$$

With this definition we have,

$$\begin{aligned} N_{a1} &= \tfrac{1}{2}N_a(1+L), & N_{a2} &= \tfrac{1}{2}N_a(1-L), \\ N_{b1} &= \tfrac{1}{2}(N_b - LN_a), & N_{b2} &= \tfrac{1}{2}(N_b + LN_a). \end{aligned} \tag{4.8-42}$$

So far everything is exact. Next we make the B–W approximation which now entails

$$N_{aa} = p\frac{N_{a1}N_{a2}}{\frac{1}{2}N}, \quad N_{bb} = p\frac{N_{b1}N_{b2}}{\frac{1}{2}N}, \quad N_{ab} = p\left(\frac{N_{a1}N_{b2}}{\frac{1}{2}N} + \frac{N_{a2}N_{b1}}{\frac{1}{2}N}\right). \tag{4.8-43}$$

The energy, expressed in these quantities, reads

$$\mathcal{E} = \tfrac{1}{2}pN\left(\varepsilon_1 c_a^2 + \varepsilon_2 c_b^2 + 2\varepsilon_{12}c_a c_b - c_a^2 \hat{\varepsilon} L^2\right), \tag{4.8-44}$$

where

$$\hat{\varepsilon} = \varepsilon_1 + \varepsilon_2 - 2\varepsilon_{12}. \tag{4.8-45}$$

The number of accessible quantum states $\Delta\Gamma(L)$ and the density of states is now

$$\Delta\Gamma(L) = \chi(L)\Delta L = \frac{(\frac{1}{2}N)!}{N_{a1}!N_{b1}!}\frac{(\frac{1}{2}N)!}{N_{a2}!N_{b2}!}, \tag{4.8-46}$$

from which we find,

$$\begin{aligned} \ln\chi(L) &\sim N_{a1}\ln\frac{N}{2N_{a1}} + N_{a2}\ln\frac{N}{2N_{a2}} + N_{b1}\ln\frac{N}{2N_{b1}} + N_{b2}\ln\frac{N}{2N_{b2}} \\ &= \tfrac{1}{2}N\{c_a(1+L)\ln[c_a(1+L)] + c_a(1-L)\ln[c_a(1-L)] \\ &\quad + (c_b - Lc_a)\ln[c_b - Lc_a] + (c_b + Lc_a)\ln[c_b + Lc_a]\}. \end{aligned} \tag{4.8-47}$$

This yields for the canonical partition function,

$$\mathcal{Z}_N = \int dL \exp\left\{\ln\chi(L) - \beta\left[\tfrac{1}{2}pN\left(\varepsilon_1 c_a^2 + \varepsilon_2 c_b^2 + 2\varepsilon_{12}c_a c_b - c_a^2\hat{\varepsilon}L^2\right)\right]\right\}. \tag{4.8-48}$$

We call the factor in{ } again $\beta\Theta(L)$, and seek to maximize it. Setting its derivative to L equal to zero, the following transcendental equation for L_0 is obtained,

$$2c_a p\hat{\varepsilon}L_0 = k_B T \ln\left[\frac{(1+L_0)(c_b + L_0 c_a)}{(1-L_0)(c_b - L_0 c_a)}\right]. \tag{4.8-49}$$

For low temperatures and $\hat{\varepsilon} > 0$ there are three solutions: a trivial solution $L_{0,1} = 0$ and two nontrivial solutions, $\pm L_{0,2}$. The trivial solution yields a maximum for $\Theta(L)$ and a minimum for the Helmholtz free energy if $T > T_c = p\hat{\varepsilon}c_a c_b / k_B$. For $c_a = c_b = \frac{1}{2}$ Eq. (4.8-49) reduces to

$$p\hat{\varepsilon}L_0 = 2k_B T \ln\frac{1+L_0}{1-L_0} \to L_0 = \tanh(\tfrac{1}{4}\beta p\hat{\varepsilon}L_0). \tag{4.8-50}$$

Clearly, the situation is entirely analogous to the Ising ferromagnet, cf. Eqs. (4.8-22) and (4.8-23). The order parameter as a function of *T* follows the law (4.8-10). It can be shown that these conclusions are valid, irrespective of the B–W approximation.

In order to obtain the equation of state $P = P(T, 1/\hat{v})$ it is desirable to evaluate the grand-canonical partition function in terms of the activity *z* introduced before. For the lattice gas – an 'alloy' consisting of filled and empty sites the pressure is found to be,

$$\begin{aligned} P &= \gamma^{-1}\{B - \tfrac{1}{8}p\varepsilon_0(1+\hat{L}^2) - \tfrac{1}{2}k_B T \ln[\tfrac{1}{4}(1-\hat{L}^2)]\}, \\ 1/\hat{v} &= \gamma^{-1}\tfrac{1}{2}(1+\hat{L}), \end{aligned} \tag{4.8-51}$$

where $\varepsilon_0 \to \hat{\varepsilon}$ and where γ is the volume of the unit cell. Further, $\hat{L}$ is the order parameter, determined by the analogue of (4.8-23), *i.e.*,

$$\hat{L} = \tanh[\beta(\tfrac{1}{4}p\varepsilon_0\hat{L} + B)], \tag{4.8-52}$$

where *B* is a free parameter which arises from the fugacity, $B = \frac{1}{4}p\varepsilon_0 - (1/2\beta)\ln z$, where $z = y\lambda_{th}^{-3}$. In order to find $P(T, 1/\hat{v})$, the second equation of (4.8-51) can be substituted for $\hat{L}$ into (4.8-52), after which *B* must be eliminated from (4.8-52) and the equation for *P*. The isotherms so obtained are given in Fig. 4-16 below.[41]

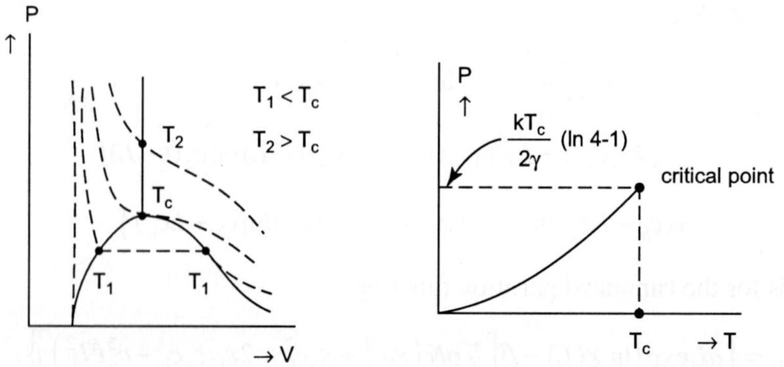

Fig. 4-16. Isotherms and *P-T* diagram for a lattice gas in the Bragg–Williams approximation. After Huang[41]. [With permission.]

[41] Cf. K. Huang, Op. Cit., 2nd Ed., Wiley, NY 1987, Fig. 14.11.

4.9 Landau–Ginzburg Theory for Phase Transitions; λ-Points

4.9.1 *General Procedure*

In the previous treatments we have used L for the long-range order parameter. For the Ising model L was found to be just m, the magnetic moment per spin. For ease of notation we shall from now on use m for the long-range order parameter, thereby noting that m is just a convenient symbol, which means something different for each type of phase transition. Thus in the binary alloy order-disorder transition $m \rightarrow n_{a1} - n_{a2}$; in other cases the parameter may be a vector or a field operator (Bose-Einstein condensation). The Ising spin model and the order-disorder transition were characterized by constant T and H (or B) in the coexisting phases, while in the gas-liquid transition T and P are held constant. In the former cases we computed the Helmholtz free energy of the system, while in the latter case the Gibbs free energy $G(T,P,N)$ is constant during the condensation process, as we showed in subsection 4.4.2.

Let us consider once more the free energy of the Ising model in the Bragg–Williams approximation, Eq. (4.8-27). For F we find the expansion[42]

$$F[N,m(H{=}0)]/N = -k_BT\ln 2 + \tfrac{1}{2}(k_BT - pJ)m^2 + \tfrac{1}{12}k_BTm^4 + \tfrac{1}{30}k_BTm^6 + \ldots \tag{4.9-1}$$

Now, assuming that the order parameter is a simple scalar as in this case, we proceed with the Landau–Ginzburg approach to phase transitions. We shall denote the free energy by $\Phi(T,m)$, leaving open whether the Helmholtz free energy F or the Gibbs free energy G applies to the problem at hand. If the free energy is analytical at the origin – which is not always the case – and if we have the usual symmetry with respect to sign reversal of m, we will have only even powers in the series. Thus we can write[43]

$$\Phi(T,m) = a_0(T) + \tfrac{1}{2}a_2(T)m^2 + \tfrac{1}{4}a_4(T)m^4 + \tfrac{1}{6}a_6(T)m^6 + \ldots \tag{4.9-2}$$

where the fractions have been added for later convenience. In addition, as exemplified by (4.9-1), the coefficient a_2 is of the nature,

$$a_2(T) = a_{20}(T)(T - T_c). \tag{4.9-3}$$

Further, for extrema we have

$$d\Phi/dm = a_2m + a_4m^3 + a_6m^5 + \ldots = 0, \quad d^2\Phi/dm^2 = a_2 + 3a_4m^2 + \ldots \tag{4.9-4}$$

---

[42] Since H=0 we must use the second line of (4.8-27) rather than the last line in which H is hidden in L_0.
[43] This expansion is *not* subject to the mean field approach. However, Landau–Ginzburg theory and mean field theory have many features in common, depending on the signs of the L–G coefficients.

These few considerations allow us to make a plot of $\Phi(T,m)$ vs. m for various T. The solution $m_0 = 0$ is viable for $T > T_c$ if $a_{20} > 0$, which gives *a fortiori* $a_2 > 0$, so that the free energy is a minimum. For $T < T_c$ we have $a_2 < 0$, giving a local maximum at $m = 0$. Further, if $a_4 > 0$ the two roots $m_0 = \pm\sqrt{-a_2 / a_4}$ are real and yield $\Phi'' = 2m_0^2 a_4$, resulting in local minima for the free energy. The equilibrium state is one of two states with broken symmetry. The situation is pictured in Fig. 4-17 below.

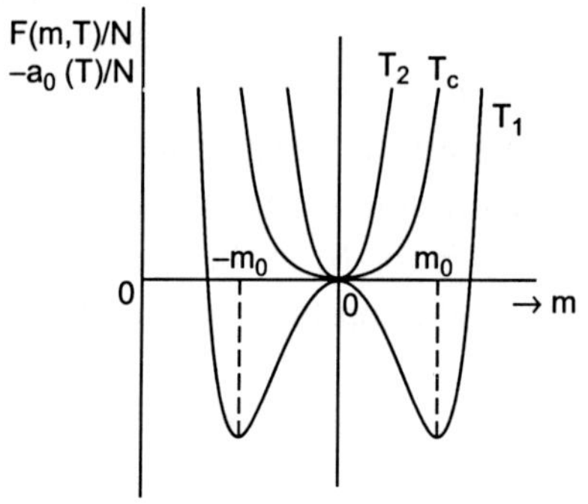

Fig. 4-17. Free energy as a function of the order parameter m in Landau–Ginzburg theory. T_2: above T_c; the equilibrium state has $m = 0$. T_1: below T_c; the equilibrium state is one of two states with broken symmetry, $m = \pm m_0$.

Employing (4.9-3) we obtain for the stable order parameter at temperatures below T_c:

$$m_0 = \sqrt{a_{20}(T)T_c / a_4}\left(1 - \frac{T}{T_c}\right)^{\frac{1}{2}}. \tag{4.9-5}$$

We have hereby tacitly assumed that the influence from the terms with $a_6, a_8, \ldots$ can be neglected. Next we find the specific heat from $S = -(\partial\Phi / \partial T)$, $C = T(\partial S / \partial T)$ or,

$$C = -T(\partial^2\Phi / \partial T^2). \tag{4.9-6}$$

We assume that all second order derivatives of the a's are zero or negligible and we also neglect da_{20} / dT and da_4 / dT. We then obtain

$$-S = \frac{\partial\Phi}{\partial T} = a_0' + \tfrac{1}{2}a_{20}m^2 + (a_2 + a_4 m^2)m\left(\frac{dm}{dT}\right), \tag{4.9-7}$$

$$\frac{\partial^2\Phi}{\partial T^2} = 2a_{20}m\frac{dm}{dT} + (a_2 + 3a_4 m^2)\left(\frac{dm}{dT}\right)^2 + (a_2 + a_4 m^2)m\left(\frac{d^2 m}{dT^2}\right). \tag{4.9-8}$$

For $T > T_c$ $m = 0$ and $dm / dT = 0$, so that $S = a_0'$ and $C = 0$. For $T < T_c$ m is given by $a_2 + a_4 m^2 = 0$. So the last terms in (4.9-7) and (4.9-8) do not contribute; the quantities dm/dT and $(dm/dT)^2$ are found from implicit differentiation of the equation for m^2. We note that at $T = T_c$ S is continuous (m being zero at both sides of T_c) while C jumps. We easily obtain,

$$\Delta C\Big|_{T_c+0}^{T_c-0} = T_c a_{20}^2 / 2a_4 \,. \tag{4.9-9}$$

The result is pictured in Fig. 4-18. The shape of this behaviour being as a "λ", we refer to the transition as a *λ-point.* This type of transition is denoted as a second-order phase transition or as a continuous phase transition[44], see next subsection.

Another situation occurs when a_2, a_4 and a_6 have a different set of signs in the neighbourhood of T_c. Then a first-order phase transition may occur, see problem 4.11 at the end of this chapter.

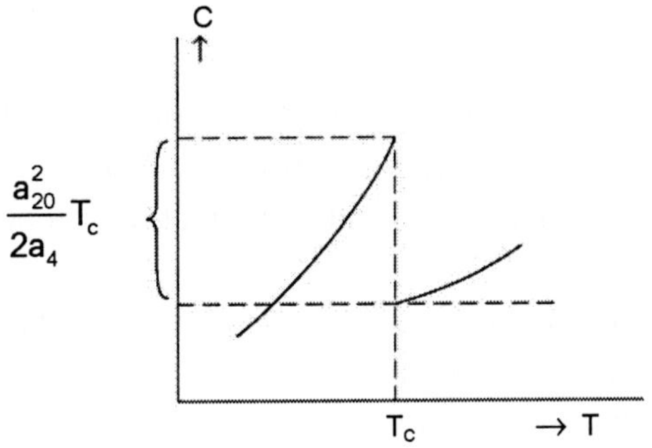

Fig. 4-18. The specific heat in the Landau–Ginzburg approach.

4.9.2 *Classification of Phase Transitions*

At this point it is expedient to recall the basic elements of changes in phase. First of all we have Gibbs' *phase rule*, taught in most elementary courses on physical chemistry. Let q be the number of independent components N^p of the system, r the number of simultaneously coexisting phases n_j and f the number of *thermodynamic degrees of freedom*, whereby we mean the number of variables that can be independently changed without affecting the equilibrium. Then the following equality holds:

$$f = q - r + 2. \tag{4.9-10}$$

In the simplest case that we deal with a single substance, manifest in two phases, we have $q = 1$, $r = 2$ so that $f = 1$. [If there are three coexisting phases, like vapour, water, ice, $f = 0$ and we have a fixed triple point.] Thus, in the free energy $\Phi(T, Q, \{N_p\})$, where Q means either V or P, we can independently change T or Q.

---

[44] The name second-order phase transition denotes that the first derivative that is discontinuous is (or in present-day terminology, has a singularity), is $\partial^2\Phi / \partial T^2$ (Ehrenfest's classification). See e.g., Moshe Gitterman and Vivian (Haim) Halpern, "Phase Transitions", World Scient. Publ. Co, Singapore, 2004.

For definiteness we consider the Gibbs free energy; then,

$$dG = -SdT + VdP + \sum_j \varsigma_j dn_j - \sum_i \bar{F}_i dA_i \,, \tag{4.9-11}$$

from which, if the external parameters are held constant, in equilibrium we have

$$(dG)\big|_{T,P,\{A_i\}} = \sum_j \varsigma_j dn_j = 0 \,. \tag{4.9-12}$$

So the chemical potentials, $\varsigma_j = (\partial G / \partial n_j)\big|_{T,P,\{A_i\}}, (j = 1,2)$ must be equal, a fact we used in the derivation of the Clausius-Clapeyron equation. However, no restriction is placed on the derivatives $V = (\partial G / \partial P)_T$ and $S = -(\partial G / \partial T)_P$.

If these derivatives are discontinuous at the transition, we speak of a first-order phase transition. Obviously, higher order derivatives will then jump as well. The situation is pictured in Fig. 4-19; the various 'cartoons' are easily drawn, noting only that the Gibbs free energy must be concave, both as a function of P and T. For an example, compare Figs. 4-19(a) and 4-4(b) , showing that the vapour-liquid transition is of first-order.

If, on the other hand, the first order derivatives are continuous at the critical temperature, then we speak of a continuous phase transition, see Fig. 4-20 (a) through (c). Now the higher order derivatives may also be continuous, or not. In Ehrenfest's classification continuous phase transitions are further distinguished by the behaviour of the second-order and higher-order derivatives. In curve 1 of Fig. 4-20(b) S is smooth and differentiable, while in curve 2 S has a cusp and different left-side and right-side derivatives at T_c. In the latter case we have a second-order phase transition. The compressibility and the heat capacity, being second-order derivatives, may have a peak at the critical temperature T_c, or they may jump, see curves 3 and 4 of Fig. 4-20(c), curve 4 being a λ-point. In the case of a peak, it may be rounded or have a cusp, i.e., the third derivative jumps, indicative of a third-order phase transition. The behaviour as in the 'true curve' for the Ising model, given by the dashed line in Fig. 4-14, is a realistic possibility for a continuous phase transition, in which no jumps to any order occur. [This is, however, contradicted for the 2D Ising model, see Fig.11-5, in which the specific heat has an 'infinite cusp'.]

Many 'modern' theorists only distinguish between first-order and continuous phase transitions cf. L.E. Reichl, Op. Cit. The reason is that Ehrenfest's classification oversimplifies the situation in that it is presumed that the discontinuities are simple jumps in the value of a derivative of the nth order, whereas in reality the discontinuities involve singularities, such as infinite jumps between left-side and right-side derivatives. Even so, Ehrenfest's classification can be maintained with more sophisticated armour.[44] A consistent older theory for second-order phase transitions has been given by Tisza; it is discussed in Callen's book (Op. Cit., § 9.8).

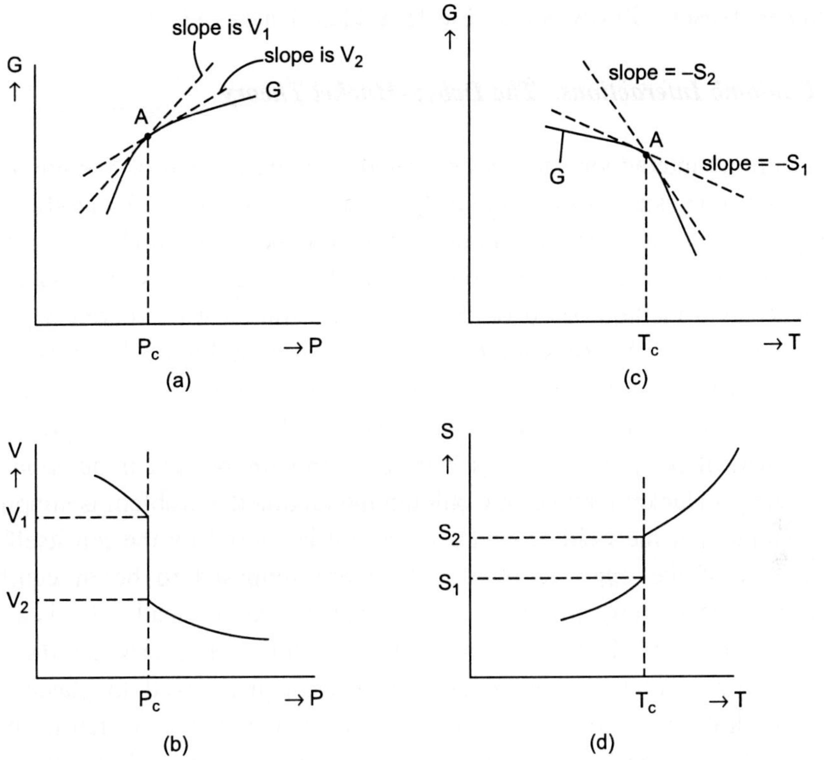

Fig. 4-19. Behaviour for a first order phase transition. (a) Gibbs free energy vs. P; (b) $V = \partial G / \partial P$ jumps at critical point; (c) Gibbs free energy vs. T; (d) $S = -\partial G / \partial T$ jumps at critical point.

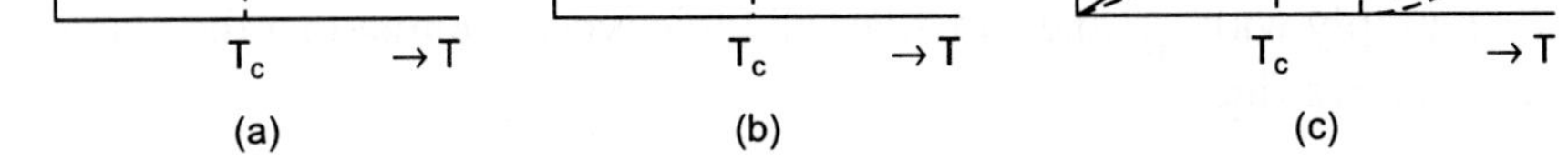

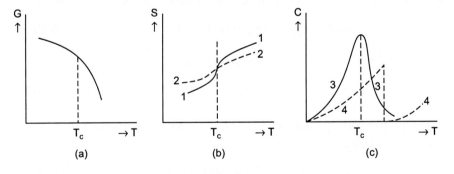

Fig. 4-20. Behaviour for a continuous phase transition. (a) G vs. T, continuous at T_c; (b) $S = -\partial G / \partial T$, continuous without break in slope (1) or with break in slope (2); (c) $C = T\partial S / \partial T$ vs. T, with peaked behaviour (3) or λ-point (4).

4.10 Ionized Gases – Plasmas – or Electron-Hole Gases in Condensed Matter

4.10.1 *Coulomb Interactions. The Debye–Hückel Theory*

The Coulomb interaction can not be treated with the method of Section 4.4 since the forces are long range, so that the integrals do not converge. A one-dimensional ionized gas (or plasma) with Coulomb interactions between neighbouring pairs of equal or opposite charges has been solved exactly by Lenard and independently by Prager, whereas a method based on functional integration was worked out later by Edwards and Lenard. A discussion of these papers is found in the book by Lieb and Mattis, cited earlier.[45] Since the potential energy in one dimension is proportional to $|x_i - x_j|$, as follows from the Green's function of the Laplacian, the problem is a nice mathematical exercise, but has little in common with the 3D computation.

In the Debye–Hückel method of Coulomb interaction the problem is simplified in that we assume that the field around a given ion is caused by the ion itself plus a cloud of ions of the opposite types, which are supposed to be in equilibrium independent of interactions with similar constellations. So, basically, this is kind of a mean field theory based on self-consistent evaluation. The original theory was developed for electrolytes.[46] It applies, however, equally well to plasmas or to electrons and holes in an ordinary or degenerate semiconductor, which have global charge neutrality. The Debye–Hückel method is not very rigorous, but will be briefly considered here because it is so simple.[47]

Let $\Phi(\mathbf{r})$ be the effective field around a given ion. Let $n_a(\mathbf{r})$ be the density of ions of type 'a' in the cloud surrounding this ion and let q_a be their charge. According to the Boltzmann distribution this density adjusts itself so that on average

$$n_a(\mathbf{r}) \approx n_{a0} \exp[-q_a \Phi(\mathbf{r}) / k_B T], \tag{4.10-1}$$

where n_{a0} is the density for $r \to \infty$. Expanding we have

$$n_a(\mathbf{r}) \approx n_{a0} - n_{a0} q_a \Phi(\mathbf{r}) / k_B T. \tag{4.10-2}$$

We let $a \to b$, multiply with q_b and sum over all ionic types, noting that the plasma as a whole is neutral; we find,

$$\sum_b q_b n_b(\mathbf{r}) = -\sum_b n_{b0} q_b^2 \Phi(\mathbf{r}) / k_B T. \tag{4.10-3}$$

But, by Poisson's law the lhs is just $-\varepsilon \nabla^2 \Phi(\mathbf{r})$, where ε is the dielectric constant.

---

[45] A. Lenard, J. Math Physics **2**, 682 (1961); S. Prager, "Advances in Chemical Physics" Vol. IV, Wiley 1962, p. 201; S. Edwards and A. Lenard, J. Math. Physics **3**, 778 (1962).

[46] P. Debye and G. Hückel, Physik. Zeitschrift, **24**, 305 (1923).

[47] See also Landau and Lifshitz, Op. Cit., § 74.

Whence, the self-consistent potential satisfies the Helmholtz equation

$$\nabla^2\Phi(\mathbf{r}) - \kappa^2\Phi(\mathbf{r}) = 0, \quad (|\mathbf{r}|>0). \tag{4.10-4}$$

Here $1/\kappa$ is the Debye length, with κ given by

$$\kappa = \left(\frac{1}{\varepsilon k_B T} \sum_b \mathcal{q}_b^2 n_{b0} \right)^{\frac{1}{2}}. \tag{4.10-5}$$

The solution for the 3D potential surrounding a charge $\mathcal{q}_a$ is

$$\Phi(\mathbf{r}) = \frac{\mathcal{q}_a e^{-\kappa r}}{4\pi\varepsilon r}, \tag{4.10-6}$$

where the constant associated with the homogeneous solution has been adjusted, such that for $r \to 0$ the field follows Coulomb's law. We note that the self-consistent potential is a *screened Coulomb potential* stemming from the cloud with ions of opposite charge. The Debye length is usually quite small. For $n \approx 10^{16}/cm^3$, $\varepsilon = 10$ and $k_B T = 0.025\ eV$, we find screening lengths of 30 Å. Thus the shielding of the ions is appreciable.

Another form for the Debye length is quite revealing. For a classical gas the Einstein relation between mobility $\hat{\mu}$ and diffusivity is

$$|\mathcal{q}_a| D_a = \hat{\mu}_a k_B T. \tag{4.10-7}$$

The conductivity is $\sigma = \sum_b n_b |\mathcal{q}_b| \hat{\mu}_b$ and the dielectric relaxation time is

$$\tau_\Omega = \varepsilon/\sigma. \tag{4.10-8}$$

If all the D's can be replaced by an average $\bar{D}$, we obtain the alternate form

$$1/\kappa = \sqrt{\bar{D}\tau_\Omega} \tag{4.10-9}$$

i.e., the diffusion length associated with the dielectric relaxation time. The screening length is therefore smaller, the higher the conductivity. For metals, however, quantum screening due to the Thomas-Fermi model may preponderate. Also, in that case we need Fermi-Dirac statistics for the degenerate electron gas. We shall later find the Einstein relation for a degenerate electron gas (cf. Section 15.2). The form (4.10-9) retains, however, its validity.

Using Fermi-Dirac statistics for the charge in the cloud we have instead of (4.10-1)

$$n_a(\mathbf{r}) \propto \left\{ e^{[\varepsilon_a - \varsigma_a - \mathcal{q}_a\Phi(\mathbf{r})]/k_B T} + 1 \right\}^{-1}, \tag{4.10-10}$$

where we inserted the electrochemical potential, see (1.3-8). We can expand the F–D function as follows,

$$f = f_0 + (\partial f / \partial q_a \Phi) q_a \Phi = f_0 - q_a \Phi (\partial f / \partial \varepsilon_a) . \tag{4.10-11}$$

Thus for small Φ we have

$$n_a(\mathbf{r}) \approx n_{a0} - q_a \Phi(\mathbf{r}) (\partial n_{a0} / \partial \varepsilon_a) . \tag{4.10-12}$$

Comparing with (4.10-2), we note that $(n_{a0} / k_B T) \rightarrow (\partial n_{a0} / \partial \varepsilon_a)$; we shall later on see that this differential is a ratio of Fermi integrals. Repeating now the same procedure, we still find a screened potential (4.10-6) with κ now given as

$$\kappa = \left(\varepsilon^{-1} \sum_b q_b^2 (\partial n_{b0} / \partial \varepsilon_b \right)^{\frac{1}{2}} . \tag{4.10-13}$$

Next we must find the Coulomb energy $\mathscr{E}_C$ and the free energy F. The Coulomb energy is given by

$$\mathscr{E}_C / V = \tfrac{1}{2} \sum_b n_{b0} q_b (\Delta \Phi_b), \tag{4.10-14}$$

where $\Delta \Phi_a$ is the field acting on a charge q_a due to the surrounding charge cloud. Expanding (4.10-6) and omitting the field from the ion itself, one obtains

$$\Delta \Phi_a \approx -q_a \kappa / 4\pi\varepsilon . \tag{4.10-15}$$

Hence we find

$$\mathscr{E}_C / V = -\frac{1}{8\pi\varepsilon} \sum_b n_{b0} q_b^2 \kappa . \tag{4.10-16}$$

For the classical gas we obtain, using (4.10-5):

$$\left(\mathscr{E}_C / V \right)_{class} = -\frac{1}{8\pi\varepsilon} \left(\frac{1}{\varepsilon k_B T} \right)^{\frac{1}{2}} \left(\sum_b q_b^2 n_{b0} \right)^{\frac{3}{2}} . \tag{4.10-17}$$

For the Fermi gas likewise, from (4.10-13):

$$\left(\mathscr{E}_C / V \right)_{Fermi} = -\frac{1}{8\pi\varepsilon^{3/2}} \left(\sum_b n_b q_b^2 \right) \left(\sum_c q_c^2 \frac{\partial n_c}{\partial \varepsilon_c} \right)^{\frac{1}{2}} . \tag{4.10-18}$$

In order to find the free energy we must use some thermodynamic relationships. For a classical gas we simply integrate

$$\left(\frac{\partial (F_q / T)}{\partial T} \right)_V = -\frac{\mathscr{E}_C}{T^2} , \tag{4.10-19}$$

where F_q is the free energy due to the interactions only. We further set $q_a = z_a e$. The integration of (4.10-19) is immediate,

$$F/V=(F_{pg}/V)-\frac{1}{12\pi}\left(\frac{e^2}{\varepsilon}\right)^{\frac{3}{2}}\left(\frac{1}{k_BT}\right)^{\frac{1}{2}}\left(\sum_b n_b z_b^2\right)^{\frac{3}{2}}. \tag{4.10-20}$$

Here F_{pg} is the free energy of a perfect gas; it enters here as an integration constant, since for $T\to\infty$ only F_{pg} remains. The connection with $\ln F_p$ (momentum part of the partition function) is

$$F_{pg}=F_p-k_BT\ln\left(V^{\Sigma_b N_b}/\prod_b N_b!\right). \tag{4.10-21}$$

The pressure, by Dalton's law, is the sum of the partial pressures[48] $P=\Sigma_b P_b$. From (4.10-20) we have for the interaction free energy per particle of type 'a',

$$\hat{f}_{q,a}=F_{q,a}/N_a=-\frac{1}{12\pi}\left(\frac{e^2}{\varepsilon}\right)^{\frac{3}{2}}\left(\frac{1}{k_BT}\right)^{\frac{1}{2}}\sqrt{n_a}\,z_a^3. \tag{4.10-22}$$

The associated partial pressure is

$$P_a=-(\partial\hat{f}_a/\partial\hat{v}_a)_T=n_a^2(\partial\hat{f}_a/\partial n_a)_T, \tag{4.10-23}$$

which gives for P, including the perfect gas contribution from (4.10-21):

$$P=\frac{k_BT}{V}\sum_b N_b+\frac{1}{3V}\mathscr{E}_C, \tag{4.10-24}$$

or in the thermodynamic limit,

$$P=k_BT\sum_b n_{b0}+\frac{1}{3}u_C. \tag{4.10-25}$$

This form is very different from the virial result for a non-ionized dense gas, obtained before.

We shall now do this for a degenerate gas with F–D statistics. Let λ be a parameter in the energy of the system. We have the Gibbs relations,

$$\begin{aligned} d\mathscr{E}&=TdS-PdV+\Sigma_b\mu_b dN_b+\Sigma_i\Lambda_i d\lambda_i,\\ dF&=-SdT-PdV+\Sigma_b\mu_b dN_b+\Sigma_i\Lambda_i d\lambda_i, \end{aligned} \tag{4.10-26}$$

from which

$$\langle\partial\mathscr{H}/\partial\lambda\rangle\rangle=(\partial\mathscr{E}_C/\partial\lambda)_{S,V\{n_{b0}\}}=(\partial F/\partial\lambda)_{T,V,\{n_{b0}\}}. \tag{4.10-27}$$

---

[48] In Dutch there is a rhyme for Dalton's law:

"In de ruimte doet elk gas,
Alsof 't alleen aanwezig was!"

While (4.10-19) can not be directly integrated because of the Fermi integrals in $(\partial n_{b0}/\partial \varepsilon_b)$, (4.10-27) does not suffer from this objection. We set $\lambda \to e^2$ and note that

$$(\partial \mathscr{E}_{C,Fermi}/\partial e^2)_{V,T,\{n_{b0}\}} = e^{-2}\mathscr{E}_{C,Fermi}. \tag{4.10-28}$$

So,

$$(\partial F_{q,Fermi}/\partial e^2)_{T,V,\{n_{b0}\}} = e^{-2}\mathscr{E}_{C,Fermi}(e^2). \tag{4.10-29}$$

Integration yields

$$F_{Fermi}/V = (F_{pg}/V) - \frac{e^3}{12\pi\varepsilon^{3/2}}\left(\sum_b n_{b0}z_b^2\right)\left(\sum_c z_c^2 \frac{\partial n_{c0}}{\partial \varepsilon_c}\right)^{\frac{1}{2}}. \tag{4.10-30}$$

It is left up to the reader to obtain the partial pressures and the total pressure. This will all be done a bit neater and more systematically in the next subsection.

4.10.2 *Coulomb Interactions via the Pair-Distribution Function; the BBGKY Hierarchy*

The previous theory was simple and straightforward, but it cannot be extended to higher order corrections. We now follow a method based on the pair distribution function. This concept is quite important in statistics of interacting particle systems and the method is illustrative for a number of applications in this field. For the problem under discussion, see Landau and Lifshitz, Op. Cit., Bogoliubov[49] or Yvon.[50]

The modified pair distribution function[51] (or two-particle distribution function) is part of an hierarchy to which the names Bogoliubov, Born, Green, Kirkwood and Yvon are attached; the hierarchy was originally developed for kinetic equations in non-equilibrium transport processes. We confine ourselves to a classical plasma or to a nondegenerate electron-hole gas in semiconductors. The canonical partition function will be written as $\mathscr{Q}_N = \mathscr{Q}_{pg}\mathscr{Q}_q$, where $\mathscr{Q}_{pg}$ is the perfect gas contribution as before, while $\mathscr{Q}_q$ is given as

$$\mathscr{Q}_q = \frac{1}{V^N}\int e^{-\mathscr{V}/k_BT}\prod_{i,s} d^3q_i^s, \tag{4.10-31}$$

with

$$\mathscr{V} = \sum_{all\ pairs}\ \sum_{p,q} u(\mathbf{r}_1^p, \mathbf{r}_2^q), \tag{4.10-32}$$

$$u(\mathbf{r}_1^p, \mathbf{r}_2^q) = z^p z^q e^2 / 4\pi\varepsilon |\mathbf{r}_1^p - \mathbf{r}_2^q|. \tag{4.10-33}$$

[49] N.N. Bogoliubov, J. Phys. USSR **10**, 256 (1946)

[50] J. Yvon, “Les corrélations dans un plasma en équilibre”, J. de Physique et le Radium **19**, 733 (1958).

[51] We use the name ‘modified’ distribution function, to distinguish it from the usual distribution functions introduced in the next section.

Note that the various species are now indicated by superscripts. The density function for the coordinates only is found to be

$$\rho(\{\mathbf{r}_i^s\}) = \int \rho(\{\mathbf{p}^s, \mathbf{r}^s\}) \prod_{i,s} d^3 p_i^s = \frac{1}{V^N \mathcal{Q}_q} e^{-\mathcal{V}/k_BT}. \tag{4.10-34}$$

We introduce the modified m-particle distribution function by

$$w_m = w_m(\mathbf{r}_1^p, \mathbf{r}_2^q, \ldots \mathbf{r}_m^{\cdot\cdot}) = \frac{1}{V^{N-m}\mathcal{Q}_q} \int e^{-\mathcal{V}/k_BT} (d^3r)^{N-m}, \tag{4.10-35}$$

where the integration is over all coordinates *not* occurring in w_m. Substituting in (4.10-35) for $\exp(-\mathcal{V}/k_BT)$ from (4.10-34) we note that the normalization is to V^m:

$$\begin{aligned} \int w_m (d^3r)^m &= \frac{1}{V^{N-m}\mathcal{Q}_q} \int (d^3r)^m V^N \mathcal{Q}_q \int \rho(\{\mathbf{r}_i^s\} (d^3r)^{N-m} \\ &= V^m \int \rho (d^3r)^N = V^m. \end{aligned} \tag{4.10-36}$$

This normalization indicates that for large inter-particle distances w_m approaches unity. In fact, we shall set

$$w_m = 1 + \omega_m. \tag{4.10-37}$$

We seek to find $w_2(\mathbf{r}_1^p, \mathbf{r}_2^q)$. Explicitly we have

$$w_2^{pq} \equiv w_2(\mathbf{r}_1^p, \mathbf{r}_2^q) = \frac{1}{V^{N-2}\mathcal{Q}_q} \int (d^3r)^{N-2} \exp\left\{ - \sum_{pairs;\, r,s} u^{rs}(\mathbf{r}^r, \mathbf{r}^s)/k_BT \right\}. \tag{4.10-38}$$

The Coulomb energy then is given by

$$\mathcal{E}_C = \frac{1}{2V^2} \sum_{p,q} N^p N^q \iint u^{pq} w_2^{pq} d^3 r_1^p d^3 r_2^q. \tag{4.10-39}$$

From the Coulomb energy we find $\mathcal{Q}_q$ by integration of the relationship

$$kT_B^2 \partial(\ln \mathcal{Q}_q)/\partial T = \mathcal{E}_C. \tag{4.10-39}$$

We now indicate the hierarchy of the functions $w_1, w_2, \ldots$ By differentiation to $\mathbf{r}^p$ we obtain,

$$\frac{\partial w_1^p(\mathbf{r}_1^p)}{\partial \mathbf{r}_1^p} = -\frac{1}{k_BT}\left(\frac{1}{V} \sum_q \int d^3 r^q N^q w_2^{pq} \frac{\partial u^{pq}}{\partial \mathbf{r}^p} \right), \tag{4.10-40}$$

$$\frac{\partial w_2^{pq}}{\partial \mathbf{r}^p} = -\frac{1}{k_BT}\left(w_2^{pq} \frac{\partial u^{pq}}{\partial \mathbf{r}^p} + \frac{1}{V} \sum_s \int d^3 r^s N^s w_3^{pqs} \frac{\partial u^{ps}}{\partial \mathbf{r}^p} \right), \tag{4.10-41}$$

etc. Note that $\partial/\partial\mathbf{r} \equiv \text{grad}_{\mathbf{r}}$. The two terms in (4.10-41) stem from the fact that when in (4.10-38) the supers r and s are equal to p and q we have a function of two variables only under the integral, but if either of them is different we have a function of three particle coordinates. Thus, we obtain a *hierarchy of coupled integro-differential equations.*

The literature on truncation procedures is voluminous, common methods being based on neglect of higher-order covariances, the ring approximation, *etc.* Clearly, a detailed discussion is outside the scope of this section. For our purpose we shall close with equation (4.10-41), by imposing Kirkwood's *superposition approximation*:

$$w_3^{pqs} = w_2^{pq} w_2^{ps} w_2^{qs}. \tag{4.10-42}$$

(This is the third-order analogue of the assumption of "molecular chaos", made by Boltzmann in his famous transport equation, $w_2^{pq} = w_1^p w_1^q$.) Employing further (4.10-37) and retaining first-order terms only, this yields

$$w_3^{pqs} = 1 + \omega_2^{pq} + \omega_2^{ps} + \omega_2^{qs}. \tag{4.10-43}$$

Also, in the first term on the right of (4.10-41) we replace w_2 by 1. Finally we note that the gas is isotropic and that grad u is an odd function; thus, upon insertion of (4.10-43) in the integral of (4.10-41), only the term ω_2^{qs} survives. Hence we obtain,

$$\frac{\partial \omega_2^{pq}}{\partial \mathbf{r}^p} = -\frac{1}{k_B T}\left(\frac{\partial u^{pq}}{\partial \mathbf{r}^p} + \sum_s \int d^3 r^s n_0^s \omega_2^{qs} \frac{\partial u^{ps}}{\partial \mathbf{r}^p} \right). \tag{4.10-44}$$

With u as in (4.10-33) this gives

$$\frac{\partial \omega_2^{pq}}{\partial \mathbf{r}^p} = -\frac{e^2}{k_B T 4\pi\varepsilon}\Bigg(z^p z^q \text{grad}_{\mathbf{r}^p} \frac{1}{|\mathbf{r}^p - \mathbf{r}^q|} + z^p \sum_s z^s n_0^s \int d^3 r^s \omega_2^{qs} (\mathbf{r}^q - \mathbf{r}^s) \text{grad}_{\mathbf{r}^p} \frac{1}{|\mathbf{r}^p - \mathbf{r}^s|} \Bigg). \tag{4.10-45}$$

We operate on both sides with the divergence and remember that $\nabla^2(1/r) = -4\pi\delta(\mathbf{r})$. Let $\mathbf{r} = \mathbf{r}^p - \mathbf{r}^q$. We then find, noting that ω_2 is even in $\mathbf{r}$,

$$\nabla^2 \omega_2^{pq}(\mathbf{r}) = \frac{e^2}{k_B T \varepsilon}\left(z^p z^q \delta(\mathbf{r}) + \sum_s z^p z^s n_0^s \omega_2^{qs}(\mathbf{r}) \right). \tag{4.10-46}$$

This system of equations is solved at once by setting

$$\nabla^2 \omega_2^{pq}(\mathbf{r}) = z^p z^q \omega(\mathbf{r}). \tag{4.10-47}$$

We obtain

$$\nabla^2 \omega(\mathbf{r}) - \kappa^2 \omega(\mathbf{r}) = (e^2 / k_B T \varepsilon)\delta(\mathbf{r}), \tag{4.10-48}$$

where

$$\kappa^2 = (e^2/k_B T\varepsilon)\sum_s (z^s)^2 n_0^s. \tag{4.10-49}$$

Apart from the constant in front of the delta function, (4.10-48) is just the defining equation for the Green's function of the Helmholtz equation if we let $\kappa \to ik$. Thus we have[52]

$$\omega(\mathbf{r}) = -\frac{e^2}{k_B T\varepsilon}\frac{e^{-\kappa r}}{4\pi r}. \tag{4.10-50}$$

For w_2^{pq} we obtain

$$w_2^{pq}(\mathbf{r}) = 1 - z^p z^q \frac{e^2}{k_B T\varepsilon}\frac{e^{-\kappa r}}{4\pi r}. \tag{4.10-51}$$

We now find the Coulomb energy of the system. The contribution from the '1' includes the divergent self-energy and is to be omitted. For the integration of the ω-part we need the simple integral

$$\int_{r\neq 0} (e^{-\kappa r}/r^2)\, d^3r = 4\pi \int_{0+}^{\infty} e^{-\kappa r} dr = 4\pi/\kappa. \tag{4.10-52}$$

This yields

$$\mathscr{E}_C/V = -\frac{1}{8\pi}\left(\frac{e^2}{\varepsilon}\right)^{\frac{3}{2}}\left(\frac{1}{k_B T}\right)^{\frac{1}{2}}\left(\sum_s n_0^s (z^s)^2\right)^{\frac{3}{2}}, \tag{4.10-53}$$

in accord with (4.10-17). The partition function is found from (4.10-39), which yields

$$\ln \mathscr{Q} = \ln \mathscr{Q}_{pg} + V\frac{1}{12\pi}\left(\frac{e^2}{\varepsilon k_B T}\right)^{\frac{3}{2}}\left(\sum_s n_0^s (z^s)^2\right)^{\frac{3}{2}}. \tag{4.10-54}$$

The ensuing free energy is in complete accord with (4.10-20). The equation of state for the pressure is again as in Eqs. (4.10-24) and (4.10-25). These results are now much better founded since we used the standard formalism of the canonical ensemble, rather than the *ad hoc* approach of Debye and Hückel.

4.11 Distribution Functions, Correlation Functions and Covariance Functions

We shall indicate the connection of w_m with the more usual description of density correlations. A complication here is that it is necessary to consider several species (at least two) for a neutral plasma. The definitions and results below will, however, be

---

[52] See Morse and Feshbach Op. Cit., pp. 804-810.

given so that they hold for any system. The canonical partition function is not factored as $\mathcal{Q}_{pg}\mathcal{Q}_q$ as in the previous Section 4.10, but as we did for the virial expansion in Section 4.4, i.e., as $\mathcal{Q}_p\mathcal{Q}_c$. The configurational partition function is more generally defined as:

$$\mathcal{Q}_c = \frac{1}{N^1!N^2!\ldots}\int e^{-\mathcal{V}/k_BT}\prod_{i,s} d^3q_i^s \,. \tag{4.11-1}$$

The density function for the coordinates only is now

$$\rho_N(\{\mathbf{r}_i^s\}) = \int \rho(p,q)\prod_{i,s} d^3p_i^s = \frac{1}{N^1!N^2!\ldots\mathcal{Q}_c} e^{-\mathcal{V}/k_BT} \,. \tag{4.11-2}$$

The *m-particle distribution function* ($m = \Sigma m^s$) for m^1 particles of kind 1, m^2 particles of kind 2, *etc.* is defined as

$$\begin{aligned} g_{m^1m^2m^3\ldots} &= \frac{N^1!}{(N^1-m^1)!}\frac{N^2!}{(N^2-m^2)!}\ldots\int (d^3r)^{N-\Sigma m^p}\rho_N(\{\mathbf{r}_i^s\}) \\ &= \frac{1}{(N^1-m^1)!}\frac{1}{(N^2-m^2)!}\ldots\frac{1}{\mathcal{Q}_c}\int (d^3r)^{N-\Sigma m^p} e^{-\mathcal{V}/k_BT} \,. \end{aligned} \tag{4.11-3}$$

The integration is over all coordinates *not* occurring in $g_{m^1m^2m^3\ldots}$.The normalization is

$$\begin{aligned} \int g_{m^1m^2m^3\ldots}(d^3r)^{\Sigma m^p} &= N^1(N^1-1)\ldots(N^1-m^1+1) \\ &\quad \times N^2(N^2-1)\ldots(N^2-m^2+1)\times\ldots \end{aligned} \tag{4.11-4}$$

The one-particle distribution function is normalized to N^p, the two-particle distribution function for two particles of the same kind is normalized to $N^p(N^p-1)$ since there are N^p ways to choose the first particle and N^p-1 ways to choose the second particle; for two particles of different species, however, the normalization is to N^pN^q. Some examples:

$$g_1^p(\mathbf{r}) = N^p\int (d^3r)^{N-1}\rho_N(\{\mathbf{r}\}) \,, \tag{4.11-5a}$$

$$g_2^{pp}(\mathbf{r}_1,\mathbf{r}_2) = N^p(N^p-1)\int (d^3r)^{N-2}\rho_N(\{\mathbf{r}\}) \,, \tag{4.11-5b}$$

$$g_2^{pq}(\mathbf{r}_1,\mathbf{r}_2) = N^pN^q\int (d^3r)^{N-2}\rho_N(\{\mathbf{r}\}) \,, \tag{4.11-5c}$$

$$g_3^{ppp}(\mathbf{r}_1,\mathbf{r}_2,\mathbf{r}_3) = N^p(N^p-1)(N^p-2)\int (d^3r)^{N-3}\rho_N(\{\mathbf{r}\}) \,, \tag{4.11-5d}$$

$$g_3^{ppq}(\mathbf{r}_1,\mathbf{r}_2,\mathbf{r}_3) = N^p(N^p-1)N^q\int (d^3r)^{N-3}\rho_N(\{\mathbf{r}\}) \,, \tag{4.11-5e}$$

$$g_3^{pqr}(\mathbf{r}_1,\mathbf{r}_2,\mathbf{r}_3) = N^pN^qN^r\int (d^3r)^{N-3}\rho_N(\{\mathbf{r}\}) \,. \tag{4.11-5f}$$

Next we define the m-point *correlation function*, with again $m = \Sigma m^s$. We write[53]

$$G_{m^1 m^2 m^3 \ldots} = \langle n^1(\mathbf{r}_1^1) n^1(\mathbf{r}_2^1) \ldots n^1(\mathbf{r}_{m^1}^1) \times n^2(\mathbf{r}_1^2) n^2(\mathbf{r}_2^2) \ldots n^2(\mathbf{r}_{m^2}^2) \times \ldots \rangle, \quad (4.11\text{-}6)$$

where the averaging is over the spatial density function $\rho_N(\{\mathbf{r}_i^s\})$. The notation is again much clearer when we write out $(p, q, r$ different or equal$)$:

$$G_1^p(\mathbf{r}) = \langle n^p(\mathbf{r}) \rangle, \quad (4.11\text{-}7a)$$

$$G_2^{pq}(\mathbf{r}_1, \mathbf{r}_2) = \langle n^p(\mathbf{r}_1) n^q(\mathbf{r}_2) \rangle, \quad (4.11\text{-}7b)$$

$$G_3^{pqr}(\mathbf{r}_1, \mathbf{r}_2, \mathbf{r}_3) = \langle n^p(\mathbf{r}_1) n^q(\mathbf{r}_2) n^r(\mathbf{r}_3) \rangle. \quad (4.11\text{-}7c)$$

The first function involves no correlations, being simply the density function. It is normalized to N^p, so clearly,[54]

$$G_1^p(\mathbf{r}) = g_1^p(\mathbf{r}). \quad (4.11\text{-}8)$$

As to the binary correlations, one easily sees that

$$G_2^{pq}(\mathbf{r}_1, \mathbf{r}_2) = g_2^{pq}(\mathbf{r}_1, \mathbf{r}_2), \quad p \neq q. \quad (4.11\text{-}9)$$

This no longer holds when dealing with alike particles, as is already born out by their different normalizations. We solve this by associating with the binary correlation (pp) the dynamical variable $\delta(\mathbf{r}_1^p - \mathbf{r})\delta(\mathbf{r}_2^p - \mathbf{r}')$ so that $G_2^{pp} \propto \delta(\mathbf{r}_1^p - \mathbf{r})\delta(\mathbf{r}_2^p - \mathbf{r}')$. Or, explicitly,

$$\begin{aligned} G_2^{pp}(\mathbf{r}_1, \mathbf{r}_2) &= \iint g_2^{pp}(\mathbf{r}, \mathbf{r}') \delta(\mathbf{r}_1^p - \mathbf{r}) \delta(\mathbf{r}_2^p - \mathbf{r}') d^3r d^3r' \\ &= g_2^{pp}(\mathbf{r}_1, \mathbf{r}_2), \quad \mathbf{r}_1 \neq \mathbf{r}_2. \end{aligned} \quad (4.11\text{-}10)$$

But,

$$\begin{aligned} \lim_{\mathbf{r}_1 \to \mathbf{r}_2} G_2^{pp}(\mathbf{r}_1, \mathbf{r}_2) &= \int g_1^p(\mathbf{r}) \delta(\mathbf{r}_1 - \mathbf{r}) \delta(\mathbf{r}_2 - \mathbf{r}) d^3r \\ &= g_1^p(\mathbf{r}_1) \delta(\mathbf{r}_1 - \mathbf{r}_2). \end{aligned} \quad (4.11\text{-}11)$$

The delta function takes care of the confluence of two position coordinates. The results of the last three equations can be summarised by

$$G_2^{pq}(\mathbf{r}_1, \mathbf{r}_2) = g_2^{pq}(\mathbf{r}_1, \mathbf{r}_2) + \delta^{pq} g_1^p(\mathbf{r}_1) \delta(\mathbf{r}_1 - \mathbf{r}_2). \quad (4.11\text{-}12)$$

---

[53] The use of the term 'correlation function' here is different from the mathematician's usage. Also, with physicists our terminology is not standard. Some authors refer to the functions g_m as correlation functions. The functions G_m, which differ from g_m at confluent points, should then be called the 'full correlation functions'.

[54] For globally homogeneous systems g_1 is the average density, being a constant. Henceforth global homogeneity will be implied.

The normalization is consistent, being always $N^p N^q$, p,q different or the same. The three-point (ternary) correlation functions are likewise obtained $(p \neq q \neq r)$:

$$G_3^{pqr} = g_3^{pqr}(\mathbf{r}_1, \mathbf{r}_2, \mathbf{r}_3), \tag{4.11-13a}$$

$$G_3^{ppq} = g_3^{ppq}(\mathbf{r}_1, \mathbf{r}_2, \mathbf{r}_3) + g_2^{pq}(\mathbf{r}_2, \mathbf{r}_3)\delta(\mathbf{r}_1 - \mathbf{r}_2), \tag{4.11-13b}$$

$$G_3^{ppp} = g_3^{ppp}(\mathbf{r}_1, \mathbf{r}_2, \mathbf{r}_3) + \sum_{(ijk)} g_2^{pp}(\mathbf{r}_j, \mathbf{r}_k)\delta(\mathbf{r}_i - \mathbf{r}_j)$$
$$+ g_1^p(\mathbf{r}_1)\delta(\mathbf{r}_1 - \mathbf{r}_2)\delta(\mathbf{r}_1 - \mathbf{r}_3). \tag{4.11-13c}$$

Here $\sum_{(ijk)}$ means that the three indices (i, j, k) are to be changed cyclically from (1,2,3).

We also give another definition for the correlation functions G_k – considering for simplicity only the case of one kind of particles,

$$G_k(\mathbf{r}_1, \mathbf{r}_2, \ldots \mathbf{r}_k) = \sum_{\{i_\alpha\}}^{(N)} \int (d^3r')^N \prod_{i=1}^{k} \delta(\mathbf{r}_i - \mathbf{r}_{i_\alpha}')\rho_N(\{\mathbf{r}'\})$$
$$= \sum_{i_1, i_2, \ldots, i_k}^{(N)} \langle \delta(\mathbf{r}_1 - \mathbf{r}_{i_1}')\delta(\mathbf{r}_2 - \mathbf{r}_{i_2}')\ldots\delta(\mathbf{r}_k - \mathbf{r}_{i_k}')\rangle. \tag{4.11-14}$$

[This is the definition that is used by Feynman in his theory of liquid ^{4}He, cf. Section 12.2.] For example, we have for the two-point (binary) correlation function

$$G_2(\mathbf{r}_1, \mathbf{r}_2) = \sum_{i,j} \int (d^3r')^N \delta(\mathbf{r}_1 - \mathbf{r}_i')\delta(\mathbf{r}_2 - \mathbf{r}_j')\rho_N(\{\mathbf{r}'\})$$
$$= \sum_i \int (d^3r')^N \delta(\mathbf{r}_1 - \mathbf{r}_i')\delta(\mathbf{r}_2 - \mathbf{r}_i')\rho_N(\{\mathbf{r}'\}) + \sum_{i \neq j} \int (d^3r')^N \delta(\mathbf{r}_1 - \mathbf{r}_i')\delta(\mathbf{r}_2 - \mathbf{r}_j')\rho_N(\{\mathbf{r}'\})$$
$$= N\delta(\mathbf{r}_1 - \mathbf{r}_2)\int (d^3r')^{N-1} \rho_N(\{\mathbf{r}'\}) + N(N-1)\int (d^3r')^{N-2} \rho_N(\{\mathbf{r}'\})$$
$$= g_1(\mathbf{r}_1)\delta(\mathbf{r}_1 - \mathbf{r}_2) + g_2(\mathbf{r}_1, \mathbf{r}_2). \tag{4.11-15}$$

This result is the same as we found in (4.11-12). The reader is encouraged to also verify (4.11-13c).[55]

Lastly, we introduce the m-point *fluctuation-correlation functions* (or covariance functions) by:

$$\Gamma_{m^1 m^2 m^3 \ldots} = \langle \Delta n^1(\mathbf{r}_1^1)\Delta n^1(\mathbf{r}_2^1)\ldots\Delta n^1(\mathbf{r}_{m^1}^1)$$
$$\times \Delta n^2(\mathbf{r}_1^2)\Delta n^2(\mathbf{r}_2^2)\ldots\Delta n^2(\mathbf{r}_{m^2}^2)\ldots\rangle. \tag{4.11-16}$$

where $\Delta n^p(\mathbf{r}) = n^p(\mathbf{r}) - \langle n^p(\mathbf{r})\rangle$. The following rules are evident (all p, q, r):

$$\Gamma_2^{pq}(\mathbf{r}_1, \mathbf{r}_2) = G_2^{pq}(\mathbf{r}_1, \mathbf{r}_2) - G_1^p(\mathbf{r}_1)G_1^p(\mathbf{r}_2), \tag{4.11-17}$$

---

[55] Some authors, e.g., Plischke and Bergersen, Op.Cit., Section 5.2, employ the form (4.11-14) for what they call the 'reduced distribution functions'; however, in the delta function product sum, they state explicitly that the spatial positions labelled with $1\ldots k$ should all be unequal, which leads to a different normalization, similarly as for our functions $g_k(\mathbf{r}_1, \mathbf{r}_2, \ldots \mathbf{r}_k)$.

$$\Gamma_3^{pqr}(\mathbf{r}_1,\mathbf{r}_2,\mathbf{r}_3) = G_3^{pqr}(\mathbf{r}_1,\mathbf{r}_2,\mathbf{r}_3) - \sum_{(ijk)} G_2^{(qr)}(\mathbf{r}_j,\mathbf{r}_k) G_1^{(p)}(\mathbf{r}_i)$$
$$+2G_1^p(\mathbf{r}_1)G_1^q(\mathbf{r}_2)G_1^r(\mathbf{r}_3), \qquad (4.11\text{-}18)$$

where the superscripts ( ) must be changed cyclically along with (i, j, k).

We return to the subject of truncation procedures. A hierarchy of equations is most easily terminated if covariance functions of higher order are successively neglected versus quantities of similar dimensions, but having one Δ less. But we must be careful with the self-variance functions, which blow up if $|\mathbf{r}-\mathbf{r}'|\to 0$. The necessity of these *dimensional singularities* is easily demonstrated.[56] For instance, the two-point covariance function is of order $\mathcal{O}(\langle n\rangle/\delta V)$:

$$\int_{\delta V}\int_{\delta V} \langle \Delta n(\mathbf{r})\Delta n(\mathbf{r}')\rangle d^3r d^3r' = \langle \Delta N_{\delta V}\rangle^2 \sim \langle N_{\delta V}\rangle = \langle n\rangle \delta V, \qquad (4.11\text{-}19)$$

or

$$\frac{\langle \Delta n(\mathbf{r})\Delta n(\mathbf{r}')\rangle}{\langle n(\mathbf{r})\rangle\langle n(\mathbf{r}')\rangle} = \begin{cases} \mathcal{O}(1/\langle n\rangle\delta V) & \text{if } |\mathbf{r}-\mathbf{r}'| \in \delta V, \\ 0 \quad \text{if } |\mathbf{r}-\mathbf{r}'|\to\infty. \end{cases} \qquad (4.11\text{-}20)$$

For $\delta V \to 0$ we write as usual $1/\delta V \to \delta(\mathbf{r}-\mathbf{r}')$. Thus, in the first order approximation we assume that there are no two-point correlations in the covariance, except for those which exist on dimensional grounds; therefore,

$$\Gamma_2^{pq} = \langle \Delta n^p(\mathbf{r}_1)\Delta n^q(\mathbf{r}_2)\rangle \to \delta^{pq}\langle n^p(\mathbf{r}_1)\rangle \delta(\mathbf{r}_1-\mathbf{r}_2). \qquad (4.11\text{-}21)$$

Or also from (4.11-17) and (4.11-21)

$$G_2^{pq}(\mathbf{r}_1,\mathbf{r}_2) \approx G_1^p(\mathbf{r}_1)G_1^q(\mathbf{r}_2) + \delta^{pq}\langle n^p(\mathbf{r}_1)\rangle\delta(\mathbf{r}_1-\mathbf{r}_2). \qquad (4.11\text{-}22)$$

Comparing with (4.11-12) this means that we implied the truncation

$$g_2^{pq}(\mathbf{r}_1,\mathbf{r}_2) \approx g_1^p(\mathbf{r}_1)g_1^q(\mathbf{r}_2), \qquad (4.11\text{-}23)$$

which is Boltzmann's assumption of molecular chaos. The dimensional singularities of the triple covariance function can easily be surmised $(p \neq q \neq r)$[57]

$$\Gamma_3^{pqr}(\mathbf{r}_1,\mathbf{r}_2,\mathbf{r}_3) \to 0, \qquad (4.11\text{-}24a)$$

$$\Gamma_3^{ppq} \to g_2^{pq}(\mathbf{r}_2,\mathbf{r}_3)\delta(\mathbf{r}_1-\mathbf{r}_2), \qquad (4.11\text{-}24b)$$

$$\Gamma_3^{ppp} \to \sum_{(ijk)} [g_2^{pp}(\mathbf{r}_j,\mathbf{r}_k) - g_1^p(\mathbf{r}_j)g_1^p(\mathbf{r}_k)]\delta(\mathbf{r}_i-\mathbf{r}_j) + g_1^p(\mathbf{r}_1)\delta(\mathbf{r}_1-\mathbf{r}_2)\delta(\mathbf{r}_1-\mathbf{r}_3). \qquad (4.11\text{-}24c)$$

---

[56] The definition of density *in a point* is already a mathematical abstraction and the density-density correlation even more so.

[57] The dimensional singularities obtained are the same as those in Eqs. (1.6-35) and (1.6-36) for $V\to\infty$.

Using (4.11-18) and comparing with Eqs. (4.11-13a,b,c) we find the second-order truncation rule (all p, q, r):

$$g_3^{pqr} \approx \sum_{(ijk)} g_1^{(p)}(\mathbf{r}_i) g_2^{(qr)}(\mathbf{r}_j, \mathbf{r}_k) - 2 g_1^p(\mathbf{r}_1) g_1^q(\mathbf{r}_2) g_1^r(\mathbf{r}_3) . \tag{4.11-25}$$

In the previous section we used the modified distribution function w_3^{pqr}. For large N^p and for globally homogeneous systems we have with $g_1^p = \langle n^p(\mathbf{r})\rangle = N^p/V$,

$$\begin{aligned} w_2^{pq}(\mathbf{r}_1,\mathbf{r}_2) &= \frac{V^2}{N^p N^q} g_2^{pq}(\mathbf{r}_1,\mathbf{r}_2) = g_2^{pq}(\mathbf{r}_1,\mathbf{r}_2)/g_1^p(\mathbf{r}_1) g_1^p(\mathbf{r}_2) , \\ w_3^{pqr}(\mathbf{r}_1,\mathbf{r}_2,\mathbf{r}_3) &= \frac{V^3}{N^p N^q N^r} g_3^{pqr}(\mathbf{r}_1,\mathbf{r}_2,\mathbf{r}_3) = g_3^{pqr}(\mathbf{r}_1,\mathbf{r}_2,\mathbf{r}_3)/g_1^p(\mathbf{r}_1) g_1^q(\mathbf{r}_2) g_1^r(\mathbf{r}_3) . \end{aligned} \tag{4.11-26}$$

Equation (4.11-25) now results in

$$w_3^{pqr} \approx w_2^{pq} + w_2^{pr} + w_2^{qr} - 2 = \omega_2^{pq} + \omega_2^{pr} + \omega_2^{qr} + 1 . \tag{4.11-27}$$

This is just the truncation rule used in the previous section. As it turns out, the further assumption $\omega_2 \ll 1$, made by Landau and Lifshitz, is unnecessary.

Finally we shall obtain g_2 and Γ_2 for the Coulomb gas explicitly. From (4.10-51), (4.10-49) and (4.11-26) we find

$$g_2^{pq}(\mathbf{r}_1,\mathbf{r}_2) = g_1^p(\mathbf{r}_1) g_1^p(\mathbf{r}_2) \left(1 - \frac{z^p z^q}{\Sigma_s n_0^s (z^s)^2} \frac{\kappa^2 e^{-\kappa|\mathbf{r}_1-\mathbf{r}_2|}}{4\pi|\mathbf{r}_1-\mathbf{r}_2|} \right) . \tag{4.11-28}$$

Using (4.11-12) and (4.11-17) we find for the covariance function

$$\langle \Delta n^p(\mathbf{r}_1) \Delta n^p(\mathbf{r}_2) \rangle = \delta^{pq} n_0^p(\mathbf{r}_1) \delta(\mathbf{r}_1 - \mathbf{r}_2) - \frac{n_0^p n_0^q z^p z^q}{\Sigma_s n_0^s (z^s)^2} \frac{\kappa^2 e^{-\kappa|\mathbf{r}_1-\mathbf{r}_2|}}{4\pi|\mathbf{r}_1-\mathbf{r}_2|} . \tag{4.11-29}$$

In particular, we consider a two-component plasma or a semiconductor with electrons $(n^p = n,\ z^p = -1)$ and holes $(n^q = p,\ z^q = 1)$. We denote further,

$$D(\mathbf{R}) = \kappa^2 e^{-\kappa R}/4\pi R . \tag{4.11-30}$$

We then obtain

$$\begin{aligned} \langle \Delta n(\mathbf{r}_1) \Delta n(\mathbf{r}_2) \rangle &= n_0(\mathbf{r}) \delta(\mathbf{r}_1 - \mathbf{r}_2) - \frac{n_0^2}{n_0 + p_0} D(\mathbf{r}_1 - \mathbf{r}_2) , \\ \langle \Delta p(\mathbf{r}_1) \Delta p(\mathbf{r}_2) \rangle &= p_0(\mathbf{r}) \delta(\mathbf{r}_1 - \mathbf{r}_2) - \frac{p_0^2}{n_0 + p_0} D(\mathbf{r}_1 - \mathbf{r}_2) , \end{aligned} \tag{4.11-31}$$

and for the cross correlation

$$\langle \Delta n(\mathbf{r}_1) \Delta p(\mathbf{r}_2) \rangle = \frac{n_0 p_0}{n_0 + p_0} D(\mathbf{r}_1 - \mathbf{r}_2) . \tag{4.11-32}$$

These results were first given in this form by Lax and Mengert.[58] An entirely different derivation, based on a transport equation for $\Gamma_2(\mathbf{r}_1,\mathbf{r}_2)$ developed by the author,[59] was given by us in 1971.[60]

We still consider the case that the interaction is of very short range. So let $\kappa \to \infty$. It is then easily shown that

$$D(\mathbf{R}) \to \delta(\mathbf{R}). \tag{4.11-33}$$

Physically the condition $\kappa \to \infty$ implies that besides charge neutrality for the system as a whole – *global* charge neutrality – we also have *local* neutrality $(\Sigma_s n_s(\mathbf{r}) z^s = 0)$. With (4.11-33), Eqs. (4.11-31) and (4.11-32) yield

$$\begin{aligned} \langle \Delta n(\mathbf{r}_1)\Delta n(\mathbf{r}_2)\rangle &= \langle \Delta p(\mathbf{r}_1)\Delta p(\mathbf{r}_2)\rangle \\ &= \langle \Delta n(\mathbf{r}_1)\Delta p(\mathbf{r}_2)\rangle = \frac{n_0 p_0}{n_0 + p_0}\delta(\mathbf{r}_1 - \mathbf{r}_2), \end{aligned} \tag{4.11-34}$$

a result well-known in the theory of carrier-density fluctuations in solids.

4.12 Problems to Chapter IV

4.1 For a quantum gas of independent particles the grand-canonical partition function was given by

$$\ln \mathcal{Z} = \mp \sum\nolimits_\alpha \ln\left[1 \mp e^{-(\varepsilon_\alpha - \zeta)/k_B T}\right], \tag{1}$$

with the upper (lower) signs applying to B–E (F–D) statistics.

(a) For the entropy show that

$$S = k_B \left(\frac{\partial (T \ln \mathcal{Z})}{\partial T}\right)_{\zeta/T}. \tag{2}$$

(b) Carry out the differentiation and establish by rearrangement the important result

$$S = -k_B \sum\nolimits_\alpha \left\{\langle n_\alpha\rangle \ln\langle n_\alpha\rangle \mp [1 \pm \langle n_\alpha\rangle] \ln[1 \pm \langle n_\alpha\rangle]\right\}. \tag{3}$$

4.2 (a) For systems with a constant number of particles, the connection between C_P and C_V is given by the well-known result

[58] M. Lax and P. Mengert, J. Phys. Chem. Solids **14**, 248 (1960).

[59] This equation is the 'lambda theorem', discussed in Section 19.6. For the original developments, see Chapter XIX, Ref. 20.

[60] K.M. Van Vliet, Physica status solidi **41**, K131 (1970).

$$C_P - C_V = -T\left(\frac{\partial V}{\partial T}\right)_P^2 \left(\frac{\partial P}{\partial V}\right)_T, \tag{4}$$

see e.g. Landau and Lifshitz, Op. Cit. §16. Express this in response functions (volume expansion coefficient α and compressibility K_T) and apply the result to a perfect gas.

(b) For a system with variable number of particles obtain a similar thermodynamic relationship between $C_{V,\varsigma/T}$ and $C_{V,N}$. Apply the result to a perfect gas and confirm Eq. (4.2-37).

(c) Apply Eq. (4) to a van der Waals gas.

4.3 In the 1D Ising model solve for the eigenvalues $\lambda_\pm$ and confirm Eq. (4.3-27); also confirm the result for the magnetization, Eq. (4.3-30). Finally, find the susceptibility χ_m near $H = 0$.

4.4 In Ornstein's method for a dense classical gas, assume that the configurational canonical partition function has been obtained, *i.e.*, start with Eq. (4.4-7).

(a) Establish *explicitly* the result for the critical temperature,

$$T_c = 4W_0/27k_B, \quad \text{where } W_0(\delta/\Delta) = \Sigma_{\alpha'} W_{\alpha\alpha'}. \tag{5}$$

(b) Compare Ornstein's form for van der Waals' equation with the standard form, Eq. (4.4-21) and identify the quantities a and b; set $n = 1$.

(c) Show that the critical temperature (inflexion point) is given by $T_c = 8a/27bR$ and show that this result is the same as obtained in part (a).

4.5 (a) Using Fourier integrals to solve the integral equation (4.4-31), show in detail that the result for the covariance function for a gas with two-body interactions in Ornstein's analysis is given by Eq. (4.4-39).

(b) Evaluate the integral (4.4-41) by contour integration, obtaining the result (4.4-44).

4.6 Employing integration by parts, show that the second virial coefficient can also be evaluated as

$$B(T) = -\frac{2\pi}{3k_B T}\int_0^\infty \left(\frac{d\varphi}{dr}\right) e^{-\varphi(r)/k_B T} r^3 dr. \tag{6}$$

(a) Let $\varphi(r) = \alpha r^{-n}, n > 3$. Establish that $B(T) = (2\pi/3)(\alpha/k_B T)^{3/n}\Gamma(1-3/n)$.

(b) Consider the Lennard-Jones or 'six-twelve potential'

$$\varphi(r)=\alpha\left[\left(\frac{r_0}{r}\right)^{12}-\left(\frac{r_0}{r}\right)^{6}\right]. \tag{7}$$

Sketch $\varphi(r)$. Obtain $B(T)$ and $P(T,\hat{v})$. Hint: Expand the term $\exp[\alpha\beta(r_0/r)^6]$ in a power series and use the same method as for (a) to obtain,

$$B(T)=-\frac{\pi}{6}r_0^3\sum_{n=0}^{\infty}\frac{1}{n!}(\alpha\beta)^{\frac{1}{4}(2n+1)}\Gamma\left(\frac{2n-1}{4}\right). \tag{8}$$

(c) Does the form for $P(T,\hat{v})$ resemble van der Waals' law? (Cf.[60a] for details.)

4.7 Consider hard-core interaction, with $\varphi(r)=\infty$, $r\le a$, and $\varphi(r)=0$ elsewhere, where ½ a is the radius of a molecule. Show that the second and third virial coefficients are[61]

$$B=\tfrac{2}{3}\pi a^3;\quad C=\tfrac{5}{8}B^2. \tag{9}$$

4.8 Consider the square well potential,

$$\varphi(r)=\begin{cases}\infty, & r<a,\\ -\phi, & a<r<b,\\ 0, & r>b.\end{cases} \tag{10}$$

(a) Sketch this potential and discuss its feasibility for real gases;
(b) Show that the second virial coefficient is given by

$$B(T)=\frac{2\pi a^3}{3}\left(1-(R^3-1)(e^{\beta\phi}-1)\right),\quad R=(b/a). \tag{11}$$

N.B. The third-order virial coefficient was obtained by Kihara.[62]

4.9 As stated in the text, one can also dispense with the procedure involving irreducible cluster integrals and simply *iterate* the equations for the pressure P [(4.7-9)] and the density n [(4.7-10)], eliminating z to successive orders. Employing this procedure, obtain the first three terms of the virial expansion,

$$P=\frac{k_BT}{\hat{v}}\left(1-\frac{b_2}{\hat{v}}+\frac{4b_2^2-2b_3}{\hat{v}^2}+\ldots\right). \tag{12}$$

[60a] A.E. Sherwood and J.M. Prausnitz, J. Chem. Phys. **41**, 413 and 429 (1964).

[61] For an elegant method see L.E. Reichl, 2nd Ed., Op. Cit., p.516-517.

[62] T. Kihara, Rev. Modern Phys. **25**, 831 ((1953); ibid. **27**, 412 (1955).

4.10 Show that the order parameter L_0 in the Bragg–Williams approximation has the temperature dependence

$$L_0 \approx \begin{cases} \sqrt{3(1-T/T_c)}\,, & T \lesssim T_c\,, \\ 1-2e^{-2T_c/T}\,, & T \gtrsim 0\,. \end{cases} \tag{13}$$

4.11 In the Landau–Ginzburg expansion (4.9-2) assume that for all temperatures in a range near T_c $a_2(T)>0$, $a_4(T)<0$ and $a_6(T)>0$. Show that a first-order phase transition takes place and obtain an expression for the latent heat of transition, $\Delta W = T\Delta S$. Hint. Above T_c $m=0$ and below T_c $m=m_0$, with m_0 determined by $(d\Phi/dm)|_{m_0}=0$, together with $\Phi(m_0)=\Phi(0)$, leading to $m_0=\sqrt{-3a_4(T_c)/4a_6(T_c)}$ and $a_2=3a_4^2/16a_6$. From the jump in order-parameter obtain $T\Delta S$.

4.12 Obtain the first three terms of the asymptotic expansion of the Gamma function, Eq. (4.6-17), cf. Morse and Feshbach I, p.442. N.B. Some printings of M–F have the wrong value for a_0.

4.13 Further to Ornstein's treatment for the pair correlation function of a classical gas, cf. Section 4.4.2, Ornstein and Zernike[63] assumed that this function could be split in a short-range part $C(r)$, called the *direct correlation function,* and a long-range part, $p(r)-C(r)$, whereby $C(r)$ satisfies the integral equation:

$$p(|\mathbf{r}_1-\mathbf{r}_2|)=C(|\mathbf{r}_1-\mathbf{r}_2|)+\langle n\rangle\int d^3r_3\, C(|\mathbf{r}_3-\mathbf{r}_2|)p(|\mathbf{r}_1-\mathbf{r}_3|). \tag{14}$$

With the convolution theorem for Fourier transforms, this leads for the structure factor as defined at the end of subsection 4.4.2 to

$$S(\mathbf{k})=\tilde{C}(\mathbf{k})+\langle n\rangle S(\mathbf{k})\tilde{C}(\mathbf{k})\,. \tag{15}$$

We now introduce the following expansion,

$$\exp[\beta\varphi(|\mathbf{r}_1-\mathbf{r}_2|)]w_2(|\mathbf{r}_1-\mathbf{r}_2|)=1+\sum\nolimits_{j=1}^{\infty} y_j(|\mathbf{r}_1-\mathbf{r}_2|)\langle n\rangle^j\,, \tag{16}$$

where the y_j can be obtained as cluster integrals similarly as in the text.
(a) From the *definition* of w_2 establish that the first order in the density term is given by

$$y_1(|\mathbf{r}_1-\mathbf{r}_2|)=\int d^3r_3 f(|\mathbf{r}_1-\mathbf{r}_3|)f(|\mathbf{r}_2-\mathbf{r}_3|)\,, \tag{17}$$

[63] L.S. Ornstein and F. Zernike, Versl. Koninkl. Ned. Akad. v. Wetenschappen **17**, 793 (1914); see also "The equilibrium Theory of Classical Fluids" (H.L. Frisch and J.L. Lebowitz, Eds.), W.A. Benjamin, NY 1964.

where $f(r)$ is the Mayer function. Noting further from (16) that

$$p(r) = f(r) + \sum_{j=1}^{\infty} y_j(r)[1 + f(r)]\langle n\rangle^j \,, \tag{18}$$

and setting $C(r) = \sum_{j=0}^{\infty} C_j(r)\langle n\rangle^j$, obtain $C_0(r)$ and $C_1(r)$. Also, show that to first order in the density

$$C(r) = [1 - \exp(\beta\varphi(r)] w_2(r) \,. \tag{19}$$

(b) For *arbitrary* density, replace $p(r)$ by $p_{PY}(r)$, substitute into the Ornstein–Zernike equation (14) and obtain the nonlinear Percus–Yevick integral equation:[64]

$$[p_{PY}(r_{12}) + 1]e^{\beta\varphi(r_{12})} = 1 + \langle n\rangle \int d^3r_3[1 - e^{\beta\varphi(r_{23})}][p_{PY}(r_{23}) + 1]\, p_{PY}(r_{13}) \,. \qquad \textbf{(20)}$$

N.B. The Percus–Yevick equation yields remarkably good results for the structure factor of many liquids, *employing a simple hard-sphere potential* for which it can be solved analytically. In Fig. 4-21(a) we reproduce $p(r)+1$, based on data by Alder and Hecht[65] and in Fig. 4-21(b) the structure factor $S(\mathbf{k})$, as computed by Verlet.[66]

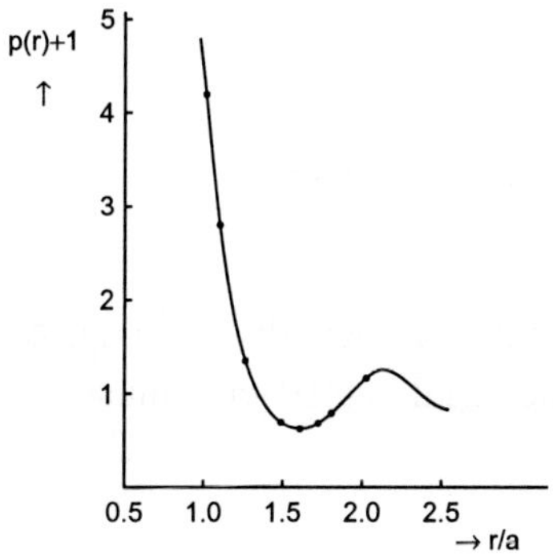

Fig. 4-21(a) Pair distribution function $p(r)$+1 based on computations from the PY Eq. with hard-sphere model, $\rho\pi a^3/6 = 0.463$ (dots); Solid line: based on molecular dynamics. Data for dots Alder and Hecht [65]. [With permission.]

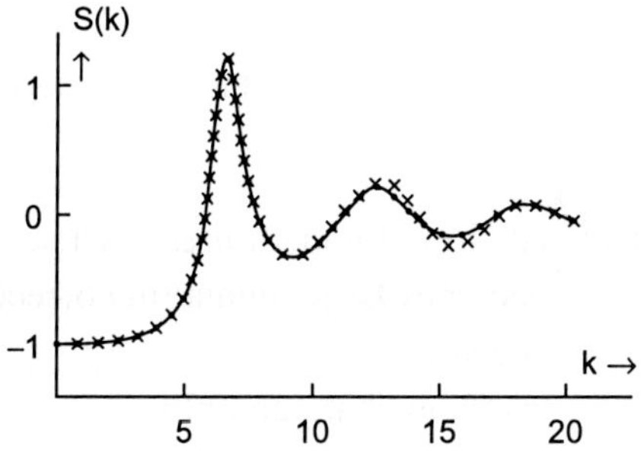

Fig. 4-21(b) Structure factor $S(\mathbf{k})$ vs. k. Solid line: PY Eq. with hard-sphere model, $T = 0.827K$, $\rho = 0.817$, $a = 1.03$ Å; crosses: neutron scattering data for ^{84}Kr, after Verlet [66]. [Reproduced by permission.]

4.14 The Gamma function has many integral representations besides the standard form. In particular, $\Gamma(z)$ can be analytically extended in the entire complex

---

[64] J.K. Percus and G.J. Yevick, Phys. Rev. **110**, 1 (1958); a clear presentation can be found in R. Balescu, "Equilibrium and Nonequilibrium Statistical Mechanics",Wiley & Sons, NY and London 1975.

[65] B.J. Alder and C.E. Hecht, J. Chemic. Physics **50**, 2032 (1969).

[66] L. Verlet, Phys. Rev. **165**, 201 (1968).

plane except at the poles $z = -1, -2, \ldots$

(a) Show that

$$\Gamma(z) = \frac{1}{2i \sin \pi z} \int_C d\zeta\, \zeta^{z-1} e^{\zeta}, \tag{21}$$

where the contour C is sketched below;[67]

(b) Replace z by $1-z$ and use the *mirror formula* $\Gamma(z)\Gamma(1-z) = \pi / \sin \pi z$, to obtain Hankel's form (in essence already given by Laplace in *Théorie Analytique des Probabilités*, 1812):

$$\frac{1}{\Gamma(z)} = \frac{1}{2\pi i} \int_C \frac{e^{\zeta}}{\zeta^z} d\zeta. \tag{22}$$

(c) Consider z to be real; expand the integrand logarithmically about the saddle point and confirm Stirling's formula.

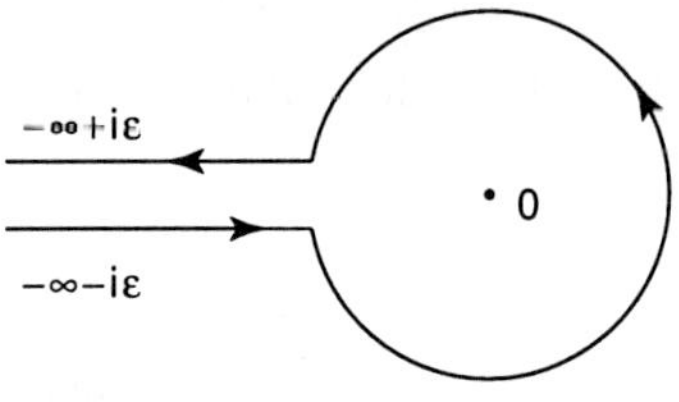

Fig. 4-22. Integration contour for problem 4.14.

4.15 At low temperatures, when the thermal wavelength $\lambda_{th}^{-1} = \sqrt{2\pi m k_B T / h^2}$ becomes large, quantum corrections to the virial coefficients must be taken into account.

(a) Show that generally

$$B(T) = B_0(T) - 2^{3/2} \lambda_{th}^3 \sum_i e^{-\beta \varepsilon_i}, \tag{23}$$

where $B_0(T)$ is the semi-classical virial coefficient and the sum is over all bound states and continuum states of the interaction potential;

(b) Assume that the wave equation for the interaction potential $\varphi(r)$ is solved by

$$\psi(r,\theta) = \sum_{\ell=0}^{\infty} \alpha_\ell (2\ell+1) P_\ell(\cos\theta) R_\ell(kr), \tag{24}$$

where $P_\ell(\cos\theta)$ are Legendre polynomials and $R_l(kr)$ satisfies the radial wave equation; for bosons ℓ must be even and for fermions ℓ must be odd. For the continuum states with $kr \gg 1$, show that

[67] E.T. Whittaker and G.N. Watson, "A Course of Modern Analysis", 4th Ed., Cambridge U. Press, 1958, p. 244ff.

$$R_\ell(kr) \sim (A_\ell / kr)\sin[kr - \tfrac{1}{2}\pi\ell + \delta_\ell(k)], \tag{25}$$

where $\delta_\ell(k)$ is the phase shift.

(c) Obtain the following final result for the second virial coefficient:

$$B(T) = B_0(T) - 2^{3/2}\lambda_{th}^3 \sum_{i,bound} e^{-\beta\varepsilon_i} - \frac{2^{3/2}\lambda_{th}^3}{\pi} \sum_{\substack{\ell\ even,\\ \text{or } odd}} (2\ell+1)\int_0^\infty dk \frac{d\delta_\ell}{dk} e^{-\beta\hbar^2 k^2/2\mu}, \tag{26}$$

where $\mu = \tfrac{1}{2}m$ is the reduced mass of two interacting bosons or fermions.

4.16 Compute the quantum correction to the second virial coefficient for hard-core bosons with radii r_1. Note that there are no bound state contributions for this case.

Chapter V

Generalized Canonical Ensembles

5.1 Formal Results

5.1.1 *The Generalized Ensemble Probability*

In this chapter we shall consider ensembles which are open with respect to the energy and to other extensive variables. The case that N is variable has already been considered in the grand-canonical ensemble. So here we have in mind ensembles with variable volume, charge, magnetization, etc. These ensembles are called *generalized canonical ensembles.* Few books pay attention to these ensembles. Exceptions are the older books of Guggenheim and of Münster, cited earlier but given once more here.[1] These ensembles are also considered in some papers by Callen and coworkers.[2] Some ensembles are really useful, like the pressure ensemble, in which $\mathscr{E}$ and V are variable, and the 'open' ensemble, in which $\mathscr{E}$, V and N are variable.

In order to derive the density operator for these ensembles, we shall not go back to the microcanonical ensemble as we did in Chapter IV; rather, we shall take the canonical ensemble as point of departure, adding the other reservoirs to the system to form a partitioned composite system. The physical set-up is sketched in Fig. 5-1.

The basic problem involving these ensembles lies in the specification of a suitable Hamiltonian. Our point of departure is akin to that of linear response theory – treated in the nonequilibrium part of this text. Thus, we shall consider the intensive variables of the system to be c-numbers, while the extensive variables are construed as quantum operators of the system. This is similar to our considerations for the grand-canonical ensemble, in which the chemical potential was a c-number, while the particle number is taken to be an operator which commutes with the Hamiltonian. Presently, however, the system operators which are allowed to vary generally do not commute with the Hamiltonian. For instance, in the charge-ensemble charge exchange involves the 'response Hamiltonian' $\mathbf{E}\cdot\Sigma_i(\mathscr{q}_i\delta\mathbf{q}_i)$, where the $\mathbf{q}_i$ are the position operators of the charged particles; for variable magnetization in a system of

---

[1] E. Guggenheim, "Statistical Thermodynamics", Cambridge University Press, 1953, p. 253ff. ; A. Münster, " Statistische Thermodynamik", Springer Verlag 1956.

[2] R.F. Greene, and H.B. Callen, Phys. Rev. **83**, 1231 (1951).

paramagnetic dipoles the corresponding term is $\mathbf{H}\cdot\Sigma_i\boldsymbol{\mu}_i$, where $\boldsymbol{\mu}_i$ are the magnetic (orbital or spin) moments. In case the volume is variable, it is not at all lucid where the operator V enters into the Hamiltonian, except for the obvious case of free particles in a box ("box quantization"). In other cases we must assume that the volume involves a term associated with the compressibility of the molecular volume through the interparticle interaction potential as considered before.[3] Notwithstanding these difficulties, we presume that by some adroit formalism a basis of eigenstates $\{|\eta\rangle\}$ of the Hamiltonian $\mathcal{H}$, with eigenvalues $\{\mathcal{E}_\eta\}$ can be constructed; the exchangeable extensive variables $A_1...A_m$ in a state $|\eta\rangle$ are characterized by their expectation values $A_{i\eta}\equiv\langle\eta|A_i|\eta\rangle$. We shall be content to find an explicit form for the diagonal part of the density operator, i.e., the ensemble probability $p_\eta(T^r,F_1^r...F_m^r)$, in which the super '*r*' stands for the reservoir values.

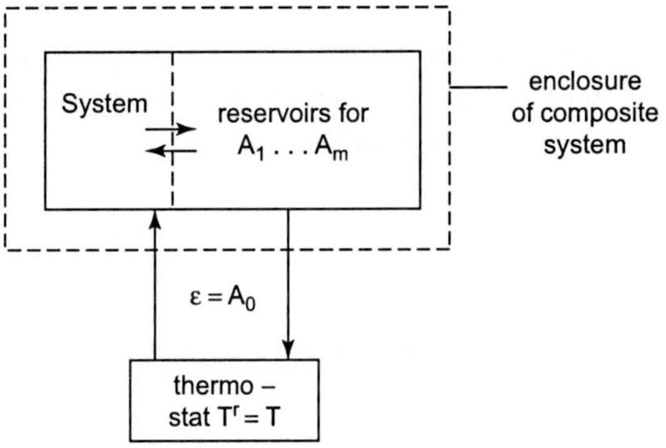

Fig. 5-1. Energy and other-variable exchange for a generalized canonical ensemble.

For the composite system, comprising the system and the reservoirs for $A_1...A_m$, in equilibrium with the heat bath of temperature $T^r\equiv T$, we have the canonical density operator

$$\rho_c = e^{(\Psi-\mathcal{H}_c)/k_BT}, \tag{5.1-1}$$

cf. Eq. (4.1-8). We take the trace over the reservoir states, which we label as $|\beta\rangle^{res}$, to obtain

$$p_\eta(T,F_1^r...F_m^r) = e^{\Psi/k_BT}e^{-\mathcal{E}_\eta(\{A_\eta\})/k_BT}\Sigma_{\beta res}e^{-\mathcal{E}_{\beta res}(\{A_{\beta res}\})/k_BT}. \tag{5.1-2}$$

Now Ψ is the free energy[4] of the composite system, $\mathcal{F}^c(T,V,...A_m)$, while $\Sigma_{\beta res}$ denotes the partition function of the reservoir, related to the nonequilibrium free energy function of the reservoir by $\exp[-\mathcal{F}^{res}(T,V^r...A_m^r)/k_BT]$. We thus obtain

---

[3] Be it noted here that in classical mechanics the volume constitutes a non-holonomic constraint, which can not be expressed by a simple differential form in the coordinates of the constituents.

[4] The free energy is denoted by script $\mathcal{F}$, to avoid confusion with the conjugate forces $\{F_i\}$.

$$p_\eta(T,F_1^r...F_m^r)=e^{[\mathcal{F}^c(T,...A_m)-\mathcal{F}^{res}(T,...A_m^r)]/k_BT}e^{-\mathcal{E}_\eta(\{A_\eta\})/k_BT}. \tag{5.1-3}$$

Next, let the reservoirs for the $A_1...A_m$ become very large; the intensive variables then assume fixed values. Since $\mathcal{F}$ is a function of state we can formally write

$$\begin{aligned}\mathcal{F}^c &= \mathcal{F}^c(T,A_1^r+A_1,...A_m^r+A_m;A_{m+1}^r+A_{m+1}...A_s^r+A_s)\\ &= \mathcal{F}^{res}(T,A_1^r+A_1,...A_m^r+A_m;A_{m+1}...A_s)+[\Delta\mathcal{F}(A_{m+1}...A_s)]_{T,A_1^r...A_m^r}.\end{aligned} \tag{5.1-4}$$

Noting that $\mathcal{F}$ is differentiable only with respect to exchangeable variables, we have

$$\begin{aligned}&\mathcal{F}^c(T,...A_m)-\mathcal{F}^{res}(T,...A_m^r)\\ &=\sum_{i=1}^m A_{i\eta}\left(\partial\mathcal{F}^{res}(T,...A_m^r)/\partial A_i^r\right)_{A_{m+1}^r...A_s^r}+[\Delta\mathcal{F}(A_{m+1}...A_s)]_{T,A_1^r...A_m^r}.\end{aligned} \tag{5.1-5}$$

The derivatives are of course the reservoir forces $F_i^r \to F_i$ (superscript 'r' dropped). The quantity $(\Delta\mathcal{F})_{A_1...A_m}$ will be called Φ (note that, if the A_i^r are kept constant, the A_i of the system are kept constant). The result is now

$$p_\eta(T,F_1...F_m)=e^{\Phi/k_BT}e^{-(\mathcal{E}_\eta+\sum_{i=1}^m F_i A_{i\eta}/k_BT)}. \tag{5.1-6}$$

The generalized canonical partition function will be denoted by Ξ. Clearly we have

$$\Phi=-k_BT\ln\Xi. \tag{5.1-7}$$

The ensemble probability p_η now takes the form,

$$p_\eta(T,F_1...F_m)=(1/\Xi)\exp\left(-[\mathcal{E}_\eta+\sum_{i=1}^m F_i A_{i\eta}]/k_BT\right). \tag{\textbf{5.1-8}}$$

In particular we have for the pressure ensemble, the 'open' ensemble and the magnetic ensemble – to mention a few – writing out the expectation values in detail,

$$\begin{aligned}p_\eta(T,P;N...)&=\Xi_p^{-1}\exp[-(\mathcal{E}_\eta+P\langle\eta|V|\eta\rangle)/k_BT],\\ p_\eta(T,P,\varsigma;...)&=\Xi_o^{-1}\exp[-(\mathcal{E}_\eta+P\langle\eta|V|\eta\rangle-\varsigma N_\eta)/k_BT],\\ p_\eta(T,H;V,N...)&=\Xi_m^{-1}\exp[-(\mathcal{E}_\eta-H\langle\eta|(\Sigma\mu)|\eta\rangle)/k_BT].\end{aligned} \tag{5.1-9}$$

At this point we shall also give the more common expression based on the entropic forces, conjugate to the exchangeable extensive variables, with $\mathcal{E}\equiv A_0$ and $Q_0=1/T$, $Q_i=F_i/T$, $(i=1...m)$. Eq. (5.1-8) then also reads (restoring the super 'r'):

$$p_\eta(T,Q_1^r...Q_m^r)=(1/\Xi)\exp\left(-\sum_{i=0}^m Q_i^r A_{i\eta}/k_B\right). \tag{\textbf{5.1-10}}$$

5.1.2 *Thermodynamic Functions*

To find the thermodynamic meaning of Ξ we will proceed somewhat differently than before. Suppose that at first there are no constraints, so that *all* A_i $(i = 0...s)$ are freely variable. We would then have had

$$\mathcal{F}^c - \mathcal{F}^{res} = -\sum_{i=1}^{s} TQ_i^r A_{i\eta} \,. \tag{5.1-11}$$

The resulting ensemble probability would be the *normalized* result,

$$p_\eta(T, Q_1^r ... Q_s^r) = e^{-\sum_{i=0}^{s} Q_i^r A_{i\eta} / k_B} \,. \tag{5.1-12}$$

We now impose constraints for $A_{m+1}...A_s$ *a posteriori* so that the corresponding forces in the system are 'frozen' to the equilibrium values, $Q_{m+1}^r ... Q_s^r \to Q_{m+1}^0 ... Q_s^0$. Hence we find

$$p_\eta(T, Q_1^r ... Q_m^r) = e^{-\sum_{i=0}^{m} Q_i^r A_{i\eta} / k_B - \sum_{i=m+1}^{s} Q_i^0 A_{i\eta}^0 / k_B} \,. \tag{5.1-13}$$

Employing the homogeneous form for the entropy we have

$$S(\{A_{i\eta}\}) = \underbrace{(\mathcal{E}_\eta / T) + \sum_{i=1}^{m} Q_i^0 A_{i\eta}^0}_{\sum_{i=0}^{m} Q_i^0 A_{i\eta}^0} + \sum_{i=m+1}^{s} Q_i^0 A_{i\eta}^0 . \tag{5.1-14}$$

We eliminate the last summand in (5.1-13) with the aid of (5.1-14); this then yields

$$p_\eta(T, Q_1^r ... Q_m^r) = e^{-[S(\{A_{i\eta}^0\}) - \sum_{i=0}^{m} Q_i^0 A_{i\eta}^0] / k_B} \, e^{-\sum_{i=0}^{m} Q_i^r A_{i\eta} / k_B} \,. \tag{5.1-15}$$

Comparing with (5.1-10) we obtained,

$$k_B \ln \Xi = S(\{A_{i\eta}^0\}) - \sum_{i=0}^{m} Q_i^0 A_{i\eta}^0 \equiv S_m(Q_0^0 ... Q_m^0; A_{m+1,\eta} ... A_{s\eta}), \tag{5.1-16}$$

where the last rhs denotes the generalized Massieu function, cf. subsection 1.3.1. There is therefore a great symmetry for all these ensembles: *The exponent of the distribution function in the customary form* (5.1-8) *contains the Legendre transform of the energy with respect to Gibbs' canonical ensemble; the resulting state function* $k_B \ln \Xi$ *is the Legendre transform or generalized Massieu function of the entropy.* The pertinent Massieu functions can also be replaced by energy Legendre transforms, in particular the free energies $\mathcal{F}$ and G, and the Gibbs function or grand potential Ω:

$$S_1 \equiv S_M(1/T; V, N, ...) = -T^{-1}\mathcal{F}(T; V, N...),$$
$$S_2 \equiv S_{Pl}(1/T, P/T; N, ...) = -T^{-1}G(T, P; N...), \tag{5.1-17}$$
$$S_3(1/T, P/T, -\varsigma/T; ...) = -T^{-1}\Omega(T, P, \varsigma; ...).$$

A survey of the various ensembles, following Guggenheim, Op. Cit., is given in Table 5-1.

Table 5-1. Survey of ensembles.

name	constraints	reservoirs	ensemble distr. (op.)	thermod. function
microcanonical	$\mathcal{E}, V, N, \{A_i\}$	none	$1/\Delta\Gamma(\mathcal{E}, V, N...)$	$S = k_B \ln \Delta\Gamma$
canonical	$V, N, \{A_i\}$	T	$\mathcal{Z}^{-1} e^{-\mathcal{H}/k_B T}$	$\mathcal{F} = -k_B T \ln \mathcal{Z}$
grand-canonical	$V, \{A_i\}$	T, ς	$\mathcal{Z}^{-1} e^{-(\mathcal{H}-\varsigma N)/k_B T}$	$K = -k_B T \ln \mathcal{Z}$
pressure ens.	$N, \{A_i\}$	T, P	$\Xi_p^{-1} e^{-(\mathcal{E}_\eta + P\langle\eta\|V\|\eta\rangle)/k_B T}$	$G = -k_B T \ln \Xi_p$
open ensemble	none	$T, P, \varsigma, \{F_i\}$	$e^{-(\mathcal{E}_\eta - \varsigma N_\eta + \Sigma F_i \langle\eta\|A_i\|\eta\rangle)/k_B T}$	- - -

5.1.3 *The Macroscopic Thermodynamic Distribution*

We will now find an expression for the probability density function $W(A_0...A_m)$ under the conditions of the ensemble, analogous to the Boltzmann-Einstein result of Section (3.7). We multiply (5.1-8) with the number of accessible quantum states $\Delta\Gamma(A_{0\eta}, A_{1\eta}, ..., A_{m\eta})$. Hence we have, writing $F_i \to TQ_i^r$

$$W(A_0...A_m) = (1/\Xi)\,\chi(A_{0\eta}, A_{1\eta}, ..., A_{m\eta}) \times \exp\left(-[(\mathcal{E}_\eta/T) + \sum_{i=1}^{m} Q_i^r A_{i\eta}]/k_B\right), \tag{5.1-18}$$

With χ we associate a nonequilibrium entropy function $S(A_{0\eta}...A_{m\eta})$:

$$\chi(A_{0\eta}, A_{1\eta}, ..., A_{m\eta}) = \hat{c}\exp[S(A_{0\eta}...A_{m\eta})/k_B], \tag{5.1-19}$$

where $\hat{c}$ is a constant of order ln N introduced for dimensional consistency. Also, we rewrite $(1/\Xi)$ as, using (5.1-16)

$$(1/\Xi) = \exp\left(-[S^0 - \sum_{i=0}^{m} Q_i^r A_i^0]/k_B\right), \tag{5.1-20}$$

where we noted that the equilibrium thermodynamic forces Q^0 equal the reservoir values Q^r and where we omitted the subs η since we consider the variables A_i to be *reasonably coarse grained* (see also the argument in the next section). We now combine (5.1-18) – (5.1-20), to obtain

$$W(A_0...A_m) = \hat{c}\exp\left(\left[S(A_0...A_m) - S^0 - \sum_{i=0}^{m} Q_i^r (A_i - A_i^0)\right]/k_B\right). \tag{5.1-21}$$

This is the desired result for the thermodynamic form of the probability density for the A_i. We shall use it later.

5.2 Transformation Theory Using the Fowler Generating Function

We give once more the ensemble probability for a system with reservoirs for the extensive variables $A_1...A_m$ to be in the state $|\eta\rangle$; the $\{A_i\}$ may concern the volume, the particle number of a given species, the total charge of the constituents, etc. We found:

$$p_\eta = \Xi^{-1} e^{-(\mathcal{E}_\eta + \sum_{i=1}^{m} F_i A_{i\eta})/k_B T}. \qquad [\textit{see } (5.1\text{-}8)]$$

The corresponding density operator, $\rho(T,\{F_i\}) = \Sigma_\eta |\eta\rangle p_\eta \langle\eta|$ can generally not be obtained in closed form since the operators $\{A_i\}$ do not commute with the Hamiltonian; in particular, the often purported form $\rho \propto \exp[-(\mathcal{H} + \Sigma F_i A_i)/k_B T]$ *is incorrect*, although this expression can serve to obtain a classical equivalent $\rho(p,q,T,\{F_i\})$, see Problem 5.1. The partition function Ξ is available, however, as a 'Zustandsomme' – sum over states:

$$\Xi = \sum_{\{\eta\}} e^{-(\mathcal{E}_\eta + \Sigma_i F_i A_{i\eta})/k_B T}. \tag{5.2-1}$$

However, we must bear in mind that the counting of the states for infinite domains of the $\{A_i\}$ cannot be achieved in any precise way, since no state space comprising all operators can be surmised. So we resort to coarse-graining, i.e., we define energy cells $\delta\mathcal{E}^J$, within which the A_i and $\mathcal{H}$ quasi-commute, similarly as in Section 1.5. Then for each cell we can define the accessible number of quantum states within $(\mathcal{E}^J, \mathcal{E}^J + \delta\mathcal{E}^J)$, denoted as usual by $\Delta\Gamma(\mathcal{E}^J, A_1,...A_m)$. Again, let $\mathcal{E}^J = A_0$ and let the generalized fugacities be defined as

$$y_i = e^{-F_i/k_BT}; \text{ or}$$
$$y_0 = e^{-1/k_BT},\ y_1 = e^{-P/k_BT},\ y_2 = e^{\varsigma/k_BT},\ etc. \tag{5.2-2}$$

The partition function can then be evaluated as

$$\Xi(y_0, y_1, \ldots y_m) = \sum_J \sum_{A_1,\ldots A_m} \left(\prod_{j=0}^{m} y_j^{A_j} \right) \Delta\Gamma(\mathscr{E}^J, A_1, \ldots A_m)\,. \tag{5.2-3}$$

We note: *The generalized canonical partition function, when expressed in the fugacities, is just the multiple generating function according to Fowler of the partition function in the microcanonical ensemble. The same applies with respect to the relationship between* $\Xi_{(m)}$ *and a less open ensemble with partition function* $\Xi_{(m-\ell)}$.

As an example we give the relations between the grand-canonical ensemble and the microcanonical ensemble and between the grand-canonical ensemble and the canonical ensemble [see also Eqs. (4.2-13) and (4.2-14)]:

$$\mathscr{Z}(y_0, y_2; V, ..) = \sum_{\mathscr{E}} \sum_N y_0^{\mathscr{E}} y_2^N \Delta\Gamma(\mathscr{E}, N; V, A_3 \ldots)\,,$$
$$\mathscr{Z}(T, y_2; V, \ldots) = \sum_N y_2^N \mathscr{Q}(T, N; V, A_3 \ldots)\,. \tag{5.2-4}$$

Thus, in statistical mechanics it is customary to evaluate the partition function for an ensemble that is sufficiently open, such that constraints imposed on the system can be circumvented – as we did for the quantum distributions of a perfect gas in subsection 4.3.1. We can then, if required by the problems at hand, go back to a lower ensemble, using the inverse Fowler transforms discussed in section 1.7.3. We give below the inverses for Eqs. (5.2-3) and (5.2-4), in which the variables $\{y_i\}$ have been extended to the variables $\{\zeta_i\}$ in the complex plane:

$$\Delta\Gamma(\mathscr{E}^J, A_1, \ldots A_m) = \left(\frac{1}{2\pi i}\right)^{m+1} \oint \ldots \oint \frac{\Xi_{(m)}(\zeta_0, \zeta_1, \ldots \zeta_m)}{\prod_{j=0..m} \zeta_j^{A_j+1}} \prod_j d\zeta_j\,. \tag{5.2-5}$$

$$\Delta\Gamma(\mathscr{E}, N; V, A_3 \ldots) = \left(\frac{1}{2\pi i}\right)^2 \oint\oint \frac{\mathscr{Z}(\zeta_0, \zeta_2; V, ..)}{\zeta_0^{\mathscr{E}+1} \zeta_2^{N+1}} d\zeta_0 d\zeta_2\,, \tag{5.2-6}$$

$$\mathscr{Q}(T, N; V, A_3 \ldots) = \left(\frac{1}{2\pi i}\right) \oint \frac{\mathscr{Z}(T, \zeta_2; V, \ldots)}{\zeta_2^{N+1}} d\zeta_2\,. \tag{5.2-7}$$

In many textbooks it is shown that the thermodynamic relationships are the same, no matter what ensemble has been used; i.e., the first order moments, $\langle A_i \rangle$ are the same in each ensemble. However, this is no longer true for higher order moments. Thus, relations involving fluctuations, variances and covariances, as well as higher order correlations, are subject to the particular constraints imposed. We will give examples for carrier density fluctuations in charge-neutral solids in Section 5.4. It is then imperative to evaluate the inversions given above. Usually this requires little computation, except judicious application of the method of steepest descent.

Before going into the problem of fluctuations in a 'lower' (less open) ensemble, in the next section we shall develop some general rules for the fluctuations in the generalized canonical ensembles at hand. These developments follow Gibbs' original ideas, statements in Fowler's book and in the cited paper by Greene and Callen.

5.3 Fluctuations: General Results

We consider once more a generalized canonical ensemble with reservoirs for $A_0 = \mathscr{E}, A_1, ... A_m$ and constraints for $A_{m+1}, ... A_s$. We are interested in some general relationships for the first order moments $\langle A_i \rangle$ and $\langle F_i \rangle$ and for the higher order moments $\langle A_i A_j \rangle$, $\langle A_i A_j A_k \rangle$, as well as the fluctuations $\langle \Delta A_i \Delta A_j \rangle$, $\langle \Delta F_i \Delta F_j \rangle$, *etc. under the conditions of the ensemble.* The basic 'trick' to obtain these moments stems from Gibbs: one either differentiates the normalization integral or the partition function to the various parameters which define the ensemble, viz., $T, F_1^r ... F_m^r$ and $A_{m+1} ... A_s$. There are, however, some problems with this method, foreseen already by Gibbs, since in some cases we obtain irrelevant or wrong results due to the non-analyticity of some derivatives, such as the pressure at the point of contact of a rigid wall. We shall discuss these problems and their remedy, based on thermodynamic arguments. We will start, however, with the statistical mechanical description, using the Hamiltonian and energy-derived forces F_i.

5.3.1 *Extensive Variables*

We repeat the expression for the partition function, Eq. (5.2-1):

$$\Xi = \sum_{\{\eta\}} \exp[-(\mathscr{E}_\eta + \sum_{i=1}^{m} F_i^r A_{i\eta}) / k_B T]. \qquad [\textit{see}\ (5.2\text{-}1)]$$

We differentiate $\ln \Xi$ to F_i^r, i.e., the ensemble undergoes a virtual change to one with slightly different reservoir parameters. Then we have for $i = 1...m$,

$$-k_BT\left(\frac{\partial \ln \Xi}{\partial F_i^r}\right)_{T,\,other\,F's}=\frac{1}{\Xi}\sum_{\{\eta\}}A_{i\eta}\exp[-(\mathcal{E}_\eta+\sum_{i=1}^{m}F_i^rA_{i\eta})/k_BT]=\langle A_i\rangle. \tag{5.3-1}$$

Likewise:

$$\begin{aligned} k_BT^2\left(\frac{\partial \ln \Xi}{\partial T}\right)_{F_1\ldots F_m} &=\frac{1}{\Xi}\sum_{\{\eta\}}(\mathcal{E}_\eta+\Sigma_iF_i^rA_{i\eta})\exp[-(\mathcal{E}_\eta+\sum_{i=1}^{m}F_i^rA_{i\eta})/k_BT] \\ &=\langle\mathcal{E}\rangle+\sum_i F_i^r\langle A_i\rangle, \end{aligned} \tag{5.3-2}$$

from which,

$$\langle\mathcal{E}\rangle=k_BT^2\left(\frac{\partial \ln \Xi}{\partial T}\right)_{F_1\ldots F_m}+k_BT\sum_{i=1}^{m}F_i^r\left(\frac{\partial \ln \Xi}{\partial F_i^r}\right)_{T,\,other\,F's}. \tag{5.3-3}$$

Differentiating (5.2-1) once more, we find with the usual algebra the second-order cumulant,

$$\langle\Delta A_i\Delta A_j\rangle=(k_BT)^2\left(\frac{\partial^2 \ln \Xi}{\partial F_i^r\partial F_j^r}\right)_{T,\,other\,F's}\qquad (i,j=1\ldots m). \tag{5.3-4}$$

An alternative result is

$$\langle\Delta A_i\Delta A_j\rangle=-k_BT\left(\frac{\partial\langle A_i\rangle}{\partial F_j^r}\right)_{T,F^r\neq F_j^r}=-k_BT\left(\frac{\partial\langle A_j\rangle}{\partial F_i^r}\right)_{T,F^r\neq F_i^r}. \tag{5.3-5}$$

By differentiating (5.3-2) once more to T or to F_i^r, similar results can be obtained for $\langle\Delta\mathcal{E}^2\rangle$ and for $\langle\Delta\mathcal{E}\Delta A_i\rangle$.

The lack of symmetry is caused by the canonical formalism. As we have seen, more properly Ξ is a function of the thermodynamic Q-forces, $1/T$ and F_i^r/T, or the fugacities $y_0,\ldots y_m$. Thus, the following, more symmetric results are easily verified:

$$\begin{aligned} \langle\Delta A_i\Delta A_j\rangle &=-k_B\left(\frac{\partial\langle A_i\rangle}{\partial(F_j^r/T)}\right)_{T,F^r\neq F_j^r}=-k_B\left(\frac{\partial\langle A_j\rangle}{\partial(F_i^r/T)}\right)_{T,F^r\neq F_i^r}, \\ \langle\Delta A_i\Delta\mathcal{E}\rangle &=-k_B\left(\frac{\partial\langle A_i\rangle}{\partial(1/T)}\right)_{F^r/T}=-k_B\left(\frac{\partial\langle\mathcal{E}\rangle}{\partial(F_i^r/T)}\right)_{T,F^r\neq F_i^r}, \\ \langle\Delta\mathcal{E}^2\rangle &=-k_B\left(\frac{\partial\langle\mathcal{E}\rangle}{\partial(1/T)}\right)_{F^r/T}=k_BT^2\,n\,C_{F_1^r/T\ldots F_m^r/T}. \end{aligned} \tag{5.3-6}$$

(N.B. "F^r" means all $\{F_i^r\}$). The quantity C is the molar heat capacity under the specified conditions and n is the number of moles in the system. E.g., for an ideal gas in the pressure ensemble, $\langle\mathcal{E}\rangle = \langle W\rangle = \tfrac{3}{2}RT+PV=\tfrac{5}{2}RT$, so that $C_{P/T}= C_P=\tfrac{5}{2}R$.

Another result for the covariances of the extensive fluctuating variables will also be derived. We write (using overhead bars rather than angular brackets), cf., Ref 2,

$$\left(\frac{\partial \bar{A}_i}{\partial F_j^r}\right)_{T,F^r\neq F_j^r} = \frac{\partial(\bar{A}_i, F_1^r,\ldots,F_{j-1}^r,F_{j+1}^r,\ldots,F_m^r)}{\partial(F_j^r, F_1^r,\ldots,F_{j-1}^r,F_{j+1}^r,\ldots,F_m^r)}\Bigg|_T = \frac{\partial(F_1^r,\ldots,F_{j-1}^r,\bar{A}_i,F_{j+1}^r,\ldots,F_m^r)}{\partial(F_1^r,\ldots,F_{j-1}^r,F_j^r,F_{j+1}^r,\ldots,F_m^r)}\Bigg|_T$$

$$= \frac{\partial(\bar{A}_1,\ldots,\bar{A}_m)}{\partial(F_1^r,\ldots,F_m^r)}\Bigg|_T \bullet \frac{\partial(F_1^r,\ldots,F_{j-1}^r,\bar{A}_i,F_{j+1}^r,\ldots,F_m^r)}{\partial(\bar{A}_1,\ldots,\bar{A}_{j-1},\bar{A}_j,\bar{A}_{j+1},\ldots,\bar{A}_m)}\Bigg|_T . \tag{5.3-7}$$

We showed previously (equilibrium conditions):

$$F_i^r = \bar{F}_i = -\left(\frac{\partial\mathcal{E}}{\partial A_i}\right)_{S,\,A_i=\bar{A}_i} = -\left(\frac{\partial\mathcal{F}}{\partial A_i}\right)_{A_i=\bar{A}_i} = -\left(\frac{\partial\mathcal{F}^0}{\partial\bar{A}_i}\right), \tag{5.3-8}$$

where $\mathcal{F}^0$ is the equilibrium free energy, being a function of the equilibrium values $\{\bar{A}_i\}$. The first Jacobian in (5.3-7) is therefore $(-1)^m / |\partial^2\mathcal{F}^0/\partial\bar{A}_i\partial\bar{A}_j|$. The second Jacobian is easily shown to be $(-1)^{m-1+i+j}M_{ij}$, where M_{ij} is the minor of the element $\partial^2\mathcal{F}^0/\partial\bar{A}_i\partial\bar{A}_j$ in the matrix $\left(\partial^2\mathcal{F}^0/\partial\bar{A}\partial\bar{A}\right)$. We thus arrive at

$$\langle\Delta A_i\Delta A_j\rangle = k_B T\left[\frac{\partial^2\mathcal{F}^0}{\partial\bar{A}\partial\bar{A}}\right]^{-1}_{ij}, \tag{5.3-9}$$

where the elements on the rhs involve the reciprocal matrix. We alluded earlier to this result [see Eq. (3.7-11)], if the a_i of that treatment become the fluctuating A_i of the present ensemble.[5]

5.3.2 *Intensive Variables*

The other fixed variables in the ensemble are $A_{m+1}\ldots A_s$. The corresponding intensive variables are subject to fluctuations. The new 'trick' is to differentiate the partition function to the parameters $A_i, i = m+1\ldots s$. We find

$$-k_B T\frac{\partial\ln\Xi}{\partial A_p} = \langle\frac{\partial\mathcal{E}}{\partial A_p}\rangle = -\bar{F}_p, \quad p = m+1\ldots s, \tag{5.3-10}$$

which is simply a confirmation of the definition of generalized force in the energy description, cf. Eq. (1.3-14). Differentiating once more, we arrive at

---

[5] Apparently the result (5.3-9) is *not* dependent on the Gaussian approximation to Eq. (3.7-11); this is important for the theory of critical fluctuations, cf., M. Klein and L. Tisza, Phys. Rev. **76**, 1861 (1949).

$$\begin{aligned}\overline{\Delta F_p \Delta F_q} &= (k_B T)^2 \frac{\partial^2 \ln \Xi}{\partial A_p \partial A_q} + k_B T \overline{\frac{\partial^2 \mathcal{E}}{\partial A_p \partial A_q}} \\ &= k_B T \frac{\partial \bar{F}_p}{\partial A_q} - k_B T \overline{\frac{\partial F_p}{\partial A_q}} \approx 0. \quad (p,q = m+1 \ldots s)\end{aligned} \tag{5.3-11}$$

Gibbs attributed the failure of this result to the fact that F_p varies very rapidly, so that $\partial F_p / \partial A_q$ is not a meaningful derivative. The situation will be remedied by incorporating also a time average over times that involve a large number of collisions; we shall denote the double average by $\tilde{F}$. According to ergodic theory, this average is equal to the average in a microcanonical ensemble. The idea is therefore to find the equations of state for a microcanonical ensemble, and subsequently to consider fluctuations if $\mathcal{E}, V, N$, *etc.* are left to vary. Thus, only the generalized forces Q_i of the microcanonical ensemble are suitable forces to do the job. We set therefore,

$$\tilde{F}_i = T Q_i = \frac{\partial S / \partial A_i}{\partial S / \partial \mathcal{E}}. \tag{5.3-12}$$

Clearly, then, the Q_i are coarser than the F_i and we can expect the derivatives $\partial Q_i / \partial A_i$ to be meaningful.

Suppose we have found the equations of state, i.e., the rhs of (5.3-12) is written as

$$\tilde{F}_i = f_i(\mathcal{E} = A_0, A_1 \ldots A_s). \tag{5.3-13}$$

The fluctuations are now – noting that $A_{m+1} \ldots A_s$ are held constant,

$$\langle \Delta \tilde{F}_i \Delta \tilde{F}_j \rangle = \sum_{k,l=0}^{m} \frac{\partial f_i}{\partial A_k} \frac{\partial f_j}{\partial A_l} \langle \Delta A_k \Delta A_l \rangle, \quad (i,j = 1..s). \tag{(5.3-14)}$$

To obtain a definitive result, we will rewrite the ensemble distribution for $W(A_1 \ldots A_m)$, Eq. (5.1-21), in the more precise form[6,7]

$$\begin{aligned}&W(A_0 \ldots A_m \,|\, Q_0^r \ldots Q_m^r; A_{m+1} \ldots A_s) \\ &\quad = \hat{c} \exp\{[S(A_0 \ldots A_m; A_{m+1} \ldots A_s) - S^0(\{A_i^0\}) - \textstyle\sum_{i=0}^{m} Q_i^r (A_i - A_i^0)]/k_B\}.\end{aligned} \tag{5.3-15}$$

We differentiate $\ln W$ to A_i $(i = 0 \ldots m)$:

$$\frac{\partial \ln W}{\partial A_i} = \frac{1}{k_B}\left(\frac{\partial S}{\partial A_i} - Q_i^r\right) = \frac{1}{k_B} \Delta Q_i. \tag{5.3-16}$$

---

[6] The vertical bar in W means that this is a *conditional probability*, subject to the specification to the right of the bar, which states the conditions of the ensemble.

[7] K.M. Van Vliet, Comptes Rendus du 7e Congrès Intern. sur la physique des semiconducteurs, Dunod, Paris 1964, p.831.

Hence,

$$\langle \Delta A_i \Delta Q_j \rangle = k_B \int ... \int \Pi_k dA_k \, \Delta A_i (\partial \ln W / \partial A_j) W$$
$$= k_B \int .. \int dA_0 ... dA_{j-1} dA_{j+1} \int dA_j \Delta A_i (\partial W / \partial A_j) . \tag{5.3-17}$$

Integrating the latter integral by parts, we obtain

$$\langle \Delta A_i \Delta Q_j \rangle = -k_B \delta_{ij} . \tag{5.3-18}$$

Whence finally,

$$\langle \Delta Q_i \Delta Q_j \rangle = \sum_{k=0}^{m} \langle \Delta Q_i \Delta A_k \rangle \left(\frac{\partial Q_j}{\partial A_k} \right)_{eq} = -k_B \left(\frac{\partial Q_j^r}{\partial \bar{A}_i} \right), \tag{5.3-19}$$

where we wrote the equilibrium derivatives in terms of the reservoir forces. In particular, we have for the temperature fluctuations

$$\langle \Delta (1/T)^2 \rangle = -k_B \partial (1/T) / \partial \mathscr{E}, \quad \text{or} \quad \langle \Delta T^2 \rangle = k_B T^2 / \mathcal{n} C, \tag{5.3-20}$$

where C is the molar heat capacity. From (5.3-12) this now yields a form for the covariances of the time-averaged energy-derived forces; with some algebra we obtain,

$$\langle \Delta \tilde{F}_i \Delta \tilde{F}_j \rangle = -k_B T^2 \left(\frac{\partial (F_j^r / T)}{\partial \bar{A}_i} \right) + k_B T^2 F_j^r \frac{\partial (F_i^r / T)}{\partial \mathscr{E}} + k_B T^2 F_i^r \frac{\partial (F_j^r / T)}{\partial \mathscr{E}} + k_B F_i^r F_j^r \frac{\partial T}{\partial \mathscr{E}} . \tag{5.3-21}$$

This expression is dimension-wise as (5.3-11), but yields realistic results, as we will show below.

5.3.3 *Examples and Conclusions*

Although some formulas are of practical importance, it is clear that the formalism is rather strained. First of all, the canonical approach singles out the energy as the basic variable, whereas, secondly, the information is too fine-grained. Even a fluctuation contains a large number of microscopic events; variables as e.g. $\Delta\mathscr{E}$ and ΔP are basically macroscopic (or at least "mesoscopic", see section 1.5). This is borne out by the fact that such fluctuations – commonly called noise – can be measured with sufficiently sensitive instruments. Noise will be described in detail in part F, dealing with stochastic processes and their spectral resolution. So here we just give just a few examples, all of them pertaining to *equilibrium* fluctuations.

Consider a capacitor in equilibrium with a charge reservoir (a large metal electrode will do). Then from (5.3-6) for the charge fluctuations:

$$\langle \Delta Q^2 \rangle = -k_B T \partial \langle Q \rangle / \partial(-\Phi^r) = k_B T C_{el}, \tag{5.3-22}$$

where C_{el} is the electrical capacitance; this result is in accord with our earlier result (1.6-67), obtained from the central limit theorem.

Electron fluctuations in an electron gas, in contact with an infinite reservoir, are also directly obtained:

$$\langle \Delta N_i^2 \rangle = -k_B T \, \partial \langle N_i \rangle / \partial \mu_i^r, \tag{5.3-23}$$

where 'i' refers to the species or band index, etc. [We note that this requires differentiation to the electro-chemical potential, whereas in subsection 4.3.1 we required a virtual variation of the energy levels.] Consider e.g. the number of electrons per cm^3 in a parabolic band. Then we have

$$\begin{gathered} \langle N \rangle = \mathcal{N}_c \mathcal{F}_{1/2}(\mu / kT); \quad \langle \Delta N^2 \rangle = \xi \langle N \rangle, \\ \xi = \mathcal{F}_{-1/2}(\mu / kT) / \mathcal{F}_{1/2}(\mu / kT), \end{gathered} \tag{5.3-24}$$

where $\mathcal{N}_c$ is the density of states, $\mu \, (= \mu^r)$ is measured with respect to the bottom of the band, and $\mathcal{F}_k$ is a Fermi integral (see footnote 15).

Next, we consider black-body radiation. Let the mode density be $Z(\nu) = 8\pi\nu^2/c^3$ and the photon density

$$q(\nu) = (8\pi\nu^2 / c^3)/(e^{h\nu / k_B T} - 1). \tag{5.3-25}$$

The energy in a volume V and in a frequency interval $d\nu$ is $\mathcal{E} = h\nu q(\nu) V d\nu \equiv Nh\nu$. For the variance we have

$$\langle \Delta N^2 \rangle = -(1/h\nu)^2 \partial \mathcal{E} / \partial(1/k_B T). \tag{5.3-26}$$

Hence,

$$\langle \Delta N^2 \rangle = \langle N \rangle [1 + \langle N \rangle / ZVd\nu] = \langle N \rangle (1 + B), \tag{5.3-27}$$

where B is the boson factor

$$B = 1/(e^{h\nu / k_B T} - 1). \tag{5.3-28}$$

This result was first obtained by Einstein.[8]

Finally, we consider an example of fluctuations of an intensive variable, viz., pressure fluctuations in a perfect mono-atomic gas. They can be measured , in principle, with a low-noise transducer and amplifying equipment. For an ensemble, open with respect to V, we have from (5.3-21) ($\overline{P} = P^r$):

$$\langle \Delta P^2 \rangle = -k_B T^2 \frac{\partial(\overline{P}/T)}{\partial V} + 2k_B T^2 \frac{\partial(\overline{P}/T)}{\partial \mathcal{E}} + k_B \overline{P}^2 \frac{\partial T}{\partial \mathcal{E}}. \tag{5.3-29}$$

[8] A. Einstein, Berliner Berichte p. 261 (1924); ib. p.3. (1925).

Now $\overline{P}/T = k_B N/V$, $T = 2\mathcal{E}/5k_B N$. These are the equations of state. We find,

$$\frac{\langle \Delta P^2 \rangle}{\overline{P}^2} = \frac{7}{5N}. \tag{5.3-30}$$

One may now repeat this for a canonical ensemble, closed with respect to *V*. According to (5.3-14) $\langle \Delta P^2 \rangle = (\partial P/\partial \mathcal{E})^2 \langle \Delta \mathcal{E}^2 \rangle$. This yields $\langle \Delta P^2 \rangle / \overline{P}^2 = 2/(3N)$ [see e.g., ter Haar[9]], which is substantially smaller.

5.4 Carrier-Density Fluctuations in a Solid

5.4.1 *Microscopic Occupancy Fluctuations*

In a solid there is usually total charge neutrality. If the Debye screening length $\to$ 0, we have also *local* charge neutrality as we saw in Section 4.9. Therefore, in any volume δV the number of charge carriers, distributed microscopically over all available quantum states, or, macroscopically distributed over the relevant electron bands and impurity levels, is subject to a particle constraint. We must therefore consider the statistics for electron gases in a canonical ensemble, with fixed total particle number. In subsection 4.3.1 we presented the usual derivation for Fermi-Dirac statistics via the grand-canonical ensemble. This result is *not* applicable to fluctuations in the electron gas in the conduction band (or hole gas in the valence band) of solids. However, we can readily obtain the canonical partition function from the grand-canonical partition function with the results of transformation theory. The necessary connection was given by the second line of Eq. (5.2-4) .

Let the one-electron states be indicated by Greek indices, as in subsection 4.3.1: $\{|\alpha)\}$ denotes a basis of states, ε_α is a one-particle energy and each state can have an occupancy $n_\alpha = 0$ or 1. [Spin, when relevant, is included in the designation α.] The grand-canonical partition function was found to be, cf. (4.3-14):

$$\mathcal{Z} = \prod_\alpha [1 + e^{-(\varepsilon_\alpha - \mu^r)/k_B T}], \tag{5.4-1}$$

where we used the electro-chemical potential μ^r since the particles (electrons or holes) are charged.[10] The particle constraint is

$$\sum_\alpha n_\alpha \varepsilon_\alpha = N, \tag{5.4-2}$$

[9] D. ter Haar, Op. Cit., p.133, Eq. (5.912).

[10] When dealing with holes, like in the valence band of a semiconductor, we will still primarily consider the electron occupancies of the available states, to maintain the universal particle constraint (5.4-2) above. The hole occupancy is $p_\alpha = 1 - n_\alpha$ and for macroscopic occupancies, discussed later, it suffices that $\Delta P_i = -\Delta N_i$.

where N is the total number of electrons in the volume considered, occupying all "relevant" bands and possible impurity levels.[11] The canonical partition function is obtained as

$$\mathcal{Z}(T,V,N) = \frac{1}{2\pi i}\oint d\zeta\, \zeta^{-N-1} \prod_{\alpha} [1+\zeta e^{-\varepsilon_\alpha / k_B T}] \,. \tag{5.4-3}$$

The evaluation of (5.4-3) is accomplished with the method of steepest descent. The integrand has a minimum on the real axis at $\zeta = x_1^*$ and a maximum in the direction parallel to the y-axis through x_1^*. So we integrate along $\zeta = x_1^* + iy$. With the integrand being denoted as $\exp[g(y)]$ and with the usual Gaussian expansion about the saddle point, we arrive at

$$\mathcal{Z}(T,V,N) = \frac{1}{\sqrt{2\pi g''(x_1^*)}} \prod_{\alpha} [1+x_1^* e^{-\varepsilon_\alpha / k_B T}] . \tag{5.4-4}$$

In $\ln \mathcal{Z}$ we can neglect $\ln g'' = \mathcal{O}(\ln N)$. The average occupancies and the covariances follow similar rules as in the grand canonical ensemble, but with now $\mathcal{Z} \to \mathcal{Z}$. Hence we have

$$\langle n_\alpha \rangle = (1/\mathcal{Z}) \sum_{\{n_\kappa\}} n_\alpha e^{-\Sigma_\gamma n_\gamma \varepsilon_\gamma / k_B T} = -k_B T \frac{\partial \ln \mathcal{Z}}{\partial \varepsilon_\alpha}, \tag{5.4-5}$$

$$\langle \Delta n_\alpha \Delta n_\beta \rangle = (k_B T)^2 \frac{\partial^2 \ln \mathcal{Z}}{\partial \varepsilon_\alpha \partial \varepsilon_\beta} = -k_B T \frac{\partial \langle n_\alpha \rangle}{\partial \varepsilon_\beta} . \tag{5.4-6}$$

From (5.4-4) and (5.4-5) we find

$$\langle n_\alpha \rangle = \frac{1}{(x_1^*)^{-1} e^{\varepsilon_\alpha / k_B T} + 1}, \tag{5.4-7}$$

The quantity x_1^* follows from the particle constraint, which determines the position of the Fermi level. So we set $x_1^* = \exp(\varepsilon_F / k_B T)$, where ε_F is now the average electrochemical potential $\langle \mu \rangle$ of the system, and *not* of the reservoir, which was discarded. For a degenerate Fermi gas (metals) the Fermi level is in the conduction band, while in nondegenerate semiconductors it is located somewhere in the band gap below the conduction band. The F–D distribution (5.4-7) now looks the same as before. However, the covariances of the occupancies are different; in fact we have from (5.4-6) and the form of x_1^*

$$\langle \Delta n_\alpha \Delta n_\beta \rangle = \langle n_\alpha \rangle [1 - \langle n_\alpha \rangle](\delta_{\alpha\beta} - \partial \varepsilon_F / \partial \varepsilon_\beta). \tag{5.4-8}$$

Note that the last term in the (..) of (5.4-8) is by no means zero; the position of the

[11] We exclude inner bands, for which the population is, for all practical purposes, frozen; also we ignore at present the particular spin and multiple-valley degeneracy factors that may modify the Fermi function.

Fermi level depends on the location of all levels $\{\varepsilon_\kappa\}$. The derivative in question involves a *virtual variation* of the level ε_β. As we have stated repeatedly, the first order moments are independent of the type of ensemble employed. However, the second order moments and other cumulants depend strongly on the constraints for the ensemble; the correction factor in (5.4-8) was known to Fowler.[12]

5.4.2 *Macroscopic Occupancy Fluctuations*[13]

We now apply these results to fluctuations of macroscopic electron occupancies in solids. Suppose we have Z_k levels of approximately the same energy. The occupancy is[14]

$$\langle n_k \rangle = \sum_{\alpha=1}^{Z_k} \langle n_\alpha \rangle = Z_k \langle n_\alpha \rangle . \tag{5.4-9}$$

The variance is given by

$$\begin{aligned}\langle \Delta n_k^2 \rangle &= \sum_{\alpha,\beta=1}^{Z_k} \langle \Delta n_\alpha \Delta n_\beta \rangle = \sum_{\alpha,\beta=1}^{Z_k} \langle n_\alpha \rangle [1-\langle n_\alpha \rangle][\delta_{\alpha\beta} - (\partial\varepsilon_F / \partial\varepsilon_\beta)] \\ &= \sum_{\alpha=1}^{Z_k} \langle n_\alpha \rangle [1-\langle n_\alpha \rangle][1-(\partial\varepsilon_F/\partial\varepsilon_\beta)] - 2\sum_{\alpha>\beta}^{Z_k} \langle n_\alpha \rangle [1-\langle n_\alpha \rangle](\partial\varepsilon_F/\partial\varepsilon_\beta).\end{aligned} \tag{5.4-10}$$

Since $\partial\varepsilon_F / \partial\varepsilon_\alpha = \partial\varepsilon_F / \partial\varepsilon_\beta$ this gives with ½ $Z_k(Z_k-1)$ terms for the second sum,

$$\begin{aligned}\langle \Delta n_k^2 \rangle &= Z_k \langle n_\alpha \rangle [1-\langle n_\alpha \rangle] \left[1 - \frac{\partial\varepsilon_F}{\partial\varepsilon_\alpha} - (Z_k - 1)\frac{\partial\varepsilon_F}{\partial\varepsilon_\alpha}\right] \\ &= Z_k \langle n_\alpha \rangle [1-\langle n_\alpha \rangle] \left[1 - Z_k \frac{\partial\varepsilon_F}{\partial\varepsilon_\alpha}\right].\end{aligned} \tag{5.4-11}$$

The last factor is somewhat misleading, since it looks to be of order Z_k. The reason is that the ε_α, after the grouping procedure, have lost their identity in the differentiation of the electrochemical potential. Thus, with ε_k being the representative energy for the group, we have according to the chain rule

$$\frac{\partial\varepsilon_F(\varepsilon_{\alpha_1} \ldots \varepsilon_{\alpha_Z})}{\partial\varepsilon_k} = \sum_{\alpha_i} \frac{\partial\varepsilon_F}{\partial\varepsilon_{\alpha_i}} \frac{\partial\varepsilon_{\alpha_i}}{\partial\varepsilon_k} \to Z_k \frac{\partial\varepsilon_F}{\partial\varepsilon_\alpha} . \tag{5.4-12}$$

---

[12] R.H. Fowler, "Statistical Mechanics", Second Ed., Cambridge Univ. Press, 1936 (reprinted 1966), Eqs. (2114) and (2115).

[13] For this subsection and the next, see a recent paper, C.M. Van Vliet, J. Applied Physics **93**, 6068 (2003).

[14] While we used Greek symbols to characterize single electron states, we shall employ Roman symbols to denote groups of states.

We therefore can rewrite (5.4-11) as

$$\langle \Delta n_k^2 \rangle = \langle n_k \rangle \left(1 - \frac{\langle n_k \rangle}{Z_k}\right)\left[1 - \frac{\partial \varepsilon_F}{\partial \varepsilon_k}\right]. \tag{5.4-13}$$

The term in [..] is the *canonical correction factor.*

This result can also be obtained directly by computing the grand-canonical partition function for states that can accommodate Z_k electrons (so-called para-statistics). This task is easily accomplished, using the binomial theorem. By inversion, one then obtains the canonical partition function, similarly as done here. The covariance (second order cumulant) then follows. This procedure is left as a problem at the end of the chapter.

5.4.3 *Examples for Nondegenerate and Degenerate Semiconductors*

We first consider electron fluctuations in a semiconductor with two sets of energy levels, ε_1 and ε_2, of abundance N_1 and N_2 and occupancy n_1 and n_2, respectively. The canonical constraint determines the Fermi level, i.e., $n_1 + n_2 = C$. For such impurity levels the occupancy is given by the "alternate" Fermi function

$$f_i = \frac{n_{i0}}{N_i} = \frac{1}{g_i e^{(\varepsilon_i - \varepsilon_F)/k_B T} + 1}, \quad (i = 1, 2). \tag{5.4-14}$$

Here the sub zero refers to the equilibrium occupancy, while g is a degeneracy factor, which depends on the nature of the defect states, donors or acceptors, on the spin, and on the number of valleys in the conduction band; see Section 5.6 at the end of this chapter for details. The constraint is therefore expressed as

$$\frac{N_1}{g_1 e^{(\varepsilon_1 - \varepsilon_F)/k_B T} + 1} + \frac{N_2}{g_2 e^{(\varepsilon_2 - \varepsilon_F)/k_B T} + 1} = C. \tag{5.4-15}$$

We now differentiate this expression implicitly to ε_1. A useful rule is – easily verifiable –

$$\frac{\partial f}{\partial \varepsilon} = -\frac{\partial f}{\partial \varepsilon_F} = -\frac{1}{k_B T} f(1-f), \tag{5.4-16}$$

which holds regardless of the factor g (so, *a fortiori*, for $g = 1$). Therefore we obtain

$$-\frac{1}{k_B T}\frac{n_{10}(N_1 - n_{10})}{N_1}\left(1 - \frac{\partial \varepsilon_F}{\partial \varepsilon_1}\right) - \frac{1}{k_B T}\frac{n_{20}(N_2 - n_{20})}{N_2}\left(-\frac{\partial \varepsilon_F}{\partial \varepsilon_1}\right) = 0. \tag{5.4-17}$$

This yields

$$\frac{\partial \varepsilon_F}{\partial \varepsilon_1} = \frac{n_{10}(1 - n_{10}/N_1)}{n_{10}(1 - n_{10}/N_1) + n_{20}(1 - n_{20}/N_2)}. \tag{5.4-18}$$

The fluctuations are therefore

$$\langle \Delta n_1^2 \rangle = \langle \Delta n_2^2 \rangle = \frac{n_{10}(1-n_{10}/N_1)\, n_{20}(1-n_{20}/N_2)}{n_{10}(1-n_{10}/N_1) + n_{20}(1-n_{20}/N_2)}. \tag{5.4-19}$$

We note for later use that this can also be written in the form

$$\begin{aligned} \frac{1}{\langle \Delta n_1^2 \rangle} = \frac{1}{\langle \Delta n_2^2 \rangle} &= \frac{1}{N_1 f_1(1-f_1)} + \frac{1}{N_2 f_2(1-f_2)} \\ &= \frac{1}{\langle \Delta n_1^2 \rangle_\infty} + \frac{1}{\langle \Delta n_2^2 \rangle_\infty}, \end{aligned} \tag{5.4-20}$$

where the variances with subscript '∞' indicate the corresponding results in a grand-canonical ensemble, in which each set of levels is in equilibrium with an infinite reservoir of chemical potential $\mu^r = \varepsilon_F$. The canonical correction factor is certainly not negligible! For the case that both terms above contribute equally, the fluctuations are reduced by 50% with respect to those in an ensemble open to *N*. We still note that for such an ensemble we simply have

$$\langle \Delta n_1^2 \rangle_\infty = N_1 f_1(1-f_1), \tag{5.4-21}$$

a result that is often 'postulated' from binomial statistics, see Eq. (1.6-7). However, this result does not apply to actual electron gases in semiconductors with a neutrality constraint – although the number of papers on electron device noise that use this erroneous result runs in the hundreds...

As a second example we consider the fluctuations in the free carrier density of a nondegenerate conduction band, which exchanges carriers with $s-1$ sets of traps, located at energies $\varepsilon_1 ... \varepsilon_{s-1}$ below the conduction band; the Fermi level is at ε_F with respect to the conduction band. For the number of electrons in the conduction band, assumed to be parabolic, we have the well-known result,

$$n_0 = N_C \exp[(\varepsilon_F - \varepsilon_c)/k_B T], \tag{5.4-21}$$

where $N_C = 2(2\pi m^* k_B T/h^2)^{3/2}$ is the statistical weight of the conduction band and $\mathscr{E}_c$ is the bottom of the band. The neutrality requirement leads to the constraint

$$n_0 + \sum_{i=1}^{s-1} n_{i0} = donor - acceptor\ excess = C. \tag{5.4-22}$$

This constraint is expressed in ε_F via (5.4-14) and (5.4-21). Implicit differentiation of (5.4-22) to ε_i then yields:

$$\frac{\partial \varepsilon_F}{\partial \varepsilon_i} = \frac{n_{i0}(N_i - n_{i0})/N_i}{\sum_k^{s-1}[n_{k0}(N_k - n_{k0})/N_k] + n_0}. \tag{5.4-23}$$

The final result for the free carrier fluctuations is

$$\langle \Delta n^2 \rangle = \sum_i^{s-1} \langle \Delta n_i^2 \rangle$$
$$= \sum_i^{s-1} N_i f_i (1-f_i) \left\{ \frac{\sum_k^{s-1} N_k f_k (1-f_k) - N_i f_i (1-f_i) + n_0}{\sum_k^{s-1} N_k f_k (1-f_k) + n_0} \right\}. \tag{5.4-24}$$

The term in { } is the correction factor with respect to the grand-canonical result.

Finally, we consider a near-intrinsic semiconductor with a small band gap, such as InSb, with fully ionized donors and acceptors. The constraint is $n_0 - p_0 = const.$, where n_0 is the number of electrons in the conduction band and p_0 is the number of holes in the valence band. Since both the electron gas and the hole gas are likely degenerate because of the proximity of the Fermi level – which may be in either band – the carrier constraint is given by the Fermi integrals[15]

$$N_C \mathcal{F}_{1/2}[(\varepsilon_F - \varepsilon_c)/k_B T] - N_V \mathcal{F}_{1/2}[(\varepsilon_v - \varepsilon_F)/k_B T] = const, \tag{5.4-25}$$

where N_C and N_V are the statistical weights of both bands. Differentiating to ε_C we obtain

$$n_0 \xi_n \left(\frac{\partial \varepsilon_F}{\partial \varepsilon_c} - 1 \right) + p_0 \xi_p \left(\frac{\partial \varepsilon_F}{\partial \varepsilon_c} \right) = 0, \tag{5.4-27}$$

where

$$\xi_n = \frac{\mathcal{F}_{-1/2}[(\varepsilon_F - \varepsilon_c)/k_B T]}{\mathcal{F}_{1/2}[(\varepsilon_F - \varepsilon_c)/k_B T]}, \quad \xi_p = \frac{\mathcal{F}_{-1/2}[(\varepsilon_v - \varepsilon_F)/k_B T]}{\mathcal{F}_{1/2}[(\varepsilon_v - \varepsilon_F)/k_B T]}. \tag{5.4-28}$$

From this we obtain $\partial \varepsilon_F / \partial \varepsilon_c$, which is the relevant differential for the variance $\langle \Delta n^2 \rangle$, as we will now show.

Generally we have for the number fluctuations of the electron gas in a band:

$$\langle \Delta n^2 \rangle = \sum_{\alpha,\beta \in \, cond.band} \langle n_\alpha \rangle [1 - \langle n_\alpha \rangle] \left(\delta_{\alpha\beta} - \frac{\partial \varepsilon_F}{\partial \varepsilon_\beta} \right). \tag{5.4-29}$$

The double sum will be converted in a double integral, using the density of states

$$Z(\varepsilon) d\varepsilon = 4\pi (2m_e^* / h^2)^{3/2} \sqrt{\varepsilon - \varepsilon_c}\, d\varepsilon \equiv C d\varepsilon, \tag{5.4-30}$$

whereas $\delta_{\alpha\beta} \to \delta(\varepsilon - \varepsilon')/Z(\varepsilon)$. Further, $\Sigma_\beta \partial \varepsilon_F / \partial \varepsilon_\beta \to \partial \varepsilon_F / \partial \varepsilon_c$, since only a virtual variation in the bottom of the conduction band has an effect on ε_F. Thus the double

---

[15] Fermi integrals are considered in detail in Chapter VIII. Here we just mention their definition and main property:

$$\mathcal{F}_k(\eta) = \frac{1}{\Gamma(k+1)} \int_0^\infty \frac{x^k}{e^{x-\eta} + 1} dx; \quad \frac{d}{d\eta} \mathcal{F}_k(\eta) = \mathcal{F}_{k-1}(\eta). \tag{5.4-26}$$

sum (5.4-29) becomes

$$\langle \Delta n^2 \rangle = C\int_{\varepsilon_c}^{\infty}\int_{\varepsilon_c}^{\infty} d\varepsilon d\varepsilon' \frac{e^{(\varepsilon-\varepsilon_F)/k_BT}}{[e^{(\varepsilon-\varepsilon_F)/k_BT}+1]^2}\delta(\varepsilon-\varepsilon')\sqrt{\varepsilon-\varepsilon_c}$$
$$= C\int_{\varepsilon_c}^{\infty} d\varepsilon \frac{e^{(\varepsilon-\varepsilon_F)/k_BT}}{[e^{(\varepsilon-\varepsilon_F)/k_BT}+1]^2}\left(\frac{\partial \varepsilon_F}{\partial \varepsilon_c}\right)\sqrt{\varepsilon-\varepsilon_c}. \tag{5.4-31}$$

From integration by parts we find[16]

$$\langle \Delta n^2 \rangle = N_C \mathcal{F}_{-\frac{1}{2}}[(\varepsilon-\varepsilon_c)/k_BT]\left(1-\frac{\partial \varepsilon_F}{\partial \varepsilon_c}\right) = \xi_n n_0 \left(1-\frac{\partial \varepsilon_F}{\partial \varepsilon_c}\right). \tag{5.4-32}$$

Substituting for the differential from (5.4-27) we find as final result

$$\langle \Delta n^2 \rangle = \frac{\xi_n \xi_p n_0 p_0}{\xi_n n_0 + \xi_p p_0}, \tag{5.4-33}$$

a result first obtained by Oliver.[17] For a nondegenerate two-band semiconductor the ξ's are unity and we re-obtain the earlier stated result (4.9-34).

5.5 Fluctuations in Systems Interacting with Finite Reservoirs

In Fig. 5-2 we describe a situation in which two systems can interact with each other, i.e., exchange variables $A_0 ... A_m$. We also indicate a large reservoir $\mathcal{R}$, with fixed thermodynamic forces $T, F_1^r ... F_m^r$. The joint system, comprising both 1 and 2 will be denoted by the superscript '*c*'. Since the free energies are additive in a canonical ensemble, $\mathcal{F}^c = \mathcal{F}^1 + \mathcal{F}^2$. However, the second-order derivatives are also additive, as we now show:

$$\frac{\partial^2 \mathcal{F}^1}{\partial A_i^1 \partial A_j^1} + \frac{\partial^2 \mathcal{F}^2}{\partial A_i^2 \partial A_j^2} = \frac{\partial^2 \mathcal{F}^1}{\partial A_i^1 \partial A_j^1} + \frac{\partial^2 \mathcal{F}^2}{\partial (A_i^c - A_i^1)\partial (A_j^c - A_j^1)}$$
$$= \frac{\partial^2 \mathcal{F}^1}{\partial A_i^1 \partial A_j^1} + \frac{\partial^2 \mathcal{F}^2}{\partial A_i^1 \partial A_j^1} = \frac{\partial^2 \mathcal{F}^c}{\partial A_i^1 \partial A_j^1}. \tag{5.5-1}$$

From (5.3-9) the first and second term on the lhs are ($1/k_BT$) times the reciprocal covariance-matrix element *ij* if each system interacted freely on its own with the reservoir $\mathcal{R}$. The extreme rhs, on the other hand, is ($1/k_BT$) times the covariance for a

[16] Since N_C and N_V are the statistical weights per unit volume, we have tacitly set $V = 1\ cm^3$, to avoid the need to distinguish between the carrier occupancies and the carrier densities.

[17] D.J. Oliver, Proc. Phys. Soc. **B 70**, 244 (1957).

system interacting with a finite reservoir. We thus arrive at

$$[\langle \Delta\mathbf{A}\Delta\mathbf{A}\rangle]_{ij}^{-1} = \left[\langle \Delta\mathbf{A}^1\Delta\mathbf{A}^1\rangle_\infty\right]_{ij}^{-1} + \left[\langle \Delta\mathbf{A}^2\Delta\mathbf{A}^2\rangle_\infty\right]_{ij}^{-1}. \tag{5.5-2}$$

The subscript '∞' again indicates the covariance for interaction of each system by itself with $\mathcal{R}$. Clearly, we can now also consider system (1+2) as another system, interacting with a system 3. The same rule would apply. By induction we could obtain a general rule for K finite systems interacting with each other. However, it would have little practical use. Even Eq. (5.5-2) is useful only if the matrix $[\langle \Delta\mathbf{A}\Delta\mathbf{A}\rangle_\infty]$ is diagonal.

As an example we consider free electrons of a nondegenerate band interacting with impurity levels of abundance N_i and average occupancy n_{i0}. If each 'band' interacted with a large reservoir by itself, we have the grand-canonical variances

$$\begin{aligned} \langle \Delta n^2\rangle_\infty &= n_0, \\ \langle \Delta n_i^2\rangle_\infty &= N_i f_i(1-f_i) = n_{i0}(1-n_{i0}/N_i). \end{aligned} \tag{5.5-3}$$

The variance in a system for which the free carriers interact with the finite reservoir of the impurity carriers is therefore,

$$\frac{1}{\langle \Delta n^2\rangle} = \frac{1}{n_0} + \frac{N_i}{n_{i0}(N_i - n_{i0})} = \frac{1}{n_0} + \frac{N_i}{(N_i - n_0)n_0}, \tag{5.5-4}$$

where we used that $n_0 + n_{i0} = N_i$. We thus obtain

$$\langle \Delta n^2\rangle = \frac{n_0(N_i - n_0)}{2N_i - n_0}. \tag{5.5-5}$$

This result was first obtained by Burgess.[18]

As another example we consider heat exchange between two finite bodies, connected to each other by a heat-conducting bar, having heat capacities C_1 and C_2, see Fig. 5-3. If each of them would be separately connected to an infinite heat reservoir, then we have as shown before,

$$\langle \Delta\mathcal{E}_i^2\rangle_\infty = k_B T^2 C, \quad (i = 1,2). \tag{5.5-6}$$

Thus, when connected, we find

$$\langle \Delta\mathcal{E}^2\rangle = k_B T^2 \frac{C_1 C_2}{C_1 + C_2}, \tag{5.5-7}$$

a result first obtained by Mrs. de Haas-Lorentz.[19]

---

[18] R.E. Burgess, Physica **20**.1007 (1954); also, Proc. Phys. Soc. **B 69**, 1020 (1956)

[19] G.L. de Haas-Lorentz, "Die Brownsche Bewegung und einige verwandte Erscheinungen" Vieweg, Braunschweig , 1912.

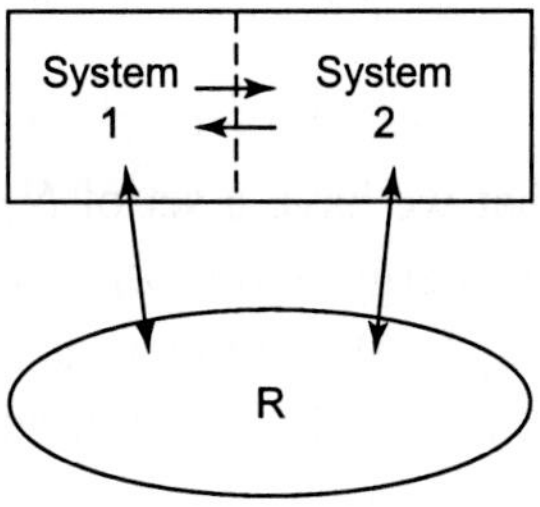

Fig. 5-2. Exchange of system 1 with a finite reservoir, system 2.

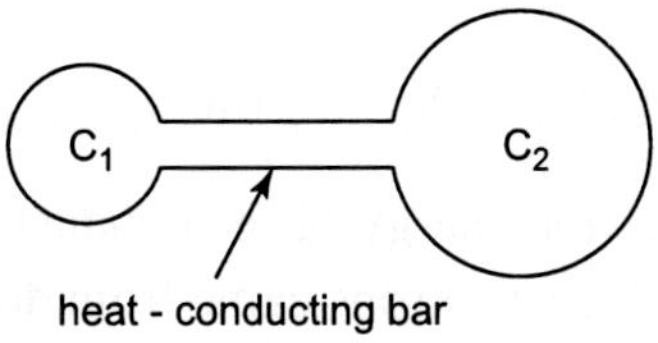

Fig. 5-3. Heat exchange between two finite bodies with heat capacitances C_1 and C_2.

5.6 Alternate Fermi–Dirac Distributions

It is usually assumed that the electrons in an energy band of a solid, when treated within a one-electron Hartree-Fock model, constitute an ideal Fermi gas. The occupancy of the energy levels is then given by the regular Fermi–Dirac function. Whereas this picture is generally correct for metals or the conduction and valence band of semiconductors, it must be amended for states arising from impurity levels associated with donors, acceptors, or crystal vacancies and other unintentional imperfections. This state of affairs was first clearly pointed out by Landsberg (1952)[20], although correct treatments were already available in the pre-world-war II books by Mott and Gurney[21] and by Wilson.[22]. Also, there was a Soviet publication in 1943 by Shifrin.[23] We think it to be proper to finish this chapter on generalized canonical ensembles and the implicated thermodynamic relationships by presenting some of these 'alternate' Fermi distributions here. While Shifrin's derivations are based on the grand-canonical ensemble – a task left as a problem to the student – all other derivations cited here, are obtained from straightforward thermodynamic

[20] P.T. Landsberg, Proc. Phys. Soc. **A 65**, 604 (1952).

[21] N.F. Mott and R. Gurney, "Electronic Processes in Ionic Crystals", Oxford, 1940.

[22] A.H. Wilson, "The Theory of Metals", Cambridge Univ. Press, 1936. (2nd Ed., N.Y., 1965).

[23] K.S. Shifrin, J. Techn. Physics USSR **14**, 43 (1943).

arguments. Below we shall follow our own discussion in a 1964 paper, already alluded to in subsection 5.3.2.[24]

First we consider the case that we have a set of N_d donors (per cm^3), occupied by n_d electrons and giving rise to energy levels ε_d below the conduction band. The entropy is as usual found from $k_B \ln \Delta\Gamma$, but there is a configurational part as well as a spin part. Since the donors can normally accommodate electrons of either spin, the spin multiplicity is 2^{n_d}. We thus have

$$\Delta\Gamma = \frac{N_d!}{(N_d - n_d)!n_d!} 2^{n_d}. \tag{5.6-1}$$

For the free energy function of the donors we obtain

$$\mathcal{F}(T, n_d) = n_d \varepsilon_d - k_B T[N_d \ln N_d - (N_d - n_d)\ln(N_d - n_d) - n_d \ln n_d + n_d \ln 2]. \tag{5.6-2}$$

We could now add to this the free energy of the conduction electrons and impose the constraint $n_d + n = N_d$ *if* all free electrons came from donor sites. However, there is no need to make this assumption and we need not impose a canonical constraint. Rather, we shall assume that the donors are in equilibrium with a large reservoir of electro-chemical potential μ^r and we set $(\partial\mathcal{F}/\partial n_d)_{equil} = \mu_r \doteq \varepsilon_F$; *if* the other electron bands in the solid are in equilibrium with the donors, then the last equality likewise applies, so that we connect to the Fermi level. Carrying out the differentiation, we find

$$\left(\frac{\partial\mathcal{F}}{\partial n_d}\right)_{n_d = n_{d0}} = \varepsilon_F = \varepsilon_d + k_B T \ln \frac{n_{d0}}{2(N_d - n_{d0})}. \tag{5.6-3}$$

From this we readily obtain

$$n_{d0} = \frac{N_d}{\frac{1}{2} e^{(\varepsilon_d - \varepsilon_F)/k_B T} + 1}. \tag{5.6-4}$$

This differs from regular Fermi-Dirac statistics by the factor ½ in front of the exponential.

For acceptor states the story is analogous. We assume that the acceptors are *s*-states, which can have two paired electrons. If n_a acceptors are ionized (i.e., *fully occupied*) then the multiplicity of the "empty" acceptors (i.e., those which contain *only one electron*) is $2^{N_a - n_a}$. Hence,

$$\Delta\Gamma = \frac{N_a!}{(N_a - n_a)!n_a!} 2^{N_a - n_a}. \tag{5.6-5}$$

The nonequilibrium free energy function becomes

---

[24] K.M. Van Vliet, this chapter, Ref. 7.

$$\mathcal{F}(T,n_a) = n_a\varepsilon_a - k_BT[N_a \ln N_a - (N_a - n_a)\ln(N_a - n_a) - n_a \ln n_a + (N_a - n_a)\ln 2]. \tag{5.6-6}$$

Again we set $(\partial\mathcal{F}/\partial n_a)_{equil.} = \mu^r \doteq \varepsilon_F$. We easily obtain,

$$n_{a0} = \frac{N_a}{2e^{(\varepsilon_a - \varepsilon_F)/k_BT} + 1}. \tag{5.6-7}$$

Other degeneracies also may play a role, such as multiplicity due to the number of minima (valleys) in the conduction band, being six for silicon. This makes for a *g*-factor of (1/12) in n-type silicon. Thus, generally, we shall write for the equilibrium occupancy of impurity levels in a solid

$$n_{i0} = \frac{N_i}{ge^{(\varepsilon_i - \varepsilon_F)/k_BT} + 1}, \tag{5.6-8}$$

where g is a small rational number, determined by the centres *and* the nature of the conduction and valence bands in the solid.

Another modification occurs if we take into account that hydrogenic impurity centres have a spectrum of excited states – similar as for hydrogen which has a ground state of -13.6 *eV* and a first excited state of -3.4 *eV*, *etc.* So, let there be n_1 electrons in the energy level ε_1 which has a degeneracy g_1, n_2 in ε_2 which has a degeneracy g_2, *etc.* We assume that there are N centres with a total occupancy $\Sigma_i n_i = n$. For the number of accessible quantum states we now have

$$\begin{aligned}\Delta\Gamma(n) &= \sum_{\{n_i\}}{}' \frac{N!}{n_1!(N-n_1)!} g_1^{n_1} \cdot \frac{(N-n_1)!}{n_1!(N-n_1-n_2)!} g_2^{n_2} \cdot \ldots \\ &= \sum_{\{n_i\}}{}' \frac{N!}{n_1!n_2!\ldots(N-n)!} g_1^{n_1} g_2^{n_2} \ldots \\ &= \frac{N!}{n!(N-n)!} \sum_{\{n_i\}}{}' \frac{n!}{n_1!n_2!\ldots} g_1^{n_1} g_2^{n_2} \ldots .\end{aligned} \tag{5.6-9}$$

The Σ' indicates that we have a restricted sum, subject to $\Sigma_i n_i = n$ and $\Sigma_i n_i \varepsilon_i = \mathcal{E}$. The latter condition can be dropped, however when we switch to the free energy function $\mathcal{F}(T,(\{n_i\})$. Using the multinomial expansion theorem we readily obtain

$$\begin{aligned}\mathcal{F}(T,n) &= \sum_i \varepsilon_i n_i - k_BT \ln \Delta\Gamma = -k_BT\left(\ln \prod_i e^{-(\varepsilon_i/k_BT)n_i} + \ln \Delta\Gamma\right) \\ &= -k_BT \ln\left\{\frac{N!}{n!(N-n)!}\left(\sum_i g_i e^{-\varepsilon_i/kT}\right)^n\right\}.\end{aligned} \tag{5.6-10}$$

With $(\partial\mathcal{F}/\partial n)_{n=n_0} = \mu^r \doteq \varepsilon_F$ the result is

$$n_0 = \frac{N}{G^{-1} e^{(\varepsilon_1 - \varepsilon_F)/k_B T} + 1}, \tag{5.6-11}$$

where

$$G = \sum_{j=1}^{\infty} g_j e^{(\varepsilon_1 - \varepsilon_j)/k_B T}. \tag{5.6-12}$$

We thus managed to again obtain a form like (5.6-8), although G is temperature dependent. The relevant energy ε_1 is the ground-state energy of the impurity centre, i.e., the usual energy associated with the donor or acceptor. Our result is in full harmony with that reported by Shifrin.

Finally, we endeavour to obtain the statistics for impurities that can trap several electrons; as an example we consider double acceptors, such as Ni or Zn in germanium. We shall denote the single electron energies by $\in_1$ and $\in_2$ and the energies of the charged centres by ε_{11}, ε_{12} and ε_2. The complication here is that the single electron energies depend on the charge already present in the centre (just as the energy levels of He are different from those of He^+). The situation is depicted in Fig. 5-4. The relevant variables are now N^{**} (neutral centres), N^{-*} and N^{*-} (centres with one electron) and N^{--} (centres with two electrons). The charge constraint for the sample as a whole reads:

$$N^{-*} + N^{*-} + 2N^{--} + \textit{other electrons} = \textit{const}. \tag{5.6-13}$$

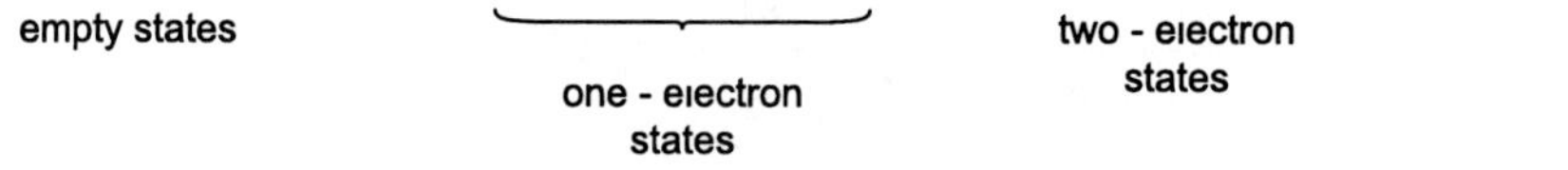

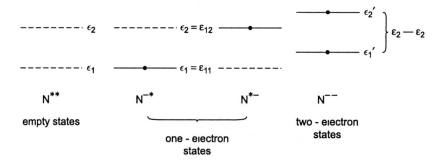

Fig. 5-4. Energy diagrams for centres that can accommodate two electrons.

Considering exchange of electrons, we shall find it expedient to define electro-chemical potentials such that

$$\left(\frac{\partial \mathcal{F}}{\partial N^{*-}}\right)_{eq} = \mu_{12}^r, \quad \left(\frac{\partial \mathcal{F}}{\partial N^{-*}}\right)_{eq} = \mu_{11}^r, \quad \left(\frac{\partial \mathcal{F}}{\partial 2N^{--}}\right)_{eq} = \mu_2^r, \tag{5.6-14}$$

For the number of accessible states we have the tetranomial coefficient

$$\Delta\Gamma(N^{*-}, N^{-*}, N^{--}) = \frac{N!}{N^{**}!N^{*-}!N^{-*}!N^{--}!} g_{11}^{N^{-*}} g_{12}^{N^{*-}} g_{22}^{N^{--}}, \tag{5.6-15}$$

where the g's are the appropriate degeneracies. For unconstrained equilibrium, we set all μ^r equal to ε_F. Defining,

$$A_{11} = g_{11}e^{(\varepsilon_F-\varepsilon_{11})/k_BT}, \quad A_{12} = g_{12}e^{(\varepsilon_F-\varepsilon_{12})/k_BT}, \quad A_2 = g_2 e^{(2\varepsilon_F-\varepsilon_2)/k_BT}, \tag{5.6-16}$$

we obtain

$$N_0^{-*} = NA_{11}/(1+\Sigma A), \quad N_0^{*-} = NA_{12}/(1+\Sigma A), \quad N_0^{--} = NA_2(1+\Sigma A). \tag{5.6-17}$$

In some problems it is useful to consider the variables $n_{10} = N_0^{-*} + N_0^{--}$ for the number in the lower energy states and $n_{20} = N_0^{*-} + N_0^{**}$ for the number in the upper energy states. (The fact that $\epsilon_1' \neq \epsilon_1$ is rather irrelevant since, when the upper acceptor level is ionized, the remaining electron falls back immediately to the level ϵ_1; it is to be noted, however, that the electrons n_1 are not all identical.) For n_1 we obtain

$$n_{10} = N\frac{g_{11}e^{(\varepsilon_F-\varepsilon_{11})/k_BT} + g_2 e^{(2\varepsilon_F-\varepsilon_2)/k_BT}}{g_{11}e^{(\varepsilon_F-\varepsilon_{11})/k_BT} + g_{12}e^{(\varepsilon_F-\varepsilon_{12})/k_BT} + g_2 e^{(2\varepsilon_F-\varepsilon_2)/k_BT} + 1}, \tag{5.6-18}$$

with a similar result for n_{20}. This differs greatly from the usual Fermi–Dirac distribution!

5.7 Problems to Chapter V

5.1 Consider the pressure ensemble for a classical gas. Explain that the semi-classical distribution function can be written as,

$$\rho(p,q;T,P,N,A_3...A_s) = \Xi_p^{-1} e^{-\beta[\mathcal{H}(p,q;V,N)+PV]}, \quad \beta = (1/k_BT), \tag{1}$$

where Ξ_p is the phase integral [cf. (5.2-3)]:

$$\Xi_p(T,P,N,A_3...A_s) = \int_{V_{min}} dV\, e^{-\beta PV} \int d\Gamma_p d\Omega_q\, e^{-\beta\mathcal{H}(p,q,V,N)}, \tag{2}$$

with $d\Gamma_p = d\Omega_p / h^{3N} N!$

(a) If the compressibility is included in the Hamiltonian, should we set V_{min}=0?

(b) For a two-body potential that incorporates a hard core for molecules of diameter a what is $V_{\min}$?

5.2 Consider the 1D classical gas with two-body interactions of Tonks and Takahashi, cf., Section 4.3.4. The volume V is now replaced by the length L, which is measured from the centre of the first molecule to the centre of the Nth molecule.

(a) Write down the configurational partition function (phase integral) of the pressure ensemble $\Xi_{p,c}$ as a function of L and N, analogous to (4.3-98);

(b) Transform to the subregion $L \geq x_N \geq x_{N-1} ... \geq x_1$ and introduce 'local'

variables ℓ and y_i, cf., (4.3-99). The partition function $\Xi_{p,c}$ now assumes the form of a Laplace-type convolution integral;
(c) Evaluate $\Xi_{p,c}$ and the Gibbs free energy per particle, $\hat{g} = G/N$, with the thermodynamic limit being implied;
(d) With $n = N/L$ being the density, obtain the pressure for Tonks' hard-core gas $(\phi \equiv 0)$:

$$P = k_B T n/(1 - na). \tag{3}$$

5.3 (a) Express the volume fluctuations $\langle \Delta V^2 \rangle$ of the pressure ensemble in terms of the isothermal compressibility K_T.
(b) Consider an open ensemble with variable energy, volume and particle number. Obtain the volume fluctuations $\langle \Delta V^2 \rangle$ for an arbitrary system and compare with the result under (a).
(c) Obtain the number fluctuations $\langle \Delta N^2 \rangle$ and compare with the result (4.2-30) of the grand-canonical ensemble.

5.4 Employing transformation theory, obtain the occupancy $\langle n_\alpha \rangle$ of a state $|\alpha)$, having energy ε_α, for a perfect Fermi gas in a microcanonical ensemble; also, obtain the covariance $\langle \Delta n_\alpha \Delta n_\beta \rangle$ and compare with (5.4-8).

5.5 In order to find the mean occupancy of electrons in donor levels that can accommodate electrons of either spin, Wilson[22,25] considered the number of accessible states $\Delta\Gamma(n\uparrow, n\downarrow \in N_d)$ as the trinomial distribution for up-spin, down-spin and vacancies, assigning equal electrochemical potentials to both 'species'.
(a) Obtain the Helmholtz free energy and the mean occupancies $\langle n\uparrow \rangle$ and $\langle n\downarrow \rangle$ and their sum n_{d0}. Show that the result is the same as in (5.6-4).
(b) Obtain the same result from summing $\sum_{n\uparrow + n\downarrow = n_{d0}} \Delta\Gamma(n\uparrow, n\downarrow \in N_d)$.

5.6 Employing mean field theory, consider the magnetization M of a system of N paramagnetic ions of spin ½ in an external magnetic field H below the Curie temperature. Instead of with a canonical ensemble, the system is described with a generalized canonical ensemble, in which the magnetization is variable and the reservoir force $F^r = -H$ is fixed.
(a) Obtain the magnetization fluctuations $\langle \Delta M^2 \rangle$ as a function of T and H. Assume $V = 1\ cm^3$. What happens when $H \to 0$ and $T \to T_c$? [Show $\langle \Delta M^2 \rangle \to \infty$.]
(b) Above T_c establish $\langle \Delta M^2 \rangle = kT\chi_{dc}$, where χ_{dc} is the differential *dc* susceptibility. Again, let $T \to T_c$. N.B. *Critical fluctuations* will be considered in Chapter IX.

---

[25] A.H. Wilson, "The Theory of Metals", Cambridge Univ. Press, 2nd Ed., (1965), Appendix A.1.

5.7 The formula for the variance of the electron occupancy of a group of Z_k states with energy ε_k can also be found from *para-statistics*. Employing an approach similar to that of subsection 4.3.1, obtain the grand-canonical partition function $\mathcal{Z}$ if the one-particle state $|k)$ can accommodate $1, 2, \ldots Z_k$ electrons. Then obtain the canonical partition function $\mathcal{Q}$ as an inverse Fowler transform, using the method of steepest descent. From $\ln \mathcal{Q}$ obtain $\langle n_k \rangle$ and $\langle \Delta n_k^2 \rangle$; confirm Eq. (5.4-13). [N.B. The method of the text, in which macroscopic occupancy fluctuations are obtained from microscopic occupancy fluctuations, is more founded.]

PART B

CLASSICAL AND QUANTUM FORMALISMS. THE BOLTZMANN GAS, THE PERFECT BOSE GAS AND FERMI GAS

Chapter VI

The Boltzmann Distribution and Chemical Applications

6.1 Aspects of Molecular Distributions

In this Part we consider systems of weakly interacting particles. Although some form of interaction is always necessary in order that equilibrium can be established, the interaction energies will be neglected in the Hamiltonian. Thus we write

$$\mathcal{H} = \sum_i h_i, \tag{6.1-1}$$

where h_i is the one-particle Hamiltonian, having r degrees of freedom, so that its generalized coordinates and momenta are $q_1 \ldots q_r$ and $p_1 \ldots p_r$. For a classical description these attributes are pictured in the $2r$-dimensional molecule space or μ-space.

Suppose that we divide up the accessible volume of the μ-space into a number of cells having volumes $\Delta\omega_1, \Delta\omega_2, \ldots, \Delta\omega_k, \ldots$. The number of particles whose motion is represented by a point in these volumes is denoted by $N_1, N_2, \ldots, N_k, \ldots$. We then call the set

$$\gamma_c = \gamma_c(\{N_i\}), \quad N_i \in \Delta\omega_i, \tag{6.1-2}$$

a *configuration* in the μ-space; the subscript 'c' stands for coarse-grained. It will be omitted in the limit that $\Delta\omega_i \to d\omega$. In this limit we write $N_i \to n_i(p,q)d\omega$ and we denote the configuration by the functional

$$\gamma = \gamma[n(p,q)]. \tag{6.1-3}$$

In the quantum description there is no μ-space. Let the one-particle states of the molecules (or constituents) be $\{|\alpha_i)\}$ where we can choose these states to be the eigenstates of the one-particle Hamiltonian, although any complete basis in the Hilbert space of h will suffice. Suppose that we lump these states into groups of Z_i, having all the approximate energy ε_i. We now consider a coarse-grained configuration to be a specification of the numbers of particles N_i that have an eigenstate in Z_i . Hence, the quantum equivalent of (6.1-2) is

$$\gamma_c = \gamma_c(\{N_i\}), \quad N_i \in Z_i. \tag{6.1-4}$$

What if we do not form groups? We then have a microscopic quantum configuration, which has a very real meaning. Remember that for non-interacting particles the many-body state will simply be $|\phi\rangle = \Pi_i[\otimes|\alpha_i\rangle]$, i.e., we have a tensor product of one-particle states. Since in a quantum view particles are indistinguishable, it suffices to simply count the number of particles n_α in each state $|\alpha\rangle$. We shall later on present the *occupation-number representation* in which the many-body state is represented by these numbers $\{n_\alpha\}$, i.e., a microscopic quantum configuration *is* the many-body state,

$$\gamma = \gamma(\{n_\alpha\}) \equiv |\gamma\rangle = |n_1, n_2, ... n_k, ...\rangle, \tag{6.1-5}$$

$|\gamma\rangle$ being the state of the system in occupation-number form. In this chapter we shall not invoke the necessary symmetrization or antisymmetrization of the tensor product $\Pi_i[\otimes|\alpha_i\rangle]$ of one-particle states; thus, we only employ a semi-quantal description, although the factor $N!$ will be introduced for indistinguishable particles, so as to obtain a proper quantum limit. The resulting "Boltzmann statistics", while useful for many systems at room temperature, fails to describe any system at low temperatures; in other words: a Boltzmann gas basically does not exist in nature.

We consider a microcanonical ensemble and we seek the appropriate Boltzmann distribution. While many texts maximize the configuration, we shall find the *average* configuration.[1] The point is that we will *not* assume that the numbers $\{N_i\}$ in a configuration are asymptotically large; we do assume, however, that the thermodynamic limit will be taken for the total number N, so that Stirling's formula applies to $\ln N!$. The average configurations will be denoted by $\overline{\gamma}_c$ for (6.1-2) and (6.1-4) and by $\overline{\gamma}$ for the microscopic specifications (6.1-3) and (6.1-5). The coarse-grained Boltzmann distribution is thus obtained as

$$\overline{\gamma}_c = \gamma_c(\{\overline{N}_i\}), \text{ with } \overline{N}_i = \sum_{\gamma_c}{}' N_i W(\gamma_c) = \sum_{\{N_k\}}{}' N_i W(N_1, N_2, ..., N_k, ...), \tag{6.1-6}$$

where the prime on the sum indicates the constraints $\Sigma_i N_i = N$ and $\Sigma_i N_i \varepsilon_i = \mathcal{E}$. These constraints will be handled with the Darwin–Fowler procedure, which is based on the method of the steepest descent.[2] For a microscopic configuration we have likewise

$$\overline{\gamma} = \gamma(\{\overline{n}_i\}), \text{ with } \overline{n}_i = \sum_{\gamma}{}' n_i p(\gamma) = \sum_{\{n_k\}}{}' n_i p(n_1, n_2, ... n_k, ...). \tag{6.1-7}$$

We must now find $W(\gamma_c)$ and $p(\gamma)$. Dealing with the classical picture first, we note that the correspondence Γ-space $\leftrightarrow$ μ-space is somewhat complex. The following can be said:

---

[1] We follow the developments as set forth in E. Schrödinger, "Statistical Thermodynamics", Cambridge Univ. Press, 1960.

[2] R.H. Fowler, "Statistical Mechanics", Cambridge Univ. Press, 2nd Ed. 1960.

1) With one point in Γ-space corresponds one configuration γ_c in μ-space. This is clear since the point in phase space fixes the coordinates of all particles;
2) With a given configuration γ_c in μ-space corresponds a large, bizarre, volume in Γ-space, called Ehrenfest's Z-star. This is illustrated in Fig. 6-1 for a system of two particles with one generalized coordinate only (read the caption of that figure).

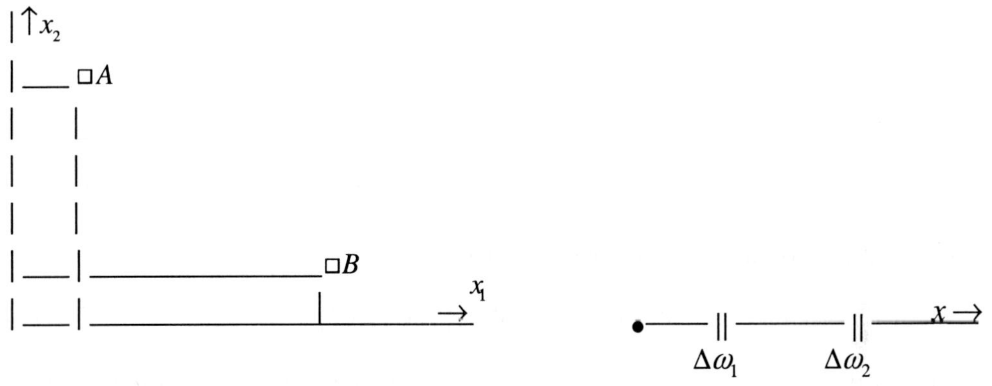

Fig. 6-1. Example of Γ-space – μ-space correspondence. The μ-space has one axis, the Γ-space two. For particle 1 in the first slot||, $\Delta\omega_1$, and particle 2 in the second slot||, $\Delta\omega_2$, the 'volume' in Γ-space is the square at A; for particle 1 in the second slot and particle 2 in the first slot, the 'volume' in Γ-space is the square at B. Clearly, both squares belong to the same configuration. Together they form Ehrenfest's Z-star. Obviously, with many axes and many particles, the Z-star becomes a large sprawling volume in the Γ-space.

It is easily seen that the volume in Γ-space $\Delta\Omega$ is given by

$$\Delta\Omega = \frac{N!}{N_1!N_2!\ldots N_k!\ldots}(\Delta\omega_1)^{N_1}(\Delta\omega_2)^{N_2}\ldots(\Delta\omega_k)^{N_k}\ldots . \tag{6.1-8}$$

The probability sought for is thus $W(\gamma_c) = C'\Delta\Omega$, which is a multinomial distribution, satisfying the particle constraint, but not yet the energy constraint. To ensure a fluid connection with the quantum case, we divide $\Delta\Omega$ by $N!h^{rN}$. But, as noted by ter Haar[3], we want to retain the multinomial coefficient *as is*. We therefore write

$$1/N!h^{rN} \sim (1/\sqrt{2\pi N})\left(\frac{e}{Nh^r}\right)^{\Sigma_k N_k} . \tag{6.1-9}$$

Consequently we have

$$W(\gamma_c) = C\delta_{\mathcal{E},\Sigma_i N_i \varepsilon_i}\frac{N!}{\prod_k N_k!}\prod_k\left(\frac{e\Delta\omega_k}{Nh^r}\right)^{N_k} . \tag{6.1-10}$$

---

[3] D. ter Haar, "Elements of Thermostatistics" (2nd Ed.) Rinehart, NY. 1966, pp. 95-96, 99.

The quantum mechanical case proceeds similarly. For indistinguishable particles we simply replace $\Delta\omega_k / h^r$ by Z_k. But we may also employ Boltzmann statistics for distinguishable objects, such as a set of oscillators; in that case we omit the factor e / Nh^r, stemming from the microcell. Thus, generally we have

$$W(\gamma_c) = C\delta_{\mathscr{E},\Sigma_i N_i \varepsilon_i} \frac{N!}{\prod_k N_k!} \prod_k (\kappa_k)^{N_k}, \tag{6.1-11}$$

where we set

$$\kappa_k = \begin{cases} e\Delta\omega_k / Nh^r, & (\textit{semi-classical}) \\ eZ_k / N \quad , & (\textit{quantum, indist}) \\ Z_k \quad , & (\textit{distinguishable}). \end{cases} \tag{6.1-12}$$

The sought for averages are then obtained from

$$\bar{N}_i = \kappa_i \partial \ln\left[\Sigma_{\gamma_c} W[(\gamma_c)/C]\right] / \partial\kappa_i. \tag{6.1-13}$$

We recall that *W/C* is also what we called the "thermodynamic probability". It is related to the equilibrium entropy, see Eq. (3.4-4). Hence, once we have obtained $\Sigma(W/C)$, we also have the entropy

$$S = k_B \ln \sum_{\gamma_c} [W(\gamma_c)/C] = k_B \ln \sum_{\gamma_c} \tilde{W}(\gamma_c). \tag{6.1-14}$$

6.2 The Darwin–Fowler Procedure

In order to evaluate $\Sigma_{\gamma_c}(\tilde{W})$ we take the generating function of Fowler with respect to $\mathscr{E}$,

$$\begin{aligned} \Phi(z;\{\kappa_i\}) &= \sum_{\mathscr{E}} z^{\mathscr{E}} \sum_{\{N_i\}}{}' \delta_{\mathscr{E},\Sigma_i \varepsilon_i N_i} \frac{N!}{N_1!N_2!\ldots N_k!\ldots} \prod_k \kappa_k^{N_k} \\ &= \sum_{\{N_i\}} \frac{N!}{N_1!N_2!\ldots N_k!\ldots} \prod_k z^{\varepsilon_k} \kappa^{N_k} = [f(z)]^N, \end{aligned} \tag{6.2-1}$$

where

$$f(z) = \kappa_1 z^{\varepsilon_1} + \kappa_2 z^{\varepsilon_2} + \ldots = \sum_k \kappa_k z^{\varepsilon_k}. \tag{6.2-2}$$

The inverse Fowler transform yields

$$\sum_{\gamma_c} \tilde{W}(\gamma_c) = \frac{1}{2\pi i} \oint_{\mathscr{C}} [f(\zeta)]^N \zeta^{-(\mathscr{E}+1)}, \tag{6.2-3}$$

where $\mathscr{C}$ encircles the origin ccw and $\mathscr{E}$ has been expressed in sufficiently fine units, cf. our discussion in subsection 1.7.3.

The integrand, denoted as $\exp g(\zeta)$, is the product of an ascending function and a descending function; it has a minimum at the real axis at the saddle point $\zeta = \zeta_0 + iy|_{y=0}$. Integrating along an axis // y-axis through the saddle point, we have by the method of steepest descent

$$\sum_{\gamma_c} \tilde{W}(\gamma_c) = \exp[g(\zeta_0)] / \sqrt{2\pi g_0''(\zeta_0)}, \tag{6.2-4}$$

where

$$g(\zeta_0) = N \ln f(\zeta_0) - \mathscr{E} \ln \zeta_0, \tag{6.2-5}$$

$$g'(\zeta_0) = N\big(f'(\zeta_0)/f(\zeta_0)\big) - \mathscr{E}/\zeta_0 = 0, \tag{6.2-6}$$

$$g''(\zeta_0) = \frac{\mathscr{E}}{\zeta_0^2} + N\left\{\frac{f''(\zeta_0)}{f(\zeta_0)} - \left[\frac{f'(\zeta_0)}{f(\zeta_0)}\right]^2\right\}. \tag{6.2-7}$$

Eq. (6.2-6) is the saddle-point condition. We note that in the thermodynamic limit $\ln g''(\zeta_0) = \mathscr{O}(\ln N)$ can be neglected. From (6.1-13) and these results we find

$$\bar{N}_i = \kappa_i \frac{\partial \ln \Sigma_{\gamma_c} \tilde{W}}{\partial \kappa_i} = \frac{N\kappa_i \zeta_0^{\varepsilon_i}}{\Sigma_k \kappa_k \zeta_0^{\varepsilon_k}} + \kappa_i \left[\frac{Nf'(\zeta_0)}{f(\zeta_0)} - \frac{\mathscr{E}}{\zeta_0}\right] \frac{\partial \zeta_0}{\partial \kappa_i}. \tag{6.2-8}$$

The latter term is zero because of the definition of the saddle point, (6.2-6). This result takes on a more familiar form if we set

$$\beta = -\ln \zeta_0 . \tag{6.2-9}$$

With the partition function in μ-space defined as

$$\mathcal{z}_\mu \equiv \sum_k \kappa_k \zeta_0^{\varepsilon_k} = \sum_k \kappa_k e^{-\beta \varepsilon_k}, \tag{6.2-10}$$

we arrive at the Boltzmann distribution

$$\bar{N}_i = (N/\mathcal{z}_\mu)\kappa_i e^{-\beta \varepsilon_i}, \tag{6.2-11}$$

whereby we note that the distribution is normalized to N: $\Sigma_i \bar{N}_i = N$.

Finally, we investigate the saddle-point condition (6.2-6). With (6.2-9) and (6.2-11) this takes the form

$$N\sum\nolimits_k \kappa_k \varepsilon_k e^{-\beta\varepsilon_k}/\mathcal{z}_\mu - \mathscr{E} = 0 \quad \text{or} \sum\nolimits_k \bar{N}_k \varepsilon_k - \mathscr{E} = 0. \tag{6.2-12}$$

This confirms the general notion that *the saddle-point condition is always just the constraint equation that we had to tackle!*

An explicit result for ζ_0 or for β is easily obtained. For the entropy we find from

Eq. (6.1-14)

$$S = k_B N \ln f(\zeta_0) - k_B \mathscr{E} \ln \zeta_0 = k_B N \ln \sum_k \kappa_k e^{-\beta \varepsilon_k} + k_B \beta \mathscr{E}, \tag{6.2-13}$$

or,

$$\frac{\partial S}{\partial \mathscr{E}} = \frac{1}{T} = k_B \beta + \left[\frac{kNf'(\zeta_0)}{f(\zeta_0)} - \frac{k\mathscr{E}}{\zeta_0} \right] \frac{\partial \zeta_0}{\partial \mathscr{E}}. \tag{6.2-14}$$

The last term does not contribute because of (6.2-6). Hence, $\beta = 1/k_B T$.

6.3 Thermodynamic Functions and Standard Forms

From (6.2-13) it is seen that the Helmholtz free energy is given by

$$F = -Nk_B T \ln \mathscr{z}_\mu, \tag{6.3-1}$$

which is entirely analogous to the result for any system in a canonical ensemble.[4] The energy is also expressible in $\mathscr{z}_\mu$. We find

$$\mathscr{E} = N\overline{\overline{\varepsilon}} = N \frac{\Sigma_i \varepsilon_i \kappa_i e^{-\beta \varepsilon_i}}{\Sigma_i \kappa_i e^{-\beta \varepsilon_i}} = -N \frac{\partial \ln \mathscr{z}_\mu}{\partial \beta} = Nk_B T^2 \frac{\partial \ln \mathscr{z}_\mu}{\partial T}; \tag{6.3-2}$$

here $\overline{\overline{\varepsilon}}$ is a molecular average. Generally we shall denote any molecular average $(1/N)\Sigma\ldots$ by a double overhead bar. For the entropy we have in terms of $\mathscr{z}_\mu$:

$$S = (\mathscr{E} - F)/T = k_B N \frac{\partial (T \ln \mathscr{z}_\mu)}{\partial T}. \tag{6.3-3}$$

We also can express the entropy as a molecular average. We then have

$$S = \mathscr{E}/T + k_B N \ln \mathscr{z}_\mu = -k_B \sum_i N_i \ln(Ne^{-\beta \varepsilon_i} / \mathscr{z}_\mu) = -k_B N \overline{\overline{\ln(\overline{N}_i / \kappa_i)}}, \tag{6.3-4}$$

where we neglected a term $\mathscr{O}(\ln N)$. This resembles the canonical expression $S = -k_B \overline{\ln \rho}$. For a quantum gas a small change appears. Since then $\kappa = eZ/N$ we have

$$S = -k_B \sum_i N_i \ln(e^{-\beta \varepsilon_i} / \mathscr{z}_\mu) = -k_B \sum_i N_i \ln(\overline{N}_i / \kappa_i N)$$

$$= -k_B \sum_i N_i \ln(\overline{N}_i / eZ_i) = -k_B \left[\sum_i N_i \ln(\overline{N}_i / Z_i) - \sum_i N_i \right]$$

[4] Thus, $\ln \mathscr{Z}_N = N \ln \mathscr{z}_\mu + \mathscr{O}(\ln N)$. But, as is easily verified, $\mathscr{Z}_N = \hat{c}(\mathscr{z}_\mu)^N$, with $\hat{c} = 1$ for distinguishable objects, while $\hat{c} = 1/\sqrt{2\pi N}$ for non-distinguishable entities.

$$= -k_B\left[N\overline{\overline{\ln(\overline{N}_i/Z_i)}} - N\right]. \tag{6.3-5}$$

The correction '$-N$' in the final expression is important for Boltzmann's H-theorem in quantum statistics, see Chapter XIII.

Standard forms

For the semi-classical case $\kappa = e\Delta\omega/h^r N$. Let $\overline{N}_i = \overline{n}(p,q)\Delta\omega_i$. For infinitesimal cells, we find

$$\overline{n}(p,q)d\omega = \frac{ed\omega}{h^r \gamma_\mu} e^{-\beta h(p,q)}, \tag{6.3-6}$$

where $h(p,q)$ is the single particle Hamiltonian. The partition function in μ-space becomes the phase integral

$$\gamma_\mu = \sum_k \kappa_k e^{-\beta\varepsilon_k} \to (e/h^r N)\int d\omega e^{-\beta h(p,q)}. \tag{6.3-7}$$

Combining (6.3-6) and (6.3-7) we have of course the simple form,

$$\overline{n}(p,q)d\omega = Ne^{-\beta h(p,q)} / \int d\omega e^{-\beta h(p,q)}. \tag{6.3-8}$$

However, the precise form of γ_μ is required for the correct thermodynamic functions.

For the quantum case with indistinguishable particles, $\kappa = eZ/N$, we have the formulae

$$\overline{N}_i = (eZ_i/\gamma_\mu)e^{-\beta\varepsilon_i}, \quad \gamma_\mu = \sum_k (eZ_k/N)e^{-\beta\varepsilon_k}, \tag{6.3-8}$$

$$\overline{N}_i = NZ_i e^{-\beta\varepsilon_i} / \Sigma_k Z_k e^{-\beta\varepsilon_k}. \tag{6.3-9}$$

In case $Z_i = g_i$, where g_i is the degeneracy of the level ε_i we obtain the occupancy n_i of this level; if $Z_i = 1$ we obtain the occupancy n_α of a state $|\alpha\rangle$. Hence, for these cases,[5]

$$\gamma_\mu = (e/N)\sum_k g_k e^{-\beta\varepsilon_k}, \quad \overline{n}_i = g_i N e^{-\beta\varepsilon_i} / \Sigma_k g_k e^{-\beta\varepsilon_k}, \tag{6.3-10}$$

$$\gamma_\mu = (e/N)\sum_\alpha e^{-\beta\varepsilon_\alpha}, \quad \overline{n}_\alpha = Ne^{-\beta\varepsilon_\alpha} / \Sigma_\alpha e^{-\beta\varepsilon_\alpha}. \tag{6.3-11}$$

Finally, for a system of distinguishable objects with $\kappa_i = Z_i = g_i$, we have the forms

$$\gamma_\mu = \sum_k g_k e^{-\beta\varepsilon_k}, \quad \overline{N}_i = Ng_i e^{-\beta\varepsilon_i} / \Sigma_k g_k e^{-\beta\varepsilon_k}. \tag{6.3-12}$$

---

[5] For these cases the $\overline{N}_i \to \overline{n}_i$ are very small numbers. The "W_{max} - method" would have required Stirling's formula for the $\overline{N}_i$, which is of course not applicable. *So the use of the Darwin-Fowler method is essential.*

6.4 Fluctuations of the Distribution Function

Since we showed that the Boltzmann distribution is the *average* distribution, there are fluctuations about the average distribution, which can be computed by the same procedure. We leave it to the reader to show the general formula, valid for any i, j:

$$\overline{\Delta N_i \Delta N_j} = \kappa_i \frac{\partial}{\partial \kappa_i}\left(\kappa_j \frac{\partial \ln \Sigma \tilde{W}}{\partial \kappa_j}\right) = \kappa_i \frac{\partial \bar{N}_j}{\partial \kappa_i} = \kappa_j \frac{\partial \bar{N}_i}{\partial \kappa_j}. \tag{6.4-1}$$

Using (6.2-11) and the saddle-point condition, straightforward evaluation yields,

$$\overline{\Delta N_i \Delta N_j} = \bar{N}_i \left[\delta_{ij} - \frac{\bar{N}_j}{N} - \kappa_j \frac{\partial \beta}{\partial \kappa_j}\left(\varepsilon_i - \frac{\mathcal{E}}{N}\right)\right]. \tag{6.4-2}$$

The first term is the classical term that would occur in a grand-canonical ensemble. The other terms are of order $(1/N)$ and vanish in an infinite system; they stem from the particle and energy constraints.

For the computation we follow Schrödinger, Op. Cit. We introduce the molecular sums

$$s_p = \sum_k \varepsilon_k^p \kappa_k e^{-\beta \varepsilon_k}. \tag{6.4-3}$$

Here p is an integer. In particular,

$$s_0 = \gamma_\mu, \quad s_p = \gamma_\mu \overline{\overline{\varepsilon^p}}. \tag{6.4-4}$$

Differentiating the first equality in (6.2-12) to κ_j we have

$$\frac{\partial}{\partial \kappa_j}\left(\frac{s_1}{s_0}\right) = 0, \text{ or } \frac{\partial s_1}{\partial \kappa_j} s_0 = \frac{\partial s_0}{\partial \kappa_j} s_1. \tag{6.4-5}$$

But from (6.4-3):

$$\begin{aligned} \partial s_1 / \partial \kappa_j &= \varepsilon_j e^{-\beta \varepsilon_j} - s_2 \partial \beta / \partial \kappa_j, \\ \partial s_0 / \partial \kappa_j &= e^{-\beta \varepsilon_j} - s_1 \partial \beta / \partial \kappa_j. \end{aligned} \tag{6.4-6}$$

Substituting into (6.4-5), one obtains

$$\frac{\partial \beta}{\partial \kappa_j} = \frac{s_0 \varepsilon_j e^{-\beta \varepsilon_j} - s_1 e^{-\beta \varepsilon_j}}{s_0 s_2 - s_1^2} = \frac{(\varepsilon_j - \overline{\overline{\varepsilon}}) e^{-\beta \varepsilon_j}}{[\overline{\overline{\varepsilon^2}} - (\overline{\overline{\varepsilon}})^2]\gamma_\mu}, \tag{6.4-7}$$

$$\kappa_j \frac{\partial \beta}{\partial \kappa_j} = \frac{\bar{N}_j}{N} \frac{(\varepsilon_j - \overline{\overline{\varepsilon}})}{\operatorname{var} \varepsilon}. \tag{6.4-8}$$

This finally yields the symmetrical form

$$\overline{\Delta N_i \Delta N_j} = \overline{N}_{i \text{ or } j}\, \delta_{ij} - \frac{\overline{N}_i \overline{N}_j}{N}\left(1 + \frac{(\varepsilon_i - \overline{\overline{\varepsilon}})(\varepsilon_j - \overline{\overline{\varepsilon}})}{\operatorname{var} \varepsilon}\right). \tag{6.4-9}$$

While the correction terms go to zero in an infinite system, they are by no means negligible in a finite system. Let us consider the energy fluctuations in a macroscopic subsystem $V' \subset V$. Let the energy of this system be

$$\mathcal{E}' = \sum_{k=1}^{N'} N_k \varepsilon_k . \tag{6.4-10}$$

We remember that the $\{\kappa_k\}$ are volumes in the μ-space. We want to sum over those cells $\kappa_1 \ldots \kappa_{N'}$ whose projections on the q-space lie within V', see Fig. 6-2.

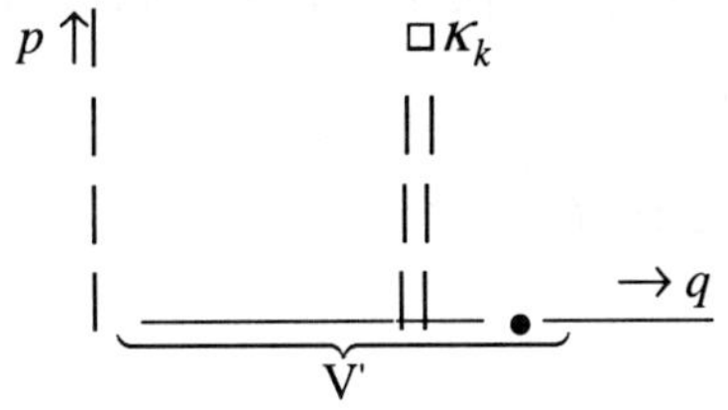

Fig. 6-2. μ-space projection on coordinate space.

For the fluctuations of $\mathcal{E}'$ we have

$$\operatorname{var} \mathcal{E}' = \sum_{i=1}^{N'} \sum_{j=1}^{N'} \overline{\Delta N_i \Delta N_j}\, \varepsilon_i \varepsilon_j . \tag{6.4-11}$$

Substituting (6.4-9) we find

$$\operatorname{var} \mathcal{E}' = \sum_{k=1}^{N'} \overline{N}_k \varepsilon_k^2 - \frac{1}{N}\left[\sum_{k=1}^{N'} \overline{N}_k \varepsilon_k\right]^2 - \frac{1}{N \operatorname{var} \varepsilon}\left[\sum_{k=1}^{N'} \overline{N}_k (\varepsilon_k - \overline{\overline{\varepsilon}}) \varepsilon_k\right]^2 . \tag{6.4-12}$$

In the volume V' all energies still occur since the sub-volume is of macroscopic size. Thus, molecular averages are the same as for the entire system. Consequently,

$$\sum_{k=1}^{N'} \overline{N}_k \varepsilon_k^p = N' \overline{\overline{\varepsilon^p}} . \tag{6.4-13}$$

We now easily obtain

$$\begin{aligned} \operatorname{var} \mathcal{E}' &= N' \overline{\overline{\varepsilon^2}} - \frac{N'^2}{N} (\overline{\overline{\varepsilon}})^2 - \frac{N'^2}{N} \operatorname{var} \varepsilon \\ &= N (\overline{\overline{\varepsilon}})^2 \frac{N' N''}{N^2} + (N \operatorname{var} \varepsilon) \frac{N' N''}{N^2} . \end{aligned} \tag{6.4-14}$$

Here we set $N'' = N - N'$. The first term represents *partition fluctuations,* giving a partial variance in accord with binomial statistics; the second term represents energy fluctuations due to *heat transfer* between the two finite systems and is just $[C'C''/(C'+C'')]k_B T^2$, conform to Eq.(5.5-7), obtained in a more general context previously.

6.5 Illustrations

6.5.1 *Effect of a Magnetic Field; Bohr–van Leeuwen Theorem*

In subsection 1.6.4 the Maxwell–Boltzmann distribution was derived from heuristic arguments. We can of course obtain it right away from the Boltzmann distribution. With $d^3p = m^3 d^3v$ we find from (6.3-8) by integrating over all positions $dq_1 \ldots dq_r$ and all non-translational momenta $dp_4 \ldots dp_r$ the M–B distribution

$$\bar{n}(\mathbf{v})d^3v = \frac{N e^{-mv^2/2k_BT} d^3v}{\int\!\!\int\!\!\int_{-\infty}^{\infty} e^{-mv^2/2k_BT} dv_1 dv_2 dv_3} = N\left(\frac{m}{2\pi k_B T}\right)^{3/2} e^{-mv^2/2k_BT} d^3v, \quad (6.5\text{-}1)$$

in accord with (1.6-72). By integrating over the polar angles in velocity space, the distribution for the velocity magnitudes, Eq. (1.6-74), is likewise obtained.

In the presence of a magnetic field the connection between p and v is more complicated. The canonical momenta are defined as $p_i = \partial\mathcal{L}/\partial\dot{q}_i$, where $\mathcal{L}$ is the Lagrangian

$$\mathcal{L} = \sum_i [\tfrac{1}{2} m\dot{q}_i^2 + (\mathcal{q}/c)\mathbf{A}(\{q_i\})\cdot\dot{\mathbf{q}}_i - \mathcal{q}\Phi(\{q_i\})], \qquad (6.5\text{-}2)$$

where $\mathbf{A}$ and Φ are the vector and scalar potential, respectively. If $\boldsymbol{\pi} = m\mathbf{v}$ is the 'kinetic momentum' and $\mathbf{p}$ is the canonical momentum, the connection between the (p,q)-space and the (v,q)-space is

$$p_i - (\mathcal{q}/c)A_i = \pi_i = mv_i, \quad q_i \to q_i,$$

$$\left|\frac{\partial(p_1, p_2, \ldots q_r)}{\partial(v_1, v_2, \ldots q_r)}\right| = m^3. \qquad (6.5\text{-}2')$$

Clearly, the Jacobian is still the same as before! We thus find the same Maxwell-Boltzmann distribution, given by (6.5-1).[6] The magnetic moment for a collection of moving charges is proportional to $\int \mathbf{r}\times\rho\mathbf{v}\,d^3v$. Since the mean velocity $\bar{\bar{\mathbf{v}}} = 0$ for any $\mathbf{A}$, we see that there is no diamagnetic moment induced by the magnetic field in a

[6] However, the path of the electrons between collisions is curved due to the Lorentz force.

classical gas. This paradox is a form of the Bohr–van Leeuwen theorem. Diamagnetism of free electrons is therefore a quantum effect; we will deal with such effects in Chapter XVII.

An electric field does, however, have an effect. The (v,q)-space distribution is then given by

$$\bar{n}(\mathbf{v},\mathbf{q})d\omega = N\left(\frac{m}{2\pi k_B T}\right)^{3/2}\left[\int e^{-q\Phi(\mathbf{q})/k_B T}\right]^{-1} e^{-mv^2/2k_B T} e^{-q\Phi(\mathbf{q})/k_B T} d\omega. \quad (6.5\text{-}3)$$

6.5.2 *Generalized Sackur–Tetrode Formula and the Equations of State*

We consider a mono-atomic perfect gas in a gravitational field. From (6.3-7) the partition function is

$$\begin{aligned} z_\mu &= \frac{e}{Nh^3}\iiint_{(-\infty,\infty)} e^{-(p_1^2+p_2^2+p_3^2)/2mk_B T} dp_1 dp_2 dp_3 \int_0^L e^{-mgz/k_B T} dz \int_{(A)} dx dy \\ &= \frac{eA}{N}\left(\frac{2\pi m k_B T}{h^2}\right)^{3/2} \frac{k_B T}{mg}\left(1 - e^{-mgL/k_B T}\right). \end{aligned} \quad (6.5\text{-}4)$$

From (6.3-1)-(6.3-3) the following results for the thermodynamic functions are obtained

$$F = -Nk_B T \ln\left[\frac{Ae}{N}\left(\frac{2\pi m k_B T}{h^2}\right)^{3/2} \frac{k_B T}{mg}\left(1 - e^{-mgL/k_B T}\right)\right], \quad (6.5\text{-}5)$$

$$\mathcal{E} = Nk_B T\left[\frac{5}{2} - \frac{(mgL/k_B T)e^{-mgL/k_B T}}{1 - e^{-mgL/k_B T}}\right], \quad (6.5\text{-}6)$$

$$S = Nk_B \ln\left[\frac{Ae^{7/2}}{N}\left(\frac{2\pi m k_B T}{h^2}\right)^{3/2} \frac{k_B T}{mg}\left(1 - e^{-mgL/k_B T}\right)\right] - \frac{NmgL}{T}\frac{e^{-mgL/k_B T}}{1 - e^{-mgL/k_B T}}. \quad (6.5\text{-}7)$$

This is the generalization of the Sackur-Tetrode formula. For small L we find the ordinary result [Eq. (3.5-1)]:

$$S = k_B N \ln\left[e^{5/2}\frac{V}{N}\left(\frac{2\pi m k_B T}{h^2}\right)^{3/2}\right]. \quad (6.5\text{-}8)$$

For large L (6.5-6) shows that $\mathcal{E} = \frac{5}{2}Nk_B T$ [providing T does not change with altitude], while for small L we recover the equipartition result $\mathcal{E} = \frac{3}{2}Nk_B T$.

We finish this subsection by stating the equations of state for the entropic intensive variables. First,

$$T^{-1} = \frac{3}{2}(k_B N/\mathcal{E}). \quad (6.5\text{-}9)$$

The entropy is

$$S = S[(\mathcal{E}(T,V,N),V,N] \equiv S(T,V,N). \tag{6.5-10}$$

This yields for the derivatives to V and to N (chain rule):

$$\left(\frac{\partial S}{\partial \mathcal{E}}\right)_{V,N}\left(\frac{\partial \mathcal{E}}{\partial V}\right)_{T,N}+\left(\frac{\partial S}{\partial V}\right)_{\mathcal{E},N}=\left(\frac{\partial S}{\partial V}\right)_{T,N}, \tag{6.5-11}$$

$$\left(\frac{\partial S}{\partial \mathcal{E}}\right)_{V,N}\left(\frac{\partial \mathcal{E}}{\partial N}\right)_{T,V}+\left(\frac{\partial S}{\partial N}\right)_{\mathcal{E},V}=\left(\frac{\partial S}{\partial N}\right)_{T,V}. \tag{6.5-12}$$

The entropic forces are the last derivatives on the lhs; further, $\partial\mathcal{E}/\partial V = 0$ and $\partial S/\partial\mathcal{E} = T^{-1}$. We thus find, for a gas of n grammolecules (*moles*)

$$\frac{P}{T}=\left(\frac{\partial S}{\partial V}\right)_{T,N}=\frac{Nk_B}{V}=\frac{nR}{V}, \tag{6.5-13}$$

the ideal gas law, while for the chemical potential we have

$$-\frac{\zeta}{T}=\left(\frac{\partial S}{\partial N}\right)_{T,V}-\frac{1}{T}\left(\frac{\partial \mathcal{E}}{\partial N}\right)_{T,V}=k_B \ln\left[\frac{V}{N}\left(\frac{2\pi m k_B T}{h^2}\right)^{3/2}\right]. \tag{6.5-14}$$

If we deal with a mixture of gases, the entropies are additive. The partial pressures are defined by $P_i/P = N_i/N$. Hence, for each component we have the law of Dalton, discussed already in subsection 4.10.1,

$$P_i = N_i k_B T/V. \tag{6.5-15}$$

We also have for the chemical potential of each component

$$\varsigma_i = k_B T \ln P_i - \xi_i(T) \tag{6.5-16}$$

with

$$\xi_i(T) = k_B T \ln\left[k_B T\left(\frac{2\pi m_i k_B T}{h^2}\right)^{3/2}\right] \equiv \frac{5}{2}k_B T \ln k_B T + k_B T \eta_i, \tag{6.5-17}$$

where η_i is called the chemical constant of the gas. In previous chapters we called the factor in parentheses $\lambda_{th,i}^{-3}$, where $\lambda_{th,i}$ is the thermal wavelength for component *i*. So, $\eta_i = -\frac{3}{2}\ln(kT\lambda_{th,i}^2)$.

One can easily show that (6.5-16) has general validity, also for poly-atomic gases. To find the applicable form for $\xi_i(T)$ we use the alternate result for the chemical potential,

$$\varsigma_i = (\partial F/\partial N_i)_{T,V}, \tag{6.5-18}$$

consistent with the middle member of (6.5-14). We write

$$\mathscr{z}_\mu = e\frac{V}{N}\left(\frac{2\pi m k_B T}{h^2}\right)^{3/2}\overset{*}{\mathscr{z}}_\mu\,, \tag{6.5-19}$$

where $\overset{*}{\mathscr{z}}_\mu$ stems from internal molecular contributions, including rotations and vibrations. For $\xi_i(T)$ we now have

$$\xi_i(T) = k_B T \ln\left[k_B T\left(\frac{2\pi m_i k_B T}{h^2}\right)^{3/2}\right] + k_B T\frac{\partial}{\partial N_i}\left(N_i \ln \overset{*}{\mathscr{z}}_{\mu,i}\right). \tag{6.5-20}$$

For the computation of $\overset{*}{\mathscr{z}}_{\mu,i}$ we must use quantum models.

6.6 Oscillators

6.6.1 *The Planck Oscillator*

The Planck oscillator is any device with energy levels

$$\varepsilon_n = (n+\tfrac{1}{2})\hbar\omega, \quad n = 0,1,\ldots \tag{6.6-1}$$

This may refer to an ordinary pendulum, a vibrating molecule, or to any normal mode of a quantized boson field with operators $(a^\dagger a + \frac{1}{2})\hbar\omega$, see next chapter. For the partition function we have from (6.3-12)

$$\begin{aligned}\mathscr{z}_\mu &= \sum_{n=0}^{\infty} e^{-(n+\frac{1}{2})\hbar\omega/k_B T} \\ &= e^{-\frac{1}{2}\hbar\omega/k_B T}\frac{1}{1-e^{-\hbar\omega/k_B T}} = \frac{1}{2}\operatorname{csch}(\hbar\omega/2k_B T).\end{aligned} \tag{6.6-2}$$

The mean energy per oscillator, including zero-point energy, is

$$\begin{aligned}\overline{\overline{\varepsilon}} &= k_B T^2 \frac{\partial \ln \mathscr{z}_\mu}{\partial T} \\ &= \frac{\hbar\omega}{2} + \frac{\hbar\omega}{e^{\hbar\omega/k_B T}-1} = \frac{\hbar\omega}{2}\coth\frac{\hbar\omega}{2k_B T}.\end{aligned} \tag{6.6-3}$$

The main thermodynamic functions are, besides $\mathscr{E} = N\overline{\overline{\varepsilon}}$,

$$F = Nk_B T \ln[2\sinh(\hbar\omega/2k_B T)], \tag{6.6-4}$$

$$S = (N\hbar\omega/2)\coth(\hbar\omega/2k_B T) - Nk_B \ln[2\sinh(\hbar\omega/2k_B T)]. \tag{6.6-5}$$

We leave it to the reader to show agreement with the result for $\Delta\Gamma$ of Eq. (2.5-29).

6.6.2 *The Fermi Oscillator*

The Fermi oscillator is a rather hypothetical device with two energy levels only:

$$\varepsilon = 0\ ,\ \varepsilon_0, \quad \text{or also} \quad \varepsilon = -\varepsilon_0/2,\ +\varepsilon_0/2. \tag{6.6-6}$$

It arises of course in spin one-half problems, or in quantized Fermi fields as eigenvalues of operators $c^\dagger c\,\varepsilon_0$. The partition function is

$$\mathcal{Z}_\mu = 1 + e^{-\varepsilon_0/k_B T} \quad \text{or} \quad 2\cosh(\varepsilon_0/2k_B T). \tag{6.6-7}$$

The mean energy is

$$\overline{\overline{\varepsilon}} = \frac{\varepsilon_0}{e^{\varepsilon_0/k_B T} + 1} \quad \text{or} \quad \frac{\varepsilon_0}{2}\tanh(\varepsilon_0/2k_B T). \tag{6.6-8}$$

We note the similarities between the two forms for $\overline{\overline{\varepsilon}}$ in (6.6-3) and the two forms of (6.6-8). Finally, for the free energy and the entropy, applicable to the second form of E.V. $\pm\varepsilon_0/2$, we obtain

$$F = -Nk_B T \ln[2\cosh(\varepsilon_0/2k_B T)], \tag{6.6-9}$$

$$S = (N\varepsilon_0/2)\tanh(\varepsilon_0/2k_B T) + Nk_B \ln[2\cosh(\varepsilon_0/2k_B T)]. \tag{6.6-10}$$

In the next few chapters we shall be concerned with the perfect Bose and the perfect Fermi gas. The oscillators in this section prepare the way for these two distinct quantum gases. It remains one of the *curious facts in theoretical physics* that we can either look at oscillators as distinguishable entities with quantized energies – the route taken by Planck in 1900 – or look at indistinguishable particles or quasi-particles, with (anti-) symmetrized wave functions that are the excitations of the associated quantum fields – as implied by Einstein's approach in 1905 to the photo-electric effect.

Many ideal gases are diatomic: H_2, O_2, N_2, etc. The one-particle Hamiltonian will then look like

$$h(p_r, J, r) = \frac{p_r^2}{2\mu} + \frac{J^2}{2\mu r^2} + V(r)\ , \tag{6.6-11}$$

where the centre of mass is taken to be the origin, μ is the reduced mass and J is the angular momentum. Besides translational energy of the centre of mass [not included in (6.6-11)], rotations and vibrations contribute to the specific heat of gases. For the translational energy we can use the classical model and $C_{V,\,trans} = \frac{3}{2}R$. The interaction potential $V(r)$ is usually approximated by any of the standard forms: the Morse potential, the Lennard-Jones potential, or the exponential potential, respectively given by

$$V(r) = A\left(e^{-2ar} - 2e^{-ar}\right); \qquad \text{(Morse)}, \tag{6.6-12}$$

$$V(r) = A/r^n - B/r^m, \quad A,B>0, \quad n>m, \quad \text{(L–J)}, \tag{6.6-13}$$

$$V(r) = Ae^{-\alpha r} - B/r^m, \quad A,B,\alpha>0. \quad \text{(exp pot)}. \tag{6.6-14}$$

Expanding up to second order, the first and third term in (6.6-11) are then equivalent to a harmonic oscillator, of which the frequency ω_{vib}, depending on the constants in the above equations, is easily obtained. The vibrational specific heat follows from $\mathcal{E} = n\overline{\overline{\varepsilon}}$ in (6.6-3) above. The characteristic temperature for vibrations $\theta_{vib} = \hbar\omega_{vib}/k_B$ is quite high, e.g., 3340 K for H_2. The middle term on the right in (6.6-11) gives rise to rotations, to be considered in detail in the next section; the characteristic temperature will be denoted by θ_{rot}. For $\theta_{rot} \ll T \ll \theta_{vib}$ rotations are the dominant factor in the specific heat of diatomic and polyatomic gases.

6.7 Rotators

A general rigid rotator has three moments of inertia I_1, I_2 and I_3 along its principal axes, denoted as ξ, η and ζ. For a symmetrical rotator $I_1 = I_2$. We consider a coordinate system moving with these axes in a general composite state of rotation, as e.g., found in Landau and Lifshitz.[7] The other characterization by Euler angles with respect to a resting coordinate system is more known, but quite laborious.[8]

We shall, however, indicate the main ideas of this method. The Euler angles for a general solid body are depicted in Fig. 6-3(a). For a symmetric rotator with $I_1 = I_2 = I$ and $I_3 = I'$ the Hamiltonian is

$$\mathcal{H} = \frac{p_\vartheta^2}{2I} + \frac{(p_\varphi - p_\chi \cos\vartheta)^2}{2I\sin^2\vartheta} + \frac{p_\chi^2}{2I'}, \tag{6.7-1}$$

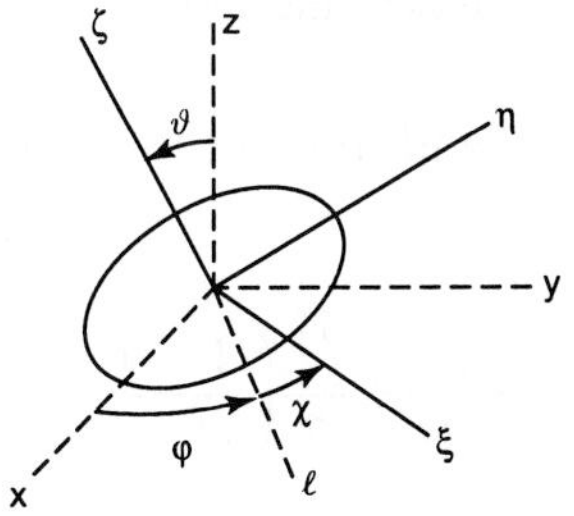

Fig. 6-3(a). Rigid rotator with Euler angles φ, ϑ, χ. ℓ is the line of nodes, i.e. the intersection of the x-y and ζ-η planes.

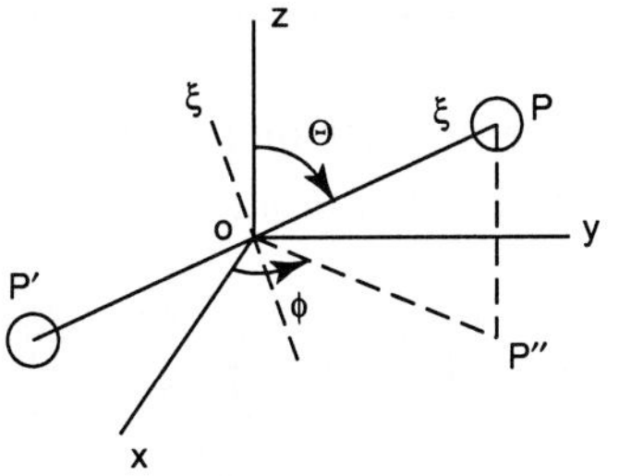

Fig. 6-3(b). A diatomic molecule is an example of a rigid rotator without axial spin. The moment of inertia I is with respect to the line $\xi \perp PP'$.

---

[7] L.D. Landau and E.M. Lifshitz, "Quantum Mechanics", Addison-Wesley 1958, p. 373ff.

[8] R.H. Fowler, Statistical Mechanics", Cambridge University Press 1936; reprinted 1966, p.21; see also, H. Goldstein, "Classical Mechanics", 2nd. Ed., Addison-Wesley, 1981, p. 130ff.

with

$$p_\vartheta = I\dot{\vartheta}, \quad p_\varphi = I\dot{\varphi}\sin^2\vartheta + I'\cos\vartheta(\dot{\chi} + \dot{\varphi}\cos\vartheta), \quad p_\chi = I'(\dot{\chi} + \dot{\varphi}\cos\vartheta). \tag{6.7-2}$$

For a symmetric rotator with no axial spin we can use polar coordinates ϕ and θ, see Fig. 6-3(b). The Hamiltonian is then

$$\mathcal{H} = \frac{1}{2I}\left(p_\theta^2 + \frac{1}{\sin^2\theta}p_\phi^2\right) = \frac{J^2}{2I}, \tag{6.7-3}$$

with

$$p_\theta = I\dot{\theta}, \quad p_\phi = I\dot{\phi}\sin^2\theta. \tag{6.7-4}$$

In (6.7-3) **J** is the angular momentum. Hermitean operators are found by expressing the momenta in Cartesian coordinates, or by proper use of the Jacobian of the volume element. If the latter is D (e.g., $D = r^2 \sin\theta$ for polar coordinates), then generally

$$p_i F = \frac{\hbar}{i}\frac{1}{\sqrt{D}}\frac{\partial}{\partial q_i}\left(\sqrt{D}F\right), \tag{6.7-5}$$

cf. Pauli.[9] Sometimes anomalies occur for p_i^2. For the momenta of the symmetrical rotator with spin we find

$$\begin{aligned} &p_\varphi = (\hbar/i)\partial/\partial\varphi, \quad p_\chi = (\hbar/i)\partial/\partial\chi, \quad p_\vartheta = (\hbar/i)(\partial/\partial\vartheta + \tfrac{1}{2}\cot\vartheta), \\ &p_\varphi^2 = (p_\varphi)^2, \quad p_\chi^2 = (p_\chi)^2, \quad p_\vartheta^2 = -\hbar^2\left[\frac{1}{\sin\vartheta}\frac{\partial}{\partial\vartheta}\left(\sin\vartheta\frac{\partial}{\partial\vartheta}\right)\right]. \end{aligned} \tag{6.7-6}$$

For the symmetric rotator without spin the operators are

$$\begin{aligned} &p_\phi = (\hbar/i)\partial/\partial\phi, \qquad p_\theta = (\hbar/i)[1/\sqrt{\sin\theta}](\partial/\partial\theta)[\sqrt{\sin\theta}], \\ &p_\phi^2 = (p_\phi)^2 = -\hbar^2\frac{\partial^2}{\partial\phi^2}, \quad p_\theta^2 = -\hbar^2\left[\frac{1}{\sin\theta}\frac{\partial}{\partial\theta}\left(\sin\theta\frac{\partial}{\partial\theta}\right)\right]. \end{aligned} \tag{6.7-7}$$

These operators give the required wave equations, as found in Fowler's treatment.

In terms of the angular momenta $\hat{J}_\xi, \hat{J}_\eta$ and $\hat{J}_\zeta$ with respect to the rotating body axes we have for the Hamiltonian

$$\mathcal{H} = \frac{1}{2}\mathsf{I}^{-1}:\hat{\mathbf{J}}\hat{\mathbf{J}} = \frac{1}{2}\left(\frac{\hat{J}_\xi^2}{I_1} + \frac{\hat{J}_\eta^2}{I_2} + \frac{\hat{J}_\zeta^2}{I_3}\right) \to \frac{\hat{J}^2}{2I} + \frac{\hat{J}_\zeta^2}{2}\left(\frac{1}{I'} - \frac{1}{I}\right), \tag{6.7-8}$$

where the last expression applies for a symmetrical rotator. For a symmetrical rotator without spin, we have the simple result, see Fig. 6-3(b):

$$\mathcal{H} = \hat{J}^2/2I. \tag{6.7-9}$$

[9] W. Pauli, "Encyclopaedia of Physics", Springer 1956, pp. 39,40; also, H. Podolski, Phys. Rev. **32**, 812-816, (1928).

The commutation rules for these moving angular momenta are slightly different from those in a fixed frame. They are obtained as follows. Let **a**(t) and **b**(t) be two vectors that commute with each other and with **J**. From the general rule for angular momenta $\mathbf{J}\times\mathbf{J} = i\mathbf{J}$ one easily derives[10]

$$[\mathbf{J}\cdot\mathbf{a}, \mathbf{J}\cdot\mathbf{b}] = i\mathbf{J}\cdot\mathbf{a}\times\mathbf{b}\,. \tag{6.7-10}$$

Thus, choosing **a**(t) and **b**(t) to be unit vectors along the moving ξ and η axes, we have $-\mathbf{J}\cdot\mathbf{a}(t) = \hat{J}_\xi$, $-\mathbf{J}\cdot\mathbf{b}(t) = \hat{J}_\eta$ and $-\mathbf{J}\cdot\mathbf{a}\times\mathbf{b} = \hat{J}_\zeta$; the minus signs stem from the fact that the rotation is reversed, looking from the moving system. Whence we have

$$\hat{J}_\xi\hat{J}_\eta - \hat{J}_\eta\hat{J}_\xi = -i\hat{J}_\varsigma \text{ or } \hat{\mathbf{J}}\times\hat{\mathbf{J}} = -i\hat{\mathbf{J}}\,. \tag{6.7-11}$$

This differs from the usual result in the minus sign at the rhs. The eigenvalues of these rotating angular momenta are of course the same, i.e.,

$$|\hat{J}| = \hbar\sqrt{j(j+1)}, \quad \hat{J}_\varsigma = \hat{m}_j\hbar, \quad \hat{m}_j = -j ... +j\,, \tag{6.7-12}$$

but the eigenstates are replaced by their duals $|\,j,m_j\rangle \rightarrow \langle \hat{m}_j j\,|$.

For a rotator with an axis of symmetry and a transverse moment of inertia $I_1 = I_2 = I$ without axial spin $\hat{J}_\zeta = 0$, the Hamiltonian was given in (6.7-9). We thus have the E.V.,

$$\varepsilon_j = (\hbar^2/2I)\,j(j+1)\,. \tag{6.7-13}$$

Although the azymuthal quantum number m_j is zero, *there is still a* (2*j*+1)*-fold degeneracy with respect to a fixed axis in space,* since the ς-axis turns about the *z*-axis, see Fig. 6-3(b). The partition function is therefore

$$\mathfrak{z}_\mu = \sum_j (2j+1)\, e^{-\hbar^2 j(j+1)/2I\,k_B T}\,. \tag{6.7-14}$$

We shall introduce the characteristic temperature θ_{rot} by $k_B\theta_{rot} = \hbar^2/2I$ and the parameter σ by $\sigma = \theta_{rot}/T$. For large σ we may approximate the sum in (6.7-14) by an integral,

$$\mathfrak{z}_\mu \sim \int_0^\infty dx(2x+1)\, e^{-\hbar^2 x(x+1)/2I\,k_B T} = 2I\,k_B T/\hbar^2\,. \tag{6.7-15}$$

The mean energy is found to be k_BT, as expected from the equipartition theorem, since there are two rotational degrees of freedom. The molar specific heat at high temperatures (but not high enough for vibrational modes to contribute) is thus $\frac{5}{2}R$. For small σ ($\leq \frac{1}{2}$) we will find from (6.7-14) another series that converges more rapidly. Generally, rotator partition functions can be expressed in Jacobi theta-

[10] A. Messiah, "Quantum Mechanics", North Holland Publ. Co. Vol. II, Eq. XIII.4 .

functions.

We need the following definitions[11]

$$\vartheta_2(z\,|\,\tau) = \vartheta_1(z+\tfrac{1}{2}\pi\,|\,\tau) = 2\sum_{n=0}^{\infty} \exp[i\pi\tau(n+\tfrac{1}{2})^2]\cos(2n+1)z,$$
$$\vartheta_3(z\,|\,\tau) = \vartheta_4(z+\tfrac{1}{2}\pi\,|\,\tau) = 1+2\sum_{n=1}^{\infty} \exp[i\pi\tau n^2]\cos 2nz. \tag{6.7-16}$$

Further we need Jacobi's transformation formula

$$\vartheta_2(z\,|\,\tau) = (-i\tau)^{-1/2}\exp(-iz^2/\pi\tau)\,\vartheta_4(-\tfrac{z}{\tau}\,|\,-\tfrac{1}{\tau})\,. \tag{6.7-17}$$

From the definition of ϑ_2 we have, taking $\tau = i\sigma/\pi$ and letting $z \to \pi z$,

$$\frac{\partial\vartheta_2(\pi z\,|\,i\sigma/\pi)}{\partial z} = -2\sum_{n=1}^{\infty} e^{-\sigma[n(n+1)+\frac{1}{4}]}(2n+1)\pi\sin[(2n+1)\pi z]\,, \tag{6.7-18}$$

from which we obtain

$$-\frac{1}{\pi^2}e^{\sigma/4}\int_0^{\infty}\frac{\partial\vartheta_2(\pi z\,|\,i\sigma/\pi)}{\partial z}\frac{dz}{z} = \frac{2}{\pi}\sum_{n=1}^{\infty} e^{-\sigma n(n+1)}(2n+1)\int_0^{\infty}\frac{\sin[(2n+1)\pi z]}{z}dz\,. \tag{6.7-19}$$

The latter (standard) integral is equal to $\tfrac{1}{2}\pi$. The rhs of (6.7-19) is just the partition function (6.7-14). We thus obtained

$$\mathfrak{z}_\mu = -\frac{1}{\pi^2}e^{\sigma/4}\int_0^{\infty}\frac{\partial\vartheta_2(\pi z\,|\,i\sigma/\pi)}{\partial z}\frac{dz}{z}\,. \tag{6.7-20}$$

Using now Jacobi's transformation formula and the definition of ϑ_4 we have

$$\vartheta_2(\pi z\,|\,i\sigma/\pi) = \sqrt{\pi/\sigma}\sum_{n=-\infty}^{\infty}(-1)^n e^{-\pi^2(z-n)^2/\sigma}\,. \tag{6.7-21}$$

This, then, yields for $\mathfrak{z}_\mu$

$$\mathfrak{z}_\mu = (1/\sigma)e^{\sigma/4}\sqrt{\pi/\sigma}\sum_{n=-\infty}^{\infty}(-1)^n\,\mathcal{P}\!\int_{-\infty}^{\infty}\frac{z-n}{z}e^{-\pi^2(z-n)^2/\sigma}dz\,, \tag{6.7-22}$$

where $\mathcal{P}$ denotes the Cauchy principal value. The substitution $z = (t/\pi)+n$ finally gives

---

[11] Cf. M. Abramowitz and Irene A. Stegun, "Handbook of Mathematical Functions", National Bureau of Standards, 1964, Chapter 16; also, E.T. Whittaker and G.N. Watson, "A course of Modern Analysis" Cambridge Univ. Press, 4th Ed. 1958, Chapter 21; see also A. Sommerfeld, "Statistical Mechanics and Thermodynamics", Acad. Press, N.Y. 1956 (Vol. V), problem IV.6, p. 384ff.

$$\mathcal{z}_\mu = \frac{1}{\sigma} e^{\sigma/4} (\pi\sigma)^{-½} \mathcal{P} \int_{-\infty}^{\infty} dt\, e^{-t^2/\sigma} \sum_{n=-\infty}^{\infty} (-1)^n \frac{t}{t+n\pi} . \tag{6.7-23}$$

The latter sum is just the partial fraction expansion of the meromorphic function $t/\sin t$, which has poles for $t = \pm n\pi,\ n \neq 0$. To evaluate the integral we use the series expansion

$$t\csc t = 1 + \frac{1}{6}t^2 + \frac{7}{360}t^4 + \frac{31}{15120}t^6 + ... \tag{6.7-24}$$

For σ small enough the Gaussian integrals can be carried out, since most of the contributions come from the neighbourhood of $t = 0$. Using (1.6-51) we obtain

$$\mathcal{z}_\mu = \frac{1}{\sigma} e^{\sigma/4} \left[1 + \frac{1}{12}\sigma + \frac{7}{480}\sigma^2 + \frac{31}{8064}\sigma^3 + ... \right]. \tag{6.7-25}$$

We also expand exp(σ/4) and the series are multiplied term by term. The final results are

$$\mathcal{z}_\mu = \frac{1}{\sigma}\left(1 + \frac{1}{3}\sigma + \frac{1}{15}\sigma^2 + \frac{4}{315}\sigma^3 + ... \right), \tag{6.7-26}$$

$$\ln \mathcal{z}_\mu = -\ln\sigma + \frac{1}{3}\sigma + \frac{1}{90}\sigma^2 + \frac{8}{2835}\sigma^3 + ... \tag{6.7-27}$$

We remember that σ was defined as θ_{rot}/T. For the mean energy we find

$$\overline{\overline{\varepsilon}} = k_B T \left(1 - \frac{\sigma}{3} - \frac{\sigma^2}{45} - \frac{8\sigma^3}{945} - ... \right). \tag{6.7-28}$$

For the molar specific heat this yields

$$C_V = R\left(1 + \frac{1}{45}\sigma^2 + \frac{16}{945}\sigma^3 + ... \right). \tag{6.7-29}$$

It is noticed that the classical value $C_V \sim R$ is reached *from above*. Consequently, the rotational specific heat exhibits a maximum. We give a figure for the rotational specific heat of an XY molecule in Fig 6-4(a); we also give a sketch for the vibrational specific heat in Fig. 6-4(b). In Fig. 6-4(c) we give the specific heat of deuterium, observed and calculated, cf. Fowler, Op. Cit.; the agreement is excellent.

We make a few remarks on rotation with axial spin. The Hamiltonian was already given in (6.7-8). The E.V. are immediately seen to be

$$\varepsilon_{j,\hat{m}_j} = \frac{\hbar^2}{2I} j(j+1) + \frac{\hbar^2}{2}\hat{m}_j^2 \left(\frac{1}{I'} - \frac{1}{I} \right). \tag{6.7-30}$$

The energy is $(2j+1)$ - fold degenerate with respect to the *possible directions* of the

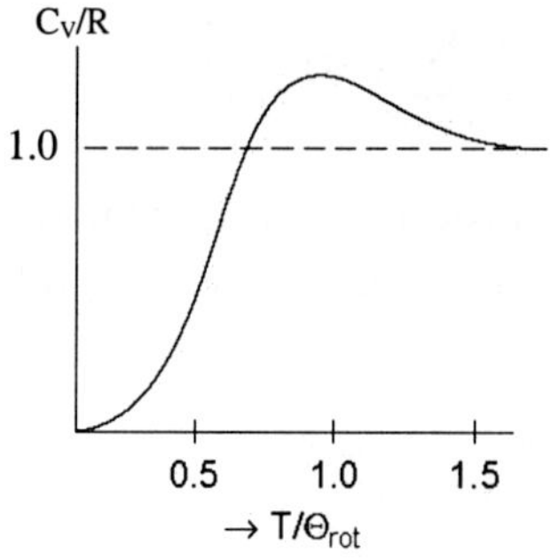

Fig. 6-4(a) Sketch of rotational specific heat vs. T/θ_{rot}.

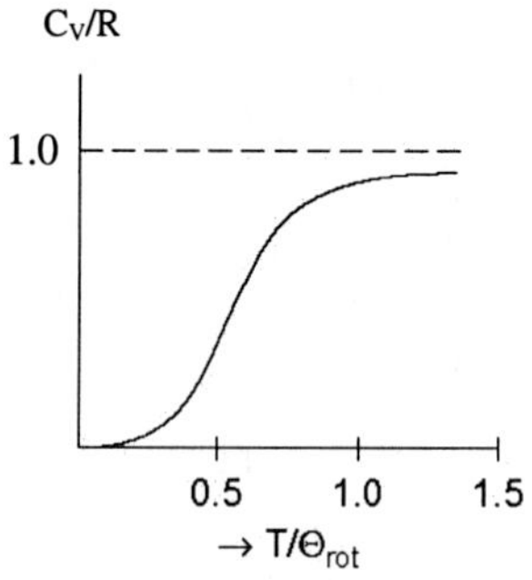

Fig. 6-4(b) Sketch of vibrational specific heat vs. T/θ_{vib}.

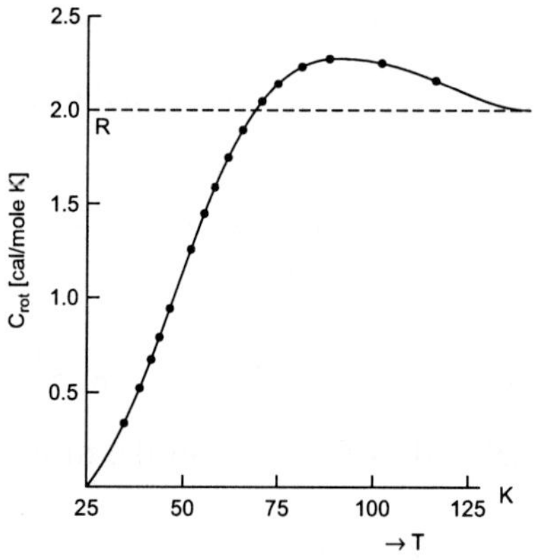

Fig. 6-4(c) Specific heat of D_2 in the range 25-125K. Solid curve computed by Fowler, Ref. 8, Fig 3.2. [With permission.] Measured points by Clusius and Bartholomé.[12]

angular momentum in space; the degeneracy with respect to the *ζ-axis is low* (only 2 for $\hat{m}_j = \pm$). However, if $I' = I$, the last term in (6.7-30) is absent and the latter degeneracy is also $2j+1$. [In the Eulerian treatment the degeneracy $(2j+1)^2$ appears explicitly, stemming both from φ and χ.] We thus find the partition function for a partially symmetrical rotator ($I \neq I'$)

$$\mathfrak{z}_\mu = \sum_{j=0}^{\infty} 2(2j+1) \sum_{\hat{m}_j=-j}^{j} e^{-j(j+1)\sigma - \hat{m}_j^2 \sigma'}, \tag{6.7-31}$$

where $\sigma' = \sigma I (I'^{-1} - I^{-1})$. For a completely symmetrical rotator, $I' = I$, the result is

$$\mathfrak{z}_\mu = \sum_{j=0}^{\infty} (2j+1)^2 e^{-j(j+1)\sigma}. \tag{6.7-32}$$

[12] Clusius and Bartholomé, Zeitschr. für Elektrochemie **40**, 524 (1934).

Both partition functions can be expressed in ϑ-functions. For both cases the high temperature limit of the mean energy is $\frac{3}{2}k_BT$, as expected from the equipartition theorem.

6.8 Dielectric and Paramagnetic Dipoles

A classical example of Boltzmann statistics – in the literary and technical sense of the word – is the calculation of the contribution to the dielectric constant due to the orientational polarization by Langevin in 1905 and by Debye in 1912.[13] Let μ be the electric dipole moment of the molecules and let $\mathbf{E}$ be the external electric field. We assume that the orientation of the dipole with respect to a set of axes is as in Fig. 6-2(b). The one-molecule Hamiltonian is, in accord with (6.7-3),

$$h(\theta,\phi,p_\theta,p_\phi)=\frac{1}{2I}\left(p_\theta^2+\frac{p_\phi^2}{\sin^2\theta}\right)-\mu E\cos\theta. \tag{6.8-1}$$

The partition function is the phase integral

$$\begin{aligned}\gamma_\mu&=\frac{e}{N\hbar^2}\int_{-\infty}^{\infty}dp_\theta\int_{-\infty}^{\infty}dp_\phi\int_0^{2\pi}d\phi\int_0^{\pi}d\theta\, e^{-h(\theta,\phi,p_\theta,p_\phi)/k_BT}\\&=\frac{4\pi^2 Iek_BT}{N\hbar^2}\int_0^{\pi}e^{\mu E\cos\theta/k_BT}\sin\theta d\theta=\frac{8\pi^2 Iek_B^2T^2}{N\hbar^2\mu E}\sinh\left(\frac{\mu E}{k_BT}\right).\end{aligned} \tag{6.8-2}$$

For the free energy in the Gibbsian sense [subsection 1.3.2] $F(T,V,N,E)$ we obtain

$$F(T,V,N,E)=-Nk_BT\ln\left[(T^2/E)\sinh(\mu E/k_BT)\right]+const. \tag{6.8-3}$$

By differentiation we find the polarization

$$\mathcal{P}=-\frac{\partial(F/V)}{\partial E}=nk_BT\frac{\partial}{\partial E}\ln\gamma_\mu=n\mu\mathcal{L}(\mu E/k_BT), \tag{6.8-4}$$

where we introduced the Langevin function

$$\mathcal{L}(x)=\coth x-\frac{1}{x}. \tag{6.8-5}$$

The electric susceptibility χ_e is given by $\partial\mathcal{P}/\partial E$. It is independent of E for low values of the electric field. For large E, $\mathcal{L}(x)\to 1$ and $\mathcal{P}\sim n\mu$, indicating that all dipoles are lined up with the field.

---

[13] P. Langevin, J. de Physique **4**, 678 (1905); P. Debye, Physik. Zeitschrift **13**, 97 (1912).

The quantum case occurs for paramagnetism of *localised* atomic moments stemming from states with total spin[14] $s \neq 0$, or, as the case may be, from orbital motion. We have the energies

$$\varepsilon_{m_s} = -m_s g \rho_B H, \quad m_s = -s, -s+1, \ldots + s. \tag{6.8-6}$$

Here g is the Landé factor and ρ_B is the Bohrmagneton. For orbital motion replace s by j and m_s by m_j. The above energies are responsible for the splitting of the spectral lines in a magnetic field (Zeeman effect). The partition function is the sum of a geometrical progression; with (6.3-12) and $Z_k = 1$,

$$\mathcal{Z}_\mu = \sum_{m_s=-s}^{s} e^{m_s g \rho_B H / k_B T} = \frac{\sinh[g \rho_B H (2s+1)/2k_B T]}{\sinh(g \rho_B H / 2k_B T)}. \tag{6.8-7}$$

From this we find the free energy F, the energy $\mathcal{E}$ and the magnetization $M = -\partial(F/V)/\partial H$. Differentiating once more, we obtain the magnetic susceptibility χ_m and the specific heat C_V. Some results are

$$\mathcal{E} = -N g \rho_B s H \mathcal{B}_s(g \rho_B H s / k_B T), \tag{6.8-8}$$

$$M = n g \rho_B s \mathcal{B}_s(g \rho_B H s / k_B T), \tag{6.8-9}$$

where $\mathcal{B}_s$ is the Brillouin function

$$\mathcal{B}_s(y) = \frac{1}{2s}\left[(2s+1)\coth\frac{y(2s+1)}{2s} - \coth\frac{y}{2s}\right]. \tag{6.8-10}$$

For small y series expansion gives

$$\mathcal{B}_s(y) \approx (s+1)y/3s. \tag{6.8-11}$$

We still note that for $s = \frac{1}{2}$ we can use the identity $2\coth 2y = \coth y + \tanh y$, in order to see that M is in agreement with the results of (4.8-23) and (4.8-31). For small H we obtain for the specific heat and the susceptibility

$$C_V = N g^2 \rho_B^2 H^2 s(s+1)/3k_B T^2, \tag{6.8-12}$$

$$\chi_m = C/T, \quad C = n g^2 \rho_B^2 s(s+1)/3k_B = n\mu^2/3k_B. \tag{6.8-13}$$

The susceptibility follows the inverse temperature or Curie's law. From the observed Curie constant C one can calculate $p_{eff} \equiv |\mu|/\rho_B$, the effective number of Bohrmagnetons of the magnetic moment and compare it with $p_{theor} = g\sqrt{s(s+1)}$ or $g\sqrt{j(j+1)}$. A comparison for the iron group is given in the Table 6-1. Generally, the agreement is good. Discrepancies can be caused by crystal

[14] We use lower case symbols for the total spin quantum number, as in most current textbooks, reserving S for the spin operator.

fields, orbital motion and nuclear spin contributions.[15]

Table 6-1. Comparison of theoretical and experimental magnetic moments for the iron group. After C. Kittel[15].

ion	config.	basic level	p (theor.) $g\sqrt{j(j+1)}$	p (theor.) $g\sqrt{s(s+1)}$	experiment
Ti^{3+} V^{4+}	$3d^1$	$^2D_{3/2}$	1.55	1.73	1.8
V^{3+}	$3d^2$	$^3F_2$	1.63	2.83	2.8
Cr^{3+} V^{2+}	$3d^3$	$^4F_{3/2}$	0.77	3.87	3.8
Mn^{3+} Cr^{2+}	$3d^4$	$^5D_0$	0.00	4.90	4.9
Fe^{3+} Mn^{2+}	$3d^5$	$^6S_{5/2}$	5.92	5.92	5.9
Fe^{2+}	$3d^6$	$^5D_4$	6.70	4.90	5.4
Co^{2+}	$3d^7$	$^4F_{9/2}$	6.64	3.87	4.8
Ni^{2+}	$3d^8$	$^3F_4$	5.59	2.83	3.2
Cu^{2+}	$3d^9$	$^2D_{5/2}$	3.55	1.73	1.9

6.9 Chemical Equilibrium and the Mass-Action Law

Consider a chemical reaction

$$\sum_i \nu_i A_i \to \sum_i \nu_i' A_i', \tag{6.9-1}$$

where A_i are the chemical symbols for the reactants and A_i' of the products and ν_i, ν_i' stands for the coefficients in the reaction equation. E.g., for $2H_2 + O_2 \to 2H_2O$, $\nu_1 = 2$, $\nu_2 = 1$, $\nu_1' = 2$. After some time the opposite reaction must become important and we reach chemical equilibrium

$$\sum_i \nu_i A_i \rightleftarrows \sum_i \nu_i' A_i'. \tag{6.9-2}$$

[15] C. Kittel, "Introduction to Solid State Physics", 3rd Ed., John Wiley, NY (1968), p.438. [Also later Editions: 8th Ed., John Wiley, NY (2005), p.308.]

In an adiabatic equilibrium system $\partial S/\partial N_i = 0$. In most chemical reactions it is more realistic to consider an isothermal, isobaric system, for which $\partial G/\partial N_i = 0$. However, since

$$\left(\frac{\partial G}{\partial N_i}\right)_{P,T} = -T\left(\frac{\partial S}{\partial N_i}\right)_{\mathcal{E},V} \tag{6.9-3}$$

the same conditions apply. For a chemical reaction generally the constituents are conserved (law of Lavoisier). Thus in (6.9-3) the derivative with respect to N_i is a *total* derivative. So we have

$$\left(\frac{\partial G}{\partial N_i}\right)_{total} = \left(\frac{\partial G}{\partial N_i}\right)_{\{N\}\neq N_i} + \sum_j{}' \left(\frac{\partial G}{\partial N_j}\right)_{\{N\}\neq N_j} \frac{dN_j}{dN_i} = 0\,, \tag{6.9-4}$$

where the summation is over all $j \neq i$. The latter differentials are the number ratios ν_j/ν_i for all species on the lhs of (6.9-2) and $\nu_j{}'/\nu_i{}'$ for the species on the rhs. Thus (6.9-4) yields

$$\sum_i \nu_i \varsigma_i = \sum_i \nu_i{}'\varsigma_i{}'. \tag{6.9-5}$$

For the chemical potentials we substitute (6.5-16). The result is

$$\prod_i (P_i)^{\nu_i} \Big/ \prod_i (P_i{}')^{\nu_i{}'} = \exp\left\{\frac{1}{k_B T}\left[\sum_i \nu_i \xi_i(T) - \sum_i \nu_i{}'\xi_i{}'(T)\right]\right\} \equiv K_P(T). \tag{6.9-6}$$

This is the *mass-action law* of Guldberg and Waage. The rhs is the reaction constant. We can also introduce the concentrations in *moles/litre*, $C_i = N_i/22.4N_A$. Using Dalton's law we thus have

$$\prod_i (C_i)^{\nu_i} \Big/ \prod_i (C_i{}')^{\nu_i{}'} = RT^{(\Sigma_i \nu_i{}' - \Sigma_i \nu_i)} K_P \equiv K_C. \tag{6.9-7}$$

We must now say more about the ξ's as given by (6.5-20). A peculiar but important contribution to $\gamma^*_{\mu,i}$ arises from the ground-state energy $\varepsilon_{0,i}$. In the partition function this gives a common factor $\exp(-\varepsilon_{0,i}/k_B T)$ in all terms, which can be ignored in all equations of state, except in ς_i. The reason is that the term $-N_i k_B T \ln(\exp - \varepsilon_{0,i}/k_B T) = N_i \varepsilon_{0,i}$ has a finite derivative to N_i. Let γ^{**}_{μ} represent the internal partition function of energies with reference to the ground-state level $\varepsilon_{0,i}$ for species i. Then, instead of (6.5-20) we write

$$\xi_i = -\varepsilon_{0,i} + k_B T \ln\left[\left(\frac{2\pi m_i k_B T}{h^2}\right)^{3/2} k_B T\right] + k_B T \frac{\partial}{\partial N_i}\left(N_i \ln \gamma^{**}_{\mu,i}\right). \tag{6.9-8}$$

Employing these ξ's in the reaction 'constants', we find

$$K_C(T) = K_{C0} \exp\left[-\frac{1}{k_B T}\left(\sum_i \nu_i \varepsilon_{0,i} - \sum_i \nu_i' \varepsilon_{0,i}'\right)\right], \tag{6.9-9}$$

with K_{C0} given by

$$K_{C0} = \frac{\Pi_i (2\pi m_i k_B T / h^2)^{3\nu_i/2} (N_A)^{-\nu_i}}{\Pi_i (2\pi m_i' k_B T / h^2)^{3\nu_i'/2} (N_A)^{-\nu_i'}}$$

$$\times \exp\left[\sum_i \nu_i \frac{\partial}{\partial N_i}(N_i \ln \gamma_{\mu,i}^{**}) - \sum_i \nu_i' \frac{\partial}{\partial N_i'}(N_i' \ln \gamma_{\mu,i}^{**})\right]. \tag{6.9-10}$$

We note that the main temperature dependence in $K_C(T)$ observed in (6.9-9) stems from the difference in ground-state levels on both sides of the reaction. This difference is also the reaction heat at absolute zero. For the case of the water-dissociation equilibrium the energy difference is $2\varepsilon_{0,\mathrm{H}_2} + \varepsilon_{0,\mathrm{O}_2} - 2\varepsilon_{0,\mathrm{H}_2\mathrm{O}} = 4.9 eV$, or 56.9 *Kcal/mole* for the formation of $2H_2O$. The resulting temperature factor is $\exp(-57{,}000/T)$, which is extremely small. Thus the equilibrium lies to the right and the dissociation of water into its constituents is minute at ordinary temperatures.[16]

In solutions the same considerations apply *grosso motto*. The mass-action law holds equally well for electrolytic reactions. The chemical potential for solutions can be written as

$$\varsigma_i = k_B T \ln C_i + \psi_i(T). \tag{6.9-11}$$

Finally, some words on the *kinetic approach* to the mass-action law. In this, more common approach, the "gain rate" $g(t)$ is balanced with the "retrieval rate" (or recombination rate) $r(t)$. We clearly have the following forms

$$g(t) = \alpha (C_1')^{\nu_1'} (C_2')^{\nu_2'} \ldots, \quad r(t) = \beta C_1^{\nu_1} C_2^{\nu_2} \ldots, \tag{6.9-12}$$

where α and β are constants (depending on T). In equilibrium $\overline{g} = \overline{r}$, from which the mass-action law (6.9-7) follows; however, the statistical mechanical nature of K_C remains unknown, although an *ad hoc* exponential temperature dependence is usually suggested. If a number of chemical reactions between the constituents are going on simultaneously we have

$$dC_i'/dt = \sum_{i=1}^{M} [g_i(t) - r_i(t)] \doteq 0, \tag{6.9-13}$$

where the latter equality holds in the steady state. However, the mass-action law, as derived before, holds for each set of rates separately. Therefore, in thermal

[16] We can of course enhance the dissociation, outside equilibrium, through the absorption of ultraviolet light, $h\nu \geq 4.9 eV$. This photo-chemical process may have been responsible for the oxygen in earth's primitive atmosphere, prior to plant photo-synthesis of O_2.

equilibrium, each term in the summand is *separately zero.* This is a manifestation of the *principle of detailed balance*,

$$\overline{g_i} = \overline{r_i} \,. \tag{6.9-14}$$

We shall be more generally concerned with this in Chapter XIII of non-equilibrium statistical mechanics, when we deal with "Boltzmann's H-theorem".

6.10 Problems to Chapter VI

6.1 Establish in detail the result for the fluctuations of the Boltzmann distribution, Eq. (6.4-2). Hint: Consider separately the cases $i \neq j$ and $i = j$.

6.2 Consider a 'generalized Fermi oscillator', with eigen-energies $\varepsilon_\alpha = 0, \varepsilon, 2\varepsilon, \ldots, p\varepsilon$. Obtain the partition function, Helmholtz free energy and mean energy.

6.3 Fill in the details for the rotator treatment, Eqs.(6.7-14) to (6.7-26). In particular *show* that the sum in (6.7-23) is the partial fraction expansion of the meromorphic function $t/\sin t$, employing Mittag–Leffler's theorem.

6.4 With less rigour, the result for the symmetrical rotator can be obtained from Euler's summation formula:

$$\sum_{n=0}^{\infty} f(n) = \int_0^{\infty} f(x)dx + \frac{1}{2} f(0) - \frac{1}{12} f'(0) + \frac{1}{720} f^{(3)}(0) - \frac{1}{30240} f^{(5)}(0) + \ldots \tag{1}$$

Obtain an approximation to the partition function and compare with the results of the text.

6.5 Consider the chemical equilibrium

$$D_2 + H_2 \rightleftarrows 2HD \,. \tag{2}$$

The temperature is sufficiently high, so that the rotational motion can be described classically. Evaluate the equilibrium constant.

6.6 In most cases the intermolecular potential has a profile that is lopsided, rather than parabolic. Often, the anharmonicity can be described by a formula

$$\varepsilon_n = (n + \tfrac{1}{2})\hbar\omega - x(n + \tfrac{1}{2})^2 \hbar\omega, \quad 0 < x < 1. \tag{3}$$

Show that to first order in x the partition function is given by

$$\ln \gamma_\mu(T) = \ln \gamma_{\mu 0}(T) + 2xue^{-u}/(1-e^{-u})^2, \quad u = \hbar\omega / k_B T. \tag{4}$$

Establish the following results for the vibrational specific heat per mole:

$$C_{vib} = R\frac{u^2 e^u}{(e^u - 1)^2}\left\{1 + \frac{2x}{1-e^{-u}}\left[-2 + u(1+3e^{-u} + 6e^{-2u})\right]\right\}; \tag{5}$$

$$C_{vib}(T \to \infty) = R\left\{\left(1 - \tfrac{1}{12}u^2 + \tfrac{1}{240}u^4\right) + 4x\left((1/u) + \tfrac{1}{80}u^3\right)\right\}. \tag{6}$$

Explain why the main correction term $4xk_B T/\hbar\omega$ increases with temperature.

6.7 For a Boltzmann gas in a gravitational field, the entropy and energy are given by the generalized Tetrode formulae (6.5-6) and (6.5-7). Let $L = V/A$, where A is constant.
(a) Find $S(T,V,N)$ and $\mathcal{E}(T,V,N)$;
(b) Using the chain rules (6.5-11) and (6.5-12), find the entropic forces P/T and $-\varsigma/T$.
(c) Is the ideal gas law still satisfied?
(d) Let $L \to \infty$ and find again P/T and $-\varsigma/T$. Is the ideal gas law satisfied?

6.8 Consider a set of symmetrical rotators with axial spin, the Hamiltonian being expressed in Euler angles φ, ϑ and χ and associated canonical momenta p_φ, p_ϑ and p_χ. Obtain the classical approximation to the partition function and the specific heat. Hint: Carry out the integrations in the order $dp_\vartheta dp_\varphi dp_\chi$ $d\varphi\, d\chi\, d\vartheta$.

6.9 Consider a system of paramagnetic spins with quantum number s. The spins are at first independent, so that Boltzmann statistics for the orientational energies applies.
(a) Express the mean magnetization in the Brillouin function $\mathcal{B}_s(y)$;
(b) Now below the Curie temperature let there be a molecular field $H_m = \gamma\langle M\rangle$. Write down a self-consistent equation for $\langle M\rangle$ in terms of $\mathcal{B}_s(y)$.
(c) Apply your result for the case that $s = ½$, $g = 2$ and obtain the mean-field expressions of Section 4.8. Identify γ.

Chapter VII

The Occupation-Number State Formalism; Spin and Statistics

7.1 Symmetrization of States

One of the basic tenets of quantum theory is that identical, non-localised particles are indistinguishable. Therefore, all many-body quantum states must be symmetrized. We also say: identical particles cannot be "painted". Thus a state with particle 1 having momenta and coordinates p_1, q_1 and particle 2 having momenta and coordinates p_2, q_2 is basically identical to a state in which particle 1 has momenta and coordinates p_2, q_2 and particle 2 has the momenta and coordinates p_1, q_1. Symmetrization means that the dynamical many-body state must lie in an appropriate subspace of the state space otherwise available. We shall now formulate this in a quantitative way.

Suppose that $\{|\alpha^k)\}$ is a complete set of single particle eigenstates. It is not necessary that $|\alpha)$ is an eigenstate of the one-particle Hamiltonian *h*, but we do assume that these states span the state space of *h*. Then the system can be characterized by the many-body states $|\varphi\rangle$ such that

$$|\varphi\rangle = |\alpha_1^a) \otimes |\alpha_2^b) ... \otimes |\alpha_N^\nu), \tag{7.1-1}$$

where particle 1 is in the state $|\alpha^a)$, particle 2 in state $|\alpha^b)$ ... and particle *N* in state $|\alpha^\nu)$. If there are no interactions, and *if* the $|\alpha)$'s are eigenstates of *h*, then $|\varphi\rangle$ is an eigenstate of the many-body Hamiltonian $\mathcal{H}$. But even if there are strong interactions, $|\varphi\rangle$ is still a state that spans the Hilbert space of $\mathcal{H}$. Most often, however, $|\varphi\rangle$ is an eigenstate of an appropriate sub-Hamiltonian $\mathcal{H}_0$, related to $\mathcal{H}$ by $\mathcal{H} = \mathcal{H}_0 + \lambda\mathcal{V}$ where the latter term represents the interparticle interactions. Now let *P* be an operator that permutes the particles; clearly $P|\varphi\rangle$ is an identical eigenstate. Since the states are normalized, we have

$$\langle\varphi|P^\dagger P|\varphi\rangle = \langle\varphi|\varphi\rangle = 1, \tag{7.1-2}$$

from which follows $P^\dagger P = I$, where *I* is the identity operator. The eigenvalues of *P* therefore lie on the unit circle in the complex plane.

Before we go into details of *P*, we consider the simpler transposition operator p_{ij} which interchanges particles *i* and *j*. Since two successive interchanges restore the original situation, we have $(p_{ij})^2 = 1$. Thus the eigenvalues are $\lambda_{ij} = \pm 1$. Consider next the transposition that interchanges particles 1 and 2. It is easily verified that

$$p_{ij} = p_{1i} p_{2j} p_{12} p_{2j} p_{1i} \,. \tag{7.1-3}$$

Whence,

$$\lambda_{ij} = (\lambda_{1i})^2 (\lambda_{2j})^2 \lambda_{12} = \lambda_{12} \,. \tag{7.1-4}$$

As a consequence, the eigenvalues of p are independent of the particular transposition envisaged. We make a spectral decomposition into projectors. Let s be the projector for the eigenvalue $+1$ and a for the eigenvalue -1; then

$$p = s - a \,. \tag{7.1-5}$$

Closure requires[1]

$$1 = s + a \,. \tag{7.1-6}$$

Thus the symmetrization operator s and the antisymmetrization operator a are given by

$$s = \tfrac{1}{2}(1 + p), \quad a = \tfrac{1}{2}(1 - p) \,. \tag{7.1-7}$$

Returning to a general permutation, we note that each permutation can be obtained as a product of transpositions. Let the eigenvalues of P be $\{\mu\}$. Then $P|\varphi\rangle = \mu|\varphi\rangle$. Let p be the number of transpositions necessary to obtain P. In combinatorial analysis p is called the parity; it is also equal to the number of inversions (larger number preceding smaller number) invoked by P, modulo 2. E.g., take the arrangement 3,1,4,2 compared to 1,2,3,4. This permutation is obtainable by the transpositions $p_{31} p_{32} p_{34}$; the number of inversions is also three. Of the p transpositions let p' have the eigenvalue $+1$ and let p'' have the eigenvalue -1. Then

$$\mu = (+1)^{p'} (-1)^{p''}, \quad p' + p'' = p \,. \tag{7.1-8}$$

The manifold of the states $P|\varphi\rangle$ gives rise to "exchange degeneracy". This degeneracy is removed by the *symmetrization postulate,* which states that the allowed transpositions must be all of the same nature, i.e., the states are either symmetrical or antisymmetrical with respect to permutation; moreover, in a given system, only one of these possibilities applies. Thus we have $p'' = 0$ or p only, so that

$$\mu = \begin{cases} +1 & (\textit{systems with symmetrical states}) \,, \\ (-1)^p & (\textit{systems with antisymmetrical states}) \,. \end{cases} \tag{7.1-9}$$

In order to obtain the symmetrized or antisymmetrized state, we must specify the projectors associated with this choice, denoted by S and A, respectively. In contrast to

---

[1] The closure property in this form is also called the "decomposition of unity".

the case of transposition operators, we have $S + A \leq 1$, since partially symmetrizing projectors have been discarded. The following forms are, however, easily surmised:

$$S = \frac{1}{N!}\sum_P P, \qquad A = \frac{1}{N!}\sum_P (-1)^p P. \tag{7.1-10}$$

Accordingly, the state space must be limited to the subspace of the projectors *S* or *A*. This subspace is the space of the *occupation-number states*, which are obtained as follows. In the states $S\,|\,\varphi\rangle$ or $A\,|\,\varphi\rangle$ it does not matter which particle is in the state $|\,\alpha)$. The state is therefore characterized by the number of particles $n_a, n_b, n_c, \ldots$ occupying the one-particle states $|\,\alpha^a), |\,\alpha^b), |\,\alpha^c)\ldots$. However, to obtain a complete set of occupation numbers, we must consider the occupancies of a complete basis $\{|\,\alpha^k)\}$ and not just those of the original state $|\,\varphi\rangle$ as given by (7.1-1), even though most states of this denumerably infinite set will be empty. Let the occupation numbers be $\{n_k\}$. The occupation-number many-body states, denoted by $|\,\gamma\rangle$, then are

$$|\,\gamma\rangle \equiv |n_1 n_2 \ldots n_k \ldots\rangle = \begin{cases} C_S S\,|\,\varphi\rangle \\ C_A A\,|\,\varphi\rangle \end{cases} \tag{7.1-11}$$

where the *C*'s are normalization factors. For a symmetrized state we have, observing (7.1-10)

$$\langle\varphi\,|\,S^\dagger S\,|\,\varphi\rangle = |\,C_S\,|^{-2} = (N!)^{-2}\sum_{P,P'}\langle\varphi\,|\,P'^\dagger\, P\,|\,\varphi\rangle. \tag{7.1-12}$$

Now when $P = P'$ there are *N*! terms $\langle\varphi\,|\,\varphi\rangle = 1$. When $P' \neq P$, there is no contribution, unless P' permutes the particles n_k occupying the same single particle state $|\,\alpha_k)$ (the effect of P' is then the same as that of *P*). Therefore,

$$\sum_{P,P'} \to N!\,n_1!\,n_2!\ldots n_k!\ldots \tag{7.1-13}$$

This yields for (7.1-12),

$$|\,C_S\,|^{-2} = \frac{N!\,n_1!\,n_2!\ldots n_k!\ldots}{(N!)^2}. \tag{7.1-14}$$

Omitting a possible phase factor we obtain for the normalized state

$$|\,\gamma\rangle_S \equiv |n_1 n_2 \ldots n_k \ldots\rangle = \left[\frac{N!}{n_1!\,n_2!\ldots n_k!\ldots}\right]^{1/2} S\,|\,\varphi\rangle. \tag{7.1-15}$$

For the antisymmetrized state the same normalization constant is formed. However, the numbers n_k are now 0 or 1 only, since application of *A* requires at least one transposition. For the latter, if $n_k \geq 2$,

$$|\alpha_1^k)|\alpha_2^k) = |\alpha_2^k)|\alpha_1^k) = \tfrac{1}{2}(1+p)|\alpha_1^k)|\alpha_2^k),$$
$$a|\alpha_1^k)|\alpha_2^k) = \tfrac{1}{4}(1-p^2)|\alpha_1^k)|\alpha_2^k) = 0, \tag{7.1-16}$$

since $p^2 = (\pm 1)^2 = 1$. The antisymmetrized state is thus the *Slater determinant*

$$|\gamma\rangle_A \equiv |n_1 n_2 ... n_k ...\rangle = (N!)^{\frac{1}{2}} A|\varphi\rangle = (N!)^{-\frac{1}{2}} \sum_P (-1)^p P|\varphi\rangle$$

$$= \frac{1}{\sqrt{N!}} \begin{vmatrix} |\alpha_1^a) & |\alpha_2^a) & ... & |\alpha_N^a) \\ |\alpha_1^b) & |\alpha_2^b) & ... & |\alpha_N^b) \\ . & & & \\ |\alpha_1^\nu) & |\alpha_2^\nu) & ... & |\alpha_N^\nu) \end{vmatrix} . \tag{7.1-17}$$

From this form *Pauli's exclusion principle* follows once more. If two particles would occupy the same state $|\alpha^k)$, then two rows would be equal $(a = b)$ and the state $|\gamma\rangle_A$ would identically vanish; hence, the occupancies are only given as $n_k = 0$ or 1. In order that the states $|\gamma\rangle_A$ are unique, it is necessary that the $\{|\alpha^k)\}$ are ordered according to a *standard arrangement.* The set of occupation numbers $n_1, n_2, ... n_k, ...$ is then also unique and $|\gamma\rangle_A = |n_1 n_2 ...\rangle$ is an *ordered state.*

We finally mention that, since we constructed different subspaces of the original tensor-product space for symmetrized and antisymmetrized states – these subspaces being orthogonal to each other – all physical systems are *delineated* into Bose-Einstein and Fermi-Dirac systems, respectively.

7.2 Systems of Bosons; Creation and Annihilation Operators

We shall outline the main tenets of a method, first devised by Dirac, which is applicable to many-body systems and to quantum fields. In these problems the number of variables is generally very large, sometimes denumerably infinite and sometimes continuum infinite. In a many-body system we are concerned with the occupancies of a basis of single particle states and in a quantum field with the occupancies of its normal modes, of which we shall see examples later. In all cases the occupation-number formalism, sometimes called 'second quantization' [2], is by far closer to classical physics than the standard quantum approach of e.g., wave mechanics. It may *a prima vista* seem to be bothersome to specify an infinite set of occupation numbers; yet its convenience lies in the fact that the numbers $n_1, n_2, ..., n_k, ...$ can be seen as eigenvalues of the number operators, which we shall denote by bold symbols $\mathbf{n}_1, \mathbf{n}_2, ..., \mathbf{n}_k, ...$. [3] We have

---

[2] The origin of the name 'second quantization' will be elucidated in Section 7.4.

[3] Dirac used regular symbols n_k for the number operators and primed numbers n_k' for the eigenvalues; see P.A.M. Dirac "Quantum Mechanics", 4th Ed., Oxford 1958.

$$\mathbf{n}_k \,|\, n_1 n_2 ... n_k ...\rangle = n_k \,|\, n_1 n_2 ... n_k ...\rangle,$$
$$\mathbf{N} \,|\, n_1 n_2 ... n_k ...\rangle = \sum_k \mathbf{n}_k \,|\, n_1 n_2 ... n_k ...\rangle = \sum_k n_k \,|\, n_1 n_2 ... n_k ...\rangle. \tag{7.2-1}$$

Clearly, all $\mathbf{n}_k$ commute with each other and with $\mathbf{N}$. The occupancy operators are the natural dynamical variables of the system; they form a complete set of commuting operators, which define the Hilbert space, known as the *Fock space*. This space is completely isomorphic with the state space of $S\,|\,\varphi\rangle$ or $A\,|\,\varphi\rangle$, as the case may be.

The fact that the number of particles in each state are observables, which determine the possible motions of the system, is an acceptable statement. How to associate operators with these observables is another problem. For bosons this was accomplished by Dirac by analogy with the quantization of generalized harmonic oscillators via 'ladder operators'. The equivalence is shown in Dirac's book by means of identical transformation properties for the system states. Rather than being concerned with the historical procedure underlying these ideas, in particular the connection of the ladder operators with the p_i and q_i of the system, we shall turn to the results of that process and *define* boson creation and annihilation operators by the commutation rules:

$$[a_k, a_l] = a_k a_l - a_l a_k = 0, \tag{7.2-2}$$

$$[a_k^\dagger, a_l^\dagger] = a_k^\dagger a_l^\dagger - a_l^\dagger a_k^\dagger = 0, \tag{7.2-3}$$

$$[a_k, a_l^\dagger] = a_k a_l^\dagger - a_l^\dagger a_k = \delta_{kl}\,. \tag{7.2-4}$$

We also define the operator $\rho_k \equiv a_k^\dagger a_k$. We compute the commutator $[\rho_k, \rho_l]$, using the standard commutator rules (which are identical to those for Poisson brackets). Hence,

$$[\rho_k, \rho_l] = [a_k^\dagger a_k, a_l^\dagger a_l] = a_k^\dagger \underbrace{[a_k, a_l^\dagger]}_{\delta_{kl}} a_l + \underbrace{[a_k^\dagger, a_l^\dagger]}_{0} a_k a_l$$
$$+ a_l^\dagger \underbrace{[a_k^\dagger, a_l]}_{-\delta_{kl}} a_k + a_l^\dagger a_k^\dagger \underbrace{[a_k, a_l]}_{0} = (a_k^\dagger a_l - a_l^\dagger a_k)\delta_{kl} = 0 \; always\,. \tag{7.2-5}$$

The operators ρ_k and ρ_l thus commute for any k and l. They have a common set of eigenstates (ES) denoted by $|\,\rho_1\rho_2...\rho_k...\rangle$ or shortly $|\,\rho\rangle$. Thus,

$$\rho_k \,|\, \rho_1\rho_2...\rangle = \rho_k{}' \,|\, \rho_1\rho_2...\rangle, \tag{7.2-6}$$

where the prime denotes the eigenvalue (EV), see footnote [3]. We need two more commutation rules obtained by considering the Hilbert space transformations $\rho_k a_k \,|\, \rho\rangle$ and $\rho_k a_k^\dagger \,|\, \rho\rangle$. We have

$$\rho_k a_k - a_k \rho_k = a_k^\dagger a_k a_k - a_k a_k^\dagger a_k = (a_k a_k^\dagger - 1)a_k - a_k a_k^\dagger a_k = -a_k, \qquad (7.2\text{-}7)$$

$$\rho_k a_k^\dagger - a_k^\dagger \rho_k = a_k^\dagger a_k a_k^\dagger - a_k^\dagger a_k^\dagger a_k = a_k^\dagger (a_k a_k^\dagger - a_k^\dagger a_k) = a_k^\dagger, \qquad (7.2\text{-}8)$$

where we used (7.2-4) for $k = l$. Thus from (7.2-7):

$$\rho_k a_k \,|\, \rho_1 \rho_2 ...\rangle = (a_k \rho_k - a_k) \,|\, \rho_1 \rho_2 ...\rangle = (\rho_k{}' - 1) a_k \,|\, \rho_1 \rho_2 ...\rangle. \qquad (7.2\text{-}9)$$

Hence, $a_k \,|\, \rho\rangle$ is an ES of ρ_k with EV $\rho_k{}' - 1$, unless $a_k \,|\, \rho\rangle$ is the null-state. To investigate this possibility, we consider its norm

$$\langle \rho \,|\, a_k^\dagger a_k \,|\, \rho\rangle = \rho_k{}' \underbrace{\langle \rho_k \,|\, \rho_k\rangle}_{>0}. \qquad (7.2\text{-}10)$$

So, for $a_k \,|\, \rho\rangle$ to be the null-state, we must have $\rho_k{}' = 0$. If, however this is not the case, then we consider the transformation $\rho_k a_k \{a_k \,|\, \rho\rangle\}$. Repeating the procedure, we find that $(a_k)^2 \,|\, \rho\rangle$ is again an ES of $|\, \rho\rangle$ with EV $\rho_k{}' - 2$, unless this state is the null-state; etc. We therefore generate the eigenvalues

$$\rho_k{}', \; \rho_k{}' - 1, \; \rho_k{}' - 2, \ldots, 0. \qquad (7.2\text{-}11)$$

Doing now the same for the transformation $\rho_k a_k^\dagger$ and applying (7.2-8) we obtain

$$\rho_k a_k^\dagger \,|\, \rho\rangle = (a_k^\dagger \rho + a_k^\dagger) \,|\, \rho\rangle = (\rho_k{}' + 1) a_k^\dagger \,|\, \rho\rangle. \qquad (7.2\text{-}12)$$

Therefore, we have a higher EV $\rho_k{}' + 1$, unless $a_k^\dagger \,|\, \rho\rangle$ is the null-state. However, its norm is

$$\langle \rho \,|\, a_k a_k^\dagger \,|\, \rho\rangle = \langle \rho \,|\, 1 + a_k^\dagger a_k) \,|\, \rho\rangle = (\rho_k{}' + 1)\langle \rho \,|\, \rho\rangle. \qquad (7.2\text{-}13)$$

This is never zero since we saw above that $\rho_k{}' \geq 0$. We thus generate the eigenvalues

$$\rho_k{}', \; \rho_k{}' + 1, \; \rho_k{}' + 2, \ldots, \textit{ad inf}. \qquad (7.2\text{-}14)$$

Accordingly, all positive integers, including zero, complete the eigenvalues of any ρ_k. We can therefore identify the ρ_k-operators with the $\mathbf{n}_k$-operators and the states $|\, \rho\rangle$ with the occupation-number states $|\, n\rangle$. Hence we found the forms

$$\begin{aligned} \mathbf{n}_k &= a_k^\dagger a_k, \\ \mathbf{n}_k \,|\, n_1 n_2 ...\rangle &= a_k^\dagger a_k \,|\, n_1 n_2 ...\rangle = n_k \,|\, n_1 n_2 ...\rangle. \end{aligned} \qquad (7.2\text{-}15)$$

The eigenstates of $\mathbf{n}_k$ are, starting from the vacuum state $|\, 0\rangle \equiv |\, 00...\rangle$:

$$|\, 0\rangle, \quad a_k^\dagger \,|\, 0\rangle / \alpha_0, \quad (a_k^\dagger)^2 \,|\, 0\rangle / \alpha_0 \alpha_1, \quad \textit{etc}. \qquad (7.2\text{-}16)$$

where the α's are normalization constants. From (7.2-13) we see

$$|\, \alpha_0 \,|^2 = 1, \quad |\, \alpha_1 \,|^2 = 2, \quad |\, \alpha_2 \,|^2 = 3, \quad \textit{etc}. \qquad (7.2\text{-}17)$$

Omitting insignificant phase factors $e^{i\phi}$, the normalized eigenstates become

$$|0\rangle, \quad a_k^\dagger|0\rangle, \quad \dots \quad (\nu!)^{-\frac{1}{2}}(a_k^\dagger)^\nu|0\rangle. \tag{7.2-18}$$

The general state is

$$\prod_k (n_k!)^{-\frac{1}{2}}(a_k^\dagger)^{n_k}|0\rangle = |n_1 n_2 ... n_k ...\rangle. \tag{7.2-19}$$

The connection between adjacent states (differing in ± for the occupancy) is more directly found from (7.2-9) and (7.2-12), noting the normalizations in (7.2-10) and (7.2-13):

$$a_k|n_1 n_2 ... n_k ...\rangle = \sqrt{n_k}\,|n_1 n_2 ... n_k - 1 ...\rangle, \tag{7.2-20}$$

$$a_k^\dagger|n_1 n_2 ... n_k ...\rangle = \sqrt{n_k + 1}\,|n_1 n_2 ... n_k + 1 ...\rangle. \tag{7.2-21}$$

These are the two principal results. The names creation operator and annihilation operator have hereby gained their meaning. (Other names are raising and lowering operators.)

7.3 Many-Body Boson Operators

7.3.1 *Expressions in Terms of* $a_k, a_k^\dagger$

We now ask: How can a system operator F act on an occupation-number state? We shall first consider the simplest many-body operator, $F = \Sigma_i f_i$, where f_i is a one-particle operator, involving the coordinates and momenta p_i and q_i of one particle. Thus consider

$$\begin{aligned} &F\left(\frac{N!}{n_1!n_2!...n_k!...}\right)^{\frac{1}{2}} S|\alpha_1^{\ell_1}\alpha_2^{\ell_2}...\alpha_k^{\ell_k}...\rangle \\ &= \left(\frac{N!}{n_1!n_2!...n_k!...}\right)^{\frac{1}{2}} \sum_i f_i\, S|\alpha_1^{\ell_1}\rangle|\alpha_2^{\ell_2}\rangle...|\alpha_k^{\ell_k}\rangle... \end{aligned} \tag{7.3-1}$$

Notice that we have slightly changed our notation: $|\alpha_1^{\ell_1}\rangle$ is the quantum state for particle 1, etc. For the quantum state of particle i we also write $|\alpha_i^{\ell_i}\rangle \equiv |\ell_i\rangle$. We further note the closure property in the space of particle i, viz., $\Sigma_k |k\rangle\langle k| = 1$, where $|k\rangle$ is a one-particle state.[4] Then, inserting this into (7.3-1) we also have

$$F|n_1 n_2 ...\rangle = \left(\frac{N!}{n_1!n_2!...n_k!...}\right)^{\frac{1}{2}} \sum_{ki} S|\alpha_1^{\ell_1}\rangle|\alpha_2^{\ell_2}\rangle...|k\rangle\langle k|f_i|\ell_i\rangle. \tag{7.3-2}$$

[4] In this section and the following ones, we only use angular brackets, since we will need one-particle states, two-particle states, etc.

Since the entire expression is symmetric in all particles – both the state and the operator F being symmetrical – we also write this simply as

$$F\,|\,n_1 n_2 \ldots\rangle = \left(\frac{N!}{n_1!n_2!\ldots n_k!\ldots}\right)^{1/2} \sum_{k\ell} n_\ell\, S\,|\,\alpha_1^{\ell_1}\alpha_2^{\ell_2}\ldots\alpha^k\ldots\rangle\langle k\,|f\,|\,\ell\rangle, \tag{7.3-3}$$

where the n_ℓ stems from the fact that n_ℓ particles have the state $|\,\ell\rangle$. Note that in (7.3-2) i sums particles and k sums states, whereas in (7.3-3) both k and ℓ sum states. The state $S\,|\,\alpha_1^{\ell_1}\alpha_2^{\ell_2}\ldots\alpha^k\ldots\rangle$ is now a many-body state in which the one-particle state $|\,\alpha^\ell\rangle = |\,\ell\rangle$ of the original state in (7.3-1) has been replaced by $|\,\alpha^k\rangle = |\,k\rangle$. Accordingly, we have

$$S\,|\,\alpha_1^{\ell_1}\alpha_2^{\ell_2}\ldots\alpha^k\ldots\rangle \propto |\,n_1, n_2, \ldots, n_k+1, n_\ell-1, \ldots\rangle. \tag{7.3-4}$$

To make the normalization correct we notice

$$\left(\frac{N!}{n_1!n_2!\ldots n_k!\ldots}\right)^{1/2} = \left(\frac{N!}{n_1!n_2!\ldots(n_k+1)!(n_\ell-1)!\ldots}\right)^{1/2}\sqrt{\frac{n_k+1}{n_\ell}}\,. \tag{7.3-5}$$

Hence, (7.3-3) finally gives

$$F\,|\,n_1 n_2\ldots\rangle = \sum_{k\ell}\sqrt{(n_k+1)n_\ell}\,|\,n_1, n_2, \ldots, n_k+1, n_\ell-1, \ldots\rangle\langle k\,|\,f\,|\,\ell\rangle. \tag{7.3-6}$$

Comparing with (7.2-20) and (7.2-21), we conclude that the action of F is the same as that of the Boson operators $a_k^\dagger a_\ell$. Explicitly we obtain

$$F = \sum_{k\ell} a_k^\dagger a_l \langle k\,|\,f\,|\,\ell\rangle. \tag{\textbf{7.3-7}}$$

Thus, any operator of the form $F = \Sigma f$ has now been expressed so as to give a transformation in the Fock space for the occupation-number states.

The computations can be repeated for more general operators like the two-particle operator

$$V = \sum_{ij} v_{ij}(\{p_i, q_i\}, \{p_j, q_j\}), \tag{7.3-8}$$

where v_{ij} acts in the Hilbert space of two particles on $|\,\alpha_i^{\ell_i}\alpha_j^{\ell_j}\rangle$. We further have, using the projector property $\Sigma_{k\ell}\,|\,k\ell\rangle\langle k\ell\,| = 1$,

$$\begin{aligned} V\,|\,n_1 n_2\ldots\rangle &= \left(\frac{N!}{n_1!n_2!\ldots n_k!\ldots}\right)^{1/2}\sum_{ij} v_{ij} S\,|\,\alpha_1^{\ell_1}\alpha_2^{\ell_2}\ldots\rangle \\ &= \left(\frac{N!}{n_1!n_2!\ldots n_k!\ldots}\right)^{1/2}\sum_{k\ell}\sum_{ij} S\,|\,\alpha_1^{\ell_1}\alpha_2^{\ell_2}\ldots k\ell\ldots\rangle\langle k\ell\,|\,v_{ij}\,|\,m_i n_j\rangle \end{aligned}$$

$$= \left(\frac{N!}{n_1! n_2! \ldots n_k! \ldots} \right)^{1/2} \sum_{k\ell mn} n_m n_n \, S \,| \alpha_1^{\ell_1} \alpha_2^{\ell_2} \ldots k\ell \ldots \rangle \langle k\ell \,|\, v \,|\, mn \rangle . \qquad (7.3\text{-}9)$$

In the second equality i and j sum particles, while k and ℓ sum states; in the final equality all indices sum over states. We now notice that in the state $S\,|\alpha_1^{\ell_1}\alpha_2^{\ell_2}\ldots k\ell\ldots\rangle$ we have a lowering of the states denoted by m and n, while we have a raising of the states k and ℓ. Renormalizing the final state, we find

$$V\,|n_1 n_2 \ldots\rangle = \sum_{k\ell mn} \sqrt{(n_k+1)(n_\ell+1) n_m n_n} \times | n_1, n_2, \ldots, n_k+1, n_\ell+1, n_m-1, n_n-1, \ldots \rangle \langle k\ell \,|\, v \,|\, mn\rangle , \qquad (7.3\text{-}10)$$

from which we conclude

$$V = \sum_{k\ell mn} a_k^\dagger a_\ell^\dagger \langle k\ell \,|\, v \,|\, mn \rangle a_m a_n . \qquad \mathbf{(7.3\text{-}11)}$$

This expression has an immediate interpretation. Most often, two-body operators depend only on the q's and represent two-body interactions, which, classically viewed, cause scattering from two initial states to two final states. In Fig. 7-1 below we indicate the scattering of two α-particles, following hyperbolae as in the Kepler problem. In quantum mechanical form we create two particles with states $|k\rangle$ and $|\ell\rangle$ and we annihilate two particles with states $|m\rangle$ and $|n\rangle$.[5] The corresponding Feynman diagram is given in Fig. 7-2.

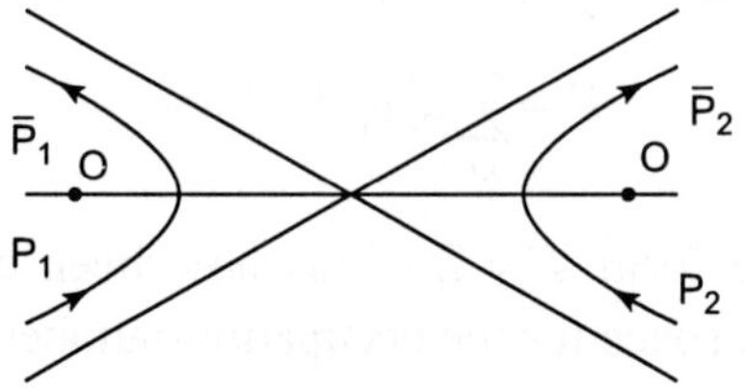

Fig. 7-1. Scattering of two α-particles, viewed classically. The path is a hyperbola.

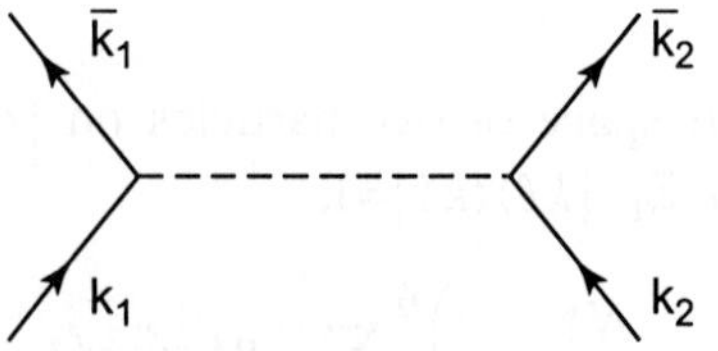

Fig. 7-2. Feynman diagram for two-body interaction.

[5] Or, less dramatic and epistemologically sound, we have a *transition* of two particles from (mn) to $(k\ell)$.

Although there are differences as to the interpretation of the quantum result, see the footnote, as a whole the quantum picture in the present formalism is much closer to the classical picture than say, wave-mechanical scattering theory.

We can of course repeat the procedure for higher order operators, like those involving three particles. Also, we must consider the case that one of the constituents is a quasi-particle, as in electron-phonon interaction. We shall deal with those problems in the specific context in which it is needed. For now, let us take a Hamiltonian of the form

$$\mathcal{H} = \sum_i h_i^0 + \tfrac{1}{2}\sum_{ij} v_{ij}^{(2)} + \tfrac{1}{6}\sum_{ijk} v_{ijk}^{(3)} + \ldots \tag{7.3-12}$$

where the fractions in front of the terms are added to ensure that interactions between like particles are counted only once. The appropriate operator in Fock space is

$$\begin{aligned}\mathcal{H} &= \sum_{k\ell} a_k^\dagger \langle k\,|\,h^0\,|\,\ell\rangle a_\ell + \tfrac{1}{2}\sum_{k\ell mn} a_k^\dagger a_\ell^\dagger \langle k\ell\,|\,v^{(2)}\,|\,mn\rangle a_m a_n \\ &+ \tfrac{1}{6}\sum_{k\ell mnpq} a_k^\dagger a_\ell^\dagger a_m^\dagger \langle k\ell m\,|\,v^{(3)}\,|\,npq\rangle a_n a_p a_q + \ldots\end{aligned} \tag{7.3-13}$$

Often, we choose as a basis $|\alpha\rangle$ the eigenstates of the unperturbed one-particle Hamiltonian h^0. Then $\langle k\,|\,h^0\,|\,\ell\rangle = \varepsilon_k^0 \delta_{k\ell}$, so we have

$$\mathcal{H} = \sum_k \mathbf{n}_k \varepsilon_k^0 + \tfrac{1}{2}\sum_{k\ell mn} a_k^\dagger a_\ell^\dagger \langle k\ell\,|\,v^{(2)}\,|\,mn\rangle a_m a_n + \ldots \tag{7.3-14}$$

From the above it is evident what an essential role is played by the boson operators $a_k, a_k^\dagger$. For the rest we need only compute the matrix elements $\langle k\,|\,f\,|\,\ell\rangle$, $\langle k\ell|v\,|\,mn\rangle$, *etc.* These can be computed in an $\mathcal{L}^2$ function-space for the one- and two-particle states involved.

The final result is that we always get a power series in the operators a_k and $a_k^\dagger$; although not necessary for bosons, we prefer to put these operators in the form of an *ordered product.* Thus, creation operators precede annihilation operators and, if the states are ordered, so must be the a's. The equivalence with generalized oscillators, mentioned before, is obtained if we formally introduce p's and q's via

$$\begin{aligned} q_k &= (a_k + a_k^\dagger)\sqrt{\hbar/2}, \\ p_k &= -i(a_k - a_k^\dagger)\sqrt{\hbar/2}. \end{aligned} \tag{7.3-15}$$

One easily confirms the commutation rule

$$[p_k, q_\ell] = p_k q_\ell - q_\ell p_k = (\hbar/i)\delta_{k\ell}. \tag{7.3-16}$$

Thus, p_k and q_k are canonical operators, satisfying the *quantum conditions.* Solving from (7.3-15) for the a's, and substituting into (7.3-13), a polynomial in p,q results, indeed representing a generalized oscillator.

7.3.2 *Field Operators and Local Observables*

Above we dealt with one form of the second quantization procedure. In the following we introduce the *field-operator form.* So far we have not put any spatial dependence in the picture. We shall do so by choosing a wave mechanical ($\mathscr{L}^2$-space) realization for the eigenstates, i.e., $|\alpha_i^a\rangle \to \alpha_a(\mathbf{r}_i)$; the state suffix is now a subscript, while the particle attribute occurs only in the position coordinate. The field operators are introduced by

$$\Psi(\mathbf{r}) = \sum_k \alpha_k(\mathbf{r}) a_k, \quad \Psi^\dagger(\mathbf{r}) = \sum_k \alpha_k^*(\mathbf{r}) a_k^\dagger. \tag{7.3-17}$$

In these expressions the Ψ's and the a's are operators, while the wave functions are c-numbers; to indicate that the a's do not operate on these at all, we put them to the right. The quantities Ψ and $\Psi^\dagger$ depend on position through the functions $\alpha_k(\mathbf{r})$, so that these are functional operators. They satisfy the commutation rules

$$\Psi(\mathbf{r})\Psi(\mathbf{r}') - \Psi(\mathbf{r}')\Psi(\mathbf{r}) = \sum_{k\ell} \alpha_k(\mathbf{r})\alpha_\ell(\mathbf{r}')[a_k, a_\ell] = 0, \tag{7.3-18}$$

or,

$$[\Psi(\mathbf{r}), \Psi(\mathbf{r}')] = [\Psi^\dagger(\mathbf{r}), \Psi^\dagger(\mathbf{r}')] = 0. \tag{7.3-19}$$

Also,

$$\Psi(\mathbf{r})\Psi^\dagger(\mathbf{r}') - \Psi^\dagger(\mathbf{r}')\Psi(\mathbf{r}) = \sum_{k\ell} \alpha_k(\mathbf{r})\alpha_\ell^*(\mathbf{r}')\underbrace{[a_k, a_\ell^\dagger]}_{\delta_{k\ell}}. \tag{7.3-20}$$

For the wave functions we have the closure relation

$$\sum\nolimits_k \alpha_k(\mathbf{r})\alpha_k^*(\mathbf{r}') = \delta(\mathbf{r} - \mathbf{r}'). \tag{7.3-21}$$

Whence, we have as basic third commutation rule

$$[\Psi(\mathbf{r}), \Psi^\dagger(\mathbf{r}')] = \delta(\mathbf{r} - \mathbf{r}'). \tag{7.3-22}$$

The operators Ψ and $\Psi^\dagger$ act on occupation-number states. E.g., we have

$$\int d^3r \Psi^\dagger(\mathbf{r})\alpha_k(\mathbf{r})|n_1 n_2 ...\rangle = \sqrt{n_k + 1}\,|n_1, ..., n_k + 1, ...\rangle. \tag{7.3-23}$$

We now turn to the forms needed for the system operators. First we consider $f = \Sigma_i f_i$ as before. We denote the wave mechanical form by $f_i(\mathbf{p},\mathbf{r})$, where as usual $\mathbf{p} = (\hbar / i)\nabla$. The matrix element has the wave mechanical form

$$\int d^3r\,\alpha_k^*(\mathbf{r})f(\mathbf{p},\mathbf{r})\alpha_\ell(\mathbf{r}) = \langle k|f|\ell\rangle. \tag{7.3-24}$$

The result (7.3-7) still holds for the Schrödinger form. Multiplying (7.3-24) with $a_k^\dagger a_\ell$, summing and using the definitions (7.3-17), we obtain at once

$$F = \int \Psi^\dagger(\mathbf{r})f(\mathbf{p},\mathbf{r})\Psi(\mathbf{r})d^3r. \tag{7.3-25}$$

For the two-particle operator $V = \frac{1}{2}\Sigma_{ij}v(\mathbf{r}_i,\mathbf{r}_j)$ the procedure is the same. The two-particle matrix element is defined as

$$\iint d^3r d^3r'\alpha_k^*(\mathbf{r})\alpha_\ell^*(\mathbf{r}')v(\mathbf{r},\mathbf{r}')\alpha_m(\mathbf{r})\alpha_n(\mathbf{r}') = \langle k\ell|v|mn\rangle. \tag{7.3-26}$$

One easily verifies that this leads to

$$V = \tfrac{1}{2}\iint \Psi^\dagger(\mathbf{r})\Psi^\dagger(\mathbf{r}')v(\mathbf{r},\mathbf{r}')\Psi(\mathbf{r})\Psi(\mathbf{r}')d^3rd^3r'. \tag{7.3-27}$$

Finally, consider a symmetrical N-body operator,

$$A = \frac{1}{N!}\sum_{i_1i_2\ldots i_N} a(\mathbf{r}_{i_1},\mathbf{r}_{i_2},\ldots\mathbf{r}_{i_N}). \tag{7.3-28}$$

The needed second quantization form is

$$A = \frac{1}{N!}\int d^{3N}r\prod_{j=1}^{N}\Psi^\dagger(\mathbf{r}_j)a(\mathbf{r}_1,\mathbf{r}_2,\ldots\mathbf{r}_N)\Psi(\mathbf{r}_j). \tag{7.3-29}$$

(This remains valid if a also depends on the momenta.)

Local observables

The field operators permit us to find expressions for local observables. First, with the particle density we associate the operator

$$\rho(\mathbf{r}) = \sum\nolimits_i \delta(\mathbf{r}-\mathbf{r}_i). \tag{7.3-30}$$

This is an operator of the form F. Hence, from (7.3-25)

$$\rho(\mathbf{r}) = \Psi^\dagger(\mathbf{r})\Psi(\mathbf{r}). \tag{7.3-31}$$

This looks like the probability density in "regular" wave mechanics, but notice, firstly, that the Ψ's are operators and, secondly, that $\rho(\mathbf{r})$ *is normalized to the total particle operator* $\mathbf{N}$; from (7.3-17):

$$\int d^3r\Psi^\dagger(\mathbf{r})\Psi(\mathbf{r}) = \sum_{k\ell}\int d^3r\,\alpha_\ell^*(\mathbf{r})\alpha_k(\mathbf{r})a_\ell^\dagger a_k$$

$$= \sum_{k\ell}\delta_{k\ell}a_\ell^\dagger a_k = \sum_k a_k^\dagger a_k = \mathbf{N}. \tag{7.3-32}$$

One can also define the number of particles in a volume V, N_V, as the integral of $\Psi^\dagger\Psi$ over that volume. It is easily shown that

$$[\mathbf{N}_{V_1}, \mathbf{N}_{V_2}] = 0, \tag{7.3-33}$$

which is true whether or not the volumes overlap. This means that we can associate a definite number of particles with each volume; however, this holds only for non-relativistic systems.

We now consider the Hamiltonian. Generally,

$$\mathcal{H} = -\frac{\hbar^2}{2m}\sum_i \nabla_i^2 + \sum_i v^{(1)}(\mathbf{r}_i) + \frac{1}{2}\sum_{ij} v^{(2)}(\mathbf{r}_i, \mathbf{r}_j) + \ldots \tag{7.3-34}$$

We use the rules (7.3-25) and (7.3-27) and apply integration by parts to the first term; then

$$\begin{aligned} \mathcal{H} &= \int \frac{\hbar^2}{2m} \nabla\Psi^\dagger(\mathbf{r}) \cdot \nabla\Psi(\mathbf{r}) d^3r + \int \Psi^\dagger(\mathbf{r}) v^{(1)}(\mathbf{r}) \Psi(\mathbf{r}) d^3r \\ &+ \frac{1}{2}\iint \Psi^\dagger(\mathbf{r})\Psi^\dagger(\mathbf{r}') v^{(2)}(\mathbf{r},\mathbf{r}')\Psi(\mathbf{r})\Psi(\mathbf{r}') d^3r d^3r' + \ldots \end{aligned} \tag{7.3-35}$$

For non-interacting systems, the first line only is retained; we have

$$\mathcal{H}_{\text{non-interact}} = \int \left[\frac{\hbar^2}{2m} \nabla\Psi^\dagger(\mathbf{r}) \cdot \nabla\Psi(\mathbf{r}) + \Psi^\dagger(\mathbf{r}) v^{(1)}(\mathbf{r}) \Psi(\mathbf{r}) \right] d^3r. \tag{7.3-36}$$

The integrand is the local energy-density operator. Finally, we write down the total momentum operator in the field language: $\mathbf{P} = (\hbar/i)\Sigma_i \nabla_i$. We find

$$\mathbf{P} = (\hbar/i)\int \Psi^\dagger(\mathbf{r})\nabla\Psi(\mathbf{r}) d^3r = (\hbar i)\int [\nabla\Psi^\dagger(\mathbf{r})]\Psi(\mathbf{r}) d^3r, \tag{7.3-37}$$

where in the latter expression we used integration by parts. The integrand in these expressions is the local momentum density. This result can also be obtained in another way. We know that the translation operator is given by[6]

$$T(\mathbf{R}) = \exp(-i\mathbf{P}\cdot\mathbf{R}/\hbar), \tag{7.3-38}$$

where $\mathbf{P}$ is again the total momentum. Thus, for the transformation $T\Psi^\dagger T^{-1}$:

$$e^{-i\mathbf{P}\cdot\mathbf{R}/\hbar}\, \Psi^\dagger(\mathbf{r})\, e^{i\mathbf{P}\cdot\mathbf{R}/\hbar} = \Psi^\dagger(\mathbf{r}+\mathbf{R}). \tag{7.3-39}$$

If we differentiate this to $\mathbf{R}$ and afterwards let $\mathbf{R} \to 0$, we find

$$[\mathbf{P}, \Psi^\dagger] = -\hbar i \nabla\Psi^\dagger, \tag{7.3-40a}$$

[6] A. Messiah, Op. Cit., Chapter XV.

and

$$[\mathbf{P}, \Psi] = \hbar i \nabla \Psi, \tag{7.3-40b}$$

concordant with (7.3-37).

The commutation rules (7.3-40) are like the regular commutation rule for $[p, f(q)]$ in a single particle system. As a matter of fact, $\Psi(\mathbf{r})$ behaves entirely like the amplitude operator of a classical wave, $\Psi(\mathbf{r})$ being the continuum analogue of an atomic system with positions $\mathbf{r}_1, \mathbf{r}_2 ... \mathbf{r}_N$. In the next section we shall show that (7.3-35) and (7.3-36) are obtainable from *direct* field quantization.

Although the field point of view is attractive, *a warning must be made*. When we first introduced a_k and $a_k^\dagger$ it was clear that we had many-particle systems in mind, in particular in view of application to statistical mechanics, where the thermodynamic limit reigns! However, some popular books on quantum field theory impose no restrictions on N; even $N = 1$ is considered admissible. For a one-particle "many-body state" $|1\rangle = |0_1, 0_2, ... 1_i, ...\rangle$ we easily find the expectation value of the density in this state:

$$\langle 1 | \rho(\mathbf{r}) | 1 \rangle = \langle 1 | \sum_{k\ell} \alpha_k^*(\mathbf{r}) \alpha_\ell(\mathbf{r}) a_k^\dagger a_\ell | 1 \rangle = | \alpha_i(\mathbf{r}) |^2 . \tag{7.3-41}$$

Since the wave function extends over the entire volume – and even outside the volume – this expectation value would indicate that the particle in indeed "smeared out" over all space![7] Even stronger, this appears as follows. The total number of particles is always an eigenvalue of the integrated field operator, i.e., *the particles are the excitations of the field.* From (7.3-31) we have

$$\mathbf{N} | n \rangle = \int \Psi^\dagger \Psi | n \rangle d^3 r = \int \rho(\mathbf{r}) | n \rangle d^3 r . \tag{7.3-42}$$

It is now tempting to conclude that $|n\rangle$ is also an eigenstate of the density operator. This is, indeed, formally correct, for it is easily shown that $[\mathbf{N}, \rho(\mathbf{r})] = 0$. Also, it is easily shown that $[\rho(\mathbf{r}), \rho(\mathbf{r}')] = 0$. Taking again the extreme case $N = 1$, it follows that we can simultaneously measure the density due to one particle at any point in space!

Before we submit to metaphysics, we should realize that the field equations, as set forth again in the next section, are *homogeneous* equations. The normalization of $\Psi^\dagger \Psi$ to $\mathbf{N}$ makes sense for a many-particle system. For $N = 1$, $\Psi^\dagger \Psi$ is an operator, representing the squared norm of the wave function, $\psi \psi^*$. *This author* adheres to the Bohr–Kramers–Slater interpretation[8] (called by others the Born interpretation[9]) of

---

[7] For this view see e.g., Henley and Thirring, "Elementary Quantum Field Theory", McGraw Hill, NY 1962.

[8] N. Bohr, H.A.Kramers and J.C. Slater, Zeitschr. für Physik **32**, 69 (1924); also Phil. Mag. **47**, 785 (1924).

quantum mechanics, according to which $\psi\psi^*$ is the *probability density* to find a particle at **r**. *If* the field formulation is to be applied to the case that $N = 1$, the field density $\Psi^\dagger\Psi$ clearly should be normalized to the unit operator; this implies that only probability densities can be simultaneously measured at any point – a non-objectionable statement.

On the contrary, for large N we have a *physical wave*, with a particle presence measurable in any point of space – depending on the time resolution of the set-up.[10]

7.4 On the Quantization of Fields

Taking a different route, we shall look upon a many-body system as an example of a 'particle field'. The amplitude of the field will be $\psi(\mathbf{r},t)$ and the intensity of the field $\psi^*\psi$. Thus, the particle density is

$$\rho(\mathbf{r},t) = \psi^*(\mathbf{r},t)\psi(\mathbf{r},t)\,, \tag{7.4-1}$$

with $\int \rho d^3r = N$. The form of such a field is semi-classically[11] given by

$$\psi(\mathbf{r},t) \equiv A(\mathbf{r},t)e^{iS(\mathbf{r},t)/\hbar} = A(\mathbf{r},t)e^{i[W(\mathbf{r})-\mathcal{E}t]/\hbar} = A(\mathbf{r},t)e^{iW(\mathbf{r})/\hbar}e^{-i\omega t}\,. \tag{7.4-2}$$

Here S and W are the Hamilton–Jacobi and Hamilton's principal function, respectively. Hence

$$\rho(\mathbf{r},t) = [A(\mathbf{r},t)]^2. \tag{7.4-3}$$

The density function must satisfy the continuity equation

$$\frac{\partial\rho(\mathbf{r},t)}{\partial t} + \mathrm{div}(\rho\mathbf{v}) = \frac{\partial[A(\mathbf{r},t)]^2}{\partial t} + \mathrm{div}\left[\frac{[A(\mathbf{r},t)]^2}{m}\nabla S\right] = 0\,, \tag{7.4-4}$$

where we used that $\mathbf{p} = \nabla S$, as follows from the constitutive equations of S. For systems without interparticle interactions, we have the Hamilton-Jacobi equation

---

[9] M. Born, Zeitschr. für Physik **38**, 803 (1926).

[10] After the Como conference of 1927, Bohr created the "Copenhagen interpretation", which endows the wave function with both mathematical and physical reality. But it is usually grossly overlooked that there is no evidence that *one* electron passing through a two-slit interferometer produces the typical interference spectrum with the main peak at the dead space midway behind the slits. For, if the source intensity is low enough, so that the effect of each passing electron is resolved, the electron produces a blob somewhere on the luminescent screen behind the slits. After many repeats, the superposition of these blobs, with weighted probabilities as observed, gives the interference spectrum. Only for high source intensity do we observe wave interference on the screen from the partial waves through the slits.

[11] W.K.B. approximation.

$$\frac{(\nabla S)^2}{2m}+\mathcal{V}+\frac{\partial S}{\partial t}=0, \tag{7.4-5}$$

where $\mathcal{V}$ is the energy in an external potential field, while (7.4-4) can be simplified to

$$m\frac{\partial A}{\partial t}+\nabla A\cdot\nabla S+\frac{1}{2}A\nabla^2 S=0. \tag{7.4-6}$$

Eqs. (7.4-5) and (7.4-6) are two equations for A and S, which together constitute the wave. We now argue that the Hamilton–Jacobi equation (7.4-5) is incomplete, since no coupling occurs with the amplitude A; a wave description requires such a coupling to allow for the possibility of interference. We therefore add a term $-(\hbar^2/2m)(\nabla^2 A)$; in the classical limit $\hbar \to 0$ this term vanishes. We then obtain

$$\begin{aligned}\frac{A(\nabla S)^2}{2m}+A\mathcal{V}+A\frac{\partial S}{\partial t}-\frac{\hbar^2}{2m}\nabla^2 A=0,\\ m\frac{\partial A}{\partial t}+\nabla A\cdot\nabla S+\frac{1}{2}A\nabla^2 S=0.\end{aligned} \tag{7.4-7}$$

We multiply the second equation by $-\hbar i/m$ and add the first equation, which yields,

$$\frac{A(\nabla S)^2}{2m}+A\mathcal{V}+A\frac{\partial S}{\partial t}-\frac{\hbar^2}{2m}\nabla^2 A-\hbar i\frac{\partial A}{\partial t}-\frac{\hbar i}{m}\nabla A\cdot\nabla S-\frac{\hbar i}{2m}A\nabla^2 S=0. \tag{7.4-8}$$

This is still multiplied by $\exp(iS/\hbar)$. After some algebra this results in[12]

$$-\frac{\hbar^2}{2m}\nabla^2\psi+\mathcal{V}\psi-\hbar i\frac{\partial\psi}{\partial t}=0. \tag{7.4-9}$$

This is the Schrödinger equation, but it appears here as a *particle-field equation*, with normalization $\int\psi^*\psi\, d^3r=N$. The norm is conserved, since by multiplying (7.4-9) with ψ^* and adding the similar complex conjugate result, we find upon integrating,

$$\frac{\partial N}{\partial t}=\frac{\hbar i}{2m}\int d^3r\left[\psi^*\nabla^2\psi-\psi\nabla^2\psi^*\right]=\frac{\hbar i}{2m}\oiint d\mathbf{S}\cdot\left[\psi^*\nabla\psi-\psi\nabla\psi^*\right]=0, \tag{7.4-10}$$

where we used Green's theorem.

Although (7.4-9) is a quantum equation, the wave under consideration *is not yet quantized.* We therefore seek to obtain the operators $\Psi^\dagger$ and Ψ, which represent the field ψ^*,ψ. This procedure has been named 'second quantization' – yet, in our view nothing in nature is quantized twice! There are two approaches to obtain the

[12] Conversely, when we substitute the first equality of (7.4-2) for ψ and separate into real and imaginary parts, both equations (7.4-7) are recovered. Thus the single result (7.4-9) is fully equivalent with the two equations of departure; see e.g., Landau and Lifshitz, "Quantum Mechanics, Addison-Wesley, 1958, p.49.

quantum forms. First, we can expand the field in normal modes. Secondly, we may find the Lagrangian of the field and the associated canonical coordinates and momenta. We shall briefly elucidate both methods.

The wave equation is of the form $(\mathcal{F}+\partial/\partial t)\Psi=0$. We expand in the EF of $\mathcal{F}$. For the present case $\mathcal{F}\to h^0$, the one-particle Hamiltonian, with eigenfunctions $\varphi_k(\mathbf{r})$. So we have

$$\left(h^0-\hbar i\partial/\partial t\right)\Psi=0, \tag{7.4-11}$$

and we expand

$$\Psi(\mathbf{r},t)=\sum\nolimits_k f_k(t)\varphi_k(\mathbf{r}), \qquad \Psi^\dagger(\mathbf{r},t)=\sum\nolimits_k f_k^\dagger(t)\varphi_k^*(\mathbf{r}). \tag{7.4-12}$$

Note that the f_k are operators. Substituting into (7.4-11) and using $h^0\varphi_k=\varepsilon_k\varphi_k$, we find from the usual procedure

$$\begin{aligned}\varepsilon_\ell f_\ell &= \hbar i\, df_\ell/dt,\\ \varepsilon_\ell f_\ell^\dagger &= -\hbar i\, df_\ell^\dagger/dt.\end{aligned} \tag{7.4-13}$$

From these two equations we find

$$\hbar i\frac{d}{dt}[f_k,f_\ell^\dagger]=(\varepsilon_k-\varepsilon_\ell)[f_k,f_\ell^\dagger], \tag{7.4-14}$$

which integrates to

$$[f_k,f_\ell^\dagger]=[f_k(0),f_\ell^\dagger(0)]\exp[-i(\varepsilon_k-\varepsilon_\ell)/\hbar]. \tag{7.4-15}$$

Now for $k\neq\ell$, noting that f_k and f_ℓ are independent operator expansion coefficients, we choose the initial commutator to be zero; it is then zero for all times. Taking $k=\ell$, we find that the commutator is constant; we set this constant equal to 1. Thus, $[f_k,f_\ell^\dagger]=\delta_{k\ell}$. Likewise it is established that $[f_k,f_\ell]=[f_k^\dagger,f_\ell^\dagger]=0$. Consequently, *we have satisfied the boson commutation rules.* Henceforth, we set $f_k=a_k$.

Recapitulating, we obtained the following results. The field operators Ψ and $\Psi^\dagger$ are expanded in the EF of h^0, the expansion coefficients being boson operators,

$$\Psi(\mathbf{r},t)=\sum\nolimits_k a_k(t)\varphi_k(\mathbf{r}), \quad \Psi^\dagger(\mathbf{r},t)=\sum\nolimits_k a_k^\dagger(t)\varphi_k^*(\mathbf{r}). \tag{7.4-16}$$

This is entirely analogous to the field-operator forms. (7.3-17), *but presently we work in the Heisenberg picture*, cf. Appendix A. We consider the commutators

$$[a_k,a_k^\dagger a_k\varepsilon_k]=[a_k,a_k^\dagger]a_k\varepsilon_k+a_k^\dagger[a_k,a_k]\varepsilon_k=a_k\varepsilon_k=\hbar i\,\dot a_k, \tag{7.4-17}$$

where we used (7.4-13). Likewise,

$$[a_k^\dagger,a_k^\dagger a_k\varepsilon_k]=a_k^\dagger\varepsilon_k=-\hbar i\,\dot a_k^\dagger. \tag{7.4.18}$$

These equations are the Heisenberg equations of motion for the operators a_k and $a_k^\dagger$, with the partial *mode Hamiltonian* $h_k^0 = \varepsilon_k a_k^\dagger a_k = \varepsilon_k \mathbf{n}_k$. This, in turn, yields the Heisenberg equation of motion for Ψ,

$$\begin{aligned}\frac{\partial \Psi(\mathbf{r},t)}{\partial t} &= \sum_k \dot{a}_k \varphi_k(\mathbf{r}) = \frac{1}{\hbar i}\sum_k [a_k\varphi_k, a_k^\dagger a_k \varepsilon_k] \\ &= \frac{1}{\hbar i}\sum_{k\ell}[a_k\varphi_k, a_\ell^\dagger a_\ell \varepsilon_\ell] = \frac{1}{\hbar i}[\Psi, \mathcal{H}^0].\end{aligned} \tag{7.4-19}$$

A similar result is obtained for $\Psi^\dagger$. For a system with interactions, the Heisenberg equation of motion applies equally well; just replace $\mathcal{H}^0$ by $\mathcal{H}$.

In the Lagrangian formalism one tries to first find the classical Lagrangian and Hamiltonian for the field. Actually, we need the Lagrangian and Hamiltonian densities, i.e., we write

$$\begin{aligned} L &= \int \mathcal{L} d^3 r, \quad \mathcal{L} = \mathcal{L}(\psi, \partial\psi/\partial\mathbf{r}, \dot{\psi}), \\ H &= \int \mathcal{H} d^3 r, \quad \mathcal{H} = \mathcal{H}(\psi, \Pi) = -\mathcal{L} + \Pi\dot{\psi}, \end{aligned} \tag{7.4-20}$$

where $\Pi = \delta L/\delta\dot{\psi}$ is the canonical momentum; the δ denotes the functional derivative. It is easily checked that the Lagrangian density

$$\mathcal{L} = \hbar i \psi^* \dot{\psi} - (\hbar^2/2m)\nabla\psi^* \cdot \nabla\psi - \mathcal{V}\psi^*\psi \tag{7.4-21}$$

has as Euler–Lagrange equation (7.4-9); the canonical momentum is found to be $\Pi = \hbar i \psi^*$. The quantized field Hamiltonian requires the mere substitutions $\psi \to \Psi$, $\psi^* \to \Psi^\dagger$. The first two terms of (7.3-35) are then recovered, but with Ψ and $\Psi^\dagger$ in the Heisenberg picture.

Other field equations

The field formalism offers clearly nothing new for particles whose Hamiltonian and interactions can be specified *a priori*. The situation is different for the relativistic Klein-Gordon field or for the electromagnetic field. The Klein-Gordon equation is

$$(1/c^2)\partial^2\Phi/\partial t^2 - \nabla^2\Phi + (mc/\hbar)^2\Phi = 0. \tag{7.4-22}$$

Resolving this equation into normal modes, one can easily express Φ in operators which satisfy boson commutation rules. In this case the field operators do not follow Eqs. (7.3-17), but instead we find

$$\Phi(\mathbf{r},t) = \sum\nolimits_k (2\omega_k)^{-1/2}\{a_k(t)\gamma_k(\mathbf{r}) + a_k^\dagger(t)\gamma_k^*(\mathbf{r})\}, \tag{7.4-23}$$

and

$$\Pi(\mathbf{r},t) = \sum\nolimits_k -i(\omega_k/2)^{1/2}\{a_k(t)\gamma_k(\mathbf{r}) - a_k^\dagger(t)\gamma_k^*(\mathbf{r})\}, \tag{7.4-24}$$

where for this case $\Pi = (1/c)\partial\Phi/\partial t$ and the $\{\gamma_k(\mathbf{r})\}$ are EF of the ∇^2 operator, cf. problem 7.4. The properties of this field are entirely different. The field Hamiltonian, density operator, etc., are found in standards texts on quantum field theory.

Finally, we make some remarks on the electromagnetic field. The field equations are

$$\nabla^2\mathbf{E} - \frac{1}{c^2}\frac{\partial^2\mathbf{E}}{\partial t^2} = 0, \qquad \nabla^2\mathbf{H} - \frac{1}{c^2}\frac{\partial^2\mathbf{H}}{\partial t^2} = 0. \tag{7.4-25}$$

The field Hamiltonian is

$$\mathcal{H} = \frac{1}{2}\int d^3r\left[\mathbf{E}^2 + (\mathrm{curl}\,\mathbf{A})^2\right], \tag{7.4-26}$$

where **E** and **A**/*c* (electric field and vector potential) are the canonically conjugate field quantities (corresponding to Φ and Π in the previous example). The field equations can be obtained from (7.4-26) by Hamilton's principle. The quantization is most easily accomplished, however, by the normal mode expansions

$$\begin{aligned} \mathbf{A}(\mathbf{r},t) &= \sum_{\mathbf{q},\lambda} c(\hbar/2\omega_{\mathbf{q}}\Omega)^{1/2}\left[a_{\mathbf{q}\lambda}\mathbf{u}_{\mathbf{q}\lambda}(\mathbf{r})e^{-i\omega_{\mathbf{q}}t} + a^\dagger_{\mathbf{q}\lambda}\mathbf{u}^*_{\mathbf{q}\lambda}(\mathbf{r})e^{i\omega_{\mathbf{q}}t}\right], \\ \mathbf{E}(\mathbf{r},t) &= \sum_{\mathbf{q},\lambda} i(\hbar\omega_{\mathbf{q}}/2\Omega)^{1/2}\left[a_{\mathbf{q}\lambda}\mathbf{u}_{\mathbf{q}\lambda}(\mathbf{r})e^{-i\omega_{\mathbf{q}}t} - a^\dagger_{\mathbf{q}\lambda}\mathbf{u}^*_{\mathbf{q}\lambda}(\mathbf{r})e^{i\omega_{\mathbf{q}}t}\right]. \end{aligned} \tag{7.4-27}$$

Here the a's are photon creation and annihilation operators, satisfying the boson commutation rules, λ is the polarization index, Ω is the volume and the $\mathbf{u}_{\mathbf{q}\lambda}$ are the eigenvector solutions of the vector Helmholtz equation $(\nabla^2 + q^2)\mathbf{u}_{\mathbf{q}\lambda} = 0$, subject to the transversality condition $\mathrm{div}\,\mathbf{u}_{\mathbf{q}\lambda} = 0$. The field Hamiltonian in terms of the normal modes becomes

$$\mathcal{H} = \sum_{\mathbf{q},\lambda} \hbar\omega_{\mathbf{q}}\, a^\dagger_{\mathbf{q}\lambda}a_{\mathbf{q}\lambda} + \textit{zero-point energy}. \tag{7.4-28}$$

We refer to books on quantum electrodynamics for more details.

7.5 Fermion Operators; Anticommutation Rules

Systems of fermions are described by the antisymmetrized state (7.1-17). The occupation-number states $|\gamma\rangle \equiv |n_1 n_2 ... n_k ...\rangle$ are obtained from antisymmetrized tensor products of one-particle states $\{|\alpha^k\rangle\}$. For fermions the inclusion of spin is essential. It will be appropriate to let the $\{|\alpha^k\rangle\}$ be ES of S_z. Explicitly we write $|\alpha^k\rangle = |\beta^{\mathbf{k}}\eta\rangle$, where $\eta = \pm\frac{1}{2}$ labels the spin. Note that we employ the same state symbol in the overall state and in the orbital parts. The latter can be written as spin projections, $\langle\eta|\alpha^k\rangle \to |\beta^{\mathbf{k}}\rangle_\eta$. These two projections can be arranged in a spinor or column matrix if desired. Generally, the label 'k' will *include* the spin, i.e. $k = (\mathbf{k},\eta)$.

Creation and annihilation operators are defined by anticommutation rules:

$$
\begin{aligned}
c_k c_l^\dagger + c_l^\dagger c_k &= [c_k, c_l^\dagger]_+ = \delta_{kl}, \\
c_k c_l + c_l c_k &= [c_k, c_l]_+ = 0, \\
c_k^\dagger c_l^\dagger + c_l^\dagger c_k^\dagger &= [c_k^\dagger, c_l^\dagger]_+ = 0.
\end{aligned}
\tag{7.5-1}
$$

These operators bear some relation to spin operators, but we shall not pursue this here; see, however, Chapter XI. In particular, we have from the above that $c_k^2 = 0$, $(c_k^\dagger)^2 = 0$. Further we have,

$$(c_k c_k^\dagger)(c_k c_k^\dagger) = c_k (c_k^\dagger c_k) c_k^\dagger = c_k (1 - c_k c_k^\dagger) c_k^\dagger = (c_k c_k^\dagger). \tag{7.5-2}$$

It follows that $(c_k c_k^\dagger)$ has the eigenvalues 0 and 1; it is therefore identical to the number operator $\mathbf{n}_k$ or the hole operator $\mathbf{p}_k = 1 - \mathbf{n}_k$, which have the same eigenvalue spectrum. Choosing the latter, we established:

$$1 - \mathbf{n}_k = \mathbf{p}_k = c_k c_k^\dagger, \quad \mathbf{n}_k = c_k^\dagger c_k. \tag{7.5-3}$$

Consider now the eigenvalue equation

$$\mathbf{n}_k \,|\, n_1 n_2 ... n_k ... \rangle = n_k \,|\, n_1 n_2 ... n_k ... \rangle. \tag{7.5-4}$$

We substitute (7.5-3) and dot into the bra $\langle n_1 n_2 ... n_k ... |$. Then we have

$$\langle n_1 n_2 ... n_k ... | \, c_k^\dagger c_k \, | n_1 n_2 ... n_k ... \rangle = n_k \langle n_1 n_2 ... n_k ... | n_1 n_2 ... n_k ... \rangle = n_k, \tag{7.5-5}$$

and also

$$\langle n_1 n_2 ... n_k ... | \, c_k c_k^\dagger \, | n_1 n_2 ... n_k ... \rangle = p_k \langle n_1 n_2 ... n_k ... | n_1 n_2 ... n_k ... \rangle = 1 - n_k. \tag{7.5-6}$$

Since n_k and p_k only have the values 0 and 1, it follows that

$$c_k \,|\, n_1 n_2 ... n_k ... \rangle = 0 \quad \text{if} \quad \mathrm{n_k} = 0, \tag{7.5-7a}$$

$$c_k^\dagger \,|\, n_1 n_2 ... n_k ... \rangle = 0 \quad \text{if} \quad p_k = 0 \quad \text{or} \quad \mathrm{n_k} = 1, \tag{7.5-7b}$$

and

$$\| c_k \,|\, n_1 n_2 ... n_k ... \rangle \| = 1 \text{ if } n_k = 1, \tag{7.5-8a}$$

$$\| c_k^\dagger \,|\, n_1 n_2 ... n_k ... \rangle \| = 1 \text{ if } p_k = 1 \text{ or } n_k = 0. \tag{7.5-8b}$$

To come any further, we consider the anticommutators

$$\mathbf{n}_k c_k + c_k \mathbf{n}_k = c_k^\dagger c_k c_k + c_k c_k^\dagger c_k = (c_k^\dagger c_k + c_k c_k^\dagger) c_k = c_k, \tag{7.5-9a}$$

and

$$\mathbf{n}_k c_k^\dagger + c_k^\dagger \mathbf{n}_k = c_k^\dagger c_k c_k^\dagger + c_k^\dagger c_k^\dagger c_k = c_k^\dagger (c_k c_k^\dagger + c_k^\dagger c_k) = c_k^\dagger. \tag{7.5-9b}$$

Whence from (7.5-9a):

$$\mathbf{n}_k c_k \,|\, n_1 n_2 ... n_k ...\rangle = (c_k - c_k \mathbf{n}_k) \,|\, n_1 n_2 ... n_k ...\rangle = (1 - n_k) c_k \,|\, n_1 n_2 ... n_k ...\rangle , \quad (7.5\text{-}10a)$$

Likewise from (7.5-9b)

$$\mathbf{n}_k c_k^\dagger \,|\, n_1 n_2 ... n_k ...\rangle = (c_k^\dagger - c_k^\dagger \mathbf{n}_k) \,|\, n_1 n_2 ... n_k ...\rangle = (1 - n_k) c_k^\dagger \,|\, n_1 n_2 ... n_k ...\rangle . \quad (7.5\text{-}10b)$$

The first one of these equations, (7.5-10a), says that $c_k \,|n\rangle$ is the null-state, or an ES of $\mathbf{n}_k$ with EV $1 - n_k$, for $n_k = 1$ or 0, respectively. In the latter case the state $c_k \,|n\rangle$ is equal to $|n_1 n_2 ... 1 - n_k ...\rangle$, except for a phase factor, which in view of (7.5-8a) is taken to be ± 1. Likewise, $c_k^\dagger \,|n\rangle$ is the null-state, or an ES of $\mathbf{n}_k$ with EV $1 - n_k$, for $n_k = 1$ or 0, respectively. Again, in the latter case the state $c_k^\dagger \,|n\rangle$ is equal to $|n_1 n_2 ... 1 - n_k ...\rangle$, except [in view of (7.5-8b)] for a factor ± 1. All possibilities are properly addressed by the combined statements – noticing the normalizations as given in (7.5-5) and (7.5-6) ,

$$c_k^\dagger \,|\, n_1 n_2 ... n_k ...\rangle = (-1)^{\varsigma_k} \sqrt{1 - n_k} \,|\, n_1 n_2 ... 1 - n_k ... \rangle , \quad \mathbf{(7.5\text{-}11)}$$

$$c_k \,|\, n_1 n_2 ... n_k ...\rangle = (-1)^{\varsigma_k} \sqrt{n_k} \,|\, n_1 n_2 ... 1 - n_k ... \rangle . \quad \mathbf{(7.5\text{-}12)}$$

Since $(n_k)^2 = n_k$ for both eigenvalues, the square root signs are usually omitted in quantum mechanics. However, *in statistical mechanics* $0 \le \langle n_k \rangle \le 1$, the square root signs should be retained. Also, for the above given forms we stress the analogy with the boson forms (7.2-20) and (7.2-21).[13]

We also give the notation found in many books. Recalling that for raising n_k must be equal to 0 and for lowering equal to 1, we have

$$\begin{aligned} &c_k^\dagger \,|\, n_1 n_2 ... 0_k ...\rangle = (-1)^{\varsigma_k} \,|\, n_1 n_2 ... 1_k ...\rangle , \\ &c_k^\dagger \,|\, n_1 n_2 ... 1_k ...\rangle = 0 , \\ &c_k \,|\, n_1 n_2 ... 1_k ...\rangle = (-1)^{\varsigma_k} \,|\, n_1 n_2 ... 0_k ...\rangle , \\ &c_k \,|\, n_1 n_2 ... 0_k ...\rangle = 0 . \end{aligned} \quad (7.5\text{-}13)$$

These equations clearly indicate that $c_k^\dagger$ is a creation operator and c_k is an annihilation operator. We must still fix the sign. We note that the anticommutation rules (7.5-1) have only been used for $k = l$. For $k \ne l$ these rules require that ς_k *be fixed such that it counts the occupation of all states preceding the state k*:

$$\varsigma(c_k) = \varsigma(c_k^\dagger) = n_1 + n_2 + ... + n_{k-1} . \quad (7.5\text{-}14)$$

---

[13] These formulas will also be employed in this form for the derivation of the Boltzmann equation in Section 17.1.

For the proof, consider a non-zero state $c_l^\dagger c_k^\dagger |n\rangle$. For this repeated operation the sign is $(-1)^{\varsigma(c_l^\dagger c_k^\dagger)}$, with

$$\begin{aligned} \varsigma(c_l^\dagger c_k^\dagger) &= 2(n_1 + n_2 + ... n_{k-1}) + 1 + n_{k+1} + ... n_{l-1}, \quad (l > k), \\ \varsigma(c_l^\dagger c_k^\dagger) &= 2(n_1 + n_2 + ... n_{l-1}) + 0 + n_{l+1} + ... n_{k-1}, \quad (l < k). \end{aligned} \tag{7.5-15}$$

In the first line the +1 stems from the value of n_k in the k-th state after raising. In the second line the 0 stems from the value of n_l prior to its raising. When we interchange k and l, the two lines of (7.5-15) are also interchanged. As a result we find

$$\varsigma(c_l^\dagger c_k^\dagger) = \varsigma(c_k^\dagger c_l^\dagger) \pm 1. \tag{7.5-16}$$

This is true only if the third line of (7.5-1) is satisfied for $k \neq l$. For the other rules consider $\varsigma(c_l c_k^\dagger)$ and $\varsigma(c_l c_k)$.

We shall derive a few more results. Let $|0\rangle = |0_1 0_2 ... 0_k ...\rangle$ be again the vacuum state. Let $k_1 k_2 ... k_r$ be a permutation of the ordered set $l_1, l_2, ..., l_r$.[14] All k's are different. Then

$$c_{k_1}^\dagger c_{k_2}^\dagger ... c_{k_r}^\dagger |0\rangle = (-1)^{p(k_1 k_2 ... k_r)} |1_{k_1} 1_{k_2} ... 1_{k_r}\rangle \tag{7.5-17}$$

where p is the parity of the permutation. The proof is evident; according to the permutation rule

$$c_{k_1}^\dagger c_{k_2}^\dagger ... c_{k_r}^\dagger = (-1)^{p(k_1 k_2 ... k_r)} c_{l_1}^\dagger c_{l_2}^\dagger ... c_{l_r}^\dagger . \tag{7.5-18}$$

Further, for the *ordered product*:

$$c_{l_1}^\dagger c_{l_2}^\dagger ... c_{l_r}^\dagger |0\rangle = |1_{l_1} 1_{l_2} ... 1_{l_r}\rangle, \tag{7.5-19}$$

the ς's being zero, for we create a succession of states $|1_r\rangle, |1_{r-1} 1_r\rangle$, *etc.*, with the occupancies of preceding states still being zero. Equations (7.5-17) and (7.5-19) indicate how the occupancy states are obtained from the vacuum state; this is the analogue of the boson expression (7.2-19). If both creation operators and annihilation operators are present, we usually try to get the latter to the right of the creation operators. We then speak of a *normal ordered product*. It is denoted as $:\prod_{i,j} c_{k_i}^\dagger c_{k_j}:$ With (7.5-17) any product can be put in the normal form. For instance, we have

$$c_{k_1}^\dagger c_{k_2}^\dagger ... c_{k_r}^\dagger c_k |0_{k_1} 0_{k_2} ... 1_k ... 0_{k_r}\rangle = \delta_{k,k_r} (-1)^{p(k_1 k_2 ... k_r)} |1_{k_1} 1_{k_2} ... 1_{k_r}\rangle. \tag{7.5-20}$$

The matrix elements of the c's are obtained from Eqs (7.5-7). The only non-vanishing possibilities are:

---

[14] If l just represents *one* quantum number it is natural to take the ordering as $l_1 < l_2 < ... < l_r$.

$$\langle n_1 n_2 ... 0_k ... | c_k | n_1 n_2 ... 1_k ... \rangle = (-1)^{(n_1 + ... + n_{k-1})} ,$$
$$\langle n_1 n_2 ... 1_k ... | c_k^\dagger | n_1 n_2 ... 0_k ... \rangle = (-1)^{(n_1 + ... + n_{k-1})} . \tag{7.5-21}$$

These equations justify that c_k and $c_k^\dagger$ are, indeed, adjoint pairs. We finally give some results for the repeated operator $c_k^\dagger c_l$. We obtain directly,

$$c_k^\dagger c_l | n_1 n_2 ... 0_k, 1_l ... \rangle = (-1)^{\varsigma(c_k^\dagger c_l)} | n_1 n_2 ... 1_k, 0_l ... \rangle, \tag{7.5-22}$$

with as only non-vanishing matrix elements

$$\langle n_1 n_2 ... 1_k, 0_l ... | c_k^\dagger c_l | n_1 n_2 ... 0_k, 1_l ... \rangle = (-1)^{\varsigma(c_k^\dagger c_l)} , \tag{7.5-23}$$

where – omitting an even sum of numbers,

$$\begin{aligned} \varsigma(c_k^\dagger c_l) &= n_{k+1} + ... n_{l-1}, \quad (k < l-1), \\ \varsigma(c_k^\dagger c_l) &= n_{l+1} + ... n_{k-1}, \quad (l < k-1), \\ \varsigma(c_k^\dagger c_l) &= 0, \qquad\qquad l = k \pm 1. \end{aligned} \tag{7.5-24}$$

7.6 Many-Body Fermion Operators

7.6.1 *Expressions in Terms of* $c_k, c_k^\dagger$.

We first consider a many-body operator which is symmetrical in all particle attributes, viz., $F = \Sigma_i f_i$. We act with F on an antisymmetrized *ordered* s-particle state. We have, cf. (7.1-17):

$$\begin{aligned} F\sqrt{N!}A | \alpha_1^{\ell_1} \alpha_2^{\ell_2} ... \alpha_s^{\ell_s} \rangle &= \sum_i f_i \sqrt{N!}A | \alpha_1^{\ell_1} \alpha_2^{\ell_2} ... \alpha_s^{\ell_s} \rangle \\ &= \sum_{k,i} \sqrt{N!}A | \alpha_1^{\ell_1} \alpha_2^{\ell_2} ... \alpha^k ... \alpha_s^{\ell_s} \rangle \langle k | f_i | \ell_i \rangle, \end{aligned} \tag{7.6-1}$$

where we used the decomposition of unity for the projectors, $\Sigma_k | k \rangle \langle k | = 1$. Note that we also used $AF = FA$. The lhs represents the many-body state $| 1_{\ell_1} 1_{\ell_2} ... 1_{\ell_s} \rangle$, which we rewrite with (7.5-19); we then obtain

$$F c_{\ell_1}^\dagger c_{\ell_2}^\dagger ... c_{\ell_s}^\dagger | 0 \rangle = \sum_{k,i} c_{\ell_1}^\dagger c_{\ell_2}^\dagger ... c_k^\dagger ... c_{\ell_s}^\dagger | 0 \rangle \langle k | f_i | \ell_i \rangle, \tag{7.6-2}$$

where $c_k^\dagger$ must occur in the same place where $c_{\ell_i}^\dagger$ stood, so that the order is preserved. Now, with Dirac[15] we bring $c_k^\dagger$ to the front, thus creating a permutation; also, we use the operator $(c_{\ell_i}^\dagger)^{-1}$ to denote that the term $c_{\ell_i}^\dagger$ of the ordered product

[15] P.A.M. Dirac, Op. Cit., § 65. Eq. (26').

must be cancelled out. We thus obtain

$$\begin{aligned} Fc_{\ell_1}^\dagger c_{\ell_2}^\dagger ...c_{\ell_s}^\dagger |0\rangle &= \sum_{k,i} (-1)^{i-1} c_k^\dagger (c_{\ell_i}^\dagger)^{-1} c_{\ell_1}^\dagger c_{\ell_2}^\dagger ...c_{\ell_s}^\dagger |0\rangle\langle k| f_i |\ell_i\rangle \\ &= \sum_{k\ell} c_k^\dagger \sum_i (-1)^{i-1} (c_{\ell_i}^\dagger)^{-1} c_{\ell_1}^\dagger c_{\ell_2}^\dagger ...c_{\ell_s}^\dagger |0\rangle \delta_{\ell,\ell_i} \langle k| f |\ell\rangle . \end{aligned} \tag{7.6-3}$$

We now note that the cancellation operator, in fact, represents an annihilation, providing a particle is present in the envisaged state; so we have

$$c_\ell c_{\ell_1}^\dagger c_{\ell_2}^\dagger ...c_{\ell_s}^\dagger |0\rangle = \sum_i (-1)^{i-1} (c_{\ell_i}^\dagger)^{-1} c_{\ell_1}^\dagger c_{\ell_2}^\dagger ...c_{\ell_s}^\dagger |0\rangle \delta_{\ell,\ell_i} . \tag{7.6-4}$$

The parity sign involving the cancellation of the particle in ℓ_i is accounted for [notice $\varsigma(c_{\ell_i}) = \varsigma(c_{\ell_i}^\dagger) = i-1$]. Substituting (7.6-4) into (7.6-5), we arrive at

$$F\{c_{\ell_1}^\dagger c_{\ell_2}^\dagger ...c_{\ell_s}^\dagger |0\rangle\} = \sum_{k\ell} c_k^\dagger c_\ell \{c_{\ell_1}^\dagger c_{\ell_2}^\dagger ...c_{\ell_s}^\dagger |0\rangle\} \langle k| f |\ell\rangle . \tag{7.6-5}$$

The set of states { } in the above expression can be made complete by using all sets $\{\ell_i\}$. So we find for F

$$F = \sum_{k\ell} c_k^\dagger c_\ell \langle k| f |\ell\rangle . \tag{7.6-6}$$

This result is the same as the corresponding formula for bosons, Eq. (7.3-7).

We now repeat this for a sum of two-body operators, $V = \Sigma v_{ij}$. Employing the projection property for an *ordered* two-particle state, $\Sigma |k\ell\rangle\langle k\ell| = 1$, where $|k\rangle$ replaces the state with suffix ℓ_i and $|\ell\rangle$ replaces the state with suffix ℓ_j, we arrive at

$$Vc_{\ell_1}^\dagger c_{\ell_2}^\dagger ...c_{\ell_s}^\dagger |0\rangle = \sum_{k\ell,ij} c_{\ell_1}^\dagger c_{\ell_2}^\dagger ...c_k^\dagger c_\ell^\dagger ...c_{\ell_s}^\dagger |0\rangle\langle k\ell| v_{ij} |\ell_i \ell_j\rangle . \tag{7.6-7}$$

Again, we bring $c_k^\dagger c_\ell^\dagger$ to the front, so that

$$\begin{aligned} Vc_{\ell_1}^\dagger c_{\ell_2}^\dagger ...c_{\ell_s}^\dagger |0\rangle &= \sum_{k\ell,ij} (-1)^{i+j} c_k^\dagger c_\ell^\dagger (c_{\ell_i}^\dagger)^{-1} (c_{\ell_j}^\dagger)^{-1} c_{\ell_1}^\dagger c_{\ell_2}^\dagger ...c_{\ell_s}^\dagger |0\rangle\langle k\ell| v_{ij} |\ell_i \ell_j\rangle \\ &= \sum_{k\ell mn} c_k^\dagger c_\ell^\dagger \sum_{ij} (-1)^{i+j} (c_{\ell_i}^\dagger)^{-1} (c_{\ell_j}^\dagger)^{-1} c_{\ell_1}^\dagger c_{\ell_2}^\dagger ...c_{\ell_s}^\dagger |0\rangle \delta_{m,\ell_i} \delta_{n,\ell_j} \langle k\ell| v |mn\rangle . \end{aligned} \tag{7.6-8}$$

But we have the following equivalence

$$c_n c_m c_{\ell_1}^\dagger c_{\ell_2}^\dagger ...c_{\ell_s}^\dagger |0\rangle = \sum_{ij} (-1)^{i+j} (c_{\ell_i}^\dagger)^{-1} (c_{\ell_j}^\dagger)^{-1} c_{\ell_1}^\dagger c_{\ell_2}^\dagger ...c_{\ell_s}^\dagger |0\rangle \delta_{m,\ell_i} \delta_{n,\ell_j} . \tag{7.6-9}$$

We note that first $m \to \ell_i$ must be cancelled and next $n \to \ell_j$. When we substitute (7.6-9) into (7.6-8), the result is

$$V\{c_{\ell_1}^\dagger c_{\ell_2}^\dagger ...c_{\ell_s}^\dagger |0\rangle\} = \sum_{k\ell mn} c_k^\dagger c_\ell^\dagger c_n c_m \{c_{\ell_1}^\dagger c_{\ell_2}^\dagger ...c_{\ell_s}^\dagger |0\rangle\} \langle k\ell| v |mn\rangle . \tag{7.6-10}$$

This can be extended to all sets $\{\ell_i\}$. We therefore obtain the occupation-number operator form sought for:

$$V = \sum_{k\ell mn} c_k^\dagger c_\ell^\dagger \langle k\ell | v | mn\rangle c_n c_m \,. \qquad \textbf{(7.6-11)}$$

This is the analogue of the boson result (7.3-11). The only difference is here that $|mn\rangle$ is an ordered two-particle state; the operators c_n and c_m must appear in the order shown. Interchange will add a minus sign, since the operators anticommute.

7.6.2 *Field Operators; Spin*

We will use an $\mathscr{L}^2$-space for the orbital part of the state; the total state space is then the tensor product of the $\mathscr{L}^2$-space and the spin space. The one-particle states in this section will then be denoted by

$$\alpha_a(\mathbf{r}_i,\eta) \equiv \langle \mathbf{r}_i \,|\, \gamma^a \eta\rangle = \gamma_a(\mathbf{r}_i)\,|\,\eta\rangle \,. \qquad (7.6\text{-}12)$$

The only explicit representation of $|\eta\rangle$ is the matrix representation (ℓ^2space), so that

$$\alpha_a(\mathbf{r}_i,+) = \gamma_a(\mathbf{r}_i)\begin{pmatrix}1\\0\end{pmatrix}, \quad \alpha_a(\mathbf{r}_i,-) = \gamma_a(\mathbf{r}_i)\begin{pmatrix}0\\1\end{pmatrix}. \qquad (7.6\text{-}13)$$

As fermion field operators we introduce

$$\begin{aligned} \Psi(\mathbf{r}) &= \sum_k \alpha_k c_k = \sum_{\mathbf{k}\eta} \alpha_{\mathbf{k}}(\mathbf{r},\eta) c_{\mathbf{k}\eta} \,, \\ \Psi^\dagger(\mathbf{r}) &= \sum_k \alpha_k^* c_k^\dagger = \sum_{\mathbf{k}\eta} \alpha_{\mathbf{k}}^*(\mathbf{r},\eta) c_{\mathbf{k}\eta}^\dagger \,. \end{aligned} \qquad (7.6\text{-}14)$$

In the first sums k includes all labels, and in the second sums we wrote the spin label separately. It is also advantageous to introduce the spin projections of Ψ and $\Psi^\dagger$, *i.e.*, the field is represented by the spinors (where we now use $\gamma_{\mathbf{k}}$ for the spatial part):

$$\Psi_\eta(\mathbf{r}) = \langle \eta \,|\, \Psi\rangle = \sum_{\mathbf{k}} \gamma_{\mathbf{k}}(\mathbf{r}) c_{\mathbf{k}\eta} \,, \quad \Psi_\eta^\dagger(\mathbf{r}) = \langle \eta \,|\, \Psi^\dagger\rangle = \sum_{\mathbf{k}} \gamma_{\mathbf{k}}^*(\mathbf{r}) c_{\mathbf{k}\eta}^\dagger \,. \qquad (7.6\text{-}15)$$

Hence,

$$\Psi^{(\dagger)} = \sum_\eta \Psi_\eta^{(\dagger)} |\eta\rangle = \Psi_+^{(\dagger)}\begin{pmatrix}1\\0\end{pmatrix} + \Psi_-^{(\dagger)}\begin{pmatrix}0\\1\end{pmatrix} = \begin{pmatrix}\Psi_+^{(\dagger)}\\ \Psi_-^{(\dagger)}\end{pmatrix}. \qquad (7.6\text{-}16)$$

The anticommutation rules are found to be

$$\left[\Psi_\lambda(\mathbf{r}), \Psi_\mu^\dagger(\mathbf{r}')\right]_+ = \delta(\mathbf{r}-\mathbf{r}')\delta_{\lambda\mu}\,, \quad \text{or} \quad \left[\Psi(\mathbf{r}), \Psi^\dagger(\mathbf{r})\right]_+ = \delta(\mathbf{r}-\mathbf{r}')\,. \qquad (7.6\text{-}17)$$

Also, we have

$$\left[\Psi_\lambda^{(\dagger)}(\mathbf{r}),\Psi_\mu^{(\dagger)}(\mathbf{r}')\right]_+=0,\quad \text{or}\quad \left[\Psi^{(\dagger)}(\mathbf{r}),\Psi^{(\dagger)}(\mathbf{r}')\right]_+=0. \tag{7.6-18}$$

We now consider an operator $F=\Sigma_i f_i(\mathbf{p},\mathbf{r})$, which does not act on the spin states. For the matrix element we have

$$\begin{aligned}\langle\mathbf{k}\lambda\,|\,f\,|\,\underset{\sim}{\ell}\mu\rangle &= \int \alpha_{\mathbf{k}}^*(\mathbf{r},\lambda) f(\mathbf{p},\mathbf{r})\,\alpha_{\underset{\sim}{\ell}}(\mathbf{r},\mu)\,d^3r \\ &= \langle\lambda\,|\,\mu\rangle \int \gamma_{\mathbf{k}}^*(\mathbf{r}) f(\mathbf{p},\mathbf{r})\,\gamma_{\underset{\sim}{\ell}}(\mathbf{r})\,d^3r = \delta_{\lambda\mu}\langle\mathbf{k}\,|\,f\,|\,\underset{\sim}{\ell}\rangle,\end{aligned} \tag{7.6-19}$$

(where $\underset{\sim}{\ell}=$ bold ℓ). From (7.6-6) and the above we thus obtain,

$$F=\sum_{\mathbf{k}\underset{\sim}{\ell},\lambda\mu} c_{\mathbf{k}\lambda}^\dagger c_{\underset{\sim}{\ell}\mu}\langle\mathbf{k}\lambda\,|\,f\,|\,\underset{\sim}{\ell}\mu\rangle=\sum_{\mathbf{k}\underset{\sim}{\ell},\eta} c_{\mathbf{k}\eta}^\dagger c_{\underset{\sim}{\ell}\eta}\langle\mathbf{k}\,|\,f\,|\,\underset{\sim}{\ell}\rangle. \tag{7.6-20}$$

Hence, employing (7.6-15) we establish:

$$F=\sum_\eta \int \Psi_\eta^\dagger(\mathbf{r}) f(\mathbf{p},\mathbf{r})\Psi_\eta(\mathbf{r})\,d^3r. \tag{7.6-21}$$

Exactly the same can be done for an operator $V=\Sigma_{ij}v(\mathbf{r}_i,\mathbf{r}_j')$. We then obtain,

$$V=\sum_{\lambda\mu}\int\!\!\int \Psi_\lambda^\dagger(\mathbf{r})\Psi_\mu^\dagger(\mathbf{r}')v(\mathbf{r},\mathbf{r}')\Psi_\mu(\mathbf{r}')\Psi_\lambda(\mathbf{r})\,d^3r d^3r'. \tag{7.6-22}$$

For the Hamiltonian this gives in particular:

$$\begin{aligned}\mathcal{H} &= \sum_\eta \int \left[(\hbar^2/2m)\nabla\Psi_\eta^\dagger(\mathbf{r})\cdot\nabla\Psi_\eta(\mathbf{r})+\Psi_\eta^\dagger(\mathbf{r})v^{(1)}(\mathbf{r})\Psi_\eta(\mathbf{r})\right]d^3r \\ &+\tfrac{1}{2}\sum_{\lambda\mu}\int\!\!\int \Psi_\lambda^\dagger(\mathbf{r})\Psi_\mu^\dagger(\mathbf{r}')v^{(2)}(\mathbf{r},\mathbf{r}')\Psi_\mu(\mathbf{r}')\Psi_\lambda(\mathbf{r})\,d^3r d^3r'+\ldots\end{aligned} \tag{7.6-23}$$

We could give here several applications. However, these will occur in conjunction with statistical physics problems. As an exception we treat in Section 7.8 the derivation of the Hartree-Fock equation, which underlies the one-electron wave functions in an electron gas in solids in the presence of electron-electron interactions. We could also consider the quantization of fermion many-particle systems from the field point of view. For spin ½ particles we must then consider the Dirac equation as the field equation. The quantization of this field is treated in a number of standard texts. A concise, readable version is found in the older book of Schiff.[16]

7.7 The Boson-Fermion Dichotomy: General Remarks

A fundamental question, still to be answered, is which particles behave as bosons and which behave as fermions. This was first studied by Pauli. The question is

[16] L. Schiff, "Quantum Mechanics", Second Ed., McGraw Hill, NY. 1955, Section 47.

intimately linked with the relativistic description of many particle systems. From the irreducible representations of the restricted Lorentz group, together with other considerations, it can be shown that particles with integral nuclear spin ($I = 0, 1, 2...$) are bosons, whereas particles with half integral spin ($I = \frac{1}{2}, \frac{3}{2}, ...$) are fermions. Elements of the proof are given in Appendix B on spin and statistics. All nuclei composed of an even number of protons plus neutrons have integral spin and are therefore bosons. Examples are He^4 or α-particles. All nuclei composed of an odd number of baryons (like Li or He^3) are fermions.

As to the fundamental particles, the following can be said. According to the standard model, *all four types of interactions,* the strong (nuclear) interaction, the weak interaction, the electromagnetic interaction and the gravitational interaction *are mediated by bosons, which 'originally' were massless.* The strong interaction is mediated by massless gluons, being spin-one bosons. The weak interaction is mediated by the $W^{\pm}$ and Z^0 spin-one vector bosons, which have acquired mass by the Higgs' mechanism (masses are 80.4 and 91.2 GeV/c^2). The Higgs vector boson ($\sim 1\,TeV/c^2$) has not yet been found. The electromagnetic interaction is mediated by photons, the well known quasi-particles with spin one. Gravitational interaction is mediated by gravitons with spin two; these quasi-particles, predicted by general relativity, have not yet been observed, although many experiments are underway.

All leptons – light particles which are not subject to the strong interaction force – are fundamental spin ½ particles: electrons, muons and tau-leptons with their neutrinos; also, their anti-particles. These are all fermions.

Hadrons – particles subject to the strong interaction – are not fundamental particles as they once were thought to be. They are composed of a combination of the six existing quarks, which are the fundamental units giving rise to the strong interaction. Mesons have a two-quark content; since all quarks are spin ½ fermions, mesons have a spin equal to zero or one, thus being bosons. Examples are the $\pi^{\pm}, \pi^0, K^{\pm}$ and η^0 mesons.

Baryons include the proton and neutron with their anti-particles, as well as the heavier hyperons, $\Lambda^0, \Sigma^{\pm}, \Sigma^0, \Omega^-$, etc. They consists of three quarks and mostly have spin ½ , although Ω^- and Δ^{++} have spin $\frac{3}{2}$; all are fermions.

In summary: *all fundamental building blocks, the six quarks and the six leptons with their anti-particles, are fermions. All particles that mediate the four interactions – with mass (obtained via the Higgs mechanism) or massless – are bosons.*

7.8 The Hartree–Fock Equation*

In a solid we are dealing with an electron gas, a system of fermions, in which electron-electron interaction is largely accounted for by an averaged field contribution, contained in the Hartree-Fock Hamiltonian and the subsequent effective one-electron eigenvalue equation. The Hartree-Fock equation, being the basis of the electron-band theory of solids, is most easily derived with the second quantization procedure, which we have just presented in the preceding sections.[17,18]

We consider an electron gas with Hamiltonian

$$H=\sum_i\left[\frac{p_i^2}{2m}+V(\mathbf{r}_i)\right]+\frac{1}{2}\sum_{ij}\frac{e^2}{4\pi|\mathbf{r}_i-\mathbf{r}_j|}\,. \tag{7.8-1}$$

Here $V(\mathbf{r})$ is the lattice potential and the last term is the electron-electron interaction in rationalized Gaussian units. The many-body field Hamiltonian is

$$\begin{aligned}\mathcal{H}_{field}&=\sum_\sigma\int d^3r\Psi_\sigma^\dagger(\mathbf{r})\left(\frac{p_i^2}{2m}+V(\mathbf{r}_i)\right)\Psi_\sigma(\mathbf{r})\\&+\frac{1}{2}\sum_{\sigma\sigma'}\int\int d^3rd^3r'\Psi_\sigma^\dagger(\mathbf{r})\Psi_{\sigma'}^\dagger(\mathbf{r}')\frac{e^2}{4\pi|\mathbf{r}_i-\mathbf{r}_j|}\Psi_{\sigma'}(\mathbf{r}')\Psi_\sigma(\mathbf{r})\,.\end{aligned} \tag{7.8-2}$$

Here σ denotes the spin quantum number.

We now consider the ground state. For the $\alpha_{\mathbf{k}}(\mathbf{r})$ we choose the one-particle wave functions (as yet unknown. but labelled with the momentum **k**) that minimize the expectation value of the energy. We write

$$\alpha_{\mathbf{k}}(\mathbf{r})=\varphi_{\mathbf{k}}(\mathbf{r})|\sigma\rangle\equiv\varphi_{\mathbf{k}\sigma}(\mathbf{r})=\begin{pmatrix}\varphi_{\mathbf{k}+}(\mathbf{r})\\\varphi_{\mathbf{k}-}(\mathbf{r})\end{pmatrix}. \tag{7.8-3}$$

The one-particle wave functions are spinors, as expected. For simplicity we shall use the subscript 'k' to denote both **k** and σ. For the ground state containing N particles we write

$$|\psi_N\rangle=\prod_k^N c_k^\dagger|0\rangle\,. \tag{7.8-4}$$

According to the variational principle we must have

$$\mathcal{E}_N=\min\frac{\langle\psi_N|\mathcal{H}|\psi_N\rangle}{\langle\psi_N|\psi_N\rangle}\,. \tag{7.8-5}$$

Assuming that $|\psi_N\rangle$ is normalized, we require $\langle\psi_N|\mathcal{H}|\delta\psi_N\rangle=0$, or, in view of (7.8-4),

$$\langle 0|\prod_n^N c_n\mathcal{H}\delta c_k^\dagger\prod_{n\neq k}^N c_n^\dagger|0\rangle\,, \tag{7.8-6}$$

[17] C. Kittel, "Quantum Theory of Solids", John Wiley and Sons, NY. 1964 pp. 80-83 (second printing 1986);

[18] P. W. Anderson, "Concepts in Solids", Benjamin , NY. and Amsterdam 1963, pp. 15-28.

where k refers to any of the electron states. Now, the only way that we can have a variation $\delta c_k^\dagger$ (symbolic!) is that an electron from the state φ_k is put into a state φ_M, with $M > N$. Thus, (7.8-6) means $\delta c_k^\dagger \to c_M^\dagger$:

$$\langle 0 | \prod_n^N c_n \mathcal{H} c_M^\dagger \prod_{n \neq k}^N c_n^\dagger | 0 \rangle = 0. \tag{7.8-7}$$

We write $\mathcal{H}$ in terms of the c-operators, following the recipe of (7.6-6) and (7.6-11); so with $\mathcal{H} = \Sigma_i h_i^{(1)} + \frac{1}{2}\Sigma_{ij} v_{ij}^{(2)}$, we have

$$\mathcal{H} = \sum_{pq} c_p^\dagger \langle p | h^{(1)} | q \rangle c_q + \tfrac{1}{2} \sum_{pqrs} c_p^\dagger c_q^\dagger \langle pq | v^{(2)} | rs \rangle c_s c_r, \tag{7.8-8}$$

where

$$\langle p | h^{(1)} | q \rangle = \int d^3 r \varphi^*_{\mathbf{p},\sigma_\mathbf{p}} \left(\frac{\hbar^2}{2m} \nabla^2 + V(\mathbf{r}) \right) \varphi_{\mathbf{q},\sigma_\mathbf{q}}(\mathbf{r}) \delta_{\sigma_\mathbf{p}\sigma_\mathbf{q}},$$

$$\langle pq | v^{(2)} | rs \rangle = \frac{e^2}{4\pi} \int\int d^3 r d^3 r' \frac{\varphi^*_{\mathbf{p},\sigma_\mathbf{p}}(\mathbf{r}) \varphi^*_{\mathbf{q},\sigma_\mathbf{q}}(\mathbf{r}') \varphi_{\mathbf{r},\sigma_\mathbf{r}}(\mathbf{r}) \varphi_{\mathbf{s},\sigma_\mathbf{s}}(\mathbf{r}')}{|\mathbf{r} - \mathbf{r}'|} \delta_{\sigma_\mathbf{p}\sigma_\mathbf{s}}. \tag{7.8-9}$$

We substitute (7.8-8) into (7.8-7), to obtain

$$\sum_{pq} \langle 0 | \prod_n^N c_n c_p^\dagger c_q c_M^\dagger \prod_{n \neq k}^N c_n^\dagger | 0 \rangle \langle p | h^{(1)} | q \rangle$$
$$+ \tfrac{1}{2} \sum_{pqrs} \langle 0 | \prod_n^N c_n c_p^\dagger c_q^\dagger c_s c_r c_M^\dagger \prod_{n \neq k}^N c_n^\dagger | 0 \rangle \langle pq | v^{(2)} | rs \rangle = 0. \tag{7.8-10}$$

The evaluation of these expressions is straightforward, but a bit lengthy. In the first expression one sees immediately that q must be equal to M in order to undo the raising by $c_M^\dagger$; if then p is equal to k, the raising products in the ket and bra match, so that the whole sequence reduces to $\langle 0 | 0 \rangle = 1$. With some more reflection, one obtains

$$\langle k | h^{(1)} | M \rangle + \sum_m^N \Big[\langle km | v^{(2)} | mM \rangle - \langle mk | v^{(2)} | mM \rangle \Big] = 0. \tag{7.8-11}$$

We also use the complex conjugate result

$$\langle M | h^{(1)} | k \rangle + \sum_m^N \Big[\langle mM | v^{(2)} | km \rangle - \langle mM | v^{(2)} | mk \rangle \Big] = 0. \tag{7.8-12}$$

Note that in both expressions $k, m \in N$, while $M > N$. Now note that the self-consistent eigenvalue ε_n of a one-body state φ_n satisfies

$$\sum_k^N \langle k | h^{(1)} | n \rangle c_k^\dagger + \sum_k^N \sum_m^N \Big[\langle km | v^{(2)} | mn \rangle - \langle mk | v^{(2)} | mn \rangle \Big] c_k^\dagger = \varepsilon_n c_n^\dagger. \tag{7.8-13}$$

Further, by adding similar contributions from (7.8-12) – noting $M > N$ – we find

$$\sum_{all\ k} \langle k | h^{(1)} | n \rangle c_k^\dagger + \sum_{all\ k}^{N} \sum_{m}^{N} \Big[\langle km | v^{(2)} | mn \rangle - \langle mk | v^{(2)} | mn \rangle \Big] c_k^\dagger = \varepsilon_n c_n^\dagger . \quad (7.8\text{-}14)$$

If this is multiplied by c_n and summed over all n, the rhs gives $\Sigma_n \varepsilon_n n_n = \mathcal{H}_{field}$. The first term on the left is the standard term, while the second term is the 'smeared out' interaction energy, taking into account exchange effects.

The Hartree-Fock equation is now just around the corner. We will obtain the field form by noting

$$c_k^\dagger = \int d^3 r \Psi^\dagger_{\sigma_\mathbf{k}}(\mathbf{r}) \varphi^*_{\mathbf{k}\sigma_\mathbf{k}}(\mathbf{r}) . \quad (7.8\text{-}15)$$

We write out the matrix elements in (7.8-14) and use the closure relation

$$\sum_{\mathbf{k},\sigma} \varphi^*_{\mathbf{k},\sigma}(\mathbf{r}) \varphi_{\mathbf{k},\sigma}(\mathbf{r}') = \delta(\mathbf{r} - \mathbf{r}') . \quad (7.8\text{-}16)$$

From (7.8-14) we then easily establish

$$\int d^3 r \Psi^\dagger_{\sigma_\mathbf{k}}(\mathbf{r}) \Bigg[h^{(1)} \varphi_{\mathbf{n},\sigma_\mathbf{n}}(\mathbf{r}) + \sum_{\mathbf{m},\sigma_\mathbf{m}}^{N} \int d^3 r' \frac{e^2}{4\pi |\mathbf{r} - \mathbf{r}'|} |\varphi_{\mathbf{m},\sigma_\mathbf{m}}(\mathbf{r}')|^2 \varphi_{\mathbf{n},\sigma_\mathbf{n}}(\mathbf{r})$$

$$- \sum_{\mathbf{m},\sigma_\mathbf{m}}^{N} \int d^3 r' \frac{e^2}{4\pi |\mathbf{r} - \mathbf{r}'|} \varphi^*_{\mathbf{m},\sigma_\mathbf{m}}(\mathbf{r}') \varphi_{\mathbf{n},\sigma_\mathbf{n}}(\mathbf{r}') \varphi_{\mathbf{m},\sigma_\mathbf{m}}(\mathbf{r}) \Bigg] = \int d^3 r \varepsilon_{\mathbf{n},\sigma_\mathbf{n}} \Psi^\dagger_{\sigma_\mathbf{k}} \varphi_{\mathbf{n},\sigma_\mathbf{n}}(\mathbf{r}) .$$

$$(7.8\text{-}17)$$

This holds for any $\Psi^\dagger_{\sigma_\mathbf{k}}$. Thus equating the integrands and substituting for $h^{(1)}$, we finally obtain

$$\left\{ \left[-\frac{\hbar^2}{2m} \nabla^2 + V(\mathbf{r}) \right] + \sum_{\mathbf{m},\sigma_\mathbf{m}}^{N} \int d^3 r' \frac{e^2}{4\pi |\mathbf{r} - \mathbf{r}'|} \left| \varphi_{\mathbf{m},\sigma_\mathbf{m}}(\mathbf{r}') \right|^2 \right\} \varphi_{\mathbf{n},\sigma_\mathbf{n}}(\mathbf{r})$$

$$- \sum_{\mathbf{m},\sigma_\mathbf{m}}^{N} \int d^3 r' \frac{e^2}{4\pi |\mathbf{r} - \mathbf{r}'|} \varphi^*_{\mathbf{m},\sigma_\mathbf{m}}(\mathbf{r}') \varphi_{\mathbf{m},\sigma_\mathbf{m}}(\mathbf{r}) \varphi_{\mathbf{n},\sigma_\mathbf{n}}(\mathbf{r}') = \varepsilon_{\mathbf{n},\sigma_\mathbf{n}} \varphi_{\mathbf{n},\sigma_\mathbf{n}}(\mathbf{r}) . \quad (7.8\text{-}18)$$

This integro-differential equation is the Hartree-Fock equation. The first term is the one-electron term without any interaction. The second term is the effect of all other electrons, being at $\mathbf{r}'$, with probability density $|\varphi(\mathbf{r}')|^2$. The third term is the exchange energy. Summarising the result by $\mathcal{H}_{H-F} \varphi_\mathbf{n}(\mathbf{r}) = \varepsilon_\mathbf{n} \varphi_\mathbf{n}$, we find that the Hartree-Fock Hamiltonian is given by

$$\mathcal{H}_{H-F} = -\frac{\hbar^2}{2m} \nabla^2 + V(\mathbf{r}) + \sum_{\mathbf{m},\sigma_\mathbf{m}}^{N} \int d^3 r' \frac{e^2}{4\pi |\mathbf{r} - \mathbf{r}'|} \left| \varphi_{\mathbf{m},\sigma_\mathbf{m}}(\mathbf{r}') \right|^2$$

$$-\sum_{\mathbf{m},\sigma_{\mathbf{m}}}^{N}\int d^3r'\frac{e^2}{4\pi|\mathbf{r}-\mathbf{r}'|}\varphi^*_{\mathbf{m},\sigma_{\mathbf{m}}}(\mathbf{r}')\varphi_{\mathbf{m},\sigma_{\mathbf{m}}}(\mathbf{r})\int d^3r\,\delta(\mathbf{r}-\mathbf{r}'). \quad (7.8\text{-}19)$$

Clearly, the solutions for the Bloch functions in electron bands in condensed matter require a great deal of numerical computation! We still note that the Hartree-Fock equation can also be obtained without the field formalism; however, the derivation requires many assumptions, is less rigorous and is quite complex also.

7.9 Problems to Chapter VII

7.1 Given is the *ordered* many-body fermion state $|1_a 1_b 0_c 1_d 0_e\rangle$. The following operation is performed:

$$c_e c_e^\dagger c_a^\dagger c_c^\dagger c_a c_d\,|1_a 1_b 0_c 1_d 0_e\rangle. \quad (1)$$

(a) Find the final state, including sign;
(b) Write the creation and annihilation operators as an ordered product, using the anti-commutation rules; find once more the final state.

7.2 A spin one particle has three states with energies $-\varepsilon$, 0, ε. We have three particles divided over these states; they obey Bose–Einstein statistics. The particles are in equilibrium with a thermal bath of temperature T.
(a) What is the number of possible many-body states? [Remember the problem of distributing N balls over Z bags with no limit to the number of balls per bag.]
(b) Write down the possible many-body occupation-number states $|\gamma\rangle = |n_a n_b n_c\rangle$, their energies $\mathcal{E}_\gamma$ and their degeneracies g_γ.
(c) Find the canonical partition function $\mathcal{Z}$ and the Helmholtz free energy F.
(d) In another universe these particles obey Fermi-Dirac statistics. Find again the partition function $\mathcal{Z}$ and the Helmholtz free energy F.

7.3 An electron gas is described with normalized plane waves, $|\mathbf{k}\rangle \to V_0^{-1/2}e^{i\mathbf{k}\cdot\mathbf{r}}$, where V_0 denotes the volume. The two-body interaction stems from a screened Coulomb potential,

$$V = \tfrac{1}{2}\Sigma_{ij}\frac{q^2 e^{-\kappa|\mathbf{r}_i-\mathbf{r}_j|}}{4\pi|\mathbf{r}_i-\mathbf{r}_j|}. \quad (2)$$

(a) Write V in second quantization form (operators $c, c^\dagger$), using eigenstates $|\mathbf{k}_1\rangle, |\mathbf{k}_2\rangle, |\bar{\mathbf{k}}_1\rangle$ and $|\bar{\mathbf{k}}_2\rangle$.
(b) Change the spatial variables to $\mathbf{R}=\mathbf{r}_1-\mathbf{r}_2$ and $\mathbf{r}_2$ and use the Fourier result $\int d^3r\exp(i\mathbf{K}\cdot\mathbf{r}) = V_0\delta_{\mathbf{K},0}$. Set $\mathbf{k}_1-\bar{\mathbf{k}}_1=\mathbf{q}$. To evaluate the remaining

integral over d^3R, use $\mathbf{q}$ as polar axis.
(c) Is the momentum conserved? Write your final operators in terms of the vectors $\mathbf{k}_1, \mathbf{k}_2$ and $\mathbf{q}$. (This problem is important in the BCS theory of superconductivity.)

7.4 A system of relativistic bosons satisfies the Klein-Gordon wave equation:

$$\square\Phi + m^2\Phi \equiv (\partial^2/\partial t^2)\Phi^2 - \nabla^2\Phi + m^2\Phi = 0, \tag{3}$$

where $\square$ is the d'Alembertian and where we set $c = \hbar = 1$. Consider the E.F. of the ∇^2 operator, $(\nabla^2 + \lambda_k^2)\gamma_k(\mathbf{r}) = 0$, and expand the field operator as usual,

$$\Phi(\mathbf{r},t) = \Sigma_k f_k(t)\gamma_k(\mathbf{r}). \tag{4}$$

Establish the normal mode expansion stated in Section 7.4.:

$$\Phi(\mathbf{r},t) = \Sigma_k (2\omega_k)^{-1/2}[a_k(t)\gamma_k(\mathbf{r}) + a_k^\dagger(t)\gamma_k^*(\mathbf{r})], \tag{5}$$

where $\omega_k = (\lambda_k^2 + m^2)^{1/2}$ and show that $f_k \equiv a_k$ and $f_k^\dagger \equiv a_k^\dagger$ are boson operators.

7.5 Given is the Hermitean Hamiltonian for a boson system

$$\mathcal{H} = \varepsilon_0 a^\dagger a + \varepsilon_1(a^\dagger a^\dagger + aa). \tag{6}$$

Diagonalize this Hamiltonian, using suitable linear combinations of $a^\dagger$ and a.

7.6 For an ordinary, non-relativistic boson system the operator for the number of particles in a volume V_i, denoted by $\mathbf{N}_i$, is given by $\mathbf{N}_i = \int_{V_i} d^3r \Psi^\dagger(\mathbf{r})\Psi(\mathbf{r})$.
(a) Show that $[\mathbf{N}_1, \mathbf{N}_2] = 0$, even if the volumes V_1 and V_2 overlap.
(b) In a relativistic boson system described by the Klein-Gordon equation, let $j^\mu = (\rho, \mathbf{J})$ be the current four-vector, with ρ denoting the particle density operator. From the continuity equation $\partial_\mu j^\mu = 0$, show that ρ is given as

$$\rho(\mathbf{r},t) = (i/2m)\left(\Psi^\dagger(\mathbf{r},t)\Pi(\mathbf{r},t) - \Pi^\dagger(\mathbf{r},t)\Psi(\mathbf{r},t)\right), \quad \Pi = \partial\Psi/\partial t, \tag{7}$$

and establish that ρ is not positive definite.
(c) Compute again $[\mathbf{N}_1, \mathbf{N}_2]$ and show that it is generally non-zero.

Chapter VIII

Ideal Quantum Gases and Elementary Excitations in Solids

8.1 Bose–Einstein Statistics for Zero-Restmass Particles; Blackbody Radiation and Photons

8.1.1 *Planck's Law; Original Considerations*

Electromagnetic radiation in the form of "heat radiation", given off by a black body, was the subject of many studies and laws established in the late 19th century. In particular, the law of Stefan-Boltzmann for the total radiation, encompassing all wavelengths, $\langle I \rangle = \sigma T^4$, and the displacements laws of Wien were a pleasant confirmation of Maxwell's basis for electromagnetic waves. The classical spectral distribution of the radiation was embodied in the radiation law of Rayleigh-Jeans, based on the mean energy per oscillator, $k_B T$ per mode, with two perpendicular transversal polarizations. This law worked alright for high wavelengths, but was in flagrant disagreement with observations for small wavelengths. This was known as the "ultra-violet catastrophe." The break-through came when Max Planck (born 1858) ushered in the quantum age with a new theory proposed in 1900, by assuming that the energy content of the vibrational modes was not continuous but discrete, i.e., $\varepsilon = n\,h\nu$, where h is a proposed new constant, to be found through comparison with experiment. Briefly, his derivation was as follows. He considered the radiation given off by a cavity with a pinhole ('Hohlraum') in which the electromagnetic field caused standing waves, whose energy is transferred via the walls to the aperture of the pinhole. [Such a cavity is the best realization of a black body, since virtually no radiation escapes from the pinhole due to multiple reflections, so the absorption coefficient $\alpha \to 1$ and the pinhole is 'black'.] The number of possible modes in the frequency interval $(\nu, \nu + d\nu)$ was known to be $(8\pi / c^3)\, V\nu^2 d\nu$, where V is the volume of the cavity. However, the energy of the oscillators is no longer $Nk_B T$ in this new view. Knowing Boltzmann's connection between entropy and the amount of disorder of a macroscopic state, Planck computed the number of 'complexions' R that the oscillator apportionments $\varepsilon_1 / h\nu$, $\varepsilon_2 / h\nu, \ldots \varepsilon_N / h\nu$, could assume, analogous to our method of (2.5-29), the result being $R = (N + P - 1)!/(P - 1)!N!$ where $P = \mathcal{E} / h\nu$. He then connected with the temperature – the only parameter that played a role in the observed universal spectral curves – by setting $1/T = k_B\, d \ln R_0 / d\mathcal{E}$. [Planck also introduced for the first time the symbol k for Boltzmann's constant, which was at that time only written as the gas constant divided by Avogadro's number.] Then from rearrangement similar as in our (3.5-13) he obtained the average energy per oscillator minus zero-point energy,

$$\langle \varepsilon_\nu \rangle = (\mathscr{E}/N) = \frac{h\nu}{e^{h\nu/k_BT} - 1}. \tag{8.1-1}$$

Multiplying with the number of modes as given above, the radiated energy density in the interval $(\nu, \nu + d\nu)$ then follows immediately[1]

$$\langle u(\nu) \rangle d\nu = \frac{8\pi h\nu^3}{c^3} \frac{d\nu}{e^{h\nu/k_BT} - 1}, \tag{8.1-2}$$

which is Planck's law. Agreement with experimental data was excellent. Of course, the quantum hypothesis was at that time entirely *ad hoc.* Tragically, Planck spent much of his later life to find a classical origin for the basic tenets of his approach. It was not until Einstein in 1905 proposed that light was *actually* made up of energy packets or quanta of magnitude $h\nu$, that a rational basis for Planck's assumption was given and a radical breach with classical physics was immanent.[2]

We briefly come back to one particular argument of the old physics school, to which this author was still exposed as a student: it was assumed that the EM field in the cavity gave rise to *standing* waves.[3,4] Assuming the cavity to be a cube of dimension $L^3 = V$, the possible wavelengths that 'fitted' these dimensions required wave numbers $\sigma = (1/\lambda)$ of the form

$$\sigma_x = n_x/2L, \quad \sigma_y = n_y/2L, \quad \sigma_z = n_z/2L, \quad n_x, n_y, n_z = 1, 2, \ldots \tag{8.1-3}$$

Note that there is a maximum wavelength $\frac{1}{2}\lambda_{max} = L$, corresponding to a minimum wave number $1/2L$ for each direction. Since $\Delta\sigma_i = 1/2L$, the mode volume in σ-space is $1/8V$ and the number of modes between σ and $\sigma + d\sigma$ $[\sigma = \sqrt{\sigma_x^2 + \sigma_y^2 + \sigma_z^2}]$ is $2 \times \frac{1}{8} 4\pi\sigma^2 d\sigma/[1/8V] = 8\pi V \sigma^2 d\sigma$, where the factor two accounts for the two transverse modes. With $\sigma = \nu/c$, the number of modes in $(\nu, \nu + d\nu)$ is found to be as given above. However, there is a minimum value of ν, associated with the dimensions of the cavity, as we indicated above, whereas Planck's law is valid for $\nu \to 0$. The remedy is indicated below: basically, the cavity is to be placed outside

---

[1] Max Planck, "Zur Theorie des Gesetzes der Energieverteilung im Normalspectrum", Verhandlungen der Deutschen Physikalischen Gesellschaft **2**, 237-245 (1900). English translation by D. ter Haar, in " The Old Quantum Theory" Oxford Univ. Press, 1967.

[2] A. Einstein, Ann. der Physik (4) **17**, 132 (1905). [Einstein still uses R/N_A for Boltzmann's constant.]

[3] R. Kronig (Ed.) "Leerboek der Natuurkunde", vol. I and II, Scheltema and Holkema, Amsterdam 1946, Chapter VIIC. Later translated into English as "Textbook of Physics", 2nd Ed., Pergamon, 1959.

[4] R. Becker, "Theorie der Elektrizität", Vol. II: "Elektronentheorie", 6th Ed., B.G. Teubner, Leipzig 1933. Translated into English , J.W. Edwards, Ann Arbor, 1946.

the space where blackbody radiation is observed and the quantization takes place in the vacuum with the electromagnetic field.

8.1.2 *Quantization of the Electromagnetic Field*

Some results from quantum electrodynamics (QED) were discussed in Section 7.4. After elimination of the longitudinal component, the field is described by running-wave modes with two linear perpendicular polarizations, cf. Eq. (7.4-27). Briefly,

$$\left.\begin{matrix}\mathbf{A}(\mathbf{r},t)\\ \mathbf{E}(\mathbf{r},t)\end{matrix}\right\} \Leftrightarrow a_{\mathbf{q},\lambda}\mathbf{u}_{\mathbf{q},\lambda}(\mathbf{r})e^{-i\omega_{\mathbf{q}}t} \pm a^{\dagger}_{\mathbf{q},\lambda}\mathbf{u}^{*}_{\mathbf{q},\lambda}(\mathbf{r})e^{i\omega_{\mathbf{q}}t}, \tag{8.1-4}$$

where the **u**'s are eigenvector solutions of the Helmholtz equation $(\nabla^2 + \mathbf{q}^2)\mathbf{u}_{\mathbf{q},\lambda} = 0$ and λ denotes the polarization. Choosing plane waves, we have $\mathbf{u}_{\mathbf{q},\lambda} = \mathbf{e}_{\lambda}\exp(\pm i\mathbf{q}\cdot\mathbf{r})$, **q** being the wave vector. It is also possible to take linear combinations of the **e**'s, $\frac{1}{2}\sqrt{2}(\mathbf{e}_1 \pm i\mathbf{e}_2)$, such that we have two circularly polarized waves; thus the photons, which are the excitations of the quantized field, have angular momentum in the direction of propagation with spin component ± 1.

We now come back to the mode density. Let V be any mathematical volume in the space of the vacuum field. Periodic boundary conditions require that **q** and so $\omega_{\mathbf{q}} = c\,|\mathbf{q}|$ are quantized,

$$\begin{aligned} &q_x = 2\pi n_x / L_x, \quad q_y = 2\pi n_y / L_y, \quad q_z = 2\pi n_z / L_z, \\ &n_x, n_y, n_z = 0, \pm 1, \pm 2, \ldots \quad L_x L_y L_z = V. \end{aligned} \tag{8.1-5}$$

The volume in **q**-space associated with one mode is $8\pi^3/V$, so the density in **q**-space is $Z(\mathbf{q})dq = 4\pi q^2 dq/(8\pi^3/V) = q^2 V dq/2\pi^2$. For the density of ω-modes we need only to multiply with a factor two for the separate polarizations and divide by $d\omega/dq$; hence, letting $z(\omega)$ denote the mode density per unit volume of space [V can afterwards $\to \infty$], we have

$$z(\omega)d\omega = \omega^2 d\omega/\pi^2 c^3, \text{ or } z(\nu)d\nu = 8\pi\nu^2 d\nu/c^3. \tag{8.1-6}$$

This is the same result as found previously, but based on a different argument!

We should now derive the statistical distribution for zero restmass particles, *i.c.*, quanta $\hbar\omega$. There is no particle conservation since each quantum is an energy packet. We can thus employ the canonical ensemble. Analogous to (4.3-1) we have for the canonical partition function

$$\mathcal{Q} = \sum_{\{n_{\mathbf{q},\lambda}\}} \exp\left[-\sum_{\mathbf{q},\lambda} n_{\mathbf{q},\lambda}\hbar\omega_{\mathbf{q}}/k_B T\right] = \prod_{\mathbf{q},\lambda}\sum_{n_{\mathbf{q},\lambda}} \exp\left[-n_{\mathbf{q},\lambda}\hbar\omega_{\mathbf{q}}/k_B T\right]. \tag{8.1-7}$$

where the n's are zero or integers. The result is immediate

$$\mathcal{Z} = \prod_{\mathbf{q},\lambda} \frac{1}{1 - e^{-\hbar\omega_{\mathbf{q}}/k_B T}}. \tag{8.1-8}$$

From this we find $\ln \mathcal{Z}$ and for the mean occupancy we readily obtain

$$\langle n_{\mathbf{q},\lambda} \rangle = -k_B T \frac{\partial \ln \mathcal{Z}_\lambda}{\partial \hbar\omega_{\mathbf{q}}} = \frac{1}{e^{\hbar\omega_{\mathbf{q}}/k_B T} - 1}. \tag{8.1-9}$$

We note that the occupancy for given $\mathbf{q}$ is the same for both polarizations. We now arrive once more at Planck's law. Multiplying (8.1-9) by $z(\omega_{\mathbf{q}})d\omega_{\mathbf{q}}\hbar\omega_{\mathbf{q}}$, as given in (8.1-6), we obtain for the mean energy density in the interval $(\omega, \omega + d\omega)$

$$\langle u(\omega) \rangle d\omega = \frac{\hbar\omega^3}{\pi^2 c^3} \frac{d\omega}{e^{\hbar\omega/k_B T} - 1}. \tag{8.1-10}$$

We shall establish a few more common results about blackbody radiation from elementary considerations. Let $K(\nu, T)d\nu$ be the radiation coming off 1 cm^2 of wall per second, emitted within one steradian and with frequency within $(\nu, \nu + d\nu)$. One refers to $K(\nu,T)$ as the *specific intensity* (R. Becker, Op. Cit., §67). It is further assumed that the radiation is isotropic. Then the energy radiated per second through an element of surface $d\mathbf{S}$ in a cone of solid angle $d\Omega$, whose principal axis makes an angle θ with $d\mathbf{S}$ is $KdS\cos\theta d\Omega$ (Lambert's law). We also introduce the *intensity* $S(\nu,T)$, emitted per second per cm^2 from one side of a surface. [By detailed balance, this is also the radiation absorbed by the surface per second per cm^2.]. We have, if ϕ is the azymuthal angle and $d\Omega = \sin\theta d\phi d\theta$,

$$S(\nu, T) = K(\nu, T) \int_0^{2\pi} d\phi \int_0^{\pi} \cos\theta \sin\theta d\theta = \pi K(\nu, T). \tag{8.1-11}$$

The *radiation pressure* is easily found. The momentum transfer of a photon striking the wall is $2mc\cos\theta$, with $m = h\nu/c^2$. The number of photons striking 1 cm^2 per second within the directional element $d\Omega$ is $(K/h\nu)\cos\theta d\Omega$. Whence,

$$\begin{aligned} P(\nu, T) &= 2mc[K(\nu, T)/h\nu] \int_0^{2\pi} d\phi \int_0^{\pi} \cos^2\theta \sin\theta d\theta \\ &= 4\pi mc\, K(\nu, T)/3h\nu = 4\pi K(\nu, T)/3c, \end{aligned} \tag{8.1-12}$$

Finally, we need the connection between $K(\nu, T)$ and $u(\nu, T)$. We observe a small volume ΔV inside a cavity and look at the radiation coming from an element dS of its wall, see Fig. 8-1. We consider the amount of energy radiated into ΔV by dS within a cone of solid angle $d\Omega$, whose axis makes an angle θ with $d\mathbf{S}$ as before. This cone cuts out of ΔV a prism of cross section da and length ℓ. Of the energy $KdS\cos\theta d\Omega$, radiated per second, the fraction ℓ/c is within ΔV. This amounts therefore to $KdS\cos\theta(da/r^2)(\ell/c)$. The energy intercepted by all elements da

stemming from the entire cavity wall is

$$u\Delta V = \oiint \frac{KdS\cos\theta}{r^2 c}\Sigma(\ell da) = \oiint \frac{KdS\cos\theta}{r^2 c}\Delta V. \tag{8.1-13}$$

Now $dS\cos\theta / r^2$ is nothing but the solid angle $d\Omega_P$, subtended by the surface element dS with respect to a point P in ΔV; so the result is

$$u(\nu,T) = 4\pi K(\nu,T)/c. \tag{8.1-14}$$

Note also $P = u/3$. From these results one finds the alternate form of Planck's law

$$\langle S(\nu,T)\rangle = \frac{1}{4}c\langle u(\nu,T)\rangle = \frac{2\pi h\nu^3}{c^2}\frac{1}{e^{h\nu/k_BT}-1}. \tag{8.1-15}$$

For the *photon intensity* – number emitted per *sec* per cm^2 of wall) – we have

$$\langle J(\nu,T)\rangle = \langle S(\nu,T)\rangle / h\nu = \frac{2\pi\nu^2}{c^2}\frac{1}{e^{h\nu/k_BT}-1}. \tag{8.1-16}$$

Lastly, the total emitted radiation per *sec* per cm^2 of wall is found by integration over all frequencies:

$$\langle I(T)\rangle = \int_0^\infty \langle S(\nu,T)\rangle d\nu = \frac{2\pi k_B{}^4T^4}{h^3c^2}\int_0^\infty \frac{x^3}{e^x-1}dx = \frac{2\pi k_B{}^4T^4}{h^3c^2}\zeta(4)\Gamma(4), \tag{8.1-17}$$

where $\zeta(\nu)$ is the Riemann zeta function; in particular $\Gamma(4) = 3!$, $\zeta(4) = \pi^4/90$. We thus obtained the Stefan-Boltzmann formula $\langle I\rangle = \sigma T^4$, with

$$\sigma = 2\pi^5 k^4 / 15c^2h^3 = 5.77\times 10^{-12}\, watt\ cm^{-2}K^{-4}. \tag{8.1-18}$$

The fact that $\langle I\rangle \propto T^4$ can also be derived from thermodynamics, but the value of σ requires a statistical basis.

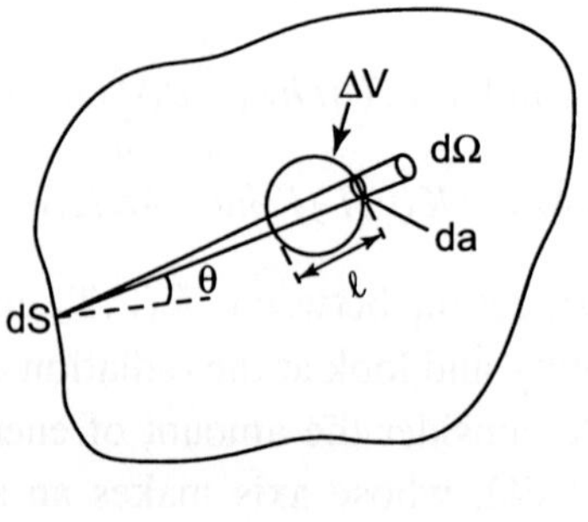

Fig. 8-1. Radiation in a cavity.

We finish with a note on the integrals

$$\mathcal{I} = \int_0^\infty x^n dx / (e^x - 1) . \tag{8.1-19}$$

We write

$$\frac{1}{e^x - 1} = \frac{e^{-x}}{1 - e^{-x}} = \sum_{\nu=1}^{\infty} e^{-\nu x} . \tag{8.1-20}$$

Hence

$$\mathcal{I} = \sum_{\nu=1}^{\infty} \int_0^\infty x^n e^{-\nu x} dx = \sum_{\nu=1}^{\infty} \frac{1}{\nu^{n+1}} \int_0^\infty t^n e^{-t} dt = \Gamma(n+1) \zeta(n+1) . \tag{8.1-21}$$

Some examples of the ζ-function are: $\zeta(2) = \pi^2/6$, $\zeta(4) = \pi^4/90$, $\zeta(6) = \pi^6/945$. Results for fractional arguments will appear in the next section.

8.2 The Perfect Bose Gas

In this section and the next one we will consider a gas of non-interacting bosons. We derive the equations of state and show the occurrence of Bose–Einstein condensation. This is a phase transition in momentum space, which can be derived exactly. With some hand-waving, this can be applied to the phase transition involving superfluidity of helium-four (^{4}He) at 2.19K. The correct theory of strongly interacting bosons will be considered in Chapter XII, dealing with quantum fluids.

The equations of state in the thermodynamic limit

The grand potential or Gibbs function $\Omega(T,V,\varsigma)$ was obtained in Section 4.3.1, see Eq. (4.3-6). We shall assume that there are no extensive variables other than $\mathcal{E}, V$ and N. Then, according to the homogeneous form for the entropy, Eq. (3.5-21),

$$\Omega = \mathcal{E} - TS - \varsigma N = -PV . \tag{8.2-1}$$

Summing over energies rather than over one-particle states, the analogue of (4.3-6) is

$$\Omega(T,V,\varsigma) = k_B T \sum_i g_i \ln\left[1 - e^{(\varsigma - \varepsilon_i)/k_B T}\right], \tag{8.2-2}$$

where g_i is the degeneracy of ε_i. Since Ω depends on one extensive and two intensive variables, the equation of state is explicit only for the intensive variable P. We have

$$P = -\Omega(T, V\varsigma)/V . \tag{8.2-3}$$

The equations of state for T and ς come in 'inverted' form:

$$N = -\partial\Omega(T,V,\varsigma)/\partial\varsigma , \tag{8.2-4}$$

$$S = -\partial\Omega(T,V,\varsigma)/\partial T. \tag{8.2-5}$$

Because of the Gibbs-Duhem relation, one of these results is superfluous, so we discard (8.2-5). We must study

$$P = \lim_{V\to\infty}\left[-\frac{\Omega(T,V,\varsigma)}{V}\right], \tag{8.2-6}$$

$$\frac{1}{\hat{v}} = \frac{N}{V} = \lim_{V\to\infty}\left[-\frac{1}{V}\frac{\partial\Omega(T,V,\varsigma}{\partial\varsigma}\right]. \tag{8.2-7}$$

The quantity $\hat{v}$ is the specific or molecular volume [not to be confused with the molar volume $\tilde{V}$].

The eigenfunctions of the one particle Schrödinger equation are $(1/\sqrt{V})e^{i\mathbf{k}\cdot\mathbf{r}}$, subject to periodic boundary conditions. Thus we have

$$\begin{aligned} &k_x = 2\pi n_x / L_x, \quad k_y = 2\pi n_y / L_y, \quad k_z = 2\pi n_z / L_z, \\ &n_x, n_y, n_z = 0, \pm 1, \pm 2, \ldots \quad L_x L_y L_z = V. \end{aligned} \tag{8.2-8}$$

The energies are

$$\varepsilon(\mathbf{k}) = \frac{\hbar^2 k^2}{2m} \simeq \frac{2\pi^2\hbar^2}{mV^{2/3}}(n_x^2 + n_y^2 + n_x^2). \tag{8.2-9}$$

Counting the number of modes between two spheres separated by *dk*, we find

$$Z(\mathbf{k})dk = \frac{4\pi k^2 dk}{8\pi^3 / V} = \frac{V}{2\pi^2}k^2 dk. \tag{8.2-10}$$

For sufficiently large quantum numbers n_x, n_y, n_z we can introduce the density of states

$$Z(\varepsilon)d\varepsilon = Z(k)\,|\,dk/d\varepsilon\,|\,d\varepsilon = 2\pi V(2m/h^2)^{3/2}\sqrt{\varepsilon}d\varepsilon. \tag{8.2-11}$$

The ground state energy is zero, and occurs for $n_x = n_y = n_z = 0$.

For large energy the sum in (8.2-2) can be approximated by an integral; for low energies, the sum remains discrete. Thus we have

$$\begin{aligned} \Omega(T,V,\varsigma)/V &= (k_BT/V)\sum_{i=0}^{I} g_i \ln\left[1 - e^{(\varsigma-\varepsilon_i)/k_BT}\right] \\ &+ \int_{\varepsilon_I}^{\infty} 2\pi(2m/h^2)^{3/2}\sqrt{\varepsilon}\ln\left[1 - e^{(\varsigma-\varepsilon_i)/k_BT}\right]d\varepsilon. \end{aligned} \tag{8.2-12}$$

We first investigate the finite sum. The first term is

$$k_BTV^{-1}\ln(1-y), \quad y = e^{\varsigma/k_BT}. \tag{8.2-13}$$

Here y is the fugacity; it is restricted to $0 \le y < 1$. We shall include, however, the value $y = 1$, requiring

$$\lim_{y\to 1, V\to\infty} V^{-1}\ln(1-y)=0\,, \tag{8.2-14}$$

which seems a reasonable proposition, since the linear ∞ 'beats' the logarithmic ∞. So the first term goes to zero. For the next terms we note from (8.2-9) that $\lim_{V\to\infty}\varepsilon_i = 0+$, indicating that they go likewise to zero. Thus,

$$\begin{aligned}\kappa = \lim_{V\to\infty}\frac{\Omega}{V} &= k_B T\int_{0+}^{\infty}\left(\frac{2m}{h^2}\right)^{3/2} 2\pi\sqrt{\varepsilon}\ln\left(1-e^{(\varsigma-\varepsilon)/k_BT}\right)d\varepsilon \\ &= -\frac{4\pi}{3}\left(\frac{2m}{h^2}\right)^{3/2}\int_0^{\infty}\frac{\varepsilon^{3/2}}{y^{-1}e^{\varepsilon/k_BT}-1}d\varepsilon,\end{aligned} \tag{8.2-15}$$

where the integral is an improper Riemann integral; in the second line we integrated by parts. This yields for P in the thermodynamic limit from (8.2-6)

$$P=\frac{4\pi}{3}\left(\frac{2m}{h^2}\right)^{3/2}\int_0^{\infty}\frac{\varepsilon^{3/2}}{y^{-1}e^{\varepsilon/k_BT}-1}d\varepsilon. \tag{8.2-16}$$

Now we must deal with (8.2-7). From (8.2-13) we have

$$-\frac{1}{V}\frac{\partial\Omega}{\partial\varsigma}=\sum_{i=0}^{I}\frac{g_i}{V}\frac{1}{y^{-1}e^{\varepsilon_i/k_BT}-1}+2\pi\left(\frac{2m}{h^2}\right)^{3/2}\int_{\varepsilon_I}^{\infty}\frac{\sqrt{\varepsilon}}{y^{-1}e^{\varepsilon/k_BT}-1}d\varepsilon. \tag{8.2-17}$$

For the first term, $i=0,\ \varepsilon_0=0,$ we have

$$\lim_{V\to\infty}\frac{1}{V}\frac{1}{y^{-1}-1}=\frac{\langle n_0\rangle}{V}, \tag{8.2-18}$$

where $\langle n_0\rangle$ is the number of particles in the ground state. At sufficiently low temperatures $\langle n_0\rangle \sim N$ and $N/V=1/\hat{v}$, so this term must certainly be retained. For the next term we obtain likewise $\langle n_1\rangle/N$. However, we now have

$$\frac{\langle n_1\rangle}{V}=\frac{1}{V}\frac{g_1}{y^{-1}e^{\varepsilon_1/k_BT}-1}\le\frac{g_1}{V[e^{\varepsilon_1/k_BT}-1]}\to\frac{g_1 mV^{2/3}k_BT}{V2\pi^2\hbar^2}\to 0, \tag{8.2-19}$$

where we used (8.2-9). Hence, this term, and likewise all the following ones, go to zero in the thermodynamic limit. We therefore arrive at

$$\frac{1}{\hat{v}}=\frac{\langle n_0\rangle}{V}+2\pi\left(\frac{2m}{h^2}\right)^{3/2}\int_0^{\infty}\frac{\sqrt{\varepsilon}}{y^{-1}e^{\varepsilon/k_BT}-1}d\varepsilon\,. \tag{8.2-20}$$

These results can be written in simple forms by introducing the Bose integrals,

$$g_\ell(y)=\frac{1}{\Gamma(\ell+1)}\int_0^{\infty}\frac{x^\ell}{e^x y^{-1}-1}dx\,. \tag{8.2-21}$$

with series expansion

$$g_\ell(y) = \frac{1}{\Gamma(\ell+1)}\int_0^\infty \frac{x^\ell (e^{-x}y)}{1-e^{-x}y}dx = \frac{1}{\Gamma(\ell+1)}\sum_{j=1}^{\infty}\int_0^\infty x^\ell e^{-jx} y^j dx = \sum_{j=1}^{\infty}\frac{y^j}{j^{\ell+1}}. \quad (8.2\text{-}22)$$

Further we have the recurrence relation

$$y\, g_\ell{}'(y) = g_{\ell-1}(y). \quad (8.2\text{-}23)$$

In particular we have

$$g_\ell(y) \le g_\ell(1) = \zeta(\ell+1), \quad (8.2\text{-}24)$$

where $\zeta(\ell+1)$ is once more the Riemann zeta function. We will need the following values

$$\zeta(\tfrac{3}{2}) = 2.612.., \quad \zeta(\tfrac{5}{2}) = 1.341.., \quad \Gamma(\tfrac{5}{2}) = \tfrac{3}{4}\sqrt{\pi}, \quad \Gamma(\tfrac{3}{2}) = \tfrac{1}{2}\sqrt{\pi}. \quad (8.2\text{-}25)$$

We now obtain with minor algebra the following results. The Gibbs function per unit volume is given as

$$\kappa = -\frac{k_B T}{\hat{v}_c(T)}\frac{g_{3/2}(y)}{\zeta(\frac{3}{2})}, \quad (8.2\text{-}26)$$

where

$$1/\hat{v}_c(T) \equiv (2m\pi k_B T/h^2)^{3/2}\zeta(\tfrac{3}{2}). \quad (8.2\text{-}27)$$

The equations of state follow from

$$P = \frac{k_B T}{\hat{v}_c(T)}\frac{g_{3/2}(y)}{\zeta(\frac{3}{2})}, \quad (8.2\text{-}28)$$

$$\frac{1}{\hat{v}} = \frac{\langle n_0\rangle}{V} + \frac{1}{\hat{v}_c(T)}\frac{g_{1/2}(y)}{\zeta(\frac{3}{2})}. \quad (8.2\text{-}29)$$

The function $g_{1/2}(y)$ is sketched in Fig. 8-2 below.

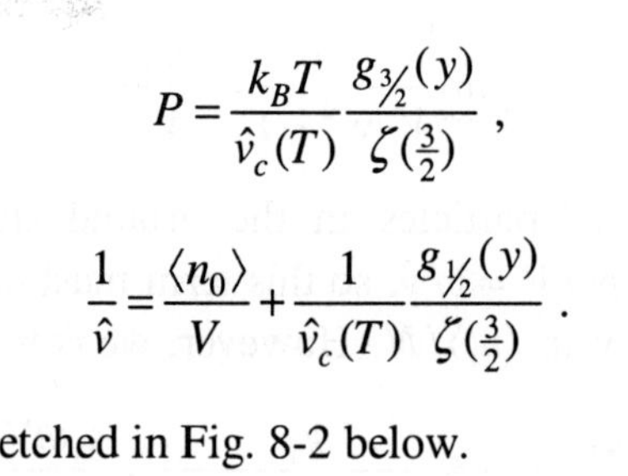

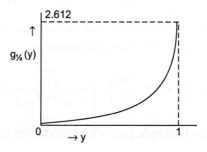

Fig. 8-2. The Bose function $g_{1/2}(y)$ vs. y.

At $y = 1$ the function is finite, but its derivative, $g_{-\frac{1}{2}}(1)$ diverges, so the tangent is vertical. The density of particles in the ground state $\langle n_0 \rangle / V$ is positive if $g_{\frac{1}{2}}(y)/\hat{v}_c(T) < g_{\frac{1}{2}}(1)/v$, which happens if T is sufficiently low, see (8.2-27). The region for which there is a sizeable fraction of particles in the ground state is called the *condensation region.*

8.3 Bose–Einstein Condensation

8.3.1 *The $P-\hat{v}$ and the $P-T$ Diagrams*

Equation (8.2- 29) will be put in the form, substituting for $\langle n_0 \rangle / V$ the B–E result,

$$\frac{\hat{v}_c(T)}{\hat{v}} = \frac{\hat{v}_c(T)y}{V(1-y)} + \frac{g_{\frac{1}{2}}(y)}{g_{\frac{1}{2}}(1)}, \tag{8.3-1}$$

which will be solved graphically.[5] In Fig. 8-3 the $-\cdot-\cdot-$ line is the total right-hand side of (8.3-1). The larger V, the steeper is the curve near $y = 1$. Therefore, if the left-hand side $\hat{v}_c / \hat{v}$ is larger than unity, the intercept point P, in the limit $V \to \infty$, is at $y = 1$.

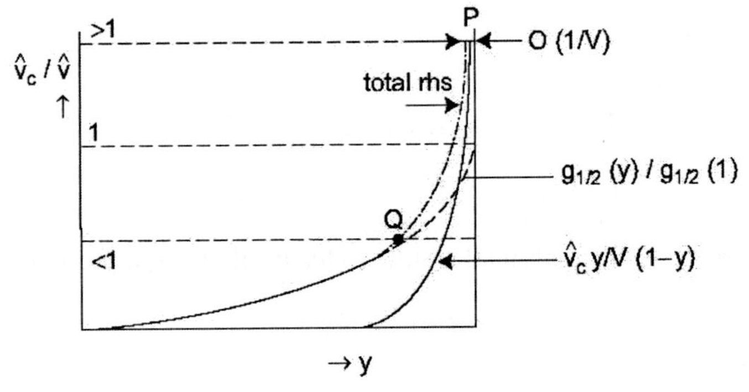

Fig. 8-3. Graphical solution of Eq. (8.3-1).

If, however, $\hat{v}_c / \hat{v}$ is smaller than unity, the intercept point Q is approximately equal to that of the curve $g_{\frac{1}{2}}(y)/g_{\frac{1}{2}}(1)$ alone. Thus the inversion of (8.2-7) leads to

$$y = \begin{cases} 1 + \mathcal{O}(1/V), & \hat{v} \le \hat{v}_c(T), \\ \text{root of } g_{\frac{1}{2}}(y)/g_{\frac{1}{2}}(1) = \hat{v}_c(T)/\hat{v}, & \hat{v} \ge \hat{v}_c(T). \end{cases} \tag{8.3-2}$$

We note that y, and therefore the chemical potential ς, is continuous at $\hat{v} = \hat{v}_c$. This should now be substituted into the result for P, Eq. (8.2-28). We have

---

[5] Cf. Kerson Huang, 2nd Ed., Op. Cit. p. 288, Fig. 12.4.

$$P = \frac{k_B T}{\hat{v}_c(T)} \frac{\zeta(\frac{5}{2})}{\zeta(\frac{3}{2})}, \qquad \hat{v} \le \hat{v}_c(T),$$
$$P = \frac{k_B T}{\hat{v}_c(T)} \frac{g_{3/2}(y)}{\zeta(\frac{3}{2})}, \; y = \text{root of...} \; \hat{v} \ge \hat{v}_c(T)\,. \tag{8.3-3}$$

The top line is a constant. The bottom expression can be found from a graph of $g_{3/2}(y)$. The resulting isotherms are given in Fig. 8-4. This is clearly a two-component phase diagram. P is constant in the domain for coexisting phases, similar

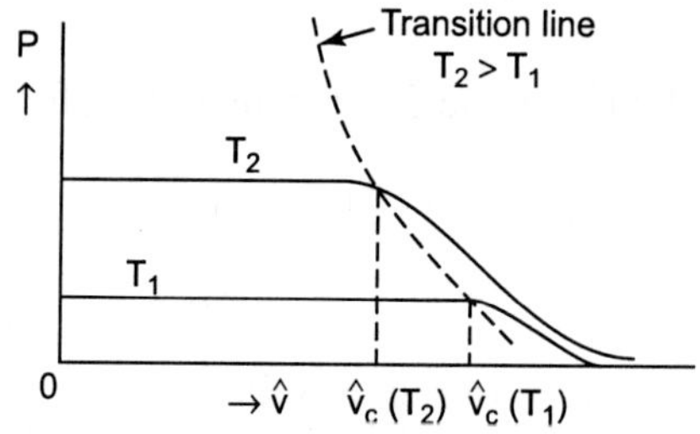

Fig. 8-4. Isotherms of perfect Bose gas. Note $T_2 > T_1$. The transition line $P_c \hat{v}_c^{5/3} = const.$ is shown.

as for a van der Waals gas. The transition line is easily found. From (8.3-3) and (8.2-27) we conclude:

$$P_c = (k_B T / \hat{v}_c)\zeta(\tfrac{5}{2}) / \zeta(\tfrac{3}{2}) = const\; T^{5/2},$$
$$\hat{v}_c = const' T^{-3/2}; \;\rightarrow\; P_c \hat{v}_c^{5/3} = C. \tag{8.3-4}$$

We now indicate some analytical results in terms of an asymptotic series in $\hat{v}$. Let

$$\alpha = (\hat{v}_c / \hat{v})\zeta(\tfrac{3}{2})\,. \tag{8.3-5}$$

Then for $\alpha < 1$,

$$g_{3/2}(y) = y + \frac{y^2}{2^{3/2}} + \frac{y^3}{3^{3/2}} + \ldots = \alpha, \tag{8.3-6}$$

which by iteration is converted to

$$y = \alpha\left\{1 - \frac{1}{2^{3/2}}\alpha + \left(\frac{1}{4} - \frac{1}{3^{3/2}}\right)\alpha^2 - \ldots\right\}. \tag{8.3-7}$$

We now use the Bose series (8.2-22) for $g_{3/2}(y)$ as well as (8.3-7); we then find for the second line of (8.3-3)

$$P = \frac{k_B T}{\hat{v}}\left[1 - \alpha 2^{-5/2} - \alpha^2(2\cdot 3^{-5/2} - 2^{-3}) - \ldots\right]$$

$$= \frac{k_B T}{\hat{v}}\left[1 - 0.1768\alpha - 0.0033\alpha^2 - \ldots\right]. \tag{8.3-8}$$

Note that this is a virial expansion. For very large $\hat{v}$ or very small α the perfect gas law $P = k_B T/\hat{v} = nk_B T$ is satisfied. Since y (or ς) is continuous at $\hat{v} = \hat{v}_c$ and the entropy and particle density become an admixture of two phases – see below – the transition is a first-order phase transition. [However, the specific heat is continuous, but its derivative jumps – see below – so that from that perspective we have a third-order phase transition.] The *real* Bose gas does not show such simple behaviour. Below we show the $P - \hat{v}$ diagram for that case.

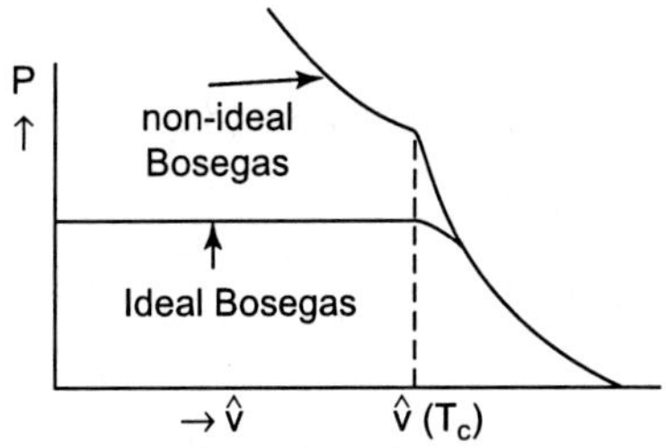

Fig. 8-5. Comparison between the real Bose gas and the perfect Bose gas.

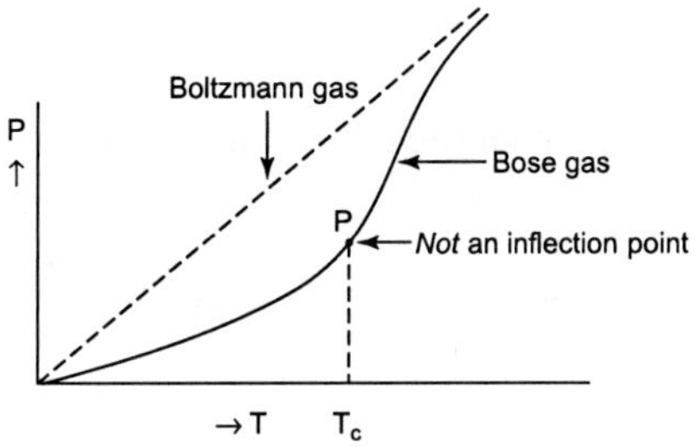

Fig. 8-6. The *P-T* diagram.

We now turn to the *P-T* diagram for fixed $\hat{v}$. From (8.2-27) we infer

$$T_c(\hat{v}) = h^2/2\pi m k_B \left(\hat{v}\zeta(\tfrac{3}{2})\right)^{2/3}. \tag{8.3-9}$$

Note that $\hat{v}_c$ and T_c are related by

$$\hat{v}_c/\hat{v} = (T_c/T)^{3/2}. \tag{8.3-10}$$

The 'translation' of Eqs. (8.3-2) and (8.3-3) is thus easy. We find

$$y = \begin{cases} 1 + \mathcal{O}(1/V), & T \le T_c\,, \\ \text{root of } g_{3/2}(y)/g_{3/2}(1) = (T_c/T)^{3/2}, & T \ge T_c\,. \end{cases} \tag{8.3-11}$$

Likewise,

$$\begin{aligned} P &= \frac{k_B T}{\hat{v}}\left(\frac{T}{T_c}\right)^{3/2} \frac{\zeta(\frac{5}{2})}{\zeta(\frac{3}{2})}, && T \le T_c\,, \\ P &= \frac{k_B T}{\hat{v}}\left(\frac{T}{T_c}\right)^{3/2} \frac{g_{5/2}(y)}{\zeta(\frac{3}{2})}, \quad y = \text{root of...} && T \ge T_c\,. \end{aligned} \tag{8.3-12}$$

The *P-T* diagram is given in Fig. 8-6. At $T = T_c$ the pressure P is continuous; yet the point P in Fig. 8-6 is *not* an inflection point, for the second derivative jumps.

8.3.2 *Coexisting Phases and Thermodynamic Functions*

In the condensation region, $y = 1$, Eq. (8.2-29) is rewritten as

$$\frac{\langle n_0 \rangle}{V} = \frac{1}{\hat{v}} - \frac{1}{\hat{v}_c} = \frac{N}{V}\left(1 - \frac{\hat{v}}{\hat{v}_c}\right), \tag{8.3-13}$$

which leads to two other interesting forms. From the first equality, with $n = 1/\hat{v}$ being the total particle *density* of the fluid, $n_n = 1/\hat{v}_c = (2\pi m k_B T/h^2)^{3/2}\zeta(\frac{3}{2})$ – see (8.2-27) – being the normal fluid density and $n_0 = \langle n_0 \rangle / V$ being the condensed state density, we have

$$n = n_n(T) + n_0(T), \quad T \leq T_c. \tag{8.3-14}$$

Secondly, from the latter equality in (8.3-13) we note, employing also (8.3-10),

$$n_0(T) \equiv \frac{\langle n_0 \rangle_T}{V} = n\left[1 - \left(\frac{T}{T_c}\right)^{3/2}\right]. \tag{8.3-15}$$

At $T = 0$ all particles are in the ground state. For $T \to T_c$ from below, the density in the condensed state slowly approaches zero, as sketched in Fig. 8-7.

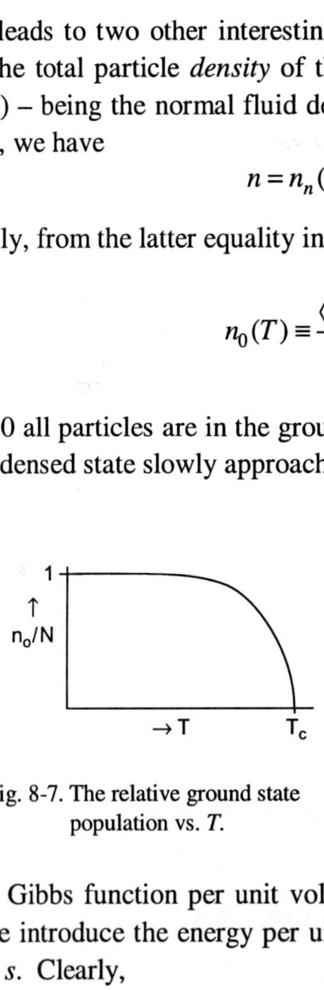

Fig. 8-7. The relative ground state population vs. *T*.

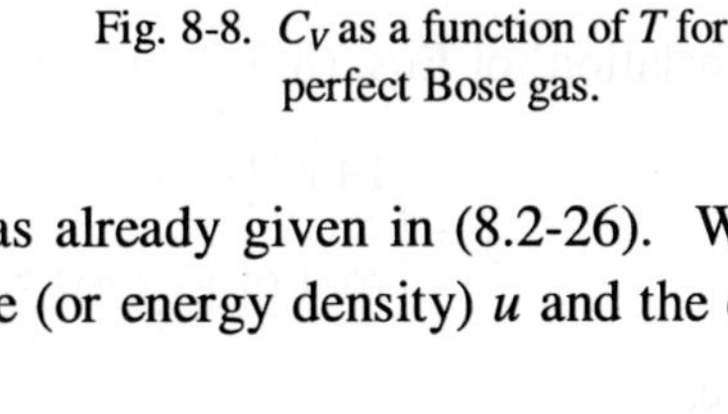

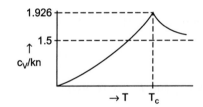

Fig. 8-8. C_V as a function of *T* for a perfect Bose gas.

The Gibbs function per unit volume κ was already given in (8.2-26). We shall likewise introduce the energy per unit volume (or energy density) *u* and the entropy density *s*. Clearly,

$$\kappa = u - Ts - \varsigma/\hat{v}. \tag{8.3-16}$$

The energy density *u* is related to the grand-canonical partition function, and thereby to κ, by [see Eq. (4.2-32) or (4.3-19)]

$$u = \left(\frac{\partial \kappa / k_B T}{\partial (1/k_B T)}\right)_{\varsigma/T} = -T^2\left(\frac{\partial \kappa / T}{\partial T}\right)_y. \tag{8.3-17}$$

We thus obtain

$$u = \frac{3}{2}\frac{k_B T}{\hat{v}}\left(\frac{T}{T_c}\right)^{3/2}\frac{\zeta(\frac{5}{2})}{\zeta(\frac{3}{2})}, \qquad T \le T_c,$$
$$u = \frac{3}{2}\frac{k_B T}{\hat{v}}\left(\frac{T}{T_c}\right)^{3/2}\frac{g_{5/2}(y)}{\zeta(\frac{3}{2})}, \qquad T \ge T_c. \tag{8.3-18}$$

From (8.2-26) , (8.3-16) and (8.3-18) we find for s:

$$s = \frac{5}{2}\frac{k_B}{\hat{v}}\left(\frac{T}{T_c}\right)^{3/2}\frac{\zeta(\frac{5}{2})}{\zeta(\frac{3}{2})}, \qquad T \le T_c,$$
$$s = \frac{5}{2}\frac{k_B}{\hat{v}}\left(\frac{T}{T_c}\right)^{3/2}\frac{g_{5/2}(y)}{\zeta(\frac{3}{2})} - \frac{k_B}{\hat{v}}\ln y, \qquad T \ge T_c. \tag{8.3-19}$$

Note that the entropy is continuous at T_c. From (8.3-18) and (8.3-12) we find

$$u = 3/2\, P, \tag{8.3-20}$$

which is the same relation as for a Boltzmann gas. For the entropy we remark that Nernst's law is satisfied – contrary to the Sackur-Tetrode formula for a Boltzmann gas – in that for $T \to 0$ the entropy becomes zero.

Next we consider the molecular or specific entropy, denoted in each phase by $\hat{s}_i$; we have $\hat{s}_i = S_i/N_i = (S_i/V)(V/N_i) = s_i v_i$. So from (8.3-19) in the condensation region,

$$\hat{s}_i = \frac{5}{2}\frac{\hat{v}_i}{\hat{v}}k_B\left(\frac{T}{T_c}\right)^{3/2}\frac{\zeta(\frac{5}{2})}{\zeta(\frac{3}{2})}. \tag{8.3-21}$$

Using further the Clausius-Clapeyron equation (3.6-15) and employing (8.3-12) for dP/dT we obtain the relationship

$$\frac{\hat{s}_n - \hat{s}_0}{\hat{v}_n - \hat{v}_0} = \frac{5}{2}\frac{k_B}{\hat{v}}\left(\frac{T}{T_c}\right)^{3/2}\frac{\zeta(\frac{5}{2})}{\zeta(\frac{3}{2})}. \tag{8.3-22}$$

This is consistent with (8.3-21), indicating once more that we have two coexisting phases in the condensation region. While, strictly speaking, these results say nothing about the relative magnitudes of $\hat{s}_n$ and $\hat{s}_0$, from the isotherms in Fig. 8-4 we see that $\hat{v}_0$ is zero, or very small; therefore, $\hat{s}_0$ is likewise zero or very small. This is confirmed by the behaviour of liquid helium II, the condensate of ^{4}He below the transition temperature of 2.19K. The theory given here would give $T_c = 3.2\,K$.

Finally, we consider the behaviour of the specific heat per unit volume, c_V. Below T_c we have from (8.3-18)

$$c_V = \frac{15}{4}\frac{k_B}{\hat{v}}\left(\frac{T}{T_c}\right)^{3/2}\frac{\zeta(\frac{5}{2})}{\zeta(\frac{3}{2})}, \qquad T \le T_c. \tag{8.3-23}$$

In the high temperature region we have

$$c_V = \frac{15}{4}\frac{k_B}{\hat{v}}\left(\frac{T}{T_c}\right)^{3/2}\frac{g_{5/2}(y)}{\zeta(\frac{3}{2})} + \frac{3}{2}\frac{k_B T}{\hat{v}}\left(\frac{T}{T_c}\right)^{3/2}\frac{g_{3/2}(y)}{\zeta(\frac{3}{2})y}\frac{dy}{dT}, \quad T \geq T_c. \tag{8.3-24}$$

The derivative dy/dT we find by implicit differentiation of the second line of (8.3-11); the result is

$$\frac{dy}{dT}\frac{1}{\zeta(\frac{3}{2})y} = -\frac{3}{2}\frac{1}{g_{-1/2}(y)T}\left(\frac{T_c}{T}\right)^{3/2}. \tag{8.3-25}$$

Substituting into (8.3-24) yields

$$c_V = \frac{15}{4}\frac{k_B}{\hat{v}}\left(\frac{T}{T_c}\right)^{3/2}\frac{g_{5/2}(y)}{\zeta(\frac{3}{2})} - \frac{9}{4}\frac{k_B}{\hat{v}}\frac{g_{3/2}(y)}{g_{-1/2}(y)}. \tag{8.3-26}$$

Since $g_{-1/2}(y)$ becomes ∞ for $y = 1$, the second term does not contribute for $T = T_c$, so that c_V is continuous; however, the derivative dc_V / dT jumps. The behaviour is pictured in Fig. 8-8. The peak value is at $1.926\, k_B n$. In actual liquid helium II the peak is at $12\, k_B n$ and the temperature law is $\propto T^3$, rather than $\propto T^{3/2}$; for details, see Section 12.4.

Lastly, we discuss the order parameter for this two component system. In subsection 7.3.2 we discussed the field representation for a system of bosons. The local particle density operator was shown to be $\rho(\mathbf{r}) = \Psi^\dagger(\mathbf{r})\Psi(\mathbf{r})$. We were dealing, however, with quantum systems having 'pure' states. Presently, the system is part of a statistical ensemble, governed by the density operator ρ_{gcan}, where we added the subscript 'grand-canonical' to avoid confusion with $\rho(\mathbf{r})$. Thus, the local density becomes

$$\rho(\mathbf{r}) = \mathrm{Tr}\{\rho_{gcan}\Psi^\dagger(\mathbf{r})\Psi(\mathbf{r})\} = \langle\Psi^\dagger(\mathbf{r})\Psi(\mathbf{r})\rangle. \tag{8.3-27}$$

Likewise, we have the local density 'matrix' [better kernel]:

$$\rho(\mathbf{r},\mathbf{r}') \equiv \rho(\mathbf{r}-\mathbf{r}') = \langle\Psi^\dagger(\mathbf{r})\Psi(\mathbf{r}')\rangle. \tag{8.3-28}$$

Employing (7.3-17) and noticing $\langle a_\mathbf{k}^\dagger a_{\mathbf{k}'}\rangle = \langle n_\mathbf{k}\rangle\delta_{\mathbf{k}\mathbf{k}'}$, the following result obtains:

$$\rho(\mathbf{r},\mathbf{r}') = \sum_\mathbf{k} \alpha_\mathbf{k}^*(\mathbf{r})\alpha_\mathbf{k}(\mathbf{r}')\langle n_\mathbf{k}\rangle. \tag{8.3-29}$$

For the perfect Bose gas the $\alpha_\mathbf{k}$'s are momentum eigenstates; hence

$$\rho(\mathbf{r},\mathbf{r}') = (1/V)\sum_\mathbf{k} e^{-i\mathbf{k}\cdot(\mathbf{r}-\mathbf{r}')}\langle n_\mathbf{k}\rangle. \tag{8.3-30}$$

Consider first the gas for $T > T_c$. Then the ground state is empty and all the $\langle n_\mathbf{k}\rangle$ are small numbers. In the thermodynamic limit,

$$\lim_{|\mathbf{r}-\mathbf{r}'|\to\infty} \rho(\mathbf{r},\mathbf{r}') = 0. \tag{8.3-31}$$

Therefore, the off-diagonal elements tend to zero. On the contrary, for $T < T_c$ the ground state with $k_x = k_y = k_z = 0$ has $\langle n_0 \rangle / V = n_0$ is finite; consequently,

$$\lim_{|\mathbf{r}-\mathbf{r}'|\to\infty} \rho(\mathbf{r},\mathbf{r}') = n_0 . \tag{8.3-32}$$

Therefore, the condensate shows "ODLRO": off-diagonal long-range order, or cooperative behaviour, like the spin-spin correlation $\langle \sigma_i^z \sigma_j^z \rangle$ in a ferromagnet (see Chapter XI); the order parameter itself is $\langle \Psi(\mathbf{r}) \rangle$ (see Chapter IX) .

For the "real" Bose gas with interactions, this conclusion remains true. While the $\alpha_{\mathbf{k}}(\mathbf{r})$ are no longer momentum ES, $\rho(\mathbf{r},\mathbf{r}')$ shows the behaviour as expressed by (8.3-32).

In this section we have discussed the 3D ideal Bose gas. It is left to the problems to show that for a 1D or 2D perfect Bose gas no Bose–Einstein condensation occurs.

8.4 The Perfect Fermi Gas

The perfect Fermi gas is much simpler than the Bose gas since no condensation occurs. Although the theory is applicable to all substances with an odd number of nucleons (half integral spin), the most fruitful application is to a perfect degenerate electron gas like in alkali metals. It will be mainly with this application in mind that the theory is presented here.

The grand potential $\Omega(T,V,\mu)$, where μ is the electrochemical potential or Fermi level, was obtained in Section 4.2. and subsection 4.3.1 from the grand-canonical partition function. We assume again that there are no other external thermodynamic variables than $\mathcal{E}, V$ and N.

The one-particle wave equation is the Hartree-Fock equation, discussed in Section 7.8. The solutions are Bloch functions

$$\alpha_{\mathbf{k}}(\mathbf{r}) = (1/\sqrt{V}) e^{i\mathbf{k}\cdot\mathbf{r}} u_{\mathbf{k}}(\mathbf{r}), \tag{8.4-1}$$

where the $u_{\mathbf{k}}(\mathbf{r})$ are periodic functions on the lattice. However, this section is not meant to be a treatise on energy bands in condensed matter. Anyway, the effect of the band structure is usually sufficiently represented by replacing the free electron mass by an effective mass m^*, which in general is anisotropic, though for cubic crystals an isotropic 'spherical band value' will do. In the alkali metals, moreover, Sommerfeld's free electron theory is quite acceptable, i.e., the eigenfunctions are taken to be plane waves and the free electron mass m can be used. This, then, will be the picture for the perfect Fermi gas presented here.

The density of states $Z(k)dk$ and $Z(\varepsilon)d\varepsilon$ is found as for the Bose gas. However, we take the lowest one-particle energy to be ε_C, being the bottom of the conduction band and we must multiply by a factor two due to two possibilities for states of spin ±½. Thus,

$$Z(\varepsilon)d\varepsilon = 4\pi V(2m/h^2)^{3/2}\sqrt{\varepsilon-\varepsilon_C}\,d\varepsilon. \tag{8.4-2}$$

Substituting this for the sum of (4.3-15), with ς replaced by μ, we now obtain the Riemann integral

$$\kappa(T,\mu) = \lim_{V\to\infty}\frac{\Omega(T,V,\mu)}{V} = -k_BT\int_{\varepsilon_C+0}^{\infty} 4\pi\left(\frac{2m}{h^2}\right)^{3/2}\sqrt{\varepsilon-\varepsilon_C}\ln\left[1+e^{(\mu-\varepsilon)/k_BT}\right]d\varepsilon. \tag{8.4-3}$$

We integrate by parts, to obtain κ as well as the pressure P,

$$P = -\kappa = \frac{8\pi}{3}\left(\frac{2m}{h^2}\right)^{3/2}\int_0^\infty \frac{\varepsilon'^{3/2}}{e^{(\varepsilon'-\Delta\mu)/k_BT}+1}d\varepsilon', \tag{8.4-4}$$

where $\Delta\mu = \mu - \varepsilon_C$. As a second equation of state we need the inverted result $n(T,\Delta\mu) = 1/\hat{v}$:

$$n = \lim_{V\to\infty}\frac{\partial(\Omega/V)}{\partial\mu} = -\frac{\partial\kappa}{\partial\mu} = 4\pi\left(\frac{2m}{h^2}\right)^{3/2}\int_0^\infty\frac{\sqrt{\varepsilon'}}{e^{(\varepsilon'-\Delta\mu)/k_BT}+1}d\varepsilon'. \tag{8.4-5}$$

We note that $n = \Sigma_i\langle n_i\rangle/V$, with $\langle n_i\rangle$ given by Fermi–Dirac statistics.

We define the Fermi integrals,

$$\mathcal{F}_k(\eta) = \frac{1}{\Gamma(k+1)}\int_0^\infty\frac{x^k}{e^{x-\eta}+1}dx, \tag{8.4-6}$$

with recurrent relation

$$\mathcal{F}_k'(\eta) = \mathcal{F}_{k-1}(\eta). \tag{8.4-7}$$

Eqs. (8.4-4) and (8.4-5) take the standard forms

$$P = k_BTN_c\mathcal{F}_{3/2}(\Delta\mu/k_BT), \tag{8.4-8}$$

$$n = N_c\mathcal{F}_{1/2}(\Delta\mu/k_BT), \tag{8.4-9}$$

with $N_c = 2(2\pi mk_BT/h^2)^{3/2}$ being the statistical weight of the conduction band. For very high temperatures $\eta = \Delta\mu/k_BT \to 0$ and $\mathcal{F}_k \to 1$. Thus, $P \to nk_BT$ and the perfect gas law applies. For very low temperatures, on the contrary,

$$\mathcal{F}_k(\eta) = \frac{1}{\Gamma(k+1)}\int_0^\infty x^k[1-\Theta(x-\eta)]dx = \frac{\eta^{k+1}}{\Gamma(k+2)}, \quad T\text{ small}, \tag{8.4-10}$$

where $\Theta(z)$ is the Heaviside function. In that approximation

$$P = \frac{16\pi}{15}\left(\frac{2m}{h^2}\right)^{3/2} (\Delta\mu)^{5/2}, \quad n = \frac{8\pi}{3}\left(\frac{2m}{h^2}\right)^{3/2} (\Delta\mu)^{3/2}. \tag{8.4-11}$$

The equations of state proper (in terms of extensive variables) then become

$$P = \frac{8\pi}{15}\frac{h^2}{m}\left(\frac{3n}{8\pi}\right)^{5/3}, \quad \Delta\mu = \frac{h^2}{2m}\left(\frac{3n}{8\pi}\right)^{2/3}. \tag{8.4-12}$$

We also have the relationship $P = (2/5)\,n\,\Delta\mu$. Note that the thermal energy $k_B T$ plays no role at low temperatures.

The Fermi wave vector k_F is defined by $\Delta\mu = \hbar^2 k_F^2/2m$. The Fermi velocity is related to this by $v_F = \hbar k_F/m$. Also, let $k_B T_F \equiv \Delta\mu$. For the alkali metals and some others the electrons form a near-perfect Fermi gas. The following values have been calculated (after C. Kittel[6]), see Table 8-1.

Table 8-1. Fermi energy and derived quantities for the alkali metals.

metal	$n\,(cm^{-3})$	$k_F\,(cm^{-1})$	$v_F\,(cm/s)$	$\Delta\mu\,(eV)$	$T_F\,(K)$
Li	4.6×10^{22}	1.10×10^{8}	1.30×10^{8}	4.7	5.5×10^{4}
Na	2.50	0.90	1.10	3.1	3.7
K	1.34	0.73	0.85	2.1	2.4
Rb	1.08	0.68	0.79	1.8	2.1
Cs	0.86	0.63	0.73	1.5	1.8
Cu	8.50	1.35	1.56	7.0	8.2
Ag	5.76	1.19	1.38	5.5	6.4
Au	5.90	1.20	1.39	5.5	6.4

For higher temperatures Eq. (8.4-10) is not sufficiently accurate. We shall develop an asymptotic series for $\mathcal{F}_k(\eta)$ in descending powers of η.

We first integrate (8.4-6) by parts.

$$\begin{aligned}\Gamma(k+1)\mathcal{F}_k(\eta) &= \int_0^\infty \frac{x^{k+1}}{(k+1)(e^{x-\eta}+1)(e^{\eta-x}+1)}dx \\ &= \int_{-\eta}^\infty \frac{(\eta+z)^{k+1}}{(k+1)(e^z+1)(e^{-z}+1)}dz.\end{aligned} \tag{8.4-13}$$

[6] C. Kittel, "Introduction to Solid State Physics", John Wiley, 3rd Ed., NY 1968, p.208. Also later Eds.

Using the binomial expansion for $(\eta + z)^{k+1}$ we find

$$\Gamma(k+1)\mathcal{F}_k(\eta) = \sum_{j=0}^{\infty} \frac{1}{k+1}\binom{k+1}{j}\eta^{k+1-j} \int_{-\infty}^{\infty} \frac{z^j}{(e^z+1)(e^{-z}+1)} dz + \mathcal{O}(e^{-\eta}) \tag{8.4-14}$$

For $j = 0$ we have

$$\int_{-\infty}^{\infty} \frac{dz}{(e^z+1)(e^{-z}+1)} = 1\,. \tag{8.4-15}$$

For $j > 0$ we use once more integration by parts (in the opposite sense); we have

$$\int_{-\infty}^{\infty} \frac{z^j}{(e^z+1)(e^{-z}+1)} dz = [1 + (-1)^j\,]\,j \int_0^{\infty} \frac{z^{j-1}}{e^z+1} dz\,. \tag{8.4-16}$$

We note that all terms with j odd cancel. For the remaining even integrals we use the identity

$$\frac{1}{e^z+1} = \frac{1}{e^z-1} - \frac{2}{e^{2z}-1}, \tag{8.4-17}$$

so that the integrals are reduced to Bose integrals. The result is

$$\int_{-\infty}^{\infty} \frac{z^j}{(e^z+1)(e^{-z}+1)} dz = 2(2n\,!)(1-2^{1-2n})\zeta(2n), \quad j = 2n,\ n = 1, 2, \ldots \tag{8.4-18}$$

So, finally, we obtain

$$\mathcal{F}_k(\eta) = \frac{\eta^{k+1}}{\Gamma(k+2)} + \sum_{n=1}^{\infty} \frac{2}{\Gamma(k+2)}\binom{k+1}{2n}(2n)!\,(1-2^{1-2n})\,\zeta(2n)\,\eta^{k+1-2n}\,. \tag{8.4-19}$$

We note that the leading term in this asymptotic expansion is our previous result (8.4-10).

As an example we compute the Fermi level at temperatures $T \ll T_F$. Taking two terms we find for (8.4-9)

$$\begin{aligned} \frac{n}{N_c} &= \frac{1}{\Gamma(\frac{5}{2})}\left[\eta^{3/2} + 2\binom{3/2}{2} 2!\,(1-2^{-1})\zeta(2)\eta^{-1/2}\right] \\ &= \frac{2}{\sqrt{\pi}}\left(\frac{2}{3}\eta^{3/2} + \frac{\pi^2}{12}\frac{1}{\eta^{1/2}}\right), \qquad \eta = \Delta\mu / k_B T. \end{aligned} \tag{8.4-20}$$

Raising both sides to the power 2/3 and substituting the second equality of (8.4-12) we find

$$\Delta\mu(0) = \left\{[\Delta\mu(T)]^{3/2} + \frac{3\pi^2}{24}(k_B T)^2[\Delta\mu(T)]^{-1/2}\right\}^{2/3} \simeq \Delta\mu(T)\left\{1 + \frac{\pi^2}{12}\left[\frac{k_B T}{\Delta\mu(0)}\right]^2\right\}, \tag{8.4-21}$$

or also,

$$\Delta\mu(T) \simeq \Delta\mu(0)\left\{1-\frac{\pi^2}{12}\left[\frac{k_BT}{\Delta\mu(0)}\right]^2\right\}. \tag{8.4-22}$$

Likewise, taking two terms of the asymptotic series one easily confirms

$$P(T) \simeq P(0)\left\{1+\frac{5\pi^2}{12}\left[\frac{k_BT}{\Delta\mu(0)}\right]^2\right\}. \tag{8.4-23}$$

Note that $\Delta\mu$ decreases (comes closer to the band edge), while P increases.

The above all pertains to $\Delta\mu > 0$, as is the case for metals and n-type degenerate semiconductors. For nondegenerate semiconductors the Fermi level is below the conduction band edge. Then with $\eta << -1$, we have

$$\mathcal{F}_k(\eta \ll -1) \sim e^{\eta}, \tag{8.4-24}$$

so that Boltzmann statistics applies. We refer to books on semiconductors.

Returning to the degenerate electron gas in metals, we complete the list of thermodynamic functions, as we did for the Bose gas. We noted already

$$\kappa = u - Ts - \mu n = -k_BTN_c(T)\mathcal{F}_{3/2}(\Delta\mu / k_BT). \tag{8.4-25}$$

The energy density is

$$u = -k_BT^2\left[\left(\partial\kappa(T,\Delta\mu/k_BT)/k_BT\right)/\partial T\right]_{\Delta\mu/T} = \tfrac{3}{2}k_BTN_c(T)\mathcal{F}_{3/2}(\Delta\mu/k_BT). \tag{8.4-26}$$

Comparing with (8.4-8) we find

$$u = 3/2\, P, \tag{8.4-27}$$

which is the same as for a Bose gas, cf., (8.3-20). From (8.4-23) we get

$$u(T) = u(0)\left\{1+\frac{5\pi^2}{12}\left[\frac{k_BT}{\Delta\mu(0)}\right]^2\right\}. \tag{8.4-28}$$

This yields for the specific heat per unit volume

$$c_V = du/dT = \tfrac{1}{2}\pi^2 k_B n\left[kT_B/\Delta\mu(0)\right]. \tag{8.4-29}$$

The strong reduction compared to the 'usual' result $\frac{3}{2}k_Bn$ is due to the fact that only a strip of electrons of width ~ $2k_BT$ about the Fermi level contributes, all other states being fully occupied or empty, see Fig. 8-9a and Fig. 8-9b. Finally, given κ, u and n, we find for the entropy density s,

$$s = \tfrac{5}{2}k_BN_c(T)\mathcal{F}_{3/2}(\Delta\mu/k_BT) - (\Delta\mu/T)N_c(T)\mathcal{F}_{1/2}(\Delta\mu/k_BT). \tag{8.4-30}$$

Using two terms in the series for the Fermi integrals, one easily obtains

$$s = \frac{4\pi^3}{3}\left(\frac{2m^*}{h^2}\right)^{3/2} k_B^2 T [\Delta\mu(0)]^{1/2}. \tag{8.4-31}$$

This is linear in T and reaches zero at $T = 0$, in accord with our interpretation of Nernst's law.

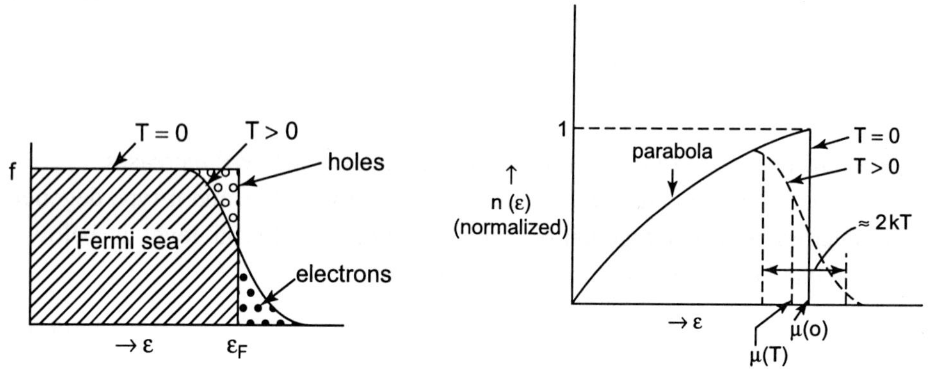

Fig. 8-9a. The Fermi function at $T = 0$ and $T > 0$. Fig. 8-9b. Electron occupancy in a parabolic band.

8.5 Lattice Vibrations; Phonons [7]

8.5.1 *Continuum Description. Einstein and Debye Specific Heat*

The specific heat of a solid stems from the lattice vibrations and from the electron gas. The latter contribution was shown to be very small in the previous section. So in this section we shall investigate the contribution from the lattice. The classical law of Dulong and Petit ascribed an amount of $k_B T$ to the energy of the vibrations. So, for N 3D oscillators, the specific heat should be $3R$ or about 6 cal per mole, a rule which is fairly well satisfied at room temperature. At lower temperatures we should use the mean energy of a quantized oscillator, as stated in (6.6-3). The zero-point energy can be omitted, since it is temperature independent. So,

$$\mathcal{E} = 3N\hbar\omega_E /(e^{\hbar\omega_E / k_B T} - 1), \tag{8.5-1}$$

where ω_E is the Einstein angular frequency. Let us define θ_E by $k_B \theta_E = \hbar\omega_E$. The specific heat is then

$$C_V = \frac{d\mathcal{E}}{dT} = 3R\left(\frac{\theta_E}{T}\right)^2 \frac{e^{\theta_E / T}}{(e^{\theta_E / T} - 1)^2}. \tag{8.5-2}$$

[7] Sections 8.5 and 8.6 have been included since the results are often needed. However, it is assumed that most students have had a course in Solid State on the level of Ashcroft and Merwin (Thomson 1976).

This yields the Dulong-Petit result for $T \gg \theta_E$ and goes to zero for temperatures below θ_E.

However, it does not correspond to the observed behaviour, which shows that at low temperatures $C_V \propto T^3$. The picture is grossly oversimplified and, particularly, a single vibrating frequency ω_E has no physical justification.

Much better is the Debye theory of specific heat, although it is still simplistic in that dispersion is neglected and that the discrete array of atoms is replaced by a continuum. To briefly establish its tenets we must find the vibrational modes for a 3D continuous elastic medium. This can be done semi-classically or completely quantum mechanically, using field quantization, see e.g., Kittel.[8] However, the field point of view adds nothing new over the semi-classical approach, which we will follow here.

The equation for the vibrations of an elastic medium is quite generally[9]:

$$\rho\frac{\partial^2\mathbf{u}(\mathbf{r},t)}{\partial t^2} = c_{11}\text{grad div }\mathbf{u} - c_{44}\text{curl curl }\mathbf{u}\,; \tag{8.5-3}$$

here $\mathbf{u}$ is the displacement at t of a point at $\mathbf{r}$ in the medium, ρ is the density and c_{11} and c_{44} are tensile and shear coefficients of elasticity. A general displacement can be split into two parts, $\mathbf{u} = \mathbf{u}^{(1)} + \mathbf{u}^{(2)}$ with

$$\begin{aligned} &\text{div}\,\mathbf{u}^{(1)} \neq 0, \quad \text{curl}\,\mathbf{u}^{(1)} = 0, \\ &\text{div}\,\mathbf{u}^{(2)} = 0, \quad \text{curl}\,\mathbf{u}^{(2)} \neq 0. \end{aligned} \tag{8.5-4}$$

This simply indicates that any vector field can be constructed from sources and vortices. For the separate parts we have

$$\begin{aligned} \rho\frac{\partial^2\mathbf{u}^{(1)}(\mathbf{r},t)}{\partial t^2} &= c_{11}\text{grad div }\mathbf{u}^{(1)}\,, \\ \rho\frac{\partial^2\mathbf{u}^{(2)}(\mathbf{r},t)}{\partial t^2} &= -c_{44}\text{curl curl }\mathbf{u}^{(2)}\,. \end{aligned} \tag{8.5-5}$$

The first wave is longitudinal since it causes volume compression and rarefaction $(dV/V = \text{div}\,\mathbf{u}^{(1)} \neq 0)$; the second wave causes rotation $(\mathbf{\Omega} = \frac{1}{2}\text{curl}\,\mathbf{u}^{(2)} \neq 0)$; it can be decomposed into two transverse waves. Both equations can be put into the form

$$\rho\frac{\partial^2\mathbf{u}(\mathbf{r},t)}{\partial t^2} = c\nabla^2\mathbf{u}(\mathbf{r},t)\,, \qquad c = c_{11} \text{ or } c_{44}\,. \tag{8.5-6}$$

(we used the identity $\nabla^2 = \text{grad div} - \text{curl curl}$). The solutions are built up of plane

8 C. Kittel, "The Quantum Theory of Solids", McGraw Hill, NY 1963; 2nd printing , Wiley N.Y. (1986).

9 See e.g., Lord Rayleigh, "The Theory of Sound", Vol. 1 and 2, 1894; Dover reprints 1945.

waves, $\mathbf{u}(\mathbf{r},t) = \exp(i\mathbf{q}\cdot\mathbf{r} - i\omega t)$. The quantization of $\mathbf{q}$ in a box of volume V with periodic boundary conditions is as in Section 8.1. Thus, analogous to (8.1-6)

$$Z(\nu)d\nu = 4\pi V\left(\frac{1}{v_\ell^3} + \frac{2}{v_t^3}\right)\nu^2 d\nu, \tag{8.5-7}$$

where $v_\ell = \sqrt{c_{11}/\rho}$ is the longitudinal velocity of sound and $v_t = \sqrt{c_{44}/\rho}$ is the transverse velocity of sound. A failure of this treatment is that no upper limit to ν occurs. This is remedied by imposing the Debye cut-off frequency, such that the number of integrated modes is $3N$. Hence,

$$\nu_D^3 = \frac{9N}{4\pi V}\left(\frac{1}{v_\ell^3} + \frac{2}{v_t^3}\right)^{-1}. \tag{8.5-8}$$

Substituting back into (8.5-7) this yields

$$Z(\nu)d\nu = (9N/\nu_D^3)\nu^2 d\nu, \quad \nu \le \nu_D. \tag{8.5-9}$$

The total vibrational energy is now

$$\mathcal{E} = \int_0^{\nu_D} \frac{9N}{\nu_D^3}\nu^2 \frac{h\nu}{e^{h\nu/k_BT} - 1} d\nu. \tag{8.5-10}$$

We introduce the Debye temperature θ_D by $k_B\theta_D = h\nu_D$. We then obtain

$$\mathcal{E} = 9Nk_BT(T/\theta_D)^3 \int_0^{\theta_D/T} x^3 dx/(e^x - 1). \tag{8.5-11}$$

For low temperatures the upper limit $\to \infty$ and the integral is $\Gamma(4)\zeta(4) = \pi^4/15$, see (8.1-21). Hence for the low temperature molar specific heat we find

$$C_V = (12\pi^4/5)(T/\theta_D)^3 R, \tag{8.5-12}$$

the renown Debye T^3 law.

8.5.2 *Normal Modes; Running-Wave Boson Operators*

We shall consider the atomic vibrations of a discrete lattice in some detail, in order to obtain the vibrational modes, of which the excitations are the acoustical or optical phonons. So, as for photons, we have two equivalent pictures. We start with the Hamiltonian associated with the mechanical vibrations in the crystal; we end up with the field Hamiltonian for the phonon gas of the various polarization branches, expressed in appropriate boson operators.

Let $\mathbf{u}_{\mathbf{l},\mathbf{b}}$ be the deviation from the equilibrium position of an atom, whereby $\mathbf{l}$ labels the unit cell to which it belongs and $\mathbf{b}$ indicates its position within the unit cell. Cartesian components, where necessary, will be labelled with a Greek subscript α, whereas the lattice (affine) components of $\mathbf{l}$ will be labelled with the Latin subscript i. Thus,

$$\mathbf{l} = \Sigma_i l_i \mathbf{a}_i = l_1\mathbf{a}_1 + l_2\mathbf{a}_2 + l_3\mathbf{a}_3 \,, \tag{8.5-13}$$

where the $\mathbf{a}$'s are the vectors of the unit cell. For the $\mathbf{b}$ we will find it easier to use a scalar label; if there are κ atoms in the unit cell, we write $\mathbf{b} = \{b^1, b^2, ... b^{3\kappa}\}$. We further have $l_i = 0,1,...L_i - 1$, with $L_1L_2L_3 = N$, N being the number of unit cells. Using periodic b.c., it follows that (l_1, l_2, l_3) and $(l_1 + L_1, l_2, l_3)$ refer to the same lattice vector; similar for the other affine directions.

The Hamiltonian is $\mathcal{H}(\mathbf{u}_{\mathbf{l},\mathbf{b}}, \mathbf{p}_{\mathbf{l},\mathbf{b}})$. We shall not write it out, but it contains the kinetic energy $\Sigma\, p_{\mathbf{l},\mathbf{b}}^2 / 2m_b$ and the coupling energy in the harmonic approximation (cubic and higher terms omitted), i.e., a quadratic form in $\mathbf{u}_{\mathbf{l},\mathbf{b}}\mathbf{u}_{\mathbf{l}\pm\mathbf{a}_i,\mathbf{b}'}$. The main tenet is that $\mathcal{H}$ is invariant against a translation, i.e., $T\mathcal{H} = \mathcal{H}T$. Thus, for the ES $|\eta\rangle$ of $\mathcal{H}$ we have

$$\mathcal{H}T(\mathbf{l}')|\eta\rangle = T(\mathbf{l}')\mathcal{H}|\eta\rangle = \mathcal{E}_\eta T(\mathbf{l}')|\eta\rangle \,. \tag{8.5-14}$$

It follows that $T(\mathbf{l}')|\eta\rangle$ is an ES with eigenvalue $\mathcal{E}_\eta$. Writing the ES also as $\eta(\{\mathbf{u}_{\mathbf{l},\mathbf{b}}\})$ we have,[10]

$$T(\mathbf{l}')\eta(\{\mathbf{u}_{\mathbf{l},\mathbf{b}}\}) = e^{i\varphi}\eta(\{\mathbf{u}_{\mathbf{l},\mathbf{b}}\}) \,. \tag{8.5-15}$$

We must fix φ in a suitable manner. For a translation over a lattice vector $\mathbf{a}_i$, let $\varphi = 2\pi q_i$. So with $\mathbf{l}' = l_1'\mathbf{a}_1 + l_2'\mathbf{a}_2 + l_3'\mathbf{a}_3$ we will have

$$T(\mathbf{l}')\eta(\{\mathbf{u}_{\mathbf{l},\mathbf{b}}\}) = e^{2\pi i(q_1 l_1' + q_2 l_2' + q_3 l_3')}\eta(\{\mathbf{u}_{\mathbf{l},\mathbf{b}}\}) \,. \tag{8.5-16}$$

The exponent can be made a scalar product if we introduce the vectors $\mathbf{q}$ of the reciprocal lattice, such that

$$\mathbf{q} = \Sigma_i q_i \mathbf{c}_i = q_1\mathbf{c}_1 + q_2\mathbf{c}_2 + q_3\mathbf{c}_3 \,, \tag{8.5-17}$$

where the reciprocal lattice vectors are defined by the usual relations $\mathbf{a}_i \cdot \mathbf{c}_j = 2\pi\delta_{ij}$. Then it follows from (8.5-13) and (8.5-17) that

$$\mathbf{q}\cdot\mathbf{l}' = 2\pi(q_1 l_1' + q_2 l_2' + q_3 l_3') \,. \tag{8.5-18}$$

Eq. (8.5-16) can now be written as

$$T(\mathbf{l}')|\eta\rangle = e^{i\mathbf{q}\cdot\mathbf{l}'}|\eta\rangle \,, \tag{8.5-19}$$

---

[10] In case of degeneracy we must make a unitary transformation for the states $|\eta, \lambda\rangle$ in order to make the argument rigorous.

where we reverted to the Dirac notation. This relationship is referred to as *Floquet's theorem*. Now if **g** is any integral vector of the reciprocal lattice, one easily sees that $\exp(i\mathbf{g}\cdot\mathbf{l}') = 1$. Therefore **q** should be limited to a reciprocal lattice unit cell. We prefer to have the components q_i both positive and negative; the unit cell is therefore taken to be the Wigner-Seitz (WS) unit cell, obtained by considering the space enclosed by all planes that bisect the distances from a given lattice point to all neighbouring lattice points.[11] *Explicitly we have:*

$$\mathbf{q} = \frac{n_1}{L_1}\mathbf{c}_1 + \frac{n_2}{L_2}\mathbf{c}_2 + \frac{n_3}{L_3}\mathbf{c}_3 , \quad \mathbf{q} \in \text{WS unit cell}, \quad n_i = 0, \pm 1, \pm 2, \ldots \pm n_{i,max} . \tag{8.5-20}$$

We note that a translation over $L_1\mathbf{a}_1$ now leads to $\exp i[L_1\mathbf{a}_1 \cdot (n_1/L_1)\mathbf{c}_1] = \exp 2\pi i n_1 = 1$, so that the periodic b.c. are satisfied. The total number of **q**-vectors is $L_1L_2L_3 = N$. The density in **q**-space follows from the spacing of vectors in (8.5-20). Clearly,

$$Z(\mathbf{q}) = L_1L_2L_3 / \mathbf{c}_1 \times \mathbf{c}_2 \cdot \mathbf{c}_3 = N\mathbf{a}_1 \times \mathbf{a}_2 \cdot \mathbf{a}_3 / 8\pi^3 = V/8\pi^3 , \tag{8.5-21}$$

where V is the macroscopic volume of the sample. Note that the density is the same as for simple cubic box quantization! The WS cell of the reciprocal lattice is also called the 'first Brillouin zone'.

We mentioned before (subsection 7.3.2) that the translation operator has the quantum form

$$T(\mathbf{l}') = e^{i\mathbf{P}_{tot}\cdot\mathbf{l}'/\hbar} , \tag{8.5-22}$$

where $\mathbf{P}_{tot}$ is the total momentum. Comparing with (8.5-19) we note that the EV of $\mathbf{P}_{tot}$ are given by

$$\mathbf{P}_{tot} = \hbar(\mathbf{q} + \mathbf{g}) . \tag{8.5-23}$$

The vector **q**, often called the 'crystal momentum', determines $\mathbf{P}_{tot}$ up to a reciprocal lattice vector.

Normal coordinates

New coordinates are introduced by the coefficients of Fourier series on the lattice:

$$\mathbf{u}_{\mathbf{l,b}} = (1/\sqrt{N})\sum_{\mathbf{q}} \mathbf{Q}_{\mathbf{q,b}} e^{-i\mathbf{q}\cdot\mathbf{l}}, \quad \mathbf{p}_{\mathbf{l,b}} = (1/\sqrt{N})\sum_{\mathbf{q}} \mathbf{P}_{\mathbf{q,b}} e^{i\mathbf{q}\cdot\mathbf{l}}, \tag{8.5-24}$$

with inversion

$$\mathbf{Q}_{\mathbf{q,b}} = (1/\sqrt{N})\sum_{\mathbf{l}} \mathbf{u}_{\mathbf{l,b}} e^{i\mathbf{q}\cdot\mathbf{l}}, \quad \mathbf{P}_{\mathbf{q,b}} = (1/\sqrt{N})\sum_{\mathbf{l}} \mathbf{p}_{\mathbf{l,b}} e^{-i\mathbf{q}\cdot\mathbf{l}}. \tag{8.5-25}$$

---

[11] The WS unit cell is the closest resemblance to a sphere. For a bcc reciprocal lattice it is the well-known truncated octahedron. Also note that WS unit cells fill the entire reciprocal lattice without gaps!

The inversion is proven with the lemmas

$$(A):\qquad \sum_{\mathbf{l}\in V} e^{\pm i\mathbf{q}\cdot\mathbf{l}} = \begin{cases} N \text{ if } \mathbf{q}=0, \\ 0 \text{ if } \mathbf{q}\neq 0. \end{cases} \tag{8.5-26a}$$

$$(B):\quad \sum_{\mathbf{q}\in WS\,cell} e^{\pm i\mathbf{q}\cdot\mathbf{l}} = \begin{cases} N \text{ if } \mathbf{l}=0, \\ 0 \text{ if } \mathbf{l}\neq 0. \end{cases} \tag{8.5-26b}$$

Using these normal coordinates the Hamiltonian becomes largely uncoupled; one finds[12]

$$\mathcal{H} = \sum_{\mathbf{q}} \left\{ \sum_{\mathbf{b}} \mathbf{P}_{\mathbf{q,b}} \cdot \mathbf{P}^\dagger_{\mathbf{q,b}} / 2m_{\mathbf{b}} + \tfrac{1}{2} \sum_{\mathbf{bb'}} \mathsf{E}^{\mathbf{bb'}}(\mathbf{q}) : \mathbf{Q}_{\mathbf{q,b}} \mathbf{Q}^\dagger_{\mathbf{q,b'}} \sqrt{m_{\mathbf{b}} m_{\mathbf{b'}}} \right\}. \tag{8.5-27}$$

The : denotes a tensor contraction with the vectors $\mathbf{QQ}^\dagger$. The latter sum has $(3\kappa)^2$ terms.

While there is no longer coupling of adjacent unit cells, the indices **b** representing different atoms within the cell, are still admixed. We must thus further diagonalize the Hermitean tensor matrix E; the result leads to the secular determinant

$$|\mathsf{E}(\mathbf{q}) - \omega^2_{\mathbf{q},p}\mathsf{I}| = 0, \quad p = 1,2,\ldots 3\kappa. \tag{8.5-28}$$

The index 'p' stands for polarization branch. After a unitary transformation in the 3κ-dimensional vector space and a rotation such that E for every $\mathbf{q}$ is on its principal axes, whereby $\{\mathbf{Q}^{(\dagger)}_{\mathbf{q,b}}, \mathbf{P}^{(\dagger)}_{\mathbf{q,b}}\} \to \{Q^{(\dagger)}_{\mathbf{q},p}\mathbf{e}_{\mathbf{q}p}, P^{(\dagger)}_{\mathbf{q},p}\mathbf{e}_{\mathbf{q}p}\}$ with the $\mathbf{e}$'s being unit vectors, the Hamiltonian becomes a sum of independent modes, which are either longitudinal or transversal:

$$\mathcal{H} = \frac{1}{2} \sum_{\mathbf{q}}^{N} \sum_{p=1}^{3\kappa} \left\{ P_{\mathbf{q},p} P^\dagger_{\mathbf{q},p} / m_p + \omega^2_{\mathbf{q},p} m_p Q_{\mathbf{q},p} Q^\dagger_{\mathbf{q},p} \right\}. \tag{8.5-29}$$

We note that $Q^\dagger_{\mathbf{q}p} = Q_{-\mathbf{q}p}$, $P^\dagger_{\mathbf{q}p} = P_{-\mathbf{q}p}$, $\omega_{-\mathbf{q}p} = \omega_{\mathbf{q}p}$. Boson operators are now introduced in a similar way as ladder operators for a single harmonic oscillator; only we must also try to decouple the $\pm\mathbf{q}$ components; hence we set,

$$\begin{aligned} a_{\mathbf{q}p} &= (2\hbar m_p \omega_{\mathbf{q}p})^{-1/2} \left(m_p \omega_{\mathbf{q}p} Q_{\mathbf{q}p} + i P_{-\mathbf{q}p} \right), \\ a^\dagger_{\mathbf{q}p} &= (2\hbar m_p \omega_{\mathbf{q}p})^{-1/2} \left(m_p \omega_{\mathbf{q}p} Q_{-\mathbf{q}p} - i P_{\mathbf{q}p} \right). \end{aligned} \tag{8.5-30}$$

Clearly $a_{\mathbf{q}p} \propto \exp(i\mathbf{q}\cdot\mathbf{l})$ and $a^\dagger_{\mathbf{q}p} \propto \exp(-i\mathbf{q}\cdot\mathbf{l})$. Moreover, we easily verify

$$[a_{\mathbf{q}p}, a^\dagger_{\mathbf{q}'p'}] = [\sqrt{\frac{m_p\omega_{\mathbf{q}p}}{2\hbar}} Q_{\mathbf{q}p} + i\sqrt{\frac{1}{2\hbar m_p \omega_{\mathbf{q}p}}} P^\dagger_{\mathbf{q}p}, \sqrt{\frac{m_{p'}\omega_{\mathbf{q}'p'}}{2\hbar}} Q^\dagger_{\mathbf{q}'p'} - i\sqrt{\frac{1}{2\hbar m_{p'}\omega_{\mathbf{q}'p'}}} P_{\mathbf{q}'p'}]$$

[12] J. Ziman, "Electrons and Phonons", Oxford University Press, 1960.

$$=\frac{1}{2\hbar}\frac{\hbar}{-i}(-i)\delta_{\mathbf{qq}'}\delta_{pp'}+\frac{1}{2\hbar}\frac{\hbar}{i}i\delta_{\mathbf{qq}'}\delta_{pp'}=\delta_{\mathbf{qq}'}\delta_{pp'}\,. \tag{8.5-31}$$

The $a_{\mathbf{q}p}$ and $a^\dagger_{\mathbf{q}p}$ are running-wave boson operators. With the lemmas given above, one can invert the relations (8.5-30) and express P and Q in $a, a^\dagger$. The Hamiltonian is then found to take the basic form for the phonon field,

$$\mathcal{H}=\sum_{\mathbf{q}}^{N}\sum_{p}^{3\kappa}\left(a^\dagger_{\mathbf{q}p}a_{\mathbf{q}p}+\tfrac{1}{2}\right)\hbar\omega_{\mathbf{q}p}\,. \tag{8.5-32}$$

The energy is accordingly

$$\mathcal{E}_{phonon\text{-}field}=\sum_{\mathbf{q}p}(n_{\mathbf{q}p}+\tfrac{1}{2})\hbar\omega_{\mathbf{q}p}\,. \tag{8.5-33}$$

We now come to the meaning of the polarization branches. If $\kappa = 1$ (one atom per unit cell), there are three *acoustical* branches. These branches pass through the origin of **q**-space and for small $|\mathbf{q}|$, $\omega_{\mathbf{q}p} \propto |\mathbf{q}|$. Since the coupling tensor was rotated to be on its main axes, one of the branches has a polarization parallel to **q**, while the other branches have a polarization perpendicular to **q**. If $\kappa = 2$, as for silicon and germanium, we have also *optical modes,* in which the two atoms move in opposition to each other. We now have six branches: one LA (longitudinal acoustical), two TA (transverse acoustical), one LO (longitudinal optical) and two TO (transverse optical). These branches are pictured in Fig. 8-10.

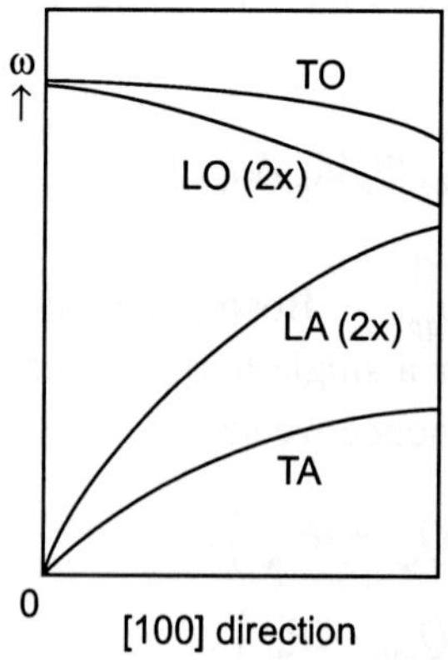

Fig. 8-10. Phonon branches in germanium [after E. Conwell, Proc. IRE (1957)]. Note that the LA and LO branches are two-fold degenerate in the [100] direction.

Density of modes and specific heat

The density of modes in **q**-space is simply $V/8\pi^3$, as we saw in (8.5-21). The frequency density is, however, more complicated. With reference to Fig. 8-11 we have for branch p

$$Z_p(\omega_p)d\omega_p = \oiint Z(\mathbf{q})dS_{\omega_p}d\omega_p / |\nabla_{\mathbf{q}}\omega_p(\mathbf{q})|$$
$$= (V/8\pi^3)\oiint dS_{\omega_p}d\omega_p / |\nabla_{\mathbf{q}}\omega_p(\mathbf{q})|, \qquad (8.5\text{-}34)$$

where the surface integral is over a constant ω-surface in $\mathbf{q}$-space. The normalized total frequency-density spectrum is therefore

$$D(\omega) = (V/8\pi^3 N)\sum_p \oiint dS_{\omega_p} / |\nabla_{\mathbf{q}}\omega_p(\mathbf{q}). \qquad (8.5\text{-}35)$$

The phonon spectrum for NaCl is given in Fig. 8-12.

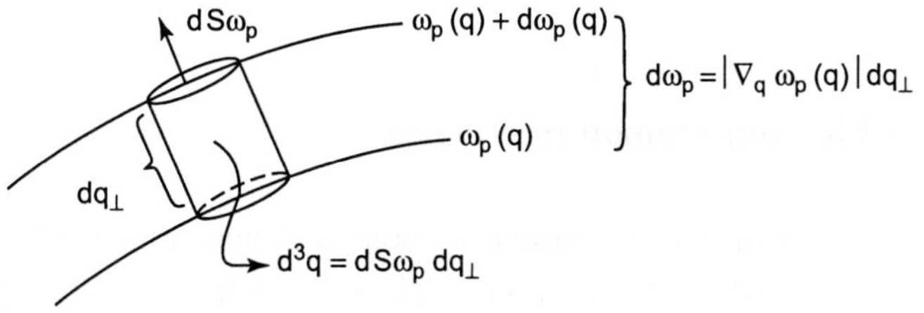

Fig. 8-11. Number of modes between two constant ω-surfaces.

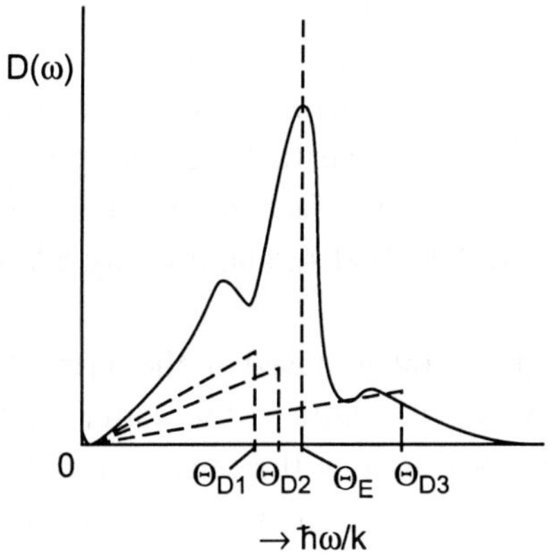

Fig. 8-12. Phonon spectrum for NaCl. The result can reasonably be presented by three Debye terms for the acoustical modes (dashed lines) and a spike-Einstein term for the optical modes. [J. Ziman, Op. Cit., Fig. 20.]

From (8.5-33) we have for the mean energy of the phonon field

$$\langle \mathscr{E} \rangle = \sum_{\mathbf{q}p} \frac{\hbar\omega_p(\mathbf{q})}{e^{\hbar\omega_p(\mathbf{q})/k_BT} - 1}, \qquad (8.5\text{-}36)$$

where we omitted again the zero-point energy. By differentiation to T we find the specific heat. For the acoustical branches we often approximate $|\nabla_{\mathbf{q}}\omega_p(\mathbf{q})| = v_p$, so that $Z_p(\omega_p) = 3\omega_p^2 N/\omega_{D,p}^3$, where $\omega_{D,p}^3 = 3Nv_p^3/4\pi V$ defines the Debye frequency for those modes. For the optical branches we write $Z_p(\omega_p) = N\delta(\omega - \omega_{E,p})$, where

$\omega_{E,p}$ is an Einstein frequency. The specific heat appears then as a sum of Debye and Einstein contributions:

$$C_V = k_B N \sum_{p=1}^{3} \left(\frac{T}{\theta_{D,p}} \right)^3 \int_0^{\theta_{D,p}/T} \frac{x^4 e^x}{(e^x - 1)^2} dx$$

$$+ k_B N \sum_{p'} \left(\frac{\theta_{E,p'}}{T} \right)^2 \frac{e^y}{(e^y - 1)^2}, \quad y \equiv \theta_{E,p'}/T. \qquad (8.5\text{-}37)$$

For the phonon spectrum of NaCl (Fig. 8-12) as inferred from experiments, the three Debye modes and one three-fold degenerate Einstein mode have been entered in. The agreement is reasonable.

8.6 Elements of Electron-Phonon Interaction

In the previous section we encountered acoustical and optical phonon branches. We shall commence with the interaction of electrons with acoustical phonons. We recall that in the absence of lattice vibrations, electrons move unhindered by the periodic array of ion cores, except that in most cases the free electron mass is to be replaced by an effective electron mass, m_e. Their mobility in that case is infinite! On average electrons move, however, with a drift velocity that is proportional to the electric field, due to collisions, either with impurities or with 'lattice vibrations'; more properly, we speak of electron-phonon collisions. The fact that the periodic array of ions has nothing to do with this is borne out by the fact that the mean free path associated with electron-phonon collisions is 100-10,000 Å, and not, say 5Å, the separation between adjacent ions.

In the present section we shall give a simplified treatment, based on the interaction of one electron with a given type of phonons in the lattice, the latter in the occupation-number representation. The Hilbert space is thus the tensor product of the phonon-Fock space and the one-electron state-space. The situation is summarised below.

	LATTICE	ELECTRON	COMPOSITE
Eigenstates	$\lvert\{n_{\mathbf{q}p}\}\rangle$	$\lvert\mathbf{k})$ or $\phi_{\mathbf{k}}(\mathbf{r})$	$\lvert\{n_{\mathbf{q},p}\}\rangle\otimes\lvert\mathbf{k})$
Eigenvalues	$\sum_{\mathbf{q},p}(n_{\mathbf{q}p}+\tfrac{1}{2})\hbar\omega_{\mathbf{q}p}$	$\varepsilon(\mathbf{k})$	$\sum_{\mathbf{q},p}(n_{\mathbf{q}p}+\tfrac{1}{2})\hbar\omega_{\mathbf{q}p}+\varepsilon(\mathbf{k})$

In the usual *adiabatic approximation* it is assumed that the phonon states are mainly unperturbed by the vibrations which deform the periodicity, as sketched in Fig. 8-13.

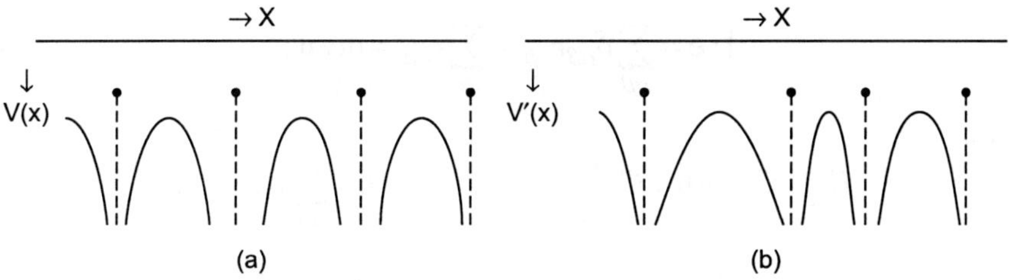

Fig. 8-13. The deformation potential. (a) Unperturbed lattice. (b) Deformed lattice due to vibrations.

As to the electrons, which are described by the one-electron Hartree-Fock Hamiltonian, cf. Section 7.8, in particular Eq. (7.8-19), the lattice potential $V(\mathbf{r})$ must now be replaced by $V[\mathbf{r}(\{\mathbf{u}_l\})]$, in which the position vector $\mathbf{r}$ is affected by the displacements $\mathbf{u}_l$ of the ions in the lattice. Therefore, the perturbation Hamiltonian of lattice plus band electron is approximately given as

$$H^1 = V[\mathbf{r}\{\mathbf{u}_l\})] - V(\mathbf{r}). \tag{8.6-1}$$

According to the Wannier-Slater theorem a perturbation Hamiltonian affects the system *as if* the band structure were changed from $\varepsilon(\mathbf{k})$ to $\varepsilon'(\mathbf{k})$, where $\mathbf{k}$ is the operator $-i\nabla$.[13]

Acoustical phonons

In this case the perturbation is due to strain associated with the displacements $\mathbf{u}_{\mathbf{l,b}}$, which for all ions '**b**' in a unit cell swing in phase. We recall the definition of the strain tensor

$$e_{\alpha\beta} = \partial u_\alpha / \partial x_\beta = \partial_\alpha u_\beta, \quad \text{or} \quad \mathsf{e} = \text{Grad}\,\mathbf{u}. \tag{8.6-2}$$

Bardeen and Shockley showed that the band-structure change due to strain is given by[14]

$$\varepsilon'(\mathbf{k}) - \varepsilon(\mathbf{k}) = \sum_{\alpha\beta} c_{\alpha\beta} e_{\alpha\beta} + c'_{\alpha\beta} k_\alpha k_\beta e_{\alpha\beta}, \tag{8.6-3}$$

which is a second-order Taylor expansion. Separating off the diagonal terms and using a relation between c and c', the result becomes, with e now being the strain operator

$$H^1 = \varepsilon'(\mathbf{k}) - \varepsilon(\mathbf{k}) = C_1 \mathsf{I} : \mathsf{e} + C_2 (\hat{\mathbf{k}}\hat{\mathbf{k}} : \mathsf{e} - \tfrac{1}{3} \mathsf{I} : \mathsf{e}), \tag{8.6-4}$$

---

[13] G.H. Wannier, Phys. Rev. **52** 191 (1937); J.C. Slater, Phys. Rev. **76**, 1592 (1949); see also, C.M. Van Vliet and A.H. Marshak, Phys. Rev. **B26**, 6734 (1982).

[14] J. Bardeen and W. S. Shockley, Phys. Rev. **80**, 72 (1950).

where $\hat{\mathbf{k}}$ is a unit vector in the direction of **k**. Now,

$$\mathsf{I}:\mathsf{e} = \sum_{\alpha\beta} \delta_{\alpha\beta} e_{\alpha\beta} = \sum_{\alpha} e_{\alpha\alpha} = \operatorname{div}\mathbf{u}\,; \tag{8.6-5}$$

therefore,

$$H^1 = C_1 \operatorname{div}\mathbf{u} + C_2 \sum_{\alpha} (\hat{k}_\alpha^2 - \tfrac{1}{3}) e_{\alpha\alpha} + C_2 \sum_{\alpha \neq \beta} \hat{k}_\alpha \hat{k}_\beta e_{\alpha\beta}. \tag{8.6-6}$$

For properly chosen axes, $\hat{k}_\alpha^2 = \frac{1}{3}\hat{k}^2 = \frac{1}{3}$, so that the second term vanishes. The last term is only important for non-cubic crystals, when there is a shear strain; in most cases it can be omitted. Hence we have for the perturbation Hamiltonian

$$H^1 = C_1 \operatorname{div}\mathbf{u}\,, \tag{8.6-7}$$

where C_1 is called the *deformation potential.* For metals it can be shown to be $C_1 \approx \frac{2}{3}\varepsilon_F$. For semiconductors, such as Ge and Si it is best taken to be an empirical constant of order ~100 *eV*.

We need to find div **u**. We take the results (8.5-30) in which we replace $\mathbf{q} \to -\mathbf{q}$ in the second line. Adding and subtracting the two parts of (8.5-30) so obtained, we obtain $P_{\mathbf{q}p}$ and $Q_{\mathbf{q}p}$. Upon rotation and a unitary transformation among the polarization branches we then obtain the original normal mode vectors $P_{\mathbf{q},\mathbf{b}}$ and $Q_{\mathbf{q},\mathbf{b}}$, from which $\mathbf{u}_{\mathbf{l},\mathbf{b}}$ is obtained by (8.5-24). Explicit results are only simple if there is only one ion in the unit cell of the (direct) lattice. Then, with $m_p \to M$, the coupling tensor matrix $\mathsf{E}(\mathbf{q})$ is diagonal; one easily obtains

$$\mathbf{u}_{\mathbf{l},\mathbf{b}} = \sum_{\mathbf{q},p} \sqrt{\frac{\hbar}{MN\omega_{\mathbf{q},p}}}\, \mathbf{e}_{\mathbf{q}p} [a_{\mathbf{q}p} e^{i\mathbf{q}\cdot\mathbf{l}} + a_{\mathbf{q}p}^\dagger e^{-i\mathbf{q}\cdot\mathbf{l}}]. \tag{8.6-8}$$

For the divergence we find

$$\operatorname{div}\mathbf{u}_{\mathbf{l},\mathbf{b}} = i \sum_{\mathbf{q},p} \sqrt{\frac{\hbar}{MN\omega_{\mathbf{q},p}}}\, (\mathbf{e}_{\mathbf{q}p} \cdot \mathbf{q}) [a_{\mathbf{q}p} e^{i\mathbf{q}\cdot\mathbf{l}} - a_{\mathbf{q}p}^\dagger e^{-i\mathbf{q}\cdot\mathbf{l}}]. \tag{8.6-9}$$

In most cases of acoustical phonon interaction the coupling with longitudinal phonons (rarefaction and dilatation) prevails; then $\mathbf{e}_{\mathbf{q}p} \cdot \mathbf{q} = q$.

Lastly, we replace the discrete variable **l** by the continuous position variable **r** and we sum over the effect of all electrons, using the Fermion operators $c_{\mathbf{k}}$ and $c_{\mathbf{k}}^\dagger$ [see (7.6-6)]; the result is

$$\begin{aligned} \mathcal{H}^1 &= \sum_{\mathbf{k},\mathbf{k}'} c_{\mathbf{k}'}^\dagger c_{\mathbf{k}} (\mathbf{k}'|H^1|\mathbf{k}) \\ &= i \sum_{\mathbf{k},\mathbf{k}',\mathbf{q}} \mathcal{F}(\mathbf{q}) \left\{ c_{\mathbf{k}'}^\dagger c_{\mathbf{k}} a_{\mathbf{q}} (\mathbf{k}'|e^{i\mathbf{q}\cdot\mathbf{r}}|\mathbf{k}) - c_{\mathbf{k}'}^\dagger c_{\mathbf{k}} a_{\mathbf{q}}^\dagger (\mathbf{k}'|e^{-i\mathbf{q}\cdot\mathbf{r}}|\mathbf{k}) \right\}, \end{aligned} \tag{8.6-10}$$

where

$$|\mathcal{F}(\mathbf{q})|^2 = \left(\frac{\hbar}{2MN\omega_\mathbf{q}}\right) q^2 C_1^2 \approx \left(\frac{\hbar q}{2V_0 \rho u_0}\right) C_1^2 , \tag{8.6-11}$$

ρ is the density and u_0 is the velocity of sound. For Bloch electrons, neglecting umklapp processes [involving addition of a reciprocal lattice vector] we have

$$(\mathbf{k}'|e^{\pm i\mathbf{q}\cdot\mathbf{r}}|\mathbf{k}) = \delta_{\mathbf{k}',\mathbf{k}\pm\mathbf{q}}. \tag{8.6-12}$$

Thus (8.6-10) becomes

$$\mathcal{H}^1 = i\sum_{\mathbf{k},\mathbf{q}} \mathcal{F}(\mathbf{q})[c^\dagger_{\mathbf{k}+\mathbf{q}} c_\mathbf{k} a_\mathbf{q} - c^\dagger_{\mathbf{k}-\mathbf{q}} c_\mathbf{k} a^\dagger_\mathbf{q}]. \tag{8.6-13}$$

The first term represents phonon absorption, while the second term denotes phonon emission; the momentum is conserved, as expected. The Feynman diagrams for these elementary processes are sketched in Fig. 8-14.

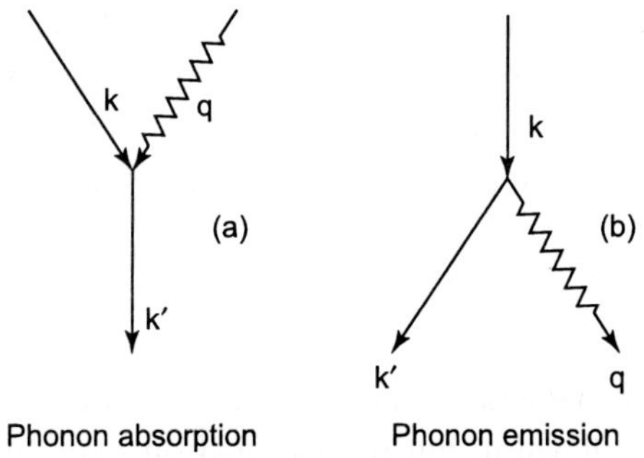

Fig. 8-14. Feynman diagrams for (a) phonon absorption and (b) phonon emission.

The many-body transitions are given by Fermi's "golden rule". Let $|\gamma\rangle$ denote a many-body state; then

$$W_{\gamma'\gamma''} = (2\pi/\hbar)|\langle\gamma'|\mathcal{H}^1|\gamma''\rangle|^2\, \delta(\mathcal{E}_{\gamma''} - \mathcal{E}_{\gamma'}). \tag{8.6-14}$$

In the nonequilibrium part these results will be fully discussed. But we can at once see the emerging structure for the possible transitions. The energy conservation, as indicated by the delta function, entails that

$$\mathcal{E}_{\gamma''} - \mathcal{E}_{\gamma'} = \varepsilon_{\mathbf{k}'} - \varepsilon_\mathbf{k} \pm \hbar\omega_\mathbf{q} . \tag{8.6-15}$$

For phonon absorption we obtain the transition rate

$$Q(\mathbf{k},\mathbf{q} \to \mathbf{k}') = (2\pi/\hbar)|\mathcal{F}(\mathbf{q})|^2\, \delta(\varepsilon_{\mathbf{k}'} - \varepsilon_\mathbf{k} - \hbar\omega_\mathbf{q})\delta_{\mathbf{k}',\mathbf{k}+\mathbf{q}} . \tag{8.6-16}$$

Likewise, for phonon emission we have

$$Q(\mathbf{k} \to \mathbf{k}',\mathbf{q}) = (2\pi/\hbar)\,|\mathcal{F}(\mathbf{q})|^2\,\delta(\varepsilon_{\mathbf{k}'} - \varepsilon_{\mathbf{k}} + \hbar\omega_{\mathbf{q}})\delta_{\mathbf{k}',\mathbf{k}-\mathbf{q}}\,. \tag{8.6-17}$$

There are other models for interaction with acoustical phonons, like the rigid ion approximation (L. Nordheim) and the deformable ion model (F. Bloch); these will not be discussed, the final results being similar to the results presented here; only the details for the coupling function $\mathcal{F}(\mathbf{q})$ are different.

Optical phonons

Optical lattice vibrations occur in lattices with at least two atoms per unit cell. Moreover, we discuss here the 'pure' case that no dipole moment is formed, which requires that the atoms are of the same type. A typical case is that of Si and Ge, with two atoms per fcc unit cell, located at (0,0,0) and $(\frac{1}{4},\frac{1}{4},\frac{1}{4})$, respectively. [The lattice is often – though wrongly – described as two 'interpenetrating fcc lattices, displaced with respect to each other by one quarter of the body diagonal.'[15]] In the optical vibration mode the two atoms swing against each other, thereby causing optical strain. The effect is particularly severe if the band-energy minima are located along the line connecting the two partners, like in Ge, for which the band valleys occur along the [111] direction. Using the Wannier-Slater theorem and following a simple model, it is assumed that the perturbation Hamiltonian can be written as

$$H^1 = \varepsilon'(\mathbf{k}) - \varepsilon(\mathbf{k}) = \sum_{p(opt)} \mathbf{D}_p \cdot \mathbf{u}_{lp}\,, \tag{8.6-18}$$

where $\mathbf{u}$ is a positive displacement vector of one of the atoms. Equation (8.6-8), obtained previously, is still valid, but the summation over 'p' now involves optical modes only. In these modes $\omega_{\mathbf{q}p}$ is nearly constant in the Brillouin zone as we noted before; moreover, it is mostly the longitudinal mode that contributes, so we will replace it by ω_0. We then have the result

$$\begin{aligned} H^1 &= \sum_{\mathbf{q},p} \sqrt{\frac{\hbar}{2V_0\rho\omega_{0,p}}}\,\mathbf{D}_p \cdot \mathbf{e}_{\mathbf{q}p}[a_{\mathbf{q}p}e^{i\mathbf{q}\cdot\mathbf{r}} + a^\dagger_{\mathbf{q}p}e^{-i\mathbf{q}\cdot\mathbf{r}}] \\ &\approx \sum_{\mathbf{q}} \sqrt{\frac{\hbar}{2V_0\rho\omega_0}}\,\mathbf{D}_0 \cdot \mathbf{e}_{\mathbf{q},long}[a_{\mathbf{q}}e^{i\mathbf{q}\cdot\mathbf{r}} + a^\dagger_{\mathbf{q}}e^{-i\mathbf{q}\cdot\mathbf{r}}]\,, \end{aligned} \tag{8.6-19}$$

where $\mathbf{D}_0$ is the optical coupling constant. The transition rates for absorption and emission have the same forms as before, cf. Eqs. (8.6-16) and (8.6-17). The coupling function is now

$$|\mathcal{F}(\mathbf{q})|^2 = \hbar D_0^2 / 2V_0\rho\omega_0 = constant. \tag{8.6-20}$$

[15] Properly: The space-lattice (or Bravais lattice) is face-centered cubic (fcc), with base two.

Piezo-electric phonons ('polar acoustical phonons')

We consider materials which have a dipole moment, such as II-VI compounds (CdS, ZnO, etc.), or III-V compounds (GaAs, InSb, GaN, etc.). We assume that there are two (or possibly more) atoms per unit cell. The total number of atoms be N. For the case of acoustical vibrations, neighbouring atoms vibrate in phase see fig. 8-15;

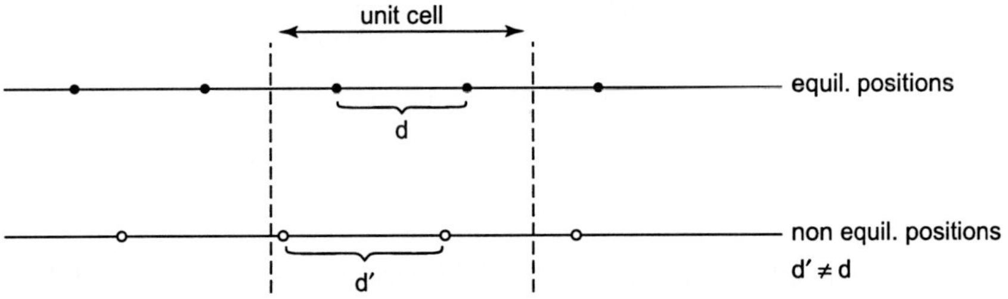

Fig. 8-15. Piezo-electric vibrations.

however, they do not have the same amplitude, so a dipole moment is produced.

Vibrations associated with two atoms per unit cell are treated in most solid-state texts. Simplified models yield, if M_1 and M_2 are the masses of positive and negative ions

$$\mathbf{u}_+ \equiv \mathbf{u}_\mathbf{l} = \sum_{\mathbf{q},p} \mathbf{e}_{\mathbf{q}p} \left(\frac{\hbar}{N(M_1+M_2)\omega_{\mathbf{q}p}} \right)^{1/2} [a_{\mathbf{q}p} e^{i\mathbf{q}\cdot\mathbf{l}} + a^\dagger_{\mathbf{q}p} e^{-i\mathbf{q}\cdot\mathbf{l}}]$$

$$\mathbf{u}_- \equiv \mathbf{u}_{\mathbf{l}+\mathbf{d}} = \sum_{\mathbf{q},p} \mathbf{e}_{\mathbf{q}p} \left(\frac{\hbar}{N(M_1+M_2)\omega_{\mathbf{q}p}} \right)^{1/2} [a_{\mathbf{q}p} e^{i\mathbf{q}\cdot(\mathbf{l}+\mathbf{d})} + a^\dagger_{\mathbf{q}p} e^{-i\mathbf{q}\cdot(\mathbf{l}+\mathbf{d})}]. \qquad (8.6\text{-}21)$$

By first order Taylor expansion and replacing **l** by **r** we find

$$\mathbf{u}_+ - \mathbf{u}_- = \sum_{\mathbf{q},p} \mathbf{e}_{\mathbf{q}p} \left(\frac{\hbar}{N(M_1+M_2)\omega_{\mathbf{q}p}} \right)^{1/2} [i\mathbf{d}\cdot\mathbf{q}\, a_{\mathbf{q}p} e^{i\mathbf{q}\cdot\mathbf{l}} - i\mathbf{d}\cdot\mathbf{q}\, a^\dagger_{\mathbf{q}p} e^{-i\mathbf{q}\cdot\mathbf{l}}]. \qquad (8.6\text{-}22)$$

The number of pairs is ½ N. Thus the polarization per unit volume becomes, e* being the effective charge, $\mathbf{P} = (Ne^*/2V_0)(\mathbf{u}_+ - \mathbf{u}_-)$. From this we obtain

$$\nabla\cdot\mathbf{P} = -\frac{Ne^*}{2V_0} \sum_{\mathbf{q},p} (\mathbf{q}\cdot\mathbf{e}_{\mathbf{q}p}) \left(\frac{\hbar}{N(M_1+M_2)\omega_{\mathbf{q}p}} \right)^{1/2} (\mathbf{d}\cdot\mathbf{q}) [a_{\mathbf{q}p} e^{i\mathbf{q}\cdot\mathbf{r}} + a^\dagger_{\mathbf{q}p} e^{-i\mathbf{q}\cdot\mathbf{r}}]. \qquad (8.6\text{-}23)$$

The electrostatic potential is found from Poisson's equation, $\nabla^2\Phi = -\rho_{bound} = \nabla\cdot\mathbf{P}$. Integrating twice this yields

$$H^1 = -e\Phi = \sum_{\mathbf{q},p} \frac{Nee^*}{2V_0 q^2} \left(\frac{\hbar}{N(M_1+M_2)\omega_{\mathbf{q}p}}\right)^{1/2} (\mathbf{d}\cdot\mathbf{q})(\mathbf{q}\cdot\mathbf{e}_{\mathbf{q}p})[a_{\mathbf{q}p}e^{i\mathbf{q}\cdot\mathbf{r}} + a^\dagger_{\mathbf{q}p}e^{-i\mathbf{q}\cdot\mathbf{r}}].$$

(8.6-24)

From this $\mathcal{H}^1$ is found. Restricting ourselves again to longitudinal phonons, $\mathbf{d}\cdot\mathbf{q}=dq$ and $\mathbf{q}\cdot\mathbf{e}_{\mathbf{q}p}=q$, we have

$$|\mathcal{F}(\mathbf{q})|^2 = \left(\frac{Nee^*d}{2V_0}\right)^2 \left(\frac{\hbar}{\rho V_0 u_0 q}\right), \tag{8.6-25}$$

where u_0 is again the velocity of sound. We note that this goes as $1/q$.

Polar optical phonons

We now look at optical vibrations of neighbouring ions in polar crystals. Since the ions vibrate in opposite directions a large dipole field is produced. For the optical displacements we have

$$\mathbf{u}_1 = \sum_{\mathbf{q},p} \mathbf{e}_{\mathbf{q}p} \frac{1}{M_1}\left(\frac{\hbar M_r}{N\omega_{0,p}}\right)^{1/2} [a_{\mathbf{q}p}e^{i\mathbf{q}\cdot\mathbf{l}} + a^\dagger_{\mathbf{q}p}e^{-i\mathbf{q}\cdot\mathbf{l}}],$$

$$\mathbf{u}_2 = -\sum_{\mathbf{q},p} \mathbf{e}_{\mathbf{q}p} \frac{1}{M_2}\left(\frac{\hbar M_r}{N\omega_{0,p}}\right)^{1/2} [a_{\mathbf{q}p}e^{i\mathbf{q}\cdot(\mathbf{l}+\mathbf{d})} + a^\dagger_{\mathbf{q}p}e^{-i\mathbf{q}\cdot(\mathbf{l}+\mathbf{d})}]; \tag{8.6-26}$$

here $M_r = M_1M_2/(M_1+M_2)$ is the reduced mass. Using zero-order Taylor expansion and replacing $\mathbf{l}$ by $\mathbf{r}$, this yields for the polarization,

$$\mathbf{P} = \frac{Ne^*}{2V_0}\sum_{\mathbf{q},p}\left(\frac{\hbar}{NM_r\omega_{0,p}}\right)^{1/2} \mathbf{e}_{\mathbf{q}p}[a_{\mathbf{q}p}e^{i\mathbf{q}\cdot\mathbf{r}} + a^\dagger_{\mathbf{q}p}e^{-i\mathbf{q}\cdot\mathbf{r}}]. \tag{8.6-27}$$

Next, $\nabla\cdot\mathbf{P}$ and Φ are found as in the previous case. The final results are

$$H^1 = \frac{iNee^*}{2V_0}\sum_{\mathbf{q},p}\left(\frac{\hbar}{NM_r\omega_{0,p}}\right)^{1/2} \frac{\mathbf{q}\cdot\mathbf{e}_{\mathbf{q}p}}{q^2}[a_{\mathbf{q}p}e^{i\mathbf{q}\cdot\mathbf{r}} - a^\dagger_{\mathbf{q}p}e^{-i\mathbf{q}\cdot\mathbf{r}}]. \tag{8.6-28}$$

From this $\mathcal{H}^1$ is again found to have the form (8.6-10), while – taking only the longitudinal optical phonons contribution – we obtain Fröhlich's renowned result:

$$|\mathcal{F}(\mathbf{q})|^2 = e^2\left(\frac{\hbar\omega_{0,long}}{2V_0}\right)^{1/2}\left(\frac{\varepsilon_s-\varepsilon_\infty}{\varepsilon_s\varepsilon_\infty}\right)\frac{1}{q^2}, \tag{8.6-29}$$

in which the following result for the effective charge has been employed

$$e^* = \left(\frac{2V_0 M_r}{N}\right)^{1/2} \left(\frac{\varepsilon_s - \varepsilon_\infty}{\varepsilon_s \varepsilon_\infty}\right)^{1/2} \omega_{0,long} , \tag{8.6-30}$$

cf. Nag[16], where ε_s and ε_∞ are the static and dynamic dielectric constant, respectively.

In summary, the dependence of the coupling constant squared is as follows:

$$\begin{aligned} &\text{acoustical phonons} &&: \ |\mathcal{F}(\mathbf{q})|^2 \propto q \ , \\ &\text{optical phonons} &&: \ |\mathcal{F}(\mathbf{q})|^2 \propto q^0 \ , \\ &\text{piezo-electric phonons} &&: \ |\mathcal{F}(\mathbf{q})|^2 \propto q^{-1} , \\ &\text{polar optical phonons} &&: \ |\mathcal{F}(\mathbf{q})|^2 \propto q^{-2} . \end{aligned} \tag{8.6-31}$$

In all above cases we assumed that the contributions from transverse modes was negligible. Also, we did not deal with umklapp processes, nor with intervalley processes. The above results will be used extensively when dealing with classical and quantum transport in the nonequilibrium part of this text.

8.7 Problems to Chapter VIII

8.1 Compute the entropy per photon $\hat{s}$ for blackbody radiation in three-dimensional space and show that it is independent of T:

$$\hat{s} = 4k_B \zeta(4) / \zeta(3) \ ; \tag{1}$$

here ζ denotes the Riemann zeta-function. Also, find a similar expression in a d-dimensional space.

8.2 For a perfect Bose gas for $\hat{v} > \hat{v}_c$ or $T > T_c$ establish in detail the virial expansion (8.3-8).

8.3 As shown in Fig. 8-8, the specific heat of a perfect Bose gas has a cusp at $T = T_c$. Establish that

$$\frac{\partial}{\partial T}\left(\frac{c_V}{kn}\right)\Bigg|_{T_c-0}^{T_c+0} = -\frac{3.66}{T_c} . \tag{2}$$

Hint: You will need the derivative dy/dT, obtained from implicit differentiation of (8.3-11), see (8.3-25). Also you will need $\zeta(3/2) = 2.612$ and the limit (obtainable from series expansion)

---

[16] B.R. Nag, "Theory of Electrical transport in Semiconductors", Pergamon Press 1970.

$$\lim_{y\to 1} g_{-3/2}(y)/[g_{-1/2}(y)]^3 = 0.159\,. \tag{3}$$

8.4 Consider a perfect Bose gas in two dimensions. Compute the grand-canonical partition function $\mathcal{Z}(T,V,y)$ and obtain the equations of state. Show that there is *no* Bose–Einstein condensation for this case.

8.5 Consider an electron gas at very low temperatures. Obtain the Helmholtz free energy per particle and show

$$\hat{f} = \tfrac{3}{5}\Delta\mu(0)\left[1 - \frac{5\pi^2}{12}\left(\frac{k_B T}{\Delta\mu(0)}\right)^2\right]. \tag{4}$$

8.6 An ideal Fermi gas has a density of states $Z(\varepsilon) = D\Theta(\varepsilon)$, where Θ is the Heaviside function and where D is a constant.
(a). Obtain the Fermi energy;
(b). For low temperatures, obtain the thermodynamic functions, κ, s, u and the pressure P, as well as the specific heat per unit volume, c_V.

8.7 Consider an ideal 3D Bose gas with density of states $Z(\varepsilon) = F\varepsilon^2\Theta(\varepsilon)$, where F is a constant. Compute the critical temperature for Bose–Einstein condensation to take place.

8.8 Equations (8.6-7) and (8.6-9) show that one electron in exchange with acoustical phonons has an interaction Hamiltonian

$$H^1 = iC_1\sum_{\mathbf{q}}\left(\hbar/2V_0\rho\omega_{\mathbf{q}}\right)^{1/2} |\mathbf{q}|[a_{\mathbf{q}}e^{i\mathbf{q}\cdot\mathbf{r}} - a_{\mathbf{q}}^{\dagger}e^{-i\mathbf{q}\cdot\mathbf{r}}]\,. \tag{5}$$

Clearly, an electron gas with ES $|\mathbf{k}0\rangle$ with no phonons excited, is not a realistic state of the system. From first order perturbation theory show that the *polaron*, being an electron 'dressed' with a virtual phonon cloud, is represented by the state

$$|\mathbf{k}0\rangle^{(1)} = |\mathbf{k}0\rangle + \sum_{\mathbf{q}} |\mathbf{k}-\mathbf{q},1_{\mathbf{q}}\rangle \frac{\langle \mathbf{k}-\mathbf{q},1_{\mathbf{q}}|\mathcal{H}^1|\mathbf{k}0\rangle}{\varepsilon_{\mathbf{k}} - \varepsilon_{\mathbf{k}-\mathbf{q}} - \hbar\omega_{\mathbf{q}}}\,, \tag{6}$$

where $\mathcal{H}^1$ is the full perturbation Hamiltonian. Compute the total number of virtual (= emitted and reabsorbed) phonons in the cloud, by acting with $\Sigma_{\mathbf{q}} a_{\mathbf{q}}$ on the state $|\mathbf{k}0\rangle^{(1)}$ and summing the squared norms of the admixture of states; show that

$$\langle N\rangle = \sum_{\mathbf{q}} C_1^2 \left(\frac{\hbar q}{2V_0 \rho u_0}\right)\left(\frac{1}{\varepsilon_{\mathbf{k}} - \varepsilon_{\mathbf{k}-\mathbf{q}} - \hbar\omega_{\mathbf{q}}}\right)^2 . \quad (7)$$

With appropriate approximations (spherical Brillouin zone with cut-off at q_D, the Debye wave vector, and $k \ll q$) show that

$$\langle N\rangle \approx \left(m^2 C_1^2 / \pi^2 \hbar^3 \rho u_0\right) \ln(q_D / q_c), \quad (8)$$

with q_c being the 'Compton wave vector' in the phonon field, $q_c = 2mu_0/\hbar$. [For reasonable numerical values, this yields $\langle N\rangle \sim 0.02$.]

8.9 If we do not make the assumption that k is negligible with respect to q, obtain the more general result:

$$\langle N\rangle = \frac{mC_1^2}{4\pi^2 \rho u_0 \hbar^3 k}\left\{(q_c - 2k)\ln\frac{q_c - 2k}{q_D + q_c - 2k} + (q_c + 2k)\ln\frac{q_c + 2k}{q_D + q_c + 2k}\right\}. \quad (9)$$

PART C

QUANTUM SYSTEMS WITH STRONG INTERACTIONS

Chapter IX

Critical Phenomena: General Features of Phase Transitions

9.1 Introductory Remarks

In the preceding chapters we have already encountered a great deal about phase transitions. The condensation of classical gases was dealt with in Chapter IV. Also, in that chapter we introduced several versions of mean field theory, the Landau-Ginzburg approach to phase transitions, as well as the classification of phase transitions, see Section 4.9. However, phase transitions in true quantum systems have been reserved for the last two chapters of this part (Part C). In Chapter XI we shall treat the two-dimensional Ising model – the historical challenge posed by Onsager – in detail, while Bose–Einstein condensation in real quantum liquids like ^{4}He and electron-pairing in superconducting electron gases or nuclear pairing in other Fermi liquids will be examined in Chapter XII. The reader will note that each and every treatment largely has its own approach; it seems fair to us to state that the diversity stands out above what is common, as is confirmed by a glance through the many books entitled "Phase Transitions", which, besides a short part on general aspects, offer a 'smorgasbord' of the great variety of systems for which the statistical mechanical description of the change in phase has been successful.[1,2,3,4]

In order to bring some unity to the presentation, we have seen to it, however, that the quantum systems presented in this part have at least one common aspect, viz. the (grand)-Hamiltonian of departure. The Ising model, after a brief description of the spin gas properties, has been translated to an interacting fermion system, following the work of Schultz, Mattis and Lieb (Chapter XI, Ref. 5.). Likewise, the Bogoliubov Hamiltonian for ^{4}He, the Fröhlich Hamiltonian for Cooper pairs, the BCS Hamiltonian for superconductors and the Balian–Werthamer Hamiltonian for condensed ^{3}He, have all been formulated in terms of interactions of quasi-particles of

[1] R. Brout, "Phase Transitions." W. A. Benjamin, New York and Amsterdam, (1965).

[2] H. E. Stanley, Phase Transitions and Critical Phenomena", Oxford Univ. Press, NY and Oxford, 1971.

[3] B. Widom, "Kritieke Verschijnselen" (Critical Phenomena), Lecture Notes Institute for Theoretical. Physics, Univ. of Amsterdam, 1972.

[4] M. Gitterman and Vivian (Haim) Halpern, "Phase Transitions", World Scient. Publ. Co. New Jersey and Singapore, 2004.

the quantum liquid, expressed in second quantization form. 'Ordinary' methods, as well as diagrammatical approaches, will be presented.

First, however, in this chapter and the next one, we shall furnish a description of the great forward strides of the second half of the twentieth century, involving critical fluctuations, critical exponents and universality, as well as aspects pertaining to renormalization. We shall be mostly brief: whole textbooks have been devoted to these subjects and it is clearly not wise to expound these here. Moreover, these general aspects have not generally led to 'easy' formulations and theories for particular systems. However, they have given a more unified outlook through renormalization theory and they do provide valuable checks on the results obtained, since the constraints for the critical exponents should be obeyed by the theory, as well as being confirmed by experiment.

9.2 Critical Fluctuations

9.2.1 *Elements of functional theory*

Although generally we do not elaborate on mathematical procedures, we shall make an exception for the concepts of functional differentiation and integration – both of which are standard ingredients of Landau theory in statistical mechanics, besides being useful for many branches of physics, in particular quantum-field theory, in which field-density Lagrangians and Hamiltonians are common place. As references we mention Van Doren et alii[5] and Beran.[6]

In principle, any function that depends on all values of another function is a functional (Volterra); e.g., $\Phi = \psi[p(\mathbf{r})]$, where we assume $\mathbf{r}$ to be a spatial variable in R^3 space. However, for the ordinary physicist these are just composite (or compound) functions which can be handled with substitutions and the chain rule. So, here we have in mind *density functionals*, in which the functional is an integral over the compound density $\psi[p(\mathbf{r})]$:

$$F[p(\mathbf{r})] = \int_\Omega d^3r\, \psi[p(\mathbf{r})]. \qquad (9.2\text{-}1)$$

For the simplest case that ψ linearly depends on $p(\mathbf{r})$, i.e., $\psi[p(\mathbf{r})] = \varphi(\mathbf{r})p(\mathbf{r})$, $F[p(\mathbf{r})]$ is a linear functional ,

$$F[\alpha p_1(\mathbf{r}) + \beta p_2(\mathbf{r})] = \alpha F[p_1(\mathbf{r})] + \beta F[p_2(\mathbf{r})]. \qquad (9.2\text{-}2)$$

---

[5] V.E. Van Doren, C. Van Alsenoy and V. Van Doren (Eds.) "Density Functional Theory and its Application to Materials", Addison-Wesley, 2000.

[6] Mark J. Beran, "Statistical Continuum Theories", Interscience Publ. N.Y. (1968); also, M.J. Beran, "Functional Theory" (1967), unpublished.

We shall now define the functional derivative, first the mathematician's way and then the physicist's way. Let $\theta(\mathbf{r})$ be a function that is positive definite and bounded in an interval $\Delta^3 r'$ centred on $\mathbf{r}'$ and zero elsewhere. The integral will be replaced by a Riemann sum with intervals of like width as $\Delta^3 r'$. The functional derivative, $\delta F / \delta p(\mathbf{r})$ in a point $\mathbf{r}'$, is then defined as[7]

$$\left.\frac{\delta F}{\delta p(\mathbf{r})}\right|_{\mathbf{r}=\mathbf{r}'} = \lim_{\Delta^3 r_i \to 0} \sum_i \frac{\Delta^3 r_i \{\psi[p(\mathbf{r}_i) + \theta(\mathbf{r}_i)] - \psi[p(\mathbf{r}_i)]\}}{\theta(\mathbf{r}_i)\Delta^3 r'}. \tag{9.2-3}$$

Now,

$$\psi[p(\mathbf{r}_i) + \theta(\mathbf{r}_i)] = \begin{cases} \psi[p(\mathbf{r}_i)] + \psi'[p(\mathbf{r}_i)]\theta(\mathbf{r}_i) + \mathcal{O}(\theta^2), & \mathbf{r}_i \in \Delta^3 r' \\ \psi[p(\mathbf{r}_i)] & \mathbf{r}_i \notin \Delta^3 r'. \end{cases} \tag{9.2-4}$$

Hence,

$$\left.\frac{\delta F}{\delta p(\mathbf{r})}\right|_{\mathbf{r}=\mathbf{r}'} = \lim_{\Delta^3 r' \to 0} \{\psi'[p(\mathbf{r} \in \Delta^3 r')] + \mathcal{O}(\theta)\} = \psi'[p(\mathbf{r}')]. \tag{9.2-5}$$

The physicist's definition is straightforward, incorporating the same idea, but employing Dirac's delta function. Let $p(\mathbf{r}) \to p(\mathbf{r}) + \kappa(\mathbf{r})\delta(\mathbf{r}-\mathbf{r}')$. The functional derivative is then defined as

$$\left.\frac{\delta F}{\delta p(\mathbf{r})}\right|_{\mathbf{r}=\mathbf{r}'} = \lim_{\kappa(\mathbf{r}) \to 0} \int_\Omega d^3 r \frac{\{\psi[p(\mathbf{r}) + \kappa(\mathbf{r})\delta(\mathbf{r}-\mathbf{r}')] - \psi[p(\mathbf{r})]\}}{\kappa(\mathbf{r})}. \tag{9.2-6}$$

[The keen reader will notice that all we did is replacing $\theta(\mathbf{r}) \equiv \bar{\theta}(\mathbf{r})/\Delta^3 r'$ by $\kappa(\mathbf{r})\delta(\mathbf{r}-\mathbf{r}')$ and avoid the Riemann sum.] Using Taylor expansion to first order and carrying out the delta integration, we arrive once more at

$$\left.\frac{\delta F}{\delta p(\mathbf{r})}\right|_{\mathbf{r}=\mathbf{r}'} = \psi'[p(\mathbf{r}')]. \tag{9.2-7}$$

The specification of the point $\mathbf{r}'$ is of course superfluous in most cases. So we simply have

$$\textit{If} \quad F[p(\mathbf{r})] = \int_\Omega d^3 r \psi[p(\mathbf{r})], \quad \textit{then} \quad \delta F / \delta p(\mathbf{r}) = \psi'[p(\mathbf{r})]. \tag{9.2-8}$$

Examples follow in the next subsection and in Section 12.7.

We will also meet the concept of functional integration, $\int \psi[\rho(\mathbf{r})]\, D[\rho(\mathbf{r})]$; this is far more involved, since a function space is infinite dimensional. Formally this integration will be defined as follows. Let $\rho(\mathbf{r})$ be sampled at a denumerable set of points $\mathbf{r}_1 ... \mathbf{r}_m$. Let again Ω be some volume in R^3. These points are usually chosen so as to be spaced in intervals of size Ω/m. We then define

---

[7] More precisely, the denominator should be $\sigma = \int_{\Delta^3 r'} d^3 r \theta(\mathbf{r})$.

$$\int \psi[\rho(\mathbf{r})]\, D[\rho(\mathbf{r})] = \lim_{m\to\infty} \int_\Omega d\rho(\mathbf{r}_m) \int_\Omega d\rho(\mathbf{r}_{m-1}) \ldots \int_\Omega d\rho(\mathbf{r}_1)\, \psi[\rho(\mathbf{r}_1 \ldots \mathbf{r}_m)]$$
$$= \lim_{m\to\infty} \prod_k \int_\Omega d\rho(\mathbf{r}_k)\, \psi[\rho(\mathbf{r}_1 \ldots \mathbf{r}_m)], \qquad (9.2\text{-}9)$$

where each integral is a Riemann-Stieltjes integral. The existence of the limit must of course be investigated for each case separately. Fortunately, often we do not *need* the answer to (9.2-9), but we can perform differentiations under the integral in order to obtain moments for continuous distributions, etc. The only functional integral that can be integrated exactly is the Gaussian integral,

$$I = \int (D\varphi) \exp\{\int dx[-\tfrac{1}{2}(\nabla\varphi(x))^2 - \tfrac{1}{2}\alpha(\varphi(x))^2 + \varphi(x)\eta(x)]\}, \qquad (9.2\text{-}10)$$

which occurs in certain formulations of the Landau theory. The argument of the exponential can be written as the scalar product $-\frac{1}{2}(\varphi, K\varphi) + (\eta, \varphi)$, $K = \alpha - \nabla^2$. One may show that this yields

$$I = (const)\exp[(\eta, K^{-1}\eta)/2], \qquad (9.2\text{-}10')$$

where K^{-1} is a Green's operator. Fourier transformation is also useful for evaluation.

9.2.2 *Landau–Ginzburg Density Functionals*

As in Section 4.9 the order parameter will be denoted by m. This symbol reminds us of course of the spontaneous magnetization in ferromagnets, but – here as well as in Section 4.9 – the order parameter denotes the appropriate quantum operator whose mean value characterizes the cooperative behaviour, e.g., the magnetization per particle m or the spin σ_i^z in the magnetic transition, the quantum-field operator $\Psi^{(\dagger)}(\mathbf{r})$ in condensed ^{4}He and the gap $\Delta_\mathbf{k}$ or the pair-operator $b_\mathbf{k}^{(\dagger)}$ in superconductors, all of which show long-range correlations. Prior to the onset of cooperative behaviour, as approached from $T > T_c$, there will be critical fluctuations, with a variance that approaches infinity when $T = T_c$. The description of these fluctuations is most easily given with a slightly expanded version of the Landau–Ginzburg method presented previously for the order parameter m; like there, the main drawback of the method is that there is no *a priori* justification for a power series expansion, which from the outset ignores possible terms like $\ln m$.

We shall consider a generalized canonical ensemble in which $m(\mathbf{r})$ is variable, while its conjugate variable H is held constant; note that the order parameter is *locally defined* in the present case. Further, this entails that we consider a free energy which is a Legendre transform of $\mathcal{E}$ with respect to the entropy, the parameter $M = \langle \int d^3r\, m(\mathbf{r})\rangle$ and possible other variables like the volume; it is denoted by Φ. More explicitly, we assume that this Legendre transform is engendered by the 'response Hamiltonian' which involves a linear coupling to $m(\mathbf{r})$,

$$\mathcal{H}_{total} = \mathcal{H} - \int d^3r\, m(\mathbf{r})h(\mathbf{r}); \qquad (9.2\text{-}11)$$

for $h(\mathbf{r}) \to 0$ this acquiesces to the Hamiltonian of the equilibrium system for which we seek to describe the fluctuations $\Delta m(\mathbf{r})$. Generally we now write

$$\Phi[T, m(\mathbf{r})] = \int d^3r \left\{ a_0(T) + \tfrac{1}{2}a_2[m(\mathbf{r})]^2 + \tfrac{1}{4}a_4[m(\mathbf{r})]^4 + \ldots \tfrac{1}{2}b[\nabla m(\mathbf{r})]^2 \right\}, \qquad (9.2\text{-}12)$$

in which the term [grad m]^2 has been added over and above the ansatz (4.9.2); we have $b > 0$, since Φ is a minimum if $\nabla m = 0$. The conjugate local force follows from functional differentiation,

$$h(\mathbf{r}) = \delta\Phi / \delta m(\mathbf{r}) = a_2 m(\mathbf{r}) + a_4[m(\mathbf{r})]^3 - b\nabla^2 m(\mathbf{r}), \qquad (9.2\text{-}13)$$

where the last term in (9.2-3) was obtained with Green's theorem, assuming that the surface integral vanishes. For $T > T_c$ the order parameter vanishes when $h = 0$ and $\nabla m = 0$, as expected; for $T < T_c$ the order parameter satisfies $m_0^2 = -a_2 / a_4$, $a_2 = a_{20}(T - T_c)$, if the coefficients are such that a second-order phase transition occurs.

We will now find the response function $\varphi(\mathbf{r}) = \delta m(\mathbf{r}) / \delta h(\mathbf{r})$ and the generalized susceptibility $\chi = \int d^3r\, \varphi(\mathbf{r})$. For the functional differentiation for this case note that we can write

$$m(\mathbf{r}) = \int d^3r' \delta(\mathbf{r} - \mathbf{r}') F[h(\mathbf{r}')],$$

so that by the rules for functional derivative

$$\varphi(\mathbf{r}) = \delta(\mathbf{r} - \mathbf{r}')[dF / dh]_{\mathbf{r}'}.$$

The implicit differentiation of (9.2-13) is thus carried out by assuming a perturbation centred on $\mathbf{r}'$, i.e. we set $h(\mathbf{r}) = h_0(\mathbf{r}) + h_1\delta(\mathbf{r} - \mathbf{r}')$ and $m(\mathbf{r}) = m_0(\mathbf{r}) + h_1\varphi(\mathbf{r})$. Substituting into (9.2-13) and retaining only terms linear in φ one easily obtains

$$\nabla^2\varphi - \xi^{-2}\varphi = -b^{-1}\delta(\mathbf{r} - \mathbf{r}'), \quad \xi = \begin{cases} \sqrt{b / a_{20}(T - T_c)}, & T > T_c, \\ \sqrt{-b / 2a_{20}(T - T_c)}, & T < T_c. \end{cases} \qquad (9.2\text{-}14)$$

The solution is well-known; remembering $\nabla^2(1/r) = -4\pi\delta(\mathbf{r})$, we find

$$\varphi(\mathbf{r} - \mathbf{r}') = e^{-|\mathbf{r}-\mathbf{r}'|/\xi} / 4\pi b |\mathbf{r} - \mathbf{r}'| \quad \text{or} \quad \varphi(\mathbf{R}) = e^{-R/\xi} / 4\pi b R. \qquad (9.2\text{-}15)$$

The quantity ξ is the *correlation length.* We note that it goes to infinity when we approach T_c, whether from above or below. For the case that the fluctuations were non-local, we saw in Chapter V that the variance $\langle \Delta M^2 \rangle$ was k_BT times the generalized susceptibility, cf. Eq. (5.3-5). The local character that we consider here leads therefore to the spatial correlation function; this will be shown more clearly in the next subsection.

9.2.3 *Spatial Correlation Function*

The two-point fluctuation-correlation function is defined as usual, cf. (4.11-17),

$$\Gamma(\mathbf{r}) \equiv \langle \Delta m(\mathbf{r}) \Delta m(0) \rangle = \langle m(\mathbf{r}) m(0) \rangle - \langle m(\mathbf{r}) \rangle \langle m(0) \rangle . \tag{9.2-16}$$

This function measures the persistence of spatial order. The average is over a grand-canonical ensemble, for the case that $h(\mathbf{r})$ in (9.2-11) $\to 0$. Assuming the system to be invariant against a spatial translation, $\langle m(\mathbf{r}) \rangle = \langle m(0) \rangle \equiv \langle m \rangle$. We will seek the Fourier transform of (9.2-16). Denoting transforms by a tilde, we have

$$m(\mathbf{r}) = (2\pi)^{-3} \int d^3k \, e^{i\mathbf{k}\cdot\mathbf{r}} \tilde{m}(\mathbf{k}) , \qquad \tilde{m}(\mathbf{k}) = \int d^3r \, e^{-i\mathbf{k}\cdot\mathbf{r}} m(\mathbf{r}) . \tag{9.2-17}$$

We note that $\tilde{m}(-\mathbf{k}) = \tilde{m}^*(\mathbf{k})$ since $m(\mathbf{r})$ is real. For the Fourier transform of Γ this yields

$$\begin{aligned} \tilde{\Gamma}(\mathbf{k}) &= \int d^3r \, e^{-i\mathbf{k}\cdot\mathbf{r}} \left\{ \langle m(\mathbf{r}) m(0) \rangle - \langle m \rangle^2 \right\} \\ &= \langle \tilde{m}(\mathbf{k}) m(0) \rangle - (2\pi)^3 \langle m \rangle^2 \delta(\mathbf{k}) \\ &\overset{\mathbf{k}\neq 0}{=} (2\pi)^{-3} \int d^3p \, \langle \tilde{m}(\mathbf{k}) \tilde{m}(\mathbf{p}) \rangle . \end{aligned} \tag{9.2-18}$$

For the correlation in **k**-space, occurring in the last line, we shall obtain a general expression, using equipartition of energy. First of all, we have

$$\begin{aligned} \langle \tilde{m}(\mathbf{k}) \tilde{m}(\mathbf{p}) \rangle &= \iint d^3r d^3r' e^{-i\mathbf{k}\cdot\mathbf{r}} e^{-i\mathbf{p}\cdot\mathbf{r}'} \langle m(\mathbf{r}) m(\mathbf{r}') \rangle \\ &\overset{\mathbf{r}-\mathbf{r}'=\mathbf{R}}{=} \iint d^3R d^3r' e^{-i(\mathbf{k}+\mathbf{p})\cdot\mathbf{r}'} e^{-i\mathbf{k}\cdot\mathbf{R}} \underbrace{\langle m(\mathbf{r}) m(\mathbf{r}') \rangle}_{F(\mathbf{R})} \\ &= (2\pi)^3 \delta(\mathbf{k}+\mathbf{p}) \int d^3R \, e^{-i\mathbf{k}\cdot\mathbf{R}} F(\mathbf{R}) . \end{aligned} \tag{9.2-19}$$

The main purpose of this exercise was to show that the Fourier coefficients for different **k**'s are not correlated, except for $\mathbf{p} = -\mathbf{k}$. We thus have

$$\langle \tilde{m}(\mathbf{k}) \tilde{m}(\mathbf{p}) \rangle = (2\pi)^3 V^{-1} \delta(\mathbf{k}+\mathbf{p}) \langle \tilde{m}(\mathbf{k}) \tilde{m}^*(\mathbf{k}) \rangle , \tag{9.2-20}$$

where we added the inverse volume factor, since otherwise there is a dimensional failure on the rhs; the problem is alleviated if we coarse-grain the **k**-space, so that the modes are discrete in V and we have a Kronecker delta. This result into (9.2-18) gives

$$\tilde{\Gamma}(\mathbf{k}) = \langle \tilde{m}(\mathbf{k}) \tilde{m}^*(\mathbf{k}) \rangle / V, \quad \mathbf{k} \neq 0 . \tag{9.2-21}$$

We now return to the free energy of the previous subsection. First consider $T > T_c$. The leading terms in the free energy expansion are

$$\Phi[T,m(\mathbf{r})]=\int d^3r\left\{\tfrac{1}{2}a_2[m(\mathbf{r})]^2-\tfrac{1}{2}bm(\mathbf{r})\nabla^2 m(\mathbf{r})\right\}, \tag{9.2-22}$$

where we integrated the last term by parts (or applied Green's theorem) and omitted the term a_0, since the curve for Φ starts at $m = 0$, see Fig. 4-17, temperature T_2; note also that $m_0=\langle m\rangle=0$. For the spatial Fourier transforms we then have from Parseval's theorem

$$\int d^3r\left\{\tfrac{1}{2}a_2[m(\mathbf{r})]^2-\tfrac{1}{2}bm(\mathbf{r})\nabla^2 m(\mathbf{r})\right\}=(2\pi)^{-3}\int d^3k\left\{(\tfrac{1}{2}a_2+\tfrac{1}{2}bk^2)\tilde{m}(\mathbf{k})\tilde{m}^*(\mathbf{k})\right\}$$
$$=\sum\nolimits_{\mathbf{k}>0}\left\{(a_2+bk^2)[\tilde{m}(\mathbf{k})\tilde{m}^*(\mathbf{k})]/V\right\}. \tag{9.2-23}$$

The energy residing in one Fourier mode is that of a classical oscillator

$$\langle\mathcal{E}(\mathbf{k})\rangle=(a_2+bk^2)\langle\tilde{m}(\mathbf{k})\tilde{m}^*(\mathbf{k})\rangle/V=k_BT. \tag{9.2-24}$$

From (9.2-21) we find

$$\tilde{\Gamma}(\mathbf{k})=\frac{k_BT}{a_2+bk^2}=\frac{k_BT/b}{k^2+\xi^{-2}}. \tag{9.2-25}$$

where $\xi=\sqrt{b/a_2}$ as before, cf. (9.2-14), first line. This has exactly the same form as the Ornstein–Zernike correlation function we studied in Section 4.4, esp. Eq. (4.4-41). We thus obtain for the order parameter fluctuation-correlation function,

$$\Gamma(\mathbf{r}-\mathbf{r}')=\langle\Delta m(\mathbf{r})\Delta m(\mathbf{r}')\rangle=k_BT\frac{e^{-R/\xi}}{4\pi bR},\qquad R=|\mathbf{r}-\mathbf{r}'|. \tag{9.2-26}$$

Thus we found that the correlation function decreases with the same correlation length as the response function; this is a general result, in stochastic theory denoted as *Onsager's principle*, where it is usually applied in the time domain. So, specifically,

$$\varphi(\mathbf{R})=(1/k_BT)\Gamma(\mathbf{R})\quad\text{and}\quad\chi=(1/k_BT)\int d^3R\,\Gamma(\mathbf{R}). \tag{9.2-27}$$

This is a particular form of the fluctuation-dissipation theorem, which links the correlation function to the response function and the generalized susceptibility.

For $T < T_c$ the leading terms in the expansion for Φ are different, as can be seen from the curve for T_1 in Fig. 4-17. The second derivative [see below Eq. (4.9-4)] is now given by $\Phi''=2m_0^2a_4=-2a_2$. So by Taylor expansion

$$\Phi[T,m(\mathbf{r})]=\int d^3r\left\{a_0-a_2[m(\mathbf{r})]^2-\tfrac{1}{2}bm(\mathbf{r})\nabla^2 m(\mathbf{r})\right\}. \tag{9.2-28}$$

The constant is of no importance since it only contributes at $\mathbf{k} = 0$. From Parseval's theorem for the Fourier transforms we now find that the energy in a mode $\mathbf{k}$ is

$$\langle\mathcal{E}(\mathbf{k})\rangle=(-2a_2+bk^2)\langle\tilde{m}(\mathbf{k})\tilde{m}^*(\mathbf{k})\rangle/V=k_BT, \tag{9.2-29}$$

from which

$$\tilde{\Gamma}(\mathbf{k}) = \frac{k_B T}{-2a_2 + bk^2} = \frac{k_B T / b}{k^2 + \xi^{-2}}, \tag{9.2-30}$$

where now $\xi = \sqrt{-b/2a_2}$, in entire harmony with (9.2-14), second line. The response function is again related to the fluctuation-correlation function by (9.2-27). For either case the correlation length depends on T near T_c as

$$\xi \propto |T - T_c|^{-1/2}. \tag{9.2-31}$$

For χ, integration gives for this model $\chi = \xi^2 / b k_B T$. So the dependence near T_c is as in mean-field theory,

$$\chi \propto |T - T_c|^{-1}. \tag{9.2-32}$$

Finally, we give another derivation of the same results, encountered in some recent texts.[8,9] The Hamiltonian (9.2-11) is exponentiated in order to define the partition function

$$\Xi = \mathrm{Tr}\left\{\exp\left(-\beta\left[\mathcal{H} - \int d^3r' m(\mathbf{r}')h(\mathbf{r}')\right]\right)\right\}. \tag{9.2-33}$$

A primary functional differentiation of ln Ξ yields $\langle m(\mathbf{r})\rangle$,

$$\langle m(\mathbf{r})\rangle = \beta^{-1}\frac{\delta \ln \Xi}{\delta h(\mathbf{r})} = \frac{\mathrm{Tr}\left\{m(\mathbf{r})\exp\left(-\beta\left[\mathcal{H} - \int d^3r' m(\mathbf{r}')h(\mathbf{r}')\right]\right)\right\}}{\mathrm{Tr}\left\{\exp\left(-\beta\left[\mathcal{H} - \int d^3r' m(\mathbf{r}')h(\mathbf{r}')\right]\right)\right\}}. \tag{9.2-34}$$

The second functional differentiation is more tricky; however, for a translationally invariant system, the factor $m(\mathbf{r})$ behind the trace in the numerator can be replaced by $m(0)$ and the differentiation goes smoothly. One easily obtains for $\varphi(\mathbf{r})$:

$$\begin{aligned}
&\frac{\delta\langle m(\mathbf{r})\rangle}{\delta h(\mathbf{r})} \\
&= \beta\left(\frac{\mathrm{Tr}\{m(\mathbf{r})m(0)e^{-\beta[\mathcal{H}-\int d^3r'...]}\}}{\mathrm{Tr}\{e^{-\beta[\mathcal{H}-\int d^3r'...]}\}} - \frac{\mathrm{Tr}\{m(\mathbf{r})e^{-\beta[\mathcal{H}-\int d^3r'...]}\}\times\mathrm{Tr}\{m(0)e^{-\beta[\mathcal{H}-\int d^3r'...]}\}}{[\mathrm{Tr}\{e^{-\beta[\mathcal{H}-\int d^3r'...]}\}]^2}\right) \\
&= \beta\big(\langle m(\mathbf{r})m(0)\rangle - \langle m(\mathbf{r})\rangle\langle m(0)\rangle\big) = \beta\langle \Delta m(\mathbf{r})\Delta m(0)\rangle \equiv \beta\Gamma(\mathbf{r}).
\end{aligned} \tag{9.2-35}$$

There are two (minor) problems with this procedure. Firstly, the two parts of the Hamiltonian (9.2-11) are not likely to commute – m being an operator and h a c-number – so that the partition function in this form only has symbolic significance.

---

[8] K. Huang, "Statistical Mechanics" Second Ed., Wiley and Sons, NY. 1987, Section 16.2.

[9] M. Plischke and B. Bergersen, 2nd Ed., World Scient. Publ. Co. 1994, Section 3.10.

Secondly, the essence of linear response theory is that it couples a response parameter to the *equilibrium fluctuations* in the system. Thus, in (9.2-35) we must *a posteriori* let $h(\mathbf{r}) \to 0$ and evaluate the fluctuations for the original Hamiltonian. A treatment that avoids the exponentiation of the Hamiltonian (9.2-11) requires time-dependent perturbation theory and Heisenberg operators; this will be the approach that we shall follow in Chapter XVI.

Before we go on we shall give the approximate solution for Γ in d dimensions. Let

$$|x|^2 = \sum_{i=1}^{d} x_i^2; \tag{9.2-36}$$

then Γ satisfies

$$\nabla^2\Gamma - \xi^{-2}\Gamma = -k_B T b^{-1}\delta^d(x). \tag{9.2-37}$$

For large $|x|$, $\nabla^2 \sim d^2/d|x|^2$ in all dimensions, so that $\Gamma \propto \exp(-|x|/\xi)$. At $T = T_c$ $\xi \to \infty$ and we must solve $\nabla^2\Gamma = -k_B T b^{-1}\delta^d(x)$. Integrating over a small d-dimensional sphere of radius ℓ we have

$$\int d^d x \nabla \cdot \nabla\Gamma = \oint\!\!\oint_{S^d} \nabla\Gamma \cdot d^d S = (d\Gamma/d\ell) 2(\pi)^{d/2}[\Gamma(\tfrac{1}{2}d)]^{-1}\ell^{d-1} = -k_B T b^{-1}, \tag{9.2-38}$$

where the constants are from Eqs. (2.5-3) and (2.5-4). Hence, $d\Gamma/d\ell \propto \ell^{-(d-1)}$. Combining with the asymptotic result, we have the approximate solution

$$\Gamma(|x|) \approx k_B T\, e^{-|x|/\xi}/b|x|^{d-2}, \quad d \neq 2. \tag{9.2-39}$$

The case $d = 2$ will be considered in subsection 9.7.4. The above is only correct either *at* T_c or far away from T_c; very near T_c we have the scaling laws discussed below.

9.3 Critical Exponents and Scaling Relations

We will now take a closer look at the behaviour of response parameters and the equations of state near the critical point. Generally, these thermodynamic quantities can be split into a 'regular' part and a singular part near T_c. The regular part is analytic except, perhaps, for a finite jump at the critical temperature; the singular part, on the contrary, is divergent or has divergent derivatives. Thus, let us set $t = (T - T_c)/T_c$. The thermodynamic quantities which we have in mind have a divergent part which follows a power law $x = |t|^z$, where z is usually a fractional exponent. The literature of the late sixties and early seventies abound with experimental data for the specific heat, magnetization, correlation length (obtainable from neutron or X-ray scattering data) and other quantities. At the same time many theories about scaling and universality saw the light and the available physical models for phase transitions, such as the Ornstein–Zernike description of vapour-liquid condensation, mean-field theory for (anti)ferromagnetic behaviour and order-disorder transitions, the Ising model (1D, 2D or 3D) and the Heisenberg spin-coupling model were analyzed to compare their predictions for these exponents. At one time it was hoped that 'really universal' exponents would emerge, both experimentally and theoretically, but that appeared to be illusionary. For instance, it seemed that the order parameter $M = \langle \int d^3r\, m(\mathbf{r}) \rangle$ would have a critical exponent of $\frac{1}{3}$ when plotted as a function of $|t|$, but very precise measurements by Ho and Litster indicated a value of 0.368 ± 0.005 for the insulating ferromagnetic material $CrBr_3$. Their data are partially reproduced in Fig. 9-1.[10] However, both experiment and theory indicated that several exact relations existed between critical exponents.

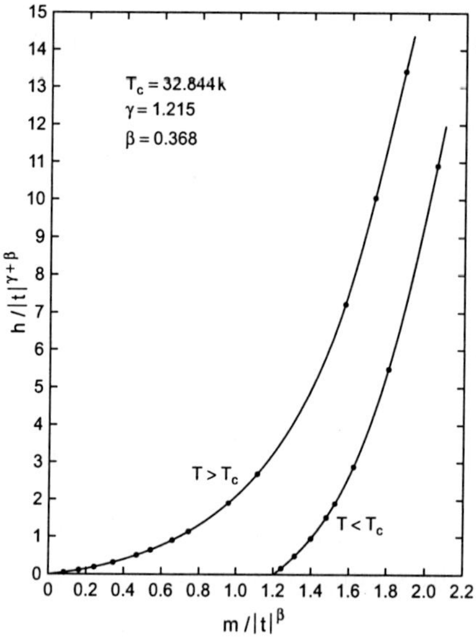

Fig. 9-1. Magnetization of $CrBr_3$ vs. reduced temperature $|T - T_c|/T_c$. After Ho and Litster.[7] [With permission.]

Besides t we introduce the relative variables $m = (M - M_c)$ and $h = (H - H_c)$. The first four critical exponents are defined as follows:

$$\begin{aligned}
&\text{heat capacity:} && C_H \sim |t|^{-\alpha} \quad , \\
&\text{susceptibility:} && \chi_T \sim |t|^{-\gamma} \quad , \\
&\text{order parameter:} && m \sim (-t)^{\beta} \quad , \\
&\text{equation of state } (t=0): && h \sim m^{\delta}\, \text{sgn}\, m \ .
\end{aligned} \tag{9.3-1}$$

The $\sim$ sign refers to the singular part of the quantities involved and means 'proportional to'. We note that we have not introduced separate symbols for power laws below T_c (sometimes denoted by $\alpha' ... \gamma'$) and above T_c, since it has been shown theoretically and verified by experiment that these coefficients are the same. While we have repeatedly stressed that our symbols are 'general', targeting any type of

---

[10] J.T. Ho and J.D. Litster, Phys. Rev. Lett. **22**, 603 (1969).

phase transition, it is clear that the symbols have been 'borrowed' from magnetic phenomena; we shall therefore – with Stanley, Op. Cit. – repeat the power laws for the classical vapour-liquid transition. The specific heat clearly means C_V and the order parameter is now $(\rho_\ell - \rho_g)/\rho_c$, where the symbol ρ denotes the mass density. Also, let p denote $(P-P_c)/P_c$. We have

$$\begin{aligned}
&\text{heat capacity:} && C_V \sim |t|^{-\alpha} \quad , \\
&\text{isothermal compressibility:} && K_T \sim |t|^{-\gamma} \quad , \\
&\text{order parameter:} && (\rho_\ell - \rho_g)/\rho_c \sim (-t)^{\beta} \quad , \\
&\text{critical isotherm } (t=0): && p \sim [(\rho_\ell - \rho_g)/\rho_c]^{\delta}\,\mathrm{sgn}[(\rho_\ell - \rho_g)/\rho_c] \,.
\end{aligned} \tag{9.3-2}$$

In the next few sections these symbols and the analogy between the 'magnetic' case and the classical case will become more clear.

Two more power laws will be introduced; they pertain to the two-point fluctuation-correlation function, or the response function, these functions being proportional to each other, as we saw before; these are

$$\begin{aligned}
&\text{correlation length:} && \xi \sim |t|^{-\nu} \quad , \\
&\text{power-law decay } r^{-p} \ (t=0): && p = d-2+\eta \,.
\end{aligned} \tag{9.3-3}$$

Here d is the dimensionality. All these power laws share two important features. First, widely different systems with critical temperatures that are orders of magnitude apart, have exponents that are close to equal; secondly, there are four relations between these coefficients, which are exactly obeyed in all models and experimentally to within measured error. The *scaling relations* involving the first four exponents are:

$$\begin{aligned}
\alpha + 2\beta + \gamma &= 2, && \text{(Rushbrooke)} \\
\beta(\delta - 1) &= \gamma. && \text{(Widom)}
\end{aligned} \tag{9.3-4}$$

In addition, there are two rules that connect the above exponents with ν, η and d (hyper scaling relations):

$$\begin{aligned}
\alpha &= 2 - \nu d \quad , && \text{(Josephson)} \\
\gamma &= \nu(2-\eta) \,. && \text{(Fisher)}
\end{aligned} \tag{9.3-5}$$

The corresponding *inequalities* are easily explained from thermodynamics. The equalities can be obtained from simple dimensional analysis and 'scaling hypotheses'. In Table 9-1 we summarise experimental data for fluids and magnetic materials and we give the various exponents for soluble models and approximate computations.[11]

Table 9-1. Critical exponents; experimental data and theoretical models

system	α	β	γ	δ	ν	η	Th.	Exp.
fluids								
CO_2	~0.1	0.34	1.35	4.2	–	–		
Xe	~0.2	0.35	1.3	4.4	0.57	–		
magnetic								
Ni	0	0.42	1.35	4.22	–	–		
EuS	0.05	0.33	–	–	–	–		
$CrBr_3$	–	0.368	–	4.3	–	–		
soluble models								
mean-field, Landau–G	0	½	1	3	½	0		
Ising 2D	0	⅛	7/4	15	1	¼		
approximate models								
Ising 3D	⅛	5/16	5/4	5	0.64	0.05		
Heisenberg 3D	1/16	5/16	21/16	–	0.7	0.04		
$\alpha + 2\beta + \gamma$							2	2.00 ± 0.01
$(\beta\delta - \gamma)/\beta$							1	0.93 ± 0.08
$(2-\eta)\nu/\gamma$							1	1.02 ± 0.05
$(2-\alpha)/\nu d$							1 or $4/d$ (*mft*)	–

To demonstrate the *universality* (within a class) of these exponents, in Fig. 9-2 we plotted experimental data for the vapour-liquid coexistence region for various substances.[12] The abscissa is $(\rho_\ell - \rho_g)/\rho_c$ while the ordinate is T/T_c. Clearly $(\rho_\ell - \rho_g)/\rho_c$ is a universal function of T/T_c. The empirical fit gives the following relations

$$\frac{\rho_\ell + \rho_g}{2\rho_c} = 1 + \frac{3}{4}\left(1 - \frac{T}{T_c}\right), \qquad \frac{\rho_\ell - \rho_g}{\rho_c} = \frac{7}{2}\left(1 - \frac{T}{T_c}\right)^{1/3}. \tag{9.3-6}$$

The second relation shows that $\beta = \frac{1}{3}$. We note that the van der Waals equation yields $\beta = ½$ as will be discussed in Problem 9.6. The values of $\alpha \ldots \delta$ for mean field theory (*mft*) are the subject of Problem 9.4.

---

[11] H. Eugene Stanley, Op. Cit., p. 47. See also, Kerson Huang, 2nd Ed., Op. Cit., p. 398.
[12] E.A. Guggenheim, J. Chemical Phys. **13**, 253 (1945).

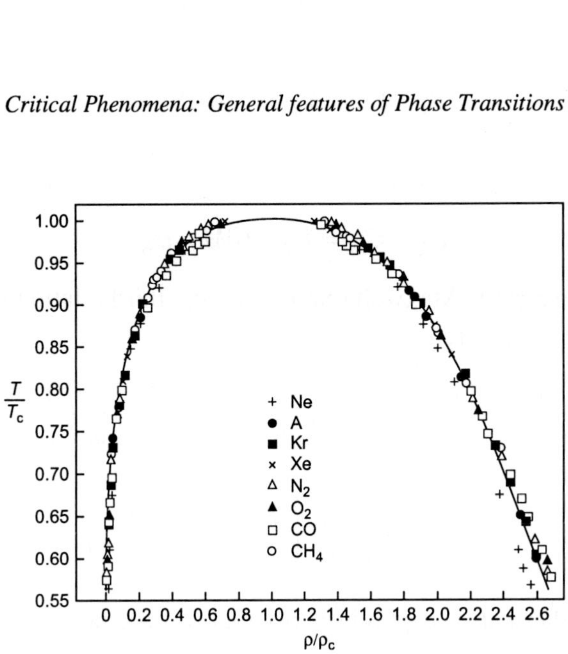

Fig. 9-2. Data for reduced vapour-liquid parameters in the coexistence region for various substances; after Ref. 12. [With permission.]

9.4 Thermodynamic Inequalities

9.4.1 *Magnetic Systems; the Curie-Point Transition*

Before the scaling equalities were fully established and experimentally verified, a number of inequalities were derived, based on the thermodynamic stability criteria and the minimum principle of the free energy, cf. Section 3.6. The most well-known is the Rushbrooke inequality.[13] Let us consider the 'magnetic Gibbs free energy'

$$G_M(T,N,H) = \mathcal{E} - TS - MH, \qquad dG_M = -SdT + \varsigma dN - MdH, \tag{9.4-1}$$

where we set $V = 1\ cm^3$, MH being an energy per unit volume. There are two forms for the specific heat, viz.,

$$C_H = T(\partial S/\partial T)_{N,H}, \qquad C_M = T(\partial S/\partial T)_{N,M}, \tag{9.4-2}$$

analogous to C_P and C_V, respectively (see, however, below). The subscript 'N' will further be suppressed. The relationship between them follows the usual pattern, see Problem 4.2. We have[14]

$$S = S[T,N,M(H,T)] \tag{9.4-3}$$

$$\left(\frac{\partial S}{\partial T}\right)_H = \left(\frac{\partial S}{\partial T}\right)_M + \left(\frac{\partial S}{\partial M}\right)_T \left(\frac{\partial M}{\partial T}\right)_H . \tag{9.4-4}$$

---

[13] G.S. Rushbrooke, J. Chemical Phys. **43**, 3439 (1963).

[14] In this chapter and the next one we mainly follow Callen's description of thermodynamic functions.

Setting $\alpha_H \equiv (\partial M / \partial T)_H$, we get

$$C_H - C_M = T(\partial S / \partial M)_T \alpha_H \,. \tag{9.4-5}$$

Now we need the magnetic Maxwell relation [cf. the third relation of (3.5-31), noting $P \to -H$] and the '-1 theorem',

$$\left(\frac{\partial S}{\partial M}\right)_T = -\left(\frac{\partial H}{\partial T}\right)_M, \qquad \left(\frac{\partial H}{\partial T}\right)_M \left(\frac{\partial M}{\partial H}\right)_T \left(\frac{\partial T}{\partial M}\right)_H = -1, \tag{9.4-6}$$

from which

$$-\left(\frac{\partial H}{\partial T}\right)_M = \left(\frac{\partial H}{\partial M}\right)_T \left(\frac{\partial M}{\partial T}\right)_H = \frac{1}{\chi_T}\alpha_H \,, \tag{9.4-7}$$

where χ_T is the isothermal susceptibility. So, (9.4-5) becomes

$$C_H - C_M = T\alpha_H^2 / \chi_T \,. \tag{9.4-8}$$

The response quantities C_H and C_M must be positive definite. Hence,

$$C_H \geq T\alpha_H^2 / \chi_T \,. \tag{9.4-9}$$

We now employ Stanley's lemma 3.[15] Consider

$$f(x) \leq g(x) \;\; \text{with} \;\; f(x) \sim x^{\lambda}, \quad g(x) \sim x^{\mu}\,; \tag{9.4-10}$$

For sufficiently small positive values of the argument, $0 < x < 1$ we have $\ln x < 0$; thus,

$$[\ln f(x) / \ln x] \geq [\ln g(x) / \ln x], \quad \text{or} \quad \lambda \geq \mu \,. \tag{9.4-11}$$

This is applied to the inequality (9.4-9), observing the scaling laws (9.3-1). We then obtain

$$2(\beta - 1) + \gamma \geq -\alpha \quad \text{or} \quad \alpha + 2\beta + \gamma \geq 2\,, \tag{9.4-12}$$

which is Rushbrooke's inequality.

This result can also be formulated for the gas-liquid transition. However, despite the correspondence $P \to -H$ mentioned above, stemming from the use of Callen's Gibbs' free energy, in the usual statistical mechanical specification H is held constant in magnetic experiments and V is specified in the gas-fluid case, cf. our discussion in subsection 3.1.2. Therefore the appropriate specific heat is C_V. With the scaling laws (9.3-2) Rushbrooke's inequality follows again, as discussed in the next subsection.

Many more inequalities have been derived, in particular by Griffiths.[16] We only mention two, viz.,

---

[15] H.E. Stanley, Op. Cit., p. 52.

[16] R.B. Griffiths, Phys. Rev. Lts. **14**, 623 (1965); J. Chemical Phys. **43**, 1958 (1965); "Lectures on Critical Phenomena", Banff Summer Institute, 1968 (unpublished).

$$\alpha + \beta(1+\delta) \geq 2\,, \tag{9.4-13}$$

$$\gamma \geq \beta(\delta - 1)\,. \tag{9.4-14}$$

The reader recognizes that all these inequalities turned out to be equalities, see (9.3-4) and (9.3-5). The derivations of the equalities require, however, additional assumptions.

9.4.2 *Vapour-Liquid Transition and the Coexistence Region*

We shall consider the behaviour along an isotherm in the coexistence region. The molecular volume at a point D of this isotherm is denoted by $\hat{v}_D$ and the molar volume by $\tilde{V}_D$, whereby we follow the notation of chapter IV; the connection is $\tilde{V}_D = N_A \hat{v}_D$, where N_A is Avogadro's number. The isotherm is sketched in Fig. 9-3; when decreasing $\tilde{V}$, point B is the end of the gas phase and point A marks the beginning of the liquid phase, while the stretch $\overline{AB}$ is in the *coexistence region A'-C-B'*. Note that $\tilde{V}_B = \tilde{V}_g$ and $\tilde{V}_A = \tilde{V}_\ell$. The pressure is constant in the coexistence region, as discussed previously. The volume jump, $\Delta\tilde{V} = (\partial G_g / \partial P)_T - (\partial G_\ell / \partial P)_T$, indicates that we have a first-order phase transition.

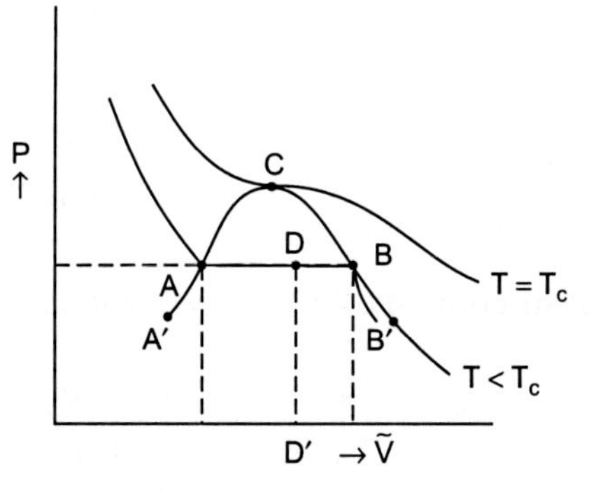

Fig. 9-3. Isotherm P vs. molar volume $\tilde{V}$ in the coexistence region.

Let x_g and x_ℓ be the molar fractions at D; then, for the volume at D: $\tilde{V} = x_g \tilde{V}_g + x_\ell \tilde{V}_\ell$. Since also $x_g + x_\ell = 1$, we have $\tilde{V}(x_g + x_\ell) = x_g \tilde{V}_g + x_\ell \tilde{V}_\ell$, from which we obtain the 'lever rule',

$$\frac{x_\ell}{x_g} = \frac{\tilde{V}_g - \tilde{V}}{\tilde{V} - \tilde{V}_\ell} = \frac{\tilde{V}_B - \tilde{V}}{\tilde{V} - \tilde{V}_A}\,. \tag{9.4-15}$$

For the molar entropy we have

$$\tilde{S} = x_g(\tilde{V},T)\tilde{S}_g[T,\tilde{V}_g(T)] + x_\ell(\tilde{V},T)\tilde{S}_\ell[T,\tilde{V}_\ell(T)]\,. \tag{9.4-16}$$

This gives for the molar specific heat at constant $\tilde{V}$:

$$C_V = x_g(\tilde{V},T)\,T\left(\frac{\partial \tilde{S}_g}{\partial T}\right)_{coex} + x_\ell(\tilde{V},T)\,T\left(\frac{\partial \tilde{S}_\ell}{\partial T}\right)_{coex} + T\left(\tilde{S}_\ell - \tilde{S}_g\right)\left(\frac{\partial x_\ell}{\partial T}\right)_{\tilde{V}}, \quad (9.4\text{-}17)$$

where we noticed that $dx_g = -dx_\ell$ and where $(\partial/\partial T)_{coex}$ means differentiation along either end of the coexistence curve (AA' or BB') in Fig. 9-3. From the chain rule we further have

$$T\left(\frac{\partial \tilde{S}_g}{\partial T}\right)_{coex} = T\left(\frac{\partial \tilde{S}_g}{\partial T}\right)_{\tilde{V}_g} + T\left(\frac{\partial \tilde{S}_g}{\partial \tilde{V}_g}\right)_T\left(\frac{\partial \tilde{V}_g}{\partial T}\right)_{coex} = C_{V_g} + T\left(\frac{\partial \tilde{S}_g}{\partial \tilde{V}_g}\right)_T\left(\frac{\partial \tilde{V}_g}{\partial T}\right)_{coex}, \quad (9.4\text{-}18)$$

with a similar relationship for the liquid phase. For the last term in (9.4-17) we use the Clausius-Clapeyron equation, cf. (3.6-16):

$$\tilde{S}_\ell - \tilde{S}_g = -\left(\tilde{V}_g - \tilde{V}_\ell\right)\left(\frac{dP}{dT}\right). \quad (9.4\text{-}19)$$

Also, from differentiation of $\tilde{V} = x_g\tilde{V}_g + x_\ell\tilde{V}_\ell$ to T along $\overline{DD'}$ we have, noting that $(d\tilde{V}/dT) = 0$,

$$\left(\tilde{V}_\ell - \tilde{V}_g\right)\left(\frac{\partial x_\ell}{\partial T}\right)_{\tilde{V}} + x_g\left(\frac{\partial \tilde{V}_g}{\partial T}\right)_{coex} + x_\ell\left(\frac{\partial \tilde{V}_\ell}{\partial T}\right)_{coex} = 0, \quad \text{or,} \quad (9.4\text{-}20)$$

$$\left(\frac{\partial x_\ell}{\partial T}\right)_{\tilde{V}} = \frac{1}{\tilde{V}_g - \tilde{V}_\ell}\left[x_g\left(\frac{\partial \tilde{V}_g}{\partial T}\right)_{coex} + x_\ell\left(\frac{\partial \tilde{V}_\ell}{\partial T}\right)_{coex}\right]. \quad (9.4\text{-}21)$$

Finally, collecting and substituting into (9.4-17) we obtain,

$$C_V = x_g C_{V_g} + x_\ell C_{V_\ell} + x_g\left(\frac{\partial \tilde{V}_g}{\partial T}\right)_{coex}\left[T\left(\frac{\partial \tilde{S}_\ell}{\partial \tilde{V}_\ell}\right)_T - T\frac{dP}{dT}\right]$$
$$+ x_\ell\left(\frac{\partial \tilde{V}_\ell}{\partial T}\right)_{coex}\left[T\left(\frac{\partial \tilde{S}_\ell}{\partial \tilde{V}_\ell}\right)_T - T\frac{dP}{dT}\right]. \quad (9.4\text{-}22)$$

Two more modifications must be made. First we use the Maxwell relation (3.5-31)

$$\left(\frac{\partial \tilde{S}_g}{\partial \tilde{V}_g}\right)_T = \left(\frac{\partial P}{\partial T}\right)_{\tilde{V}_g}, \quad \left(\frac{\partial \tilde{S}_\ell}{\partial \tilde{V}_\ell}\right)_T = \left(\frac{\partial P}{\partial T}\right)_{\tilde{V}_\ell}. \quad (9.4\text{-}23)$$

Secondly, since $P = P[T, \tilde{V}(T)]$ we have, applying this at either end of the isobar $\overline{AB}$,

$$\frac{dP}{dT} = \left(\frac{\partial P}{\partial T}\right)_{\tilde{V}_g} + \left(\frac{\partial P}{\partial \tilde{V}_g}\right)_T\left(\frac{\partial \tilde{V}_g}{\partial T}\right)_{coex} = \left(\frac{\partial P}{\partial T}\right)_{\tilde{V}_\ell} + \left(\frac{\partial P}{\partial \tilde{V}_\ell}\right)_T\left(\frac{\partial \tilde{V}_\ell}{\partial T}\right)_{coex}. \quad (9.4\text{-}24)$$

Combining the last two results gives

$$\left(\frac{\partial \tilde{S}_g}{\partial \tilde{V}_g}\right)_T = \frac{dP}{dT} - \left(\frac{\partial \tilde{V}_g}{\partial T}\right)_{coex}\left(\frac{\partial P}{\partial \tilde{V}_g}\right)_T, \quad \left(\frac{\partial \tilde{S}_\ell}{\partial \tilde{V}_\ell}\right)_T = \frac{dP}{dT} - \left(\frac{\partial \tilde{V}_\ell}{\partial T}\right)_{coex}\left(\frac{\partial P}{\partial \tilde{V}_\ell}\right)_T. \tag{9.4-25}$$

With this in (9.4-22), the result for C_V becomes

$$C_V = x_g\left[C_{V_g} - T\left(\frac{\partial \tilde{V}_g}{\partial T}\right)^2_{coex}\left(\frac{\partial P}{\partial \tilde{V}_g}\right)_T\right] + x_\ell\left[C_{V_g} - T\left(\frac{\partial \tilde{V}_\ell}{\partial T}\right)^2_{coex}\left(\frac{\partial P}{\partial \tilde{V}_\ell}\right)_T\right]. \tag{9.4-26}$$

Note that this has the familiar form (Problem 4.2), although C_P does not occur, being ∞ in the coexistence region. Lastly, we change to the mass density, $\rho_i = M_i/\tilde{V}_i$ $(i = g, \ell)$, where M is the molar mass and we introduce the isothermal compressibilities. With c_V denoting the specific heat per gram, we have as final expression

$$c_V = x_g c_{V_g} + x_\ell c_{V_\ell} + \frac{x_g T}{K_{T,g}\rho_g^3}\left(\frac{\partial \rho_g}{\partial T}\right)^2_{coex} + \frac{x_\ell T}{K_{T,\ell}\rho_\ell^3}\left(\frac{\partial \rho_\ell}{\partial T}\right)^2_{coex}. \tag{9.4-27}$$

In order to study the divergence near the critical temperature, we must let $T \to T_c$. We then have

$$c_V \geq \frac{\frac{1}{2}T}{K_{T,c}\rho_c^3}\left[\left(\frac{\partial \rho_g}{\partial T}\right)^2_{coex} + \left(\frac{\partial \rho_\ell}{\partial T}\right)^2_{coex}\right], \tag{9.4-28}$$

where we noted that $\rho_g \approx \rho_\ell \approx \rho_c$, so that $x_g \approx x_\ell \approx \frac{1}{2}$. For the remainder we write

$$\frac{\partial \rho_g}{\partial T} = \frac{\partial \rho_c}{\partial T} - \frac{\partial(\rho_c - \rho_g)}{\partial T}, \quad \frac{\partial \rho_\ell}{\partial T} = \frac{\partial \rho_c}{\partial T} + \frac{\partial(\rho_\ell - \rho_c)}{\partial T}, \tag{9.4-29}$$

in which $\rho_c = \lim_{T\to T_c}\rho_D$. From the lever rule and $x_g \approx x_\ell \to \frac{1}{2}$ we see that the last differentials in the two above expressions are near-equal. This, then, yields

$$\left(\frac{\partial \rho_g}{\partial T}\right)^2 + \left(\frac{\partial \rho_\ell}{\partial T}\right)^2 = \left[\left(\frac{\partial \rho_g}{\partial T}\right) + \left(\frac{\partial \rho_\ell}{\partial T}\right)\right]^2 - 2\left(\frac{\partial \rho_g}{\partial T}\right)\left(\frac{\partial \rho_\ell}{\partial T}\right)$$

$$= A - 2\left(\frac{\partial \rho_c}{\partial T}\right)^2 + 2\left(\frac{\partial(\rho_c - \rho_g)}{\partial T}\right)\left(\frac{\partial(\rho_\ell - \rho_c)}{\partial T}\right) \approx A - 2\left(\frac{\partial \rho_c}{\partial T}\right)^2 + \frac{1}{2}\left(\frac{\partial(\rho_\ell - \rho_g)}{\partial T}\right)^2, \tag{9.4-30}$$

where A is the term in square brackets at the first rhs.[17] Whence from (9.4-28) and

[17] The exercise can be refined if we make the switch $\tilde{V} \to M/\rho$ *after* similar manipulations are applied to (9.4-26).

(9.4-30),

$$c_V \geq \frac{T}{2\rho_c^3}\frac{1}{K_{T,c}}\left[A - 2\left(\frac{\partial\rho_c}{\partial T}\right)^2 + \frac{1}{2}\left(\frac{\partial(\rho_\ell - \rho_g)}{\partial T}\right)^2\right]. \tag{9.4-31}$$

The terms that scale divergently are c_V, K_T and $\rho_\ell - \rho_g$. As to *A*, as exemplified by the *empirical* result (9.3-6), first statement, it is expected to be finite at T_c and approach $2(\partial\rho_c/\partial T)^2$. With the scaling laws (9.3-2) and the earlier given lemma, we find

$$-\alpha \leq \gamma + 2(\beta - 1), \tag{9.4-32}$$

which yields Rushbrooke's inequality, being the same as for the magnetic transition.

9.5 Dimensional Analysis

Simple dimensional considerations support the scaling relations. We consider again the magnetic case. Let *L* be the unit of length. Further, the magnetic Gibbs' free energy, divided by k_BT and per unit 'volume', will be denoted as g_M and has the dimension L^{-d}:

$$[g_M] \equiv [G_M / k_BTV] = L^{-d}. \tag{9.5-1}$$

The heat capacity follows by differentiation to *T*. For the correlation function we have from (9.3-3)

$$[\Gamma] = L^{2-d-\eta}. \tag{9.5-2}$$

Since this is also the dimension of $\langle\Delta m(\mathbf{r})\Delta m(0)\rangle$ we have

$$[m] = L^{(2-d-\eta)/2}. \tag{9.5-3}$$

Now let us rewrite (9.2-17) for *d* dimensions. We then note

$$[\chi k_BT] = L^{2-\eta}. \tag{9.5-4}$$

For the conjugate force we know that *HM* has the dimension of energy per unit 'volume', while $H = \chi M$. Hence $H^2/\chi k_BT$ has the dimension L^d. With $H = hk_BT$ it follows that

$$[H] = L^{(d+2-\eta)/2}, \qquad [hk_BT] = L^{(d+2-\eta)/2}. \tag{9.5-5}$$

Now we make the assumption (scaling hypothesis) that any length scales with the correlation length ξ, which itself near the critical temperature scales with $|t|^{-\nu}$. Employing also (9.3-1) or (9.3-2) we then establish the following identities:

$$\begin{aligned}
&[m]\rightarrow && -\nu(2-d-\eta)/2=\beta\ ,\\
&[\chi k_BT]\rightarrow && -\nu(2-\eta)=-\gamma\,,\\
&[hk_BT]\rightarrow && \nu(d+2-\eta)/2=\delta\beta\,,\\
&[C_V/k_B]=(d^2/d\,|t\,|^2)[g_M]\rightarrow && (\nu d-2)=-\alpha\,.
\end{aligned}\tag{9.5-6}$$

The second relationship is Fisher's law, while the fourth relationship is Josephson's law. By addition and subtraction one easily obtains Widom's and Rushbrooke's equalities. Of course, the basic assumption that all lengths should scale as ξ has not been proven with this argument. More sophisticated hypotheses will be put forward in the next few sections.

9.6 Other Scaling Hypotheses

9.6.1 *Widom Scaling*

Homogeneous functions of first order were met in subsection 3.5.2. Homogeneous functions of any integral order and their corresponding Euler relations can be found in many a text on function theory. Let us consider

$$F(\mu x_1,\mu x_2,\ldots,\mu x_s)=\mu^r F(x_1,x_2,\ldots x_s)\,. \tag{9.6-1}$$

By differentiation to μ, setting afterwards $\mu = 1$, one obtains Euler's relation

$$\textstyle\sum_k \partial F/\partial x_k = r\,F(x_1,x_2,\ldots x_s)\,. \tag{9.6-2}$$

With the present developments in view, we shall allow r to be a fractional number. Let $\mu=\lambda^{1/r}$. We then also have

$$F(\lambda^{1/r}x_1,\lambda^{1/r}x_2,\ldots,\lambda^{1/r}x_s)=\lambda F(x_1,x_2,\ldots x_s)\,. \tag{9.6-3}$$

This is still a homogeneous function of order r. Take now $\lambda = x_i^{-r}$ and let $x_j = 0,\ i\neq j$. Then we get

$$F(0,0,\ldots,1,\ldots,0)=x_i^{-r}F(0,0,\ldots,x_i,\ldots,0)\,. \tag{9.6-4}$$

Since the lhs is a constant, all $F(x_i)$ scale as x_i^r.

In scaling theory we need functions of several variables which we will call 'generalized homogeneous' functions of fractional order, i.e., they scale as in (9.6-4) but with a different exponent for different variables. In 1965 Widom suggested that the (magnetic) Gibbs free energy should have a regular and a singular part, which near T_c has a particular scaling form.[18] So, for the free energy per particle and in units k_BT, we should have

---

[18] B. Widom, J. Chemical Phys. **43**, 3898 (1965).

$$\hat{g}(T,H) = \hat{g}_{reg}(T,H) + \hat{g}_s(t,h), \tag{9.6-5}$$

with

$$\hat{g}_s(\lambda^p t, \lambda^q h) = \lambda \hat{g}_s(t,h). \tag{9.6-6}$$

As before, $t = (T - T_c)/T_c$, $h = H/k_B T$ and p and q are fractional exponents. We shall show that the critical exponents $\alpha \ldots \delta$ can all be expressed in p and q; consequently, the scaling relations should follow if Widom's hypothesis is correct. The magnetization per spin m below T_c is as always obtained from differentiation to h. Thus, from (9-6-6), starting with the rhs,

$$\lambda m(-t,h) = -\lambda \left(\frac{\partial \hat{g}_s(-t,h)}{\partial h} \right)_t = -\lambda^q \left(\frac{\partial \hat{g}_s(-\lambda^p t, \lambda^q h)}{\partial h} \right)_t = \lambda^q m(-\lambda^p t, \lambda^q h). \tag{9.6-7}$$

Now let $\lambda = (-t)^{-1/p}$ and let $h = 0$. We then have

$$(-t)^{-1/p} m(-t,0) = (-t)^{-q/p} m(1,0). \tag{9.6-8}$$

The quantity $m(1,0)$ is a mere constant. So we obtain for $m(-t,0)$

$$m \sim (-t)^{(1-q)/p} \quad \rightarrow \beta = (1-q)/p. \tag{9.6-9}$$

As to the susceptibility, since it is the second derivative of g or the first derivative of m, we must differentiate (9.6-7) with respect to h; this yields, both above and below T_c,

$$\lambda \chi(|t|,h) = \lambda^{2q} \chi(\lambda^p |t|, \lambda^q h). \tag{9.6-10}$$

Now we set $h = 0$ and we take $\lambda = |t|^{-1/p}$. This gives us

$$\chi(|t|,0) = |t|^{(1-2q)/p} \chi(1,0) \qquad \rightarrow \gamma = (2q-1)/p. \tag{9.6-11}$$

Next, the heat capacity follows from $C_H = -t(\partial^2 \hat{g}/\partial t)^2$. So, (9.6-6) is differentiated twice to t. Starting with the rhs, this yields

$$\frac{\lambda}{t} C_H(t,h) = -\lambda \left(\frac{\partial^2 \hat{g}_s(t,h)}{\partial t^2} \right)_h = -\lambda^{2p} \left(\frac{\partial^2 \hat{g}_s(\lambda^p t, \lambda^q h)}{\partial t^2} \right)_h = \frac{\lambda^{2p}}{t} C_H(\lambda^p t, \lambda^q h). \tag{9.6-12}$$

Again. afterwards we should set $\lambda = t^{-1/p}$ and $h = 0$; also replace t by $|t|$. This gives

$$C_H \sim |t|^{(1-2p)/p} \qquad \rightarrow \alpha = (2p-1)/p. \tag{9.6-13}$$

Finally, we will find the equation of state. In (9.6-6) we set $t = 0$ and $\lambda = h^{-1/q}$. This leads to

$$\hat{g}_s(0,1) = h^{-1/q} \hat{g}_s(0,h), \text{ or } \hat{g}_s(0,h) = h^{1/q} \hat{g}_s(0,1). \tag{9.6-14}$$

Differentiating to h we find

$$m \sim h^{(1-q)/q} \qquad \rightarrow \delta = q/(1-q)\,. \tag{9.6-15}$$

It is now a matter of trivial algebra to obtain the two scaling relations. From (9.6-9), (9.6-11) and (9.6-13) one finds right away the Rushbrooke equality $\alpha + 2\beta + \gamma = 2$, while from (9.6-9), (9.6-11) and (9.6-15) one obtains Widom's relation $\beta(\delta - 1) = \gamma$. The hyperscaling relations, which link the critical exponents to the dimensionality of the system, will be discussed in the next subsection.

9.6.2 *Kadanoff Scaling*

In the second part of the nineteen-sixties, Kadanoff et al. wrote an extensive review paper which summarised the many experimental studies and theoretical findings of the time; the paper is well worth reading still today, even though many questions from that era have now been settled.[19] After reviewing the Landau–Ginzburg approach and the derivations of the response and correlation function for critical fluctuations, already discussed by us in the preceding sections, they present an original outlook on the scaling problem, notably for the Ising model. The basic idea is that in any cooperative phenomenon the system can be divided into subvolumes or 'blocks', which sufficiently close to the critical point will themselves behave as the individual spins do on a microscopic scale. And, naturally, these blocks, in turn, could organize themselves into superblocks, which also behave collectively when we get still closer to the critical point, and so on. The reason is, of course, that the correlation length increases indefinitely when $T \rightarrow T_c$ until at $T = T_c$ $\xi = \infty$, so that the system becomes scale-invariant. In Fig. 9-4 the situation is sketched for a square 2D Ising spin lattice. Kerson Huang has illumined his review with the poem of Jonathan Swift (and an appropriate illustration)[20] [clearly, Swift used the opposite scaling order…]:

> "Great fleas have lesser fleas
> Upon their backs to bite 'em
> And lesser fleas have lesser still
> And so ad infinitum."

So, to be definite, let a_0 be the lattice dimension and let the block length be $K = La_0$, where $a_0 \ll K \ll \xi$. The number of spin sites per block is L^d, where d is the dimensionality. Each spin site i is as usual endowed with the spin quantum

[19] L.P. Kadanoff, W. Götze, D. Hamblin, R. Hecht, E.A.S. Lewis, V.V. Palciauskas, M. Rail, J. Swift, D. Aspnes and J. Kane, Revs. Modern Phys. **39**, 395 (1967).

[20] K. Huang, 2nd Ed., Op. Cit., p.404.

number σ_I and likewise, when the block J is near-coherent, it will be assigned a spin number σ_J. The Ising Hamiltonian and the canonical partition function can now be written in terms of spin-spin coupling involving the σ_J's; the details are considered in the next chapter.

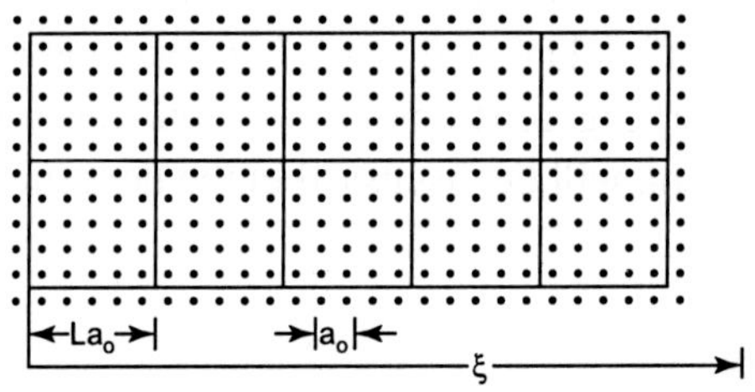

Fig. 9-4. Division of square 2D Ising lattice into blocks of 5×5 spins. After Kadanoff et al. (Ref. 15).

Once we have the partition function, we can connect again with the free energy, which – for consistency reason in this chapter – we shall call the magnetic Gibbs free energy per spin, $\hat{g}(t,h)$. For L to be $\gg 1$ we must rescale t, such that $\bar{t}=tL^x$, where x is as yet unknown, but should be positive definite, since the blocks are farther from criticality than the original spins. Likewise, we make the replacement $\bar{h}=hL^y$, $y>0$. The following scaling function is then apparent:

$$\hat{g}_s(L^x t, L^y h) = L^d \hat{g}_s(t,h), \tag{9.6-16}$$

where 'sub s' denotes again the singular part. The factor L^d stems from the fact that in the lhs we deal with the free energy per block spin – the individual spins having lost their significance – while on the rhs we have the free energy per individual spin; yet, the functional dependence on t and h should be the same. Also, we can write down a scaling function for the correlation length in terms of L. The following relationship presents itself as plausible:

$$\xi(L^x t, L^y h) = L^{-1}\xi(t,h). \tag{9.6-17}$$

We now make the connection with the previous Widom scaling function for $\hat{g}_s$. Clearly, (9.6-16) is a special form of (9.6-6); in fact, if we set $\lambda = L^d$ we have the same scaling form, providing

$$x = pd, \quad y = qd. \tag{9.6-18}$$

In (9.6-17) we replace t by $|t|$, since this form is valid on both sides of the critical point, we let $L=|t|^{-1/x}$ and we set $h=0$. This gives

$$\xi(t,h) = |t|^{-1/x}\xi(1,0) \quad \rightarrow \xi \sim |t|^{-1/x}, \tag{9.6-19}$$

so that with (9.6-19) and the definition of the coefficient ν in (9.3-3) we find

$$x = pd = 1/\nu. \tag{9.6-20}$$

Further, from (9.6-13) we have $p = 1/(2-\alpha)$. Combining this with the above, we obtained the hyperscaling relation,

$$\nu d = 2-\alpha, \tag{9.6-21}$$

which is the Josephson equality.

To establish the Fisher relationship, we must consider the order-parameter, i.e., the spin-spin fluctuation-correlation function. As before we introduce local fields $h(\mathbf{r})$. The correlation function is given as the second functional derivative

$$\Gamma(r) = k_B T\, \delta^2 \hat{g}_s / [\delta h(\mathbf{r})]^2 . \tag{9.6-22}$$

Now,

$$\delta^2 \hat{g}_s (L^x t, L^y h(\mathbf{r})) = L^{2d} \delta^2 \hat{g}_s (t, h(\mathbf{r})) . \tag{9.6-23}$$

so that

$$\Gamma[(r/L), L^x t, L^y h(\mathbf{r})][\delta \bar{h}(\mathbf{r})]^2 = L^{2d} \Gamma[(r), t, h(\mathbf{r})][\delta h(\mathbf{r})]^2 . \tag{9.6-24}$$

Also, we have $\delta h(\mathbf{r}) / \delta \bar{h}(\mathbf{r}) = L^{-y}$. We thus obtain

$$\Gamma[(r/L), L^x t, L^y h(\mathbf{r})] = L^{2(d-y)} \Gamma[r, t, h(\mathbf{r})] . \tag{9.6-25}$$

Next we set $L = r$ and $t = 0$. This gives

$$\Gamma[1, 0, r^y h(\mathbf{r})] = r^{2(d-y)} \Gamma[r, t, h(\mathbf{r})] \qquad \rightarrow \Gamma \sim r^{-2(d-y)} . \tag{9.6-26}$$

Comparing with the general scaling law $\Gamma \sim r^{-(d-2+\eta)}$ and using $y = qd$, we obtain

$$2d(1-q) = d-2+\eta; \tag{9.6-27}$$

with q given by (9.6-15) this leads to

$$2-\eta = d\frac{\delta-1}{\delta+1}. \tag{9.6-28}$$

From Widom's and Rushbrooke's relationships the coefficient β is eliminated to yield

$$\frac{\delta-1}{\delta+1} = \frac{\gamma}{2-\alpha}. \tag{9.6-29}$$

Further, we employ Josephson's equality (9.6-21); this, then, gives the final hyperscaling law

$$2-\eta = \gamma/\nu, \tag{9.6-30}$$

which establishes Fisher's relationship.

9.7 Other Topics in Phase Transitions

9.7.1 *Symmetry Breaking and Order Parameter*

The phenomenon of cooperative behaviour is at once an experimental fact and a baffling theoretical mystery. Let us consider first again the case of the magnetic transition at the Curie point. At elevated temperatures a spin lattice possesses rotational symmetry in all directions; notably, if we introduce a set of axes, the total spin angular momentum $|\mathbf{S}|$ and the components S_α ($\alpha = x, y$ or z) are conserved, i.e., they commute with the Hamiltonian of the system (although not with each other). However, below the Curie point the spins line up along a given direction, which we may have selected by introducing a later vanishing magnetic field, or which is spontaneously chosen; in either case we shall consider this to be the z-direction. While there is still a twofold symmetry (and consequent degeneracy) in that $\langle\mathbf{m}\rangle = \pm\langle m\rangle\hat{\mathbf{z}}$ are both possible, rotations about the system axes are absent, indicating that the symmetry of the condensed state is greatly reduced. We speak of 'spontaneous symmetry breaking'; the situation has already been depicted in the Landau–Ginzburg results for a second-order phase transition, see Fig. 4-17.

Reverting to the case that we did not apply a later vanishing field, one may argue that the symmetry is still always there, but that the other possible ground states have not been realised. Like the considerations by many with regard to the uncertainty principle, this is an *epistemological* point of view. In fact, the system, once oriented along the z-direction, will not spontaneously shift to another direction, unless one waits for a Poincaré cycle ("waiting for Godot"...). Our *ontological* viewpoint is therefore that the symmetry *has been* reduced. In the case at hand, this means that S_x and S_y are no longer associated with generators for rotations. In fact, as we show subsequently, this will give us a method for uniquely determining the order parameter for a given phase-transition process.

We return to the general question of symmetry in nature. Both in classical and quantum mechanics symmetries involve a group of transformations which derive from a generating function or a generating operator associated with an observable Q; the corresponding quantum operator we shall denote by $\hat{Q}$ and we have $[\hat{Q}, \mathcal{H}] = 0$. The most common symmetries in statistical mechanics involve rotations, translations and global or local gauge invariance; the generators for these groups are angular momentum, linear momentum and particle number, respectively. Examples of phase transitions in which symmetries are spontaneously broken are:

a) Paramagnetic – magnetic transition; rotational symmetry reduced;
b) Order – disorder transition; translational and point symmetry reduced;
c) Superfluid transition; global gauge invariance relinquished;
d) Pairing in superconductors; local gauge invariance relinquished.

The cases a) and c) will be considered in more detail.

If J_α is a component of the total angular momentum, rotations are generated by $\hat{Q} = \exp[iJ_\alpha \theta / \hbar]$. For the magnetic transition the spin angular momenta commute at all times with the microscopic Hamiltonian. Now let us suppose that we describe the system with the canonical ensemble having the density operator $\rho \propto \exp[-\beta\mathcal{H}]$. Below the Curie temperature the spins along the x and y directions will no longer commute with the exponentiated Hamiltonian $\exp[-\beta\mathcal{H}]$ due to bifurcation (*cf* Chapter XI). Thus,

$$[\rho_{can}, S_{x,t}] \neq 0, \qquad [\rho_{can}, S_{y,t}] \neq 0 \tag{9.7-1}$$

(the sub 't' denotes that this is the total spin momentum). Now consider the average

$$\mathrm{Tr}\{\rho_{can} S_{z,t}\} = \frac{1}{\hbar i}\mathrm{Tr}\{\rho_{can}[S_{x,t}, S_{y,t}]\} = \frac{1}{\hbar i}\mathrm{Tr}\{[\rho_{can}, S_{x,t}]S_{y,t}\} = -\frac{1}{\hbar i}\mathrm{Tr}\{[\rho_{can}, S_{y,t}]S_{x,t}\}; \tag{9.7-2}$$

note that we used the commutation rule for components of the angular momentum $[\mathbf{S}\times\mathbf{S} = \hbar i\,\mathbf{S}]$ and the cyclic property of the trace. Above the Curie temperature the average on the lhs is zero, but below T_c it is finite because of (9.7-1). The order parameter must therefore be related to this average. In fact, for the magnetization per spin m we have

$$\langle m\rangle = (\rho_B / N)\mathrm{Tr}\{\rho_{can} S_{z,t}\}, \tag{9.7-3}$$

where ρ_B is the Bohrmagneton. We conclude: *the order parameter is a linear functional of the non-vanishing commutators associated with the broken symmetries.*

Next, we consider the broken global gauge invariance of a superfluid. Due to Bose–Einstein condensation in the ground state, the particle number operator, **N**, no longer commutes with the Hamiltonian. We recall that in the field-operator formalism

$$\mathbf{N} = \int d^3r \Psi^\dagger(\mathbf{r})\Psi(\mathbf{r}). \tag{9.7-4}$$

When **N** is conserved, global gauge invariance entails that the transformation

$$\Psi(\mathbf{r}) \to \Psi(\mathbf{r})e^{i\gamma}, \tag{9.7-5}$$

with arbitrary but constant phase, leaves the system invariant. The average of the field operator $\langle\Psi(\mathbf{r})\rangle$ is then zero, since a phase γ occurs as often as a phase $(\gamma + \pi)$, so that the corresponding values of Ψ cancel each other. Below T_c there is still a local phase and we can write

$$\Psi(\mathbf{r}) = Ae^{i\gamma(\mathbf{r})} \qquad (A > 0). \tag{9.7-6}$$

As we shall see in Section 12.1, the superfluid velocity is given by the gradient $\nabla\gamma(r)$. Since **N** is not conserved, we must resort to the grand-canonical ensemble. Now we look at the consequences of $[\mathbf{N}, \mathcal{H}_{grand}] \neq 0$. Recalling the commutation rules (7.3-18) and (7.3-22), we easily establish

$$[\Psi(\mathbf{r}),\mathbf{N}] = \Psi(\mathbf{r}), \qquad [\Psi^\dagger(\mathbf{r}),\mathbf{N}] = -\Psi^\dagger(\mathbf{r}). \tag{9.7-7}$$

We thus have

$$\begin{aligned}\langle\Psi(\mathbf{r})\rangle &= \mathrm{Tr}\{\rho_{gcan}\Psi(\mathbf{r})\}\\ &= \mathrm{Tr}\{\rho_{gcan}[\Psi(\mathbf{r}),\mathbf{N}]\} = -\mathrm{Tr}\{[\rho_{gcan},\mathbf{N}]\Psi(\mathbf{r})\},\end{aligned} \tag{9.7-8a}$$

$$\langle\Psi^\dagger(\mathbf{r})\rangle = \mathrm{Tr}\{\rho_{gcan}[\mathbf{N},\Psi^\dagger(\mathbf{r})]\} = \mathrm{Tr}\{[\rho_{gcan},\mathbf{N}]\,\Psi^\dagger(\mathbf{r})\}. \tag{9.7-8b}$$

Below the lambda-point the commutators on the rhs are non-zero; *the order parameter, being a linear functional of these commutators, can be taken to be either* $\Psi(\mathbf{r})$ or $\Psi^\dagger(\mathbf{r})$.

These examples have made abundantly clear that broken symmetries play an essential role in the ordering of the condensed phase. The identification of the operators involved in the symmetry breaking leads to identification of the order operator(s).

Finally, a word about *Goldstone excitations*. The condensed state is a "ground state at temperature T"; there are long wavelength fluctuations, causing phonon-like excitations. In the case of magnetism these are spin waves, to be treated in Chapter XI. In superfluids there are sound waves, rotons and vortices. If the condensed system is coupled to an electromagnetic field, "massive photons" may appear (Higgs photons). The Goldstone excitations then provide the longitudinal degree of freedom for this electromagnetic wave. This is exemplified by the Meissner effect in type I superconductors, the dynamical photon mass being inversely proportional to the penetration depth.[21] We shall refrain from treating these details in this text, considering only statistical mechanical aspects.

9.7.2 *The Tricritical Point*

Anyone, who ever studied in related fields, like material science or chemistry – as we did – knows that physical systems show much more complexity than the usual physicist considers; in particular, in phase transitions there are variables other than the temperature which underlie a transition of phase: we think of colloidal suspensions and binary mixtures in which a phase separation occurs at a critical ratio, order-disorder in percolation systems for which a critical cluster size may cause the system to coalesce, etc. While these phenomena may occur at any temperature, usually the temperature remains a critical variable, so that phase transitions may be triggered by both a change in T and of another critical variable in the system, which we shall denote by x. Phase diagrams of such systems are often very complex with

[21] A concise description is found in Jackson's book: J.D. Jackson, "Classical Electrodynamics", 3rd Ed., Wiley, N.Y., 1999, pp. 603-605.

several phases present and various coexistence regions. Systems in which temperature and molar concentration are the pertinent variables occur most often and are amenable to the Landau–Ginzburg approach. Such systems may show a tricritical point (Griffiths), being a point in T, x - space where two distinct transitions of phase meet each other.

The most well known system, much studied by physicists and chemists alike, involves mixtures of ^{3}He and ^{4}He. At low molar ^{3}He concentration, the mixture behaves as a regular substance, showing a continuous, second-order phase transition with T_c depending on the molar content x of ^{3}He, decreasing as x grows; we thus have a line of lambda points. For x increasing to a critical concentration x_c of 0.67, the superfluid transition will no longer appear, but the mixture will undergo a phase separation into a ^{4}He-rich superfluid and a ^{3}He-rich normal fluid; observation shows that this is a first-order transition. The phase diagram is given in Fig. 9-5 below, the various regions being self-evident. The point where the two phase transitions meet is a tricritical point. A quantitative model was developed by Blume, Emery and Griffiths and is referred to as the BEV-model.[22]

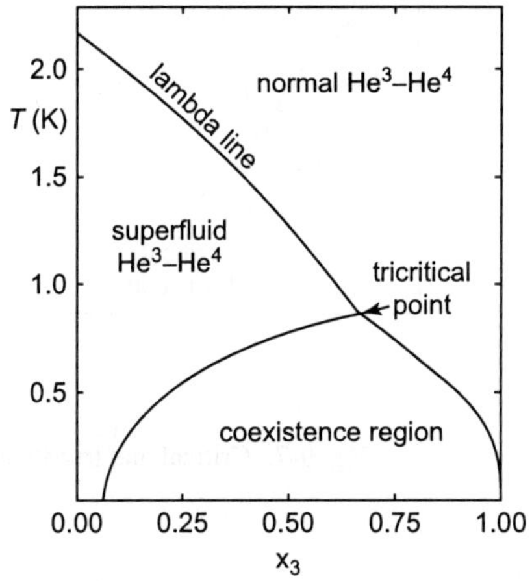

Fig. 9-5. Phase diagram for a binary mixture of ^{4}He and ^{3}He with tricritical point

A number of other substances show a tricritical point. We mention the solid ammonium chloride, whose orientational phase transition changes from first order to second order as a function of pressure; ferrochloride, which undergoes a continuous phase transition at low magnetic fields, giving a line of lambda points, and separates into a mixed phase of coexisting ferromagnetic and antiferromagnetic components at sufficiently high magnetic fields; a variety of many-phase liquid crystals and of nematic crystals show very interesting behaviour; etc. Really, the physics and chemistry of phase transitions is extremely rich!

We shall now endeavour to briefly describe the tricritical behaviour in terms of Landau–Ginzburg theory. Let m still denote the order parameter [$\Psi(\mathbf{r})$ for the case of

---

22 H. Blume, V.J. Emery and R.B. Griffiths, Phys. Rev. **A4**, 1071 (1971).

liquid He] and h its conjugate field; as auxiliary order parameter we take the molar concentration x of one component of the binary mixture [^{3}He], with the associated force being Δ [the difference of the chemical potentials $\varsigma_3 - \varsigma_4$ for the above case]. We then have for the Gibbs free energy per particle

$$\hat{g}(T,m,\Delta) = a_0(T,\Delta) + \tfrac{1}{2}a_2(T,\Delta)m^2 + \tfrac{1}{4}a_4(T,\Delta)m^4 + \tfrac{1}{6}a_6(T,\Delta)m^6 + \dots \quad (9.7\text{-}9)$$

When $a_2(T,\Delta) = 0$ we have a critical transition and, assuming $a_4(T,\Delta) > 0$, this transition is continuous, giving rise to a line of lambda points – a curve in the T-Δ plane. When the temperature is steadily lowered, we may encounter a temperature T_t at which $a_4(T,\Delta)$ becomes zero as well, and then becomes < 0 while $a_6(T,\Delta)$ remains > 0; this possibility was examined in Problem 4.11. We then have a first order phase transition, giving another line in the T-Δ plane. Both lines are sketched in Fig. 9-6. The two lines meet at the tricritical point T_t in a *smooth* manner, as we presently show.

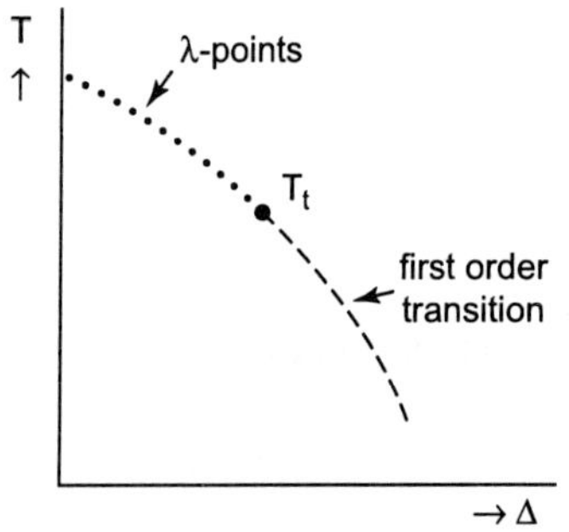

Fig. 9-6. Behaviour near tricritical point.
Dotted line: set of lambda points;
dashed line: first-order transitions.

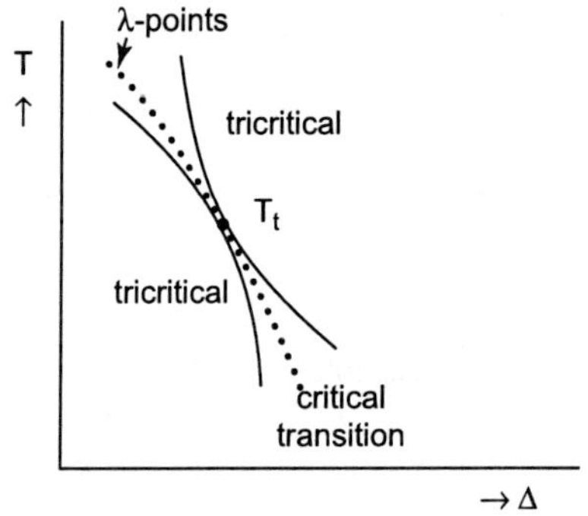

Fig. 9-7. Critical and tricritical regions.

The line $a_2(T,\Delta) = 0$ has a tangent with direction cosines

$$da_2 \to (\partial a_2/\partial T)_\Delta \text{ and } (\partial a_2/\partial\Delta)_T. \quad (9.7\text{-}10)$$

The first-order transition line is given by (see Problem 4.11) $f(T,\Delta) \equiv a_2 - 3a_4^2/16a_6 = 0$. The direction cosines are

$$(\partial a_2/\partial T)_\Delta - (\partial[3a_4^2/16a_6]/\partial T)_\Delta \text{ and } (\partial a_2/\partial\Delta)_T - (\partial[3a_4^2/16a_6]/\partial\Delta)_T. \quad (9.7\text{-}11)$$

The derivatives of the square brackets are clearly proportional to a_4. Thus, when $a_4 \to 0$, the two tangents become equal and the lines smoothly meet at the tricritical point.

We now take a look at the critical exponents. The transition lines have $(\partial\hat{g}/\partial m)_{T,\Delta} = 0$, which has as non-trivial solution away from the tricritical point

$$m^2 = -\frac{a_4}{2a_6} + \sqrt{\frac{a_4^2}{4a_6^2} - \frac{a_2}{a_6}}. \tag{9.7-12}$$

In the ordered state, $\Delta > \Delta_t$, $|a_2|$ grows faster than $|a_4|$; in fact, we expect that $|a_2/a_6| \gg |a_4^2/4a_6^2|$. Then $m^2 \approx \sqrt{a_{20}(T,\Delta)(T_t - T)/a_6}$. This yields the exponent $\beta_t = \frac{1}{4}$. For the susceptibility we have

$$\chi^{-1} \propto d^2\hat{g}/dm^2 = a_2 + 3a_4m^2 + 5a_4m^4 \ldots \tag{9.7-13}$$

In the tricritical region a_2 and a_6m^4 go as $(T_t - T)$; as $a_4 \propto (T_t - T)$, a_4m^2 goes as $(T_t - T)^{3/2}$. Since the slowest terms survive, this entails $\gamma_t = 1$. For the specific heat we note

$$\left(\frac{\partial \hat{g}}{\partial T}\right)_\Delta = \frac{d\hat{g}}{dm^2}\left(\frac{\partial m^2}{\partial T}\right)_\Delta = \frac{1}{2}a_{20}(T_t - T)\left(\frac{\partial m^2}{\partial T}\right)_\Delta + \ldots$$

$$C = -T\left(\frac{\partial^2 \hat{g}}{\partial T^2}\right)_\Delta = \frac{1}{2}Ta_{20}\left(\frac{\partial m^2}{\partial T}\right)_\Delta - \frac{1}{2}Ta_{20}(T_t - T)\left(\frac{\partial^2 m^2}{\partial T^2}\right)_\Delta + \ldots \tag{9.7-14}$$

The first derivative of m^2 goes as $(T_t - T)^{-1/2}$ and the second derivative as $(T_t - T)^{-3/2}$; so all terms go as $(T_t - T)^{-1/2}$. We thus arrive at:

$$\alpha_t = \tfrac{1}{2}, \qquad \beta_t = \tfrac{1}{4}, \qquad \gamma_t = 1. \tag{9.7-15}$$

Rushbrooke's equality is found to hold. We invite the reader to obtain the following coefficients along a critical line, noting that there is no jump in the specific heat in this model:

$$\alpha_u = -1, \qquad \beta_u = \tfrac{1}{2}, \qquad \gamma_u = 2, \tag{9.7-16}$$

where sub 'u' denotes a path along a critical line, see Fig. 9-7. Rushbrooke's relationship is likewise satisfied.

All these considerations have been more or less phenomenological. A microscopic model should be able to furnish details for the critical parameters, *in casu*, T_t and x_t. The BEV model does exactly that. The Hamiltonian of departure is an Ising-like Hamiltonian, except that the spins have eigenvalues 0 and ± 1. We refer to their paper. The result $x_t = \frac{2}{3}$ is in very good agreement with the data for the ^{3,4}He mixture ($x_t = 0.67$); the value $T_t(x_t)/T_c(0) = \frac{1}{3}$ is somewhat lower than the observed value of 0.40. An interesting concise summary of the model is found in Plischke and Bergersen.[23]

[23] M. Plischke and B. Bergersen, 2nd Ed., Op. Cit., p. 100-102.

9.7.3 *The Ginzburg Criterion*

V.L. Ginzburg derived an argument to show when Landau–Ginzburg theory would fail. His main point is that the predictions of L–G theory are reliable only when the fluctuations of the order parameter are much smaller than the mean value. This self-consistency criterion is known as the *Ginzburg criterion*[24]. If the criterion fails, the results of L–G theory are in the best case unreliable and in the worst case scenario, flatly wrong. To quote Kadanoff [25]: "the argument uses L–G theory to predict its own demise."

We recall the form of the fluctuation-correlation function

$$\Gamma(|x|) = \langle \Delta m(|x|)\Delta m(0)\rangle = e^{-|x|/\xi}/b|x|^{d-2}, \tag{9.7-17}$$

cf., Eq. (9.2-39). Let $M = \int d^d x m(|x|)$; then the relative rms fluctuation must be less than 1, cf. Section 3.8.4, i.e., $[\langle \Delta M^2\rangle]^{1/2} \ll \langle M\rangle$. In the same vein, let Ω_d be a volume (d-dimensional sphere) with radius $|x| = \xi$. Then we require, with $\Gamma(|x|) \approx |x|^{2-d}$,

$$\int_{\Omega_d} d^d x |x|^{2-d} = A_d \int_0^{\xi} d|x||x|^{d-1}|x|^{2-d} \ll \int_{\Omega_d} d^d x \langle m^2\rangle = A_d \int_0^{\xi} d|x||x|^{d-1}\langle m^2\rangle,$$

or, with the area factor A_d cancelling out and performing the integrations,

$$\tfrac{1}{2}\xi^2 \ll (1/d)\xi^d \langle m^2\rangle, \quad \to \quad |t|^{-2\nu} \ll |t|^{2\beta-\nu d}. \tag{9.7-18}$$

For $|t| < 1$ this amounts to

$$|t|^{\nu d - 2\nu - 2\beta} \ll 1, \quad \text{or} \quad \nu d - 2\nu - 2\beta \geq 0. \tag{9.7-19}$$

This places a limit on the dimension for which the theory is consistent:

$$d \geq 2 + (2\beta/\nu). \tag{9.7-20}$$

The equal sign defines the so-called 'upper critical dimension'; for lower dimensions the theory may fail due to the extent of fluctuations.

We computed already the coefficient β in the Landau–Ginzburg approach, both for a critical and tricritical process. As to the coherence length ξ, it is reasonable to assume that it scales similarly as the thermal wave length we met at many occasions, $\lambda_{th}^{-1} = [mk_B(T - T_c)/2\pi\hbar^2]^{1/2}$, where we adjusted the result so that $\lambda_{th} = \infty$ at $T = T_c$, in keeping with our observations for critical fluctuations at T_c. So, $\xi \propto \lambda_{th} \sim |t|^{-1/2}$. This entails that in any dimension $\nu = \frac{1}{2}$, in accord with (9.2-31).

---

[24] V.L. Ginzburg, Fiz. Tverd. Tela **2**, 2031 (1960) [Sov. Phys.-Solid State **2**, 1824 (1960)]; also, L.P. Kadanoff et al. Ref. 15, Section IIF.

[25] L.P. Kadanoff "Statistical Physics: Statics, Dynamics, and Renormalization" World Scient. Publ. Co., Singapore and N. J., 2000, p. 242.

Also, in Landau–Ginzburg $\eta = 0$. For the usual critical process in Landau–Ginzburg theory, $\beta = ½$, so that the upper critical dimension is $d = 4$. For a tricritical process $\beta = ¼$, which results in $d = 3$. These dimensions play a role in the theory of renormalization, discussed in the next chapter.

9.7.4 *The Kosterlitz–Thouless Transition*

Lastly, we will discuss an other type of phase transition, which occurs in 2D systems and was first described by Kosterlitz and Thouless.[26] Mathematically, two-dimensional systems are singled out, since the Green's function for the spatial covariance is not a power law but has a logarithmic singularity. Physically, many 2D systems have a finite temperature phase transition, but there is no long-range order below T_c. Examples are found in thin planar magnets, films of liquid ^{4}He, liquid films or monolayers and certain adsorbed gaseous layers on planar crystalline substrates, among others. A number of cases are treated in the cited papers by Kosterlitz and Thouless, in which references to earlier studies are given. In most cases the excitations above the ground state involve besides the usual phonons or, as the case may be, spin waves (magnons), a distribution of vortices *which are closely bound in clusters or pairs* below the critical temperature where they stabilise the condensed state while at higher temperatures they are set free. We will start with planar magnets.

The exchange Hamiltonian is as for the X-Y model (Problem 9.7), but we assume that the spins are 'classical', having a full range of orientations. Thus, we write

$$\mathcal{H} = -J \sum_{(\mathbf{l},\mathbf{l}')} \mathbf{S}_\mathbf{l} \cdot \mathbf{S}_{\mathbf{l}'} = -JS^2 \sum_{(\mathbf{l},\mathbf{l}')} \cos(\phi_{\mathbf{l}'} - \phi_\mathbf{l}), \tag{9.7-21}$$

where the orientation ϕ is measured with respect to some fixed axis. For temperatures not too far below the critical point we can restrict ourselves at first to nearest neighbour coupling, the angular differences being small; hence we expand the cosine,

$$\mathcal{H} = -JS^2 \sum_{\langle \mathbf{l},\mathbf{l}'\rangle} \cos(\phi_{\mathbf{l}'} - \phi_\mathbf{l}) = -\tfrac{1}{2} pNJS^2 + \tfrac{1}{2} JS^2 \sum_{\langle \mathbf{l},\mathbf{l}'\rangle} (\phi_{\mathbf{l}'} - \phi_\mathbf{l})^2 , \tag{9.7-22}$$

where p is the coordination number for the site. In the absence of any vortices the spin-spin correlation function is the straightforward average

$$\langle \mathbf{S}_\mathbf{l} \cdot \mathbf{S}_{\mathbf{l}'} \rangle = S^2 \langle \exp\{i(\phi_{\mathbf{l}'} - \phi_\mathbf{l})\}\rangle \to S^2 \langle \exp\{i[\phi(\mathbf{r}') - \phi(\mathbf{r})]\}\rangle , \tag{9.7-23}$$

where in the last transition we ignored the discrete lattice aspects. With a similar approximation for the Hamiltonian we can write for (9.7-22)

$$\mathcal{H} = \mathcal{E}_0 I + \tfrac{1}{2} JS^2 \sum_{\langle \mathbf{l},\mathbf{l}'\rangle} (\phi_{\mathbf{l}'} - \phi_\mathbf{l})^2 \approx \mathcal{E}_0 I + \tfrac{1}{2} JS^2 \int d^2 r \nabla\phi(\mathbf{r}) \cdot \nabla\phi(\mathbf{r}) , \tag{9.7-24}$$

---

[26] J.M. Kosterlitz and D.J. Thouless, J. Phys. C **5**, L 123 (1972); ibid. J. Phys. C **6**, 1181 (1973).

where $\mathscr{E}_0$ is the ground-state energy. For the present case the angle $\phi(\mathbf{r})$ with respect to a given axis is the order parameter in the system, whose (non-equilibrium) free energy in the Landau–Ginzburg form (9.2-12) will have a term $\frac{1}{2}JS^2\int d^2r\nabla\phi\cdot\nabla\phi$, indicating that $b\to JS^2$. Taking a conditional average over an ensemble with fixed $\phi(\mathbf{r})$, we find [cf. Eq. (9.2-37)] that $\langle\Delta\phi(\mathbf{r}')\Delta\phi(\mathbf{r})\rangle_{\phi(\mathbf{r})}=-[k_BT/JS^2]\,g(\mathbf{r}'-\mathbf{r})$, where $g(\mathbf{r}'-\mathbf{r})$ is the 2D Laplacian Green's function, so that

$$\langle\Delta\phi(\mathbf{r}')\Delta\phi(\mathbf{r})\rangle_{\phi(\mathbf{r})}=-[k_BT/2\pi JS^2]\ln[|\mathbf{r}-\mathbf{r}'|/r_0], \qquad (9.7\text{-}25)$$

where r_0 is a suitable cut-off parameter of the order of the spacing. Taking next an average over all $\phi(\mathbf{r}')$ with $R=|\mathbf{r}'-\mathbf{r}|$, we find for the spin-spin correlation function

$$\begin{aligned}\langle\mathbf{S_l}\cdot\mathbf{S_{l'}}\rangle &= S^2\{1-\tfrac{1}{2}S^2\langle[\phi(\mathbf{r}')-\phi(\mathbf{r})]^2\rangle\}\\ &\simeq S^2[1+\langle\langle\Delta\phi(\mathbf{r}')\Delta\phi(\mathbf{r})\rangle_{\phi(\mathbf{r})}\rangle]=S^2[1-(k_BT/2\pi JS^2)\ln(R/r_0)], \qquad (9.7\text{-}26)\end{aligned}$$

with a more detailed argument being given by Berenzinskii.[27] Now we might assume that with growing order the nearest neighbour coupling would spread, so that (9.7-26) engenders a fuller spin-spin ordering, thus giving for (9.7-23) and any pair ($\mathbf{r}'$,$\mathbf{r}$), being M neighbours apart,

$$\begin{aligned}\langle\mathbf{S_l}\cdot\mathbf{S_{l'}}\rangle &= S^2\langle\exp\{i\textstyle\sum_{k=1}^{M}[\phi(\mathbf{r}^k_{neigh})-\phi(\mathbf{r}_k)]\}\rangle\\ &\simeq S^2[1-M^{-1}(k_BT/2\pi JS^2)\ln(R/r_0)]^M\\ &\sim S^2\exp\{-(k_BT/2\pi JS^2)\ln(R/r_0)\}=S^2(R/r_0)^{-k_BT/2\pi JS^2}. \qquad (9.7\text{-}27)\end{aligned}$$

This indicates an algebraic roll-off with a critical exponent $\eta=\eta(T)$; clearly, we do *not* obtain a regular ordered state below the critical temperature T_c with fixed η; this is entirely due to the logarithmic form of the 2D Green's function.[28]

The essence of the theory is the presence of vortices. The result $\langle[\phi(\mathbf{r}')-\phi(\mathbf{r})]^2\rangle \propto \ln|\mathbf{r}'-\mathbf{r}|$ remains valid, but it goes to show that for large separations the spins will have gone through several revolutions relative to each other. The vortices form a pattern of gradually rotating spins when following a contour about their centre. It is now useful to split the phase changes $\delta\phi$ in a spin-wave part $\delta\psi$ and a vortice part $\delta\varphi$, the two being uncorrelated whereby along any closed contour about the vortice centre

$$\textstyle\sum\!\!\!\!\!\bigcirc\ \delta\psi=\oint\nabla\psi\cdot d\ell=0, \quad \text{while } \oint\nabla\varphi\cdot d\ell=2\pi q, \quad q=\pm1,\pm2,\ldots \qquad (9.7\text{-}28)$$

The vortex distribution is given as $\rho(\mathbf{r})=\sum_q q\,\delta(\mathbf{r}-\mathbf{r}_q)$, where $\mathbf{r}_q$ is the vortex centre. Clearly $\varphi(\mathbf{r})$ behaves like a scalar magnetic potential with a singular charge density stemming from a mathematical point dipole.[29] It thus satisfies Poisson's equation,

---

[27] V.L. Berenzinskii, Sov. Phys. JETP **32**, 493 (1970).

[28] Note that in three dimensions we would have $\langle\mathbf{S}_i\cdot\mathbf{S}_j\rangle/S^2=\exp[-(k_BT/4\pi JS^2R^{1+\eta})]\to1$ if $R\to\infty$.

[29] Cf. J.R. Reitz and F.J. Milford, "Foundations of Electromagn. Theory", Addison–Wesley 1960, § 8-8.

$$\nabla^2\varphi = 2\pi\sum_q q\,\delta(\mathbf{r}-\mathbf{r}_q). \tag{9.7-29}$$

The energy of the spin waves is the same as before, cf. (9.7-24), with now $\phi \to \psi$; see also Chapter XI, Section 11.5. They still form a low temperature 'critical phase', but without long-range order. We must now show that the vortices lead to a stable condensed state below T_c and a fully disordered state above T_c. Let at first there be only one vortex. Its energy is most easily found directly from (9.7-28); we have

$$\mathscr{E}_{\text{vortex}} = \tfrac{1}{2}JS^2\int d^2r\,\nabla\varphi(\mathbf{r})\cdot\nabla\varphi(\mathbf{r}) = \pi q^2 JS^2\int_{r_0}^{R} dr/r = \pi q^2 JS^2 \ln(R/r_0). \tag{9.7-30}$$

Noting the two possible rotations, the vortex entropy is found to be $k_B \ln(R/r_0)^2$, so that the change in free energy is $\Delta\Phi = (\pi q^2 JS^2 - 2k_B T)\ln(R/r_0)$, which is positive for low temperatures; clearly, single vortices do not occur in the ordered state. Better fare pairs of vortices; their energy is obtained from (9.7-29). We easily find

$$\begin{aligned}\mathscr{E}_{\text{vortex pair}}(\mathbf{r},\mathbf{r}') &= -4\pi^2 q_1 q_2 JS^2 \iint d^2r_1\, d^2r_2\, \delta(\mathbf{r}-\mathbf{r}_1)\delta(\mathbf{r}'-\mathbf{r}_2) g(\mathbf{r}_1-\mathbf{r}_2)\\ &= -2\pi q_1 q_2 JS^2 \ln[|\mathbf{r}-\mathbf{r}'|/r_0].\end{aligned} \tag{9.7-31}$$

This results in a decrease in free energy, thereby stabilising the condensed state. The total vorticity of this state is zero. At temperatures above T_c the pairs unwind and free vortices exist until all order vanishes. For T_c one obtains $\pi J/k_B T_c \approx 1.12$.

The final form for the vortex-pair energy is entirely analogous to the Coulomb energy of a 2D gas, which is used as a primary model in the Kosterlitz–Thouless papers. The partition function for a planar plasma with two types of opposite charges can be computed similarly as in Section 4.10. Pairs of opposite charges form a bound state below T_c, causing the conductivity σ to be zero. Above T_c the pairs dissociate and σ is finite. The dielectric constant has a singularity as in mean field theory.

The case of a planar liquid helium film is quite similar to the magnetic model considered above. In this case there are *material* vortices of the superfluid, discussed in Section 12.1. The gradient $\nabla\varphi$ is given by the velocity of the film relative to the substrate, cf. Eqs. (12.1-13) and (12.1-14). The critical temperature is given by the upper limit $k_B T_c \le \pi\hbar^2\rho/2m$, where ρ is the planar density and m the effective particle mass. Below T_c these vortices are bound in clusters of zero vorticity.

Finally, we mention the case of a 2D liquid film, in which the transition to a crystalline lattice often has no long range order of the usual type, this being destroyed by long wavelength phonon modes. The displacements can be written as the sum $\mathbf{u}(\mathbf{r}) = \mathbf{v}(\mathbf{r}) + \mathbf{w}(\mathbf{r})$, where $\mathbf{v}(\mathbf{r})$ represents lattice vibrations and $\mathbf{w}(\mathbf{r})$ the effect of dislocations. The latter satisfy $\oint \mathbf{w}(\mathbf{r})\,d\ell = \mathbf{b}$, where $\mathbf{b}$ is the Burger's vector. The condensed state is characterized by the formation of dislocation pairs with oppositely directed Burger's vectors. Although this behaviour has been observed in some liquid films (as apparent from non-sharp Bragg diffraction peaks indicative of limited ordering), most melting is a regular first-order transition.

9.8 Problems to Chapter IX

9.1 Some liquid crystals undergo a phase transition from isotropic to nematic. The Maier–Saupe model can be 'translated' into a Landau–Ginzburg description that has a cubic term in the expansion for the free energy, i.e.,

$$\Phi(T,m) = a_0 + \tfrac{1}{2}a_2 m^2 + \tfrac{1}{3}a_3 m^3 + \tfrac{1}{4}a_4 m^4 + \ldots \qquad (1)$$

For stability we assume that $a_4 > 0$, while we take a_3 to be < 0. For a sketch of Φ vs. m see Fig.9.8. Find the order parameter m_0 and obtain $a_2(T_c)$. Show that there is a first-order phase transition; find the jump $\Delta\Phi = T_c \Delta S$ at the transition.

9.2 In d dimensions, let the Landau–Ginzburg expansion be

$$\Phi[T,m(r)] = \int d^d\rho\{a_0 + \tfrac{1}{2}a_2[m(\rho)]^2 - \tfrac{1}{2}b(d^2/d\rho^2)m(\rho)\}, \qquad (2)$$

where $\rho = |x|$. Use equipartition and the Fourier transform in d dimensions to obtain a result like (9.2-25). Find the inverse Fourier transform for any $d \neq 2$ and show that

$$\Gamma(\rho) = k_B T b^{-1} \rho^{2-d} K_p(\rho/\xi), \qquad p = (d-2)/2, \qquad (3)$$

where K_p is the modified Bessel function of order p. Use the asymptotic expansion to affirm (9.2-39).

9.3 Derive Griffiths' thermodynamic inequalities (9.4-13) and (9.4-14).
N.B. For (9.4-13) employ the 'magnetic' Helmholtz free energy $F(T,M)$, which is concave with respect to variations in T and convex with respect to variations in M, cf. Problem 3.3. So the point $T = T_c$, $M = 0$ is a saddle point. Define new functions

$$F^*(T,M) \equiv [F(T,M) - F_c] + (T - T_c)S_c, \qquad S^*(T,M) \equiv S(T,M) - S_c. \qquad (4)$$

Here, $F_c = F(T_c,M)$, etc. Note that $S^*(T,M) = -(\partial F^*/\partial T)_M$ and that F^* has the same type of saddle point structure as F. In Fig. 9-9 we give *a 3D sketch* of F^* (z-axis) and the *T-M* saddle-point plane. Choose a fixed value T_1 on the saddle point surface and consider $F^*(T, M_1)$. Draw the tangent to this curve at a point T_1 and notice that all points on the surface fall below this tangent line, since F^* for constant M is concave. Accordingly,

$$F^*(T_c,M_1) \le F^*(T_1,M_1) - (T_c - T_1)S^*(T_1,M_1). \qquad (5)$$

Show that *a fortiori*

$$F^*(T_c,M_1) \le -(T_c - T_1)S^*(T_1,0). \qquad (6)$$

Now $F^* \sim M_1^{\delta+1} \sim (-t)^{\beta(\delta+1)}$, with a similar statement applying to the rhs of (6).
(a) Obtain the result $\beta(\delta+1) \geq 2-\alpha$ and fill in all the details.
(b) Give a similar proof for the relationship $\gamma \geq \beta(\delta-1)$.

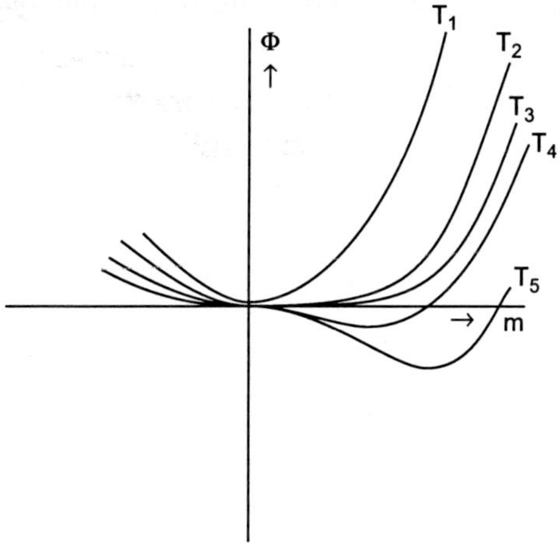

Fig. 9-8. Free energy vs. m when there is a cubic term.

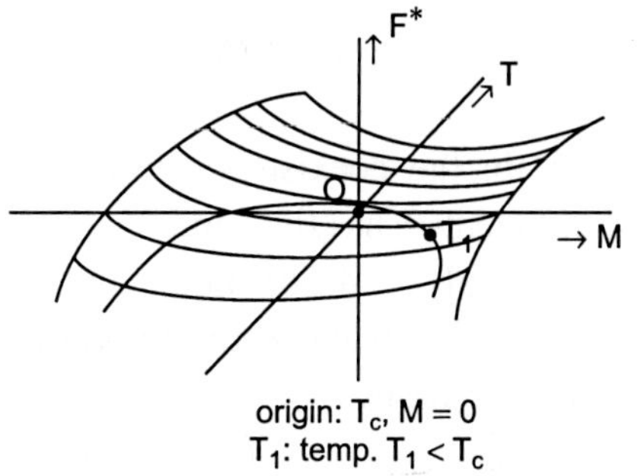

Fig. 9-9. Saddle-point surface of F^* for $T = T_c$ and $M = 0$.

9.4 (a) The easiest form of mean field theory (*mft*) was given in subsection 4.8.1. From the developments there conclude that for *mft* $\beta = \frac{1}{2}, \gamma = 1$ and $\alpha = 0$. To obtain δ, set $m \to \overline{m}\rho_B$, $(\rho_B / k_B T)H = \overline{h}$ and express $\overline{h} + \overline{h}_m$ in terms of $\overline{m}$ by inverting (4.8-5) and expanding the logarithm. Show that

$$\overline{h} = \overline{m}[(1-(T_c/T)] + \tfrac{1}{3}\overline{m}^3 + \ldots \tag{7}$$

and conclude that for the critical isotherm $\delta = 3$. [Note: $\overline{h}_c = \overline{m}_c = 0$.]
(b) Based on subsection 4.8.2, obtain similar results for $\alpha \ldots \delta$ from the Bragg–Williams approximation by expanding the free energy for the general case that $H \neq 0$.

9.5 Consider a classical gas which obeys van der Waals' equation. In reduced variables the equation is given in (4.4-25). Set $t = T'-1$, $\omega = V'-1$, $\pi = P'-1$. Solve for π:

$$\pi = \frac{8t + 16t\omega + 8t\omega^2 - 3\omega^3}{2 + 7\omega + 8\omega^2 + 3\omega^3}. \tag{8}$$

(a) Define the critical isotherm and show that $\delta = 3$;
(b) Find the isothermal compressibility and show that $\gamma = 1$.

9.6 Van der Waals gas, continued. Near $t = 0$ we have

$$\pi = 4t - 6t\omega + 9t\omega^2 - \tfrac{3}{2}\omega^3 + \dots \tag{9}$$

(a) In the coexistence region, write $\omega_\ell = -\bar{\omega}_\ell$ and $\omega_g = \bar{\omega}_g$ where the barred quantities are positive. The corresponding density parameter is $\bar{\rho}_\ell + \bar{\rho}_g$. Require that $\pi(t, -\bar{\omega}_\ell) = \pi(t, \bar{\omega}_g)$ and that the Maxwell construction is applied,

$$\int_{-\bar{\omega}_\ell}^{\bar{\omega}_g} (\omega + 1)\, d\pi = 0, \qquad d\pi = \underbrace{(\partial\pi/\partial t)_\omega dt}_{0} + (\partial\pi/\partial\omega)_t\, d\omega. \tag{10}$$

Obtain the coexistence law near T_c [see (9.3-2)] and show that the critical exponent β has the value ½.

(b) Apply (9.4-26) along an isochor near $t = 0$. Show that $(\partial\tilde{V}_{g,\ell}/\partial T)_{coex} = \pm|t|^{-\frac{1}{2}}$, evaluate the pressure differentials and obtain the jump in specific heat per mole,

$$C_{V_c}(T_c - 0) - C_{V_c}(T_c + 0) = (9/2)R. \tag{11}$$

(c) What is the value of the exponent α, consistent with Rushbrooke's relationship? Explain.

9.7 *The X-Y Model of Ferromagnetism.* In the X-Y model it is assumed that only x and y spin-components couple, i.e., the Hamiltonian has nearest neighbour interaction

$$\mathcal{H}_{XY} = -(J_\perp/2\hbar^2)\sum_{\mathbf{l},\mathbf{a}}(S_{\mathbf{l}}^+ S_{\mathbf{l+a}}^- + S_{\mathbf{l}}^- S_{\mathbf{l+a}}^+). \tag{12}$$

The model is worked out in Mattis' book[30] and in other texts.[31] The problem is closely related to 'Bethe's ansatz' for the Heisenberg linear chain.[32]

From the Heisenberg Hamiltonian, obtain the Hamiltonian (12) assuming that there is no exchange integral J_z. Set $\beta J_\perp/4\hbar^2 = K_0$. As basis states choose the eigenstates of S_z. The linear chain with s = ½ can be solved exactly. Assume that the ground state with $\mathcal{E} = 0$ (zero spin deviations) has all spins down: $|0\rangle = |-\tfrac{1}{2}\rangle_1 \dots |-\tfrac{1}{2}\rangle_N$. From translational invariance, one may guess that the 'one-particle states' with (N – 1) spins down and one up are given by

$$|\psi_q\rangle = (1/\sqrt{N})\sum_\ell e^{iq\ell} S_\ell^+ |0\rangle. \tag{13}$$

(a) Show that this is an ES with EV $\mathcal{E}_q = (4K_0/\beta)\cos q$; ℓ denotes a lattice site.

(b) Show that a state with any number $M < N$ spins up is given by the determinantal state

---

[30] D.C. Mattis, "The Theory of Magnetism", Harper & Row, NY and London, 1965.

[31] R.K. Pathria, "Statistical Mechanics", 2nd Ed. Elsevier, Amsterdam 1996, p. 316 and 404.

[32] H. Bethe, Z. für Physik, **71**, 205 (1931).

$$|\psi_{q_1q_2\ldots q_M}\rangle = C\sum_{\ell_1\ldots\ell_M} F^{\ell_1\ell_2\ldots\ell_M}_{q_1q_2\ldots q_M} S^+_{\ell_1}S^+_{\ell_2}\ldots S^+_{\ell_M}\,|0\rangle, \tag{14}$$

where C is a normalization constant and where F is the determinant

$$F^{\ell_1\ell_2\ldots\ell_M}_{q_1q_2\ldots q_M} = (-1)^p \begin{pmatrix} e^{iq_1\ell_1} & \ldots & e^{iq_M\ell_1} \\ \vdots & \ddots & \vdots \\ e^{iq_1\ell_M} & \ldots & e^{iq_M\ell_M} \end{pmatrix}, \tag{15}$$

with p being the parity of the subset of the N ordered spin sites and with each q satisfying anti-periodic b.c., $q = [\pi(2n+1)/N]$. Note that all q's must be different.

(c) Show that the energy EV is given by $\mathscr{E}_{q_1q_2\ldots} = (4K_0/\beta)\Sigma_i \cos q_i$.

N.B. The ground state, in terms of the wave numbers q, means that all states in the ranges

$$-\pi \le q \le -\tfrac{1}{2}\pi \qquad \tfrac{1}{2}\pi \le q \le \pi \tag{16}$$

are occupied, while all other states are empty. The situation is pictured in Fig. 9-10. The XY model is *naturally* described by a fermion-state analogue. [This procedure will fully be utilized in the 2D Ising description of Chapter XI.]

9.8 The partition function and free energy for the XY model, with magnetic energy $-g\rho_B H\Sigma_\ell S^z_\ell$ added, has been computed by Katsura[33]

$$(F/Nk_BT) = -\frac{1}{\pi}\int_0^\pi dt \ln[2\cosh 2(h - K_0\cos t)], \tag{17}$$

where $h = \beta g \rho_B H/2$. Obtain expressions for the mean energy, the specific heat $(h=0)$, the magnetization and the susceptibility. [For plots of these quantities, obtained by numerical integration, see Katsura, loc. cit.]

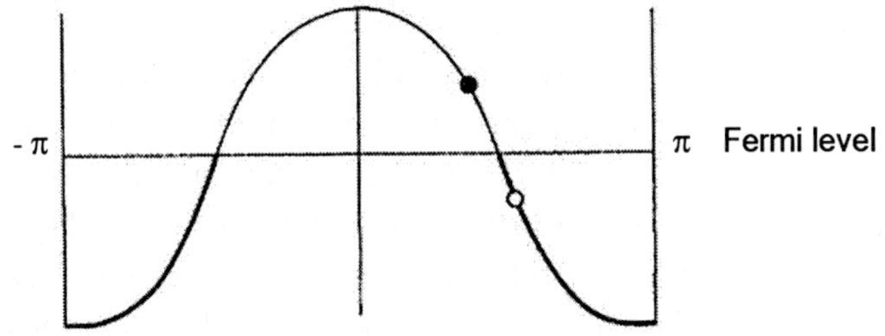

Fig. 9-10. X-Y model, showing occupied and unoccupied states. Open circle and dot indicate a 'quasi-hole' and 'quasi-particle', created in an elementary excitation. [After Mattis, Op. Cit.. with permission.]

[33] S. Katsura, Phys. Rev. **127**, 1508 (1962).

Chapter X

Renormalization: Theory and Examples

10.1 Objective of Renormalization

Leo Kadanoff's scaling with block spins has indicated that the microscopic variables in the Hamiltonian are not strictly necessary in order to obtain the Gibbs' free energy of the system. In fact, we can perform a continued coarse-graining procedure, relinquishing many degrees of freedom, until the linear dimension of the n-th block becomes comparable to the correlation length ξ; also, the closer we are to the critical point, the more rescaling operations we can perform, since $\xi \to \infty$ when $T \to T_c$. This qualitative idea was quantitatively worked out by Ken Wilson in two milestone papers in the Physical Review of 1971.[1] Successive transformations give rise to the *renormalization group* (R), whereby the coupling constants, denoted by the vector $\mathbf{K}$, continually change, with $\mathbf{K}^{(n)}$ approaching a 'fixed point' $\mathbf{K}^*$ at the critical temperature T_c. While no Hamiltonian is diagonalized, the infinitesimal transformations of the group near the fixed point reveal the proper scaling form for the Gibbs' free energy and thereby allow us to obtain the critical exponents for a given universality class. This, indeed, is a very great achievement which, moreover, lends credence to the earlier *ad hoc* introduced scaling functions of Widom and Kadanoff. However, as we will see, the procedure is not an easy exercise, since a lot of formalism as well as computational work has to be absorbed. While many of the early computations were carried out for known systems – such as the 2D Ising model – presently much more refined computations for new systems are as reliable as very accurate series-expansion results.

In this chapter we shall first exemplify the renormalization idea with the linear chain Ising model, which has no phase transition (except, so to speak, at absolute zero); the treatment will indeed reveal two fixed points, viz. at $T = 0$ and $T = \infty$. Next, we shall present the normalization group in its generality, following largely the treatment by Th. Niemeijer and J. van Leeuwen.[2] For other reviews see Wilson and Kogut[3] and Michael Fisher.[4] As application we shall obtain the critical exponents for

---

[1] K.G. Wilson, Phys. Rev. **B4**, 3174 and 3184 (1971).

[2] Th. Niemeijer and J.M.J. van Leeuwen in "Phase Transitions and Critical Phenomena" Vol. 6 (C. Domb and M.S. Green, Eds.), Academic Press, N.Y. !976, pp. 425-505.

[3] K.G. Wilson and J. Kogut, Phys. Rep. **12**, 75–199 (1974).

[4] M.E. Fisher, Rev. Mod. Phys. **4**, 597 (1974).

a 2D triangular spin system, also originally presented by Niemeijer and van Leeuwen.[5] This makes up the first part of this chapter.

In the second part we will present the essential ingredients of some quite ambitious applications of the renormalization group, involving the 'classical spin' Gaussian model, the S^4 model and the 'epsilon-expansion'.[3] This becomes rather involved since, at this moment, we shall refrain from using diagrammatic techniques (see Chapter XII, Sections 12.9-12.13) although these techniques are presently widely used.[6] For more elegant presentations using quantum field theory we refer to Brezin et al.[7] We also note that a very readable and more detailed chapter on this topic can be found in Plischke and Bergersen.[8]

10.2 The Linear Spin-Chain Revisited

The 1D Ising spin Hamiltonian with nearest neighbour interaction only was given in (4.3-21), which we rewrite as

$$-\beta\mathcal{H} = K\sum_{i=1}^{N}\sigma_i\sigma_{i+1} + h\sum_{i=1}^{N}\sigma_i, \tag{10.2-1}$$

where $K = \beta J_0$ and $h = \beta\rho_B H$; further, σ_i is the spin operator in the z-direction represented, if so desired, by the Pauli spin matrix $\boldsymbol{\sigma}^z$. The eigenvalues are denoted by $\sigma_i' = \pm 1$, satisfying $\sigma_i|\sigma_i\rangle = \sigma_i'|\sigma_i\rangle$. The canonical partition function, after applying periodic b.c., was previously found to be [subsection 4.3.2]:

$$\begin{aligned}\mathcal{Z} &= \sum_{\{\sigma_i'=\pm1\}} \langle\{\sigma\}|\left\{\exp\sum_{i=1}^{N}\left[K\sigma_i\sigma_{i+1} + \tfrac{1}{2}h(\sigma_i+\sigma_{i+1})\right]\right\}|\{\sigma\}\rangle \\ &= \sum_{\{\sigma_i'=\pm1\}} \langle\{\sigma\}|\prod_{i=1}^{N}\left\{\exp\left[K\sigma_i\sigma_{i+1} + \tfrac{1}{2}h(\sigma_i+\sigma_{i+1})\right]\right\}|\{\sigma\}\rangle \\ &= \sum_{\{\sigma_i'=\pm1\}}\prod_{i=1}^{N}\left[\langle\sigma_i|\left\{\exp\left[K\sigma_i\sigma_{i+1} + \tfrac{1}{2}h(\sigma_i+\sigma_{i+1})\right]\right\}|\sigma_{i+1}\rangle\right].\end{aligned} \tag{10.2-2}$$

In the last line we have put the bra $\langle\{\sigma\}| = \Pi_i\langle\sigma_i|$ [and the ket $|\{\sigma\}\rangle$] inside the product, distributing the states over the appropriate operators $\{\ldots\}$, noting that they are orthogonal to all others. Each factor under the product sign is a matrix of identical

---

[5] Th. Niemeijer and J.M.J. van Leeuwen, Phys. Rev. Lett. **31**, 1411 (1973); also Physica **71**, 17 (1974).

[6] Sheng-Keng Ma, "Modern Theory of Critical Phenomena", Benjamin, NY. and Amsterdam, (1976).

[7] E. Brezin, J.D. Le Guillou and J. Zinn-Justin, in "Phase Transitions and Critical Phenomena" Vol. 6 (C. Domb and M.S. Green, Eds.), Academic Press, NY. !976.

[8] M. Plischke and B. Bergersen, "Equilibrium Statistical Physics",2rd Ed., World Scientific Publ. Co. Singapore, 1994, Chapter 6. [Also, 3rd Ed., World Scientific Publ. Co. Singapore 2006, Chapter 7.]

structure, as we noted in subsection 4.3.2; the internal sums denote matrix multiplication. With only the bra $\langle\sigma_1|$ and the ket $|\sigma_{N+1}\rangle=|\sigma_1\rangle$ remaining, the result is $\mathrm{Tr}[\ldots]^N$.

Presently, our idea is to thin out degrees of freedom by arranging the spins in blocks of two, see Fig. 10-1. This is very easily done by writing

$$\mathcal{Z}=\mathrm{Tr}\Big[\langle\sigma_i|\exp\{K\sigma_i\sigma_{i+1}+\tfrac{1}{2}h(\sigma_i+\sigma_{i+1})\}|\sigma_{i+1}\rangle\Big]^N$$
$$=\mathrm{Tr}\Big[\Big(\langle\sigma_i|\exp\{K\sigma_i\sigma_{i+1}+\tfrac{1}{2}h(\sigma_i+\sigma_{i+1})\}|\sigma_{i+1}\rangle\Big)^2\Big]^{N/2}. \qquad (10.2\text{-}3)$$

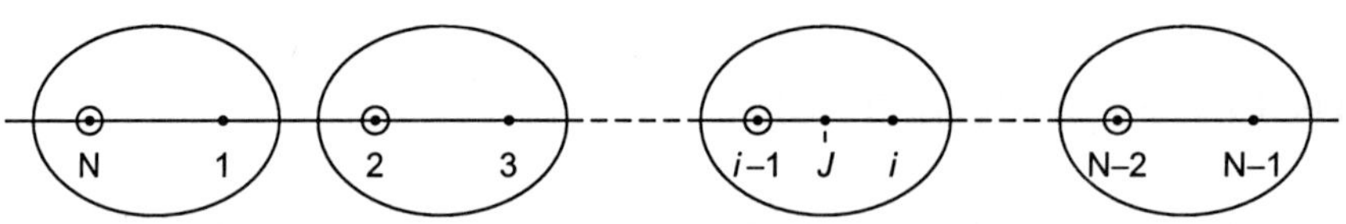

Fig. 10-1. N spins are grouped in blocks of two. The new leaders have been circled. We assumed that N is even and i is odd.

Now we work our way backward to obtain the analogue of the result of departure for the partition function. Let the even sites be the leaders in the spin blocks; they are circled in Fig. 10-1. Summing over the connected spins of block J we have

$$f_J(\sigma_{i-1},\sigma_i)\equiv\langle\sigma_{i-1}|\sum_{\sigma_i'=\pm1}e^{K\sigma_{i-1}'\sigma_i'+(h/2)(\sigma_{i-1}'+\sigma_i')}e^{K\sigma_i'\sigma_{i+1}'+(h/2)(\sigma_i'+\sigma_{i+1}')}|\sigma_{i+1}\rangle$$
$$=\langle\sigma_{i-1}|e^{(h/2)(\sigma_{i-1}'+\sigma_{i+1}')}\sum_{\sigma_i'=\pm1}e^{K\sigma_i'(\sigma_{i-1}'+\sigma_{i+1}')+h\sigma_i'}|\sigma_{i+1}\rangle$$
$$=\langle\sigma_{i-1}|e^{(h/2)(\sigma_{i-1}+\sigma_{i+1})}2\cosh[K(\sigma_{i-1}+\sigma_{i+1})+h]|\sigma_{i+1}\rangle, \qquad (10.2\text{-}4)$$

where in the last line we rewrote the σ_{i-1} and σ_{i+1} as operators.

We now demand that the operator in (10.2-4) be written as a block-spin interaction operator

$$2e^{h(\sigma_J+\sigma_{J+1})/2}\cosh[K(\sigma_J+\sigma_{J+1})+h]=\exp\{-2g^0+K'\sigma_J\sigma_{J+1}+\tfrac{1}{2}h'(\sigma_J+\sigma_{J+1})\}, \qquad (10.2\text{-}5)$$

where we wrote $\sigma_{i-1}\equiv\sigma_J$ and $\sigma_{i+1}\equiv\sigma_{J+1}$. These *even*-numbered operators now interact with other even spins only, as expressed above.[9] To obtain the new interaction constant and the new magnetic field, use $\cosh(A+B)=\cosh A\cosh B+\sinh A\sinh B$ and expand the cosh and the sinh in a Taylor series, noting that the

[9] Of course, neither the physical picture, nor the mathematical structure, changes if the block spins are thought of as acting in the middle of each block.

squared matrix $\sigma^2 = 1$, so that

$$\cosh K\sigma = \sum_{n=0}^{\infty} \frac{K^{2n}\sigma^{2n}}{(2n)!} = \sum_{n=0}^{\infty} \frac{K^{2n}}{(2n)!} = \cosh K, \tag{10.2-6}$$

and likewise

$$\sinh K\sigma = \sigma \sinh K. \tag{10.2-7}$$

With some algebra one obtains:

$$\begin{aligned} K' &= \tfrac{1}{4}\ln\{\cosh(2K+h)\cosh(2K-h)/\cosh^2 h\}, \\ h' &= h + \tfrac{1}{2}\ln\{\cosh(2K+h)/\cosh(2K-h)\}, \\ g^0 &= -\tfrac{1}{2}\ln 2 - \tfrac{1}{8}\ln\{\cosh(2K+h)\cosh(2K-h)\cosh^2 h\}. \end{aligned} \tag{10.2-8}$$

The results (10.2-4) and (10.2-5) are substituted into (10.2-3); this yields

$$\mathcal{Z} = [\exp(-Ng^0)] \sum_{\{\sigma_J'\}} \langle\{\sigma\}| \left\{ \exp \sum_{J=1}^{N/2} \left[K'\sigma_J\sigma_{J+1} + \tfrac{1}{2}h'(\sigma_J + \sigma_{J+1}) \right] \right\} |\{\sigma\}\rangle. \tag{10.2-9}$$

For the 'magnetic Gibbs free energy' per spin in units k_BT we thus find,

$$\hat{g}(K,h) = -\ln\left[\mathcal{Z}(K,h)\right]^{1/N} = g^0(K,h) - \ln\left[\mathcal{Z}(K',h')\right]^{2/N}. \tag{10.2-10}$$

This can of course be continued, the result being

$$\hat{g}(K,h) = \sum_{n=0}^{\infty} \left(\frac{1}{2}\right)^n g^{(n)}(K_n, h_n). \tag{10.2-11}$$

Since the Gibbs free energy is finite, the sum can be expected to converge. We also note that we defined $K = J_0/k_BT$; hence the form $\hat{g}(K,h)$ implies the form $\hat{g}(T,h)$.

The salient feature of this approach is that the functional connections $K_n \to K_{n+1}$ and $h_n \to h_{n+1}$ are the same for all n. Making K and h a two-component vector $\mathbf{K}$, we write

$$\mathbf{K}^{(n+1)} = R[\mathbf{K}^{(n)}], \tag{10.2-12}$$

where R denotes the renormalization group. The group property is expressed by: if R_1 and R_2 are transformations of $\mathbf{K}$, then R_1R_2 is also a transformation of $\mathbf{K}$; this is easily verified for the example under discussion. However, the group has no inverse (spins cannot be 'unblocked' *ad infinitum*), so the designation 'group' is a misnomer. We may now ask: if $n \to \infty$ and $\mathbf{K} \to \mathbf{K}^*$, under what conditions do we have a fixed point, $\mathbf{K} = \mathbf{K}^*$? For the present case, let at first $h = 0$. We then have from (10.2-8):

$$K_{n+1} = \tfrac{1}{2}\ln\cosh 2K_n \le K_n. \tag{10.2-13}$$

Thus in a 'flow-diagram' K decreases in each step of iteration. There are two fixed points, $K = K^*$, viz.

$$K^* = 0 \ (T = \infty) \quad \text{and} \quad K^* = \infty \ \ (T = 0)\,. \tag{10.2-14}$$

The fixed point at T = 0 is a critical point, having infinite correlation length (see below), but it cannot be realised and is unstable. In the other case the system is non-interacting and trivial. For both cases the Hamiltonian is scale-invariant, having $\xi = \infty$ or $\xi = 0$. In all instances we can do the rescaling of Kadanoff's picture until the block-spin separation is of the order of the correlation length. If this stage is reached after n iterations, then the free energy becomes

$$\hat{g}(K,0) = \sum_{k=0}^{n} (\tfrac{1}{2})^k g^{(k)}(K_k,0) - (\tfrac{1}{2})^{n+1} \ln 2\,. \tag{10.2-15}$$

The last term is the negative entropy (in units k_B) of the random spins of the last block.

For finite h some new features arise. From the second line in (10.2-8) we see that

$$h_{n+1} \geq h_n. \tag{10.2-16}$$

The equal sign only holds for K = 0 or for $K = \infty$; so the fixed points apply to *both* components of the earlier introduced vector **K**. But, while K decreases with increasing n, h increases. It is expedient to indicate the flow of **K** on the interaction surface (plane in this case). For the 1D Ising model the flow diagram is pictured in Fig. 10-2.

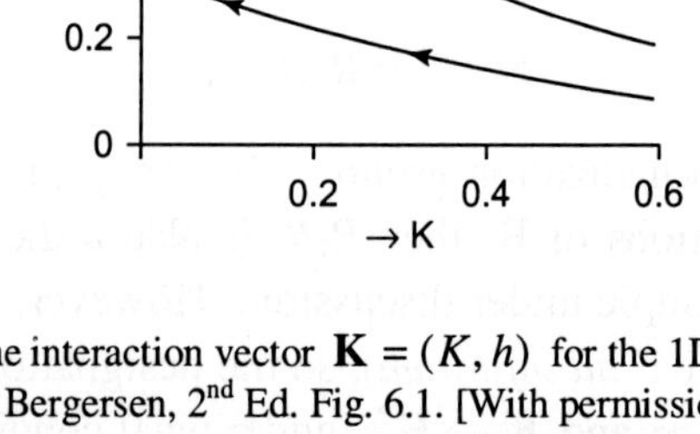

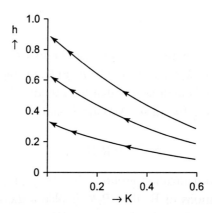

Fig. 10-2. Flow diagram of the interaction vector $\mathbf{K} = (K, h)$ for the 1D Ising model. After Plischke and Bergersen, 2nd Ed. Fig. 6.1. [With permission.]

Finally, we shall obtain an explicit form for the spin-spin correlation function, denoted as $G(i,j) = G(j-i) = \langle \sigma_i \sigma_j \rangle$. The connection with the correlation-fluctuation function is as usual,

$$\Gamma(i,j) = \langle \Delta\sigma_i \Delta\sigma_j \rangle = \langle \sigma_i \sigma_j \rangle - \langle \sigma_i \rangle \langle \sigma_j \rangle = G(i,j) - \langle \sigma \rangle^2. \qquad (10.2\text{-}17)$$

The latter equality supposes that the system is homogeneous. For $T < T_c$ one can expect that $\Gamma \to 0$ if $|i-j| \to \infty$. Then, the magnetization in Bohrmagnetons is given by

$$m_0^2 = \lim_{|i-j| \to \infty} \langle \sigma_i \sigma_j \rangle, \qquad (10.2\text{-}18)$$

a formula we will meet in the Onsager 2D Ising problem. For the 1D chain, $\langle \sigma \rangle = 0$ at any finite temperature. For the correlation function in the absence of a magnetic field we have from the foregoing for an open-ended chain of N spins (no periodic b.c.)[10]

$$\langle \sigma_i \sigma_j \rangle = \mathcal{Z}_N^{-1} \,\mathrm{Tr}\left\{ (\sigma_i \sigma_j) \exp\left[\sum_{\ell=1}^{N-1} K_\ell \sigma_\ell \sigma_{\ell+1} \right] \right\}, \quad \mathcal{Z}_N = 2^N \prod_{\ell=1}^{N-1} \cosh K_\ell \,, \qquad (10.2\text{-}19)$$

where at first we will allow for a position-dependent interaction K. Since $\sigma^2 = 1$, we can write ($j > i$)

$$\sigma_i \sigma_j = (\sigma_i \sigma_{i+1})(\sigma_{i+1} \sigma_{i+2}) \ldots (\sigma_{j-1} \sigma_j). \qquad (10.2\text{-}20)$$

One easily verifies:

$$\langle \sigma_i \sigma_j \rangle = \mathcal{Z}_N^{-1} \left. \frac{\partial^{j-i} \mathcal{Z}_N (K_1 \ldots K_{N-1})}{\partial K_i \partial K_{i+1} \ldots \partial K_{j-1}} \right|_{\{K_\ell\} = K} = \left(\tanh K \right)^{j-i} \equiv \exp[-(j-i)/\xi], \qquad (10.2\text{-}20')$$

where ξ is the correlation length. Explicitly, $\xi = -[\ln \tanh K]^{-1}$. At $T = 0$, $\xi \to \infty$, while if $T \to \infty$, $\xi \to 0$. For finite low temperatures, $\xi \approx \frac{1}{2} e^{2K}$.

10.3 The Renormalization Group

10.3.1 *Fixed Points, Infinitesimal Transformations and Scaling Fields*

The rescaling of the linear chain was a straightforward procedure. The extension to rescaling of multidimensional systems of spins is not immediate, however. In fact, one can easily see that Kadanoff's original procedure is beset with difficulties. Take for instance a 2D square lattice (Fig. 10-3). One can then define block spins for every square composed of four spins. Suppose that the original spins only have nearest neighbour interaction, involving the lattice spacing a_0. After the grouping of spins, one finds that there are now interactions of block A's upper right-corner spin

[10] The partition function for the finite chain follows from summation over the last spin; independently of the value for σ_{N-1}' one has $\Sigma_{\sigma_N' = \pm 1} \exp(K_{N-1} \sigma_{N-1}' \sigma_N') = 2 \cosh K_{N-1}$. This yields the recurrent relation $\mathcal{Z}_N = (2 \cosh K_{N-1}) \mathcal{Z}_{N-1}$, which is solved by the expression in (10.2-19).

with block B's lower left-corner spin. These are placed 'kiddie-corner' with respect to each other, separated by a distance $a_0\sqrt{2}$. Thus, after rescaling, next to nearest neighbour interactions play a role. There can be no isomorphy between the block-spin interactions and the original spin interactions, unless from the beginning these next-to nearest neighbour interactions are taken into account. Therefore, the simple Ising picture must be revised in order that the renormalization group works out. Or, in other words, the interaction parameters which make up the rescaling vector **K** must be enlarged and the surface of criticality, composed of the components of this vector or of appropriately chosen 'scaling fields', has a higher dimensionality than the simple Ising model would have led us to believe.

The correct procedure is as follows. Let $\{\sigma_i\}$ denote the site spins in a d-dimensional lattice. The interactions of these spins do not only involve a sufficiently large set of n-th order neighbours, but also clusters of spins as required by symmetry. For instance, for a triangular lattice there could be three-spin interactions $K_{123}\sigma_1\sigma_2\sigma_3$. Generally, let α denote a subset of all sites and let $\kappa_\alpha = \Pi_{i\in\alpha}\sigma_i$ denote the relevant two-spin, three spin, etc. interactions, leading to a Hamiltonian (in units k_BT) of the form

$$-\mathcal{H}[\{K_\alpha\},\{\sigma_i\},N] = \sum_\alpha K_\alpha \kappa_\alpha(\sigma_i), \tag{10.3-1}$$

where the sum extends over all non-empty sets. In practice we may have the following interactions

$$\kappa_1 = \sum_{\langle i,j\rangle} \sigma_i\sigma_j, \quad \kappa_2 = \sum_{\langle\langle i,j\rangle\rangle} \sigma_i\sigma_j, \quad \kappa_3 = \sum_{\langle i,j,k\rangle} \sigma_i\sigma_j\sigma_k, \quad \kappa_4 = \sum_i \sigma_i, \quad etc. \tag{10.3-1'}$$

with their attending interaction parameters $K_1 ... K_4, etc.$ Here $\langle i, j\rangle$ stands for nearest neighbour spins, $\langle\langle i, j\rangle\rangle$ for next nearest neighbour spins, $\langle i, j, k\rangle$ for triple sites, etc. As in the 1D Ising chain, the simplest one-spin 'interaction' stands for the magnetic energy with $K_4 = h$. All relevant interaction parameters K_α make up the interaction vector **K**.

The next thorny problem is how to do the mapping for the block spins (called 'cell-spins' by Niemeijer and van Leeuwen). Let $\sigma_I' \equiv \{\sigma_i\}_I$ represent the block spin for the collection of spins in the block. The weight function will be denoted by $P_I(\sigma_I',\{\sigma_i\}_I)$ and the overall weight by $P = \Pi_I P_I$. There are various rules for stipulating the weight P_I, e.g., the 'majority rule'. Basically, P_I is a projection operator, mapping $\{\sigma_i\}$ into $\{\sigma_I'\}$, such that the form of the partition function is preserved. There are two requirements for the P_I:

$$\begin{aligned} &(i) \quad P_I(\sigma_I',\{\sigma_i\}_I) > 0; \\ &(ii) \quad \sum_{\{\sigma'_I\}} P_I(\sigma_I',\{\sigma_i\}_I) = 1. \end{aligned} \tag{10.3-2}$$

The new Hamiltonian now must satisfy

$$\exp[-\mathcal{H}'(\{K'_\alpha\},\{\sigma_I'\},N')+N\gamma(\mathbf{K})] = \mathop{\mathrm{Tr}}_{\{\sigma_i\in\sigma_I'\}} \prod_I [P_I(\sigma_I',\{\sigma_i\})\, e^{-\mathcal{H}(\{K_\alpha\},\{\sigma_i\},N)}], \tag{10.3-3}$$

where $\gamma(\mathbf{K})$ is a spin-independent function. If we now sum the lhs over all block-spins, we obtain, employing (10.3-2),

$$\mathop{\mathrm{Tr}}_{\{\sigma_I'\}}\left\{e^{-\mathcal{H}'[\{K'_\alpha\},\{\sigma_I'\},N']+N\gamma(\mathbf{K})}\right\} = \mathop{\mathrm{Tr}}_{\{\sigma_i\},\{\sigma_I'\}}\left\{\prod_I P_I(\sigma_I',\{\sigma_i\})\, e^{-\mathcal{H}[\{K_\alpha\},\{\sigma_i\},N]}\right\}$$

$$= \mathop{\mathrm{Tr}}_{\{\sigma_i\}}\{\underbrace{\prod_I \sum_{\sigma_I'}\left(P_I(\sigma_I',\{\sigma_i\})\right)}_{1} e^{-\mathcal{H}[\{K_\alpha\},\{\sigma_i\},N]}\} = \mathop{\mathrm{Tr}}_{\{\sigma_i\}}\left\{e^{-\mathcal{H}[\{K_\alpha\},\{\sigma_i\},N]}\right\}. \tag{10.3-4}$$

The contemplated block-spin Hamiltonian, being isomorphic with the original Hamiltonian, has again the form (10.3-1), apart from $\gamma(\mathbf{K})$, i.e.,

$$-\mathcal{H}'[\{K'_\alpha\},\{\sigma_I'\},N'] = \sum_\alpha K'_\alpha \kappa_\alpha(\sigma_I') - N\gamma(\mathbf{K}); \tag{10.3-5}$$

note that the κ's have *not* been primed, since they must be the same functions of the σ_I' in $\mathcal{H}'$ as they were of the σ_i in $\mathcal{H}$. The partition function (10.3-4) yields the connection for the free energies per spin $[\hat{g} = (-\ln \mathcal{Z})/\text{spin}]$; noting that $N' = N/b^d$, where ba_0 is the linear dimension of the block, we obtain

$$\hat{g}(\mathbf{K}) = \gamma(\mathbf{K}) + b^{-d}\hat{g}(\mathbf{K}'). \tag{10.3-6}$$

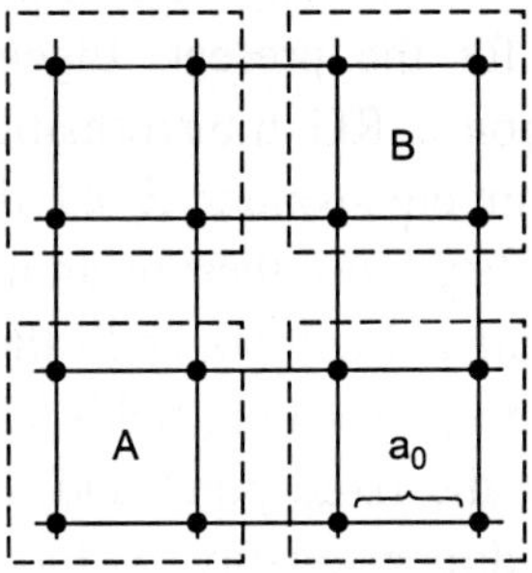

Fig. 10-3. Square lattice with blocks of 4 ×4 spins.

The rescaling of the vector $\mathbf{K}$ is implied by (10.3-3). We shall write $\mathbf{K}' = R(\mathbf{K})$, where R denotes the renormalization group; after n rescalings we get

$$R(\mathbf{K}^{(n)}) = \mathbf{K}^{(n+1)}. \tag{10.3-7}$$

Let for $n \to \infty$ a fixed point be reached; it has the property that

$$R(\mathbf{K}^*) = \mathbf{K}^*. \tag{10.3-8}$$

Subtraction of the last two equations gives [we write $\mathbf{K}^{(n+1)} = \overline{\mathbf{K}}'$, $\mathbf{K}^{(n)} = \overline{\mathbf{K}}$]:

$$\overline{\mathbf{K}}' - \mathbf{K}^* \equiv \delta\mathbf{K}' = R(\overline{\mathbf{K}}) - R(\mathbf{K}^*) = \mathsf{T}\delta\mathbf{K} + \mathcal{O}(\delta K^2)\,, \tag{10.3-9}$$

where $\mathsf{T} = \{T_{\alpha\beta}\}$, with $T_{\alpha\beta} = \partial R_\alpha / \partial K_\beta$ denoting the group of infinitesimal transformations near the fixed point $\mathbf{K}^*$. The matrix $\mathsf{T}(\mathbf{K}^*)$ is generally not symmetrical, so it has different right and left eigenvectors which, moreover, are not orthogonal to each other. Even so, the left eigenvectors [row matrices $(\varphi_1\ \varphi_2 \ldots)$] will be employed to define a set of affine axes for the tangent hypersurface at $\mathbf{K}^*$ in K_α-space. We have

$$\varphi\mathsf{T} = \lambda\varphi, \quad \text{or} \quad \Sigma_\alpha \varphi_\alpha^i T_{\alpha\beta} = \lambda_i \varphi_\beta^i, \tag{10.3-10}$$

where 'i' labels the different eigenvectors and eigenvalues. The usefulness of these eigenvectors lies in the fact that they do not mix with one another under a RG transformation. The proof is direct. We shall introduce 'normal coordinates' by the linear combination

$$u_i = \Sigma_\alpha \varphi_\alpha^i \delta K_\alpha. \tag{10.3-11}$$

We then have

$$\begin{aligned} u_i' &= \Sigma_\alpha \varphi_\alpha^i (\delta K_\alpha') = \Sigma_{\alpha\beta} \varphi_\alpha^i T_{\alpha\beta} \delta K_\beta \\ &= \lambda_i \Sigma_\beta \varphi_\beta^i (\delta K_\beta) = \lambda_i \Sigma_\alpha \varphi_\alpha^i (\delta K_\alpha) = \lambda_i u_i\,. \end{aligned} \tag{10.3-12}$$

The $u_i(\mathbf{K})$ are called *scaling fields*. Although we cannot be sure that all eigenvalues are real, we will assume so for the present. Eigenvalues $\lambda_i > 1$ will be called 'relevant'; they increase u_i upon a RG transformation and move away from the critical point, while for irrelevant eigenvalues $\lambda_i < 1$ the u_i decrease, i.e., they move toward the critical point where they loose their meaning (being identically zero). The case that $\lambda_i = 1$ is called marginal and will not be further discussed here. At the fixed point all δK_α are zero and, thence, all scaling fields take the value zero. Conversely, the equations $u_i(\mathbf{K}) = 0$ *define* the fixed point and the tangent hypersurface in its neighbourhood. Let n be the dimensionality of the $\mathbf{K}$-space and let there be m relevant scaling fields. The critical surface (and its tangent surface at the fixed point) will then have the dimension $n - m$. In Fig. 10-4a we have made an attempt to picture the critical hypersurface and a scaling field. We get a better grasp of the situation for the case that the dimension of the $\mathbf{K}$-space is two, shown in Fig. 10-4b. We have pictured the case that there is a line of transition points and an isolated fixed point at T_c. Let K_1 be J_1/k_BT; the points to the right of the line indicate high temperatures (disordered state) and those to the left low temperatures (ordered state). This aspect does not change by thinning out degrees of freedom, so the transformations do not cross this line. The flow in the regions $T \gtrless T_t$ is therefore as indicated in Fig. 4b. Points in the region A move toward $T = \infty$, while points in B move to $T = 0$. The critical point is hyperbolic, having one EV larger than zero and the other EV less than zero. The critical 'surface' is one-dimensional, indicated by the line $u_1(\mathbf{K})$. The critical temperature of the isolated fixed point is determined by the equation of the scaling field $u_1(\mathbf{K}) = 0$.[11] More generally, for any dimension [see (10.3-11)]:

$$k_B T_c(J) = \sum_\alpha \varphi_\alpha^T / \sum_\alpha \varphi_\alpha^T K_\alpha^* . \tag{10.3-13}$$

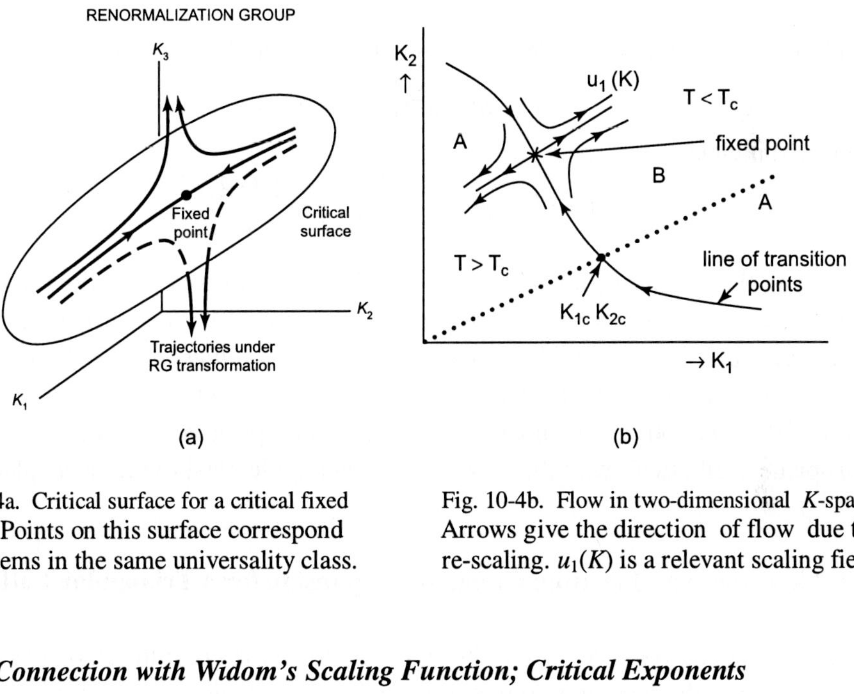

Fig. 10-4a. Critical surface for a critical fixed point. Points on this surface correspond to systems in the same universality class.

Fig. 10-4b. Flow in two-dimensional K-space. Arrows give the direction of flow due to re-scaling. $u_1(K)$ is a relevant scaling field.

10.3.2 *Connection with Widom's Scaling Function; Critical Exponents*

We return to the form for the Gibbs free energy per spin, as given in (10.3-6). From what has been said before, we can expect $\gamma(\mathbf{K})$ to be a regular function of the interaction components K_α. The other part will be identified as the singular part (subscript 's'). In the preceding subsection we have seen that the interaction parameters are best represented by the scaling fields, $u_i(\mathbf{K})$. Hence, the free energy can be written as a function of these scaling fields, whereby Eq.(10.3-12) gives the behaviour of its singular part upon a renormalization transformation. Accordingly

[11] It is possible to go beyond the linear approximation implied by (10.3-9); then a quadratic equation gives a more accurate estimate of the critical temperature. See Niemeijer and van Leeuwen, Eq. (2.19).

we have, solving from (10.3-11) for $K_\alpha(\{u_i\})$

$$\hat{g}_s(u_1,u_2,...) = b^{-d}\hat{g}_s(\lambda_1 u_1,\lambda_2 u_2,...). \tag{10.3-14}$$

The u_i are now identified with the physical scaling parameters of a given problem. Generally, since at the critical point all u_i are zero, one finds the critical behaviour for a given u, say u_1, *near* the critical point by setting $u_2,u_3,... = 0$. Thus,

$$\hat{g}_s(u_1,0,0,...) = b^{-d}\hat{g}_s(\lambda_1 u_1,0,0,...), \tag{10.3-15}$$

which gives the scaling behaviour for u_1. Now let only $K = K(t)$ and h be the variables of interest. Then (10.3-14) gives

$$\hat{g}_s(t,h) = b^{-d}\hat{g}_s(\lambda_1 t,\lambda_2 h), \tag{10.3-16}$$

which is to be compared with Widom's scaling function (9.6-6) which reads

$$\hat{g}_s(t,h) = \lambda^{-1}\hat{g}_s(\lambda^p t,\lambda^q h). \tag{10.3-17}$$

Comparison yields

$$\lambda = b^d, \quad p = \frac{1}{d}\frac{\ln\lambda_1}{\ln b}, \quad q = \frac{1}{d}\frac{\ln\lambda_2}{\ln b}. \tag{10.3-18}$$

Employing the developments of Section 9.6, all critical exponents follow from p and q. We thus obtained one of the most gratifying results of the renormalization theory: *the critical exponents are determined by the eigenvalues of the linearized (or infinitesimal) transformation relations near the critical point.*

For (10.3-17) both eigenvalues will turn out to be positive; the fixed point in the (K,h)-plane is elliptical, with the tangent plane being identical to the (K,h)-plane.

10.4 Niemeijer–van Leeuwen Cumulant Expansion for a Triangular Lattice

For a triangular (or face-centred hexagonal) lattice the grouping in blocks of three spins is natural; the resulting groups of triangles form again a triangular lattice, as is pictured in Fig. 10-5. The spins in a block of three need no 'tiebreaker rule': their sum is always up or down *majoritaire*. As projectors we choose:

$$\begin{aligned} &P_I(\sigma_I{}',\sigma_{1I},\sigma_{2I},\sigma_{3I}) = \delta_{\sigma_I{}',\varphi_I}, \\ &\varphi_I(\{\sigma_i\}) = \tfrac{1}{2}(\sigma_{1I}+\sigma_{2I}+\sigma_{3I}-\sigma_{1I}\sigma_{2I}\sigma_{3I}). \end{aligned} \tag{10.4-1}$$

One easily finds that the block spin is +1 if two or more spins are up and −1 if two or more spins are down. Also, the possible projectors sum to unity. Now the mapping (10.3-3) must be effectuated. Whether or not the results of (10.3-3) and (10.4-1) will lead to analytical transformation equations cannot be foreseen with certainty.

Considerations by Griffiths and Pearce suggest non-analytic results.[12] However, appropriate approximations will produce analytical recursion relations. Cumulant expansions as well as cluster approximations, such as we met in the treatment of classical gases, have been the most widely used approximation schemes. Here we employ the former, following Refs. 5 and 2.

To begin with, the Hamiltonian of departure will be Ising-like, but we must go beyond nearest neighbours, i.e.,

$$-\mathcal{H} = \sum_{i,j} K_{ij}\sigma_i\sigma_j + h\sum_i \sigma_i\,, \tag{10.4-2}$$

where $\mathcal{H}$ is again in units k_BT. Let us divide the Hamiltonian in two parts, $\mathcal{H}_0 + V$, whereby $\mathcal{H}_0$ will contain all intra-block couplings and V the inter-block couplings,

$$-\mathcal{H}^0 = \sum_I K_1(\sigma_{1I}\sigma_{2I} + \sigma_{1I}\sigma_{3I} + \sigma_{2I}\sigma_{3I}) + h\sum_I(\sigma_{1I} + \sigma_{2I} + \sigma_{3I})\,, \tag{10.4-3}$$

$$-V = \sum_{I,J,n} K_n \sum_{i,j} \sigma_{iI}\sigma_{jJ}\,, \tag{10.4-4}$$

where K_n is the n-th nearest neighbour coupling constant and i and j are labels in block I and J, respectively. Using the rhs of (10.3-3), the partition function can now be evaluated as[13]

$$\mathcal{Z} = \operatorname*{Tr}_{\{\sigma\}}\left\{\prod_I \delta_{\sigma_I',\varphi_I} e^{-\mathcal{H}_0} e^{-V}\right\} = \mathcal{Z}_0\langle e^{-V}\rangle_0\,, \tag{10.4-5}$$

where the zero-order average is

$$\langle A\rangle_0 = \mathcal{Z}_0^{-1} \operatorname*{Tr}_{\{\sigma\}}\left\{\prod_I \delta_{\sigma_I',\varphi_I} e^{-\mathcal{H}_0} A\right\}. \tag{10.4-6}$$

We recall that the characteristic function was defined as $\Phi_V(iu) = \langle \exp(iuV)\rangle$, which has a Taylor expansion in terms of the moments $m_n = \langle V^n\rangle$. The corresponding expansion for $\ln\Phi$ is the cumulant expansion $\Sigma_{n=1}^{\infty} c_n (iu)^n / n!$, cf. Eq. (1.7-51). Presently we set $iu = -1$, which leads to

$$\langle e^{-V}\rangle_0 = [\Phi_V(u)]_{iu=-1}$$
$$= \exp\left(\sum_{n=1}^{\infty} \frac{c_n(-1)^n}{n!}\right) = e^{-\langle V\rangle_0 + \frac{1}{2}[\langle V^2\rangle_0 - \langle V\rangle_0^2] - \frac{1}{6}[\langle V^3\rangle_0 - 3\langle V^2\rangle_0\langle V\rangle_0 + 2\langle V\rangle_0^3] + \ldots} \tag{10.4-7}$$

The cumulant expansion can be truncated at any order.

---

[12] R.B. Griffiths and P.A. Pearce, Phys. Rev. Lett. **41**, 917 (1978).

[13] We note that H_0 and V commute, since they refer to different spin interactions, so that the exponential can be factored.

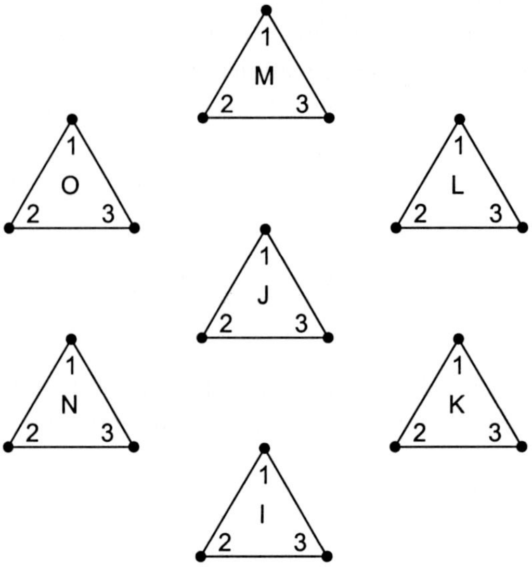

Fig. 10-5. Triangular lattice, rearranged in triangle-block spins.

10.4.1 *First-order Results*

Within a block only nearest neighbour interaction can occur (excluding triple spin interactions), so we set $K_1 = K$ and all other $K_n = 0$. In Eq. (10.4-5) we can interchange the sum and the product signs; so, the partition function $\mathcal{Z}_0$ is a product of single block partition functions z_0. Hence, let us configure

$$\mathcal{Z}_0 = \prod_I z_{I0}, \qquad z_{I0} = \mathop{\mathrm{Tr}}_{\sigma}\left\{\delta_{\sigma_I',\varphi}\, e^{-\mathcal{H}_0}\right\} \equiv e^{P+Q\sigma_I'}\,, \tag{10.4-8}$$

with $\mathcal{H}_0$ given by (10.4-3). There are four spin configurations in each block, viz.,

$$\begin{array}{cccc} (a)\ \uparrow\uparrow\uparrow, & (b)\ \uparrow\uparrow\downarrow, & (c)\ \uparrow\downarrow\downarrow, & (d)\ \downarrow\downarrow\downarrow\,. \\ 1 & 3 & 3 & 1 \end{array} \tag{10.4-9}$$

The numbers underneath indicate their relative abundances, the total number of configurations being $2^3 = 8$. The weights P give the block spin $\sigma_I' = 1$ for (a) and (b) and $\sigma_I' = -1$ for the cases (c) and (d). The single block Hamiltonians are

$$\begin{array}{ll} -h_0(a) = 3K + 3h, & -h_0(b) = -K + h, \\ -h_0(c) = -K - h, & -h_0(d) = 3K - 3h. \end{array} \tag{10.4-10}$$

Evaluating separately the contributions for $\sigma_I' = \pm 1$, we have from (10.4-8) - (10.4-10):

$$e^{P+Q}=e^{3K+3h}+3e^{-K+h}, \qquad e^{P-Q}=e^{3K-3h}+3e^{-K-h}. \tag{10.4-11}$$

This yields

$$\begin{aligned} P&=\tfrac{1}{2}\ln[(e^{3K+3h}+3e^{-K+h})(e^{3K-3h}+3e^{-K-h})], \\ Q&=\tfrac{1}{2}\ln[(e^{3K+3h}+3e^{-K+h})/(e^{3K-3h}+3e^{-K-h})]. \end{aligned} \tag{10.4-12}$$

To obtain exp[$\langle -V\rangle_0$] we need $\langle \sigma_{iI}\sigma_{jJ}\rangle_0$. However, spins in different blocks are not correlated in order $\mathfrak{z}_0$, so we can write $\langle \sigma_{iI}\sigma_{jJ}\rangle_0=\langle \sigma_{iI}\rangle_0\langle \sigma_{jJ}\rangle_0$. Further, because of this independence, we have with

$$\langle \sigma_{iI}\rangle_0 \equiv R+S\sigma_I', \tag{10.4-13}$$

$$\begin{aligned} e^{-\langle V\rangle_0} &= \prod_{\langle I,J\rangle} e^{K\Sigma_{i,j}\langle \sigma_{iI}\rangle_0\langle \sigma_{jJ}\rangle_0} = \prod_{\langle I,J\rangle} e^{2K\langle \sigma_{iI}\rangle_0\langle \sigma_{jJ}\rangle_0} \\ &= \prod_{\langle I,J\rangle} e^{2K(R+S\sigma_I')(R+S\sigma_J')}, \end{aligned} \tag{10.4-14}$$

where we noted that a given block has two nearest neighbour spin-couplings with an adjacent block – see Fig. 10-5 – and we perceived that the average over $\mathfrak{z}_0$ depends on the sign of the block-spin considered. For $\sigma_{iI}=+1,-1$ in (10.4-13) we obtain, respectively,

$$R+S=\frac{e^{3K+3h}+e^{-K+h}}{e^{3K+3h}+3e^{-K+h}}\equiv U, \qquad R-S=-\frac{e^{3K-3h}+e^{-K-h}}{e^{3K-3h}+3e^{-K-h}}\equiv -W, \tag{10.4-15}$$

from which

$$R=\tfrac{1}{2}(U-W), \qquad S=\tfrac{1}{2}(U+W). \tag{10.4-16}$$

Now from (10.3-3), (10.4-5), (10.4-8), (10.4-13) and 10.4-14) we established the connection

$$\begin{aligned} -\mathcal{H}'[K',h',\tfrac{1}{3}N,\{\sigma_I'\}]+N\gamma(K,h) &= \ln \mathcal{Z}_0+\ln\langle e^{-V}\rangle_0 \\ \approx \ln \mathcal{Z}_0-\langle V\rangle_0 &= \sum_I [P+Q\sigma_I']+2K\sum_{\langle I,J\rangle}[(R+S\sigma_I')(R+S\sigma_J')]. \end{aligned} \tag{10.4-17}$$

We must equate this with the intended form

$$-\mathcal{H}'[K',h',\tfrac{1}{3}N,\{\sigma_I'\}]+N\gamma(K,h)=K'\sum_{\langle I,J\rangle}\sigma_I'\sigma_J'+h'\sum_I \sigma_I'+N\gamma(K,h). \tag{10.4-18}$$

Noting that the sum of nearest pairs $\Sigma_{\langle I,J\rangle}$ is $(N/3)\,p$, where $p=3$ is the coordination number (i.e., the number of adjacent triangles), we finally obtain the recursion relations

$$\begin{aligned}\gamma &= \tfrac{1}{3}P(K,h) + 2K[R(K,h)]^2, \\ K' &= 2K[S(K,h)]^2, \\ h' &= Q(K,h) + 12K\,R(K,h)\,S(K,h).\end{aligned} \tag{10.4-19}$$

For the phase transition paramagnetic → ferromagnetic we set $h = 0$. The relation for K then becomes simple:

$$K' = 2K\left(\frac{e^{3K} + e^{-K}}{e^{3K} + 3e^{-K}}\right)^2. \tag{10.4-20}$$

For $K \ll 0$, $K' \approx \frac{1}{2}K$ while for $K \gg 0$, $K' \approx 2K$. Thus, at some finite value of K the flow reverses itself. The same can be shown to hold for h' in the neighbourhood of $h = 0$. There is therefore an elliptic critical point at $K = K^*$ and $h^* = 0$. From (10.4-20), setting $K' = K \to K^*$ we find that K^* is determined by

$$\frac{e^{3K^*} + e^{-K^*}}{e^{3K^*} + 3e^{-K^*}} = \frac{e^{4K^*} + 1}{e^{4K^*} + 3} = \frac{1}{2}\sqrt{2}, \tag{10.4-21}$$

from which

$$K^* = \tfrac{1}{4}\ln(1 + 2\sqrt{2}) = 0.3356. \tag{10.4-22}$$

This yields for the critical temperature $(1/K^*) = k_B T_c / J = 2.980$. The 2D triangular spin lattice can also be solved exactly (as in Chapter XI) giving $k_B T_c / J = [\frac{1}{4}\ln 3]^{-1}$ $= 3.641$. For small K the flow is toward the non-interacting high-temperature fixed point, while for large K the flow is toward the ground-state fixed point at $T = 0$. The K-h flow diagram is shown in Fig. 10-6.

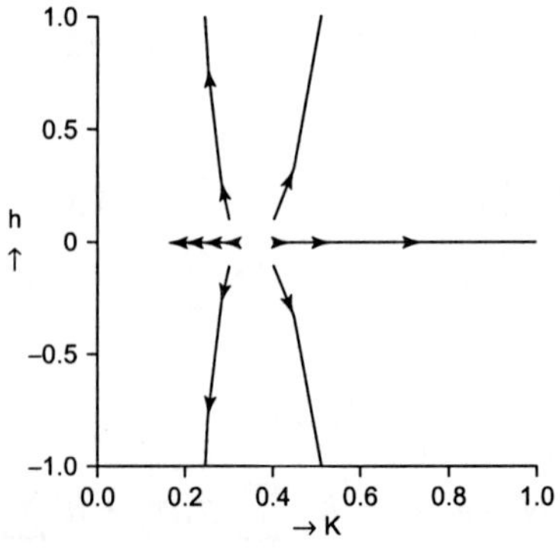

Fig. 10-6. *K-h* flow diagram with elliptical fixed point for triangular lattice in first-order cumulant approximation. After Plischke and Bergersen, Op. Cit. 2nd Ed., Fig. 6.4. [With permission.]

Finally, we must find the infinitesimal recursion relations near $K = K^*$, $h^* = 0$. We easily verify $\partial h'/\partial K|_{K^*,h^*} = 0$, $\partial K'/\partial h|_{K^*,h^*} = 0$. The diagonal elements of the **T**-

matrix are straightforward. We obtain

$$\left(\frac{\partial K'}{\partial K}\right)_{K^*,h=0} = \lambda_K = 1 + 8\sqrt{2}K^* \frac{e^{3K^*}}{(e^{3K^*}+e^{-K^*})^2} = 1.623,$$
$$\left(\frac{\partial h'}{\partial h}\right)_{K^*,h=0} = \lambda_h = \left(\frac{e^{3K^*}+e^{-K^*}}{e^{3K^*}+3e^{-K^*}}\right)\left(3 + 48K^* \frac{e^{2K^*}}{(e^{3K^*}+3e^{-K^*})^2}\right) = 2.78. \qquad (10.4\text{-}23)$$

From (10.3-18), with $b = \sqrt{3}$, we thus find:

$$p = \frac{\ln \lambda_K}{2\ln\sqrt{3}} = 0.441, \qquad q = \frac{\ln \lambda_h}{2\ln\sqrt{3}} = 0.931. \qquad (10.4\text{-}24)$$

The exact solution for the 2D triangular lattice yields $\lambda_{K,exact} = 1.73$, $\lambda_{h,exact} = 2.80$. Clearly, even the first-order cumulant treatment is not too far off. The critical exponents for the values obtained in (10.4-24) are:

$$\alpha = 2 - (1/p) = -0.27, \qquad \beta = (1-q)/p = 0.16,$$
$$\gamma = (2q-1)/p = 1.95, \qquad \delta = q/(1-q) = 13.5. \qquad (10.4\text{-}25)$$

10.4.2 *Higher-Order Results*

When higher-order cumulant terms are retained, the intracell couplings remain the same, notably contained in $\mathfrak{z}_0$, but the intercell couplings must be extended to higher order neighbours. In Fig. 10-5 *I* and *J* are first-order neighbours, *I* and *L* are second-order neighbours, while *I* and *M* are third-order neighbours. In what follows we shall set h = 0,[14] in order to simplify the treatment, and we shall denote the *n*-th order neighbour interaction constants (following N-vL) by *K*, *L* and *M* (n = 1,2,3). Yet, only binary interactions will be taken into account and higher order products, involving three or more spins will be excluded. The functions $Q(K, h = 0)$ and $R(K, h = 0)$ are zero and play no longer a role. For the cumulant terms we will only need three averages, viz.,

$$\langle\sigma_{iI}\rangle_0 = \frac{e^{3K}+e^{-K}}{e^{3K}+3e^{-K}}\sigma_I' = S_0\sigma_I', \qquad \langle\sigma_{iI}^2\rangle_0 = \frac{e^{3K}+e^{-K}}{e^{3K}+3e^{-K}} = S_0,$$
$$\langle\sigma_{iI}\sigma_{jI}\rangle_0 = \frac{e^{3K}-e^{-K}}{e^{3K}+3e^{-K}} = T_0, \qquad (10.4\text{-}26)$$

where we averaged over all configurations (10.4-9); we note that the latter two results are valid for either sign of the cell spin σ_I'. We seek to obtain the recursion relations for K', L' and M' in terms of *K*,*L* and *M*. The ordering will be in powers of *K* and,

[14] For the complete treatment, see Niemeijer and van Leeuwen, (1974), loc. cit..

for the second cumulant approximation, the L and M contributions will only be incorporated into the first cumulant.

Starting with K', the first-order cumulant contribution, associated with nearest cell interactions, was found before:

$$K'|_{n.n.,1} = 2K\,S_0^2. \tag{10.4-27}$$

To obtain the second cumulant contribution, we reiterate the two types of nearest-neighbour cell transitions:

$$\begin{aligned} V_{IJ} &= K(\sigma_{1I}\sigma_{2J} + \sigma_{1I}\sigma_{3J}), \\ V_{IK} &= K(\sigma_{1I}\sigma_{2K} + \sigma_{3I}\sigma_{2K}). \end{aligned} \tag{10.4-28}$$

For the first-type interaction the second cumulant contribution is

$$\begin{aligned} &\frac{1}{2!}K^2\Big(\langle(\sigma_{1I}\sigma_{2J} + \sigma_{1I}\sigma_{3J})^2\rangle - \langle(\sigma_{1I}\sigma_{2J} + \sigma_{1I}\sigma_{3J})\rangle^2\Big) \\ &= \frac{1}{2!}K^2\Big(\langle\sigma_{1I}^2\sigma_{2J}^2\rangle + \langle\sigma_{1I}^2\sigma_{3J}^2\rangle + 2\langle\sigma_{1I}\sigma_{2J}\sigma_{1I}\sigma_{3J}\rangle \\ &-\langle\sigma_{1I}\sigma_{2J}\rangle^2 - \langle\sigma_{1I}\sigma_{3J}\rangle^2 - 2\langle\sigma_{1I}\sigma_{2J}\rangle\langle\sigma_{1I}\sigma_{3J}\rangle\Big) \\ &= \frac{1}{2!}K^2\Big[2\langle\sigma^2\rangle^2 + 2\langle\sigma_{1I}\rangle\langle\sigma_{2J}\sigma_{3J}\rangle\langle\sigma_{1I}\rangle - 4\langle\sigma_I\rangle^2\langle\sigma_J\rangle^2\Big] \\ &= \frac{1}{2!}K^2\Big[2S_0^2 + 2S_0^2T_0 - 4S_0^4\Big] \xrightarrow{\times 4} 2K^2\Big[2S_0^2 + 2S_0^2T_0 - 4S_0^4\Big]. \end{aligned} \tag{10.4-29}$$

All we used here is that spins in different cells are uncorrelated. The second type of interaction in (10.4-28) yields the same result. Further we noted that there are two of each type of interactions, so that the result must be multiplied by four. We now have

$$K'|_{n.n.,2} = 2K^2\Big[2S_0^2 + 2S_0^2T_0 - 4S_0^4\Big]. \tag{10.4-30}$$

Notice that the second cumulant contribution goes with K^2. For the second and third nearest neighbour interactions we get, to first cumulant approximation, contributions similar in form as (10.4-27). In the *I-L* connection there are three internal spins that do the coupling, whereas in the *I-M* connection there are two internal spins that couple. We thus find forthwith

$$K'|_{sec.n.,1} = 3LS_0^2, \qquad K'|_{third\,n.,1} = 2MS_0^2. \tag{10.4-31}$$

Collecting we obtain,

$$K' = 2K\,S_0^2 + 2K^2\Big[2S_0^2 + 2S_0^2T_0 - 4S_0^4\Big] + 3LS_0^2 + 2MS_0^2. \tag{10.4-32}$$

Let us next consider the transformed third neighbour interaction, M'. The effect due to nearest neighbour interactions K necessarily involves two steps: cell $I \leftrightarrow$ cell $J \leftrightarrow$ cell M ; this being proportional to K^2, there is no first cumulant contribution. For the second cumulant contribution we have two product terms

$$\begin{aligned} M'|_{n.n.,2} &= (1/2!)2K^2\left(\langle\sigma_{1I}(\sigma_{2J}+\sigma_{3J})\sigma_{1J}(\sigma_{2M}+\sigma_{3M})\rangle\right. \\ &\quad \left.-\langle\sigma_{1I}(\sigma_{2J}+\sigma_{3J})\rangle\langle\sigma_{1J}(\sigma_{2M}+\sigma_{3M})\rangle\right) \\ &= K^2\left(4\langle\sigma_I\rangle\langle\sigma_{iJ}\sigma_{jJ}\rangle\langle\sigma_M\rangle - 4\langle\sigma_I\rangle\langle\sigma_{iJ}\rangle\langle\sigma_{jJ}\rangle\langle\sigma_M\rangle\right) = 4K^2\left(S_0^2T_0 - S_0^4\right)[\sigma_I'\sigma_M']. \end{aligned} \tag{10.4-33}$$

While it seems that the rescaled coupling constant for this case depends on the cell spins, this is not a real possibility near a critical point, since the spins increasingly align themselves; we shall therefore set the final bracket [..] equal to unity. According to the N-vL scheme, there are no original second or third order interactions to be included in the rescaled third neighbour effect. Hence, (10.4-33) is the full result for M'.

The last function needed is the rescaled second neighbour interaction $L'(K,L,M)$. The second cumulant result associated with nearest neighbour effects K^2 involves 'double hopping': cell $I \rightarrow$ cell $K \rightarrow$ cell L or cell $I \rightarrow$ cell $J \rightarrow$ cell L; the details are left as a problem. The full result for the rescaling equations is

$$\begin{aligned} K' &= 2KS_0^2 + 2K^2\left[2S_0^2 + 2S_0^2T_0 - 4S_0^4\right] + 3LS_0^2 + 2MS_0^2, \\ L' &= K^2(7S_0^2T_0 + S_0^2 - 8S_0^4) + MS_0^2, \\ M' &= 4K^2\left(S_0^2T_0 - S_0^4\right). \end{aligned} \tag{10.4-34}$$

The critical point is obtained from the intersection of the critical surface with the K-axis. This yields for the critical temperature $1/K^* = 3.997$, in better agreement with the actual value of $k_BT_c/J = 3.641$. The T-matrix is found to be (N-vL, Op. Cit.):

$$\mathrm{T} = \begin{pmatrix} 1.8966 & 1.3446 & 0.8964 \\ -0.0403 & 0 & 0.4482 \\ -0.0782 & 0 & 0 \end{pmatrix}. \tag{10.4-35}$$

The eigenvalues are: $\lambda_1 = 1.7835$, $\lambda_2 = 0.2286$, $\lambda_3 = -0.1156$. Only the first one is relevant, leading to the critical point $\lambda_K = 1.7835$, closer to the exact value $\lambda_{K,exact} = 1.73$. Altogether, we have demonstrated that higher-order cumulant evaluations lead to much improved results. Clearly, the computational enthusiast has here a fertile terrain! Moreover, a combination with cluster-expansion methods, also initiated by Niemeijer and van Leeuwen and carried to greater perfection by many others, is useful and leads to faster convergent results.

10.5 The 'Classical Spin' Gaussian Model

"Exact statistical mechanics" aims at solving precisely defined Hamiltonian systems, such as the 1D linear spin chain and the 2D triangular lattice which we have presented. However, often more is accomplished with approximate formalisms that encompass a much larger variety of systems, such as the mean-field approach or the Landau–Ginzburg theory which we have explored at several occasions. It is in that vein that Wilson, in the papers already cited, developed a general "spin-field" approach, known as the S^4 model or the Landau–Ginzburg–Wilson Hamiltonian model. For concreteness we shall be fixed on a d-dimensional spin lattice, with order-parameter $\mathbf{S}(\mathbf{r})$; other formulations are in closer harmony with the general Landau–Ginzburg form. The Hamiltonian of departure in units k_BT will be

$$\mathcal{H} = -\sum_{\mathbf{l},\mathbf{a}} K(\mathbf{a})\mathbf{S}_\mathbf{l}\cdot\mathbf{S}_{\mathbf{l}+\mathbf{a}} - \tilde{h}\sum_{\mathbf{l}} S_\mathbf{l}^i\,, \qquad (10.5\text{-}1)$$

where $\mathbf{l}$ denotes a lattice position, $\mathbf{a}$ labels the nearest neighbours and i is the magnetic field coordinate of the n-component spins. The spins will be dimensionless classical quantities with a continuum of values in R_n. As such they may be compared with the lattice vectors $\mathbf{u}_\mathbf{l}$ in our treatment of discrete lattice vibrations, cf., subsection 8.5.2; we may, however, extend these vectors to all position space, with $\mathbf{S}_\mathbf{l} \to \mathbf{S}(\mathbf{r})$ being a *spin-field*. Whereas quantum spins have a fixed norm, for classical spins in the Landau–Ginzburg–Wilson model the following weight-function is introduced:

$$W(\mathbf{S}_\mathbf{l}) = e^{-(b/2)\mathbf{S}_\mathbf{l}\cdot\mathbf{S}_\mathbf{l} - c(\mathbf{S}_\mathbf{l}\cdot\mathbf{S}_\mathbf{l})^2}\,, \qquad (10.5\text{-}2)$$

where b and c are parameters. For the case of a one-component spin and with $b = -4c$ the function (apart from a constant) becomes $W = \exp[-c(S_\mathbf{l}^2 - 1)^2]$, which represents a fairly good approximation to the Ising model since it is peaked at $S_\mathbf{l} = \pm 1$. The aetiology of the function $W(\mathbf{S}_\mathbf{l})$ is depicted in Fig. 10-7. The 'true' partition function now reads

$$\mathcal{Z} = \prod_{\mathbf{l}} \int W(\mathbf{S}_\mathbf{l})\, d^n S_\mathbf{l}\, e^{-\mathcal{H}}\,. \qquad (10.5\text{-}3)$$

However, we shall find it more convenient to define an effective Hamiltonian by[15]

$$\hat{\mathcal{H}} = \mathcal{H} - \sum_{\mathbf{l}} \ln W(\mathbf{S}_\mathbf{l}) = \sum_{\mathbf{l}}\left[-\sum_{\mathbf{a}}\left(K(\mathbf{a})\mathbf{S}_\mathbf{l}\cdot\mathbf{S}_{\mathbf{l}+\mathbf{a}}\right) - \tilde{h}S_\mathbf{l}^i + \tfrac{1}{2}b\mathbf{S}_\mathbf{l}\cdot\mathbf{S}_\mathbf{l} + c(\mathbf{S}_\mathbf{l}\cdot\mathbf{S}_\mathbf{l})^2\right]$$

$$= \frac{1}{2}K\sum_{\mathbf{l}}\left[\sum_{\mathbf{a}}(\mathbf{S}_{\mathbf{l}+\mathbf{a}} - \mathbf{S}_\mathbf{l})\cdot(\mathbf{S}_{\mathbf{l}+\mathbf{a}} - \mathbf{S}_\mathbf{l}) - \frac{2\tilde{h}}{K}S_\mathbf{l}^i + \left(\frac{b}{K} - 2d\right)\mathbf{S}_\mathbf{l}\cdot\mathbf{S}_\mathbf{l} + \frac{2c}{K}(\mathbf{S}_\mathbf{l}\cdot\mathbf{S}_\mathbf{l})^2\right]. \qquad (10.5\text{-}4)$$

---

[15] K.G. Wilson and J. Kogut, loc. cit. Eq. (3.12) ff.

We assumed that K does not depend on $\mathbf{a}$ and used that the number of one-sided nearest neighbours (noting that afterwards we sum over all $\mathbf{l}$) of the simple cubic lattice is d. This permits us to write down an ersatz partition function

$$\mathscr{Z} = \prod_{\mathbf{l}} \int d^n S_{\mathbf{l}}\, e^{-\hat{\mathscr{H}}}. \tag{10.5-5}$$

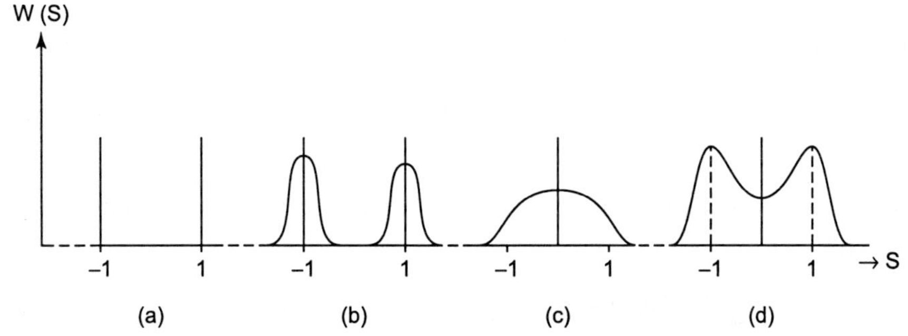

Fig. 10-7. Aetiology of $W(\mathbf{S}_{\mathbf{l}})$. (a). Ising spins; (b) Broadening; (c) Gaussian model; (d) Quartic model with $b = -4c$. [After Wilson and Kogut, Fig. 3.1 and Reichl 2nd Ed. Fig. 8.3.]

Next we introduce a Fourier series on the reciprocal lattice; employing (temporarily) the standard amplitudes of condensed-matter physics (chapter VIII), we write

$$\mathbf{S}_{\mathbf{l}} = (1/\sqrt{N})\sum_{\mathbf{q}} \bar{\mathbf{S}}_{\mathbf{q}} e^{i\mathbf{q}\cdot\mathbf{l}}, \qquad \bar{\mathbf{S}}_{\mathbf{q}} = (1/\sqrt{N})\sum_{\mathbf{q}} \mathbf{S}_{\mathbf{l}} e^{-i\mathbf{q}\cdot\mathbf{l}}. \tag{10.5-6}$$

For the conversion to these new variables we need the usual lemmas, Eqs. (8.5-26):

$$\sum_{\mathbf{l}\in V} e^{\pm i\mathbf{q}\cdot\mathbf{l}} = N\delta_{\mathbf{q},0}, \qquad \sum_{\mathbf{q}\in BZ} e^{\pm i\mathbf{q}\cdot\mathbf{l}} = N\delta_{\mathbf{l},0}, \tag{10.5-7}$$

where BZ denotes the simple cubic Brillouin zone $-\pi/a_0 \le q_i \le \pi/a_0$, $i = 1...d$. There is, however, a problem with this notation in that N here is a constant *that does not scale.*[16] In addition, $\mathbf{q}$ is usually taken to be continuous. The changes needed are simple.[17]

(a) We redefine the Fourier representation by – a_0 being the lattice constant –

$$\begin{aligned} &\mathbf{S}_{\mathbf{q}} = a_0^d \sum_{\mathbf{l}} \mathbf{S}_{\mathbf{l}} e^{-i\mathbf{q}\cdot\mathbf{l}}, \\ &\mathbf{S}_{\mathbf{l}} = (1/Na_0^d)\sum_{\mathbf{q}} \mathbf{S}_{\mathbf{q}} e^{i\mathbf{q}\cdot\mathbf{l}} = V^{-1}\sum_{\mathbf{q}} \mathbf{S}_{\mathbf{q}} e^{i\mathbf{q}\cdot\mathbf{l}} = [1/(2\pi)^d]\int d^d q\, \mathbf{S}_{\mathbf{q}} e^{i\mathbf{q}\cdot\mathbf{l}}. \end{aligned} \tag{10.5-8}$$

---

[16] Cf. the magnetic field contribution to the Hamiltonian in Plischke and Bergersen (Ref.8), Eqs. (7.108) - (7.114).

[17] R.E. Reichl, 2nd Ed. p 448; note that her lemma (8.105) is correct, but in (8.106) we think that a factor $(2\pi)^d$ should be incorporated on the right-hand side.

(b) For the lemmas we need the connection $\delta_{\mathbf{q},0} \to [(2\pi)^d / V]\delta(\mathbf{q})$; so for (10.5-7)

$$\frac{a_0^d}{V} \sum_{\mathbf{q} \in BZ} e^{\pm i\mathbf{q}\cdot\mathbf{l}} = \delta_{\mathbf{l},0}, \quad \text{or} \quad \left(\frac{a_0}{2\pi}\right)^d \int_{BZ} d^d q\, e^{\pm i\mathbf{q}\cdot\mathbf{l}} = \delta_{\mathbf{l},0},$$
$$a_0^d \sum_{\mathbf{l} \in V} e^{\pm i\mathbf{q}\cdot\mathbf{l}} = V\delta_{\mathbf{q},0}, \quad \text{or} \quad a_0^d \sum_{\mathbf{l} \in V} e^{\pm i\mathbf{q}\cdot\mathbf{l}} = (2\pi)^d \delta(\mathbf{q}). \tag{10.5-9}$$

The conversion of (10.5-4) is straightforward. For the first double sum one obtains

$$\sum_{\mathbf{l}} \sum_{\mathbf{a}} (\mathbf{S}_{\mathbf{l}+\mathbf{a}} - \mathbf{S}_{\mathbf{l}}) \cdot (\mathbf{S}_{\mathbf{l}+\mathbf{a}} - \mathbf{S}_{\mathbf{l}}) = \left(\frac{1}{2\pi a_0}\right)^d \int_{BZ} d^d q\, S_{\mathbf{q}} S_{-\mathbf{q}} \sum_{\mathbf{a}}^{d} [2 - 2\cos(\mathbf{q}\cdot\mathbf{a})]. \tag{10.5-10}$$

Since only long-wavelength fluctuations play a role near a critical point, we expand for small $|\mathbf{q}|$:

$$\sum_{\mathbf{a}}^{d} [2 - 2\cos(\mathbf{q}\cdot\mathbf{a})] = q_1^2 a_0^2 + q_2^2 a_0^2 ... + q_d^2 a_0^2 = q^2 a_0^2. \tag{10.5-11}$$

The fourth order term goes likewise. The full result becomes

$$\hat{\mathscr{H}} = \frac{1}{2} K a_0^{2-d} \left\{ \left(\frac{1}{2\pi}\right)^d \int_{BZ} d^d q\, \mathbf{S}_{\mathbf{q}} \cdot \mathbf{S}_{-\mathbf{q}} \left(q^2 + \frac{b - 2Kd}{K a_0^2} \right) - \frac{2\tilde{h}}{K a_0^2} S^i_{\mathbf{q}=0} \right.$$
$$\left. + \frac{2c}{K a_0^2} \left(\frac{1}{2\pi}\right)^{3d} \iiiint_{BZ} d^d q_1 ... d^d q_4\, (\mathbf{S}_{\mathbf{q}_1} \cdot \mathbf{S}_{\mathbf{q}_2})(\mathbf{S}_{\mathbf{q}_3} \cdot \mathbf{S}_{\mathbf{q}_4})\, \delta(\mathbf{q}_1 + \mathbf{q}_2 + \mathbf{q}_3 + \mathbf{q}_4) \right\}. \tag{10.5-12}$$

We change the Fourier coefficients such that $\mathbf{S}_{\mathbf{q}} = \mathbf{S}(\mathbf{q}) / a_0^{1-d/2} \sqrt{K}$ and set

$$r = \frac{b - 2Kd}{K a_0^2}, \qquad h = \tilde{h} \frac{1}{\sqrt{a_0^{2+d} K}}, \qquad u = \frac{c}{a_0^{4-d} K^2}; \tag{10.5-13}$$

The effective Hamiltonian then reads – note the coefficient ½ of $(r + q^2)$ –

$$\hat{\mathscr{H}}[\{\mathbf{S}(\mathbf{q})\}] = \left(\frac{1}{2\pi}\right)^d \int_{BZ} d^d q\, \mathbf{S}(\mathbf{q}) \cdot \mathbf{S}(-\mathbf{q}) \frac{1}{2} \left(r + q^2\right) - h S^i(0)$$
$$+ u \left(\frac{1}{2\pi}\right)^{3d} \iiiint_{BZ} d^d q_1 ... d^d q_4\, [\mathbf{S}(\mathbf{q}_1) \cdot \mathbf{S}(\mathbf{q}_2)][\mathbf{S}(\mathbf{q}_3) \cdot \mathbf{S}(\mathbf{q}_4)]\, \delta(\mathbf{q}_1 + \mathbf{q}_2 + \mathbf{q}_3 + \mathbf{q}_4). \tag{10.5-14}$$

In what follows we restrict ourselves to the Gaussian model, setting $u = 0$, $b > 0$. We shall assume that the localized spins are replaced by a spin field, $\mathbf{l} \to \mathbf{r}$. The Brillouin zone for the wave vectors $\mathbf{q}$ will be approximated by a hypersphere of approximate radius π / a_0. The partition function is then given as

$$\mathscr{Z}=\prod_{q<\pi/a_0}\int D^n S(\mathbf{q})\exp\left[-\left(\frac{1}{2\pi}\right)^d\int_{q<\pi/a_0} d^d q\mathbf{S}(\mathbf{q})\cdot\mathbf{S}(-\mathbf{q})\frac{1}{2}\left(r+q^2\right)+hS^i(0)\right], \quad (10.5\text{-}15)$$

which is a multiple Gaussian-type functional integral. As previously noted, it can be carried out exactly; however, our present goal is to obtain the scaling relations upon renormalization. So let ℓa_0 be a block length with $\ell>1$. The new cut-off in **q**-space is $q_\ell=\pi/a_0\ell$. Being interested mainly in long wavelength (small **q**) modes, let the functional integration over all modes **S(q)** with **q** within the shell $q_\ell<q<\pi/a_0$ be carried out. This leaves for the partition function the result

$$\mathscr{Z}=const\prod_{q<\pi/a_0\ell}\int D^n S(\mathbf{q})\exp\left[-\left(\frac{1}{2\pi}\right)^d\int_{q<\pi/a_0\ell} d^d q\mathbf{S}(\mathbf{q})\cdot\mathbf{S}(-\mathbf{q})\frac{1}{2}\left(r+q^2\right)+hS^i(0)\right]. \quad (10.5\text{-}16)$$

Next, we rescale to the original values, i.e., we set $\mathbf{q}=\mathbf{q}'/\ell$. The exponent in (10.5-16) then becomes

$$\hat{\mathscr{H}}'[\{\mathbf{S}(\mathbf{q}'/\ell)\}]=\left(\frac{1}{2\pi}\right)^d\int_{q'<\pi/a_0} d^d q'\mathbf{S}(\mathbf{q}'/\ell)\cdot\mathbf{S}(-\mathbf{q}'/\ell)\ell^{-d}\frac{1}{2}\left(r+\frac{q'^2}{\ell^2}\right)-hS^i(0'), \quad (10.5\text{-}17)$$

where the factor ℓ^{-d} compensates for the enlargement to the original Brillouin zone. The coefficient of q'^2 should remain ½, so we require

$$\mathbf{S}(\mathbf{q}'/\ell)=\varsigma(\ell)\mathbf{S}'(\mathbf{q}') \text{ with } \varsigma(\ell)=\ell^{1+d/2}. \quad (10.5\text{-}18)$$

We then have

$$\hat{\mathscr{H}}'[\{\mathbf{S}'(\mathbf{q}')\}]=\left(\frac{1}{2\pi}\right)^d\int_{q'<\pi/a_0} d^d q'\mathbf{S}'(\mathbf{q}')\cdot\mathbf{S}'(-\mathbf{q}')\frac{1}{2}\left(r\ell^2+q'^2\right)-h\ell^{1+d/2}S'^i(0). \quad (10.5\text{-}19)$$

This is now back into the original form

$$(1/2\pi)^d\int d^d q\tfrac{1}{2}(r'+q^2)\mathbf{S}'(\mathbf{q})\cdot\mathbf{S}'(-\mathbf{q})-h'S'^i(0)$$

providing the following renormalization (recursion) relations apply:

$$r'=r\ell^2, \qquad h'=h\ell^{1+d/2}. \quad (10.5\text{-}20)$$

There are three fixed points: $r=\infty\rightarrow T=\infty$; $r=-\infty\rightarrow T=0$; $r=0\rightarrow T=T_c$. Let us make the plausible assumption that r is a temperature-like variable; the eigenvalues for the above relations then are

$$\partial r'/dr|_{T_c}=\lambda_T=\ell^2, \qquad \partial h'/\partial h|_{T_c}=\lambda_h=\ell^{1+d/2}. \quad (10.5\text{-}21)$$

From (10.3-18) [with $b=\ell$] we find: $p=2/d$, $q=(1/d)+\tfrac{1}{2}$. This yields the following critical exponents,

$$\alpha=2-\tfrac{1}{2}d,\quad \beta=\tfrac{1}{4}d-\tfrac{1}{2},\quad \gamma=1,\quad \delta=(d+2)/(d-2). \qquad (10.5\text{-}22)$$

We note that Rushbrooke's and Widom's scaling relations are satisfied. For $d = 4$ the exponents are identical with the mean-field and the usual Landau–Ginzburg results.

Finally we shall obtain the correlation function, defined as

$$\Gamma(\mathbf{x})=\frac{1}{N}\sum_{\mathbf{l}}\langle S_{\mathbf{l}}^k S_{\mathbf{l}+\mathbf{x}}^k\rangle = A\int d^d q\langle S^k(\mathbf{q})S^k(-\mathbf{q})\rangle e^{-i\mathbf{q}\cdot\mathbf{x}} \equiv A(2\pi)^d\int d^d q\tilde{\Gamma}(\mathbf{q})e^{-i\mathbf{q}\cdot\mathbf{x}}, \qquad (10.5\text{-}23)$$

where A is a constant. With the previously given effective Hamiltonian we then have

$$\tilde{\Gamma}(\mathbf{q})=\left(\frac{1}{2\pi}\right)^d\frac{\int\int DS^k(\mathbf{q})DS^k(-\mathbf{q})\,S^k(\mathbf{q})S^k(-\mathbf{q})\exp\left[-(r+q^2)S^k(\mathbf{q})S^k(-\mathbf{q})\right]}{\int\int DS^k(\mathbf{q})DS^k(-\mathbf{q})\exp\left[-(r+q^2)S^k(\mathbf{q})S^k(-\mathbf{q})\right]}, \qquad (10.5\text{-}24)$$

where we set $h = 0$ and where we integrated out all other variables, $S^j(\mathbf{q})$, $j\neq k$, in the partition function, noting, moreover, that the term $S^k(\mathbf{q})S^k(-\mathbf{q})$ of the exponent occurs twice, thereby cancelling the factor ½. Let us now write $S^k(\pm\mathbf{q})=x\pm iy$, where x and y are functions on **q**-space. Functional integrals were defined in (9.2-9), requiring a sequence of integrations at various points. For the present case we can simply consider a given point (x_p, y_p) at the time and integrate for this value; all other integrals in the numerator and denominator cancel. So we obtain,

$$\tilde{\Gamma}(\mathbf{q})=\left(\frac{1}{2\pi}\right)^d\frac{\int_{-\infty}^{\infty}\int_{-\infty}^{\infty}dx_p dy_p(x_p^2+y_p^2)\exp\left[-(r+q^2)(x_p^2+y_p^2)\right]}{\int_{-\infty}^{\infty}\int_{-\infty}^{\infty}dx_p dy_p\exp\left[-(r+q^2)(x_p^2+y_p^2)\right]}$$

$$=\left(\frac{1}{2\pi}\right)^d\frac{2\pi\int_0^{\infty}d\rho_p\rho_p^3\exp\left[-(r+q^2)\rho_p^2\right]}{2\pi\int_0^{\infty}d\rho_p\rho_p\exp\left[-(r+q^2)\rho_p^2\right]}=\left(\frac{1}{2\pi}\right)^d\frac{1}{(r+q^2)}, \qquad (10.5\text{-}25)$$

where we converted to polar coordinates and used (1.6-54). Identifying r with the reciprocal correlation length squared, $r=\xi^{-2}$, we have exactly the form that we met before, Eq. (9.2-15). Its inversion for a d-dimensional structure was given in (9.2-39); hence we find

$$\Gamma(\mathbf{x})=e^{-|x|\sqrt{r}}/(2\pi)^d\,|x|^{d-2}. \qquad (10.5\text{-}26)$$

Since $r\sim|t|$, ξ scales with $|t|^{-\frac{1}{2}}$, indicating that $\nu=\tfrac{1}{2}$, as in Landau–Ginzburg theory; also, the other hyperscaling exponent apparently satisfies $\eta=0$, as expected.

10.6 Elements of the S^4 Model and the Epsilon Expansion

The parameter u will be restored, so that we deal with the effective Hamiltonian (10.5-14); in some treatments a sixth-order spin term is added as well. We shall follow a cumulant-expansion, similar to what we did in Section 10.4. Thus, we write $\hat{\mathcal{H}} = \hat{\mathcal{H}}_0 + \hat{\mathcal{H}}_1$, with

$$\hat{\mathcal{H}}_0 = \left(\frac{1}{2\pi}\right)^d \int_{|\mathbf{q}|<\pi/a_0} d^d q \mathbf{S}(\mathbf{q}) \cdot \mathbf{S}(-\mathbf{q}) \frac{1}{2}\left(r + q^2\right) - h S^i(0), \tag{10.6-1}$$

$$\hat{\mathcal{H}}_1 = u\left(\frac{1}{2\pi}\right)^{3d} \iiiint_{|\mathbf{q}|<\pi/a_0} d^d q_1 \ldots d^d q_4 \; [\mathbf{S}(\mathbf{q}_1)\cdot\mathbf{S}(\mathbf{q}_2)][\mathbf{S}(\mathbf{q}_3)\cdot\mathbf{S}(\mathbf{q}_4)] \, \delta(\mathbf{q}_1 + \mathbf{q}_2 + \mathbf{q}_3 + \mathbf{q}_4). \tag{10.6-2}$$

Let us now have a block of ℓ spins. Then $\mathbf{q}$ shrinks to $q < \pi / a_0 \ell$. We must integrate over the short wavelength modes in the *outside* shell $\pi / a_0 \ell < q < \pi / a_0$ and retain the new essential modes within the sphere $q < \pi / a_0 \ell$. [This is the momentum-space equivalent of summing over the spins *inside* a block spin.] The procedure, while similar to that in the Gaussian model, is more involved because of the quartic term. Let us write down the partition function in the cumulant approximation form:

$$\mathcal{Z} = \prod_{q<\pi/a_0\ell} \int D^n S(\mathbf{q}) \prod_{\pi/a_0\ell<q<\pi/a_0} \int D^n S(\mathbf{q}) e^{-\hat{\mathcal{H}}_0} \langle e^{-\hat{\mathcal{H}}_1} \rangle, \tag{10.6-3}$$

where the average now is defined as

$$\langle P \rangle = \frac{\prod_{\pi/a_0\ell<q<\pi/a_0} \int D^n S(\mathbf{q}) e^{-\hat{\mathcal{H}}_0} P}{\prod_{\pi/a_0\ell<q<\pi/a_0} \int D^n S(\mathbf{q}) e^{-\hat{\mathcal{H}}_0}}. \tag{10.6-4}$$

We note that the denominator of (10.6-4) and the second product part in (10.6-3) cancel in as far as the spins of $\hat{\mathcal{H}}_0$ in the outer shell are concerned. Further, with the cumulant expansion up to second-order, we have

$$\langle e^{-\hat{\mathcal{H}}_1} \rangle = e^{-\langle \hat{\mathcal{H}}_1 \rangle + \frac{1}{2}[\langle \hat{\mathcal{H}}_1^2 \rangle - \langle \hat{\mathcal{H}}_1 \rangle^2]}. \tag{10.6-5}$$

Rewriting the partition function as

$$\mathcal{Z} = \prod_{q<\pi/a_0\ell} \int D^n S(\mathbf{q}) \exp\left(-\mathcal{H}'[\{\mathbf{S}(\mathbf{q})\}]\right), \tag{10.6-5'}$$

the primed Hamiltonian is

$$\mathcal{H}'[\{\mathbf{S}(\mathbf{q})\}] = \hat{\mathcal{H}}_0 + \langle\hat{\mathcal{H}}_1\rangle - \frac{1}{2}\left(\langle\hat{\mathcal{H}}_1^2\rangle - \langle\hat{\mathcal{H}}_1\rangle^2\right). \tag{10.6-6}$$

We proceed with the first cumulant evaluation. Since the averages concern only the spins in the outer shell, there is a piece in $\langle\hat{\mathcal{H}}_1\rangle$ that is not integrated and remains *as is*; ditto for the non-cancelled spins in $\hat{\mathcal{H}}_0$. Therefore, there is a part of $\mathcal{H}'$, associated with all spins having $\mathbf{q}$ in the sphere,

$$\hat{\mathcal{H}}'[\{\mathbf{S}(\mathbf{q})\}]\Big|_{\mathbf{q}\in\, sphere} = \left(\frac{1}{2\pi}\right)^d \int\limits_{q<\pi/a_0\ell} d^d q\, \mathbf{S}(\mathbf{q})\mathbf{S}(-\mathbf{q})\frac{1}{2}(r+q^2) - hS^i(0)$$

$$+u\left(\frac{1}{2\pi}\right)^{3d} \iiiint\limits_{q<\pi/a_0\ell} d^d q_1..d^d q_4\, \mathbf{S}(\mathbf{q}_1)\cdot\mathbf{S}(\mathbf{q}_2)\mathbf{S}(\mathbf{q}_3)\cdot\mathbf{S}(\mathbf{q}_4)\,\delta(\mathbf{q}_1+..+\mathbf{q}_4)\ . \tag{10.6-7}$$

Next, there are four other contributions to be considered.

(a) All $\mathbf{q}$'s are in the outer shell. Then all spin variables are integrated out, adding an insignificant constant to the partition function;
(b) One or three of the $\mathbf{q}$-vectors lie in the outer shell. The resulting averaged spin product is zero because of inversion symmetry in $\mathbf{q}$-space;
(c) Two $\mathbf{q}$-vectors are in the sphere, say $\mathbf{q}_1$ and $\mathbf{q}_2$, while the other ones, $\mathbf{q}_3$ and $\mathbf{q}_4$, are in the outer shell. Let $\mathbf{q}_1 = -\mathbf{q}_2, \mathbf{q}_3 = -\mathbf{q}_4$. This yields the following contribution,

$$\text{quartic term}\Big|_{(c)}$$

$$= u\left(\frac{1}{2\pi}\right)^{3d} \int\limits_{q<\pi/a_0\ell} d^d q_1 \sum_{i=1}^{n}\sum_{j=1}^{n} S^i(\mathbf{q}_1)S^i(-\mathbf{q}_1) \int\limits_{shell} d^d q_3 \langle S^j(\mathbf{q}_3)S^j(-\mathbf{q}_3)\rangle .$$

The sum over j yields $(2\pi)^d n\tilde{\Gamma}(\mathbf{q})$, where n is the number of components of the spin and $\tilde{\Gamma}(\mathbf{q})$ is the correlation function we met before; with the sum over i this reconstitutes the dot product, yielding $(2\pi)^{-2d} 2nu\int d^d q_1\, \mathbf{S}(\mathbf{q}_1)\cdot\mathbf{S}(-\mathbf{q}_1)\int d^d q_3\, \tilde{\Gamma}(\mathbf{q}_3)$; the factor 2 stems from the interchangeability of $\mathbf{q}_1$ and $\mathbf{q}_2$.
(d) The final possibility is that we have a quartic spin arrangement similar to case (c), but with the shell spins not stemming from the same scalar product. The $\mathbf{q}$-permutations give a factor 2! 2! = 4. Hence,

$$\text{quartic term}\Big|_{(d)}$$

$$= 4u\left(\frac{1}{2\pi}\right)^{3d} \int\limits_{q<\pi/a_0\ell} d^d q_1 \sum_{i=1}^{n}\sum_{j=1}^{n} S^i(\mathbf{q}_1)S^j(-\mathbf{q}_1) \int\limits_{shell} d^d q_3 \underbrace{\langle S^i(\mathbf{q}_3)S^j(-\mathbf{q}_3)\rangle}_{2\pi\tilde{\Gamma}(\mathbf{q}_3)\delta_{ij}} .$$

We now collect the results to obtain

$$\mathcal{H}'[\{\mathbf{S}(\mathbf{q})\}]=\left(\frac{1}{2\pi}\right)^d \int_{q<\pi/a_0\ell} d^d q\mathbf{S}(\mathbf{q})\mathbf{S}(-\mathbf{q})$$

$$\times\left(\frac{r+q^2}{2}+u(2n+4)\left(\frac{1}{2\pi}\right)^d \int_{\pi/a_0\ell<q'<\pi/a_0} d^d q'\tilde{\Gamma}(\mathbf{q}')\right)-hS^i(0)$$

$$+u\left(\frac{1}{2\pi}\right)^{3d} \iiiint_{q<\pi/a_0\ell} d^d q_1..d^d q_4\mathbf{S}(\mathbf{q}_1)\cdot\mathbf{S}(\mathbf{q}_2)\mathbf{S}(\mathbf{q}_3)\cdot\mathbf{S}(\mathbf{q}_4)\ \delta(\mathbf{q}_1+..\mathbf{q}_4)\ . \quad (10.6\text{-}8)$$

As in the Gaussian model we scale back to the original domain for $\mathbf{q}$, i.e., we set $\mathbf{q}=\mathbf{q}'/\ell$ and $\mathbf{S}(\mathbf{q}'/\ell)=\ell^{1+d/2}\mathbf{S}'(\mathbf{q}')$, cf.(10.5-18). This yields, noting $\delta(\mathbf{q}/\ell)=\ell^d\delta(\mathbf{q})$,

$$\mathcal{H}'[\{\mathbf{S}'(\mathbf{q}')\}]=\left(\frac{1}{2\pi}\right)^d \int_{q'<\pi/a_0} d^d q'\mathbf{S}'(\mathbf{q}')\mathbf{S}'(-\mathbf{q}')$$

$$\times\left(\frac{r\ell^2+q'^2}{2}+u(2n+4)\ell^2\left(\frac{1}{2\pi}\right)^d \int_{\pi/a_0\ell<\bar{q}<\pi/a_0} d^d\bar{q}\tilde{\Gamma}(\bar{\mathbf{q}})\right)-h\ell^{1+d/2}S'^i(0)$$

$$+u\ell^{4-d}\left(\frac{1}{2\pi}\right)^{3d} \iiiint_{q'<\pi/a_0} d^d q_1'..d^d q_4'[\mathbf{S}'(\mathbf{q}_1')\cdot\mathbf{S}'(\mathbf{q}_2')][\mathbf{S}'(\mathbf{q}_3')\cdot\mathbf{S}'(\mathbf{q}_4')]\,\delta(\mathbf{q}_1'+..\mathbf{q}_4') .$$

(10.6-9)

[Note: we did not rescale the correlation function, since this involves a definite integral, denoted as I.] The power for rescaling u is $4-d\equiv\varepsilon$, being the first-order term of the renowned Wilson–Fisher *epsilon expansion*. The recursion relations are found to be

$$r'=r\ell^2+4(n+2)\ell^2uI, \qquad u'=u\ell^{4-d}, \qquad h'=h\ell^{1+\frac{1}{2}d}. \qquad (10.6\text{-}10)$$

The integral will be evaluated later; here we need only that $I>0$. We note that for $d>4$ $u'<u$, i.e. in this case the Gaussian fixed point is slightly shifted but still stable. However, for $d<4$ $u'>u$, meaning that the fourth order term grows and that another fixed point may come in. Indeed, let us suppose that a more complete computation will lead to

$$u'=u\ell^{4-d}[1-\varphi(r,\ell)u]. \qquad (10.6\text{-}11)$$

Besides the Gaussian fixed point $u^*=0$, there will then be another fixed point (u^*,r^*) with

$$u^*=\frac{1-\ell^{-\varepsilon}}{\varphi(l,r^*)}\simeq\varepsilon\ln\ell/\varphi(l,r^*), \qquad (10.6\text{-}12)$$

which will be stable for appropriate r^* and $\varphi(r,\ell)$.

We shall proceed to the computation of the second cumulant term, $-\frac{1}{2}(\langle\hat{\mathcal{H}}_1^2\rangle-\langle\hat{\mathcal{H}}_1\rangle^2)$, keeping only powers of first order in ε. In general we have, after scaling to blocks $a_0\ell$,

$$-\tfrac{1}{2}(\langle\hat{\mathcal{H}}_1'^2\rangle-\langle\hat{\mathcal{H}}_1'\rangle^2)$$

$$=-\frac{u^2}{2}\left(\frac{1}{2\pi}\right)^{6d}\int_{q<\pi/a_0\ell} d^dq \int_{shell} d^dq[\mathbf{S}(\mathbf{q}_1)\cdot\mathbf{S}(\mathbf{q}_2)][\mathbf{S}(\mathbf{q}_3)\cdot\mathbf{S}(\mathbf{q}_4)]\delta(\mathbf{q}_1+\mathbf{q}_2+\mathbf{q}_3+\mathbf{q}_4)$$

$$\times[\mathbf{S}(\mathbf{q}_5)\cdot\mathbf{S}(\mathbf{q}_6)][\mathbf{S}(\mathbf{q}_7)\cdot\mathbf{S}(\mathbf{q}_8)]\delta(\mathbf{q}_5+\mathbf{q}_6+\mathbf{q}_7+\mathbf{q}_8)+\tfrac{1}{2}\langle\hat{\mathcal{H}}_1'\rangle^2, \tag{10.6-13}$$

where we have not yet incorporated the averaging bars for the integrand, since they need to be placed around those spins which are in the shell. If all eight spins are in the shell, they are simply integrated out, giving a constant to the partition function. Next, if all four spins of either sequence $\mathbf{q}_1\ldots\mathbf{q}_4$ or $\mathbf{q}_5\ldots\mathbf{q}_8$ are in the outer shell, their product being averaged, the result will be cancelled by the squared averages of the last term, $[\frac{1}{2}\langle\hat{\mathcal{H}}_1'\rangle^2]$, in (10.6-13). To see this we need to apply Wick's theorem to the averaged four-spin product, which states that it can be replaced by pairwise two-spin product averages [Wick's theorem will be considered in Chapter XII]. Finally we note that any sequence with an odd number of spins in the shell will vanish because of symmetry considerations. Rests therefore as only viable possibility that *two of the spins* $\mathbf{S}(\mathbf{q}_1)..\mathbf{S}(\mathbf{q}_4)$ *and two of the spins* $\mathbf{S}(\mathbf{q}_5)...\mathbf{S}(\mathbf{q}_8)$ *have their momenta in the outer shell.* Now there are still some distinct cases to be considered.

(a) The two momentum vectors in the outer shell for each of the two sets stem from the same scalar product. E.g., let $\mathbf{q}_3 = -\mathbf{q}_7$ and $\mathbf{q}_4 = -\mathbf{q}_8$ be the outer shell momenta. There will be eight such cases, giving a contribution

$$2^{nd}\,\text{cum}\big|_{(a)}=-4u^2\left(\frac{1}{2\pi}\right)^{6d}\int_{\substack{\mathbf{q}_1,\mathbf{q}_2,\mathbf{q}_5,\mathbf{q}_6\\ \in\, sphere}} d^dq \int_{\substack{\mathbf{q}_3,\mathbf{q}_4\\ \in\, shell}} d^dq[\mathbf{S}(\mathbf{q}_1)\cdot\mathbf{S}(\mathbf{q}_2)][\mathbf{S}(\mathbf{q}_5)\cdot\mathbf{S}(\mathbf{q}_6)]$$

$$\times\sum_{i,j=1}^{n}\langle S^i(\mathbf{q}_3)S^j(-\mathbf{q}_3)S^i(\mathbf{q}_4)S^j(-\mathbf{q}_4)\rangle\delta(\mathbf{q}_1+\mathbf{q}_2+\mathbf{q}_3+\mathbf{q}_4)\delta(\mathbf{q}_5+\mathbf{q}_6-\mathbf{q}_3-\mathbf{q}_4)$$

$$=-4u^2\left(\frac{1}{2\pi}\right)^{6d}\int_{\substack{\mathbf{q}_1,\mathbf{q}_2,\mathbf{q}_5,\mathbf{q}_6\\ \in\, sphere}} d^dq[\mathbf{S}(\mathbf{q}_1)\cdot\mathbf{S}(\mathbf{q}_2)][\mathbf{S}(\mathbf{q}_5)\cdot\mathbf{S}(\mathbf{q}_6)]\delta(\mathbf{q}_1+\mathbf{q}_2+\mathbf{q}_5+\mathbf{q}_6)$$

$$\times\int_{\substack{\mathbf{q}_3,\mathbf{q}_4\\ \in\, shell}} d^dq\sum_{i,j=1}^{n}\underbrace{\langle S^i(\mathbf{q}_3)S^j(-\mathbf{q}_3)\rangle}_{(2\pi)^d\tilde{\Gamma}(\mathbf{q}_3)\delta_{ij}}\underbrace{\langle S^i(\mathbf{q}_4)S^j(-\mathbf{q}_4)\rangle}_{(2\pi)^d\tilde{\Gamma}(\mathbf{q}_4)\delta_{ij}}\,\delta(\mathbf{q}_5+\mathbf{q}_6-\mathbf{q}_3-\mathbf{q}_4)$$

$$= -4nu^2 \left(\frac{1}{2\pi}\right)^{3d} \int\limits_{\substack{\mathbf{q}_1,\mathbf{q}_2,\mathbf{q}_5,\mathbf{q}_6 \\ \in\, sphere}} d^d q [\mathbf{S}(\mathbf{q}_1)\cdot\mathbf{S}(\mathbf{q}_2)][\mathbf{S}(\mathbf{q}_5)\cdot\mathbf{S}(\mathbf{q}_6)]\delta(\mathbf{q}_1+\mathbf{q}_2+\mathbf{q}_5+\mathbf{q}_6)$$

$$\times \left(\frac{1}{2\pi}\right)^{d} \int\limits_{\substack{\mathbf{q}_3,\mathbf{q}_4 \\ \in\, shell}} d^d q \tilde{\Gamma}(\mathbf{q}_3)\tilde{\Gamma}(\mathbf{q}_4)\delta(\mathbf{q}_3+\mathbf{q}_4-\mathbf{q}_5-\mathbf{q}_6). \qquad (10.6\text{-}14)$$

Note that we applied Wick's theorem in the middle expression and that we used the definition of the two-point correlation function in the final result.

(b) The total numbers of pairing being $2\times(4!/2!2!)^2 = 72$, there remain 64 pairings for which the spin pairs stem from arbitrary combinations of each of the four-products. Labelling the spin components with superscripts i, j, k, l we get, taking again as example the pairs $\mathbf{q}_3 = -\mathbf{q}_7$ and $\mathbf{q}_4 = -\mathbf{q}_8$ placed in the outer shell,

$$2^{nd}\ \text{cum}\big|_{(b)}$$

$$= -32u^2 \left(\frac{1}{2\pi}\right)^{6d} \int\limits_{\substack{\mathbf{q}_1,\mathbf{q}_2,\mathbf{q}_5,\mathbf{q}_6 \\ \in\, sphere}} d^d q \sum_{i,j,k,l} S^i(\mathbf{q}_1)S^j(\mathbf{q}_2)S^k(\mathbf{q}_5)S^l(\mathbf{q}_6)\delta(\mathbf{q}_1+\mathbf{q}_2+\mathbf{q}_3+\mathbf{q}_4)$$

$$\times \int\limits_{\substack{\mathbf{q}_3,\mathbf{q}_4 \\ \in\, shell}} d^d q\, \underbrace{\langle S^i(\mathbf{q}_3)S^k(-\mathbf{q}_3)\rangle}_{(2\pi)^d\tilde{\Gamma}(\mathbf{q}_3)\delta_{ik}} \underbrace{\langle S^j(\mathbf{q}_4)S^l(-\mathbf{q}_4)\rangle}_{(2\pi)^d\tilde{\Gamma}(\mathbf{q}_4)\delta_{jl}} \delta(\mathbf{q}_5-\mathbf{q}_3+\mathbf{q}_6-\mathbf{q}_4)$$

$$= -32u^2 \left(\frac{1}{2\pi}\right)^{3d} \int\limits_{\substack{\mathbf{q}_1,\mathbf{q}_2,\mathbf{q}_5,\mathbf{q}_6 \\ \in\, sphere}} d^d q [\mathbf{S}(\mathbf{q}_1)\cdot\mathbf{S}(\mathbf{q}_2)][\mathbf{S}(\mathbf{q}_5)\cdot\mathbf{S}(\mathbf{q}_6)]\delta(\mathbf{q}_1+\mathbf{q}_2+\mathbf{q}_5+\mathbf{q}_6)$$

$$\times \left(\frac{1}{2\pi}\right)^{d} \int\limits_{\substack{\mathbf{q}_3,\mathbf{q}_4 \\ \in\, shell}} d^d q \tilde{\Gamma}(\mathbf{q}_3)\tilde{\Gamma}(\mathbf{q}_4)\delta(\mathbf{q}_3+\mathbf{q}_4-\mathbf{q}_5-\mathbf{q}_6). \qquad (10.6\text{-}15)$$

(c) There are a few more cases (e.g., $\mathbf{q}_3 = -\mathbf{q}_4 = \mathbf{q}_7 = -\mathbf{q}_8$ are all in the outer shell), but their integration over the shell yields vanishing small results.

The shell integrals in (10.6-14) and (10.6-15) will be denoted by *J*. Both *I* and *J* will be evaluated below. We now add the above results for the second cumulant and we do the usual rescaling, $\mathbf{q} = \mathbf{q}'/\ell$ and $\mathbf{S}(\mathbf{q}'/\ell) = \ell^{1+d/2}\mathbf{S}'(\mathbf{q}')$. We relabel some of the $\mathbf{q}$'s and we set (with Wilson and Kogut) $\pi/a_0 = 1$. The total contribution then becomes

$$2^{nd}\,\mathrm{cum}\Big|_{rescaled}$$

$$= -4u^2\ell^{4-d}(n+8)\left(\frac{1}{2\pi}\right)^{3d}\int_{q_1,q_2,q_3,q_4<1} d^dq[\mathbf{S}(\mathbf{q}_1)\cdot\mathbf{S}(\mathbf{q}_2)][\mathbf{S}(\mathbf{q}_3)\cdot\mathbf{S}(\mathbf{q}_4)]$$

$$\times\delta(\mathbf{q}_1+\mathbf{q}_2+\mathbf{q}_3+\mathbf{q}_4)\left(\frac{1}{2\pi}\right)^{d}\int_{\substack{q,q'\\ \in\, shell}} d^dq\tilde{\Gamma}(\mathbf{q})\tilde{\Gamma}(\mathbf{q}')\delta(\mathbf{q}+\mathbf{q}'-\mathbf{q}_3-\mathbf{q}_4). \quad (10.6\text{-}16)$$

The total rescaled Hamiltonian is obtained by adding (10.6-9) to (10.6-16). The following recursion relations are read from these results:

$$\begin{aligned} r' &= r\ell^2 + 4(n+2)u\ell^2 I, \\ u' &= u\ell^{4-d}[1-4(n+8)uJ], \\ h' &= h\ell^{1+\frac{1}{2}d}. \end{aligned} \quad (10.6\text{-}17)$$

We proceed to the evaluation of the integrals. The rescaling parameter for quasi-continuous rescaling near a critical point will be of the form $\ell = 1+\delta$, where δ is a small quantity. The integral I then becomes

$$I = (2\pi)^{-d}\int_{1/(1+\delta)<q<1} d^dq\tilde{\Gamma}(q) = [A_d/(2\pi)^{2d}]\delta\frac{1}{r+1} \equiv C\delta\frac{1}{r+1}, \quad (10.6\text{-}18)$$

where A_d is the area of the d-dimensional hypersphere with unit radius, cf. Eq. (2.5-4). The other integral goes likewise but requires an *ad hoc* approximation; we have

$$\begin{aligned} J &= (2\pi)^{-d}\iint_{1/(1+\delta)<q<1} d^dqd^dq'\tilde{\Gamma}(q)\tilde{\Gamma}(q')\delta(\mathbf{q}+\mathbf{q}'-\mathbf{q}_3-\mathbf{q}_4) \\ &= (2\pi)^{-d}\int_{1/(1+\delta)<q<1} d^dq\tilde{\Gamma}(q)\tilde{\Gamma}(|\mathbf{q}-\mathbf{q}_3-\mathbf{q}_4|) \\ &\approx (2\pi)^{-d}\int_{1/(1+\delta)<q<1} d^dq[\tilde{\Gamma}(q)]^2 = C\delta\left(\frac{1}{r+1}\right)^2. \end{aligned} \quad (10.6\text{-}19)$$

Finally we are in a position to get the fruit of all this labour. The fixed point discussed before follows from $r = r' = r^*$ and $u = u' = u^*$. Keeping lowest orders in ε and δ one finds

$$r^* = -\frac{(n+2)}{2(n+8)}\varepsilon, \quad u^* = \frac{1}{4C(n+8)}\varepsilon. \quad (10.6\text{-}20)$$

We substitute the results for the two integrals into (10.6-17); this yields

$$r' = r\ell^2 + [4(n+2)u\ell^2 C\delta]/(r+1)\,,$$
$$u' = u\ell^{\varepsilon}[1-4(n+8)uC\delta/(r+1)^2]\,, \qquad (10.6\text{-}21)$$
$$h' = h\ell^{1+\frac{1}{2}d}\,.$$

Next we need the matrix associated with the infinitesimal transformations near the critical point. The following results are easily established,

$$\frac{\partial r'}{\partial r} = \ell^2\left[1-\frac{4(n+2)u^* C\delta}{(r^*+1)^2}\right] = \ell^2 - \frac{n+2}{n+8}\frac{(1+\delta)^2\delta\varepsilon}{(r^*+1)^2} \approx \ell^2 - \frac{n+2}{n+8}\delta\varepsilon + \mathcal{O}(\varepsilon^2)\,,$$

$$\frac{\partial r'}{\partial u} = \frac{4(n+2)\ell^2 C}{(r^*+1)}\delta \approx 4(n+2)C\delta + \mathcal{O}(\varepsilon^2)\,,$$

$$\frac{\partial u'}{\partial r} = u^{*2}\ell^{\varepsilon}\frac{8(n+8)C\delta}{(r^*+1)^3} = \mathcal{O}(\varepsilon^2)\,,$$

$$\frac{\partial u'}{\partial u} = \ell^{\varepsilon}\left[1-\frac{8(n+8)u^* C\delta}{(r^*+1)^2}\right] = (1+\delta)^{\varepsilon}\left[1-\frac{2\varepsilon\delta}{(r^*+1)^2}\right]$$

$$\approx (1+\delta)^{\varepsilon}(1+\delta)^{-2\varepsilon} = (1+\delta)^{-\varepsilon}\,. \qquad (10.6\text{-}22)$$

In the approximations denoted with '$\approx$', terms of order δ^2 have been neglected and/or the assumption $r^* \ll 1$ was made. The transformation matrix (excluding the h-rescaling) now reads

$$\mathsf{T} = \begin{pmatrix} (1+\delta)^2 - \frac{n+2}{n+8}\delta\varepsilon & 4(n+2)C\delta \\ 0 & (1+\delta)^{-\varepsilon} \end{pmatrix}. \qquad (10.6\text{-}23)$$

Had we added the rescaling for the magnetic field, the matrix would have been 3×3, there being three eigenvalues; hence,

$$\lambda_1 = (1+\delta)^2 - \frac{n+2}{n+8}\delta\varepsilon, \quad \lambda_2 = (1+\delta)^{-\varepsilon}, \quad \lambda_h = \ell^{1+\frac{1}{2}d} = \ell^{3-\frac{1}{2}\varepsilon}. \qquad (10.6\text{-}24)$$

Altogether, there are three scaling fields, u_1, u_2 and u_h. The free energy scaling function is

$$\hat{g}_s(u_1,u_2,u_h) = \ell^{-d}\,\hat{g}_s(u_1\ell^{pd}, u_2\ell^{sd}, u_h\ell^{qd})\,, \qquad (10.6\text{-}25)$$

see subsection 10.3.2, where the factors $\ell^{pd}, \ell^{sd}, \ell^{qd}$ are the eigenvalues, λ_1, λ_2 and λ_h, respectively. Previously we associated λ_1 with temperature scaling, whence the use of the exponent p. The eigenvalue λ_h is associated with magnetic-field scaling, having the exponent q. As to λ_2, it is not associated with any critical exponent, being *irrelevant* $\lambda_2 < 1$. We thus obtained

$$p = \frac{\ln \lambda_1}{d \ln(1+\delta)} \approx \frac{2}{d} - \frac{n+2}{n+8}\frac{\varepsilon}{d}, \qquad s = -\frac{\varepsilon}{d}, \qquad q = \frac{3}{d} - \frac{\varepsilon}{2d}. \tag{10.6-26}$$

The results can also be expressed in n and ε only by using $d = 4 - \varepsilon$. Usually the exponents are further approximated since, in the S^4 model as presented, we have not considered any terms of order ε^2 or higher. For example, one easily finds

$$\frac{1}{p} \approx 2 + \varepsilon\left(\frac{n+2}{n+8} - \frac{1}{2}\right), \qquad \forall \quad \frac{(n+2)^2}{4(n+8)^2}\varepsilon^2 \ll 1. \tag{10.6-27}$$

For the Heisenberg 3D spin Hamiltonian with $n = 3$ and $\varepsilon = 1$ the error is 5%; but often the error is substantial and one should bear in mind that in the real 3D world, the epsilon expansion does not converge! One may hope, however, that the S^4 model gives *insight* into critical processes by allowing us to consider, instead, a dimensionality *slightly above* $d = 3$, i.e., $\varepsilon = 1 - \xi$, $\xi > 0$. Most investigators consider the expansion under those conditions to be an asymptotic expansion, for which a small number of powers in ε will yield a meaningful description. With this admonition in mind, we present the usual first-order results for the critical exponents as found in the literature, related to p and q according to the formulas derived in subsection 9.6.1. The following then are easily verified

$$\alpha = \left[\frac{1}{2} - \frac{n+2}{n+8}\right]\varepsilon + \mathcal{O}(\varepsilon^2), \quad \beta = \frac{1}{2} - \frac{3\varepsilon}{2(n+8)} + \mathcal{O}(\varepsilon^2),$$
$$\gamma = 1 + \frac{(n+2)\varepsilon}{2(n+8)} + \mathcal{O}(\varepsilon^2), \qquad \delta = 3 + \varepsilon + \mathcal{O}(\varepsilon^2). \tag{10.6-28}$$

Both Rushbrooke's and Widom's scaling relationships are satisfied. In Table 10-1 we give the results for the various exponents for $\varepsilon = 1$ and $n = 1$, 2 or 3, representing the 3D Ising model, the X-Y model and the Heisenberg Hamiltonian, respectively.

Table 10-1. Critical exponents from the S^4 model for the Ising, X-Y and Heisenberg Hamiltonians.

Exp.	Ising 3D	X-Y model	Heisenberg	series expansions
α	1/6	1/10	1/22	
β	1/3	17/20	9/11	
γ	7/6	6/5	27/22	[1.24, 1.33, 1.44]
δ	4	4	4	

For comparison with other theoretical computations and experimental values, see Table 9-1 in the preceding chapter. One particular aspect of the S^4 model stands out: It clearly demonstrates the principle of *universality*: critical exponents for all Hamiltonians with given d and n are the same.

Finally we must comment on the "flow" of the scaling-fields parameters. From the incorporation of the quartic term and computation up to the second cumulant, we found that a second non-trivial fixed point arose. When $d > 4$ or $\varepsilon < 0$, the second fixed point becomes unstable , i.e., becomes unphysical; the system then settles in the Gaussian fixed point. For $\varepsilon > 0$ the system moves away from the Gaussian fixed point to the non-trivial fixed point along the line $(r^* - r)/(u^* - u) = -2(n+2)/C$. The flow diagram in the two-parameter (u,r) coupling constant space is sketched in Fig. 10-8. The magnetic field parameter h could be drawn as a third axis ⊥ the plane.

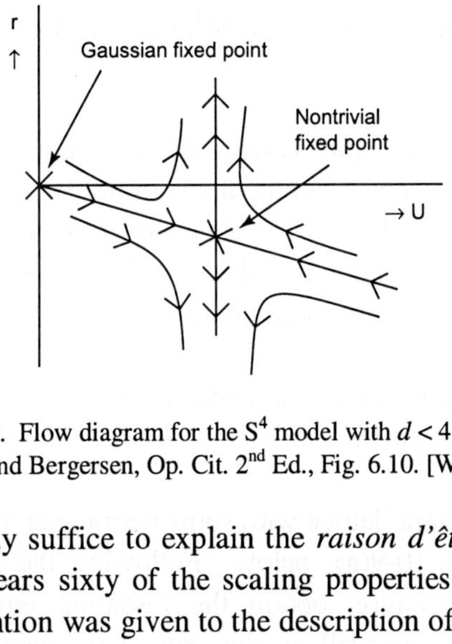

Fig. 10-8. Flow diagram for the S^4 model with $d < 4$ ($\varepsilon > 0$).
After Plischke and Bergersen, Op. Cit. 2nd Ed., Fig. 6.10. [With permission.]

A final paragraph may suffice to explain the *raison d'être* of this chapter. After the discoveries in the years sixty of the scaling properties and the various critical exponents, renewed attention was given to the description of phase transitions from a more general vantage point. It was soon recognized that Widom's and Kadanoff's rescaling procedures were deficient and in need of a more general framework. This was found in terms of the renormalization group, initiated by Wilson, a colleague of Widom at Cornell University, in the early seventies. Our chapters IX and X are to be seen together, with the latter chapter being the closure of the ideas developed in the former. While the ultimate models that we discussed may have seemed to be somewhat esoteric, they have opened up a much wider view than usually befits a physical theory. In fact, the underlying ideas are probably more of a philosophical nature than 'just physics'. We are reminded – once again – that (a few centuries ago) physics was called *natural philosophy*. It is not surprising that renormalization theory has found applications in other domains of physics, in particular for condensed matter problems, dynamical systems, as well as in the socio-economic sciences.

10.7 Problems to Chapter X

10.1 Consider again the linear spin chain rescaling of Section 10.2. Show that after n rescalings – thinning out by ½ – the number of remaining spins is $N(\frac{1}{2})^{n+1}$.
(a) What is the correlation length if this 'n' is the last possible rescaling procedure?
(b) What is the entropy per spin, $\hat{s}$, associated with the remaining spins? [*answer*: $(\frac{1}{2})^{n+1} k_B \ln 2.$]
(c) From the rescaling relationships (10.2-8) find the infinitesimal scaling matrix

$$\mathsf{T} = \begin{pmatrix} \partial K'/\partial K & \partial K'/\partial h \\ \partial h'/\partial K & \partial h'/\partial h \end{pmatrix}_{K^*=\infty, h^*=0} . \tag{1}$$

Show that the eigenvalues are *marginal*, i.e., the system is unstable at that fixed point ($T = 0$).

10.2 For the triangular 2D spin lattice the Niemeijer–van Leeuwen rescaling equations, up to the second cumulant expansion, were listed in Eqs. (10.4-34). Review the results for K' and M' obtained in the text [what is the origin and meaning of each contribution?] and *derive* the result for the coupling constant of the rescaled second-nearest neighbours, L'.

Problems 10.3-10.5 require a small computer program for numerical solutions or a software program like *Mathematica*.

10.3 Consider the 2D square lattice with spins rearranged into square blocks of nine spins each, see Fig. 10-9(a) below. Following the Niemeijer–van Leeuwen renormalization procedure, obtain the rescaling equations up to the first cumulant expansion, as well as the critical temperature T_c and the scaling coefficients α, β, γ and δ. Compare with the results in Eqs. (10.4-25).

10.4 For the same 2D square lattice, rearrange the spins into blocks of five spins each, as indicated in Fig. 10-9(b) below. Consider separate subspaces for the case that the central spin is up or down. Obtain again the rescaling equations and critical exponents in the first cumulant approximation.

10.5 Extend the results of Problem 10.4 up to the second cumulant expansion.
(a) Obtain the fixed point value K^* and the critical temperature T_c;
(b) Find numerical values for the T-matrix and obtain the eigenvalues;
(c) Obtain the critical exponents and compare with the exact square-lattice Ising values from Chapter XI: $k_B T_c / J = 2.269$; α=0, β=0.125, γ=1.750, δ=15.

10.6 For Wilson's S^4 model do *not* set $\ell = 1+\delta$ and let $\pi / a_0 = \rho$. Obtain more exact results for the fixed points r^*, u^* and h^* and find the T-matrix up to order ε, making *reasonable* approximations where necessary. Compare with the results obtained in the text.

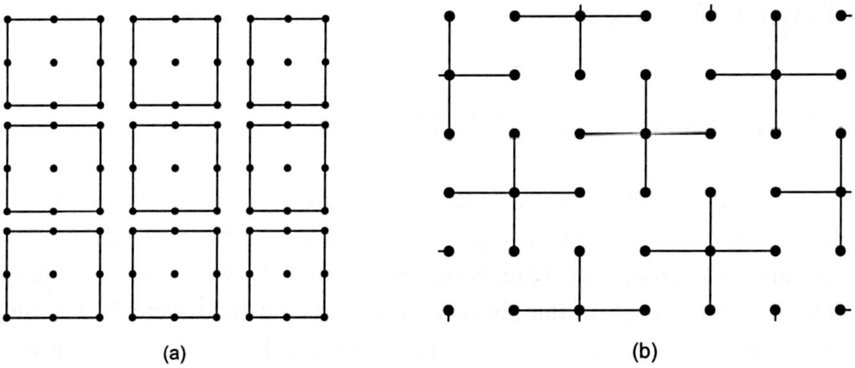

Fig. 10-9. 2D square spin lattice. (a): grouping into blocks of nine; (b): grouping into blocks of five. Discern nearest neighbour block spins, second neighbour block spins, etc.

Chapter XI

The Two-Dimensional Ising Model and Spin Waves

11.1 Historical Notes. Review of the 1D Model

The 2D Ising model of ferromagnetism and anti-ferromagnetism is one of the rare problems in statistical mechanics that shows a phase transition, paramagnetic state → ferro (or anti-ferro) magnetic state, which is *exactly solvable.* The canonical partition function and free energy in the absence of a magnetic field were first computed by Lars Onsager, who used the theory of Lie algebras. The solution was presented by Onsager as a "discussion remark", following a paper by Gregory Wannier at a meeting of the New York Academy of Sciences in February 1942; his statement was published two years later, in the Physical review of 1944.[1] The critical temperature is obtained from the condition

$$\sinh\frac{2J_1}{k_B T_c}\sinh\frac{2J_2}{k_B T_c}=1, \tag{11.1-1}$$

where J_1 and J_2 are the exchange integrals in the two directions.

Even more challenging was the subsequent computation of the free energy in an external field and the spontaneous magnetization per spin, $m_0(T)$, in units ρ_B (Bohrmagneton); it is given by

$$m_0(T)=-\lim_{H\to 0}[\partial\hat{f}(H,T)/\partial H]=(1-x^{-2})^{1/8},\quad T<T_c, \tag{11.1-2}$$

with

$$x=\sinh\frac{2J_1}{k_B T}\sinh\frac{2J_2}{k_B T}. \tag{11.1-3}$$

From (11.1-1) and (11.1-2), when $T\to T_c$ from below, the limiting form of the magnetization is

$$m_0(T)\approx(1-T/T_c)^{\beta},\quad \beta=1/8. \tag{11.1-4}$$

Onsager stated this result at a conference on the theory of phase transitions at Cornell University in 1948, following a paper by Laszlo Tisza and repeated it at the first post-war IUPAP meeting in Florence, Italy later that year; it was finally

---

[1] L. Onsager, Phys. Rev. **65**, 117 (1944).

presented for the general public as a discussion remark in 1949.[2] Yet, Onsager never published the derivation of his result! The puzzle was solved by C.N. Yang in a very complicated paper, which appeared in 1952.[3] Subsequent simpler solutions were presented by Newell and Montroll[4], and by various others. So much for the history... It would take another twelve years before Schultz, Mattis and Lieb[5] published a difficult, but readable account, which made the topic accessible to less specialized physicists and to graduate textbooks for students. A treatment based on that paper is found in Mattis' book[6] and in the more recent book of Plischke and Bergersen[7]. Our treatment here will basically follow the papers by Schultz et al.[5] and by Montroll, Potts and Ward in the "Onsager celebration issue" of 1963 [8].

We begin with a review of the 1D model, differing from the standard treatment as presented in section 4.3.3, but based on the "transfer-matrix method". We recall the general Heisenberg Hamiltonian for interacting spins in a magnetic field:

$$\mathcal{H} = -\frac{1}{\hbar^2}\sum_{\mathbf{l},\mathbf{l}'}{}' J_{\mathbf{l},\mathbf{l}'}\mathbf{S}_\mathbf{l}\cdot\mathbf{S}_{\mathbf{l}'} - gH\frac{\rho_B}{\hbar}\sum_\mathbf{l} S^z_\mathbf{l}. \tag{11.1-5}$$

Here $\rho_B = e\hbar/2m_e$ is the Bohrmagneton; the Landé factor g will be taken to be 2.0 . In the Ising model it is assumed that the x-y interaction $S^x_\mathbf{l}S^x_{\mathbf{l}'} + S^y_\mathbf{l}S^y_{\mathbf{l}'}$ averages to zero; the so-obtained Ising Hamiltonian only contains S^z. Writing further $S^z = \sigma_z\hbar/2$ where σ_z is the spin operator with EV ± 1, $J_0 = J_{\mathbf{l},\mathbf{l+a}}/4$ (nearest neighbour interaction only) and $h = (g\rho_B/2)H = \rho_B H$, we obtain the more pleasant form,

$$\mathcal{H} = -J_0\sum_{\mathbf{l},\mathbf{a}}\sigma_\mathbf{l}\sigma_{\mathbf{l+a}} - h\sum_\mathbf{l}\sigma_\mathbf{l}, \quad (\sigma\equiv\sigma_z), \tag{11.1-6}$$

where $\mathbf{l}$ labels all sites in the lattice of any dimension.

We return to the 1D case (linear chain), for which the canonical partition function, based on (11.1-6), is found to be

$$\mathcal{Z} = \mathrm{Tr}\left\{\left(e^{\beta h\sigma_1}e^{K\sigma_1\sigma_2}\right)\left(e^{\beta h\sigma_2}e^{K\sigma_2\sigma_3}\right)\ldots\left(e^{\beta h\sigma_N}e^{K\sigma_N\sigma_1}\right)\right\}, \tag{11.1-7}$$

where as usual $\beta = 1/k_BT$, $K = \beta J_0$, and where we used periodic boundary conditions

---

[2] L. Onsager, Nuovo Cimento, Suppl. **6**, 261 (1949).

[3] C.N. Yang, Phys. Rev **85**, 809 (1952); ibid. Rev. Mod. Physics **34**, 694 (1962).

[4] G.F. Newell and E.W. Montroll, Rev. Mod. Physics **25**, 353 (1952).

[5] T. Schultz, D. Mattis and E. Lieb, Rev. Mod. Physics **36**, 856-871 (1964).

[6] D.C. Mattis, "The Theory of Magnetism", Harper and Row, 1965. [There is presently a newer version of this book, entitled "ibid. – Statics and Dynamics", Springer-Verlag, Berlin, 1981.]

[7] M. Plischke and B. Bergersen, "Equilibrium Statistical Physics" 2nd Ed., World Scientific, 1994 Chapter 5.

[8] E.W. Montroll, R.B. Potts and J.C. Ward, J. Math. Physics **4**, 308-322 (1963).

(b.c.) $\sigma_{i+N} = \sigma_i$. In (11.1-7) we grouped the terms together differently than in subsection 4.3.3, in order to get a more lucid form for the transfer matrix, which can easily be extended to two or more dimensions. As basis states we use the ES of the spin operator σ_z

$$\sigma_z\,|+1\rangle = |+1\rangle, \quad \sigma_z\,|-1\rangle = -|-1\rangle, \tag{11.1-8}$$

indicating the EV ± 1. In a matrix *realization* of the spin space [Hilbert space $\ell^2(R_2)$] we have

$$|+1\rangle = \begin{pmatrix} 1 \\ 0 \end{pmatrix}, \quad |-1\rangle = \begin{pmatrix} 0 \\ 1 \end{pmatrix}. \tag{11.1-9}$$

The pertinent Pauli spin operators in this space are

$$\sigma_z = \begin{pmatrix} 1 & 0 \\ 0 & -1 \end{pmatrix}, \quad \sigma^+ = \begin{pmatrix} 0 & 1 \\ 0 & 0 \end{pmatrix}, \quad \sigma^- = \begin{pmatrix} 0 & 0 \\ 1 & 0 \end{pmatrix}, \quad \mathbf{1} = \begin{pmatrix} 1 & 0 \\ 0 & 1 \end{pmatrix}, \tag{11.1.10}$$

where we added the unit matrix.[9] We also used the rules $\sigma^{\pm} = \frac{1}{2}(\sigma_x \pm i\sigma_y)$, which imply

$$\sigma_x = \begin{pmatrix} 0 & 1 \\ 1 & 0 \end{pmatrix}, \quad \sigma_y = \begin{pmatrix} 0 & -i \\ i & 0 \end{pmatrix}. \tag{11.1-11}$$

Both matrices are Hermitean but have off-diagonal elements only. We also review some well-known rules.[10] Each matrix squares to the unit matrix:

$$\sigma_x^2 = \sigma_y^2 = \sigma_z^2 = \mathbf{1}. \tag{11.1-12}$$

From the angular momentum rule, valid for **L** or for **S**, $\boldsymbol{\sigma}\times\boldsymbol{\sigma} = 2i\boldsymbol{\sigma}$, we have

$$\begin{aligned}[\sigma_x, \sigma_y]_+ &= \sigma_x\sigma_y + \sigma_y\sigma_x = (\sigma_x\sigma_y - \sigma_y\sigma_x) + 2\sigma_y\sigma_x \\ &= 2(i\sigma_z + \sigma_y\sigma_x) = 0,\end{aligned} \tag{11.1-13}$$

and similarly for $[\sigma_x, \sigma_z]_+$ and $[\sigma_y, \sigma_z]_+$. Thus the spin operators anticommute. For $\sigma^{\pm}$ one obtains

$$[\sigma^+, \sigma^-]_+ = \sigma^+\sigma^- + \sigma^-\sigma^+ = \mathbf{1}. \tag{11.1-14}$$

These rules are interconnected and one of them entails the other two.

We return to the partition function (11.1-7). The two exponential operators in each bracket will be denoted by V_1 and V_2, respectively. Clearly, V_1 is diagonal in the space of σ_z

$$\langle \pm 1 | V_1 | \pm 1 \rangle = e^{\pm\beta h}, \quad \langle \pm 1 | V_1 | \mp 1 \rangle = 0. \tag{11.1-15}$$

---

[9] In the relativistic description the 4-vector of the spin is $\sigma = (\mathbf{1}, \sigma_x, \sigma_y, \sigma_z)$, cf. (B.3-7), Appendix B.

[10] P.A.M. Dirac, Op. Cit., p. 149; A. Messiah, Op. Cit., Vol. II, p. 545.

Hence $V_1 = \exp(\beta h \sigma_z)$, in which V_1 and σ_z may be taken to be either as operators or as eigenvalues. For the matrix V_2 the product $\sigma_i \sigma_{i+1}$ has the four values $1, -1, -1, 1$, so that

$$\langle "\pm 1" | V_2 | "\pm 1" \rangle = e^K, \quad \langle "\pm 1" | V_2 | "\mp 1" \rangle = e^{-K}. \tag{11.1-16}$$

We used the quotes to indicate that these are product spin states. The matrix V_2 becomes

$$\begin{pmatrix} e^K & e^{-K} \\ e^{-K} & e^K \end{pmatrix} = \begin{pmatrix} e^K & 0 \\ 0 & e^K \end{pmatrix} + \begin{pmatrix} 0 & e^{-K} \\ e^{-K} & 0 \end{pmatrix}, \tag{11.1-17}$$

so that

$$V_2 = e^K \mathbf{1} + e^{-K} \sigma_x \equiv A \exp(K^* \sigma_x), \tag{11.1-18}$$

where the extreme rhs *defines* the scalar functions A and K^*. To see what these are, we write the exponential operator in terms of hyperbolic operators, both being defined by series expansion; hence, employing the relation $(\sigma_x)^{2n} = \mathbf{1}$ for any integer n, we have

$$\begin{aligned} e^{K^* \sigma_x} &= \cosh(K^* \sigma_x) + \sinh(K^* \sigma_x) = \sum_{n=0}^{\infty} \frac{(K^* \sigma_x)^{2n}}{(2n)!} + \sum_{n=0}^{\infty} \frac{(K^* \sigma_x)^{2n+1}}{(2n+1)!} \\ &= \sum_{n=0}^{\infty} \frac{(K^*)^{2n}}{(2n)!} \mathbf{1} + \sum_{n=0}^{\infty} \frac{(K^*)^{2n+1}}{(2n+1)!} \sigma_x = (\cosh K^*) \mathbf{1} + (\sinh K^*) \sigma_x . \end{aligned} \tag{11.1-19}$$

Therefore, the proposed relationship (11.1-18) entails

$$\begin{aligned} A \cosh K^* &= e^K , \\ A \sinh K^* &= e^{-K} . \end{aligned} \tag{11.1-20}$$

Squaring plus subtracting and dividing these two equations yields:

$$\left. \begin{aligned} A &= \sqrt{2 \sinh 2K} \\ \tanh K^* &= \exp(-2K) \end{aligned} \right\} . \tag{11.1-21}$$

We are now able to write the partition function in a conducive form; we have

$$\mathcal{Z} = \mathrm{Tr}\{(V_1 V_2)^N\} = \mathrm{Tr}\{(V_1 V_2^{1/2} V_2^{1/2})^N\}. \tag{11.1-22}$$

Since the trace is invariant under cyclic permutation, Tr(ABC)=Tr(CAB)=Tr(BCA), we obtain

$$\mathcal{Z} = \mathrm{Tr}\{(V_2^{1/2} V_1 V_2^{1/2})^N\} = \lambda_1^N + \lambda_2^N, \tag{11.1-23}$$

where λ_1 and λ_2 are EV of the Hermitean matrix[11]

[11] Note that $V_1 V_2$ is non-Hermitean: $(V_1 V_2)^\dagger = V_2^\dagger V_1^\dagger = V_2 V_1$, but $V^\dagger = (V_2^{1/2} V_1 V_2^{1/2})^\dagger = V$.

$$V = V_2^{1/2} V_1 V_2^{1/2} = \sqrt{2\sinh 2K}\, e^{K^*\sigma_x/2} e^{\beta h \sigma_z} e^{K^*\sigma_x/2}. \qquad (11.1\text{-}24)$$

We leave it up to the problems to show that the EV are the same as in our previous treatment,

$$\lambda_{1,2} = e^K \cosh\beta h \pm \sqrt{e^{2K}\sinh^2\beta h + e^{-2K}}\,, \qquad (11.1\text{-}25)$$

cf. Eq. (4.3-27). In the absence of an external field, we have the simpler result

$$\lambda_{1,2}(h=0) = e^K \pm e^{-K} = A\exp(\pm K^*). \qquad (11.1\text{-}26)$$

The free energy, given previously, is likewise recovered.

11.2 The Transfer Matrix for the Rectangular Lattice

11.2.1 *Procedure*

We shall consider a rectangular lattice with exchange integrals J_1 and J_2 in the two directions. We assume that there are M columns and N rows. Ultimately, we will take the thermodynamic limit, by which we mean $M, N \to \infty$, keeping the ratio M/N fixed. The spins are now labelled with a row index n and a column index m, i.e., we have spins $\{\sigma_{nm,z}\}$. For now the subscript 'z' will be suppressed. The full Hamiltonian then reads, [cf., Mattis[6], Op. Cit., p. 257]

$$\mathcal{H} = -\sum_{n,m}\left[J_1\sigma_{n,m}\sigma_{n+1,m} + J_2\sigma_{n,m}\sigma_{n,m+1}\right] - h\sum_{n,m}\sigma_{n,m}\,. \qquad (11.2\text{-}1)$$

So J_1 denotes row-coupling within a given column and J_2 denotes column-coupling within a given row. The periodic b.c. are

$$\sigma_{n+N,m} = \sigma_{n,m} \text{ and } \sigma_{n,m+M} = \sigma_{n,m}\,. \qquad (11.2\text{-}2)$$

The b.c. can physically be realised if the lattice is bent into a toroid. In this section we shall be concerned with the partition function in the absence of a magnetic field; thus, we consider the simpler basic Hamiltonian

$$\mathcal{H}_0 = -\sum_{n,m}\left[J_1\sigma_{n,m}\sigma_{n+1,m} + J_2\sigma_{n,m}\sigma_{n,m+1}\right]. \qquad (11.2\text{-}3)$$

We shall concentrate on the M columns to define a basis of spin states for the lattice. Accordingly, we introduce the 2^M many-spin states

$$|\{\sigma\}\rangle \equiv |\sigma_1\rangle \otimes |\sigma_2\rangle \otimes \ldots \otimes |\sigma_M\rangle = |\sigma_1\sigma_2\ldots\sigma_M\rangle\,. \qquad (11.2\text{-}4)$$

Clearly, the Hilbert space is the tensor product of the M individual spin-spaces. In a single spin space we had

$$\begin{aligned}
&\sigma_z \,|\pm 1\rangle = \pm\,|\pm 1\rangle, \\
&\sigma^+ \,|+1\rangle = 0, \qquad \sigma^+ \,|-1\rangle = |+1\rangle, \\
&\sigma^- \,|+1\rangle = |-1\rangle, \quad \sigma^- \,|-1\rangle = 0 .
\end{aligned} \tag{11.2-5}$$

The first statement is obvious. For the latter statements, just write out the results in the ℓ_2 space, e.g.,

$$\sigma^+ \,|+1\rangle = \begin{pmatrix} 0 & 1 \\ 0 & 0 \end{pmatrix}\begin{pmatrix} 1 \\ 0 \end{pmatrix} = 0, \qquad \sigma^+ \,|-1\rangle = \begin{pmatrix} 0 & 1 \\ 0 & 0 \end{pmatrix}\begin{pmatrix} 0 \\ 1 \end{pmatrix} = |+1\rangle . \tag{11.2-6}$$

Therefore, the following rules now apply:

$$\begin{aligned}
&\sigma_{j,z} \,|\,\sigma_1\sigma_2 ... \sigma_M \rangle = \sigma_j \,|\,\sigma_1\sigma_2 ... \sigma_j ... \sigma_M \rangle , \\
&\sigma_j^+ \,|\,\sigma_1\sigma_2 ... \sigma_M \rangle = \delta_{\sigma_j,-1} \,|\,\sigma_1\sigma_2 ... \sigma_j + 2 ... \sigma_M \rangle , \\
&\sigma_j^- \,|\,\sigma_1\sigma_2 ... \sigma_M \rangle = \delta_{\sigma_j,+1} \,|\,\sigma_1\sigma_2 ... \sigma_j - 2 ... \sigma_M \rangle .
\end{aligned} \tag{11.2-7}$$

The commutators $[\sigma_{j\alpha}, \sigma_{k\beta}]$ are zero for $j \neq k$, with α and β referring to Cartesian components; for $j = k$ the usual Pauli spin anti- and regular commutation rules apply.

We now come to the partition function. We introduce $K_i = \beta J_i$, $(i = 1, 2)$. We have

$$\mathcal{Z} = \mathrm{Tr}\left\{ \exp\left(K_1 \sum_{n,m} \sigma_{n,m}\sigma_{n+1,m} + K_2 \sum_{n,m} \sigma_{n,m}\sigma_{n,m+1} \right) \right\}. \tag{11.2-8}$$

For the purpose of finding the trace, we can factor the exponential[12], since in the spin-space representation the exponential yields c-numbers. Further, we note that the two parts in the exponential are in many respects the same as the two parts we considered for the 1D case. With the basis as chosen, the second part is diagonal in the spin space while the first part is non-diagonal. So we introduce for given n:

$$V_2 = \exp\left(K_2 \sum_m \sigma_{n,m}\sigma_{n,m+1} \right) = \exp\left(K_2 \sum_{j=1}^{M} \sigma_{j,z}\sigma_{j+1,z} \right), \tag{11.2-9}$$

$$V_1 = \exp\left(K_1 \sum_m \sigma_{n,m}\sigma_{n+1,m} \right) = \prod_{j=1}^{M} \left[(\exp K_1)\mathbf{1} + (\exp(-K_1))\sigma_{j,x} \right], \tag{11.2-10}$$

where we used the same decomposition as in Eqs. (11.1-16) – (11.1-18). With the

[12] The reader is reminded that, in general, for non-commuting operators, exp(A+B) ≠ exp(A) exp(B); see e.g., W. Louisell, "Radiation and Noise in Quantum Electronics", McGraw-Hill 1964, Section 3.2. Generally one can show: $e^{A+B} = e^A e^B e^{-\frac{1}{2}[A,B]} = e^B e^A e^{\frac{1}{2}[A,B]}$ if $[A,[A,B]] = [B,[A,B]] = 0$.

subsequent arguments given there we arrive at

$$V_1 = (2\sinh 2K_1)^{M/2} \exp\left(K_1^* \sum_j \sigma_{j,x}\right), \tag{11.2-11}$$

with

$$\tanh K_1^* = \exp(-2K_1). \tag{11.2-12}$$

The partition function thus becomes

$$\mathcal{Z} = \mathrm{Tr}\left\{(\mathrm{V}_2 V_1)^N\right\} = \mathrm{Tr}\left\{(\mathrm{V}_1^{1/2} \mathrm{V}_2 V_1^{1/2})^N\right\}, \tag{11.2-13}$$

where, explicitly,

$$V = \mathrm{V}_1^{1/2} \mathrm{V}_2 V_1^{1/2}$$

$$= (2\sinh 2K_1)^{M/2} \exp\left(\frac{K_1^*}{2} \sum_{j=1}^{M} \sigma_{j,x}\right) \exp\left[K_2 \sum_{j=1}^{M} \sigma_{j,z}\sigma_{j+1,z}\right] \exp\left(\frac{K_1^*}{2} \sum_{j=1}^{M} \sigma_{j,x}\right). \tag{11.2-14}$$

The challenge is now to find the eigenvalues of this transfer matrix, which is Hermitean, but has a non-trivial structurc!

11.2.2 *Transformation to an Interacting Fermion Problem*

First we perform a ccw rotation of the spin operators about the y-axis, Fig. 11-1. The rotation leaves the EV invariant, but we have $\sigma_{j,z} \to \sigma_{j,x}$, $\sigma_{j,x} \to \sigma_{j,-z}$ for all j.

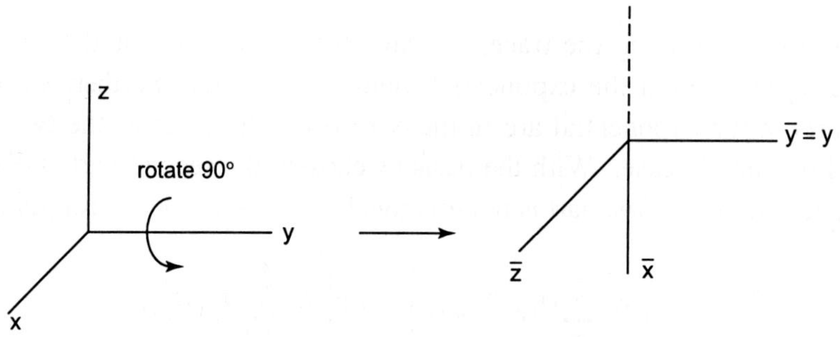

Fig. 11-1. Rotation about the y-axis through 90 degrees counterclockwise.

The resulting expressions will be expressed in σ^+ and σ^- only, employing

$$\begin{aligned} \sigma_{j,x} &= (\sigma_j^+ + \sigma_j^-), \\ \sigma_{j,-z} &= -\sigma_{j,z} = -(2\sigma_j^+\sigma_j^- - \mathbf{1}). \end{aligned} \tag{11.2-15}$$

This yields

$$V_2 = \exp\left(K_2 \sum_{j=1}^{M} (\sigma_j^+ + \sigma_j^-)(\sigma_{j+1}^+ + \sigma_{j+1}^-) \right), \tag{11.2-16}$$

$$V_1 = (2\sinh 2K_1)^{M/2} \exp\left(-2K_1^* \sum_{j=1}^{M} (\sigma_j^+ \sigma_j^- - \tfrac{1}{2}\mathbf{1}) \right). \tag{11.2-17}$$

Let us look at the last two rules of (11.2-7). We note that σ^+ can only act on a $|-1\rangle$ state and then flips it to $|+1\rangle$; *vice versa* for σ^-. Pictorially

$$\sigma^+ \downarrow = \uparrow, \qquad \sigma^- \uparrow = \downarrow .$$

Had we labelled the spin states as $|0\rangle$ and $|1\rangle$, then $\sigma^+|0\rangle \to |1\rangle$ and $\sigma^-|1\rangle \to |0\rangle$. The new spin rules would read

$$\begin{aligned} \sigma_j^+ \,|\, \sigma_1\sigma_2 \ldots \sigma_M \rangle &= \delta_{\sigma_j,0} \,|\, \sigma_1\sigma_2 \ldots, \sigma_j + 1, \ldots \sigma_M \rangle , \\ \sigma_j^- \,|\, \sigma_1\sigma_2 \ldots \sigma_M \rangle &= \delta_{\sigma_j,1} \,|\, \sigma_1\sigma_2 \ldots, \sigma_j - 1, \ldots \sigma_M \rangle . \end{aligned} \tag{11.2-18}$$

Clearly, there is an *isomorphism* between spin-states and fermion states. The only difference involves the sign on the rhs of (11.2-18), which for fermion operators c_j and $c_j^\dagger$ contains the factor $(-1)^\xi$, where ξ counts the occupancies of all ordered states prior to position '*j*'. The relevant factor is therefore

$$(-1)^\xi = (-1)^{\sum_{m=1}^{j-1} c_m^\dagger c_m} = \exp\left(\pi i \sum_{m=1}^{j-1} c_m^\dagger c_m \right). \tag{11.2-19}$$

The isomorphism therefore is expressible by the *Jordan–Wigner transformation* of the Pauli spin matrices, or their *generic spin operators* into *fermion operators*:

$$\begin{aligned} \sigma_j^+ &= e^{\pi i \sum_m^{j-1} c_m^\dagger c_m} c_j^\dagger , \\ \sigma_j^- &= c_j e^{-\pi i \sum_m^{j-1} c_m^\dagger c_m} = e^{\pi i \sum_m^{j-1} c_m^\dagger c_m} c_j . \end{aligned} \tag{11.2-20}$$

(the latter form recognizes that c_j and $c_m^\dagger c_m$ commute, while we added $2\pi i$ to the exponent).

To verify the previously stated spin commutation rules, we compute the result from (11.2-20) for $n > j$:

$$[\sigma_j^-, \sigma_n^+] = \exp\left\{ \pi i \sum_{m=j+1}^{n-1} c_m^\dagger c_m \right\} \left(c_j e^{\pi i c_j^\dagger c_j} c_n^\dagger - c_n^\dagger e^{\pi i c_j^\dagger c_j} c_j \right). \tag{11.2-21}$$

Now we have

$$\left(e^{\pi i c_j^\dagger c_j}\right)c_j = \sum_{k=0}^{\infty}\frac{(\pi i c_j^\dagger c_j)^k}{k!}c_j = c_j \qquad \begin{bmatrix}\text{term with } k=0 \text{ only}\\ \text{since } c_j^2=0.\end{bmatrix},$$
$$c_j\left(e^{\pi i c_j^\dagger c_j}\right) = c_j\left(e^{\pi i(1-c_j c_j^\dagger)}\right) = -c_j \qquad \text{[similar argument]}. \tag{11.2-22}$$

With these relations the last factor of (11.2-21) becomes $-c_j c_n^\dagger - c_n^\dagger c_j = -\delta_{jn} \to 0$. Also, for $j = n$, one easily confirms that the on-site anticommutator is given by $[\sigma_j^+, \sigma_j^-]_+ = [c_j^\dagger, c_j]_+ = 1$, in accord with (11.1-14).

We continue to express the Gibbs factors V_1 and V_2 into fermion operators. The result for V_1 is direct

$$V_1 = (2\sinh 2K_1)^{M/2}\exp\left\{-2K_1^* \sum_{j=1}^{M}(c_j^\dagger c_j - \tfrac{1}{2}\mathsf{I})\right\}, \tag{11.2-23}$$

where I is the idem- or identity operator; if the result is clear it will be suppressed. For V_2 there is a problem due to the periodic boundary conditions. One arrives at

$$V_2 = \exp\left\{K_2 \sum_{j=1}^{M}(c_j^\dagger - c_j)(c_{j+1}^\dagger + c_{j+1})\right\}. \tag{11.2-24}$$

This expression involves the total number operator $\mathbf{n} = \Sigma_{j=1}^{M} c_j^\dagger c_j$. The b.c. become

$$\begin{aligned} c_{M+1} &= -c_1, \quad c_{M+1}^\dagger = -c_1^\dagger, \quad \text{for } n \textit{ even},\\ c_{M+1} &= c_1, \quad c_{M+1}^\dagger = c_1^\dagger, \quad \text{for } n \textit{ odd}. \end{aligned} \tag{11.2-25}$$

With this choice of b.c., the operators V_1 and V_2 are translationally invariant, as required.

Digression: details on the conversion

(a) *For V_1*. We have

$$\sigma_j^+\sigma_j^- - \tfrac{1}{2}\mathbf{1} = e^{\pi i \sum_{m=1}^{j-1} c_m^\dagger c_m}\, c_j^\dagger c_j\, e^{-\pi i \sum_{m=1}^{j-1} c_m^\dagger c_m} - \tfrac{1}{2}\mathsf{I}. \tag{11.2-26}$$

All operators on the rhs are number operators, which commute with each other. We can therefore bring the factor $c_j^\dagger c_j$ to the right; the result (11.2-23) is then immanent.

(b) *For V_2*. When $j \neq M$ one easily finds [expand $\exp(c_j^\dagger c_j)$] :

$$\begin{aligned}(\sigma_j^+ + \sigma_j^-)(\sigma_{j+1}^+ + \sigma_{j+1}^-) &= c_j^\dagger c_{j+1}^\dagger + c_j^\dagger c_{j+1} + c_{j+1}^\dagger c_j + c_{j+1} c_j\\ &= c_j^\dagger c_{j+1}^\dagger + c_j^\dagger c_{j+1} - c_j c_{j+1}^\dagger - c_j c_{j+1} = (c_j^\dagger - c_j)(c_{j+1}^\dagger + c_{j+1}).\end{aligned} \tag{11.2-27}$$

This yields (11.2-24). The case $j = M$ needs special investigation. We then have

$$(\sigma_M^+ + \sigma_M^-)(\sigma_1^+ + \sigma_1^-)$$

$$= \exp\left(\pi i \sum_{k=1}^{M-1} c_k^\dagger c_k\right) c_M^\dagger (c_1^\dagger + c_1) + \exp\left(\pi i \sum_{k=1}^{M-1} c_k^\dagger c_k\right) c_M (c_1^\dagger + c_1)$$

$$= \exp\left(\pi i \sum_{k=1}^{M} c_k^\dagger c_k\right) e^{-\pi i c_M^\dagger c_M} (c_M^\dagger + c_M)(c_1^\dagger + c_1) = (-1)^{\mathbf{n}} (-c_M^\dagger + c_M)(c_1^\dagger + c_1), \quad (11.2\text{-}28)$$

where we expanded the exponential $\exp(-\pi i c_M^\dagger c_M)$, using $(c_M^\dagger)^2 = c_M^2 = 0$ as in (11.2-22); further $\mathbf{n}$ is the *total* fermion occupancy operator $\Sigma_1^M c_k^\dagger c_k$. While $\mathbf{n}$ does commute with V_1, it does not so with V_2; however, the factor $(-1)^{\mathbf{n}}$ commutes with V_2 if the factors in that operator change the occupancy by 0 or ±2. Thus the problem requires that we separately consider subspaces with even and with odd total number of fermions. The general result (11.2-24) can be maintained for $j = M$, providing the b.c. are stipulated as in (11.2-25). N.B. The total fermion occupancy corresponds to the total number of "up-spins", which can be even or odd.

11.2.3 *Running-Wave Fermion Operators*

In order to uncouple the quadratic expression in V_2, we make a canonical transformation to running-wave fermion operators on the lattice, analogous to the phonon-mode boson operators for vibrations in solids, see Section 8.5. For our operators c_q and $c_q^\dagger$ we follow the definition by Schultz et al.[5] (see also Mattis[6] p.264):

$$c_q = \frac{e^{i\pi/4}}{\sqrt{M}} \sum_{j=1}^{M} c_j e^{-iqj}, \quad c_q^\dagger = \frac{e^{-i\pi/4}}{\sqrt{M}} \sum_{j=1}^{M} c_j^\dagger e^{iqj}. \quad (11.2\text{-}29)$$

Note that the exponential is a wave on the lattice. Had we replaced j by the spin position vector $\mathbf{l} = j\mathbf{d}$, where $\mathbf{d}$ is the horizontal spin-lattice constant, then the exponentials would have been $\exp(\pm i\mathbf{q}\cdot\mathbf{l})$, as for the phonon gas. To invert the relationships (11.2-29) we use two lemmas analogous to Eqs. (8.5-27a and b):

$$(A): \quad \sum\nolimits_{j=1}^{M} e^{\pm iqj} = M\delta_{q,0}, \qquad (B): \quad \sum\nolimits_{all\ q} e^{\pm iqj} = M\delta_{j,0}. \quad (11.2\text{-}30)$$

The range of the q's will be established shortly, but suffice it here to say that the transformation can only be canonical if there are M q-modes. For the inversion of the first statement of (11.2-29), multiply with $\exp(iqj')$ and sum over all q. We then have

$$\sum_q c_q e^{iqj'} = \frac{e^{i\pi/4}}{\sqrt{M}} \sum_q \sum_j c_j e^{iq(j'-j)}$$

$$= \frac{e^{i\pi/4}}{\sqrt{M}} \sum_j \sum_q c_j e^{iq(j'-j)} \overset{(B)}{=} \frac{e^{i\pi/4}}{\sqrt{M}} \sum_j M c_j \delta_{j,j'} = \sqrt{M} e^{i\pi/4} c_{j'}. \quad (11.2\text{-}31)$$

Hence we have, changing $j' \to j$,

$$c_j = \frac{e^{-i\pi/4}}{\sqrt{M}} \sum_q c_q e^{iqj}; \quad \text{likewise} \quad c_j^\dagger = \frac{e^{i\pi/4}}{\sqrt{M}} \sum_q c_q^\dagger e^{-iqj}. \tag{11.2-32}$$

The same lemmas can be used to show that the c_q's satisfy the fermion anticommutation rule

$$[c_q, c_{q'}^\dagger]_+ = \delta_{q,q'}; \tag{11.2-33}$$

the proof is left to the reader.

We must fix the store of q-values, such that the b.c. (11.2-25) are satisfied. We will set,

$$q = k\pi / M, \text{ with } \begin{cases} k = \pm 1, \pm 3, \pm 5, \ldots \pm (M-1), & \textit{for n even}, \\ k = \ \ 0, \pm 2, \pm 4, \ldots + M \quad , & \textit{for n odd}. \end{cases} \tag{11.2-34}$$

(we assumed without undue restriction that M is even). In each case there are M values of q, like in the first Brillouin zone of a solid; note that for n odd we *omitted* the value $k = -M$, corresponding to $q = -\pi$.

We must now transform the operators V_1 and V_2. The exponential in V_2 gives

$$\sum_j (c_j^\dagger - c_j)(c_{j+1}^\dagger + c_{j+1})$$

$$= \frac{1}{M} \sum_j \sum_{q,q'} \left(e^{i\pi/4} c_q^\dagger e^{-iqj} - e^{-i\pi/4} c_q e^{iqj} \right) \left(e^{i\pi/4} c_{q'}^\dagger e^{-iq'j} e^{-iq'} + e^{-i\pi/4} c_{q'} e^{iq'j} e^{iq'} \right)$$

$$= \frac{1}{M} \sum_j \sum_{q,q'} \left\{ i c_q^\dagger c_{q'}^\dagger \underbrace{e^{-i(q+q')j}}_{M\delta_{q',-q}} e^{-iq'} + c_q^\dagger c_{q'} \underbrace{e^{-i(q-q')}}_{M\delta_{q',q}} e^{iq'} - c_q c_{q'}^\dagger \underbrace{e^{i(q-q')j}}_{M\delta_{q',q}} e^{-iq'} + i c_q c_{q'} \underbrace{e^{i(q+q')j}}_{M\delta_{q',-q}} e^{iq'} \right\}$$

$$= \sum_q \left(i c_q^\dagger c_{-q}^\dagger e^{iq} + c_q^\dagger c_q e^{iq} - c_q c_q^\dagger e^{-iq} + i c_q c_{-q} e^{-iq} \right); \tag{11.2-35}$$

we used here Lemma (A). We now split into two sums, $\Sigma_{q>0}$ and $\Sigma_{q<0}$; in the latter we change $q \to -q$. [The terms with $q = 0$ and $q = \pi$, left out in this procedure for odd n, will be dealt with later.] We thus obtain

$$\begin{aligned}(11.2-35) \to \sum_{q>0} \Big(& i c_q^\dagger c_{-q}^\dagger e^{iq} + c_q^\dagger c_q e^{iq} - c_q c_q^\dagger e^{-iq} + i c_q c_{-q} e^{-iq} \\ & + i c_{-q}^\dagger c_q^\dagger e^{-iq} + c_{-q}^\dagger c_{-q} e^{-iq} - c_{-q} c_{-q}^\dagger e^{iq} + i c_{-q} c_q e^{iq} \Big). \end{aligned} \tag{11.2-36}$$

Some terms will be changed, using the anticommutation rules; this yields[13]

[13] Note that the extra terms $-\Sigma_{q>0}(e^{iq} + e^{-iq}) = -\Sigma_{all\, q} \exp(iq)$ are zero according to Lemma (B).

$$(11.2-36) \to \sum_{q>0}\Big(ic_q^\dagger c_{-q}^\dagger e^{iq} + c_q^\dagger c_q e^{iq} + c_q^\dagger c_q e^{-iq} - e^{-iq} + ic_q c_{-q} e^{-iq}$$

$$- ic_q^\dagger c_{-q}^\dagger e^{-iq} + c_{-q}^\dagger c_{-q} e^{-iq} + c_{-q}^\dagger c_{-q} e^{iq} - e^{iq} - ic_q c_{-q} e^{iq}\Big)$$

$$= 2\sum_{q>0}\Big[\cos q\,\big(c_q^\dagger c_q + c_{-q}^\dagger c_{-q}\big) + \sin q\,\big(c_{-q}^\dagger c_q^\dagger + c_q c_{-q}\big)\Big]. \qquad (11.2\text{-}37)$$

Hence we find

$$V_2 = \prod_{q>0} V_{2q}, \text{ with}$$

$$V_{2q} = \exp\Big\{2K_2\Big[\cos q\,\big(c_q^\dagger c_q + c_{-q}^\dagger c_{-q}\big) + \sin q\,\big(c_{-q}^\dagger c_q^\dagger + c_q c_{-q}\big)\Big]\Big\}. \qquad (11.2\text{-}38)$$

The c-operators are now uncoupled, except for the mixing of q and $-q$. We leave it to the reader to obtain the result for V_1,

$$V_1 = (2\sinh 2K_1)^{M/2} \prod_{q>0} V_{1q}, \text{ with}$$

$$V_{1q} = \exp\big\{-2K_1^*(c_q^\dagger c_q + c_{-q}^\dagger c_{-q} - 1)\big\}. \qquad (11.2\text{-}39)$$

For the case of odd n we still also need V_{1q} and V_{2q} for $q = 0$ and $q = \pi$. These operators are given by

$$V_{10} = \exp[2K_1^*(c_0^\dagger c_0 - \tfrac{1}{2})], \quad V_{1\pi} = \exp[2K_1^*(c_\pi^\dagger c_\pi - \tfrac{1}{2})],$$
$$V_{20} = \exp(2K_2 c_0^\dagger c_0), \qquad V_{2\pi} = \exp(-2K_2 c_\pi^\dagger c_\pi). \qquad (11.2\text{-}40)$$

They are already in diagonal form and offer no problems.

To compute the partition function we need to evaluate – see Eq. (11.2-13) :

$$\mathcal{Z}_{(even)} = \mathrm{Tr}\Big\{\Big[(2\sinh 2K_1)^{M/2} \prod\nolimits_{q>0} V_q\Big]^N\Big\},$$

$$V_q = (V_{1q})^{1/2} V_{2q} (V_{1q})^{1/2}, \qquad (11.2\text{-}41)$$

where the q's are chosen according to the upper line of (11.2-34). For a state with odd total occupancy the effect of the diagonal operators (11.2-40) must be added on to the above product; to keep their joint occupancy odd, we assume the 0-state is occupied while the π-state is left empty. Moreover, the wave-vectors of the lower line of (11.2-34) will be formally assigned the wave vector $\bar{q}$. We thus have

$$\mathcal{Z}_{(odd)} = \mathrm{Tr}\Big\{\Big[(2\sinh 2K_1)^{M/2}\Big(\prod\nolimits_{\bar{q}>0} V_{\bar{q}}\Big) V_0 V_\pi\Big]^N\Big\},$$

$$V_{\bar{q}} = (V_{1\bar{q}})^{1/2} V_{2\bar{q}} (V_{1\bar{q}})^{1/2}, \quad V_{0\ or\ \pi} = (V_{10\ or\ 1\pi})^{1/2} V_{20\ or\ 2\pi} (V_{10\ or\ 1\pi})^{1/2}. \qquad (11.2\text{-}42)$$

For now all wave-vectors will for convenience still be denoted by 'q'. For a given q the occupation-number space involves the four basic states

$$\begin{aligned} &|0_{-q}0_q\rangle, && |0_{-q}1_q\rangle = c_q^\dagger\,|0_{-q}0_q\rangle, \\ &|1_{-q}0_q\rangle = c_{-q}^\dagger|0_{-q}0_q\rangle, && |1_{-q}1_q\rangle = c_{-q}^\dagger c_q^\dagger\,|0_{-q}0_q\rangle. \end{aligned} \tag{11.2-43}$$

For simplicity of notation we omitted in the kets all q' with $q' \neq q$ or $-q$; also, without loss of generality, we assumed that the total preceding occupancy is even so that there is no factor $(-1)^\xi$ to account for. [NB. The state $|0_{-q}0_q\rangle$ above is *not* the ground state, which will be defined later.] Since V_{1q} only depends on the diagonal operators $\mathbf{n}_q$ and $\mathbf{n}_{-q}$, their action on $|0_{-q}1_q\rangle$ and $|1_{-q}0_q\rangle$ leaves the states unaltered, as is easily verified. As to V_{2q}, its action on these states yields the null-state: the cos part is diagonal and the sin part would involve $c_{\pm q}^{\dagger 2}$ or $c_{\pm q}^2 \to 0$. These states can therefore be deleted from the bi-spinor (11.2-43), leaving as basis states $|1_{-q}1_q\rangle$ and $|0_{-q}0_q\rangle$.[14] Now we have $V_{1q}^{1/2}|1_{-q}1_q\rangle = \exp(-K_1^*)|1_{-q}1_q\rangle$ and $V_{1q}^{1/2}|0_{-q}0_q\rangle = \exp(K_1^*) \times |0_{-q}0_q\rangle$. Hence, the matrix $V_{1q}^{1/2}$ is given by

$$V_{1q}^{1/2} = \begin{pmatrix} \langle 1_{-q}1_q\,|V_{1q}^{1/2}\,|1_{-q}1_q\rangle & \langle 1_{-q}1_q\,|V_{1q}^{1/2}\,|0_{-q}0_q\rangle \\ \langle 0_{-q}0_q\,|V_{1q}^{1/2}\,|1_{-q}1_q\rangle & \langle 0_{-q}0_q\,|V_{1q}^{1/2}\,|0_{-q}0_q\rangle \end{pmatrix} = \begin{pmatrix} e^{-K_1^*} & 0 \\ 0 & e^{K_1^*} \end{pmatrix}. \tag{11.2-44}$$

To obtain the matrix for V_{2q} we must employ more armour. Since both pseudo-spinor states involve two occupancies, Schultz et al. introduce pair operators $b_q = c_q c_{-q}$ and $b_q^\dagger = c_{-q}^\dagger c_q^\dagger$, similar to the Cooper-pair operators in BCS theory, discussed in the next chapter. Clearly, $c_{-q}^\dagger c_q^\dagger + c_{-q}c_q = b_q^\dagger + b_q$. For the diagonal operators note that, as far as operations on pair states are concerned, we can set $c_q^\dagger c_q = c_{-q}^\dagger c_{-q} c_q^\dagger c_q = b_q^\dagger b_q$ and similarly for $c_{-q}^\dagger c_{-q}$. Now as indicated in connection with BCS theory [Eq. (12-6-2')], the pair operators have commutation properties identical to spin lowering and raising operators, *without* the sign-change problem of the Jordan-Wigner transformation. Thus $b_q \to \beta_q^-$, $b_q^\dagger \to \beta_q^+$. In *this* (isomorphic) spin-space we have a representation by Pauli spin matrices such that

$$c_q^\dagger c_q + c_{-q}^\dagger c_{-q} = 2b_q^\dagger b_q \to 2\beta_q^+\beta_q^- \equiv \beta_{q,z} + 1, \tag{11.2-45}$$

$$c_{-q}^\dagger c_q^\dagger + c_q c_{-q} = b_q^\dagger + b_q \to \beta_q^+ + \beta_q^- \equiv \beta_{q,x}. \tag{11.2-46}$$

We still need another operator

$$\overline{\beta}_q \equiv \beta_{q,z}\cos q + \beta_{q,x}\sin q, \tag{11.2-47}$$

which has the important property that $(\overline{\beta}_q)^2 = 1$ since $[\beta_{q,z}, \beta_{q,x}]_+ = 0$ [analogous to (11.1-13)]. For V_{2q} this yields

---

[14] These states can be thought of as a pseudo-spinor, arranged in a column matrix.

$$V_{2q}=\exp\left\{2K_2[(\beta_{q,z}+1)\cos q+\beta_{qx}\sin q]\right\}$$
$$=\left[\exp(2K_2\cos q)\right]\left[\exp(2K_2\overline{\beta}_q)\right]$$
$$=\left[\exp(2K_2\cos q)\right]\left((\cosh 2K_2)\mathbf{1}+(\sinh 2K_2)\overline{\beta}_q\right), \qquad (11.2\text{-}47)$$

where we used a procedure similar to (11.1-19). Translating back into the pair operators, this gives

$$V_{2q}=\left[\exp(2K_2\cos q)\right]$$
$$\times\left\{(\cosh 2K_2)\mathrm{I}+(\sinh 2K_2)[(2b_q^\dagger b_q-1)\cos q+(b_q^\dagger+b_q)\sin q]\right\}. \qquad (11.2\text{-}48)$$

The resulting matrix elements $\langle\cdot\cdot|V_{2q}|\cdot\cdot\rangle$ are found from application of the usual rules $b_q^\dagger|0_{-q}0_q\rangle=|1_{-q}1_q\rangle$, $b_q^\dagger b_q|1_{-q}1_q\rangle=|1_{-q}1_q\rangle$, etc. We thus easily obtain

$$V_{2q}=e^{2K_2\cos q}\begin{pmatrix}\cosh 2K_2+\sinh 2K_2\cos q & \sinh 2K_2\sin q\\ \sinh 2K_2\sin q & \cosh 2K_2-\sinh 2K_2\cos q\end{pmatrix}. \qquad (11.2\text{-}49)$$

Employing also (11.2-44), the full transfer matrix V_q becomes

$$V_q=\begin{pmatrix}e^{-K_1^*} & 0\\ 0 & e^{K_1^*}\end{pmatrix}\left(V_{2q}\right)\begin{pmatrix}e^{-K_1^*} & 0\\ 0 & e^{K_1^*}\end{pmatrix}. \qquad (11.2\text{-}50)$$

or,

$$V_q=e^{2K_2\cos q}\begin{pmatrix}A_q & C_q\\ C_q & B_q\end{pmatrix}, \qquad (11.2\text{-}51)$$

where[15]

$$A_q=(\cosh 2K_2+\cos q\sinh 2K_2)e^{-2K_1^*},$$
$$B_q=(\cosh 2K_2-\cos q\sinh 2K_2)\,e^{2K_1^*}, \qquad (11.2\text{-}52)$$
$$C_q=\sin q\sinh 2K_2\,.$$

The eigenvalues are given by

$$\lambda_\pm=e^{2K_2\cos q}\left\{\tfrac{1}{2}(A_q+B_q)\pm\left[\left(\tfrac{1}{2}(A_q-B_q)\right)^2+C_q^2\right]^{1/2}\right\}. \qquad (11.2\text{-}53)$$

---

[15] The matrix given here is considerably simpler than that of Schultz *et alii*, since they computed V_q as $V_{2q}^{1/2}V_{1q}V_{2q}^{1/2}$. However, the EV must be independent of the procedure chosen. Our result (11.2-56) is the result of the original literature[5,8] and of most books (e.g., Mattis, Huang). We note that some other treatments[7] have a similar result, but with the change $\cos q\to-\cos q$. This has no effect on the free energy but gives problems for the spin-spin correlation function, see Sec. 11.4.

While this can be "grinded out" we proceed as follows. Write the EV as

$$\lambda_\pm = e^{2K_2\cos q}\mu_\pm, \text{ where } \mu_\pm = e^{\pm\varepsilon(q)}. \tag{11.2-54}$$

We then have for half the sum of the roots,

$$\tfrac{1}{2}(\mu_+ + \mu_-) = \cosh\varepsilon(q) = \tfrac{1}{2}(A_q + B_q). \tag{11.2-55}$$

From the expressions for A_q and B_q given above, one immediately obtains

$$\cosh\varepsilon(q) = \cosh 2K_2 \cosh 2K_1^* - \cos q \sinh 2K_2 \sinh 2K_1^*. \tag{11.2-56}$$

Likewise,

$$\tfrac{1}{2}(\mu_+ - \mu_-) = \sinh\varepsilon(q) = \left[\left(\tfrac{1}{2}(A_q - B_q)\right)^2 + C_q^2\right]^{1/2}. \tag{11.2-57}$$

If, in fact, (11.2-54) is the solution, we must verify that $\cosh^2\varepsilon(q) - \sinh^2\varepsilon(q) = 1$, so that

$$\left[\tfrac{1}{2}(A_q + B_q)\right]^2 - \left[\left(\tfrac{1}{2}(A_q - B_q)\right)^2 + C_q^2\right] = 1 \text{ or } A_q B_q - C_q^2 = 1. \tag{11.2-58}$$

This equality is shown to be correct by elementary algebra. We thus found

$$\lambda_\pm = e^{2K_2\cos q \pm \varepsilon(q)}. \tag{11.2-59}$$

11.2.4 *Bogoliubov–Valatin Transformation* [16]

The final uncoupling of q and $-q$ operators can also be accomplished by a rotation over an angle ϕ_q. New fermion operators are introduced by the canonical transformation

$$\begin{aligned} \eta_q &= \cos\phi_q\, c_q + \sin\phi_q\, c_{-q}^\dagger, \quad \eta_q^\dagger = h.c., \\ \eta_{-q} &= \cos\phi_q c_{-q} - \sin\phi_q\, c_q^\dagger, \quad \eta_{-q}^\dagger = h.c., \end{aligned} \tag{11.2-60}$$

where *h.c.* stands for Hermitean conjugate. Its inversion is

$$\begin{aligned} c_q &= \cos\phi_q\, \eta_q - \sin\phi_q\, \eta_{-q}^\dagger, \quad c_q^\dagger = h.c., \\ c_{-q} &= \cos\phi_q\, \eta_{-q} + \sin\phi_q\, \eta_q^\dagger, \quad c_{-q}^\dagger = h.c.. \end{aligned} \tag{11.2-61}$$

We shall not carry out the computations but indicate the method. The operator V_{1q} – see Eq. (11.2-39) – becomes

[16] N.N. Bogoliubov, J. Phys. USSR **11**, 23 (1947); ibid. Nuovo Cimento **7**, 794 (1958). J.G. Valatin, Nuovo Cimento **7**, 843 (1958).

$$V_{1q} = \exp\{K_1^*[\cos 2\phi_q (\eta_q^\dagger \eta_q + \eta_{-q}^\dagger \eta_{-q} - 1) + \sin 2\phi_q [\eta_{-q}^\dagger \eta_q^\dagger + h.c.]\}. \qquad (11.2\text{-}62)$$

(note: $h.c. = \eta_q \eta_{-q}$). The matrix elements are easily found. Doing similarly for V_{2q}, the new matrix V_q is obtained and ϕ_q is chosen such that the matrix is diagonal, which then is re-exponentiated. The resulting diagonal operator is (note that $\bar{\mathbf{n}}_q = \eta_q^\dagger \eta_q$ and $\bar{\mathbf{n}}_{-q} = \eta_{-q}^\dagger \eta_{-q}$ commute):

$$V_q = \exp(2K_2 \cos q)\exp[-\varepsilon(q)(\bar{\mathbf{n}}_q + \bar{\mathbf{n}}_{-q} - 1)]. \qquad (11.2\text{-}63)$$

For ϕ_q one obtains: $\tan\phi_q = C_q/[\exp\varepsilon(q) - A_q]$, or by some manipulation more succinctly,

$$\tan 2\phi_q = 2C_q/(B_q - A_q), \quad \operatorname{sgn} 2\phi_q = \operatorname{sgn} q. \qquad (11.2\text{-}64)$$

where A_q, B_q and C_q are as given previously. The new eigenstates are

$$|\Psi_0\rangle = \sin\phi_q |1_{-q}1_q\rangle + \cos\phi_q |0_{-q}0_q\rangle,$$
$$|\Psi_1\rangle = \cos\phi_q |1_{-q}1_q\rangle - \sin\phi_q |0_{-q}0_q\rangle, \qquad (11.2\text{-}65)$$

being a rotation of the previous eigenvectors.[17] From (11.2-63) the eigenvalues $\lambda_\pm$ follow once more. The largest EV λ_+ corresponds to $|\Psi_0\rangle \equiv |\bar{0}_{-q}\bar{0}_q\rangle$, called the "*ground state*" at T.

11.3 The Critical Temperature and the Thermodynamic Functions

In this section we shall obtain the free energy, being as usual given as $F = -\ln \mathcal{Q}/\beta$. In the thermodynamic limit we have for the free energy per spin

$$\hat{f} = \lim_{M,N\to\infty} (-\ln \mathcal{Q})/\beta MN. \qquad (11.3\text{-}1)$$

It suffices to use the largest eigenvalue only in the trace. Thus from (11.2-41) and (11-2-59) we have

$$\Lambda_{even} = (2\sinh 2K_1)^{M/2} \prod_{q>0} \lambda_+(q) = (2\sinh 2K_1)^{M/2} \exp\left\{\sum_{q>0} [2K_2 \cos q + \varepsilon(q)]\right\}$$

$$= (2\sinh 2K_1)^{M/2} \exp\left\{K_2 \sum_{all\, q} e^{iq} + \tfrac{1}{2}\sum_{all\, q} \varepsilon(q)\right\} = (2\sinh 2K_1)^{M/2} \exp\left\{\tfrac{1}{2}\sum_q \varepsilon(q)\right\}. \qquad (11.3\text{-}2)$$

---

[17] These eigenvectors could also have been found directly in the previous subsection in the standard way by requiring

$$V_q \begin{pmatrix} \psi_1 \\ \psi_2 \end{pmatrix} = \lambda_\pm \begin{pmatrix} \psi_1 \\ \psi_2 \end{pmatrix}, \qquad (11.2\text{-}66)$$

the values for $\cos\phi_q$ and $\sin\phi_q$ being commensurate with Eq. (11.2-64).

[See footnote 13 of this chapter.] We note that $\varepsilon(q)$ is an even function in q; also, despite the sinh in its constitutive equation, (11.2-56), it is easily seen that $\varepsilon(q) \geq 0$. The minimum value occurs for $\cos q = -1$ or $q = \pi$. For that case $\varepsilon(q \to \pi) = 2|K_2 - K_1^*|$.[18]

We now turn to the case of odd total fermion occupancy. We need the additional EV from (11.2-40), remembering that the 0-state is occupied and the π-state is empty,

$$\lambda_+[V_{10}^{1/2} V_{20} V_{10}^{1/2}] = e^{2K_2 + K_1^*}, \qquad \lambda_+[V_{1\pi}^{1/2} V_{2\pi} V_{1\pi}^{1/2}] = e^{-K_1^*} \quad . \tag{11.3-3}$$

From (11.2-42) and the above,

$$\Lambda_{odd} = (2\sinh 2K_1)^{M/2} \exp\left[2K_2 + \tfrac{1}{2}\sum_{\bar{q}\neq 0,\pi} \varepsilon(\bar{q})\right]. \tag{11.3-4}$$

For the exponent we have

$$2K_2 + \tfrac{1}{2}\sum_{\bar{q}\neq 0,\pi} \varepsilon(\bar{q})$$

$$= \tfrac{1}{2}\sum_{all\,\bar{q}} \varepsilon(\bar{q}) + K_2 - K_1^* - |K_2 - K_1^*| = \begin{cases} \tfrac{1}{2}\Sigma_{\bar{q}}\varepsilon(\bar{q}) & K_2 > K_1^*, \\ 2(K_2 - K_1^*) + \tfrac{1}{2}\Sigma_{\bar{q}}\varepsilon(\bar{q}), & K_2 < K_1^*. \end{cases} \tag{11.3-5}$$

We recall that $K_2 = \beta J_2 = J_2 / k_B T$; thus, K_2 is large at low temperatures. We also have $\tan K_1^* = \exp(-2J_1 / k_B T)$, which is small at low temperatures. Since we look for cooperative effects that occur at low temperatures, the second possibility in (11.3-5) will be discarded. The result is then

$$\Lambda_{odd} = (2\sinh 2K_1)^{M/2} \exp\left[\tfrac{1}{2}\sum_{\bar{q}} \varepsilon(\bar{q})\right]. \tag{11.3-6}$$

We saw before that there are M distinct even and odd wave vectors, given in (11.2-34). In the limit that $M \to \infty$ these sequences of wave vectors become identical and we can set $\Sigma_{\bar{q}}\varepsilon(\bar{q}) \to \Sigma_q \varepsilon(q)$. Thus, the EV Λ_{even} and Λ_{odd} are two-fold degenerate below a critical temperature T_c, which is determined by $K_2 = K_1^*$. Now from (11.2-12) we have $\tanh K_1^* = e^{-2K_1}$, $\coth K_1^* = e^{2K_1}$, from which by subtraction

$$\sinh 2K_1 = \tfrac{1}{2}(\coth K_1^* - \tanh K_1^*) = \frac{\cosh^2 K_1^* - \sinh^2 K_1^*}{2\sinh K_1^* \cosh K_1^*} = \frac{1}{\sinh 2K_1^*}. \tag{11.3-7}$$

Since $K_2(T_c) = K_1^*(T_c)$ we find that T_c is determined by

---

[18] $\cosh 2K_2 \cosh 2K_1^* - \sinh 2K_2 \sinh 2K_1^* = \cosh 2(K_2 - K_1^*)$. Likewise we find $\varepsilon(q \to 0) = 2(K_2 + K_1^*)$.

$$\sinh(2J_1/k_BT_c)\sinh(2J_2/k_BT_c)=1. \tag{11.3-8}$$

For the special case of a square lattice with $J_1 = J_2 = J$, this yields

$$k_BT_c/J = 2.269185... \tag{11.3-9}$$

We still note that the degeneracy of the EV adds a term $-k_BT\ln 2$ to the free energy, which, in the thermodynamic limit can be neglected.

The two-fold degeneracy of the ground state indicates that we have *bifurcation*; there are two branch solutions with long-range order: either all spins are up or all spins are down. Although we have not included an external magnetic field, we indicate in Fig.11-2 the limiting behaviour of the $m-h$ curve when $h\to 0$.

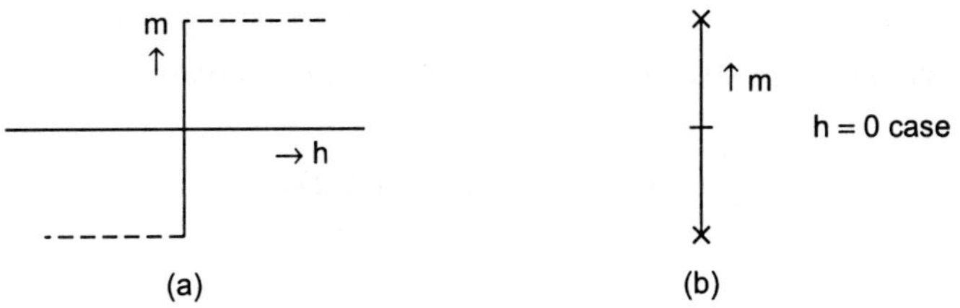

Fig. 11-2. (a) m vs. h curve. (b) m for $h\to 0$.

Thermodynamic functions

The free energy is obtained as $F=-\beta^{-1}\ln[\Lambda_+^N+\Lambda_-^N]\to-\beta^{-1}N\ln\Lambda_+$, so that from (11.3-2) [or (11.3-6)] and (11.3-1):

$$\beta\hat{f}(0,T)=-(1/2)\ln(2\sinh 2K_1)-\lim_M(1/2M)\textstyle\sum_q\varepsilon(q). \tag{11.3-10}$$

[Here the '(0,*T*)' indicates that the magnetic field h = 0.] Since for $M\to\infty$ the modes become continuous, the sum can be replaced with a Riemann integral, noting that the mode interval is $(\Delta q)=2\pi/M$. Hence,

$$\beta\hat{f}(0,T)=-\frac{1}{2}\ln(2\sinh 2K_1)-\frac{1}{4\pi}\int_{-\pi}^{\pi}\varepsilon(q)dq. \tag{11.3-11}$$

Although this result is correct, it is highly non-symmetric in the row and column parameters due to our choice to use the column spin states as basis; the ultimate *challenge* is now to obtain a symmetrical result. This was historically achieved with 'Onsager's identity' which we will meet below.

We shall modify the defining equation for $\varepsilon(q)$. Firstly we note that (11.3-7) holds for both variables, so that

$$\sinh 2K_i^* \sinh 2K_i = 1, \quad (i=1,2). \tag{11.3-12a}$$

Secondly, it follows that

$$\cosh 2K_i^* = \sqrt{1+\mathrm{csch}^2 2K_i} = \coth 2K_i \quad (i=1,2). \tag{11.3-12b}$$

Consequently, we have the alternate expression

$$\cosh \varepsilon(q) = \frac{\sinh 2K_2}{\sinh 2K_1}\left(\cosh 2K_1 \coth 2K_2 - \cos q\right). \tag{11.3-13}$$

The integration involved in (11.3-11) is carried out in several of the original articles of the cited open literature. Readable accounts are also found in Huang[19] and in Plischke and Bergersen[20], but both treatments assume a priori a square lattice, so some essential elements of the method are lost. We need *Onsager's identity*,

$$\frac{1}{2\pi}\int_0^{2\pi} dt \ln(2\cosh x \pm 2\cos t) = |x|. \tag{11.3-14}$$

To prove this, differentiate to x and set

$$f(x) = (1/2\pi)\int_0^{2\pi} dt \frac{\sinh x}{\cosh x \pm \cos t}. \tag{11.3-15}$$

To change into a contour integral we make the usual substitution $z = \exp(it)$. One easily finds $f(x) = \mathrm{sgn}\, x$. By integration we find (11.3-14).

From (11.3-13) and (11.3-14) we now have, setting $x = \varepsilon(q)$,[21]

$$\begin{aligned}\varepsilon(q) &= \frac{1}{2\pi}\int_0^{2\pi} dt \ln[2\cosh\varepsilon(q) \pm 2\cos t] \\ &= \frac{1}{\pi}\int_0^{\pi} dt \left\{ \ln\left[\frac{2}{\sinh 2K_1}\left(\cosh 2K_1 \cosh 2K_2 - \cos q \sinh 2K_2\right) - 2\cos t\right]\right\}. \end{aligned}\tag{11.3-16}$$

We need the integral $I = -(1/4\pi)\int_{-\pi}^{\pi} dq\, \varepsilon(q) = -(1/2\pi)\int_0^{\pi} dq\, \varepsilon(q)$. Note $\varepsilon(q) = \varepsilon(-q)$. From (11.3-16) we have,

$$I = -\frac{1}{2\pi^2}\int_0^{\pi} dq \int_0^{\pi} dt \left\{ \ln\left[\frac{2}{\sinh 2K_1}(\cosh 2K_1 \cosh 2K_2 - \cos q \sinh 2K_2) - 2\cos t\right]\right\}$$

---

[19] Kerson Huang, 2nd Ed., Op. Cit., pp. 386 - 391.

[20] M. Plischke and B. Bergersen, 2nd Ed. p. 174ff.

[21] In the integral for dt we choose the − sign on the interval (0,π) and the + sign on (π,2π) changing $t \to t + \pi$.

$$= -\frac{1}{2\pi^2}\int_0^\pi dt'\int_0^\pi dt\{\ln 2 + \ln[\cosh 2K_1 \cosh 2K_2 - \cos t'\sinh 2K_2 - \cos t \sinh 2K_1]$$
$$-\ln \sinh 2K_1\}, \tag{11.3-17}$$

where we set $t'=q$. Now going back to the free energy (11.3-11), the first term, when added under the integral above, cancels the last ln sinh term; so the final result is,

$$\beta\hat{f}(0,T) = -\frac{1}{2\pi^2}\int_0^\pi dt'\int_0^\pi dt\{\ln 2+\ln[\cosh 2K_1\cosh 2K_2 - \cos t'\sinh 2K_2 - \cos t \sinh 2K_1]\}, \tag{11.3-18}$$

which is symmetrical in K_1 and K_2, in accord with Mattis, Op. Cit, Eq. (57), p. 267.

To obtain a result in terms of tabulated integrals, we consider further only the case of a square lattice, for which $K_1 = K_2 = K$. The expression (11.3-13) for $\cosh \varepsilon(q)$ now simplifies considerably; from Onsager's identity we obtain

$$\varepsilon(q) = \frac{1}{\pi}\int_0^\pi dt\{\ln[2\cosh 2K \coth 2K - 2(\cos q + \cos t)]$$
$$= \frac{1}{\pi}\int_0^\pi dt\left\{\ln\left[2\cosh 2K \coth 2K - 4\cos\frac{q+t}{2}\cos\frac{q-t}{2}\right]\right\}. \tag{11.3-19}$$

We choose new variables, $\omega_1 = \frac{1}{2}(q-t)$, $\omega_2 = \frac{1}{2}(q+t)$. The Jacobian is 2 and the new area of integration is indicated in Fig. 11-3 by the striped square. Obviously, nothing changes if instead we integrate over the dashed rectangle. Hence for the integral over $\varepsilon(q)$

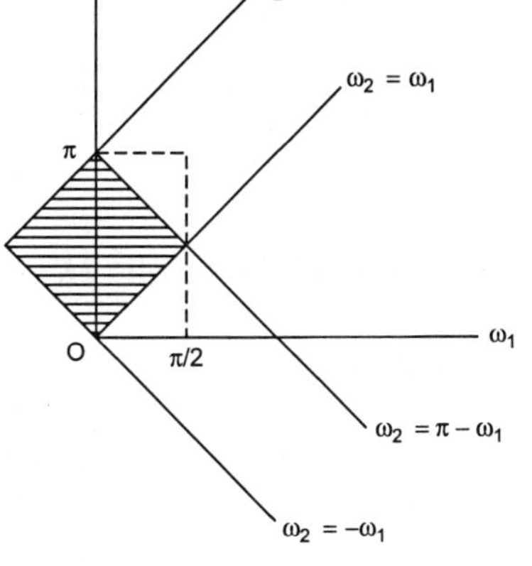

Fig. 11-3. Diagram for change of variables.

$$I = -\frac{1}{\pi^2}\int_0^{\pi} d\omega_2 \int_0^{\pi/2} d\omega_1 \{\ln[2\cosh 2K \coth 2K - 4\cos\omega_1 \cos\omega_2]\}. \quad (11.3\text{-}20)$$

This is modified to read

$$I = -\frac{1}{\pi^2}\int_0^{\pi} d\omega_2 \int_0^{\pi/2} d\omega_1 \ln(2\cos 2\omega_1)$$
$$-\frac{1}{\pi^2}\int_0^{\pi} d\omega_2 \int_0^{\pi/2} d\omega_1 \ln\left[2\frac{\cosh 2K \coth 2K}{2\cos\omega_1} - 2\cos\omega_2\right]. \quad (11.3\text{-}21)$$

Setting $I = I_1 + I_2$, the second integral can be evaluated with Onsager's identity, yielding

$$I_2 = -\frac{1}{\pi}\int_0^{\pi/2} \mathrm{d}\omega_1 \operatorname{arcosh}\left[\frac{\cosh 2K \coth 2K}{2\cos\omega_1}\right]. \quad (11.3\text{-}22)$$

For the first integral we find

$$I_1 = -\frac{1}{\pi}\int_0^{\pi/2} d\omega_1 \ln(2\cos\omega_1). \quad (11.3\text{-}23)$$

We employ $\operatorname{arcosh} x = \ln(x + \sqrt{x^2 - 1})$, set $\omega_1 = \pi/2 - \theta$ and combine the integrals again; hence

$$I = -\frac{1}{\pi}\int_0^{\pi/2} d\theta \ln\left\{\cosh 2K \coth 2K + \sqrt{\cosh^2 2K \coth^2 2K - 4\sin^2\theta}\right\}$$
$$= -\frac{1}{\pi}\int_0^{\pi/2} d\theta \ln\left\{2\cosh 2K \coth 2K\left(\frac{1+\sqrt{1-\Psi^2\sin^2\theta}}{2}\right)\right\}$$
$$= -\frac{1}{2}\ln(2\cosh 2K \coth 2K) - \frac{1}{\pi}\int_0^{\pi/2} d\theta \ln\left(\frac{1+\sqrt{1-\Psi^2\sin^2\theta}}{2}\right), \quad (11.3\text{-}24)$$

with

$$\Psi(K) = 2\sinh 2K / \cosh^2 2K. \quad (11.3\text{-}25)$$

We finally substitute into (11.3-11) to obtain

$$\beta\hat{f}(0,T) = -[\ln(2\cosh 2K)] - \frac{1}{\pi}\int_0^{\pi/2} d\theta \ln\left(\frac{1+\sqrt{1-[\Psi(K)]^2\sin^2\theta}}{2}\right). \quad (11.3\text{-}26)$$

The function $\Psi(K)$ is pictured in Fig. 11-4. The maximum is $\Psi = 1$, occurring for $\sinh^2 2K = 1$ or $T = T_c$.

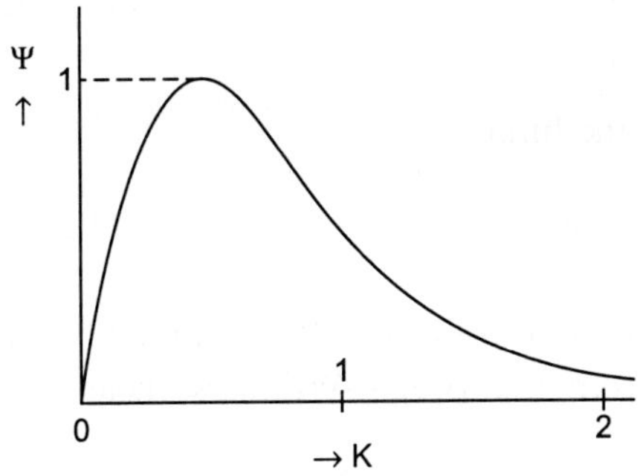

Fig. 11-4. The function $\Psi(K)$.

The energy per spin is given by

$$\hat{u}(T) = (d/d\beta)[\hat{f}(0,T)] = -J\coth 2K\left[1+\frac{2}{\pi}(2\tanh^2 2K-1)\mathcal{K}_1(\Psi)\right], \quad (11.3\text{-}27)$$

where

$$\mathcal{K}_1(\Psi) = \int_0^{\pi/2} \frac{d\phi}{\sqrt{1-\Psi^2\sin^2\phi}} \quad (11.3\text{-}28)$$

is the complete elliptic integral of the first kind. When $\Psi \to 1$, i.e., at $T = T_c$ the term $2\tanh^2 2K - 1 \to 0$, so that the energy per spin is continuous at the transition. To obtain the specific heat, we must differentiate once more. This is left to the problems. The result is

$$C(T)/k_B = \frac{4}{\pi}(K\coth 2K)^2\{\mathcal{K}_1(\Psi) - \mathcal{E}_1(\Psi) + (\tanh^2 2K - 1)[\frac{\pi}{2} + \left(2\tanh^2 2K - 1\right)\mathcal{K}_1(\Psi)\}, \quad (11.3\text{-}29)$$

where

$$\mathcal{E}_1(\Psi) = \int_0^{\pi/2} d\phi\sqrt{1-\Psi^2\sin^2\phi} \quad (11.3\text{-}30)$$

is the complete elliptic integral of the second kind. Near T_c the specific heat can be approximated by

$$C(T)/k_B = -\frac{2}{\pi}\left(\frac{2J}{k_B T_c}\right)^2 \ln\left|1-\frac{T}{T_c}\right| + const. \quad (11.3\text{-}31)$$

Thus, at $T = T_c$ there is not a discontinuity but a logarithmic divergence. According to scaling theory, the behaviour should follow the law

$$C(T) \propto \left| 1 - \frac{T}{T_c} \right|^{-\alpha}, \tag{11.3-32}$$

where α is the critical exponent. Now we have the limit

$$\lim_{\alpha \to 0} \alpha^{-1}\left(z^{-\alpha} - 1\right) = -\ln z. \tag{11.3-33}$$

The present result therefore amounts to a scaling law with $\alpha \to 0$. In Fig. 11-5 we give a sketch of $C(T)$ obtained here and from the Bragg–Williams mean-field approximation. The difference is striking.

We also give some results for the 3D Ising model. It has been definitively shown that the corresponding Hamiltonian cannot be diagonalized by any of the mentioned procedures. However, highly accurate series expansions for this problem have been carried out. The specific heat has a logarithmic singularity, several theoretical arguments to the contrary notwithstanding. The singularity is asymmetrical, resembling more the Curie-Weiss behaviour, giving some re-assurance for the possible validity of the molecular field approach. Results are shown in Fig. 11.6.

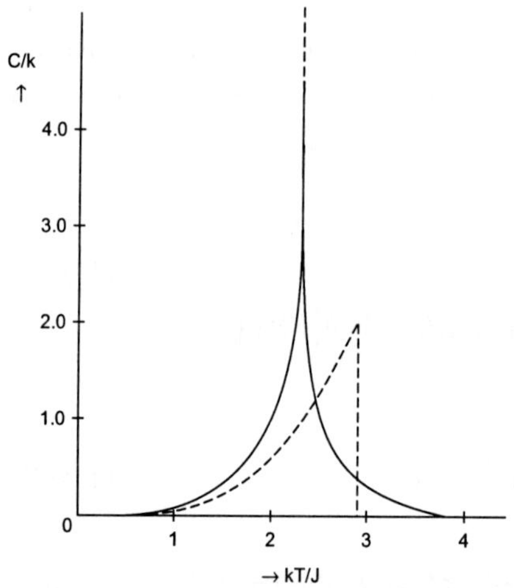

Fig. 11-5. $C(T)$ for the 2D Ising model: solid curve: exact specific heat from 2D results; dashed curve: mean-field approximation. After Mattis, Op. Cit., Fig. 9.2. [With permission.]

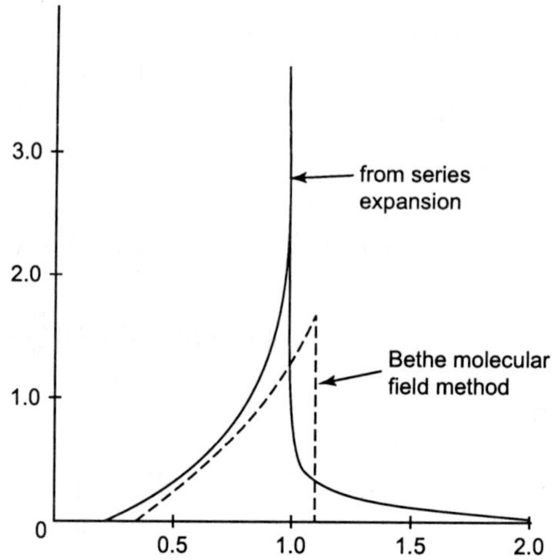

Fig. 11-6. Specific heat vs. k_BT/J for the 3D Ising model from series expansion (solid curve) and molecular-field results (dashed curve). After Mattis, Op. Cit., Fig. 9.3. [With permission.]

11.4 The Spontaneous Magnetization; Onsager's result

When the interaction energy with a magnetic field is added, as at the onset of our discussion, the complexity is greatly increased, see Yang [3]. We need another transfer operator, V_3

$$V_3 = \exp(\beta h \sum_j^M \sigma_{j,z}). \tag{11.4-1}$$

Due to the magnetic field energy the degeneracy between the two ground states below T_c is now lifted, much like in the Zeeman effect; the new states give rise to a macroscopic magnetization parallel to h. The partition function is formally given by

$$\mathcal{Z} = \mathrm{Tr}\{(V_1 V_2 V_3)^N\}. \tag{11.4-2}$$

The transfer matrix must be Hermitized. Noticing that V_2 and V_3 commute with each other, but not with V_1, we should set

$$\mathcal{Z} = \mathrm{Tr}\left\{\left[V_1^{1/2}\left(\frac{V_2V_3 + V_3V_2}{2}\right)V_1^{1/2}\right]^N\right\}. \tag{11.4-3}$$

The spontaneous magnetization in units ρ_B per spin is obtained from

$$m_0(T) = -\lim_{h\to 0}(\partial/\partial h)\hat{f}(h,T), \tag{11.4-4}$$

where as usual $\hat{f}(h,T) = -\ln \mathcal{Z}/\beta MN$. The procedure is *very complex* and will not be considered. Instead, we shall obtain $m_0(T)$ from the spin-spin correlation function.[5,8]

11.4.1 *Spin-Spin Correlation Function*

As an alternative to the method used by Yang, it is suggested by the mean-field theory approach discussed in a previous chapter, that spontaneous magnetization occurring for zero external field is associated with long-range spin ordering, given by the spin-spin correlation function. For the 2D rectangular lattice the magnetization per spin in Bohrmagnetons, $m_0(T)$, should be defined by [(see (10.2-18)]:

$$[m_0(T)]^2 = \lim_{M,N\to\infty} \langle(\Sigma_i \sigma_{\mathbf{R}_i})^2\rangle_{eq}/MN = \lim_{\mathbf{R}_{ij}\to\infty} \langle \sigma_{\mathbf{R}_i}\sigma_{\mathbf{R}_j}\rangle_{eq}. \tag{11.4-5}$$

Here $\mathbf{R}_i$ and $\mathbf{R}_j$ are spin positions and '*eq*' means an equilibrium average, based on the canonical ensemble used in this chapter. In the thermodynamic limit as applied, the short-range correlations disappear and the long-range order is expressed by the far rhs of (11.4-5). It is clear that $m_0(T)$ cannot exceed unity, meaning that all spins are 'lined up'. Our challenge is now to obtain the result stated in (11.1-2) and (11.1-

3). While we will never know what method (or methods?) Onsager used when he wrote this landmark formula on a blackboard for a broad audience at Cornell University in 1948, we do know that Onsager was familiar with certain properties of *Toeplitz determinants*, which were communicated to the Yale mathematician S. Katani, and which formed the subject of several papers by G. Szego[22] and by Marc Kac.[23] In the present section, which will be based on Section IV of the treatment by Schultz, Mattis and Lieb[5] that we have followed so far, as well as on the earlier basic paper by Elliott Montroll, Potts and Ward[24], we shall see that the evaluation of (11.4-5) will lead to a Toeplitz determinant, that can be evaluated by the powerful Kac-Szego theorem, to be discussed. We *may therefore guess* that Onsager presumably employed a method analogous to the approach of this subsection, and not the extremely complex Yang method, in which the magnetic field must be explicitly introduced and then later on let go to zero. *Anyway*, we believe that the method of Eq. (11.4-5) is the only sensible way to present the Onsager result in a textbook for students or fellow-researchers.

Without loss of generality it suffices to obtain the spin-spin correlation function along the elements of a row in the rectangular lattice. The basis spin space is the same as before, but we rotated about the y-axis, see Fig. 11-1, so that we must evaluate

$$\langle\sigma_{i,x}\sigma_{j,x}\rangle_{eq} = \mathrm{Tr}\{\sigma_{i,x}\sigma_{j,x}e^{-\beta\mathcal{H}}\}/\mathrm{Tr}e^{-\beta\mathcal{H}} = \mathrm{Tr}\{\sigma_{i,x}\sigma_{j,x}V^N\}/\mathrm{Tr}V^N, \quad (11.4\text{-}6)$$

where i and j represent spins in a row separated by $(j-i)d$ and where $V = \Pi_q V_q$ is the transfer operator, *properly chosen*, see below. Denoting the ES as Ψ_α ($\alpha = 0,1$), we have

$$Tr\{\sigma_{i,x}\sigma_{j,x}V^N\} = \sum_{\alpha,\beta}\langle\Psi_\alpha|\sigma_{i,x}\sigma_{j,x}|\Psi_\beta\rangle\underbrace{\langle\Psi_\beta|V^N|\Psi_\alpha\rangle}_{\Lambda_\alpha\delta_{\alpha\beta}}$$

$$= \sum_\alpha\langle\Psi_\alpha|\sigma_{i,x}\sigma_{j,x}|\Psi_\alpha\rangle\Lambda_\alpha^N. \quad (11.4\text{-}7)$$

Keeping only the largest EV, this gives,

$$\langle\sigma_{i,x}\sigma_{j,x}\rangle_{eq} = \langle\Psi_0|(\sigma_i^+ + \sigma_i^-)(\sigma_j^+ + \sigma_j^-)|\Psi_0\rangle. \quad (11.4\text{-}8)$$

---

[22] G. Szego, Communications du séminaire mathématique de l'Université de Lund, tome supplementaire (1952) (dedicated to M. Riesz) p. 228-238; also V. Grenander and G. Szego, "Toeplitz forms and their applications", U. of California Press, Berkeley 1958.

[23] M. Kac, Duke Math. Journal, p.21 (1954); M. Kac, "Probability and Related Topics in Physical Sciences", Interscience, N.Y., 1959.

[24] See Ref. 8 of this chapter; we will not dwell on their main treatment, which involves the use of Pfaffians to obtain the transfer operator; however, their paper, Section 8, has the details on the Kac-Szego theorem and the evaluation of the Toeplitz determinant.

Let us now consider the expression $\mathrm{Tr}\{\sigma_{i,x}\sigma_{j,x}V^N\}$ in more detail. The operator V^N can be Hermitized in two ways. First, we can write

$$\begin{aligned}\mathrm{Tr}\{\sigma_{i,x}\sigma_{j,x}e^{-\beta\mathcal{H}}\} &= \mathrm{Tr}\{\sigma_{i,x}\sigma_{j,x}V_1^N V_2^{\frac{1}{2}N} V_2^{\frac{1}{2}N}\} \\ &= \mathrm{Tr}\{V_2^{\frac{1}{2}N}\sigma_{i,x}\sigma_{j,x}V_1^N V_2^{\frac{1}{2}N}\} = \mathrm{Tr}\{\sigma_{i,x}\sigma_{j,x}V_2^{\frac{1}{2}N}V_1^N V_2^{\frac{1}{2}N}\},\end{aligned} \tag{11.4-9}$$

where we used the fact that $(\sigma_{i,x}\sigma_{j,x})$ and V_2 [as given in original form,(11.2-15), (11.2 -16)] commute. So, introducing $\bar{V} = V_2^{\frac{1}{2}}V_1V_2^{\frac{1}{2}}$, we have $\langle\sigma_{i,x}\sigma_{j,x}\rangle = \mathrm{Tr}\{\sigma_{i,x}\sigma_{j,x}\bar{V}^N\}$. Secondly, with our previous Hermitized form one can show that we end up with

$$\mathrm{Tr}\{\sigma_{i,x}\sigma_{j,x}e^{-\beta\mathcal{H}}\} = \mathrm{Tr}\{(V_1^{-\frac{1}{2}}\sigma_{i,x}\sigma_{j,x}V_1^{\frac{1}{2}})V_1^{\frac{1}{2}N}V_2^N V_1^{\frac{1}{2}N}\}, \tag{11.4-10}$$

The evaluation of this expression is a whole new ballgame, see SML[5], Eq. (4.25); it will not be pursued.

In this section we must therefore consider the more complex transfer operator $\bar{V}$. The main new ingredient is the form for $V_{2q}^{\frac{1}{2}}$. The original exponential form was in (11.2-38). Taking the square root means the replacement $2K_2 \to K_2$. We can thus employ the previous pair-operator procedure, leading to a matrix analogous to (11.2-49):

$$V_{2q}^{\frac{1}{2}} = e^{K_2\cos q}\begin{pmatrix}\cosh K_2 + \sinh K_2\cos q & \sinh K_2 \sin q \\ \sinh K_2 \sin q & \cosh K_2 - \sinh K_2 \cos q\end{pmatrix}. \tag{11.4-11}$$

The transfer matrix has again the form (11.2-51), but with new coefficients:

$$\begin{aligned}A_q &= e^{-2K_1^*}[\cosh^2 K_2 + 2\sinh K_2 \cosh K_2 \cos q + \sinh^2 K_2 \cos^2 q] \\ &\quad + e^{2K_1^*}(\sinh^2 K_2 \sin^2 q),\end{aligned} \tag{11.4-12a}$$

$$\begin{aligned}B_q &= e^{2K_1^*}[\cosh^2 K_2 - 2\sinh K_2 \cosh K_2 \cos q + \sinh^2 K_2 \cos^2 q] \\ &\quad + e^{-2K_1^*}(\sinh^2 K_2 \sin^2 q),\end{aligned} \tag{11.4-12b}$$

$$C_q = 2\sinh K_2 \sin q[\cosh 2K_1^* \cosh K_2 - \sinh 2K_1^* \sinh K_2 \cos q] \tag{11.4-12c}$$

We now translate to fermion operators. Because i and j are widely separated lattice points on a row, the results will *not* be bilinear in the c's and $c^\dagger$'s! Generally we have

$$\begin{aligned}&\sigma_i^+ = Q_i c_i^\dagger, \quad \sigma_i^- = c_i Q_i^\dagger, \quad \text{where} \\ &Q_i^{(\dagger)} = \exp[\pm\pi i \sum_{k<i} c_k^\dagger c_k] = \prod_{k<i}(-1)^{\mathbf{n}_k}.\end{aligned} \tag{11.4-13}$$

Now consider

$$(c_k^\dagger + c_k)(c_k^\dagger - c_k) = (c_k^\dagger)^2 - c_k^\dagger c_k + c_k c_k^\dagger - (c_k)^2 = 1 - 2\mathbf{n}_k. \tag{11.4-14}$$

Acting on a state $|n_1...n_k...\rangle$ with $n_k = 0,1$, we note that $(1-2\mathrm{n}_k)|n_1..n_k..\rangle = (-1)^{n_k}|n_1..n_k..\rangle = (-1)^{\mathbf{n}_k}\,|n_1..n_k..\rangle$. Therefore, we have the operator equivalence

$$\prod_k (-1)^{\mathbf{n}_k} = \prod_k (c_k^\dagger + c_k)(c_k^\dagger - c_k). \tag{11.4-15}$$

Returning to the product operators in (11.4-8), noting that Q_i and $c_i^{(\dagger)}$ commute, we encounter terms like

$$Q_i^{(\dagger)} Q_j^{(\dagger)} = \prod_{k=i+1}^{j-1} (-1)^{\mathbf{n}_k}\, e^{\pm \pi i c_i^\dagger c_i}, \tag{11.4-16}$$

$$c_i^\dagger e^{\pm \pi i c_i^\dagger c_i} = c_i^\dagger, \quad c_i e^{\pm \pi i c_i^\dagger c_i} = -c_i\,. \tag{11.4-17}$$

[see also (11.2-22)]. From (11.4-13) through (11.4-17) we arrive at the main result:

$$\langle \sigma_{i,x}\sigma_{j,x}\rangle_{eq}$$
$$= \langle \Psi_0 | (c_i^\dagger - c_i)(c_{i+1}^\dagger + c_{i+1})(c_{i+1}^\dagger - c_{i+1}) ... (c_{j-1}^\dagger + c_{j-1})(c_{j-1}^\dagger - c_{j-1})(c_j^\dagger + c_j) | \Psi_0 \rangle. \tag{11.4-18}$$

To evaluate this expectation value, we employ Wick's theorem of many-body physics, which states[25] : "Associate the operators in all possible pair products, replace each pair by its ground-state expectation value and provide each product with $(-1)^p$ to take care of the number of transpositions p to attain each permutation $\mathscr{P}$ of the operators; finally sum over all permutations." The result thus obtained clearly defines a determinant.

The contractions which occur are of three types, two of which vanish,

$$\langle \Psi_0 | (c_m^\dagger + c_m)(c_n^\dagger + c_n) | \Psi_0 \rangle = 0, \tag{11.4-19a}$$

$$\langle \Psi_0 | (c_m^\dagger - c_m)(c_n^\dagger - c_n) | \Psi_0 \rangle = 0; \tag{11.4-19b}$$

remains

$$\langle \Psi_0 | (c_m^\dagger - c_m)(c_n^\dagger + c_n) | \Psi_0 \rangle \equiv a_{m,n-1}\,. \tag{11.4-20}$$

To obtain the elements $a_{m,n-1}$ we transform to running-wave fermion operators, c_q and $c_q^\dagger$ and subsequently with the Bogoliubov-Valatin transformation to the operators η_q and $\eta_q^\dagger$. We thus write

$$c_m = M^{-1/2} e^{-i\pi/4} \sum_{q'} e^{iq'm} (\eta_{q'} \cos\phi_{q'} - \eta_{-q'}^\dagger \sin\phi_{q'})\ ,$$

$$c_m^\dagger = M^{-1/2} e^{i\pi/4} \sum_{q'} e^{-iq'm} (\eta_{q'}^\dagger \cos\phi_{q'} - \eta_{-q'} \sin\phi_{q'})\ ,$$

---

[25] See e.g., A.L. Fetter and J.D. Walecka. "Quantum Theory of Many-Particle Systems", McGraw Hill 1971, p. 83ff; also Chapter XII, subsection 12.9.2.

$$c_{n+1} = M^{-\frac{1}{2}} e^{-i\pi/4} \sum_q e^{iqn} e^{iq} (\eta_q \cos\phi_q - \eta_{-q}^\dagger \sin\phi_q) \ ,$$

$$c_{n+1}^\dagger = M^{-\frac{1}{2}} e^{i\pi/4} \sum_q e^{-iqn} e^{-iq} (\eta_q^\dagger \cos\phi_q - \eta_{-q} \sin\phi_q). \qquad (11.4\text{-}21)$$

For the daggered operators we change the summation variables $q' \to -q'$, $q \to -q$, to get

$$c_m^\dagger - c_m = M^{-\frac{1}{2}} e^{-i\pi/4} \sum_{q'} e^{iq'm} (-\eta_{q'} + i\eta_{-q'}^\dagger)(\cos\phi_{q'} - i\sin\phi_{q'}) \ ,$$

$$c_{n+1}^\dagger + c_{n+1} = M^{-\frac{1}{2}} e^{-i\pi/4} \sum_q e^{iqn} e^{iq} (\eta_q + i\eta_{-q}^\dagger)(\cos\phi_q + i\sin\phi_q) \ . \qquad (11.4\text{-}22)$$

Now $(\eta_q + i\eta_{-q}^\dagger) | \bar{0}_{-q} \bar{0}_q \rangle = i | \bar{1}_{-q} \bar{0}_q \rangle$. Next, acting on this with $(-\eta_{q'} + i\eta_{-q'}^\dagger)$ we see that q' must be equal to $-q$, in order to recreate the ground state; the operator $i\eta_{-q'}^\dagger = i\eta_q^\dagger$ has no effect since it creates the state $| \bar{1}_{-q} \bar{1}_q \rangle$, which is orthogonal to $| \bar{0}_{-q} \bar{0}_q \rangle$. Therefore, the result is [26]:

$$\langle \Psi_0 | (c_m^\dagger - c_m)(c_{n+1}^\dagger + c_{n+1}) | \Psi_0 \rangle$$

$$= M^{-1} e^{-i\pi/2} \sum_q e^{iq(n-m)} e^{iq} e^{-i\pi/2} (\cos\phi_q + i\sin\phi_q)^2 .$$

$$= M^{-1} \sum_q e^{iq(n-m)} e^{i(q+2\phi_q)} \overset{q \to -q}{=} M^{-1} \sum_q e^{-iq(n-m)} e^{-i(q+2\phi_q)} = a_{m,n} \ , \qquad (11.4\text{-}23)$$

where in the last transition we used $\operatorname{sgn}\phi_q = \operatorname{sgn} q$. Summing all pair products with their appropriate parity we obtain the Toeplitz determinant [27]

$$\langle \sigma_{m,x} \sigma_{m',x} \rangle_{eq} = \begin{vmatrix} a_{m,m} & a_{m,m+1} & \cdots & a_{m,m'-1} \\ a_{m+1,m} & \cdot & & \cdot \\ \vdots & \vdots & & \vdots \\ a_{m'-1,m} & \cdot & \cdots & a_{m'-1,m'-1} \end{vmatrix} \to \begin{vmatrix} a_0 & a_1 & \cdots & a_{m'-1} \\ a_{-1} & \cdot & & \cdot \\ \vdots & \vdots & & \vdots \\ a_{-m'+1} & a_{-m'+2} & \cdots & a_0 \end{vmatrix} .$$

(11.4-24)

In the second form, used by MPW[8], the correlation runs from the site m' to site 0, the rank of the determinant being m'. The Toeplitz determinant, denoted as $T(F)$, is closely related to the cyclic determinant $C(F)$, as will now be indicated.[28]

---

[26] Since ϕ_q is defined modulo $\pi/2$ [see (11.2-64)] we absorb the factor $\exp(-i\pi)$ into $\exp(-2i\phi_q)$.

[27] In a Toeplitz determinant the element a_{ij} depends only on $|i-j|$.

[28] For a cyclic determinant $a_{m+M,m'+M} = a_{m,m'}$; for a Toeplitz determinant of rank M, $a_{m,m'} = 0$ for $m, m' > M$. For an example see Mattis, Op. Cit., p. 270.

Let the element $a_{m,n} \equiv a(n-m)$ have a Fourier series in q-space, i.e. we introduce

$$F(q)=\sum_{p=1}^{M} e^{iqp}a(p)\,, \quad \text{with} \quad a(n-m)=\frac{1}{M}\sum_{q} e^{-iq(n-m)}F(q)\,. \tag{11.4-25}$$

In the limit of large M the cyclic determinant is easily shown to be

$$C(F)=\prod_{q} F(q) \sim \exp\left\{M\int_{-\pi}^{\pi} \ln F(q)dq/2\pi\right\}. \tag{11.4-26}$$

The Kac-Szego theorem yields the Toeplitz determinant $T(F)$ in terms of $C(F)$, see Ref. 23,

$$T(F)=C(F)\exp\left\{\sum_{n=1}^{\infty} nk_n k_{-n}\right\}, \tag{11.4-27}$$

where the k's are determined by the Fourier expansion of $\ln F$:

$$\ln F(q) = \sum_{n=-\infty}^{\infty} k_n e^{inq}\,. \tag{11.4-28}$$

11.4.2 *Evaluation of the Toeplitz Determinant for the Correlation*

We return to the Toeplitz determinant under consideration. We can read the Fourier coefficients out of (11.4-23):

$$F(q)=\exp[-i(2\phi_q+q)]\,. \tag{11.4-29}$$

Since $\operatorname{sgn}\phi_q=\operatorname{sgn} q$, $\ln F(q)$ is an odd function and the integral (11.4-26) vanishes prior to taking the thermodynamic limit; consequently, $C=1$.[29] Lastly, we need a form for $\exp(-2i\phi_q)$; this can be found from $\tan 2\phi_q$, as given in (11.2-64). With a lengthy (but straightforward) computation one obtains a final beautiful result:

$$\begin{aligned} e^{-2i\phi_q} &= \sqrt{\frac{1-i\tan 2\phi_q}{1+i\tan 2\phi_q}} \\ &= e^{iq}\left[\frac{e^{iq}\tanh K_1^*\tanh K_2+e^{-iq}\tanh K_1^*\coth K_2-(1+\tanh^2 K_1^*)}{e^{-iq}\tanh K_1^*\tanh K_2+e^{iq}\tanh K_1^*\coth K_2-(1+\tanh^2 K_1^*)}\right]^{1/2}. \end{aligned} \tag{11.4-30}$$

The rhs can be factored, as is confirmed by minor algebra; the result for $F(q)$ becomes

---

[29] Mathematical justification for the order of the limit procedure is found in the last section of Schultz *et alii.*

$$F(q) = e^{-i(2\phi_q + q)} = \left[\frac{(1 - x_1^{-1}e^{iq})(1 - x_2 e^{-iq})}{(1 - x_1^{-1}e^{-iq})(1 - x_2 e^{iq})}\right]^{1/2}, \quad T < T_c, \tag{11.4-31}$$

where

$$\begin{aligned} x_1^{-1} &= \tanh K_1^* \tanh K_2 < 1, \\ x_2 &= \tanh K_1^* / \tanh K_2 < 1, \quad (T < T_c). \end{aligned} \tag{11.4-32}$$

Next, we expand $\ln F(q)$. To that end we shall revert to the notation of Montroll et al.[8], who first evaluated the Toeplitz determinant. So let $x_1^{-1} = z_1^* z_2$, $x_2 = z_1^*/z_2$, from which

$$z_2 = \sqrt{1/x_1 x_2} = \tanh K_2, \quad z_1^* = \sqrt{x_2/x_1} = \tanh K_1^*. \tag{11.4-33}$$

From $\tanh K_1^* = \exp(-2K_1)$ also

$$z_1^* = (1 - z_1)/(1 + z_1), \quad z_1 = \tanh K_1. \tag{11.4-34}$$

Equation (11.4-31) now yields

$$\begin{aligned} \ln F(q) &= \ln\left\{\frac{(1 - z_1^* z_2 e^{iq})[1 - (z_1^*/z_2)e^{-iq}]}{(1 - z_1^* z_2 e^{-iq})[1 - (z_1^*/z_2)e^{iq}]}\right\}^{1/2} \\ &= \frac{1}{2}\left\{\ln(1 - z_1^* z_2 e^{iq}) + \ln[1 - (z_1^*/z_2)e^{-iq}] - \ln(1 - z_1^* z_2 e^{-iq}) - \ln[1 - (z_1^*/z_2)e^{iq}]\right\} \\ &= \frac{1}{2}\sum_{n=1}^{\infty}\left\{\frac{1}{n}\left[-(z_1^* z_2)^n + (z_1^*/z_2)^n\right]e^{inq} + \frac{1}{n}\left[(z_1^* z_2)^n - (z_1^*/z_2)^n\right]e^{-inq}\right\}. \end{aligned} \tag{11.4-35}$$

Whence, the Fourier coefficients are

$$k_n = -k_{-n} = \frac{1}{2n}\left[-(z_1^* z_2)^n + (z_1^*/z_2)^n\right]. \tag{11.4-36}$$

We must now perform the sum, see (11.4-27),

$$\begin{aligned} \sum_{n=1}^{\infty} n k_n k_{-n} &= \frac{1}{4}\sum_{n=1}^{\infty}\left\{\frac{1}{n}\left[-(z_1^* z_2)^{2n} + 2(z_1^*)^{2n} - (z_1^*/z_2)^{2n}\right]\right\} \\ &= \frac{1}{4}\ln\left\{\frac{[1 - (z_1^* z_2)^2][1 - (z_1^*/z_2)^2]}{(1 - z_1^{*2})^2}\right\} = \frac{1}{4}\ln\left\{1 - \frac{(1 - z_2^2)^2 z_1^{*2}}{(1 - z_1^*)^2 z_2^2}\right\}, \end{aligned} \tag{11.4-37}$$

We eliminate z_1^* with the aid of (11.4-34), to obtain for $T(F)$:

$$T(F) = \left[1 - \frac{(1 - z_1^2)^2 (1 - z_2^2)^2}{16 z_1^2 z_2^2}\right]^{1/4}. \tag{11.4-38}$$

Thus, from (11.4-24) and (11.4-38) we obtain for the magnetization per spin, (11.4-5)

$$m_0(T) = \left[\lim_{|m'-m|\to\infty} \langle \sigma_{m,x} \sigma_{m',x} \rangle_{eq} \right]^{1/2} = \left[1 - \frac{(1-z_1^2)^2(1-z_2^2)^2}{16 z_1^2 z_2^2} \right]^{1/8} . \quad (11.4\text{-}39)$$

With $z_i = \tanh K_i$, $(i = 1,2)$ we finally arrive at

$$m_0(T) = [1 - (\sinh 2K_1 \sinh 2K_2)^{-2}]^{1/8} , \quad (11.4\text{-}40)$$

which is *Onsager's cryptogram!* Employing (11.3-8) one easily establishes the limiting form for the magnetization with the temperature approaching T_c from below,

$$m_0(T) \approx [1 - (T/T_c)]^{1/8} , \quad T \le T_c . \quad (11.4\text{-}41)$$

A similar computation can also be carried out for $T > T_c$; we recall that this merely requires an interchange of K and K^*, or in the present context, of z and z^*. The corresponding result for the Fourier coefficients k_n and k_{-n} is then found to be

$$k_n = -k_{-n} = \frac{1}{2n}[-(z_2 z_1^*)^n - (z_2/z_1^*)^n + 2(-1)^n] . \quad (11.4\text{-}42)$$

This yields

$$\sum_{n=1}^{\infty} n k_n k_{-n} = \sum_{n=1}^{\infty} \left\{ -\frac{1}{n} + \frac{(-1)^n}{n}\left[(z_2/z_1^*)^n + (z_2 z_1^*)^n \right] - \frac{1}{4n}\left[(z_2/z_1^*)^n + (z_2 z_1^*)^n \right]^2 \right\} = -\infty , \quad (11.4\text{-}43)$$

since $\Sigma_n (1/n)$ diverges. Therefore $T(F) = \exp(-\infty) = 0$ and there is no spontaneous magnetization. The Ising magnet above T_c is simply paramagnetic.

With another approach based on an exact series expansion, Fisher[30] has obtained the following result for the paramagnetic susceptibility

$$\chi(T) = \frac{N \rho_B^2}{k_B T_c}\left(1 - \frac{T_c}{T}\right)^{-\gamma} , \gamma = 7/4 , \quad T > T_c . \quad (11.4\text{-}44)$$

This, then, is the final part of the *Ising saga!* For those who believe that this was 'all very difficult', we invite them to read the older presentations, based on Lie algebras, Pfaffians, the 'bathroom-tile lattice', – see Kasteleyn's contribution in the same issue of J. Math. Physics[31] – or some original observations by Kaufman and Onsager[32] and many others. Yet, progress in exact methods in statistical physics is being made continually. Mattis, Op. Cit. p.257, states that "perhaps, 'a generation hence' the 2D Ising magnet will be seen as 'a trivial' problem!"

---

[30] M.E. Fisher, J. Math Physics **4**, 278 (1963).
[31] P.W. Kasteleyn, J. Math. Phys. **4**, 287 (1963).
[32] B. Kaufman and L. Onsager, Phys. Rev. **76**, 1244 (1949)

11.5 Ferromagnetism as Excitations of Magnons[33]

In the Ising model as presented we assumed that the x- and y- components averaged out to zero and we assumed *a priori* that the cooperative state was ferromagnetic ($J_1, J_2 > 0$), having spin $s = \frac{1}{2}$. We now turn to the more general problem in which the spin operator **S** can be characterized by $s = \frac{1}{2}, 1, \frac{3}{2}, \ldots$ The ES are as usual given as $|\, s, m_s\rangle$, with $m_s = -s, -s+1, \ldots, s-1, s$. We shall omit effects of magnetostatic dipole-dipole coupling, spin-orbit coupling and similar effects. Spin lowering and raising operators are defined in the usual way (**l** = bold ℓ)

$$S_{\mathbf{l}}^{+} = S_{\mathbf{l}}^{x} + iS_{\mathbf{l}}^{y}, \quad S_{\mathbf{l}}^{-} = S_{\mathbf{l}}^{x} - iS_{\mathbf{l}}^{y}. \tag{11.5-1}$$

The Heisenberg Hamiltonian becomes

$$\mathcal{H} = -\frac{1}{\hbar^2} \sum_{\mathbf{l},\mathbf{l}'}{}' J_{\mathbf{l},\mathbf{l}'} \left\{ S_{\mathbf{l}}^{z} S_{\mathbf{l}'}^{z} + \tfrac{1}{2}\left(S_{\mathbf{l}}^{+} S_{\mathbf{l}'}^{-} + S_{\mathbf{l}}^{-} S_{\mathbf{l}'}^{+} \right) \right\} - gH \frac{\rho_B}{\hbar} \sum_{\mathbf{l}} S_{\mathbf{l}}^{z}. \tag{11.5-2}$$

In the spin-wave procedure we seek to split the Hamiltonian in a ground state Hamiltonian, $\mathcal{H}_0$, and a quadratic part which is interpreted as a sum of harmonic oscillators, giving rise to spin waves or excitations of magnons, with Hamiltonian $\mathcal{H}_{s.w.}$. The uncoupling of the quadratic part is obtained with normal modes, similarly as for phonons, cf. Section 8.5. The approximations for this 'linear' spin-wave treatment are rather severe, however, and some results boast more vigour than rigour!

We proceed to the task. The ground state will be taken the state in which all magnetic quantum numbers are equal to s ("vacuum-state", no spin-deviations), thus minimizing the magnetic energy term, with all N moments parallel to **H**:

$$|0\rangle = |s\rangle_1 |s\rangle_2 \ldots |s\rangle_N. \tag{11.5-3}$$

Since $S_{\mathbf{l}}^{+} |0\rangle = S_{\mathbf{l}'}^{+} |0\rangle = 0$, the second term in (11.5-2) does not contribute, hence

$$\mathcal{H}_0 |0\rangle = -\sum_{\mathbf{l},\mathbf{l}'}{}' J_{\mathbf{l},\mathbf{l}'} s^2 |0\rangle - sNg\rho_B H |0\rangle. \tag{11.5-4}$$

This implies the ground-state Hamiltonian and ground-state energy

$$\mathcal{H}_0 = -\sum_{\mathbf{l},\mathbf{l}'}{}' J_{\mathbf{l},\mathbf{l}'} S_{\mathbf{l}}^{z} S_{\mathbf{l}'}^{z} - sNg\rho_B H\, I, \tag{11.5-5a}$$

$$\mathcal{E}_0 = -\sum_{\mathbf{l},\mathbf{l}'}{}' J_{\mathbf{l},\mathbf{l}'} s^2 - sNg\rho_B H, \tag{11.5-5b}$$

---

[33] Cf. D.C. Mattis, Op. Cit., Chapter 6; C. Kittel, "The Quantum Theory of Solids", John Wiley and Sons, N.Y., Chapter 4; 2nd printing 1986; J. Van Kranendonk and J.H. Van Vleck, Rev. Mod. Physics **30**,1 (1958).

where I is the identity operator. This is the exchange magnetic energy with all the spins lined up. We notice that s^2 occurs rather than $s(s+1)$, which indicate that the old semi-classical vector model yields the correct results.

For other states we introduce the *spin-deviation number*

$$n_1 = s - m_{s,1}\,. \qquad (11.5\text{-}6)$$

We recall the angular momenta rules[34]

$$S_1^- \,|\, s,m_{s,1}\rangle = \hbar\sqrt{(s-m_{s,1}+1)(s+m_{s,1})}\,|\, s,m_{s,1}-1\rangle \to \hbar\sqrt{(n_1+1)(2s-n_1)}\,|\, s,n_1+1\rangle.$$
$$S_1^+ \,|\, s,m_{s,1}\rangle = \hbar\sqrt{(s-m_{s,1})(s+m_{s,1}+1)}\,|\, s,m_{s,1}+1\rangle \to \hbar\sqrt{n_1(2s-n_1+1)}\,|\, s,n_1-1\rangle, \qquad (11.5\text{-}7)$$

We now make the approximation $2s \gg n_1$.[35] We can then introduce the boson operators[36]

$$a_1 = S_1^+/\hbar\sqrt{2s}, \quad a_1^\dagger = S_1^-/\hbar\sqrt{2s}\,. \qquad (11.5\text{-}8)$$

From (11.5-7) we have, introducing the spin-deviation number states for the entire lattice,

$$a_1\,|\, n_1 n_2 \ldots n_1 \ldots\rangle = \left(\frac{2s-n_1+1}{2s}\right)^{1/2}\sqrt{n_1}\,|\, n_1 n_2 \ldots n_1 - 1 \ldots\rangle \approx \sqrt{n_1}\,|\, n_1 n_2 \ldots n_1 \ldots\rangle,$$

$$a_1^\dagger\,|\, n_1 n_2 \ldots n_1 \ldots\rangle = \left(\frac{2s-n_1}{2s}\right)^{1/2}\sqrt{n_1+1}\,|\, n_1 n_2 \ldots n_1 + 1 \ldots\rangle \approx \sqrt{n_1+1}\,|\, n_1 n_2 \ldots n_1 + 1 \ldots\rangle. \qquad (11.5\text{-}9)$$

For the number operator we obtain

$$\begin{aligned} \mathbf{n}_1 &= a_1^\dagger a_1 = S_1^- S_1^+/2s\hbar^2 = (S_1^x - iS_1^y)(S_1^x + iS_1^y)/2s\hbar^2 \\ &= [(S^2 - S_1^{z2}) + i\underbrace{(S_1^x S_1^y - S_1^y S_1^x)}_{\hbar i S_1^z}]/2s\hbar^2\,, \end{aligned} \qquad (11.5\text{-}10)$$

Operating on a spin-deviation number state, we have

$$\begin{aligned} \mathbf{n}_1\,|\, n_1 n_2 \ldots n_1 \ldots\rangle &= \{[s(s+1) - m_{s,1}^2 - m_{s,1}]/2s\}\,|\, n_1 n_2 \ldots n_1 \ldots\rangle = [(s-m_{s,1})(s+m_{s,1}+1)/2s] \\ &\times|\, n_1 n_2 \ldots n_1 \ldots\rangle = [n_1(2s-n_1+1)/2s]\,|\, n_1 n_2 \ldots n_1 \ldots\rangle \approx n_1\,|\, n_1 n_2 \ldots n_1 \ldots\rangle, \end{aligned}$$
$$(11.5\text{-}11)$$

which is the expected result, valid under the same approximations. One easily confirms the boson commutation rules,

---

[34] A. Messiah, Op. Cit., Vol. II Chapter XIII p. 514 ff.

[35] Actually, it can be shown that the milder condition for the *sum of all spins* $2Ns \gg \Sigma_1 n_1$ suffices.

[36] An *exact* spin-boson equivalent is expressed by the Holstein-Primakoff transformation, see Eq. (7) in Problem 11.10; cf. T. Holstein and H. Primakoff, Phys. Rec. **58**, 1098 (1940).

$$[a_{\mathbf{l}}, a_{\mathbf{l}'}^{\dagger}] \approx \delta_{\mathbf{l},\mathbf{l}'}, \quad [a_{\mathbf{l}}, a_{\mathbf{l}'}] = [a_{\mathbf{l}}^{\dagger}, a_{\mathbf{l}'}^{\dagger}] = 0. \tag{11.5-12}$$

We proceed to find the spin-wave Hamiltonian. We write $S_{\mathbf{l}}^{z} = \hbar(sI - a_{\mathbf{l}}^{\dagger} a_{\mathbf{l}})$ with I again the identity operator. Substituting this and (11.5-8) in the Hamiltonian (11.5-2) we get:

$$\begin{aligned}\mathcal{H} &= -\sum_{\mathbf{l},\mathbf{l}'}{}' J_{\mathbf{l},\mathbf{l}'} \{(sI - a_{\mathbf{l}}^{\dagger} a_{\mathbf{l}})(sI - a_{\mathbf{l}'}^{\dagger} a_{\mathbf{l}'}) + s(a_{\mathbf{l}}^{\dagger} a_{\mathbf{l}'} + a_{\mathbf{l}} a_{\mathbf{l}'}^{\dagger})\} - gH\rho_B \sum_{\mathbf{l}} (sI - a_{\mathbf{l}}^{\dagger} a_{\mathbf{l}}) \\ &= \underbrace{\{-\sum_{\mathbf{l},\mathbf{l}'}{}' J_{\mathbf{l},\mathbf{l}'} s^2 - sNg\rho_B H\} I}_{\mathcal{E}_0} \underbrace{- \sum_{\mathbf{l},\mathbf{l}'} J_{\mathbf{l},\mathbf{l}'} s (a_{\mathbf{l}}^{\dagger} a_{\mathbf{l}'} + a_{\mathbf{l}} a_{\mathbf{l}'}^{\dagger} - a_{\mathbf{l}}^{\dagger} a_{\mathbf{l}} - a_{\mathbf{l}'}^{\dagger} a_{\mathbf{l}'}) + gH\rho_B \sum_{\mathbf{l}} a_{\mathbf{l}}^{\dagger} a_{\mathbf{l}}}_{\mathcal{H}_{s.w.}} .\end{aligned} \tag{11.5-13}$$

Terms of order $\mathcal{O}(a^4)$ have been omitted. To uncouple the quadratic terms, we transform to running-wave boson operators,

$$a_{\mathbf{q}} = N^{-1/2} \sum_{\mathbf{l}} a_{\mathbf{l}} e^{-i\mathbf{q}\cdot\mathbf{l}}, \quad a_{\mathbf{q}}^{\dagger} = N^{-1/2} \sum_{\mathbf{l}} a_{\mathbf{l}}^{\dagger} e^{i\mathbf{q}\cdot\mathbf{l}}, \tag{11.5-14}$$

with inversion

$$a_{\mathbf{l}} = N^{-1/2} \sum_{\mathbf{q}} a_{\mathbf{q}} e^{i\mathbf{q}\cdot\mathbf{l}}, \quad a_{\mathbf{l}}^{\dagger} = N^{-1/2} \sum_{\mathbf{q}} a_{\mathbf{q}}^{\dagger} e^{-i\mathbf{q}\cdot\mathbf{l}}. \tag{11.5-15}$$

Clearly, we also need the lemmas Eqs. (8.5-26), employed in the treatment of phonons. The $\mathbf{q}$'s are equidistantly spaced about the origin of $\mathbf{q}$-space, such that there are N values in a reciprocal lattice unit cell. The transformation is straightforward; we find:

$$\mathcal{H}_{s.w.} = \sum_{\mathbf{q}} \left\{ 2s \sum_{\mathbf{L}} J(\mathbf{L}) a_{\mathbf{q}}^{\dagger} a_{\mathbf{q}} (1 - \cos \mathbf{q}\cdot\mathbf{L}) + a_{\mathbf{q}}^{\dagger} a_{\mathbf{q}} g\rho_B H \right\}, \tag{11.5-16}$$

where $J(\mathbf{L}) = J_{\mathbf{l},\mathbf{l}+\mathbf{L}}$ is the exchange energy for sites which are a vector $\mathbf{L}$ apart. In what follows we set $H = 0$. The spin-wave Hamiltonian then becomes

$$\mathcal{H}_{s.w.}^{0} = \sum_{\mathbf{q}} \mathbf{n}_{\mathbf{q}} \hbar\omega_{\mathbf{q}}, \quad \hbar\omega_{\mathbf{q}} = 2s \sum_{\mathbf{L}} J(\mathbf{L})(1 - \cos \mathbf{q}\cdot\mathbf{L}), \tag{11.5-17}$$

where the last part is the dispersion relation. For small $\mathbf{q}$ and nearest-neighbour interaction only in a simple cubic lattice we have the approximate result

$$\begin{aligned}\sum_{\mathbf{L}=\mathbf{a}_i} J(\mathbf{L})(1 - \cos \mathbf{q}\cdot\mathbf{L}) &= 2J_0[(1 - \cos q_x a) + (1 - \cos q_y a) + (1 - \cos q_z a)] \\ &\approx 2J_0\left(\tfrac{1}{2} q_x^2 a^2 + \tfrac{1}{2} q_y^2 a^2 + \tfrac{1}{2} q_z^2 a^2\right) = J_0 q^2 a^2 .\end{aligned} \tag{11.5-18}$$

The magnetization will be computed in the next section.

11.6 Bose–Einstein Statistics for Magnons

The number of magnons in a mode $\omega_{\mathbf{q}}$ is given by

$$\langle n_{\mathbf{q}} \rangle = \frac{1}{e^{\hbar\omega_{\mathbf{q}}/k_B T} - 1}. \tag{11.6-1}$$

The total spin deviation equals the number of magnons excited $[\Sigma_l n_l = \Sigma_{\mathbf{q}} n_{\mathbf{q}}]$; hence denoting by M_0 the ground-state magnetization, we have

$$(M_0 - M)/M_0 = \Sigma_{\mathbf{q}} n_{\mathbf{q}} / sN, \tag{11.6-2}$$

or for the relative magnetization

$$m(T) \equiv \langle M \rangle / M_0 = 1 - \Sigma_{\mathbf{q}} \langle n_{\mathbf{q}} \rangle / sN. \tag{11.6-3}$$

The density in **q**-space is $V/(2\pi)^d$, where d is the dimensionality. Thus,

$$\Sigma_{\mathbf{q}} \langle n_{\mathbf{q}} \rangle = \frac{V}{(2\pi)^d} \int_{Brill.zone} d^d q \frac{1}{e^{\hbar\omega_{\mathbf{q}}/k_B T} - 1}. \tag{11.6-4}$$

At very low temperatures only small q-values contribute and we may use (11.5-18). We then obtain

$$\Sigma_{\mathbf{q}} \langle n_{\mathbf{q}} \rangle = \frac{V}{(2\pi)^d} \int_{\mathbf{q}-space} d^d q \frac{1}{e^{2sJ_0 q^2 a^2 / k_B T} - 1} = V \left(\frac{k_B T}{8\pi^2 a^2 s J_0} \right)^{d/2} \int_{all\, x} \frac{d^d x}{e^{x^2} - 1}. \tag{11.6-5}$$

The integral diverges in one and in two dimensions. The latter fact contradicts the results of the 2D Ising model. However, a nonlinear 2D treatment shows the existence of bound states, see Mattis, Op. Cit. For three dimensions, noting $V/a^3 = N$, setting $x^2 = y$ and employing the integral (8.1-19)-(8.1-21) we arrive at

$$\Sigma_{\mathbf{q}} \langle n_{\mathbf{q}} \rangle = \frac{N}{4\pi^2} \left(\frac{k_B T}{2sJ_0} \right)^{3/2} \zeta(\tfrac{3}{2}) \Gamma(\tfrac{3}{2}). \tag{11.6-6}$$

This yields for the relative magnetization

$$m(T) = 1 - \frac{2.31}{4\pi^2 s^{5/2}} \left(\frac{k_B T}{2J_0} \right)^{3/2}, \tag{11.6-7}$$

which is the celebrated "Bloch 3/2 law". One usually puts it in the form

$$m(T) = 1 - (T/T_0)^{3/2}. \tag{11.6-8}$$

However, we may not identify T_0 with the Curie temperature T_c, for the result (11.6-7) usually does not hold in that temperature region.

So generally for a s.c. lattice we must use $\hbar\omega_{\mathbf{q}} = 6sJ_0(1 - \frac{1}{3}\cos q_x a - \frac{1}{3}\cos q_y a - \frac{1}{3}\cos q_z a)$. With a series expansion for $1/(e^y - 1)$ we then have

$$m(T) = 1 - \frac{1}{s}\left(\frac{a}{2\pi}\right)^3 \sum_{p=1}^{\infty} \iiint_{-\pi/a}^{\pi/a} dq_x dq_y dq_z e^{-p\hbar\omega_{\mathbf{q}}/k_B T}. \quad (11\text{-}6\text{-}9)$$

Recalling the definite integral

$$\frac{1}{2\pi}\int_{-\pi}^{\pi} d\theta e^{z\cos\theta} \cos p\theta = I_p(z), \quad (11.6\text{-}10)$$

where $I_p(z)$ is the modified Bessel function of the first kind, we obtain

$$m(T) = 1 - \frac{1}{s}\left\{\sum_{p=1}^{\infty} e^{-2psJ_0/k_B T} I_0(2psJ_0/k_B T)\right\}^3. \quad (11.6\text{-}11)$$

With the asymptotic expansion $I_0(z) \sim (1/\sqrt{2\pi z})\exp(-z)$ the leading term for low temperatures is once more the Bloch 3/2 law. For high temperatures $\hbar\omega_{\mathbf{q}} < k_B T$ another approximation is useful; keeping two terms in the series expansion we note

$$\frac{1}{e^{\hbar\omega_{\mathbf{q}}/k_B T} - 1} \approx \frac{k_B T}{\hbar\omega_{\mathbf{q}}} - \frac{1}{2}. \quad (11.6\text{-}12)$$

This yields

$$m(T) \sim 1 + \frac{1}{2s} - \frac{k_B T}{6s^2 J_0}\left\{\left(\frac{a}{2\pi}\right)^3 \iiint_{-\pi/a}^{\pi/a} \frac{dq_x dq_y dq_z}{1 - \frac{1}{3}\cos q_x a - \frac{1}{3}\cos q_y a - \frac{1}{3}\cos q_z a}\right\}. \quad (11.6\text{-}13)$$

This integral has, amazingly, been evaluated by Watson, cf. Mattis, Op. Cit., p. 147. Its value: $W = 1.516386$. So finally

$$m(T) = 1 + \frac{1}{2s} - \frac{k_B T W}{6s^2 J_0}. \quad (11.6\text{-}14)$$

The Curie temperature follows from $m(T_c) = 0$. Hence we have

$$k_B T_c = 3.96 s^2 J_0[1 + (1/2s)]. \quad (11.6\text{-}15)$$

The reader is encouraged to also compute the result for T_c from mean field theory for the Heisenberg Hamiltonian for arbitrary s, to obtain

$$k_B T_c = 4s^2 J_0[1 + (1/s)]. \quad (11.6\text{-}16)$$

Closer scrutiny shows that the results of (linear) spin-wave theory are off by ~ 20%. Refinements, giving agreement to within a few percent are found in Mattis, Op. Cit.

11.7 The Heisenberg Antiferromagnet

We now proceed to the spin-wave treatment for the anti-ferromagnetic state. We consider the basic case that we can arrange the lattice into two sublattices with direct overlap integrals of negative magnitude, such that adjacent spins couple to have opposite orientation; see Fig 11-7 for the exchange integral in the iron group [^{24}Cr – ^{28}Ni]. Such is the case e.g., for Cr or Mn; in Cr the structure is body-centered cubic (b.c.c.), which can be viewed as two interpenetrating s.c. lattices with opposite spins. More complicated cases, including those involving super-exchange, can be found in texts on condensed matter physics. The preferred spin orientation of each sublattice will be characterized by the spin number $\sigma_{\mathbf{l}}$, which is $\pm$ for $\uparrow\downarrow$, respectively. The label '**l**' will denote *all* sites.

The basic difficulty is that the ground state is not known a priori and is *not* an eigenstate of the Heisenberg Hamiltonian. Yet, we assume that a ground state exists. The spin-deviation numbers are once more defined as deviations from this state, i.e.,

$$n_{\mathbf{l}} = s - \sigma_{\mathbf{l}} m_{s,\mathbf{l}}. \tag{11.7-1}$$

The spin-deviation number state is denoted, as previously, by $|n_1 n_2 ... n_{\mathbf{l}} ...\rangle$. The spin raising and lowering operators still satisfy the basic angular momenta rules, that now translate to

$$\begin{aligned}
(1/\hbar\sqrt{2s})S^{+}|n_1 n_2 ... n_{\mathbf{l}} ...\rangle &= (1/\sqrt{2s})\sqrt{(s-\sigma s+\sigma n)(s+\sigma s-\sigma n+1)}\,|n_1 n_2 ... n_{\mathbf{l}} -1 ...\rangle,\\
(1/\hbar\sqrt{2s})S^{-}|n_1 n_2 ... n_{\mathbf{l}} ...\rangle &= (1/\sqrt{2s})\sqrt{(s-\sigma s+\sigma n+1)(s+\sigma s-\sigma n)}\,|n_1 n_2 ... n_{\mathbf{l}} +1 ...\rangle.
\end{aligned} \tag{11.7-2}$$

Boson operators are defined somewhat differently:

$$a_{\mathbf{l}} = \frac{S_{\mathbf{l}}^{x} + i\sigma_{\mathbf{l}} S_{\mathbf{l}}^{y}}{\hbar\sqrt{2s}} = \frac{1}{\hbar\sqrt{2s}}\begin{cases} \overset{\sigma_{\mathbf{l}}=1}{\longrightarrow} S_{\mathbf{l}}^{+} \\ \overset{\sigma_{\mathbf{l}}=-1}{\longrightarrow} S_{\mathbf{l}}^{-} \end{cases}, \qquad a_{\mathbf{l}}^{\dagger} = \frac{S_{\mathbf{l}}^{x} - i\sigma_{\mathbf{l}} S_{\mathbf{l}}^{y}}{\hbar\sqrt{2s}} = \frac{1}{\hbar\sqrt{2s}}\begin{cases} \overset{\sigma_{\mathbf{l}}=1}{\longrightarrow} S_{\mathbf{l}}^{-} \\ \overset{\sigma_{\mathbf{l}}=-1}{\longrightarrow} S_{\mathbf{l}}^{+} \end{cases}. \tag{11.7-3}$$

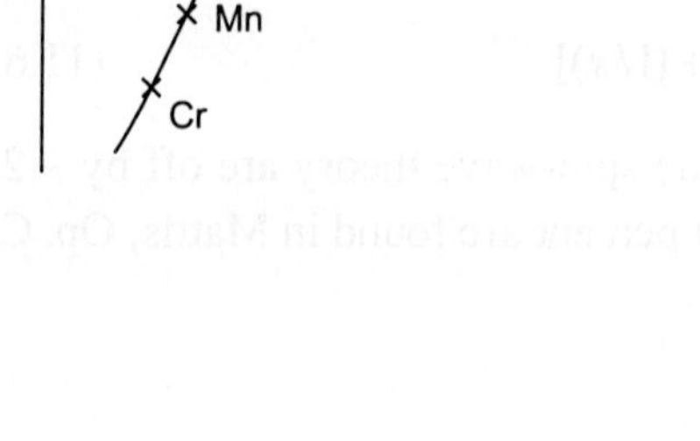

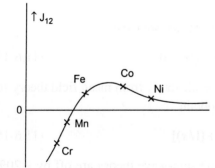

Fig. 11-7. Exchange integrals in the iron group.

It is readily shown that these boson operators still satisfy the boson commutation rules and that the deviation-number operator is $a_{\mathbf{l}}^\dagger a_{\mathbf{l}}$ as before. From (11.7-3) we solve for $S_{\mathbf{l}}^x$ and $S_{\mathbf{l}}^y$, while for $S_{\mathbf{l}}^z$ we have $S_{\mathbf{l}}^z = \hbar\sigma_{\mathbf{l}}(sI - a_{\mathbf{l}}^\dagger a_{\mathbf{l}})$. Substitution then yields for the Heisenberg Hamiltonian:

$$\mathcal{H} = -\sum_{\mathbf{l},\mathbf{l}'}{}' J_{\mathbf{l},\mathbf{l}'}\left\{\tfrac{1}{2}s(a_{\mathbf{l}}a_{\mathbf{l}'} + a_{\mathbf{l}}^\dagger a_{\mathbf{l}'}^\dagger)(1-\sigma_{\mathbf{l}}\sigma_{\mathbf{l}'}) + \tfrac{1}{2}s(a_{\mathbf{l}}^\dagger a_{\mathbf{l}'} + a_{\mathbf{l}}a_{\mathbf{l}'}^\dagger)(1+\sigma_{\mathbf{l}}\sigma_{\mathbf{l}'}) \right.$$
$$\left. +\sigma_{\mathbf{l}}\sigma_{\mathbf{l}'}s^2 - \sigma_{\mathbf{l}}\sigma_{\mathbf{l}'}s(a_{\mathbf{l}}^\dagger a_{\mathbf{l}} + a_{\mathbf{l}'}^\dagger a_{\mathbf{l}'})\right\} - g\rho_B H\sum_{\mathbf{l}}\sigma_{\mathbf{l}}(sI - a_{\mathbf{l}}^\dagger a_{\mathbf{l}}), \qquad (11.7\text{-}4)$$

where as before terms of order $\mathcal{O}(a^4)$ have been dropped. It may not be concluded that terms not containing $a_{\mathbf{l}}$ or $a_{\mathbf{l}}^\dagger$ pertain to the ground state; however, we can drop $\Sigma_{\mathbf{l}}\sigma_{\mathbf{l}}s$ which is zero for a fully compensated anti-ferromagnet. The part $-g\rho_B H\Sigma_{\mathbf{l}}a_{\mathbf{l}}^\dagger a_{\mathbf{l}}$ relates to the susceptibility in the cooperative state; however, it gives complications and will not be considered. As before, we write $J_{\mathbf{l},\mathbf{l}+\mathbf{L}} = J(\mathbf{L})$. For nearest neighbour interaction , involving spin-spin coupling in different sublattices, we have $\mathbf{L}\to\mathbf{L}_0$ with $\mathrm{L}_0 = a\sqrt{2}$ for s.c. sublattices. While $J(\mathbf{L}_0)\equiv -J_0 < 0$, further neighbours may have a positive exchange integral, though of smaller magnitude. We also write $\sigma_{\mathbf{l}}\sigma_{\mathbf{l}+\mathbf{L}} = \sigma(\mathbf{L})$, where the latter is independent of **l** and denotes the sign of the product with respect to $\sigma(0) = \sigma^2 = 1$. We now transform to running-wave boson operators, $a_{\mathbf{q}}$ and $a_{\mathbf{q}}^\dagger$ to obtain:

$$\mathcal{H} = -\sum_{\mathbf{L}} J(\mathbf{L})N\sigma(\mathbf{L})s^2 - \tfrac{1}{2}s\sum_{\mathbf{q}}\sum_{\mathbf{L}} J(\mathbf{L})\left\{[a_{\mathbf{q}}a_{-\mathbf{q}}e^{i\mathbf{q}\cdot\mathbf{L}} + a_{\mathbf{q}}^\dagger a_{-\mathbf{q}}^\dagger e^{-i\mathbf{q}\cdot\mathbf{L}}]\right.$$
$$\left. \times[1-\sigma(\mathbf{L})] + [a_{\mathbf{q}}^\dagger a_{\mathbf{q}}e^{-i\mathbf{q}\cdot\mathbf{L}} + a_{\mathbf{q}}a_{\mathbf{q}}^\dagger e^{i\mathbf{q}\cdot\mathbf{L}}][1+\sigma(\mathbf{L})] - 4\sigma(\mathbf{L})a_{\mathbf{q}}^\dagger a_{\mathbf{q}}\right\}. \qquad (11.7\text{-}5)$$

The result must be even in **L**, so, $\exp(\pm i\mathbf{q}\cdot\mathbf{L}) \to \cos(\mathbf{q}\cdot\mathbf{L})$. Employing further $[a_{\mathbf{q}}, a_{\mathbf{q}}^\dagger] = 1$ to symmetrize the last term, we arrive at

$$\mathcal{H} = -\sum_{\mathbf{L}} J(\mathbf{L})N\sigma(\mathbf{L})s(s+1) - \tfrac{1}{2}s\sum_{\mathbf{q}}\sum_{\mathbf{L}} J(\mathbf{L})\left\{[a_{\mathbf{q}}a_{-\mathbf{q}} + a_{\mathbf{q}}^\dagger a_{-\mathbf{q}}^\dagger]\right.$$
$$\left. \times[1-\sigma(\mathbf{L})]\cos\mathbf{q}\cdot\mathbf{L} + [a_{\mathbf{q}}^\dagger a_{\mathbf{q}} + a_{\mathbf{q}}a_{\mathbf{q}}^\dagger][\big(1+\sigma(\mathbf{L})\big)\cos\mathbf{q}\cdot\mathbf{L} - 2\sigma(\mathbf{L})]\right\}. \qquad (11.7\text{-}6)$$

This Hamiltonian is not yet diagonal, due to the fact that the translational symmetry in an antiferromagnet is reduced, e.g., from b.c.c. to s.c. ; the states for **q** and –**q** are now mixed. The problem is alleviated with the Bogoliubov transformation, which involves a rotation over an imaginary angle $iu_{\mathbf{q}}$. So we set

$$\left.\begin{aligned} a_{\mathbf{q}} &= (\cosh u_{\mathbf{q}})b_{\mathbf{q}} + (\sinh u_{\mathbf{q}})b_{-\mathbf{q}}^\dagger \\ a_{\mathbf{q}}^\dagger &= (\sinh u_{\mathbf{q}})b_{-\mathbf{q}} + (\cosh u_{\mathbf{q}})b_{\mathbf{q}}^\dagger \end{aligned}\right\} \text{all pos. and neg.} \{q_\alpha\}, \qquad (11.7\text{-}7)$$

with $u_{\mathbf{q}} = u_{-\mathbf{q}}$. One easily checks that these new operators satisfy the boson commutator rules. The quantity $u_{\mathbf{q}}$ is to be chosen such that in the new Hamiltonian the off-diagonal part is identically zero. Though this can be done for the form (11.7-6), we shall only consider the case that nearest neighbour interaction is predominant; then

$$\begin{aligned}&\sum_{\mathbf{L}} J(\mathbf{L})\sigma(\mathbf{L}) \approx pJ_0,\\&\sum_{\mathbf{L}} J(\mathbf{L})[1-\sigma(\mathbf{L})]f(\mathbf{L}) \approx -2J_0\sum_{\mathbf{L}_0} f(\mathbf{L}_0),\\&\sum_{\mathbf{L}} J(\mathbf{L})[1+\sigma(\mathbf{L})]f(\mathbf{L}) \approx 0.\end{aligned} \tag{11.7-8}$$

Here p is the coordination number (number of nearest neighbours). Carrying out the Bogoliubov transformation for the simplified Hamiltonian, the off-diagonal terms require that

$$(b_{\mathbf{q}}b_{-\mathbf{q}} + b^\dagger_{-\mathbf{q}}b^\dagger_{\mathbf{q}})[(\cosh^2 u_{\mathbf{q}} + \sinh^2 u_{\mathbf{q}})(\sum_{\mathbf{L}_0}\tfrac{1}{p}\cos\mathbf{q}\cdot\mathbf{L}_0) + 2\sinh u_{\mathbf{q}}\cosh u_{\mathbf{q}}] = 0, \tag{11.7-9}$$

$$or,\ \tanh 2u_{\mathbf{q}} = -\tfrac{1}{p}\sum_{\mathbf{L}_0}\cos\mathbf{q}\cdot\mathbf{L}_0\,. \tag{11.7-10}$$

The Hamiltonian is now diagonal,

$$\mathcal{H} = -pNJ_0 s(s+1) + 2psJ_0\sum_{\mathbf{q}}(\bar{\mathbf{n}}_{\mathbf{q}} + \frac{1}{2})\sqrt{1-\tanh^2 2u_{\mathbf{q}}}\,, \tag{11.7-11}$$

where $\bar{\mathbf{n}}_{\mathbf{q}} = b^\dagger_{\mathbf{q}}b_{\mathbf{q}}$ is the new number operator with EV $\bar{n}_{\mathbf{q}} = 0,1,2,\ldots$.

Finally, the ground-state energy for $\bar{n}_{\mathbf{q}} = 0$ can be found:

$$\mathcal{E}_0 = -pNJ_0 s(s+1) + psJ_0\sum_{\mathbf{q}}(\sqrt{1-\tanh^2 2u_{\mathbf{q}}} = -pNJ_0 s^2\left(1+\frac{\gamma}{ps}\right), \tag{11.7-12}$$

where γ is a number between zero and one. The following results are found:

linear chain	$p=2$	$\gamma = 0.726$
square	$p=4$	$\gamma = 0.632$
simple cubic	$p=6$	$\gamma = 0.58$
b.c.c.	$p=8$	$\gamma = 0.58$

In all cases the ground-state energy is larger than the energy from which we started, i.e., the Ising energy, see the first term in (11.7-5). Substituting (11.7-10) into (11.7-11), we can write the Hamiltonian as

$$\mathcal{H} = -pNJ_0 s(s+1) + \sum_{\mathbf{q}}(\bar{\mathbf{n}}_{\mathbf{q}} + \frac{1}{2})\hbar\omega_{\mathbf{q}}\,, \tag{11.7-13}$$

where the magnon spectrum is given by

$$\hbar\omega_{\mathbf{q}} = 2J_0\sqrt{p^2 - (\sum_{\mathbf{L}_0} \cos\mathbf{q}\cdot\mathbf{L}_0)^2} \xrightarrow{small\ q} 2pJ_0 q\,. \qquad (11.7\text{-}14)$$

We note that for small $|\mathbf{q}|$ the spectrum is linear in q and not quadratic as previously, cf. Eq. (11.5-18). Furthermore, the Hamiltonian (11.7-13) indicates that we must associate zero-point energy with anti-ferromagnetic spin waves, in contrast to the ferromagnetic case.

We further consider the anti-ferromagnetic state as a function of temperature. Obviously, the average relative magnetization $(M - M_0)/M_0 = \Sigma_{\mathbf{l}}\sigma_{\mathbf{l}}\langle n_{\mathbf{l}}\rangle / sN$ vanishes. There are fluctuations, however, about this equilibrium state. The average spin deviation is

$$\begin{aligned}\mathbf{n}_{\mathbf{l}} &= \frac{1}{N}\sum_{\mathbf{q}} a_{\mathbf{q}}^{\dagger}a_{\mathbf{q}} = \frac{1}{N}\sum_{\mathbf{q}}[(\cosh^2 u_{\mathbf{q}})b_{\mathbf{q}}^{\dagger}b_{\mathbf{q}} + (\sinh^2 u_{\mathbf{q}})b_{\mathbf{q}}b_{\mathbf{q}}^{\dagger}] \\ &= \frac{1}{N}\sum_{\mathbf{q}}\left[(\cosh 2u_{\mathbf{q}})(\overline{\mathbf{n}}_{\mathbf{q}} + \frac{1}{2}) - \frac{1}{2}\right]. \qquad (11.7\text{-}15)\end{aligned}$$

Now

$$\cosh 2u_{\mathbf{q}} = 1/\sqrt{1 - \tanh^2 2u_{\mathbf{q}}} = 2pJ_0/\hbar\omega_{\mathbf{q}}\,, \qquad (11.7\text{-}16)$$

where we used (11.7-10) and (11.7-14). Therefore, from Bose–Einstein statistics including the zero-point energy,

$$\begin{aligned}\langle n_{\mathbf{l}}\rangle &= \frac{1}{N}\sum_{\mathbf{q}}\left\{\left(\frac{1}{e^{\hbar\omega_{\mathbf{q}}/k_BT} - 1} + \frac{1}{2}\right)\frac{2pJ_0}{\hbar\omega_{\mathbf{q}}} - \frac{1}{2}\right\} = \left(\frac{1}{2\pi}\right)^d \\ &\times \int_{\mathbf{q}\in Brill.zone} d^d q\left\{\left(\frac{1}{e^{\hbar\omega_{\mathbf{q}}/k_BT} - 1} + \frac{1}{2}\right)\frac{2pJ_0}{\hbar\omega_{\mathbf{q}}} - \frac{1}{2}\right\} \\ &\approx \left(\frac{1}{2\pi}\right)^d \int_{\mathbf{q}-space} \frac{d^d q}{Cq}\frac{1}{e^{2pCqJ_0/k_BT} - 1}. \qquad (11.7\text{-}17)\end{aligned}$$

where C is a geometry factor. The approximation on the far rhs blows up in one and two dimensions, but is finite for the 3D case. For $\langle n_{\mathbf{l}}\rangle = s$ we have an estimate for the Néel temperature T_N. However, we must use the full integral for this temperature region, so T_N can only be obtained after numerical integration.

Concluding remarks

In this chapter we have presented two radically different approaches to the cooperative ferromagnetic or anti-ferromagnetic state. In the Ising approach the

Hamiltonian was transformed to that for an interacting fermion system. In the spin-wave approach, on the other hand, the problem is transformed to a system of bosonic excitations, called magnons. These two approaches make us mindful of the *ambivalence* of systems of many-body angular momenta. While the spin momenta are defined by the commutation property $\mathbf{S}\times\mathbf{S} = \hbar i\mathbf{S}$, for the spin operator $\mathbf{S}$ with $s = ½$ the anti-commutation properties stand out above the commutation rules. Yet, in no way can we obtain a straightforward equivalence with either a boson or a fermion gas. So, the computations need much adaptation and in each case are complex. We have treated the various approaches with the purpose of statistical mechanics in view. However, texts on many-body physics or condensed matter physics will give many more details for *real* magnetic systems.

11.8 Problems to Chapter XI

11.1 For the 1D Ising model, obtain the eigenvalues (11.1-25) by the following method:

(a) Establish that

$$\begin{aligned} V_2^{1/2} &= \sqrt{A}\exp(\tfrac{1}{2}K^*\sigma_x) = \sqrt{A}[(\cosh\tfrac{1}{2}K^*)\mathbf{1} + (\sinh\tfrac{1}{2}K^*)\sigma_x], \\ V_1 &= \exp(\beta h\sigma_z) = (\cosh\beta h)\mathbf{1} + (\sinh\beta h)\sigma_z . \end{aligned} \tag{1}$$

(b) Using the Pauli matrices, obtain the full transfer matrix $V = V_2^{1/2}V_1V_2^{1/2}$ and find its eigenvalues.

11.2 For the 2D model the transfer operator V_1 (in terms of spin operators) reads

$$V_1 = (2\sinh 2K_1)^{M/2}\exp\left(-2K_1^*\sum_{j=1}^{M}(\sigma_j^+\sigma_j^- - \tfrac{1}{2}\mathbf{1})\right). \tag{2}$$

(a) Convert to fermion operators $c_j^\dagger$ and c_j. *Show details.*

(b) Convert to running-wave fermion operators $c_q^\dagger$ and c_q. Obtain Eq.(11.2-39) of the text.

11.3. Obtain the eigenvalues of the diagonal operators listed in (11.2-40).

11.4. As indicated in subsection 11.2.4 the final transfer operator $V_q = V_{1q}^{1/2}V_{2q}V_{1q}^{1/2}$ can be brought in diagonal form by a rotation known as the Bogoliubov–Valatin transformation. The eigenstates appertaining to the new V_q are stated in Eq. (11.2-65), where $|\Psi_0\rangle$ is the "ground state" at temperature T, belonging to λ_+, while $|\Psi_1\rangle$ corresponds to λ_-. Obtain this result directly from the eigenvector procedure outlined in footnote 17, where V_q is the *original*, nondiagonal operator (11.2-51). In particular, find $\tan\phi_q = C_q/[\exp\varepsilon(q) - A_q]$ and obtain the form (11.2-64):

$$\tan 2\phi_q = 2C_q/(B_q - A_q), \quad \operatorname{sgn} 2\phi_q = \operatorname{sgn} q. \tag{3}$$

11.5. Prove *Onsager's identity* by carrying out the contour-integral procedure, outlined in Eq. (11.3-15) and subsequent phrases.

11.6. Onsager's identity is also valid when $\sin t$ is substituted for $\cos t$, i.e.,

$$\int_0^{2\pi} dt \ln(2\cosh x - 2\sin t) = 2\pi x, \quad x > 0. \tag{4}$$

Differentiate this to x and establish Eq. (4) by contour integration.

11.7 The matrix representation of V_{2q} can also be obtained without the use of pair operators by the following ansatz:

$$\begin{aligned} V_{2q}\,|1_{-q}1_q\rangle &= \alpha(K_2)\,|1_{-q}1_q\rangle + \beta(K_2)\,|0_{-q}0_q\rangle, \\ V_{2q}\,|0_{-q}0_q\rangle &= \gamma(K_2)\,|1_{-q}1_q\rangle + \delta(K_2)\,|0_{-q}0_q\rangle. \end{aligned} \tag{5}$$

Differentiate these equations to K_2 and express the results in terms of the two pair-states; comparison with (5) above gives two sets of coupled differential equations. Solve these equations by quadrature and obtain the result stated in (11.2-49).

11.8 Starting from the free energy for the square-lattice 2D Ising model, Eq. (11.3-26), obtain the energy per spin, (11.3-27) and the ensuing specific heat. Also confirm the logarithmic expansion near the critical temperature, (11.3-31).
Hints: (i) Use the following connection between the two complete elliptic integrals:

$$\frac{d\mathcal{K}_1(z)}{dz} = \frac{\mathcal{E}_1(z)}{z(1-z^2)} - \frac{\mathcal{K}_1(z)}{z}. \tag{6}$$

(ii) To obtain the form (11.3-31) first prove

$$\Psi^2 = 1 - \delta^2, \text{ where } \delta^2 = \alpha^2(1 - T/T_c)^2 \text{ or } \delta = \alpha\,|1 - T/T_c\,|, \quad \alpha > 0. \tag{7}$$

(iii) The logarithmic divergence stems solely from $\mathcal{K}_1(\Psi)$; the other terms are continuous and contribute to 'const.'. Substitute (7) into $\mathcal{K}_1(\Psi)$ and pull out the logarithmic contribution, before employing Taylor expansion about $\delta = 0$.

11.9 Consider a gas of magnons in a ferromagnet with dispersion relation $\hbar\omega_{\mathbf{q}} = Dq^2$, where $D = 2sJ_0a^2$, cf. (11.5-18). From the spin-wave Hamiltonian (11.5-17) obtain the free energy density, the energy density and the compressibility.

11.10 In nonlinear spin-wave theory quartic terms are retained but terms of order $\mathcal{O}(a^{(\dagger)6})$ are neglected. To convert the Heisenberg Hamiltonian to a boson form, one makes first a Holstein–Primakoff transformation and next a similarity transformation, (set $\hbar = 1$) viz.,

$$S_{\mathbf{l}}^{+} = \sqrt{2s}\sqrt{1 - a_{\mathbf{l}}^{\dagger}a_{\mathbf{l}}/2s}\, a_{\mathbf{l}}, \quad S_{\mathbf{l}}^{-} = \sqrt{2s}\, a_{\mathbf{l}}^{\dagger}\sqrt{1 - a_{\mathbf{l}}^{\dagger}a_{\mathbf{l}}/2s}, \tag{8}$$

$$a_{\mathbf{l}} \to 1/\sqrt{1 - a_{\mathbf{l}}^{\dagger}a_{\mathbf{l}}/2s}\, a_{\mathbf{l}} \quad a_{\mathbf{l}}^{\dagger} \to a_{\mathbf{l}}^{\dagger}\sqrt{1 - a_{\mathbf{l}}^{\dagger}a_{\mathbf{l}}/2s}\ . \tag{9}$$

(a) Denoting the ground-state Hamiltonian as $\mathcal{E}_0 I$, show that the following form is found

$$\mathcal{H} = \mathcal{E}_0 I - 2s\sum_{\mathbf{l},\mathbf{l}'}{}' J_{\mathbf{l}\mathbf{l}'}(a_{\mathbf{l}}^{\dagger} - a_{\mathbf{l}'}^{\dagger})a_{\mathbf{l}'}\left(1 - \frac{a_{\mathbf{l}}^{\dagger}a_{\mathbf{l}}}{2}\right) + g\rho_B H \sum_{\mathbf{l}} \mathbf{n}_{\mathbf{l}}\,. \tag{10}$$

(b) Convert to running-wave boson operators, keeping only diagonal terms $f(a_{\mathbf{q}}^{\dagger}a_{\mathbf{q}})$ and quartic terms that conserve momentum, $(\mathbf{q}_1,\mathbf{q}_3) = (\mathbf{q}_2,\mathbf{q}_4)$ or $(\mathbf{q}_4,\mathbf{q}_2)$. With $J(\mathbf{q}) = (1/2N)\Sigma_{\mathbf{l},\mathbf{l}'}\exp[i\mathbf{q}\cdot(\mathbf{l}-\mathbf{l}')]J_{\mathbf{l}\mathbf{l}'}$ being the Fourier transform, establish the following useful nonlinear spin-wave Hamiltonian:

$$\begin{aligned}\mathcal{H}_{nonl.s.w.} &= \sum\nolimits_{\mathbf{q}}[2sJ(0) - 2sJ(\mathbf{q}) + g\rho_B H]n_{\mathbf{q}} \\ &\quad - (1/N)\sum\nolimits_{\mathbf{q},\mathbf{q}'}[J(0) + J(\mathbf{q}-\mathbf{q}') - J(\mathbf{q}) - J(\mathbf{q}')]n_{\mathbf{q}}n_{\mathbf{q}'}\,.\end{aligned} \tag{11}$$

Chapter XII

Aspects of Quantum Fluids

1. VARIOUS SPECIAL THEORIES

12.1 Superfluidity; General Features

Despite the fact that an ideal Bose gas shows a phase transition at $3.2K$ – not too far from the observed transition point of $2.17K$ for ^{4}He – it can readily be shown that the ideal gas condensate cannot account for superfluidity. Rather, the latter is only possible if *short-range repulsive interactions between the particles are taken into account.* The elementary argument demonstrating this fact is found notably in Landau and Lifshitz.[1]

Consider laminar flow through a conduit, in which particles exchange energy with the walls. Suppose that a momentum $\mathbf{p}$ is imparted to one of the particles. Noting that for the condensed state $p = \hbar k = 0$, the gas experiences a "single particle excitation" with $p \neq 0$. Let $\mathbf{v}$ be the velocity of the fluid. In a frame in which the fluid is initially at rest, the acquired energy for all N particles is

$$\Delta\mathscr{E} = (N-1)\frac{mv^2}{2} + \frac{(\mathbf{p}+m\mathbf{v})^2}{2m} - N\frac{mv^2}{2}, \tag{12.1-1}$$

which is a decrease, $\Delta\mathscr{E} < 0$, if

$$\frac{p^2}{2m} + \mathbf{p}\cdot\mathbf{v} < 0. \tag{12.1-2}$$

Suppose that the excitation momentum is antiparallel to $\mathbf{v}$, i.e., $\mathbf{p}\cdot\mathbf{v} = -pv$, then the above condition merely requires $p < 2mv$. Since p will be close to the condensed state with zero momentum, such excitation momenta are always available. Therefore, collisions with the wall lead to a loss of energy, appearing as frictional heat, so that superfluidity will not occur: the ideal Bose condensate is *not* a superfluid. From a more a stringent point of view, the reason is that particle-like excitations, with a parabolic dispersion curve $\varepsilon(\mathbf{k}) = \hbar^2k^2/2m$, yield a zero tangent $d\varepsilon/d\mathbf{k}$ near $\mathbf{k} = 0$. In fact, all excitations of a Bose quantum fluid, whether

---

[1] L.D. Landau and E.M. Lifshitz, "Statistical Physics", Addison–Wesley 1958, § 66 and 67.

exhibiting superfluidity or not, never involve single particles but consist of *elementary excitations or quasi-particles.*

For any system of strongly interacting particles, described by the name quantum fluids, only quantum states for the system as a whole have quantum mechanical meaning. Nevertheless, we may assume that the interactions are adiabatically turned on, so that a ^{4}He quantum liquid still has integral spin, as expected from its atomic composition. At $T = 0$ only the ground-state is occupied. We now consider weakly excited states, characterized by elementary excitations involving creation or annihilation of a *single* entity, carrying along one integral unit $\hbar$ of angular momentum, as dictated by the quantum system. The quasi-particles therefore are bosons and the elementary excitations, corresponding to normal sound waves in the liquid, will be denoted as 'phonons'. Near $k = 0$ they have a linear energy spectrum $\varepsilon(k) = u\hbar k$, where u is the velocity of sound; we note that $\hbar k$ is a *real momentum* and not a quasi-momentum as in a solid. The usual treatment of this type is due to Landau and was proposed in 1941.[2] Later justification was given by Feynman.[3] Yet, the first (largely unnoticed) paper was presented by Bijl in 1940 during WWII.[4]

If $\varepsilon(k)$ for small k is known, we can find the thermodynamic quantities of the quantum system as in Section 8.5 for phonons in condensed matter. The main difference is that these phonons are only longitudinal. So, all expressions must be divided by a factor three. We thus obtain for the free energy

$$F = F_0 - V\frac{\pi^2 (k_B T)^4}{90(\hbar u)^3}, \tag{12.1-3}$$

where F_0 is the free energy of the quantum liquid at $T = 0$. For the energy likewise,

$$\mathscr{E} = \mathscr{E}_0 + V\frac{\pi^2 (k_B T)^4}{30(\hbar u)^3}, \tag{12.1-4}$$

while for the specific heat we have the modified Debye T^3 law

$$C_v = \frac{2\pi^2 V k_B}{15(\hbar u)^3}(k_B T)^3 . \tag{12.1-5}$$

Clearly, the linear dispersion law can only be valid for small k or long wavelengths. Thereafter $\varepsilon(k)$ will decrease with k, till a minimum is reached. Experimentally the excitation spectrum can be obtained from inelastic neutron scattering data. Measurements for ^{4}He by Cowley and Woods[5] closely follow the behaviour shown in

---

[2] L.D. Landau, J. Phys. USSR **5**, 71 (1941); ibid. **11**, 91 (1947).
[3] R.P. Feynman, Phys. Rev. **94**, 262 (1954).
[4] A. Bijl, Physica **7**, 860 (1940).
[5] R.A. Cowley and A.D.B. Woods, Can. J. Physics **49**, 177 (1971).

Fig. 12-1. They fit a formula of the type

$$\varepsilon(k) = \Delta + \hbar^2 (k - k_0)^2 / 2\mu, \tag{12.1-6}$$

where $\Delta / k_B = 8.7K$, $k_0 = 0.19 nm^{-1}$ and $\mu = 0.19 m_{\rm He}$. Excitations near the minimum are historically called 'rotons', an unfortunate name since these have nothing to do with vortices, to be discussed later.

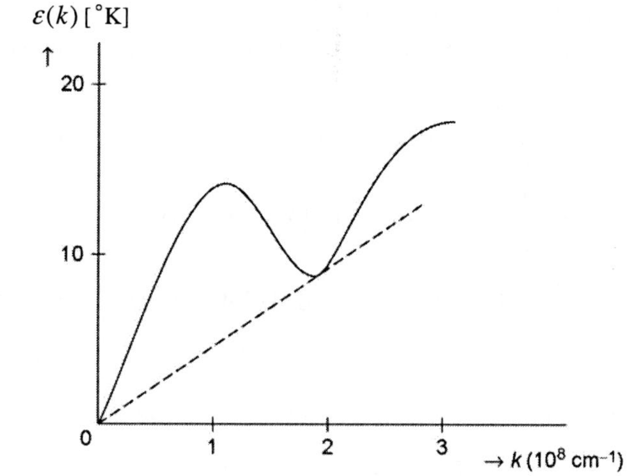

Fig. 12-1. Excitation spectrum of a superfluid, modelled after Landau[1] and data by Cowley and Woods, Ref. 5, Fig. 20. [With permission.]

We now indicate that a spectrum as in Fig.12-1 is conducive to superfluidity. Let us consider the conduit through which the fluid moves to be a classical object with mass M and momentum $\mathbf{P} = M\mathbf{v}$ and relative momentum $\mathbf{P} - \hbar\mathbf{k}$ with respect to a quasi-particle of momentum $\hbar\mathbf{k}$. In the rest frame of the fluid the condition for creating excitations is

$$P^2/2M - (\mathbf{P} - \hbar\mathbf{k})^2/2M = \varepsilon(k). \tag{12.1-7}$$

Neglecting the small term $\hbar^2 k^2 / 2M$, and with θ denoting the angle between $\mathbf{P}$ and $\mathbf{k}$, we have $|\mathbf{P}/M| = |\mathbf{v}| = \varepsilon(k)/\hbar k \cos\theta$, or $v_{\rm min} = \varepsilon(k)/\hbar k$. Thus, for $v > \varepsilon(k)/\hbar k$ – see the tangent to the curve (dashed line) in Fig.12-1 – energy can be converted to heat; for smaller v superfluidity sets in. For the data given, v_{min} = 60 *m/s*; in practice it is much less. The crucial element is therefore that the excitation spectrum exhibits a minimum *not* occurring at $k = 0$ and above the k-axis.

We will discuss a number of properties associated with the superfluid, but refer to specialized texts for details. For the particle density we had below T_c [see (8.3-14)]: $n = n_s + n_n$, where sub s refers to the superfluid (condensate) and sub n to the normal fluid. Likewise for the fluid mass-density $(\rho = nm)$,

$$\rho = \rho_s + \rho_n. \tag{12.1-8}$$

Let $\chi^{(\dagger)}(\mathbf{r}) = m^{1/2}\Psi^{(\dagger)}(\mathbf{r})$; the long-range order is then determined by

$$m \lim_{|\mathbf{r}-\mathbf{r}'|\to\infty} \langle \Psi^\dagger(\mathbf{r})\Psi(\mathbf{r}')\rangle = \lim_{|\mathbf{r}-\mathbf{r}'|\to\infty} \langle \chi^\dagger(\mathbf{r})\chi(\mathbf{r}')\rangle, \tag{12.1-9}$$

similarly as for the Ising magnet; the average is over a grand-canonical ensemble. The single particle classical mass-current density of the superfluid is given by $m\mathbf{u}_{s,i}\delta(\mathbf{r}-\mathbf{r}_i) \equiv \mathbf{j}_i(\mathbf{r})$, $i \in N$. In the quantum picture $\mathbf{u}_s = (\hbar/mi)\nabla$. Hermitizing the product $\delta(\mathbf{r}-\mathbf{r}')\nabla$ and employing Green's theorem[6] we find for the many-body mass flux

$$\begin{aligned}\langle \mathbf{J}_s(\mathbf{r})\rangle &= \frac{\hbar}{2im}\Big\langle \int d^3r' \chi^\dagger(\mathbf{r}')[\delta(\mathbf{r}-\mathbf{r}')\nabla_{\mathbf{r}'} + \nabla_{\mathbf{r}'}\delta(\mathbf{r}-\mathbf{r}')]\chi(\mathbf{r}')\Big\rangle \\ &= \frac{\hbar}{2im}\Big[\langle \chi^\dagger(\mathbf{r})\nabla\chi(\mathbf{r}) - \big(\nabla\chi^\dagger(\mathbf{r})\big)\chi(\mathbf{r})\rangle\Big].\end{aligned} \tag{12.1-10}$$

We now write $\chi^{(\dagger)} = A^{(*)}\exp[\pm i\gamma(r)]$; the superfluid density is given by

$$\rho_s = \langle \chi^\dagger \chi\rangle = \langle A^* A\rangle, \tag{12.1-11}$$

analogous to the square of a classical amplitude. Further we have

$$\langle \mathbf{J}_s(\mathbf{r})\rangle = (\hbar\rho_s/m)\nabla\gamma(\mathbf{r}), \tag{12.1-12}$$

implying,

$$\mathbf{u}_s^0 \equiv \langle \mathbf{u}_s\rangle = (\hbar/m)\nabla\gamma(\mathbf{r}). \tag{12.1-13}$$

Since $\mathbf{u}_s^0$ is proportional to the gradient of $\gamma(\mathbf{r})$, $\mathrm{curl}\,\mathbf{u}_s^0 = 0$. So, one expects that the circulation $\oint \mathbf{u}_s^0 \cdot d\ell$ will be zero. However, $\gamma(\mathbf{r})$ is only defined modulo 2π. So $\oint \nabla\gamma \cdot d\ell$ can satisfy any of the values $2\pi n$, where n is an integer, taken to be positive or zero. Hence for the circulation

$$\oint \mathbf{u}_s^0 \cdot d\ell = nh/m. \tag{12.1-14}$$

The physical interpretation of all this is that, besides laminar flow, there are quantized vortices. *Such vortices are realized when superfluid circulates about a core of normal fluid.*

Experimentally this is verified by the "rotating cylinder experiment", carried out by Osborne[7] and in a slightly different form by others.[8] The point is that in a rotating cylinder filled with an ordinary liquid the meniscus has a depth z, which is related to

---

[6] The operator $i\nabla$ is self-adjoint; by Green's theorem for the scalar product $(i\nabla\alpha,\beta) - (\alpha, i\nabla\beta) = C[\alpha,\beta]$, where C is the bilinear concomitant, a vanishing surface integral.

[7] D. Osborne, Proc. Phys. Soc. (London) **A 63**, 909 (1950).

[8] E.L. Andronokashvili and Yu. G. Mamaladze, Rev. Mod. Physics **3**, 567 (1966); see also K. Huang, Op. Cit., p. 381.

the radial distance r by the simple relation $z(r) = \omega^2 r^2 / 2g$ (i.e., the shape is that of a paraboloid of revolution) based on balancing of centripetal and gravitational forces.[9] For a quantum liquid it could be expected that only the normal component would rotate, so that the centripetal force involves ρ_n while the gravitation force contains $\rho_n + \rho_s$:

$$z(r) = \rho_n \omega^2 r^2 / 2\rho g\,. \tag{12.1-15}$$

However, the classical result was observed; the usual laminar flow of the superfluid component is destroyed through the excitation of vortices.

Another special feature, peculiar to superfluids, is the *fountain effect.* It is realised by the experiment of Fig.12-2. Upon heating the superfluid as shown, a 'fountain' is created.

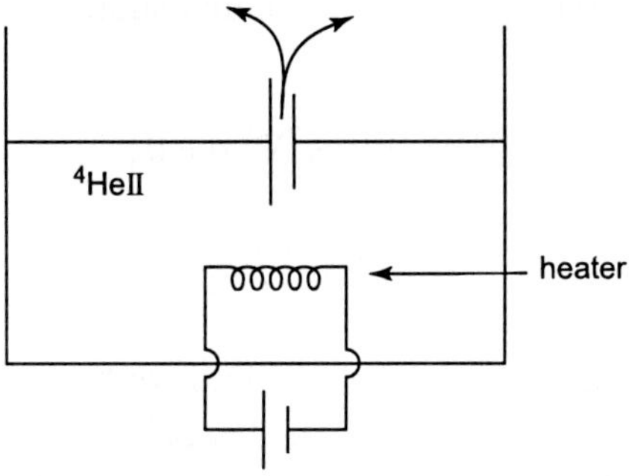

Fig. 12-2. The fountain effect.

The explanation rests upon the fact that the condensate has zero entropy, as shown in Section 8.3. Now consider the Gibbs-Duhem relation,

$$N d\varsigma = -S dT + V dP. \tag{12.1-16}$$

Let Δ denote the difference for quantities in the tube and in the container. As in any particle equilibrium, $\Delta\varsigma = 0$. However, there is no heat flow, so the temperatures in the tube and the container can be different. If the tube is clogged for the normal fluid, a pressure differential can likewise build up. From (12.1-16) it then follows that

$$\frac{\Delta T}{\Delta P} = \frac{V}{S} = \frac{\hat{v}_n}{\hat{s}_n}, \tag{12.1-17}$$

since $\hat{s}_s = \hat{v}_s = 0$. Thus, an imposed temperature differential causes a pressure differential (and *vice versa*), which explains the fountain effect. The inverse

[9] For a derivation see Osborne, loc. cit.; also, see Sears and Zemansky, "University Physics" (11th Ed.), Pearson/Addison Wesley 2004, Chapter XIV, Problem 14.76, based on the equations of fluid dynamics.

phenomenon is known as the 'mechano-caloric effect'.

We finally come to yet another phenomenon, the occurrence of *second sound.* We begin with the hydrodynamic conservation laws – to be considered in detail in the non-equilibrium part of this text – for mass and momentum balance:

$$\partial\rho/\partial t+\nabla\cdot\mathbf{J}=0,\qquad \partial\mathbf{J}/dt+\nabla\cdot\mathsf{P}=0, \tag{12.1-18}$$

where $\mathsf{P}=\mathsf{I}P$ is the pressure tensor in an isotropic fluid. Note that $\nabla\cdot(\mathsf{I}P)=\nabla P$. Pressure fluctuations in a sound wave are sufficiently fast, so that they can be taken to be adiabatic; hence

$$\delta P=(\partial P/\partial\rho)_S\,\delta\rho. \tag{12.1-19}$$

We now consider again the two equations of (12.1-18); operate with $\partial/\partial t$ on the first equation and with $\nabla\cdot$ on the second equation and subtract. Using also (12.1-19) the result is the wave equation

$$\partial^2\rho/\partial t^2-u^2\nabla^2\rho=0, \tag{12.1-20}$$

where

$$u=\sqrt{(\partial P/\partial\rho)_S} \tag{12.1-21}$$

is the velocity of sound. Since in a quantum fluid we have coexisting superfluid and normal fluid components (which are *not* associated with physically different constituents!) there are two types of motion, with sound velocities u_1 and u_2. In *first sound* the normal and superfluid components are in phase with each other, as in ordinary sound. In *second sound* the normal and superfluid components may move against each other, so as to preserve the total mass density. Since the superfluid has no entropy, second sound can be seen as a thermal wave.

More precisely, we should think of ^{4}He as a quantum liquid, for which the normal component can be identified with collective phonon-like excitations of the ground state, representing first sound. Next, second sound can be seen as a sound wave in the 'gas of excitations'. [As such, it should also exist for the phonon gas of very pure solids, for which it was indeed observed by de Gennes.[10]]

This interpretation allowed Landau to make some quantitative predictions for very low temperatures. Let $n_\mathbf{k}$ be the number of excitations of energy $\hbar\omega_\mathbf{k}=\hbar u_1 k$, where u_1 is the velocity of first sound, so that the energy carried by this mode is $n_\mathbf{k}\hbar u_1 k$. The number of modes in a volume V is as usual $(V/8\pi^3)d^3k$. Hence the free energy of the gas of phonon-like excitations is

$$F_{exc}=(Vk_BT/2\pi^2u_1^3)\int_0^\infty d\omega\,\omega^2\ln\left(1-e^{-\beta\hbar\omega}\right). \tag{12.1-22}$$

---

[10] P.G. de Gennes, "The Physics of Liquid Crystals", Oxford, Clarendon Press, 1974.

Note that integration by parts, observing (8.1-21), yields our previously stated result (12.1-3). With $P = -\partial F / \partial V$, $\mathcal{E} = \partial(\beta F)/\partial \beta$, this yields the equation of state

$$PV = \tfrac{1}{3}\mathcal{E}, \tag{12.1-23}$$

which, by the way, is similar as for an ideal Bose gas $[P = \tfrac{2}{3}u]$, although we deal here with the gas of excitations and not at all with the particles! Defining the 'mass' of the quasi-particles as $\hbar k / u_1 = \hbar\omega_{\mathbf{k}} / u_1^2$, the density for this gas is

$$\rho_{quasi} = \mathcal{E} / u_1^2 V = 3P / u_1^2, \tag{12.1-24}$$

where we used (12.1-23). For the velocity of second sound we then find, employing (12.1-21):

$$u_2 = \sqrt{\partial P / \partial \rho_{quasi}} = u_1 / \sqrt{3}. \tag{12.1-25}$$

At very low temperatures, $T < 1K$, Landau's prediction is closely obeyed; in the range between $1K$ and $2K$, however, $u_1 \approx 240 m/s$ while $u_2 \approx 20 m/s$.

12.2 Elements of Feynman's Theory[11]

12.2.1 *The Ground State and Single-Quantum Excited State*

Feynman's analysis throws far more light on the problem than Landau's theory, which was mainly guided by data on the specific heat $C_V(T)$. His theory shows how to get explicit results when almost nothing is known about the system. The article quoted, cf. Ref.3, is a continuation of two earlier studies cited therein. Feynman raises the question: what sort of excitations can we expect to occur in a system of tightly bound particles being in the ground state. Several possibilities are proposed, of which we will only mention two.

Let the ground state be described by $\Psi_0(\mathbf{r}_1, ..., \mathbf{r}_N)$. Since the Hamiltonian is a Hermitean operator, we can always choose this wave function to be real; moreover, since the Schrödinger equation is a second-order partial differential equation with solutions that must be symmetric in all coordinates, the ground-state function for a boson system has no zeros and can be chosen to be positive definite.[12] On the contrary, the first excited state Ψ_1 must be orthogonal to the ground state, i.e.,

$$(\Psi_1, \Psi_0) = \int d^{3N} r \Psi_1(\{\mathbf{r}\}) \Psi_0(\{\mathbf{r}\}) = 0; \tag{12.2-1}$$

---

[11] See also, K. Huang, 1st Ed., Op. Cit., p.383ff (not in Huang's second edition) [Huang's density functions ρ_m are *not* those used by Feynman and by us]; also, K. Huang in "Studies in Statistical Mechanics" II (J. de Boer and G. Uhlenbeck, Eds.) 1964, p. 1-24.

[12] Cf., O. Penrose and L. Onsager, Phys. Rev. **104**, 576 (1956).

therefore, it should be positive for some configurations $(\mathbf{r}_1,\mathbf{r}_2,...\mathbf{r}_N)$ and negative for others. Thus, as a first possibility, suppose that an excitation represents a rotation of a small ring of atoms about a given atom. The ring should be the smallest ring that can turn easily as an independent unit in view of the interatomic forces. This is pictured in Fig. 12-3, the ring being chosen to comprise six atoms. Interchange of two atoms does not change the dynamical state. Thus, if a rotation by 60 degrees (see the α-positions outlined by heavy circles) has Ψ_1 being positive, a rotation by 30 degrees (β-positions indicated by the intermediate light circles) should have Ψ_1 being negative. While nothing has been said so far about the effect on other atoms in the system, the wave function should drop to zero if two atoms overlap, just as for the ground state. Assigning a factor f_i to each atom in the excited state, we must have $f_i = 1/6$ for each α-position, $f_i = -1/6$ for each β-position and $f_i \approx 0$ for all other atoms. Thus Feynman proposes that $\Psi_1 = \Sigma_i f_i \Psi_0$ is a good representation for the described excited state; note that the summation extends over all atoms since no part of the liquid can be dissociated from the others. More likely, in our opinion, is another example in which the α-positions and β-positions are distributed throughout the liquid. In order for the kinetic energy to be low, these positions should be as far as possible removed from each other, however. This is schematically sketched in Fig. 12-4. An excitation would be a change from occupied α-positions to occupied β-positions. While this is not clearly readable from Fig. 12-4, one must in addition assume that the two types of configurations are more preponderant in different parts of the system, so that the excitation represents a *density fluctuation* in the system. More than the rotation of atoms in a ring, this suggests already that we here have a sound wave, driven by the zero-point energy of this state. So, whatever, the precise nature of the excitation, these and other examples given by Feynman suggests that we should have

$$\Psi_1(\mathbf{r}_1,\mathbf{r}_2,...\mathbf{r}_N) = (\Sigma_i f_i)\Psi_0(\mathbf{r}_1,\mathbf{r}_2,...\mathbf{r}_N). \tag{12.2-2}$$

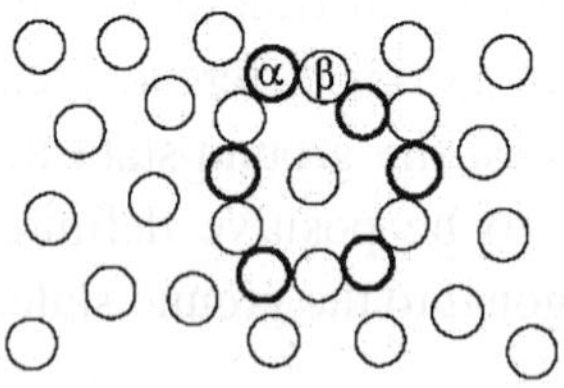

Fig. 12-3. Excitation involving rotation of a ring of atoms, involving α- and β-sites. [After Feynman, Op. Cit., Fig. 1.]

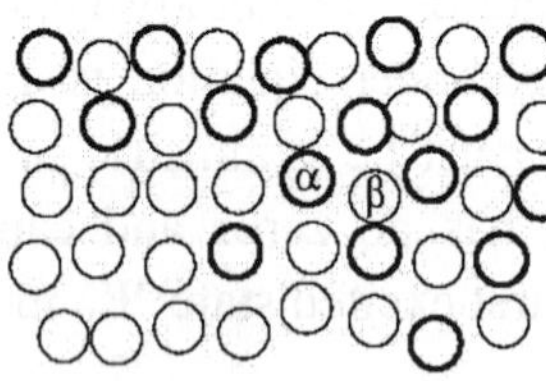

Fig. 12-4. Excitation with α- and β-sites distributed throughout the system, but not uniformly. [After Feynman, Op. Cit., Fig.2.]

12.2.2 *The Excitation Spectrum for T > 0*

Having decided that Eq. (12.2-2) should be a good form for the first excited state, we must decide on the form of f_i; this will be accomplished by using the variational principle. The Hamiltonian of the system, resulting in energies above the ground state is

$$H = \mathcal{H} - \mathcal{E}_0 I = -\frac{\hbar^2}{2m}\sum_i \nabla_i^2 + \sum_{i<j}\varphi_{ij} - \mathcal{E}_0 I, \tag{12.2-3}$$

where I is the identity operator and $\mathcal{E}_0$ is the ground state energy. Further we denote by F the sum over f_i, $F \equiv \Sigma_i f_i$. We thus have $\Psi_1 = F\Psi_0$. We shall minimize $(H\Psi_1, \Psi_1)$ – or in Dirac notation $\langle \Psi_1 | H | \Psi_1 \rangle$ – subject to $(\Psi_1, \Psi_1) = 1$. We have

$$H\Psi_1 = HF\Psi_0 = -\frac{\hbar^2}{2m}\sum_i [(\nabla_i^2 F)\Psi_0 + 2(\nabla_i F)\cdot(\nabla_i \Psi_0)] + \underbrace{(H\Psi_0)}_{0} F$$

$$= \frac{1}{\Psi_0}\left[-\frac{\hbar^2}{2m}\sum_i \nabla_i \cdot \Psi_0^2 \nabla_i F\right]. \tag{12.2-4}$$

The quantity Ψ_0^2 is the probability density for the particle distribution, $\rho_N(\mathbf{r}_1, \mathbf{r}_2, .., \mathbf{r}_N)$; note that we are dealing here with a *pure state*. Accordingly we have

$$\mathcal{E} = (H\Psi_1, \Psi_1) = \left[-\frac{\hbar^2}{2m}\int d^{3N}r\, F^* \sum_i \nabla_i \cdot \rho_N \nabla_i F\right]$$

$$= \frac{\hbar^2}{2m}\sum_i \int d^{3N}r (\nabla_i F^*)\cdot(\nabla_i F)\rho_N = \frac{\hbar^2}{2m}\sum_i \int d^{3N}r\, \rho_N\, \nabla_i f^*(\mathbf{r}_i)\cdot \nabla_i f(\mathbf{r}_i), \tag{12.2-5}$$

where we used Green's theorem. Since ρ is a symmetrical function of the positions, we can write

$$\mathcal{E} = \frac{\hbar^2 N}{2m}\int d^{3N}r\, \rho_N\, \nabla_i f^*(\mathbf{r}_i)\cdot\nabla_i f(\mathbf{r}_i) = \frac{\hbar^2 N \rho_1}{2m}\int d^3 r\, \rho_N\, \nabla_i f^*(\mathbf{r}_i)\cdot\nabla_i f(\mathbf{r}_i). \tag{12.2-6}$$

Here, $\rho_1 = \langle n \rangle$ is the first of the hierarchy of m-point density functions defined in Feynman's paper,

$$\rho_m(\mathbf{r}_1, \mathbf{r}_2, ..., \mathbf{r}_m)$$
$$= \sum_{i_1}\sum_{i_2}...\sum_{i_m}\int d^{3N}r'\,\delta(\mathbf{r}_1 - \mathbf{r}_{i_1}')\delta(\mathbf{r}_2 - \mathbf{r}_{i_2}')...\delta(\mathbf{r}_m - \mathbf{r}_{i_m}')\rho_N(\{\mathbf{r}'\}). \tag{12.2-7}$$

The first two density functions are

$$\rho_1(\mathbf{r}_1) = \Sigma_i \int d^{3N}r'\,\delta(\mathbf{r}_1 - \mathbf{r}_i')\rho_N(\{\mathbf{r}_i'\}) = N\int d^{3(N-1)}r'\rho(\mathbf{r}_1, \mathbf{r}_2' ... \mathbf{r}_N') = \langle n(\mathbf{r}_1)\rangle = N/V,$$

$$\rho_2(\mathbf{r}_1,\mathbf{r}_2)=\Sigma_i\Sigma_j\int d^{3N}r'\delta(\mathbf{r}_1-\mathbf{r}_i')\delta(\mathbf{r}_2-\mathbf{r}_j')\rho_N(\{\mathbf{r}'\})=\sum_i\int d^{3(N-1)}r'..+\sum_{i\neq j}\int d^{3(N-2)}r'..$$
$$=N\delta(\mathbf{r}_1-\mathbf{r}_2)\int d^{3(N-1)}r'\rho_N(\mathbf{r}_1,\{\mathbf{r}'\})+N(N-1)\int d^{3(N-2)}r'\rho_N(\mathbf{r}_1,\mathbf{r}_2,\{\mathbf{r}'\}). \quad (12.2\text{-}8)$$

Next we compute the normalization integral

$$(\Psi_1,\Psi_1)=\sum_i\sum_j\int d^{3N}r\rho_N(\{\mathbf{r}\})f_i^*(\mathbf{r}_i)f_j(\mathbf{r}_j)$$
$$=N\int d^{3N}r\rho_N\, f^*(\mathbf{r}_i)f(\mathbf{r}_i)+N(N-1)\int d^{3N}r\rho_N\, f^*(\mathbf{r}_i)f(\mathbf{r}_j). \quad (12.2\text{-}9)$$

Comparing with (12.2-8) we obtain

$$\mathcal{J}=(\Psi_1,\Psi_1)=\iint d^3r_1d^3r_2\, f_1^*(\mathbf{r}_1)f(\mathbf{r}_2)\rho_2(\mathbf{r}_1,\mathbf{r}_2). \quad (12.2\text{-}10)$$

We proceed to minimize $\mathcal{E}/N$ as given in (12.2-6) subject to $\mathcal{J}$ satisfying (12.2-10). With $\hbar\omega$ being a Lagrangian multiplier, we must set $\delta[(\mathcal{E}/N)-\hbar\omega\mathcal{J}]=0$. Or,

$$\delta\int d^3r_1f^*(\mathbf{r}_1)\left\{-\frac{\hbar^2\rho_1}{2m}\left[\nabla^2f(\mathbf{r}_1)\right]-\hbar\omega\int d^3r_2f(\mathbf{r}_2)\rho_2(\mathbf{r}_1,\mathbf{r}_2)\right\}=0. \quad (12.2\text{-}11)$$

We now introduce $D(\mathbf{r}_1,\mathbf{r}_2)=\rho_2(\mathbf{r}_1,\mathbf{r}_2)/\rho_1(\mathbf{r}_2)$. We note that ρ_2 is a density-density correlation, which can be expected to depend only on the distance separating the particles; the same applies to D: $D(|\mathbf{r}_1-\mathbf{r}_2|)=\rho_2(|\mathbf{r}_1-\mathbf{r}_2|)/\rho_1$.[13] The quantity $D(r)(V/N)-1\equiv p(r)$ is the pair-correlation function, the same function that was introduced in subsection 4.4.1 for a van der Waals gas. Writing in (12.2-11) $\rho_2=\rho_1D(r)$, we see that the variation is zero if

$$-\frac{\hbar^2}{2m}\nabla^2f(\mathbf{r}_1)-\hbar\omega\int d^3r_2f(\mathbf{r}_2)D(|\mathbf{r}_1-\mathbf{r}_2|)=0. \quad (12.2\text{-}12)$$

This integro-differential equation is solved by $f(\mathbf{r})=\exp(i\mathbf{k}\cdot\mathbf{r})$. The corresponding value of the Lagrangian multiplier is $\hbar\omega_\mathbf{k}$ and satisfies

$$\frac{\hbar^2k^2}{2m}=\hbar\omega_\mathbf{k}\int d^3r e^{i\mathbf{k}\cdot\mathbf{r}}D(|\mathbf{r}|). \quad (12.2\text{-}13)$$

Now the structure factor $S(\mathbf{k})$ is defined as the Fourier transform of $(N/V)p(|\mathbf{r}|)$ $=D(|\mathbf{r}|)-N/V$. However, we note that $(N/V)\int d^3re^{i\mathbf{k}\cdot\mathbf{r}}=N(8\pi^3/V)\delta(\mathbf{k})\approx N\delta_{\mathbf{k},0}$. But all excitations involve $\mathbf{k}>0$, so this term is of no consequence and in (12.2-13) we can replace $D(|\mathbf{r}|)$ by $D(|\mathbf{r}|)-N/V$ with impunity. Further, it is clear that $S(\mathbf{k})$ is only a function of $k=|\mathbf{k}|$. So, (12.2-13) entails

[13] From this definition $D(\mathbf{r}_1,\mathbf{r}_2)$ is the conditional probability density to find a particle at $\mathbf{r}_1$, given that there is another at $\mathbf{r}_2$ (Feynman); however, this 'probability density' must be normalized to N.

$$\hbar\omega_{\mathbf{k}} = \hbar^2 k^2 / 2mS(k). \qquad \textbf{(12.2-14)}$$

This is the main result, generally referred to as the *Bijl-Feynman formula*. Further, the eigenvalue belonging to $\Psi_{1\mathbf{k}} \propto \Sigma_i \exp(i\mathbf{k}\cdot\mathbf{r}_i)\Psi_0$ is

$$\mathcal{E}_{\mathbf{k}} = \mathcal{E}_0 + \hbar\omega_{\mathbf{k}}. \qquad \textbf{(12.2-15)}$$

We finally note that the total momentum operator **P**, acting on $\Psi_{1\mathbf{k}}$ yields

$$\mathbf{P}\Psi_{1\mathbf{k}} = (\hbar / i)\Sigma_i \nabla_i \Psi_{1\mathbf{k}} = \hbar\mathbf{k}\,\Psi_{1\mathbf{k}}, \qquad (12.2\text{-}16)$$

whereby we used that $\mathbf{P}\Psi_0 = 0$. Thus,

$$\hbar\mathbf{k}(\Psi_{1\mathbf{k}},\Psi_0) = (P\Psi_{1\mathbf{k}},\Psi_0) = (\Psi_{1\mathbf{k}},P\Psi_0) = 0. \qquad (12.2\text{-}17)$$

Noting $\mathbf{k} \neq 0$, this confirms that $\Psi_{1\mathbf{k}}$ is orthogonal to Ψ_0, which was the point of departure. Also, the excited states for different **k**'s are orthogonal to each other, since they belong to different EV of the momentum operator.

In a subsequent section Feynman discusses the possible shape of the energy spectrum and compares the deductions with experimentally obtained data of the structure factor from neutron diffraction. Clearly, these measurements are performed at low but finite temperature. While Feynman's paper only deals with the density distributions for the pure state Ψ_0, we believe that now some statistical mechanics should be brought in, an average over a canonical (or grand-canonical) ensemble being appropriate. Quoting Feynman: *"Since the structure of the liquid ought to be more or less like that in a classical liquid..."*, a density function $\rho_N(\{\mathbf{r}\})$, obtained by integrating over the fluctuating momenta should replace Ψ_0^2 for $T > 0$. The density functions $\rho_m(\mathbf{r}_1...\mathbf{r}_m)$ then are replaced by the correlation functions $G_m(\mathbf{r}_1...\mathbf{r}_m)$, as is noted from their identical definitions (4.11-14) and (12.2-7).

The shape of $D(|\mathbf{r}|)$ and of the Fourier transform $S(\mathbf{k})$ can be qualitatively predicted from simple arguments. If $\mathbf{r} = 0$, i.e., the positions $\mathbf{r}_1$ and $\mathbf{r}_2$ coincide, $D(|\mathbf{r}|) = \delta(\mathbf{r})$; this indicates that the Fourier transform for large **k** has as asymptotic value $S(\mathbf{k}) \to 1$, while also $S(k) \to 1$.[14] The behaviour for small k depends on the variations of $D(r)$ over large distances. One may expect $D(r)$ to have a maximum at the nearest neighbour spacing causing $S(k)$ to have a maximum near $k = 2\pi / a$. For larger r, $D(r)$ will fall again and after some other weaker maxima and minima approach a constant value . The form of $S(k)$ will therefore initially be determined by the density fluctuations in the liquid. Representing the density by

$$\rho(\mathbf{r}) = \Sigma_i \delta(\mathbf{r} - \mathbf{r}_i), \qquad (12.2\text{-}18)$$

we have for its Fourier transform

---

[14] We have $S(k) = 4\pi \int D(r)[\sin(kr)/kr]\, r^2 dr$; $\delta(\mathbf{r}) = \delta(r)/4\pi r^2$, so $S(k) \to \int \delta(r)[\sin(kr)/kr]dr = 1$.

$$a_{\mathbf{k}} = \int \rho(\mathbf{r}) e^{i\mathbf{k}\cdot\mathbf{r}} d^3 r = \Sigma_i e^{i\mathbf{k}\cdot\mathbf{r}_i}. \qquad (12.2\text{-}19)$$

Apparently, $S(\mathbf{k})$ is the expectation value of $|a_{\mathbf{k}}|^2$. For long wave length sound $a_{\mathbf{k}}$ is the coordinate of a normal mode with a mean potential energy that is half of the total energy, i.e., $\frac{1}{2}\hbar\omega_{\mathbf{k}}$. This implies $S(k) = \hbar k / 2mu$ for small k, u being the velocity of sound. So (12.2-14) yields $\hbar\omega_{\mathbf{k}} = \hbar k u$. Altogether, $S(k)$ starts out linearly, then reaches a maximum at $k = 2\pi / a$, after which it falls with some minor oscillations to the asymptotic value 1.

These predictions agree well with the measured structure factor for liquid ^{4}He.[15] The excitation spectrum computed from their data is shown in Fig. 12-5. The roton part of the excitation spectrum is slightly higher than in Landau's curve, which was based on data for the specific heat $C_v(T)$.

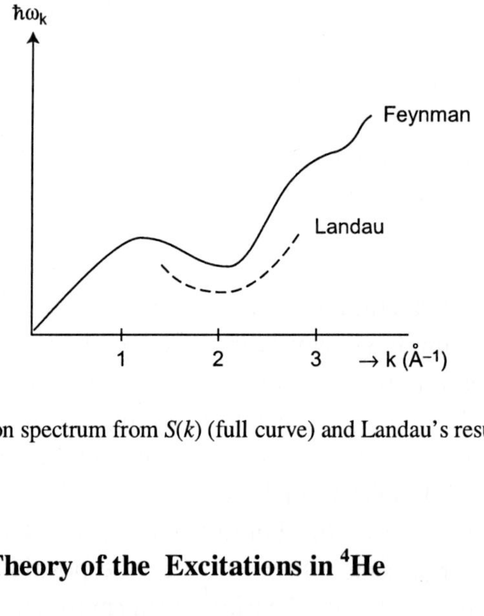

Fig. 12-5. Excitation spectrum from $S(k)$ (full curve) and Landau's results (dashed curve).

12.3 Bogoliubov's Theory of the Excitations in ^{4}He

12.3.1 *The Grand Hamiltonian*

While both Landau's and Feynman's theories consider the behaviour of quantum liquids from very broad and general quantum mechanical principles, one would expect that any fundamental theory would look in more detail at the interactions and use a second quantization procedure. So we now briefly discuss one of the oldest approaches involving such a description with a crude potential for the binary interactions, given in the late forties by Bogoliubov,[16] having as purpose to obtain the phonon-like excitation spectrum in the long wave length or small k limit.

---

[15] D.G. Henshaw and D.G. Hurst, Phys. Rev. **91**, 1222 (1953).

[16] N.N. Bogoliubov, J. Phys. USSR **11**,23 (1947).

We start with the 'grand Hamiltonian' using momentum eigenstates $\{|\mathbf{k})\}$ as basis for the unperturbed Hamiltonian, with binary interactions $v(\mathbf{r}_i,\mathbf{r}_j)$ in the second quantization form. We thus have

$$\mathcal{H}_{grand} = \mathcal{H} - \varsigma\mathbf{N} = \sum_{\mathbf{k}}[\varepsilon(\mathbf{k}) - \varsigma]a_{\mathbf{k}}^{\dagger}a_{\mathbf{k}} + \frac{1}{2}\sum_{\bar{\mathbf{k}},\bar{\mathbf{k}}',\mathbf{k},\mathbf{k}'} a_{\bar{\mathbf{k}}}^{\dagger}a_{\bar{\mathbf{k}}'}^{\dagger}(\bar{\mathbf{k}}\bar{\mathbf{k}}'|v(\mathbf{r},\mathbf{r}')|\mathbf{k}\mathbf{k}')a_{\mathbf{k}'}a_{\mathbf{k}} , \tag{12.3-1}$$

where the a's are boson operators, $\mathbf{N}$ is the number operator, ς is the chemical potential and $\varepsilon(\mathbf{k}) = \hbar^2k^2/2m$ is the free particle energy. With V being the volume, the momentum ES are $|\mathbf{k}) \to V^{-1/2}\exp(i\mathbf{k}\cdot\mathbf{r})$. Further we assume that the binary potential involves a central force,

$$v(\mathbf{r},\mathbf{r}') = v(\mathbf{r}-\mathbf{r}') = v(\mathbf{R}) , \tag{12.3-2}$$

with Fourier series $\Sigma_{\mathbf{q}}V_{\mathbf{q}}\exp(i\mathbf{q}\cdot\mathbf{R})$,

$$V_{\mathbf{q}} = V^{-1}\int_V d^3R\, e^{-i\mathbf{q}\cdot\mathbf{R}}v(\mathbf{R}) . \tag{12.3-3}$$

The second term of (12.3-1) becomes $a_{\bar{\mathbf{k}}}^{\dagger}a_{\bar{\mathbf{k}}'}^{\dagger}a_{\mathbf{k}'}a_{\mathbf{k}}$ times

$$\frac{1}{2V^2}\sum_{\bar{\mathbf{k}},\bar{\mathbf{k}}',\mathbf{k},\mathbf{k}'}\sum_{\mathbf{q}}V_{\mathbf{q}}\underbrace{\int d^3r e^{-i\mathbf{r}\cdot(\bar{\mathbf{k}}-\mathbf{k}-\mathbf{q})}}_{V\delta_{\bar{\mathbf{k}},\mathbf{k}+\mathbf{q}}}\underbrace{\int d^3r' e^{-i\mathbf{r}'\cdot(\bar{\mathbf{k}}'-\mathbf{k}'+\mathbf{q})}}_{V\delta_{\bar{\mathbf{k}}',\mathbf{k}'-\mathbf{q}}} , \tag{12.3-4}$$

where the Kronecker deltas follow from the theory of Fourier series. We can now carry out the sum over $\bar{\mathbf{k}}$ and $\bar{\mathbf{k}}'$; the remaining sum is

$$\text{sum} = \frac{1}{2}\sum_{\mathbf{k},\mathbf{k}',\mathbf{q}} V_{\mathbf{q}}a_{\mathbf{k}+\mathbf{q}}^{\dagger}a_{\mathbf{k}'-\mathbf{q}}^{\dagger}a_{\mathbf{k}'}a_{\mathbf{k}} . \tag{12.3-5}$$

To obtain an approximate result, we consider a finite core $v(\mathbf{R}) = \varphi_0/V_{core}$, $|\mathbf{R}| \le a$. Then,

$$V_{\mathbf{q}} = V^{-1}\int_0^a dr\int dS_r(\varphi_0/V_{core})e^{i\mathbf{q}\cdot\mathbf{r}} = 4\pi(\varphi_0/V\,V_{core})\int_0^a \frac{\sin qr}{qr}r^2dr. \tag{12.3-6}$$

$V_{\mathbf{q}}$ is appreciable only for $|\mathbf{q}| < 1/a$ so we replace it by $V_0 = \varphi_0/V$; we note that for repulsive forces $V_0 > 0$. This, then, yields for the grand Hamiltonian

$$\mathcal{H}_{grand} = \sum_{\mathbf{k}}[\varepsilon(\mathbf{k}) - \varsigma]a_{\mathbf{k}}^{\dagger}a_{\mathbf{k}} + \frac{\varphi_0}{2V}\sum_{\mathbf{k},\mathbf{k}',\mathbf{q}} a_{\mathbf{k}+\mathbf{q}}^{\dagger}a_{\mathbf{k}'-\mathbf{q}}^{\dagger}a_{\mathbf{k}'}a_{\mathbf{k}} . \tag{12.3-7}$$

We assume that we are well below the λ-point, so that an appreciable number of particles n_0 is in the state $\mathbf{k} = 0$. The many-body state is as usual denoted by the occupation-number state $|n_0 \ldots n_{\mathbf{k}} \ldots\rangle$, with the ground state in the absence of interactions being $|N,0,0....\rangle$. Now *with* interactions these many-body states still form a complete basis, although they are not eigenstates of the Hamiltonian for the interacting system. Yet it can be shown that a large fraction $n_0 = pN$, $(p<1)$, is condensed, having $\mathbf{k}=0$. We denote the ground state of $\mathcal{H}_{grand}$ by $|\text{O}\rangle \equiv |n_0\rangle$. We further make the following approximations for the creation and annihilation operators employed in this theory:

$$\begin{aligned} a_0^\dagger\,|n_0\rangle &= \sqrt{n_0+1}\,|n_0+1\rangle \approx \sqrt{n_0}\,|n_0\rangle, \\ a_0\,|n_0\rangle &= \sqrt{n_0}\,|n_0-1\rangle \approx \sqrt{n_0}\,|n_0\rangle. \end{aligned} \tag{12.3-8}$$

From the triple sum in (12.3-7) we select the terms for which one or more operators act on $|n_0\rangle$. First of all there is the term with $\mathbf{k}=\mathbf{k}'=\mathbf{q}=0$; it yields $(\varphi_0/2V)\mathbf{n}_0^2$. Then there are the terms of order n_0; terms yielding $\sqrt{n_0}$ will be dropped. So, besides the first term there are six combinations to be kept; denoting the operators by Ω we have

$$\begin{aligned}
&(a)\;\; \mathbf{k}=\mathbf{k}'=0,\; \mathbf{q}\neq 0 \rightarrow \mathbf{\Omega}_1\,|n_0\rangle = \mathbf{n}_0 a_\mathbf{q}^\dagger a_{-\mathbf{q}}^\dagger\,|n_0\rangle, \\
&(b)\;\; \mathbf{k}=0,\; \mathbf{q}=-\mathbf{k} \quad\;\; \rightarrow \mathbf{\Omega}_2\,|n_0\rangle = \mathbf{n}_0 a_{\mathbf{k}'}^\dagger a_{\mathbf{k}'}|n_0\rangle, \\
&(c)\;\; \mathbf{k}=0,\; \mathbf{q}=\mathbf{k}' \quad\;\;\; \rightarrow \mathbf{\Omega}_3\,|n_0\rangle = \mathbf{n}_0 a_{\mathbf{k}'}^\dagger a_{\mathbf{k}'}|n_0\rangle, \\
&(d)\;\; \mathbf{k}=-\mathbf{k}'=-\mathbf{q} \quad\; \rightarrow \mathbf{\Omega}_4\,|n_0\rangle \approx \mathbf{n}_0 a_\mathbf{k} a_{-\mathbf{k}}\,|n_0\rangle, \\
&(e)\;\; \mathbf{k}=\mathbf{q}=0,\; \mathbf{k}'\neq 0 \rightarrow \mathbf{\Omega}_5\,|n_0\rangle = \mathbf{n}_0 a_{\mathbf{k}'}^\dagger a_{\mathbf{k}'}|n_0\rangle, \\
&(f)\;\; \mathbf{k}'=\mathbf{q}=0,\; \mathbf{k}\neq 0 \rightarrow \mathbf{\Omega}_6\,|n_0\rangle = \mathbf{n}_0 a_\mathbf{k}^\dagger a_\mathbf{k}\;\,|n_0\rangle.
\end{aligned} \tag{12.3-9}$$

We just verify one such result; for *(b)*: $\mathbf{\Omega}_2|n_0\rangle = a_0^\dagger a_{\mathbf{k}'}^\dagger a_{\mathbf{k}'} a_0\,|n_0\rangle = \mathbf{n}_0 a_{\mathbf{k}'}^\dagger a_{\mathbf{k}'}|n_0\rangle$, where we used the fact that boson operators for different subscripts commute.[17] Note also that terms *(b)* and *(c)* can be combined. Separating off the terms with $\mathbf{k}=0$ the grand Hamiltonian becomes

$$\begin{aligned} \mathcal{H}_{grand} &= (\varepsilon_0-\varsigma)\mathbf{n}_0 + \frac{\varphi_0}{2V}\mathbf{n}_0^2 + \sum_\mathbf{k}{}' \left[\varepsilon(\mathbf{k}) - \varsigma + \frac{\varphi_0}{V}\mathbf{n}_0\right] a_\mathbf{k}^\dagger a_\mathbf{k} \\ &+ \frac{\varphi_0}{2V}\mathbf{n}_0 \sum_\mathbf{k}{}' \left(a_\mathbf{k}^\dagger a_{-\mathbf{k}}^\dagger + a_{-\mathbf{k}} a_\mathbf{k} + a_\mathbf{k}^\dagger a_\mathbf{k} + a_{-\mathbf{k}}^\dagger a_{-\mathbf{k}}\right), \end{aligned} \tag{12.3-10}$$

where Σ' means that $\mathbf{k}=0$ is to be excluded; note that in the last term the subscript $\mathbf{k}'$ was changed to $-\mathbf{k}$. Next, we introduce the new operators, cf. Fetter and Walecka,[18]

---

[17] See also G.D. Mahan, "Many-Particle Physics", 3rd Ed., Kluwer Academic/Plenum Publ., N.Y., 2000.

[18] A.L. Fetter and J.D. Walecka, "Quantum Theory of Many Particle Systems", McGraw-Hill, N.Y. 1971.

$$\xi_0^\dagger = V^{-1/2} a_0^\dagger, \quad \xi_0 = V^{-1/2} a_0, \tag{12.3-11}$$

with commutator $[\xi_0, \xi_0^\dagger] = 1/V \to 0$ in the thermodynamic limit. The ξ's can thus be treated as *c*-numbers. We then have $\langle O|\xi_0^\dagger \xi_0|O\rangle = \rho_0$, where ρ_0 is the density of particles in $|O\rangle$. Lastly, we must eliminate the chemical potential ς by minimizing the grand potential (or Gibbs' function) $-k_B T \ln \mathcal{Z}$ with respect to the variational parameter ρ_0. This is carried out in Fetter and Walecka (p. 206ff) ; we shall briefly outline the derivation at the end of this section. The result is

$$\varsigma = \varphi_0 \rho_0 \,. \tag{12.3-12}$$

Upon substitution this yields

$$\mathcal{H}_{grand} = -\frac{1}{2}\varphi_0\rho_0 n_0 + \frac{1}{2}\sum_{\mathbf{k}}{}' \Big\{ \varepsilon(\mathbf{k})\left(a_\mathbf{k}^\dagger a_\mathbf{k} + a_{-\mathbf{k}}^\dagger a_{-\mathbf{k}}\right)$$
$$+\varphi_0\rho_0\left(a_\mathbf{k}^\dagger a_{-\mathbf{k}}^\dagger + a_{-\mathbf{k}} a_\mathbf{k} + a_\mathbf{k}^\dagger a_\mathbf{k} + a_{-\mathbf{k}}^\dagger a_{-\mathbf{k}}\right)\Big\}, \tag{12.3-13}$$

which consists of a diagonal part and a nondiagonal part. In particular, the terms with $\mathbf{k}$ and $-\mathbf{k}$ must be unravelled. Note also that we set $\varepsilon_0 = 0$.

The entire grand Hamiltonian can be fully diagonalized with the Bogoliubov transformation (rotation over an imaginary angle) which we met before, see Eq. (11.7-7). We introduce the new boson operators $\eta_\mathbf{k}$ and $\eta_\mathbf{k}^\dagger$ slightly differently, however, by $(\theta_\mathbf{k} = \theta_{-\mathbf{k}})$:

$$\begin{aligned} a_\mathbf{k} &= \eta_\mathbf{k} \cosh\theta_\mathbf{k} - \eta_{-\mathbf{k}}^\dagger \sinh\theta_\mathbf{k}, \\ a_{-\mathbf{k}} &= \eta_{-\mathbf{k}} \cosh\theta_\mathbf{k} - \eta_\mathbf{k}^\dagger \sinh\theta_\mathbf{k}, \end{aligned} \tag{12.3-14}$$

and similarly for $a_\mathbf{k}^\dagger$ and $a_{-\mathbf{k}}^\dagger$; it is easily verified that $[\eta_\mathbf{k}, \eta_\mathbf{k}^\dagger] = [\eta_{-\mathbf{k}}, \eta_{-\mathbf{k}}^\dagger] = 1$. Doing the substitution and using the above commutation rules, one easily obtains

$$\begin{aligned} \mathcal{H}_{grand} &= -\tfrac{1}{2}\varphi_0\rho_0 n_0 + \tfrac{1}{2}\sum{}'_{\mathbf{k}} \{[\varepsilon(\mathbf{k}) + \varphi_0\rho_0] \\ &\times[(\eta_\mathbf{k}^\dagger\eta_\mathbf{k} + \eta_{-\mathbf{k}}^\dagger\eta_{-\mathbf{k}} + 1)\cosh 2\theta_\mathbf{k} - (\eta_\mathbf{k}^\dagger\eta_{-\mathbf{k}}^\dagger + \eta_\mathbf{k}\eta_{-\mathbf{k}})\sinh 2\theta_\mathbf{k} - 1] \\ &+\varphi_0\rho_0[(\eta_\mathbf{k}^\dagger\eta_{-\mathbf{k}}^\dagger + \eta_\mathbf{k}\eta_{-\mathbf{k}})\cosh 2\theta_\mathbf{k} - (\eta_\mathbf{k}^\dagger\eta_\mathbf{k} + \eta_{-\mathbf{k}}^\dagger\eta_{-\mathbf{k}} + 1)\sinh 2\theta_\mathbf{k}]\}. \end{aligned} \tag{12.3-15}$$

We set the new nondiagonal part equal to zero:

$$\begin{aligned} \mathcal{H}_{grand,nd} &= \tfrac{1}{2}\sum{}'_{\mathbf{k}} \{-[\varepsilon(\mathbf{k}) + \varphi_0\rho_0](\eta_\mathbf{k}^\dagger\eta_{-\mathbf{k}}^\dagger + \eta_\mathbf{k}\eta_{-\mathbf{k}})\sinh 2\theta_\mathbf{k} \\ &+\varphi_0\rho_0(\eta_\mathbf{k}^\dagger\eta_{-\mathbf{k}}^\dagger + \eta_\mathbf{k}\eta_{-\mathbf{k}})\cosh 2\theta_\mathbf{k}\} = 0, \end{aligned} \tag{12.3-16}$$

from which

$$\tanh 2\theta_\mathbf{k} = \frac{\varphi_0\rho_0}{\varepsilon(\mathbf{k}) + \varphi_0\rho_0}. \tag{12.3-17}$$

This is substituted into the rest of $\mathcal{H}_{grand}$, which now is diagonal; with $\eta_{\mathbf{k}}^{\dagger}\eta_{\mathbf{k}} = \bar{\mathbf{n}}_{\mathbf{k}}$ we now have

$$\mathcal{H}_{grand} = -\frac{1}{2}\varphi_0\rho_0 n_0 + \frac{1}{2}\sum_{\mathbf{k}}{}'\{(\bar{\mathbf{n}}_{\mathbf{k}} + \bar{\mathbf{n}}_{-\mathbf{k}} + 1)[\varepsilon(\mathbf{k}) + \varphi_0\rho_0]$$
$$\times\cosh 2\theta_{\mathbf{k}} - \varphi_0\rho_0 \sinh 2\theta_{\mathbf{k}}\} - \frac{1}{2}\sum_{\mathbf{k}}{}'[\varepsilon(\mathbf{k}) + \varphi_0\rho_0]. \qquad (12.3\text{-}18)$$

The following results are to be inserted,

$$\tfrac{1}{2}\sum_{\mathbf{k}}{}'(\bar{\mathbf{n}}_{\mathbf{k}} + \bar{\mathbf{n}}_{-\mathbf{k}} + 1) = \sum_{\mathbf{k}}{}'(\bar{\mathbf{n}}_{\mathbf{k}} + \tfrac{1}{2}), \qquad (12.3\text{-}19)$$

$$\cosh 2\theta_{\mathbf{k}} = \frac{1}{\sqrt{1-\tanh^2 2\theta_{\mathbf{k}}}} = \frac{\varepsilon(\mathbf{k}) + \varphi_0\rho_0}{\sqrt{\varepsilon(\mathbf{k})[\varepsilon(\mathbf{k}) + 2\varphi_0\rho_0]}},$$
$$\sinh 2\theta_{\mathbf{k}} = \frac{1}{\sqrt{\coth^2 2\theta_{\mathbf{k}} - 1}} = \frac{\varphi_0\rho_0}{\sqrt{\varepsilon(\mathbf{k})[\varepsilon(\mathbf{k}) + 2\varphi_0\rho_0]}}. \qquad (12.3\text{-}20)$$

This finally yields

$$\mathcal{H}_{grand} = -\frac{1}{2}\varphi_0\rho_0 n_0 + \sum_{\mathbf{k}}{}'(\bar{\mathbf{n}}_{\mathbf{k}} + \frac{1}{2})\sqrt{\varepsilon(\mathbf{k})[\varepsilon(\mathbf{k}) + 2\varphi_0\rho_0]} - \frac{1}{2}\sum_{\mathbf{k}}{}'[\varepsilon(\mathbf{k}) + \varphi_0\rho_0]. \qquad (12.3\text{-}21)$$

In this result the third term represents a shift in ground-state energy which, in fact, diverges (a consequence of replacing all $V_{\mathbf{q}}$ by V_0) ; it will be omitted. Since the theory supposed that $|\mathbf{k}|$ is small, in the second term, which adds the zero-point energy of the excitations to the ground-state energy, we shall neglect $\varepsilon(\mathbf{k})$ with respect to $\varphi_0\rho_0$. We then have, over and above the ground-state energy,

$$\bar{\mathcal{H}}_{grand} \approx \sum_{\mathbf{k}}{}'(\bar{\mathbf{n}}_{\mathbf{k}} + \frac{1}{2})\sqrt{2\varepsilon(\mathbf{k})\varphi_0\rho_0} = \sum_{\mathbf{k}}{}'(\bar{\mathbf{n}}_{\mathbf{k}} + \frac{1}{2})\underbrace{\hbar k\sqrt{\varphi_0\rho_0/m}}_{\hbar\omega_{\mathbf{k}}}. \qquad (12.3\text{-}22)$$

This explains the linear k-dependence of the phonon-like excitations. From the above and with the connection $\varphi_0 = V(0)$, where $V(0)$ is the Fourier transform of $v(\mathbf{R})$ for $\mathbf{q} \to 0$, we have

$$\hbar\omega_{\mathbf{k}} = \hbar k\sqrt{\rho_0 V(0)/m} \equiv \hbar k\bar{u}. \qquad (12\text{-}3\text{-}23)$$

The present calculation does not allow us to immediately identify $\bar{u}$ with the speed of sound, since basically the approach is equivalent to the use of a single-particle Green's function, rather than the density correlation function. However, a more detailed calculation, carried out in Fetter and Walecka, shows $\bar{u} \to u$.[19]

---

[19] A.L. Fetter and J.D. Walecka, Op. Cit., Section 22.

12.3.2 *The Chemical Potential*

The rendering of Bogoliubov's theory so far has not clarified why we employed the grand Hamiltonian. The main reason is that the number operator, after splitting off the $\mathbf{k}=0$ state population n_0, is no longer a constant of motion. We must therefore resort to the grand-canonical ensemble, in which N is variable. The number operator then becomes

$$N = n_0 + \sum_{\mathbf{k}}{}' \langle a_{\mathbf{k}}^{\dagger} a_{\mathbf{k}} \rangle . \tag{12.3-24}$$

Since the chemical potential is specified, we must determine n_0 in such a manner that it minimizes the grand potential

$$\Omega(T,V,\varsigma) = -(1/\beta)\ln \mathcal{Z}, \quad \mathcal{Z} = \mathrm{Tr} e^{-\beta\mathcal{H}_{grand}} . \tag{12.3-25}$$

Hence,

$$\delta\Omega = (\partial\Omega/\partial n_0)_{T,V,\varsigma(n_0)}\delta n_0 = 0 . \tag{12.3-26}$$

Since $\mathcal{H} = \mathcal{H}_0 + \mathcal{V}$, where $\mathcal{H}_0$ is kinetic energy and $\mathcal{V}$ is the interaction energy, we also have

$$\mathcal{H}_{grand} = \mathcal{H}_0 - \varsigma N + \mathcal{V} = \mathcal{H}_{grand}^0 + \mathcal{V}. \tag{12.3-27}$$

From (12.3-24) through (12.3-27) it follows that we must require

$$\frac{\partial\Omega}{\partial n_0} = -\frac{1}{\beta\mathcal{Z}}\frac{\partial}{\partial n_0}\mathrm{Tr}\left\{e^{-\beta[\mathcal{H}_0 - \varsigma(n_0 + \Sigma_{\mathbf{k}}{}' a_{\mathbf{k}}^{\dagger} a_{\mathbf{k}}) + \mathcal{V}]}\right\} = 0, \tag{12.3-28}$$

or

$$\varsigma\,\mathrm{Tr}\left\{e^{-\beta[\mathcal{H}_0 - \varsigma N + \mathcal{V}]}\right\} = \mathrm{Tr}\left\{(\partial\mathcal{V}/\partial n_0)\, e^{-\beta[\mathcal{H}_0 - \varsigma N + \mathcal{V}]}\right\}. \tag{12.3-29}$$

The reader will realise that the exponential, $\exp[-\beta(\mathcal{H}_{grand}^0 + \mathcal{V})]$, is a *symbolic* expressions, since $\mathcal{H}_{grand}^0$ and $\mathcal{V}$ do not commute, cf., Chapter 11, footnote [12]. This can be resolved by either using an infinite series of repeated commutators in the exponential, or by using an infinite perturbation expansion involving time-ordered integrals – a procedure borrowed from time-dependent perturbation theory in the interaction picture. Employing the latter, in Section 12.9 we show that

$$e^{-\beta(\mathcal{H}_{grand}^0 + \mathcal{V})} = \sum_{n=0}^{\infty}\frac{(-1)^n}{n!}\int_0^{\beta} d\tau_n \int_0^{\beta} d\tau_{n-1} \ldots \int_0^{\beta} d\tau_1\, e^{-\beta\mathcal{H}_{grand}^0}\, \mathcal{T}\left[\tilde{V}(\tau_n)\tilde{V}(\tau_{n-1})\ldots\tilde{V}(\tau_1)\right], \tag{12.3-30}$$

where

$$\tilde{V}(\tau) = e^{\tau\mathcal{H}_{grand}^0}\,\mathcal{V}\,e^{-\tau\mathcal{H}_{grand}^0} \tag{12.3-31}$$

and in which $\mathcal{T}$ is the time-ordering operator.

As first noted by Felix Bloch the operator $\tilde{V}(\tau)$ is *not* Hermitean, unless one considers it to be an interaction operator, extended in the complex plane for imaginary time with $\tau = i\hbar^{-1}t$.[20] This is the view which we espouse in this text, so that $\tilde{V}(\tau) = \mathcal{V}_I(-i\hbar\tau)$. Time ordering then involves tau-ordering along the imaginary axis, so that $\beta > \tau_n > \tau_{n-1} > ... > \tau_1$.

We are now ready to obtain the chemical potential as given by (12.3-29). The trace will be evaluated in the basis of eigenstates appertaining to $\mathcal{H}^0_{grand}$. For temperatures near $T = 0$ it suffices to consider only the matrix elements involving the lowest state, the ground state of $\mathcal{H}^0_{grand}$, denoted as $|\mathrm{O}\rangle$. We thus find ς as

$$\varsigma = \frac{\sum_{n=0}^{\infty} \frac{(-1)^n}{n!} \int_0^\beta d\tau_n \int_0^\beta d\tau_{n-1} .. \int_0^\beta d\tau_1 \langle \mathrm{O}|(\partial\mathcal{V}/\partial n_0) e^{-\beta\mathcal{H}^0_{grand}} \mathcal{T}\left[\tilde{V}(\tau_n)\tilde{V}(\tau_{n-1})..\tilde{V}(\tau_1)\right]|\mathrm{O}\rangle}{\sum_{n=0}^{\infty} \frac{(-1)^n}{n!} \int_0^\beta d\tau_n \int_0^\beta d\tau_{n-1} .. \int_0^\beta d\tau_1 \langle \mathrm{O}| e^{-\beta\mathcal{H}^0_{grand}} \mathcal{T}\left[\tilde{V}(\tau_n)\tilde{V}(\tau_{n-1})...\tilde{V}(\tau_1)\right]|\mathrm{O}\rangle}. \tag{12.3-32}$$

In this we need to substitute (12.3-31). For $T \to 0$ or $\beta \to \infty$, the lowest order contribution, stemming from the term $n = 0$, suffices; thus we have

$$\varsigma = \langle \mathrm{O}| \frac{\partial \mathcal{V}}{\partial n_0} |\mathrm{O}\rangle. \tag{12.3-33}$$

We now rewrite (12.3-10), using the operator notation of (12.3-9). We then have

$$\mathcal{H}_{grand} = (\varepsilon_0 - \varsigma)\mathbf{n}_0 + \sum_{\mathbf{k}}{}' [\varepsilon(\mathbf{k}) - \varsigma] a^\dagger_{\mathbf{k}} a_{\mathbf{k}} + \frac{\varphi_0}{2V}\mathbf{n}_0^2 + \frac{\varphi_0}{2V}\sum_{i=1}^{6}\sum_{\mathbf{k}}{}' \mathbf{\Omega}_i. \tag{12.3-34}$$

The first two terms clearly make up $\mathcal{H}^0_{grand}$, so the second part corresponds – in the approximations made – to the potential energy operator. Hence,

$$\mathcal{V}_{(op)} = \frac{\varphi_0}{2V}\mathbf{n}_0^2 + \frac{\varphi_0}{2V}\sum_{i=1}^{6}\sum_{\mathbf{k}}{}' \mathbf{\Omega}_i . \tag{12.3-35}$$

The $\mathbf{\Omega}_i$ operators which contain $a^{(\dagger)}_{\pm\mathbf{k}} a^{(\dagger)}_{\pm\mathbf{k}}$, $\mathbf{k} \neq 0$, destroy the ground state, so they do not contribute. The remaining term yields

$$\varsigma = \frac{\partial}{\partial n_0}\langle \mathrm{O}|\mathcal{V}|\mathrm{O}\rangle = \frac{\partial}{\partial n_0}\left(\frac{\varphi_0}{2V}n_0^2\right) = \varphi_0\rho_0, \tag{12.3-36}$$

QED — cf. (12.3-12).

---

[20] F. Bloch, Zeitschr. für Physik, **74**, 295 (1932).

12.4 Gaseous Atomic Bose–Einstein Condensates

Prior to the mid-nineties ^{4}He was the only known substance exhibiting Bose–Einstein condensation (BEC) and it was only in the late fifties that superfluidity and the two-component model of London and Tisza attributed the observed properties to BEC. Landau's brilliant 1941 theory, discussed previously, aimed solely at understanding the properties of ^{4}He and nowhere mentioned BEC. The reason is that to fully understand the properties of strongly interacting condensed bosons, as pertaining to *liquid* helium, the concept of broken symmetry and associated field-theoretic methods had to be developed, as was done in the period from 1957 to 1965.

In the mean time, research was started on the quest to observe the BEC state in gaseous materials and in spin-polarized hydrogen, $H\uparrow$. A gas of $H\uparrow$ atoms is stable till $T = 0$, since no bound state between two spin polarized hydrogen atoms exists, so that no liquid or solid state should form even at very low temperatures. Experimental groups, using cryogenic cooling and magnetic traps to bring this gas to the BEC threshold, were started in the late seventies at MIT and at the University of Amsterdam. Their efforts finally paid off 20 years later, see [21]. However, researchers looking for BEC in magnetically trapped alkali atoms, in particular ^{87}Rb and later ^{23}Na, were more successful. We have the first report of a condensate of ~ 2000 atoms of gaseous ^{87}Rb in a magnetic trap by a group in Boulder, Colorado [Andersen, Wieman and Cornell and coworkers] in 1995.[22] The reader may be astonished that these materials have bosonic properties. The point is that these atoms have both a nuclear spin **I** with quantum number $I = \frac{3}{2}$ and an electronic spin **S** with $S = \frac{1}{2}$; thus the total spin operator $\mathbf{F} = \mathbf{I} + \mathbf{S}$ has as quantum number $F = 1$ or 2, leading to eight possible atomic hyperfine-structure states. In a magnetic trap the total spin magnetic moment adjusts in the field [cf. Eqs. (6.8-6) and (6.8-7)], whereby the relevant atomic states are bosons, the number of neutrons being even in all alkali atoms and having an integral spin. Moreover, a metastable gaseous state can be maintained down to nano-Kelvins. Such temperatures are reached by initial laser cooling and subsequent 'evaporative' magnetic cooling; by lowering the trap potential sufficiently slow, one allows the hottest fraction of atoms, which are near the periphery of the trap to escape, after which the colder atoms re-thermalise. At a temperature for which the thermal de Broglie wavelength

$$\lambda_{th} \equiv \left(h^2/2\pi m k_B T\right)^{1/2} \tag{12.4-1}$$

is such that $n\lambda_{th}^3$ is of order unity [or, more precisely $\zeta(3/2) = 2.612$], where n is the

[21] Physics Today, October 1998.

[22] M.H. Andersen, J.R. Ensher, M.R. Matthews, C.E. Wieman and E.A. Cornell, Science **269**,198 (1995); see also Physics Today, August 1995.

density of the atoms, BEC sets in, see Section 8.3. Thus, the critical temperature is given as

$$k_B T_c \approx (h^2 / 2\pi m) n^{2/3}, \tag{12.4-2}$$

cf. Eq. (8.3-9). Confinement traps with a radius of the orders of microns are employed in order to increase the density, such that manageable values for T_c are obtained.

12.4.1 *Quantum Equations for the Near-Perfect B–E Gas*

We shall briefly describe the static and dynamic properties of these new Bose Einstein gases. Since the literature on this topic has exploded into many thousands of papers after its discovery, we shall largely follow a recent survey article by Allan Griffin of the University of Toronto.[23]

The magnetic confinement traps currently in use are well described by a harmonic potential, usually anisotropic. The trap potential is thus given as

$$V_{tr}(\mathbf{r}) = \begin{cases} \frac{1}{2} m\omega_0^2 r^2 & \text{(isotropic)} \\ \frac{1}{2} m(\omega_{0x}^2 + \omega_{0y}^2 + \omega_{0z}^2) & \text{(anisotr.)} \end{cases} \tag{12.4-3}$$

with $\omega_0 / 2\pi$ being of order 100 Hz. The energy levels are

$$\varepsilon_i = \varepsilon_{n_x, n_y, n_z} = (n_x + n_y + n_z + \tfrac{3}{2})\hbar\omega_0 . \tag{12.4-4}$$

At first we treat the condensate as an ideal B–E gas; the ground-state single particle wave function is then

$$\phi_0(\mathbf{r}) \sim e^{-r^2/2\alpha^2}, \tag{12.4-5}$$

where the oscillator length $\alpha = [\hbar / m\omega_0]^{1/2}$ is of order 1 μm in present-day traps. The size of the condensate in this description is determined by α, the density being $n_c(\mathbf{r}) = |\phi_0(\mathbf{r})|^2$. For the non-condensate density $\tilde{n}_0(\mathbf{r})$, called the 'thermal cloud', we can use the semi-classical limit, since the thermal energy $k_B T$ is much larger than the energy spacing of the oscillator levels; hence,

$$\tilde{n}_0(\mathbf{r}) \sim e^{-V_{tr}(\mathbf{r})/k_B T} = e^{-r^2/2R_T^2}, \tag{12.4-6}$$

where

$$R_T = \sqrt{k_B T / m\omega_0^2} = \alpha\sqrt{(k_B T / \hbar\omega_0)} . \tag{12.4-7}$$

Clearly, $R_T >> \alpha$; the size of the thermal cloud is thus much larger than the size of the

[23] A. Griffin in "*Theoretical Physics at the end of the Twentieth Century*", CRM series in Mathematical Physics (Yvan Saint-Aubin and Luc Vinet, Eds.), Springer 2002, pp. 277-305.

condensate. The onset of BEC is therefore marked by the appearing of a sharp high density peak at the centre of the condensate, usually at some hundred nano-*Kelvins*. This transition temperature is easily estimated for a harmonic trap. With N_c and $\tilde{N}_0$ being the condensate and thermal cloud numbers, respectively, we have $N = N_c + \tilde{N}_0$; in the continuum approximation and noting that the chemical potential equals $\frac{3}{2}\hbar\omega_0$, we find

$$N_c = \int d^3r \,|\phi_0(\mathbf{r})|^2 \quad ,$$
$$\tilde{N}_0 = \int_0^\infty dn_x \int_0^\infty dn_y \int_0^\infty dn_z \frac{1}{e^{\hbar\omega_0(n_x+n_y+n_z)/k_BT} - 1} = \left(\frac{k_BT}{\hbar\omega_0}\right)^3 \zeta(3) \,. \tag{12.4-8}$$

For the integral we used

$$\int_0^\infty dx \int_0^\infty dy \int_0^\infty dz \sum_{n=1}^\infty e^{-n(x+y+z)} = \sum_{n=1}^\infty \frac{1}{n^3} = \zeta(3) = 1.202 \,. \tag{12.4-9}$$

Since $N_c = 0+$ when $T \to T_c$ from below, we obtain from (12.4-8):

$$\frac{N_c(T)}{N} = \left(1 - \frac{T}{T_c}\right)^3 , \quad \text{with } T_c = 0.94 N^{1/3} \hbar\omega_0 / k_B \,. \tag{12.4.10}$$

However, soon after the discovery of atomic Bose–Einstein condensates it was realised that even in these dilute gases interactions, although weak, play a basic role, transforming the subject into a nontrivial many-body problem. As in subsection 12.3.1 we assume that the two-body potential $v(\mathbf{R})$ has a repulsive hard core and a weak attractive tail. Usually we can approximate the potential by that of an *s*-wave scattering event, represented by the pseudo potential[24]

$$v(\mathbf{R}) \approx (h^2/m) a\delta(\mathbf{R}) \equiv g\delta(\mathbf{R}) \,, \tag{12.4-11}$$

where a is the *s*-wave scattering length, being much less than the average atomic distance, i.e., $na^3 << 1$. For the alkali atoms $v(\mathbf{R})$ almost has a bound state and a is large, viz. 58 Å for ^{87}Rb and 28 Å for ^{23}Na. By varying the magnetic field the value of a can be changed and even be reversed (near a 'Feshbach resonance'), from repulsive $(a > 0)$ to attractive $(a < 0)$.

Due to the interactions, N is no longer a constant of motion and, as in ^{4}He, the number of condensed particles N_c fluctuates. The description requires the use of local, time-dependent field operators, $\Psi(\mathbf{r},t)$ and $\Psi^\dagger(\mathbf{r},t)$. There is now a *macroscopic* wave function,

$$\Phi(\mathbf{r},t) = \langle \Psi(\mathbf{r},t) \rangle = \sqrt{n_c(\mathbf{r},t)}\, e^{i\theta(\mathbf{r},t)} \,. \tag{12.4-12}$$

---

[24] J. Dalibard in "Bose–Einstein Condensation in Atomic Gases" (M.I. Inguscio, S. Stringari and C.E. Wieman Eds.) [Proc. Intern. School of Physics Enrico Fermi], IOS Press, Amsterdam 1999.

For the superfluid motion we have

$$\nabla\theta = m\mathbf{u}_s/\hbar, \tag{12.4-13}$$

entirely analogous to (12.1-13). The basic equation for the macroscopic wave function for a condensate at $T = 0$ has a Hartree-type form and was proposed by Gross and Pitaevskii in 1961[25]; it reads:

$$\left[-\frac{\hbar^2}{2m}\nabla^2 + V_{tr}(\mathbf{r}) + g n_c(\mathbf{r},t)\right]\Phi(\mathbf{r},t) = \hbar i\frac{\partial\Phi(\mathbf{r},t)}{\partial t}, \quad n_c(\mathbf{r},t) = |\Phi(\mathbf{r},t)|^2. \tag{12.4-14}$$

This nonlinear equation describes the motion of the condensate in a self-consistent Hartree potential produced by the condensate,

$$\begin{aligned} V_H(\mathbf{r},t) &= \int d^3r\, v(\mathbf{r}-\mathbf{r}')n_c(\mathbf{r}',t) \\ &= \int d^3r\, g\delta(\mathbf{r}-\mathbf{r}')n_c(\mathbf{r}',t) = g n_c(\mathbf{r},t). \end{aligned} \tag{12.4-15}$$

In the next subsection we shall consider the quantum-field properties and the collective modes, for which we shall find a spectrum similar to that of Bogoliubov's theory for ^{4}He.

12.4.2 *Properties and Solutions of the Gross-Pitaevskii Equation*

The G–P equation is a quantum field equation for the many-particle system in the sense of the field description in Section 7.4. Whereas it has the appearance of a Hartree equation, it is based on the Heisenberg equation of motion for the operators $\Psi(\mathbf{r},t)$ and $\Psi^\dagger(\mathbf{r},t)$. The complete Hamiltonian is written as $\mathcal{H} + \mathcal{H}_{SB}$, where $\mathcal{H}_{SB}$ is a symmetry-breaking perturbation, which allows $\langle\Psi^{(\dagger)}\rangle$ to be finite.[26] For $\mathcal{H}$ we have the quantum-field expression (7.3-35), in which $v^{(1)}$ is to be taken to be the potential $V_{tr}(\mathbf{r})$ and $v^{(2)}(\mathbf{r},\mathbf{r}')$ the interaction potential $v(\mathbf{R})$. The commutator bracket $[\Psi,\mathcal{H}]$ is easily obtained from the standard commutation rules for the Ψ's; e.g., for the kinetic energy term,

$$\begin{aligned} &[\Psi(\mathbf{r},t), \int d^3r'\{(\hbar^2/2m)\nabla\Psi^\dagger(\mathbf{r}',t)\cdot\nabla\Psi(\mathbf{r}',t)\}] \\ &= \frac{\hbar^2}{2m}\int d^3r' \underbrace{[\Psi(\mathbf{r},t),\nabla\Psi^\dagger(\mathbf{r}',t)]}_{\nabla\delta(\mathbf{r}-\mathbf{r}')}\cdot\nabla\Psi(\mathbf{r}',t) \\ &+ \frac{\hbar^2}{2m}\int d^3r'\nabla\Psi^\dagger(\mathbf{r}',t)\cdot\underbrace{[\Psi(\mathbf{r},t),\nabla\Psi(\mathbf{r}',t)]}_{0} = -\frac{\hbar^2}{2m}\nabla^2\Psi(\mathbf{r},t), \end{aligned} \tag{12.4-16}$$

---

[25] E.P. Gross, Nuovo Cimento **20**, 454 (1961); L.P. Pitaevskii, Soviet Physics JETP **13**, 451 (1961).
[26] P.W. Andersen, Rev. Mod. Phys. **38**, 298 (1966).

which looks like the 'Schrödinger' term. For the symmetry-breaking part one has,

$$\mathcal{H}_{SB} = \delta V(\mathbf{r},t) + \int d^3r[\eta(\mathbf{r},t)\Psi^\dagger(\mathbf{r},t) + h.c.], \tag{12.4-17}$$

where $\delta V(\mathbf{r},t)$ is a small driving potential causing the response expressed by $\eta(\mathbf{r},t)$. The complete equation of motion is easily shown to be

$$\frac{\partial \Psi(\mathbf{r},t)}{\partial t} = \frac{1}{\hbar i}\left[-\frac{\hbar^2}{2m}\nabla^2 + V_{tr}(\mathbf{r}) + \delta V(\mathbf{r},t)\right]\Psi(\mathbf{r},t)$$

$$+\eta(\mathbf{r},t) + g\Psi^\dagger(\mathbf{r},t)\Psi(\mathbf{r},t)\Psi(\mathbf{r},t). \tag{12.4-18}$$

Following Beliaev[27] the condensate part is separated off:

$$\Psi(\mathbf{r},t) = \langle\Psi(\mathbf{r},t)\rangle + \tilde{\Psi}(\mathbf{r},t) \equiv \Phi(\mathbf{r},t) + \tilde{\Psi}(\mathbf{r},t). \tag{12.4-19}$$

At temperatures $T \ll T_c$ we can neglect $\tilde{\Psi}$ and after $\delta V, \eta \to 0$ the G–P equation follows. We still note that Φ, as given by (12.4-12), is a two-variable order parameter; while N_c fluctuates, its conjugate canonical variable, being the phase θ, is fixed so that the macroscopic wave function $\Phi(\mathbf{r},t)$ represents a *coherent wave.*

The stationary G–P equation is obtained by setting $\Phi(\mathbf{r},t) = \Phi_s(\mathbf{r})\exp(-i\bar{\varsigma}t/\hbar)$. This yields

$$\left[-\frac{\hbar^2}{2m}\nabla^2 - \bar{\varsigma} + V_{tr}(\mathbf{r}) + g\,|\Phi_s(\mathbf{r})|^2\right]\Phi_s(\mathbf{r}) = 0. \tag{12.4-20}$$

The expression in [ ] is easily recognized to be the grand Hamiltonian for this problem, in which the number of condensate atoms N_c is variable, but the chemical potential $\varsigma = \bar{\varsigma}$ is fixed; henceforth, the overhead bar will be omitted. To obtain an estimate of the density profile, the kinetic energy contribution will at first be neglected – called the 'Thomas-Fermi approximation' in the BEC literature. We obtain

$$n_{c0}(\mathbf{r}) = g^{-1}[\varsigma - V_{tr}(\mathbf{r})] = g^{-1}(\varsigma - \tfrac{1}{2}m\omega_0^2 r^2). \tag{12.4-21}$$

Denoting the 'radius' of the condensate by R_{TF}, by integration one finds ς in terms of N_c. Since also $\varsigma \approx \frac{1}{2}m\omega_0^2 R_{TF}^2$, the following result obtains:

$$R_{TF} \approx \alpha\left(15N_c a/\alpha\right)^{1/5}, \tag{12.4-22}$$

which generally is much larger than the original harmonic oscillator width α. Therefore, the interactions, although weak, have caused the density profile to become significantly broader than for the perfect B–E gas, thereby decreasing the peak at the centre of the trap. We clearly witness here that the neighbouring bosons repel!

---

[27] S.T. Beliaev, Soviet Physics JETP **7**, 289 (1958).

In order to obtain *collective mode excitations* we linearize the G–P equation around the stationary wave function, i.e. we write

$$\Phi(\mathbf{r},t) = e^{-i\varsigma t/\hbar}[\Phi_s(\mathbf{r},t) + \delta\Phi(\mathbf{r},t)]\,. \qquad (12.4\text{-}23)$$

Substituting in the G–P equation (12.4-14), we obtain

$$\hbar i\frac{\partial\delta\Phi(\mathbf{r},t)}{\partial t} = \left[-\frac{\hbar^2}{2m}\nabla^2 + V_{tr}(\mathbf{r}) - \varsigma + g\left(\Phi_s\Phi_s^* + \Phi_s^*\delta\Phi + \Phi_s\delta\Phi^*\right)\right][\Phi_s + \delta\Phi]\,. \qquad (12.4\text{-}24)$$

Retaining only terms linear in $\delta\Phi$ and employing (12.4-20) we find

$$\hbar i\frac{\partial\delta\Phi(\mathbf{r},t)}{\partial t} = \left[-\frac{\hbar^2}{2m}\nabla^2 + V_{tr}(\mathbf{r}) - \varsigma + 2g\,|\Phi_s|^2\right]\delta\Phi(\mathbf{r},t) + g\Phi_s^2\delta\Phi^*(\mathbf{r},t)\,. \qquad (12.4\text{-}25)$$

There is a similar equation for $\partial\delta\Phi^*/\partial t$. These equations are solved with the ansatz

$$\delta\Phi(\mathbf{r},t) = u(\mathbf{r})e^{-i\omega t} + w(\mathbf{r})e^{i\omega t}\,. \qquad (12.4\text{-}26)$$

This yields two coupled Bogoliubov equations:

$$\begin{aligned} &\left[-\frac{\hbar^2}{2m}\nabla^2 + V_{tr}(\mathbf{r}) - \varsigma + 2gn_{c0}(\mathbf{r})\right]u(\mathbf{r}) + gn_{c0}(\mathbf{r})w(\mathbf{r}) = \mathcal{E}_i u(\mathbf{r}),\\ &\left[-\frac{\hbar^2}{2m}\nabla^2 + V_{tr}(\mathbf{r}) - \varsigma + 2gn_{c0}(\mathbf{r})\right]w(\mathbf{r}) + gn_{c0}(\mathbf{r})u(\mathbf{r}) = -\mathcal{E}_i w(\mathbf{r})\,. \end{aligned} \qquad (12.4\text{-}27)$$

Herein $\mathcal{E}_i = \hbar\omega$ are the excitation energies of the condensate. These two equations have been solved numerically by several groups; the results are in good agreement with the observed frequency spectrum.

We may get a feeling for the possible results, if we consider a very shallow trap, so that we may assume that the condensate profile is nearly uniform. We then set

$$u(\mathbf{r}) = u_0 e^{i\mathbf{k}\cdot\mathbf{r}}, \quad w(\mathbf{r}) = w_0 e^{i\mathbf{k}\cdot\mathbf{r}}\,. \qquad (12.4\text{-}28)$$

This gives two homogeneous equations for the amplitudes u_0 and w_0. Neglecting $V_{tr}(\mathbf{r})$, the determinant is

$$\begin{vmatrix} \dfrac{\hbar^2k^2}{2m} - \varsigma + 2gn_{c0} - \hbar\omega & gn_{c0} \\ gn_{c0} & \dfrac{\hbar^2k^2}{2m} - \varsigma + 2gn_{c0} + \hbar\omega \end{vmatrix} = 0\,. \qquad (12.4\text{-}29)$$

This yields

$$\left(\frac{\hbar^2 k^2}{2m} - \varsigma + 2gn_{c0}\right)^2 - (\hbar\omega)^2 = (gn_{c0})^2 . \tag{12.4-30}$$

We need to know the chemical potential. It follows from (12.4-21), with the same assumption that $V_{tr}(\mathbf{r})$ is negligible. We thus have

$$\varsigma \approx gn_{c0} . \tag{12.4-31}$$

The reader will notice that this is the same result as found in Section 12.3 – see (12.3-36). Equation (12.4-30) now gives

$$\hbar\omega = \sqrt{\varepsilon(\mathbf{k})[\varepsilon(\mathbf{k}) + 2gn_{c0}]}, \quad \varepsilon(\mathbf{k}) = \hbar^2 k^2 / 2m , \tag{12.4-32}$$

which is the same expression as in Bogoliubov's theory for liquid helium! For long wavelengths $\varepsilon(\mathbf{k}) << 2gn_{c0}$, we have 'phonon-like' excitations,

$$\omega = ku_B, \quad u_B = \sqrt{gn_{c0}/m} . \tag{12.4-33}$$

The cross-over from particle excitation to collective phonon-like excitations occurs for

$$(\hbar^2 k_o^2 / 2m) = 2gn_{c0} \text{ or } k_0 = \sqrt{4mn_{c0}g} / \hbar . \tag{12.4-34}$$

This shows that even weak interactions drastically alter the behaviour from that of a perfect B–E gas. It can also be shown that the interactions stabilize the superfluid motion.

The oscillations of the condensate for $T > 0$ also involve the excitations of the non-condensate part of the gas. From linear response arguments it can be shown that the non-condensate density fluctuations have the same spectrum as $\delta\Phi(\mathbf{r},t)$. We refer to the literature.[28] A particularly interesting excitation is the 'dipole mode', representing a rigid oscillation of the centre of mass of the condensate profile. We then have

$$n_c(\mathbf{r},t) = n_{c0}\left(\mathbf{r} - \boldsymbol{\chi}(t)\right), \tag{12.4-35}$$

where $\dot{\boldsymbol{\chi}} = \mathbf{v}_c$. The time-dependent centre of mass satisfies

$$d^2\boldsymbol{\chi}/dt^2 + \omega_0^2 \boldsymbol{\chi}(t) = 0 . \tag{12.4-36}$$

This mode at the trap frequency of a parabolic trap is called the Kohn mode, after a similar effect for interacting fermions. This 'sloshing' mode provides one of the best methods to measure the natural frequency of the trap.

[28] A. Griffin, "Excitations in a Bose-condensed Liquid", Cambridge Univ. Press , 1993.

12.5 Superconductivity

12.5.1 *The Fröhlich Hamiltonian*

Superconductivity – the phenomenon of electrical currents, which in certain metals and metal-like materials experience no resistance below a critical temperature T_c but persist forever – was discovered by Heike Kamerlingh Onnes in 1911. In ordinary or 'low T_c' superconductors the critical temperature is below $10K$, e.g., Pb(7.3K), Hg(4.1K), Zn(0.8K). Superconductivity in a Fermi electron gas is the counterpart of superfluidity in a Bose gas like ^{4}He. However, the usual perturbation approaches, besides explaining the formation of Cooper pairs treated in this section, fail to predict quantitatively viable results. A successful theory, known as the BCS theory after the three Nobel laureates, Bardeen, Cooper and Schrieffer [29], was finally proposed in the late fifties; we will briefly consider some tenets in the next section.

First, however, we go back to the origin of the phenomenon. For many metals it was found that the critical temperature was related to the isotopic mass by the relationship $M^{1/2}T_c = const.$; this led Fröhlich to believe that lattice phonons were involved in the apparent bonding of electrons.[30] More specifically, he assumed that zero-point phonons cause an attractive interaction of two electrons near the Fermi surface of the material. Below a certain temperature, this attraction would cancel out or over-compensate the usual screened electron repulsion $(e^2/4\pi r)\exp(-r/\lambda_0)$ [where λ_0 is the Thomas-Fermi screening length], the net attraction giving rise to pairing, i.e., the formation of what is now known as *Cooper pairs*.

Assuming for a moment that the above suggestions can be substantiated, we can readily write down the Hamiltonian – analogous to that of the superfluid, see (12.3-7):

$$\mathcal{H} = \mathcal{H}_F + \sum_{|\mathbf{k}|>k_F;\,\sigma} \varepsilon(\mathbf{k})\, c^\dagger_{\mathbf{k},\sigma} c_{\mathbf{k},\sigma} + \frac{1}{2}\sum_{\mathbf{k},\mathbf{k}',\mathbf{q},\sigma,\sigma'} V(\mathbf{q})\, c^\dagger_{\mathbf{k}+\mathbf{q},\sigma} c^\dagger_{\mathbf{k}'-\mathbf{q},\sigma'} c_{\mathbf{k}',\sigma'} c_{\mathbf{k},\sigma}\,. \tag{12.5-1}$$

Here $\mathcal{H}_F$ is the energy of the filled Fermi sea, σ represents the spin, $V(\mathbf{q})$ is the Fourier coefficient of $v(\mathbf{R})$; it will be replaced by $V(0) \equiv -\upsilon_0$, where now $\upsilon_0 > 0$. The computation below shows that both $\mathbf{k}$ and $\mathbf{k}'$ in the second summand must lay near the Fermi surface.

We return to the origin of the indirect electron-electron interaction, mediated by the phonon field. We recall the form of the Hamiltonian for electron-phonon interaction, given in (8.6-13):

---

[29] J. Bardeen, L.N. Cooper and J.R. Schrieffer, Phys. Rev. **108**, 1175 (1957).

[30] H. Fröhlich, Phys. Rev. **79**, 845 (1950).

$$\mathcal{H}^1 = i \sum_{\mathbf{k},\mathbf{q},\sigma} \mathcal{F}(\mathbf{q})[c^\dagger_{\mathbf{k}+\mathbf{q},\sigma} c_{\mathbf{k},\sigma} a_{\mathbf{q}} - c^\dagger_{\mathbf{k}-\mathbf{q},\sigma} c_{\mathbf{k},\sigma} a^\dagger_{\mathbf{q}}], \tag{12.5-2}$$

where $\mathcal{F}(\mathbf{q})$ is the coupling strength. In first order, $\mathcal{H}^1$ leads to scattering and electrical resistance; in second order it leads to self-energy and to coupling between electrons. In a physical sense, one electron polarizes the lattice and the other one reacts to the polarization. The following perturbation arguments are a modified form of those found in Kittel.[31]. The total Hamiltonian is obtained by adding to (12.5-2)

$$\mathcal{H}^0 = \sum_{\mathbf{k},\sigma} \varepsilon(\mathbf{k}) c^\dagger_{\mathbf{k},\sigma} c_{\mathbf{k},\sigma} + \sum_{\mathbf{q}} \hbar\omega_{\mathbf{q}} (a^\dagger_{\mathbf{q}} a_{\mathbf{q}} + \tfrac{1}{2}) . \tag{12.5-3}$$

We now consider $\mathcal{H} = \mathcal{H}^0 + \lambda \mathcal{H}^1$ and we let

$$\hat{H} = e^{iS} \mathcal{H} e^{-iS}, \quad S = I + \lambda S^1, \tag{12.5-4}$$

(I is the identity operator) be a transformation that is unitary and such that $\hat{H}$ is diagonal up to order $\mathcal{O}(\lambda^2)$. From series expansion of the exponentials we find

$$\hat{H} = \mathcal{H} - i[\mathcal{H}, S] - \tfrac{1}{2}[[\mathcal{H}, S], S] - i(\tfrac{1}{6})... \tag{12.5-5}$$

Substituting for $\mathcal{H}$ and S we have

$$\begin{aligned} \hat{H} &= \mathcal{H}^0 + \lambda\mathcal{H}^1 - i[\mathcal{H}^0, \lambda S^1] + \mathcal{O}(\lambda^2) \\ &= \mathcal{H}^0 + \lambda\mathcal{H}^1 - i[\mathcal{H}^0, S] + \mathcal{O}(\lambda^2) . \end{aligned} \tag{12.5-6}$$

Clearly the first order terms should be zero, so S is determined by[32]

$$\lambda \mathcal{H}^1 = i[\mathcal{H}^0, S]. \tag{12.5-7}$$

In the representation of $\mathcal{H}^0$, $\hat{H}$ has no off-diagonal terms up to $\mathcal{O}(\lambda^2)$. Using this S we can iterate a step further; hence,

$$\begin{aligned} \hat{H} = e^{iS}\mathcal{H}e^{-iS} &= \mathcal{H}^0 + \underbrace{\lambda\mathcal{H}^1 - i[\mathcal{H}^0, S]}_{0} - i\lambda[\mathcal{H}^1, S] \\ &- \tfrac{1}{2}[[\mathcal{H}^0, S], S] + \mathcal{O}(\lambda^3) = \mathcal{H}^0 - \tfrac{1}{2} i\lambda[\mathcal{H}^1, S] + \mathcal{O}(\lambda^3). \end{aligned} \tag{12.5-8}$$

In (12.5-7) we take the matrix elements in the states of $\mathcal{H}^0$, denoted as $|\{n_k\},\{N_q\}\rangle$ or shortly $|n, N\rangle$; we also use the convention that $k = (\mathbf{k}, \sigma)$. We obtain

$$\langle n, N | S | \bar{n}, \bar{N} \rangle = i\lambda \frac{\langle n, N | \mathcal{H}^1 | \bar{n}, \bar{N} \rangle}{\mathcal{E}_{\bar{n},\bar{N}} - \mathcal{E}_{n,N}} . \tag{12.5-9}$$

---

[31] C. Kittel, "The Quantum Theory of Solids", 2nd printing, Wiley, N.Y. 1986.

[32] Since $\mathcal{H}^1$ is Hermitean, S is Hermitean – as required from the outset – the commutator of Hermitean operators being anti-Hermitean.

In the problem at hand, we set $\lambda = 1$; then $\mathcal{F}(\mathbf{q})$ serves as a strength parameter, $\hat{H}$ being diagonal up to $\mathcal{O}(\mathcal{F}^2)$.

Into (12.5-9) we substitute (12.5-2). To obtain the effective electron-electron interaction, we only take the matrix elements for the phonon part of the states. Near $T = 0$ it suffices to consider the matrix elements for the phonon states $|1_\mathbf{q}\rangle$ and $|0_\mathbf{q}\rangle$. This yields

$$\begin{aligned}\langle 1_\mathbf{q} | S | 0_\mathbf{q} \rangle &= \mathcal{F}(\mathbf{q}) \sum_k c^\dagger_{\mathbf{k}-\mathbf{q},\sigma} c_{\mathbf{k},\sigma} \frac{1}{\varepsilon_\mathbf{k} - \varepsilon_{\mathbf{k}-\mathbf{q}} - \hbar\omega_\mathbf{q}}, \\ \langle 0_\mathbf{q} | S | 1_\mathbf{q} \rangle &= -\mathcal{F}(\mathbf{q}) \sum_{k'} c^\dagger_{\mathbf{k}'+\mathbf{q},\sigma} c_{\mathbf{k}',\sigma} \frac{1}{\varepsilon_{\mathbf{k}'} - \varepsilon_{\mathbf{k}'+\mathbf{q}} + \hbar\omega_\mathbf{q}},\end{aligned} \tag{12.5-10}$$

Note that the rhs is still an operator in the electron Fock space. The corresponding operator in (12.5-8) is denoted as $\hat{H}_e$. We then arrive at

$$\hat{H}_e = \mathcal{H}_e^0 + \frac{1}{2}\mathcal{F}^2 \sum_\mathbf{q} \sum_{k,k'} c^\dagger_{\mathbf{k}'+\mathbf{q},\sigma'} c_{\mathbf{k}',\sigma'} c^\dagger_{\mathbf{k}-\mathbf{q},\sigma} c_{\mathbf{k},\sigma} \times \left(\frac{1}{\varepsilon_\mathbf{k} - \varepsilon_{\mathbf{k}-\mathbf{q}} - \hbar\omega_\mathbf{q}} - \frac{1}{\varepsilon_{\mathbf{k}'} - \varepsilon_{\mathbf{k}'+\mathbf{q}} + \hbar\omega_\mathbf{q}} \right). \tag{12.5-11}$$

Let us regroup the phonon terms with $\mathbf{q}$ and $-\mathbf{q}$. The latter give a contribution, noticing that $\omega_\mathbf{q} = \omega_{-\mathbf{q}}$,

$$\sum_{k,k'} c^\dagger_{\mathbf{k}'-\mathbf{q},\sigma'} c_{\mathbf{k}',\sigma'} c^\dagger_{\mathbf{k}+\mathbf{q},\sigma} c_{\mathbf{k},\sigma} \left(\frac{1}{\varepsilon_\mathbf{k} - \varepsilon_{\mathbf{k}+\mathbf{q}} - \hbar\omega_\mathbf{q}} - \frac{1}{\varepsilon_{\mathbf{k}'} - \varepsilon_{\mathbf{k}'-\mathbf{q}} + \hbar\omega_\mathbf{q}} \right). \tag{12.5-12}$$

Next we interchange k and k' and we re-order the c-operators; the sum of the $\mathbf{q}$ and $-\mathbf{q}$ terms then gives the final result

$$\hat{H}_e = \mathcal{H}_e^0 + \mathcal{H}_e^1, \tag{12.5-13}$$

with

$$\mathcal{H}_e^1 = \frac{1}{2}\mathcal{F}^2 \sum_\mathbf{q} \sum_{k,k'} c^\dagger_{\mathbf{k}'+\mathbf{q},\sigma'} c_{\mathbf{k}',\sigma'} c^\dagger_{\mathbf{k}-\mathbf{q},\sigma} c_{\mathbf{k},\sigma} \left(\frac{2\hbar\omega_\mathbf{q}}{(\varepsilon_\mathbf{k} - \varepsilon_{\mathbf{k}-\mathbf{q}})^2 - \hbar^2\omega_\mathbf{q}^2} \right). \tag{12.5-14}$$

The diagram for the interaction is shown in Figs. 12-6(a) and 12-6(b). We note that the exchange is mediated by *virtual phonons*: a phonon is emitted and re-absorbed, so that the phonon-line is between two vertices. Thus, in Fig.(a) we put together the two diagrams of Fig. 8-14; the final diagram is then that of Fig.(b), having only external electron lines. The interaction is attractive (negative) for excitation energies $|\varepsilon_{\mathbf{k}\pm\mathbf{q}} - \varepsilon_\mathbf{k}| < \hbar\omega_\mathbf{q}$; it is repulsive otherwise. Even in the attractive region, we still should add the opposing screened Coulomb potential. However, for sufficiently large strength of the coupling constant, it may be neglected. It is usually supposed that the phonons are near the Debye limit, so the condition for $\mathcal{H}_e^1$ to be attractive can be written as

$$\varepsilon_F - \hbar\omega_D < \varepsilon_{\mathbf{k}}, \varepsilon_{\mathbf{k}\pm\mathbf{q}} < \varepsilon_F + \hbar\omega_D \,. \qquad (12.5\text{-}15)$$

Dropping all other terms in the summation over **q**, denoting the remaining terms of the bracket ( ) in (12.5-14) by $V(\mathbf{q})$ and separating off the **k**-vectors in the Fermi sea comprised in $\mathcal{H}_F$, the Hamiltonian of (12.5-1) has been duly obtained.

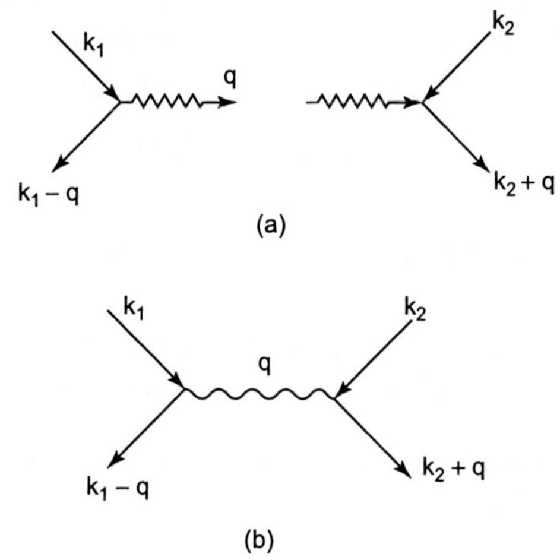

Fig. 12-6. Electron-electron interaction mediated by lattice phonons. Fig.(a): phonon emission and phonon absorption; Fig.(b): actual Feynman diagram involving a virtual phonon.

12.5.2 *Cooper Pairs*

We shall shift the energy scale, such that $\varepsilon_F = 0$. The ground state, or 'quasi-vacuum state', representing the filled Fermi sea, is then, see Fig. 12-7,

$$|\mathrm{O}\rangle = \prod_{|\mathbf{k}|\le k_F,\,\sigma} c^\dagger_{\mathbf{k},\sigma}\,|0\rangle \,. \qquad (12.5\text{-}16)$$

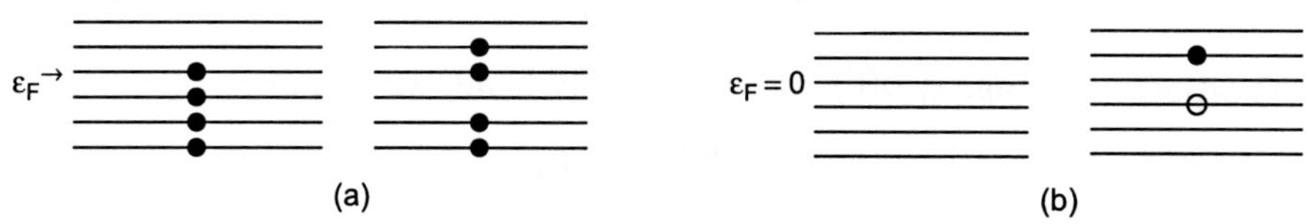

Fig. 12-7. Unperturbed Fermi system. (a) Ordinary picture of Fermi sea and excited state; (b): Particle-hole picture.

We will endeavour to find a bound state $|\Psi\rangle$ with energy $\mathcal{E}_\Psi$ such that $\mathcal{E}_\Psi < 0$. We try the following form for the state involving one Cooper pair,

$$|\Psi\rangle = \sum_{\mathbf{k}} \alpha(\mathbf{k}) c^\dagger_{\mathbf{k},\frac{1}{2}} c^\dagger_{-\mathbf{k},-\frac{1}{2}} |\text{O}\rangle, \tag{12.5-17}$$

in which two hole-electron excitations are created, with the paired electrons outside the Fermi sphere having conserved momentum and spin. We now require that the new ground state energy satisfy the Schrödinger equation $(\overline{\mathcal{H}} - \mathcal{E}_\Psi)|\Psi\rangle = 0$, with $\overline{\mathcal{H}}$ being the Fröhlich Hamiltonian of (12.5-1) minus $\mathcal{H}_F$; we then obtain the following relations for the set of variational parameters $\{\alpha(\mathbf{k})\}$ (see Problem 12-5):

$$[2\varepsilon(\mathbf{k}) - \mathcal{E}_\Psi]\alpha(\mathbf{k}) = \upsilon_0 \sum_{\mathbf{q}} \alpha(\mathbf{k}-\mathbf{q}). \tag{12.5-18}$$

Using the density of states $Z(\mathbf{q})$ in $\mathbf{q}$-space, we also write

$$\Lambda \equiv \sum_{\mathbf{q}} \alpha(\mathbf{k}-\mathbf{q}) \approx \int_{\mathbf{q}\in\Delta\mathbf{k}} \alpha(\mathbf{k}-\mathbf{q}) Z(\mathbf{q}) d^3q = \int_{\varepsilon_F=0}^{\hbar\omega_D} \alpha(\varepsilon)\rho(\varepsilon)d\varepsilon, \tag{12.5-19}$$

where $\rho(\varepsilon)d\varepsilon$ is the number of states in $(\varepsilon, \varepsilon + d\varepsilon)$. Returning to (12.5-18), we have, dividing both sides by $[2\varepsilon - \mathcal{E}]$ and integrating over the energy shell,

$$\Lambda \approx \upsilon_0 \Lambda \int_0^{\hbar\omega_D} \frac{\rho(\varepsilon)}{2\varepsilon - \mathcal{E}} d\varepsilon \approx \frac{\Lambda}{2} \upsilon_0 \rho(0) \ln\left(\frac{2\hbar\omega_D - \mathcal{E}}{-\mathcal{E}}\right), \tag{12.5-20}$$

where we pulled $\rho(\varepsilon)$ in front, evaluating it at the Fermi level; also note that we need no absolute sign for the argument of the logarithm, as long as we confirm that $\mathcal{E}$ is a negative quantity. This follows indeed from (12.5-20), which yields $\mathcal{E}$ in terms of the coupling strength:

$$\mathcal{E} = \frac{-2\hbar\omega_D}{\exp[2/\upsilon_0\rho(0)] - 1}. \tag{12.5-21}$$

This result is in accord with Schrieffer[33], obtained by a quite different, very precise analysis. For rather weak coupling, $\upsilon_0\rho(0) < 1$, this can be approximated as

$$\mathcal{E} = -2\hbar\omega_D e^{-2/\upsilon_0\rho(0)}. \tag{12.5-22a}$$

For very strong coupling, $\upsilon_0\rho(0) \sim 1$, on the other hand, we obtain

$$\mathcal{E} = -\hbar\omega_D \upsilon_0 \rho(0). \tag{12.5-22b}$$

These results indicate that even for weak interactions – providing they buck out the repulsive screened Coulomb energy – there always exists a bound state for a Cooper

---

[33] J. Robert Schrieffer, "Theory of superconductivity", Benjamin, NY 1964, p. 31.

pair with energy less than ε_F. Or, to put it differently, the Fermi sea is unstable against the formation of Cooper pairs, which form a condensate separated from the nearest single particle state by a gap Δ, to be specified in the next section.

It can also be shown that a state as considered above, with zero centre-of-mass momentum and opposite spins (singlet state), is the lowest paired electron state.

In Schrieffer's more exact analysis the eigenvalue problem is treated by an expansion in spherical polynomials Y_ℓ^m. Instead of our Λ Schrieffer arrives at a function $\Phi(\mathscr{E}_{\ell m})$:

$$\Phi(\mathscr{E}_{\ell m}) = \sum_{\mathbf{k}} |\, w_k^\ell \,|^2 \frac{1}{\mathscr{E}_{\ell m} - 2\varepsilon(\mathbf{k})}, \tag{12.5-23}$$

where w_k^ℓ is unity inside the Debye shell and zero elsewhere. A plot of $\Phi(\mathscr{E})$ is given in Fig. 12-8. It shows clearly how bound states are formed.

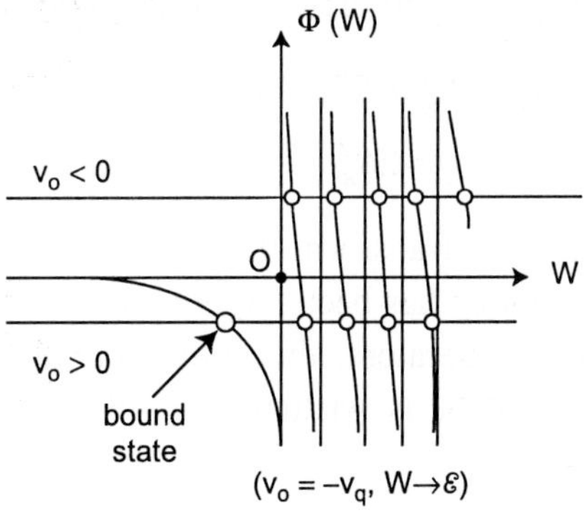

Fig. 12-8. Plot of $\Phi(\mathscr{E})$ *vs* $\mathscr{E}$. For repulsive interaction, $V_{\mathbf{q}} > 0$, all states are in the continuum; for attractive interaction, $V_{\mathbf{q}} < 0$, a bound state is split off. After Schrieffer, Ref. 33, Fig. 2-2. [With permission.]

12.6 The BCS Hamiltonian

12.6.1 *The Ground State*

In the previous section we considered a Fermi gas and one singlet Cooper pair $(s = 0)$; from the outset we excluded triplet pairs (parallel spins giving $s = 1$) and interactions below the Debye shell. In this section we briefly describe the BCS Hamiltonian, which assumes the same features, but allows for a condensate of many Cooper pairs[34]. As was the case for superfluids, the number of particles, *N*, will not

[34] For the ground state problem we follow Schrieffer, Op. Cit.. For the finite temperature theory, we use elements of the original BCS article and of the simplification thereof by Bogoliubov and Valatin, see Chapter XI ref.16. The Hamiltonian in BCS and in our rendering thereof is *not* a mean-field Hamiltonian [as often claimed in view of the self-consistent gap equation], the interaction potential being short-range.

be a constant of motion and the number operator **N** does not commute with the Hamiltonian. We shall therefore work with the grand-Hamiltonian to allow for number fluctuations. But first we introduce the pair operators

$$b_{\mathbf{k}}^{\dagger} = c_{\mathbf{k},\frac{1}{2}}^{\dagger} c_{-\mathbf{k},-\frac{1}{2}}^{\dagger}, \quad b_{\mathbf{k}} = c_{-\mathbf{k},-\frac{1}{2}} c_{\mathbf{k},\frac{1}{2}} . \tag{12.6-1}$$

The commutation rules are

$$\begin{aligned} &[b_{\mathbf{k}}^{\dagger}, b_{\mathbf{k}'}^{\dagger}] = 0, \quad [b_{\mathbf{k}}, b_{\mathbf{k}'}] = 0, \\ &[b_{\mathbf{k}}, b_{\mathbf{k}'}^{\dagger}] = \delta_{\mathbf{k}\mathbf{k}'}(1 - \mathbf{n}_{\mathbf{k}} - \mathbf{n}_{-\mathbf{k}}). \end{aligned} \tag{12.6-2}$$

The first two rules and the third one for $\mathbf{k} \neq \mathbf{k}'$ are as for boson operators; however, the pairs are "*bosons, not quite*". Actually, for $\mathbf{k} = \mathbf{k}'$ the exclusion principle reigns: $b_{\mathbf{k}}^{\dagger 2} = b_{\mathbf{k}}^2 = 0$. The reader will recognise that the commutation rules are similar to those of spin operators for spin ½ particles, cf., (11.5-1); we have

$$\begin{aligned} &b_{\mathbf{k}}^{\dagger} = S_{\mathbf{k}}^x + iS_{\mathbf{k}}^y = S_{\mathbf{k}}^+, \quad b_{\mathbf{k}} = S_{\mathbf{k}}^x - iS_{\mathbf{k}}^y = S_{\mathbf{k}}^-, \\ &1 - \mathbf{n}_{\mathbf{k}} - \mathbf{n}_{-\mathbf{k}} = 2S_{\mathbf{k}}^z . \end{aligned} \tag{12.6-2'}$$

Some older treatments therefore transform the problem to that of a spin gas, e.g. in Kittel, Op. Cit., but this technique has been largely abandoned at present.

In terms of the original c-operators, the grand Hamiltonian of the system, that forms the starting point of BCS, is written as [with $k = (\mathbf{k}, \sigma)$ and μ being the electrochemical potential] [35]

$$\begin{aligned} \mathcal{H}_{grand} = &\sum_{|\mathbf{k}|<k_F,\sigma} |\varepsilon_{\mathbf{k}} - \mu|(1 - c_k^{\dagger} c_k) + \sum_{|\mathbf{k}|>k_F,\sigma} (\varepsilon_{\mathbf{k}} - \mu) c_k^{\dagger} c_k \\ &+ \frac{1}{2} \sum_{k,k',\mathbf{q}} c_{\mathbf{k}'+\mathbf{q},\sigma'}^{\dagger} c_{\mathbf{k}',\sigma'} c_{\mathbf{k}-\mathbf{q},\sigma}^{\dagger} c_{\mathbf{k},\sigma} \left(\frac{e^2}{|\mathbf{q}|^2 + \kappa^2} + \frac{2\hbar\omega_{\mathbf{q}} |\mathcal{F}|^2}{(\varepsilon_{\mathbf{k}} - \varepsilon_{\mathbf{k}-\mathbf{q}})^2 - \hbar^2 \omega_q^2} \right). \end{aligned} \tag{12.6-3}$$

The first term is the pool of holes below ε_F while the second term contains the electron energies outside the Fermi surface (see Fig. 8.9a); the third term is the screened Coulomb energy (see problem 7.3) and the final term represents the electron-electron interactions mediated by virtual phonons, computed before [note that the $\varepsilon_{\mathbf{k}}$ are renormalized Bloch energies]. The interaction term is of paramount importance in the shell of energies $(\varepsilon_F - \hbar\omega_D, \varepsilon_F + \hbar\omega_D)$, see (12.5-15). Bardeen, Cooper and Schrieffer now argue that only long wavelength phonons with $\mathbf{q} \approx 0$ contribute and that of the combination of c-operators only those representing singlet pairs with net zero momentum and zero spin, $(\mathbf{k}\uparrow, -\mathbf{k}\downarrow)$, need to be retained. They introduce the "reduced grand Hamiltonian", written as

[35] The μ is not in the BCS article, but the necessity for its incorporation is found in Schrieffer Op. Cit., who introduces it as a Lagrangian multiplier, cf. his Eq. (2-27).

$$\mathcal{H}_{grand,red.} = 2\sum_{|\mathbf{k}|<k_F} |\varepsilon(\mathbf{k})-\mu| b_{\mathbf{k}} b_{\mathbf{k}}^{\dagger} + 2\sum_{|\mathbf{k}|>k_F} [\varepsilon(\mathbf{k})-\mu] b_{\mathbf{k}}^{\dagger} b_{\mathbf{k}} + \sum_{\mathbf{k},\mathbf{k}'} V_{\mathbf{k}\mathbf{k}'} b_{\mathbf{k}}^{\dagger} b_{\mathbf{k}'} . \quad (12.6\text{-}4)$$

The interaction terms as well as the Fermi terms include only singlet pairs, since triplet states in the electron gas do not contribute to excitations.

For the ground state that includes the condensate Schrieffer employs the trial state, familiar in the polaron problem,

$$|\Psi_0\rangle = C\prod_{\mathbf{k}} e^{g_{\mathbf{k}} b_{\mathbf{k}}^{\dagger}} |0\rangle = C\prod_{\mathbf{k}} (1+g_{\mathbf{k}} b_{\mathbf{k}}^{\dagger})|0\rangle , \quad (12.6\text{-}5)$$

where $|0\rangle$ is the vacuum state. Normalization gives $C^{-1} = \Pi_{\mathbf{k}}\sqrt{1+g_{\mathbf{k}}^2}$. The variational principle requires that upon variation of $g_{\mathbf{k}}$

$$\delta W = \delta\langle\Psi_0|\mathcal{H}_{gr,red.}|\Psi_0\rangle = 0 . \quad (12.6\text{-}6)$$

For the normal state $g_{\mathbf{k}} = \infty$, $|\mathbf{k}| < k_F$ and $g_{\mathbf{k}} = 0$, $|\mathbf{k}| > k_F$. Indeed,

$$\prod_{\mathbf{k}} b_{\mathbf{k}}^{\dagger}|0\rangle = \prod_{\mathbf{k}} c_{\mathbf{k}\uparrow}^{\dagger} c_{-\mathbf{k}\downarrow}^{\dagger}|0\rangle = \prod_{\mathbf{k},\sigma} c_{\mathbf{k},\sigma}^{\dagger}|0\rangle \equiv |O\rangle . \quad (12.6\text{-}7)$$

The BCS theory is most easily handled if the following quantities are introduced

$$u_{\mathbf{k}} = \frac{1}{\sqrt{1+g_{\mathbf{k}}^2}}, \qquad v_{\mathbf{k}} = \frac{g_{\mathbf{k}}}{\sqrt{1+g_{\mathbf{k}}^2}}, \quad (12.6\text{-}8)$$

with $u_{\mathbf{k}}^2 + v_{\mathbf{k}}^2 = 1$. This means that we must minimize the variational quantity W:

$$W = 2\sum_{\mathbf{k}} [\varepsilon(\mathbf{k})-\mu] v_{\mathbf{k}}^2 + \sum_{\mathbf{k},\mathbf{k}'} V_{\mathbf{k}\mathbf{k}'} u_{\mathbf{k}} v_{\mathbf{k}} u_{\mathbf{k}'} v_{\mathbf{k}'} , \quad (12.6\text{-}9)$$

$$\delta W = \delta\left\{2\sum_{\mathbf{k}} [\varepsilon(\mathbf{k})-\mu] v_{\mathbf{k}}^2 + \sum_{\mathbf{k},\mathbf{k}'} V_{\mathbf{k}\mathbf{k}'} u_{\mathbf{k}} v_{\mathbf{k}} u_{\mathbf{k}'} v_{\mathbf{k}'}\right\} = 0 . \quad (12.6\text{-}10)$$

When carrying out the variation we can eliminate δu since $u\delta u + v\delta v = 0$. Accordingly,

$$\delta W = 2\sum_{\mathbf{k}} \left\{ 2[\varepsilon(\mathbf{k})-\mu] v_{\mathbf{k}} + \sum_{\mathbf{k}'} V_{\mathbf{k}\mathbf{k}'} u_{\mathbf{k}'} v_{\mathbf{k}'} \frac{u_{\mathbf{k}}^2 - v_{\mathbf{k}}^2}{u_{\mathbf{k}}} \right\} \delta v_{\mathbf{k}} = 0 . \quad (12.6\text{-}11)$$

Since the $\delta v_{\mathbf{k}}$ are independent, the expression in the curly brackets is zero:

$$2[\varepsilon(\mathbf{k})-\mu] u_{\mathbf{k}} v_{\mathbf{k}} + \sum_{\mathbf{k}'} V_{\mathbf{k}\mathbf{k}'} u_{\mathbf{k}'} v_{\mathbf{k}'} (u_{\mathbf{k}}^2 - v_{\mathbf{k}}^2) = 0 . \quad (12.6\text{-}12)$$

Now two new parameters are introduced: the gap parameter $\Delta_{\mathbf{k}}$ and the energy $\mathcal{E}_{\mathbf{k}}$, defined through

$$u_{\mathbf{k}}^2 = \frac{1}{2}\left(1 + \frac{\varepsilon(\mathbf{k}) - \mu}{\mathscr{E}_{\mathbf{k}}}\right), \quad v_{\mathbf{k}}^2 = \frac{1}{2}\left(1 - \frac{\varepsilon(\mathbf{k}) - \mu}{\mathscr{E}_{\mathbf{k}}}\right), \tag{12.6-13}$$

$$u_{\mathbf{k}} v_{\mathbf{k}} = \frac{\Delta_{\mathbf{k}}}{2\mathscr{E}_{\mathbf{k}}}, \tag{12.6-14}$$

which satisfies $u_{\mathbf{k}}^2 + v_{\mathbf{k}}^2 = 1$, while $u_{\mathbf{k}}^2 - v_{\mathbf{k}}^2 = [\varepsilon(\mathbf{k}) - \mu]/\mathscr{E}_{\mathbf{k}}$. Substituting into (12.6-12), we find the simple 'integral equation', referred to as *the gap equation*:

$$\Delta_{\mathbf{k}} + \sum_{\mathbf{k}'} V_{\mathbf{k}\mathbf{k}'}(\Delta_{\mathbf{k}'}/2\mathscr{E}_{\mathbf{k}'}) = 0. \tag{12.6-15}$$

The energies $\varepsilon(\mathbf{k}) \equiv \varepsilon_{\mathbf{k}}$ will be measured with respect to the Fermi level ε_F of the normal state, which is taken to be zero, as in the previous section. Then μ is the shift of the electrochemical potential between the normal (N) and the superconducting state (S). In a system possessing electron-hole symmetry, one finds that this shift is very small, so we can set $\mu \approx 0$, where necessary; however, to retain the proper 'image' of the theory, we will formally employ

$$\hat{\varepsilon}_{\mathbf{k}} = \varepsilon_{\mathbf{k}} - \mu. \tag{12.6-15'}$$

In order to solve (12.6-15) we must express $\mathscr{E}_{\mathbf{k}}$ in $\Delta_{\mathbf{k}}$. We do some elementary algebra. From (12.6-14) we have $\mathscr{E}_{\mathbf{k}} = \Delta_{\mathbf{k}}/2u_{\mathbf{k}}v_{\mathbf{k}}$; using the statements in the line below (12.6-14) we have

$$\mathscr{E}_{\mathbf{k}} = \frac{\Delta_{\mathbf{k}}}{2u_{\mathbf{k}}v_{\mathbf{k}}} = \frac{\Delta_{\mathbf{k}}}{\sqrt{(u_{\mathbf{k}}^2 + v_{\mathbf{k}}^2)^2 - (u_{\mathbf{k}}^2 - v_{\mathbf{k}}^2)^2}} = \frac{\Delta_{\mathbf{k}}}{\sqrt{1 - \hat{\varepsilon}_{\mathbf{k}}^2/\mathscr{E}_{\mathbf{k}}^2}},$$

from which we obtain

$$\mathscr{E}_{\mathbf{k}} = \sqrt{\hat{\varepsilon}_{\mathbf{k}}^2 + \Delta_{\mathbf{k}}^2}. \tag{12.6-16}$$

In the next subsection we will show that $\mathscr{E}_{\mathbf{k}}$ is the energy needed to create a quasi-particle of momentum **k** in the superconducting state.

Assuming that the energies as well as the Fermi surface are spheres in **k**-space, the interaction involves only the shell defined by (12.5-15); see Fig. 12- 9 below. We set

$$V_{\mathbf{k}\mathbf{k}'} = \begin{cases} -v_0, & \mathbf{k}, \mathbf{k}' \in \text{shell } \delta\varepsilon, \\ 0 & \text{elsewhere.} \end{cases} \tag{12.6-17}$$

Likewise, we replace $\Delta_{\mathbf{k}}$ by Δ_0 within the shell. Then (12.6-15) can be written as an energy integral

$$\Delta_0 = v_0 \int_{-\hbar\omega_D}^{\hbar\omega_D} d\varepsilon\, \rho(\varepsilon) \frac{\Delta_0}{2\sqrt{\hat{\varepsilon}^2 + \Delta_0^2}} \approx \Delta_0 v_0 \rho(0) \operatorname{arsinh}\left(\frac{\hbar\omega_D}{\Delta_0}\right). \tag{12.6-18}$$

From this we obtain for the gap

$$\Delta_0 = \frac{\hbar\omega_D}{\sinh[1/\upsilon_0\rho(0)]} \approx 2\hbar\omega_D \exp\left[-\frac{1}{\upsilon_0\rho(0)}\right]. \qquad (12.6\text{-}19)$$

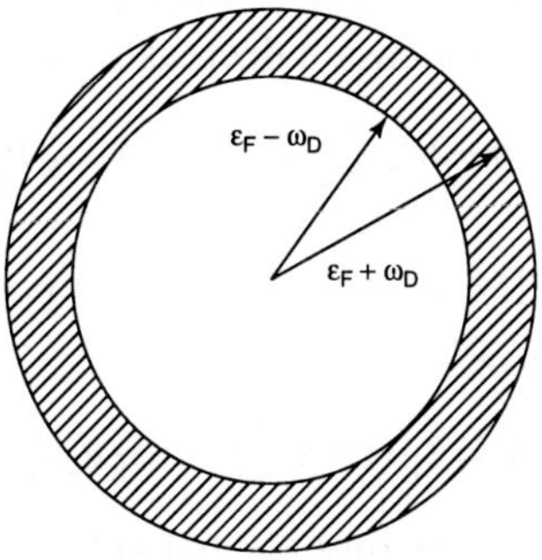

Fig. 12-9. Shell of one-electron states (shaded) in forming the BCS ground state.

Finally, we must show that the superconducting state has a lower energy than the normal state. So we go back to the expression for the energy W of (12.6-9), which for the solution obtained will be denoted as W_S. We must subtract the ground state energy for the normal state, i.e., $2\Sigma_{\mathbf{k}}[\varepsilon(\mathbf{k})-\mu]$. One obtains

$$W_S - W_N = -\frac{1}{2}\rho(0)\Delta_0^2 \approx -2\rho(0)(\hbar\omega_D)^2 \exp\left[-\frac{2}{\upsilon_0\rho(0)}\right], \qquad (12.6\text{-}20)$$

where the last rhs holds for weak coupling, $\upsilon_0\rho(0)<1$; for strong coupling, $\upsilon_0\rho(0)\sim 1$,

$$W_S - W_N \approx -\tfrac{1}{2}\upsilon_0[\rho(0)\hbar\omega_D]^2 . \qquad (12.6\text{-}20')$$

In all cases the new ground state has a lower energy than the normal state. It is to be noted that the energy of the condensate for weak and intermediate coupling is not a simple power of the coupling constant υ_0, so that a perturbation treatment could not have given this result – unless an infinite number of selected diagrams are summed.[36]

The gap is responsible for the absence of dissipation in current flow. The superconducting state can be destroyed, however, by a magnetic field **H** of magnitude

$$\tfrac{1}{2}H^2 \geq W_N - W_S \text{ , or } H_{crit} = \Delta\sqrt{\rho(0)} . \qquad (12.6\text{-}21)$$

Using experimental values for Δ and $\rho(0)$, this relationship is reasonably well satisfied.

[36] L.P. Kadanoff and P.C. Martin, Phys. Rev. **124**, 670 (1961).

12.6.2 *Finite Temperature Results*

Away from the ground state there will be quasi-particle excitations. Thus, an electron, say $\bar{\mathbf{k}}\uparrow$, from the pair $(\bar{\mathbf{k}}\uparrow,-\bar{\mathbf{k}}\downarrow)$ may be excited to join the normal electron pool, leaving a hole $-\bar{\mathbf{k}}\downarrow$ behind; these excitations sometimes are denoted as 'bogolons'. The energy of the condensate decreases, its change being [cf.(12.6-9)],

$$\Delta W^{(1)} = -2\hat{\varepsilon}_{\bar{\mathbf{k}}} v_{\bar{\mathbf{k}}}^2 - 2\sum_{\mathbf{k}'} V_{\bar{\mathbf{k}}\mathbf{k}'} u_{\mathbf{k}'} v_{\mathbf{k}'} u_{\bar{\mathbf{k}}} v_{\bar{\mathbf{k}}} \,. \qquad (12.6\text{-}22)$$

To this we must add the gain for the normal electron pool, $\Delta W^{(2)} = \hat{\varepsilon}_{\bar{\mathbf{k}}}$. Employing (12.6-15) and (12.6-14) with $\mathbf{k}\to\mathbf{k}'$, this yields

$$\Delta W = \hat{\varepsilon}_{\bar{\mathbf{k}}}(1-2v_{\bar{\mathbf{k}}}^2) + 2\Delta_{\bar{\mathbf{k}}} u_{\bar{\mathbf{k}}} v_{\bar{\mathbf{k}}} \,. \qquad (12.6\text{-}22')$$

We substitute for $v_{\bar{\mathbf{k}}}^2$ from (12.6-13), use (12.6-14) for $\mathbf{k}\to\bar{\mathbf{k}}$ and (12.6-16) to obtain

$$W_{\bar{\mathbf{k}}\uparrow} - W_0 = \frac{\hat{\varepsilon}_{\bar{\mathbf{k}}}^2}{\mathcal{E}_{\bar{\mathbf{k}}}} + \frac{\Delta_{\bar{\mathbf{k}}}^2}{\mathcal{E}_{\bar{\mathbf{k}}}} = \mathcal{E}_{\bar{\mathbf{k}}} \,, \qquad (12.6\text{-}23)$$

$\mathcal{E}_{\mathbf{k}}$ being the energy of the quasi-particles. Its minimum value occurs for $\hat{\varepsilon}_{\mathbf{k}} = \varepsilon_F = 0$, being just Δ_F, the gap at the Fermi level, which in the usual weak coupling limit is of order $10^{-4}\times(\varepsilon_F - \varepsilon_C)$, where ε_C is the bottom of the conduction band. This gives us a rough estimate of the critical temperature; with $k_B T_c \sim \Delta_F$ we arrive at temperatures of $10K$ or less.

We must proceed to a more serious computation. We shall work in the grand-canonical ensemble, which allows for fluctuations of energy and particle number. The quantities $u_{\mathbf{k}}$ and $v_{\mathbf{k}}$, still satisfying (12.6-13) and (12.6-14) for fixed electrochemical potential – remember that μ is actually the potential of the reservoir μ^r – are subject to fluctuations. We must therefore take an average involving the grand-canonical density operator; i.e., for each operator Ω we have

$$\langle\Omega\rangle_{th} = \mathrm{Tr}\{\Omega\exp(-\beta\mathcal{H}_{grand})\}/\mathrm{Tr}\exp(-\beta\mathcal{H}_{grand}) \,, \qquad (12.6\text{-}24)$$

where β is as usual $1/k_B T$ and where the subscript "*th*" means that this is a 'thermal' average and not an expectation value in a pure state, as was the case for the ground-state considerations; we shall, however drop this subscript, the implications being understood from the context. The gap equation, see (12.6-15) and (12.6-14), allows us now to define a temperature-dependent gap by analogy,

$$\langle\Delta_{\mathbf{k}}(T)\rangle = -\sum_{\mathbf{k}'} V_{\mathbf{k}\mathbf{k}'}\langle u_{\mathbf{k}'} v_{\mathbf{k}'}\rangle = -\sum_{\mathbf{k}'} V_{\mathbf{k}\mathbf{k}'}\langle b_{\mathbf{k}'}^{(\dagger)}\rangle. \qquad (12.6\text{-}25)$$

The last rhs follows from the temperature analogue of $\langle b_{\mathbf{k}}^{(\dagger)}\rangle_{\Psi_0} = g_{\mathbf{k}}/(1+g_{\mathbf{k}}^2) = u_{\mathbf{k}} v_{\mathbf{k}}$.

We now take a look at the reduced grand Hamiltonian, which with the shift in energy scale reads:

$$H_{grand,red.} = \sum_{|\mathbf{k}|>k_F} 2\hat{\varepsilon}_{\mathbf{k}} b^\dagger_{\mathbf{k}} b_{\mathbf{k}} + \sum_{\mathbf{k},\mathbf{k}'} V_{\mathbf{kk}'} b^\dagger_{\mathbf{k}} b_{\mathbf{k}'}, \tag{12.6-26}$$

where for simplicity we omitted the energies below k_F. Both parts will be rewritten in terms of the original fermion operators; for the first term we have

$$\sum_{\mathbf{k},\sigma} \hat{\varepsilon}_{\mathbf{k}} c^\dagger_{\mathbf{k},\sigma} c_{\mathbf{k},\sigma} = \sum_{\mathbf{k}} \hat{\varepsilon}_{\mathbf{k}} (c^\dagger_{\mathbf{k},1/2} c_{\mathbf{k},1/2} + c^\dagger_{\mathbf{k},-1/2} c_{\mathbf{k},-1/2}) = \sum_{\mathbf{k}} \hat{\varepsilon}_{\mathbf{k}} (c^\dagger_{\mathbf{k},1/2} c_{\mathbf{k},1/2} + c^\dagger_{-\mathbf{k},-1/2} c_{-\mathbf{k},-1/2}).$$

We add the second part of (12.6-26), expressing the b's into c's and we symmetrize the double sum by interchanging $\mathbf{k}$ and $\mathbf{k}'$ and taking half the sum. Thus we obtain

$$\begin{aligned}\mathcal{H}_{grand,red.} &= \sum_{\mathbf{k}} \hat{\varepsilon}_{\mathbf{k}} \left(c^\dagger_{\mathbf{k},1/2} c_{\mathbf{k},1/2} + c^\dagger_{-\mathbf{k},-1/2} c_{-\mathbf{k},-1/2} \right) \\ &+ \frac{1}{2} \sum_{\mathbf{k},\mathbf{k}'} V_{\mathbf{kk}'} \left\{ \left(c^\dagger_{\mathbf{k},1/2} c^\dagger_{-\mathbf{k},-1/2} c_{-\mathbf{k}',-1/2} c_{\mathbf{k}',1/2} + c^\dagger_{\mathbf{k}',1/2} c^\dagger_{-\mathbf{k}',-1/2} c_{-\mathbf{k},-1/2} c_{\mathbf{k},1/2} \right) \right\}. \end{aligned} \tag{12.6-27}$$

In order to exponentiate the grand Hamiltonian, as required for the grand partition function, the expression should first be diagonalized. This cannot readily be done, however, since we have a quartic combination of the c's, which, by the way, is the Fourier transform of the pair-correlation function.[37] There are two ways out. First, we can pass on to a thermodynamic description, which yields the grand potential; this is the route followed in the original BCS article (they call it the free energy). Secondly, we can seek to convert the double sum to a near-equivalent bilinear form and employ a Bogoliubov–Valatin transformation to diagonalize the remaining bilinear expression. To attain this objective, we must transfer to the Heisenberg picture with $b_{\mathbf{k}}(t) = U^\dagger(t,0) b_{\mathbf{k}} U(t,0)$ and, following Schrieffer, linearize the equations of motion, although, in contrast to Schrieffer's procedure[38], we have no explicit need for the commutators $[b_k^{(\dagger)}, \mathcal{H}_{gr\ red}]$ occurring in the Heisenberg equations. Let us take one of the terms of the double sum and write it in the pair-operator form. We then have for the derivative:

$$\begin{aligned}(d/dt)\Lambda_{\mathbf{kk}'}(t) &\equiv \tfrac{1}{2}(d/dt)\left\{ V_{\mathbf{kk}'} [b^\dagger_{\mathbf{k}}(t) b_{\mathbf{k}'}(t) + b^\dagger_{\mathbf{k}'}(t) b_{\mathbf{k}}(t)] \right\} \\ &= \tfrac{1}{2} V_{\mathbf{kk}'} \left\{ \left[b^\dagger_{\mathbf{k}}(t) \dot{b}_{\mathbf{k}'} + \dot{b}^\dagger_{\mathbf{k}} b_{\mathbf{k}'}(t) \right] + \underbrace{\left[b^\dagger_{\mathbf{k}'}(t) \dot{b}_{\mathbf{k}} + \dot{b}^\dagger_{\mathbf{k}'} b_{\mathbf{k}}(t) \right]}_{\mathbf{k}\leftrightarrow\mathbf{k}'} \right\} \\ &= V_{\mathbf{kk}'} \left[b^\dagger_{\mathbf{k}}(t) \dot{b}_{\mathbf{k}'} + \dot{b}^\dagger_{\mathbf{k}} b_{\mathbf{k}'}(t) \right] \approx V_{\mathbf{kk}'} \left\{ \langle b^\dagger_{\mathbf{k}} \rangle \dot{b}_{\mathbf{k}'} + \dot{b}^\dagger_{\mathbf{k}} \langle b_{\mathbf{k}'} \rangle \right\}. \end{aligned} \tag{12.6-28}$$

In the second step we presumed the $\Sigma_{\mathbf{kk}'}$ to be applied; in the last step we linearized

---

[37] J.G. Valatin, Op. Cit, Eq. (20a): $\rho_{\sigma\sigma'} = \langle \Psi^\dagger_\sigma(\mathbf{r}) \Psi^\dagger_{\sigma'}(\mathbf{r}') \Psi_{\sigma'}(\mathbf{r}') \Psi_\sigma(\mathbf{r}) \rangle$, with $\Psi^{(\dagger)}_\sigma(\mathbf{r}) = \Sigma_{\mathbf{k}} \psi^{(*)}_{\mathbf{k}}(\mathbf{r}) c^{(\dagger)}_{\mathbf{k}\sigma}$, where the $\psi_{\mathbf{k}}$ are taken to be plane waves.

[38] For the linearization of the Heisenberg equations of motion, Schrieffer obtains the eigenvalues of the commutators $[b^{(\dagger)}_{\mathbf{k}}, \mathcal{H}_{red}]$ and proceeds accordingly; cf., J.R. Schrieffer, Op. Cit p. 49.

the motion. Since the averages are numbers independent of time, we can integrate to obtain [39]

$$\Lambda_{\mathbf{kk}'}(t) = V_{\mathbf{kk}'}\left\{\langle b_{\mathbf{k}}^{\dagger}\rangle b_{\mathbf{k}'}(t) + b_{\mathbf{k}}^{\dagger}(t)\langle b_{\mathbf{k}'}\rangle\right\}. \qquad (12.6\text{-}28')$$

The unitary transformation $U(t,0)\Lambda_{\mathbf{kk}'}(t)U^{\dagger}(t,0)$ brings us back to the Schrödinger picture. The linearization of the Heisenberg expression has given us a factor two for the interaction sum.[40] Returning to the c-operators, we thus established:

$$\begin{aligned}\mathcal{H}_{grand,red.} &= \sum_{\mathbf{k}} \hat{\varepsilon}_{\mathbf{k}} \left(c_{\mathbf{k},\frac{1}{2}}^{\dagger} c_{\mathbf{k},\frac{1}{2}} + c_{-\mathbf{k},-\frac{1}{2}}^{\dagger} c_{-\mathbf{k},-\frac{1}{2}}\right) \\ &+ \sum_{\mathbf{k},\mathbf{k}'} V_{\mathbf{kk}'} \left\{\left(c_{\mathbf{k},\frac{1}{2}}^{\dagger} c_{-\mathbf{k},-\frac{1}{2}}^{\dagger} \langle b_{\mathbf{k}'}\rangle + \langle b_{\mathbf{k}'}^{\dagger}\rangle c_{-\mathbf{k},-\frac{1}{2}} c_{\mathbf{k},\frac{1}{2}}\right)\right\} \\ &= \sum_{\mathbf{k}} \left\{\hat{\varepsilon}_{\mathbf{k}} \left(c_{\mathbf{k},\frac{1}{2}}^{\dagger} c_{\mathbf{k},\frac{1}{2}} + c_{-\mathbf{k},-\frac{1}{2}}^{\dagger} c_{-\mathbf{k},-\frac{1}{2}}\right) - \Delta_k \left(c_{\mathbf{k},\frac{1}{2}}^{\dagger} c_{-\mathbf{k},-\frac{1}{2}}^{\dagger} + c_{-\mathbf{k},-\frac{1}{2}} c_{\mathbf{k},\frac{1}{2}}\right)\right\}, \qquad (12.6\text{-}29)\end{aligned}$$

where we used (12.6-25). We note that, as always, the nondiagonal terms are Hermitean conjugates.

The Bogoliubov–Valatin transformation, a rotation in **k**-space, was already stated in the 2D Ising problem, see Eqs. (11.2-60) and (11.2-61); we repeat the latter equations here. Let η_k and $\eta_k^{\dagger}$ be new fermion operators; we use $k = (\mathbf{k}, \sigma)$. We have:

$$\begin{aligned} c_k &= \cos\phi_k\, \eta_k - \sin\phi_k\, \eta_{-k}^{\dagger}, \quad c_k^{\dagger} = h.c. \,, \\ c_{-k} &= \cos\phi_k\, \eta_{-k} + \sin\phi_k\, \eta_k^{\dagger}, \quad c_{-k}^{\dagger} = h.c. \,. \end{aligned} \qquad (12.6\text{-}30)$$

Note that $\operatorname{sgn}\phi_k = \operatorname{sgn} k$. Substituting into (12.6-29) one readily obtains the diagonal and nondiagonal parts:

$$\mathcal{H}_{gr\text{-}red,d} = \sum_{\mathbf{k}} \{(\overline{\mathbf{n}}_k + \overline{\mathbf{n}}_{-k} - 1)(\hat{\varepsilon}_{\mathbf{k}} \cos 2\phi_k - \Delta_{\mathbf{k}} \sin 2\phi_k) + \hat{\varepsilon}_{\mathbf{k}}\}, \qquad (12.6\text{-}31)$$

$$\mathcal{H}_{gr\text{-}red,nd} = -\tfrac{1}{2}\sum_{\mathbf{k}} \left\{(\eta_k^{\dagger}\eta_{-k}^{\dagger} + \eta_{-k}\eta_k)(\hat{\varepsilon}_{\mathbf{k}} \sin 2\phi_k + \Delta_{\mathbf{k}} \cos 2\phi_k)\right\}. \qquad (12.6\text{-}32)$$

The nondiagonal part is zero for

---

[39] This result can also be established by writing the interaction Hamiltonian as a time-ordered perturbation series analogous to (12.3-30), so that $\mathcal{Z}/\mathcal{Z}_0 = \mathcal{T}\{\exp[-\int_0^{\beta} \mathcal{H}_I(-i\hbar\tau)d\tau]\}$, where $\mathcal{T}$ is the time-ordering operator. Evaluation of the path integral then leads to the above result with the averages representing the Cooper pair field; see the article on high T_c superconductivity in YBCO by A. Widom, T. Yuan, H. Jiang and C. Vittoria, Physica **A 210**, 496 (1994).

[40] The problem discussed here is also found with other authors. In Reichl's treatment the factor two simply 'drops out of the blue' (2nd Ed. Wiley 1998, p.408); Plischke and Bergersen admit to 'over-counting' and deduct the double average $V_{\mathbf{kk}'}\langle b_{\mathbf{k}}^{\dagger}\rangle\langle b_{\mathbf{k}}\rangle$ (2nd Ed. World Scient. Publ. Co, 1994, p. 366), appealing to an *ad hoc* mean-field argument.

$$\tan 2\phi_k = -\Delta_{\mathbf{k}} / \hat{\varepsilon}_{\mathbf{k}} . \tag{12.6-33}$$

Since $\Delta_{\mathbf{k}} / \hat{\varepsilon}_{\mathbf{k}} \ll 1$, the angle $2\phi_k$ must lay in the fourth quadrant; so $\cos 2\phi_k = \hat{\varepsilon}_{\mathbf{k}} / \mathcal{E}_{\mathbf{k}}$ and $\sin 2\phi_k = -\Delta_{\mathbf{k}} / \mathcal{E}_{\mathbf{k}}$. The resulting reduced grand Hamiltonian becomes

$$\mathcal{H}_{grand,red.} = \sum_{\mathbf{k}} \left\{ \mathcal{E}_{\mathbf{k}} (\bar{\mathbf{n}}_{\mathbf{k},\frac{1}{2}} + \bar{\mathbf{n}}_{-\mathbf{k},-\frac{1}{2}} - 1) + \hat{\varepsilon}_{\mathbf{k}} \right\} = \sum_{\mathbf{k},\sigma} \left\{ \mathcal{E}_{\mathbf{k}} \bar{\mathbf{n}}_{k,\sigma} + \tfrac{1}{2} (\hat{\varepsilon}_{\mathbf{k}} - \mathcal{E}_{\mathbf{k}}) \right\}. \tag{12.6-34}$$

We note that for $\Delta_{\mathbf{k}} = 0$ this reduces to $\Sigma_{\mathbf{k},\sigma} \hat{\varepsilon}_{\mathbf{k}} \bar{\mathbf{n}}_{\mathbf{k},\sigma}$, which is the first term of the reduced grand Hamiltonian. For $\Delta_{\mathbf{k}} \neq 0$, we have

$$\mathcal{H}_{grand,red.} = \sum_{\mathbf{k},\sigma} \mathcal{E}_{\mathbf{k}} \bar{\mathbf{n}}_{\mathbf{k},\sigma} + const = 2 \sum_{\mathbf{k}} \mathcal{E}_{\mathbf{k}} \bar{\mathbf{n}}_{\mathbf{k},\sigma} + const . \tag{12.6-35}$$

This result is completely consistent with our definition of a quasi-particle, being an excitation of an electron from a singlet pair, leading to a state of either spin, $\mathbf{k}\uparrow$ or $-\mathbf{k}\downarrow$. The constant is insignificant, since it will cancel out in expressions like (12.6-24), as well as in the many expressions where we deal with derivatives of the logarithm of the grand partition function, $\ln \mathcal{Z}$. We conclude that $\mathcal{H}_{grand,red.}$ represents the shifted energies $\mathcal{E}_{\mathbf{k}} - \mu$ of excitations of the condensate, the quasi-particles being independent as in a Hartree-Fock treatment. We can therefore formally apply the thermodynamic results for a perfect Fermi gas, obtained in section 8.4, with the replacement $\hat{\varepsilon}_{\mathbf{k}} \to \mathcal{E}_{\mathbf{k}}$.

We start with the computation of the gap $\Delta_{\mathbf{k}}(T)$ from the self-consistent gap equation. Taking now the remaining average for $\bar{\mathbf{n}}_{\mathbf{k},\sigma} = \eta^\dagger_{\mathbf{k},\sigma} \eta_{\mathbf{k},\sigma}$, we obtain the Fermi–Dirac result

$$f_{\mathbf{k}} \equiv \langle \bar{\mathbf{n}}_{\mathbf{k},\sigma} \rangle = 1/[\exp(\beta \mathcal{E}_{\mathbf{k}}) + 1] , \tag{12.6-36}$$

where the overhead bar is further omitted. We express $b_{\mathbf{k}}$ (or $b^\dagger_{\mathbf{k}}$) in terms of the new fermion operators, dropping the nondiagonal part, which has zero trace. Hence,

$$\begin{aligned} \langle b_{\mathbf{k}} \rangle = \langle b^\dagger_{\mathbf{k}} \rangle &= \tfrac{1}{2} \langle \mathbf{n}_k + \mathbf{n}_{-k} - 1 \rangle \sin 2\phi_k \\ &= (\Delta_{\mathbf{k}} / \mathcal{E}_{\mathbf{k}}) \left(\tfrac{1}{2} - \langle \mathbf{n}_k \rangle \right) = (\Delta_{\mathbf{k}} / 2\mathcal{E}_{\mathbf{k}}) \tanh(\tfrac{1}{2} \beta \mathcal{E}_{\mathbf{k}}) . \end{aligned} \tag{12.6-37}$$

This is substituted into (12.6-25), yielding

$$\Delta_{\mathbf{k}}(T) = -\sum_{\mathbf{k}'} V_{\mathbf{k}\mathbf{k}'} \frac{\Delta_{\mathbf{k}'}(T)}{2\mathcal{E}_{\mathbf{k}'}} \tanh\left(\frac{\beta \mathcal{E}_{\mathbf{k}'}}{2} \right). \tag{12.6-38}$$

Next, we make the assumption (12.6-17). Moreover, approximating the sum by an integral as we did previously, we obtain

$$1 \approx v_0 \rho(0) \int_{-\hbar\omega_D}^{\hbar\omega_D} d\varepsilon \frac{\tanh[\tfrac{1}{2} \beta \sqrt{\varepsilon^2 + \Delta_0^2(T)}]}{2\sqrt{\varepsilon^2 + \Delta_0^2(T)}} . \tag{12.6-39}$$

This is the generalization of (12.6-18) for the ground state. The critical temperature, at which the gap – and thus the superconducting state – disappears is subject to the integral equation

$$\frac{1}{\upsilon_0\rho(0)} = \int_0^{\hbar\omega_D} \frac{d\varepsilon}{\varepsilon}\tanh\left(\frac{\varepsilon}{2k_B T_c}\right). \tag{12.6-40}$$

The integral has been computed in the weak coupling limit; the result yields

$$k_B T_c = 1.14\hbar\omega_D \exp\left[-\frac{1}{\upsilon_0\rho(0)}\right]. \tag{12.6-41}$$

The gap which is experimentally measured is $2\Delta_0$, since it takes an energy Δ_0 to create each quasi-particle in a one-electron transition.[41] Comparing (12.6-19) and (12.6-41), we find

$$2\Delta_0(T=0)/k_B T_c = 3.52. \tag{12.6-42}$$

This ratio is in good agreement with experiment for most metals, an exception being lead and mercury, in which temperature-dependent damping effects play a role.[42] From (12.6-39) the temperature dependence of the gap can be found. The approximate result is

$$\Delta_0(T)/\Delta_0(0) = 1.74[1-(T/T_c)]^{1/2}. \tag{12.6-43}$$

It would carry us too far to enter into the details of the thermodynamic properties that follow from the BCS theory; we refer to a survey article by Bardeen and Schrieffer.[43] The grand-canonical partition function is clearly as in (4.3-18)

$$\ln\mathcal{Z} = \sum_{\mathbf{k}} \ln\left[1+e^{-\beta\mathcal{E}_{\mathbf{k}}}\right], \tag{12.6-44}$$

but we cannot convert to Fermi integrals, since $\mathcal{E}_{\mathbf{k}}$ is not a simple function of ε, which occurs in the density of states. The average energy can be written down, however, immediately:

$$\langle W\rangle = 2\sum_{\mathbf{k}}\langle n_{\mathbf{k}}\rangle\mathcal{E}_{\mathbf{k}} = 2\sum_{\mathbf{k}} f_{\mathbf{k}}\mathcal{E}_{\mathbf{k}}. \tag{12.6-45}$$

The electronic specific heat is found by differentiating both $f_{\mathbf{k}}$ and $\mathcal{E}_{\mathbf{k}}$ to T. The following result, neglecting a small term, is easily verified,

---

[41] J.R. Schrieffer, Op. Cit., p.45, Fig. 2-2; G.D. Mahan, Op. Cit., p.439.

[42] H. Suhl, Bull. Am, Phys, Soc., **6** , 119 (1961); Y. Wada, Rev. Mod. Physics **36**, 253 (1964).

[43] J. Bardeen and J.R. Schrieffer, "Progress in Low Temperature Physics" Vol. III, North Holland Publ. Co. Amsterdam and NY 1961.

$$C_{el} = 2k_B \beta^2 \sum_{\mathbf{k}} f_{\mathbf{k}}(1-f_{\mathbf{k}}) \left[\mathscr{E}_{\mathbf{k}}^2 + \frac{\beta}{2} \frac{d\Delta_{\mathbf{k}}^2}{d\beta} \right]. \tag{12.6-46}$$

It can be shown that very close to T_c $\Delta_{\mathbf{k}}^2 \propto (T_c - T)$. Note $\beta(d/d\beta) = -T(d/dT)$, so that there is a contribution from the last term, while the other part smoothly approaches that of the normal state. There is thus a jump in specific heat or, in Ehrenfest's terminology [Chapter IV, ref. 43], we have a second order phase transition. A sketch of $C_{el}(T)$ is given in Fig. 12-10.

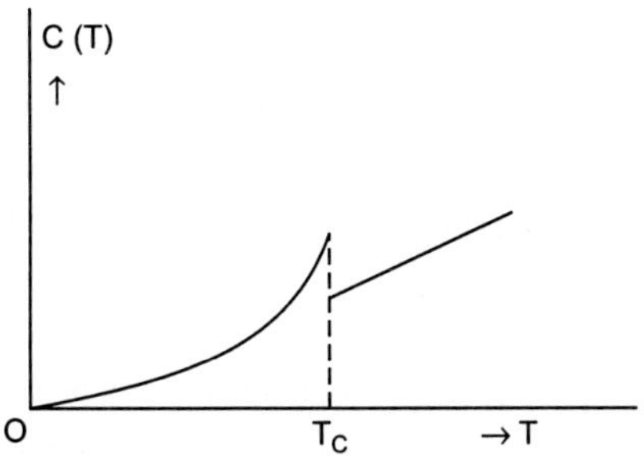

Fig. 12-10. Electronic specific heat vs. *T*.

12.7 Excitations in Fermi liquids; ^{3}He

12.7.1 *Original Fermi Liquid Theory (Landau) and Some Empirical Data*

Natural helium consists of two isotopes, ^{4}He and ^{3}He. The latter is a Fermi gas, with an abundance of only 1 part in a million. It is usually obtained through β-decay of ^{3}H, or tritium. It was first liquefied in 1948 by Sydoriack et al.[44] Since its availability in large enough quantities was assured, it has been abundantly studied, both theoretically and experimentally. From ideal Fermi gas theory, cf. Eq. (4.8-12), with the mass and density now referring to ^{3}He as given in Table 12-1 below, its Fermi energy $\varepsilon_F = \varsigma$ [the latter denoting the chemical potential of the pseudo-particles] is found to be $4.29 \times 10^{-4} eV$ or $4.97K$. The actual degeneracy energy is about two-thirds of this value. Fermi liquid theory, dealing with excitations in the liquid small compared to ε_F, should not be used above around 0.1-0.2*K*; it is then useful for two decades of lower temperatures, until the onset of the superfluid phases commences in the milli-*Kelvin* range.

The original Fermi-liquid theory was proposed by Landau in 1956; in fact, the essential elements can already be found in Landau and Lifshitz.[45] Whereas so far we

---

[44] S.G. Sydoriack, E.R. Grilly and E.F. Hammel, Phys Rev. **75**, 303 (1949).

[45] L. Landau and E.M. Lifshitz, Op. Cit. § 68. See also L. Landau, Soviet Physics JETP **3**, 920 (1956).

have only focussed on the electron gas in condensed matter to discuss Fermi systems, ^{3}He is a far more prominent example, since here we deal with interacting fermions, without the influence of a lattice and the Coulomb interactions, which we studied in Sections 7.8 and 8.4. Clearly, at the temperatures envisaged, the atoms form part of a liquid of strongly coupled constituents, and we cannot expect that the usual Fermi-Dirac distribution would directly apply. Let $n_{\mathbf{p}}$ denote the average distribution over momentum states ($\mathbf{p} = \hbar\mathbf{k}$), where – as for Bose quantum fluids – $\mathbf{p}$ is a true momentum and not a quasi-momentum as in a solid lattice; for convenience the usual averaging brackets $\langle..\rangle$ will be omitted in this section. The distribution $n_{\mathbf{p}}$ can be inferred from neutron-scattering experiments. In Fig. 12-11 we have reproduced the momentum distribution for ^{3}He at 0.37*K* as obtained by Mook.[46] The results can be fitted by a Fermi function, but with a temperature $T_{eff} = 1.8K$, which signifies that the distribution is grossly 'smeared out' compared to the F-D distribution in gases, as expected for strongly repulsive particles.

Table 12-1 Some properties of liquid helium; pressure 1 *bar*.

parameter	^{4}He	^{3}He
m	6.65×10^{-27} kg	5.01×10^{-27} kg
n	2.18×10^{28} atoms/m^3	1.64×10^{28} atoms/m^3
$u_1(1K)$	238 m/s	182 m/s
$\mathcal{E}/N$	$-7.20K$/atom	$-2.52K$/atom

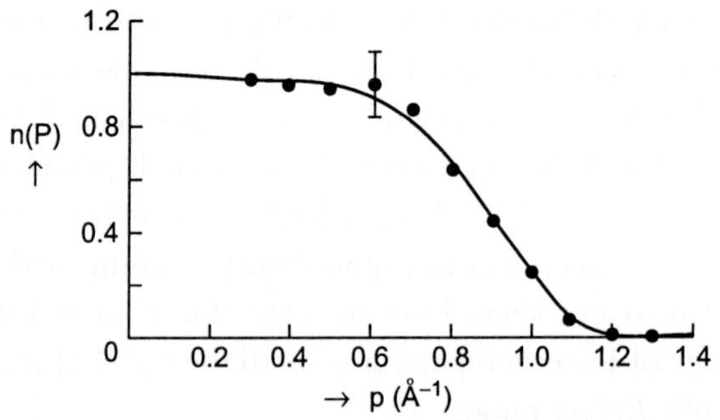

Fig. 12-11. Momentum distribution deduced from neutron scattering on ^{3}He at 0.37K. The solid line is a least-squares fit to the Fermi distribution with an effective temperature of 1.8K. After Mook.[46] [With permission.]

[46] H.A. Mook, Phys. Rev. Lett. **55**, 2452 (1985).

Due to the exclusion principle and its concomitant need for antisymmetrized wave functions, the theory of Fermi liquids is in many respects more complicated than the Bose-type systems we considered previously. There are a number of readable surveys from which the student may profit. We mention the older treatment of Pines and Nozières (1966[47]), modern reviews by Baym and Pethick (1978[48]) or Leggett (1975[49]) and a useful section in Mahan's book (2000[50]). Detailed considerations need a diagrammatic treatment, as discussed in later sections. Here we just mention that the imaginary part of the self-energy, $\mathrm{Im}\,\Sigma$, can be shown to be small, so that the spectral function $A(\mathbf{p},\omega)$ is peaked and resembles a delta function[51]; we can thus for given $\hbar\omega$ associate a momentum $\mathbf{p}$ with the constituents, which we shall call *quasi-particles*. Basically, we may think of these as 'dressed' ^{3}He atoms, thereby acquiring an effective mass m^* (Leggett[49]). Landau[45] describes this perhaps more succinctly: "a quasi-particle in the Fermi-type spectrum can, in a sense, be regarded as an atom in the self-consistent field of the surrounding atoms." Experimental data suggest that $m^* = 2.76m$, where m is the bare mass of the atom. We can also surmise this as follows. The energy $\varepsilon_\mathbf{p}$ should be zero for $\mathbf{p} = 0$. A formal expansion about $\mathbf{p} = 0$ will not have any odd terms – in contrast to a Bose liquid – because of particle-hole symmetry; the first important term is then $\varepsilon_\mathbf{p} \propto p^2$, which defines the effective mass with $\varepsilon_\mathbf{p} = p^2/2m^*$. The distribution $n_\mathbf{p}$ will at all times refer to these quasi-particles. In accordance with Landau's viewpoint we shall also introduce a self-consistent energy $\tilde{\varepsilon}_\mathbf{p}$ which is more than $\varepsilon_\mathbf{p}$, see below.

Landau's theory, as for the case of ^{4}He, was guided by experimental results on the specific heat and the compressibility, required for the velocity of sound; it is therefore to be regarded as a phenomenological theory. The following basic assumptions will be made and are upheld by comparison of the theory with the experimental data. First of all, it must be assumed that the interactions can in some sense be 'adiabatically turned on', so that the number of quasi-particles remains the same as the number of separate atoms in the gaseous state. Secondly and as a consequence of this supposition, the quasi-particles or excitations[51a] obey Fermi-Dirac statistics, *with some qualification.* Fermi liquid theory aims primarily at describing low-lying excited states which naturally involve particle-hole pairs; these excitations are fermions. In addition, there are two other types of excitations. Density fluctuations or zero-sound waves can be viewed as collective resonances of

---

[47] D. Pines and P. Nozières, "The Theory of Quantum Liquids", Benjamin 1966.

[48] G. Baym and C. Pethick in "The Physics of Liquid and Solid Helium," Part 2 (J.B. Ketterson and K. Benneman, Eds.), J. Wiley, N.Y. 1971, Chapter 1.

[49] A.J. Leggett, Rev. Modern Physics **47**, 331-414 (1975).

[50] G.D. Mahan, Op. Cit., 3rd Ed., Kluwer Acad./Plenum, 2000, pp.713-742.

[51] W. Jones and N.H. March, "Theoretical Solid State Physics", Wiley & Sons, London, p. 140 ff.

[51a] Following Landau, the denotations quasi-particles and excitations will be taken to be synonymous.

the primary particle-hole fluid and are carried by phonons.[51b] Further there are spin waves carried by paramagnons; they represent spin fluctuations involving pairs of opposite spin. They are inadequately accounted for by the original Fermi liquid theory, but a number of modern approaches, using quantum-field Hamiltonians of the types we studied before, have been developed. And, last but not least, we must mention that at very low temperatures pairing can take place, resulting in a boson-like superfluid. Contrary to Cooper pairs in BCS theory, these pairs usually have the triplet state, with $S = 1$.[52] Superfluidity in ^{3}He was discovered in 1972 by Osheroff, Richardson and Lee.[53] They identified two phases in the liquid, designated *A* and *B*; a Nobel prize followed. Curiously, the theory for this phenomenon did exist already, since Balian and Werthamer[54] developed the triplet pairing theory as an alternative for the BSC theory in 1963, in order to (possibly) explain some anomalous results in superconducting materials like Sn, Hg and others. All these developments will be briefly discussed in the next few sections.

We now return to Fermi liquid theory, which is our basic quest in this section. The ground state is a 'quasi-vacuum' – as for an electron gas – and will be designated as $|O\rangle$, its energy being $\mathscr{E}_0$. At $T=0$ the state is mainly filled up to the Fermi radius p_F, although there is some fuzziness for the distribution of quasi-particles in systems with interactions, cf. Ref. 51, loc. cit. Fig. 2.5; the distribution will be denoted by $n_{\mathbf{p}}^0$. Clearly, this is a fictitious concept, except near the Fermi radius. The ground-state energy will be written as $\mathscr{E}_0 = const + \sum_{\mathbf{p},\sigma} n_{\mathbf{p}}^0 \varepsilon_{\mathbf{p}}$, where σ denotes the spin; the 'constant' needs no discussion. At nonzero temperatures the exclusion principle forces many excitations in states outside p_F; we shall denote their occupancy by $n_{\mathbf{p}}$. We will assume that the average occupancy (Heaviside-like behaviour or step-down function) can still be represented by the ordinary F–D distribution,

$$n_{\mathbf{p}}^0 = n_F(\varepsilon_{\mathbf{p}} - \varsigma) \equiv \lim_{T\to 0}\left\{1/[e^{\beta(\varepsilon_{\mathbf{p}}-\varsigma)} + 1]\right\}; \tag{12.7-1}$$

note that the Fermi-Dirac function will be denoted by the middle member for the present considerations. More important is the 'difference distribution'; setting $\delta n_{\mathbf{p}} = n_{\mathbf{p}} - n_{\mathbf{p}}^0$, we have for the energy with respect to the ground state

$$\mathscr{E} = \mathscr{E}_0 + \sum_{\mathbf{p},\sigma} \varepsilon_{\mathbf{p}} \delta n_{\mathbf{p}} = \mathscr{E}_0 + \int \Delta_{p\sigma}^3 \varepsilon_{\mathbf{p}} \delta n_{\mathbf{p}}\,. \tag{12.7-2}$$

where $\Delta_{p\sigma}^3$ is a shortcut notation for integration over p-space including the density of states $1/8\pi^3\hbar^3$ [we set V_0 = unity] and summation over the spin $\sigma = \pm 1$.

It must now be born in mind that the total number of quasi-particles in the ground

---

[51b] The sound-wave phonons in ^{3}He are the equivalent of plasmons in an electron gas, cf. Chapter XVI.

[52] It is customary to use the symbols S, s and σ (instead of I) for the nuclear spin of the quasi-particles.

[53] D.D. Osheroff, R.C. Richardson and D.M. Lee, Phys. Rev. Lett. **28**, 885 (1972).

[54] R. Balian and N.R. Werthamer, Phys. Rev. **131**, 1553 (1963).

state, N_0, is not a constant of motion, nor is the number N above the ground state. Due to their fluctuations we need an ensemble with specified chemical potential, such as we used in previous sections. Fortunately, for the Fermi gas ς is hardly affected by $\Sigma\delta n_{\mathbf{p}}$ and we can use the ground state value. Instead of $\mathscr{E}$ as in (12.7-2) we should consider the grand-ensemble Legendre transform

$$\mathscr{E} - \varsigma N \equiv \hat{\mathscr{E}}, \quad N = N_0 + \Sigma\delta n_{\mathbf{p}}, \tag{12.7-3}$$

where we shall call $\hat{\mathscr{E}}$ the grand energy. Denoting $\mathscr{E}_0 - \varsigma N_0$ by $\hat{\mathscr{E}}_0$, trivial algebra shows

$$\hat{\mathscr{E}} = \hat{\mathscr{E}}_0 + \int \Delta^3_{p\sigma}(\varepsilon_{\mathbf{p}} - \varsigma)\,\delta n_{\mathbf{p}}. \tag{12.7-4}$$

To be noted here is that $|\mathbf{p}|$ is always very near the Fermi value p_F, so that the integrand is actually of second order smallness, i.e., $\propto (\delta n_{\mathbf{p}})^2$.

Landau now recognised that, after all, Eq. (12.7-4) cannot be the full story, since in the above each quasi-particle contributes independently to the grand energy. Therefore, he added a binary interaction term, which redefines the character of each excitation. Let $\varphi_{\mathbf{p}\sigma,\mathbf{p}\sigma'}$ be a binary interaction energy; the grand energy up to terms $\mathscr{O}(\delta n_{\mathbf{p}})^3$ should then be modified from (12.7-4) to read:

$$\hat{\mathscr{E}} = \hat{\mathscr{E}}_0 + \int \Delta^3_{p\sigma}(\varepsilon_{\mathbf{p}} - \varsigma)\delta n_{\mathbf{p}\sigma} + \frac{1}{2}\iint \Delta^3_{p\sigma}\Delta^3_{p'\sigma'}\varphi_{\mathbf{p}\sigma,\mathbf{p}'\sigma'}\delta n_{\mathbf{p}\sigma}\delta n_{\mathbf{p}'\sigma'}. \tag{12.7-5}$$

The interaction terms are spin-dependent, but they do not represent ordinary dipole-dipole coupling, which is negligibly small. Rather, the exchange hole around each quasi-particle causes an exchange energy which is spin-dependent analogous to the spin-spin coupling in a Heisenberg Hamiltonian. A full discussion is found in Leggett, Op. Cit. Generally, $\varphi_{\mathbf{p}\sigma,\mathbf{p}'\sigma'}$ contains a spin-symmetric and asymmetric contribution and we have

$$\varphi_{\mathbf{p}\sigma,\mathbf{p}'\sigma'} = \varphi^s_{\mathbf{pp}'} + (\boldsymbol{\sigma}\cdot\boldsymbol{\sigma}')\varphi^a_{\mathbf{pp}'}, \tag{12.7-6}$$

where $\boldsymbol{\sigma} = (\sigma_x, \sigma_y, \sigma_z)$ refers to the Pauli spin matrices. As to the two momenta of each term, we note that both have magnitudes which for all practical purposes are equal to p_F ; therefore only their mutual angle γ is relevant. Consequently, an expansion in Legendre polynomials is a natural representation. It is customary to remove a factor ρ_F, being the density of states at the Fermi surface, from the factors φ; we have $\rho_F = m^* p_F / \pi^2\hbar^3$. Thus we employ the expansions

$$\varphi^{s,a}_{\mathbf{pp}'}(\gamma) = (1/\rho_F)\sum_{\ell=0}^{\infty} F^{s,a}_{\ell} P_{\ell}(\cos\gamma). \tag{12.7-7}$$

We note that the coefficients F_l^s and F_l^a are also called F_l and Z_l in many papers. Generally only the coefficients for $l = 0,1$ are important. These, then, are the four dimensionless Fermi liquid parameters which relate to experimental data and which can be further explored by microscopic theories.

We briefly return to the system energy, implied by (12.7-5):

$$\mathscr{E} = \mathscr{E}_0 + \int \Delta^3_{p\sigma} \varepsilon_{\mathbf{p}} \delta n_{\mathbf{p}\sigma} + \frac{1}{2} \iint \Delta^3_{p\sigma} \Delta^3_{p'\sigma'} \varphi_{\mathbf{p}\sigma,\mathbf{p}'\sigma'} \delta n_{\mathbf{p}\sigma} \delta n_{\mathbf{p}'\sigma'} . \tag{12.7-8}$$

The renormalized quasi-particle energy is now obtained as the functional derivative of $\mathscr{E}$ with respect to $\delta n_{\mathbf{p}\sigma}$:

$$\tilde{\varepsilon}_{\mathbf{p}} = \frac{\delta \mathscr{E}}{\delta(\delta n_{\mathbf{p}\sigma})} = \varepsilon_{\mathbf{p}} + \int \Delta^3_{p'\sigma'} \varphi_{\mathbf{p}\sigma,\mathbf{p}'\sigma'} \delta n_{\mathbf{p}'\sigma'} . \tag{12.7-9}$$

In terms of the new energies the ground-state occupancies will be rewritten as $\tilde{n}^0_{p\sigma} = n_F(\tilde{\varepsilon}_{\mathbf{p}} - \varsigma)$. For the excitations above the ground state we find:

$$\delta\tilde{n}_{\mathbf{p}\sigma} = n_{\mathbf{p}\sigma} - n_F(\tilde{\varepsilon}_{\mathbf{p}} - \varsigma) = n_{\mathbf{p}\sigma} - n_F(\varepsilon_{\mathbf{p}} - \varsigma + \tilde{\varepsilon}_{\mathbf{p}} - \varepsilon_{\mathbf{p}}) . \tag{12.7-10}$$

From first order Taylor expansion we have

$$n_F(\varepsilon_{\mathbf{p}} - \varsigma + \tilde{\varepsilon}_{\mathbf{p}} - \varepsilon_{\mathbf{p}}) = n_F(\varepsilon_{\mathbf{p}} - \varsigma) + (\tilde{\varepsilon}_{\mathbf{p}} - \varepsilon_{\mathbf{p}}) \frac{dn_F(\varepsilon_{\mathbf{p}} - \varsigma)}{d\varepsilon_{\mathbf{p}}} . \tag{12.7-11}$$

From the last two equations we obtain the connection

$$\delta\tilde{n}_{\mathbf{p}\sigma} = \delta n_{\mathbf{p}\sigma} - (\tilde{\varepsilon}_{\mathbf{p}} - \varepsilon_{\mathbf{p}}) \frac{dn_F(\varepsilon_{\mathbf{p}} - \varsigma)}{d\varepsilon_{\mathbf{p}}} . \tag{12.7-12}$$

Finally, substitution from (12.7-9) yields the relationship

$$\delta\tilde{n}_{\mathbf{p}\sigma} = \delta n_{\mathbf{p}\sigma} - \frac{dn_F(\varepsilon_{\mathbf{p}} - \varsigma)}{d\varepsilon_{\mathbf{p}}} \int \Delta^3_{p'\sigma'} \varphi_{\mathbf{p}\sigma,\mathbf{p}'\sigma'} \delta n_{\mathbf{p}'\sigma'} . \tag{12.7-13}$$

The factor in front of the integral resembles a delta function, $-dn_F / d\varepsilon_{\mathbf{p}} \approx \delta(\varepsilon_{\mathbf{p}} - \varsigma)$. The spin dependence means that there is a symmetric and antisymmetric contribution. Accordingly we write,

$$\delta n_{\mathbf{p}\pm} = \delta n^s_{\mathbf{p}} \pm \delta n^a_{\mathbf{p}} , \tag{12.7-14a}$$

$$\delta\tilde{n}_{\mathbf{p}\pm} = \delta\tilde{n}^s_{\mathbf{p}} \pm \delta\tilde{n}^a_{\mathbf{p}} . \tag{12.7-14b}$$

The momentum **p** is expressed in polar coordinates, with $p \approx p_F$ and the expressions (12.7-14a and b) are expanded in spherical polynomials $Y_{lm}(\theta,\phi)$. Hence we have

$$\delta n^{s,a}_{\mathbf{p}} = \sum_{\ell m} \delta(\varepsilon_{\mathbf{p}} - \varsigma) \delta n^{s,a}_{\ell m} Y_{\ell m}(\theta,\phi) , \tag{12.7-15}$$

$$\delta\tilde{n}_{\mathbf{p}}^{s,a} = \sum_{\ell m} \delta(\varepsilon_{\mathbf{p}} - \varsigma)\, \delta\tilde{n}_{\ell m}^{s,a} Y_{\ell m}(\theta,\phi) \,, \tag{12.7-16}$$

where θ and ϕ are the polar angles of **p**. Likewise, let θ' and ϕ' are the polar angles of $\mathbf{p}'$. Equation (12.7-7) is rewritten using the addition theorem for spherical polynomials[55],

$$\varphi_{\mathbf{pp}'}^{s,a}(\gamma) = (1/\rho_F) \sum_{\ell m} F_\ell^{s,a} \frac{4\pi}{2\ell+1} Y_{\ell m}(\theta,\phi) Y_{\ell m}^*(\theta',\phi') \,. \tag{12.7-17}$$

This is substituted into (12.7-13); dividing out the delta functions, we get

$$\sum_{\ell m} \delta\tilde{n}_{\ell m}^{s,a} Y_{\ell m}(\theta,\phi) = \sum_{\ell m} \delta n_{\ell m}^{s,a} Y_{\ell m}(\theta,\phi)$$
$$+ (1/\rho_F) \int \delta(\varepsilon_{\mathbf{p}'} - \varsigma) \Delta^3 p' \sum_{\ell m} F_\ell^{s,a} \frac{4\pi}{2\ell+1} Y_{\ell m}(\theta,\phi) Y_{\ell m}^*(\theta',\phi') \sum_{\ell' m'} \delta n_{\ell' m'}^{s,a} Y_{\ell' m'}(\theta',\phi'). \tag{12.7-18}$$

Now $\Delta^3 p' \to 2(1/8\pi^3\hbar^3) p'^2 dp' d\Omega_{p'} = (m^*/4\pi^3\hbar^3) p' d\varepsilon_{\mathbf{p}'} d\Omega_{p'}$; the delta function is then integrated and the integral over $d\Omega_{p'}$ is carried out, using orthogonality of the spherical polynomials. Comparing the coefficients of $Y_{\ell m}(\theta,\phi)$ on both sides, we have

$$\delta\tilde{n}_{\ell m}^{s,a} = \left(1 + F_\ell^{s,a} \frac{1}{2\ell+1}\right) \delta n_{\ell m}^{s,a} . \tag{12.7-19}$$

The four quantities F_0^s, F_1^s, F_0^a and F_1^a can be related to experimental data, as we now show.

First of all, we consider the specific heat, C_V. The free particle value is analogous to the result (8.4-29), with the mass now being $m(^3\text{He})$; employing also the analogue of (8.4-11) one easily finds $C_{V0} = (\pi^2/3) k_B^2 T \rho_{F0}$, where ρ_{F0} is the density of states at the Fermi level, $\rho_{F0} = mk_F/\pi^2\hbar^2$. For the Fermi liquid the interactions have no effect on the specific heat, which is entirely determined by a small band of excitations near the Fermi energy; hence, $C_V/C_{V0} = \rho_F/\rho_{F0} = m^*/m$. For the current density one may surmise that either of the two expressions below is applicable:

$$J = \sum_{\mathbf{p},\sigma} (p/m^*) \delta\tilde{n}_{\mathbf{p}\sigma} = \sum_{\mathbf{p},\sigma} (p/m) \delta n_{\mathbf{p}\sigma} . \tag{12.7-20}$$

The current operators are nonzero if the distributions $\delta\tilde{n}_{\mathbf{p}\sigma}$ and $\delta n_{\mathbf{p}\sigma}$ have a nonzero component of angular momentum, $\ell = 1$, $m = 0$. Hence, from (12.7-19) and (12.7-20) we find the ratio m^*/m, yielding

$$C_V/C_{V0} = m^*/m = 1 + \tfrac{1}{3} F_1^s . \tag{12.7-21}$$

[55] See e.g., J.D. Jackson, "Classical Electrodynamics", 3rd Ed., John Wiley & Sons, NY, 1999, p. 110.

With $m^*/m \approx 2.76$, the value of F_1^s is found to be $F_1^s = 5.27$.

We proceed to compute the compressibility and the velocity of sound. As to the former, from (4.2-29) and (4.2-25) we have

$$K^{-1} = N\rho(\partial\varsigma/\partial N)_T. \tag{12.7-22}$$

(NB. The isothermal and adiabatic compressibility are the same near $T=0$). Following Pines and Nozières, Op. Cit., let the Fermi surface be increased by $d\varepsilon_F = d\varsigma$. Considering two points A and B, relating to the original and increased Fermi surface, we have $\varepsilon_B(\varsigma + d\varsigma) - \varepsilon_A(\varsigma) = d\varsigma$, see Fig. 12-12; the change of the normal is dp_F.

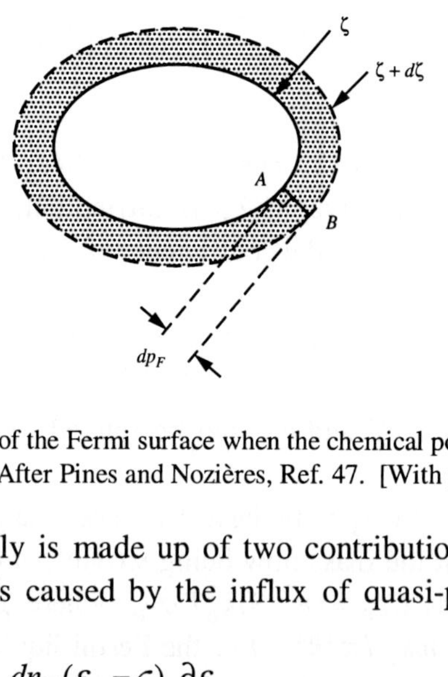

Fig. 12-12. Deformation of the Fermi surface when the chemical potential is changed from ς to $\varsigma + d\varsigma$. After Pines and Nozières, Ref. 47. [With permission.]

The energy change actually is made up of two contributions. The direct change is $v_{\mathbf{p}} dp_F$. Another change is caused by the influx of quasi-particles into the shaded region. For this we have

$$\delta n_{\mathbf{p}\sigma} = -\frac{dn_F(\varepsilon_{\mathbf{p}} - \varsigma)}{d\varepsilon_{\mathbf{p}}} \frac{\partial \varepsilon_{\mathbf{p}}}{\partial p} dp_F \approx \delta(\varepsilon_{\mathbf{p}} - \varsigma) v_{\mathbf{p}} dp_F. \tag{12.7-23}$$

We insert this into the above; dividing by $d\varsigma$, we obtain the integral equation

$$v_{\mathbf{p}}(\partial p_F/\partial\varsigma) + \int \Delta^3 p' \delta(\varepsilon_{\mathbf{p}'} - \varsigma)\varphi_{\mathbf{pp}'} v_{\mathbf{p}'}(\partial p'_F/\partial\varsigma) = 1. \tag{12.7-24}$$

Finally we need $dN = \Sigma_{\mathbf{p}\sigma}\delta n_{\mathbf{p}} \approx \Sigma_{\mathbf{p}\sigma}\delta(\varepsilon_{\mathbf{p}} - \varsigma) v_{\mathbf{p}} dp_F$. Hence,

$$\partial N/\partial\varsigma \approx \sum_{\mathbf{p},\sigma} \delta(\varepsilon_{\mathbf{p}} - \varsigma) v_{\mathbf{p}} (\partial p_F/\partial\varsigma). \tag{12.7-25}$$

For an isotropic system $\partial p_F/\partial\varsigma$ is independent of direction. We then expand $\varphi_{\mathbf{pp}'}$ in Legendre polynomials, cf. (12.7-7); the term with $\ell = 0$ suffices. Returning to (12.7-22), we easily obtain – using the previous procedure for the integral part of (12.7-24):

$$\frac{K}{K_0} = \frac{1+F_0^s}{1+\frac{1}{3}F_1^s}, \qquad K_0 = \frac{\rho_{F0}}{n_0^2}. \tag{12.7-26}$$

In this result K_0 is the compressibility for a free particle gas, which is easily found from the analogues of the results in (8.4-11). This also yields the velocity of first sound, as given by (12.1-21); hence,

$$\frac{\partial P}{\partial \rho} = \frac{1}{m^*}\frac{\partial P}{\partial(1/\hat{v})} = -\frac{\hat{v}}{m^*}\hat{v}\frac{\partial P}{\partial \hat{v}} = \frac{1}{\rho K}. \tag{12.7-27}$$

So we find

$$\frac{u}{u_0} = \frac{1}{\sqrt{1+F_0^s}}, \qquad u_0 = \frac{p_F}{m\sqrt{3}}. \tag{12.7-28}$$

Lastly, we shall find an expression for the spin susceptibility. It will yield F_0^a, the isotropic asymmetric spin-dependent part of the interaction. In a magnetic field **H** the Fermi surface splits into two surfaces with the same chemical potential (being in equilibrium). The new local energies are $\tilde{\varepsilon}_{\mathbf{p}} - g\rho_B(\frac{1}{2})\sigma H$, where $\sigma = \pm 1$ and we set $g = 2$. With Taylor expansion we have

$$\delta\tilde{n}_{\mathbf{p}\sigma}(H) = -\rho_B \sigma H \frac{dn_F(\tilde{\varepsilon}_{\mathbf{p}} - \varsigma)}{d\tilde{\varepsilon}_{\mathbf{p}}} \approx -\rho_B \sigma H \frac{dn_F}{d\varepsilon_{\mathbf{p}}}. \tag{12.7-29}$$

In order to compute the magnetization $M = \Sigma_{\mathbf{p}\sigma}\rho_B \sigma \delta n_{\mathbf{p}\sigma}$ we must pass from $\delta\tilde{n}_{\mathbf{p}\sigma}$ to $\delta n_{\mathbf{p}\sigma}$. This is easily done if we assume that the system is isotropic; $\delta\tilde{n}_{\mathbf{p}\sigma}$ is then isotropic and spin-asymmetric (changes sign when the spin is flipped). From (12.7-19) with $\ell = 0$ one finds

$$\delta n_{\mathbf{p}\sigma} = \frac{\delta\tilde{n}_{\mathbf{p}\sigma}}{1+F_0^a} = -\frac{\rho_B \sigma H}{1+F_0^a}\frac{dn_F}{d\varepsilon_{\mathbf{p}}}, \tag{12.7-30}$$

which yields

$$M = -\int \Delta^3 p \frac{\rho_B^2 H}{1+F_0^a}\frac{dn_F}{d\varepsilon_{\mathbf{p}}}, \tag{12.7-31}$$

where we replaced the sum by an integral involving the density of states and we noted that $\sigma^2 = 1$ for either spin direction. Setting once more $-dn_F/d\varepsilon_{\mathbf{p}}$ $\approx \delta(\varepsilon_{\mathbf{p}} - \varsigma)$ one easily obtains

$$M = \frac{\rho_F \rho_B^2 H}{1+F_0^a}. \tag{12.7-32}$$

The magnetization is modified by the exchange-interaction parameter F_0^a. For the susceptibility it follows that

$$\frac{\chi}{\chi_0} = \frac{1+\frac{1}{3}F_1^a}{1+F_0^a}, \qquad \chi_0 = \rho_F \rho_B^2. \tag{12.7-33}$$

The results for the various parameters, obtained from comparison with experimental data, are listed in Table 12-2[56] We also give there some other parameters, to be discussed shortly.

Table 12-2. Fermi liquid parameters; $A_\ell = F_\ell /[1 + F_\ell /(2\ell + 1)]$.

Parameter	Experimental data	Parameter	Experimental data
F_0^s	9.15	A_0^s	0.90
F_1^s	5.27	A_1^s	1.91
F_0^a	− 0.70	A_0^a	− 2.33
F_1^a	− 0.55	A_1^a	− 0.67

The above factors $F_{0,1}^{s,a}$ were derived from the interaction of 'bare' quasi-particles, see (12.7-5) and the resulting energy expression (12.7-9). In order to understand the spin dependence in pairing that leads to superfluidity, it is useful to consider the interaction of dressed particles. Instead of (12.7-9) we then write

$$\tilde{\varepsilon}_{\mathbf{p}} - \varepsilon_{\mathbf{p}} = \int \Delta^3_{p',\sigma'} A_{\mathbf{p}\sigma,\mathbf{p}'\sigma'} \delta\tilde{n}_{\mathbf{p}'\sigma'}. \tag{12.7-34}$$

Similar to (12.7-6) we explicit the spin dependence by setting

$$A_{\mathbf{p}\sigma,\mathbf{p}'\sigma'} = A^s_{\mathbf{pp}'} + (\boldsymbol{\sigma}\cdot\boldsymbol{\sigma}')A^a_{\mathbf{pp}'}, \tag{12.7-35}$$

with an expansion in spherical polynomials as in (12.7-7)

$$A^{s,a}_{\mathbf{pp}'} = (1/\rho_F)\sum_{\ell=0}^{\infty} A_\ell^{s,a} P_\ell(\cos\theta). \tag{12.7-36}$$

With the expansions in spherical polynomials previously given for the distributions δn and $\delta\tilde{n}$ one easily obtains

[56] D.S. Greywall, Phys. Rev. **B27**, 2747 (1983).

$$A_\ell^{s,a} = \frac{F_\ell^{s,a}}{1 + F_\ell^{s,a}/(2\ell+1)}. \tag{12.7-37}$$

Comparing the parameters for the bare-particle interaction (F's) and those for the dressed-particle interaction (A's), we notice that the zero order spin-independent interaction is strongly reduced for the dressed particles: $A_0^s \ll F_0^s$. This is the analogue of electron screening in metals. The spin-dependent interaction A_0^a, on the other hand, is strongly enhanced and attractive if the two spins are aligned. This is a strong indication for the possibility of triplet pairing, discussed in subsection 12.8.2.

12.7.2 *The Ground State and Pair-Correlation Function*

The ground state is more complicated than for a boson liquid such as ^{4}He, since we must make provisions to satisfy the antisymmetric character which is responsible for the Pauli exclusion principle. So, let x_i represent position and spin, $x_i = (\mathbf{r}_i, \sigma)$; we must ensure that

$$\Psi_0(x_1, x_2, \ldots x_i, x_j, \ldots x_1) = -\Psi_0(x_1, x_2, \ldots x_j, x_i, \ldots x_1). \tag{12.7-38}$$

The problem has been much investigated by Feenberg and coworkers.[57] The ground state of choice is taken to be a product of a bosonic state, composed of *correlated basic functions* [CBS's or BDF's after Bijl[4]–Dingle[58]–Jastrow[59]] and a Slater determinant as discussed in Section 7.1. Accordingly we set

$$\Psi_0 = A_N \exp\left[-\sum_{i<j} u(\mathbf{r}_i - \mathbf{r}_j)\right](N!)^{-1/2} \det\left(\alpha_{\mathbf{p}_1}^1 \alpha_{\mathbf{p}_2}^2 \ldots \alpha_{\mathbf{p}_N}^N\right), \tag{12.7-39}$$

where the u's are two-body interaction functions similar as in classical fluids, whereas the single particle functions in the determinant are taken to be plane waves of momentum $\mathbf{p}_i$. The 'boson-part', will also be denoted as Ψ_{B0}. Usually one takes a repulsive potential of the form

$$\Psi_{B0} = A_N \exp\left[-\sum_{i<j}\left(d/|\mathbf{r}_i - \mathbf{r}_j|\right)^l\right], \tag{12.7-40}$$

where l is often taken to be five (see Massey) or ten (Massey and Woo).[60] The ground-state energy is obtained as $(\mathcal{H}\Psi_0, \Psi_0)/(\Psi_0, \Psi_0)$, where $\mathcal{H}$ contains the usual kinetic energy part and binary potential energies $\Sigma_{i<j}\varphi_{ij}(\mathbf{r}_{ij})$. For the latter one often

---

[57] E. Feenberg, "Theory of Quantum Fluids" Academic Press NY, 1969.

[58] R.M. Dingle, Philos. Mag. **40**, 573 (1949).

[59] R. Jastrow, Phys. Rev. **98**, 1479 (1955).

[60] W.E. Massey, Phys. Rev. **151**, 253 (1966); W.E. Massey and Chia-Wei Woo, Phys. Rev. **164**, 256 (1967).

uses the classical 6-12 Lennard-Jones potential $\varphi(r) = 4\varepsilon[(d/r)^{12} - (d/r)^6]$, with the de Boer–Michels values $\varepsilon = 10.22K$ (in units $k_B T$) and $d = 2.556$ Å.[61] The pair-correlation function is obtained from the second-order BBGKY equation, (4.10-41), with as closure condition Kirkwood's superposition principle, (4.10-42). Both cumulant expansions and extensive numerical techniques for solving these equations are employed. One first determines the values for a hypothetical ^{3}He Bose gas and adds the Slater determinantal plane wave functions to incorporate the Fermi 'correction'. Readable accounts are found in Massey and Massey and Woo, Op. Cit. For full details we refer to Feenberg's book. From the pair-correlation function the liquid structure factor $S(k)$, being the adjusted Fourier transform

$$S(k) = 1 + \langle n \rangle \int p(r) e^{i\mathbf{k}\cdot\mathbf{r}} d^3 r \tag{12.7-41}$$

is then obtained. As for ^{4}He this function can be compared with experimental data obtained from X-ray or neutron scattering. In Fig. 12-13 we give the computed data obtained by Massey and Woo, together with X-ray data[62] by Achter and Meyer and by Hallock. We note that the Massey-Woo calculations were carried out *prior* to the reported experimental data; the results are clearly remarkably good! Somewhat less satisfactory is the ground-state energy per atom, for which M–W found $-1.35K$; experimental data suggest the value $-2.52K$. We note that of the theoretical value an amount $\Delta\mathscr{E}_0 = -1.15K$ stems from the modification by the Slater determinant. Clearly, the Bose liquid value is too small to allow accurate computation. A more likely Bose liquid value is reported by Schiff and Verlett, viz. -1.27 K.[63]

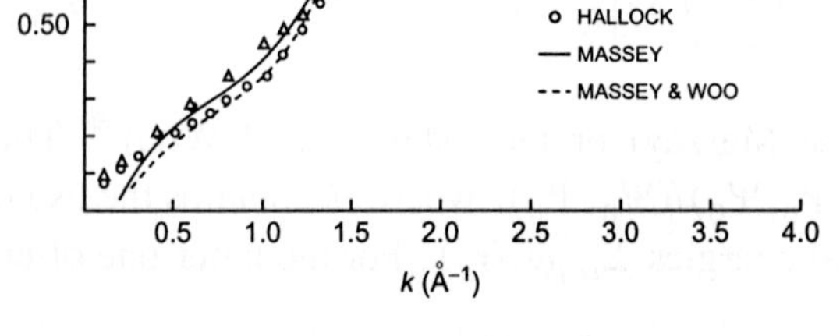

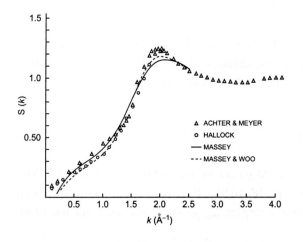

Fig. 12-13. The structure factor $S(k)$ for liquid ^{3}He. Solid line: computed by Massey and Woo[60]; Δ : from X-ray scattering data by Achter and Meyer[62]; O : from X-ray scattering data by Hallock[62]. [With permission.]

---

[61] J. de Boer and A. Michels, Physica **5**, 945 (1938).

[62] E.K. Achter and L. Meyer, Phys. Rev. **188,** 291 (1969); R.B. Hallock, J. Low Temp. Phys. **9**, 109 (1972).

[63] D. Schiff and L. Verlett, Phys. Rev.**160**, 208 (1967).

12.8 Modern Developments of ^{3}He

12.8.1 *Other Excitations*

Neutron-scattering data were first reported by Scherm et al. at the Laue–Langevin Institute in Grenoble.[64] A neutron probe reveals much of the dynamics of ^{3}He since it couples both to the density fluctuations of the quasi-particles and to the spin fluctuations of the nuclear spin system. The first spectra at $0.63K$ only showed a broad peak of inelastically scattered neutrons, corresponding to energy transfers giving excitations of single electron-hole pairs near the Fermi energy, without any evidence of a collective density mode involving 'zero sound', predicted by Pines in the sixties[47], or of possible spin fluctuations. Subsequent measurements by Sköld and others at Argonne National Laboratories at a temperature of $15mK$, summarised by Sköld and Pelizzari[65], showed two peaks, one at 0.2 meV and one at approximately 1.0 meV. The low energy peak was within the particle-hole band and explained as being caused by spin fluctuations, whereas the ~1.0 meV peak was attributed to scattering with longitudinal density fluctuations, thus evidencing the presence of zero sound. The results are compared with theoretical considerations based on the 'random phase approximation' (RPA).[66]

The excitations targeted in these measurements are density fluctuations or spin fluctuations, represented by a spectral density function $\mathbb{S}_d(\mathbf{q},\omega)$ or $\mathbb{S}_s(\mathbf{q},\omega)$, respectively. This function is two times the spatial and temporal Fourier transform of the retarded density-density (or spin-spin) correlation function of the process envisaged. Turning first to density fluctuations: if the particle density operator is denoted by the Heisenberg form $\rho(\mathbf{r},t) \equiv \Psi^\dagger(\mathbf{r},t)\Psi(\mathbf{r},t)$, its spatial transform is $\overline{\rho}(\mathbf{q},t)$ and the *dynamical structure factor* is defined as

$$S_d(\mathbf{q},\omega) = \int_{-\infty}^{\infty} dt\, e^{-i\omega t} \langle \overline{\rho}(\mathbf{q},t)\overline{\rho}^\dagger(\mathbf{q},0)\rangle, \tag{12.8-1}$$

where the average is over a grand-canonical ensemble. The connection with the spectral density function is given by

$$\mathbb{S}_d(\mathbf{q},\omega) = [1+e^{-\beta\hbar\omega}]S_d(\mathbf{q},\omega), \text{ where } \beta = 1/k_B T. \tag{12.8-2}$$

Although there are attempts in the literature to obtain the retarded density-density correlation function from an appropriate Hamiltonian[67], the function is most easily

---

[64] R. Scherm, W.G. Stirling, A.D.B. Woods, R.A. Cowley and G.I. Coombs, J. Phys. **C 9**, L341 (1974).

[65] K. Sköld and C.A. Pelizzari, Phil. Trans. Royal Soc. London, **B 290**, 606 (1980).

[66] The RPA is a procedure by Bohm and Pines in which certain position-dependent terms in $\mathcal{H}_{grand}$ are ignored; D. Bohm and D. Pines, Phys. Rev. **92**, 609 (1953). In diagrammatic representations, a large class of diagrams is omitted and only ring diagrams are summed, see Sect. 12.13.2.

[67] G.D. Mahan, Op. Cit., Eq. (11.191).

obtained from linear response theory, discussed in the non-equilibrium part, Sections 16.5 and 16.6. In that event we consider the perturbation of the system by an external Hamiltonian, which couples to the density variation; the response is characterized by a generalized susceptibility $\chi_d(\mathbf{q},\omega)$. For the present case, we need to take into account that there is a collective change in the environment of all pseudo-particles – the equivalent of polarization in an electron gas – so we need the 'dressed' susceptibility $\chi_d^{dr}(\mathbf{q},\omega)$. From the fluctuation-dissipation theorem it follows that we have the connection

$$S_d(\mathbf{q},\omega) = -2\hbar V_0 \,\mathrm{Im}\,\chi_d^{dr}(\mathbf{q},\omega). \tag{12.8-3}$$

The dressed susceptibility has the form

$$\chi_d^{dr}(\mathbf{q},\omega) = \frac{\chi_d(\mathbf{q},\omega)}{1 - C(\mathbf{q})\chi_d(\mathbf{q},\omega)}, \tag{12.8-4}$$

where C is a coupling function. In the RPA and for $|\omega| \gg qv_F$ this takes the form

$$\chi_d^{dr}(\mathbf{q},\omega) = \frac{P(\mathbf{q},\omega)}{1 - (F_0^s / \rho_F)P(\mathbf{q},\omega)}, \tag{12.8-5}$$

where P is part of the polarization energy; it has the form of the bare-mass Lindhard function, discussed in subsection 16.3.3. The dynamical structure factor $S_d(\mathbf{q},\omega)$ so obtained pertains to *zero sound.*

For spin fluctuations the theory is less developed; however, analogous to (12.8-4) we have

$$\chi_s^{dr}(\mathbf{q},\omega) = \frac{\chi_s(\mathbf{q},\omega)}{1 - C'(\mathbf{q})\chi_s(\mathbf{q},\omega)}. \tag{12.8-6}$$

Two types of spin fluctuations can be expected. First, it has been suggested that collective excitations might occur, giving rise to "spin sound". It is, however, readily shown that the eigenfrequency would be given by $\omega_0 = qv_F\sqrt{F_0^a/3}$, which is an imaginary number; such modes are therefore totally damped and do not exist. The other possibility is that spin *pairs* of opposite angular momentum would be excited. Fluctuations associated with such spin pairs are the subject of *paramagnon* theory. Some models consider a Hamiltonian with an interaction term

$$\mathcal{V}_{inter} = I \sum_{\mathbf{q}} \sum_{\mathbf{k},\mathbf{k}'} c^\dagger_{\mathbf{k}'+\mathbf{q}\uparrow} c_{\mathbf{k}'\uparrow} c^\dagger_{\mathbf{k}-\mathbf{q}\downarrow} c_{\mathbf{k}\downarrow} \overset{\mathbf{q}\to 0}{\Rightarrow} I \sum_{\mathbf{k}} b^\dagger_{\mathbf{k}} b_{\mathbf{k}}, \tag{12.8-7}$$

where $b_{\mathbf{k}}$ and $b_{\mathbf{k}}^\dagger$ are pair operators and $I > 0$ is the paramagnon energy; note that the coupling involves a vertex with four pair-wise opposite spins. The connection with the spin susceptibility is found in the article by Sköld and Pelizzari; they show that $C'(\mathbf{q}) \to -I$, while $\chi_s(\mathbf{q},\omega)$ is the Lindhard function for the bare mass; this is further discussed in subsection 16.6.3. In the RPA and long wavelength limit, Leggett[49] proposed the expression

$$\chi_s^{dr}(\mathbf{q},\omega) = \frac{\overline{P}(\mathbf{q},\omega)}{1-(F_0^a/\rho_F)\overline{P}(\mathbf{q},\omega)}, \quad F_0^a < 0. \tag{12.8-8}$$

The spin polarization $\overline{P}$ can be shown to be

$$\overline{P}(\mathbf{q},\omega) = -\rho_F\left[1-\frac{i\pi}{2}\left(\frac{\omega}{qv_F}\right)\Theta(qv_F-|\omega|)\right]. \tag{12.8-9}$$

Simple algebra gives

$$\mathrm{Im}\,\chi_s^{dr} = \rho_F\frac{1}{\left|F_0^a(1+F_0^a)\right|}\frac{\omega\omega_a}{\omega^2+\omega_a^2}, \quad |\omega| < qv_F, \tag{12.8-10}$$

where

$$\omega_a = 2qv_F/\pi|A_0^a|. \tag{12.8-11}$$

There are no real poles and the dynamical structure factor will show a peaked behaviour, with a rather strong coupling since $|F_0^a(1+F_0^a)|^{-1} = 4.76$.

Sköld and Pelizzari give detailed measurements for the dynamical structure factor, which in their case is a weighted combination of $S_d(\mathbf{q},\omega)$ and $S_s(\mathbf{q},\omega)$. In Fig 12-14 we reproduce one of their figures of the measured structure factor which shows three contributions: a ~ 0.2 meV peak associated with spin fluctuations, a 1.3 meV peak attributed to zero-sound and a broad spectrum associated with multiple pair excitations.

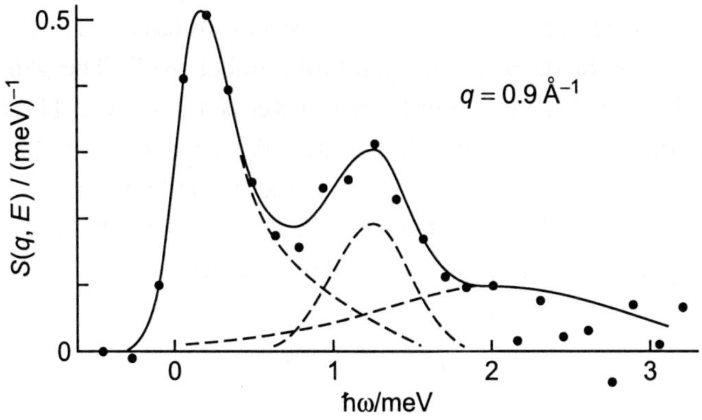

Fig. 12-14. The dynamical structure factor vs. $\hbar\omega$ at $q = 0.9$ Å^{-1}. The three contributions are explained in the text. After Sköld and Pelizzari.[65] [With permission.]

12.8.2 *Balian–Werthamer (B–W) Hamiltonian for the Superfluid Phases*

The phase diagram for liquid ^{3}He below three milli-*Kelvin* is shown in Fig. 12-15. Two superfluid phases have been found; the most common *B*-phase prevails at pressures below 20 *Bar*, whereas the *A*-phase is found in a narrow temperature range around 2.5 milli-*Kelvin* at pressures above 20 Bar, but before solidification occurs. In the *B*-phase triplet pairing explains most of the features. The *A*-phase requires additional assumptions.

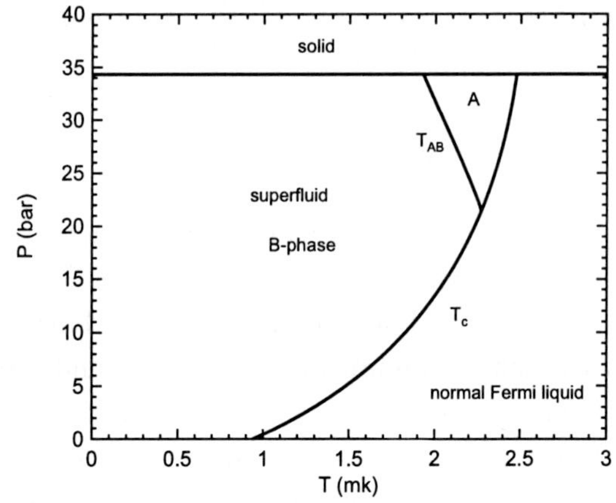

Fig. 12-15. Phase diagram for ^{3}He below 3.0 mK. After Greywall.[68]
[With permission.]

The triplet-pairing state has $S = 1$, but the total angular momentum has values $|L - S| \ldots L + S$. With the relative angular momentum having $L = 1$, this makes for many possible values of the Zeeman quantum number m_J.[69] The gap equation now becomes a matrix equation and the order-parameter is a 7-spinor. There is therefore a rich variety of ordering in the superfluid state. We shall give a brief description of the B–W results. A matrix description is desirable from the onset.

Balian and Werthamer[54] start with a reduced Hamiltonian similar as in BCS theory but avoid the explicit use of pair operators. The interaction Hamiltonian thus reads

$$\mathcal{V} = \frac{1}{2} \sum_{\mathbf{kk}'\sigma\sigma'} V_{\mathbf{kk}'} c^\dagger_{-\mathbf{k}\sigma'} c^\dagger_{\mathbf{k}\sigma} c_{\mathbf{k}'\sigma} c_{-\mathbf{k}\sigma'}, \tag{12.8-12}$$

where the $c, c^\dagger$ are the usual fermion operators and σ denotes the z-component of the spin. Note that we couple quasi-particle states of opposite momenta and both same

---

[68] D.S. Greywall, Phys. Rev. **B 33**, 7520 (1986).

[69] However, one cannot combine the angular momenta with Clebsch-Gordon coefficients as for a He_2 molecule, since the pairs are a collective property.

or opposite spins. To (12.8-12) we must add the free quasi-particle contribution; as in the BCS theory the chemical potential ς will be chosen to be zero and all energies will be shifted accordingly. For symmetry purposes a 'rotation' in the spin-space is performed:

$$c_{\mathbf{k}\sigma} = \sum_{\sigma'}(u^{\mathbf{k}}_{\sigma\sigma'}\alpha_{\mathbf{k}\sigma'} + v^{\mathbf{k}}_{\sigma\sigma'}\alpha^{\dagger}_{-\mathbf{k}\sigma'}), \quad c^{\dagger}_{\mathbf{k}\sigma} = adjoint, \tag{12.8-13}$$

where the transformation matrix must be unitary in order to preserve the anti-commutation rules. The following four-component matrices will be employed

$$c^{\mathbf{k}} \equiv \{c_{\mathbf{k}\uparrow}\ c_{\mathbf{k}\downarrow}\ c^{\dagger}_{-\mathbf{k}\uparrow}\ c^{\dagger}_{-\mathbf{k}\downarrow}\}, \quad \alpha^{\mathbf{k}} \equiv \{\alpha_{\mathbf{k}+}\ \alpha_{\mathbf{k}-}\ \alpha^{\dagger}_{-\mathbf{k}+}\ \alpha^{\dagger}_{-\mathbf{k}-}\}. \tag{12.8-14}$$

The transformation is now written as

$$c^{\mathbf{k}} = \begin{pmatrix} u^{\mathbf{k}} & v^{\mathbf{k}} \\ v^{-\mathbf{k}*} & u^{-\mathbf{k}*} \end{pmatrix}\alpha^{\mathbf{k}} \equiv U^{\mathbf{k}}\alpha^{\mathbf{k}}, \quad U^{\mathbf{k}}U^{\mathbf{k}\dagger} = 1; \tag{12.8-15}$$

note that the elements $u^{\mathbf{k}}, v^{\mathbf{k}}$, *etc.*, are each two by two matrices (subs σ, σ'), making $U^{\mathbf{k}}$ a partitioned matrix. Moreover, $U^{\mathbf{k}}$ and $U^{-\mathbf{k}}$ are not independent but related by – see their definition (12.8-15) – $U^{\mathbf{k}} = \bar{J}U^{-\mathbf{k}*}\bar{J}$, where $\bar{J}$ is the symplectic matrix $\bar{J} = \left(\begin{smallmatrix} 0 & 1 \\ 1 & 0 \end{smallmatrix}\right)$.

For the ensemble the following exponentiated diagonal bilinear form is chosen as density operator:

$$\rho = A\exp[-\beta\sum_{\mathbf{k}\sigma}\mathcal{E}_{\mathbf{k}\sigma}\alpha^{\dagger}_{\mathbf{k}\sigma}\alpha_{\mathbf{k}\sigma}], \quad \mathrm{Tr}\rho = 1. \tag{12.8-16}$$

As an alternative to the Bogoliubov–Valatin transformation that was used in our version of the BCS theory we shall, with the authors, resort to a thermodynamic argument. Thus, we must choose the $U^{\mathbf{k}}$ and $\mathcal{E}^{\mathbf{k}}$ such that the 'free energy' associated with ρ will be a minimum. In order to do the variational problem we also consider averages of the original independent variables. Thus, we define the 4×4 matrix

$$\langle c^{\mathbf{k}}c^{\mathbf{k}\dagger}\rangle = \mathrm{Tr}\{\rho c^{\mathbf{k}}c^{\mathbf{k}\dagger}\} = \frac{1}{2}\begin{pmatrix} 1+w^{\mathbf{k}} & x^{\mathbf{k}} \\ -x^{-\mathbf{k}*} & 1-w^{-\mathbf{k}*} \end{pmatrix} \equiv \frac{1}{2}W^{\mathbf{k}}, \tag{12.8-17}$$

or explicitly,

$$\langle c_{\mathbf{k}\sigma}c^{\dagger}_{\mathbf{k}\sigma'}\rangle = \tfrac{1}{2}(\delta_{\sigma\sigma'} + w^{\mathbf{k}}_{\sigma\sigma'}), \quad \langle c_{\mathbf{k}\sigma}c_{-\mathbf{k}\sigma'}\rangle = \tfrac{1}{2}x^{\mathbf{k}}_{\sigma\sigma'}. \tag{12.8-18}$$

Again we find that $W^{\mathbf{k}}$ and $W^{-\mathbf{k}}$ are related by $W^{\mathbf{k}} = W^{\mathbf{k}\dagger} = -\bar{J}W^{-\mathbf{k}*}\bar{J}$. Furthermore, $W^{\mathbf{k}}$ is closely related to $U^{\mathbf{k}}$ and $\mathcal{E}_{\mathbf{k}\sigma}$.

The diagonal (grand-) canonical density operator yields for the hole-occupancies the standard form

$$\langle \alpha^{\mathbf{k}} \alpha^{\mathbf{k}\dagger} \rangle = \begin{pmatrix} 1-f(\mathcal{E}_{\mathbf{k}+}) & & & 0 \\ & 1-f(\mathcal{E}_{\mathbf{k}-}) & & \\ & & 1-f(\mathcal{E}_{-\mathbf{k}+}) & \\ 0 & & & 1-f(\mathcal{E}_{-\mathbf{k}-}) \end{pmatrix} = \frac{1}{2}(1+\tanh\frac{1}{2}\beta\mathcal{E}^{\mathbf{k}}), \quad (12.8\text{-}19)$$

where f is the usual Fermi–Dirac function and $\mathcal{E}^{\mathbf{k}}$ is the diagonal matrix of signature $[\mathcal{E}_{\mathbf{k}+}, \mathcal{E}_{\mathbf{k}-}, -\mathcal{E}_{-\mathbf{k}+}, -\mathcal{E}_{-\mathbf{k}+}]$. Comparison of Eqs. (12.8-17) and (12.8-19), using (11.8-15) shows that

$$W^{\mathbf{k}} = \tanh(\tfrac{1}{2}\beta U^{\mathbf{k}} \mathcal{E}^{\mathbf{k}} U^{\mathbf{k}\dagger}). \quad (12.8\text{-}20)$$

Hence, $U^{\mathbf{k}}$ is the unitary transformation that diagonalizes the Hermitean matrix $W^{\mathbf{k}}$, whose real eigenvalues are functions of $\mathcal{E}_{\mathbf{k}\sigma}$; therefore, the $W^{\mathbf{k}}$ can be used as variational parameters in lieu of the $U^{\mathbf{k}}$ plus $\mathcal{E}^{\mathbf{k}}$. This is a great simplification in view of the fact that the subsidiary conditions on the $W^{\mathbf{k}}$ are linear $[W^{\mathbf{k}} = W^{\mathbf{k}\dagger} = -JW^{-\mathbf{k}*}J\,]$, in contrast to the conditions on the $U^{\mathbf{k}}$.

Now we look at the Gibbs' function, or "free energy" in the B–W paper, $F = \langle \mathcal{H}_{gr\ red.} \rangle - TS$, to be expressed in the $W^{\mathbf{k}}$. Denoting by $\hat{\varepsilon}_{\mathbf{k}} = \varepsilon_{\mathbf{k}} - \varsigma$ the reduced quasi-particle energies, the grand Hamiltonian becomes

$$\mathcal{H}_{gr\ red.} = \sum\nolimits_{\mathbf{k}\sigma} \hat{\varepsilon}_{\mathbf{k}} c^{\dagger}_{\mathbf{k}\sigma} c_{\mathbf{k}\sigma} + \mathcal{V}, \quad (12.8\text{-}21)$$

where $\mathcal{V}$ is given by the quartic expression (12.8-12). From Wick's theorem and (12.8-16) we then easily obtain

$$\langle \mathcal{H}_{gr\ red.} \rangle = \tfrac{1}{2}\mathrm{tr}\left\{ \sum\nolimits_{\mathbf{k}} \hat{\varepsilon}_{\mathbf{k}} (1 - w^{\mathbf{k}}) + \tfrac{1}{4} \sum\nolimits_{\mathbf{k}\mathbf{k}'} V_{\mathbf{k}\mathbf{k}'} x^{\mathbf{k}\dagger} x^{\mathbf{k}'} \right\}, \quad (12.8\text{-}22)$$

where 'tr' denotes the trace in the matrix space employed. For the entropy one finds

$$\begin{aligned} S &= -k_B\, \mathrm{tr}\{\langle \rho \ln \rho \rangle\} \\ &= -k_B \sum\nolimits_{\mathbf{k}\sigma} \{ f(\mathcal{E}_{\mathbf{k}\sigma}) \ln f(\mathcal{E}_{\mathbf{k}\sigma}) + [1 - f(\mathcal{E}_{\mathbf{k}\sigma})] \ln[1 - f(\mathcal{E}_{\mathbf{k}\sigma})] \} \\ &= -k_B \sum_{\mathbf{k}} \mathrm{tr}\{ \tfrac{1}{2}(1+W^{\mathbf{k}}) \ln \tfrac{1}{2}(1+W^{\mathbf{k}}) \} = -k_B \sum_{\mathbf{k}} \mathrm{tr}\{ \tfrac{1}{2}(1-W^{\mathbf{k}}) \ln \tfrac{1}{2}(1-W^{\mathbf{k}}) \}. \end{aligned} \quad (12.8\text{-}23)$$

This yields the following variational form:

$$\delta F = -\frac{1}{4} \mathrm{tr} \sum\nolimits_{\mathbf{k}} \delta W^{\mathbf{k}} \left[\hat{\mathcal{E}}^{\mathbf{k}} - \frac{1}{\beta} \ln \frac{1+W^{\mathbf{k}}}{1-W^{\mathbf{k}}} \right], \quad (12.8\text{-}24)$$

where now $\hat{\mathcal{E}}^{\mathbf{k}}$ is a partitioned matrix and the gap equation is a two by two matrix:

$$\hat{\mathcal{E}}^{\mathbf{k}} \equiv \begin{pmatrix} \hat{\varepsilon}_{\mathbf{k}} \mathbf{1} & \Delta^{\mathbf{k}} \\ -\Delta^{-\mathbf{k}*} & -\hat{\varepsilon}_{\mathbf{k}} \mathbf{1} \end{pmatrix}, \quad \Delta^{\mathbf{k}} = -\frac{1}{2} \sum\nolimits_{\mathbf{k}'} V_{\mathbf{k}\mathbf{k}'} x^{\mathbf{k}'} = -\tilde{\Delta}^{-\mathbf{k}}. \quad (12.8\text{-}25)$$

Since the brackets [ ] in (12.8-24) must be zero, we finally find – see also (12.8-20),

$$W^{\mathbf{k}} = \tanh\tfrac{1}{2}\beta\hat{\mathscr{E}}^{\mathbf{k}}, \quad \hat{\mathscr{E}}^{\mathbf{k}} = U^{\mathbf{k}}\mathscr{E}^{\mathbf{k}}U^{\mathbf{k}\dagger}; \tag{12.8-26}$$

so that $U^{\mathbf{k}}$ also diagonalizes $\mathscr{E}^{\mathbf{k}}$.

Next, Balian and Werthamer study the possible couplings and resulting gap components. There are clearly a large variety of possible conditions that render the free energy stationary. They restrict themselves to a class of solutions for which the gap matrix is proportional to a unitary matrix, so that

$$\Delta^{\mathbf{k}}\Delta^{\mathbf{k}\dagger} = \Delta^{\mathbf{k}}\Delta^{\mathbf{k}*} = |\Delta_{\mathbf{k}}|^2 . \tag{12.8-27}$$

This, in turn, implies that $\hat{\mathscr{E}}^{\mathbf{k}}$ is proportional to the unit matrix, see (12.8-25); with (12.8-26b) we obtain

$$(\hat{\mathscr{E}}^{\mathbf{k}})^2 = (\mathscr{E}^{\mathbf{k}})^2 = \varepsilon_{\mathbf{k}}^2 + |\Delta_{\mathbf{k}}|^2 \equiv \mathscr{E}_{\mathbf{k}}^2 . \tag{12.8-28}$$

As usual, the coupling kernel only acts within a shell $|\hat{\varepsilon}_{\mathbf{k}}| \le \hbar\omega_0$, so $V_{\mathbf{kk}'}$ is replaced by a constant value at the Fermi energy. With an expansion in spherical harmonics we have for the first two terms

$$V_{kk'} = V_0 - 3V_1(k,k')\hat{\mathbf{k}}\cdot\hat{\mathbf{k}}'. \tag{12.8-29}$$

where $\hat{\mathbf{k}}$ is a unit vector and $V_1 > 0$, resulting in an attractive potential. Balian and Werthamer show that the 'equal spin pairing' (ESP) for which $\sigma = \sigma' = \pm 1$ and $m_S = 1$ or -1, studied previously by Anderson and Morel[70], leads to an anisotropic gap, which is not energetically favourable, nor is such an anisotropy observed. However, there is also the $m_S = 0$ triplet state associated with the p-wave. In this configuration spin and orbital momenta are aligned along the same axes, resulting in

$$\Delta_0^{\mathbf{k}} = \Delta_k \begin{pmatrix} -\hat{k}_x + i\hat{k}_y & \hat{k}_z \\ \hat{k}_z & \hat{k}_x + i\hat{k}_y \end{pmatrix} = \Delta_k \left(\frac{4\pi}{3}\right)^{1/2} \begin{pmatrix} Y_{11}^*(\hat{k})\sqrt{2} & Y_{10}^*(\hat{k}) \\ Y_{10}^*(\hat{k}) & Y_{1,-1}^*(\hat{k})\sqrt{2} \end{pmatrix}. \tag{12.8-30}$$

where we noted that $\hat{k}_x \pm i\hat{k}_y = \sin\theta_k \exp(\pm i\phi_k)$, $\hat{k}_z = \cos\theta_k$, giving the above spherical polynomials. This finally entails the equation for the applicable gap,

$$\Delta_k = \tfrac{1}{2}\sum\nolimits_{\mathbf{k}'} V_1(k,k')(\Delta_{k'}/\mathscr{E}_{k'})\tanh(\tfrac{1}{2}\beta\mathscr{E}_{k'}) . \tag{12.8-31}$$

This result is entirely similar to what we obtained in the BCS theory, see Eq. (12.6-38). The physical difference is that the radial part $V(k,k')$ of the p-wave potential replaces the s-wave interaction potential of the BCS theory for superconductivity.

The above dealt with the major aspects of the superfluid B-state in ^{3}He. Many further calculations, in particular for the spin susceptibility and the response functions, are meritorious; the reader is encouraged to peruse the abundant literature.

---

[70] P.W. Anderson and P. Morel, Phys. Rev. **123**, 1911 (1961).

2. Formal Theory; Diagrammatic Methods

12.9 Perturbation Expansion of the Grand-Canonical Partition Function

In this section we shall obtain an expansion for the grand-canonical partition function. To ease on the notation we denote the grand Hamiltonian by the symbol K and employ the symbol Ω for the Gibbs function or grand potential. We then have

$$\Omega(T,V,\varsigma) = -\beta^{-1}\ln\mathcal{Z}, \quad \mathcal{Z} = \mathrm{Tr}\{\exp-\beta K\} = \mathrm{Tr}\{\exp[-\beta(K^0 + \lambda\mathcal{V})]\}. \quad (12.9\text{-}1)$$

The term K^0 involves the unperturbed energy of the particles, typically $\Sigma_k(\varepsilon_k - \varsigma)\hat{e}_k^\dagger\hat{e}_k$, where $\hat{e}_k^\dagger$ and $\hat{e}_k$ will be either boson or fermion creation and annihilation operators; the designation 'k' usually stands for momentum states, i.e., for bosons $\hat{e}_k, \hat{e}_k^\dagger \to a_{\mathbf{k}}, a_{\mathbf{k}}^\dagger$, but will include spin for the case of fermions, $\hat{e}_k, \hat{e}_k^\dagger \to c_{\mathbf{k}\sigma}, c_{\mathbf{k}\sigma}^\dagger$. The potential energy will be expressed in a series of normally ordered $\hat{e}$-products; while the ordering is unimportant for a basis of boson states, for the sake of uniformity of presentation in this chapter, we shall assume that both the boson and fermion bases are ordered. The reader will realize that the far rhs of Eq. (12.9-1) is a *symbolic* expression, since the operators K^0 and $\lambda\mathcal{V}$ do not commute. We will also formally write

$$\mathcal{Z} = \mathrm{Tr}\{e^{-\beta K^0} U_I(-i\hbar\beta)\}, \quad (12.9\text{-}2)$$

where $U_I(t)$ is the interaction operator

$$U_I(t,t_0) \equiv U_I(t-t_0) = U^{(0)\dagger}(t-t_0)e^{-iK(t-t_0)/\hbar} = e^{iK^0(t-t_0)/\hbar}e^{-i(K^0+\lambda\mathcal{V})(t-t_0)/\hbar}. \quad (12.9\text{-}3)$$

In particular, for imaginary time, $t - t_0 \to -i\hbar\beta$,

$$U_I(-i\hbar\beta) = e^{\beta K^0}e^{-\beta(K^0+\lambda\mathcal{V})}. \quad (12.9\text{-}4)$$

Clearly, (12.9-2) and (12.9-4) reconstitute the expression of departure, (12.9-1). To obtain useful results, we must give the perturbation expansion of the interaction operator $U_I(t)$.

12.9.1 *The Interaction Picture; Expansion of the Evolution Operator*

We review some basic results, which we have been further elaborated in Appendix A. The evolution operator $U(t,t_0) \equiv \exp[-iK(t-t_0)/\hbar]$ satisfies the differential equation [71]

---

[71] A. Messiah, Op. Cit., Vol. II, p. 722*ff*. Note that in the usual quantum context the grand Hamiltonian K is replaced by the ordinary Hamiltonian $\mathcal{H}$ and K^0 by the unperturbed Hamiltonian $\mathcal{H}^0$.

$$\hbar i \frac{dU(t,t_0)}{dt} = KU(t,t_0), \quad K = K^0 + \lambda \mathcal{V}. \tag{12.9-5}$$

By $U^0(t,t_0) = \exp[-iK^0(t-t_0)/\hbar]$ we denote the evolution operator in the absence of perturbations; it satisfies

$$\hbar i \frac{dU^0(t,t_0)}{dt} = K^0 U^0(t,t_0). \tag{12.9-6}$$

Next we introduce the interaction operators,

$$U_I(t,t_0) = U^{0\dagger}(t,t_0)U(t,t_0) = e^{itK^0/\hbar}e^{-i(t-t_0)K/\hbar}e^{-it_0K^0/\hbar}, \tag{12.9-7}$$

$$\mathcal{V}_I(t) = U^{0\dagger}(t,t_0)\mathcal{V}U^0(t,t_0) = e^{i(t-t_0)K^0/\hbar}\mathcal{V}e^{-i(t-t_0)K^0/\hbar}, \tag{12.9-8}$$

where the non-subscripted operator $\mathcal{V}$ is the Schrödinger operator.[72] From the last four equations one easily obtains

$$\frac{dU_I(t,t_0)}{dt} = \frac{1}{\hbar i} U^{0\dagger}(t,t_0)\lambda\mathcal{V}(t)\underbrace{U^0(t,t_0)U^{0\dagger}(t,t_0)}_{inserted=1}U(t,t_0) = \frac{1}{\hbar i}\lambda\mathcal{V}_I(t)U_I(t,t_0), \tag{12.9-9}$$

which has as formal solution the integral equation

$$U_I(t,t_0) = 1 - i(\lambda/\hbar)\int_{t_0}^{t} \mathcal{V}_I(t')U_I(t',t_0)dt'. \tag{12.9-10}$$

This is further evaluated by iteration. In zeroth order $U_I^{(0)} = 1$, while in first order

$$U_I^1(t,t_0) = 1 - i(\lambda/\hbar)\int_{t_0}^{t} \mathcal{V}_I(t')U_I^{(0)}(t',t_0)dt' = 1 - i(\lambda/\hbar)\int_{t_0}^{t}\mathcal{V}_I(t')dt' \equiv 1 + U_I^{(1)}(t,t_0). \tag{12.9-11}$$

Generally we obtain

$$U_I(t,t_0) = 1 + \sum_{n=1}^{\infty} U_I^{(n)}(t,t_0), \tag{12.9-12}$$

where,

$$U_I^{(n)}(t,t_0) = (\lambda/\hbar i)^n \int_{t_0}^{t} dt_n \int_{t_0}^{t_n} dt_{n-1} \int_{t_0}^{t_{n-1}} dt_{n-2} \ldots \int_{t_0}^{t_2} dt_1 \mathcal{V}_I(t_n)\mathcal{V}_I(t_{n-1})\mathcal{V}_I(t_{n-2})\ldots\mathcal{V}_I(t_1). \tag{12.9-13}$$

Note that we have here a *time-ordered integral* as discussed in section 2.5.2, which is pre-eminently amenable to Laplace transformation – a property encountered in non-

---

[72] This operator may depend on t in time-dependent perturbation theory. As to the final rhs of (12.9-7), although K and K^0 do not commute, the factor $\exp(-it_0K^0/\hbar)$ can be placed to the right, since it leaves the norm of $U_I(t,t_0)$ unchanged while still satisfying Eq. (12.9-9).

equilibrium theory. Further, inverting (12.9-7)

$$U(t,t_0) = U^0(t,t_0)U_I(t,t_0), \tag{12.9-14}$$

and employing the composition property

$$U^0(t,t_0)U^{0\dagger}(t',t_0) = e^{-i(t-t_0)K^0/\hbar}e^{i(t'-t_0)K^0/\hbar} = e^{-i(t-t')K^0/\hbar} = U^0(t,t'), \tag{12.9-15}$$

we obtain the perturbation series for the evolution operator

$$U(t,t_0) = U^0(t,t_0) + \sum_{n=1}^{\infty} U^{(n)}(t,t_0), \tag{12.9-16}$$

where

$$U^{(n)}(t,t_0) = (\lambda/\hbar i)^n \int_{t_0}^{t} dt_n \int_{t_0}^{t_n} dt_{n-1} \int_{t_0}^{t_{n-1}} dt_{n-2} \ldots \int_{t_0}^{t_2} dt_1$$

$$\times U^0(t,t_n)\mathcal{V}(t_n)U^0(t_n,t_{n-1})\mathcal{V}(t_{n-1})\ldots U^0(t_2,t_1)\mathcal{V}(t_1)U^0(t_1,t_0). \tag{12.9-17}$$

We return to Eq. (12.9-13), needed for the grand-canonical partition function as given by (12.9-2). As usual $\mathcal{V}$ is a sum of two-body potentials, $\mathcal{V} = \frac{1}{2}\Sigma_{i,j}v_{ij}$. If exchange energy plays no role or can be neglected (Hartree approximation) so that no symmetrizing or antisymmetrizing of the operators $\mathcal{V}$ in (12.9-13) is necessary, we can write – see Section 4.5 ,

$$U_I^{(n)}(t,t_0) = \frac{(\lambda/\hbar i)^n}{n!} \int_{t_0}^{t} dt_n \int_{t_0}^{t} dt_{n-1} \int_{t_0}^{t} dt_{n-2} \ldots \int_{t_0}^{t} dt_1 \mathcal{T}\{\mathcal{V}_I(t_n)\mathcal{V}_I(t_{n-1})\mathcal{V}_I(t_{n-2})\ldots\mathcal{V}_I(t_1)\}, \tag{12.9-18}$$

where $\mathcal{T}$ designates time-ordering: $t_n > t_{n-1} > t_{n-2} > \ldots t_1 > t_0$. Usually we take $t_0 = 0$. We now have the appealing simple forms,

$$U_I(t) = \mathcal{T}\left\{\exp\left[-(i/\hbar)\int_0^t \lambda\mathcal{V}_I(t')dt'\right]\right\}, \tag{12.9-19}$$

and

$$U_I(-i\hbar\beta) = \mathcal{T}\left\{\exp\left[-\int_0^\beta \lambda\mathcal{V}_I(-i\hbar\tau)d\tau\right]\right\}, \tag{12.9-20}$$

where we set $\tau = i\hbar^{-1}t'$. The time-ordering operator now denotes tau-ordering along the imaginary axis. We will set $\tilde{V}(\tau) \equiv \mathcal{V}_I(-i\hbar\tau)$; then we have the usual definition[73]

$$\tilde{V}(\tau) = \exp(\tau K^0)\mathcal{V}\exp(-\tau K^0). \tag{12.9-21}$$

---

[73] L.E. Reichl, "A Modern Course in Statistical Physics", first Ed., U. of Texas Press, p. 401ff. [This material is not in the second edition.]

Since exchange energy usually *is* important, we will now give a general method. From (12.9-2) and (12.9-12)+(12.9-13), with the substitutions $t \to -i\hbar\beta,\ t_i \to -i\hbar\tau_i$, we obtain

$$\mathcal{Z} = \sum_{n=0}^{\infty} (-\lambda)^n \int_0^{\beta} d\tau_n \int_0^{\tau_n} d\tau_{n-1} \dots \int_0^{\tau_2} d\tau_1 \mathrm{Tr}\left\{ e^{-\beta K^0} \tilde{V}(\tau_n)\tilde{V}(\tau_{n-1})\dots\tilde{V}(\tau_1)\right\}. \quad (12.9\text{-}22)$$

The chemical potential ς, contained in K^0, depends on λ, so that (12.9-22) is *not* a simple power expansion in orders of perturbation. We have already noted

$$K^0 = \sum_k (\varepsilon_k - \varsigma)\hat{e}_k^\dagger \hat{e}_k, \quad (12.9\text{-}23)$$

while for $\tilde{V}$ we write

$$\tilde{V} = \frac{1}{2} \sum_{k_1,k_2,k_3,k_4} \hat{e}_{k_1}^\dagger \hat{e}_{k_2}^\dagger \langle k_1 k_2 | \tilde{v} | k_3 k_4 \rangle \hat{e}_{k_4} \hat{e}_{k_3}, \quad k = (\mathbf{k}) \text{ or } k = (\mathbf{k}, \sigma). \quad (12.9\text{-}24)$$

It is now customary to add Heaviside functions to the $\tilde{V}$-products; e.g., for the second-order term:

$$2^{nd}\,order\ term = (-\lambda)^2 \int_0^{\beta} d\tau_2 \int_0^{\beta} d\tau_1 \mathrm{Tr}\left\{ e^{-\beta K^0} \tilde{V}(\tau_2)\tilde{V}(\tau_1)\Theta(\tau_2 - \tau_1)\right\}$$
$$= \frac{(-\lambda)^2}{2!} \int_0^{\beta} d\tau_2 \int_0^{\beta} d\tau_1 \mathrm{Tr}\left\{ e^{-\beta K^0} \left[\tilde{V}(\tau_2)\tilde{V}(\tau_1)\Theta(\tau_2 - \tau_1) + \varepsilon^P \tilde{V}(\tau_1)\tilde{V}(\tau_2)\Theta(\tau_1 - \tau_2)\right]\right\}$$
$$= \frac{(-\lambda)^2}{2!} \int_0^{\beta} d\tau_2 \int_0^{\beta} d\tau_1 \mathrm{Tr}\left\{ e^{-\beta K^0} \mathcal{T}\left[\tilde{V}(\tau_2)\tilde{V}(\tau_1)\right]\right\}, \quad (12.9\text{-}25)$$

where

$$\varepsilon = \begin{cases} +1 \text{ for bosons} \\ -1 \text{ for fermions,} \end{cases} \quad (12.9\text{-}26)$$

and where $\mathcal{T}$ is a time-ordering as well as (anti)symmetrizing operator, providing every permutation with a parity *p*; otherwise stated, *$\mathcal{T}$ denotes the sum of all n! time-ordered products of $\tilde{V}$-operators*. With these measures, the grand partition function takes the form

$$\mathcal{Z} = \sum_{n=0}^{\infty} \frac{(-\lambda)^n}{n!} \int_0^{\beta} d\tau_n \int_0^{\beta} d\tau_{n-1} \dots \int_0^{\beta} d\tau_1 \mathrm{Tr}\left\{ e^{-\beta K^0} \mathcal{T}\left[\tilde{V}(\tau_n)\tilde{V}(\tau_{n-1})\dots\tilde{V}(\tau_1)\right]\right\}$$
$$= \mathrm{Tr}\left\{ \exp(-\beta K^0) \mathcal{T}\left[\exp(-\lambda \int_0^{\beta} \tilde{V}(\tau) d\tau \right]\right\}, \quad (12.9\text{-}27)$$

in harmony with our previous results (12.9-2) and (12.9-20,21). We note once more that (12.9-21) is not Hermitean *unless* τ is an imaginary time $it/\hbar$; also – in our view – τ^{-1} has no thermodynamic meaning as a temperature, unlike $1/\beta$ $(=k_BT)$.

Finally, we give an often used alternate form. Let

$$\mathcal{Z}/\mathcal{Z}_0 = \sum_{n=0}^{\infty} \lambda^n W_n(T,V,\varsigma), \quad \mathcal{Z}_0 = \mathrm{Tr}\{\exp(-\beta K^0)\}. \tag{12.9-28}$$

We thus have

$$W_n = \left(\mathrm{Tr}\{e^{-\beta K^0}\}\right)^{-1} \frac{(-1)^n}{n!} \int_0^\beta d\tau_n \int_0^\beta d\tau_{n-1} \ldots \int_0^\beta d\tau_1 \mathrm{Tr}\left\{e^{-\beta K^0} \mathcal{T}\left[\tilde{V}(\tau_n)\tilde{V}(\tau_{n-1})\ldots\tilde{V}(\tau_1)\right]\right\}. \tag{12.9-29}$$

This leads to a cumulant expansion

$$\ln(\mathcal{Z}/\mathcal{Z}_0) = \sum_{m=1}^{\infty} \lambda^m C_m(T,V,\varsigma); \tag{12.9-30}$$

The cumulants are given by

$$\begin{aligned} C_1 &= W_1, \\ C_2 &= W_2 - \tfrac{1}{2} W_1^2, \\ C_3 &= W_3 - W_1 W_2 + \tfrac{1}{3} W_1^3, \ldots, etc. \end{aligned} \tag{12.9-31}$$

The grand potential then follows from

$$\Omega(T,V,\varsigma) = -(\beta)^{-1} \ln \mathrm{Tr}\{\exp[-\beta K^0(T,V,\varsigma)]\} - (\beta)^{-1} \sum\nolimits_{m=1}^{\infty} \lambda^m C_m(T,V,\varsigma). \tag{12.9-32}$$

The problem now has been reduced to evaluating the trace in (12.9-29), i.e., finding the averaged $\tilde{V}$-operator products at a temperature *T*; this can be done with Wick's theorem.

12.9.2 *Generalized Wick's Theorem*

Wick's original theorem provides for the computation of normally ordered products of operators – either involving creation and annihilation operators or field operators $\Psi^\dagger, \Psi$ in the many-body ground state.[74] Later it was generalized to give the expectation value of statistical product operators in an ensemble at a temperature *T*.[75] Basically, for our purpose the theorem states:

---

[74] G.C. Wick, Phys. Rev. **80**, 268 (1950).

[75] C. Bloch and C. De Dominicis, Nucl. Phys. **7**, 459 (1958); M. Gaudin, ibid. **15**, 89 (1960).

The average of an operator product, involving creation and annihilation operators, is obtained by associating the product into pairs of such operators, for which the expectation value is evaluated, supplying each such pair-expression product with its parity sign and summing over all possible pair-expression products.

For a proof of Wick's original theorem, see Fetter and Walecka, Op. Cit. For the generalized theorem, read the above cited literature. [Product averages involving two creation operators or two annihilation operators are generally zero; they must be added, however, for Cooper-pairs, cf. e.g., Eq. (12.6-29).]

In the present context we will need the product averages of (imaginary) time-dependent operators. We therefore first take a look at the interaction-operator averages $\langle \tilde{e}_k^\dagger(\tau_2)\tilde{e}_{k'}(\tau_1)\rangle$. We recall the following results for the (interaction) creation and annihilation from the Heisenberg equation of motion, see Eqs. (7.4-17) and (7.4-18):

$$\begin{aligned} \hbar i(d/dt)\tilde{e}_k(t) &= [\tilde{e}_k(t), K^0] = (\varepsilon_k - \varsigma)\tilde{e}_k(t), \\ -\hbar i(d/dt)\tilde{e}_k^\dagger(t) &= [\tilde{e}_k^\dagger(t), K^0] = (\varepsilon_k - \varsigma)\tilde{e}_k^\dagger(t). \end{aligned} \tag{12.9-33}$$

Letting $t \rightarrow -i\hbar\tau$ we also have

$$(d/d\tau)\tilde{e}_k(\tau) = -[\tilde{e}_k(\tau), K^0] = -(\varepsilon_k - \varsigma)\tilde{e}_k(\tau), \tag{12.9-34}$$

$$(d/d\tau)\tilde{e}_k^\dagger(\tau) = [\tilde{e}_k^\dagger(\tau), K^0] = (\varepsilon_k - \varsigma)\tilde{e}_k^\dagger(\tau), \tag{12.9-35}$$

with solutions

$$\tilde{e}_k(\tau) = e^{-\tau(\varepsilon_k - \varsigma)}\hat{e}_k, \qquad \tilde{e}_k^\dagger(\tau) = e^{\tau(\varepsilon_k - \varsigma)}\hat{e}_k^\dagger. \tag{12.9-36}$$

For $\tau \rightarrow \beta$ (12.9-36) reads

$$\hat{e}_k e^{-\beta(\varepsilon_k - \varsigma)} = \tilde{e}_k(\beta) = e^{\beta K^0}\hat{e}_k e^{-\beta K^0} \text{ or } e^{-\beta K^0}\hat{e}_k e^{-\beta(\varepsilon_k - \varsigma)} = \hat{e}_k e^{-\beta K^0}. \tag{12.9-37}$$

Let $\rho_0 = \exp(-\beta K^0)/\mathcal{Z}_0$ be the density operator when the interactions are turned off. We then have the averages at temperature T:

$$\begin{aligned} \langle \hat{e}_{k'}^\dagger \hat{e}_k \rangle_0 &= \mathrm{Tr}\{\rho_0 \hat{e}_{k'}^\dagger \hat{e}_k\} = \mp\delta_{k,k'} \pm \mathrm{Tr}\{\rho_0 \hat{e}_k \hat{e}_{k'}^\dagger\} \\ &= \mp\delta_{k,k'} \pm e^{\beta(\varepsilon_k - \varsigma)}\mathrm{Tr}\{(e^{-\beta K^0}\hat{e}_k e^{-\beta(\varepsilon_k - \varsigma)})\hat{e}_{k'}^\dagger\}/\mathcal{Z}_0 \\ &\overset{(12.9\text{-}37)}{=} \mp\delta_{k,k'} \pm e^{\beta(\varepsilon_k - \varsigma)}\mathrm{Tr}\{\hat{e}_k e^{-\beta K^0}\hat{e}_{k'}^\dagger\}/\mathcal{Z}_0 = \mp\delta_{k,k'} \pm e^{\beta(\varepsilon_k - \varsigma)}\langle \hat{e}_{k'}^\dagger \hat{e}_k\rangle_0, \end{aligned} \tag{12.9-38}$$

where we used the cyclic invariance of the trace; note that the upper signs pertain to bosons and the lower ones to fermions. Rearranging we find

$$\langle \hat{e}_{k'}^\dagger \hat{e}_k \rangle_0 = \langle n_k \rangle_0 \delta_{k,k'} = \frac{\delta_{k,k'}}{e^{\beta(\varepsilon_k - \varsigma)} \mp 1}, \tag{12.9-39}$$

which is the usual Bose–Einstein or Fermi–Dirac distribution. With Wick's theorem we can easily obtain averages for other normal products, the simplest one being

$$\mathrm{Tr}\{\rho_0 \hat{e}^\dagger_{k_1}\hat{e}^\dagger_{k_2}\hat{e}_{k_4}\hat{e}_{k_3}\} = \langle\hat{e}^\dagger_{k_1}\hat{e}_{k_3}\rangle_0\langle\hat{e}^\dagger_{k_2}\hat{e}_{k_4}\rangle_0 \pm \langle\hat{e}^\dagger_{k_1}\hat{e}_{k_4}\rangle_0\langle\hat{e}^\dagger_{k_2}\hat{e}_{k_3}\rangle_0\,; \qquad (12.9\text{-}40)$$

the $\langle..\rangle_0$ are averages over the free quasi-particle operator ρ_0. Note that a 6-product will have $3!$ terms; the complexity rapidly grows. However, the main complication is that we will need time-ordered products, which require free propagators or *Matsubara Green's functions.* Thus, consider the two-time product,

$$\mathcal{T}[\tilde{e}_k(\tau_2)\tilde{e}^\dagger_{k'}(\tau_1)] = \tilde{e}_k(\tau_2)\tilde{e}^\dagger_{k'}(\tau_1)\Theta(\tau_2-\tau_1) \pm \tilde{e}^\dagger_{k'}(\tau_1)\tilde{e}_k(\tau_2)\Theta(\tau_1-\tau_2). \quad (12.9\text{-}41)$$

We define a contraction to be the average of a time-ordered product pair. In particular

$$\mathrm{Tr}\{\rho_0\mathcal{T}[\tilde{e}_k(\tau_2)\tilde{e}^\dagger_{k'}(\tau_1)]\} = \langle\mathcal{T}[\tilde{e}_k(\tau_2)\tilde{e}^\dagger_{k'}(\tau_1)]\rangle_0 \equiv -G_0(k,k',\tau_2-\tau_1). \quad (12.9\text{-}42)$$

Using the rules (12.9-35) and writing $\langle n_k\rangle_0 = n_0(k)$, one easily obtains

$$G_0(k,k',\tau_2-\tau_1) = \delta_{kk'}[\mp n_0(k)\Theta(\tau_1-\tau_2) - (1 \pm n_0(k)\Theta(\tau_2-\tau_1)]e^{-(\varepsilon_k-\varsigma)(\tau_2-\tau_1)}. \quad (12.9\text{-}43)$$

Obviously, only one of the two terms applies. For the equal time Green's function we have $G_0(k,k',0-) = \mp\delta_{k,k'}n_0(k)$. For the four-time product we find from Wick's theorem in a straightforward manner

$$\begin{aligned}&\langle\mathcal{T}[\tilde{e}^\dagger_{k_1}(\tau_4)\tilde{e}^\dagger_{k_2}(\tau_3)\tilde{e}_{k_3}(\tau_2)\tilde{e}_{k_4}(\tau_1)]\rangle_0\\ &\quad= \delta_{k_1,k_4}\delta_{k_2,k_3}G_0(k_1,k_4,\tau_4-\tau_1)G_0(k_2,k_3,\tau_3-\tau_2)\\ &\quad\pm \delta_{k_1,k_3}\delta_{k_2,k_4}G_0(k_1,k_3,\tau_3-\tau_1)G_0(k_2,k_4,\tau_4-\tau_2).\end{aligned} \qquad (12.9\text{-}44)$$

Finally, we return to the form for the terms W_n, see (12.9-29). With our present notation we have

$$W_n = \frac{(-1)^n}{n!}\int_0^\beta d\tau_n\int_0^\beta d\tau_{n-1}\ldots\int_0^\beta d\tau_1 \mathrm{Tr}\Big\{\rho_0\mathcal{T}\Big[e^{\tau_n K^0}\mathcal{V}e^{-\tau_n K^0}\,e^{\tau_{n-1}K^0}\mathcal{V}e^{-\tau_{n-1}K^0}\ldots e^{\tau_1 K^0}\mathcal{V}e^{-\tau_1 K^0}\Big]\Big\}$$

$$= \frac{(-1)^n}{n!2^n}\sum_{\{\bar{k}_i',\bar{k}_i,k_i',k_i\}}\int_0^\beta d\tau_n\int_0^\beta d\tau_{n-1}..\int_0^\beta d\tau_1\langle\tilde{\mathcal{T}}[\tilde{e}^\dagger_{\bar{k}'_n}(\tau_n)\tilde{e}^\dagger_{\bar{k}_n}(\tau_n)\langle\bar{k}'_n\bar{k}_n|v|k'_n k_n\rangle\tilde{e}_{k_n}(\tau_n)\tilde{e}_{k'_n}(\tau_n)$$

$$\times\tilde{e}^\dagger_{\bar{k}'_{n-1}}(\tau_{n-1})\tilde{e}^\dagger_{\bar{k}_{n-1}}(\tau_{n-1})\,\langle\bar{k}'_{n-1}\bar{k}_{n-1}|v|k'_{n-1}k_{n-1}\rangle\tilde{e}_{k_{n-1}}(\tau_{n-1})\tilde{e}_{k'_{n-1}}(\tau_{n-1})\ \ldots$$

$$\times\tilde{e}^\dagger_{\bar{k}'_1}(\tau_1)\tilde{e}^\dagger_{\bar{k}_1}(\tau_1)\langle\bar{k}'_1\bar{k}_1|v|k'_1k_1\rangle\langle\tilde{e}_{k_1}(\tau_1)\tilde{e}_{k'_1}(\tau_1)]\rangle_0. \qquad (12.9\text{-}45)$$

The creation and annihilation operators can, of course, all be put to the front and all the matrix elements assembled at the end. Yet this becomes rapidly very nasty; so in order to keep track of all the terms, a diagrammatic representation will be followed.

12.10 Momentum-Space Diagrams

We shall study a few terms of the expansion for the grand-canonical partition function, Eqs. (12.9-29) and (12.9-45), in detail. For W_1 we have

$$W_1 = \frac{-1}{2}\int_0^\beta d\tau_1 \sum_{k_1k_2k_3k_4} \langle \mathcal{T}[\tilde{e}^\dagger_{k_1}(\tau_1)\tilde{e}^\dagger_{k_2}(\tau_1)\tilde{e}_{k_4}(\tau_1)\tilde{e}_{k_3}(\tau_1)]\rangle\langle k_1k_2|v|k_3k_4\rangle. \quad (12.10\text{-}1)$$

The Green's functions will be simplified, in view of the Kronecker deltas which occur in the free propagators:

$$G_0(k,k',\tau_2-\tau_1)\delta_{k,k'} \equiv G_0(k,\tau_2-\tau_1). \quad (12.10\text{-}2)$$

From Wicks theorem we obtain for W_1

$$W_1 = \frac{-1}{2}\left\{\beta\sum_{k_1k_2} G_0(k_1,0)G_0(k_2,0)\langle k_1k_2|v|k_1k_2\rangle + \varepsilon\sum_{k_1k_2} G_0(k_1,0)G_0(k_2,0)\langle k_1k_2|v|k_2k_1\rangle\right\}. \quad (12.10\text{-}3)$$

where $G_0(k,0) \equiv G(k,0-) = \mp n_0(k)$. The first term represents the direct interaction, while the second one represents the exchange energy. In a similar way we find for W_2 with the help of Wick's theorem[76]

$$W_2 = \frac{1}{8}\left[\sum_{k_1k_2}\int_0^\beta d\tau_1 G_0(k_1,0)G_0(k_2,0)\langle k_1k_2|v|k_1k_2\rangle\right]^2 \quad (A)$$

$$+\varepsilon\frac{1}{4}\sum_{k_1\ldots k_4}\int_0^\beta d\tau_2\int_0^\beta d\tau_1 G_0(k_1,0)G_0(k_2,0)G_0(k_3,0)G_0(k_4,0)$$

$$\times\langle k_1k_2|v|k_1k_2\rangle\langle k_3k_4|v|k_4k_3\rangle \quad (B)$$

$$+\frac{1}{8}\left[\sum_{k_1k_2}\int_0^\beta d\tau_1 G_0(k_1,0)G_0(k_2,0)\langle k_1k_2|v|k_2k_1\rangle\right]^2 \quad (C)$$

$$+\varepsilon\frac{1}{2}\sum_{k_1..k_4}\int_0^\beta d\tau_2\int_0^\beta d\tau_1 G_0(k_2,\tau_2-\tau_1)G_0(k_3,\tau_1-\tau_2)G_0(k_1,0)G_0(k_4,0)$$

$$\times\langle k_1k_3|v|k_1k_2\rangle\langle k_4k_2|v|k_4k_3\rangle \quad (D)$$

---

[76]This result is obtained from the $4! = 24$ terms stemming from Wick's theorem; terms can be combined by interchanging τ_1 and τ_2, relabelling the summation variables and using Eq.(12.10-6) below.

$$+\sum_{k_1..k_4}\int_0^\beta d\tau_2\int_0^\beta d\tau_1 G_0(k_2,\tau_2-\tau_1)G_0(k_3,\tau_1-\tau_2)G_0(k_1,0)G_0(k_4,0)$$
$$\times\langle k_1k_3|v|k_2k_1\rangle\langle k_4k_2|v|k_4k_3\rangle \qquad (E)$$
$$+\varepsilon\frac{1}{2}\sum_{k_1..k_4}\int_0^\beta d\tau_2\int_0^\beta d\tau_1 G_0(k_2,\tau_2-\tau_1)G_0(k_3,\tau_1-\tau_2)G_0(k_1,0)G_0(k_4,0)$$
$$\times\langle k_1k_3|v|k_2k_1\rangle\langle k_4k_2|v|k_3k_4\rangle \qquad (F)$$
$$+\varepsilon\frac{1}{4}\sum_{k_1..k_4}\int_0^\beta d\tau_2\int_0^\beta d\tau_1 G_0(k_1,\tau_2-\tau_1)G_0(k_2,\tau_2-\tau_1)G_0(k_3,\tau_1-\tau_2)G_0(k_4,\tau_1-\tau_2)$$
$$\times\langle k_1k_2|v|k_4k_3\rangle\langle k_3k_4|v|k_1k_2\rangle \qquad (G)$$
$$+\frac{1}{4}\sum_{k_1..k_4}\int_0^\beta d\tau_2\int_0^\beta d\tau_1 G_0(k_1,\tau_2-\tau_1)G_0(k_2,\tau_2-\tau_1)G_0(k_3,\tau_1-\tau_2)G_0(k_4,\tau_1-\tau_2)$$
$$\times\langle k_1k_2|v|k_3k_4\rangle\langle k_3k_4|v|k_1k_2\rangle. \qquad (H)$$
$$(12.10\text{-}4)$$

For W_1 we have two terms and for W_2 eight terms. This number rapidly grows!

12.10.1 *Feynman Diagrams*

Clearly it is advisable to devise a pattern that represent each term by established rules. Feynman diagrams were first developed by Feynman in his work on quantum electrodynamics.[77] The connection with quantum field theory was established by Dyson.[78] For those unfamiliar with these diagrams we refer to Fetter and Walecka, Op. Cit., and to other standard works[79]; of these the treatise by Negele and Orland is most applicable to the present material.

The rules for associating diagrams with each term and the weighting factors to be assigned are mostly straightforward. We list them here as follows.[80]

---

[77] R.P. Feynman, Phys. Rev.**76**, 749 and 769 (1949).
[78] F.J. Dyson, Phys. Rev. **75**, 486 and 1736 (1949).
[79] R.D. Mattuch, "A guide to Feynman diagrams in the many-body problem," 2nd Ed., McGraw-Hill, NY, 1976; A.A. Abrikosov, L.P. Gorkov and I.E. Dzyaloshinski, "Methods of Quantum Field Theory in Statistical Physics," Prentice Hall, Englewood Cliffs, NJ, 1963; G.D. Mahan, Op. Cit., Chapter 3; J.W. Negele and H. Orland, "Quantum Many-Particle Systems," Addison-Wesley Publ. Co., NY, 1988.
[80] The rules will here be stated only for two-body interactions; for the effect of an external field (one-body interactions) or when two particles and a massless quasi-particle interact (electron-phonon or –photon collisions) the rules must be modified – see e.g., G.D. Mahan, Op. Cit.

(i) An interaction is represented by a horizontal wavy line or 'spring' (most older books) or a dashed line (Negele & Orland and this text) with end points that are labelled L and R; together they are called 'the vertex'. For two-body interactions, two lines each emanate from L and R. The wavy or dashed lines are labelled with ordered times $t_n ... t_1$ or in 'temperature diagrams' with imaginary times $\tau_n ... \tau_1$. Hence,

$$\mathcal{V}(t) \text{ or } \hat{V}(\tau) \quad \longrightarrow$$

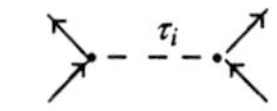

(ii) Solid lines $\rightarrow$ represent states; each line is labelled with a state momentum k_i or (k_i, η_i), where η_i denotes the spin part. Diagrams may be closed, in which case there are loops. If external lines occur, label them $t' \xrightarrow{k'}$ and $\xrightarrow{k} t$ for time-ordered diagrams and $\tau' \xrightarrow{k'}$ and $\xrightarrow{k} \tau$ in 'temperature diagrams'.

(iii) Assign a product term $-\langle k_1 k_2 | v | k_3 k_4 \rangle$ to each vertex, where k_1 is the left outgoing line, k_2 is the right outgoing line, k_3 is the left incoming line and k_4 is the right incoming line.

(iv) Assign a Green's function to each directed line. For a line representing the state k_i going from t_j to t_i, or in the present situation from τ_j or τ_i, write $G_0(k_i, \tau_j - \tau_i)$. Sum over all internal state labels and integrate $[\int_{t_0}^{t} dt_n .. dt_1]$ or $[\int_0^{\beta} d\tau_n .. d\tau_1]$ all internal variables.

Next comes the problem of 'weight'. Then the following applies.

(v) For each fermion loop in a diagram we need a factor (−), so that the diagram is to be provided with $(-1)^F$, where F is the number of loops; no factor is needed for bosons.

(vi) Each diagram has a symmetry factor S, for which a factor $1/S$ is given to the weight of the diagram. Technically S is the number of transformations, composed of time permutations and vertices-extremity exchanges or a combination thereof, which transform the diagram into a deformation of itself. While often this can be decided 'at a glance' visual assignment is fallible. It helps therefore to know the sum-rule: $\Sigma(1/S) = (2n-1)!!$ This provides a check on the result for all n-th order diagrams.

We finally arrive at the desired result for our purpose:

$$W_n = \sum \textit{all weighted topologically-distinct n-th order diagrams}. \qquad (12.10\text{-}5)$$

We shall illustrate these rules to obtain the diagrammatic representation for the two and eight Feynman diagrams for W_1 and W_2, respectively. The diagrams for W_1 are in Fig. 12-16. They clearly show the effect of exchange. The first diagram has two loops, so its Fermi factor is one; the second one has one loop through the vertex points, so the Fermi factor is ε. Both have a symmetry factor of two, giving rise to a weight of ½ ; the left diagram can be rotated about a vertical axis leaving it invariant, while the right diagram must be successively horizontally and vertically rotated in order to remain invariant. Needless to say that the labels are of no importance and are omitted in most diagrammatic representations.

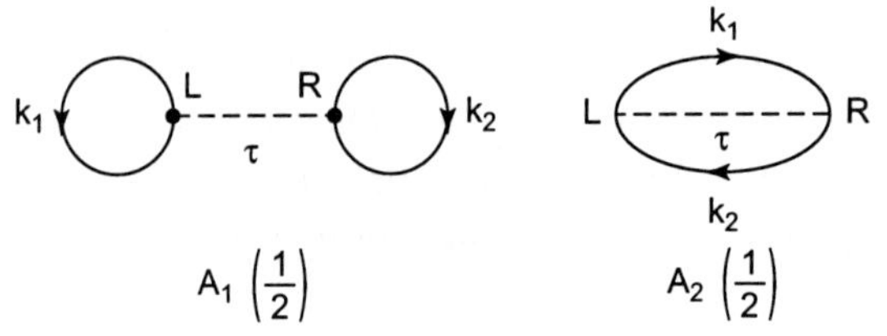

Fig. 12-16. Feynman diagrams for W_1.

For W_2 we have eight distinct diagrams, pictured in Fig. 12-17. They are labelled (*A*) to (*H*); these labels have been entered in the various parts of Eq. (12.10-6). The reader is encouraged to verify the matrix elements and Green's functions associated with each diagram. Again, we have added all the states and the *L*-*R* positions of the vertices, as well as the tau's; given that the state labels as well as the tau's are dummy variables, they can be freely assigned and are usually not shown. The weights, being the reciprocal symmetry factors $1/S$, are given in parentheses for each diagram; we note that $\Sigma(1/S) = (2n-1)!! \xrightarrow{n=2} 3$. The number of fermion loops varies from four in (*A*) to one in (*F*) and (*G*). We note that these particular sets of Feynman diagrams have no external lines but are all closed; the initial and final states $(0, \beta)$ only enter as integration points.

An explanation is still in order with respect to the statement 'topologically distinct'. First of all we notice the following equality for two-body interactions,

$$\langle \alpha\beta \,|\, v \,|\, \gamma\delta \rangle = \langle \beta\alpha \,|\, v \,|\, \delta\gamma \rangle. \tag{12.10-6}$$

This entails that we can interchange the extremities denoted as *L* and *R*, as we now show. Following Negele and Orland[81] we consider the diagram Γ in which four states with directed lines $\alpha ... \delta$ connect one vertex with the rest of the diagram, called *F*. Also, we consider the diagram Γ' for which the labels *L* and *R* are interchanged; both diagrams are depicted in Fig. 12-18. We readily write down the expressions:

[81] Negele and Orland, Op. Cit., p.82-83.

$$\Gamma = \tfrac{1}{2}\int_0^\beta d\tau \sum_{\alpha..\delta} G_0(\alpha,\tau_\alpha-\tau)G_0(\gamma,\tau-\tau_\gamma)G_0(\beta,\tau_\beta-\tau)G_0(\delta,\tau-\tau_\delta)F\langle\alpha\beta|v|\gamma\delta\rangle,$$

$$\Gamma' = \tfrac{1}{2}\int_0^\beta d\tau \sum_{\alpha..\delta} G_0(\alpha,\tau_\alpha-\tau)G_0(\gamma,\tau-\tau_\gamma)G_0(\beta,\tau_\beta-\tau)G_0(\delta,\tau-\tau_\delta)F\langle\beta\alpha|v|\delta\gamma\rangle.$$

(12.10-7)

Because of (12.10-6) the two diagrams are identical. In Fig. 12-19 we give a number of useful equivalent non-labelled Feynman diagrams.

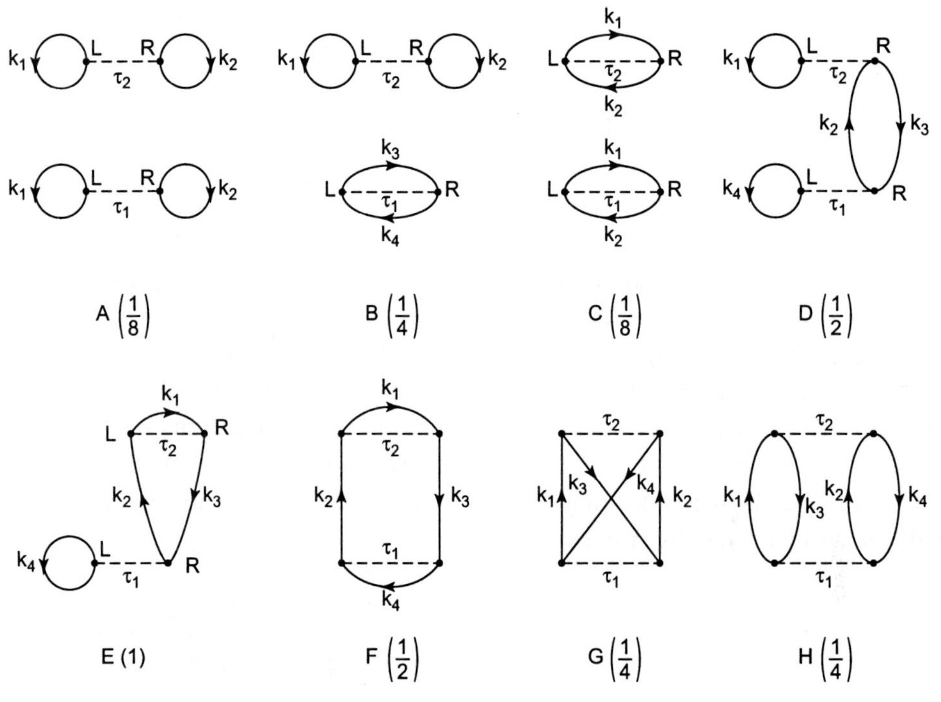

Fig. 12-17. Feynman diagrams for W_2.

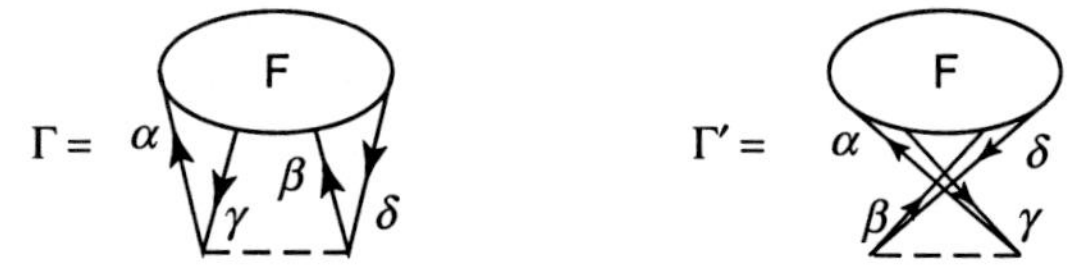

Fig. 12-18. Effect of extremity interchange. After Negele and Orland (see Ref. 79). [With permission.]

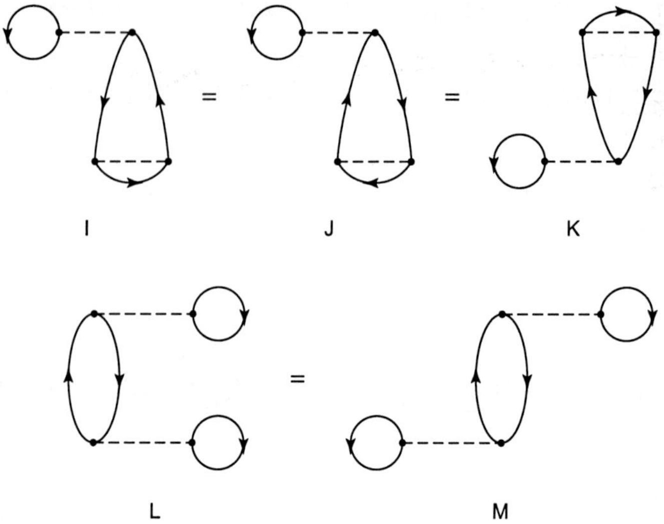

Fig. 12-19. Some common second order diagrams which are topologically indistinct.

In higher orders the direct ring diagrams are of particular importance, since they are used in the "ring approximation". In Fig. 12-20 we give the third order ring diagram. Equivalent deformations are obtained by the permutations of (τ_1, τ_2, τ_3), combined with appropriate interchanges of the extremities. For this diagram $S = 3! = 6$. For higher order ring diagrams, $n > 3$, also shown in Fig. 12-20, one easily shows that generally $S = 2n$.

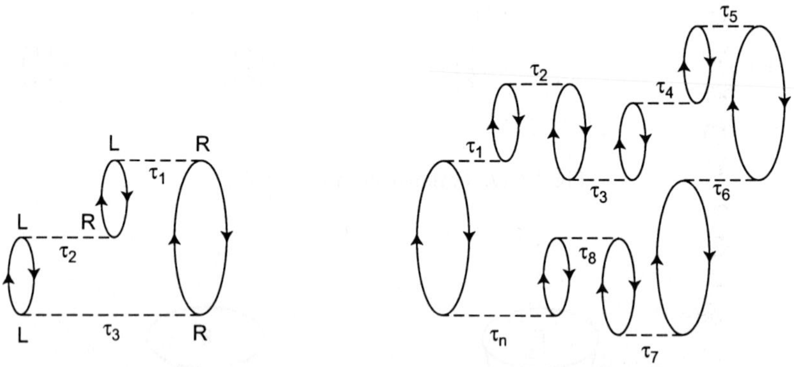

Fig. 12-20. Third order direct ring diagram (left) and general direct ring diagram (right).

We leave it to the problems to obtain the following compact result for the sum of the first order exchange graph and all direct ring diagrams:

$$\textit{Ring sum} = \sum_{n=1}^{\infty} \frac{(-\varepsilon)^n}{2n} \,'\mathrm{tr}'\{[G_0 G_0 v]^n\} = -\frac{1}{2} \,'\mathrm{tr}'\{\ln(1+\varepsilon v G_0 G_0)\}$$

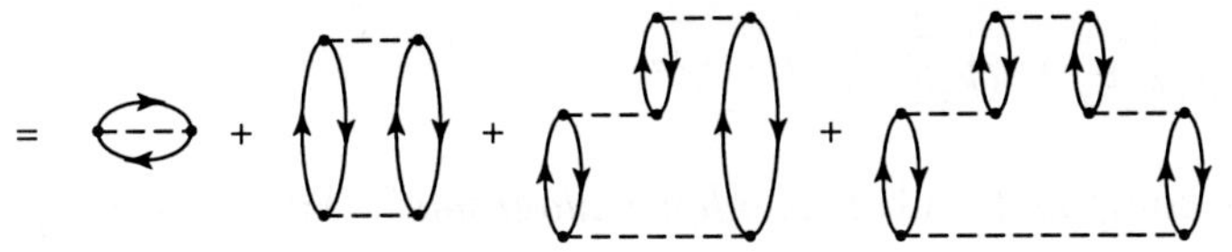

$$+\cdots \quad (12.10\text{-}8)$$

[The meaning of 'tr' is explained in problem 12-7.]

12.10.2 *Hugenholtz Diagrams*

For many purposes it is unnecessary to distinguish between direct and exchange contributions. We can then work with the summary matrix elements $\langle k_1 k_2 | v | k_3 k_4 \rangle^{s,a}$:

$$\mathcal{V} = \frac{1}{4} \sum_{k_1,k_2,k_3,k_4} \hat{e}^{\dagger}_{k_1} \hat{e}^{\dagger}_{k_2} \langle k_1 k_2 | v | k_3 k_4 \rangle^{s,a} \hat{e}_{k_4} \hat{e}_{k_3}, \quad k = (\mathbf{k}) \text{ or } k = (\mathbf{k}, \sigma), \quad (12.10\text{-}9)$$

where the (anti)symmetrized matrix elements denote

$$\langle k_1 k_2 | v | k_3 k_4 \rangle^{s,a} = \langle k_1 k_2 | v | k_3 k_4 \rangle + \varepsilon \langle k_1 k_2 | v | k_4 k_3 \rangle . \quad (12.10\text{-}10)$$

Many contractions can now be combined, as illustrated in the relationship below:

$$\sum_{\gamma\delta} \hat{e}^{\dagger}_{\alpha} \hat{e}^{\dagger}_{\beta} \hat{e}_{\delta} \hat{e}_{\gamma} \ldots \hat{e}^{\dagger}_{\alpha'} \hat{e}^{\dagger}_{\beta'} \langle \alpha\beta | v | \gamma\delta \rangle^{s,a} = \sum_{\gamma\delta} \hat{e}^{\dagger}_{\alpha} \hat{e}^{\dagger}_{\beta} \hat{e}_{\delta} \hat{e}_{\gamma} \ldots \hat{e}^{\dagger}_{\alpha'} \hat{e}^{\dagger}_{\beta'} \langle \alpha\beta | v | \gamma\delta \rangle^{s,a} . \quad (12.10\text{-}11)$$

For W_1 and W_2 we now obtain the shorter forms

$$W_1 = \frac{-1}{2} \beta \sum_{k_1 k_2} G_0(k_1, 0) G_0(k_2, 0) \langle k_1 k_2 | v | k_1 k_2 \rangle^{s,a} , \quad (12.10\text{-}12)$$

$$W_2 = \frac{1}{8} \left[\sum_{k_1 k_2} \int_0^{\beta} d\tau_1 G_0(k_1,0) G_0(k_2,0) \langle k_1 k_2 | v | k_1 k_2 \rangle^{s,a} \right]^2$$

$$+ \varepsilon \frac{1}{2} \sum_{k_1..k_4} \int_0^{\beta} d\tau_2 \int_0^{\beta} d\tau_1 G_0(k_2, \tau_2 - \tau_1) G_0(k_3, \tau_1 - \tau_2) G_0(k_1, 0) G_0(k_4, 0)$$

$$\times \langle k_1 k_3 | v | k_1 k_2 \rangle^{s,a} \langle k_4 k_2 | v | k_4 k_3 \rangle^{s,a}$$

$$+\frac{1}{8}\sum_{k_1..k_4}\int_0^\beta d\tau_2\int_0^\beta d\tau_1 G_0(k_1,\tau_2-\tau_1)G_0(k_2,\tau_2-\tau_1)G_0(k_3,\tau_1-\tau_2)G_0(k_4,\tau_1-\tau_2)$$

$$\times\langle k_1k_2\,|\,v\,|\,k_3k_4\rangle^{s,a}\langle k_3k_4\,|\,v\,|\,k_1k_2\rangle^{s,a}\,. \qquad (12.10\text{-}13)$$

All W_n can be computed with *Hugenholtz diagrams*.[82] The main simplification is that we need not specify the order of ingoing or outgoing lines at a vertex. The vertex is therefore represented by a single dot to which the time label τ_i is given; four lines of directed states emanate from the vertex, two ingoing and two outgoing ones:

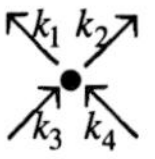

$$=-\langle k_1k_2\,|\,v\,|\,k_3k_4\rangle^{s,a} \qquad (12.10\text{-}14)$$

To the directed lines representing states we assign propagators $G_0(k,\tau_j-\tau_i)$ similarly as before. There are now new features, however, for the Fermi factor and the symmetry factor S_H. First we consider the Fermi factor. To obtain the number of Fermi loops, we must 'pull the vertex apart' to give it a left and right side and we count the number of Fermi loops as before. At first glance it seems that this could give ambiguous results, since a Hugenholtz diagram can be associated with different Feynman diagrams. We give an illustration in Fig. 12-21. A second order *H*-diagram is represented by two different Feynman diagrams when the dot-vertex is resolved in an *L-R* vertex. The diagram to the left has two Fermi loops, whereas the one at the right has a single Fermi loop, thus resulting in factors ε^2 and ε, respectively. The matrix elements are also different, viz.

$$left \rightarrow \varepsilon^2\langle\alpha\beta\,|\,v\,|\,\gamma\delta\rangle^{s,a}\langle\gamma\delta\,|\,v\,|\,\alpha\beta\rangle^{s,a}\,,$$

while

$$right \rightarrow \varepsilon\langle\alpha\beta\,|\,v\,|\,\gamma\delta\rangle^{s,a}\langle\gamma\delta\,|\,v\,|\,\beta\alpha\rangle^{s,a} = \varepsilon^2\langle\alpha\beta\,|\,v\,|\,\gamma\delta\rangle^{s,a}\langle\gamma\delta\,|\,v\,|\,\alpha\beta\rangle^{s,a}\,,$$

so that in the end both choices give the same result.

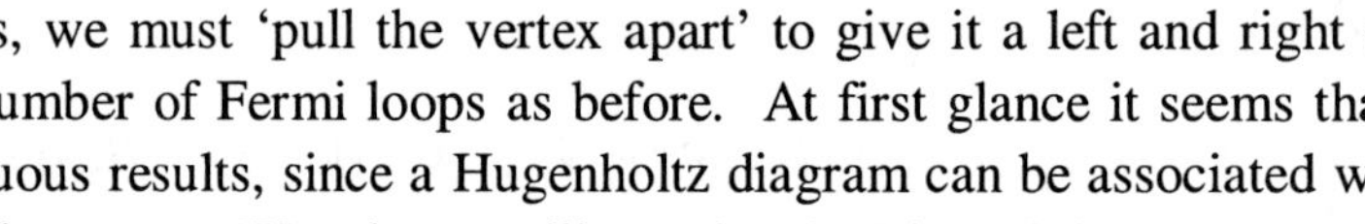

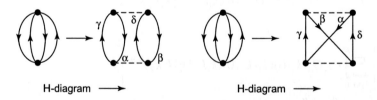

Fig. 12-21. A second-order *H*-diagram has been resolved into two different Feynman diagrams.

---

[82] N.M. Hugenholtz, Physica **23**, 481 (1957); N.M. Hugenholtz and D. Pines, Phys. Rev. **116**, 489 (1959). Some authors refer to these diagrams as zero-diagrams (0-diagrams) and one-diagrams (1-diagrams).

The symmetry factor, S_H, and its concomitant weight, $1/S_H$, are quite different from the previous situation, with a given H-diagram possibly representing a large number of Feynman diagrams. This factor is obtained as follows. Since vertices only differ from each other by the tau's assigned to them, we count the number of vertex exchanges that leave the diagram invariant, being S_V. Next we observe that pairs of equally directed lines, going both from a given vertex to the same vertex (which may be a different vertex or the original one), can be interchanged. Denoting this number by n_e, we have the result

$$S_H = S_V 2^{n_e}, \qquad 1/S_H = 1/S_V 2^{n_e} . \tag{12.10-15}$$

To verify the results for all n^{th} order diagrams we have the sum-rule:

$$\sum \frac{1}{S_H} = \sum \frac{1}{S_V 2^{n_e}} = \frac{(2n)!}{n! 2^{2n}} = \frac{(2n-1)!!}{2^n} . \tag{12.10-16}$$

We proceed to give the diagrams of first, second and third order. For W_1 we have one diagram,

$$W_1 = \text{OO} = \frac{1}{1\times 2} \sum_{k_1 k_2} \int_0^\beta d\tau\, G_0(k_1,0) G_0(k_2,0) \left[-\langle k_1 k_2 | v | k_1 k_2 \rangle^{s,a} \right], \tag{12.10-17}$$

where we noted that for the one vertex $S_V = 1$ and that that there is one equivalent pair of lines. The H-diagrams for W_2 are given in Fig. 12-22, together with the weights as stated in (12.10-15). For W_3 the diagrams are in Fig. 12-23. There are eight terms, labelled (A) through (H); the weights are easily verified. To be noted is that W_2 has one disconnected diagram, while W_3 has three disconnected diagrams.

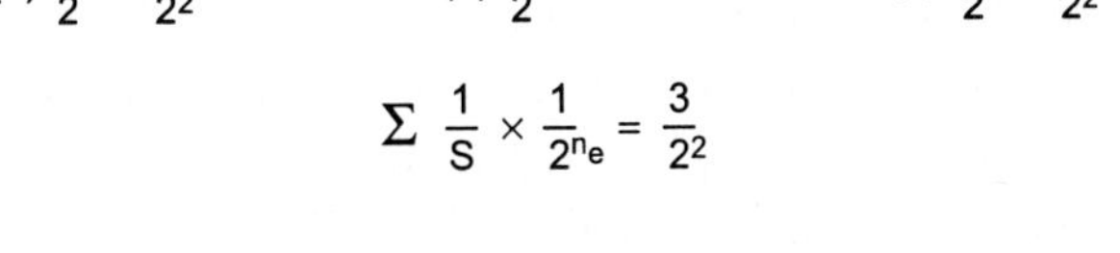

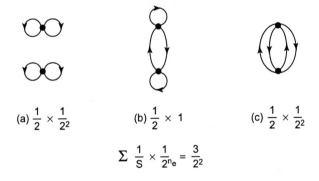

Fig. 12-22. Hugenholtz diagrams for W_2.

We finally return to the meaning of all this for statistical mechanics. While the grand-canonical partition function is fully given by the sum of all W_n's, for the grand potential we need the cumulants, stated in Eq. (12.9-32). When the recipe there is carried out – easily doable up to order C_3 – one arrives at the following

Cluster linkage theorem: Only connected diagrams contribute to the cumulants[83]

or,

$$\Omega(T,V,\varsigma)-\Omega_0(T,V,\varsigma)=-\beta^{-1}\sum(\textit{all connected diagrams}). \quad (12.10\text{-}18)$$

Going up to order λ^3, we find that there are just eight H-diagrams, with one, two or three vertices, that contribute.

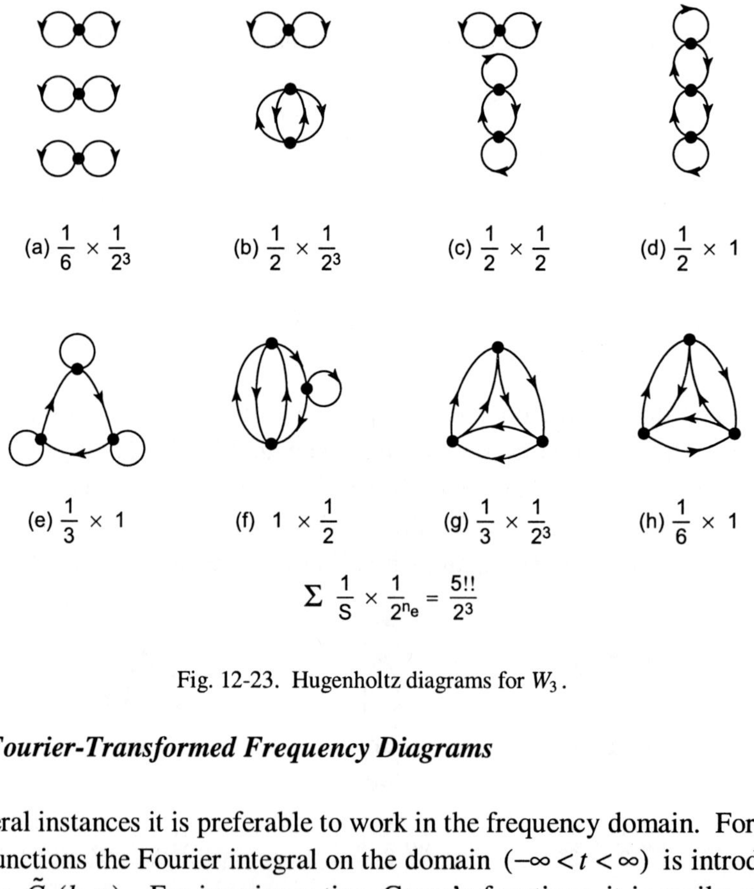

Fig. 12-23. Hugenholtz diagrams for W_3.

12.10.3 *Fourier-Transformed Frequency Diagrams*

In several instances it is preferable to work in the frequency domain. For real time Green's functions the Fourier integral on the domain $(-\infty < t < \infty)$ is introduced and denoted by $\tilde{G}_0(k,\omega)$. For imaginary time Green's functions, it is easily established that the domain of validity is $(-\beta < \tau < \beta)$. Outside an interval $0 < |\tau| < \beta$ the function is extended periodically or anti-periodically, as required for bosons and fermions, respectively. For fermions the proof is as follows. We have for $-\beta < \tau < 0$,

$$G_0(k,\tau)=(1/\mathcal{Z}_0)\mathrm{Tr}\{e^{-\beta K^0}\tilde{e}_k^\dagger(0)\tilde{e}_k(\tau)\}=(1/\mathcal{Z}_0)\mathrm{Tr}\{\tilde{e}_k(\tau)e^{-\beta K^0}\tilde{e}_k^\dagger(0)\}$$

---

[83] For a general proof, see Negele and Orland, Op. Cit., p. 96-97.

$$= (1/\mathcal{Z}_0)\mathrm{Tr}\{e^{-\beta K^0}\underbrace{e^{\beta K^0}\tilde{e}_k(\tau)e^{-\beta K^0}}_{\hat{e}_k(\tau+\beta)}\tilde{e}_k^\dagger(0)\} = -G^0(k,\tau+\beta)\,. \qquad (12.10\text{-}19)$$

Similarly we find $G_0(k,\tau>0) = -G_0(k,\tau-\beta)$. Likewise, for bosons the first equality in (12.10-19) carries a minus sign, so that the final rhs has a plus sign. Altogether we can write for appropriate τ,

$$G_0(k,\tau\pm\beta) = \varepsilon G_0(k,\tau)\,. \qquad (12.10\text{-}20)$$

Note that in any case $G_0(k,\tau\pm 2\beta) = G_0(k,\tau)$, indicating periodicity over a period 2β.

We now introduce the Fourier coefficients and Fourier series as follows,

$$\bar{G}_0(k,\omega_n) = \tfrac{1}{2}\int_{-\beta}^{\beta} d\tau\, e^{i\omega_n\tau}G_0(k,\tau), \quad G_0(k,\tau) = \beta^{-1}\sum_{n=-\infty}^{\infty} e^{-i\omega_n\tau}\bar{G}_0(k,\omega_n)\,, \qquad (12.10\text{-}21)$$

where ω_n is differently determined for periodic and anti-periodic functions, so that we have

$$\omega_n = \frac{2n\pi}{\beta} \quad (\text{bosons})\,, \qquad \omega_n = \frac{(2n+1)\pi}{\beta} \quad (\text{fermions})\,. \qquad (12.10\text{-}22)$$

Further, breaking up the interval and employing (12.19-20) and (12.10-22),

$$\begin{aligned}\bar{G}_0(k,\omega_n) &= \tfrac{1}{2}\int_{-\beta}^{0} d\tau\, e^{i\omega_n\tau}G_0(k,\tau) + \tfrac{1}{2}\int_0^{\beta} d\tau\, e^{i\omega_n\tau}G_0(k,\tau)\\ &= \varepsilon\tfrac{1}{2}\int_{-\beta}^{0} d\tau\, e^{i\omega_n\tau}G_0(k,\tau+\beta) + \tfrac{1}{2}\int_0^{\beta} d\tau\, e^{i\omega_n\tau}G_0(k,\tau)\\ &= \tfrac{1}{2}\Big[(1+\varepsilon\underbrace{e^{-i\omega_n\beta}}_{\varepsilon}\Big]\int_0^{\beta} d\tau\, e^{i\omega_n\tau}G_0(k,\tau) = \int_0^{\beta} d\tau\, e^{i\omega_n\tau}G_0(k,\tau)\,, \end{aligned} \qquad (12.10\text{-}23)$$

where we set $\tau' = \tau+\beta$ for the first integral of the middle line. From (12.9-43) we obtain

$$\bar{G}_0(k,\omega_n) = -[1+\varepsilon n_0(k)]\int_0^{\beta} d\tau\, e^{(i\omega_n-\hat{\varepsilon}_k)\tau} = \frac{1}{i\omega_n - \hat{\varepsilon}_k}\,, \qquad (12.10\text{-}24)$$

where $\hat{\varepsilon}_k = \varepsilon_k - \varsigma$, as in Eq. (12.6-15'). We note that this result is *identical* for bosons and fermions. For the time domain propagator we then find

$$G_0(k,\tau) = (1/\beta)\sum\nolimits_n e^{-i\omega_n\tau}\frac{1}{i\omega_n - \hat{\varepsilon}_k}\,. \qquad (12.10\text{-}25)$$

Integrating over τ from 0 to β, at every vertex 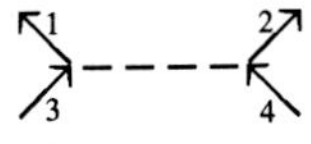 we have a factor

$$\int_0^{\beta} d\tau \exp[(i\omega_{n_1} + i\omega_{n_2} - i\omega_{n_3} - i\omega_{n_4})\tau] = \beta\delta_{\omega_{n_1}+\omega_{n_2},\,\omega_{n_3}+\omega_{n_4}}\,. \qquad (12.10\text{-}26)$$

So the rules for the frequency diagrams are a bit modified:

(i) With an interaction we associate a vertex and assign $-\delta_{n_1+n_2,n_3+n_4}\langle k_1k_2|v|k_3k_4\rangle^{[s,a]}$, where the [*s*,*a*] is omitted for Feynman diagrams and applied for Hugenholtz diagrams;

(ii) With each directed line we associate a state $(k_i,n_i)\equiv(k_i,\omega_{n_i})$ and assign $\bar{G}_0(k_i,\omega_{n_i})$;

(iii) For each line that passes through one vertex [corresponding to $G_0(k,0-)$], add a convergence factor $e^{i\omega_n\delta}$, with $\delta\to 0+$;

(iv) We are to sum over all internal state variables $k_1...k_p$ and frequencies $\omega_{n_1}...\omega_{n_p}$.

(v) Multiply the total with $1/\beta^n$, where *n* is the number of interactions.

With these rules we can easily write down the terms associated with a given diagram. E.g., for the Hugenholtz diagram (b) of W_2 [see Fig. 12-21] we have

$$(b)=-\frac{1}{2}\varepsilon\frac{1}{\beta^2}\sum_{k_1...k_4}\lim_{\delta_1\to 0}\lim_{\delta_4\to 0}\sum_{n_1..n_4}\prod_{\ell=1}^{4}\frac{1}{i\omega_{n_\ell}-\hat{\varepsilon}_{k_\ell}}e^{i\omega_{n_1}\delta_1}e^{i\omega_{n_4}\delta_4}\delta_{v_1}\delta_{v_2}\langle k_1k_2|v|k_1k_3\rangle^{s,a}$$
$$\times\langle k_3k_4|v|k_2k_4\rangle^{s,a}=-\frac{1}{2}\varepsilon\frac{1}{\beta^2}\sum_{k_1...k_4}\lim_{\delta_1\to 0}\lim_{\delta_4\to 0}\prod_{\ell=1}^{4}\sum_{n_\ell}\frac{1}{i\omega_{n_\ell}-\hat{\varepsilon}_{k_\ell}}e^{i\omega_{n_1}\delta_1}e^{i\omega_{n_4}\delta_4}\delta_{v_1}\delta_{v_2}$$
$$\times\langle k_1k_2|v|k_1k_3\rangle^{s,a}\langle k_3k_4|v|k_2k_4\rangle^{s,a}, \qquad (12.10\text{-}27)$$

where δ_v means the vertex restriction indicated in (12.10-26).

We shall now show how the convergence factor works out by evaluating one of the sums. We need

$$\lim_{\delta_1\to 0}\sum_{n_1=-\infty}^{\infty}e^{i\omega_{n_1}\delta_1}\frac{1}{i\omega_{n_1}-\hat{\varepsilon}_{k_1}}\equiv\lim_{\delta\to 0}\sum_{n=-\infty}^{\infty}e^{i\omega_n\delta}\frac{1}{i\omega_n-\hat{\varepsilon}_k}. \qquad (12.10\text{-}28)$$

Consider the meromorphic function $\pm\beta/[e^{\beta z}\mp 1]$, where the upper sign is for bosons and the lower sign for fermions. With the ω_n given by (12.10-22), we see that there is an infinite number of poles along the imaginary axis at $z_p=i\omega_n$, each with unit residue. So,

$$\frac{1}{\beta}\lim_{\delta\to 0}\sum_{n=-\infty}^{\infty}e^{i\omega_n\delta}\frac{1}{i\omega_n-\hat{\varepsilon}_k}=\lim_{\delta\to 0}\frac{\pm 1}{2\pi i}\oint_C dz\frac{1}{e^{\beta z}\mp 1}\frac{e^{\delta z}}{z-\hat{\varepsilon}_k}; \qquad (12.10\text{-}29)$$

the poles and the contour *C* are shown in Fig. 12-24, together with another little contour *c* around $\hat{\varepsilon}_k$ and a big circular contour Γ which for $|z|\to\infty$ encloses all poles of the rhs of (12.10-29).[84] At the big contour, if $|z| \to \infty$ along a ray with $\operatorname{Re} z > 0$, the integrand is of order $|z|^{-1} \exp[-(\beta - \delta)\operatorname{Re} z]$, while along a ray with $\operatorname{Re} z < 0$, the integrand is of order $|z|^{-1} \exp(\delta \operatorname{Re} z)$. Since $\beta > \delta > 0$, the integrand times the semi-arc length goes to zero in each case. Thus, the contour integral over the full arc Γ is asymptotically zero. Since the contour Γ is the sum of the contours C and C', we arrive at the result [85]

$$\frac{1}{\beta}\lim_{\delta\to 0}\sum_{n=-\infty}^{\infty} e^{i\omega_n\delta}\frac{1}{i\omega_n - \hat{\varepsilon}_k} = -\lim_{\delta\to 0}\frac{\pm 1}{2\pi i}\oint_{C'} dz\frac{1}{e^{\beta z}\mp 1}\frac{e^{\delta z}}{z-\hat{\varepsilon}_k}$$
$$= -\frac{1}{2\pi i}\oint_{C'} dz\frac{\pm 1}{e^{\beta z}\mp 1}\frac{1}{z-\hat{\varepsilon}_k} = \mp\frac{1}{e^{\beta\hat{\varepsilon}_k}\mp 1}, \qquad (12.10\text{-}30)$$

where the final rhs is $2\pi i$ times the residue of the single pole at $z = \hat{\varepsilon}_k$. This, then, is the full Fourier series that is to be associated with $G_0(k,0)$.

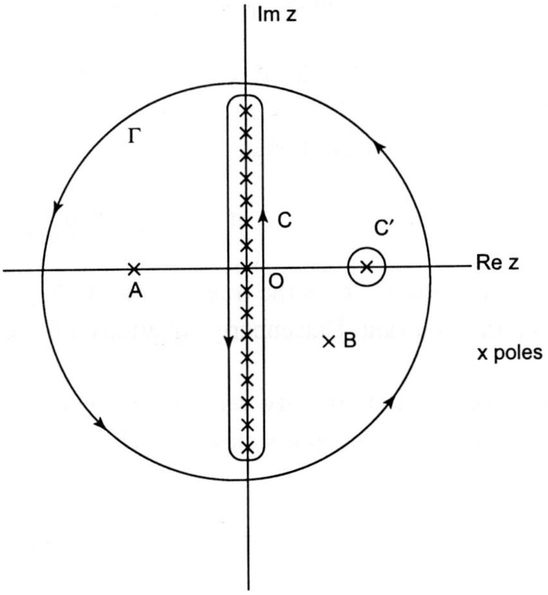

Fig. 12-24. Contours Γ, C and c used for loop evaluation; $\times$'s denote the poles. [The points A and B in the figure are for later use.]

---

[84] The original argument is found with E.W. Montroll and J.C. Ward, Phys. Fluids **1**, 55 (1958). We follow Mattuck, Op. Cit., p. 250; for a slightly different presentation, see Reichl (first Ed.) p. 412-413.
[85] Note that we first established $\oint_C .. = -\oint_c ..$; only *after* this result, the limit $\delta \to 0$ is carried out.

12.11 Full Propagators (or Green's Functions) for Normal Quantum Fluids

12.11.1 *Spatial and Momentum Forms*

The language of second quantization employs both the creation-annihilation operator form and the local field-operator form, as introduced in Chapter VII. While so far most of our presentation has been based on the former formalism, we note that most treatments of quantum liquids use the field-operator form. Accordingly, in the present section we shall first introduce the full propagators or Green's functions in the $\mathbf{r}$-t *or* $\mathbf{r}$-τ coordinates. Needless to say that this presentation can directly be carried over to relativistic quantum-field theory and quantum electrodynamics.[86] Averages will now be based on the complete grand-canonical density operator, $\rho_{gcan.} = \exp(-\beta K)/\mathcal{Z}$. The Green's function is then defined as the Heisenberg-operator average

$$G_{\lambda\mu}(\mathbf{r},\tau;\mathbf{r}',\tau') \equiv -\mathrm{Tr}\{\rho_{gcan.}\mathcal{T}[\Psi_\lambda(\mathbf{r},\tau)\Psi_\mu^\dagger(\mathbf{r}',\tau')]\}$$

$$= -\langle \mathcal{T}[\Psi_\lambda(\mathbf{r},\tau)\Psi_\mu^\dagger(\mathbf{r}',\tau')]\rangle, \qquad (12.11\text{-}1)$$

where λ and μ are spin indices, which are to be omitted for Bose systems. The Heisenberg operators $\Psi_\lambda(\mathbf{r},\tau)$ and $\Psi_\mu^\dagger(\mathbf{r},\tau)$ refer to the 'modified Heisenberg picture', in which K replaces the Hamiltonian $\mathcal{H}$; moreover, $\tau = i\hbar^{-1}t$. Accordingly we have

$$\Psi_\lambda(\mathbf{r},\tau) = e^{\tau K}\Psi_\lambda(\mathbf{r})e^{-\tau K}, \qquad \Psi_\mu^\dagger(\mathbf{r},\tau) = e^{\tau K}\Psi_\mu^\dagger(\mathbf{r})e^{-\tau K}. \qquad (12.11\text{-}2)$$

[Again, as long as τ is imaginary, $\Psi^\dagger$ is the true adjoint of Ψ.] Differentiating Eqs. (12.11-2), we obtain the relevant Heisenberg equations of motion; they will be considered below.

For a pure state the one-particle density was shown to be $\mathrm{tr}[\Psi^\dagger(\mathbf{r})\Psi(\mathbf{r})]$, where 'tr' refers to the matrix space of the spin states. For the present statistical situation we have likewise

$$\langle\rho(\mathbf{r})\rangle = \mathrm{tr}\langle\Psi^\dagger(\mathbf{r})\Psi(\mathbf{r})\rangle = \sum\nolimits_\lambda \langle\Psi_\lambda^\dagger(\mathbf{r})\Psi_\lambda(\mathbf{r})\rangle. \qquad (12.11\text{-}3)$$

This result is easily linked to the Green's function G. We have

$$\mathrm{tr}[G(\mathbf{r},\tau;\mathbf{r},\tau+0)] = \mp\sum\nolimits_\lambda \mathrm{Tr}\{\rho_{gcan.}\Psi_\lambda^\dagger(\mathbf{r},\tau)\Psi_\lambda(\mathbf{r},\tau)\}$$

$$= \mp\sum\nolimits_\lambda \mathrm{Tr}\{e^{-\beta K}e^{\tau K}\Psi_\lambda^\dagger(\mathbf{r})\underbrace{e^{-\tau K}e^{\tau K}}_{1}\Psi_\lambda(\mathbf{r})e^{-\tau K}\}/\mathcal{Z}$$

$$= \mp\sum\nolimits_\lambda \mathrm{Tr}\{\overbrace{e^{-\tau K}e^{-\beta K}}^{commute}e^{\tau K}\Psi_\lambda^\dagger(\mathbf{r})\Psi_\lambda(\mathbf{r})\}/\mathcal{Z} = \mp\mathrm{tr}\langle\Psi^\dagger(\mathbf{r})\Psi(\mathbf{r})\rangle = \mp\langle\rho(\mathbf{r})\rangle, \ (12.11\text{-}4)$$

---

[86] See e.g. A.I. Akhiezer and V.B. Berestetskii, "Quantum Electrodynamics", Interscience Publ. NY (1965).

where we used the invariance of the trace under cyclic change. In addition, we note the *general property* that averages are the same in the Schrödinger and Heisenberg pictures. Integrating (12.11-4) we find the average of the *fluctuating* particle number

$$\langle N\rangle = \mp\int d^3r\,\mathrm{tr}[G(\mathbf{r},\tau;\mathbf{r},\tau+0)]. \qquad (12.11\text{-}5)$$

While the present notation for the Green's function is the one most widely used, it be noted that for time-independent Hamiltonians – inherent in the forms of (12.11-2) – we can write

$$G(\mathbf{r},\tau;\mathbf{r}',\tau') = G(\mathbf{r},\mathbf{r}',\tau-\tau'). \qquad (12.11\text{-}6)$$

All thermodynamic functions can be derived from knowledge of $G(\mathbf{r},\tau;\mathbf{r}',\tau')$. First, consider operators $F=\Sigma_i f_i$, where f_i is a one-body operator. Analogous to the results for a pure state as dealt with in chapter VII, we now have

$$\langle F\rangle = \mathrm{tr}\int d^3r\langle\Psi^\dagger(\mathbf{r})f(\mathbf{r})\Psi(\mathbf{r})\rangle = \mathrm{tr}\int d^3r\,\mathrm{Tr}\{\rho_{gcan.}\Psi^\dagger(\mathbf{r})f(\mathbf{r})\Psi(\mathbf{r})\}$$
$$= \mathrm{tr}\int d^3r \lim_{\mathbf{r}'\to\mathbf{r}}\mathrm{Tr}\{\rho_{gcan.}f(\mathbf{r})\Psi^\dagger(\mathbf{r}',\tau)\Psi(\mathbf{r},\tau)\} = \mp\int d^3r \lim_{\mathbf{r}'\to\mathbf{r}}\lim_{\tau'\to\tau+0}\mathrm{tr}[f(\mathbf{r})G(\mathbf{r},\tau;\mathbf{r}'\tau')]. \qquad (12.11\text{-}7)$$

As an example we consider the mean kinetic energy. Then $f\to(-\hbar^2/2m)\nabla^2$, so that

$$\langle\mathcal{T}\rangle = \mp\int d^3r \lim_{\mathbf{r}'\to\mathbf{r}}\mathrm{tr}\Big[-(\hbar^2/2m)\nabla^2 G(\mathbf{r},\tau;\mathbf{r}',\tau+0)\Big]. \qquad (12.11\text{-}8)$$

For the potential energy, being a two-body operator, we must go back to the Heisenberg equations of motion. We have

$$(\partial/\partial\tau)\Psi_\lambda(\mathbf{r},\tau) = -[\Psi_\lambda,K], \qquad (\partial/\partial\tau)\Psi_\lambda^\dagger(\mathbf{r},\tau) = [\Psi_\lambda^\dagger,K]. \qquad (12.11\text{-}9)$$

The commutator is easily evaluated. One obtains

$$\frac{\partial}{\partial\tau}\Psi_\lambda(\mathbf{r},\tau) = \frac{\hbar^2}{2m}\nabla^2\Psi_\lambda(\mathbf{r},\tau) + \varsigma\Psi_\lambda(\mathbf{r},\tau)$$
$$-\sum_\eta\int d^3r''\Psi_\eta^\dagger(\mathbf{r}'',\tau)\Psi_\eta(\mathbf{r}'',\tau)v(\mathbf{r}-\mathbf{r}'')\Psi_\lambda(\mathbf{r},\tau),$$

$$\frac{\partial}{\partial\tau}\Psi_\mu^\dagger(\mathbf{r},\tau) = \frac{-\hbar^2}{2m}\nabla^2\Psi_\lambda(\mathbf{r},\tau) - \varsigma\Psi_\lambda(\mathbf{r},\tau)$$
$$+\sum_\eta\int d^3r''\Psi_\eta^\dagger(\mathbf{r}'',\tau)\Psi_\eta(\mathbf{r}'',\tau)v(\mathbf{r}-\mathbf{r}'')\Psi_\mu^\dagger(\mathbf{r},\tau). \qquad (12.11\text{-}10)$$

From the first equation we then find

$$\int d^3r \lim_{\mathbf{r}'\to\mathbf{r}} \lim_{\tau'\to\tau+0} \frac{\partial}{\partial\tau} G_{\lambda\mu}(\mathbf{r},\tau;\mathbf{r}'\tau') = \mp\mathrm{Tr}\left\{ \rho_{gcan.}\left[\Psi^\dagger_\mu(\mathbf{r},\tau)\left(\frac{\hbar^2}{2m}\nabla^2 + \varsigma\right)\Psi_\lambda(\mathbf{r},\tau)\right.\right.$$

$$\left.\left. -\sum_\eta \int d^3r \int d^3r'' \Psi^\dagger_\mu(\mathbf{r},\tau)\Psi^\dagger_\eta(\mathbf{r}'',\tau)v(\mathbf{r}-\mathbf{r}'')\Psi_\eta(\mathbf{r}'',\tau)\Psi_\lambda(\mathbf{r},\tau)\right]\right\}, \quad (12.11\text{-}11)$$

where we commuted $v(\mathbf{r}\text{-}\mathbf{r}'')$ and $\Psi^\dagger_\eta$, which is permissible if the potential is not spin-dependent. Defining $\langle \mathcal{V} \rangle$ as the finite temperature analogue of (7.6-22) [with a factor ½, see (7.6-23)], we thus obtained

$$\langle \mathcal{V} \rangle = \mp\tfrac{1}{2}\int d^3r \lim_{\mathbf{r}'\to\mathbf{r}} \lim_{\tau'\to\tau+0}\left[-\frac{\partial}{\partial\tau} + \frac{\hbar^2}{2m}\nabla^2 + \varsigma \right] \mathrm{tr}G(\mathbf{r},\tau;\mathbf{r}'\tau'). \quad (12.11\text{-}12)$$

Adding the kinetic energy, this gives for the mean energy of the system

$$\langle \mathcal{E} \rangle = \langle \mathcal{V} \rangle + \langle \mathcal{T} \rangle = \mp\tfrac{1}{2}\int d^3r \lim_{\mathbf{r}'\to\mathbf{r}} \lim_{\tau'\to\tau+0}\left[-\frac{\partial}{\partial\tau} - \frac{\hbar^2}{2m}\nabla^2 + \varsigma \right] \mathrm{tr}G(\mathbf{r},\tau;\mathbf{r}'\tau'). \quad (12.11\text{-}13)$$

Next we will obtain the grand potential $\Omega(T,V,\varsigma)$. Let us consider $K(\lambda) = K_0 + \lambda\mathcal{V}$. Differentiating the grand partition function $\mathcal{Z}_\lambda = \mathrm{Tr}\{\exp[-\beta(K^0 + \lambda\mathcal{V})]\}$ with respect to λ, we have

$$\partial\Omega_\lambda / \partial\lambda = (-\beta\mathcal{Z}_\lambda)^{-1}\partial\mathcal{Z}_\lambda / \partial\lambda. \quad (12.11\text{-}14)$$

The differentiation yields[87]

$$\partial\mathcal{Z}_\lambda / \partial\lambda = -\beta\lambda^{-1}\mathcal{Z}_\lambda\langle\lambda\mathcal{V}\rangle_\lambda. \quad (12.11\text{-}15)$$

Substituting into (12.11-14), this yields

$$\Omega(T,V,\varsigma) = \Omega_0(T,V,\varsigma) + \int_0^1 \lambda^{-1} d\lambda \langle\lambda\mathcal{V}\rangle_\lambda, \quad (12.11\text{-}16)$$

which is expressible in the Green's function via (12.11-12). This expression is useful for the quantum theory of an interacting electron gas.[88]

Finally, let us derive the '*mathematical source equation*' for the Green's function. The time derivative must be done more carefully, employing the time-ordering and (anti)-symmetrization operator $\mathcal{T}$ in a general way, as previously defined. Thus,

---

[87] This is not a trivial differentiation. The correct way is to expand the exponential, giving terms like $(1/n!)(-\beta)^n \mathrm{Tr}[(K^0 + \lambda\mathcal{V})^n]$. Upon differentiating a factor $\mathcal{V}$ will be perched somewhere between $n-1$ operators $K + \lambda\mathcal{V}$. Because of the cyclic invariance of the trace, these n factors will ultimately be identical, yielding $[1/(n-1)!](-\beta^n)\mathrm{Tr}\{(K^0 + \lambda\mathcal{V})^{n-1}\mathcal{V}\}$. The exponential is then reconstituted.

[88] See Fetter and Walecka, Op. Cit. Section 30 and this chapter, Section 12.13.2.

$$\frac{\partial}{\partial\tau}G_{\lambda\mu}(\mathbf{r},\tau;\mathbf{r}'\tau') = -\frac{\partial}{\partial\tau}\mathrm{Tr}\left\{\rho_{gcan}\mathcal{T}\left[\Psi_\lambda(\mathbf{r},\tau)\Psi_\mu^\dagger(\mathbf{r}',\tau')\right]\right\}$$

$$= -\frac{\partial}{\partial\tau}\mathrm{Tr}\left\{\rho_{gcan}\left(\Psi_\lambda(\mathbf{r},\tau)\Psi_\mu^\dagger(\mathbf{r}',\tau')\Theta(\tau-\tau') \pm \Psi_\mu^\dagger(\mathbf{r}',\tau')\Psi_\lambda(\mathbf{r},\tau)\Theta(\tau'-\tau)\right)\right\}$$

$$= -\mathrm{Tr}\left\{\rho_{gcan}\left[\Psi_\lambda(\mathbf{r},\tau),\Psi_\mu^\dagger(\mathbf{r}',\tau')\right]_\mp\right\}\delta(\tau-\tau') - \mathrm{Tr}\left\{\rho_{gcan}\mathcal{T}\left[\frac{\partial}{\partial\tau}[\Psi_\lambda(\mathbf{r},\tau)]\Psi_\mu^\dagger(\mathbf{r}',\tau')\right]\right\}. \tag{12.11-17}$$

In this expression the (anti)commutator is simply $\delta_{\lambda\mu}\delta(\mathbf{r}-\mathbf{r}')$. The derivative in the second term we computed already from the Heisenberg equation; the result is linear for the kinetic and chemical energy contributions, but leads to a quartic expression for the potential energy, which must be time-ordered and averaged. Using Wick's theorem we average in pairs and denote the result by $-\Lambda_\mathbf{r} G_{\lambda\mu}$. We then obtain,

$$\left(\frac{\partial}{\partial\tau}+\Lambda_\mathbf{r}\right)G_{\lambda\mu}(\mathbf{r},\tau;\mathbf{r}'\tau') = \delta_{\lambda\mu}\delta(\mathbf{r}-\mathbf{r}')\delta(\tau-\tau'), \tag{12.11-18}$$

where

$$\Lambda_\mathbf{r} = -\frac{\hbar^2}{2m}\nabla_\mathbf{r}^2 - \varsigma + \int d^3r'' \underbrace{\Sigma_\eta\langle\Psi_\eta^\dagger(\mathbf{r}'',\tau)\Psi_\eta(\mathbf{r}'',\tau)\rangle}_{\langle\rho(\mathbf{r}'')\rangle} v(\mathbf{r}-\mathbf{r}'') \pm exch.term; \tag{12.11-19}$$

the exchange term involves the Green's function $G_\eta(\mathbf{r}'',\tau;\mathbf{r}'\tau')$. Basically, the interaction energy has given us a Hartree-Fock-like result [cf. Section 7.8] ; of course, this result is purely formal since $\langle\rho(\mathbf{r}'')\rangle$ requires knowledge of $\mathrm{tr}\,G$.

Note that everything said here pertains to normal Bose or Fermi fluids. Green's functions for superconductors are considered in Problem 12-6. Since the BCS interaction is short-range, there is ultimately a *linear* matrix Green's function.

For the remainder of this section we shall convert to the momentum space. As in Chapter VII, cf. Eqs. (7.3-17) and (7.6-15), we expand the field operators in a basis $\{\varphi_\mathbf{k}(\mathbf{r})\}$ of single particle states:

$$\Psi_\lambda(\mathbf{r},\tau) = \sum_\mathbf{k}\varphi_\mathbf{k}(\mathbf{r})\hat{e}_{\mathbf{k}\lambda}(\tau), \quad \Psi_\mu^\dagger(\mathbf{r}',\tau') = \sum_{\mathbf{k}'}\varphi_{\mathbf{k}'}^*(\mathbf{r})\hat{e}_{\mathbf{k}'\mu}^\dagger(\tau). \tag{12.11-20}$$

As basis we shall use normalized momentum states. We then obtain

$$G_{\lambda\mu}(\mathbf{r},\tau;\mathbf{r}'\tau') = \frac{1}{V_0}\sum_{\mathbf{k},\mathbf{k}'} e^{i\mathbf{r}\cdot\mathbf{k}}e^{-i\mathbf{r}'\cdot\mathbf{k}'}G_{\lambda\mu}(\mathbf{k},\tau;\mathbf{k}',\tau'), \tag{12.11-21}$$

where V_0 is the volume and the momentum-space propagator is defined by

$$G_{\lambda\mu}(\mathbf{k},\tau;\mathbf{k}',\tau') \equiv -\mathrm{Tr}\{\rho_{gcan.}\mathcal{T}[\hat{e}_{\mathbf{k}\lambda}(\tau)\hat{e}_{\mathbf{k}'\mu}^\dagger(\tau')]\}. \tag{12.11-22}$$

Clearly, (12.11-21) is a double Fourier series (bilinear form). Its inversion reads

$$G_{\lambda\mu}(\mathbf{k},\tau;\mathbf{k}',\tau') = \frac{1}{V_0}\int_{V_0}\int_{V_0} d^3r d^3r' e^{-i\mathbf{k}\cdot\mathbf{r}} e^{i\mathbf{k}'\cdot\mathbf{r}'} G_{\lambda\mu}(\mathbf{r},\tau;\mathbf{r}',\tau'). \qquad (12.11\text{-}23)$$

While the **k**'s are discrete for a finite volume, in the thermodynamic limit we have a Fourier integral. In that event, we remove the factor $(1/V_0)$ from (12.11-23) and put $(1/V_0^2)$ in front of (12.11-21). Replacing also $\Sigma_{\mathbf{k},\mathbf{k}'}$ by $[V_0/8\pi^3]^2 \int\int d^3k\, d^3k'$, we obtain

$$G_{\lambda\mu}(\mathbf{r},\tau;\mathbf{r}'\tau') = \left(\frac{1}{8\pi^3}\right)^2 \int\int d^3k\, d^3k' G_{\lambda\mu}(\mathbf{k},\tau;\mathbf{k}',\tau'). \qquad (12.11\text{-}24)$$

From the rhs of (12.11-22) we have, writing out the $\mathcal{T}$ operator,

$$-\lim_{\mathbf{k}\to\mathbf{k}'}\lim_{\tau'\to\tau+0} \mathrm{tr}\Big(\mathrm{Tr}\{\rho_{gcan.}\mathcal{T}[\hat{e}_{\mathbf{k}\lambda}(\tau)\hat{e}^\dagger_{\mathbf{k}'\mu}(\tau')]\}\Big) = \mp\langle n_{\mathbf{k}}(\tau)\rangle\Theta(\tau'-\tau). \quad (12.11\text{-}25)$$

The equilibrium average does of course not depend on τ. So we connect the momentum distribution with the Green's function by

$$n(\mathbf{k}) \equiv \langle n_{\mathbf{k}}\rangle = \mp \lim_{\mathbf{k}'\to\mathbf{k}}\lim_{\tau'\to\tau+0} \mathrm{tr}[G_{\lambda\mu}(\mathbf{k},\tau;\mathbf{k}',\tau')]. \qquad (12.11\text{-}26)$$

Prior to taking the first limit, the expression can be interpreted as the one-particle density matrix; we therefore also have

$$(\mathbf{k}\,|\,\rho_1\,|\,\mathbf{k}') = \mp \lim_{\tau'\to\tau+0} \mathrm{tr}[G_{\lambda\mu}(\mathbf{k},\tau;\mathbf{k}',\tau')]. \qquad (12.11\text{-}27)$$

Finally, for the case of a time-independent Hamiltonian, it be noted that the Green's function only depends on the difference of the two τ's. So,

$$G_{\lambda\mu}(\mathbf{k},\tau;\mathbf{k}',\tau') = G_{\lambda\mu}(\mathbf{k},\mathbf{k}';\tau-\tau'). \qquad (12.11\text{-}28)$$

As for the free propagators we shall henceforth denote by k the vector **k** (bosons) or $(\mathbf{k},\lambda)$ (fermions). The function $G(k,k';\tau)$ is periodic over an interval $(-\beta<\tau<\beta)$, thus permitting a Fourier series $G(k,k';\tau) = \beta^{-1}\Sigma_{-\infty}^{\infty} e^{-i\omega_n\tau}\bar{G}(k,k';\omega_n)$; the frequencies ω_n are again given by (12.10-22). Analogous to (12.10-23) the Fourier coefficients are obtained as

$$\bar{G}(k,k';\omega_n) = \tfrac{1}{2}\int_{-\beta}^{\beta} d\tau e^{i\omega_n\tau} G(k,k';\tau) = \int_0^{\beta} d\tau e^{i\omega_n\tau} G(k,k';\tau). \qquad (12.11\text{-}29)$$

From (12.11-22) we have

$$\begin{aligned} G(k,k';\tau-\tau') &= -\mathrm{Tr}\Big\{\rho_{gcan.}\mathcal{T}\Big[\hat{e}_k(\tau)e^\dagger_{k'}(\tau')\Big]\Big\} \\ &= -\Big(\mathrm{Tr}\big\{\rho_{gcan.}\hat{e}_k(\tau)e^\dagger_{k'}(\tau')\big\}\Theta(\tau-\tau') \pm \mathrm{Tr}\big\{\rho_{gcan.}e^\dagger_{k'}(\tau')\hat{e}_k(\tau)\big\}\Theta(\tau'-\tau)\Big) \end{aligned} \quad (12.11\text{-}30)$$

Let $\{|\overline{\eta}_{i\lambda}\rangle\}$ be the eigenstates of $K = \mathcal{H} - \varsigma \Sigma_k \varepsilon_k \mathbf{n}_k$ and let $\{\kappa_i\}$ be the eigenvalues. Taking the trace in this representation and using

$$\hat{e}_k(\tau) = e^{\tau K}\hat{e}_k e^{-\tau K}, \qquad \hat{e}^{\dagger}_{k'}(\tau') = e^{\tau' K}\hat{e}_{k'} e^{-\tau' K}, \tag{12.11-31}$$

we easily find

$$G(k,k';\tau-\tau') = -\sum_{i,j,\lambda,\mu}\Big\{\langle\overline{\eta}_{i\lambda}|\hat{e}_k|\overline{\eta}_{j\mu}\rangle\langle\overline{\eta}_{j\mu}|\hat{e}^{\dagger}_{k'}|\overline{\eta}_{i\lambda}\rangle\Theta(\tau-\tau')e^{(\tau-\tau')(\kappa_i-\kappa_j)}$$
$$\pm\langle\overline{\eta}_{i\lambda}|\hat{e}^{\dagger}_{k'}|\overline{\eta}_{j\mu}\rangle\langle\overline{\eta}_{j\mu}|\hat{e}_k|\overline{\eta}_{i\lambda}\rangle\Theta(\tau'-\tau)e^{-(\tau-\tau')(\kappa_i-\kappa_j)}\Big\}. \tag{12.11-32}$$

For the integral (12.11-29) this yields with $\tau - \tau' \to \tau$

$$\overline{G}(k,k';\omega_n) = \sum_{i,j,\lambda,\mu}\frac{\langle\overline{\eta}_{i\lambda}|\hat{e}_k|\overline{\eta}_{j\mu}\rangle\langle\overline{\eta}_{j\mu}|\hat{e}^{\dagger}_{k'}|\overline{\eta}_{i\lambda}\rangle\Big[1\mp e^{-\beta(\kappa_j-\kappa_i)}\Big]}{i\omega_n-(\kappa_j-\kappa_i)}. \tag{12.11-33}$$

Although the ES $\{|\overline{\eta}_{i\lambda}\rangle\}$ are generally unknown, we found the important result: *the excitation energies are the poles of* $\overline{G}(k,k';\omega_n)$. In the next subsection we shall indicate how to obtain $\overline{G}(k,k';\omega_n)$ from a diagrammatic expansion.

12.11.2 *Cumulant Expansion of the Green's Function in Free Propagators*

We shall indicate the connection of $G(k,k';\tau-\tau')$ with $G_0(k,\tau-\tau')\delta_{k,k'}$. First of all we note

$$\hat{e}_k(\tau) = U^{\dagger}(\tau)\hat{e}_k U(\tau) = \underbrace{U^{\dagger}(\tau)U^0(\tau)}_{U_I^{\dagger}(\tau)}\underbrace{U^{0\dagger}(\tau)\hat{e}_k U^0(\tau)}_{\tilde{e}_k(\tau)}\underbrace{U^{0\dagger}(\tau)U(\tau)}_{U_I(\tau)}; \tag{12.11-34}$$

or,

$$\hat{e}_k(\tau) = U_I^{\dagger}(\tau)\tilde{e}_k(\tau)U_I(\tau), \qquad \hat{e}^{\dagger}_{k'}(\tau') = U_I^{\dagger}(\tau')\tilde{e}^{\dagger}_{k'}(\tau')U_I(\tau'). \tag{12.11-35}$$

Writing further $\exp(-\beta K) = \exp(-\beta K^0)U_I(\beta)$ and noting also the composition property

$$U_I(\tau)U_I^{\dagger}(\tau') = U_I(\tau,0)U_I^{\dagger}(\tau',0) = U_I(\tau,\tau'), \tag{12.11-36}$$

easily proven from the far rhs of (12.9-7), we arrive at

$$G(k,k';\tau-\tau') = -(\mathcal{Z}_0/\mathcal{Z})\mathrm{Tr}\Big\{\rho_0\mathcal{T}\Big[U_I(\beta,\tau)\tilde{e}_k(\tau)U_I(\tau,\tau')\tilde{e}^{\dagger}_{k'}(\tau')U_I(\tau',0)\Big]\Big\}. \tag{12.11-37}$$

We must now expand the three factors U_I in a perturbation series, which is as in (12.9-18) with imaginary times, or more directly as in the rhs of (12-9-27); one thus obtains a result of the following structure:

$$G(k,k';\tau-\tau') = -\frac{\mathcal{Z}_0}{\mathcal{Z}}\sum_{m,p,q}\frac{(-\lambda)^{m+p+q}}{m!p!q!}\int_\tau^\beta d\tau_m...d\tau_1\int_{\tau'}^{\tau} d\tau_p...d\tau_1'\int_0^{\tau'} d\tau_q...d\tau_1''$$
$$\times\mathrm{Tr}\left\{\rho_0\mathcal{T}\left[\tilde{V}(\tau_m)...\tilde{V}(\tau_1)\tilde{e}_k(\tau)\tilde{V}(\tau_p)...\tilde{V}(\tau_1')\tilde{e}_{k'}^\dagger\tilde{V}(\tau_q)...\tilde{V}(\tau_1'')\right]\right\}. \quad (12.11\text{-}38)$$

Suppose that we look for a term of order $n = m+p+q$; there are $n!/m!p!q!$ such terms, each term thus having a weight $W = (m!p!q!/n!)$. The time-ordering operator will further give a slew of Heaviside functions, thus allowing us to take each integral over the full interval 0 to β. Multiplying the coefficients in the above expression by W, we obtain,

$$G(k,k';\tau-\tau') = -\frac{\mathcal{Z}_0}{\mathcal{Z}}\sum_{n=0}^{\infty}\frac{(-\lambda)^n}{n!}\int_0^\beta d\tau_n\int_0^\beta d\tau_{n-1}...\int_0^\beta d\tau_1$$
$$\times\mathrm{Tr}\left\{\rho_0\mathcal{T}\left[\tilde{V}(\tau_n)\tilde{V}(\tau_{n-1})...\tilde{V}(\tau_1)\tilde{e}_k(\tau)\tilde{e}_{k'}^\dagger(\tau')\right]\right\}. \quad (12.11\text{-}39)$$

We substitute for $\mathcal{Z}/\mathcal{Z}_0$ from (12.9-28) and (12.9-29) to arrive at:

$$G(k,k';\tau-\tau') = \frac{-\sum_{n=0}^{\infty}\frac{(-\lambda)^n}{n!}\int_0^\beta d\tau_n...\int_0^\beta d\tau_1\mathrm{Tr}\left\{\rho_0\mathcal{T}\left[\tilde{V}(\tau_n)...\tilde{V}(\tau_1)\tilde{e}_k(\tau)\tilde{e}_{k'}^\dagger(\tau')\right]\right\}}{\sum_{n=0}^{\infty}\frac{(-\lambda)^n}{n!}\int_0^\beta d\tau_n...\int_0^\beta d\tau_1\mathrm{Tr}\left\{\rho_0\mathcal{T}\left[\tilde{V}(\tau_n)...\tilde{V}(\tau_1)\right]\right\}},$$

(12.11-40)

in full agreement with a similar expression in Mattuck.[89] The denominator is $\Sigma_n\lambda^n W_n$.

In order to obtain a perturbation expansion ordered in powers of λ, the denominator needs to be inverted ($1/Den$) to the appropriate order and then multiplied with the numerator. The task is not as arduous as it seems; moreover, cancellations of various terms – ($1/Den$) giving many opposite signs – lead to a result in which only the cumulants, defined by (12.9-30) and (12.9-31), remain. To not take up space, we shall demonstrate this only for the zero-order and first-order contributions; the reader is encouraged to try a few more terms by him/herself.[90] The zero-order term is immediate,

$$G^{(0)}(k,k';\tau-\tau') = -\langle\tilde{e}_k(\tau)\tilde{e}_{k'}^\dagger(\tau')\rangle_0 = G_0(k,\tau-\tau')\delta_{kk'}. \quad (12.11\text{-}41)$$

For the next order we have for the denominator: $Den = 1+\lambda W_1$, or $1/Den = 1-\lambda W_1 + \mathcal{O}(\lambda^2)$. For the numerator we get to first order

---

[89] R.D. Mattuck, Op. Cit., Appendix E, Eq. (E.17).

[90] See also L.E. Reichl, Op. Cit. (*first Edition*), Section 12E.2.

$$Num = \lambda \int_0^\beta d\tau_1 \langle \mathcal{T}[\tilde{V}(\tau_1)\tilde{e}_k(\tau)\tilde{e}_{k'}^\dagger(\tau')]\rangle_0$$

$$= \frac{\lambda}{2} \sum_{k_1k_2k_3k_4} \int_0^\beta d\tau_1 \langle k_1k_2|v|k_3k_4\rangle \langle \mathcal{T}[\tilde{e}_{k_1}^\dagger(\tau_1)\tilde{e}_{k_2}^\dagger(\tau_1)\tilde{e}_{k_4}(\tau_1)\tilde{e}_{k_3}(\tau_1)\tilde{e}_k(\tau)\tilde{e}_{k'}^\dagger(\tau')]\rangle_0. \tag{12.11-42}$$

We denote k by '5' and k' by '6' and interchange k_3 and k_4, giving a factor ε. The average is obtained with Wick's theorem. Denoting its operation by $\mathcal{W}$, we have for $\mathcal{T}[..]$:

$$\mathcal{T}[...] = \varepsilon \mathcal{W}\langle [1^\dagger 2^\dagger 3456^\dagger]\rangle_0$$

$$= \varepsilon\big\{ \varepsilon\langle 1^\dagger 3\rangle_0 \langle 2^\dagger 4\rangle_0 \langle 56^\dagger\rangle_0 + \langle 1^\dagger 4\rangle_0 \langle 2^\dagger 3\rangle_0 \langle 56^\dagger\rangle_0 + \varepsilon\langle 1^\dagger 5\rangle_0 \langle 2^\dagger 3\rangle_0 \langle 46^\dagger\rangle_0$$

$$+ \langle 1^\dagger 3\rangle_0 \langle 2^\dagger 5\rangle_0 \langle 46^\dagger\rangle_0 + \varepsilon\langle 1^\dagger 4\rangle_0 \langle 2^\dagger 5\rangle_0 \langle 36^\dagger\rangle_0 + \langle 1^\dagger 5\rangle_0 \langle 2^\dagger 4\rangle_0 \langle 36^\dagger\rangle_0 \big\}. \tag{12.11-43}$$

In the first two brackets of each term we interchange the two operators; this change has parity 2, leaving the sign as is. Next we put in the τ's and substitute the free propagators, to obtain

$$(\lambda/2)\sum_{k_1...k_4} \int_0^\beta d\tau_1 \langle 12|v|34\rangle \mathcal{T}[...]$$

$$= -\varepsilon \left\{ \frac{\lambda}{2} \sum_{k_1k_2} \beta G_0(k_1,0) G_0(k_2,0) G_0(k,\tau-\tau')\delta_{kk'} \big[\varepsilon\langle 12|v|12\rangle + \langle 12|v|21\rangle\big] \right.$$

$$\left. + \lambda \sum_{k_1} \int_0^\beta d\tau_1 G_0(k_1,0) G_0(k,\tau-\tau_1) G_0(k',\tau_1-\tau')\big[\langle 1k|v|1k'\rangle + \varepsilon\langle 1k|v|k'1\rangle\big] \right\}, \tag{12.11.44}$$

where we used (12.10-6) to combine two pairs of terms. Now the $-\lambda W_1$ term resulting from the denominator contributes after being multiplied with the zero-order numerator, Eq. (12.11-41). Substituting for W_1 from (12.10-3), one finds that it cancels the first line of (12.11-44), having two non-connected diagrams. In the remaining part we set $\lambda = 1$. The two terms of that part yield two connected Feynman diagrams, the 'bubble' and the 'open oyster'. So, let us represent $G(k,k';\tau-\tau')$ by a double arrowed line. Then the following result is obtained:

 $= \sum$ *all directed topologically distinct connected diagrams.* (12.11-45)

More fully, we give the zero-order, first order and a number of second order diagrams in the following 'equation',

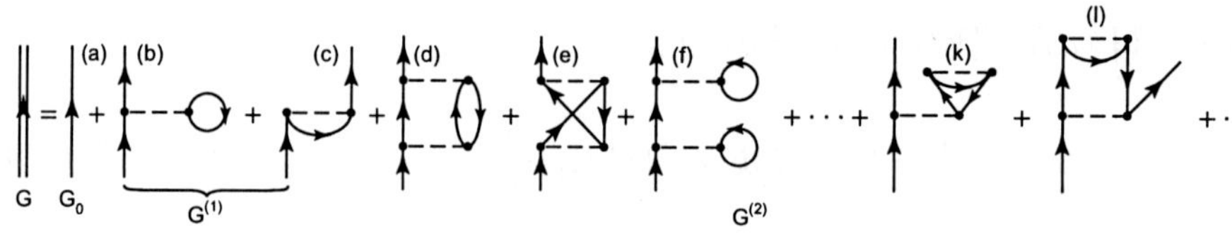

(12.11-46)

It will not have escaped the reader that the two pairs of matrix elements in (12.11-44) can be combined if one is not specifically interested in the exchange contributions. Therefore it is far more simple to use Hugenholtz diagrams, with summary matrix elements (12.10-10). Up to second order there are just five diagrams. They are given in the 'equation' below

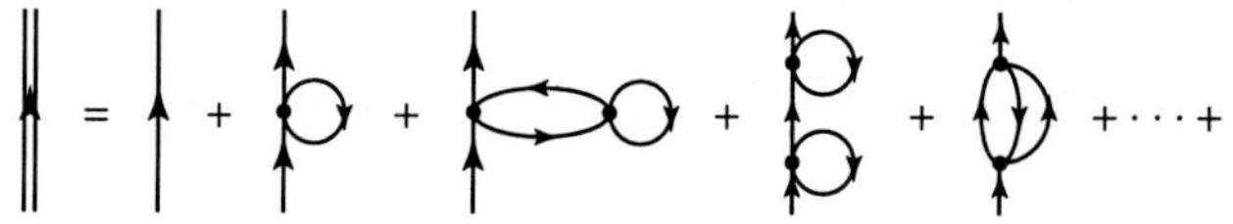

(12.11-47)

The algebraic result is easily written out with the rules previously given.

12.12 Self-energy and Dyson's Equation

To fully appreciate the concept of self-energy and Dyson's integral equation for the full Green's function, it is best to first again consider the spatial form $G_{\lambda\mu}(\mathbf{r},\tau;\mathbf{r}'\tau')$. The diagrammatic results of Eqs. (12.11-46) and (12.11-47) can be carried over nearly verbatim, with $k = (\mathbf{k},\lambda)$ now replaced by $\mathbf{r}$ and λ, while instead of summing over internal k's we now integrate over internal $\mathbf{r}$'s and still sum over the spins. The diagrammatic analysis has shown that $G_{\lambda\mu}(\mathbf{r},\tau;\mathbf{r}'\tau')$ is obtained as a sum of objects with an incoming and outgoing line attached. These objects can be collected in a single skeleton denoted by $\hat{\Sigma}$, the self-energy, see the diagram of Fig. 12-25. Formally we then have

$$G_{\lambda\mu}(\mathbf{r}\tau;\mathbf{r}',\tau') = G^0_{\lambda\mu}(\mathbf{r}\tau;\mathbf{r}',\tau')$$
$$+\sum_{\nu\rho}\int\!\!...\!\int d^3r_1 d^3\bar{r}_1 d\tau_1 d\bar{\tau}_1 G^0_{\lambda\nu}(\mathbf{r}\tau;\mathbf{r}_1,\tau_1)\hat{\Sigma}_{\nu\rho}(\mathbf{r}_1\tau_1;\bar{\mathbf{r}}_1\bar{\tau}_1)G^0_{\rho\mu}(\bar{\mathbf{r}}_1\bar{\tau}_1;\mathbf{r}',\tau')\,. \quad (12.12\text{-}1)$$

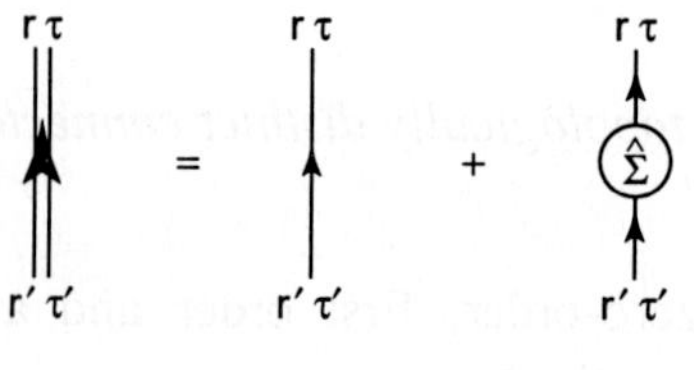

Fig. 12-25. Diagram representing Eq. (12.12-1).

However, $\hat{\Sigma}$ so obtained is not yet the *proper self-energy* Σ (or *irreducible self-energy*), the reason being that many diagrams can be converted to simpler diagrams by cutting an internal line, e.g. the double bubble diagram [sixth from left, denoted by (f)] in (12.11-46). Such diagrams lead us to the connection between $\hat{\Sigma}$ and Σ, illustrated in Fig. (12-26).

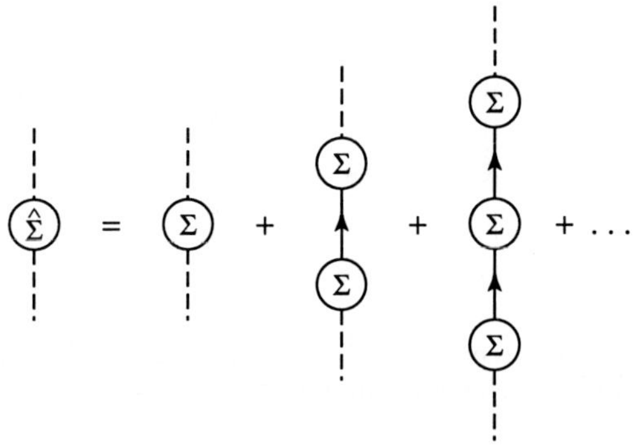

Fig. 12-26. Self-energy $\hat{\Sigma}$ as a sum of skeletons with repeated proper self-energy Σ.

Substituting this diagrammatic summation into Eq. (12.12-1), we arrive at

$$\begin{aligned}
G_{\lambda\mu}(\mathbf{r}\tau;\mathbf{r}'\tau') &= G^0_{\lambda\mu}(\mathbf{r}\tau;\mathbf{r}'\tau') \\
&+\sum_{\nu\rho}\int\ldots\int d^3r_1 d^3\bar{r}_1 d\tau_1 d\bar{\tau}_1 G^0_{\lambda\nu}(\mathbf{r}\tau;\mathbf{r}_1,\tau_1)\Sigma_{\nu\rho}(\mathbf{r}_1\tau_1;\bar{\mathbf{r}}_1\bar{\tau}_1)G^0_{\rho\mu}(\bar{\mathbf{r}}_1\bar{\tau}_1;\mathbf{r}'\tau') \\
&+\sum_{\nu_1\nu_2\rho_1\rho_2}\int\ldots\int d^3r_1 d^3\bar{r}_1 d\tau_1 d\bar{\tau}_1 \int\ldots\int d^3r_2 d^3\bar{r}_2 d\tau_2 d\bar{\tau}_2 G^0_{\lambda\nu_1}(\mathbf{r}\tau;\mathbf{r}_1\tau_1) \\
&\times\Sigma_{\nu_1\rho_1}(\mathbf{r}_1\tau_1;\bar{\mathbf{r}}_1\bar{\tau}_1)G^0_{\rho_1\nu_2}(\bar{\mathbf{r}}_1\bar{\tau}_1;\mathbf{r}_2,\tau_2)\Sigma_{\nu_2\rho_2}(\mathbf{r}_2\tau_2;\bar{\mathbf{r}}_2\bar{\tau}_2)G^0_{\rho_2\mu}(\bar{\mathbf{r}}_2\bar{\tau}_2;\mathbf{r}'\tau') + \ldots + \textit{ad inf}.
\end{aligned}$$

(12.12-2)

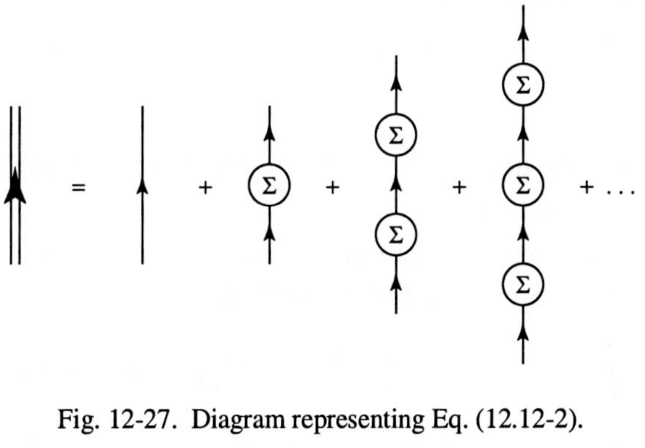

Fig. 12-27. Diagram representing Eq. (12.12-2).

This equation is solved by *Dyson's equation* [78]:

$$G_{\lambda\mu}(\mathbf{r}\tau;\mathbf{r}'\tau') = G^0_{\lambda\mu}(\mathbf{r}\tau;\mathbf{r}'\tau')$$
$$+\sum_{\nu\rho}\int ...\int d^3r_1 d^3\bar{r}_1 d\tau_1 d\bar{\tau}_1 G^0_{\lambda\nu}(\mathbf{r}\tau;\mathbf{r}_1,\tau_1)\Sigma_{\nu\rho}(\mathbf{r}_1\tau_1;\bar{\mathbf{r}}_1\bar{\tau}_1)G_{\rho\mu}(\bar{\mathbf{r}}_1\bar{\tau}_1;\mathbf{r}'\tau'). \quad \textbf{(12-12-3)}$$

To verify the result, simply iterate (12.12-3), to re-obtain (12.12-2) to any order. Dyson's equation is diagrammatically shown in Fig. 12-28.

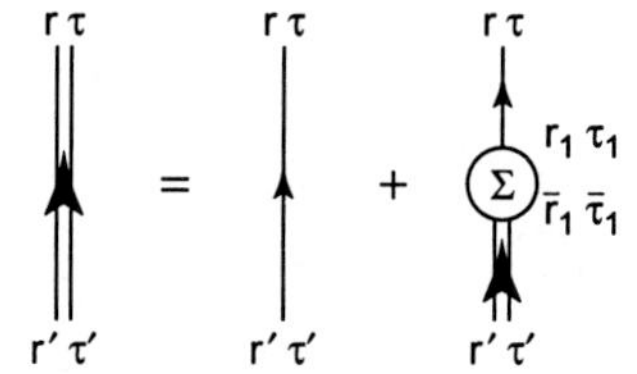

Fig. 12-28. Diagram representing Dyson's equation (12.12-3). Each entry is a matrix element with (suppressed) spin indices.

We shall now restate these results in the $(\mathbf{k},\tau)$ language and in the Fourier-series language. Let

$$\Sigma_{\lambda\mu}(\mathbf{r}\tau;\mathbf{r}'\tau') = \frac{1}{V_0}\sum_{\mathbf{k},\mathbf{k}'} e^{i\mathbf{k}\cdot\mathbf{r}} e^{-i\mathbf{k}'\cdot\mathbf{r}'}\Sigma_{\lambda\mu}(\mathbf{k}\tau;\mathbf{k}'\tau') \qquad (12.12\text{-}4a)$$

with

$$\Sigma_{\lambda\mu}(\mathbf{k}\tau;\mathbf{k}'\tau') = \frac{1}{V_0}\int\int d^3r d^3r' e^{-i\mathbf{k}\cdot\mathbf{r}} e^{i\mathbf{k}'\cdot\mathbf{r}'}\Sigma_{\lambda\mu}(\mathbf{r}\tau;\mathbf{r}'\tau'); \qquad (12.12\text{-}4b)$$

multiplying Dyson's equation with $V_0^{-1}\int d^3r \int d^3r' e^{-i\mathbf{k}\cdot\mathbf{r}} e^{i\mathbf{k}'\cdot\mathbf{r}'}$, substituting (12.12-4a) and employing also (12.11-23), one finds the momentum-space result, being another form of Dyson's equation,

$$G_{\lambda\mu}(\mathbf{k}\tau;\mathbf{k}'\tau') = G^0_{\lambda\mu}(\mathbf{k}\tau;\mathbf{k}'\tau')$$
$$+\sum_{\nu\rho}\sum_{\mathbf{k}_1\bar{\mathbf{k}}_1}\int\int d\tau_1 d\bar{\tau}_1 G^0_{\lambda\nu}(\mathbf{k}\tau;\mathbf{k}_1\tau_1)\Sigma_{\nu\rho}(\mathbf{k}_1\tau_1;\bar{\mathbf{k}}_1\bar{\tau}_1)G_{\rho\mu}(\bar{\mathbf{k}}_1\bar{\tau}_1;\mathbf{k}'\tau'). \quad \textbf{(12.12-5)}$$

We note that all functions in (12.12-5) – being equilibrium averages – only depend on the difference $\Delta\tau$ of the two τ's.

We now make the assumption that the system is *homogeneous*, i.e., invariant with regard to a translation $\mathbf{r}_0$. From (12.11-24) we find

$$G_{\lambda\mu}(\mathbf{k},\tau;\mathbf{k}',\tau') = \frac{1}{V_0}\int\int d^3r d^3r' e^{-i\mathbf{k}\cdot(\mathbf{r}+\mathbf{r}_0)} e^{i\mathbf{k}'\cdot(\mathbf{r}'+\mathbf{r}_0)} G_{\lambda\mu}(\mathbf{r}+\mathbf{r}_0,\tau;\mathbf{r}'+\mathbf{r}_0,\tau')$$

$$= \frac{1}{V_0}\int\int d^3r d^3r' e^{-i\mathbf{k}.\mathbf{r}} e^{i\mathbf{k}'\cdot\mathbf{r}'} e^{-i(\mathbf{k}-\mathbf{k}')\cdot\mathbf{r}_0} G_{\lambda\mu}(\mathbf{r},\tau;\mathbf{r}',\tau') = e^{-i(\mathbf{k}-\mathbf{k}')\cdot\mathbf{r}_0} G_{\lambda\mu}(\mathbf{k},\tau;\mathbf{k}',\tau').$$

Since this must hold for arbitrary $\mathbf{r}_0$, we clearly have $\mathbf{k} = \mathbf{k}'$, or $G_{\lambda\mu}(\mathbf{k},\tau;\mathbf{k}'\tau') = G_{\lambda\mu}(\mathbf{k},\Delta\tau)\delta_{\mathbf{k},\mathbf{k}'}$; this holds, *a fortiori*, for $G^0_{\lambda\mu}$. Next, with G being periodic over the interval $\Delta\tau \in (-\beta,\beta)$, we define frequency Fourier series with coefficients in ω_n; we then arrive at the form

$$\overline{G}_{\lambda\mu}(\mathbf{k},\omega_n) = \overline{G}^0_{\lambda\mu}(\mathbf{k},\omega_n) + \sum_{\nu,\rho} \Sigma_{\nu\rho}(\mathbf{k},\omega_n)\overline{G}_{\rho\mu}(\mathbf{k},\omega_n), \tag{12.12-6}$$

which is a simple matrix equation. With $\overline{G}_0$ given by (12.10-24), Eq.(12.12-6) yields

$$\overline{G}_{\lambda\mu}(\mathbf{k},\omega_n) = [(i\omega_n - \hat{\varepsilon}^0_k)\mathsf{I} - \Sigma(\mathbf{k},\omega_n)]^{-1}_{\lambda\mu}, \tag{12.12-7}$$

where $[..]^{-1}$ is the reciprocal matrix; we have added a super '0' to $\hat{\varepsilon}_k$, being the unperturbed single-particle energy. For the case that $\overline{G}$, $\overline{G}_0$ and Σ are diagonal in the spin indices this gives

$$\overline{G}_{\lambda\mu}(\mathbf{k},\omega_n) = \frac{1}{i\omega_n - \hat{\varepsilon}^0_k - \Sigma(\mathbf{k},\omega_n)}\delta_{\lambda\mu}. \tag{12.12-8}$$

Comparing with (12.11-33) we note that this unique result allows us to obtain the excitations of the quantum liquid from the poles of $\overline{G}_{\lambda\mu}(\mathbf{k},\omega_n)$, while their relative strengths are given by the residues.

The importance of Dyson's equation should not be underestimated. Suppose we evaluate Σ only to second order. As (12.11-46) shows, there are two diagrams in first order and ten diagrams in second order, of which only five have been shown.[91] Five of these diagrams are reducible like (f), while the other five are irreducible. So, up to second order there are in total seven irreducible diagrams to be included. However, as Fig.12-26 shows, these seven diagrams are repeated *ad infinitum*, giving rise to strings of bubble diagrams, oyster diagrams, their mixtures, etc. Thus $\Sigma^{(2)}$ sums ∞^7 diagrams! To summarize, quoting Fetter and Walecka: "this example [$\Sigma^{(2)}$] clearly demonstrates the power of Dyson's equation, because *any* approximation for Σ generates an *infinite-order* approximate series for the Green's function. Dyson's equation thus enables us to sum an infinite class of perturbation terms in a compact form."[92] In Fig. 12-29 we reproduce the self energy associated with the skeletons of diagrams (*b*), (*c*) and (*d*) of Eq. (12.11-47) in the $(\mathbf{k},\omega_n)$ language.[93] Note that in the $(\mathbf{k},\omega_n)$ language the summation is simply executed as in a geometrical progression! In the next section we will meet two very successful approximation schemes, viz. the Hartree-Fock approximation and the ring approximation for Fermi liquids.

---

[91] All ten are shown in Fetter and Walecka, Op. Cit, Fig. 9.8; they should be converted to the more modern convention in which all vertices are drawn horizontally.

[92] A.L. Fetter and J.D. Walecka, Op. Cit., p.109.

[93] R.D. Mattuck, Op. Cit., p. 180, Eq. (10.5); this equation sums the 'bubble', 'open oyster' and the 'pair-bubble', with their repetitions.

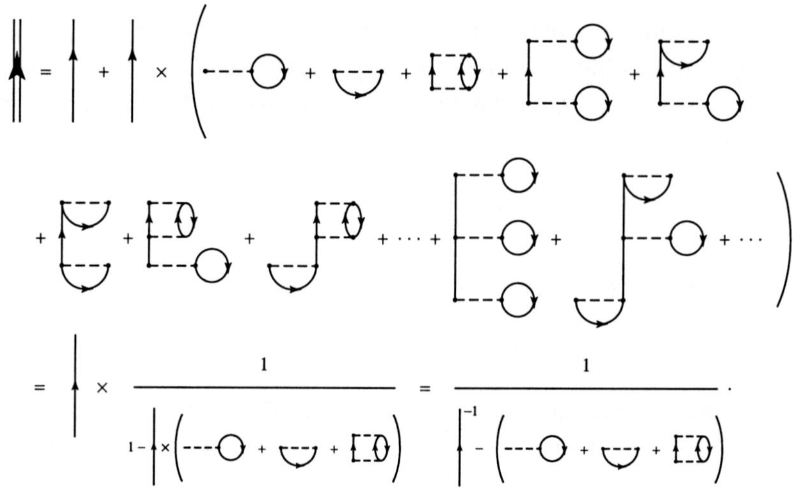

Fig. 12-29. The Green's function with Σ resulting from diagrams (*b*), (*c*) and (*d*); cf. R.D. Mattuck. [93].

Finally, we come back to the earlier given results for the average particle number $\langle N\rangle$, energy $\langle \mathscr{E}\rangle$ and the grand potential Ω, Eqs. (12.11-5), (12.11-12) and (12.11-16). The following results are easily verified:

$$\langle N\rangle = \mp\frac{2s+1}{\beta}\lim_{\delta\to 0}\sum_{\mathbf{k}}\sum_{n} e^{i\omega_n\delta}\overline{G}(\mathbf{k},\omega_n), \tag{12.12-9}$$

$$\langle \mathscr{E}\rangle = \mp\frac{2s+1}{\beta}\lim_{\delta\to 0}\sum_{\mathbf{k}}\sum_{n}\left[\hat{\varepsilon}^0_{\mathbf{k}} + \varsigma + \tfrac{1}{2}\Sigma(\mathbf{k},\omega_n)\right] e^{i\omega_n\delta}\overline{G}(\mathbf{k},\omega_n), \tag{12.12-10}$$

$$\Omega(T,V,\varsigma) = \Omega_0(T,V,\varsigma) \mp \frac{2s+1}{2\beta}\lim_{\delta\to 0}\sum_{\mathbf{k}}\sum_{n}\int_0^1 \frac{d\lambda}{\lambda}\Sigma_\lambda(\mathbf{k},\omega_n)e^{i\omega_n\delta}\overline{G}_\lambda(\mathbf{k},\omega_n). \tag{12.12-11}$$

Here s is the spin [the term $2s+1$ stems from $\mathrm{tr}\overline{G} = \Sigma_\eta \overline{G}_{\eta\eta} = 2s+1$]. For Ω we need the Green's function in an ensemble with density operator $\rho_\lambda = [\exp(K^0 + \lambda\mathscr{V})]/\mathscr{Z}_\lambda$.

12.13 Fermi Liquids Revisited

12.13.1 *The Hartree–Fock Approximation* [94]

In our previous discussion on Fermi liquids we dealt with the Landau approach, in which the constituents were dressed with the interactions in the liquid and move in the potential associated with the interactions. We will now be able to quantify this in

[94] The Hartree-Fock approximation can equally well be done for Bose liquids, with the usual sign changes applied. However, for Bose liquids the zero-temperature limit cannot be taken because of BEC.

a microscopic rather than phenomenological way. Firstly, the particle is dressed with an average potential coming from the presence of all other particles, which is just the proper self-energy. In the Hartree-Fock approximation we restrict ourselves to the first two diagrams associated with the two-body interaction and exchange potential, pictured in Fig. 12-30a. However, taken at face-value, the background particles in this approximation are themselves taken to be non-interacting, an oversimplification. Thus, secondly, the background-particles should also be endowed with their self-energies; the addition of all corresponding diagrams leads to a modified proper self energy and Hartree-Fock Green's function, depicted in Fig. 12.30b.

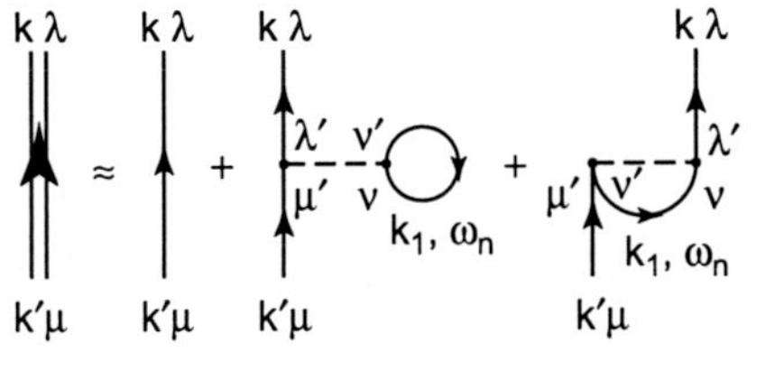

Fig. 12-30a. Lowest-order proper self-energy.

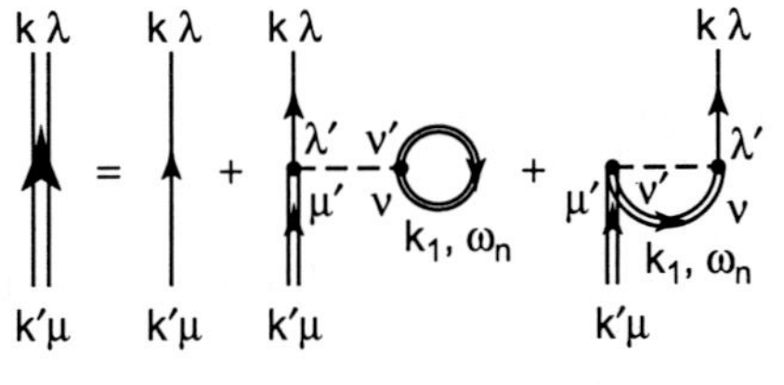

Fig. 12-30b. Hartree-Fock Green's function.

The computation of the proper self-energy implied by the diagrams of Fig. 12-30b is straightforward. In order to obtain the connection with the usual Hartree-Fock equation for the case that we deal with an electron gas (cf., Section 7.8), we should evaluate the Green's function in coordinate-space; since this is rather laborious, it has been relegated to the problems (Problem 12.8). For the general treatment pertaining to Fermi liquids the momentum-frequency method is faster and adequate. A note is in order, however, for the placing of the spin indices, given that we have double line parts in the diagrams. The reader will undoubtedly have noted that in the presentation of Dyson's equation we have refrained from using a symbol 'k' which denotes both **k** and spin η, the reason being that this does not generally apply to double-line structures [other authors use heavy lines]. Whereas for translation-invariant systems each double line can still be endowed with a momentum **k** and frequency ω_n, the spin indices are placed at the vertices; in the Green's function G representing the double line, the spin of the vertex going to is entered as first subscript, while the spin of the tail is the second subscript: $\beta \Rightarrow \alpha = G_{\alpha\beta}(\mathbf{k},\omega_n)$. For G^0 in 'temperature diagrams', the spin should be the same at both ends and a Kronecker delta is in order. As to the vertex contribution, we have the usual two-body matrix element *including spin.* If the potential is spin-independent, the vertex will have a spin part $\Sigma_{\lambda'\nu'}\langle\lambda\nu|\lambda'\nu'\rangle = \Sigma_{\lambda'\nu'}\delta_{\lambda\lambda'}\delta_{\nu\nu'}$, indicating that *both particles retain their spin.* The delta's here denote the unit matrix **1** in spinor space[95].With this in mind, we obtain for G

[95] Fetter and Walecka, Op. Cit., p. 103-104.

$$\bar{G}_{\lambda\mu}(\mathbf{k},\omega_n) = \bar{G}^0_{\lambda\mu}(\mathbf{k},\omega_n)\delta_{\lambda\mu}$$

$$+\beta^{-1}\sum_{n'}\sum_{\mathbf{k}_1,\nu}\lim_{\delta\to 0} e^{i\omega_{n'}\delta}\left\{\bar{G}^0_{\lambda\lambda}(\mathbf{k},\omega_n)\bar{G}_{\nu\nu}(\mathbf{k}_1,\omega_{n'})\bar{G}_{\lambda\mu}(\mathbf{k},\omega_n)\langle\mathbf{kk}_1|v|\mathbf{kk}_1\rangle \mathrm{tr}(1)\right.$$

$$\left.-\bar{G}^0_{\lambda\lambda}(\mathbf{k},\omega_n)\bar{G}_{\lambda\nu}(\mathbf{k}_1,\omega_{n'})\bar{G}_{\nu\mu}(\mathbf{k},\omega_n)\langle\mathbf{k}_1\mathbf{k}|v|\mathbf{kk}_1\rangle\right\}. \quad (12.13\text{-}1)$$

As usual, we write the two-body potential as a Fourier series, $v(\mathbf{r}) = \Sigma_{\mathbf{q}} e^{i\mathbf{q}\cdot\mathbf{r}} V(\mathbf{q})$; this leads to two Kronecker delta's for the vertex momenta, see Eq. (12.3-4). The two matrix elements above then become $V(0)$ and $V(\mathbf{k}_1 - \mathbf{k})$, respectively. Writing out Dyson's equation, noting $\bar{G}^0_{\lambda\nu} = \bar{G}^0_{\lambda\lambda}\delta_{\lambda\nu}$ and comparing with (12.13-1), we find

$$\Sigma^{(1)}_{\lambda\nu} = \frac{1}{\beta V_0}\sum_{n'}\sum_{\mathbf{k}_1}\lim_{\delta\to 0} e^{i\omega_{n'}\delta}\left[\bar{G}_{\nu\nu}(\mathbf{k}_1,\omega_{n'})V(0)(2s+1)\delta_{\lambda\nu} - \bar{G}_{\lambda\nu}(\mathbf{k}_1,\omega_{n'})V(\mathbf{k}_1-\mathbf{k})\right]. \quad (12.13\text{-}2)$$

Let $\bar{G}_{\lambda\nu} = \bar{G}_{HF}\delta_{\lambda\nu}$, which allows us to also set $\Sigma^{(1)}_{\lambda\nu} = \Sigma_{HF}\delta_{\lambda\nu}$; this finally yields,

$$\Sigma_{HF}(\mathbf{k};T) = \frac{1}{\beta V_0}\sum_{n'}\sum_{\mathbf{k}_1}\lim_{\delta\to 0} e^{i\omega_{n'}\delta}\bar{G}_{HF}(\mathbf{k}_1,\omega_{n'})\left[(2s+1)V(0) - V(\mathbf{k}_1-\mathbf{k})\right]. \quad (12\text{-}13\text{-}3)$$

For $\bar{G}_{HF}$ we have from (12.12-8),

$$\bar{G}_{HF}(\mathbf{k},\omega_n) = \frac{1}{i\omega_n - \hat{\varepsilon}^0_{\mathbf{k}} - \Sigma_{HF}(\mathbf{k};T)}. \quad (12.13\text{-}4)$$

The last two equations must be solved self-consistently; we note that $\bar{G}$ and Σ also depend on temperature through ω_n.

We shall find a low-temperature approximation for Σ_{HF} and apply it to obtain the specific heat for a spin-½ fermion system. From (12.13-4) we have

$$[\bar{G}_{HF}(\mathbf{k},\omega_n;T)]^{-1} = i\omega_n - \hat{\varepsilon}^0_{\mathbf{k}} - \Sigma_{HF}(\mathbf{k};T),$$
$$[\bar{G}_{HF}(\mathbf{k},\omega_n;0)]^{-1} = i\omega_n - \hat{\varepsilon}^0_{\mathbf{k}} - \Sigma_{HF}(\mathbf{k};0).$$

Subtracting and approximating the denominator $1/\bar{G}(T)\bar{G}(0) \approx 1/[\bar{G}(0)]^2$, we obtain

$$\bar{G}_{HF}(\mathbf{k},\omega_n;T) \simeq \bar{G}_{HF}(\mathbf{k},\omega_n;0) + [\Sigma_{HF}(\mathbf{k};T) - \Sigma_{HF}(\mathbf{k};0)][\bar{G}_{HF}(\mathbf{k},\omega_n;0)]^2. \quad (12.13\text{-}5)$$

This is substituted into (12.13-3), to yield in the thermodynamic limit,

$$\Sigma_{HF}(\mathbf{k};T) = (2\pi)^{-3}\beta^{-1}\sum_{n'}\int d^3k'\left[2V(0) - V(\mathbf{k}'-\mathbf{k})\right]\lim_{\delta\to 0} e^{i\omega_n\delta}$$

$$\times\left\{\frac{1}{i\omega_{n'} - \hat{\varepsilon}^0_{\mathbf{k}'} - \Sigma_{HF}(\mathbf{k}';0)} + \left[\Sigma_{HF}(\mathbf{k}';T) - \Sigma_{HF}(\mathbf{k}';0)\right]\left(\frac{1}{i\omega_{n'} - \hat{\varepsilon}^0_{\mathbf{k}'} - \Sigma_{HF}(\mathbf{k}';0)}\right)^2\right\}. \quad (12.13\text{-}6)$$

We have used here the form for $\bar{G}_{HF}(\mathbf{k},\omega_n;0)$ corresponding to (12.13-4); this should be compared to the free propagator form $\bar{G}_0(\mathbf{k},\omega_n)=1/(i\omega_n-\hat{\varepsilon}^0_{\mathbf{k}})$. We are thus led to define the new 'particle-energy-in-first-order' or the 'pseudo-particle' energy,

$$\hat{\varepsilon}^1_{\mathbf{k}}=\hat{\varepsilon}^0_{\mathbf{k}}+\Sigma_{HF}(\mathbf{k};0) \quad \text{or} \quad \varepsilon^1_{\mathbf{k}}=\varepsilon^0_{\mathbf{k}}+\Sigma_{HF}(\mathbf{k};0). \tag{12.13-7}$$

The H–F Green's function then simply reads

$$\bar{G}_{HF}(\mathbf{k},\omega_n;T)=1/(i\omega_n-\varepsilon^1_{\mathbf{k}}-\varsigma). \tag{12.13-8}$$

The sum over n' and the limit procedure is now performed as in (12.10-30). For the first term in the curly bracket we thus obtain

$$\textit{first term}=(2\pi)^{-3}\int d^3k'\left[2V(0)-V(\mathbf{k}'-\mathbf{k})\right]n_{\mathbf{k}'}(T), \tag{12.13-9}$$

where $n_{\mathbf{k}}(T)$ is the Fermi–Dirac distribution; with $\varsigma=\varepsilon_F$ we have

$$n_{\mathbf{k}}(T)=1/[e^{\beta(\varepsilon^1_{\mathbf{k}}-\varepsilon_F)}+1]. \tag{12.13-10}$$

Ergo: *In the Hartree–Fock approximation the pseudo-particles still have a Fermi–Dirac distribution and it is legitimate to still introduce a Fermi surface and to speak of a filled Fermi sea at T = 0. This is mainly due to the fact that the HF self-energy is frequency-independent.*

To perform the second sum, note that

$$\frac{1}{[i\omega_n-\varepsilon^1_{\mathbf{k}}-\varsigma]^2}=\frac{\partial}{\partial\varepsilon^1_{\mathbf{k}}}\left(\frac{1}{i\omega_n-\varepsilon^1_{\mathbf{k}}-\varsigma}\right);$$

so the result is immediate,

$$\textit{second term}=(2\pi)^{-3}\int d^3k'[2V(0)-V(\mathbf{k}'-\mathbf{k})][\Sigma_{HF}(\mathbf{k}';T)-\Sigma_{HF}(\mathbf{k}';0)]\frac{\partial n_{\mathbf{k}'}}{\partial\varepsilon^1_{\mathbf{k}'}}. \tag{12.13-11}$$

Collecting we find

$$\Sigma_{HF}(\mathbf{k};T)=(2\pi)^{-3}\int d^3k'[2V(0)-V(\mathbf{k}'-\mathbf{k})] \times\left\{n_{\mathbf{k}'}(T)+[\Sigma_{HF}(\mathbf{k}';T)-\Sigma_{HF}(\mathbf{k}';0)]\frac{\partial n_{\mathbf{k}'}}{\partial\varepsilon^1_{\mathbf{k}}}\right\}. \tag{12.13-12}$$

A fortiori,

$$\Sigma_{HF}(\mathbf{k};0)=(2\pi)^{-3}\int d^3k'[2V(0)-V(\mathbf{k}'-\mathbf{k})]\,n_{\mathbf{k}'}(0). \tag{12-13-13}$$

Subtracting we have the integral equation

$$\Sigma_{HF}(\mathbf{k};T)-\Sigma_{HF}(\mathbf{k};0)=(2\pi)^{-3}\int d^3k'[2V(0)-V(\mathbf{k}'-\mathbf{k})]$$
$$\times\left\{\left(n_{\mathbf{k}'}(T)-n_{\mathbf{k}'}(0)\right)+[\Sigma_{HF}(\mathbf{k}';T)-\Sigma_{HF}(\mathbf{k}';0)]\frac{\partial n_{\mathbf{k}'}}{\partial\varepsilon^1_{\mathbf{k}'}}\right\}. \quad (12.13\text{-}14)$$

Finally we turn to the appropriate thermodynamic functions. The grand potential was given by (12.12-11), but the integration is cumbersome. We therefore turn to the grand energy $\langle\hat{\mathcal{E}}\rangle=\langle\mathcal{E}\rangle-\varsigma\langle N\rangle$. This thermodynamic function will be called the 'grand-energy function',

$$\langle\hat{\mathcal{E}}\rangle=\mathrm{Tr}(\rho_{gcan.}K)\equiv K(S,V,\varsigma)=\mathcal{E}-\varsigma N, \quad (12.13\text{-}15)$$

with

$$dK=d\mathcal{E}-\varsigma dN-Nd\varsigma=TdS-PdV-Nd\varsigma, \quad (12.13\text{-}16)$$

where we used Gibbs' relation (1.4-1). Of course, we can also write $K=K(T,V,\varsigma)$, so that

$$\left(\frac{\partial K}{\partial T}\right)_{V,\varsigma}=\left(\frac{\partial K}{\partial S}\right)_{V,\varsigma}\left(\frac{\partial S}{\partial T}\right)_{V,\varsigma}=T\left(\frac{\partial S}{\partial T}\right)_{V,\varsigma}. \quad (12.13\text{-}17)$$

Our objective is to obtain K and through integration of the above result the entropy for given T, V and N. From (12.12-9) and (12.12-10) we have

$$\begin{aligned}\langle\hat{\mathcal{E}}\rangle&=\beta^{-1}\lim_{\delta\to0}\sum_{\mathbf{k}}\sum_n[2\hat{\varepsilon}^0_{\mathbf{k}}+\Sigma_{HF}(\mathbf{k};T)]e^{i\omega_n\delta}\overline{G}_{HF}(\mathbf{k},\omega_n;T)\\&=\beta^{-1}\lim_{\delta\to0}\sum_{\mathbf{k}}\sum_n[2(\varepsilon^1_{\mathbf{k}}-\varsigma)-\Sigma_{HF}(\mathbf{k};0)]e^{i\omega_n\delta}\overline{G}_{HF}(\mathbf{k},\omega_n;T)\\&+\beta^{-1}\lim_{\delta\to0}\sum_{\mathbf{k}}\sum_n[\Sigma_{HF}(\mathbf{k};T)-\Sigma_{HF}(\mathbf{k};0)]e^{i\omega_n\delta}\overline{G}_{HF}(\mathbf{k},\omega_n;T).\end{aligned} \quad (12.13\text{-}18)$$

We substitute (12-13-5) in both parts, but omit the term of order $[\Sigma_{HF}(T)-\Sigma_{\mathrm{HF}}(0)]^2$. The sums over n' are then carried out as before; the following result is readily obtained,

$$\lim_{th}K(T,V,\varsigma)/V_0=(2\pi)^{-3}\int d^3k\Big[2(\varepsilon^1_{\mathbf{k}}-\varsigma)-\Sigma_{HF}(\mathbf{k};0)\Big]n_{\mathbf{k}}(T)$$
$$+(2\pi)^{-3}\int d^3k\left\{n_k(T)+\Big[2(\varepsilon^1_{\mathbf{k}}-\varsigma)-\Sigma_{HF}(\mathbf{k};0)\Big]\frac{\partial n_k(T)}{\partial\varepsilon^1_k}\right\}\left(\Sigma_{HF}(\mathbf{k};T)-\Sigma_{HF}(\mathbf{k};0)\right).$$
$$(12.13\text{-}19)$$

Since K and V_0 are both extensive quantities, the thermodynamic limit can be taken; the symbol $\lim_{th}$ will henceforth be omitted but is implied. For $T=0$ the second line does not contribute. We then arrive at the result, valid near $T=0$,

$$[K(T,V,\varsigma)-K(0,V,\varsigma)]/V_0=(2\pi)^{-3}\int d^3k\,2(\varepsilon^1_{\mathbf{k}}-\varsigma)\big[n_{\mathbf{k}}(T)-n_{\mathbf{k}}(0)\big]$$
$$+(2\pi)^{-3}\int d^3k\left\{2(\varepsilon^1_{\mathbf{k}}-\varsigma)\frac{\partial n_k(0)}{\partial\varepsilon^1_{\mathbf{k}}}\right\}\big(\Sigma_{HF}(\mathbf{k};T)-\Sigma_{HF}(\mathbf{k};0)\big)$$
$$+(2\pi)^{-3}\int d^3k\,n_{\mathbf{k}}(0)\big(\Sigma_{HF}(\mathbf{k};T)-\Sigma_{HF}(\mathbf{k};0)\big)$$
$$-(2\pi)^{-3}\int d^3k\,\Sigma_{HF}(\mathbf{k};0)\left\{\big[n_{\mathbf{k}}(T)-n_{\mathbf{k}}(0)\big]+\frac{\partial n_k(0)}{\partial\varepsilon^1_k}\big(\Sigma_{HF}(\mathbf{k};T)-\Sigma_{HF}(\mathbf{k};0)\big)\right\}.$$
(12.13-20)

With $n_{\mathbf{k}}(0)=\Theta(\varsigma-\varepsilon^1_{\mathbf{k}})$ we have $\partial n_{\mathbf{k}}(0)/\partial\varepsilon^1_{\mathbf{k}}=-\delta(\varepsilon^1_{\mathbf{k}}-\varsigma)$. This causes the second term above to vanish. Furthermore, the last two terms cancel exactly if we substitute (12.13-14) in the third term and (12.13-13) in the fourth term, interchanging k and k' in the double integral. The final simple result is

$$[K(T,V,\varsigma)-K(0,V,\varsigma)]/V_0=(2\pi)^{-3}\int d^3k\,2(\varepsilon^1_{\mathbf{k}}-\varsigma)\big[n_{\mathbf{k}}(T)-n_{\mathbf{k}}(0)\big].\quad(12.13\text{-}21)$$

This result indicates that the only low temperature corrections to the grand-energy function arise from a redistribution of the pseudo-particles over the energy levels $\varepsilon_{\mathbf{k}}{}^1$, stemming from interactions in the ground state.

Next we shall obtain the entropy, employing (12.13-17). Let $s(T,\varsigma)$ = $\lim_{th}$ $S(T,V,\varsigma)/V$. In order to evaluate the entropy, we will write the distribution $n_{\mathbf{k}}$ as[96]

$$n_{\mathbf{k}}=\tfrac{1}{2}\big(1-\tanh[(\varepsilon^1_{\mathbf{k}}-\varsigma)/2k_BT]\big).\quad(12.13\text{-}22)$$

From (12.13-17) and (12.13-21) we then have

$$T\left(\frac{\partial s}{\partial T}\right)_\varsigma=\frac{\partial}{\partial T}\int\frac{d^3k}{8\pi^3}(\varepsilon^1_{\mathbf{k}}-\varsigma)\left(1-\tanh\frac{\varepsilon^1_{\mathbf{k}}-\varsigma}{2k_BT}\right)$$
$$=\frac{1}{4\pi^2k_BT^2}\int_0^\infty k^2\left(\frac{dk}{d\varepsilon^1_{\mathbf{k}}}\right)d\varepsilon^1_{\mathbf{k}}(\varepsilon^1_{\mathbf{k}}-\varsigma)^2\,\mathrm{sech}^2\frac{\varepsilon^1_{\mathbf{k}}-\varsigma}{2k_BT},\quad(12.13\text{-}23)$$

where we carried out the angular integration, assuming that $\varepsilon^1_{\mathbf{k}}$ is isotropic in **k**-space. We set $\xi=(\varepsilon^1_k-\varsigma)/2k_BT$ and note that for low temperatures $\mathrm{sech}^2\xi$ is sharply peaked about $\xi=0$; we thus arrive at

$$T\left(\frac{\partial s}{\partial T}\right)_\varsigma=k_B^2T\left(k^2\frac{dk}{d\varepsilon^1_k}\right)_{\varepsilon^1_k=\varepsilon_F}\frac{2}{\pi^2}\int_{-\infty}^{\infty}d\xi\,\xi^2\mathrm{sech}^2\xi=\tfrac{1}{3}k_B^2T\left(k^2\frac{dk}{d\varepsilon^1_k}\right)_{\varepsilon^1_k=\varepsilon_F};\quad(12.13\text{-}24)$$

---

[96] Fetter and Walecka, Op. Cit., p. 265ff.

for the integral see the tables.[97] The integration for constant ς is trivial, yielding

$$s(T,\varsigma) = \tfrac{1}{3}k_B^2 T k_F^2 \left(\frac{d\varepsilon_k^1}{dk} \right)^{-1}_{\varepsilon_k^1 = \varepsilon_F} . \qquad (12.13\text{-}25)$$

To convert to a function of the temperature and the density, we evaluate the latter at $T \to 0$,

$$n = 2(2\pi)^{-3} \int d^3k \Theta(k_F - k) = (3\pi^2)^{-1} k_F^3 . \qquad (12.13\text{-}26)$$

The chemical potential, being a function of the interactions, entails the Fermi surface

$$\varsigma = \varepsilon_F = \frac{\hbar^2 k_F^2}{2m} + \Sigma_{HF}(k_F;0) = \frac{\hbar^2 k_F^2}{2m} + \left\{ \frac{1}{8\pi^3} \int d^3k' [2V(0) - V(\mathbf{k}'-\mathbf{k}) \Theta(k_F - k') \right\}_{k=k_F} . \qquad (12.13\text{-}27)$$

The derivative at the Fermi surface defines an effective mass,

$$(d\varepsilon_k^1 / dk)_{k=k_F} \equiv \hbar^2 k_F / m^* , \qquad (12.13\text{-}28)$$

which leads to the form

$$\frac{1}{m^*} = \frac{1}{m} + \frac{1}{\hbar^2 k_F} \left(\frac{\partial \Sigma_{HF}}{\partial k} \right)_{k=k_F} . \qquad (12.13\text{-}29)$$

Substituting (12.13-28) and (12.13-26) into (12.13-25), we finally arrive at

$$s(T,n) = \tfrac{1}{3} k_B^2 T (m^* / \hbar^2)(3\pi^2 n)^{1/3} . \qquad (12.13\text{-}30)$$

For $c_V = T(\partial s/\partial T)_n$ we obtain the same value, as is easily verified. However, a more lucid result is obtained if from (12.13-26) we write $(3\pi^2 n)^{1/3} = 3\pi^2 n / k_F^2$; we then get

$$c_V = \frac{\pi^2}{2} k_B n \left(\frac{k_B T}{\hbar^2 k_F^2 / 2m^*} \right), \qquad (12.13\text{-}31)$$

which has the same form as the result for a perfect Fermi gas [analogous to Eq. (8.4-29) for an electron gas] with now m replaced by m^*. While this may seem trivial, it be noted that m^* is *fully defined in terms of the Fourier coefficients* $V(\mathbf{q})$ *of the two-body interaction potential* – rather than by Landau's phenomenological Fermi liquid parameters – cf. Eq. (12.7-21).

---

[97] I.S. Gradshteyn and I.M. Ryzhik, "Table of Integrals, Series and Products", Acad. Press, NY, 1980, § 3.527.

12.13.2 *The Ring Approximation (RPA)*

Whereas the Hartree–Fock approximation has successfully been used for many molecular and condensed matter materials, as well as for interacting Bose liquids above the BEC temperature, it has two fundamental drawbacks: it is not readily adapted to Fermi liquids with a strict hard core-repulsion or to the homogeneous electron gas as it applies to ordinary metals, unless modifications are made. The fundamental problem is that the Fourier coefficients $V(\mathbf{q})$ of the interaction potential must be bounded for all $\mathbf{q}$, which is not realisable for a hard core potential with $v(\mathbf{r}) = \infty$ for $|\mathbf{r}| < a$, or for a Coulomb potential $V(\mathbf{q}) = e^2/|\mathbf{q}|^2$ for which $V(0) = \infty$.

We first address the hard core problem. In classical gases the problem was circumvented by introducing the Mayer function f, which is far better behaved for $|\mathbf{r}| \to 0$ than the interaction potential itself. In quantum fluids there is not such a device; correlated basis functions have been successful in numerical computations as we saw before, but they have no place in diagrammatic methods. However, the ordinary Born approximation for s-wave scattering of two particles can help us out. We then have the standard result [see Chapter XIII, Eq. (13.3-13)]:

$$V(\mathbf{k}'-\mathbf{k}) = \int d^3r e^{-i(\mathbf{k}'-\mathbf{k})\cdot\mathbf{r}} v(\mathbf{r}) = -\frac{4\pi\hbar^2}{m} f_B(\mathbf{k},\mathbf{k}') a_B \overset{\mathbf{k},\mathbf{k}'\to 0}{\longrightarrow} \frac{4\pi\hbar^2}{m} a_B . \qquad (12.13\text{-}32)$$

where a_B is the scattering length; this result can be carried over *verbatim* for singular potentials, with a_B being now the range a of the hard core. The above gives $V(0)$, but the same result holds approximately also for $V(\mathbf{k}'-\mathbf{k})$; with $\int d^3k = 4\pi k_F{}^3/3$ being the volume of the Fermi sphere, we find for the proper self-energy in first approximation from (12.13-27)

$$\varsigma = \varepsilon_F = \left(\hbar^2 k_F^2/2m\right)\left(1 + (4k_F a/3\pi)\right). \qquad (12.13\text{-}33a)$$

However, we have to add the effect of the background particles on the scattering process. This requires the evaluation of the second-order ring diagram and a ladder diagram, shown in Fig. 12-31. Since the potential is represented by its Fourier coefficients the vertex evaluation yields the usual momentum conservation laws, shown for each vertex. We shall refrain from the evaluation, but state the result:

$$\varsigma = \varepsilon_F = \frac{\hbar^2 k_F^2}{2m}\left[1 + \frac{4}{3\pi}k_F a + \frac{4}{15\pi^2}(11 - 2\ln 2)(k_F a)^2\right]. \qquad (12.13\text{-}33b)$$

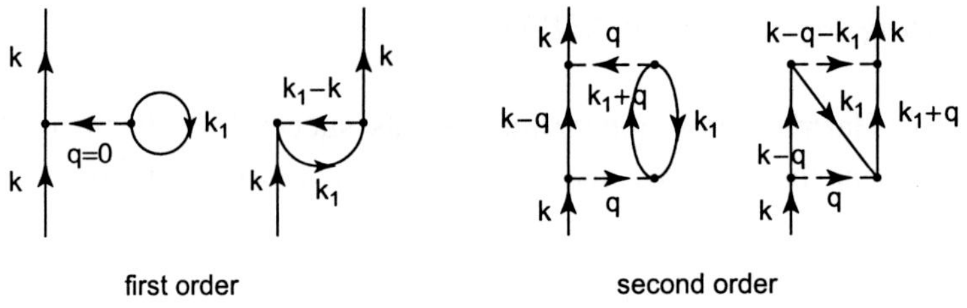

Fig. 12-31. First and second-order diagrams for evaluation of the self energy.

In the rest of this section we consider the homogeneous electron gas, having normal metals in mind. We have already mentioned the problem with the long length Coulomb tail, which gives a divergence for $|\mathbf{q}| \to 0$. More pressing is the effect of the positive ion-core background; in a neutral homogeneous electron gas, it fully cancels the direct interaction, or the contribution of all bubble diagrams (called 'tadpole' diagrams by some authors) to the self-energy, leaving only the exchange contribution, see the most right diagram in Fig. 12-30b. Obviously, such a description is insufficient, so second-order diagrams and their repetitions should be added, such as the ring diagram and the ladder diagram of Fig. 12-31, among others. In the present context, the ring contributions preponderate. So we will do the ring diagram summation, which for historical reasons is also known as the random-phase-approximation or RPA; for its original form, see the paper by Montroll and Ward, already quoted.[84] To obtain Σ we shall avail ourselves of the 'polarization insertion' Π, obtained as follows. Whenever we consider the interaction between two particles in an n-th order connected diagram, we have two states at the left side of the first interaction vertex, with the states at the right side vertex engaging in $n-1$ further interactions and two final states appearing at the last right side. The situation is pictured in Fig. 12-32. Using Fourier coefficients for the matrix elements, the total interaction is represented by a double arrowed dashed line. For simplicity we consider spin-independent potentials, so that the spin states give Kronecker deltas as before. With $V(\mathbf{q})_{\alpha\beta,\lambda\mu} = V(\mathbf{q})\delta_{\alpha\beta}\delta_{\lambda\mu}$ the result for the polarization insertion depicted in Fig.12-32 reads

$$\mathbf{V}(\mathbf{q}) = V(\mathbf{q}) + V(\mathbf{q})\hat{\Pi}(\mathbf{q})V(\mathbf{q})\,; \tag{12.13-34}$$

here we have printed a bold $\mathbf{V}$ to represent the effective interaction, while V denotes, as before, the matrix element in the unperturbed basis states (plane waves) of K^0. Eq. (12.13-34) is not the full story, since strings of polarization diagrams may occur, entirely analogous to the situation of Fig. 12-26; we thus are led to define the proper (or irreducible) polarization Π, associated with irreducible diagrams only. We then have

$$\mathbf{V}(\mathbf{q}) = V(\mathbf{q}) + V(\mathbf{q})\Pi(\mathbf{q})V(\mathbf{q}) + V(\mathbf{q})\Pi(\mathbf{q})V(\mathbf{q})\Pi(\mathbf{q})V(\mathbf{q}) + \ldots + \textit{ad inf.}$$

$$= V(\mathbf{q}) + V(\mathbf{q})\Pi(\mathbf{q})\mathbf{V}(\mathbf{q}), \tag{12.13-35}$$

where the last line reproduces the first line by iteration. The latter equation is a simple algebraic equation with solution

$$\mathbf{V}(\mathbf{q}) = \frac{V(\mathbf{q})}{1 - \Pi(\mathbf{q})V(\mathbf{q})}. \tag{12.13-36}$$

As before, we are using frequency diagrams, i.e., the lines at a vertex are characterized by a momentum and a frequency. According to the rules, both are conserved at a vertex point. For the ring approximation the polarization of the medium is easily made explicit, see Fig. 12.33. Because of the vertex restriction (12.10-26) we encounter both odd and even frequencies; they are denoted by ω_n and ν_n, respectively.

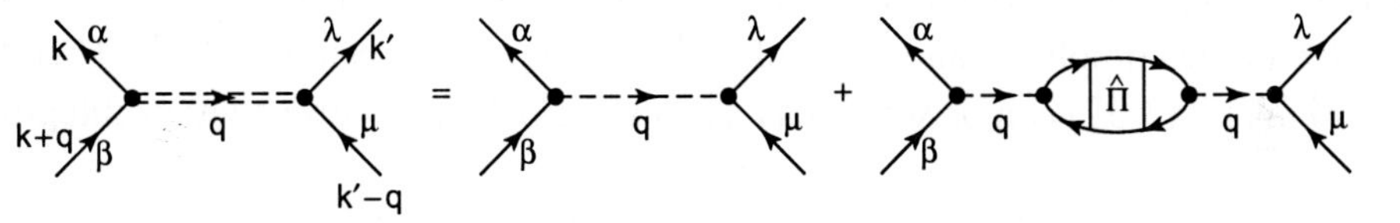

Fig. 12-32. Diagram for the polarization insertion $\hat{\Pi}$.

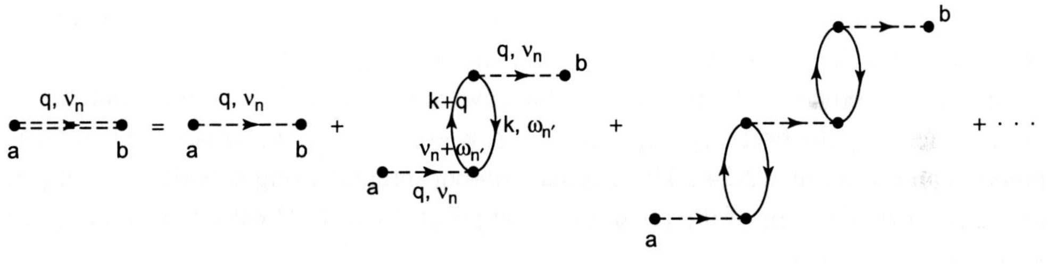

Fig. 12-33. Effective two-body interaction $\Rightarrow$ in the ring approximation.

By summing the first two contributions, we can identify the polarization Π. We easily obtain

$$\mathbf{V}_R(\mathbf{q},\nu_n) = V_R(\mathbf{q},\nu_n) + \frac{2}{\beta}[V_R(\mathbf{q},\nu_n)]^2 \sum_{\omega_{n'}} \frac{1}{(2\pi)^3} \int d^3k \bar{G}_0(\mathbf{k},\omega_{n'}) \bar{G}_0(\mathbf{k}+\mathbf{q},\omega_{n'}+\nu_n) + \ldots$$

Note that a factor $-\beta^{-1}$ arises from the extra interaction, while a factor -2 arises from the spin around the closed fermion loop. Comparing with (12.13-34), we find Π_R to be

$$\Pi_R(\mathbf{q},\nu_n) = \frac{2}{\beta}\sum_{\omega_{n'}}\frac{1}{(2\pi)^3}\int d^3k\bar{G}_0(\mathbf{k},\omega_{n'})\bar{G}_0(\mathbf{k}+\mathbf{q},\omega_{n'}+\nu_n)$$
$$= \frac{2}{\beta}\sum_{\omega_{n'}}\frac{1}{(2\pi)^3}\int d^3k\left(\frac{1}{i\omega_{n'}-(\varepsilon_{\mathbf{k}}^0-\mu)}\right)\left(\frac{1}{i(\omega_{n'}+\nu_n)-(\varepsilon_{\mathbf{k+q}}^0-\mu)}\right), \quad (12.13\text{-}37)$$

where the chemical potential ς now has been replaced by the electrochemical potential μ. As expected from the general picture in Fig. 12-32, the polarization insertion is a summed product of free propagators; however, the details depend on the second-order diagram of departure. We note that each term in (12.13-37) is of order ω_n^{-2}, so the sum converges absolutely.

It is useful to add a redundant convergence factor $\exp(i\omega_{n'}\delta)$, so that we can evaluate the sum by a contour integral, analogous to the previous procedure in Section 12.10.3. We know that the Fermi distribution $n_{\mathbf{k}}^0$, denoted as $n^0(z)$ when extended in the complex plane, has poles at $z_i = i\omega_n$, with residue $-1/\beta$. Therefore, the sum in (12.13-37) can be written as a counter clockwise (ccw) contour integral,

$$sum = -\frac{\beta}{2\pi i}\oint_C dz\, n^0(z)\left(\frac{1}{z-(\varepsilon_{\mathbf{k}}^0-\mu)}\right)\left(\frac{1}{z+i\nu_n-(\varepsilon_{\mathbf{k+q}}^0-\mu)}\right), \quad (12.13\text{-}38)$$

where C is the same contour as in Fig. 12-24, enclosing all the poles of $n^0(z)$. In addition, we consider the contour Γ, being a circle of infinite radius enclosing all poles; as shown before, this contour integral yields zero. The contour integral Γ can, however, be replaced by the ccw contour C and two small contours c_1 and c_2, enclosing the poles $z_1 = \varepsilon_{\mathbf{k}}^0 - \mu$ and $z_2 = -i\nu_n + \varepsilon_{\mathbf{k+q}} - \mu$, which are shown as points A and B in Fig. 12-24. The original contour integral along C is therefore equal to $-2\pi i$ times the sum of the residues of the poles A and B. Leaving trivial algebra aside, we find for Π_R

$$\Pi_R(\mathbf{q},\nu_n) = -2\int d^3k\frac{1}{(2\pi)^3}\frac{n_{\mathbf{k+q}}^0 - n_{\mathbf{k}}^0}{i\nu_n-(\varepsilon_{\mathbf{k+q}}^0-\varepsilon_{\mathbf{k}}^0)}. \quad (12.13\text{-}39)$$

Next, we want to find the proper self-energy associated with the ring sum. Remembering that the self-energy is a sum of skeleton-interactions, perched between an incoming and outgoing propagator, all we have to do is connect the points a and b of the potential diagrams in Fig. 12-33; the resulting sum of diagrams, as shown in Fig. 12-34, yields the self-energy. We thus obtain

$$\Sigma_R'(\mathbf{k},\omega_n) = -\frac{1}{\beta(2\pi)^3}\sum_{\nu_n}\int d^3q\,\mathbf{V}(\mathbf{q},\nu_n)G_0(\mathbf{k}-\mathbf{q},\omega_n-\nu_n)$$
$$= -\frac{1}{\beta(2\pi)^3}\sum_{\nu_n}\int d^3q\,\frac{V(\mathbf{q})}{1-\Pi(\mathbf{q},\nu_n))V(\mathbf{q})}G_0(\mathbf{k}-\mathbf{q},\omega_n-\nu_n). \quad (12\text{-}13\text{-}40)$$

The amazing fact here is that the self-energy so obtained is finite – except for a removable logarithmic divergence – whereas the separate ring diagrams are highly divergent! The power of the technique presented is evident.

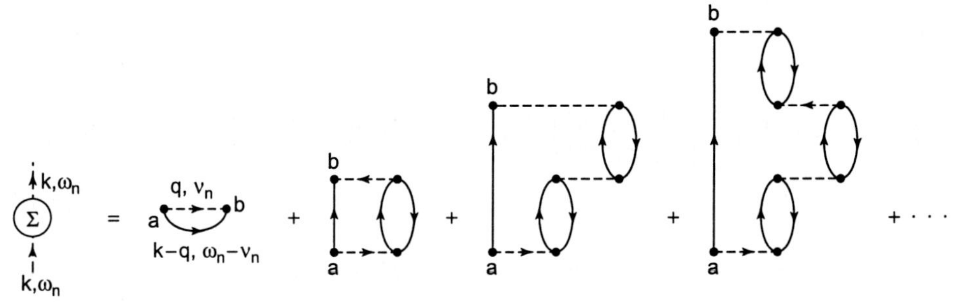

Fig. 12-34. Self-energy for the Ring Approximation.

Very little work is required now to obtain the grand potential from Eq. (12.12-11). The integrand in that formula calls for adding a factor $\overline{G}_\lambda$ to the self energy; this is equivalent to connecting each of the ends (a,b) in the diagrams of Fig. 12-34 with a propagator $\overline{G}_0$. *This yields just the closed-ring sum, depicted in Eq.* (12.10-8)! We have:

$$\Omega_R' - \Omega_0 = \int_0^1 \frac{d\lambda}{\lambda} \sum_{\omega_n} \frac{V_0}{\beta(2\pi)^3} \int d^3k \Sigma_{R\lambda} \overline{G}_{R\lambda}(\mathbf{k}, \omega_n)$$

$$\to -\int_0^1 \frac{d\lambda}{\lambda} \frac{V_0}{\beta^2 (2\pi)^6} \sum_{\omega_n} \sum_{\nu_n} \int d^3k \int d^3q \frac{\lambda V(\mathbf{q})}{1 - \lambda \Pi_R(\mathbf{q}, \nu_n) V(\mathbf{q})} \overline{G}_0(\mathbf{k} - \mathbf{q}, \omega_n - \nu_n)$$

$$\times \overline{G}_0(\mathbf{k}, \omega_n) = -\int_0^1 \frac{d\lambda}{\lambda} \frac{V_0}{2\beta(2\pi)^3} \sum_{\nu_n} \int d^3q \frac{\lambda V(\mathbf{q}) \Pi_R(\mathbf{q}, \nu_n)}{1 - \lambda \Pi_R(\mathbf{q}, \nu_n) V(\mathbf{q})}, \qquad (12.13\text{-}41)$$

where we used that $\overline{G}_0$ is invariant under change $\mathbf{q}, \nu_n \to -\mathbf{q}, -\nu_n$. The integration yields

$$\Omega_R'(T, V_0, \varsigma) - \Omega_0(T, V_0, \varsigma) = \left(V_0 / 2\beta(2\pi)^3\right) \sum_{\nu_n} \int d^3q \ln\left[1 - \Pi_R(\mathbf{q}, \nu_n) V(\mathbf{q})\right].$$

(12.13-42)

We notice that it is exactly the same as (12.10-8) for the cumulant expansion after the 'trace' is worked out – see Problem (12.7).

Yet this answer, supported by the original paper of Montroll and Ward[84], is not entirely satisfactory. The first order ring – the exchange graph – is usually omitted in the ring sum. If necessary, its self-energy and resulting Ω_1 can be added separately. The deletion of the exchange diagram modifies the result (12.13-40) slightly, in that $\mathbf{V(q)}$ is replaced by $\mathbf{V(q)} - V(\mathbf{q})$. The results for Σ_R and Ω_R then become

$$\Sigma_R(\mathbf{k},\omega_n) = -\frac{1}{\beta(2\pi)^3}\sum_{\nu_n}\int d^3q\left(\frac{V(\mathbf{q})}{1-\Pi(\mathbf{q},\nu_n))V(\mathbf{q})} - V(\mathbf{q})\right)G_0(\mathbf{k}-\mathbf{q},\omega_n-\nu_n) \tag{12.13-43}$$

and

$$\Omega_R(T,V_0,\varsigma) - \Omega_0 = \frac{V_0}{2\beta(2\pi)^3}\sum_{\nu_n}\int d^3q\{\ln[1-\Pi_R(\mathbf{q},\nu_n)V(\mathbf{q})] + V(\mathbf{q})\Pi_R(\mathbf{q},\nu_n)\}. \tag{12.13-44}$$

(a) *Classical electron gas with positive charge background*

The ring diagrams stand out from other diagrams in that they give a "feedback loop" to the original interaction. It will therefore come as no surprise that for a classical electron gas they will alter the divergent Coulomb interaction, so that shielding occurs as in the Debye–Hückel theory of an ionized plasma, treated in Section 4.8.

At high temperatures the gas can be described by a Boltzmann distribution. The density in velocity space is given in (6.5-1), which yields for the 'quantum distribution'

$$n_{\mathbf{k}}^0 = 2e^{\beta(\mu-\varepsilon_k^0)} \approx \frac{(2\pi)^3}{V_0}\frac{\partial(v_x,v_y,v_z)}{\partial(k_x k_y k_z)}N\left(\frac{m}{2\pi k_B T}\right)^{3/2} e^{-\beta\varepsilon_{\mathbf{v}}} = n\lambda_{th}^3 e^{-\beta\varepsilon_{\mathbf{k}}^0}, \tag{12.13-45}$$

where μ is the electrochemical potential, the factor 2 stems from the spin occupancy, and λ_{th} is the thermal wavelength as in Chapter IV,

$$\lambda_{th} = \left(\frac{2\pi\hbar^2}{mk_B T}\right)^{1/2}. \tag{12.13-46}$$

From (12.13-45) we have *as a first approximation* – the μ also has a part stemming from the interactions, see below,

$$e^{\beta\mu} \approx \tfrac{1}{2}n\lambda_{th}^3. \tag{12.13-47}$$

We can now proceed to the evaluation of the polarization. The boson-like discrete frequencies ν_n are given by $2\pi\ell/\beta$; below we shall argue that only $\nu_n = 0$ need to be considered. So from (12.13-39) we have

$$\Pi_R(\mathbf{q},0) = 2\int(d^3k/8\pi^3)\frac{n_{\mathbf{k+q}}^0 - n_{\mathbf{k}}^0}{\varepsilon_{\mathbf{k+q}}^0 - \varepsilon_{\mathbf{k}}^0}. \tag{12.13-48}$$

With the numerator and the denominator being zero for $\mathbf{q} = 0$, we will use the Cauchy principal value $\mathcal{P}$ for the singularity. We write as always $\varepsilon_{\mathbf{k}}^0 = \hbar^2 k^2 / 2m$; with d^3k taken in polar coordinates, the integrals are straightforward and one obtains

$$\begin{aligned}\Pi_R(\mathbf{q},0) &= \frac{2m}{(\pi\hbar)^2}\mathcal{P}\int_0^\infty k^2 dk\, n_{\mathbf{k}}^0 \int_{-1}^1 dz \frac{1}{2kqz - q^2} \\ &= \frac{m}{(\pi\hbar)^2}\mathcal{P}\int_0^\infty k^2 dk\, n_{\mathbf{k}}^0 \frac{1}{q}\ln\left|\frac{2k-q}{2k+q}\right| = -2\beta e^{\beta\mu}\lambda_{th}^{-3}\varphi(q\lambda_{th}),\end{aligned} \tag{12.13-49}$$

where

$$\varphi(x) = \frac{1}{\pi x}\int_0^\infty d\xi\, \xi e^{-\xi^2/4\pi} \ln\left|\frac{2\xi + x}{2\xi - x}\right|; \quad \varphi(0) = 1. \tag{12.13-50}$$

For high temperatures, the approximation $\varphi = \varphi(0) = 1$ is appropriate. Employing further the approximate relation for the electrochemical potential (12.13-47), we obtain

$$\Pi_R(\mathbf{q},0) = -n/k_B T = -\kappa^2/e^2, \tag{12.13-51}$$

where κ is the shielding exponent and κ^{-1} is the Debye length, as given in Eq. (4.10-5). For the effective potential $\mathbf{V(q)}$ we thus established

$$\mathbf{V}(\mathbf{q}) = \frac{e^2}{|\mathbf{q}|^2 + \kappa^2} = \int d^3r\, e^{i\mathbf{q}\cdot\mathbf{r}} \frac{e^2}{4\pi r} e^{-\kappa r}. \tag{12.13-52}$$

So, the polarization caused $\mathbf{V(q)}$ to be the Fourier transform of a screened Coulomb potential [cf. Eq. (12.6-3)] with the usual Debye screening length, which is, indeed, a remarkable result!

As to Ω, we split (12.13-44) in a term with $\nu_n = 0$ and the remainder ($\ell = 1,\ldots$). When the remainder is evaluated around $\mathbf{q} = 0$, the result is easily shown to be bounded; however, the ln term with $\nu_n = 0$ has a logarithmic divergence for $\mathbf{q} \to 0$. Since in all diagrams the most divergent ones were kept, we discard all terms but the $\nu_n = 0$ term. Employing the previously obtained form for Π_R (12.13-49), a straightforward computation yields

$$(\Omega_R - \Omega_0)/V_0 = \beta^{1/2}(e^3/4\pi^2)(2e^{\beta\mu}\lambda_{th}^{-3})^{3/2}\int_0^\infty x^2 dx\left\{\ln[1 + x^{-2}\varphi(e\alpha x)] - x^{-2}\varphi(e\alpha x)\right\},$$

where α is a constant and where the change in variables $q^2 = ne^2x^2$ was made. The main result is of order e^3; we obtain the leading contribution in the coupling parameter e^2 by setting $e = 0$ in the argument of φ. We then arrive at

$$(\Omega_R - \Omega_0)/V_0 = \beta^{1/2}(e^3/4\pi^2)(2e^{\beta\mu}\lambda_{th}^{-3})^{3/2}$$
$$\times\int_0^\infty x^2 dx\left\{\ln(1+x^{-2}) - x^{-2}\right\} = -\tfrac{1}{3}\beta^{1/2}(e^3/4\pi)(2e^{\beta\mu}\lambda_{th}^{-3})^{3/2}, \quad (12.13\text{-}53)$$

where we used integration by parts. The thermodynamic limit exists and is implied.

The equations of state now easily follow. For the pressure we have $P = -\Omega_R/V_0$ to which must be added the classical part $P_0 = -\Omega_0/V_0 = 2k_BTe^{\beta\mu}\lambda_{th}^{-3}$. The density is related to the electrochemical potential by $n = -[\partial(\Omega/V_0)/\partial\mu]_T$, which must be inverted to obtain $\mu(n)$. The results are

$$\begin{aligned} P &= 2k_BTe^{\beta\mu}\lambda_{th}^{-3} + \tfrac{1}{3}\beta^{1/2}(e^3/4\pi)(2e^{\beta\mu}\lambda_{th}^{-3})^{3/2}, \\ e^{\beta\mu} &= \tfrac{1}{2}n\lambda_{th}^3[1-(\beta^{3/2}e^3n^{1/2}/8\pi)]. \end{aligned} \quad (12.13\text{-}54)$$

Note the correction to (12.13-47). Eliminating μ, this finally yields

$$P = nk_BT\left(1 - \frac{1}{3}\frac{e^3n^{1/2}}{8\pi(k_BT)^{3/2}}\right), \quad (12.13\text{-}55)$$

in full agreement with Eqs. (4.10-17) and (4.10-25).

(b) *Quantum electron gas near $T = 0$*

The theory for the quantum gas at low temperatures is very different from the preceding classical computation. For one thing the corrections to the Fermi energy are small and we can with impunity change from the grand potential to the free energy $F(T,V,N)$ at an early stage. We will see that the main long-wavelength corrections stem from Ω_R[98], but there are also contributions from the first-order exchange diagram and from the second-order ladder (exchange) diagram [Fig. 12-31], with grand potentials Ω_1 and Ω_2, respectively. The detailed theory is in Kohn and Luttinger's paper and Fetter and Walecka.[99]. Many details are also in Mahan and a remarkable concise summary is in Reichl's book.[100] We shall only consider the contributions from Ω_R; the reader is encouraged to verify the self-energies and grand potentials for the exchange and ladder diagrams mentioned above:

$$\Omega_1(T,V_0,\mu)/V_0 = (2\pi)^{-6}\int d^3k\, d^3q\, V(\mathbf{k}-\mathbf{q})n_{\mathbf{k}}^0 n_{\mathbf{q}}^0, \quad (12.13\text{-}56)$$

---

98 M. Gel-Mann and K.A. Brueckner, Phys. Rev **106**, 364 (1957).

99 W. Kohn and J.M. Luttinger, Phys. rev. **118**, 41 (1960); Fetter and Walecka, Op. Cit., p 281 ff.

100 G.D. Mahan, Op. Cit. Section 5.1.8; L.E. Reichl, Op. Cit., First Ed., p. 426-428.

$$\Omega_2(T,V_0,\mu)/V_0 = \frac{1}{(2\pi)^9}\int d^3k d^3q d^3p V(\mathbf{q})V(\mathbf{k}+\mathbf{q}+\mathbf{p})\frac{n^0_{\mathbf{k}}n^0_{\mathbf{p}}(1-n^0_{\mathbf{k+q}})(1-n^0_{\mathbf{p+q}})}{\varepsilon^0_{\mathbf{k+q}}+\varepsilon^0_{\mathbf{p+q}}-\varepsilon^0_{\mathbf{k}}-\varepsilon^0_{\mathbf{p}}}. \tag{12.13-57}$$

Since the integrand in Π_R only has one simple pole, we shall replace the frequency sum by an integral, noting $\Delta\nu_n^{-1} = \beta/2\pi$. We thus have, writing out $V(\mathbf{q})$,

$$\Delta\Omega_R(T,V_0,\mu)/V_0 = \frac{1}{2(2\pi)^4}\int_{-\infty}^{\infty} d\nu \int d^3q\left\{\ln\left[1-(e^2/q^2)\Pi_R(\mathbf{q},\nu)\right]+(e^2/q^2)\Pi_R(\mathbf{q},\nu)\right\}. \tag{12.13-58}$$

For the polarization we have for small $\mathbf{q}$

$$\Pi_R(\mathbf{q},\nu) = -2\int\frac{d^3k}{(2\pi)^3}\frac{\mathbf{q}\cdot\nabla_{\mathbf{k}}n^0_{\mathbf{k}}}{i\nu-(\hbar^2/2m)\mathbf{q}\cdot\mathbf{k}}. \tag{12.13-59}$$

Near $T = 0$ we use the Heaviside function to describe the distribution n_k^0. So we have

$$\nabla_{\mathbf{k}}\Theta\,[(\hbar^2/2m)(k_0-k)^2] = -\hat{\mathbf{k}}\,\delta(k-k_0); \tag{12.13-60}$$

here $k_0 = \hbar^{-1}\sqrt{2m\mu(0)}$. For $\mathbf{q}\to 0$ we obtain for Π_R,

$$\Pi_R(\mathbf{q},\nu) = -\frac{k_0 m}{\pi^2\hbar^2}R(x), \quad R(x) = \int_0^1\frac{dz\,z^2}{z^2+x^2} = 1 - x\arctan\frac{1}{x}, \tag{12.13-61}$$

where $x = m\nu/\hbar q k_0$; we also introduce another dimensionless variable $y = q/k_0$. The integral for Ω_R is now split into two regions, $y < y_0 \ll 1$ and $y > y_0$, to obtain

$$\frac{\Delta\Omega_R}{V_0} = \frac{\hbar^2 k_0^5}{(2\pi)^3 m}\int_{-\infty}^{\infty}dx\int_0^{y_0}dy y^3\left\{\ln\left[1+\frac{me^2}{\hbar^2 k_0\pi^2 y^2}R(x)\right]-\frac{me^2}{\hbar^2 k_0 \pi^2 y^2}R(x)\right\}+\frac{\Omega_{R2}(y>y_0)}{V_0}, \tag{12.13-62}$$

with

$$\frac{\Omega_{R2}}{V_0} = -\frac{1}{2}\frac{\hbar^2 k_0^5}{(2\pi)^3 m}\int_{-\infty}^{\infty}dx\int_{y_0}^{\infty}dy y^3\left[\frac{e^2}{k_0^2 y^2}\Pi_R\left(k_0 y, \frac{\hbar k_0^2 xy}{m}\right)\right]^2, \tag{12.13-63}$$

where in the second part the logarithm has been expanded to retain the leading term in e^4. In (12.13-62) the integral over y can be carried out to yield

$$\int_0^{y_0} dy y^3\left\{\ln\left[1+\frac{me^2}{\hbar^2 k_0\pi^2 y^2}R(x)\right]-\frac{me^2}{\hbar^2 k_0\pi^2 y^2}R(x)\right\}$$

$$= \left(\frac{me\,R(x)}{4\hbar^2 k_0\pi^2}\right)^2\left\{\left(\frac{\hbar^2 k_0\pi^2 y_0^2}{me^2 R(x)}\right)^2\left[\ln\left(1+\frac{me^2 R(x)}{\hbar^2 k_0\pi^2 y_0^2}\right)-\frac{me^2 R(x)}{\hbar^2 k_0 \pi^2 y_0^2}\right]\right.$$

$$-\ln\left(1+\frac{\hbar^2 k_0\pi^2 y_0^2}{me^2R(x)}\right)\Bigg\} \approx \frac{1}{4}\left(\frac{me^2R(x)}{\hbar^2 k_0\pi^2}\right)^2\left\{\ln\frac{me^2R(x)}{\hbar^2 k_0\pi^2}-\frac{1}{2}-2\ln y_0\right\}+\mathcal{O}(e^6). \quad (12.13\text{-}64)$$

This must still be integrated over dx. The term $\ln y_0$ diverges for $y_0 \to 0$ and so does the contribution from Ω_{R2}, but together they give a finite result. The integral of $R^2(x)$ is most easily found by going back to its definition,

$$\int_{-\infty}^{\infty} dx[R(x)]^2 = \int_0^1 dy\int_0^1 dz\int_{-\infty}^{\infty} dx\frac{y^2z^2}{(y^2+x^2)(z^2+x^2)}$$
$$= \pi\int_0^1 dy\int_0^1 dz\frac{yz}{y+z} = \tfrac{2}{3}\pi(1-\ln 2). \quad (12.13\text{-}65)$$

By numerical integration one also obtains

$$\int_{-\infty}^{\infty} dx[R(x)]^2 \ln R(x) = 0.551\times\tfrac{2}{3}\pi(1-\ln 2). \quad (12.13.66)$$

It can be proven that the μ which follows from a full computation of the total grand potential does not differ appreciably from ε_F^0; the latter is expressible in the electron density by Eq. (8-4-11), which also gives the grand potential $\Omega_0/V_0 = -P$. So we can substitute

$$\varepsilon_F^0 = (\hbar^2k_0^2/2m) = (\hbar^2/2m)(3\pi^2 n)^{2/3}. \quad (12.13\text{-}67)$$

A Legendre transform gives us the Helmholtz free energy per unit volume. Since we consider temperatures near $T = 0$, the entropy can be neglected; so, for the energy we write $\mathcal{E} = \mathcal{E}_0 + \mathcal{E}_{corr} = \mathcal{E}_0 + (Ne^2/8\pi a_0)\varepsilon_{corr}$, where a_0 is the Bohr length $a_0 = 4\pi\hbar^2/me^2$; note that ε_{corr} is a dimensionless quantity, with $e^2/8\pi a_0$ being the Rydberg hR. From the above we have

$$\mathcal{E}_{corr}/N = n^{-1}[(\Omega_{R1}+\Omega_{R2}+\Omega_2)/V_0]. \quad (12.13\text{-}68)$$

Further, the density factor r_s is introduced as the ratio of the particle radius to the Bohr length. From the convergent parts of Ω_{R1} the correlation energy is found to be $\varepsilon_{corr} = -0.091 + 0.062\ln r_s$. The divergent part plus the contribution from Ω_{R2} contributes an amount $\delta = -0.051$ and Ω_2 can be computed exactly[101] yielding $\varepsilon_2 = \frac{1}{3}\ln 2 - (3/2\pi^2)\zeta(3) = 0.048$. So the total correlation energy is

$$\varepsilon_{corr} = -0.094 + 0.0622\ln r_s + 0.018\, r_s \ln r_s + O(r_s^2). \quad (12.13\text{-}69)$$

For metals r_s is between 1.8 and 6.0. For $r_s \approx 2.5$ the correlation energy changes sign. In Fig 12-35 we give the plot obtained by Carr and Maradudin[102]. The results

---

[101] L. Onsager, L. Mittag and M.J. Stephen, Ann Physik **18**, 71 (1966).

[102] W.J. Carr and A.A. Maradudin, Phys Rev. **A 133**, 371 (1964).

are judged to be adequate for sufficiently dense metals, $r_s < 2.5$. The polarization has resulted in an effective attractive potential, stabilizing the electron gas.

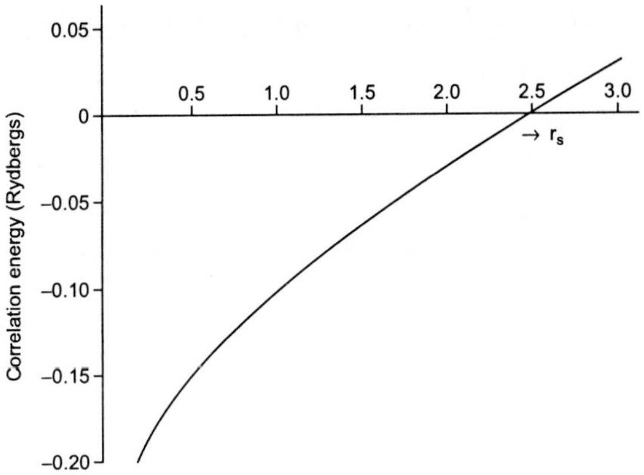

Fig. 12-35. Correlation energy of a quantum electron gas in Rydbergs. After Carr and Maradudin[102]. [With permission.]

12.14 Problems to Chapter XII

12.1 For small values of the momentum, the ε vs. p dispersion curve of a Bose superfluid is linear, indicating phonon-like excitations. We can therefore use Bose–Einstein statistics for massless quasi-particles, as in the Debye theory of a solid; the only difference is that there is only one vibrational mode, instead of one longitudinal and two transverse. Along these lines show that the Helmholtz free energy is given by

$$F = F_0 - V\frac{\pi^2 (k_B T)^4}{90(\hbar u)^3}, \tag{1}$$

where F_0 is the free energy at absolute zero and u is the velocity of sound. Next find the Debye law for the specific heat: $C = const\, T^3$ and identify the constant.

12.2 For the roton part of the dispersion curve the free energy can best be evaluated with the classical expression

$$F = -Nk_B T \ln \frac{eV}{N(2\pi\hbar)^3} \int d^3 p\, e^{-\varepsilon / k_B T}, \tag{2}$$

Since the number of rotons N is variable, you should minimize F with respect to N; show that the corresponding value of the free energy is $F_r = -k_B T N_r$, where the roton number is

$$N_r = (V/8\pi^3)\int e^{-\beta\varepsilon(k)} d^3k. \tag{3}$$

Substitute for $\varepsilon(k)$ from Eq.(12.1-6); evaluate the integral by writing $d^3k \approx 4\pi k_0^2 dk$, where k_0 is the momentum of the minimum. Also, obtain the roton contribution to the entropy and to the specific heat.

12.3 The structure factor $S(\mathbf{k})$ is defined as the Fourier transform of $\langle n\rangle p(r), [r = |\mathbf{r}|]$ where p denotes the pair-correlation function. The quantity that is measured from X-ray or neutron scattering data and which figures in the Bijl-Feynman formula is $S(k), [k = |\mathbf{k}|]$. Show that

$$S(k) = 4\pi\langle n\rangle \int_0^\infty p(r)[\sin(kr)/kr] r^2 dr. \tag{4}$$

12.4 For a Bose superfluid the grand Hamiltonian, after elimination of the chemical potential, is given by

$$\begin{aligned}\mathcal{H}_{grand} &= -\tfrac{1}{2}\varphi_0\rho_0 n_0 + \tfrac{1}{2}\sum\nolimits'_{\mathbf{k}} \Big\{\varepsilon(\mathbf{k})\left(a_{\mathbf{k}}^\dagger a_{\mathbf{k}} + a_{-\mathbf{k}}^\dagger a_{-\mathbf{k}}\right) \\ &\quad + \varphi_0\rho_0\left(a_{\mathbf{k}}^\dagger a_{-\mathbf{k}}^\dagger + a_{-\mathbf{k}} a_{\mathbf{k}} + a_{\mathbf{k}}^\dagger a_{\mathbf{k}} + a_{-\mathbf{k}}^\dagger a_{-\mathbf{k}}\right)\Big\}.\end{aligned} \tag{5}$$

Now use the Bogoliubov transformation with new boson operators $\eta_k, \eta_{-k}, \eta_k^\dagger, \eta_{-k}^\dagger$ and diagonalize the grand-Hamiltonian; show that $\theta_{\mathbf{k}}$ should be given by $\tan(2\theta_{\mathbf{k}}) = \varphi_0\rho_0/[\varepsilon(\mathbf{k}) + \varphi_0\rho_0]$, in order that the non-diagonal part vanishes. Find the new grand-Hamiltonian and confirm (12.3-22).

12.5 For the Cooper problem involving a single pair the state $|\Psi\rangle = \Sigma_{\mathbf{k}}\alpha(\mathbf{k}) c_{\mathbf{k},\frac{1}{2}}^\dagger c_{-\mathbf{k},-\frac{1}{2}}^\dagger |O\rangle$ must satisfy the Schrödinger equation $\mathcal{H}|\Psi\rangle = \mathcal{E}|\Psi\rangle$, where $\mathcal{H}$ is the Fröhlich Hamiltonian (12.5-1) minus the Fermi sea and $\alpha(\mathbf{k})$ is a variational parameter. Obtain the functional equation (12.5-18):

$$[2\varepsilon(\mathbf{k}) - \mathcal{E}]\alpha(\mathbf{k}) = \upsilon_0 \Sigma_{\mathbf{q}}\alpha(\mathbf{k} - \mathbf{q}). \tag{6}$$

Hint. In the double or triple sum involving $\mathcal{H}|\Psi\rangle$ $\mathbf{k}$ and $\mathbf{k}'$ must be chosen such that the state $|\Psi\rangle$ is recreated, in order to equal $\mathcal{E}|\Psi\rangle$.
N.B. You can also minimize $\langle\Psi|\mathcal{H}|\Psi\rangle$, subject to $\langle\Psi|\Psi\rangle = \Sigma_{\mathbf{k}}[\alpha(\mathbf{k})]^2$.

12.6 The BCS Hamiltonian for finite temperatures, after a suitable linearization procedure, was obtained in Eq. (12.6-29), first and second line. Another useful form follows when the interaction is expressed in local field operators.
(a) Show that

$$\mathcal{H}_{grand,red.}$$
$$= K_0 - \upsilon_0 \int d^3r \left\{ \Psi^\dagger_\uparrow(\mathbf{r})\Psi^\dagger_\downarrow(\mathbf{r})\langle\Psi_\downarrow(\mathbf{r})\Psi_\uparrow(\mathbf{r})\rangle + \langle\Psi^\dagger_\uparrow(\mathbf{r})\Psi^\dagger_\downarrow(\mathbf{r})\rangle\Psi_\downarrow(\mathbf{r})\Psi_\uparrow(\mathbf{r})\right\}, \quad (7)$$

cf. Fetter and Walecka, Op. Cit., p. 441, Eq. (51.7).
Hint. For $\mathbf{q} \to 0$ the phonon-mediated interaction becomes *local*, i.e., $v(\mathbf{r}-\mathbf{r}') = -\upsilon_0\delta(\mathbf{r}-\mathbf{r}')$, $\upsilon_0 > 0$. Replace $V_{\mathbf{kk}'}$ by the more general form $V_{\mathbf{k\bar{k},k'\bar{k}'}}$, where

$$V_{\mathbf{k\bar{k},k'\bar{k}'}} = -\upsilon_0 \int d^3r\, \psi^*_k(\mathbf{r})\psi^*_{\bar{k}}(\mathbf{r})\psi_{k'}(\mathbf{r})\psi_{\bar{k}'}(\mathbf{r}), \quad (8)$$

thus obtaining,

$$\mathcal{H}_{grand} = K_0 - \tfrac{1}{2}\upsilon_0 \sum_{\lambda,\mu} \int d^3r \left\{ \Psi^\dagger_\lambda(\mathbf{r})\Psi^\dagger_\mu(\mathbf{r})\langle\Psi_\mu(\mathbf{r})\Psi_\lambda(\mathbf{r})\rangle + h.c.\right\}. \quad (9)$$

Now make the BCS restriction that only pairs with zero momentum and zero spin contribute.
(b) Define the 2×2 matrix Green's function $\mathsf{G}(\mathbf{r}\tau,\mathbf{r}'\tau')$ by

$$\mathsf{G}(\mathbf{r}\tau,\mathbf{r}'\tau') \equiv -\langle\mathcal{T}[\Psi(\mathbf{r}\tau)\Psi^{(\dagger)}(\mathbf{r}'\tau')]\rangle = \begin{pmatrix} G(\mathbf{r}\tau,\mathbf{r}'\tau') & F(\mathbf{r}\tau,\mathbf{r}'\tau') \\ F^\dagger(\mathbf{r}\tau,\mathbf{r}'\tau') & -G(\mathbf{r}\tau,\mathbf{r}'\tau') \end{pmatrix}, \quad (10)$$

where G pertains to the product of $\Psi\Psi^\dagger$ with two up spins, and F refers to the product $\Psi\Psi$ with opposite spins. From the Heisenberg equations of motion, show that the matrix Green's function satisfies the usual mathematical 'source equation':

$$\mathsf{D}_{\mathbf{r}\tau}\mathsf{G}(\mathbf{r}\tau,\mathbf{r}'\tau') = \mathbf{1}\,\delta(\mathbf{r}-\mathbf{r}')\delta(\tau-\tau'), \quad (11)$$

where $\mathsf{D}_{\mathbf{r}\tau}$ is the differential operator

$$\mathsf{D}_{\mathbf{r}\tau} = \begin{pmatrix} -\frac{\partial}{\partial\tau} - \frac{1}{2m}\left(-\frac{\hbar}{i}\nabla + \frac{e}{c}\mathbf{A}\right)^2 + \mu & \Delta(\mathbf{r}) \\ \Delta^*(\mathbf{r}) & -\frac{\partial}{\partial\tau} + \frac{1}{2m}\left(-\frac{\hbar}{i}\nabla + \frac{e}{c}\mathbf{A}\right)^2 - \mu \end{pmatrix}, \quad (12)$$

in which $\Delta(\mathbf{r}) = \upsilon_0 F(\mathbf{r},\tau+0,\mathbf{r}\tau) \equiv -\upsilon_0\langle\tilde{\mathcal{T}}[\Psi_\uparrow(\mathbf{r})\Psi_\downarrow(\mathbf{r})]\rangle$ is the gap function, introduced by Abrikosov et al., Op. Cit. p. 320.

12.7 In Eq. (12-10-8) an interaction and an incoming and outgoing propagator are combined to a matrix

$$(vG_0G_0)_{\alpha\beta\tau,\alpha'\beta'\tau'} \equiv \langle\alpha\alpha'|v|\beta\beta'\rangle G_0(\alpha',\tau'-\tau)G_0(\beta',\tau-\tau'). \quad (13)$$

For an n-th order ring diagram, the diagram returns to the first vertex; denoting by ‘tr’ the matrix-trace and tau-integrations, the form (12.10-8) should be evident. Obtain and elaborate the following equivalent result

$$\textit{Ring sum} = -\tfrac{1}{2}\text{'tr'}\{\ln(1+\varepsilon v G_0 G_0)\}. \tag{14}$$

Transfer to the momentum-frequency description and confirm Eq. (12.13-42).

12.8 Obtain the Green’s function and the proper self-energy in the Hartree Fock approximation in coordinate space. Define Fourier series as usual, obtain $G^0_{HF}(\mathbf{r},\mathbf{r}',\omega_n)$ as well as $G_{HF}(\mathbf{r},\mathbf{r}',\omega_n)$. Let $\{\varphi_i(\mathbf{r})\}$ be a complete set of one-particle eigenstates with eigenvalues ε_i^1. Expand G as a bilinear series of the EF φ_i and evaluate the corresponding frequency sum in Σ_{HF}, to obtain

$$\begin{aligned}\Sigma_{HF}(\mathbf{r},\mathbf{r}') = (2s+1)\delta(\mathbf{r}-\mathbf{r}')\int d^3r_1 V(\mathbf{r}-\mathbf{r}_1)\sum\nolimits_j |\varphi_j(\mathbf{r}_1)|^2 n_j \\ \pm V(\mathbf{r}-\mathbf{r}')\sum\nolimits_j \varphi_j(\mathbf{r})\varphi_j^*(\mathbf{r}')n_j,\end{aligned} \tag{15}$$

where n_j is the B–E or F–D distribution. Now operate with $\mathscr{L}=i\omega_n+(\hbar^2/2m)\nabla^2 +\varsigma - V^{(1)}(\mathbf{r})$ on the Fourier-transformed Dyson’s equation (12.12-3), multiply with $\varphi_j(\mathbf{r}')$ and integrate over d^3r'; establish the *finite temperature* Hartree–Fock equation

$$\left[-(\hbar^2/2m)\nabla^2 + V^{(1)}(\mathbf{r})\right]\varphi_j(\mathbf{r}) + \int d^3r_1 \Sigma_{HF}(\mathbf{r},\mathbf{r}_1)\varphi_j(\mathbf{r}_1) = \varepsilon_j^1\varphi_j(\mathbf{r}). \tag{16}$$

N.B. Note that Σ_{HF} acts as a *static non-local* potential.

NON-EQUILIBRIUM STATISTICAL MECHANICS

Ab umbris et imaginibus ad veritatem!

PART D

CLASSICAL TRANSPORT THEORY

Chapter XIII

The Boltzmann Transport Equation and Boltzmann's H–Theorem

13.1 Introduction to Boltzmann Theory

Transport of electricity and heat forms one of the oldest branches of non-equilibrium statistical mechanics, with roots going back to the nineteenth century. On the many-body level the basic transport equation in Γ-space or phase space for the phase-density function $\rho(p,q,t)$ was developed by Liouville in 1838; on the one-particle level, permissible if particle interactions are weak, the Boltzmann transport equation (BTE), developed in 1871[1], describes the rate of change of the density function in μ-space $f(\mathbf{v},\mathbf{r},t)$ and the ensuing dissipative behaviour contained in Boltzmann's H-theorem[2,3]. Early in the twentieth century both approaches were endowed with quantum mechanical meaning. Replacing the Poisson bracket with the commutator bracket, the Liouville equation goes over into the von Neumann equation, as discussed in Chapter II. For solids with a periodic structure and wave functions satisfying Floquet's theorem, see Chapter VIII, Boltzmann's equation easily describes the transport for the semi-quantal occupation function $f(\mathbf{k},\mathbf{r},t)$. Be it noted from the onset that Boltzmann's transport equation is nonlinear in the occupancy function and applicable to near-equilibrium situations, as well as to transport behaviour in a steady state far from thermal equilibrium.

The most striking feature of the BTE is that it predicts irreversible behaviour for the system, *in casu* an increase in entropy (decrease in *H*) when the system approaches the equilibrium state, for which it assumes its maximum value; this notwithstanding that the derivation by Boltzmann was based on straightforward arguments from reversible Newtonian mechanics! We shall briefly repeat his arguments in the next section. While the BTE is *correct* under the assumptions stated and still is a valuable tool for many transport problems in plasma physics, hydrodynamics and condensed matter, clearly something is 'fishy', in particularly in light of the failure of his contemporary J. Willard Gibbs to establish the law of

---

[1] L. Boltzmann, Wiener Berichte **63**, 712 (1871).

[2] L. Boltzmann, "Vorlesungen über Gastheorie", J.A. Barth, Leipzig 1896, 1898. English Translation: "Lectures on Gas Theory", Univ. of California Press, 1964.

[3] Boltzmann's 'H' means *capital eta*, a symbol he employed for the negative of the entropy S.

increase in entropy from ensemble theory without coarse-graining, cf. subsections 3.8.2 and 3.8.3. So where, in essence, did Boltzmann 'err' (or 'distort') applying the tenets of classical mechanics? His contemporaries assailed a number of assumptions, in particular his assumption of *molecular chaos*, which assumes that in a two-particle collision $f_2(\mathbf{v}_1,\mathbf{v}_2,\mathbf{r},t) \approx f(\mathbf{v}_1,\mathbf{r},t)f(\mathbf{v}_2,\mathbf{r},t)$. However, it can be shown that if another closure relation is used, e.g. Kirkwood's truncation rule of subsection 4.10.2, the law of the increase in entropy follows just the same. The problem, therefore lies much deeper in that Boltzmann from the onset assumes that the mechanical forces, acting on a particle, are of a *twofold* nature: a) ponderomotive forces, derivable from a potential field; b) 'dispersion forces', associated with the momentum transfer in a collision, not derivable from a potential, although fully subject to Newton's laws. Here then, in our opinion, is the very origin of the fact that Boltzmann's *mechanical* basis led to far-reaching *thermodynamic* results! We have here, in fact, a nineteenth century precursor of what quantum mechanical perturbation theory would do some sixty years later: the (time-dependent) Hamiltonian is split into an essential part $\mathcal{H}^0$ and a perturbational part $\lambda V(t)$; there is then an agent, $\lambda V(t)$ – in Boltzmann's case the collision dynamics – which randomizes the 'motion proper', contained in $\mathcal{H}^0$.

The critique with respect to Boltzmann's results by his contemporaries has been ferocious! While the central point we stated above was not the *leitmotiv* for the attacks, a number of paradoxes made the rounds to prove the 'impossibility' of Boltzmann's results, the most well-known ones being the 'Umkehreinwand' (reversal paradox, Loschmidt, 1876) and the 'Wiederkehreinwand' (Zermelo, 1896). These 'paradoxes' will briefly be discussed later and refuted. However, Boltzmann, although admired by many in his age for his far-reaching results in *wärmetheorie* (molecular thermodynamics) became more and more an isolated and withdrawn researcher. The reader would do well to read in this regard the preface to his second volume on *gastheorie* in 1898. In 1906 Boltzmann, the pioneer, took his life.

In the next section we shall give what is essentially Boltzmann's own derivation of the BTE; as such, it is found in virtually all present-day texts on non-equilibrium kinetic theory. Clearly, however, there is more to it since the obstacle – deriving irreversibility from a mechanical basis – remains! In addition, the $\mathbf{k}$,$\mathbf{r}$-space distribution, alluded to above, cannot strictly exist because of the uncertainty principle and must be replaced by the Wigner distribution function, which is a quantum distribution amenable to a phase-space description. Altogether, it is imperative that we first operate on a many-body level, where linear operator algebra applies. Then, via the Pauli–Van Hove master equation, obtainable from the von Neumann equation with a perturbation expansion and/or projection operators, we will go back to the one-particle level and *duly derive* Boltzmann's equation, a task started by Van Hove (1955, 1960: collision terms) and completed by the author c.s. (Van Vliet 1979, 2002: streaming terms). For educational reasons this program is reserved for Part E of this book, while we start with the (unfounded) *usual* approach.

13.2 The Boltzmann Equation in Velocity-Position Space

Let $f(\mathbf{r},\mathbf{v},t)d^3r\,d^3v$ be the number of particles with position and velocity vectors within an element d^3rd^3v of μ-space[4], centred on **r** and **v**. Thus f is a density function in (**v**, **r**)-space. Integration over all velocities gives

$$n(\mathbf{r},t) = \int d^3v\, f(\mathbf{r},\mathbf{v},t), \tag{13.2-1}$$

where $n(\mathbf{r},t)$ is the density at **r** for time t. Further integration gives the normalization

$$N(t) = \int d^3r \int d^3v\, f(\mathbf{r},\mathbf{v},t). \tag{13.2-2}$$

Our aim is to find an equation for f. The distribution changes due to two causes:

(i) Streaming, i.e., the point in μ-space changes due to velocity and acceleration;

(ii) Collisions with other particles.

First we establish the effect due to streaming. Consider at time t the particles in $d\varpi = d^3rd^3v$, i.e., $f(\mathbf{r},\mathbf{v},t)d\varpi$. At time $t + dt$ we have

$$f(\mathbf{r}+d\mathbf{r},\mathbf{v}+d\mathbf{v},t+dt)d\varpi' = f(\mathbf{r}+\dot{\mathbf{r}}dt,\mathbf{v}+\dot{\mathbf{v}}dt,t+dt)d\varpi'. \tag{13.2-3}$$

Let at first there be no magnetic field. Then $\dot{\mathbf{r}} = \mathbf{v}$ and $\dot{\mathbf{v}} = \mathbf{F}/m$, where **F** is the applied force. Hence,

$$f(\mathbf{r}+d\mathbf{r},\mathbf{v}+d\mathbf{v},t+dt)d\varpi' = f[\mathbf{r}+\mathbf{v}dt,\mathbf{v}+(\mathbf{F}/m)\,dt,t+dt]d\varpi'. \tag{13.2-4}$$

We note that the element $d\varpi$ changes due to streaming as well; e.g. for a 2D μ-space a rectangle at t changes into a parallelogram at t + dt. Generally the Jacobian is

$$\frac{\partial(x',y',z',v_x',v_y',v_z')}{\partial(x,y,z,v_x,v_y,v_z)} = \begin{vmatrix} 1 & 0 & 0 & dt & 0 & 0 \\ 0 & 1 & 0 & 0 & dt & 0 \\ 0 & 0 & 1 & 0 & 0 & dt \\ \frac{\partial a_x}{\partial x}dt & & & 1 & \ddots & \end{vmatrix} = 1 + \mathcal{O}(dt^2). \tag{13.2-5}$$

Therefore, up to first order in dt we have

$$\begin{aligned}(\Delta f)_{str} &= f[\mathbf{r}+\mathbf{v}dt,\mathbf{v}+(\mathbf{F}/m)dt,t+dt] - f(\mathbf{r},\mathbf{v},t) \\ &= \left[\frac{\partial f}{\partial t} + \mathbf{v}\cdot\nabla_{\mathbf{r}} f + (\mathbf{F}/m)\cdot\nabla_{\mathbf{v}} f\right]dt \equiv (Df)dt,\end{aligned} \tag{13.2-6}$$

where D is the streaming operator, defined by the middle member of (13.2-6).

We now briefly consider a magnetic field. The Hamiltonian is then given by

---

[4] Strictly speaking, the μ-space refers to the attributes **r** and **p**; however, for the Boltzmann description it is preferable to work with **r** and **v**, in order to obtain universal results in the presence of a magnetic field.

$$\mathcal{H} = [\mathbf{p} - (q/c)\mathbf{A}(\mathbf{r},t)]^2/2m + q\Phi(\mathbf{r},t) . \qquad (13.2\text{-}7)$$

From Hamilton's equations we have

$$\dot{\mathbf{r}} = \partial\mathcal{H}/\partial\mathbf{p} = [\mathbf{p} - (q/c)\mathbf{A}]/m = \mathbf{v} , \qquad (13.2\text{-}8)$$

and

$$\begin{aligned}
m\dot{v}_x &= \dot{p}_x - (q/c)\dot{A}_x = -\partial\mathcal{H}/\partial x - (q/c)\dot{A}_x \\
&= -\frac{\partial}{\partial x}\left[\frac{[\mathbf{p}-(q/c)\mathbf{A}]^2}{2m} + q\Phi\right] - \frac{q}{c}\left[\frac{\partial A_x}{\partial t} + \frac{\partial A_x}{\partial x}\dot{x} + \frac{\partial A_x}{\partial y}\dot{y} + \frac{\partial A_x}{\partial z}\dot{z}\right] \\
&= \frac{q}{c}\left(\underbrace{v_x\frac{\partial A_x}{\partial x}}_{cancels} + v_y\frac{\partial A_y}{\partial x} + v_z\frac{\partial A_z}{\partial x}\right) - q\left(\frac{\partial\Phi}{\partial x} + \frac{1}{c}\frac{\partial A_x}{\partial t}\right) - \frac{q}{c}\left[\underbrace{\frac{\partial A_x}{\partial x}v_x}_{cancels} + \frac{\partial A_x}{\partial y}v_y + \frac{\partial A_x}{\partial z}v_z\right] \\
&= \frac{q}{c}\left[v_y\left(\frac{\partial A_y}{\partial x} - \frac{\partial A_x}{\partial y}\right) - v_z\left(\frac{\partial A_x}{\partial z} - \frac{\partial A_z}{\partial x}\right)\right] + qE_x = \frac{q}{c}[\mathbf{v}\times(\nabla\times\mathbf{A})]_x + qE_x = F_x ,
\end{aligned}$$

(13.2-9)

where F_x is the Lorentz force. Consequently, the streaming part $(\Delta f)_{str}$ has the same form as before.

Next we consider the change due to collisions. Let $R(\mathbf{r}\,\mathbf{v},\mathbf{v}')dt$ be the number of collisions in dt for which the *primary* particle changes its velocity $\mathbf{v}\rightarrow\mathbf{v}'$ and let $R(\mathbf{r}\,\mathbf{v}',\mathbf{v})dt$ be the number of inverse collisions $\mathbf{v}'\rightarrow\mathbf{v}$; note that the position of the particle does not change since the duration of the collision process is too short to be affected by the external forces. Therefore, the change due to collisions is

$$(\Delta f)_{coll} = \int d^3v\left[R(\mathbf{r}\,\mathbf{v}',\mathbf{v}) - R(\mathbf{r}\,\mathbf{v},\mathbf{v}')\right]dt \equiv (\partial f/\partial t)_{coll}\,dt . \qquad (13.2\text{-}10)$$

Balancing the loss due to streaming with the net gain due to collisions, we have, dividing by dt,

$$\left[\frac{\partial f}{\partial t} + \mathbf{v}\cdot\nabla_{\mathbf{r}}f + (\mathbf{F}/m)\cdot\nabla_{\mathbf{v}}f\right] = \int d^3v\left[R(\mathbf{r}\,\mathbf{v}',\mathbf{v}) - R(\mathbf{r}\,\mathbf{v},\mathbf{v}')\right] = \left(\frac{\partial f}{\partial t}\right)_{coll} . \qquad (13.2\text{-}11)$$

We now compute Boltzmann's form for the collision integral, assuming two-body collisions of like particles. Clearly, the result will be quadratic in f. Momentum and energy conservation require

$$\mathbf{v}_1 + \mathbf{v}_2 = \mathbf{v}_1' + \mathbf{v}_2' \equiv 2\mathbf{c} , \qquad (13.2\text{-}12)$$

$$\frac{1}{2}\left(mv_1^2 + mv_2^2\right) = \frac{1}{2}\left(mv_1'^2 + mv_2'^2\right) = \mathcal{T} , \qquad (13.2\text{-}13)$$

where primes refer to the velocities after the collision. In addition we set $\mathbf{u} = \mathbf{v}_1 - \mathbf{v}_2$. The centre of mass moves with the velocity $\mathbf{c}$; the relative velocities are

$$\mathbf{v}_1 - \mathbf{c} = \tfrac{1}{2}\mathbf{u} \text{ and } \mathbf{v}_2 - \mathbf{c} = -\tfrac{1}{2}\mathbf{u}. \tag{13.2-14}$$

The kinetic energy can now also be written as

$$\mathcal{T} = \tfrac{1}{2}m(v_1^2 + v_2^2) = mc^2 + \tfrac{1}{2}mu^2. \tag{13.2-15}$$

Since $\mathcal{T}$ and $\mathbf{c}$ are conserved, u is also conserved, i.e., $u = u'$. A pictorial view in velocity space is given in Fig. 13-1. Note that $\mathbf{u}$ and $\mathbf{u}'$ are always along the diameter of a sphere, whereas $\mathbf{c}$ is fixed from a point P to its centre O. For the scattering in the centre-of-mass system, see Fig. 13-2.

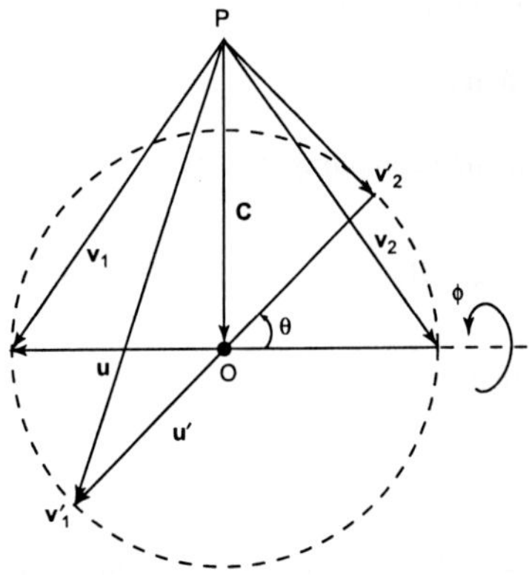

Fig. 13-1. Pictorial form of conservation laws.

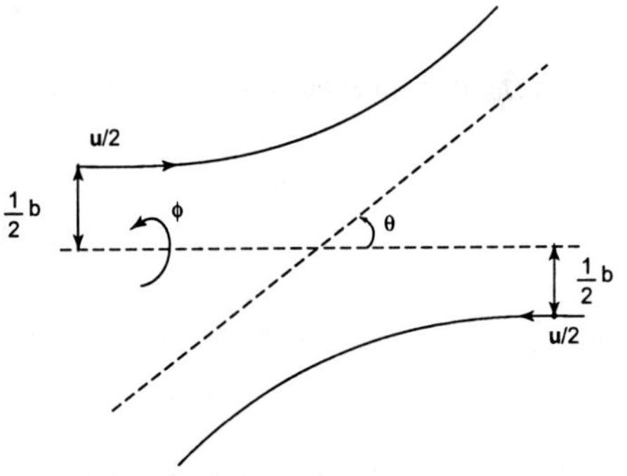

Fig. 13-2. Scattering in the centre-of-mass system.

The parameter b is called the impact parameter. The solid angle $\Omega = (\phi, \theta)$ gives the direction of $\mathbf{u}'$ with respect to $\mathbf{u}$, which is thought to be fixed. The cross section $\sigma(\Omega)$ gives the probability $\sigma(\Omega)d\Omega$ of scattering into the interval $(\Omega, \Omega + d\Omega)$ per unit incoming flux $I = |\mathbf{v}_2 - \mathbf{v}_1| = u$. Thus the transition probability per unit time is $u\sigma(\Omega)d\Omega$. To obtain the scattering rate this must be multiplied by $f_1 f_2$ and integrated, in principle, over all coordinates besides $\mathbf{v}$, the velocity of the incoming particle with the distribution f_1. From the Jacobian one easily establishes $d^3v_1'd^3v_2' = d^3u'd^3c' = u'^2 d\Omega\, du' d^3c'$. Since u' and $\mathbf{c}$ are fixed, their integration involves a delta function. We must therefore only integrate over the variables $d^3v_2 d\Omega$. So we find

$$\left(\frac{\partial f}{\partial t}\right)_{coll} = \iint d^3v_2 \sigma(\Omega) d\Omega\, u\left(f_1' f_2' - f_1 f_2\right), \quad f_1 \equiv f(\mathbf{r}, \mathbf{v}_1 = \mathbf{v}), \text{ etc.} \tag{13.2-16}$$

This is Boltzmann's famous expression for two-body collisions; see Lectures on Gas Theory, Op. Cit. p. 48.

Some remarks are still in order. We assumed reciprocity of $\sigma(\Omega)$ for inverse collisions. This point is belaboured at great length in Boltzmann's derivation. It is always true if **u** depends only on $\mathbf{v}_1$ and $\mathbf{v}_2$, which therefore completely specify the initial state. If **u** is spin-dependent, (13.2-16) will not hold, unless averaged over all spin directions. Secondly, we – like Boltzmann – made the assumption of *molecular chaos*, already mentioned in Section 13.1. This is not an essential limitation for the validity of the results.

We also briefly consider collisions between unlike particles, say electrons and ions. Let the velocities before the collision be **v** and **V** and their masses be m and M. We recall the definition of the centre-of-mass coordinate,

$$m\mathbf{r} + M\mathbf{R} = (m+M)\boldsymbol{\rho}. \tag{13.2-17}$$

Secondly, we recall the concept of reduced mass. Let $\mathbf{F}_{12}$ be a central force; then

$$\mathbf{F}_{12} = m\ddot{\mathbf{r}}, \qquad \mathbf{F}_{21} = -\mathbf{F}_{12} = M\ddot{\mathbf{R}}. \tag{13.2-18}$$

Multiplying the first equation with M and the second one with m, and subtracting we have

$$\mathbf{F}_{12} = \frac{mM}{m+M}\frac{d^2}{dt^2}(\mathbf{r}-\mathbf{R}) = \mu\frac{d^2}{dt^2}(\mathbf{r}-\mathbf{R}). \tag{13.2-19}$$

Thus, the force acts as if the reduced mass undergoes a displacement $\mathbf{r}-\mathbf{R}$.

Turning now to collisions, momentum conservation gives

$$m\mathbf{v} + M\mathbf{V} = (m+M)\mathbf{c}, \qquad \mathbf{c} = \dot{\boldsymbol{\rho}}. \tag{13.2-20}$$

Again, let the difference in velocities define the vector **u**: $\mathbf{v}-\mathbf{V}=\mathbf{u}$. In terms of **c** and **u** we have for the velocities relative to the (moving) centre-of-mass

$$\mathbf{u}_1 = \mathbf{v}-\mathbf{c} = (\mu/m)\mathbf{u}, \qquad \mathbf{u}_2 = \mathbf{V}-\mathbf{c} = -(\mu/M)\mathbf{u}. \tag{13.2-21}$$

Energy conservation is similar to the alike particle case:

$$\tfrac{1}{2}mv^2 + \tfrac{1}{2}MV^2 = \mathcal{T} = \tfrac{1}{2}(m+M)c^2 + \tfrac{1}{2}\mu u^2. \tag{13.2-22}$$

The latter equality follows from substituting (13.2-21). It expresses, of course, that the total kinetic energy is comprised of the centre-of-mass motion and the relative motion. Since $\mathcal{T}$ and **c** are conserved, it follows that **u** is conserved; so, as before, $\mathbf{u}=\mathbf{u}'$. The scattering hyperbola in the centre-of-mass system is the same as in Fig. 13-2, while in the laboratory system the lighter particle is stronger deflected than the heavier particle; for a picture see any introductory book, e.g., Reif.[5]

A more useful picture is obtained from the basal plane of the tetrahedron formed by the velocity vectors **v**,**V**,**v**' and **V**' in velocity space (the apex being the origin), given in Fig. 13.3; see also Allis.[6] The endpoint c of the c-o-m velocity $\overline{\mathrm{Oc}}$ is also in this plane. To see this, let us draw about c two spheres with radii $mu/(M+m)$ and $Mu/(M+m)$; these are the relative speeds of the two particles, both before and after the collision. The endpoints v,v ' and V,V ' of the velocity vectors lie on these spheres.

---

[5] F. Reif, "Fundamentals of Statistical and Thermal Physics", McGraw-Hill, N.Y., (1965).

[6] W.P. Allis, "Encyclopaedia of Physics Vol.21, F. Flügge, Ed., Springer Verlag 1956.

Moreover, since the directions of the relative velocities $\mathbf{v}-\mathbf{c}$ and $\mathbf{V}-\mathbf{c}$ are along $\pm\mathbf{u}$, the line $\overline{vV}$ passes through c; the same holds for the line $\overline{v'V'}$. Thus the velocity diagram of Fig. 13-3 is vindicated.

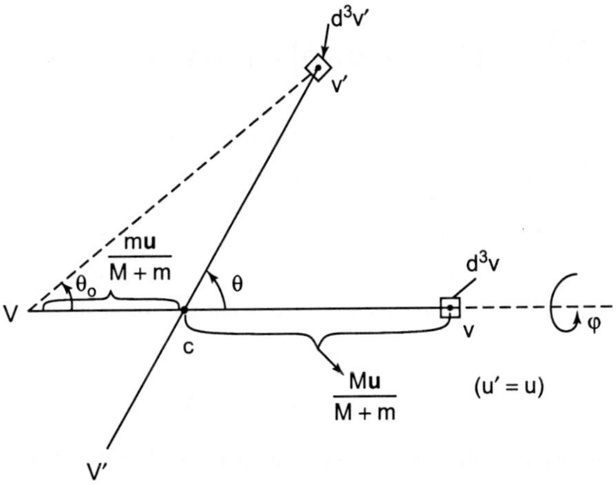

Fig. 13-3. Velocity-vector endpoints; θ and θ_0 are scattering angles in c-o-m and in the laboratory system for an observer traveling with the c-o-m velocity.

In triangle Vcv ' we have

$$\overline{v'V}^2 = \left(\frac{mu}{M+m}\right)^2 + \left(\frac{Mu}{M+m}\right)^2 + 2\frac{mMu^2}{(M+m)^2}\cos\theta , \qquad (13.2\text{-}23)$$

or

$$(\mathbf{v}'-\mathbf{V})^2 = u^2 - \frac{2mM}{(M+m)^2}u^2(1-\cos\theta) . \qquad (13.2\text{-}24)$$

If the heavy scatterer (mass M) was initially at rest, $\mathbf{V}=0$, then $\mathbf{u}=\mathbf{v}$ and (13.2-24) results in

$$v^2 - v'^2 = \frac{2Mm}{(M+m)^2}v^2(1-\cos\theta) . \qquad (13.2\text{-}25)$$

This gives the energy loss due to the recoil. For $M \gg m$ it is quite small:

$$-\frac{\Delta\varepsilon}{\varepsilon} = \frac{2Mm}{(M+m)^2}(1-\cos\theta) . \qquad (13.2\text{-}26)$$

From triangle Vcv' we can also find the relationship between θ and θ_0:

$$\sin^2\theta_0 = \frac{M^2\sin^2\theta}{M^2+m^2+2Mm\cos\theta} . \qquad (13.2\text{-}27)$$

A bit of algebra gives the standard result[7]

$$\tan\theta_0 = \frac{\sin\theta}{\cos\theta + m/M} . \qquad (13.2\text{-}28)$$

[7] H. Goldstein, "Classical Mechanics" 2nd Ed., Wiley, N.Y. 1980, Chapter 3.

Finally, we indicate the new form for the scattering integral. One easily finds that the Jacobian for the transformation $d^3v d^3V = d^3v' d^3V' = |J|\, d^3u\, d^3c$ is unity. Whence, we have analogous to (13.2-16), suppressing the position **r**:

$$\left(\frac{\partial f}{\partial t}\right)_{coll} = \int\int d^3V u\, \sigma(\Omega) d\Omega \left[f(\mathbf{v}')g(\mathbf{V}') - f(\mathbf{v})g(\mathbf{V}) \right]. \tag{13.2-29}$$

Although this is the standard expression, it is often not useful, since g(**V'**) is not easily integrated with respect to d^3V. Following Allis, Op. Cit., we replace the μ-space element d^3V by $d^3V' d^3v' / d^3v$. Although we deal with the inverse collisions, we first assume the direct process to occur and ask for the size of the volume element d^3v such that the final velocity ends up into d^3v' for given initial **V** and given θ of the scattering process. With reference to Fig. 13-3, we note that the shapes of these volume elements are similar – since **V** and all angles are fixed – but the linear dimensions differ by the ratio $\overline{vV} / \overline{v'V}$. Thus,

$$d^3v' / d^3v = \left(|\mathbf{v}' - \mathbf{V}| / |\mathbf{v} - \mathbf{V}| \right)^3 . \tag{13.2-30}$$

Since we are dealing with inverse collisions, upon changing $\mathbf{V} \to \mathbf{V}'$ in (13.2-30), Eq. (13.2-29) yields

$$\left(\frac{\partial f}{\partial t}\right)_{coll} = \int\int d^3V' u\, \sigma(\Omega) d\Omega \left(\frac{|\mathbf{v}' - \mathbf{V}'|}{|\mathbf{v} - \mathbf{V}'|} \right)^3 f(\mathbf{v}')g(\mathbf{V}') - \int\int d^3V u\, \sigma(\Omega) d\Omega\, f(\mathbf{v})g(\mathbf{V}). \tag{13.2-31}$$

This form will be employed later in computing the collision integral for plasma heating in an electromagnetic field, as e.g. occurring in the Tokamak.

13.3 The Boltzmann Equation for Solids with Extended States

In solids the electronic states are Bloch states (or simply plane waves) characterized by a crystal momentum **k**. In the absence of a magnetic field, the canonical momentum is $\hbar\mathbf{k}$ modulo a reciprocal lattice vector. Since $[p_i, r_j] = (\hbar / i)\delta_{ij}$ **r** and **k** cannot simultaneously be specified. Nevertheless it is customary to introduce a function $f(\mathbf{r}, \mathbf{k}, t)$ which measures the *occupancy* of a one-particle state $|\mathbf{k})$ at time t 'in the neighbourhood of a position **r**' (formulation of Ziman in "Electrons and Phonons" [8]). Note that for the quantum case this is a *number*, not a density as in the **v**,**r**-language. For the Fermi-Dirac case, $0 \le f \le 1$. The normalization is

$$\sum_{\mathbf{k}} f(\mathbf{r}, \mathbf{k}, t) = n(\mathbf{r}, t) \Delta^3 r, \tag{13.3-1}$$

where $n(\mathbf{r},t)$ is the density at **r** at time t and $\Delta^3 r$ is the purported neighbourhood. Further normalization gives

$$\sum_i n(\mathbf{r}_i, t) \Delta^3 r_i \approx \int n(\mathbf{r}, t) d^3 r = N(t). \tag{13.3-2}$$

---

[8] J. M. Ziman, "Electrons and Phonons", Oxford University Press, 1960, p. 264.

Although **k** is a discrete vector in the first Brillouin zone, we may consider it to be quasi-continuous and use the density of states to convert the sum to an integral. A small complication is that in some cases the spin multiplicity is also included in the density of states. In that case, we shall add a caret. Thus, $Z(\mathbf{k}) = V_0 / 8\pi^3$ and, when including spin, $\hat{Z}(\mathbf{k}) = V_0 / 4\pi^3$; here V_0 is the appropriate volume.

The BTE is easily obtained. For the change due to streaming we have

$$(\Delta f)_{str} = f(\mathbf{r} + \dot{\mathbf{r}}dt, \mathbf{k} + \dot{\mathbf{k}}dt, t + dt) - f(\mathbf{r}, \mathbf{k}, t) = \left[\frac{\partial f}{\partial t} + \dot{\mathbf{r}} \cdot \nabla_\mathbf{r} f + \dot{\mathbf{k}} \cdot \nabla_\mathbf{k} f\right] dt \,. \tag{13.3-3}$$

In the absence of a magnetic field but in a potential $\mathcal{V}(\mathbf{r},t)$ the band electrons are subject to the Wannier Hamiltonian, consisting of the band structure $\varepsilon(\mathbf{k})$, with $\mathbf{k}_{op} = -i\nabla$, on which the potential field $\mathcal{V}(\mathbf{r},t)$ is superimposed;[9] hence,

$$\mathcal{H}_W(\mathbf{r},\mathbf{k},t) = \varepsilon(\mathbf{k}) + \mathcal{V}(\mathbf{r},t) \,. \tag{13.3-4}$$

In the present method $\mathcal{H}_W$ is treated like a classical Hamiltonian. So, by Hamilton's equations

$$\dot{\mathbf{r}} = \hbar^{-1}\partial\mathcal{H}_W \partial\mathbf{k} = \hbar^{-1}\nabla_\mathbf{k}\varepsilon(\mathbf{k}) = \mathbf{v}_\mathbf{k} , \tag{13.3-5}$$

where $\mathbf{v_k}$ is the Bloch velocity, and

$$\hbar\dot{\mathbf{k}} = -\partial\mathcal{H}_W / \partial\mathbf{r} = -\nabla_\mathbf{r}\mathcal{V} = \mathbf{F} \,. \tag{13.3-6}$$

This exercise can be repeated if there is an electromagnetic field. Then $\hbar\mathbf{k} = \mathbf{p} - (q/c)\mathbf{A}(\mathbf{r},t)$. The Wannier Hamiltonian then reads

$$\mathcal{H}_W = \varepsilon\left(\hbar^{-1}[\mathbf{p} - (q/c)\mathbf{A}(\mathbf{r},t)]\right) + q\Phi(\mathbf{r},t). \tag{13.3-7}$$

It is left to the problems to show that — as in (13.3-6) — $\dot{\mathbf{k}} = \mathbf{F}/\hbar$, where **F** is the Lorentz force. Thus, finally

$$(\Delta f)_{str} = \left[\frac{\partial f}{\partial t} + \mathbf{v}_\mathbf{k} \cdot \nabla_\mathbf{r} f + (\mathbf{F}/\hbar) \cdot \nabla_\mathbf{k} f\right] dt \equiv (Df)dt \,. \tag{13.3-8}$$

Let again $R(\mathbf{r}\,\mathbf{k},\mathbf{k}')$ denote the transitions per second due to collisions $\mathbf{k} \to \mathbf{k}'$. Likewise the inverse collisions are denoted by $R(\mathbf{r}\,\mathbf{k}',\mathbf{k})$; then, whatever the precise process, formally we have

$$(\Delta f)_{coll} = \sum_{\mathbf{k}'}\left[R(\mathbf{r}\,\mathbf{k}',\mathbf{k}) - R(\mathbf{r}\,\mathbf{k},\mathbf{k}')\right] \equiv (\partial f / \partial t)_{coll}\, dt \,. \tag{13.3-9}$$

On equating this yields the transport equation

---

[9] See Reference 12 of Chapter VIII.

$$\left[\frac{\partial f}{\partial t}+\mathbf{v_k}\cdot\nabla_\mathbf{r} f+(\mathbf{F}/\hbar)\cdot\nabla_\mathbf{k} f\right]=\int Z(\mathbf{k}')d^3k'[R(\mathbf{rk}',\mathbf{k})-R(\mathbf{rk},\mathbf{k}')]=\left(\frac{\partial f}{\partial t}\right)_{coll} . \quad (13.3\text{-}10)$$

We must now be more specific about collisions. We shall distinguish three types.

(a) One-body collisions

These are collisions with (nearly) fixed obstacles. Let there be $\hat{N}$ scatterers and let $Q(\mathbf{k},\mathbf{k}')$ be the transition probability per unit time for a transition $\mathbf{k}\to\mathbf{k}'$ to occur. Let λv be the scattering potential; then, from Fermi's golden rule

$$Q(\mathbf{k},\mathbf{k}')=\frac{2\pi\lambda^2}{\hbar}|(\mathbf{k}\,|\,v\,|\,\mathbf{k}')|^2\,\delta(\varepsilon_\mathbf{k}-\varepsilon_{\mathbf{k}'}) . \quad (13.3\text{-}11)$$

We have

$$R(\mathbf{rk},\mathbf{k}')=\hat{N}Q(\mathbf{k},\mathbf{k}')f(\mathbf{r},\mathbf{k},t), \quad \text{(Boltzmann part.)}$$
$$R(\mathbf{rk},\mathbf{k}')=\hat{N}Q(\mathbf{k},\mathbf{k}')f(\mathbf{r},\mathbf{k},t)[1-f(\mathbf{r},\mathbf{k}',t)], \quad \text{(F–D particles)} \quad (13.3\text{-}12)$$
$$R(\mathbf{rk},\mathbf{k}')=\hat{N}Q(\mathbf{k},\mathbf{k}')f(\mathbf{r},\mathbf{k},t)[1+f(\mathbf{r},\mathbf{k}',t)]. \quad \text{(B–E particles)}$$

The factor $[1-f(\mathbf{r},\mathbf{k}',t)]$ reflects the exclusion principle: the state $|\mathbf{k}')$ must be available for the transition to be possible. For the B–E case we really need a many-body computation (given later in Part E). The transition probability involves the matrix elements $|\langle n_{\mathbf{k}'}+1|\,a^\dagger_{\mathbf{k}'}\,|n_{\mathbf{k}'}\rangle|^2=n_{\mathbf{k}'}+1$ with $\mathrm{Tr}\{\rho(n_{\mathbf{k}'}+1)\}=f(\mathbf{k}')+1$. To summarise we shall employ the symbol epsilon as in Eq. (12.9-26), being – 1 for Fermi–Dirac particles and + 1 for Bose–Einstein particles; in addition, we set $\varepsilon=0$ for Boltzmann particles. We also introduce the scattering kernel *pertaining to the test particles only*. For one-body collisions

$$w_{\mathbf{kk}'}=\hat{N}\,Q(\mathbf{k},\mathbf{k}') . \quad (13.3\text{-}13)$$

The form for Q, (13.3-11) indicates that there is *microscopic reversibility* $Q(\mathbf{k},\mathbf{k}')=Q(\mathbf{k}',\mathbf{k})$. Also, for one-body collisions we have $w_{\mathbf{kk}'}=w_{\mathbf{k}'\mathbf{k}}$. The above yields

$$R(\mathbf{rk},\mathbf{k}')=w_{\mathbf{kk}'}f(\mathbf{r},\mathbf{k},t)[1+\varepsilon f(\mathbf{r},\mathbf{k}',t)] . \quad (13.3\text{-}14)$$

The collision term will henceforth be indicated by $-\mathscr{B}_\mathbf{k} f(\mathbf{r},\mathbf{k},t)$. The BTE now takes the following *standard form*:

$$\frac{\partial f(\mathbf{r},\mathbf{k},t)}{\partial t}+\mathbf{v_k}\cdot\nabla_\mathbf{r} f+(\mathbf{F}/\hbar)\cdot\nabla_\mathbf{k} f+\mathscr{B}_\mathbf{k} f=0 , \quad \mathbf{(13.3\text{-}15)}$$

where $\mathscr{B}_\mathbf{k} f$ is the (generally nonlinear) Boltzmann collision operator

$$\mathscr{B}_{\mathbf{k}} f = \sum_{\mathbf{k}'} \{ f(\mathbf{r},\mathbf{k},t)[1+\varepsilon f(\mathbf{r},\mathbf{k}',t)] w_{\mathbf{k}\mathbf{k}'} - f(\mathbf{r},\mathbf{k}',t)[1+\varepsilon f(\mathbf{r},\mathbf{k},t)] w_{\mathbf{k}'\mathbf{k}} \}. \quad \textbf{(13.3-16)}$$

Eqs. (13.3-15) and (13.3-16) give the complete transport equation, being an integro-differential equation, if $\Sigma_{\mathbf{k}'}$ is replaced by $\int d^3k' \, Z(\mathbf{k}')$. As we shall see below, this form is valid for any type of collisions, but generally $w_{\mathbf{k}\mathbf{k}'} \neq w_{\mathbf{k}'\mathbf{k}}$.

(b) *Two-body collisions*

As before, the test particles have states $|\mathbf{k})$, energies $\varepsilon_{\mathbf{k}}$ and distribution $f(\mathbf{r},\mathbf{k},t)$, while the field particles have states $|\mathbf{K})$, energies $E_{\mathbf{K}}$ and distribution $F(\mathbf{r},\mathbf{K},t)$. The microscopic transition probability per second, $\mathbf{k},\mathbf{K} \to \mathbf{k}',\mathbf{K}'$ is now governed by $Q(\mathbf{kK},\mathbf{k}'\mathbf{K}')$, where by Fermi's golden rule

$$Q(\mathbf{kK},\mathbf{k}'\mathbf{K}') = \frac{2\pi\lambda^2}{\hbar} |\langle \mathbf{kK} | v^{(2)} | \mathbf{k}'\mathbf{K}' \rangle|^2 \, \delta(\varepsilon_{\mathbf{k}} + E_{\mathbf{K}} - \varepsilon_{\mathbf{k}'} - E_{\mathbf{K}'}) = Q(\mathbf{k}'\mathbf{K}',\mathbf{kK}). \tag{13.3-17}$$

The last equality again expresses the principle of microscopic reversibility. The two-body interaction is denoted by $\lambda v^{(2)}(|\mathbf{r}_1 - \mathbf{r}_2|)$; the matrix element can be computed, if so desired, by the wave-mechanical form

$$\langle \mathbf{kK} | v^{(2)} | \mathbf{k}'\mathbf{K}' \rangle = \iint d^3r_1 d^3r_2 \, \varphi_{\mathbf{k}}^*(\mathbf{r}_1) \psi_{\mathbf{K}}^*(\mathbf{r}_2) v^{(2)}(|\mathbf{r}_1 - \mathbf{r}_2|) \varphi_{\mathbf{k}'}(\mathbf{r}_1) \psi_{\mathbf{K}'}(\mathbf{r}_2). \tag{13.3-18}$$

For the direct and inverse scattering rates we find a quartic expression,

$$\begin{aligned} R(\mathbf{rk},\mathbf{k}') = \sum_{\mathbf{KK}'} & f(\mathbf{r},\mathbf{k},t)[1+\varepsilon_1 f(\mathbf{r},\mathbf{k}',t)] \\ & \times F(\mathbf{r},\mathbf{K},t)[1+\varepsilon_2 F(\mathbf{r},\mathbf{K}',t)] Q(\mathbf{kK},\mathbf{k}'\mathbf{K}'). \end{aligned} \tag{13.3-19}$$

Again, let $w_{\mathbf{k}\mathbf{k}}'$ be the transition kernel for the test particles, regardless the state of the field particles (scatterers). Clearly,

$$w_{\mathbf{k}\mathbf{k}'} = \sum_{\mathbf{KK}'} F(\mathbf{r},\mathbf{K},t)[1+\varepsilon_2 F(\mathbf{r},\mathbf{K}',t)] Q(\mathbf{kK},\mathbf{k}'\mathbf{K}'). \tag{13.3-20}$$

Although the Q's are microscopically reversible, the $w_{\mathbf{k}\mathbf{k}}'$ are not. *If* the particles with which the test particles collide are much heavier, it can often be assumed that they remain in thermal equilibrium ['adiabatic approximation' $F(\mathbf{r},\mathbf{K},t) \approx F_{eq}(\mathbf{r},\mathbf{K})$]. Replacing in (13.3-20) F by F_{eq}, one easily finds – employing energy conservation as expressed by the delta function in (13.3-17),

$$w_{\mathbf{k}\mathbf{k}'} = e^{\beta(\varepsilon_{\mathbf{k}} - \varepsilon_{\mathbf{k}'})} w_{\mathbf{k}'\mathbf{k}}, \qquad \beta = 1/k_B T. \tag{13.3-21}$$

Further, when (13.3-20) is substituted into (13.3-19), one finds that the appropriate form for the BTE is formally the same as for one-body collisions, i.e., (13.3-15) +

(13.3-16), although the expressions for the $w_{\mathbf{kk}'}$ are much more involved.

(c) *Collisions with quasi-particles, in casu electron-phonon interactions*

Collisions now involve either emission or absorption of phonons, such as described in Section 8.6. The microscopic transition rates were found in Eqs. (8.6-16) and (8.6-17). We recall them here:

$$\begin{aligned} Q(\mathbf{k},\mathbf{q}\to\mathbf{k}') &= (2\pi/\hbar)\,|\mathcal{F}(\mathbf{q})|^2\,\delta(\varepsilon_{\mathbf{k}'}-\varepsilon_{\mathbf{k}}-\hbar\omega_{\mathbf{q}})\delta_{\mathbf{k}',\mathbf{k}+\mathbf{q}}, \quad \text{(abs.)} \\ Q(\mathbf{k}\to\mathbf{k}',\mathbf{q}) &= (2\pi/\hbar)\,|\mathcal{F}(\mathbf{q})|^2\,\delta(\varepsilon_{\mathbf{k}'}-\varepsilon_{\mathbf{k}}+\hbar\omega_{\mathbf{q}})\delta_{\mathbf{k}',\mathbf{k}-\mathbf{q}}, \quad \text{(em.)} \end{aligned} \tag{13.3-22}$$

where $E_{\mathbf{q}} = \hbar\omega_{\mathbf{q}}$ and where the Kronecker delta expresses momentum conservation. The form of $\mathcal{F}(\mathbf{q})$ depends on the type of phonons involved (acoustical, optical, etc). Microscopic reversibility is expressed by

$$Q(\mathbf{k},\mathbf{q}\to\mathbf{k}') = Q(\mathbf{k}'\to\mathbf{k},\mathbf{q}). \tag{13.3.23}$$

From a many-body treatment involving creation and annihilation operators, to be discussed in Part E, one finds that the absorption rate is proportional to the occupancy $F_{\mathbf{q}}$, while the emission rate is proportional to $F_{\mathbf{q}}$ + 1, since this entails the matrix element $|\langle N_{\mathbf{q}}+1|a_{\mathbf{q}}^{\dagger}|N_{\mathbf{q}}\rangle|^2 = N_{\mathbf{q}}+1$ with $\mathrm{Tr}\{\rho(N_{\mathbf{q}}+1)\} = F_{\mathbf{q}}+1$, – as for material bosons. Therefore, one expects the plausible form

$$\begin{aligned} R(\mathbf{r}\,\mathbf{k},\mathbf{k}') &= \sum_{\mathbf{q}} f(\mathbf{r},\mathbf{k},t)[1-f(\mathbf{r},\mathbf{k}',t)] \\ &\times\left[F_{\mathbf{q}}\,Q(\mathbf{k},\mathbf{q}\to\mathbf{k}') + (F_{\mathbf{q}}+1)\,Q(\mathbf{k}\to\mathbf{k}',\mathbf{q})\right]. \end{aligned} \tag{13.3-24}$$

It behoves us to introduce a kernel for the electron transitions as caused by both phonon absorption and emission:

$$w_{\mathbf{kk}'} = \sum_{\mathbf{q}}\left[F_{\mathbf{q}}\,Q(\mathbf{k},\mathbf{q}\to\mathbf{q}') + (F_{\mathbf{q}}+1)\,Q(\mathbf{k}\to\mathbf{k}',\mathbf{q})\right]. \tag{13.3-25}$$

Usually it is assumed that the phonon distribution remains in equilibrium (no 'phonon drag'). From the equilibrium B–E distribution and energy conservation, the rule (13.3-21) is again validated. Moreover, when (13.3-25) is substituted into (13.3-24), the standard form is once more established.

In conclusion: the streaming part *and* the collision part can always be expressed by Eqs (13.3-15) and (3.13-16). However, what goes into the transition probabilities $w_{\mathbf{kk}'}$ depends in detail on the collision processes envisaged.

13.4 Connection with the Cross Section; Examples of $\sigma(\Omega)$

13.4.1 *Matrix Element Squared* $\leftrightarrow$ *Cross Section*

The connection with the cross section is simple only for one-body collisions, since in that case we have no need for the centre-of-mass coordinates, the scatterers being fixed. The microscopic transition probability per second is given by (13.3-11). It is advantageous to consider the inverse collisions, $\mathbf{k}'\to\mathbf{k}$, with the cross section being defined as the probability for scattering per second of an incoming particle in the direction Ω of $\mathbf{k}$ with respect to given $\mathbf{k}'$ per unit incoming flux, $I = v_{\mathbf{k}'}/V_0$, where V_0 is the sample volume. Thus,

$$\sigma(\Omega)d\Omega = \frac{V_0}{v_{\mathbf{k}'}}\int_{|\mathbf{k}|\,only} Z(\mathbf{k})d^3k\, Q(\mathbf{k}',\mathbf{k}) . \tag{13.4-1}$$

Assuming that spin is conserved in a collision (i.e., λv is spin-independent), we set $Z(\mathbf{k}) = V_0/8\pi^3$. The vector $\mathbf{k}$ is represented by polar coordinates, $\mathbf{k} = (k,\Omega)$, where $k = |\mathbf{k}|$. We further write

$$d^3k = \frac{d\varepsilon_{\mathbf{k}}dS_{\mathbf{k}}}{|\,\mathrm{grad}_{\mathbf{k}}\varepsilon_{\mathbf{k}}\,|} = \frac{d\varepsilon_{\mathbf{k}}k^2 d\Omega}{\hbar v_{\mathbf{k}}} . \tag{13.4-2}$$

Assuming that the energies only depend on k and substituting (13.3-11) and (13.4.2) into (13.4-1), we obtain

$$\sigma(\Omega)d\Omega = \frac{V_0^2}{8\pi^3 v_{\mathbf{k}'}}\frac{2\pi\lambda^2}{\hbar}\int_0^\infty \frac{d\varepsilon_{\mathbf{k}}k^2 d\Omega}{\hbar v_{\mathbf{k}}} |(\mathbf{k}|v|\mathbf{k}')|^2\, \delta(\varepsilon_{\mathbf{k}}-\varepsilon_{\mathbf{k}'}) . \tag{13.4-3}$$

In the effective mass approximation $k^2 = 2m^*\varepsilon_{\mathbf{k}}/\hbar^2$ and $v_{\mathbf{k}'}v_{\mathbf{k}} = v_{\mathbf{k}}^2 = 2\varepsilon_{\mathbf{k}}/m^*$. Carrying out the integration, we arrive at

$$\sigma(\Omega)d\Omega = \frac{V_0^2 m^{*2}\lambda^2}{4\pi^2\hbar^4}|(\mathbf{k}|v|\mathbf{k}')|^2\, d\Omega , \tag{13.4-4}$$

which is the desired result. Note that the states $|\mathbf{k})$ and $|\mathbf{k}')$ are $\propto V_0^{-1/2}$, so that the volume drops out.

Next we substitute the above connection into (13.3-13), using (13.3-11), to obtain

$$w_{\mathbf{k}\mathbf{k}'} = \frac{8\pi^3\hbar^3\hat{N}}{m^{*2}V_0^2}\sigma(\Omega)\delta(\varepsilon_{\mathbf{k}}-\varepsilon_{\mathbf{k}'}) . \tag{13.4-5}$$

Since for one-body collisions $w_{\mathbf{k}\mathbf{k}'} = w_{\mathbf{k}'\mathbf{k}}$, the quadratic terms in (13.3-16) cancel, so,

$$\mathscr{B}_{\mathbf{k}} f(\mathbf{r},\mathbf{k},t) = \sum_{\mathbf{k}'}[f(\mathbf{r},\mathbf{k},t)] - f(\mathbf{r},\mathbf{k}',t)]w_{\mathbf{k}\mathbf{k}'}$$

$$=\frac{\hbar^3\hat{N}}{m^{*2}V_0}\int\frac{d\varepsilon_{\mathbf{k}}}{\hbar v_{\mathbf{k}}}k^2\sigma(\Omega)d\Omega\delta(\varepsilon_{\mathbf{k}}-\varepsilon_{\mathbf{k}'})[f(\mathbf{r},\mathbf{k},t)]-f(\mathbf{r},\mathbf{k}',t)]. \quad (13.4\text{-}6)$$

Finally, carrying out the integral over $d\varepsilon_{\mathbf{k}}$, we find

$$-\mathscr{B}_{\mathbf{k}}f=\left(\frac{\partial f}{\partial t}\right)_{coll}=\frac{\hat{N}}{V_0}\int v_{\mathbf{k}}\sigma(\Omega)d\Omega\left[f(\mathbf{r},\mathbf{k}',t)-f(\mathbf{r},\mathbf{k},t)\right], \qquad (13.4\text{-}7)$$

which is the one-body **k**-space analogue of Boltzmann's collision integral (13.2-16).

It is instructive to also derive this result directly from the wave picture. The incoming and scattered wave we denote as

$$\psi=A\left[e^{ikz}+\psi_1(\mathbf{r}')\right]. \qquad (13.4\text{-}8)$$

In the first-order Born approximation

$$\psi_1(\mathbf{r}')=-\frac{A}{4\pi}\int\frac{1}{|\mathbf{r}-\mathbf{r}'|}e^{ik|\mathbf{r}-\mathbf{r}'|}e^{ikz}U(\mathbf{r})d^3r, \qquad (13.4\text{-}9)$$

where $U(\mathbf{r})=(2m^*/\hbar^2)\lambda v(\mathbf{r})$. The integration involves spherical waves emanating from all points **r**. For the far field we have the approximations, cf. Fig. 13-4,

$$\begin{aligned}&1/|\mathbf{r}-\mathbf{r}'|\approx 1/r', \qquad k'\approx k,\\ &|\mathbf{r}-\mathbf{r}'|=\overline{MP}\approx\overline{M'P}=r'-r\cos\alpha\approx r'-r\cos\vartheta=r'-\mathbf{k}'\cdot\mathbf{r}/k.\end{aligned} \qquad (13.4\text{-}10)$$

Note that $\mathbf{k}=(k_z,0,0)$ is taken as polar axis. Hence, for (13.4-9)

$$\psi_1(\mathbf{r}')=-\frac{A}{4\pi}\frac{e^{ikr'}}{r'}\frac{2m^*\lambda}{\hbar^2}\int d^3r\,e^{i\mathbf{k}\cdot\mathbf{r}}v(\mathbf{r})e^{-i\mathbf{k}'\cdot\mathbf{r}}. \qquad (13.4\text{-}11)$$

By definition of cross section, $\sigma(\theta,\varphi)=|r'\psi_1(\mathbf{r}')|^2/A^2$. Thus, (13.4-11) yields

$$\sigma(\theta,\varphi)=\frac{m^{*2}\lambda^2}{4\pi^2\hbar^4}V_0^2\,|(\mathbf{k}\,|\,v\,|\,\mathbf{k}')|^2, \qquad (13.4\text{-}12)$$

in full agreement with (13.4-4).

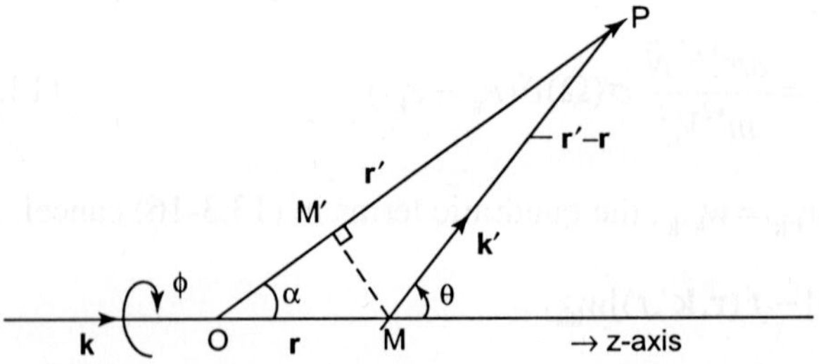

Fig. 13-4. Diagram of scattered wave

Some brief remarks on other treatments for the collision term are in order. As it stands, the collision operator is not suitable for variational manipulation. This may be seen as follows. Consider the stationary case, $\partial f / \partial t = 0$. Then, we introduce with Kohler, Sondheimer, Ziman and others[10] the function $\Phi(\mathbf{r},\mathbf{k})$ by

$$f(\mathbf{r},\mathbf{k}) = f_{eq}(\mathbf{r},\mathbf{k}) - \Phi(\mathbf{r},\mathbf{k})\partial f_{eq}(\mathbf{r},\mathbf{k}) / \partial \varepsilon_{\mathbf{k}}, \tag{13.4-13}$$

where $f_{eq}(\mathbf{r},\mathbf{k})$ is the equilibrium distribution, which depends on $\mathbf{k}$ only through $\varepsilon_{\mathbf{k}}$. We have the well-known result, valid for Boltzmann, F–D and B–E distributions,

$$-\frac{\partial f_{eq}(\mathbf{r},\mathbf{k})}{\partial \varepsilon_{\mathbf{k}}} = \frac{f_{eq}(1+\varepsilon f_{eq})}{k_B T}. \tag{13.4-14}$$

In the case of the F–D distribution $\partial f_{eq} / \partial \varepsilon$ resembles a delta function (and is so at $T = 0$). After removing this from $f(\mathbf{r},\mathbf{k})$, we expect the behaviour of $\Phi(\mathbf{r},\mathbf{k})$ to be rather smooth and suitable for variational techniques. From the above,

$$f(\mathbf{r},\mathbf{k}) = f_{eq}(\mathbf{r},\mathbf{k}) + [\Phi(\mathbf{r},\mathbf{k}) / k_B T] f_{eq}(1+\varepsilon f_{eq}). \tag{13.4-15}$$

This is substituted into the collision operator (13.4-7). We obtain,

$$\mathscr{B}_{\mathbf{k}} f = \frac{\hat{N}}{V_0 k_B T} \int v_{\mathbf{k}} \sigma(\Omega) d\Omega [\Phi(\mathbf{r},\mathbf{k}) - \Phi(\mathbf{r},\mathbf{k}')] f_{eq}(1+\varepsilon f_{eq}). \tag{13.4-16}$$

This is called the *canonical form*. Similar expressions can be formulated for two-body collisions and collisions with quasi-particles for near-equilibrium transport.

13.4.2 *Classical and Quantum Mechanical Examples of $\sigma(\Omega)$*

We present two examples of quasi-one-body scattering and two examples of two-body scattering.

(a) *Scattering of an electron (or positron) by an atom*

Generally inelastic collisions occur, in which an atom is excited from an atomic state $|\,n)$ to $|\,m)$ due to the impact. However, we shall presume that $n \approx m$. Thus for the matrix element we have assuming plane waves $e^{i\mathbf{k}\cdot\mathbf{R}}/V_0^{1/2}$ for the incoming particle

$$(\mathbf{k}n\,|\,\lambda v(n,\mathbf{R})\,|\,\mathbf{k}'n) = \int d^3R\, e^{i\mathbf{K}\cdot\mathbf{R}} U(n,\mathbf{R}) / V_0\,, \tag{13.4-17}$$

where $\mathbf{K} = \mathbf{k}' - \mathbf{k}$, $\mathbf{R} = \mathbf{r} - \mathbf{r}_{at}$. Clearly, the rhs of (13.4-17) is the Fourier transform of the potential energy stemming from the nucleus and the electron cloud. It is most

[10] J. Ziman, "Electrons and Phonons", Op. Cit. p. 275 ff.

easily found from Fourier transforming the Poisson equation associated with the charge density:

$$\nabla^2\Phi(\mathbf{R}) = -q[Z\delta(\mathbf{R}) - n(\mathbf{R})], \tag{13.4-18}$$

where Z is the atomic number. Further, $\langle U(\mathbf{R})\rangle = q\Phi(\mathbf{R})$, where $\langle U(\mathbf{R})\rangle$ is averaged with respect to the states of the atom, see below. Transforming, we have

$$-K^2\tilde{\Phi}(\mathbf{K}) = -q[Z - F(\mathbf{K})], \tag{13.4-19}$$

where $F(\mathbf{K})$ is called the atomic form factor,

$$F(\mathbf{K}) = \int d^3R\, e^{i\mathbf{K}\cdot\mathbf{R}} n(\mathbf{R}). \tag{13.4-20}$$

Hence,

$$\int d^3R\, e^{i\mathbf{K}\cdot\mathbf{R}} \langle U(\mathbf{R})\rangle / V_0 = \frac{q^2}{K^2}[Z - F(\mathbf{K})]. \tag{13.4-21}$$

Since the scattering is near-elastic, $K = |\mathbf{k}' - \mathbf{k}| = 2k\sin(\theta/2)$, where θ is the angle enclosed by $\mathbf{k}$ and $\mathbf{k}'$. Analogous to (13.4-4) we obtain for these quasi-one-body collisions

$$\begin{aligned}\sigma(\Omega)d\Omega &= \frac{V_0^2\mu^2}{4\pi^2\hbar^4}\sum_n |(\mathbf{k}n|\lambda v(n,\mathbf{R})|\mathbf{k}'n)|^2 p(n)\, d\Omega \\ &= \frac{\mu^2}{4\pi^2\hbar^4}\left|\int d^3R\, e^{i\mathbf{K}\cdot\mathbf{R}}\langle U(\mathbf{R})\rangle\right|^2 d\Omega .\end{aligned} \tag{13.4-22}$$

Substituting (13.4-21),

$$\sigma(\Omega)d\Omega = \frac{q^4\mu^2}{64\pi^2\hbar^4 k^4}\csc^4(\theta/2)\left[Z - F\left(2k\sin\frac{\theta}{2}\right)\right]^2 d\Omega . \tag{13.4-23}$$

For small K we can expand the form factor up to second order; one finds

$$Z - F(K) \simeq \frac{1}{6}K^2\int n(R)\, R^2 d^3R \tag{13.4-24}$$

(the dipole moment vanishes; the surviving term is the mean square radius in the electron cloud). For small K (low energy or small angle scattering) from (13.4-22),

$$\sigma_{small\ K}(\Omega) \approx \left[\frac{q^2\mu}{12\pi\hbar^2}\int n(R)\, R^2 d^3R\right]^2 . \tag{13.4-25}$$

The scattering is then isotropic (over the allowed range of angles). For larger values of K, we have $F(K) \sim 0$ (rapidly oscillating integrand); then from (13.4-23) we recover the renowned Rutherford formula,

$$\sigma_{Ruth}(\Omega)=\left(\frac{Ze^2}{8\pi\mu u^2}\right)^2 \csc^4(\theta/2)\,, \tag{13.4-26}$$

where we used $q = -e$ for the electronic charge and where u is the relative velocity of the incoming charge, as defined previously. More accurate results can be obtained with the Thomas–Fermi model for the electron cloud.

(b) *Electron-ionized impurity scattering*

For mobility computations in semiconductors, we will need the cross section due to ionized donor or acceptor scattering. The classical hyperbolae are depicted in Fig. (13-5), (*a*) and (*b*). In the most common treatment due to Brooks and Herring[11] a screened Coulomb potential is employed; for another treatment, based upon an unscreened Coulomb potential with impact-parameter cut-off by Conwell and Weiszkopf[12], we refer to solid-state texts. So we have

$$\lambda v(\mathbf{r}) = -\frac{(Zq)e}{4\pi\varepsilon r}e^{-\kappa r}\,, \tag{13.4-27}$$

where ε is the (relative) dielectric constant and where Zq is the charge of the impurity centre, being positive for donors and negative for acceptors, while e is the absolute value of the electronic charge as before; κ is the reciprocal Debye length, cf. subsection 4.10.1. The matrix element again has the form of a Fourier transform,

$$(\mathbf{k}'|\lambda v|\mathbf{k})V_0 = -\int d^3r\, e^{-i\mathbf{K}\cdot\mathbf{r}}\frac{(Zq)e}{4\pi\varepsilon r}e^{-\kappa r}\,, \qquad \mathbf{K}=\mathbf{k}'-\mathbf{k}\,. \tag{13.4-28}$$

This transform has already been evaluated on a number of occasions (take $\mathbf{K}$ as polar axis, carry out the angular integrations, and recognise the remaining integral as a Laplace transform, or do the contour integration]. One obtains

$$(\mathbf{k}'|\lambda v|\mathbf{k})V_0 = -\frac{(Zq)e}{\varepsilon}\frac{1}{|\mathbf{k}-\mathbf{k}'|^2+\kappa^2}\,, \qquad |\mathbf{k}-\mathbf{k}'| = k\sqrt{2(1-\cos\theta)}\,. \tag{13.4-29}$$

It follows that

$$\sigma(\Omega)d\Omega = \frac{m^{*2}Z^2e^4}{4\pi^2\hbar^4\varepsilon^2}\left[\frac{1}{2k^2(1-\cos\theta)+\kappa^2}\right]^2 d\Omega\,. \tag{13.4-30}$$

---

[11] H. Brooks in Advances in Electronics and Electron Physics (L. Martin, Ed.) Vol. **7**, 85-182, Ac. Press, NY, 1955.

[12] E.M. Conwell and V.F. Weiszkopf, Phys. Rev. **77**, 388 (1950).

If $\kappa \to 0$, with $1-\cos\theta = 2\sin^2(\theta/2)$ we obtain once more Rutherford's formula. However, if there is no screening, the mobility will diverge logarithmically, as we will see later. In metals there is a similar screening effect of one electron due to the presence of all others; the screening parameter has a different form, however, as shown in example (d). A final warning is in order. Although the hyperbolae of Fig. (13-5) are very suggestive and can be found in many textbooks, *these hyperbolae have no more realistic meaning than the orbits of the 1913 Bohr model of the hydrogen atom!* The only true state of affairs is depicted in the Feynman diagram of Fig. 7-2; in the quantum view, all that is known are the initial and final states and the interaction potential causing the transition.

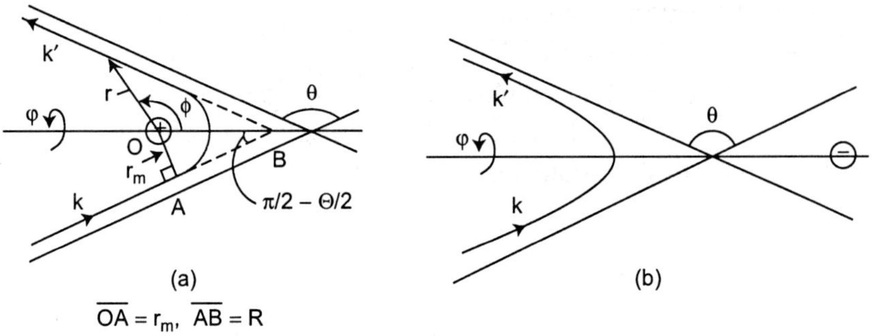

(*a*) (*b*)

Fig. 13-5. Ionized impurity scattering: (*a*) Ionized donor scattering; (*b*) Ionized acceptor scattering.

(c) *Hard sphere two-body interactions*

A simple example is at hand when we consider two hard-core spheres, with joint diameter *a*, that collide. Two pictures in the c-o-m system are shown in Fig. 13-6 (*a*) and (*b*). With $\mathbf{u} = \mathbf{v}_1 - \mathbf{v}_2$ and $\mathbf{c}$ the c-o-m velocity, the relative velocities were given by Eq. (13.2-21) with the replacement $(\mathbf{v}, \mathbf{V}) \to (\mathbf{v}_1, \mathbf{v}_2)$. From Fig. 13-6 (*a*) we see,

$$b = a\sin\vartheta, \qquad \theta = \pi - 2\vartheta, \tag{13.4-31}$$

or

$$b(\theta) = a\cos(\theta/2). \tag{13.4-32}$$

Here, b is the impact parameter. From Fig. 13-6 (*b*) it is evident that the scattering is determined by either $(b, db, d\varphi)$ or $(\Omega, d\theta, d\varphi)$. It may be helpful to consider the following equivalent definition of cross section: If I is the incident flux, then $I\sigma d\Omega$ is the number of particles crossing an area $\sigma d\Omega$ per second. From the incoming side, this is the area $bd\varphi\, db$, while at the scattered side this is the area $\sigma\sin\theta\, d\theta\, d\varphi$. Equating, we have

$$\sigma(\Omega) = \frac{b(\theta)}{\sin\theta}\left|\frac{db}{d\theta}\right|. \tag{13.4-33}$$

This relationship is *general*. For the present case, substituting (13.4-32), we find

$$\sigma(\Omega) = \frac{a\cos(\theta/2)}{\sin\theta}\frac{a}{2}\sin(\theta/2) = \frac{a^2}{4}. \tag{13.4-34}$$

The cross section is apparently isotropic. The total cross section is $4\pi(a^2/4) = \pi a^2$, as expected. We shall also introduce the momentum-loss cross section. The momentum change is $\mu(\mathbf{u}-\mathbf{u}') = \mu\mathbf{u}(1-\cos\theta)$. Hence,

$$\sigma_{total}^{mom} = \int \sigma(\Omega)\,(1-\cos\theta)d\Omega\,. \tag{13.4-35}$$

For hard sphere interaction, the result is likewise πa^2.

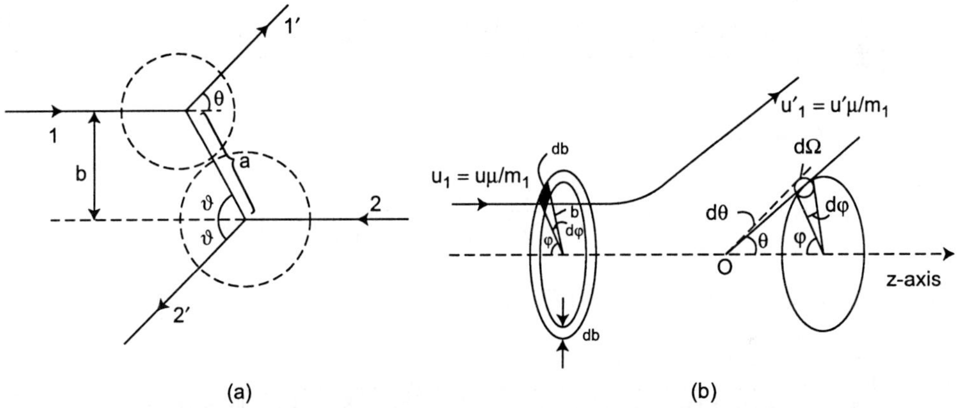

Fig. 13-6. (*a*) Hard-sphere scattering; (*b*) Scattering by a potential centred at O in the c-o-m system.

(d) *Electron-electron scattering*

In metals the Coulomb interaction can be split into a long-range effect, involving collective motion (plasma part) and a short-range effect (screened potential part). The screening in this case has no thermal basis, but arises from the exclusion principle and is obtained with the Thomas–Fermi method.[13] The resulting potential for electron-electron interaction is

$$\lambda v(\mathbf{r}_1,\mathbf{r}_2) = \frac{e^2}{4\pi}\frac{e^{-\alpha|\mathbf{r}_1-\mathbf{r}_2|}}{|\mathbf{r}_1-\mathbf{r}_2|}, \qquad \alpha^2 = e^2\rho(\varepsilon_F), \tag{13.4-36}$$

where $\rho(\varepsilon_F)$ is the density of states at the Fermi level, cf. Eq. (12.7-7). We shall

[13] J. Ziman, Op. Cit., p.159-160.

also write $\lambda v = U(\mathbf{r}_1, \mathbf{r}_2)$. In order to find the matrix element, we recognise that $U(\mathbf{R})$ is the Green's function of the Helmholtz equation,

$$(\nabla^2 - \alpha^2)U(\mathbf{r}_1, \mathbf{r}_2) = -e^2\delta(\mathbf{R}), \quad \mathbf{R} = \mathbf{r}_1 - \mathbf{r}_2. \tag{13.4-37}$$

Employing as eigenfunctions plane waves in a box V_0, the Green's function has a bilinear expansion; from the Fourier transform of (13.4-37)

$$(-K^2 - \alpha^2)\hat{U}(\mathbf{K}) = -e^2, \tag{13.4-38}$$

we find by inversion

$$U(\mathbf{R}) = \frac{e^2}{V_0}\sum_{\mathbf{K}} \frac{e^{-i\mathbf{K}\cdot(\mathbf{r}_1 - \mathbf{r}_2)}}{K^2 + \alpha^2}, \tag{13.4-39}$$

which is the sought for bilinear series. The matrix element is now

$$\langle \mathbf{k}_1'\mathbf{k}_2' | U(\mathbf{r}_1, \mathbf{r}_2) | \mathbf{k}_1\mathbf{k}_2 \rangle = \left(e^2 / V_0^3\right)$$

$$\times \sum_{\mathbf{K}} \iint d^3r_1 d^3r_2 \frac{e^{-i(\mathbf{k}_1' - \mathbf{k}_1)\cdot\mathbf{r}_1} e^{-i(\mathbf{k}_2' - \mathbf{k}_2)\cdot\mathbf{r}_2} e^{-i\mathbf{K}\cdot(\mathbf{r}_1 - \mathbf{r}_2)}}{K^2 + \alpha^2} = \frac{e^2}{V_0}\sum_{\mathbf{K}} \frac{\delta_{\mathbf{K}, \mathbf{k}_1 - \mathbf{k}_1'}\delta_{\mathbf{K}, \mathbf{k}_2' - \mathbf{k}_2}}{K^2 + \alpha^2}. \tag{13.4-40}$$

Carrying out $\Sigma_{\mathbf{K}}$ we arrive at

$$\left|\langle \mathbf{k}_1'\mathbf{k}_2' | U(\mathbf{r}_1, \mathbf{r}_2) | \mathbf{k}_1\mathbf{k}_2 \rangle\right|^2 = \frac{e^4}{V_0^2}\left[\frac{1}{|\mathbf{k}_1 - \mathbf{k}_1'|^2 + \alpha^2}\right]^2 \delta_{\mathbf{k}_1 + \mathbf{k}_2, \mathbf{k}_1' + \mathbf{k}_2'}. \tag{13.4-41}$$

This is to be substituted into (13.3-17); hence,

$$Q(\mathbf{k}_1\mathbf{k}_2, \mathbf{k}_1'\mathbf{k}_2') = \frac{2\pi e^4}{\hbar V_0^2}\left[\frac{1}{|\mathbf{k}_1 - \mathbf{k}_1'|^2 + \alpha^2}\right]^2 \delta(\varepsilon_{\mathbf{k}_1} + \varepsilon_{\mathbf{k}_2} - \varepsilon_{\mathbf{k}_1'} - \varepsilon_{\mathbf{k}_2'})\delta_{\mathbf{k}_1 + \mathbf{k}_2, \mathbf{k}_1' + \mathbf{k}_2'}. \tag{13.4-42}$$

To be noted, the transitions exhibit both energy and momentum conservation. We have excluded umklapp processes, since we used plane waves, rather than Bloch states. The above result can be further substituted in $w_{\mathbf{k}_1\mathbf{k}_1'}$, whereby the Boltzmann operator follows.

13.5 Boltzmann's H-Theorem

13.5.1 *Derivation*

We will study the trend towards equilibrium. To that purpose we introduce Boltzmann's H (capital eta), which is the negative of the entropy function S. From his transport equation Boltzmann established: $d\mathrm{H}/dt \le 0$. A simple proof, based on Boltzmann's own form for the collision integral in velocity space, can be found in a variety of books.[14] The proof can easily be given, however, in a far broader context for any Boltzmann, F–D or B–E gas using the semi-classical **r**,**k**-space description we have presented so far; we shall largely follow the treatment given by Tolman.[15] The collisions are assumed to be two-particle interactions, the test particles having a distribution $f(\mathbf{r},\mathbf{k},t)$ and the field particles having a distribution $F(\mathbf{r},\mathbf{K},t)$. One-body collisions are included, if the field particles are near-fixed obstacles, so that $\sum_{\mathbf{K}} F(1+\varepsilon_2 F)..\to \hat{N}$. For a proof involving collisions with quasi-particles, see a study by the author.[16]

We will write down the coupled Boltzmann equations for $f(\mathbf{r},\mathbf{k},t)$ and $F(\mathbf{r},\mathbf{K},t)$, employing the notation $f' = f(\mathbf{r},\mathbf{k}',t)$ and $F' = F(\mathbf{r},\mathbf{K}',t)$, etc. The transition probability $Q(\mathbf{kK},\mathbf{k}'\mathbf{K}')$ and its reverse will be denoted by Q_{if} (initial and final state); further $\int Z(\mathbf{k})d^3k$ will be denoted by $\int \Delta^3 k$ and likewise $\int Z(\mathbf{K})d^3K$ will be denoted by $\int \Delta^3 K$. Thus, Eqs. (13.3-15), (13.3-16) and (13.3-20) yield:

$$\begin{aligned}
&\frac{\partial f}{\partial t} + \mathbf{v}_{\mathbf{k}}\cdot\nabla_{\mathbf{r}} f + (\mathbf{F}_1/\hbar)\cdot\nabla_{\mathbf{k}} f = \iiint \Delta^3 k' \Delta^3 K \Delta^3 K' Q_{if} \\
&\qquad \times[f'F' - fF + \varepsilon_1 ff'(F'-F) + \varepsilon_2 FF'(f'-f)], \\
&\frac{\partial F}{\partial t} + \mathbf{v}_{\mathbf{K}}\cdot\nabla_{\mathbf{r}} F + (\mathbf{F}_2/\hbar)\cdot\nabla_{\mathbf{K}} F = \iiint \Delta^3 K' \Delta^3 k \Delta^3 k' Q_{if} \\
&\qquad \times[f'F' - fF + \varepsilon_1 ff'(F'-F) + \varepsilon_2 FF'(f'-f)].
\end{aligned} \tag{13.5-1}$$

Next, we introduce an *h*-function for each gas. Let

$$h_f \equiv \int \Delta^3 k \left[f \ln f - \varepsilon_1 (1+\varepsilon_1 f)\ln(1+\varepsilon_1 f) - f(1-\varepsilon_1^2) \right] \tag{13.5-2}$$

and likewise

$$h_F \equiv \int \Delta^3 K \left[F \ln F - \varepsilon_2 (1+\varepsilon_2 F)\ln(1+\varepsilon_2 F) - F(1-\varepsilon_2^2) \right]. \tag{13.5-3}$$

The reader will recognise that, when multiplied with $-k_B$, for $\varepsilon = 0$ these are the non-equilibrium extensions for the entropy of a Boltzmann gas [(Eq.(6.3-5)], while for $\varepsilon = \pm 1$ they are the results for the entropy of a quantum gas [see Problem 4.1(b)]. Let also $h = h_f + h_F$. Further we need the *h*-flux, denoted by ה (Hebrew *he*):

$$ה_f = \int \Delta^3 k \mathbf{v}_{\mathbf{k}} \left[f \ln f - \varepsilon_1 (1+\varepsilon_1 f)\ln(1+\varepsilon_1 f) - f(1-\varepsilon_1^2) \right], \tag{13.5-4a}$$

$$ה_F = \int \Delta^3 K \mathbf{v}_{\mathbf{K}} \left[F \ln F - \varepsilon_2 (1+\varepsilon_2 F)\ln(1+\varepsilon_2 F) - F(1-\varepsilon_2^2) \right]. \tag{13.5-4b}$$

---

[14] L.E. Reichl, "A Modern Course in Statistical Physics", 2nd Ed., John Wiley, NY, 1998 and Wiley-VCH Verlag, Weinheim, 2004, p. 180-182; K. Huang, "Statistical Mechanics", 2nd Ed., John Wiley, NY, 2000, p. 74 and p. 85-89.

[15] R.C. Tolman, "The Principles of Statistical Mechanics", Oxford Univ. Press, 1938, p. 453 ff.

[16] K.M. Van Vliet, Phys. Stat. Solidi (b) **78**, 667 (1976), Section 3.

In addition, ה = $ה_f$ + $ה_F$. The first form of Boltzmann's H-theorem which we shall meet states that

$$\frac{\partial h}{\partial t} + \text{div}\, \text{ה} \equiv \eta(r,t) \leq 0; \tag{13.5-5}$$

here $-k_B\eta = \sigma$ is the entropy production in non-equilibrium thermodynamics.[17]

To prove this, we first differentiate (13.5-2) and we substitute for $\partial f / \partial t$ from the Boltzmann's transport equation:

$$\frac{\partial h_f}{\partial t} = -\int \Delta^3 k \left[\mathbf{v_k} \cdot \nabla_{\mathbf{r}} f + (\mathbf{F}_1 / \hbar) \cdot \nabla_{\mathbf{k}} f + \mathscr{B}_{\mathbf{k}} f \right] \left[\ln f - \varepsilon_1^2 \ln(1+\varepsilon_1 f) \right]. \tag{13.5-6}$$

The integral of $\nabla_{\mathbf{k}} f$ vanishes by Gauss' theorem if f is well behaved for large $|\mathbf{k}|$. The integral of $\mathbf{v_k} \cdot \nabla_{\mathbf{r}} f$ just yields $\nabla_{\mathbf{r}} \cdot$ ה. We thus obtain

$$\eta_f = \frac{\partial h_f}{\partial t} + \text{div}\, \text{ה} = -\int \Delta^3 k (\mathscr{B}_{\mathbf{k}} f) \left[\ln f - \varepsilon_1^2 \ln(1+\varepsilon_1 f) \right]. \tag{13.5-7}$$

A similar result applies for η_F, defined analogously. We now add the results for η_f and η_F and we substitute for the collision operators from (13.5-1). The result is the 'quadruple' integral for $\eta = \eta_f + \eta_F$:

$$\begin{aligned} \eta = \iiiint \Delta^3 k \Delta^3 k' \Delta^3 K \Delta^3 K' Q_{if} & \left[f'F' - fF + \varepsilon_1 f f'(F'-F) \right. \\ & \left. + \varepsilon_2 F F'(f'-f) \right] \left[\ln(fF) - \varepsilon_1^2 \ln(1+\varepsilon_1 f) - \varepsilon_2^2 \ln(1+\varepsilon_2 F) \right]. \end{aligned} \tag{13.5-8}$$

Next, we interchange $\mathbf{k}, \mathbf{K}$ and $\mathbf{k}', \mathbf{K}'$, noticing that $Q_{if} = Q_{fi}$ and we take half the sum of (13.5-8) and the new expression to obtain (notice the change in sign of the first [ ])

$$\begin{aligned} \eta = \frac{1}{2} \iiiint \Delta^3 k \Delta^3 k' \Delta^3 K \Delta^3 K' Q_{if} & \left[f'F' - fF + \varepsilon_1 f f'(F'-F) \right. \\ & \left. + \varepsilon_2 F F'(f'-f) \right] \left[\ln(fF) - \ln(f'F') - \varepsilon_1^2 \ln(1+\varepsilon_1 f) \right. \\ & \left. - \varepsilon_2^2 \ln(1+\varepsilon_2 F) + \varepsilon_1^2 \ln(1+\varepsilon_1 f') + \varepsilon_2^2 \ln(1+\varepsilon_2 F') \right]. \end{aligned} \tag{13.5-9}$$

(a) Consider first the Boltzmann case, $\varepsilon_1 = \varepsilon_2 = 0$. Then we have the simple result,

$$\eta = \frac{1}{2} \iiiint \Delta^3 k \ldots Q_{if} [f'F' - fF][\ln(fF) - \ln(f'F')] \leq 0, \tag{13.5-10}$$

since the integrand is always negative by Klein's lemma, $(a-b)(\ln a - \ln b) \geq 0$. Moreover, $\eta = 0$ *if and only if* the integrand is zero. Defining the equilibrium state as

---

[17] S.R. de Groot and P. Mazur, "Non-equilibrium Thermodynamics", North Holland, Amsterdam, 1962 p. 22. Thus, (13.5-5) also reads $\partial s / \partial t + \text{div}\mathbf{J}_s \equiv \sigma(\mathbf{r},t) \geq 0$.

the state with zero entropy production (see also below), we see that the equilibrium state requires

$$f_{eq}F_{eq} = f'_{eq}F'_{eq}\,. \tag{13.5-11}$$

(b) The case that $\varepsilon_1 = 0$ and $\varepsilon_2 \neq 0$ is left to the reader.

(c) When ε_1 and ε_2 are both non-zero, we have collisions between quantum particles. With $\varepsilon_1^2 = \varepsilon_2^2 = 1$ (13.5-9) yields with minor algebra

$$\begin{aligned}\eta = \frac{1}{2}\int\!\!\int\!\!\int\!\!\int \Delta^3 k ... Q_{if}\,&[(f'F' + \varepsilon_1 ff'F' + \varepsilon_2 FF'f' + \varepsilon_1\varepsilon_2 ff'FF')\\ &-(fF + \varepsilon_1 ff'F + \varepsilon_2 FF'f + \varepsilon_1\varepsilon_2 ff'FF')]\{\ln[fF(1+\varepsilon_1 f')(1+\varepsilon_2 F')]\\ &-\ln[f'F'(1+\varepsilon_1 f)(1+\varepsilon_2 F)]\}.\end{aligned} \tag{13.5-12}$$

The first square bracket is immediately rewritten in its original form, viz.,

$$\{[f'F'(1+\varepsilon_1 f)(1+\varepsilon_2 F) - fF(1+\varepsilon_1 f')(1+\varepsilon_2 F')]\}\,. \tag{13.5-13}$$

Thus, the integrand is again of the form $-(a-b)(\ln a - \ln b)$ and we have $\eta \leq 0$, as claimed. Moreover, $\eta = 0$ if and only if $a = b$, or, for the equilibrium distribution the expression of (13.5-13) is zero; hence,

$$f'_{eq}(1+\varepsilon_1 f_{eq})F'_{eq}(1+\varepsilon_2 F_{eq}) = f_{eq}(1+\varepsilon_1 f'_{eq})F_{eq}(1+\varepsilon_2 F'_{eq})\,. \tag{13.5-14}$$

If we multiply the lhs with Q_{fi} and the rhs with $Q_{if} = Q_{fi}$, we obtain *a fortiori,*

$$Q_{fi}f'_{eq}(1+\varepsilon_1 f_{eq})F'_{eq}(1+\varepsilon_2 F_{eq}) = Q_{if}f_{eq}(1+\varepsilon_1 f'_{eq})F_{eq}(1+\varepsilon_2 F'_{eq})\,. \tag{13.5-14'}$$

Let us now look at the original two-body collision term as contained in Eqs.(13.3-10) and (13.3-19) [or Eqs.(13.3-16) and (13.3-20)]. We then see from (13.5-14') above that in equilibrium, not just the collision integral is zero, but the collision *integrand* as well! This is a remarkable result, indicating that we have *"detailed balance"*: in an equilibrium state, each transition rate $\mathbf{kK} \to \mathbf{k'K'}$ is equal to the inverse transition rate $\mathbf{k'K'} \to \mathbf{kK}$. For a macroscopic analogue (obtainable by multiplying with the applicable density of microscopic states): the flux of all molecules in an equilibrium gas, going from point *A* of a room to point *B* of that room, is equal to the flux of molecules going from *B* to *A*. These far-reaching conclusions all follow from the first form of the H-theorem. We also conclude from the BTE, Eq. (13.3-15), that the collision term and the streaming part are *separately* zero in equilibrium. Lastly, thermal equilibrium is vastly different from a 'mere' steady state for which $\partial f/\partial t = 0$.

We now consider a second form of Boltzmann's H-theorem. Let the H-function be defined as $\mathrm{H} = \mathrm{H}_f + \mathrm{H}_F$, where

$$\mathrm{H}_f = \int d^3r\, h_f = \iint Z(\mathbf{k}) d^3k d^3r \left[f \ln f - \varepsilon_1 (1+\varepsilon_1 f) \ln(1+\varepsilon_1 f) - f(1-\varepsilon_1^2) \right] \quad (13.5\text{-}15)$$

with a similar definition for H_F. We now *restrict ourselves to a closed system.* Since in a closed system there can be no net flux leaving the volume of the gases under consideration, integration of (13.5-5) yields upon application of Gauss' theorem,

$$\frac{d\mathrm{H}}{dt} = \int d^3r \left[\frac{\partial h}{\partial t} \right] + \oint\!\!\oint dS \cdot \pi = \int d^3\eta(\mathbf{r},t) \leq 0, \quad (13.5\text{-}16)$$

whereby we noted that the surface integral is zero. The result $d\mathrm{H}/dt \leq 0$ is the usual form of Boltzmann's H-theorem. Since in equilibrium $d/dt = 0$, H reaches its minimum, being the equilibrium state, for which the entropy production is indeed zero. The previous form is more general, however, since it holds equally well for open systems. In fact, *any* thermodynamical statement involving a closed system is formal only since no measurement can be made without interaction to the outside. The entropy formulation of the first form (13.5-5), or the standard nonequilibrium form of footnote 17, is preferable over the "S_{max}-law" for closed systems found in most texts.

13.5.2 *Further Discussion of Boltzmann's H-theorem*

Historically the most well-known objections, stated during Boltzmann's lifetime, have been the "reversal paradox" (Loschmidt's Umkehreinwand, 1876)[18] and Zermelo's "recurrence paradox" (Wiederkehreinwand, 1896)[19]. We commence with the former.

Consider a gas approaching equilibrium and going through the molecular configurations (phases) $\Gamma_1, \Gamma_2, ..., \Gamma_m$. During this process H($t$) decreases. Now, at a certain time, when the system has a configuration Γ_i, we reverse all velocities (and the magnetic field). This does not affect the Hamiltonian, so it does not affect Boltzmann's H-function. However, the system will now pass through all states of motion in reverse order, $\Gamma_i, \Gamma_{i-1}, ..., \Gamma_1$. Thus, H($t$) will increase, contrary to Boltzmann's H-theorem!

This "paradox" has been the subject of extensive discussions, both in Boltzmann's time and *today.*[20,21] The basic point to be considered here is that Boltzmann's $-k_B\mathrm{H}(t)$ is not the most microscopic entropy function for a non-equilibrium state. For a

[18] J. Loschmidt, "Über den Zustand des Wärmegleichgewichtes eines Systems von Körpern mit Rucksicht auf die Schwerkraft", Wiener Berichte **73**, 128, 366 (1876); **75**, 287 (1877); **76**, 209 (1877).

[19] E. Zermelo, "Über einen Satz der Dynamik und die mechanische Wärmetheorie", Ann. der Physik [3] **57**, 485 (1896).

[20] J.L. Lebowitz, "Boltzmann's Entropy and Time's Arrow", Physics Today **46**, 32 (September 1993).

[21] J. Gollub and D. Pine, "Microscopic Irreversibility and Chaos", Physics Today **39,** 8 (August 2006).

classical gas ($\varepsilon = 0$) Eq. (13.5-2) amounts to $S = -k_B \int (f \ln f - f) d\varpi$, where $f = \langle n(\mathbf{r}, \mathbf{v}, t) \rangle$. Obviously, the most microscopic entropy should be based on $n(\mathbf{r}, \mathbf{v}, t)$ and not on the average distribution function, since the latter as a whole fluctuates, cf. Section 6.4. This Boltzmann entropy function *fluctuates*, in contrast to the Gibbs entropy function based on ensemble theory, cf. Section 3.8. Therefore, microscopically both increases and decreases occur, although on average the H-function decreases [or $S(t)$ increases], (see also Huang, Op. Cit, Fig. 4.7). While in some sense the microscopic reversibility is reflected by the presence of *noise* as described, it still is highly unlikely that a temporary reversal of velocities will bring the system back in the configuration Γ_1, as believed by Loschmidt. Boltzmann's argument was that the non-equilibrium entropy he envisioned was proportional to the logarithm of the probability W for a specified macrostate to occur; using ergodic theory, the latter, in turn, was proportional to the volume in phase space $\Delta\Omega$ occupied by the specification. Since no system can be entirely isolated, disturbances that increase this volume, such as the collapse of an internal wall or other constraints, are more abundant than temporary decreases of the phase-space volume; in short, a reversed sequence $\Gamma_i \to \Gamma_{i-1} \to \Gamma_{i-2}$ could well occur, but a macroscopic time reversed sequence $\Gamma_i \to ... \Gamma_1$ will not be observed: the system is stable with respect to the present but not with respect to the past.

The recurrence paradox, enunciated by Zermelo, requires less dramatic actions such as reversing (by some means) all molecular velocities; on the contrary, just leave the system to itself. Then, according to a theorem of classical mechanics by Poincaré, the system must, at last, return to its original phase. However, the time required for a Poincaré cycle grows exponentially with the number of degrees of freedom of the system under consideration. For a macroscopic system, typically $t \sim t_0 \exp(N)$; clearly, with $N \approx 10^{22}$ and whatever the value of t_0, this would by far exceed the age of the universe! This 'paradox' can therefore be discarded.

Does the above historical discussion now mean that 'all is well?' We quote Lebowitz: "Given that microscopic physical laws are reversible, why do all macroscopic events have a preferred time direction? *Boltzmann's thoughts on this question have withstood the test of time*" (italics mine). While we concur with his conclusion, and while we firmly believe that Boltzmann sufficiently refuted the two stated (and other) paradoxes, still the BTE and the concomitant H-theorem beg for a fully founded derivation, such as we have alluded to in the introductory section of this chapter. Basically, the origin of randomness must be introduced on the many-body level, in which the kinetic equations are linear and the test and field particles are treated on an equal footing. It will also be clear there that no theoretical treatment will lead to irreversibility *unless* perturbations (internal or with the surroundings) are considered, possibly to all orders; this will be examined in Chapters XVI and XVII.

13.6 The Equilibrium Solutions

13.6.1 *The Classical Gas*

We shall restrict ourselves to homogeneous systems. The functional equation (13.5-11) will be solved for the case that test and field particles are from the same gas (like-particles collisions).[22] Taking the logarithm,

$$\ln f_{eq}(\mathbf{v}_1) + \ln f_{eq}(\mathbf{v}_2) = \ln f_{eq}(\mathbf{v}_1') + \ln f_{eq}(\mathbf{v}_2'). \tag{13.6-1}$$

Thus, the logarithm of f_{eq} is conserved in a collision. According to classical mechanics, only a limited number of constants of motion are relevant; for spinless particles, these are the total energy and the total momentum. The expression $\ln f_{eq}$ must linearly depend on these; we are therefore led to

$$\ln f_{eq}(\mathbf{v}) = -A(\mathbf{v}-\mathbf{w})^2 + \ln C, \text{ or } f_{eq}(\mathbf{v}) = Ce^{-A(\mathbf{v}-\mathbf{w})^2}, \tag{13.6-2}$$

where A, C and $\mathbf{w}$ are constants. Requiring that $\langle \mathbf{v} \rangle = 0$, as for any random process in the absence of barycentric flow, we find that $\mathbf{w} = 0$. Further, from normalization, $\int f_{eq}(\mathbf{v}) d^3v = n(\mathbf{r}) = n$, where n is the particle density, we obtain $C = (A/\pi)^{3/2} n$; hence,

$$f_{eq}(\mathbf{v}) = n(m\beta/2\pi)^{3/2} e^{-\beta\varepsilon}, \tag{13.6-3}$$

which is the Boltzmann distribution whereby we set $\beta = 2A/m$. Obviously, the concept of temperature does not come in from a transport equation based on mechanical concepts; however, we have already remarked – and will show so in the next section – that $-k_B \mathrm{H}_{eq} = S_{eq}$; it then easily follows that $\beta = 1/k_B T$. After integrating over the spatial volume and the polar angles in $\mathbf{v}$-space, the above result gives

$$f_{eq}(v) = 4\pi N \left(\frac{m\beta}{2\pi}\right)^{3/2} v^2 e^{-\beta m v^2/2}, \tag{13.6-4}$$

which is the Maxwell distribution, cf. (1.6-74) and (6.5-1).

13.6.2 *Quantum Gases*

We start from (13.5-14); dividing by $f_{eq} f_{eq}' F_{eq} F_{eq}'$ this yields

$$\left(\frac{1}{f_{eq}} + \varepsilon_1\right)\left(\frac{1}{F_{eq}} + \varepsilon_2\right) = \left(\frac{1}{f_{eq}'} + \varepsilon_1\right)\left(\frac{1}{F_{eq}'} + \varepsilon_2\right). \tag{13.6-5}$$

---

[22] Boltzmann considered the full collision term, involving like-particles and non-alike particles for two interacting gases; Lectures on Gas Theory, Op. Cit, p. 51 Eq. (31).

We now limit ourselves to like-particles, i.e., $f_{eq} = F_{eq}$ and $\varepsilon_1 = \varepsilon_2$. We then obtain, suppressing at this moment the position dependence **r**,

$$\left(\frac{1}{f_{eq}(\mathbf{k}_1)}+\varepsilon\right)\left(\frac{1}{f_{eq}(\mathbf{k}_2)}+\varepsilon\right)=\left(\frac{1}{f_{eq}(\mathbf{k}_1')}+\varepsilon\right)\left(\frac{1}{f_{eq}(\mathbf{k}_2')}+\varepsilon\right). \tag{13.6-6}$$

Clearly the logarithm of the term in each bracket is conserved. For a quantum gas only the Hamiltonian is a constant of motion. Denoting the one particle energy as $\varepsilon_{\mathbf{k}}$, we set

$$\ln\left(\frac{1}{f_{eq}(\mathbf{k})}+\varepsilon\right)=\beta(\varepsilon_{\mathbf{k}}-\varsigma),\quad \text{or}\quad f_{eq}(\mathbf{k})=\frac{1}{e^{\beta(\varepsilon_{\mathbf{k}}-\varsigma)}-\varepsilon}, \tag{13.6-7}$$

where β and ς are still to be determined; from the choice of symbols the reader will surmise that they will turn out to be $1/k_BT$ and the *chemical potential*, respectively. With $\varepsilon = \pm 1$, the B–E distribution (upper sign) and F–D distribution (lower sign) have been recovered.

In the general case there is also streaming, so the dependence on **r** must be taken into account. In equilibrium the streaming part of the BTE is independently zero, as we noted in the previous section. The parameter ς could therefore depend on **r**; since β will be linked to the temperature, it will remain a constant for a true equilibrium state. Formally, the full distribution therefore reads

$$f_{eq}(\mathbf{k},\mathbf{r})=\frac{1}{e^{\beta[\varepsilon_{\mathbf{k}}-\varsigma(\mathbf{r})]}-\varepsilon}. \tag{13.6-8}$$

The absence of streaming in thermal equilibrium requires

$$\nabla_{\mathbf{k}}\varepsilon_{\mathbf{k}}\cdot\nabla_{\mathbf{r}}f_{eq}+q[-\nabla_{\mathbf{r}}\Phi(\mathbf{r})+(\mathbf{v}_{\mathbf{k}}/c)\times\mathbf{B}(\mathbf{r})]\cdot\nabla_{\mathbf{k}}\varepsilon_{\mathbf{k}}(\partial f_{eq}/\partial\varepsilon_{\mathbf{k}})=0, \tag{13.6-9}$$

where we used $\hbar\mathbf{v}_{\mathbf{k}}=\nabla_{\mathbf{k}}\varepsilon_{\mathbf{k}}$. Interchanging the scalar and vector products in the last expression indicates that the magnetic force does not contribute. Straightforward algebra, using (13.4-14) and $\partial/\partial\varsigma=-\partial/\partial\varepsilon_{\mathbf{k}}$, yields

$$\nabla_{\mathbf{k}}\varepsilon_{\mathbf{k}}\cdot[\beta\nabla_{\mathbf{r}}\varsigma(\mathbf{r})f_{eq}(1+\varepsilon f_{eq})+q\nabla_{\mathbf{r}}\Phi(\mathbf{r})\beta f_{eq}(1+\varepsilon f_{eq})]=0, \tag{13.6-10}$$

which results in $\nabla_{\mathbf{r}}\varsigma+q\nabla_{\mathbf{r}}\Phi(\mathbf{r})=0$. This, in turn, integrates to

$$\varsigma(\mathbf{r})+q\Phi(\mathbf{r})=\mu, \tag{13.6-11}$$

where μ is a new constant. Comparing with Eq.(1.3-7) of Chapter I, the new constant is nothing but the *electrochemical potential*. The distribution (13.6-8) thereby becomes

$$f_{eq}(\mathbf{k},\mathbf{r})=\frac{1}{e^{\beta[\varepsilon_{\mathbf{k}}+q\Phi(\mathbf{r})-\mu]}-\varepsilon}. \tag{13.6-12}$$

Let us now consider an electron gas in a solid. For that case $\varepsilon = -1$ and $q = -e$.[23] The electrochemical potential then equals the Fermi level. Further, in the spirit of the Wannier Hamiltonian, one introduces the *total energy*, being the sum of the band energy and the superimposed potential energy. Hence, $\varepsilon_{tot}(\mathbf{k},\mathbf{r}) = \varepsilon_{\mathbf{k}} - e\Phi(\mathbf{r})$. We now have two equivalent expressions for the electron occupancy in an energy band of a solid:

a) : $$f_{eq}(\mathbf{k},\mathbf{r}) = \frac{1}{e^{\beta[\varepsilon_{\mathbf{k}} - \varsigma(\mathbf{r})]} + 1}; \qquad (13.6\text{-}13)$$

b) : $$f_{eq}(\mathbf{k},\mathbf{r}) = \frac{1}{e^{\beta[\varepsilon_{tot}(\mathbf{k},\mathbf{r}) - \mu]} + 1}. \qquad (13.6\text{-}14)$$

We note: For an electron gas, *either* use the band energy in conjunction with the chemical potential, *or* the total energy in conjunction with the electrochemical potential or Fermi level. Needless to say that this rule is 'conveniently forgotten' in many a text! As mitigating circumstances, in a metal we often have $|e\Phi| \ll \varepsilon_{\mathbf{k}}$, so that in practice $\varepsilon_{tot} \approx \varepsilon_{\mathbf{k}}$. This is, however, not true in nondegenerate semiconductors!

13.7 The Equilibrium Entropy

The h-function was introduced in (13.5-2). Reverting to a notation involving a sum over **k**-space and with $f_{eq} \equiv f(\mathbf{k})\big|_{eq}$, we have

$$\begin{aligned} h_{eq} &= \sum_{\mathbf{k}} \left[f_{eq} \ln f_{eq} - \varepsilon(1+\varepsilon f_{eq})\ln(1+\varepsilon f_{eq}) - f_{eq}(1-\varepsilon^2) \right] \\ &= \sum_{\mathbf{k}} \left\{ -f_{eq} \ln[1+\varepsilon f_{eq})^{\varepsilon^2} / f_{eq}] - \varepsilon \ln(1+\varepsilon f_{eq}) - f_{eq}(1-\varepsilon^2) \right\}. \end{aligned} \qquad (13.7\text{-}1)$$

For the quantum cases we have

$$f_{eq} = \frac{1}{e^{\beta(\varepsilon_{\mathbf{k}} - \varsigma)} - \varepsilon}, \qquad 1+\varepsilon f_{eq} = \frac{e^{\beta(\varepsilon_{\mathbf{k}} - \varsigma)}}{e^{\beta(\varepsilon_{\mathbf{k}} - \varsigma)} - \varepsilon}, \qquad (13.7\text{-}2)$$

from which

$$(1+\varepsilon f_{eq})^{\varepsilon^2} / f_{eq} = \exp[\beta(\varepsilon_{\mathbf{k}} - \varsigma)]. \qquad (13.7\text{-}3)$$

For the Boltzmann gas we obtain the same result, since

$$f_{eq}^{Bo}(\mathbf{k}) = n\left(\frac{m\beta}{2\pi}\right)^{3/2} e^{-\beta\varepsilon_{\mathbf{k}}} \frac{\partial(v_x,v_y,v_z)}{\partial(k_x,k_y,k_z)} 8\pi^3$$

---

[23] Even for holes in the valence band of a semiconductor, it is customary to consider the *electron* occupancy; thus, the above formulas still apply, with the replacement of ε_C by $-\varepsilon_V$ at appropriate places.

$$= n\left(2\pi\beta\hbar^2/m\right)^{3/2} e^{-\beta\varepsilon_{\mathbf{k}}} = e^{-\beta(\varepsilon_{\mathbf{k}}-\varsigma)}, \tag{13.7-4}$$

providing we define the chemical potential for a Boltzmann gas by[24]

$$\beta\varsigma = -\ln\left[\frac{1}{n(\mathbf{r})}\left(\frac{m}{2\pi\hbar^2\beta}\right)^{3/2}\right] - \ln g, \tag{13.7-5}$$

which agrees with (6.5-14), using the customary value for β as confirmed below. With (13.7-3) substituted into (13.7-1) we obtain for all cases

$$h_{eq} = \sum_{\mathbf{k}}\left\{f_{eq}\beta(\varsigma-\varepsilon_{\mathbf{k}}) + \varepsilon\ln\left[1-\varepsilon e^{\beta(\varsigma-\varepsilon_{\mathbf{k}})}\right] - f_{eq}(1-\varepsilon^2)\right\}. \tag{13.7-6}$$

Employing the normalization (13.3-1) this yields

$$h_{eq} = \left\{\beta\varsigma n(\mathbf{r}) - \beta u(\mathbf{r}) - n(\mathbf{r})(1-\varepsilon^2)\right\}\Delta^3 r + \varepsilon\sum_{\mathbf{k}}\ln\left[1-\varepsilon e^{\beta(\varsigma-\varepsilon_{\mathbf{k}})}\right], \tag{13.7-7}$$

where $u(\mathbf{r})$ is the energy density. For Boltzmann's H-function this gives[25] by summing over all elements $\Delta^3 r_i$,

$$\begin{aligned}
\mathrm{H}_{eq} &= \sum_i\left\{\beta\varsigma n(\mathbf{r}_i) - \beta u(\mathbf{r}_i) - n(\mathbf{r}_i)(1-\varepsilon^2)\right\}\Delta^3 r_i + \varepsilon\sum_{\mathbf{k},i}\ln\left[1-\varepsilon e^{\beta(\varsigma-\varepsilon_{\mathbf{k}})}\right] \\
&\simeq \int d^3r\left\{\beta\varsigma n(\mathbf{r}) - \beta u(\mathbf{r}) - n(\mathbf{r})(1-\varepsilon^2)\right\} + \varepsilon\sum_{\mathbf{k},i}\ln\left[1-\varepsilon e^{\beta(\varsigma-\varepsilon_{\mathbf{k}})}\right] \\
&= -\beta\mathcal{E} - N(1-\varepsilon^2-\beta\varsigma) + \varepsilon\sum_{\mathbf{k}\in\Omega_{\mathbf{k}}}\ln\left[1-\varepsilon e^{\beta(\varsigma-\varepsilon_{\mathbf{k}})}\right],
\end{aligned} \tag{13.7-8}$$

where $\Omega_{\mathbf{k}}$ is the first Brillouin zone for a sample of volume $V = V_0$.

Let us now introduce a quantity $\tilde{S}$ by $\tilde{S} = -k_B\mathrm{H}_{eq}$. For a classical gas we have with $\varepsilon = 0$,

$$\tilde{S} = k_B\beta\mathcal{E} + k_B N - k_B\beta\varsigma N. \tag{13.7-9}$$

This is consistent with the homogeneous form for the equilibrium entropy, $\tilde{S} \to S^0$,

$$TS^0 = \mathcal{E} + PV - \varsigma N = \mathcal{E} + Nk_BT - \varsigma N, \tag{13.7-10}$$

iff $\beta = 1/k_BT$ and the 'zeta' that was introduced in the equilibrium solution is indeed the chemical potential ς. Next, we consider a quantum gas with $\varepsilon^2 = 1$. We then have

---

[24] Here g is the spin degeneracy (2 for a free electron gas). For the Boltzmann gas considered here, g =1.

[25] We suppose that the system is homogeneous, so that ς is constant. If this is not the case, we should replace $\varepsilon_{\mathbf{k}}$ by ε_{tot} and use that μ is constant in the integration over d^3r; the extension is trivial.

$$\tilde{S} = k_B \beta \mathscr{E} - k_B \beta \varsigma N - \varepsilon k_B \sum_{\mathbf{k} \in \Omega_{\mathbf{k}}} \ln\left[1 - \varepsilon e^{\beta(\varsigma - \varepsilon_{\mathbf{k}})}\right]. \tag{13.7-11}$$

This is to be compared with the homogeneous form for the equilibrium entropy,

$$TS^0 = \mathscr{E} - \varsigma N - \Omega = \mathscr{E} - \varsigma N - k_B T \varepsilon \sum_{\mathbf{k} \in \Omega_{\mathbf{k}}} \ln\left[1 - \varepsilon e^{(\varsigma - \varepsilon_{\mathbf{k}})/k_B T}\right], \tag{13.7-12}$$

based on the form (4.3-18) for the grand potential of a B–E or F–D gas. Clearly, full equivalence of $\tilde{S}$ with S^0 is again obtained iff $\beta = 1/k_B T$ and the 'zeta' that we introduced is the chemical potential ς.

In other books the equivalence of $-k_B \mathrm{H}_{eq}$ and S^0 is often shown by variation of the parameters $\mathscr{E}$, N and V. This is a bit tricky, however, since the variation of V only shows up in the sum over the **k**-states via the density of states $Z(\mathbf{k})$; see, e.g. Sommerfeld.[26]

[26] A. Sommerfeld, "Thermodynamics and Statistical Mechanics" (Vol. V of "Lectures on Theoretical Physics"), Acad. Press, NY , 1956, p. 221ff.

13.8 Problems to Chapter 13

13.1 Charge carriers in the conduction band [or valence band for p-type materials], placed in an electromagnetic field, are subject to the Wannier Hamiltonian (13.3-7). Show that $\dot{\mathbf{k}} = \mathbf{F}/\hbar$, where $\mathbf{F}$ is the Lorentz force.

13.2 Consider electron-phonon collisions with transition probabilities per second $Q(\mathbf{k},\mathbf{q} \to \mathbf{k}')$ and $Q(\mathbf{k} \to \mathbf{k}',\mathbf{q})$ for absorption and emission processes, respectively. The lattice remains in equilibrium (no phonon drag), so that the phonons still obey B–E statistics. Establish the relationship

$$w_{\mathbf{kk}'} = e^{(\varepsilon_{\mathbf{k}} - \varepsilon_{\mathbf{k}'})/k_B T} w_{\mathbf{k}'\mathbf{k}} \,. \tag{1}$$

13.3 A classical gas interacts with a quantum gas. Use the negative entropy production function $\eta(\mathbf{r},t)$ defined in (13.5-5) and show that $\eta(\mathbf{r},t) \le 0$ as well as $d\mathrm{H}/dt \le 0$.

13.4 Boltzmann's H-function for a closed system depends on the energy $\mathcal{E}$, the volume V and the particle number N of the system. Assume that the system is not necessarily yet in equilibrium. By variation of the above thermodynamic variables, show that –H is proportional to the Boltzmann entropy function S, i.e., show that –H satisfies the Gibbs' relation of Section 1.3.

13.5 The relaxation time for electron-ionized impurity scattering can be shown to be given by $1/\tau = \hat{N} v_k \sigma_m$, where σ_m is the integrated momentum-loss cross section [see (13.4-35)] and $\hat{N}$ is the density of ionized impurities.

(a) Evaluate the integral, using a screened Coulomb potential and obtain the Brooks–Herring result:

$$\frac{1}{\tau(\varepsilon)} = A\left[\ln(1+2b) - \frac{2b}{1+2b}\right], \tag{2}$$

where

$$\begin{aligned} A &= 2\pi \hat{N} v_{\mathbf{k}}(\varepsilon) R^2 = \hat{N} Z^2 e^4 (2\varepsilon)^{-3/2} (m^*)^{-1/2} / 8\pi \varepsilon_d^{\,2}\,, \\ b &= 2k^2/\kappa^2 = 4m^* \varepsilon / \kappa^2 \hbar^2 \,. \end{aligned} \tag{3}$$

Here R^2 is the prefactor in the scattering formula (13.4-30), $\pm Ze$ is the charge of the ionized impurities and ε_d is the dielectric constant (not to be confused with the energy ε).

(b) In the Conwell–Weiszkopf procedure an unscreened potential is employed, but the impact parameter has a cut-off θ_m due to the fact that the impact parameter b cannot exceed half the distance r_m between two impurity ions. Show that $\sin(\theta_m/2) = 1/\lambda$, where λ is the eccentricity of the hyperbola. Also obtain $\cot(\theta_m/2) = r_m/R$. [$R = \overline{AB}$ in Fig. 13-5]. Integrating the momentum-loss cross section, establish that in this model

$$1/\tau(\varepsilon) = 2\pi\hat{N}v_{\mathbf{k}}(\varepsilon)R^2 \ln(1 + r_m^2/R^2). \tag{4}$$

(c) Which model would better describe the scattering process at (i) high impurity density, (ii) low impurity density.

13.6 In a metal the screening is associated with the exclusion principle. Consult a text in condensed matter physics and show that the screened part of the potential is given by the Thomas–Fermi expression (13.4-36).

Chapter XIV

Hydrodynamic Equations and Conservation Theorems; Barycentric Flow

14.1 Conservation Theorems

In this chapter we shall derive a number of conservation theorems, the phenomenological equations and the hydrodynamic equations. In the present section we commence the derivations in the **k**,**r** language, applicable to solids, but we soon go over to the **v**,**r** language, customary in plasma physics and hydrodynamics. The reader should be aware that all considerations in this chapter involve quasi-classical molecular arguments, based on "Boltzmann theory". Although we shall meet the Onsager–Casimir symmetry relations, no justification will be given here. It should be borne in mind that our entire Part D of non-equilibrium statistical mechanics needs a many-body quantum-theoretical foundation. It will also come as no surprise that some aspects of classical transport theory are *plainly wrong*, in particular with respect to galvanomagnetic phenomena and thermomagnetic phenomena. Many parts of such books as Ziman's "Electron's and Phonons", Nag's "Theory of Electrical Transport in Semiconductors", Harman and Honig's "Thermomagnetic Phenomena" (to mention just a few) can presently be *dismissed.*[1] Roughly, we can state that the **v**,**r** language results, as applied to plasmas and hydrodynamics, still are useful today. As to solids, the results of this chapter and the next one must be compared with the correct results based on the theory of quantum transport, found in later chapters. The demarcation line for transport theory in condensed matter was most clearly marked with the survey article by Kahn and Frederikse in Solid-State Physics, Vol. 9.[2] Unfortunately, a lot of erroneous and outdated theory persists even in 'modern' texts.

14.1.1 *Full Theorems*

We consider a number of species distinguished by the superscript 'i'. In classical systems this could refer to a mixture of gases or to plasmas with various kinds of

[1] J. Ziman, Op. Cit; B.R. Nag, "Theory of Electrical Transport in Semiconductors", Pergamon Press, Oxford (1972); T.C. Harman and J.M. Honig, "Thermoelectric and ... Effects", McGraw Hill NY, 1967.
[2] A.H. Kahn and H.P.R. Frederikse, "Oscillatory Behaviour of Magnetic Susceptibility and Electronic Conductivity" in Solid State Physics (F. Seitz and D. Turnbull, Eds.) **9**, 257-293 (1959).

ions. In solids, we might deal with electrons in various bands (conduction band, valence band, impurity bands).

We now have a Boltzmann equation for each species. The collision term will be written as $-\mathscr{B}_{\mathbf{k}^i} f^i(\mathbf{r},\mathbf{k}^i,t)$. It should be understood that there are collisions between like particles, involving four factors f^i as we saw before, and between unlike particles, involving two factors f^i and two factors f^j. In solids the former denote intraband scattering and the latter interband scattering. [In principle, we could also include electron-phonon collisions, if the phonons are one of the 'species'.] There is, however, in this section no need to specify the details of the collision terms. In particular, the results apply equally well to gases or plasmas, for which the substitution $\mathbf{k}^i \to \mathbf{v}^i$ leads to more conventional results.

We multiply each term in the Boltzmann equation for $f^i(\mathbf{r},\mathbf{k}^i,t)$ with $\Psi(\mathbf{r},\mathbf{k}^i)$ and integrate over $\Delta^3 k^i$. We then obtain

$$\int \Delta^3 k^i \Psi(\mathbf{r},\mathbf{k}^i)\left[\frac{\partial}{\partial t} + \mathbf{v}_{\mathbf{k}^i}\cdot\nabla_{\mathbf{r}} + (\mathbf{F}^i/\hbar)\cdot\nabla_{\mathbf{k}^i}\right] f^i(\mathbf{r},\mathbf{k}^i,t)$$
$$= -\int \Delta^3 k^i \Psi(\mathbf{r},\mathbf{k}^i)\mathscr{B}_{\mathbf{k}^i} f^i(\mathbf{r},\mathbf{k}^i,t). \tag{14.1-1}$$

This can be written in the form

$$\frac{\partial}{\partial t}\int \Delta^3 k^i \Psi f^i + \nabla_{\mathbf{r}}\cdot\int \Delta^3 k^i \Psi \mathbf{v}_{\mathbf{k}^i} f^i - \int \Delta^3 k^i (\nabla_{\mathbf{r}}\Psi)\cdot\mathbf{v}_{\mathbf{k}^i} f^i + \hbar^{-1}\int \Delta^3 k^i \nabla_{\mathbf{k}^i}\cdot(\Psi \mathbf{F}^i f^i)$$
$$-\hbar^{-1}\int \Delta^3 k^i (\nabla_{\mathbf{k}^i}\Psi)\cdot\mathbf{F}^i f^i - \hbar^{-1}\int \Delta^3 k^i (\nabla_{\mathbf{k}^i}\cdot\mathbf{F}^i)\Psi f^i = -\int \Delta^3 k^i \Psi \mathscr{B}_{\mathbf{k}^i} f^i. \tag{14.1-2}$$

The fourth term vanishes by Gauss' theorem if f^i goes to zero at large $\mathbf{k}$. The sixth term also vanishes if $\mathbf{F}$ is the Lorentz force [for the magnetic part use that $\nabla_{\mathbf{k}} \times \mathbf{v}_{\mathbf{k}} = \text{curl}_{\mathbf{k}}\ \text{grad}_{\mathbf{k}}(\varepsilon_{\mathbf{k}}/\hbar) = 0$].

Let now $\Psi(\mathbf{r},\mathbf{k}^i)$ be a *collisional invariant* i.e., a quantity that is conserved in a collision. Then the rhs of (14.1-2) vanishes when summed over all species. The proof is simple. For two-body collisions between species i and j (either equal or different indices) we have

$$\sum_i \int \Delta^3 k^i \Psi(\mathbf{r},\mathbf{k}^i)\mathscr{B}_{\mathbf{k}^i} f^i(\mathbf{r},\mathbf{k}^i,t) = \sum_{ij}\iiiint \Delta^3 k^i \Delta^3 k_1^i \Delta^3 k^j \Delta^3 k_1^j \Psi(\mathbf{r},\mathbf{k}^i)$$
$$\times Q(\mathbf{k}^i\mathbf{k}^j,\mathbf{k}_1^i\mathbf{k}_1^j)\left[f^i f^j - f_1^i f_1^j + \varepsilon^i f^i f_1^i (f^j - f_1^j) + \varepsilon^j f^j f_1^j (f^i - f_1^i)\right]. \tag{14.1-3}$$

We now interchange $\mathbf{k}^i$ and $\mathbf{k}^j$ with $\mathbf{k}_1^i$ and $\mathbf{k}_1^j$, noticing that Q is reciprocal and we take half the sum of the original and the changed expression. Next, we interchange the superscripts i and j and we take again half the sum. The result is

$$\sum_i \int \Delta^3 k^i \Psi(\mathbf{r},\mathbf{k}^i) \mathcal{R}_{\mathbf{k}^i} f^i(\mathbf{r},\mathbf{k}^i,t) = \frac{1}{4} \sum_{ij} \iiiint \Delta^3 k^i \Delta^3 k_1^i \Delta^3 k^j \Delta^3 k_1^j Q_{if}$$
$$\times \left[\Psi(\mathbf{r},\mathbf{k}^i) + \Psi(\mathbf{r},\mathbf{k}^j) - \Psi(\mathbf{r},\mathbf{k}_1^i) - \Psi(\mathbf{r},\mathbf{k}_1^j) \right] [......], \tag{14.1-4}$$

where the last bracket is the same as in (14.1-3); it is non-zero outside equilibrium. Since Ψ is a collisional invariant, the factor involving the four Ψ's must vanish; this completes the proof.

We now introduce averages for a given species (band or molecular averages):

$$\overline{\overline{F(\mathbf{r},t)}} = \frac{1}{n^i(\mathbf{r},t)\Delta^3 r} \int \Delta^3 k^i F(\mathbf{r},\mathbf{k}^i) f^i(\mathbf{r},\mathbf{k}^i,t), \tag{14.1-5}$$

$$J_F(\mathbf{r},t) = \frac{1}{\Delta^3 r} \int \Delta^3 k^i \mathbf{v}_{\mathbf{k}^i} F(\mathbf{r},\mathbf{k}^i) f^i(\mathbf{r},\mathbf{k}^i,t). \tag{14.1-6}$$

Note that for $F = 1$ (14.1-5) is just the normalization integral (13.3-1). We thus obtain from (14.1-2) and the vanishing of the collision sum

$$\sum_i \left\{ \frac{\partial}{\partial t} \left[n^i(\mathbf{r},t) \overline{\overline{\Psi(\mathbf{r},t)}} \right] + \nabla_{\mathbf{r}} \cdot \mathbf{J}_\Psi(\mathbf{r},t) \right.$$
$$\left. - n^i(\mathbf{r},t) \overline{\overline{\mathbf{v}_{\mathbf{k}^i} \cdot \nabla_{\mathbf{r}} \Psi(\mathbf{r},t)}} - n^i(\mathbf{r},t) \overline{\overline{(\mathbf{F}^i/\hbar) \cdot \nabla_{\mathbf{k}^i} \Psi(\mathbf{r},t)}} \right\} = 0. \tag{14.1-7}$$

This is the *Maxwell–Boltzmann conservation theorem.* We shall meet a number of invariants; for some invariants, such as mass and charge, the sum over all species can be omitted. We consider these, as well as momentum and mass conservation, in the four categories below.

(a) *Mass conservation.* Let $\Psi = m^i$. Note that $m^i n^i(\mathbf{r},t) = \rho^i(\mathbf{r},t)$ is the mass density. The average velocity is the drift velocity,

$$\mathbf{v}_d^i \equiv \overline{\overline{\mathbf{v}^i}} = \left[1/n^i(\mathbf{r},t)\Delta^3 r \right] \int \Delta^3 k^i \mathbf{v}_{\mathbf{k}^i} f^i(\mathbf{r},\mathbf{k}^i,t). \tag{14.1-8}$$

The M–B conservation theorem now yields the continuity equation

$$\frac{\partial \rho^i}{\partial t} + \mathrm{div}(\rho^i \mathbf{v}_d^i) = 0 \quad \text{or} \quad \frac{\partial \rho^i}{\partial t} + \mathrm{div}\mathbf{J}_p^i = 0, \tag{14.1-9}$$

where $\mathbf{J}_p^i$ is the particle flux.

(b) *Charge conservation.* The electrical current for species '*i*' is defined as $\mathbf{J}_e^i = \mathcal{q} n^i(\mathbf{r},t)\mathbf{v}_d^i$, with the charge density being $\rho_e^i = \mathcal{q} n^i(\mathbf{r},t)$. The continuity equation is then expected to be

$$\frac{\partial \rho_e^i}{\partial t} + \text{div}\mathbf{J}_e^i = 0. \tag{14.1-10}$$

The result is correct for metals. However, in semiconductors the charge density of *one band* is not conserved if there is generation and recombination of charge carriers. Rather than summing over i – when (14.1-10) is rigorously correct, see below – it is for this case better to separate off the collision terms involving generation and recombination (interband transitions) from the rhs of (14.1-1) and subsequent equations. Denoting by P^{ij} the number of transitions from band i (integrated over $\Delta^3 k^i$) to band j (integrated over $\Delta^3 k^j$), one easily finds that (14.1-10) is to be replaced by

$$\frac{\partial \rho_e^i}{\partial t} + \text{div}\mathbf{J}_e^i = \sum_j q\left[\langle P^{ji}\rangle - \langle P^{ij}\rangle\right]. \tag{14.1-11}$$

Summing over i one has the total balance

$$\sum_i \frac{\partial \rho_e^i}{\partial t} + \text{div}\mathbf{J}_e^i = \sum_{ij} q\left[\langle P^{ji}\rangle - \langle P^{ij}\rangle\right] = 0. \tag{14.1-12}$$

The rhs is clearly zero by interchange of i and j in one of the terms.

(c) *Momentum conservation.* For solids we put $\Psi = \hbar\mathbf{k}^i$ and we consider normal processes only (no umklapp). Then (14.1-7) yields

$$\sum_i \left\{ \frac{\partial}{\partial t}\left(n^i \overline{\overline{\hbar\mathbf{k}^i}} \right) + \nabla_{\mathbf{r}} \cdot \left(n^i \overline{\overline{\nabla_{\mathbf{k}^i}\varepsilon_{\mathbf{k}^i}\mathbf{k}^i}} \right) \right\} = \sum_i n^i \overline{\overline{\mathbf{F}^i}}. \tag{14.1-13}$$

For spherical energy surfaces, $\varepsilon_{\mathbf{k}^i} = \hbar^2 k^{i2}/2m^{*i}$, this can be written as[3]

$$\sum_i \left\{ \frac{\partial(\rho^i \mathbf{v}_d^i)}{\partial t} + \text{Div}\left[\rho^i \overline{\overline{\mathbf{v}_{\mathbf{k}^i}\mathbf{v}_{\mathbf{k}^i}}} \right] \right\} = \sum_i \rho^i \mathbf{F}_0^i / m^{*i}, \tag{14.1-14}$$

where the subscript zero on **F** denotes the band average (if **F** is velocity-dependent). There is no centre-of-mass motion for this case.[4]

For classical systems, involving a mixture of gases or ionized species like in a plasma, the same holds, whereby we omit the subscript '**k**' on **v** and use the free particle masses m^i; hence we have

[3] We use the notation $\nabla \cdot \mathrm{T} = \mathrm{Div}\,\mathrm{T}$ when T is a tensor and $\nabla\mathbf{a} = \mathrm{Grad}\,\mathbf{a}$, this quantity being a tensor.

[4] The complications of barycentric velocity are of no consequence for a solid, since in the entropy production one can replace **w** by one of the component velocities, see Ilya Prigogine, "Étude thermodynamique des phénomènes irréversibles", Desoer, Liège 1947. As a rule we choose for this the lattice, whose velocity can be set equal to zero. The lattice can further be omitted from the system.

$$\sum_i \left\{ \frac{\partial(\rho^i \mathbf{v}_d^i)}{\partial t} + \mathrm{Div}\left[\rho^i \overline{\overline{\mathbf{v}^i \mathbf{v}^i}} \right] \right\} = \sum_i \rho^i \mathbf{F}_0^i / m^i . \tag{14.1-15}$$

However, we must now introduce the statistical centre-of-mass velocity or *barycentric* velocity. Since $\dot{\mathbf{R}}\Sigma_i m^i = \Sigma_i m^i \dot{\mathbf{r}}^i$, averaging over all $\mathbf{v}^i$ gives

$$\rho \mathbf{w} = \sum\nolimits_i \rho^i \overline{\overline{\mathbf{v}^i}} = \sum\nolimits_i \rho^i \mathbf{v}_d^i , \tag{14.1-16}$$

where $\rho = \Sigma_i \rho^i$ is the total density. While the energy per particle of species i is $\varepsilon_i = \frac{1}{2} m^i v^{i2}$, the internal or 'random' energy will be defined as

$$u^i = \tfrac{1}{2} m^i (\mathbf{v}^i - \mathbf{w}) \cdot (\mathbf{v}^i - \mathbf{w}) \equiv \tfrac{1}{2} m^i (\mathbf{v}^i - \mathbf{w})^2 . \tag{14.1-17}$$

The total random energy per unit volume is given by $u = \Sigma_i \int d^3 v^i u^i f^i$. In addition, we introduce the kinematic pressure tensor P associated with momentum transfer at position $\mathbf{r}$ and the viscous pressure tensor Π, defined as

$$\mathsf{P} = \sum_i m^i \int (\mathbf{v}^i - \mathbf{w})(\mathbf{v}^i - \mathbf{w}) f^i d^3 v^i , \tag{14.1-18}$$

$$\Pi = \sum_i m^i \int \left[(\mathbf{v}^i - \mathbf{w})(\mathbf{v}^i - \mathbf{w}) - \tfrac{1}{3} (\mathbf{v}^i - \mathbf{w})^2 \mathsf{I} \right] f^i d^3 v^i , \tag{14.1-19}$$

where I is the unit tensor. Now with simple algebra

$$\mathbf{v}^i \mathbf{v}^i = (\mathbf{v}^i - \mathbf{w})(\mathbf{v}^i - \mathbf{w}) + \mathbf{v}^i \mathbf{w} + \mathbf{w} \mathbf{v}^i - \mathbf{w}\mathbf{w}, \tag{14.1-20}$$

so that

$$\begin{aligned} \sum_i \rho^i \overline{\overline{\mathbf{v}^i \mathbf{v}^i}} &= \sum_i m^i n^i \overline{\overline{\mathbf{v}^i \mathbf{v}^i}} = \sum_i m^i \int \mathbf{v}^i \mathbf{v}^i f^i d^3 v^i \\ &= \mathsf{P} + 2\rho \mathbf{w}\mathbf{w} - \rho \mathbf{w}\mathbf{w} = \mathsf{P} + \rho \mathbf{w}\mathbf{w} . \end{aligned} \tag{14.1-21}$$

Substituting into (14.1-15) and using also (14.1-16), we obtain

$$\frac{\partial(\rho \mathbf{w})}{\partial t} + \mathrm{Div}(\mathsf{P} + \rho \mathbf{w}\mathbf{w}) = \sum_i \rho^i (\mathbf{F}_0^i / m^i) . \tag{\textbf{14.1-22}}$$

This conservation theorem will be used later to obtain the Navier–Stokes equation.

(d) *Energy conservation.* First we shall consider the quantum case. Let $\Psi = \varepsilon_{\mathbf{k}^i}$. The energy density is given by $u = \sum_i (1/\Delta^3 r) \int \Delta^3 k_i f^i$.[5] The energy flux is correspondingly defined as

---

[5] Note that for an inhomogeneous electron gas as considered here, $Z(\mathbf{k}) = (2)\Delta^3 r / 8\pi^3$, where the factor (2) accounts for the spin when applicable; the arbitrary volume $\Delta^3 r$ therefore drops out.

$$\mathbf{J}_u = \sum_i \frac{1}{\Delta^3 r} \int \Delta^3 k^i \varepsilon_{\mathbf{k}^i} \mathbf{v}_{\mathbf{k}^i} f^i(\mathbf{r}, \mathbf{k}^i, t). \tag{14.1-23}$$

Lastly, the mass current density (or particle flux) is defined as

$$\mathbf{J}^i_p = \frac{1}{\Delta^3 r} \int \Delta^3 k^i m^{*i} \mathbf{v}_{\mathbf{k}^i} f^i(\mathbf{r}, \mathbf{k}^i, t). \tag{14.1-24}$$

It is now easily verified that (14.1-7) results in

$$\frac{\partial u(\mathbf{r},t)}{\partial t} + \operatorname{div} \mathbf{J}_u = \sum_i (\mathbf{F}^i / m^{*i}) \cdot \mathbf{J}^i_p . \tag{14.1-25}$$

This is the energy-balance equation. It can also be put in the form of a heat-balance equation. To that purpose the entropy current should be defined as – shown in more detail in the next chapter,

$$\mathbf{J}_s = \sum_i \frac{1}{\Delta^3 r} \int \Delta^3 k^i \frac{\varepsilon_{\mathbf{k}^i} - \varsigma^i(\mathbf{r})}{T(\mathbf{r})} \mathbf{v}_{\mathbf{k}^i} f^i(\mathbf{r}, \mathbf{k}^i, t), \tag{14.1-26}$$

where $\varsigma(\mathbf{r})$ is the local chemical potential and $T(\mathbf{r})$ is the local temperature. Clearly we have

$$\mathbf{J}_u = T\mathbf{J}_s + \sum_i (\varsigma^i / m^{*i}) \mathbf{J}^i_p . \tag{14.1-27}$$

This is substituted into (14.1-25). Noticing that the magnetic force does not contribute to the energy balance since $\mathbf{F}_{magn} \cdot \mathbf{J}^i_p = 0$, with $\mathbf{F} = -\nabla \mathcal{V}$ we easily arrive at

$$\frac{\partial u(\mathbf{r},t)}{\partial t} + \operatorname{div}(T\mathbf{J}_s) + \sum_i (\varsigma^i / m^{*i}) \operatorname{div} \mathbf{J}^i_p = -\sum_i \mathbf{J}^i_p \cdot \operatorname{grad}(\tilde{\mu}^i / m^{*i}), \tag{14.1-28}$$

where $\tilde{\mu}(\mathbf{r})$ is the 'generalized chemical potential', defined as $\tilde{\mu}(\mathbf{r}) = \varsigma(\mathbf{r}) + \mathcal{V}(\mathbf{r})$.

Turning next to classical systems with barycentric flow, we set ψ equal to the internal energy (14.1-17). The conservation theorem now yields

$$\sum_i \left\{ \frac{\partial}{\partial t}\left[n^i \overline{\overline{u^i}} \right] + \nabla_{\mathbf{r}} \cdot \left[n^i \overline{\overline{\mathbf{v}^i u^i}} \right] - \left[n^i \overline{\overline{\mathbf{v}^i \cdot \nabla_{\mathbf{r}} u^i}} \right] \right\} = \sum_i \mathbf{F}^i \cdot n^i \overline{\overline{(\mathbf{v}^i - \mathbf{w})}}. \tag{14.1-29}$$

The first term on the left is simply $\partial u / \partial t$. In the second term, replace $\mathbf{v}^i$ by $\mathbf{w} + (\mathbf{v}^i - \mathbf{w})$. For classical systems with barycentric flow the internal energy flux and the particle fluxes are defined as

$$\mathbf{J}_u = \tfrac{1}{2} \sum_i \int d^3 v^i m^i (\mathbf{v}^i - \mathbf{w})(\mathbf{v}^i - \mathbf{w})^2 f^i(\mathbf{r}, \mathbf{v}^i, t) = \tfrac{1}{2} \sum_i \rho^i \overline{\overline{(\mathbf{v}^i - \mathbf{w})(\mathbf{v}^i - \mathbf{w})^2}}, \tag{14.1-30}$$

$$\mathbf{J}_p^i = m^i \int d^3v(\mathbf{v}^i - \mathbf{w}) = \rho^i \overline{\overline{(\mathbf{v}^i - \mathbf{w})}}. \qquad (14.1\text{-}31)$$

It is customary for classical systems with flow $\mathbf{w}$ to refer to $\mathbf{J}_u$ as *heat flux* and to the $\mathbf{J}_p^i$ as *diffusion fluxes.* We will also do so here in order to not be at variance with the abundant literature on this subject; however, the subscripts 'u' and 'p' define these fluxes more clearly.[6] We substitute (14.1-30) into (14.1-29). The second term now yields $\text{div}(u\mathbf{w} + \mathbf{J}_u)$. The third term is written as

$$\sum_i \rho^i \overline{\overline{\mathbf{v}^i(\mathbf{v}^i - \mathbf{w})}} : \nabla_\mathbf{r}\mathbf{w} = \sum_i \rho^i \overline{\overline{(\mathbf{v}^i - \mathbf{w})(\mathbf{v}^i - \mathbf{w})}} : \nabla_\mathbf{r}\mathbf{w} = \mathsf{P} : \nabla_\mathbf{r}\mathbf{w}. \qquad (14.1\text{-}32)$$

Finally, the rhs of (14.1-29) can be written as $\Sigma_i(\mathbf{F}^i/m^i)\cdot\mathbf{J}_p^i$. Collecting terms, we have the energy balance

$$\frac{\partial u}{\partial t} + \text{div}(u\mathbf{w} + \mathbf{J}_u) + \mathsf{P} : \text{Grad}\,\mathbf{w} = \sum_i (\mathbf{F}^i/m^i)\cdot\mathbf{J}_p^i. \qquad \mathbf{(14.1\text{-}33)}$$

This result will be used to develop the heat-conduction equation.

14.1.2 *Zero-order or Eulerian Conservation Theorems*

The classical equilibrium distribution was obtained in subsection 13.6.1. Presently we have that $\langle \mathbf{v}^i \rangle = \mathbf{w} \neq 0$, so that $\mathbf{w}$ is maintained. Our philosophy is now that outside equilibrium we have to a first approximation still *local* equilibrium, i.e., the M–B form is still applicable, but the parameters β and $\mathbf{w}$ that were introduced become functions of position. Thus, we assume as *zero-order* solution to the Boltzmann transport equation the local displaced Maxwellian

$$f_0^i(\mathbf{r},\mathbf{v}^i) = n^i(\mathbf{r})\left[\frac{m^i}{2\pi k_B T(\mathbf{r})}\right]^{3/2} \exp\left\{-\frac{m^i[\mathbf{v}^i - \mathbf{w}(\mathbf{r})]^2}{2k_B T(\mathbf{r})}\right\}. \qquad (14.1\text{-}34)$$

We use once more the chemical potential for a classical gas given in (13.7-5). Then the above takes the form

$$f_0^i(\mathbf{r},\mathbf{v}^i) = \left(\frac{m^i}{2\pi\hbar}\right)^3 \exp\left\{-\frac{1}{k_B T(\mathbf{r})}\left(\varsigma(\mathbf{r}) - \frac{1}{2}m^i\left[\mathbf{v}^i - \mathbf{w}(\mathbf{r})\right]^2\right)\right\}. \qquad (14.1\text{-}35)$$

We shall go one step beyond this and allow for time dependence for the non-

[6] In fact, there is a Babylonian confusion of tongues when it comes to "heat flux", since any quantity that in the entropy production is conjugate to ∇T, $\nabla(1/T)$ or $(\nabla T)/T$ is called a heat flux. Our own preference is the "heat flux" $T\mathbf{J}_s$, but from time to time we will need these other fluxes. For more on this, cf. S.R. de Groot and P. Mazur, Op. Cit., Chapter XIII.

stationary case, $\partial f^i/\partial t \neq 0$. Then we assume that the solution of the BTE to zero-order is given by the zero-order *Enskog distribution*

$$f_0^i(\mathbf{r},\mathbf{v}^i,t)=\left(\frac{m^i}{2\pi\hbar}\right)^3 \exp\left\{-\frac{1}{k_BT(\mathbf{r},t)}\left(\varsigma(\mathbf{r},t)-\frac{1}{2}m^i\left[\mathbf{v}^i-\mathbf{w}(\mathbf{r},t)\right]^2\right)\right\}. \quad (14.1\text{-}36)$$

It is to be noted that $f_0^i(\mathbf{r},\mathbf{v}^i,t)$ is still an integral of the H-theorem and satisfies detailed balance; however, the streaming term $Df_0^i(\mathbf{r},\mathbf{v}^i,t)\neq 0$.

We now define averages in zero order. One finds directly

$$\overline{\overline{\mathbf{v}^i}}^{\,0}=\frac{1}{n^i(\mathbf{r},t)}\int d^3v^i\mathbf{v}^i f_0^i(\mathbf{r},\mathbf{v}^i,t)=\mathbf{w}(\mathbf{r},t), \quad (14.1\text{-}37)$$

as expected. Also we find

$$\mathbf{J}_{u,0}=\tfrac{1}{2}\overline{\overline{\rho^i(\mathbf{r},t)(\mathbf{v}^i-\mathbf{w})(\mathbf{v}^i-\mathbf{w})^2}}^{\,0}=0. \quad (14.1\text{-}38)$$

Likewise, for the diffusion fluxes we have $\mathbf{J}_{p,0}^i=0$. For the pressure tensor, with f^i replaced by f_0^i, the integral yields the hydrostatic pressure[7]

$$\mathsf{P}_0=P\mathsf{I} \text{ with } P=\sum_i n^i(\mathbf{r},t)k_BT(\mathbf{r},t). \quad (14.1\text{-}39)$$

By the same token, one can show that the two parts of the viscous pressure tensor yield opposite diagonal contributions; thus $\Pi_0=0$. Also, for the internal energy density we now have

$$u=\tfrac{3}{2}\sum\nolimits_i n^i(\mathbf{r},t)k_BT(\mathbf{r},t)=\tfrac{3}{2}P. \quad (14.1\text{-}40)$$

With these results the conservation theorems simplify to the zero-order or Eulerian conservation theorems. Eqs. (14.1-9), (14.1-22) and (14.1-33) yield[8]

$$\frac{\partial\rho^i}{\partial t}+\mathrm{div}(\rho^i\mathbf{w})=0, \quad (14.1\text{-}41)$$

$$\frac{\partial\mathbf{w}}{\partial t}+\mathbf{w}\cdot\mathrm{Grad}\,\mathbf{w}+\frac{1}{\rho}\mathrm{grad}\,P=\frac{1}{\rho}\sum_i\rho^i\left(\frac{\mathbf{F}^i}{m^i}\right), \quad (14.1\text{-}42)$$

$$\frac{\partial u}{\partial t}+\mathrm{div}(u\mathbf{w})+P\,\mathrm{div}\,\mathbf{w}=0. \quad (14.1\text{-}43)$$

---

[7] The integral follows from a general class of integrals computed in Eqs. (14.2-26) [with $g(s)$ = 1] through (14.2-29). Further we need the Gaussian integral $\int_0^\infty ds\, s^4\exp(-\alpha s^2)=(3/8)(\pi/\alpha^3)^{1/2}$.

[8] To obtain (14.1-42) we need both (14.1-22) and (14.1-41). Also, we used that $\mathrm{Div}(P\mathsf{I})=\mathrm{Grad}\,P$; similar operations apply to the other terms.

Further, we introduce the convected derivatives

$$\frac{d^i}{dt} = \frac{\partial}{\partial t} + \mathbf{v}^i \cdot \nabla_{\mathbf{r}} \,. \tag{14.1-44}$$

The Eulerian conservation theorems then take the forms

$$\frac{d^i \rho^i}{dt} = (\mathbf{v}^i - \mathbf{w}) \cdot \nabla_{\mathbf{r}} \rho^i - \rho^i \nabla_{\mathbf{r}} \cdot \mathbf{w} \,, \tag{14.1-45}$$

$$\frac{d^i \mathbf{w}}{dt} = (\mathbf{v}^i - \mathbf{w}) \cdot \nabla_{\mathbf{r}} \mathbf{w} - \frac{1}{\rho} \nabla_{\mathbf{r}} P + \frac{1}{\rho} \sum_i \rho^i \left(\frac{\mathbf{F}^i}{m^i} \right). \tag{14.1-46}$$

$$\frac{d^i u}{dt} = (\mathbf{v}^i - \mathbf{w}) \cdot \nabla_{\mathbf{r}} u - \underbrace{[P \nabla_{\mathbf{r}} \cdot \mathbf{w} + u \nabla_{\mathbf{r}} \cdot \mathbf{w}]}_{(5/3) u \nabla_{\mathbf{r}} \cdot \mathbf{w}} . \tag{14.1-47}$$

For the last equation we used (14.1-40). Further we have

$$\frac{d^i u}{dt} = \frac{3}{2} k_B \sum_i \left(\frac{\rho^i}{m^i} \right) \frac{dT}{dt} + \frac{3}{2} k_B T \sum_i \frac{1}{m^i} \frac{d^i \rho^i}{dt} \,, \tag{14.1-48}$$

$$\nabla_{\mathbf{r}} u = \frac{3}{2} k_B \sum_i \left(\frac{\rho^i}{m^i} \right) \nabla_{\mathbf{r}} T + \frac{3}{2} k_B T \sum_i \frac{1}{m^i} \nabla_{\mathbf{r}} \rho^i \,. \tag{14.1-49}$$

When this is substituted into (14.1-47) and use is made again of (14.1-45), the final result is

$$\frac{d^i T}{dt} = (\mathbf{v}^i - \mathbf{w}) \cdot \nabla_{\mathbf{r}} T - \frac{2}{3} T \nabla_{\mathbf{r}} \cdot \mathbf{w} \,. \tag{14.1-50}$$

Equations (14.1-45), (14.1-46) and (14.1-50) are the forms of the Eulerian conservation theorems we shall need in the computation of the phenomenological equations for the flow averages to first order of the Chapman–Enskog expansion in the next section.

14.2 The Phenomenological Equations in Classical Systems

14.2.1 *The Basis of the Flow Problem*

In thermal equilibrium we found that the distribution f_{eq} involved a number of constants, like $\tilde{\mu}$, β and $\mathbf{w}$. Outside equilibrium, but not too far from it, the distribution can be assumed to have the perturbation form, $f = f_0 + f_1$, where f_0 is the *local* equilibrium distribution in which the parameters μ, β and $\mathbf{w}$ become dependent on position $\mathbf{r}$, giving rise to gradients $\nabla\tilde{\mu}$, $\nabla\beta$ and $\nabla\mathbf{w}$. One can also include the time dependence for non-stationary problems, cf. the transition from (14.1-35) to (14.1-36). The functions f_0 and f_1 are then the zero-order and first-order Enskog distributions, respectively.

As for the fluxes $J_p(\mathbf{r},t)$, $J_u(\mathbf{r},t)$ and $J_s(\mathbf{r},t)$, etc., as well as for the flow tensors $\mathsf{P}(\mathbf{r},t)$ and $\Pi(\mathbf{r},t)$, we can expect that as a first approximation they will become proportional to the above-mentioned gradients, which henceforth will be called *generalized thermodynamic forces* $\psi_\alpha(\mathbf{r},t)$, while $\tilde{\mu}(\mathbf{r},t)$, $\beta(\mathbf{r},t)$ and $\mathbf{w}(\mathbf{r},t)$ [or properly chosen functions of these] will be denoted as *flow potentials.* The "phenomenological equations" will then most generally have the form

$$\mathbf{J}_\alpha(\mathbf{r},t) = \sum_\beta \int_0^{t+0} dt' \hat{\mathsf{L}}_{\alpha\beta}(\mathbf{r},t-t') \cdot \psi_\beta(\mathbf{r},t'). \tag{14.2-1}$$

The convolution form will automatically appear in linear response theory (LRT) as discussed in Part E.

In the present framework of Boltzmann theory, we assume that an adiabatic assumption $\hat{\mathsf{L}}(\mathbf{r},t) = \mathsf{L}(\mathbf{r},t)\delta(t-t')$ can be made; this involves that macroscopic time shall be larger than the characteristic times of collisions, typically of order 10^{-11} s in gases. In the present context this means that we may employ the zero-order Eulerian equations for the time dependence of $\partial\rho/\partial t$, $\partial u/\partial t$ and $\partial\mathbf{w}/\partial t$. We thus arrive at the phenomenological equations of macroscopic thermodynamics, in which memory effects of the convolution procedure have been erased,

$$J_\alpha(\mathbf{r},t) = \sum_\beta \mathsf{L}_{\alpha\beta}(\mathbf{r},t) \cdot \psi_\beta(\mathbf{r},t) \to \sum_\beta \mathsf{L}_{\alpha\beta}(\mathbf{r}) \cdot \psi_\beta(\mathbf{r},t), \tag{14.2-2}$$

where the final rhs holds in the (usual) quasi-stationary regime. The tensors L are generalized conductances. In a similar fashion, the pressure tensors $\mathsf{P}(\mathbf{r},t)$ and $\Pi(\mathbf{r},t)$ will become linearly related to the components of $\nabla\mathbf{w}$.

The choice of the flow potentials and their resultant thermodynamic forces is not unique, however, nor is the choice of the appropriate fluxes. For instance, instead of $\nabla\tilde{\mu}$, ∇T and $\nabla\mathbf{w}$, we could use other combinations, involving $\nabla\varsigma$, $\nabla(1/T)$, etc. As to the fluxes, we met already two energy fluxes, $\mathbf{J}_u$ and $T\mathbf{J}_s$, while others are possible. Thus, for (14.2-2) to be meaningful, we should make the right conjugation of fluxes and forces. If that is done, we will obtain the Onsager–Casimir *reciprocity relations,*

$$\mathsf{L}_{\alpha\beta}(\mathbf{B}) = \tilde{\mathsf{L}}_{\beta\alpha}(-\mathbf{B}), \tag{14.2-3}$$

where the tilde denotes the transpose tensor; for the relations to be valid, we must therefore transpose 'everything': the components, the current labels and the magnetic induction $\mathbf{B}$. A general proof will be given in the context of linear response theory.

Some authors have stressed the 'arbitrariness' involved in the conjugation choices. Truesdell maintains that no adequate foundation for the results (14.2-3) exists and that the reciprocity relations are a mere figment of the physicist's *desire* for symmetry.[9] We do not share his opinion. Basically, there are two procedures to obtain the proper conjugated variables. First, a macroscopic approach may be developed as in the first part of de Groot and Mazur, Op. Cit.; one then computes the entropy production of the entropy balance equation, $\sigma(\mathbf{r},t)$, which is a convex function, being quadratic in the fluxes or bilinear in the conjugated pairs, i.e. one obtains the form

$$\sigma(\mathbf{r},t) = \mathbf{J}_T \cdot \psi_T + \mathbf{J}_F \cdot \psi_F - T^{-1}\mathsf{\Pi} : \mathrm{Grad}\,\mathbf{w} - T^{-1}\sum_\ell J_\ell A_\ell\,, \qquad (14.2\text{-}4)$$

in which $\mathbf{J}_T$ is an appropriate form of the heat current with ψ_T being a specific functional of ∇T, $\mathbf{J}_F$ is an appropriate form of particle transport with ψ_F being the conjugated mechanical or electrical force stemming from $\nabla\tilde{\mu}$ or $\nabla\varsigma$, while the terms $J_\ell A_\ell$ are associated with chemical reactions. We note that the three parts, vectorial, tensorial and scalar, do not mix but contribute separately. Therefore, a computation of $\sigma(\mathbf{r},t)$ always reveals which variables are conjugated to each other and are to serve as forces and fluxes.

The second method, which we shall follow here, is more reliable; we will solve the Boltzmann equation for f_1 and compute the fluxes and the pressure tensors by integration over the distribution function. When properly done, linear relations between the **J**'s and ψ's, as well as between $\mathsf{\Pi}$ and $\nabla\mathbf{w}$ arise and the phenomenological equations (14.2-2) emerge.

Previously we found that there are no contributions to $\mathbf{J}^i_p$, $\mathbf{J}_s$ and $\mathsf{\Pi}$ in zero order. Therefore, up to first order of the Chapman-Enskog expansion, $f^i = f^i_0 + f^i_1$, we have

$$\mathbf{J}^i_p = m^i \int d^3v^i (\mathbf{v}^i - \mathbf{w}) f^i_1(\mathbf{r},\mathbf{v}^i,t)\,, \qquad (14.2\text{-}5)$$

$$\mathbf{J}_s = T^{-1}\sum_i \int d^3v^i (u^i - \varsigma^i)(\mathbf{v}^i - \mathbf{w}) f^i_1(\mathbf{r},\mathbf{v}^i,t)\,, \qquad (14.2\text{-}6)$$

$$\mathsf{P} = P\mathsf{I} + \mathsf{\Pi}, \quad \mathsf{\Pi} = \sum_i m^i \int d^3v^i (\mathbf{v}^i - \mathbf{w})(\mathbf{v}^i - \mathbf{w}) f^i(\mathbf{r},\mathbf{v}^i,t); \qquad (14.2\text{-}7)$$

for the last result to be correct, we shall show later that the second part of the viscous pressure tensor to first order vanishes,

$$\mathsf{I}\int d^3v^i (\mathbf{v}^i - \mathbf{w})^2 f^i_1 = 0\,. \qquad (14.2\text{-}7')$$

[9] C. Truesdell, "Rational Thermodynamics", McGraw Hill, NY,1969.

14.2.2 *Relaxation-Time Model*

The computation of the collision integral is the essential part in solving the Boltzmann transport equation. However, this is often done better numerically, e.g. by employing a Monte Carlo technique, than analytically. One way out is to postulate that the system relaxes to equilibrium with some velocity-dependent relaxation time $\tau(\mathbf{v})$, i.e.

$$-\mathcal{B}_{\mathbf{v}} f(\mathbf{r},\mathbf{v},t) \equiv \left(\frac{\partial f}{\partial t}\right)_{coll} = -\frac{(f-f_0)}{\tau(\mathbf{v})} = -\frac{f_1}{\tau(\mathbf{v})}. \tag{14.2-8}$$

Sometimes, such a form can be approximately justified.[10] Let us for the moment assume the validity of (14.2-8). We substitute $f = f_0 + f_1$ in the BTE; the streaming part becomes

$$Df = \frac{\partial f_0}{\partial t} + \frac{\partial f_1}{\partial t} + \mathbf{v}\cdot\nabla_{\mathbf{r}} f_0 + (\mathbf{F}/m)\cdot\nabla_{\mathbf{v}} f_0 + \mathbf{v}\cdot\nabla_{\mathbf{r}} f_1 + (\mathbf{F}/m)\cdot\nabla_{\mathbf{v}} f_1 \,. \tag{14.2-9}$$

We assume that $|\partial f_1/\partial t| \ll |\partial f_0/\partial t|$. As to the last two terms, close to equilibrium (small forces, small gradients) we assume them to be of second order; they will be omitted.[11] With the collision part given by (14.2-8), by equating the above two equations we now find,

$$f_1(\mathbf{r},\mathbf{v},t) = -\tau(\mathbf{v})\left(\frac{\partial}{\partial t} + \mathbf{v}\cdot\nabla_{\mathbf{r}} + (\mathbf{F}/m)\cdot\nabla_{\mathbf{v}}\right) f_0(\mathbf{r},\mathbf{v},t). \tag{14.2-10}$$

In most actual cases τ depends on $\mathbf{v}$ only through the energy ε_v so we set $\tau(\mathbf{v}) \to \tau(v)$. Physically $\tau(v)$ represents the time between collisions.

To elucidate such a solution we shall indicate that an *exact* relaxation time always exists when we are dealing with elastic one-body collisions. The following form then holds

$$\begin{aligned}-\mathcal{B}_{\mathbf{v}} f(\mathbf{r},\mathbf{v},t) &= \frac{N}{V_0}\int d\Omega\, v\,\sigma(\Omega)\left[f(\mathbf{r},\mathbf{v}',t) - f(\mathbf{r},\mathbf{v},t)\right] \\ &= \frac{N}{V_0}\int d\Omega\, v\,\sigma(\Omega)\left[f_1(\mathbf{r},\mathbf{v}',t) - f_1(\mathbf{r},\mathbf{v},t)\right].\end{aligned} \tag{14.2-11}$$

---

[10] The main requirement is that the collision operator is linear or can be linearized in *f*. Thus, since $\mathcal{B}_{\mathbf{v}} f_0 = 0$, we simply have $\tau(\mathbf{v}) = \mathcal{B}_{\mathbf{v}}^{-1}$. Then (14.2-10) reads more generally

$$f_1(\mathbf{r},\mathbf{v},t) = -\mathcal{B}_{\mathbf{v}}^{-1}\left[\left(\frac{\partial}{\partial t} + \mathbf{v}\cdot\nabla_{\mathbf{r}} + (\mathbf{F}/m)\cdot\nabla_{\mathbf{v}}\right) f_0\right]. \tag{14.2-10'}$$

[11] However, in the presence of a magnetic field, $(\mathbf{F}_{magn.}/m)\cdot\nabla_{\mathbf{v}} f_0$ vanishes and the second order term must be maintained (see next chapter). In this section we presume that $\mathbf{B} = 0$.

Now let us try the solution

$$f_1 = \mathbf{v}\cdot\mathbf{g}_1(\mathbf{r},\varepsilon_{\mathbf{v}})/v. \tag{14.2-12}$$

Let further χ be the angle between $\mathbf{g}_1$ and $\mathbf{v}$ while χ' be the angle between $\mathbf{g}_1$ and $\mathbf{v}'$. Then (14.2-11) gives

$$-\mathscr{R}_{\mathbf{v}} f(\mathbf{r},\mathbf{v},t) = \frac{N}{V_0}\int d\Omega\, v\,\sigma(\Omega)(\cos\chi' - \cos\chi)\, g_1(\mathbf{r},\varepsilon_{\mathbf{v}}). \tag{14.2-13}$$

Also, let θ be the angle between $\mathbf{v}$ and $\mathbf{v}'$ and ψ the corresponding azymuthal angle, see Fig. 14-1 below. The following relationship holds[12]

$$\cos\chi' = \cos\chi\cos\theta + \sin\chi\sin\theta\cos\varphi. \tag{14.2-15}$$

Substituting this into (14.2-13), we find

$$-\mathscr{R}_{\mathbf{v}} f = -(N/V_0)\int d\Omega\, v\,\sigma(\Omega)(1-\cos\theta)\cos\chi\, g_1(\mathbf{r},\varepsilon)$$
$$+(N/V_0)\int d\Omega\, v\,\sigma(\Omega)\sin\chi\sin\theta\cos\varphi\, g_1(\mathbf{r},\varepsilon); \tag{14.2-16}$$

here $d\Omega = \sin\theta d\theta d\varphi$. The second integral yields zero because of the integration over dφ. In the remaining part $\cos\chi\, g_1(\mathbf{r},\varepsilon) = f_1$ can be taken out of the integral.

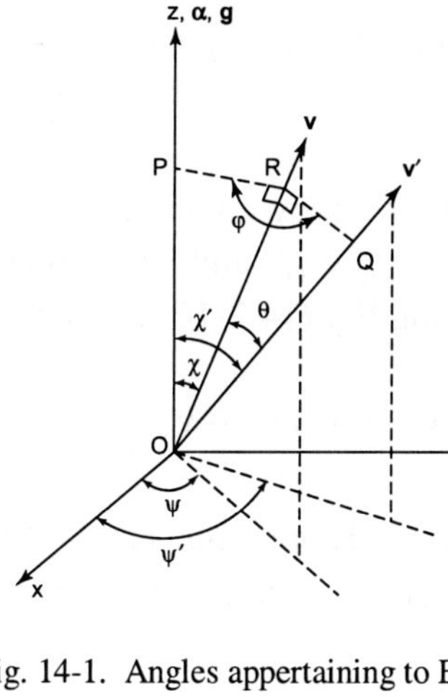

Fig. 14-1. Angles appertaining to Eq. (14.2-16).
$\theta = \angle$QOR, $\chi = \angle$POR, $\chi' = \angle$POQ, $\varphi = \angle$PRQ, $\angle$PRO = $\angle$QRO = 90 °.

---

[12] The general theorem reads (see e.g. H. Margenau and G. M. Murphy " The Mathematics of Physics and Chemistry" 2nd Ed., Van Nostrand, Princeton NJ, 1956, Section 3.7): If θ_1, φ_1 and θ_2, φ_2 denote, respectively, the polar and azymuthal angles of two lines passing through the origin, then the angle Θ between these lines is given by

$$\cos\Theta = \cos\theta_1\cos\theta_2 + \sin\theta_1\sin\theta_2\cos(\varphi_1 - \varphi_2). \tag{14.2-14}$$

For the present case, the polar axis of the theorem is the direction $\mathbf{v}$ and the z-axis and $\mathbf{v}'$ are the two lines.

Hence,

$$-\mathcal{R}_{\mathbf{v}} f(\mathbf{r},\mathbf{v},t) = -f_1(\mathbf{r},\varepsilon)/\tau(\varepsilon), \tag{14.2-17}$$

where

$$[1/\tau(\varepsilon)] = (N/V_0)\,v(\varepsilon)\int d\Omega\,\sigma(\Omega)\,(1-\cos\theta). \tag{14.2-18}$$

Clearly, there exists an *exact* relaxation time for this process! While $\tau(\varepsilon)$ is energy dependent, the mean free path $\lambda = v(\varepsilon)\tau(\varepsilon)$ is not if σ is not; in fact,

$$1/\lambda = [1/\tau(\varepsilon)] = \hat{N}\int d\Omega\,\sigma(\Omega)\,(1-\cos\theta). \tag{14.2-19}$$

We still note that for the stationary case, $\partial f_0/\partial t = 0$, with $\nabla_{\mathbf{v}} f_0 = (\partial f_0/\partial\varepsilon)m\mathbf{v}$, (14.2-10) yields

$$f_1 = -\tau(\varepsilon)\mathbf{v}\cdot[\nabla_{\mathbf{r}} f_0 + \mathbf{F}\,\partial f_0/\partial\varepsilon] \equiv -\tau(\varepsilon)\mathbf{v}\cdot\boldsymbol{\alpha}, \tag{14.2-20}$$

where

$$\boldsymbol{\alpha} = \nabla_{\mathbf{r}} f_0 + \mathbf{F}\,\partial f_0/\partial\varepsilon \tag{14.2-21}$$

is called the *streaming vector.* This vector plays an important role in the computation of transport coefficients in solids, as will be discussed in the next chapter.

14.2.3 *Computation of the Vector and Tensor Flow averages in Systems with Barycentric Flow*

We resume the consideration of many-component systems, each component having a distribution function $f^i = f_0^1 + f_1^i$ and we suppose that relaxation times $\tau^i(u^i) = \tau^i(|\mathbf{v}^i - \mathbf{w}|)$ exist. Applying Eq. (14.2-10) for each species, we have

$$f_1^i = -\tau^i\left(\frac{\partial}{\partial t} + \mathbf{v}\cdot\nabla_{\mathbf{r}} + (\mathbf{F}^i/m^1)\cdot\nabla_{\mathbf{v}}\right)f_0^i \equiv -\tau^i D^i f_0^i, \tag{14.2-22}$$

where D^i is the streaming operator. Employing (14.1-36), the various terms are easily computed. With the notation of convected derivative [Eq. (14.1-44)] one readily finds

$$D^i f_0^i = (k_BT)^{-1} f_0^i\left\{T\frac{d^i(\varsigma^i/T)}{dt} + \frac{u^i}{T}\frac{d^iT}{dt} + m^i(\mathbf{v}^i - \mathbf{w})\cdot\frac{d^i\mathbf{w}}{dt} - \mathbf{F}^i\cdot(\mathbf{v}^i - \mathbf{w})\right\}. \tag{14.2-23}$$

For the derivatives d^iT/dt and $d^i\mathbf{w}/dt$ we use the Eulerian conservation theorems. For ς^i we employ (13.7-5). Together with (14.1-50) and (14.1-45) this yields

$$\frac{d^i(\varsigma^i/T)}{dt} = \frac{k_B}{\rho^i}\frac{d^i\rho^i}{dt} - \frac{3k_B}{2T}\frac{d^iT}{dt} = (\mathbf{v}^i - \mathbf{w})\frac{k_B}{\rho^i}\cdot\nabla\rho^i - k_B\nabla\cdot\mathbf{w} - \frac{3k_B}{2T}\frac{d^iT}{dt}$$

$$= (\mathbf{v}^i - \mathbf{w}) \cdot \left[\nabla \left(\frac{\varsigma^i}{T} \right) + \frac{3k_B}{2T} \nabla T \right] - k_B \nabla \cdot \mathbf{w} - \frac{3k_B}{2T} \frac{d^i T}{dt} . \qquad (14.2\text{-}24)$$

We now substitute (14.1-46), (14.1-50) and (14.2-24) into (14.2-23); several terms cancel and the straightforward result is

$$D^i f_0^i = (k_B T)^{-1} f_0^i \left\{ (\mathbf{v}^i - \mathbf{w}) \cdot \left[\nabla \tilde{\mu}^i + m^i \rho^{-1} \sum_j \rho^j (\mathbf{F}^j / m^j) - m^i \rho^{-1} \nabla P \right] \right.$$

$$\left. + (u^i - \varsigma^i)(\mathbf{v}^i - \mathbf{w}) \cdot (\nabla T)/T + m^i \left[(\mathbf{v}^i - \mathbf{w})(\mathbf{v}^i - \mathbf{w}) - \frac{1}{3} (\mathbf{v}^i - \mathbf{w})^2 \mathsf{I} \right] : \nabla \mathbf{w} \right\} . \qquad \mathbf{(14.2\text{-}25)}$$

[We used that $\mathsf{I} : \mathrm{grad}\, \mathbf{w} = \mathrm{div}\, \mathbf{w}$.] Multiplication with $-\tau^i(u^i)$ then gives $f_1^i(\mathbf{r}, \mathbf{v}^i, t)$. This result shows that f_1^i is indeed linear in the gradients of the flow potentials $\nabla\tilde{\mu}$, ∇T and $\nabla \mathbf{w}$, so all flow averages, fluxes and the pressure tensors, will become linear in these thermodynamic forces; the term $m^i \rho^{-1} \Sigma_j \rho^j (\mathbf{F}^j / m^j) - m^i \rho^{-1} \nabla P$ is a 'nuisance term', which will be eliminated below. The result (14.2-25) can be shown to be identical to that of de Groot and Mazur.[13]

To obtain the flow averages, set $\mathbf{v}^i - \mathbf{w} = \mathbf{s}^i$. For various purposes we will need integrals of the type

$$\mathsf{T} = \int d^3 s\, \mathbf{s}\mathbf{s}\, g(s) e^{-as^2} , \qquad (14.2\text{-}26)$$

where $g(s)$ is a real function and $s = |\mathbf{s}|$. For the off-diagonal components we have

$$T_{\alpha\beta} = \int_{-\infty}^{\infty} \int_{-\infty}^{\infty} \int_{-\infty}^{\infty} ds_\alpha ds_\beta ds_\gamma s_\alpha s_\beta g[(s_\alpha^2 + s_\beta^2 + s_\gamma^2)^{1/2}] e^{-as^2} = 0, \quad (\alpha \neq \beta), \qquad (14.2\text{-}27)$$

since the integrals over ds_α and ds_β are odd quasi-Gaussian integrals which vanish. For the diagonal components we have

$$T_{\alpha\alpha} = \int_0^{2\pi} d\phi \int_0^{\pi} \sin\theta d\theta \int_0^{\infty} ds\, s^2 s_\alpha^2 g(s) e^{-as^2} = \frac{4\pi}{3} \int_0^{\infty} ds s^4 g(s) e^{-as^2} . \qquad (14.2\text{-}28)$$

Combining the two results, we have

$$\mathsf{T} = \frac{4\pi}{3} \mathsf{I} \int_0^{\infty} ds\, s^4 g(s) e^{-as^2} . \qquad (14.2\text{-}29)$$

---

[13] de Groot and Mazur, Op. Cit. P. 176. Their result is easily translated into ours. Note that their μ^i corresponds to our ς^i. Further, we need the connection

$$T d^i (\varsigma^i / T) = (d\varsigma^i)_T - (h^i / T)\, dT, \qquad (14.2\text{-}25')$$

where h^i is the specific enthalpy. For a classical gas, $h^i = \frac{5}{2} k_B T$.

For later use we also introduce the integrals[14]

$$\mathsf{M}_n = \frac{8\pi g\sqrt{2m}}{3k_BTh^3}\mathsf{I}\int_0^\infty du\,\tau(u)u^{3/2}(u-\varsigma)^n e^{(\varsigma-u)/k_BT}\,; \qquad (14.2\text{-}30)$$

these integrals can be superscripted with '*i*' if $\varsigma \to \varsigma^i$ and $\tau \to \tau^i$. By expanding $(u-\varsigma)^n$ in powers of *u*, these integrals can be expressed in Γ-functions.

Starting now with the particle flux $\mathbf{J}_p^i$, the last term of Eq. (14.2-25) does not contribute, since it leads to odd averages that are zero. The term with $\nabla\tilde{\mu}$ involves the integral

$$-(m^i/k_B)\int d^3|\,\mathbf{v}^i-\mathbf{w}\,|(\mathbf{v}^i-\mathbf{w})(\mathbf{v}^i-\mathbf{w})\tau^i(|\,\mathbf{v}^i-\mathbf{w}\,|)f_0^i$$

$$= -\frac{(m^i)^4}{k_BTh^3}\int d^3s\,\mathbf{ss}\,\tau^i(s)e^{(\varsigma^i-\frac{1}{2}m^is^2)/k_BT}$$

$$\overset{(14.2-29)}{=} -\frac{4\pi}{3}\frac{(m^i)^4}{k_BTh^3}\mathsf{I}\int_0^\infty ds\,s^4\,\tau^i(s)e^{(\varsigma^i-\frac{1}{2}m^is^2)/k_BT} \overset{s=\sqrt{2u/m^i}}{=} -m^i\mathsf{M}_0^i\,. \qquad (14.2\text{-}31)$$

In a similar way, one finds that the term with $(\nabla T)/T$ results in $-m^i\mathsf{M}_1^i$. The computation of the entropy flux, going similarly, is left to the reader. We finally obtain the results

$$\mathbf{J}_p^i = -\mathsf{L}'^{ii}\cdot[(\nabla\tilde{\mu}^i)/m^i + \rho^{-1}\textstyle\sum_j\rho^j(\mathbf{F}^j/m^j) - \rho^{-1}\nabla P]/T - \mathsf{L}'^{is}\cdot(\nabla T)/T, \quad (14.2\text{-}32)$$

$$\mathbf{J}_s = -\textstyle\sum_i\mathsf{L}'^{si}\cdot[(\nabla\tilde{\mu}^i)/m^i + \rho^{-1}\sum_j\rho^j(\mathbf{F}^j/m^j) - \rho^{-1}\nabla P]/T - \mathsf{L}'^{ss}\cdot(\nabla T)/T. \quad (14.2\text{-}33)$$

The generalized conductivities are given by

$$\begin{aligned} &\mathsf{L}'^{ii} = (m^i)^2T\,\mathsf{M}_0^i\,, \qquad &&\mathsf{L}'^{is} = m^i\,\mathsf{M}_1^i\,, \\ &\mathsf{L}'^{si} = m^i\,\mathsf{M}_1^i\,, &&\mathsf{L}'^{ss} = \textstyle\sum_i T^{-1}\,\mathsf{M}_2^i\,. \end{aligned} \qquad (14.2\text{-}34)$$

The reader will have noticed that the gradient $\nabla\tilde{\mu}^i$ is divided by m^iT in the phenomenological equations (14.2-32) and (14.2-33); the need for this will be borne out by the forms for the entropy production Section 14.4. We note that the Onsager relations $L'^{is}_{\alpha\beta} = L'^{si}_{\beta\alpha}$ resulted from our computations. For gases the conductances are generally isotropic, so it suffices to set $\mathsf{L}' = L'\mathsf{I}$ and we can write $\mathsf{L}'\cdot\nabla\tilde{\mu} = L'\nabla\tilde{\mu}$.

We now recognise that the currents are not independent. From the definition of barycentric velocity (14.1-16) it follows that $\Sigma_{j=1}^r\mathbf{J}_p^j = 0$. thus, $\mathbf{J}^r$ can be taken to be superfluous. This will be employed to eliminate the terms $\rho^{-1}\Sigma_{j=1}^r\rho^j(\mathbf{F}^j/m^j)$ $-\rho^{-1}\nabla P$. The final results then read:

---

[14] *g* is a spin factor, which in the present context is set equal to unity.

$$\mathbf{J}_p^i = -\sum_{j=1}^{r-1} L^{ij}\left\{[(\nabla\tilde{\mu}^j)/m^j] - [(\nabla\tilde{\mu}^r)/m^r]\right\}/T - L^{is}(\nabla T)/T, \quad i = 1 \ldots r-1,$$

$$\mathbf{J}_s = -\sum_{j=1}^{r-1} L^{sj}\left\{[(\nabla\tilde{\mu}^j)/m^j] - [(\nabla\tilde{\mu}^r)/m^r]\right\}/T - L^{is}(\nabla T)/T. \qquad (14.2\text{-}35)$$

The Onsager relations $L^{ij} = L^{ji}$ and $L^{is} = L^{si}$ are again satisfied. The connection with the previous coefficients L'^{ij} and L'^{is} is a matter of trivial algebra.

In addition to the fluxes we shall evaluate the kinematic and viscous pressure tensor, for which the thermodynamic tensor force $\nabla\mathbf{w}$ will appear. Let us first consider

$$\mathcal{I} \equiv \int d^3v^i (\mathbf{v}^i - \mathbf{w})^2 f_1^i\,, \qquad (14.2\text{-}36)$$

where f_1^i is based on (14.2-25). The terms with $\nabla\tilde{\mu}$ and ∇T involve $\overline{\overline{(\mathbf{v}^i - \mathbf{w})^3}}^{\,0} = 0$. So, only $\nabla\mathbf{w}$ contributes. Setting $\mathbf{v}^i - \mathbf{w} = \mathbf{s}$, we need the integral

$$\int d^3s\left[\mathbf{ss}\, s^2 e^{-as^2} - \tfrac{1}{3}\mathsf{I}\, s^4 e^{-as^2}\right]. \qquad (14.2\text{-}37)$$

Employing (14.2-29), the first part is $(4\pi/3)\mathsf{I}\int_0^\infty ds\, s^6 \exp(-as^2)$; this cancels the second integral if for the latter we write $d^3s = d\Omega\, s^2 ds$. Hence, in (14.2-36) $\mathcal{I} = 0$. This corroborates our previously stated result, cf. (14.2-7) and the next two lines.

Rests to calculate the viscous pressure tensor, cf. the second statement in (14.2-7). We thus must compute the integral

$$-(m^i)^2 (k_B T)^{-1} \int d^3s\, \tau^i(s)\mathbf{ss}\left[\mathbf{ss} - \tfrac{1}{3}s^2\mathsf{I}\right] : \nabla\mathbf{w}\, f_0^i\,. \qquad (14.2\text{-}38)$$

Pure algebra confirms the identity

$$\left[\mathbf{ss} - \tfrac{1}{3}s^2\mathsf{I}\right] : \nabla\mathbf{w} = \sum_{\alpha\beta} \Lambda_{\alpha\beta}\left(s_\alpha s_\beta - \tfrac{1}{3}s^2\delta_{\alpha\beta}\right), \qquad (14.2\text{-}39)$$

where Λ is the symmetric tensor

$$\Lambda_{\alpha\beta} = \tfrac{1}{2}\left(\frac{\partial w_\alpha}{\partial x_\beta} + \frac{\partial w_\beta}{\partial x_\alpha}\right) \equiv \tfrac{1}{2}\left(\partial_\beta w_\alpha + \partial_\alpha w_\beta\right). \qquad (14.2\text{-}40)$$

So, for Π we have, assuming that $\tau^i(s)$ can be put in front of the integral as τ_0^i,

$$\Pi_{\alpha\beta} = -\sum_i \frac{(m^i)^2 \tau_0^i}{k_B T} \sum_{\gamma\delta} \Lambda_{\gamma\delta} \int d^3s\; s_\alpha s_\beta \left(s_\gamma s_\delta - \tfrac{1}{3}s^2\delta_{\gamma\delta}\right) f_0^i\,. \qquad (14.2\text{-}41)$$

This will be evaluated as in Huang.[15] We first note that Π is a symmetric tensor of zero trace and it depends linearly on the symmetric tensor Λ. Since $\mathrm{Tr}\,\Lambda = \nabla\cdot\mathbf{w}$, Π

---

[15] K. Huang, "Statistical Mechanics", 2nd Ed., Wiley NY, 1987, Sections 5.4 and 5.5.

must be of the form

$$\Pi = -2\sum_i \vartheta^i \left[\Lambda - \tfrac{1}{3}\mathsf{I}\nabla\cdot\mathbf{w}\right], \tag{14.2-42}$$

or in component form

$$\Pi_{\alpha\beta} = -2\sum\nolimits_i \vartheta^i \left(\Lambda_{\alpha\beta} - \tfrac{1}{3}\delta_{\alpha\beta}\nabla\cdot\mathbf{w}\right). \tag{14.2-42'}$$

It remains to compute ϑ^i. For this purpose it suffices to obtain any component, e.g., Π_{12}. From (14.2-41) we have

$$\begin{aligned}\Pi_{12} &= -\sum_i \frac{(m^i)^2\tau_0^i}{k_BT}\sum_{\gamma\delta}\Lambda_{\gamma\delta}\int d^3s\; s_1s_2\left(s_\gamma s_\delta - \tfrac{1}{3}s^2\delta_{\gamma\delta}\right)f_0^i \\ &= -2\sum_i \frac{(m^i)^2\tau_0^i}{k_BT}\Lambda_{12}\int d^3s\; s_1^2s_2^2 f_0^i\,.\end{aligned} \tag{14.2-43}$$

[The last form is obtained by writing out the double sum and changing the integration variables.] Now, from (14.2-42') we also have

$$\Pi_{12} = -2\sum_i \vartheta^i\Lambda_{12}\,. \tag{14.2-44}$$

Comparing, we finally obtain,

$$\begin{aligned}\vartheta^i &= \frac{(m^i)^2\tau_0^i}{k_BT}\int d^3s\; s_1^2s_2^2 f_0^i = \frac{(m^i)^2\tau_0^i n^i}{k_BT}\left(\frac{m^i}{2\pi k_BT}\right)^{3/2}\iiint_{-\infty}^{\infty} ds_1ds_2ds_3 \\ &\quad\times s_1^2s_2^2\, e^{-m(s_1^2+s_2^2+s_3^2)/2k_BT} = \tau_0^i n^i k_BT = \tau_0^i P^i,\end{aligned} \tag{14.2-45}$$

where we evaluated the Gaussian integrals, cf. (1.6-51).

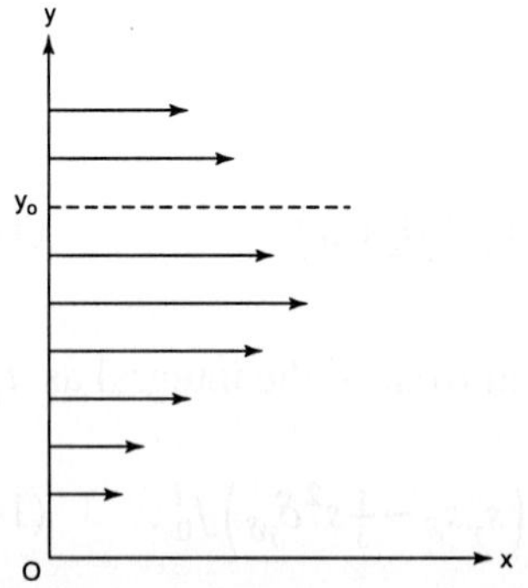

Fig. 14-2. Nonuniform flow.

It remains to show that ϑ^i is the ordinary shear viscosity as regularly defined. Consider a flow in the x-direction only with velocity differential in the y-direction, see Fig. 14-2. Let F_{xy} be the frictional force experienced by the gas at layer y_0. By definition of viscosity

$$F_{xy} = -\sum_i \vartheta^i (\partial w_x / \partial y) = -2\sum_i \vartheta^i \Lambda_{xy} \,. \qquad (14.2\text{-}46)$$

Now F_{xy} is caused by the net amount of momentum transported per second per unit area in the y-direction,

$$F_{xy} = \sum_i m^i \int d^3v^i (v_x^i - w_x)(v_y^i - w_y)(f_0^i + f_1^i) = P_{xy} = \Pi_{xy} \qquad (14.2\text{-}47)$$

(note that f_0^i does not contribute to the off-diagonal components). From the above

$$\Pi_{xy} = -2\sum_i \vartheta^i \Lambda_{xy} \,, \qquad (14.2\text{-}48)$$

in accord with (14.2-44).

The diagonal part of Π, obtained from the mathematical argument presented above, involves the trace; a simple physical picture is not easily given.

Although the formalism presented in this section is general, we still note that all computations given here are based on the Boltzmann distribution, so that there is no bulk viscosity ϑ_v^i. However, such a contribution shows up in real gases and fluids, for which Boltzmann theory does not suffice. Hydrodynamics of such systems is usually treated in specialised monographs; it will not be considered in this text.

14.3 The Hydrodynamic Equations

In Section 14.1 we developed conservation theorems for $\partial\rho/\partial t$, $\partial(\rho\mathbf{w})/\partial t$ and $\partial u/\partial t$ by integrating out the collision terms. These theorems do not imply, however, that the collision integral has no effect! The scattering mechanism, as represented by the relaxation time $\tau(v)$, still enters into the various flow averages $\mathbf{J}_p^i$, $\mathbf{J}_u$, $\mathbf{J}_s$, P and Π occurring in the conservation theorems. When the latter are expressed in the thermodynamic forces, substitution of these phenomenological equations into the conservation theorems yields the *hydrodynamic equations.*

Starting with mass conservation, we have for systems with barycentric flow

$$\frac{\partial\rho^i}{\partial t} + \operatorname{div}(\rho^i\mathbf{w}) + \operatorname{div}\mathbf{J}_p^i = 0 \,. \qquad (14.3\text{-}1)$$

In this section we shall explicitly assume that the system in '*mechanical equilibrium*' so that the centre-of-mass acceleration $\partial\mathbf{w}/\partial t$ vanishes. Also, we assume that $\nabla\mathbf{w}$ and $\nabla\mathcal{V}$ are small quantities compared to other gradients; then from (14.1-42) we have that $\Sigma_i\rho^i(\mathbf{F}_i/m_i) \approx \operatorname{grad} P$ and $\nabla\tilde{\mu}^i \approx \nabla\varsigma^i$. Eq. (14.2-25) now yields for the first-order Enskog distribution,

$$f_1^i = -\tau^i Df_0^i = -(k_BT)^{-1}\tau^i f_0^i \left\{(\mathbf{v}^i - \mathbf{w})\cdot\nabla\varsigma^i + (u^i - \varsigma^i)(\mathbf{v}^i - \mathbf{w})\cdot(\nabla T)/T\right\}. \quad (14.3\text{-}2)$$

We still write, employing (13.7-5),

$$\nabla\varsigma^i = k_B T(\nabla n^i)/n^i + terms \propto \nabla T. \tag{14.3-3}$$

To obtain the diffusion contribution, we consider isothermal conditions, $\nabla T = 0$. Using the last two equations for the diffusion fluxes $\mathbf{J}'^i_p$, Eq. (14.3-1) yields the diffusion equation

$$\frac{\partial n^i}{\partial t} + \mathbf{w}\cdot \mathrm{grad}\, n^i = \nabla\cdot[\mathsf{D}'^i\cdot\nabla n^i], \tag{\textbf{14.3-4}}$$

where D'^i is the diffusion tensor,

$$\mathsf{D}'^i = (1/n^i)\int d^3|\mathbf{v}^i - \mathbf{w}|\ \tau(|\,\mathbf{v}^i - \mathbf{w}\,|)\ (\mathbf{v}^i - \mathbf{w})(\mathbf{v}^i - \mathbf{w}) f_0^i = \overline{\overline{\tau(s)\mathbf{ss}}}^{\,0} \equiv D'\mathsf{I}, \tag{14.3-5}$$

with

$$D'^i = \frac{4\pi}{3n^i(\mathbf{r},t)}\int_0^\infty ds\tau(s)s^4 f_0^i(\mathbf{r},s,t). \tag{14.3-6}$$

Diffusion is actually a complicated process, cf., de Groot and Mazur, Op. Cit., p. 239ff. While the form (14.3-4) occurs naturally in this approach, it is possible with Prigogine's theorem and a different ansatz to obtain a different rhs for Eq. (14.3-4), viz., $D^i\nabla^2 n^i$, as will be shown in the next chapter when dealing with solids.

Next we consider momentum transport. We substitute for P from Eqs. (14.2-7) and (14.2-42) into the conservation theorem (14.1-22); this yields

$$\frac{\partial(\rho\mathbf{w})}{\partial t} + \mathrm{Div}\left[\rho\mathbf{ww} + P\mathsf{I} + \frac{2}{3}\mathsf{I}\vartheta(\nabla\cdot\mathbf{w}) - 2\vartheta\Lambda\right] = \sum_i \rho^i(\mathbf{F}^i/m^i), \tag{14.3-7}$$

where $\vartheta = \Sigma_i\vartheta^i$ is the total shear viscosity. In what follows we restrict ourselves to a one-component isotropic fluid.[16] The diffusion flux $\mathbf{J}_p$ is then zero, from which $\partial\rho/\partial t + \mathrm{div}(\rho\mathbf{w}) = 0$; this is used to eliminate the terms in ρ. We thus obtain

$$\frac{\partial\mathbf{w}}{\partial t} + \mathbf{w}\cdot\mathrm{Grad}\,\mathbf{w} = -\frac{1}{\rho}\left[\mathrm{grad}\,P + \frac{2}{3}\vartheta\,\mathrm{grad}\,(\nabla\cdot\mathbf{w}) - 2\vartheta\,\mathrm{Div}\,\Lambda\right] + \mathbf{F}/m. \tag{14.3-8}$$

For the remaining terms on the rhs we write

$$\begin{aligned}\tfrac{2}{3}\mathrm{grad}\,(\nabla\cdot\mathbf{w}) - 2\,\mathrm{Div}\,\Lambda &= -\tfrac{1}{3}\mathrm{grad}\,(\nabla\cdot\mathbf{w}) + \mathrm{grad}\,(\nabla\cdot\mathbf{w}) - 2\,\mathrm{Div}\,\Lambda \\ &= -\tfrac{1}{3}\mathrm{grad}\,(\nabla\cdot\mathbf{w}) - \nabla^2\mathbf{w},\end{aligned} \tag{14.3-9}$$

where the last equality follows from the definition of Λ. Eq. (14.3-8) now results in

[16] We have treated the "fluid" as a dilute gas; the Navier–Stokes equation holds equally well, however, for real fluids, although P and ϑ then obey more realistic equations of state.

$$\frac{\partial \mathbf{w}}{\partial t} + \mathbf{w}\cdot \mathrm{Grad}\,\mathbf{w} = \frac{1}{\rho}\left[-\mathrm{grad}\,P + \frac{1}{3}\vartheta\,\mathrm{grad}\,\mathrm{div}\,\mathbf{w} + \vartheta\nabla^2\mathbf{w}\right] + \mathbf{F}/m. \quad \mathbf{(14.3\text{-}10)}$$

This is the *Navier–Stokes* equation.

Finally, the energy transport equation will be obtained for *zero-diffusion fluxes*, i.e., the 'open-circuited' case. Under such circumstances we have from (14.1-27), (14.2-32) and (14.2-33),

$$\mathbf{J}_u = T\mathbf{J}_s = -\mathsf{\lambda}\cdot\nabla T, \quad (14.3\text{-}11)$$

where

$$\mathsf{\lambda} = \sum_i \mathsf{L}'^{si}\cdot(\mathsf{L}'^{ii})^{-1}\cdot\mathsf{L}'^{is} - \mathsf{L}'^{ss}, \quad (14.3\text{-}12)$$

or, using (14.2-34)

$$\mathsf{\lambda} = \sum_i \left[\mathsf{M}_1^i\cdot(\mathsf{M}_0^i)^{-1}\cdot\mathsf{M}_1^i - \mathsf{M}_2^i\right]/T. \quad (14.3\text{-}12')$$

Equation (14.3-11) defines the heat-conductivity tensor $\mathsf{\lambda}$. Employing moreover for P the above indicated results, substitution of (14.3-11) in the energy conservation theorem (14.1-33) yields

$$\frac{\partial u}{\partial t} + \mathrm{div}(u\mathbf{w}) + \left(P + \frac{2}{3}\vartheta\,\mathrm{div}\,\mathbf{w}\right)\mathrm{div}\,\mathbf{w} + \mathrm{div}(\mathsf{\lambda}\cdot\nabla T) - 2\vartheta\Lambda : \mathrm{Grad}\,\mathbf{w} = 0. \quad (14.3\text{-}13)$$

Next, we split $\mathrm{Grad}\,\mathbf{w}$ into a diagonal and off-diagonal part,

$$\mathrm{Grad}\,\mathbf{w} = \tfrac{1}{3}\mathsf{I}\,\mathrm{div}\,\mathbf{w} + \overset{0}{\mathrm{Grad}}\,\mathbf{w}. \quad (14.3\text{-}14)$$

The latter traceless tensor is resolved into a symmetric and antisymmetric part,

$$\overset{0}{\mathrm{Grad}}\,\mathbf{w} = (\overset{0}{\mathrm{Grad}}\,\mathbf{w})^s + (\overset{0}{\mathrm{Grad}}\,\mathbf{w})^a, \quad (14.3\text{-}15)$$

where

$$(\overset{0}{\mathrm{Grad}}\,w)^s_{\alpha\beta} = \tfrac{1}{2}(\partial_\alpha w_\beta + \partial_\beta w_\alpha) - \tfrac{1}{3}\delta_{\alpha\beta}\textstyle\sum_\gamma \partial_\gamma w_\gamma = \Lambda_{\alpha\beta} - \tfrac{1}{3}\delta_{\alpha\beta}\,\mathrm{div}\,\mathbf{w},$$

$$(\overset{0}{\mathrm{Grad}}\,w)^a_{\alpha\beta} = \tfrac{1}{2}(\partial_\alpha w_\beta - \partial_\beta w_\alpha). \quad (14.3\text{-}16)$$

Noting that the double dot product of a symmetric and antisymmetric tensor vanishes and writing $\mathsf{\lambda} = \lambda\mathsf{I}$, we obtain the energy-transport equation

$$\frac{\partial u}{\partial t} + \mathbf{w}\cdot\mathrm{grad}\,u + u\,\mathrm{div}\,\mathbf{w} + P\,\mathrm{div}\,\mathbf{w} = \lambda\nabla^2 T + 2\vartheta\Lambda : (\Lambda - \tfrac{1}{3}\mathsf{I}\,\mathrm{div}\,\mathbf{w}). \quad \mathbf{(14.3\text{-}17)}$$

[In a more complete treatment a term $-\vartheta_v(\mathrm{div}\,\mathbf{w})^2$ would still be added on the rhs.] Often, the viscous contributions on the rhs can be neglected. Writing then further $u = c_V\rho T$, where c_V is the specific heat per gram and $\rho = \Sigma_i\rho^i$, we find for the 'open-circuited' case (zero-diffusion fluxes) after some algebra

$$\frac{\partial T}{\partial t} + \mathbf{w}\cdot \operatorname{grad} T + (P/\rho c_V)\operatorname{div}\mathbf{w} = (\lambda/\rho c_V)\nabla^2 T\,, \tag{14.3-18}$$

which is the usual heat-conduction equation. It can of course be derived by much more elementary means.

14.4 Computation of the Entropy Production

Finally, in this chapter, we shall compute the entropy production and show that it has the form (14.2-4), although chemical reactions will not be considered.

We commence with a few more remarks on the Chapman–Enskog method. In general, the method seeks to obtain an iteration procedure for solving the Boltzmann equation by the series

$$f^i = f_0^i + f_1^i + f_2^i + \ldots \tag{14.4-1}$$

Since the collision integral for f_0^i vanished, we solved for f_1^i from $Df_0^i = -\mathcal{B}_v f_1^i$, where D means the streaming part. Next we put $f_0^i + f_1^i$ in the streaming terms and add f_2^i to the collision integral. Thus we must solve

$$D(f_0^i + f_1^i) = -\mathcal{B}_v(f_1^i + f_2^i),\ etc. \tag{14.4 -2}$$

If $\mathcal{B}_v$ were linear, the first terms on each side would cancel; however, that is usually not the case. The solution may be accomplished by linearizing or by employing the variational procedure mentioned in subsection 13.4.1. In each case we usually set[17]

$$f_\nu^i = f_0^i\,\varphi_\nu^i\,. \tag{14.4-3}$$

We still note that the higher-order functions are subject to three constraints. Consider the following equalities:

$$\rho^i = m^i \int d^3v^i f^i \qquad = m^i \int d^3v^i f_0^i\,, \tag{14.4-4a}$$

$$\rho\mathbf{w} = \sum\nolimits_i m^i \int d^3v^i\, \mathbf{v}^i f^i = \sum\nolimits_i m^i \int d^3v^i\, \mathbf{v}^i f_0^i\,, \tag{14.4-4b}$$

$$u = \sum\nolimits_i \int d^3v^i u^i f^i \quad = \sum\nolimits_i \int d^3v^i u^i f_0^i\,. \tag{14.4-4c}$$

In these relations the middle member is true by definition, while the last rhs follows from the form of f_0^i. Therefore, we are led to the following conditions:

$$\int d^3v^i f_\nu^i = 0, \quad (\nu \geq 1), \tag{14.4-5a}$$

[17] S.R. de Groot and P. Mazur, Op. Cit., Chapter IX, § 5.

$$\sum_i m^i \int d^3v^i\,(\mathbf{v}^i - \mathbf{w})\, f_\nu^i = 0, \quad (\nu \geq 1), \tag{14.4-5b}$$

$$\sum_i m^i \int d^3v^i\,(\mathbf{v}^i - \mathbf{w})^2\, f_\nu^i = 0, \quad (\nu \geq 1). \tag{14.4-5c}$$

We now turn to the entropy production in the first Enskog approximation. In slight deviation from the developments in Chapter XIII we set

$$s = -k_B h, \quad J_s = -k_B(\daleth - h\mathbf{w}), \quad \sigma = -k_B \eta, \tag{14.4-6}$$

with the entropy production satisfying

$$\frac{\partial s}{\partial t} + \mathrm{div}(\mathbf{J}_s + s\mathbf{w}) = \sigma. \tag{14.4-7}$$

We deal with classical gases, so in Eqs. (13.5-2) and (13.5-4a) $\varepsilon^i = 0$. This leaves us with the integrals[18]

$$\mathbf{J}_s(\mathbf{r},t) = -k_B \sum_i \int d^3v^i\,(\mathbf{v}^i - \mathbf{w})\, f^i \ln f^i, \quad s(\mathbf{r},t) = -k_B \sum_i \int d^3v^i f^i \ln f^i. \tag{14.4-8}$$

This is substituted into (14.4-7); differentiating $s(\mathbf{r},t)$ and substituting for $\partial f^i/\partial t$ from the Boltzmann equation, we find similar to (13.5-7)

$$\sigma = k_B \sum_i \int d^3v^i (\mathscr{B}_{\mathbf{v}^i} f^i) \ln f^i = -k_B \sum_i \int d^3v^i (Df_0^i) \ln(f_0^i + f_0^1). \tag{14.4-10}$$

The further computation of σ goes smoothly, since Df_0^i was evaluated in (14.2-25). Moreover, Taylor expansion gives

$$\ln(f_0^i + f_1^i) = \ln f_0^i + \varphi_i^1 = \frac{\varsigma^i - u^i}{k_B T} + \varphi_i^1 + const. \tag{14.4-11}$$

[the insignificant constant can be omitted since the zero point for entropy is not fixed in classical systems]. We thus obtain

$$\sigma = \frac{1}{T} \sum_i \int d^3v^i f_0^i \left[\frac{\varsigma^i - u^i}{k_B T}\right] \Bigg\{ -(\mathbf{v}^i - \mathbf{w}) \cdot \left[\nabla \tilde{\mu}^i + \frac{m^i}{\rho} \sum_j \rho^j \frac{\mathbf{F}_j}{m^j} - \frac{m^i}{\rho} \nabla P\right]$$
$$+ (u^i - \varsigma^i)(\mathbf{v}^i - \mathbf{w}) \cdot \frac{\nabla T}{T} + m^i \left[(\mathbf{v}^i - \mathbf{w})(\mathbf{v}^i - \mathbf{w}) - \frac{1}{3}\mathsf{I}(\mathbf{v}^i - \mathbf{w})^2 : \nabla \mathbf{w}\right] \Bigg\}$$

---

[18] It is instructive to expand $\mathbf{J}_s$ up to first order; one rapidly finds $f \ln f = (f_0 + f_1) \ln f_0 + f_1$. Noting that $\ln f_0 = (\varsigma^i - u^i)/k_B T$ and that the integral involving f_1^i yields $\Sigma_i \mathbf{J}_p^i = 0$, we established

$$\mathbf{J}_s(\mathbf{r},t) = \sum_i \int d^3v^i \frac{u^i - \varsigma^i}{T} (\mathbf{v}^i - \mathbf{w}) f_1^i, \tag{14.4-9}$$

confirming the curious fact that the chemical energy does not contribute to the entropy flow, cf. (14.2-6).

$$+\frac{1}{T}\sum_i \int d^3v^i f_1^i \left\{ -(\mathbf{v}^i-\mathbf{w})\cdot\left[\nabla\tilde{\mu}^i+\frac{m^i}{\rho}\sum_j \rho^j \frac{\mathbf{F}_j}{m^j}-\frac{m^i}{\rho}\nabla P\right]\right.$$

$$\left. +(u^i-\varsigma^i)(\mathbf{v}^i-\mathbf{w})\cdot\frac{\nabla T}{T}+m^i\left[(\mathbf{v}^i-\mathbf{w})(\mathbf{v}^i-\mathbf{w})-\frac{1}{3}\mathsf{I}(\mathbf{v}^i-\mathbf{w})^2:\nabla\mathbf{w}\right]\right\}. \quad (14.4\text{-}12)$$

Now, as to the integrals involving f_0^i: the terms involving $(\mathbf{v}^i-\mathbf{w})\cdot\nabla$.. are odd Gaussians that vanish, while the term in $\nabla\mathbf{w}$ yields $\Pi_0:\nabla\mathbf{w}=0$; there are therefore no contributions in zero-order. Regarding the integrals involving f_1^i: the terms based on $\sum_i(m^i/\rho)[\sum_j(\mathbf{F}_j/m^j)-\nabla P]\cdot(\mathbf{v}^i-\mathbf{w})$ do not contribute because of (14.4-5b); the remaining terms are easily found to corroborate the bilinear form

$$\sigma=-(1/T)\left[\sum_i(\mathbf{J}_p^i/m^i)\cdot\nabla\tilde{\mu}^i+\mathbf{J}_s\cdot\nabla T+\Pi:\nabla\mathbf{w}\right]$$

$$=-(1/T)\left[\sum_{i=1}^{r-1}\mathbf{J}_p^i\cdot\left(\frac{\nabla\tilde{\mu}^i}{m^i}-\frac{\nabla\tilde{\mu}^r}{m^r}\right)+\mathbf{J}_s\cdot\nabla T+\Pi:\nabla\mathbf{w}\right]. \quad (14.4\text{-}13)$$

[In the last expression we eliminated the superfluous variable $\mathbf{J}_p^r$.] This result shows that the independent thermodynamic forces are indeed the vector forces $(\nabla\tilde{\mu}^i/m^i)-(\nabla\tilde{\mu}^r/m^r)$, ∇T and the tensor force $\nabla\mathbf{w}$. The associated 'heat current' is $\mathbf{J}_s$. We can also find another form based on $\mathbf{J}_u=T\mathbf{J}_s+\Sigma_i(\varsigma^i/m^i)\mathbf{J}_p^i$. We note that

$$\frac{\nabla\tilde{\mu}^i}{m^i}-\frac{\varsigma^i}{m^iT}\nabla T=\frac{\nabla\varsigma^i}{m^i}-\frac{\mathbf{F}^i}{m^i}+\frac{\varsigma^i}{m^i}T\nabla\left(\frac{1}{T}\right)=T\nabla\left(\frac{\varsigma^i}{m^iT}\right)-\frac{\mathbf{F}^i}{m^i}. \quad (14.4\text{-}14)$$

We thus obtain the alternate form

$$\sigma=-\sum_i\mathbf{J}_p^i\cdot\left[\nabla\left(\frac{\varsigma^i}{m^iT}\right)-\frac{\mathbf{F}^i}{m^iT}\right]+\mathbf{J}_u\cdot\nabla\left(\frac{1}{T}\right)-\Pi:\nabla\mathbf{w}/T$$

$$=-\sum_{i=1}^{r-1}\mathbf{J}_p^i\cdot\left[\nabla\left(\frac{\varsigma^i}{m^iT}\right)-\nabla\left(\frac{\varsigma^r}{m^rT}\right)-\left(\frac{\mathbf{F}^i}{m^iT}-\frac{\mathbf{F}^r}{m^rT}\right)\right]+\mathbf{J}_u\cdot\nabla\left(\frac{1}{T}\right)-\Pi:\frac{\nabla\mathbf{w}}{T}. \quad (14.4\text{-}15)$$

Let us now write these results in the form

$$\sigma=\sum_{i=1}^{r-1}\mathbf{J}_p^i\cdot(\psi_p^i-\psi_p^r)+\mathbf{J}_T\cdot\psi_T-\Pi:\nabla\mathbf{w}/T; \quad (14.4\text{-}16)$$

then we have the choices given in Table 14-1 below.

Table 14-1. Choices of thermodynamic forces and conjugate fluxes.

ψ_T	ψ_p^i	$\mathbf{J}_T$	$\mathbf{J}_p^i$
$-\dfrac{1}{T}\nabla T$	$-\dfrac{1}{T}\nabla\left(\dfrac{\tilde{\mu}^i}{m^i}\right)$	$\mathbf{J}_s$	$\mathbf{J}_p^i$
$\nabla\left(\dfrac{1}{T}\right)$	$-\nabla\left(\dfrac{\varsigma^i}{m^i T}\right)+\dfrac{\mathbf{F}^i}{m^i T}$	$\mathbf{J}_u$	$\mathbf{J}_p^i$

Let us next insert the phenomenological equations for the fluxes in the form:

$$\mathbf{J}_p^i = \mathsf{L}^{ii}\cdot(\psi_p^i - \psi_p^r) + \mathsf{L}^{iT}\cdot\psi_T\,, \tag{14.4-17}$$

$$\mathbf{J}_T = \sum_{i=1}^{r-1}\mathsf{L}^{Ti}\cdot(\psi_p^i - \psi_p^r) + \mathsf{L}^{TT}\cdot\psi_T\,, \tag{14.4-18}$$

together with the result for Π, Eq. (14.2-42). This results in the quadratic form:

$$\sigma = \sum_{i=1}^{r-1}\mathsf{L}^{ii}:(\psi_p^i - \psi_p^r)(\psi_p^i - \psi_p^r) + \sum_{i=1}^{r-1}\mathsf{L}^{iT}:\psi_T(\psi_p^i - \psi_p^r)$$
$$+\sum_{i=1}^{r-1}\mathsf{L}^{Ti}:(\psi_p^i - \psi_p^r)\psi_T + \mathsf{L}^{TT}:\psi_T\psi_T + 2\sum_i(\vartheta^i/T)(\Lambda - \tfrac{1}{3}\mathsf{I}\nabla\cdot\mathbf{w}):\nabla\mathbf{w}\,. \tag{14.4-19}$$

We note that the vectorial and tensorial parts do not mix, as claimed before. For the former part, the expression obtained is positive semi-definite[19] only if the Onsager–Casimir relations stated in Eq. (14.2-3) hold, the present explicit form being

$$\mathsf{L}^{iT}(\mathbf{B}) = \tilde{\mathsf{L}}^{Ti}(-\mathbf{B})\,, \tag{14.4-20}$$

where the tilde denotes the transpose. We note that the magnetic field reversal is inherent in the velocity-dependent Lorentz force, i.e., in ψ_p^i, see table 14-1.[20] In isotropic gases, as well as in cubic crystals, the conductance tensors are scalars that multiply the unit tensor. In addition to (14.4-20), we should impose that the self-conductances are positive and that $-\frac{1}{4}\times$ the discriminant be semi-positive definite, similarly as in Section 3.6.

---

[19] This is a form of *Curie's principle*, cf. S.R. de Groot and P. Mazur, Op. Cit., Chapter IV, § 2.

[20] The double dot product as used here derives from the form $\mathbf{P}\cdot\mathsf{L}\cdot\mathbf{Q} \equiv \mathsf{L}:\mathbf{PQ}$ or $P_\alpha L_{\alpha\beta} Q_\beta = L_{\alpha\beta} P_\alpha Q_\beta$, where the summation convention is implied. So in order that $\mathsf{L}^{iT}:\mathbf{PQ} = \mathsf{L}^{Ti}:\mathbf{PQ}$ we need that

$$L_{\alpha\beta}^{iT} P_\alpha Q_\beta = L_{\alpha\beta}^{Ti} Q_\alpha P_\beta \overset{\alpha\leftrightarrow\beta}{=} L_{\beta\alpha}^{Ti} P_\alpha Q_\beta \quad \text{or} \quad \mathsf{L}^{iT}(\mathbf{B}) = \tilde{\mathsf{L}}^{Ti}(-\mathbf{B})\,,$$

which are the Onsager–Casimir relations (14.4-20).

14.5 Problems to Chapter XIV

14.1 Given is the local equilibrium distribution for a Boltzmann gas, (14.1-35). Consider two-body collisions $\mathbf{v}_1, \mathbf{v}_2 \to \mathbf{v}_1', \mathbf{v}_2'$, with transition probability per second $Q_{if} = Q_{fi}$. Show by *direct* computation that the distributions satisfy detailed balance,

$$f_0(\mathbf{r}, \mathbf{v}_1) f_0(\mathbf{r}, \mathbf{v}_2) Q_{if} = f_0(\mathbf{r}, \mathbf{v}_1') f_0(\mathbf{r}, \mathbf{v}_2') Q_{fi}. \tag{1}$$

14.2 Show that the fluxes $\mathbf{J}_{u,0}$ and $\mathbf{J}^{i}_{p,0}$, being averages over the zero-order Enskog distribution, are zero and show that $\mathsf{P}_0 = P\mathsf{I}$, with P being a sum of partial pressures satisfying Dalton's law.

14.3 Up to first order Enskog, the fluxes and pressure tensors are given by Eqs. (14.2-5) through (14.2-7); explain in detail. In particular, *prove* that the second part of the viscous pressure tensor, defined in (14.1-19), vanishes in the first order Enskog approximation. Hint: Use (14.2-29).

14.4 In the relaxation-time approximation the first-order Enskog distribution is related to the zero-order Enskog distribution by $f_1^i = -\tau^i D^i f_0^i$, where D^i is the streaming operator for component 'i'. *Show in detail* that $D^i f_0^i$ is given by the expression (14.2-25), which is linear in the gradients $\nabla\tilde{\mu}, \nabla T$ and $\nabla\mathbf{w}$. [This is quite lengthy, but many terms cancel.]

14.5 Consider the vectorial contribution to the entropy production σ for a one-component classical gas;
(a) Give the bilinear form involving thermodynamic forces and fluxes;
(b) Give the quadratic form involving the conductances and the forces.
(c) Express the conductances in suitable integrals [cf. (14.2-30)] and show that Onsager's relations are satisfied; further show that $L^{pp}, L^{ss} > 0$ and that $-\frac{1}{4}\times$ the discriminant, $(L^{pp}L^{ss} - L^{ps}L^{sp}) \geq 0$. [N.B. Use suitable expansions for the integrals.]

Chapter XV

Further Applications

1. NEAR-EQUILIBRIUM TRANSPORT

15.1 Electron Gas in Metals: the Perturbation Description

In this chapter-part we shall derive the conventional galvanomagnetic coefficients for the electrical fluxes and heat fluxes in a degenerate electron gas, using the elementary Sommerfeld–Drude model for metals close to the equilibrium state. The extension to the behaviour in nondegenerate semiconductors is straightforward, with appropriate sign changes being made for the hole-gas in the valence band. Where necessary, we assume that there is an isotropic effective mass m^*, which derives from the energy-band tensor, $m^*\mathsf{I} = \hbar^2[\partial^2\varepsilon(\mathbf{k})/\partial\mathbf{k}\partial\mathbf{k}]^{-1}$. Refinements involving ellipsoidal energy surfaces, degenerate surfaces, etc., are found in standard texts on solids.

We commence with the elementary perturbation description in the absence of a magnetic field, which presents itself as 'obvious', although it leads to an unconventional — or wrong? — form for the diffusion coefficient and ensuing Einstein relationship between diffusivity and mobility. This exercise in futility will prepare the reader to appreciate the 'less obvious' and far more powerful streaming-vector method, to be discussed in the next section. So, let $f(\mathbf{r},\mathbf{k}) = f_0[\mathbf{r},\varepsilon(\mathbf{k})] + f_1(\mathbf{r},\mathbf{k})$, where f_0 is the *local* equilibrium distribution, patterned after Eqs. (13.6-13) and (13.6-14),

$$\begin{aligned} f_0(\mathbf{r},\mathbf{k}) &= \left\{\exp\left[\left(\varepsilon_{tot}(\mathbf{r},\mathbf{k}) - \bar{\mu}(\mathbf{r})\right)/k_BT(\mathbf{r})\right]+1\right\}^{-1} \\ &= \left\{\exp\left[\left(\varepsilon(\mathbf{k}) - \varsigma(\mathbf{r})\right)/k_BT(\mathbf{r})\right]+1\right\}^{-1}, \end{aligned} \qquad (15.1\text{-}1)$$

where ε_{tot} is the total energy, i.e., the superposition of the band energy $\varepsilon(\mathbf{k})$ and the energy $-e\Phi(\mathbf{r})$, while $\varsigma(\mathbf{r})$ and $\bar{\mu}(\mathbf{r})$ relate to the position-dependent chemical potential and electrochemical potential, respectively; the quantity $\bar{\mu}(\mathbf{r})$ is also called the quasi-Fermi level in device-oriented literature (the overhead bar has been added since μ is used for the mobility). With $\varepsilon_C(\mathbf{r})$ being the bottom of the conduction band, it is useful to also introduce the relative band energies (kinetic energies)

$$\hat{\varepsilon}(\mathbf{k}) = \varepsilon(\mathbf{k}) - \varepsilon_C = \varepsilon_{tot}(\mathbf{r},\mathbf{k}) - \varepsilon_{tot,C}(\mathbf{r},\mathbf{k})\,, \qquad (15.1\text{-}2)$$

where $\varepsilon_{tot,C}(\mathbf{r})$ refers to the tilted bottom of the band, subject to the modulation of an external field, as dictated by the Wannier–Slater description discussed previously. For the band picture in a constant electric field, see Fig. 15-1.

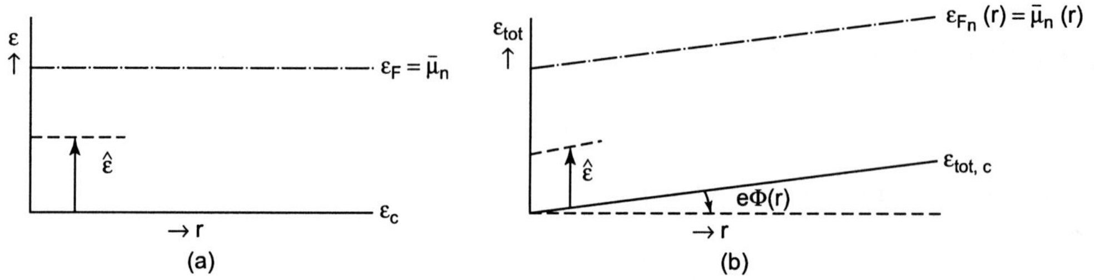

Fig. 15-1. Band-diagrams. (a) No field applied; (b) Conduction band in a constant field.

We will assume that there is a relaxation time $\tau(\mathbf{k}) = \tau[\varepsilon(\mathbf{k})]$; note that the derivation of subsection 14.2.2 can be literally carried over in the **k**,**r** formalism, so that a relaxation time always exists if the collisions are elastic, as for impurity scattering or near-elastic acoustic phonon scattering (considered later in this chapter). We thus have for the collision integral

$$-\mathcal{B}_{\mathbf{k}} f(\mathbf{r},\mathbf{k}) = -f_1(\mathbf{r},\mathbf{k})/\tau(\mathbf{k}). \tag{15.1-3}$$

For stationary flow, $\partial/\partial t = 0$, and assuming that there is no magnetic force, the perturbation approximation to the Boltzmann transport equation $\mathcal{D}_{\mathbf{k}} f_0 = -\mathcal{B}_{\mathbf{k}} f_1$ now yields

$$f_1(\mathbf{r},\mathbf{k}) = -\tau(\mathbf{k})[\mathbf{v}_{\mathbf{k}}\cdot\nabla_{\mathbf{r}} - (e\mathbf{E}/\hbar)\cdot\nabla_{\mathbf{k}}]f_0(\mathbf{r},\mathbf{k}). \tag{15.1-4}$$

For the electrical current density (current flux) we have the usual expression

$$\mathbf{J}_e = -(e/4\pi^3)\int d^3k\, \mathbf{v}_{\mathbf{k}} f_1(\mathbf{r},\mathbf{k}), \tag{15.1-5}$$

where we included a spin multiplicity of two in the density of states. Substituting (15.1-4), we see at once that the current flux contains two parts, a drift-current flux and a diffusion-current flux. We shall define local band averages by

$$\overline{\overline{\varphi(\mathbf{r})}}^{(0)} = [1/4\pi^3 n(\mathbf{r})]\int d^3k\, \varphi(\mathbf{r},\mathbf{k}) f_0(\mathbf{r},\mathbf{k}), \tag{15.1-6}$$

entirely analogous to the developments in Section (14.1). We thus obtain

$$\mathbf{J}_{e,drift} = -(e^2\mathbf{E}/4\pi^3)\cdot\int d^3k\,\tau(\mathbf{k})\mathbf{v}_{\mathbf{k}}\mathbf{v}_{\mathbf{k}}(\partial f_0/\partial\varepsilon) = [e\mu_n(\mathbf{r})n(\mathbf{r})]\cdot\mathbf{E}, \tag{15.1-7}$$

where the mobility tensor is given as

$$\mu_n(\mathbf{r}) = -e\,\overline{\overline{\tau(\mathbf{k})\mathbf{v}_{\mathbf{k}}\mathbf{v}_{\mathbf{k}}(\partial\ln f_0)/\partial\varepsilon}}^{(0)}. \tag{15.1-8}$$

Turning now to the second contribution, we find the diffusion flux to be

$$\mathbf{J}_{e,diff} = e\nabla_{\mathbf{r}} \cdot [\mathsf{D}'_n(\mathbf{r})n(\mathbf{r})]\,, \tag{15.1-9}$$

where

$$\mathsf{D}'_n(\mathbf{r}) = \overline{\overline{\tau(\mathbf{k})\mathbf{v_k}\mathbf{v_k}}}^{(0)}\,, \tag{15.1-10}$$

in which $\mathsf{D}'_n(\mathbf{r})n(\mathbf{r})$ depends on **r** through $f_0(\mathbf{r},\mathbf{k})$. For the case that the energy surfaces are spherical, the tensors are isotropic and we have $\mu = \mu\mathsf{I}$, $\mathsf{D}' = D'\mathsf{I}$.

There are several objections against the result obtained. First, the diffusion current defined here contains both *chemical* and *thermal* diffusion, since nowhere we assumed that $\nabla T(\mathbf{r})$ was zero. Secondly, the form '$\nabla[D'n]$' is different from the usual form $D\nabla n$, which prompted Landsberg to point out that the result is unconventional (at least) or erroneous.[1] Third, the results lead to an Einstein relationship that involves the details of the scattering process. To see this more clearly, we will write out the integrals for the mobility and the diffusivity, using integration by parts for the former. We then have

$$\mu_n = \frac{e}{4\pi^3 n(\mathbf{r})}\int d^3k \frac{\partial}{\partial\varepsilon}[\tau(\varepsilon)\mathbf{v_k}\mathbf{v_k}]f_0(\mathbf{r},\varepsilon)]\,, \tag{15.1-11}$$

$$\mathsf{D}_n = \frac{1}{4\pi^3 n(\mathbf{r})}\int d^3k\,\tau(\varepsilon)\mathbf{v_k}\mathbf{v_k}f_0(\mathbf{r},\varepsilon)\,. \tag{15.1-12}$$

Let us assume that the relaxation time has the energy dependence $\tau(\varepsilon) = A\hat{\varepsilon}^{\gamma}$. Both integrals are then easily expressed in Fermi integrals, cf. Eq. (8.4-6). Employing the second form of (15.1-1) the following 'generalized Einstein relationship' then follows[2]

$$e\mathsf{D}'_n = \mu_n k_B T \frac{\mathcal{F}_{\frac{3}{2}+\gamma}[(\varsigma - \varepsilon_C)/k_B T]}{\mathcal{F}_{\frac{1}{2}+\gamma}[(\varsigma - \varepsilon_C)/k_B T]}\,. \tag{15.1-13}$$

Only for a nondegenerate band (as in a wide-gap semiconductor) can we revert to Boltzmann statistics, whereby $\mathcal{F}_k(\eta \to -\infty) = e^{\eta}$ for any order *k*, so that we approach the *ordinary* Einstein relationship — which does not depend on the scattering mechanism :

$$e\mathsf{D}_n = \mu_n k_B T\,. \tag{15.1-14}$$

For further comments, see Marshak and Van Vliet.[3]

---

[1] P.T. Landsberg, "*D*grad *n* or grad(*Dn*) ?" J. Appl. Phys. **56**, 1119 (1984).

[2] A.H. Marshak and D. Asaf III, Solid State Electr. **16**, 675 (1973).

[3] A.H. Marshak and C.M. Van Vliet, Proc. IEEE **72**, 148-164 (1984).

15.2 The Streaming-Vector Method [4]

15.2.1 *Fluxes in Absence of a Magnetic Field*

Analogous to subsection 14.2.2 we make a different ansatz

$$f(\mathbf{r},\mathbf{k}) = f_0[\mathbf{r},\varepsilon(\mathbf{k})] + \underbrace{\mathbf{v_k}\cdot\mathbf{g}_1[\mathbf{r},\varepsilon(\mathbf{k})]/v_\mathbf{k}}_{f_1(\mathbf{r},\varepsilon)}, \tag{15.2-1}$$

where $\mathbf{g}_1$ is the correction in first order; since it is a vector, we have more leeway to satisfy the BTE; note that $v_\mathbf{k} = |\mathbf{v_k}|$. If χ is the angle between $\mathbf{g}_1$ and $\mathbf{v_k}$, then the added term is $g_1 \cos\chi$, indicating that, in fact, (15.2-1) is a first-order expansion in Legendre polynomials. The force acting on the electrons with charge $q = -e$ is $-e\mathbf{E} = e\nabla\Phi(\mathbf{r})$. Substitution of f_0 in the streaming terms of the Boltzmann equation yields

$$\begin{aligned}\mathcal{D}f_0 &= \mathbf{v_k}\cdot\nabla_\mathbf{r} f_0(\mathbf{r},\varepsilon) - (e\mathbf{E}/\hbar)\cdot\nabla_\mathbf{k} f_0(\mathbf{r},\varepsilon)\\ &= \mathbf{v_k}\cdot[\nabla_\mathbf{r} f_0 + e\nabla_\mathbf{r}\Phi\, \partial f_0/\partial\varepsilon] \equiv \mathbf{v_k}\cdot\boldsymbol{\alpha},\end{aligned} \tag{15.2-2}$$

where $\boldsymbol{\alpha}$ is the *streaming vector*; notice that we used $\nabla_\mathbf{k} f_0 = (\partial f_0/\partial\varepsilon)\nabla_\mathbf{k}\varepsilon(\mathbf{k})$ and $\mathbf{v_k} = \hbar^{-1}\nabla_\mathbf{k}\varepsilon(\mathbf{k})$. With the collision term being

$$-\mathcal{B}_\mathbf{k} f = -\mathbf{v_k}\cdot\mathbf{g}_1(\mathbf{r},\varepsilon)/\tau(\varepsilon)v_\mathbf{k}, \tag{15.2-3}$$

the Boltzmann equation in perturbation form, $\mathcal{D}f_0 + \mathcal{B}_\mathbf{k} f_1 = 0$, yields

$$\mathbf{v_k}\cdot[\boldsymbol{\alpha} + \mathbf{g}_1(\mathbf{r},\varepsilon)/\tau(\varepsilon)v_\mathbf{k}] = 0. \tag{15.2-4}$$

Since the streaming vector can have any direction with respect to $\mathbf{v_k}$, we must have

$$\begin{aligned}&\mathbf{g}_1(\mathbf{r},\varepsilon) = -v_\mathbf{k}\tau(\varepsilon)\boldsymbol{\alpha}, \text{ or}\\ &f_1(\mathbf{r},\varepsilon) = -\mathbf{v_k}\tau(\varepsilon)\cdot\boldsymbol{\alpha}\ .\end{aligned} \tag{15.2-5}$$

Our main objective is now to find suitable expressions for $\boldsymbol{\alpha}$. From (15.1-1) one easily obtains,

$$\begin{aligned}\boldsymbol{\alpha} &= -\frac{\partial f_0}{\partial\varepsilon}\left[\nabla\bar{\mu}_n + \frac{\varepsilon(\mathbf{k}) - \varsigma_n(\mathbf{r})}{T(\mathbf{r})}\nabla T\right]\\ &= -\frac{\partial f_0}{\partial\varepsilon}\left[\nabla\left(\frac{\bar{\mu}_n}{T}\right) - \varepsilon_{tot}(\mathbf{r},\mathbf{k})\nabla\left(\frac{1}{T}\right)\right]T(\mathbf{r})\\ &= -\frac{\partial f_0}{\partial\varepsilon}\left[\left(\nabla\left(\frac{\varsigma_n}{T}\right) - \frac{\mathbf{F}_n}{T}\right) - \varepsilon(\mathbf{k})\nabla\left(\frac{1}{T}\right)\right]T(\mathbf{r})\end{aligned}$$

[4] The next two sections are largely based on the following two articles: K.M. Van Vliet and A.H. Marshak, Physica Stat. Solidi (b) **78**, 501 (1976); K.M. Van Vliet, Phys. Stat. Solidi (b) **78**, 667 (1976).

$$= -\frac{\partial f_0}{\partial \varepsilon}\left[\left(\nabla^T \varsigma_n - \mathbf{F}_n\right) - \left(\varepsilon(\mathbf{k}) - \varsigma_n(\mathbf{r}) + T\frac{\partial \varsigma_n}{\partial T}\right)T(\mathbf{r})\nabla\left(\frac{1}{T}\right)\right]. \qquad (15.2\text{-}6)$$

Only the first form will be used at present; the other terms will be needed in Section 5.3. Further note that the symbol ∇^T here stands for the gradient, excluding temperature variations; generally, $\nabla f = \nabla^\gamma f + (\partial f / \partial \gamma)\nabla \gamma$.

We proceed to find the current flux $\mathbf{J}_e$ for the conduction band. We have

$$\mathbf{J}_e(\mathbf{r}) = -\frac{e}{4\pi^3}\int d^3k \frac{\partial f_0}{\partial \varepsilon} \mathbf{v_k v_k} \tau(\varepsilon) \cdot \left[\nabla \bar{\mu}_n + \frac{\varepsilon(\mathbf{k}) - \varsigma_n(\mathbf{r})}{T(\mathbf{r})}\nabla T\right]. \qquad (15.2\text{-}7)$$

This is clearly of the form

$$\mathbf{J}_e(\mathbf{r}) = \mathsf{L}_{ee}(\mathbf{r}) \cdot (\nabla \bar{\mu}_n / e) + \mathsf{L}_{es}(\mathbf{r}) \cdot \nabla T\,, \qquad (15.2\text{-}8)$$

where $\mathsf{L}_{ee}(\mathbf{r}) = \sigma(\mathbf{r})$ is the inhomogeneous conductivity tensor and $\mathsf{L}_{es}(\mathbf{r})$ denotes a cross tensor, related to the Seebeck effect, see below; as before, we set $\sigma = e\mu_n n$, μ_n being the mobility tensor. From these results we see that μ_n is the same band integral as in the previous section, Eq. (15.1-11). A more lucid form is obtained by writing $d^3k = dS_\mathbf{k} d\varepsilon / |\,\mathrm{grad}_\mathbf{k}\, \varepsilon(\mathbf{k})| = dS_\mathbf{k} d\varepsilon / \hbar v_\mathbf{k}$. Let further $\rho(\varepsilon)$ be the density of states for the band

$$\rho(\varepsilon) = \frac{1}{4\pi^3} \oint_{\varepsilon(\mathbf{k})=\varepsilon} dS_\mathbf{k} / \hbar v_\mathbf{k}\,; \qquad (15.2\text{-}9)$$

Local $\mathbf{k}$-space averages are then defined as

$$\overline{\overline{\varphi(\mathbf{r})}}^{(0)} = [1/n(\mathbf{r})]\int_{\varepsilon_C}^{\infty} d\varepsilon\, \rho(\varepsilon)\varphi(\mathbf{r},\varepsilon) f_0(\mathbf{r},\varepsilon)\,. \qquad (15.2\text{-}10)$$

We now obtain

$$\mu_n(\mathbf{r}) = \frac{e}{4\pi^3}\overline{\overline{\left[\frac{1}{\rho(\varepsilon)}\frac{\partial}{\partial \varepsilon}\oint \frac{dS_\mathbf{k}}{\hbar v_\mathbf{k}} \mathbf{v_k v_k}\tau(\varepsilon)\right]}}^{(0)}, \qquad (15.2\text{-}11)$$

and likewise

$$\mathsf{L}_{es}(\mathbf{r}) = \frac{en(\mathbf{r})}{4\pi^3 T(\mathbf{r})}\overline{\overline{\left[\frac{\varepsilon - \varsigma_n(\mathbf{r})}{\rho(\varepsilon)}\frac{\partial}{\partial \varepsilon}\oint \frac{dS_\mathbf{k}}{\hbar v_\mathbf{k}} \mathbf{v_k v_k}\tau(\varepsilon)\right]}}^{(0)}. \qquad (15.2\text{-}12)$$

At sufficiently low temperatures at which the electron gas is fully degenerate, we can pull the relaxation time out of the integrand as τ_F and set $\partial f_0 / \partial \varepsilon = -\delta(\varepsilon - \varepsilon_F)$. For spherical energy surfaces, noting the relationship $n(\mathbf{r}) \approx n_0 = k_F^3 / 3\pi^2$ [cf. (8.4-11)], one then readily confirms the elementary result

$$\mu_n = \mu_n \mathsf{I}, \quad \mu_n = e\tau_F / m^*\,. \qquad (15.2\text{-}13)$$

Next we shall show that the first part in (15.2-8) comprises *both drift and chemical diffusion.* With $\overline{\mu}_n / e = \varsigma_n / e - \Phi$ we have the two contributions

$$\mathbf{J}_{e,drift} = en(\mathbf{r})\mu_n \cdot (-\nabla\Phi) = en(\mathbf{r})\mu \cdot \mathbf{E}, \tag{15.2-14}$$

$$\mathbf{J}_{e,diff+} = \mu_n \cdot n(\mathbf{r})\nabla\varsigma_n = \mu_n \cdot n(\mathbf{r})[\nabla^T \varsigma_n + (\partial\varsigma_n / \partial T)\nabla T]. \tag{15.2-15}$$

The second part above refers to the Soret effect, see de Groot and Mazur, Op. Cit. The chemical diffusion proper is thus given by

$$\mathbf{J}_{e,diff} = \mu_n \cdot n(\mathbf{r})\nabla^T \varsigma_n = \mu_n \cdot (\partial\varsigma_n / \partial \ln n)_T \nabla n \equiv e\mathrm{D}_n \cdot \nabla n. \tag{15.2-16}$$

The present treatment has given us the correct form for the diffusion current, while also we established the *generalized Einstein relation*

$$e\mathrm{D}_n(\mathbf{r}) = \mu_n(\mathbf{r})(\partial\varsigma_n / \partial \ln n)_T . \tag{15.2-17}$$

To obtain some final explicit and useful results, we note the band integrals

$$n(r) = \int_{\varepsilon_C}^{\infty} d\varepsilon f(\varepsilon, \varsigma_n, T)\rho(\varepsilon), \tag{15.2-18}$$

$$\begin{aligned} [\partial n(r) / \partial\varsigma_n]_T &= \int_{\varepsilon_C}^{\infty} d\varepsilon [\partial f(\varepsilon, \varsigma_n, T) / \partial\varsigma_n]_T \rho(\varepsilon) \\ &= \int_{\varepsilon_C}^{\infty} d\varepsilon f(\varepsilon, \varsigma_n, T)[\partial\rho(\varepsilon) / \partial\varepsilon]_T , \end{aligned} \tag{15.2-19}$$

where we used that $\partial f_0 / \partial\varsigma_n = -\partial f_0 / \partial\varepsilon$ and subsequent integration by parts. Whence, with $\epsilon = \varepsilon / k_B T$,

$$e\mathrm{D}_n(\mathbf{r}) = \mu_n(\mathbf{r}) k_B T \frac{\int_{\epsilon_C}^{\infty} d\epsilon\, f_0(\epsilon)\rho(\epsilon)}{\int_{\epsilon_C}^{\infty} d\epsilon\, f_0(\epsilon)[\partial\rho(\epsilon) / \partial\epsilon]_T}. \tag{\textbf{15.2-20}}$$

The ratio of the band integrals can be computed for any band structure. For parabolic bands (spherical or ellipsoidal energy surfaces with density-of-states effective mass m^*) one has $\rho(\epsilon) = \pi^{-2}\sqrt{2}(m^*/\hbar^2)^{3/2}(k_B T)^{1/2}(\epsilon - \epsilon_C)^{1/2}$; this into (15.2-21) gives the generalized Einstein relation in the form

$$e\mathrm{D}_n(\mathbf{r}) = \mu_n(\mathbf{r}) k_B T \frac{\mathcal{F}_{1/2}[(\varsigma_n - \varepsilon_C) / k_B T]}{\mathcal{F}_{-1/2}[(\varsigma_n - \varepsilon_C) / k_B T]}. \tag{\textbf{15.2-21}}$$

The general form (15.2-20) was given by Van Vliet and Marshak[4], while the result (15.2-21) for parabolic bands was first obtained by Landsberg.[5]

---

[5] P.T. Landsberg, Proc. Roy. Soc. A **213**, 226 (1952).

A few final remarks are in order. First, the connection with the diffusion tensor of the previous section for an isothermal system is as follows. We can write

$$\mathbf{J}_{e,diff}\,|_T = e\mathsf{D}'_n \cdot \nabla^T n(\mathbf{r}) + en(\mathbf{r})\nabla^T \cdot \mathsf{D}'_n$$
$$= e\mathsf{D}'_n \cdot \nabla^T n(\mathbf{r}) + en(\mathbf{r})\frac{\partial \mathsf{D}'_n / \partial \varsigma_n}{\partial n / \partial \varsigma_n} \cdot \nabla^T n(\mathbf{r}) = e\mathsf{D}_n \cdot \nabla^T n(\mathbf{r}), \qquad (15.2\text{-}22)$$

from which

$$\mathsf{D}_n = \mathsf{D}'_n + \frac{\partial \mathsf{D}'_n / \partial \varsigma_n}{\partial \ln n / \partial \varsigma_n}. \qquad (15.2\text{-}23)$$

This can also be obtained directly from the band integral for D'_n, cf. Eq. (15.1-10) [Problem 15.2]. Secondly, we will find the connection with the Seebeck coefficient by writing $\mathsf{L}_{es} = -\sigma \cdot \mathsf{S}$. This yields

$$\mathbf{J}_e = \sigma \cdot \left[\nabla(\bar{\mu}_n / e) - \mathsf{S} \cdot \nabla T\right]. \qquad (15.2\text{-}24)$$

For an open-circuited bar ($\mathbf{J}_e = 0$), we find the thermopower

$$\Phi_2 - \Phi_1 = \frac{1}{e}\left(\varsigma_{n,2} - \varsigma_{n,1}\right) + \int_2^1 dx\, \mathsf{S}_n \cdot \nabla T. \qquad (15.2\text{-}25)$$

One must still account for the difference in chemical potentials at the two ends in order to measure the Seebeck tensor S_n. Usually, the chemical potential difference is cancelled out by an arrangement in which the bar is connected in series with a reference material, while the other terminal of the bar and the reference material are connected to a potentiometer by identical leads (see Harman and Honig, Op. Cit.).

Lastly, we shall make some comments on the results for a degenerate hole gas, as in the valence band of narrow gap *p*-type semiconductors [or for some metals with a positive Hall coefficient]. When computing the electrical flux, or mass flux for that matter, it is of course inconsequential, whether one sums the motion of all electrons in the band or of all holes in the band, since the current of a filled band is zero.[6] The formal proof is trivial:

$$\mathbf{J}_e = -e\int \Delta^3 k\, \mathbf{v}_{\mathbf{k}} f_0(\mathbf{r},\mathbf{k}) = e\int \Delta^3 k\, \mathbf{v}_{\mathbf{k}}[1 - f_0(\mathbf{r},\mathbf{k})]$$
$$= e\int \Delta^3 k\, \mathbf{v}_{-\mathbf{k}}[1 - f_0(\mathbf{r},-\mathbf{k})] = e\int \Delta^3 k\, \mathbf{v}^h_{\mathbf{k}} f^h_0(\mathbf{r},\mathbf{k})] = \mathbf{J}^h_e. \qquad (15.2\text{-}26)$$

We employed here the invariance of the Brillouin zone under inversion $\mathbf{k} \to -\mathbf{k}$ and we added the superscript '*h*' for hole attributes; in particular, note that the hole velocity is given by $\mathbf{v}^h_{\mathbf{k}} = \mathbf{v}_{-\mathbf{k}}$, while the charge sign is $+e$. The hole distribution,

---

[6] When there is gridlock in traffic, it is easier to study the motion of the fewer empty spaces!

analogous to (15.1-1) is now, employing inversion symmetry $\varepsilon(-\mathbf{k}) = \varepsilon(\mathbf{k})$,

$$f_0^h(\mathbf{r},\mathbf{k}) = \left\{\exp\left[\left(\bar{\mu}_p(\mathbf{r}) - \varepsilon_{tot}(\mathbf{r},\mathbf{k})\right)/k_B T(\mathbf{r})\right]+1\right\}^{-1}$$
$$= \left\{\exp\left[\left(\varsigma_p(\mathbf{r}) - \varepsilon(\mathbf{k})\right)/k_B T(\mathbf{r})\right]+1\right\}^{-1}. \qquad (15.2\text{-}27)$$

Now we have noted before that the band energies $\varepsilon(\mathbf{k})$ and $\varepsilon_{tot}(\mathbf{r},\mathbf{k})$ are *electron* energies. Thus, the charge entering in the potential energy is $q\Phi$ with $q = -e$, even for the valence band. The connection between the chemical and electrochemical potentials is therefore $\bar{\mu}_{n \text{ or } p} = \varsigma_{n \text{ or } p} - e\Phi$. Hence, *whereas for holes the charge +e enters into the flux and the Lorentz force, the charge –e occurs in all energy relationships.*[7] The following pertinent results are mentioned. The streaming vector is given by

$$\boldsymbol{\alpha}^h = -\frac{\partial f_0^h}{\partial \varepsilon}\left[\nabla \bar{\mu}_p + \frac{\varepsilon(\mathbf{k}) - \varsigma_p(\mathbf{r})}{T(r)}\nabla T\right]. \qquad (15.2\text{-}28)$$

This leads to the hole-current flux

$$\mathbf{J}_e^h(\mathbf{r}) = \mathsf{L}_{ee}^h \cdot \nabla(\mu_p/e) + \mathsf{L}_{es}^h \cdot \nabla T\,, \qquad (15.2\text{-}29)$$

where $\mathsf{L}_{ee}^h = \sigma_p = e\mu_p p(\mathbf{r})$. The band integrals for μ_p and D_p are similar as for the electron tensors; however, the energy integration is now effectively from $-\infty$ to ε_V. Finally, we give the generalized Einstein relations; for any arbitrary band with energy density $\rho^h(\varepsilon)$ we have the form

$$e\mathsf{D}_p(\mathbf{r}) = -\mu_p(\mathbf{r})k_B T \frac{\int_{-\infty}^{\epsilon_V} d\epsilon\, f_0^h(\epsilon)\rho^h(\epsilon)}{\int_{-\infty}^{\epsilon_V} d\epsilon\, f_0^h(\epsilon)[\partial\rho^h(\epsilon)/\partial\epsilon]_T}, \qquad (15.2\text{-}30)$$

while for parabolic bands

$$e\mathsf{D}_p(\mathbf{r}) = \mu_p(\mathbf{r})k_B T \frac{\mathcal{F}_{1/2}[(\varepsilon_V - \varsigma_p)/k_B T]}{\mathcal{F}_{-1/2}[(\varepsilon_V - \varsigma_p)/k_B T]}. \qquad (15.2\text{-}31)$$

15.2.2 *Incorporation of a Magnetic Field*

The formal incorporation of a magnetic field, as reflected by the Lorentz force, is only easily done if the energy surfaces (including the Fermi surface) are spheres. The zero-order distribution, however, gives no streaming term, since $(-e\mathbf{v}_\mathbf{k}/\hbar c)\times\mathbf{B}\cdot\nabla_\mathbf{k} f_0$

---

[7] We know of only one book in which hole energies and hole electrochemical potentials are used; see Harman and Honig, Chapter XIV, Ref. 1; the results so obtained are highly confusing and undesirable.

$=(e\mathbf{v}_\mathbf{k}/c)\times\mathbf{v}_\mathbf{k}(\partial f_0/\partial\varepsilon)\cdot\mathbf{B}=0$. We must therefore compute the streaming effect to first order; the result is straightforward and instead of (15.2-4) we now find that the Boltzmann equation for an electron gas is satisfied for

$$\mathbf{v}_\mathbf{k}\cdot[(\mathbf{g}_1/\tau v_\mathbf{k})-\boldsymbol{\omega}_c\times(\mathbf{g}_1/v_\mathbf{k})+\boldsymbol{\alpha}]=0, \tag{15.2-32}$$

where $\boldsymbol{\alpha}$ is again the streaming vector as previously given and $\boldsymbol{\omega}_c$ is the cyclotron frequency vector, with $\boldsymbol{\omega}_c=e\mathbf{B}/m^*c$. We now choose the direction of $\mathbf{g}_1$ such that the vector in square brackets has no components along $\mathbf{v}_\mathbf{k}$. Hence,

$$\mathbf{g}_1-\boldsymbol{\omega}_c\tau\times\mathbf{g}_1=-\tau v_\mathbf{k}\,\boldsymbol{\alpha}. \tag{15.2-33}$$

To obtain a solution, we represent $\mathbf{g}_1$ by a triad of non-coplanar vectors; specifically,

$$\mathbf{g}_1=\hat{g}_1\boldsymbol{\alpha}+\hat{g}_2\boldsymbol{\omega}_c+\hat{g}_3(\boldsymbol{\alpha}\times\boldsymbol{\omega}_c),$$

which also gives

$$\boldsymbol{\omega}_c\times\mathbf{g}_1=-\hat{g}_1(\boldsymbol{\alpha}\times\boldsymbol{\omega}_c)+\hat{g}_3\boldsymbol{\alpha}\omega_c^2-\hat{g}_3\boldsymbol{\omega}_c(\boldsymbol{\alpha}\cdot\boldsymbol{\omega}_c).$$

This is substituted into (15.2-33) and the triad coefficients are equated; one obtains

$$\mathbf{g}_1=-\frac{v_\mathbf{k}\tau}{1+\omega_c^2\tau^2}[\boldsymbol{\alpha}+\tau^2(\boldsymbol{\alpha}\cdot\boldsymbol{\omega}_c)\,\boldsymbol{\omega}_c-\tau(\boldsymbol{\alpha}\times\boldsymbol{\omega}_c)]. \tag{15.2-34}$$

In many treatments this is considered to be the final answer.[8] Clearly, we then obtain for the fluxes two extra terms depending on $\mathbf{B}$ and B^2. However, the result (15.2-34) can be put in a more appropriate form by introducing a tensor $\boldsymbol{\tau}$ which depends on $\mathbf{B}$.[9] Thus, we seek for $\mathbf{g}_1$ to establish the form

$$\mathbf{g}_1=-v_\mathbf{k}\boldsymbol{\tau}\cdot\boldsymbol{\alpha}\ \text{ with }\ \boldsymbol{\tau}=\boldsymbol{\tau}(\varepsilon,\mathbf{B}). \tag{15.2-35}$$

We assign Cartesian axes and choose $\mathbf{B}$ to be in the z-direction and let the x and y-directions be arbitrary in the plane perpendicular to $\mathbf{B}$. We then write (15.2-35) in its vector components. E.g., for the x-direction we have

$$g_{1x}=-v_\mathbf{k}(\boldsymbol{\tau}\cdot\boldsymbol{\alpha})_x=-v_\mathbf{k}(\tau_{xx}\alpha_x+\tau_{xy}\alpha_y+\tau_{xz}\alpha_z). \tag{15.2-36}$$

From (15.2-34) we also have

$$g_{1x}=-\frac{v_\mathbf{k}\tau}{1+\omega_c^2\tau^2}(\alpha_x-\alpha_y\omega_c\tau). \tag{15.2-36'}$$

Equating the coefficients of α_x, α_y and α_z ($\boldsymbol{\alpha}$ being arbitrary), we find $\tau_{xx}=\tau/(1+$

[8] See F.J. Blatt, "Physics of Electronic Conduction in Solids", McGraw Hill, 1968; B.R. Nag, Op.Cit.

[9] The method to be described here is well-known to plasma physicists, but seems to be obscure in condensed matter physics; see W.P. Allis, Chapter XIII, Ref. 6 and this Chapter, Van Vliet, Ref. 4.

$\omega_c^2\tau^2)$, $\tau_{xy} = -\omega_c\tau^2/(1+\omega_c^2\tau^2)$ and $\tau_{xz} = 0$. With similar comparisons for the y and z-components we obtain all nine tensor components, to wit

$$\mathsf{T}(\varepsilon,\mathbf{B}) = \begin{pmatrix} \tau/(1+\omega_c^2\tau^2) & -\omega_c\tau^2/(1+\omega_c^2\tau^2) & 0 \\ \omega_c\tau^2/(1+\omega_c^2\tau^2) & \tau/(1+\omega_c^2\tau^2) & 0 \\ 0 & 0 & \tau \end{pmatrix}. \tag{15.2-37}$$

The inverse tensor has a more simple structure:

$$\mathsf{T}^{-1}(\varepsilon,\mathbf{B}) = \begin{pmatrix} 1/\tau & \omega_c & 0 \\ -\omega_c & 1/\tau & 0 \\ 0 & 0 & 1/\tau \end{pmatrix}. \tag{15.2-38}$$

We thus showed that the streaming vector can be put in the form (15.2-35); comparing with the result in the absence of a magnetic field, we note that all we did previously remains valid, providing in all band expressions we make the change

$$\mathbf{v_k}\mathbf{v_k}\tau(\varepsilon) \to \mathbf{v_k}\mathbf{v_k}\cdot\mathsf{T}(\varepsilon,\mathbf{B}). \tag{15.2-39}$$

For example, the detailed integral for the conductivity tensor $\sigma(\mathbf{B})$ now reads

$$\sigma(\mathbf{B}) = -\frac{e^2}{4\pi}\int d^3k \frac{\partial f_0}{\partial\varepsilon}\mathbf{v_k}\mathbf{v_k}\cdot\mathsf{T}(\varepsilon,\mathbf{B}). \tag{15.2-40}$$

The reader should note that this result entails a host of galvanomagnetic effects. It should yield the zero-field conductivity, the zero-field resistivity, the longitudinal and transverse magneto-resistance, the Hall effect, etc. The tensor L_{es}, likewise, harbours a host of thermomagnetic effects. We shall meet them later, see Section 15.4.

Unfortunately, all this lofty work may not give correct results, as we indicated already in the opening statements of Section 14.1. The reason is that the application of a magnetic field gives rise to Landau 'orbits' (states). This means that there are no momentum eigenstates in three directions. Depending on the gauge, there may be still a **k**-space in the y and z-directions (Landau gauge), or in the z-direction only (symmetrical or Dingle gauge). In all circumstances, the states and/or 'orbits' in the x-y plane are quantized. So, the semi-classical description with **k** as a quasi-continuous variable is simply *fallacious*. This will abundantly be born out with specific examples that we shall present in Section 15.4. While in the absence of a magnetic field much of the semiclassical Boltzmann treatment is still correct, in the presence of a magnetic field we *must use* many-body techniques, such as linear response theory, or the quantum Boltzmann equation (QBE), developed by the author and coworkers in the nineteen-eighties; both are set forth in Part E of the non-equilibrium theory.

15.3 Entropy Production and Heat Flux

In this section we shall compute the entropy production for an electron gas, being the quantum analogue of Section 14.4. The band index will be denoted by the superscript 'i', referring usually to the conduction band and the valence band, or two overlapping conduction bands, etc., as the case may be. The entropy production is

$$\sigma(\mathbf{r},t) = \frac{\partial s}{\partial t} + \operatorname{div} \mathbf{J}_s; \quad s = -k_B h, \quad \mathbf{J}_s = -k_B \daleth\,; \tag{15.3-1}$$

the h-function is given in (13.5-2) and $\daleth$ is given in (13.5-4a), with now $\varepsilon = -1$ for a Fermi gas. We thus obtain, cf. (13.5-7),

$$\sigma(\mathbf{r},t) = \frac{k_B}{4\pi^3} \sum_i \int d^3k (\mathscr{R}^i_{\mathbf{k}} f^i)[\ln f^i - \ln(1-f^i)]\,. \tag{15.3-2}$$

With Taylor expansion about the zero-order Enskog distribution $f_0^i(\mathbf{r},\mathbf{k},t)$ we have

$$\ln\frac{f^i}{1-f^i} = \ln\frac{f_0^i}{1-f_0^i} + \frac{1}{f_0^i(1-f_0^i)}(f^i - f_0^i) = \frac{\varsigma^i(\mathbf{r},t) - \varepsilon^i(\mathbf{k})}{k_B T(\mathbf{r},t)} - \frac{f_1^i(\mathbf{r},\mathbf{k},t)}{k_B T(\mathbf{r},t)(\partial f_0^i / \partial \varepsilon^i)}\,,$$

where we used (13.4-14). The first part does not contribute since $\varsigma^i - \varepsilon^i$ is a collisional invariant. For the remaining we note from the Boltzmann equation

$$-\mathscr{R}^i_{\mathbf{k}} f^i = -\mathscr{R}^i_{\mathbf{k}} f_1^i = (\partial f_0^i / \partial t) + \mathbf{v}^i_{\mathbf{k}} \cdot \boldsymbol{\alpha}^i\,, \tag{15.3-3}$$

where $\boldsymbol{\alpha}$ is the streaming vector. From (15.2-6) we obtain the following four forms

$$\begin{aligned}
\sigma(\mathbf{r},t) &= -\frac{1}{4\pi^3}\sum_i \int d^3k f_1^i(\mathbf{r},\mathbf{k},t)\,\mathbf{v}^i_{\mathbf{k}} \cdot \left[\frac{1}{T(\mathbf{r},t)}\nabla\bar{\mu}^i + \frac{\varepsilon^i(\mathbf{k}) - \varsigma^i(\mathbf{r},t)}{[T(\mathbf{r},t)]^2}\nabla T\right] \\
&= -\frac{1}{4\pi^3}\sum_i \int d^3k f_1^i(\mathbf{r},\mathbf{k},t)\,\mathbf{v}^i_{\mathbf{k}} \cdot \left[\nabla\left(\frac{\bar{\mu}^i}{T}\right) - \varepsilon^i_{tot}(\mathbf{r},\mathbf{k})\nabla\left(\frac{1}{T}\right)\right] \\
&= -\frac{1}{4\pi^3}\sum_i \int d^3k f_1^i(\mathbf{r},\mathbf{k},t)\,\mathbf{v}^i_{\mathbf{k}} \cdot \left[\left(\nabla\left(\frac{\varsigma^i}{T}\right) - \frac{\mathbf{F}^i}{T}\right) - \varepsilon^i(\mathbf{k})\nabla\left(\frac{1}{T}\right)\right] \\
&= -\frac{1}{4\pi^3}\sum_i \int d^3k f_1^i(\mathbf{r},\mathbf{k},t)\,\mathbf{v}^i_{\mathbf{k}} \cdot \left[\left(\frac{\nabla^T \varsigma^i - \mathbf{F}^i}{T(\mathbf{r},t)}\right) - \left(\varepsilon^i(\mathbf{k}) - \varsigma^i(\mathbf{r},t) + T(\mathbf{r},t)\frac{\partial \varsigma^i}{\partial T}\right)\nabla\left(\frac{1}{T}\right)\right].
\end{aligned}$$

(15.3-4)

This result is readily expressed in terms of fluxes and conjugate forces. For the particle flux and electrical flux we have

$$\mathbf{J}_e^i(\mathbf{r},t) = e\mathbf{J}_p^i(\mathbf{r},t) = -(e/4\pi^3)\int d^3k\, \mathbf{v}_\mathbf{k}^i f_1^i(\mathbf{r},\mathbf{k},t). \quad (15.3\text{-}5)$$

For the entropy current [see (13.5-4a)] we use Taylor expansion similarly as above, to obtain

$$\mathbf{J}_s(\mathbf{r},t) = (1/4\pi^3)\textstyle\sum_i \int d^3k\, \mathbf{v}_\mathbf{k}^i \{[\varepsilon^i(\mathbf{k}) - \varsigma^i(\mathbf{r},t)]/T(\mathbf{r},t)\} f_1^i(\mathbf{r},\mathbf{k},t). \quad (15.3\text{-}6)$$

The first line of (15.3-4) is therefore

$$\sigma(\mathbf{r},t) = -[1/T(\mathbf{r},t)]\left(-\textstyle\sum_i \mathbf{J}_e^i \cdot \nabla(\bar{\mu}^i/e) + \mathbf{J}_s \cdot \nabla T\right). \quad (15.3\text{-}7)$$

However, there are several other choices. We also introduce the internal energy flux

$$\mathbf{J}_U(r,t) = (1/4\pi^3)\textstyle\sum_i \int d^3k\, \mathbf{v}_\mathbf{k}^i \varepsilon^i(\mathbf{k})\, f_1^i(\mathbf{r},\mathbf{k},t), \quad (15.3\text{-}8)$$

the total energy flux

$$\mathbf{J}_W(r,t) = (1/4\pi^3)\textstyle\sum_i \int d^3k\, \mathbf{v}_\mathbf{k}^i \varepsilon_{tot}^i(\mathbf{k})\, f_1^i(\mathbf{r},\mathbf{k},t), \quad (15.3\text{-}9)$$

as well as the heat flux

$$\mathbf{J}_Q(r,t) = (1/4\pi^3)\textstyle\sum_i \int d^3k\, \mathbf{v}_\mathbf{k}^i \left(\varepsilon^i(\mathbf{k}) - \varsigma^i(r,t) + T(r,t)(\partial\varsigma^i/\partial T\right) f_1^i(\mathbf{r},\mathbf{k},t). \quad (15.3\text{-}10)$$

Loosely speaking, $\mathbf{J}_s$, $\mathbf{J}_U$, $\mathbf{J}_W$ and $\mathbf{J}_Q$ are all 'heat fluxes'(and are designated as such in various texts). There are several relations between them, e.g.,

$$T\mathbf{J}_s = \mathbf{J}_W + \textstyle\sum_i \bar{\mu}^i \mathbf{J}_p^i, \quad \mathbf{J}_W = \mathbf{J}_U - \textstyle\sum_i \Phi \mathbf{J}_e^i. \quad (15.3\text{-}11)$$

Clearly, the next three lines of (15.3-4) involve these other heat fluxes. Thus, writing

$$\sigma(\mathbf{r},t) = \mathbf{J}_T \cdot \boldsymbol{\chi}_T + \textstyle\sum_i \mathbf{J}_F^i \cdot \boldsymbol{\chi}_F^i, \quad (15.3\text{-}12)$$

where $\boldsymbol{\chi}_T$ is a 'heat force' and $\boldsymbol{\chi}_F$ is a particle-type ponderomotive force, we have the possibilities listed in Table 15-1.

Table 15-1. Choice of heat flux and particle flux with their conjugate forces.

χ_T	χ_F^i	$\mathbf{J}_T$	$\mathbf{J}_F^i$
$-(1/T)\nabla T$	$-(1/T)\nabla(\bar{\mu}^i/e)$	$\mathbf{J}_s$	$-\mathbf{J}_e^i$
$\nabla(1/T)$	$-\nabla(\bar{\mu}^i/eT)$	$\mathbf{J}_W$	$-\mathbf{J}_e^i$
$\nabla(1/T)$	$-\nabla(\varsigma^i/T) + \mathbf{F}^i/T$	$\mathbf{J}_U$	$\mathbf{J}_p^i$
$\nabla(1/T)$	$-(1/T)(\nabla^T\varsigma^i - \mathbf{F}^i)$	$\mathbf{J}_Q$	$\mathbf{J}_p^i$

15.4 The Phenomenological Equations for Solids

15.4.1 *General Scheme*

In what follows we have mainly metals or degenerate n-type semiconductors in mind, but the results can be carried over, with some adaptation, to semiconductors of all types. For p-type materials the electron-occupancy distribution f may be replaced by the hole distribution $-f^h$, see (15.2-27). So, as fluxes we will choose $\mathbf{J}_s$ and $\mathbf{J}_e^i$, where i is the band designation. For the conjugate forces, we multiply each one with $T(\mathbf{r})$, which will not affect the Onsager-Casimir relations. We repeat the expressions for the fluxes already shown in Section 15.2 and 15.3,

$$-\mathbf{J}_e^i(\mathbf{r}) = \frac{-e}{4\pi^3}\int d\varepsilon^i f_0^i \frac{\partial}{\partial \varepsilon^i}\oint \frac{dS_{\mathbf{k}}^i}{\hbar v_{\mathbf{k}}^i} \mathbf{v_k v_k}\cdot \mathsf{T}\cdot\left[\nabla\bar{\mu} + \frac{\varepsilon^i - \varsigma^i}{T(r)}\nabla T\right]. \qquad (15.4\text{-}1)$$

$$\mathbf{J}_s(\mathbf{r}) = -\frac{1}{4\pi^3}\sum_i \int d\varepsilon^i f_0^i \frac{\partial}{\partial \varepsilon^i}\oint \frac{dS_{\mathbf{k}}^i}{\hbar v_{\mathbf{k}}^i}\frac{\varepsilon^i - \varsigma^i}{T(r)} \mathbf{v_k v_k}\cdot \mathsf{T}\cdot\left[\nabla\bar{\mu} + \frac{\varepsilon^i - \varsigma^i}{T(r)}\nabla T\right]. \qquad (15.4\text{-}2)$$

In accordance with Table 15-1 we write

$$\begin{aligned} -\mathbf{J}_e^i &= \mathsf{L}^{ii}\cdot\nabla(-\bar{\mu}^i/e) - \mathsf{L}^{is}\cdot\nabla T, \\ \mathbf{J}_s &= \textstyle\sum_i \mathsf{L}^{si}\cdot\nabla(-\bar{\mu}^i/e) - \mathsf{L}^{ss}\cdot\nabla T\,; \end{aligned} \qquad (15.4\text{-}3)$$

We now define the integrals

$$\mathsf{K}_n = \frac{1}{4\pi^3}\int d\varepsilon\, f_0 \frac{\partial}{\partial\varepsilon}\oint \frac{dS_{\mathbf{k}}}{\hbar v_{\mathbf{k}}}(\varepsilon - \varsigma)^n (\mathbf{v_k v_k})\cdot \mathsf{T}(\varepsilon, \mathbf{B}), \qquad (15.4\text{-}4)$$

which can be superscripted with i. While the definition holds for any band structure, the integrals are straightforward only when the energy surfaces in **k**-space are spherical and T can be pulled in front of the integral. In that event $\mathbf{v_k v_k} \to \frac{1}{3}v_k^2 \mathsf{I}$ and $\mathsf{T} \to \mathsf{T}_F$, so we have

$$\begin{aligned} \mathsf{K}_0 &= (1/m^*)\mathsf{T}_F N_C \mathcal{F}_{1/2}(\gamma), \\ \mathsf{K}_1 &= (\pi^2 k_B T/3m^*)\mathsf{T}_F N_C \mathcal{F}_{-1/2}(\gamma), \\ \mathsf{K}_2 &= (\pi^2 k_B^2 T^2/3m^*)\mathsf{T}_F N_C \mathcal{F}_{1/2}(\gamma), \end{aligned} \qquad (15.4\text{-}5)$$

where N_C is the density of states (for the conduction band) and $\gamma = (\varsigma - \varepsilon)/k_B T$. From (15.4-1) – (15.4-4) one easily finds

$$\begin{aligned} \mathsf{L}^{ii} &= \sigma^i = e^2 \mathsf{K}_0^i, \quad \mathsf{L}^{is} = (e/T)\mathsf{K}_1^i, \\ \mathsf{L}^{si} &= (e/T)\mathsf{K}_1^i, \quad \mathsf{L}^{ss} = (1/T^2)\textstyle\sum_i \mathsf{K}_2^i. \end{aligned} \qquad (15.4\text{-}6)$$

Since the tensor T is antisymmetric, $\mathsf{T}(\mathbf{B}) = \tilde{\mathsf{T}}(-\mathbf{B})$, we see that the Onsager-Casimir relations are valid,

$$\begin{gathered} \mathsf{L}^{is}(\mathbf{B}) = \mathsf{L}^{si}(\mathbf{B}) = \tilde{\mathsf{L}}^{si}(-\mathbf{B}), \\ \mathsf{L}^{ii}(\mathbf{B}) = \tilde{\mathsf{L}}^{ii}(-\mathbf{B}), \quad \mathsf{L}^{ss}(\mathbf{B}) = \tilde{\mathsf{L}}^{ss}(-\mathbf{B}). \end{gathered} \tag{15.4-7}$$

Besides these thermodynamic relationships, there are other microscopic relations, that cannot be predicted by non-equilibrium thermodynamics. To that end we note the following general formula for electron-band integrals: let ψ be any function that depends on $\mathbf{k}$ only via $\varepsilon(\mathbf{k})$; then

$$\int_{band} d\varepsilon\, f_0 \frac{\partial \Psi}{\partial \varepsilon} = \Psi(\varsigma) + \frac{1}{6}\pi^2 (k_B T)^2 \left(\frac{\partial^2 \Psi}{\partial \varepsilon^2} \right)_{\varepsilon=\varsigma} + \ldots \tag{15.4-8}$$

With $\Psi(\varepsilon) \to (1/4\pi^3)\oint (\mathbf{v_k v_k} \cdot \mathsf{T})(\varepsilon - \varsigma)^n dS_{\mathbf{k}} / \hbar v_{\mathbf{k}}$ this gives the results

$$\mathsf{K}_2 = \tfrac{1}{3}\pi^2 (k_B T)^2 \mathsf{K}_0 , \tag{15.4-9}$$

$$\mathsf{K}_1 = \tfrac{1}{3}\pi^2 (k_B T)^2 [\partial \mathsf{K}_0(\varsigma)/\partial \varsigma] , \tag{15.4-10}$$

which will be used below.

Next, we shall employ a hybrid version of the phenomenological equations, in which $\mathbf{J}_e^i$ and ∇T are the independent variables. For simplicity, let there be only one type of electron current, $\mathbf{J}_e$, pertaining to the conduction band. Then we write

$$\begin{gathered} \nabla(\bar{\mu}/e) = \mathsf{p} \cdot \mathbf{J}_e + \mathsf{\eta} \cdot \nabla T , \\ \mathbf{J}_s = (\mathsf{\Pi}/T) \cdot \mathbf{J}_e - (\mathsf{\lambda}/T) \cdot \nabla T . \end{gathered} \tag{15.4-11}$$

Here p is the resistivity tensor, $\mathsf{\eta}$ is the Seebeck tensor (it measures the voltage produced by a thermal gradient in an open-circuited sample), $\mathsf{\Pi}$ is the Peltier tensor and $\mathsf{\lambda}$ is the heat-conductivity tensor. One finds

$$\begin{gathered} \mathsf{p} = \sigma^{-1} = (\mathsf{L}_{ee})^{-1}, \quad \mathsf{\eta} = -(\mathsf{L}_{ee})^{-1} \cdot \mathsf{L}_{es} , \\ \mathsf{\Pi}/T = -\mathsf{L}_{se} \cdot (\mathsf{L}_{ee})^{-1}, \quad \mathsf{\lambda}/T = \mathsf{L}_{ss} - \mathsf{L}_{se} \cdot (\mathsf{L}_{ee})^{-1} \cdot \mathsf{L}_{es} . \end{gathered} \tag{15.4-12}$$

In a degenerate band (as in metals) the approximation $\mathsf{\lambda}/T = \mathsf{L}_{ss}$ suffices. Now, from (15.4-6), which in the present notation reads

$$\begin{gathered} \mathsf{L}_{ee} = \sigma = e^2 \mathsf{K}_0 , \quad \mathsf{L}_{es} = (e/T)\mathsf{K}_1 , \\ \mathsf{L}_{se} = (e/T)\mathsf{K}_1 , \quad \mathsf{L}_{ss} = (1/T^2)\mathsf{K}_2 \end{gathered} \tag{15.4-13}$$

and with (15.4-9) we find obtain the *Wiedemann–Franz* law, relating the electrical conductivity and the heat conductivity,

$$\mathsf{\lambda} = (\pi^2 k_B^2 T / 3e^2)\sigma . \tag{15.4-14}$$

Therefore, metals that are good electrical conductors, are also good heat conductors, and *vice versa.*[10] From (15.4-10) one also obtains a relation between the Seebeck coefficient and the electrical conductivity; in the absence of a magnetic field we have

$$\eta = -(\pi^2 k_B / 3e)\partial \ln \sigma(\varsigma)/\partial\varsigma . \tag{15.4-15}$$

Further, note the well-known relationship between the Seebeck and Peltier tensors

$$\boldsymbol{\eta} = \boldsymbol{\pi}/T . \tag{15.4-16}$$

15.4.2 *Galvanomagnetic and Thermomagnetic Effects*[11]

In the presence of a magnetic field, there is an overwhelming variety of effects and cross effects; the reader is advised to consult the many specialised texts in this area. Our purpose of this subsection is to consider four of such effects. We will show that two examples give a correct answer (fortuitous perhaps), while two other ones give a manifestly wrong result.

The arrangement will be such that $\Delta\varsigma = 0$, achievable with a circuit involving identical leads, so that $\nabla\bar{\mu}/e = -\nabla\Phi$, while the magnetic field will always be assumed to be in the z-direction. We can open-circuit or short-circuit with respect to $\mathbf{J}_e$ and we can impose that $\nabla T = 0$ (isothermal conditions) or that $\mathbf{J}_s = 0$ (adiabatic conditions). Since all effects involve the tensor $\boldsymbol{\tau}(\varepsilon_F, \mathbf{B})$, the transport tensors $\boldsymbol{\sigma}$ and $\boldsymbol{\rho}$ contain all galvanomagnetic effects (Hall effect, magnetoresistance) and the transport tensors $\boldsymbol{\lambda}, \boldsymbol{\pi}$ and $\boldsymbol{\eta}$ contain a host of thermomagnetic effects. A list of most effects is found in Ziman, Op. Cit., p. 497. We give a brief version in Table 15-2. In the first column we give the name of the effect and in the second column their applicable symbols according to Wilson.[12] The third and fourth column list the definition and the stated conditions. For instance, the isothermal Hall effect is based on the independent variables $\mathbf{J}_e$ and ∇T as in (15.4-11), with $\nabla T = 0$ (actually, $\nabla T_x = \nabla T_y = 0$ suffices). The measurement is $-(\nabla\Phi)_y / J_x = \rho_{yx} = -\rho_{xy}$. As another example, in the Righi–Leduc effect the independent variables are $\mathbf{J}_e$ and $\mathbf{J}_s$. If $J_{ex} = J_{ey} = 0$ and $J_{sy} = 0$, a cross-temperature gradient is set up due to the magnetic deflection of electrons in an adiabatic system. The effect measures $(\nabla T)_y / (\nabla T)_x$.

---

[10] The Wiedemann–Franz law also holds for classical semiconductors, but the coefficient ( ) in (15.4-14) is slightly altered; one finds

$$\boldsymbol{\lambda} = (\tfrac{5}{2} + \gamma)(k_B^2 T / e^2)\boldsymbol{\sigma}, \tag{15.4-14'}$$

where γ is the exponent of the scattering time, as in Section 15.1. Whilst it seems, therefore, that one cannot have a high heat conductivity in an electrical insulator, an exception is a material like sapphire, in which the heat conductivity is caused by the phonon gas. Thus, thin samples of sapphire are standard use in cryostats with a 'cold-finger' that holds the device to be cooled.

[11] In this subsection we shall employ *SI* units; *B* is in *Tesla* and the cyclotron frequency is $\omega_c = eB/m^*$.

[12] A.H. Wilson, "The Theory of Metals", Cambridge University Press, 1st Ed. 1936, 2nd Ed., 1965.

Table 15-2. Galvanomagnetic and thermomagnetic effects.

name	symbol	definition	conditions	tensor-components
Hall (isothermal)	$R_H B$	E_y / J_{ex}	$J_{ey} = \nabla T_x = \nabla T_y = 0$	$-\rho_{xy}$
Hall (adiabatic)	$R_H B$	E_y / J_{ex}	$J_{ey} = \nabla T_x = J_{sy} = 0$	$-\rho_{xy} - \eta_{xx}\pi_{xy} / \lambda_{xx}$
Ettinghausen	$A_E B$	$\nabla T_y / J_{ex}$	$J_{ey} = \nabla T_x = J_{sy} = 0$	$-\pi_{xy} / \lambda_{xx}$
Nernst	—	$\nabla T_x / J_{ex}$	$J_{ey} = J_{sy} = \nabla T_y = 0$	π_{xx} / λ_{xx}
Ettingh.–Nernst (is)	$\mathcal{B}_{EN} B$	$E_x / \nabla T_x$	$J_{ex} = J_{ey} = \nabla T_y = 0$	η_{xx}
Ettingh.–Nernst (ad)	$\mathcal{B}_{EN} B$	$E_x / \nabla T_x$	$J_{ex} = J_{ey} = J_{sy} = 0$	$\eta_{xx} + \eta_{xy}^2 / \lambda_{xx}$
Seebeck	η or Q	$E_z / \nabla T_x$	$J_{ez} = 0$ (no B)	η_{zz}
Peltier	π	TJ_{sz} / J_{ez}	$\nabla T_z = 0$ (no B)	π_{zz}
Transv. Magnetores.	$\rho_0(1+\mathcal{B}_\perp B^2)$	E_x / J_{ex}	$J_{ey} = \nabla T_x = \nabla T_y = 0$	ρ_{xx} or ρ_{yy}
Long.Magnetoresist.	$\rho_0(1+\mathcal{B}_{//} B^2)$	E_z / J_{ez}	$\nabla T_z = 0$	ρ_{zz}
Righi–Leduc	$\mathcal{B}_{RL} B$	$\nabla T_y / \nabla T_x$	$J_{ex} = J_{ey} = J_{sy} = 0$	$\lambda_{xy} / \lambda_{xx}$

(a) *The 3D isothermal Hall effect*

The Hall effect refers to the cross voltage arising from the deflection of carriers in a magnetic field. In the standard arrangement B is in the z-direction, the current is in the x-direction and the measured Hall voltage is in the y-direction, from which E_y is obtained. In the absence of a temperature gradient we have from (15.4-6) (first entry) and (15.4-5) (first line) [and $n = N_C \mathcal{F}_{1/2}(\gamma)$] the standard result $\sigma = (e^2 n \tau_F / m^*)$. The resistivity tensor is the inverse; hence,

$$E_y / J_x \Big|_{\nabla T=0} = \rho_{yx} = (m^* / e^2 n)(\tau_F^{-1})_{yx} = -(m^* / e^2 n)\,\omega_c \,. \qquad (15.4\text{-}17)$$

With $\omega_c = eB / m^*$ this yields, setting $\rho_{yx} = R_H B$, where R_H is the Hall coefficient,

$$R_H = -1/ en \,. \qquad (15.4\text{-}18)$$

For a hole band, we have seen that $\tau^h \rightarrow \tilde{\tau}$. So the Hall coefficient for a degenerate hole band is

$$R_H = 1/ ep \,. \qquad (15.4\text{-}19)$$

This is the correct result, which will be borne out by the quantum approach. [However, for nondegenerate semiconductors, the classical derivation yields a factor $3\pi/8 = 1.18$ over the above result, which is dubious.]

(b) *Transverse magnetoresistance*

The resistance is characterized by the resistivity ρ_{xx}. From τ^{-1} as given by (15.2-38) we have

$$\rho_{xx} = \rho_{yy} = m^* / e^2 n \tau_F \,, \tag{15.4-20}$$

which does not depend on B! So, there is *no* transverse magnetoresistance; this *échec* is known as Tonk's theorem.

Sondheimer and Wilson 'remedied' the situation by suggesting that magnetoresistance requires the presence of a composite conduction band, being composed of two overlapping bands, the *s* and *d*-band.[13] Their elaborate model gives indeed a finite result. Van Vliet and Avello have shown that their result also follows directly with the tensor-τ description given here.[14] We then have for the conductivity tensor of the composite band

$$\boldsymbol{\sigma} = en_1\boldsymbol{\mu}_1 + en_2\boldsymbol{\mu}_2 \,, \tag{15.4-21}$$

where the mobility tensors are $\boldsymbol{\mu}_{1,2} = e\boldsymbol{\tau}_{1,2}(\varepsilon_F)/m^*_{1,2}$. The full tensor for $\boldsymbol{\sigma}$ is easily written down; like in (15.2-37) it has a 2×2 core and edging with zeros, except for σ_{zz}. Thus we find that $\rho_{xx} = (\boldsymbol{\sigma}^{-1})_{xx} = D_{xx}/|D|$, where D_{xx} is the cofactor and $|D|$ is the determinant. This yields

$$\rho_{xx} = \frac{(e^2 n_1/m_1^*)F_1 + (e^2 n_2/m_2^*)F_2}{\left[(e^2 n_1/m_1^*)F_1 + (e^2 n_2/m_2^*)F_2\right]^2 + \left[\omega_{c1}\tau_1(e^2 n_1/m_1^*)F_1 + \omega_{c2}\tau_2(e^2 n_2/m_2^*)F_2\right]^2}; \tag{15.4-22}$$

here $F_i = \tau_i/(1+\omega_{ci}^2\tau_i^2) \equiv \tau_i/f_i$, $f_i = 1+\omega_{ci}^2\tau_i^2$ $(i = 1,2)$. With a bit of algebra this gives

$$\rho_{xx} = \frac{\sigma_1 f_2 + \sigma_2 f_1}{\sigma_1^2 f_2 + \sigma_2^2 f_1 + 2\sigma_1\sigma_2(1+\omega_{c1}\omega_{c2}\tau_1\tau_2)}. \tag{15.4-23}$$

We recognise $\rho_{xx}^0 \equiv \rho_{xx}(\mathbf{B}=0) = 1/(\sigma_1+\sigma_2)$. Thus,

$$\frac{\rho_{xx}}{\rho_{xx}^0} = 1 + \frac{-2\sigma_1\sigma_2(1+\omega_{c1}\omega_{c2}\tau_1\tau_2) + \sigma_1\sigma_2(f_1+f_2)}{\sigma_1^2 f_2 + \sigma_2^2 f_1 + 2\sigma_1\sigma_2(1+\omega_{c1}\omega_{c2}\tau_1\tau_2)}. \tag{15.4-24}$$

With minor algebra we find

$$\frac{\rho_{xx}(\mathbf{B}) - \rho_{xx}^0}{\rho_{xx}^0} = \frac{\sigma_1\sigma_2 B^2(\mu_1-\mu_2)^2}{(\sigma_1+\sigma_2)^2 + B^2(\sigma_1\mu_2+\sigma_2\mu_1)^2}. \tag{15.4-25}$$

For low fields the B^2 term in the denominator can be dismissed, yielding for $\mathscr{B}$ as defined in the Table,

$$\mathscr{B} = \sigma_1\sigma_2(\mu_1-\mu_2)^2/(\sigma_1+\sigma_2)^2 \,. \tag{15.4-26}$$

While this result is in agreement with Ziman (Op. Cit. p.494) – obtained by a

[13] E.H. Sondheimer and A.H. Wilson, Proc. Royal Soc. A **190**, 435 (1947).

[14] C.M. Van Vliet and M.Y. Avello, Phys. Stat. Solidi (b) **187**, 169 (1995).

different procedure – it differs from Sondheimer and Wilson's original result, not surprisingly! The problem with the two-band model as presented above is that it is hard to conceive a temperature range in which two degenerate bands have a common Fermi level. But, as we noted already, in the original model Sondheimer and Wilson envisaged two overlapping *s* and *d*-bands, in which the latter is a hole band. Thus,

$$\varepsilon_1 = \hbar^2 k^2 / 2m_s, \quad \varepsilon_2 = A - \hbar^2 k^2 / 2m_d,$$

where m_s is the electron effective mass of the *s*-band and m_d is the effective mass of the vacancies in the *d*-band. Since both bands are partially occupied, the Fermi level must be such that $0 \le \varepsilon_F < A$, where '0' refers to the bottom of the *s*-band. For the *d*-band the τ-tensor must be replaced by its transpose, which means that the sign in front of ω_{c2} is everywhere to be changed. Making the changes in numerator and denominator of (15.4-24), one easily arrives at the corrected result

$$\frac{\rho_{xx}(\mathbf{B}) - \rho_{xx}^0}{\rho_{xx}^0} = \frac{\sigma_1 \sigma_2 B^2 (\mu_1 + \mu_2)^2}{(\sigma_1 + \sigma_2)^2 + B^2 (\sigma_1 \mu_2 - \sigma_2 \mu_1)^2}. \tag{15.4-25'}$$

This leads to Wilson's coefficient

$$\mathscr{B} = \sigma_1 \sigma_2 (\mu_1 + \mu_2)^2 / (\sigma_1 + \sigma_2)^2, \tag{15.4-26'}$$

a far more reasonable result, being entirely in accord with Sondheimer and Wilson's original paper. Their band model is upheld by the structure for divalent materials such as the alkaline earth metals. However, the basic objections against the semi-classical **k**-space treatment remain.

(c) *The Righi–Leduc effect*

When electrons are deflected in a magnetic field under adiabatic conditions, temperature gradients arise; the Righi–Leduc effect seeks to measure $(\nabla T)_y / (\nabla T)_x$. Let us reconsider (15.4-11), second line, taking the y-component,

$$\lambda_{yx} (\nabla T)_x + \lambda_{yy} (\nabla T)_y + \lambda_{yz} (\nabla T)_z - \pi_{yz} J_{ez} = 0, \tag{15.4-27}$$

where we noted the defining conditions of Table 15.2. Because of cylindrical symmetry, the results $/\!/ \,\mathbf{B}$ and $\perp \mathbf{B}$ are separately zero. The above equation then yields for the effect in the *x*-*y* plane

$$(\nabla T)_y / (\nabla T)_x = -\lambda_{yx} / \lambda_{yy} = \lambda_{xy} / \lambda_{yy}. \tag{15.4-28}$$

For an isotropic solid we find

$$\lambda_{xy} / \lambda_{yy} = (\mathrm{\tau}_F)_{xy} / (\mathrm{\tau}_F)_{yy} = -\omega_c \tau_F = -eB\tau_F / m^*. \tag{15.4-29}$$

The effect is proportional to *B* and the Righi–Leduc coefficient is

$$\mathcal{B}_{RL} = -e\tau_F / m^* = -\mu_n \,. \tag{15.4-30}$$

For a hole band one obtains similarly

$$\mathcal{B}_{RL} = e\tau_F / m_h^* = \mu_p \,. \tag{15.4-31}$$

Note the sign change, analogous to the Hall effect.

(d) *The isothermal Nernst–Ettinghausen effect*

The thermo-voltage in the x-direction due to a temperature gradient in that direction when the sample is open-circuited is the Seebeck effect. However, when the temperature gradient in the y-direction is held zero while a magnetic field along the z-direction is applied, it is called the Nernst–Ettinghausen effect. From (15.4-11), first line, we have for the x-component, noting the conditions in Table 15.2,

$$E_x = \rho_{xz} J_{ez} + \eta_{xx} (\nabla T)_x + \eta_{xz} (\nabla T)_z \,. \tag{15.4-32}$$

Again, the behaviour in the x-y plane separates itself from that along the z-direction. Thus (15.4-32) yields

$$E_x / (\nabla T)_x = \eta_{xx} \,. \tag{15.4-33}$$

For an isotropic parabolic band we have from (15.5-12), second entry, and (15.4-13)

$$\begin{aligned} \eta_{xx} &= -(e/T)(\rho \cdot K_1)_{xx} \\ &= -\frac{e}{T}\frac{m^*}{e^2 n}\left(\tau_F^{-1} \cdot \tau_F\right)_{xx} \frac{\pi^2 k_B T}{3m^*} N_C \mathcal{F}_{-\frac{1}{2}}(\gamma) = -\frac{\pi^2 k_B}{3e} \mathcal{F}_{-\frac{1}{2}}(\gamma) / \mathcal{F}_{\frac{1}{2}}(\gamma) \,, \end{aligned} \tag{15.4-34}$$

which is independent of **B**, another *échec*. The Nernst–Ettinghausen effect is important for thermomagnetic energy conversion, see e.g., Harman and Honig, Op. Cit., p. 311ff.

We trust that these examples have been revealing: the classical treatment for the behaviour of condensed matter in a magnetic field can largely be discarded. The quantum treatment of some of these effects will come in later chapters. However, 'a lot of water must still flow under the bridge' before we arrive at that stage.

15.5 Mobility Computations *

15.5.1 *Resistivity of Metals; Bloch's Formula*

In elementary treatments it is assumed, often without any theoretical backing, that a relaxation time exists, in which case the aforementioned formulas for the mobility and other transport coefficients can be employed. However, for most metals, subject to impurity scattering and acoustical phonon scattering, such is not the case.

Generally, Matthiessen's rule applies, which says $1/\mu = (1/\mu_{imp}) + (1/\mu_{phon})$, which simply follows from the addition of the collision operators for the two scattering mechanisms. For the mobility due to impurity scattering, see problem 13.5. For the effect of phonon scattering, mainly involving longitudinal acoustical phonons, we must employ variational techniques, such as discussed briefly at the end of subsection 13.4.1; we return to these techniques now.

For the distribution function in zero and first order, we start with (13.4.13) and (13.4-14), setting $\varepsilon = -1$. Let us further set $-\mathscr{B}_{\mathbf{k}} f(\mathbf{k}) = -P\Phi(\mathbf{k})$ and $Df_0(\mathbf{k}) = X(\mathbf{k})$, where D is the streaming operator; note that we consider the stationary case and suppress the possible **r**-dependence. The scalar product will be defined as

$$(\Phi,\Psi) = 2\int \Delta^3 k\, \Phi\Psi, \quad \Delta^3 k = (V_0/8\pi^3)d^3k, \tag{15.5-1}$$

the factor 2 accounting for the spin. With methods similar to those for the H-theorem, one easily shows that P is self-adjoint, $(P\Phi,\Psi) = (\Phi, P\Psi)$. The Boltzmann equation, with the introduction of the operator P takes the symbolic form $P\Phi = X(\mathbf{k})$. Instead, we try to satisfy the weaker condition,

$$(\Phi, P\Phi) = (\Phi, X). \tag{15.5-2}$$

The variational principle in its first form states that *of all functions* Φ *that satisfy* (15.5-2), *the function* Φ_0 *which satisfies the Boltzmann equation gives* $(\Phi,P\Phi)$ *its maximum value.* The proof is simple: since $(P(\Phi-\Phi_0), P(\Phi-\Phi_0)) \geq 0$, we have

$$(P\Phi,\Phi) + (P\Phi_0,\Phi_0) - 2(P\Phi_0,\Phi) \geq 0. \tag{15.5-3}$$

Now the last term is just $-2(X,\Phi)$, which by (15.5-2) equals $-2(P\Phi,\Phi)$ so that

$$(P\Phi_0,\Phi_0) \geq (P\Phi,\Phi), \tag{15.5-4}$$

as had to be proven. From this follows the second form of the variational principle, viz. *that* $(P\Phi,\Phi)/(X,\Phi)^2$ *is a minimum for the true solution* Φ_0.

Employing a suitable set of trial functions, one sets $\Phi(\mathbf{k}) = \sum_i \eta_i \varphi_i(\mathbf{k})$, from which

$$(\Phi, X) = \sum_i X_i \eta_i, \quad (\Phi, P\Phi) = \sum_{ij} P_{ij}\eta_i\eta_j, \tag{15.5-5}$$

where X_i is the scalar product (X,φ_i) and P_{ij} is the matrix element $(\varphi_i, P\varphi_j)$.

We must now maximize $\Sigma_{ij}\eta_i\eta_j P_{ij}$ under the constraint

$$\sum_i \eta_i \left(\sum_j \eta_j P_{ij} - X_i \right) = 0. \tag{15.5-6}$$

The result is as expected,

$$\sum_j \eta_j P_{ij} = X_i. \tag{15.5-7}$$

These equations can be inverted to obtain the η_i if the set of trial functions is small.

Most of the work goes into the evaluation of the integrals for X_i and P_{ij}.

For the streaming part we have as previously shown, $X = -\mathbf{v_k}\cdot\boldsymbol{\alpha}$, where $\boldsymbol{\alpha}$ is the streaming vector, for which we take the first form in (15.2-6). This yields

$$X_i = 2\int \Delta^3 k \varphi_i(\mathbf{k})\,(\partial f_0/\partial\varepsilon)\mathbf{v_k}\cdot\left[\nabla\bar{\mu} + T^{-1}(\varepsilon-\varsigma)\nabla T\right]. \tag{15.5-8}$$

For the fluxes $\mathbf{J}_e$ and $\mathbf{J}_s$ we have the integrals based on (13.4-13),

$$\mathbf{J}_e = 2e\int \Delta^3 k\,\mathbf{v_k}\Phi(\mathbf{k})(\partial f_0/\partial\varepsilon) \equiv \sum\nolimits_i \mathbf{J}_{e,i}\eta_i, \tag{15.5-9}$$

$$\mathbf{J}_s = 2\int \Delta^3 k T^{-1}(\varepsilon-\varsigma)\mathbf{v_k}\Phi(\mathbf{k})(\partial f_0/\partial\varepsilon) \equiv \sum\nolimits_i \mathbf{J}_{s,i}\eta_i. \tag{15.5-10}$$

The 'partial' currents are defined analogously,

$$\mathbf{J}_{e,i} = 2e\int \Delta^3 k\,\mathbf{v_k}\varphi_i(\mathbf{k})(\partial f_0/\partial\varepsilon), \tag{15.5-11}$$

$$\mathbf{J}_{s,i} = 2\int \Delta^3 k T^{-1}(\varepsilon-\varsigma)\mathbf{v_k}\varphi_i(\mathbf{k})(\partial f_0/\partial\varepsilon). \tag{15.5-12}$$

Substituting into (15.5-7) we get

$$\mathbf{J}_{e,i}\cdot\nabla(\bar{\mu}/e) - \mathbf{J}_{s,i}\cdot\nabla T = \sum\nolimits_j \eta_j P_{ij}. \tag{15.5-13}$$

Before we solve this, it is instructive to multiply the result by η_i and sum over i; we then obtain

$$\mathbf{J}_e\cdot\nabla(\bar{\mu}/e) - \mathbf{J}_s\cdot\nabla T = (P\Phi,\Phi) = T\sigma_f, \tag{15.5-14}$$

where σ_f is the entropy production, indicating the thermodynamic meaning of this approach. Indeed, the dissipative nature of the lhs is clear; the first term is the Joule heat, $\mathsf{\rho}:\mathbf{J}_e\mathbf{J}_e$. Next, we solve for the η_i; this then yields for the full fluxes

$$\mathbf{J}_e = \sum_{ij}(P^{-1})_{ij}\mathbf{J}_{e,i}\left[\mathbf{J}_{e,j}\cdot\nabla(\bar{\mu}/e) - \mathbf{J}_{s,j}\cdot\nabla T\right], \tag{15.5-15}$$

$$\mathbf{J}_s = \sum_{ij}(P^{-1})_{ij}\mathbf{J}_{s,i}\left[\mathbf{J}_{e,j}\cdot\nabla(\bar{\mu}/e) - \mathbf{J}_{s,j}\cdot\nabla T\right]. \tag{15.5-16}$$

Note that all quantities on the rhs are known, once the set of trial functions is chosen judiciously. These equations thus entail the various tensors. In particularly,

$$\mathsf{\sigma} = \sum_{ij}(P^{-1})_{ij}\mathbf{J}_{e,i}\mathbf{J}_{e,j}, \quad \mathsf{\lambda}/T \simeq \sum_{ij}(P^{-1})_{ij}\mathbf{J}_{s,i}\mathbf{J}_{s,j}. \tag{15.5-17}$$

Clearly, we have somehow inverted the Boltzmann operator by this procedure!

In the Bloch approach to the resistivity in metals, only one trial function is chosen, i.e., $\Phi = \eta\varphi$. Then (15.5-17) with (15.5-11) gives for the isotropic lattice resistivity the result [note that for that case we have simply $P^{-1} = 1/(\varphi, P\varphi)$]:

$$\rho_L \approx \frac{3(\varphi, P\varphi)}{2e^2\left[\int \Delta^3 k v_{\mathbf{k}} \varphi(\mathbf{k}) \partial f_0 / \partial \varepsilon\right]^2} \equiv 3\frac{A}{B}. \tag{15.5-18}$$

For the case of one-body collisions, we have seen that $\varphi(\mathbf{k}) = const\ \mathbf{k} \cdot \boldsymbol{\alpha}$ is an exact solution. Spurred by that result, we take $\varphi(\mathbf{k}) = \mathbf{k} \cdot \mathbf{u}$, where $\mathbf{u}$ is a unit vector in the direction of the electrical field $\mathbf{E}$ (in the z-direction), assuming that there is no temperature gradient; the constant we omitted, since (15.5-18) is homogeneous in φ. The second form of the variational principle indicates that the result so obtained overestimates ρ if the chosen $\varphi(\mathbf{k})$ is not the exact solution.

Rests the evaluation of two miserable integrals. Substituting from (13.3-22) we find that phonon emission and phonon absorption contribute equally to the dissipative effect $(\varphi, P\varphi)$; hence (see footnote 16),

$$\begin{aligned} A &= (\varphi, P\varphi) \\ &= \frac{8\pi}{k_B T \hbar} \iint \Delta^3 k \Delta^3 k' f_0(\mathbf{k})[1 - f_0(\mathbf{k}')] F_0(\mathbf{q}) \varphi(\mathbf{k})[\varphi(\mathbf{k}) - \varphi(\mathbf{k}')] \,|\mathcal{F}(\mathbf{q})|^2 \delta(\varepsilon_{\mathbf{k}} - \varepsilon_{\mathbf{k}'} + \hbar\omega_{\mathbf{q}}) \\ &= \frac{4\pi}{k_B T \hbar} \iint \Delta^3 k \Delta^3 k' f_0(\mathbf{k})[1 - f_0(\mathbf{k}')] F_0(\mathbf{q})[\varphi(\mathbf{k}) - \varphi(\mathbf{k}')]^2 \,|\mathcal{F}(\mathbf{q})|^2\, \delta(\varepsilon_{\mathbf{k}} - \varepsilon_{\mathbf{k}'} + \hbar\omega_{\mathbf{q}}), \end{aligned} \tag{15.5-19}$$

where $\mathbf{q} = \mathbf{k}' - \mathbf{k}$. For the coupling strength we found previously [see (8.6-11)]

$$|\mathcal{F}(\mathbf{q})|^2 = \frac{\hbar q}{2V_0 \rho u_0} C_1^2 = \frac{\hbar C_1^2}{2V_0 \rho \omega_{\mathbf{q}}} |\mathbf{k}' - \mathbf{k}|^2, \tag{15.5-20}$$

where C_1 is the deformation potential. We note that $|\mathbf{k}' - \mathbf{k}|^2 \approx 2k_F^2(1 - \cos\theta)$ since the collisions are near-elastic and both $\mathbf{k}$-vectors hover close to the Fermi surface, so that $k \approx k' \approx k_F$. Substituting (15.5-20) and the proposed form for $\varphi(\mathbf{k})$ we have

$$\begin{aligned} A = \frac{2\pi\hbar C_1^2}{\rho V_0 k_B T} &\iint \Delta^3 k \Delta^3 k' f_0(\mathbf{k})[1 - f_0(\mathbf{k}')] F_0(\mathbf{q})[\mathbf{u} \cdot (\mathbf{k}' - \mathbf{k})]^2 \,|\mathbf{k}' - \mathbf{k}|^2 \\ &\times \delta(\varepsilon_{\mathbf{k}} - \varepsilon_{\mathbf{k}'} + \hbar\omega_{\mathbf{q}}) / \hbar\omega_{\mathbf{q}}. \end{aligned} \tag{15.5-21}$$

Next, with the usual change in $\mathbf{k}$-space $d^3k \to d\varepsilon dS_{\mathbf{k}} / \hbar v_{\mathbf{k}}$ we have

$$\int \Delta^3 k' \delta(\varepsilon_{\mathbf{k}} - \varepsilon_{\mathbf{k}'} + \hbar\omega_{\mathbf{q}}) \ldots = \frac{V_0}{8\pi^3} \int d\varepsilon' \oint \frac{dS_{\mathbf{k}'}}{\hbar v_{\mathbf{k}'}} \delta(\varepsilon_{\mathbf{k}} - \varepsilon' + \hbar\omega_{\mathbf{q}}) \ldots \approx \frac{V_0}{8\pi^3 \hbar} \oint \frac{dS'_F}{v_F} \ldots$$

where ... denotes the rest of the expression. Next, the integration over $d\varepsilon_{\mathbf{k}}$ is performed; we have

$$\int \Delta^3 k f_0(\mathbf{k})[1 - f_0(\mathbf{k}')] = \frac{V_0}{8\pi^3} \int d\varepsilon \oint \frac{dS_{\mathbf{k}}}{\hbar v_{\mathbf{k}}} \frac{1}{e^{(\varepsilon - \varsigma)/k_B T} + 1} \frac{1}{e^{-(\varepsilon + \hbar\omega_q - \varsigma)/k_B T} + 1}$$

$$= \frac{V_0 k_B T}{8\pi^3} \oint \frac{dS_F}{\hbar v_F} \int \frac{d\eta}{(e^{\eta}+1)(e^{-(\eta+z)}+1)}; \qquad (15.5\text{-}22)$$

here $z = \hbar\omega_{\mathbf{q}} / k_B T$ and the limits on η are $\approx (-[\varepsilon_F - \varepsilon_C]/k_B T, \infty)$ but may be replaced by $(-\infty, \infty)$. The integral can be found in the tables, giving $z/(1-e^{-z})$.[15] Substituting also the Bose–Einstein distribution for $F_0(\mathbf{q})$ we are left with

$$A = \left(\frac{V_0}{8\pi^3 \hbar}\right)^2 \frac{2\pi\hbar C_1^2}{\rho V_0 k_B T} \oint\oint \frac{dS_F}{v_F} \frac{dS_{F'}}{v_F} \frac{[\mathbf{u}\cdot(\mathbf{k}'-\mathbf{k})]^2 |\mathbf{k}'-\mathbf{k}|^2}{(e^{\hbar\omega_{\mathbf{q}}/k_B T}-1)(1-e^{-\hbar\omega_{\mathbf{q}}/k_B T})}. \qquad (15.5\text{-}23)$$

We must now integrate over the directions of **k** and **k'**. First note that $[\mathbf{u}\cdot(\mathbf{k}-\mathbf{k}')]^2 = q_z^2 \to \frac{1}{3}q^2$ when we average over all directions of **k**. The mutual orientation of **k** and **k'** gives more surprises, however. From $q^2 = |\mathbf{k}'-\mathbf{k}|^2 \approx 2k_F^2(1-\cos\theta)$ we have $qdq = k_F^2 \sin\theta \, d\theta$; this yields after integration over $d\phi$

$$\oint_{\phi \text{ only}} dS_{F'} = 2\pi k_F^2 \sin\theta \, d\theta = \tfrac{1}{2} S_F q dq / k_F^2, \qquad (15.5\text{-}24)$$

where we approximated the Fermi surface by a sphere. The remaining θ integration has now been replaced by an integration over all phonon wave-vector magnitudes! Sondheimer has shown that we must have $q \le 2k$ in order for momentum conservation (excluding umklapp) to take place. However, the condition that q be limited by the Debye wave vector q_D is usually the more stringent one. We therefore obtain

$$A = \left(\frac{V_0}{8\pi^3 \hbar}\right)^2 \frac{2\pi\hbar C_1^2}{\rho V_0 k_B T} \frac{S_F^2}{6k_F^2 v_F^2} \int_0^{q_D} \frac{q^5 dq}{(e^{\hbar\omega_{\mathbf{q}}/k_B T}-1)(1-e^{-\hbar\omega_{\mathbf{q}}/k_B T})}. \qquad (15.5\text{-}25)$$

We now will evaluate the second integral B; from its definition [see (15.5-18)] we readily find, approximating $\partial f_0/\partial\varepsilon$ by $-\delta(\varepsilon-\varepsilon_F)$,

$$B = \left(\frac{V_0}{8\pi^3}\right)^2 \frac{2e^2}{\hbar^2} \left[\int\oint d\varepsilon dS_{\mathbf{k}} (\mathbf{k}\cdot\mathbf{u}) \delta(\varepsilon-\varepsilon_F)\right]^2 = \frac{2}{3}\left(\frac{V_0}{8\pi^3}\right)^2 \left(\frac{eS_F k_F}{\hbar}\right)^2. \qquad (15.5\text{-}26)$$

This finally yields

$$\rho_L = \frac{3\pi\hbar C_1^2}{2e^2 mn k_B T k_F^4 v_F^2} \int_0^{q_D} \frac{q^5 dq}{(e^{\hbar\omega_{\mathbf{q}}/k_B T}-1)(1-e^{-\hbar\omega_{\mathbf{q}}/k_B T})}. \qquad (15.5\text{-}27)$$

We define the bosonic integrals (also met in the Debye theory of the specific heat):

$$\mathcal{J}_n(x) = \int_0^x \frac{z^n e^z}{(e^z-1)^2} dz \quad \text{with } \mathcal{J}_n(x) \sim \int_0^\infty \frac{nz^{n-1}}{e^z-1} dz = \Gamma(n+1)\zeta(n). \qquad (15.5\text{-}28)$$

---

[15] I.S. Gradshteyn and I.M. Ryzhik, Op. Cit., Section 3.31.

Employing the Debye temperature $\Theta_D = \hbar u_0 q_D / k_B$ we obtain as final result

$$\rho_L = \frac{3\pi\hbar C_1^2 q_D^6}{2e^2 mn\, k_B \Theta_D k_F^4 v_f^2}\left(\frac{T}{\Theta_D}\right)^5 \mathcal{J}_5\left(\frac{T}{\Theta_D}\right), \tag{15.5-29}$$

which is the renown Bloch formula.[16] It describes quite well the lattice resistivity of a pure metal. For low temperatures we have the T^5 law, similar to the Debye T^3 law of the specific heat. Also, we have the approximate result $\mathcal{J}_5(1) \approx 0.25$. With the asymptotic low temperature value $\mathcal{J}_5 \sim 5!\zeta(5) = 124.4$ we then have

$$\rho_L(T) \approx 500\,(T/\Theta_D)^5 \rho_L(\Theta_D). \tag{15.5-30}$$

15.5.2 *Acoustic Phonon Scattering in Nondegenerate Semiconductors*

Terms $1 - f(\mathbf{k})$ in the Boltzmann equation will be approximated by 1. We make the usual streaming-vector ansatz (15.2-1) and assume that there is no magnetic field. The collision integral is

$$-\mathcal{B}_{\mathbf{k}} f(\mathbf{k}) = \sum\nolimits_{\mathbf{k}'} [w_{\mathbf{k}'\mathbf{k}}\, \mathbf{v}_{\mathbf{k}'} \cdot \mathbf{g}_1(\varepsilon_{\mathbf{k}'})/v_{\mathbf{k}'} - w_{\mathbf{k}\mathbf{k}'}\, \mathbf{v}_{\mathbf{k}} \cdot \mathbf{g}_1(\varepsilon_{\mathbf{k}})/v_{\mathbf{k}}]. \tag{15.5-31}$$

We choose $\mathbf{g}_1$ to be along the z-axis and denote by χ and χ' the angles of $\mathbf{k}$ and $\mathbf{k}'$ with respect to $\mathbf{g}_1$. The pictures to be drawn are similar to Fig. 14-1, but, in addition, we put in the vector $\mathbf{q} = \mathbf{k}' - \mathbf{k}$ (absorption) or $\mathbf{q} = \mathbf{k} - \mathbf{k}'$(emission); the pictures are shown in Figs. 15-2a and 15.2b. Note that the vectors $\mathbf{g}_1$, $\mathbf{k}$ and $\mathbf{k}'$ are not coplanar, but engender an azimuthal angle φ. Employing detailed balance, $f_0(\varepsilon_{\mathbf{k}}) w_{\mathbf{k}\mathbf{k}'} = f_0(\varepsilon_{\mathbf{k}'}) w_{\mathbf{k}'\mathbf{k}}$ and $\varepsilon(\mathbf{k}) \approx \varepsilon(\mathbf{k}')$, since the collisions are near-elastic, we find

$$-\mathcal{B}_{\mathbf{k}} f(\mathbf{k}) \approx -\cos\chi\, g_1(\varepsilon_{\mathbf{k}}) \sum_{\mathbf{k}'} w_{\mathbf{k}\mathbf{k}'} [1 - \cos\chi'/\cos\chi]. \tag{15.5-32}$$

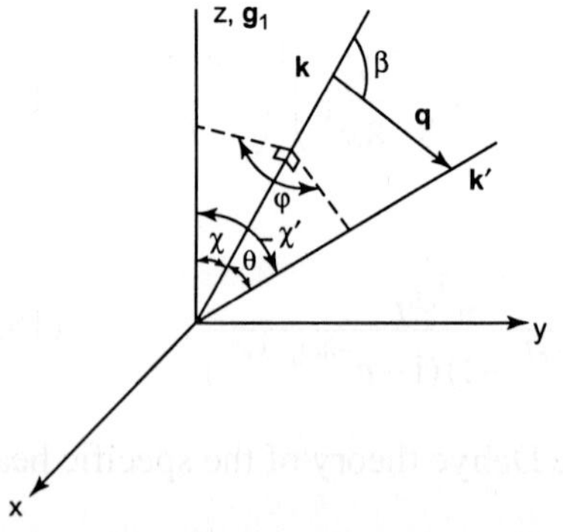

Fig. 15-2a. Pertaining to phonon absorption.

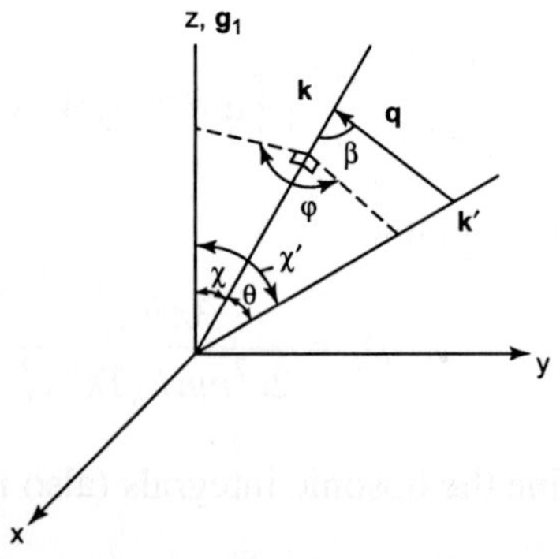

Fig. 15-2b. Pertaining to phonon emission.

---

[16] F. Bloch, Zeitschr. für Physik **59**, 208 (1930). Our result differs by a factor of two, since we took the spin into account in the streaming integral, while noting that spin is conserved in a collision.

We thus have a relaxation time

$$\frac{1}{\tau(\varepsilon_{\mathbf{k}})} = \sum_{\mathbf{k}'} w_{\mathbf{kk}'} \left[1 - \frac{\cos\chi'}{\cos\chi} \right]. \qquad (15.5\text{-}33)$$

Next, consider **k** to be the polar axis; with (14.2-15) and the sine rule $k'\sin\theta = q\sin\beta$ we find

$$\cos\chi'/\cos\chi = 1 \mp (q/k)\cos\beta + (q/k)\tan\chi \sin\beta\cos\varphi, \qquad (15.5\text{-}34)$$

where the upper sign pertains to emission and the lower sign to absorption; this fixes $1/\tau$. Next we use the results

$$w_{\mathbf{kk}'}^{em} = (V_0/4\pi^2\hbar)\int d^3q[F(\mathbf{q})+1](\hbar q C_1^2/2V_0\rho u_0)\delta(\varepsilon_{\mathbf{k}} - \varepsilon_{\mathbf{k}'} - \hbar\omega_{\mathbf{q}})\delta_{\mathbf{k}',\mathbf{k}-\mathbf{q}},$$
$$w_{\mathbf{kk}'}^{ab} = (V_0/4\pi^2\hbar)\int d^3q\, F(\mathbf{q})(\hbar q C_1^2/2V_0\rho u_0)\delta(\varepsilon_{\mathbf{k}} - \varepsilon_{\mathbf{k}'} + \hbar\omega_{\mathbf{q}})\delta_{\mathbf{k}',\mathbf{k}+\mathbf{q}}, \qquad (15.5\text{-}35)$$

where we substituted for $|\mathcal{F}(\mathbf{q})|^2$ from (8.6-11). With the spherical angles of **q** being β and φ and the transition rates depending only on |**q**| and β, the second part of (15.5-34) does not contribute because of $\int \cos\varphi\, d\varphi = 0$. Accordingly we find

$$\frac{1}{\tau(\varepsilon_k)} = \frac{C_1^2}{4\pi\rho u_0 k}\int q^4 dq \int_0^\pi d\beta \{[F(q)+1]\delta(\varepsilon_{\mathbf{k}} - \varepsilon_{\mathbf{k}-\mathbf{q}} - \hbar\omega_{\mathbf{q}})$$
$$- F(q)\delta(\varepsilon_{\mathbf{k}} - \varepsilon_{\mathbf{k}+\mathbf{q}} + \hbar\omega_{\mathbf{q}})\}\cos\beta\sin\beta, \qquad (15.5\text{-}36)$$

where q is integrated over the Brillouin zone, limits $(0, q_{BR})$. Most easily this is done as in Wilson's book.[17] Thus, expanding the argument of the delta function, we have

$$\delta(\varepsilon_{\mathbf{k}\mp\mathbf{q}} - \varepsilon_{\mathbf{k}} \pm \hbar\omega_{\mathbf{q}}) \approx \delta[\mp\mathbf{q}\cdot\nabla_{\mathbf{k}}\varepsilon_{\mathbf{k}} + \tfrac{1}{2}q^2(d^2\varepsilon_{\mathbf{k}}/dk^2) \pm \hbar\omega_{\mathbf{q}}]$$
$$= (1/q\hbar)\delta[\mp v_{\mathbf{k}}\cos\beta + (q\hbar/2m^*) \pm u_0], \qquad (15.5\text{-}37)$$

where we noted that $\delta(\alpha x) = \alpha^{-1}\delta(x)$. We introduce the new variables

$$y = \mp v_{\mathbf{k}}\cos\beta + (q\hbar/2m^*) \pm u_0, \quad Q = q\hbar/2m^*, \qquad (15.5\text{-}38)$$

with Jacobian

$$J = \frac{\partial(Q,y)}{\partial(q,\beta)} = \pm\frac{\hbar v_{\mathbf{k}}}{2m^*}\sin\beta, \qquad (15.5\text{-}39)$$

so that $dq\,d\beta \to dQ\,dy/|J|$. The domain of integration follows from the limits on q and β to be inserted in (15.5-38). The diagram is given in Figs. 15-3a and 15-3b. Since we have to integrate $\delta(y)$, only the shaded rectangles (of thickness $d \to 0$) are of importance. The upper limits on Q are then $v_{\mathbf{k}} \mp u_0 \approx v_{\mathbf{k}}$ since $u_0 \ll v_{\mathbf{k}}$. This yields

[17] A.H. Wilson, 2nd Ed., Op. Cit. Section 9.34. Wilson's treatment is more cumbersome, since he avoids the explicit delta functions, following the more exact expressions of time-dependent perturbation theory.

$$\frac{1}{\tau(\varepsilon_{\mathbf{k}})} \approx \frac{m^* C_1^2}{2\pi\rho u_0 k^2\hbar^2 v_{\mathbf{k}}}\int_0^{v_{\mathbf{k}}} Q^3 dQ \left\{[F(Q)+1]\left(\frac{2m^*}{\hbar}\right)^3\left(\frac{Q+u_0}{v_{\mathbf{k}}}\right)+F(Q)\left(\frac{2m^*}{\hbar}\right)^3\left(\frac{Q-u_0}{v_{\mathbf{k}}}\right)\right\}$$
(15.5-40)

Notice that the factors $(\pm Q+u_0)/v_{\mathbf{k}}$ stem from the factors $\cos\beta$ that follow from (15.5-38) for $y=0$. The final result is now immediate,

$$\frac{1}{\tau(\varepsilon_{\mathbf{k}})} = \frac{4m^{*5}C_1^2}{\pi\rho u_0 k^2\hbar^6 v_{\mathbf{k}}}\int_0^{v_{\mathbf{k}}} Q^4 dQ[2F(Q)+1]. \tag{15.5-41}$$

For sufficiently high temperatures, $\hbar\omega_{\mathbf{q}}/k_BT \ll 1$, we have $F(Q)\approx k_BT/\hbar\omega_{\mathbf{q}} \gg 1$. This yields

$$\frac{1}{\tau(\varepsilon_{\mathbf{k}})} = \frac{\sqrt{2}}{\pi}\frac{C_1^2 (m^*)^{3/2} k_BT\varepsilon_{\mathbf{k}}^{1/2}}{\rho u_0^2\hbar^4}. \tag{15.5-42}$$

Clearly, there is an energy-independent mean free path,

$$\ell = \frac{\pi\rho u_0^2\hbar^4}{C_1^2 m^{*2} k_BT}. \tag{15.5-43}$$

Finally, for the mobility we obtain

$$\mu_{ac} = \frac{e}{m^*}\frac{\overline{\overline{v_{\mathbf{k}}^2\tau(\varepsilon_{\mathbf{k}})}}^{(0)}}{\overline{\overline{v_{\mathbf{k}}^2}}^{(0)}} = \frac{2\sqrt{2\pi}}{3}\frac{e\rho u_0^2\hbar^4}{C_1^2 (m^*)^{5/2}}\left(\frac{1}{k_BT}\right)^{3/2}. \tag{15.5-44}$$

We thus found the well-known $T^{-3/2}$ dependence, with the mobility decreasing with increasing temperature due to the increase of the lattice vibrational energy.

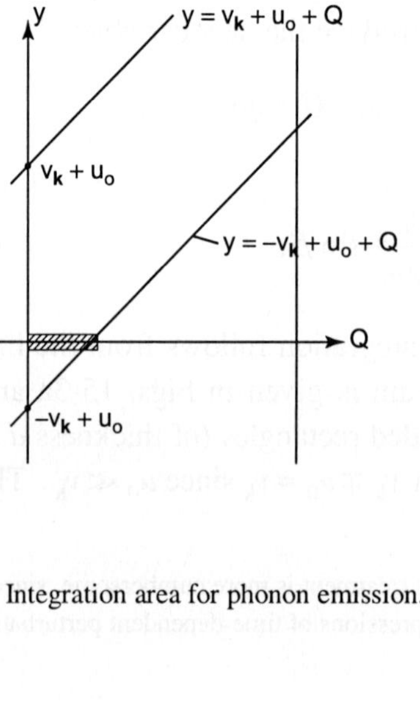

Fig. 15-3a. Integration area for phonon emission.

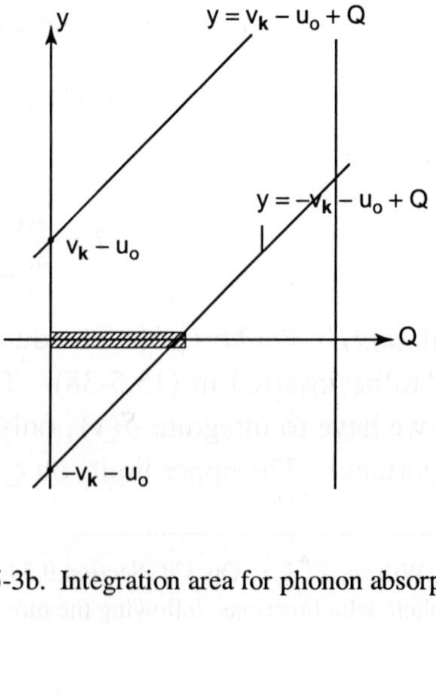

Fig. 15-3b. Integration area for phonon absorption.

2. Transport Far from Equilibrium; Steady-State Distributions and Flow

15.6 The Coupled Boltzmann Equations in the v-Language; Expansion in Spherical Polynomials

In this part we shall consider solutions to the Boltzmann equation far from equilibrium. There is nothing prohibiting us from doing so: the Boltzmann equation was based on mechanical premises and it was only Boltzmann's genius, that endowed the results with thermodynamic attributes such as temperature and entropy. The equation should be valid, however, in a stationary state with no detailed balance, although total balance of transition rates must still occur. Local equilibrium distributions, such as we considered before, will have no direct place in the solutions of the transport equation, although they may serve as 'asymptotic checks' on the solutions if we wish to approach equilibrium *a posteriori.* We will first deal with heating in plasmas – a topic of practical importance if temperatures capable of fusion in magnetically confined plasmas are to be reached in apparatus like the 'Tokamac' – and next with hot electron gases in semiconductors subject to high electric fields.

For ionized gases we employ the **v**,**r**-description and some symmetry is presumed. The imposition of boundary conditions will be left to more technically oriented treatments. Generally, we assume that the solution can be expressed by an expansion in Legendre polynomials. We thus write[18]

$$f(\mathbf{r},\mathbf{v}) = \sum_{\ell=0}^{\infty} g_\ell(\mathbf{r},\mathbf{v}) P_\ell(\cos\chi)\,, \tag{15.6-1}$$

Obviously, up to first order, this is of the same form as encountered previously,

$$f(\mathbf{r},\mathbf{v}) = g_0(\mathbf{r},\mathbf{v}) + g_1(\mathbf{r},\mathbf{v})\cos\chi = g_0(\mathbf{r},\mathbf{v}) + \mathbf{g}_1(\mathbf{r},\mathbf{v})\cdot\mathbf{v}/v\,, \tag{15.6-2}$$

where $\mathbf{g}_1$ is the first-order vector and polar axis, while χ gives the direction of **v**. The problem is now that neither g_0 nor g_1 is known; generally, we will obtain coupled equations for the series (15.6-1) which must be truncated at some desirable stage.

(a) *The streaming terms*

We first consider the spatial gradient term. To simplify matters, we assume at first that this is the only streaming term, and we take the polar axis is along $\nabla_\mathbf{r}$. Then,

$$\mathbf{v}\cdot\nabla_\mathbf{r}[g_\ell(r,v)P_\ell(\cos\chi)] = v\cos\chi P_\ell(\cos\chi)\partial g_\ell/\partial z\,. \tag{15.6-3}$$

[18] These developments are taken from class notes of my late colleague at the University of Minnesota: Hendrik J. Oskam, "Plasma Physics", University of Minnesota, 1966 (unpublished).

using the identity[19]

$$(2\ell+1)\cos\chi\, P_\ell(\cos\chi) = (\ell+1)P_{\ell+1}(\cos\chi) + \ell P_{\ell-1}(\cos\chi)\,, \qquad (15.6\text{-}4)$$

this yields

$$\mathbf{v}\cdot\nabla_{\mathbf{r}}[g_\ell(r,v)P_\ell(\cos\chi)] = v\frac{(\ell+1)P_{\ell+1}(\cos\chi) + \ell P_{\ell-1}(\cos\chi)}{2\ell+1}\frac{\partial g_\ell}{\partial z}. \qquad (15.6\text{-}5)$$

We now sum over ℓ and rearrange according to orders of P_ℓ; we thus obtain

$$\mathbf{v}\cdot\nabla_{\mathbf{r}} f(\mathbf{r},\mathbf{v}) = v\sum_{\ell=0}^{\infty}\frac{\partial}{\partial z}\left[\frac{\ell}{2\ell-1}g_{\ell-1} + \frac{\ell+1}{2\ell+3}g_{l+1}\right]P_\ell(\cos\chi)\,. \qquad (15.6\text{-}6)$$

For the first two terms this gives

$$\mathbf{v}\cdot\nabla_{\mathbf{r}} f(\mathbf{r},\mathbf{v}) = \frac{v}{3}\frac{\partial g_1}{\partial z} + v\cos\chi\frac{\partial}{\partial z}\left(g_0 + \frac{2}{5}g_2\right). \qquad (15.6\text{-}7)$$

The restriction on the choice of polar axis is now removed, noting that the scalar product $\mathbf{A}\cdot\mathbf{B}$ is invariant against rotation. Thus (15.6-7) entails the general form

$$\mathbf{v}\cdot\nabla_{\mathbf{r}} f(\mathbf{r},\mathbf{v}) = \frac{v}{3}\nabla_{\mathbf{r}}\cdot\mathbf{g}_1 + \mathbf{v}\cdot\nabla_{\mathbf{r}}\left(g_0 + \frac{2}{5}g_2\right), \qquad (15.6\text{-}8)$$

with the direction of $\mathbf{g}_1$ still being at liberty.

Next we look at the electrical force term, $\mathbf{F} = q\mathbf{E}$. taking temporarily the polar axis along $\mathbf{F}$, we have

$$(q\mathbf{E}/m)\cdot\nabla_{\mathbf{v}}[g_\ell(\mathbf{r},\mathbf{v})P_\ell(\cos\chi)] = (qE/m)(\partial/\partial v_z)[g_\ell(\mathbf{r},\mathbf{v})P_\ell(\cos\chi)]\,. \qquad (15.6\text{-}9)$$

Since $v_z = v\cos\chi$, we have

$$\begin{aligned}\frac{\partial}{\partial v_z} &= \left(\frac{\partial v}{\partial v_z}\right)_{v_x,v_y}\frac{\partial}{\partial v} + \left(\frac{\partial\cos\chi}{\partial v_z}\right)_{v_x,v_y}\frac{\partial}{\partial(\cos\chi)}\\ &= \cos\chi\frac{\partial}{\partial v} + \left(\frac{\partial v}{\partial v_z}\right)_{v_x,v_y}\left[\frac{v_z}{(v_x^2+v_y^2+v_z^2)^{1/2}}\right]\frac{\partial}{\partial(\cos\chi)}\\ &= \cos\chi\frac{\partial}{\partial v} + \frac{\sin^2\chi}{v}\frac{\partial}{\partial(\cos\chi)}. \end{aligned} \qquad (15.6\text{-}10)$$

We also use the identity[20]

$$-\sin^2\chi\frac{\partial P_\ell(\cos\chi)}{\partial(\cos\chi)} = \ell\cos\chi\, P_\ell(\cos\chi) - \ell P_{\ell-1}(\cos\chi)\,, \qquad (15.6\text{-}11)$$

---

[19] Milton Abramowitz and Irene A. Stegun, "Handbook of Mathematical Functions" National Bureau of Standards, 1964, Chapter 8.

[20] Or, $(y^2-1)\partial P_\ell(y)/\partial y = \ell y P_\ell - \ell P_{\ell-1}$.

together with (15.6-4), to give

$$(q\mathbf{E}/m)\cdot\nabla_{\mathbf{v}}[g_\ell(\mathbf{r},\mathbf{v})P_\ell(\cos\chi)] = (qE/m)$$
$$\times\left\{\left[\frac{\ell+1}{2\ell+1}P_{\ell+1}(\cos\chi)+\frac{\ell}{2\ell+1}P_{\ell-1}(\cos\chi)\right]\frac{\partial g_\ell}{\partial v}+\frac{g_\ell}{v}\frac{\ell+1}{2\ell+1}[P_{\ell-1}(\cos\chi)-P_{\ell+1}(\cos\chi)]\right\}. \tag{15.6-12}$$

Summing over ℓ and rearranging according to the order of the P_ℓ, we then find

$$(q\mathbf{E}/m)\cdot\nabla_{\mathbf{v}}f(\mathbf{r},\mathbf{v})$$
$$=(qE/m)\sum_\ell\left\{\frac{\ell}{2\ell-1}v^{\ell-1}\frac{\partial}{\partial v}\left(\frac{g_{\ell-1}}{v^{\ell-1}}\right)+\frac{\ell+1}{2\ell+3}\frac{1}{v^{\ell+2}}\frac{\partial}{\partial v}\left(v^{\ell+2}g_{\ell+1}\right)\right\}P_\ell(\cos\chi). \tag{15.6-13}$$

For the first two terms of the expansion we now have

$$(q\mathbf{E}/m)\cdot\nabla_{\mathbf{v}}f(\mathbf{r},\mathbf{v})=(qE/m)\left\{\frac{1}{3v^2}\frac{\partial}{\partial v}(v^2g_1)+\left[\frac{\partial g_0}{\partial v}+\frac{2}{5}\frac{1}{v^3}\frac{\partial}{\partial v}(v^3g_2)\right]\cos\chi\right\}. \tag{15.6-14}$$

We now drop again the particular choice of polar axis, to obtain the general result

$$(q\mathbf{E}/m)\cdot\nabla_{\mathbf{v}}f(\mathbf{r},\mathbf{v})=(q\mathbf{E}/m)\cdot\left\{\frac{1}{3v^2}\frac{\partial}{\partial v}(v^2\mathbf{g}_1)+\mathbf{v}\left[\frac{1}{v}\frac{\partial g_0}{\partial v}+\frac{2}{5}\frac{1}{v^4}\frac{\partial}{\partial v}(v^3g_2)\right]\right\}. \tag{15.6-15}$$

Finally we consider the magnetic force term, $q(\mathbf{v}/mc)\times\mathbf{B}\cdot\nabla_{\mathbf{v}}f$. Clearly, the equivalent of the second bracket of (15.6-15) gives zero. Rests the equivalent of the first term, which is found to be

$$\textit{Magn. streaming term} = q\frac{\mathbf{v}}{mc}\times\mathbf{B}\cdot\frac{1}{3v^2}\frac{\partial}{\partial v}(v^2\mathbf{g}_1)=\mathbf{v}\cdot\frac{q\mathbf{B}}{mc}\times\frac{1}{3v^2}\frac{\partial}{\partial v}(v^2\mathbf{g}_1)$$
$$=-\mathbf{v}\cdot\boldsymbol{\omega}_c\times\frac{1}{3v^2}\frac{\partial}{\partial v}[v^3(\mathbf{g}_1/v)]=-\mathbf{v}\cdot\boldsymbol{\omega}_c\times(\mathbf{g}_1/v)-\frac{1}{3}\mathbf{v}\cdot\boldsymbol{\omega}_c\times\frac{\partial\mathbf{g}_1}{\partial v}, \tag{15.6-16}$$

where we defined again the cyclotron frequency vector, $\boldsymbol{\omega}_c=-(q\mathbf{B}/mc)$.

(b) *The collision term*

The collision term for plasmas involving electron-ion collisions is linear in the electron distribution. We write down (15.6-1) both for $f(\mathbf{r},\mathbf{v})$ and $f(\mathbf{r},\mathbf{v}')$. Let as in Fig. 14-1 χ be the angle of $\mathbf{v}'$ with the polar axis and let θ_0 be the angle between $\mathbf{v}$ and $\mathbf{v}'$ in the laboratory system. Then by the addition theorem of spherical harmonics

$$P_\ell(\cos\chi')=P_\ell(\cos\chi)P_\ell(\cos\theta_0)+2\sum_{m=1}^{\ell}\frac{(\ell-m)!}{(\ell+m)!}P_\ell^m(\cos\chi)P_\ell^m(\cos\theta_0)\cos m\varphi, \tag{15.6-17}$$

where $P_\ell^m(y)$ are the associated Legendre polynomials. For the collision integral (13-2-31) we have the form

$$-\mathcal{B}_v f(\mathbf{r},\mathbf{v}) = \sum_{\ell=0}^{\infty}\left\{\iint d^3V' u\sigma(\Omega)d\Omega\left[\frac{|\mathbf{v}'-\mathbf{V}'|}{|\mathbf{v}-\mathbf{V}'|}\right]^3 g_\ell(\mathbf{r},\mathbf{v}')P_\ell(\cos\chi')F(\mathbf{V}')\right.$$

$$\left.-\iint d^3V u\sigma(\Omega)d\Omega\, g_\ell(\mathbf{r},\mathbf{v})P_\ell(\cos\chi)F(\mathbf{V})\right\}. \tag{15.6-18}$$

We substitute (15.6-17) and notice that $\int_0^{2\pi}\cos m\varphi\, d\varphi = 0$, thus being left with

$$-\mathcal{B}_v f(\mathbf{r},\mathbf{v}) = \sum_{\ell=0}^{\infty} P_\ell(\cos\chi)\left\{\iint d^3V' u\sigma(\Omega)d\Omega\left[\frac{|\mathbf{v}'-\mathbf{V}'|}{|\mathbf{v}-\mathbf{V}'|}\right]^3 g_\ell(\mathbf{r},\mathbf{v}')P_\ell(\cos\theta_0)F(\mathbf{V}')\right.$$

$$\left.-\iint d^3V u\sigma(\Omega)d\Omega\, g_\ell(\mathbf{r},\mathbf{v})\,F(\mathbf{V})\right\}; \tag{15.6-19}$$

clearly, this term also neatly expands in Legendre polynomials. If we take just two terms, we shall write

$$-\mathcal{B}_v f(\mathbf{r},\mathbf{v}) = C_0 + C_1\cos\chi = C_0 + \mathbf{v}\cdot(\mathbf{C}_1/v)\,, \tag{15.6-20}$$

where we chose $\mathbf{C}_1$ to have the same direction as $\mathbf{g}_1$. One obtains the following forms:

$$C_0 = \iint d^3V' u\sigma(\Omega)d\Omega\left[\frac{|\mathbf{v}'-\mathbf{V}'|}{|\mathbf{v}-\mathbf{V}'|}\right]^3 g_0(\mathbf{r},\mathbf{v}')\,F(\mathbf{V}')$$

$$-\iint d^3V u\sigma(\Omega)d\Omega\, g_\ell(\mathbf{r},\mathbf{v})\,F(\mathbf{V})\Big\}, \tag{15.6-21}$$

$$C_1 = \iint d^3V' u\sigma(\Omega)d\Omega\left[\frac{|\mathbf{v}'-\mathbf{V}'|}{|\mathbf{v}-\mathbf{V}'|}\right]^3 g_1(\mathbf{r},\mathbf{v}')\cos\theta_0\, F(\mathbf{V}')$$

$$-\iint d^3V u\sigma(\Omega)d\Omega\, g_1(\mathbf{r},\mathbf{v})\,F(\mathbf{V})\Big\}. \tag{15.6-22}$$

(c) *The coupled Boltzmann equations for g_0 and g_1*

We collect all terms that we computed for the Boltzmann equation, bringing as usual the magnetic streaming term to the collision-operator side.[21] We then have

$$\frac{v}{3}\nabla_{\mathbf{r}}\cdot\mathbf{g}_1 + \mathbf{v}\cdot\nabla_{\mathbf{r}}\left(g_0+\frac{2}{5}g_2\right) + \mathbf{v}\cdot\left\{\frac{q\mathbf{E}}{m}\left[\frac{1}{v}\frac{\partial g_0}{\partial v}+\frac{2}{5}\frac{1}{v^4}\frac{\partial}{\partial v}(v^3 g_2)\right]\right\}$$

$$+\frac{q\mathbf{E}}{m}\cdot\frac{1}{3v^2}\frac{\partial}{\partial v}(v^2\mathbf{g}_1) = C_0 + C_1\cos\chi + \mathbf{v}\cdot\boldsymbol{\omega}_c\times(\mathbf{g}_1/v) + \frac{1}{3}\mathbf{v}\cdot\boldsymbol{\omega}_c\times\frac{\partial\mathbf{g}_1}{\partial v}. \tag{15.6-23}$$

---

[21] The magnetic field does not contribute to streaming proper, but curves the paths between collisions.

If there is no magnetic field we choose as polar axis

$$\mathbf{z} \equiv \nabla_{\mathbf{r}}\left(g_0 + \frac{2}{5}g_2\right) + \frac{q\mathbf{E}}{m}\left[\frac{1}{v}\frac{\partial g_0}{\partial v} + \frac{2}{5}\frac{1}{v^4}\frac{\partial}{\partial v}(v^3 g_2)\right] \tag{15.6-24}$$

(the reader will recognise that this is the 'streaming vector' of the previous sections). Then (15.6-23 also reads

$$\frac{v}{3}\nabla_{\mathbf{r}} \cdot \mathbf{g}_1 + \frac{q\mathbf{E}}{m} \cdot \frac{1}{3v^2}\frac{\partial}{\partial v}(v^2 \mathbf{g}_1) + vz\cos\chi = C_0 + C_1 \cos\chi . \tag{15.6-25}$$

Equating separately the terms without direction dependence and the terms in cosχ (corresponding to the Legendre polynomials with $\ell = 1, 2$, respectively), we obtain the coupled Boltzmann equations for g_0 and $\mathbf{g}_1$:

$$\begin{aligned} &\frac{v}{3}\nabla_{\mathbf{r}} \cdot \mathbf{g}_1 + \frac{q\mathbf{E}}{m} \cdot \frac{1}{3v^2}\frac{\partial}{\partial v}(v^2 \mathbf{g}_1) = C_0 , \\ &\nabla_{\mathbf{r}}\left(g_0 + \frac{2}{5}g_2\right) + \frac{q\mathbf{E}}{m}\left[\frac{1}{v}\frac{\partial g_0}{\partial v} + \frac{2}{5}\frac{1}{v^4}\frac{\partial}{\partial v}(v^3 g_2)\right] = \mathbf{C}_1 / v . \end{aligned} \tag{15.6-26}$$

In the presence of a magnetic field the present method does not work, since the vector product in (15.6-23) requires the occurrence of associated Legendre polynomials. A faster solution for the leading terms is obtained if we split (15.6-23) into parts which are symmetric under **v**-inversion and antisymmetric (sign change) under **v**-inversion. We then easily find

$$\frac{v}{3}\nabla_{\mathbf{r}} \cdot \mathbf{g}_1 + \frac{q\mathbf{E}}{m} \cdot \frac{1}{3v^2}\frac{\partial}{\partial v}(v^2 \mathbf{g}_1) = C_0 , \tag{15.6-27}$$

$$\nabla_{\mathbf{r}}\left(g_0 + \frac{2}{5}g_2\right) + \frac{q\mathbf{E}}{m}\left[\frac{1}{v}\frac{\partial g_0}{\partial v} + \frac{2}{5}\frac{1}{v^4}\frac{\partial}{\partial v}(v^3 g_2)\right] = \frac{\mathbf{C}_1}{v} + \frac{\boldsymbol{\omega}_c \times \mathbf{g}_1}{v} + \frac{1}{3}\boldsymbol{\omega}_c \times \frac{\partial \mathbf{g}_1}{\partial v} , \tag{15.6-28}$$

where we chose the direction of $\mathbf{g}_1$ (i.e., the polar axis) such that the vector implied by (15.6-28) [*lhs* – *rhs*] has no components along **v**. We still note that the term with $\partial \mathbf{g}_1 / \partial v$ is of similar smallness as g_2. Since we do not want to involve a higher order equation yet, we will truncate the coupled equations by omitting g_2 and $\partial \mathbf{g}_1 / \partial v$. More complete and correct results may be obtained with other methods, such as Monte Carlo techniques.

15.7 The Zero-order and First-order Collision Integrals in a Binary Plasma

We shall now evaluate the zero-order collision integral (15.6-21). For infinitely heavy ions and completely elastic collisions we have $\mathbf{V}' = 0$, $\mathbf{v}' = \mathbf{u}' = \mathbf{u} = \mathbf{v}$, which gives $C_0 = 0$. In reality the ions are not infinitely heavy, however, which causes energy transfers. In addition, a certain number of inelastic collisions will occur, due to excitations of the ions or neutral atoms. We shall allow for this by letting u' be possibly different from u. The collision integral in zero-order then reads

$$C_0 = \int\int d^3V' u'\sigma(\Omega)d\Omega \left[\frac{|\mathbf{v}'-\mathbf{V}'|}{|\mathbf{v}-\mathbf{V}'|}\right]^3 g_0(\mathbf{r},\mathbf{v}')F(\mathbf{V}') - \int\int d^3V u\sigma(\Omega)d\Omega\, g_\ell(\mathbf{r},\mathbf{v})F(\mathbf{V})\}. \tag{15.7-1}$$

At first we shall neglect the thermal motion of the ions, i.e., $T_i = 0$. Then for the collisions of the first integral, involving inverse collisions $(\mathbf{v}',\mathbf{V}') \to (\mathbf{v},\mathbf{V})$, we have $\mathbf{V}'=0$, giving for the large square bracket the value (v'^3/v^3); the ions recoil, however, under impact, so $\mathbf{V} \neq 0$. The relative energy loss is given by (13.2-25):

$$v^2 - v'^2 = \frac{2Mm}{(M+m)^2} v^2(1-\cos\theta) \equiv Kv^2. \tag{15.7-2}$$

We recall that θ is the scattering angle in the centre-of-mass system, see Fig. 13-3. Now consider the quantity $u'v'^3 g_0(v')$ to be a function of v'^2. Then by first order Taylor expansion, cf. W.P. Allis, Op. Cit. (Chapter 13, Ref. 6), we have

$$u'v'^3 g_0(v') = uv^3 g_0(v) + Kv^2 (d/dv^2)[uv^3 g_0(v)]. \tag{15.7-3}$$

This is substituted into (15.7-1). We easily obtain by integration over d^3V', with $\hat{N}$ being the ion density,

$$C_0 = \hat{N}\int \sigma(\Omega)d\Omega \frac{K}{v}\frac{d}{dv^2}[uv^3 g_0(v)] = \frac{2Mm}{(M+m)^2}\frac{d}{v\,dv^2}[(\tau_p^*)^{-1} v^3 g_0(v)], \tag{15.7-4}$$

where we introduced the momentum-loss relaxation time in the centre-of-mass system, τ_p^*, defined by

$$\frac{1}{\tau_p^*} = \hat{N}\int \sigma(\Omega)\,d\Omega\, u(1-\cos\theta). \tag{15.7-5}$$

This time can be related to its equivalent in the laboratory system, τ_p – defined as the above with $\theta \to \theta_0$ – by employing (13.2-27); one easily shows

$$(1-\cos\theta_0)\,d(\cos\theta_0) \approx \frac{M}{M+m}(1-\cos\theta)\,d(\cos\theta), \text{ or, } \tau_p^* \approx \frac{M}{M+m}\tau_p. \tag{15.7-6}$$

This then yields

$$C_0 \approx \frac{2m}{M+m}\frac{d}{v\,dv^2}[\tau_p^{-1} v^3 g_0(v)]. \tag{15.7-7}$$

Next, we consider ions with a finite temperature. Although $g_0(v)$ is a non-equilibrium distribution, we make the *adiabatic assumption* that the ions stay close to equilibrium, *i.e.*, $F(\mathbf{V}')$ is a local Maxwellian with $T_i = T(\mathbf{r})$. The effect of a finite $\mathbf{V}'$ corresponds to a tolerance $\mathcal{O}(V'^2)$ in the energies of the electrons. So we set $g_0(v) \to g_0(v) + \mathcal{O}(V'^2)\, dg_0/dv^2$. This causes the collision integral to take the form

$$C_0 = \frac{2m}{M+m}\frac{d}{v\,dv^2}\left\{\tau_p^{-1}v^3\left[g_0(v) + \mathcal{O}(V'^2)\frac{dg_0}{dv^2}\right]\right\}. \tag{15.7-8}$$

The correction $\mathcal{O}(V'^2)$ will be determined by the requirement that C_0 vanishes if the electrons have an electron temperature $T_e = T(\mathbf{r})$. In that case

$$dg_0/dv^2 = -g_0 m/2k_BT_e = -g_0 m/2k_BT(\mathbf{r}), \tag{15.7-9}$$

so that $\mathcal{O}(V'^2) \approx 2k_BT(\mathbf{r})/m$. The final result for the collision integral in zeroth-order is therefore

$$C_0 = \frac{m}{M+m}\frac{1}{v^2}\frac{d}{dv}\left\{\tau_p^{-1}v^3\left[g_0(v) + \frac{k_BT(\mathbf{r})}{mv}\frac{dg_0}{dv}\right]\right\}. \tag{15.7-10}$$

For the collision integral in first order C_1 we make somewhat more crude approximations. We set the large square bracket in (15.6-22) equal to unity and we approximate $g_1(\mathbf{r},\mathbf{v}') \approx g_1(\mathbf{r},\mathbf{v})$. We then readily obtain

$$C_1 \approx -\hat{N} g_1(\mathbf{r},\mathbf{v})\int \sigma(\Omega)\, d\Omega\, u(1-\cos\theta_0) = -g_1/\tau_p\,. \tag{15.7-11}$$

15.8 Electron Heating in Plasmas: the Druyvesteyn Distribution

We proceed to solve the coupled equations for g_0 and g_1, omitting higher order terms. The equations then read

$$\frac{v}{3}\nabla_{\mathbf{r}}\cdot\mathbf{g}_1 + \frac{q\mathbf{E}}{m}\cdot\frac{1}{3v^2}\frac{d}{dv}(v^2\mathbf{g}_1) = C_0\,, \tag{15.8-1}$$

$$\nabla_{\mathbf{r}} g_0 + \frac{q\mathbf{E}}{mv}\frac{dg_0}{dv} = -\frac{1}{v}\left[\frac{\mathbf{g}_1}{\tau_p} + \mathbf{g}_1 \times \boldsymbol{\omega}_c\right]. \tag{15.8-2}$$

The latter equation has entirely the same form as (15.2-34). We can thus introduce a tensor $\boldsymbol{\tau}_p$ that has exactly the same form as in (15.2-37). Accordingly, for $\mathbf{g}_1$ we obtain

$$\mathbf{g}_1 = -v\boldsymbol{\tau}_p\cdot\left[\nabla_{\mathbf{r}} g_0 + \frac{q\mathbf{E}}{mv}\frac{dg_0}{dv}\right]. \tag{15.8-3}$$

This is substituted into (16.3-1) together with (16.2-10) and the terms involving $\nabla_{\mathbf{r}}$ are dropped, whereby we assume that the spatial dependence is sufficiently accounted for by the local field **E**(**r**) and the temperature *T*(**r**). We thus find

$$-\frac{1}{3}\left(\frac{q}{m}\right)^2\frac{d}{dv}\left[\frac{dg_0}{dv}v^2\mathsf{T}_p:\mathbf{EE}\right]=\frac{m}{M+m}\frac{d}{dv}\left[v^3\tau_p^{-1}\left(g_0+\frac{k_BT}{mv}\frac{dg_0}{dv}\right)\right] \quad (15.8\text{-}4)$$

which can at once be integrated to [with b.c. $g_0(v\to\infty)=0$]

$$-\frac{1}{3}\left(\frac{q}{m}\right)^2\frac{dg_0}{dv}\mathsf{T}_p:\mathbf{EE}=\frac{m}{M+m}\frac{v}{\tau_p}\left(g_0+\frac{k_BT}{mv}\frac{dg_0}{dv}\right). \quad (15.8\text{-}5)$$

Since T_p is antisymmetric, $\mathsf{T}_p:\mathbf{EE}=\sum_{i=x,y,z}\tau_{p,ii}E_i^2$. We now split **E** in a part $E_\perp$ perpendicular to **B** and a part $E_\parallel$ parallel to **B**; hence,

$$\mathsf{T}_p:\mathbf{EE}=\frac{\tau_p}{1+\omega_c^2\tau_p^2}E_\perp^2+\tau_pE_\parallel^2. \quad (15.8\text{-}6)$$

Equation (15.8-5) then gives the differential equation

$$\frac{dg_0}{g_0}=-\frac{\frac{1}{2}md(v^2)}{k_BT+\frac{1}{3}q^2(M+m)m^{-2}[E_\perp^2\tau_p^2(1+\omega_c^2\tau_p^2)^{-1}+E_\parallel^2\tau_p^2]}. \quad (15.8\text{-}7)$$

We will consider several cases.

(a) *Constant relaxation time*

Equation (15.8-7) integrates to

$$g_0=A_1\exp(-mv^2/2k_BT_e) \quad (15.8\text{-}8)$$

which is a Maxwellian with electron temperature

$$T_e=T+\frac{q^2}{3k_Bm}\frac{M+m}{m}\left[\frac{\tau_p^2}{1+\omega_c^2\tau_p^2}E_\perp^2+\tau_p^2E_\parallel^2\right]. \quad (15.8\text{-}9)$$

Note that for $\mathbf{E}=0$ the local equilibrium distribution $T_e=T(\mathbf{r})$ is recovered.

(b) *No magnetic field, constant mean free path*

Now $\tau_p=\lambda/v$. Assuming that *E* is large enough so that k_BT in the denominator can be neglected, the integration is straightforward:

$$g_0 = A_2 \exp\left[-\frac{3m}{M+m} \frac{\varepsilon^2}{q^2 E^2 \lambda^2} \right], \tag{15.8-10}$$

where $\varepsilon = \frac{1}{2}mv^2$. This is the Druyvesteyn distribution, an asymptotic distribution for high electric fields; note that there is no reference left to the ambient temperature.

(c) *Large B* $(\omega_c \tau_p \gg 1)$ *with* $\mathbf{E} \perp \mathbf{B}$.

The factor [ ] in the denominator of (15.8-7) becomes $E^2/\omega_c^2 = E^2 m^2 c^2 / q^2 B^2$; the time τ_p drops out. The integration yields again a Maxwellian

$$g_0 = A_3 \exp(-mv^2 / 2k_B T_e') \tag{15.8-11}$$

with electron temperature

$$T_e' = T + \tfrac{1}{3} k_B^{-1} (M+m) c^2 (E/B)^2 . \tag{15.8-12}$$

In the *H–L* units employed here, *E* and *B* have the same dimension; thus, although the expression involves the 'rest energy' $(M+m)c^2$, since we chose *B* to be very large, *E* must also be very large! This is one of the many problems in using heated plasmas to attain temperatures at which fusion can occur.[22]

15.9 Coupled Boltzmann Equations for Hot Electrons in Semiconductors

When we are in a stationary non-equilibrium state far from equilibrium, the energy of the electrons (or holes) cannot be conserved; rather heating of the electron (or hole) gas will occur. Clearly, for that case *k* is not a constant, which leads us to the choice of a slightly different ansatz

$$f(\mathbf{r},\mathbf{k}) = g_0(\mathbf{r},\varepsilon) + \mathbf{k} \cdot \vec{\mathcal{G}}_1(\mathbf{r},\varepsilon) . \tag{15.9-1}$$

This is to be substituted both in the streaming terms and the collision terms. For the zeroth-order contribution to streaming we have as usual

$$\mathbf{v_k} \cdot \nabla_\mathbf{r} g_0 + (q\mathbf{E}/\hbar) \cdot \nabla_\mathbf{k} g_0 = \mathbf{v_k} \cdot [\nabla_\mathbf{r} g_0 + q\mathbf{E}(\partial g_0 / \partial \varepsilon)] = \mathbf{v_k} \cdot \boldsymbol{\alpha} , \tag{15.9-2}$$

where $\boldsymbol{\alpha}$ is the streaming vector. For the first-order contribution without a magnetic field, using similar operations we find

$$\begin{aligned}
&\mathbf{v_k} \cdot \nabla_\mathbf{r} (\mathbf{k} \cdot \vec{\mathcal{G}}_1) + (q\mathbf{E}/\hbar) \cdot \nabla_\mathbf{k} (\mathbf{k} \cdot \vec{\mathcal{G}}_1) \\
&= \underbrace{\mathbf{v_k k}}_{\frac{1}{3} v_\mathbf{k} k \mathsf{I}} : [\nabla_\mathbf{r} \vec{\mathcal{G}}_1 + q\mathbf{E}(\partial \vec{\mathcal{G}}_1 / \partial \varepsilon)] + (q\mathbf{E}/\hbar) \cdot \vec{\mathcal{G}}_1 \\
&= \tfrac{1}{3} v_\mathbf{k} k [\nabla_\mathbf{r} \cdot \vec{\mathcal{G}}_1 + q\mathbf{E} \cdot (\partial \vec{\mathcal{G}}_1 / \partial \varepsilon)] + (q\mathbf{E}/\hbar) \cdot \vec{\mathcal{G}}_1 .
\end{aligned} \tag{15.9-3}$$

---

[22] See e.g., "Fusion", E. Teller, Ed., Vol. 1, Acad. Press, NY 1981, Chapter 3.

One may verify that the two contributions involving the electric field can be combined to give a result comparable to (15.6-14) employed for plasmas.[23] For the collision integral the formal result is $C_0 + \mathbf{k}\cdot(\mathbf{C}_1/k)$. The Boltzmann equation now reads

$$\mathbf{v_k}\cdot[\nabla_\mathbf{r} g_0 + \mathscr{q}\mathbf{E}(\partial g_0/\partial\varepsilon)] + \tfrac{1}{3}v_\mathbf{k}k[\nabla_\mathbf{r}\cdot\vec{\mathscr{G}}_1 + \mathscr{q}\mathbf{E}\cdot(\partial\vec{\mathscr{G}}_1/\partial\varepsilon)] + (\mathscr{q}\mathbf{E}/\hbar)\cdot\vec{\mathscr{G}}_1 = C_0 + \mathbf{k}\cdot(\mathbf{C}_1/k). \qquad (15.9\text{-}4)$$

Consider now inversion $\mathbf{k} \to -\mathbf{k}$. We separate the above into two coupled relations by equating separately the terms which are invariant under inversion and those which change sign under inversion; hence we obtain (note that $\mathbf{v_k}$ is parallel to $\mathbf{k}$) with $v_\mathbf{k}k = 2\varepsilon/\hbar$,

$$\frac{2\varepsilon}{3\hbar}\left\{\nabla_\mathbf{r}\cdot\vec{\mathscr{G}}_1(\varepsilon) + \mathscr{q}\mathbf{E}\cdot\left[\frac{\partial\vec{\mathscr{G}}_1}{\partial\varepsilon} + \frac{3}{2\varepsilon}\vec{\mathscr{G}}_1(\varepsilon)\right]\right\} = C_0\,, \qquad (15.9\text{-}5)$$

$$\nabla_\mathbf{r} g_0(\varepsilon) + \mathscr{q}\mathbf{E}(\partial g_0/\partial\varepsilon) = \boldsymbol{\alpha} = \mathbf{C}_1/v_\mathbf{k}\,. \qquad (15.9\text{-}6)$$

Our task is now twofold. Firstly, we must compute the collision integral up to first order; secondly, the above coupled equations must be solved simultaneously.

15.10 The Steady-State Distribution for a Hot Electron Gas

We consider acoustic phonon scattering, as discussed previously. In the collision integral the summation over $\mathbf{k}'$, involving the Kronecker deltas for momentum conservation, is carried out; this yields, suppressing the spatial variables,

$$-\mathscr{B}_\mathbf{k} f(\mathbf{k}) = (2\pi/\hbar)\sum_\mathbf{q} |\mathscr{F}(\mathbf{q})|^2$$
$$\times\{f(\mathbf{k}+\mathbf{q})[F(\mathbf{q})+1]\delta(\varepsilon_{\mathbf{k+q}} - \varepsilon_\mathbf{k} - \hbar\omega_\mathbf{q}) + f(\mathbf{k}-\mathbf{q})F(\mathbf{q})\delta(\varepsilon_{\mathbf{k-q}} - \varepsilon_\mathbf{k} + \hbar\omega_\mathbf{q})$$
$$-f(\mathbf{k})[F(\mathbf{q})+1]\delta(\varepsilon_{\mathbf{k-q}} - \varepsilon_\mathbf{k} + \hbar\omega_\mathbf{q}) - f(\mathbf{k})F(\mathbf{q})\delta(\varepsilon_{\mathbf{k+q}} - \varepsilon_\mathbf{k} - \hbar\omega_\mathbf{q})\}. \qquad (15.10\text{-}1)$$

The delta functions are combined and we write $\sum_\mathbf{q} \to (V_0/8\pi^3)\,q^2 dq \sin\vartheta\, d\varphi\, d\vartheta$, where the polar axis is taken along $\mathbf{k}$. We also substitute (8.6-11); this yields

$$-\mathscr{B}_\mathbf{k} f(\mathbf{k}) = \frac{C_1^2}{8\pi^2 u_0\rho}\int_0^{2\pi} d\varphi \int_0^{\pi} \sin\vartheta\, d\vartheta \int_0^{q_{Br}} q^3 dq$$
$$\times\Big(\{f(\mathbf{k}+\mathbf{q})[F(\mathbf{q})+1] - f(\mathbf{k})F(\mathbf{q})\}\delta(\varepsilon_{\mathbf{k+q}} - \varepsilon_\mathbf{k} - \hbar\omega_\mathbf{q})$$
$$+\{f(\mathbf{k}-\mathbf{q})F(\mathbf{q}) - f(\mathbf{k})[F(\mathbf{q})+1]\}\delta(\varepsilon_{\mathbf{k-q}} - \varepsilon_\mathbf{k} + \hbar\omega_\mathbf{q})\Big). \qquad (15.10\text{-}2)$$

---

[23] So (16.4-3) also reads *str. term in first order* $= \tfrac{1}{3}v_\mathbf{k}k\nabla_\mathbf{r}\cdot\vec{\mathscr{G}}_1 + \frac{1}{3v_\mathbf{k}}\mathscr{q}\mathbf{E}\cdot\frac{\partial}{\partial\varepsilon}[v_k^2 k\vec{\mathscr{G}}_1]$. (15.9-3')

Part of this integral can be evaluated as in subsection 15.5-2. Thus we expand the argument of the delta functions and we introduce the new variables Q and y as in (15.5-38). The domain of integration is as in Figs. 15-3a and 15-3b. The integrals then become of the form

$$\int_0^{\pi} d\vartheta \int_0^{q_{Br}} dq\, \delta[\mp v_{\mathbf{k}} \cos\vartheta + q\hbar/2m^* \pm u_0](q^2/\hbar)\sin\vartheta\, \Psi(q)$$
$$\to \int_0^{Q_{\max}} \int_{-\infty}^{\infty} dQ dy \delta(y)(2m^*/\hbar^2)^2 (Q^2/\hbar)\,|\partial(Q,y)/\partial(q,\vartheta)|^{-1}\Psi(Q);$$

the Jacobian $|\partial(Q,y)/\partial(q,\vartheta)|^{-1} = 2m^*/(\hbar v_{\mathbf{k}} \sin\vartheta)$ cancels out the factor $\sin\vartheta$ in the integral, which simplifies the problem considerably. The limits Q_{max}, following from $y = 0$, are seen to be $v_{\mathbf{k}} \pm u_0$. So, the following result obtains

$$-\mathscr{B}_{\mathbf{k}} f = \frac{C_1^2 m^{*3}}{\pi^2 u_0 \rho \hbar^4 v_{\mathbf{k}}} \left\{ \int_0^{2\pi} d\varphi \int_0^{v_{\mathbf{k}}-u_0} \frac{Q^2 dQ}{e^{\hbar\omega_Q/k_BT} - 1} \left[f(\mathbf{k}-\mathbf{q}) - f(\mathbf{k}) e^{\hbar\omega_Q/k_BT} \right] \right.$$
$$\left. + \int_0^{2\pi} d\varphi \int_0^{v_{\mathbf{k}}+u_0} \frac{Q^2 dQ}{e^{\hbar\omega_Q/k_BT} - 1} \left[f(\mathbf{k}+\mathbf{q}) e^{\hbar\omega_Q/k_BT} - f(\mathbf{k}) \right] \right\}, \tag{15.10-3}$$

where we substituted the B–E distribution, assuming, as before, that the phonons remain in equilibrium (no "phonon-drag"). This is now split into the parts $C_0 + \mathbf{k}\cdot(\mathbf{C}_1/k)$, using for $f(\mathbf{k})$ the form (15.9-1).

We first evaluate C_0. We have, carrying out the integral over $d\varphi$,

$$C_0 = \frac{2C_1^2 m^{*3}}{\pi u_0 \rho \hbar^4 v_{\mathbf{k}}} \left\{ \int_0^{v_{\mathbf{k}}-u_0} \frac{Q^2 dQ}{e^{\hbar\omega_Q/k_BT} - 1} \left[g_0(\varepsilon - \hbar\omega_Q) - g_0(\varepsilon) e^{\hbar\omega_Q/k_BT} \right] \right.$$
$$\left. + \int_0^{v_{\mathbf{k}}+u_0} \frac{Q^2 dQ}{e^{\hbar\omega_Q/k_BT} - 1} \left[g_0(\varepsilon + \hbar\omega_Q) e^{\hbar\omega_Q/k_BT} - g_0(\varepsilon) \right] \right\}. \tag{15.10-4}$$

For not too low temperatures we expand $(\hbar\omega_Q = 2m^* Q u_0)$:

$$\begin{aligned} e^{\hbar\omega_Q/k_BT} &= 1 + (2m^* Q u_0/k_BT) + \tfrac{1}{2}(2m^* Q u_0/k_BT)^2 + \ldots \\ g_0(\varepsilon \pm \hbar\omega_Q) &= g_0(\varepsilon) \pm (2m^* Q u_0) g_0'(\varepsilon) + \tfrac{1}{2}(2m^* Q u_0)^2 g_0''(\varepsilon) + \ldots \end{aligned} \tag{15.10-5}$$

This is substituted into(15.10-4); we denote the integrals by $\mathscr{I}_1$ and $\mathscr{I}_2$, respectively. So

$$\mathscr{I}_1 = \int_0^{v_{\mathbf{k}}-u_0} \frac{Q^2 dQ}{(2m^* Q u_0/k_BT) + \frac{1}{2}(2m^* Q u_0/k_BT)^2 + \ldots} \Big[g_0 - (2m^* Q u_0) g_0'$$
$$+ \tfrac{1}{2}(2m^* Q u_0)^2 g_0'' + \ldots - g_0 - (2m^* Q u_0/k_BT) g_0 - \tfrac{1}{2}(2m^* Q u_0/k_BT)^2 g_0 + \ldots \Big]$$

$$= -\int_0^{v_{\mathbf{k}}-u_0} Q^2 dQ\, g_0 + \int_0^{v_{\mathbf{k}}-u_0} \frac{Q^2 dQ\, k_B T}{2m^* u_0 Q}\left[-(2m^* Q u_0) g_0' + \tfrac{1}{2}(2m^* Q u_0)^2 g_0''\right]$$
$$\times\left[1 - \tfrac{1}{2}(2m^* Q u_0 / k_B T) + \tfrac{1}{4}(2m^* Q u_0 / k_B T)^2 - \ldots\right]$$
$$= -\int_0^{v_{\mathbf{k}}-u_0} Q^2 dQ\,[g_0 + k_B T g_0'] + \int_0^{v_{\mathbf{k}}-u_0} Q^2 dQ\left[\tfrac{1}{2}(2m^* Q u_0) g_0'\right.$$
$$\left. + \tfrac{1}{2}(2m^* Q u_0 k_B T) g_0'' + \mathcal{O}(\hbar^2 \omega_Q^2 g_0' / k_B T)\right]. \quad (15.10\text{-}6)$$

In a similar fashion the second integral is evaluated; one obtains

$$\mathcal{I}_2 = \int_0^{v_{\mathbf{k}}+u_0} Q^2 dQ\,[g_0 + k_B T g_0'] + \int_0^{v_{\mathbf{k}}+u_0} Q^2 dQ\left[\tfrac{1}{2}(2m^* Q u_0) g_0'\right.$$
$$\left. + \tfrac{1}{2}(2m^* Q u_0 k_B T) g_0'' + \mathcal{O}(\hbar^2 \omega_Q^2 g_0' / k_B T)\right]. \quad (15.10\text{-}7)$$

The integrals are now added and terms of order u_0^2 are neglected *versus* $v_{\mathbf{k}}^2$. As final result we find

$$C_0 = \frac{2C_1^2 m^{*3} v_{\mathbf{k}} k_B T}{\pi \rho \hbar^4}\left\{\frac{2g_0}{k_B T} + \left[\frac{\varepsilon}{k_B T} + 2\right] g_0' + \varepsilon g_0''\right\}. \quad (15.10\text{-}8)$$

Rests to evaluate the collision integral in first order (*c.i.f.o.*). From (16.5-3) and (15.9-1) we have, remembering that $\vec{\mathcal{B}}_1$ was taken along the streaming vector $\boldsymbol{\alpha}$,

$$c.i.f.o. = \frac{2C_1^2 m^{*3}}{\pi u_0 \rho \hbar^4 v_{\mathbf{k}}}$$
$$\times\left\{\frac{1}{2\pi}\int_0^{2\pi} d\varphi \int_0^{v_{\mathbf{k}}-u_0} \frac{Q^2 dQ}{e^{\hbar\omega_Q / k_B T} - 1}\left[\mathcal{B}_1(\varepsilon - \hbar\omega_Q)(k_\alpha - q_\alpha) - \mathcal{B}_1(\varepsilon) k_\alpha e^{\hbar\omega_Q / k_B T}\right]\right.$$
$$\left. + \frac{1}{2\pi}\int_0^{2\pi} d\varphi \int_0^{v_{\mathbf{k}}+u_0} \frac{Q^2 dQ}{e^{\hbar\omega_Q / k_B T} - 1}\left[\mathcal{B}_1(\varepsilon + \hbar\omega_Q)(k_\alpha + q_\alpha) e^{\hbar\omega_Q / k_B T} - \mathcal{B}_1(\varepsilon) k_\alpha\right]\right\}. \quad (15.10\text{-}9)$$

We make again the expansion (15.10-5) for the B–E distribution, but we further write $\mathcal{B}_1(\varepsilon \pm \hbar\omega_Q) \approx \mathcal{B}_1(\varepsilon)$, given that this is a first-order correction already. The two double integrals will be denoted by $\mathcal{I}_3$ and $\mathcal{I}_4$.[24] For the former we then have

$$\mathcal{I}_3 = \frac{1}{2\pi}\int_0^{2\pi} d\varphi \int_0^{v_{\mathbf{k}}-u_0} \frac{Q^2 dQ}{(2m^* Q u_0 / k_B T) + \tfrac{1}{2}(2m^* Q u_0 / k_B T)^2}$$

---

[24] The details of these integrals were carried out by my graduate student Alain Sirois.

$$\times\left\{(k_\alpha - q_\alpha)\mathcal{G}_1(\varepsilon) - k_\alpha\mathcal{G}_1(\varepsilon)\left[1+(2m^*Qu_0/k_BT)+\tfrac{1}{2}(2m^*Qu_0/k_BT)^2\right]\right\}, \quad (15.10\text{-}10)$$

while for $\mathcal{I}_4$

$$\mathcal{I}_4 = \frac{1}{2\pi}\int_0^{2\pi} d\varphi \int_0^{v_\mathbf{k}+u_0} \frac{Q^2 dQ}{(2m^*Qu_0/k_BT)+\tfrac{1}{2}(2m^*Qu_0/k_BT)^2}$$

$$\times\left\{(k_\alpha + q_\alpha)\mathcal{G}_1(\varepsilon)\left[1+(2m^*Qu_0/k_BT)+\tfrac{1}{2}(2m^*Qu_0/k_BT)^2\right] - k_\alpha\mathcal{G}_1(\varepsilon)\right\}. \quad (15.10\text{-}11)$$

We shall first evaluate

$$\mathcal{I} = \mp\int_0^{2\pi} d\varphi q_\alpha = \mp\int_0^{2\pi} d\varphi q \cos\Theta, \quad (15.10\text{-}12)$$

with the upper sign pertaining to emission and the lower sign to absorption; the angles are indicated in Fig. 15-4. We have the usual relation

$$\cos\Theta = \cos\chi\cos\vartheta + \sin\chi\sin\vartheta\cos(\varphi-\varphi_0).$$

From this it follows that

$$\mathcal{I} = \mp 2\pi q\cos\chi\cos\vartheta = \mp 2\pi q(k_\alpha/k)\cos\vartheta. \quad (15.10\text{-}13)$$

As previously noted, $\cos\vartheta$ follows from $y = 0$, so that $\cos\vartheta = (\pm Q + u_0)/v_\mathbf{k}$. Hence,

$$\mathcal{I} = -2\pi Q(2m^{*2}/\hbar^2)(k_\alpha/k^2)(Q \pm u_0). \quad (15.10\text{-}14)$$

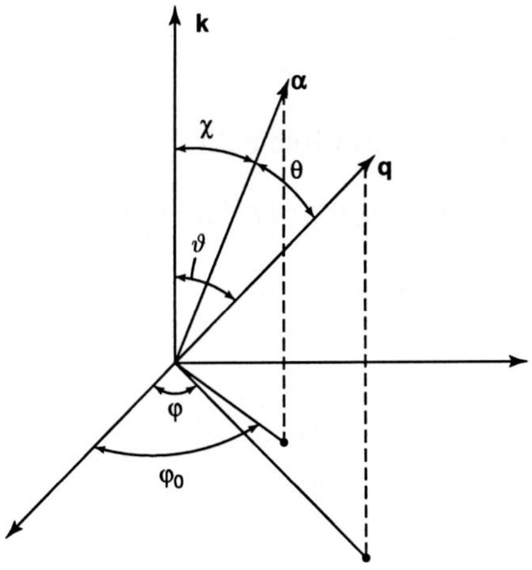

Fig. 15-4. Pertaining to the integrals $\mathcal{I}_3$ and $\mathcal{I}_4$.

Returning to (15.10-10) and (15.10-11), the integrals involving the k_α part are simple since the [ ] terms cancel the denominator below Q^2dQ. The result is therefore

$$\mathcal{I}_3 + \mathcal{I}_4 = -\left(\frac{2m^{*2}}{\hbar^2}\right)\frac{k_\alpha}{k^2}\int_0^{v_\mathbf{k}-u_0}\frac{Q^3 dQ\,(Q+u_0)\,\mathcal{G}_1(\varepsilon)}{(2m^*Qu_0/k_BT)+\frac{1}{2}(2m^*Qu_0/k_BT)^2}$$

$$-\left(\frac{2m^{*2}}{\hbar^2}\right)\frac{k_\alpha}{k^2}\int_0^{v_\mathbf{k}+u_0}\frac{Q^3 dQ\,(Q-u_0)\,\mathcal{G}_1(\varepsilon)}{(2m^*Qu_0/k_BT)+\frac{1}{2}(2m^*Qu_0/k_BT)^2}\left[1+(2m^*Qu_0/k_BT)\right.$$

$$\left.+\tfrac{1}{2}(2m^*Qu_0/k_BT)^2\right] - \int_0^{v_\mathbf{k}-u_0} Q^2 dQ\, k_\alpha \mathcal{G}_1(\varepsilon) + \int_0^{v_\mathbf{k}+u_0} Q^2 dQ\, k_\alpha \mathcal{G}_1(\varepsilon). \qquad (15.10\text{-}15)$$

We only keep the terms up to Q^3 and we neglect u_0 *vs.* $v_\mathbf{k}$ in the limits of the integrals. This, then, yields for (15.10-9):

$$c.i.f.o. = \mathbf{k}\cdot(\mathbf{C}_1/k) = -\frac{2C_1^2 m^{*3} v_\mathbf{k} k_BT}{\pi\rho\hbar^4}\left(\frac{v_\mathbf{k}}{2\hbar u_0^2}\right)\mathbf{k}\cdot(\vec{\mathcal{G}}_1/k), \qquad (15.10\text{-}16)$$

which links $\mathbf{C}_1$ to $\vec{\mathcal{G}}_1$.

The results for C_0 and $\mathbf{C}_1$ are to be substituted into the previously found coupled Boltzmann equations. Thus from (15.10-16) and (15.10-6):

$$\mathbf{k}\cdot\vec{\mathcal{G}}_1(\varepsilon) = -\frac{\pi\rho\hbar^4}{2C_1^2 m^{*3} k_BT}\left(\frac{2\hbar u_0^2}{v_\mathbf{k}}\right)\mathbf{k}\cdot\boldsymbol{\alpha} \equiv -\mathbf{k}\cdot\boldsymbol{\alpha}\,(\hbar/m^* v_\mathbf{k})\lambda, \qquad (15.10\text{-}17)$$

or, employing the standard form for $\boldsymbol{\alpha}$ we find for the vector function $\vec{\mathcal{G}}_1(\varepsilon)$

$$\vec{\mathcal{G}}_1(\varepsilon) = -\frac{\hbar\lambda}{\sqrt{2m^*\varepsilon}}\left[\nabla_\mathbf{r} g_0 + q\mathbf{E}\, g_0'\right]; \qquad (15.10\text{-}18)$$

here λ is the energy-independent mean free path

$$\lambda = (\pi\rho\hbar^4 u_0^2 / C_1^2 m^{*2} k_BT). \qquad (15.10\text{-}19)$$

Despite the entirely different treatment, this is the same as in (15.5-43). With this mean free path the collision integral in zeroth order (16.5-8) becomes

$$C_0 = \frac{2u_0^2\sqrt{2m^*\varepsilon}}{\lambda}\left\{\frac{2g_0}{k_BT} + \left[\frac{\varepsilon}{k_BT}+2\right]g_0' + \varepsilon g_0''\right\}. \qquad (15.10\text{-}20)$$

Finally, both results (16.5-18) and (16.5-20) are substituted into (16.4-5); as in our treatment of plasma heating, we drop henceforth the gradient terms, whereby we assume that the spatial dependence is sufficiently accounted for by $\mathbf{E}(\mathbf{r})$ and $T(\mathbf{r})$ [note that then $\mathcal{G}_1$ has the direction of $\mathbf{E}$]. The straightforward result is

$$(\varepsilon + pk_BT)g_0'' + \left(2 + \frac{\varepsilon}{k_BT} + \frac{pk_BT}{\varepsilon}\right)g_0' + \frac{2}{k_BT}g_0(\varepsilon) = 0, \qquad (15.10\text{-}21)$$

where

$$p=(qE\lambda)^2/6m^*u_0^2k_BT. \tag{15.10-22}$$

Equation (15.10-21) is the renowned differential equation of Yamashita and Watanabe.[25] Amazingly, this differential equation has an exact solution, the Davydov distribution,

$$g_0(\varepsilon)=A_0\left(\frac{\varepsilon}{k_BT}+p\right)^p\exp\left(-\frac{\varepsilon}{k_BT}\right). \tag{15.10-23}$$

For $E\to 0$ this simply approaches the local equilibrium distribution. For large E, however, we have $(p\gg 1)$

$$\left(\frac{\varepsilon}{k_BT}+p\right)^p=\exp\left\{p\ln\left[p\left(1+\frac{\varepsilon}{pk_BT}\right)\right]\right\}\sim\exp\left[p\ln p+\frac{\varepsilon}{k_BT}-\frac{\varepsilon^2}{2p(k_BT)^2}\right].$$

Thus,

$$g_0(\varepsilon)\sim A_0p^p\exp\left[-\frac{\varepsilon^2}{2p(k_BT)^2}\right]=A_p\exp\left[-\frac{\varepsilon^2}{2p(k_BT)^2}\right], \tag{15.10-24}$$

which is the Druyvesteyn distribution. Normalization in **k**-space requires

$$\sum_{\mathbf{k}}g_0(\varepsilon_{\mathbf{k}})=\frac{\Delta^3r}{4\pi^3}\int d^3k\,g_0(\varepsilon_{\mathbf{k}})=n^i(\mathbf{r})\Delta^3r, \tag{15.10-25}$$

so that

$$A_p=4\pi^2n^i(\mathbf{r})\hbar^3/(2m^*k_BT)^{3/2}(2p)^{3/4}\Gamma(3/4), \tag{15.10-26}$$

where $n^i=n$ or p. We can also write down the distribution for $|\mathbf{v}|=v$:

$$\mathcal{G}_0(v)=4\pi v^2(m^*/\hbar)^3[g_0(\varepsilon)/4\pi^3]=A_p'v^2\exp\left[-\frac{\varepsilon^2}{2p(k_BT)^2}\right], \tag{15.10-27}$$

with

$$A_p'=4n^i(\mathbf{r})\left(m^*/2k_BT\right)^{3/2}1/(2p)^{3/4}\Gamma(3/4). \tag{15.10-28}$$

In no case does the distribution resemble a Maxwellian, so an electron (or hole) temperature cannot be defined. The field strength necessary for an appreciable deviation from the Maxwell distribution due to field-induced heating is found by setting $p=1$. Hence from (15.10-22),

$$E_{crit}=u_0\sqrt{6m^*k_BT}/|q|\lambda. \tag{15.10-29}$$

---

[25] J. Yamashita and M. Watanabe, Progress Theor. Physics Japan, **12**, 443 (1953).

This gives 200 *V/cm* for Ge and 1000 *V/cm* for Si. In practise somewhat higher values are found.

At very low temperatures the above results are not valid, because of our expansion of the B–E distribution. Besides, other scattering often dominates, such as optical phonon emission (1 *K* or less in Ge). Oscillatory behaviour, known as 'streaming' may set in, giving a velocity-velocity correlation function that damps out only very slowly. This is shown in Monte-Carlo computations as reported by Mitin and the author.[26]

15.11 Transport in Hot Electron Systems

The current density is given by, employing (15.10-18) of the previous section,

$$\mathbf{J} = \frac{q}{4\pi^3}\int \mathbf{v_k}\mathbf{k}\cdot\vec{\mathcal{G}}_1(\varepsilon_\mathbf{k})\,d^3k$$

$$= -\frac{q}{4\pi^3}\int \frac{\hbar\lambda}{\sqrt{2m^*\varepsilon_\mathbf{k}}}\left(\frac{m^*}{\hbar}\right)\mathbf{v_k}\mathbf{v_k}\cdot\left[\nabla_\mathbf{r} g_0 + q\mathbf{E}\frac{\partial g_0}{\partial \varepsilon_\mathbf{k}}\right]d^3k. \qquad (15.11\text{-}1)$$

With $\mathbf{v_k}\mathbf{v_k} \to \frac{1}{3}v_k^2\mathsf{I}$ and $\int_\Omega [g_0(\varepsilon_\mathbf{k})/4\pi^3]d^3k = 4\pi v^2 g_0(v)\,dv$ this gives, restoring the **r**-dependence

$$\mathbf{J} = -\frac{4\pi}{3}q\lambda\mathsf{I}\cdot\int v^2\left[v\nabla_\mathbf{r} g_0(\mathbf{r},v) + \frac{q\mathbf{E}}{m^*}\frac{\partial g_0}{\partial v}\right]dv. \qquad (15.11\text{-}2)$$

Integrating the last term by parts, we obtain

$$\mathbf{J} = -\frac{4\pi}{3}q\lambda\left\{-\nabla_\mathbf{r}\left[\int v^3 g_0(\mathbf{r},v)dv + \left(\frac{2q\mathbf{E}}{m^*}\right)\int v g_0(\mathbf{r},v)\,dv\right]\right\}. \qquad (15.11\text{-}3)$$

Now for any *F*(*v*) we have the steady-state average

$$\langle F(v)\rangle_{ss} = \frac{1}{n^i(\mathbf{r})}\int 4\pi v^2 dv\,F(v)\,g_0(\mathbf{r},v) = \frac{1}{n^i(\mathbf{r})}\int dv\,F(v)\,\mathcal{G}_0(\mathbf{r},v). \quad (15.11\text{-}4)$$

Therefore. we have

$$\mathbf{J} = -qD(E)\nabla_\mathbf{r} n^i(\mathbf{r}) + |q|\mu(E)n^i(\mathbf{r})\mathbf{E}, \qquad (15.11\text{-}5)$$

where the hot-carrier diffusivity and mobility are given by

$$D(E) = \frac{1}{3}\lambda\langle v\rangle_{ss}, \quad \mu(E) = \frac{2|q|}{3m^*}\lambda\langle\frac{1}{v}\rangle. \qquad (15.11\text{-}6)$$

---

[26] V. Mitin and C.M. Van Vliet, Phys. Rev. **B41**, 5332 (1990).

Previously we established (16.5-27)

$$\mathcal{G}_0(v) = A_p' v^2 \exp\left[-\frac{m^* v^4}{8p(k_B T)^2}\right]. \qquad (15.11\text{-}7)$$

For the averages we note that

$$\int_0^\infty v^3 e^{-\alpha v^4} dv = \frac{1}{4\alpha}\int_0^\infty e^{-\alpha v^4} d(\alpha v^4) = \frac{1}{4\alpha}, \qquad \int_0^\infty v e^{-\alpha v^4} dv = \frac{1}{2\sqrt{\alpha}}\int_0^\infty e^{-y^2} dy = \frac{1}{4}\sqrt{\frac{\pi}{\alpha}}.$$

With A_p' given by (16.5-28), this yields

$$D = \tfrac{1}{3}\lambda(2k_B T/m^*)^{1/2}(2p)^{1/4}/\Gamma(3/4), \qquad (15.11\text{-}8)$$

$$\mu = \frac{4|q|\lambda}{3}\left(\frac{1}{2\pi m^* k_B T}\right)^{1/2} \frac{\pi}{2(2p)^{1/4}\Gamma(3/4)}. \qquad (15.11\text{-}9)$$

We had for the low-field mobility [see 15.5-44)]

$$\mu_0 = 4|q|\lambda/3(2\pi m^* k_B T)^{1/2}, \qquad (15.11\text{-}10)$$

and we recall the critical field strength (15.10-29)

$$E_c = \left\{\pi^{3/2}/[\Gamma(3/4)]^2\sqrt{6}\right\}(u_0/\mu_0) = 1.514(u_0/\mu_0), \qquad (15.11\text{-}11)$$

while the low-field diffusivity is related to the low-field mobility by the ordinary Einstein relation, $|q|D_0 = \mu_0 k_B T$. We thus obtain the essential results

$$\mu(E) = \mu_0 (E_c/|E|)^{1/2}, \qquad (15.11\text{-}12)$$

$$D(E) = \tfrac{3}{4}\left(\tfrac{2}{3}\right)^{1/2}\left\{\pi^{3/2}/[\Gamma(3/4)]^2\sqrt{6}\right\}D_0(|E|/E_c)^{1/2}. \qquad (15.11\text{-}13)$$

Since the sample is not in thermal equilibrium, Nyquist's formula for the noise power [at frequencies $h\nu \ll k_B T$] $P(\nu) = k_B T$, see (1.6-69), is not valid.[27] Yet, the best value for the electron temperature comes from the measured noise power under hot electron conditions, defined by $|q|D(E) = (d/dE)[E\mu(E)]k_B T_n$. From (15.11-12) and (15.11-13) we find

$$T_n(E) = 1.85T\,[|E|/E_c]. \qquad (15.11\text{-}14)$$

For measured data, which are often exacerbated by space-charge limited flow, we refer to the literature.[28,29]

[27] This is the noise power delivered to a matched load, $P(\nu) = \frac{1}{4}\langle\Delta\Phi^2\rangle_\nu/R = \frac{1}{4}S_\Phi(\nu)/R = k_B T$.

[28] A. Gisolf and R.J.J. Zijlstra, Solid State Electr. **17**, 839 (1974).

[29] G. Bosman, Ph.D Thesis, Univ. of Utrecht 1981; C.F. Whiteside, Ph.D. Thesis, Univ. of Florida 1987.

15.12 Problems to Chapter XV

15.1 Show that the conductivity of a metal with arbitrary $\varepsilon(\mathbf{k})$ near $T = 0$ is given by

$$\sigma = (e^2/12\pi^3\hbar)\tau(\varepsilon_F)v_F S_F, \tag{1}$$

where v_F is the Fermi velocity and S_F is the area of the Fermi surface.

15.2 Starting with the band integral for D'_n, obtain the relationship with D_n, stated in Eq. (15.2-24).

15.3 Give a derivation for the band-integral result, Eq. (15.4-8).

15.4 The generalized Einstein relation for any band structure was given in (15.2-21). Consider a conduction band with ellipsoidal energy surfaces, like in Si and Ge, and let m_1, m_2 and m_3 be the masses when the ellipsoids are on their main axes and M the number of equivalent minima (valleys).
(a) Show that band is parabolic, $\rho(\varepsilon) = \sqrt{2}\, m_e^{3/2}/\pi^2\hbar^3(\varepsilon - \varepsilon_C)^{1/2}$, where m_e is the density-of-states effective mass, given by $m_e = (m_1 m_2 m_3 M^2)^{1/3}$.
(b) From (15.2-21) obtain Landsberg's result, (15.2-22).

15.5 Consider a band for which the density of states can be written as the series

$$g(\varepsilon) = \sum\nolimits_\nu \alpha_\nu \hat{\varepsilon}^{\nu/2}, \tag{2}$$

where $\hat{\varepsilon}$ is the kinetic energy $\varepsilon - \varepsilon_C$.
(a) Obtain the generalized Einstein relation in terms of Fermi integrals;
(b) For the Kane band, applicable to most III-V compounds, the band structure is given by $\hat{\varepsilon}(\mathbf{k}) = (\hbar^2 k^2/2m^*)(1 - \kappa k^2)$. Show that the density of states can be represented by Eq. (2) with $\alpha_1 = \sqrt{2}(m^*)^{3/2}/\pi^2\hbar^3$, $\alpha_2 = 0$, $\alpha_3 = 5\kappa m^* \alpha_1/\hbar^2$, all higher ν being zero. Obtain the applicable generalized Einstein relation.

15.6 For nondegenerate semiconductors, approximate the F–D distribution by its classical equivalent. Assume a scattering mechanism $\tau(\hat{\varepsilon}) = A e^{\gamma\hat{\varepsilon}}$.
(a) In the effective mass approximation (spherical energy surfaces) evaluate the tensors L_{ee}, L_{es}, L_{se} and L_{ss}. You should express the results in the integrals

$$M_n = \frac{2\sqrt{2m^*}\,\tau_0}{3\pi^2 k_B T\hbar^3} e^{\Delta\varsigma/k_BT} \int_0^\infty d\hat{\varepsilon}\, \tau(\hat{\varepsilon})\hat{\varepsilon}^{3/2}(\hat{\varepsilon} - \Delta\varsigma)^n e^{-\hat{\varepsilon}/k_BT}$$
$$= \frac{2\sqrt{2m^*}\,\tau_0}{3\pi^2 k_B T\hbar^3} e^{\Delta\varsigma/k_BT} \sum_{\ell=0}^{n} \binom{n}{\ell} (-1)^{n-\ell}(\Delta\varsigma)^{n-\ell}(k_BT)^{\frac{5}{2}+\lambda+\ell}\,\Gamma(\tfrac{5}{2}+\gamma+\ell), \tag{3}$$

where $\Delta\varsigma = \varsigma - \varepsilon_C$.
(b) Obtain the Wiedemann–Franz ratio between thermal and electrical conductivity, employing (15.4-12) and confirm (15.4-14'). N.B. You will need the *full* expression (15.4-12), last entry, for the semiconductor case.

15.7 Show that the Nernst effect is given by π_{xx} / λ_{xx} and compute this effect for a degenerate electron gas.

15.8 For optical phonon scattering, with $|\mathcal{F}(\mathbf{q})|^2$ being independent of $\mathbf{q}$, the collisions are 'randomizing', i.e.,

$$w^{em}_{\mathbf{kk}'} = w^{em}_{\mathbf{k},-\mathbf{k}'} \text{ and } w^{ab}_{\mathbf{kk}'} = w^{ab}_{\mathbf{k},-\mathbf{k}'} . \tag{4}$$

Show that

$$\frac{1}{\tau(\varepsilon_{\mathbf{k}})} = \sum_{\mathbf{k}'} w_{\mathbf{kk}'} . \tag{5}$$

With a method that goes analogous to that for acoustical phonon scattering, establish that

$$\frac{1}{\tau(\varepsilon_{\mathbf{k}})} = \frac{D_0^2 m^* \sqrt{2m^*}}{2\pi\rho\omega_0\hbar^3} \left\{ (F_0 + 1)(\varepsilon_{\mathbf{k}} - \hbar\omega_0)^{1/2} \Theta(\varepsilon_{\mathbf{k}} - \hbar\omega_0) + F_0(\varepsilon_{\mathbf{k}} + \hbar\omega_0)^{1/2} \right\}, \tag{6}$$

where $\Theta(z)$ is the Heaviside function. Also show that $\mu_{opt} \propto T^{-1/2}(e^{\hbar\omega_0/k_BT} - 1)$.

15.9 Formulate the two coupled Boltzmann equations for a nondegenerate semiconductor placed in a magnetic field, assuming that this field is represented by the semi-classical Lorentz force.

15.10 (a) Verify in detail all steps and approximations leading up to the differential equation of Yamashita and Watanabe.
(b) Show that it is satisfied by the Davydov distribution and obtain the normalization constant A_p.

15.11 Sometimes a noise temperature is defined by

$$|\mathcal{q}| D(E) = \mu(E) k_B T_n'(E) . \tag{7}$$

(a) Show that $T_n'(E) = 0.927\, T(E/E_c) = \frac{1}{2} T_n(E)$.
(b) Explain why in a realistic experimental circuit with nonlinear behaviour the noise temperature T_n ,as defined in Section 15.11, is what will be measured. Hint: Find the noise power in a band interval $d\nu$, delivered by the nonlinear sample into a matched load.

15.12 Find an asymptotic approximation to the differential equation of Yamashita and Watanabe for very high electric fields. Show that it produces the Druyvesteyn distribution.

PART E

LINEAR RESPONSE THEORY AND QUANTUM TRANSPORT

Chapter XVI

Linear Response Theory, Reduced Operators and Convergent Forms

1. THE ORIGINAL KUBO–GREEN FORMALISM

16.1 Introduction to Linear Response Theory

As noted in the previous chapters, transport properties in quantum systems with strong interactions are very inadequately described within the classical or semi-quantal Boltzmann framework. So around the middle of the twentieth century alternative theories based on a many-body quantum approach, generally known as '*linear response theory*' saw the light, with the basic results for generalized conductivities and susceptibilities being encompassed in the *Kubo–Green formulas*[1,2]. Kubo's approach is the more general one, detailing many earlier results obtained by Kirkwood, Green and others. Basically, Kubo solves the von Neumann equation for a system whose Hamiltonian has been augmented with a 'response Hamiltonian' involving an applied ponderomotive stimulus $AF(t)$, in order to obtain the change in motion $\langle \Delta B(t) \rangle$ of any operator in the system representing an observable $\mathscr{B}$, in terms of a response function $\phi_{BA}(t)$ of the system. The latter is shown to be linearly related to the correlation function of the spontaneous fluctuations, $\langle \Delta B(t) \Delta A(0) \rangle$, occurring in the equilibrium state, cf. our developments in Chapter 9, subsections 9.2.2 and 9.2.3. In this manner very general expressions can be obtained for the transport coefficients.

The correlation expressions can also be Fourier transformed; the results can then be shown to be equivalent with the fluctuation-dissipation theorem, which gives the connection between the dissipative generalized conductance or susceptance of a process and the spectral density of the fluctuations in the system. We met an example of this theorem already in Nyquist's formula for the thermal noise of a system, as expressed in the dissipative part of the electrical impedance of a linear network, cf. Section 1.6, Eq. (1.6-69). The general form of the fluctuation-dissipation theorem was first obtained in a paper by Callen and Welton, using time-

---

[1] R. Kubo, (a) Can. J. Phys. **34**, 1274 (1956); ibid. (b) J. Phys. Soc. Japan **12**, 570 (1957).

[2] M.S. Green, J. Chem. Phys. **20**, 1281 (1952); ibid. **22**, 398 (1954).

dependent perturbation theory and by Callen and Greene employing thermodynamic methods[3]; see also[4]. Kubo's treatment is, however, more straightforward. Basically, *the Kubo–Green relations do in the time domain what the fluctuation-dissipation theorem does in the frequency domain.* Dealing just with the problem of electrical conduction, the theory goes back in essence to Nyquist's theorem and before that to Einstein's 1905 paper on Brownian motion and the Ph.D. thesis of de Haas–Lorentz.[5]

Yet, the Kubo–Green relations have a special appeal since they deal directly with the microscopic quantum mechanical motion of the process. Moreover, the derivation is by many believed to be 'exact', except for the linearization in the applied field, which *a prima vista* looks no worse than the similar procedure applied in the perturbation solution of the Boltzmann equation. But, in contrast to the latter, stronger interactions can in principle be included in the computations of the transport coefficients, see e.g., the papers by Chester and Thellung[6] and Verboven[7].

However, it is precisely the generality and apparent simplicity of the theory which have invited criticism. Van Kampen has emphasized that there is a vast difference between microscopic linearity and macroscopic linearity of the responses.[8] Also, for applications it is usually found that Kubo's expressions are too general, so that somewhere in the application a *randomness assumption* must be made. In this respect, the various attempts to compare the results from Kubo's theory with those of the simple Boltzmann approach, are illuminating, cf. Refs.[7,9,10,11]. These papers have motivated us to reconsider Kubo's linear response theory, in order to give the theory a firmer mathematical footing and *physical content.* Also, since a randomness assumption must be made somewhere – whatever otherwise is the meaning of fluctuations in the theory? – *it should be made on the many-body-level* and not on the Boltzmann level, where it requires new assumptions for each collision process.

Our original work in this area is found in four papers on LRT, entitled "Linear Response Theory Revisited", I – IV (1978–1984); they will be cited when the occasion arises. However, fundamental contributions in the area of quantum transport, as well as applications by the author and coworkers to a large variety of phenomena (quantum Hall effect, Aharonov–Bohm effect, cyclotron resonance, electron-phonon effects, to mention just a few) have continued to 2000 and beyond.

---

[3] H.B. Callen and T.A. Welton, Phys. Rev **83**, 34 (1951); — and R.F. Greene, Phys. Rev **86**, 702 (1952).
[4] L. Tisza and I. Manning, Phys. Rev. **105**, 2695 (1957); K.M. Van Vliet, Phys. Rev. **109**, 1021 (1958).
[5] A. Einstein, Ann. der Physik **17**, 549 (1905). Gertruida L. de Haas–Lorentz (Leiden) "Die Brownsche Bewegung und einige anverwandte Erscheinungen", Vieweg Verlag, Braunschweig, Germany, 1912.
[6] G.V. Chester and A. Thellung, Proc. Phys. Soc. **73**, 745 (1960).
[7] E. Verboven, Physica **26**, 1091 (1960).
[8] N.G. van Kampen, Physica Norvegica **5**, 279 (1971).
[9] S. Fujita and R. Abe, J. Math. Phys. **3**, 350 (1962).
[10] S.F. Edwards, Phil. Mag. **3**, 33 and 1020 (1958).
[11] D.A. Greenwood, Proc. Phys. Soc. **71**, 585 (1968).

16.2 The Response Function and the Relaxation Function

We will summarise the main aspects of linear response theory as originally developed, following in particular Kubo[1,12], Mazo[13] and Montroll.[14] Also, we follow more or less Part *A* of our exposition set forth in our first paper on the subject.[15]

Kubo considers a system with Hamiltonian

$$\mathcal{H}_{total} = \mathcal{H} - AF(t), \qquad (16.2\text{-}1)$$

where $\mathcal{H}$ is the Hamiltonian of the system proper and $AF(t)$ is the coupling with an external field; A is an operator corresponding to some observable $\mathcal{A}$ and $F(t)$ is a (complex) time function (c-number). For example, for a system in an electric field we have $AF(t) \rightarrow \mathbf{A}\cdot\mathbf{F}(t) = \sum_i q_i(\mathbf{r}_i - \mathbf{r}_{i0})\cdot\mathbf{E}(t)$ in which $\mathbf{r}_i - \mathbf{r}_{i0}$ is the shift in position of charge q_i and $\mathbf{E}(t)$ is the time-dependent electric field. Another example is that of paramagnetic dipoles in a magnetic field, $\mathbf{A}\cdot\mathbf{F}(t) = \sum_i \boldsymbol{\mu}_i\cdot\mathbf{H}(t)$. In Kubo's papers, as well as other treatments we know of, the field is supposed to be turned on at $t = -\infty$; the response of an operator B, representing an observable $\mathcal{B}$, is then sought for. However, these treatments seem to be unaware of *standard response theory*, employed in electrical engineering and other disciplines, for the response of electrical parameters in networks, following a disturbance applied at $t = 0$; the response is then computed with Laplace transforms and no convergence factor for an 'adiabatically turned on' field is necessary since the Laplace transform has better convergence properties and is tailored for this purpose. Accordingly, we seek the solution of the von Neumann equation for the Hamiltonian (16.2-1), viz.,

$$\frac{\partial\rho}{\partial t} + \left(\frac{i}{\hbar}\right)[\mathcal{H},\rho] = \left(\frac{i}{\hbar}\right)\Theta(t)F(t)[A,\rho], \qquad (16.2\text{-}2)$$

where $\Theta(t)$ is the Heaviside function, being zero for $t < 0$ and unity for $t > 0$. The formal solution of (16.2-2) is given by the Volterra integral equation

$$\rho(t) = \rho_0(t) + \left(\frac{i}{\hbar}\right)\int_0^t e^{-i\mathcal{H}(t-t')/\hbar}F(t')[A,\rho(t')]e^{i\mathcal{H}(t-t')/\hbar}dt', \quad t>0, \qquad (16.2\text{-}3)$$

where $\rho_0(t)$ is the solution of the homogeneous equation, i.e., of (16.2-2) with the rhs being zero. Obviously, (16.2-3) can be iterated any desired number of times, substituting at first $\rho(t') = \rho_0(t)$ at the rhs, then the new solution $\rho_1(t')$, and so on,

---

[12] R. Kubo in "Lectures in Theoretical Physics" Vol. I (W.E. Britten and L.G. Durham, Eds.), Boulder, CO, 1958, Interscience, NY 1959.

[13] R.M. Mazo, "Statistical Mechanics of Transport Processes", Pergamon Press, NY 1960, Chapter 10.

[14] E. Montroll in "Lectures in Theoretical Physics" Vol. III (W.E. Britten, B.W. Downs and J. Downs, Eds.) Boulder CO, 1960, Interscience, NY 1961.

[15] K.M. Van Vliet, "Linear response Theory Revisited I. The Many-Body Van Hove Limit." J. Math. Phys. **19**, 1345-1370 (1978), henceforth referred to as LRT I.

ad infinitum. For the full solution so obtained, see Kubo's article of 1957.[16] However, as recognised by Kubo, such a solution is as a rule not useful. Kubo therefore assumes that the first iteration suffices, i.e., on the rhs of (16.2-3) we replace $\rho(t')$ by $\rho_0(t)$. Equation (16.2-3) then becomes linear in the applied force $F(t)$, with the response thereby also becoming linear, hence the name "linear response theory" (LRT). As to the form of $\rho_0(t)$, Kubo assumes that prior to turning on the external disturbance, the system has thermalized an infinite time, so that $\rho_0(t)$ is the equilibrium density operator, ρ_{eq}. Needless to say perhaps that there is *no* approach to equilibrium for the von Neumann equation, cf., our discussion in subsection 3.8.2, while also the von Neumann equation strictly applies to a microcanonical ensemble. However, Kubo chooses for ρ_{eq} the canonical density operator, i.e., interactions with a heat bath are allowed for. These, then, could entail an approach to equilibrium, although the nature of the needed interactions are nowhere explicited in Kubo's papers.

We now proceed with the formal part of the theory, leaving more discussion about the method to later, see Section 16.8. For the response at $t > 0$ of an arbitrary operator in the system $\langle \Delta B(t) \rangle = \mathrm{Tr}[\rho(t)B] - \mathrm{Tr}(\rho_{eq}B)$ we find

$$\langle \Delta B(t) \rangle = (i/\hbar)\mathrm{Tr}\left\{\int_0^t dt' B e^{-i\mathcal{H}(t-t')/\hbar} F(t')[A,\rho_{eq}] e^{i\mathcal{H}(t-t')/\hbar}\right\}, \quad (16.2\text{-}4)$$

with

$$\rho_{eq} = e^{-\beta\mathcal{H}}/\mathcal{Z}, \quad \mathcal{Z} = \mathrm{Tr}\left(e^{-\beta\mathcal{H}}\right), \quad \beta = 1/k_B T. \quad (16.2\text{-}5)$$

We now use the fact that the trace is invariant under cyclic permutation, $\mathrm{Tr}\,ABC...$ $= \mathrm{Tr}\,BC...A$, etc.; we then find

$$\begin{aligned}\langle \Delta B(t) \rangle &= (i/\hbar)\mathrm{Tr}\int_0^t dt' B(t-t')\,[A,\rho_{eq}]F(t') \\ &= (1/\hbar i)\mathrm{Tr}\int_0^t dt'[A,B(t-t')]\rho_{eq}\,F(t'),\end{aligned} \quad (16.2\text{-}6)$$

where $B(t)$ is the Heisenberg operator

$$B(t) = e^{i\mathcal{H}t/\hbar} B e^{-i\mathcal{H}t/\hbar}. \quad (16.2\text{-}7)$$

For later reference we also indicate the Liouville solution. The lhs of (16.2-2) can also be written as $\partial\rho/\partial t + i\mathcal{L}\rho$, where $\mathcal{L}$ is the Liouville operator defined by

$$i\mathcal{L}C = \{C,\mathcal{H}\} \text{ (classical) or } (i/\hbar)[\mathcal{H},C] \text{ (quantum mechanical)}. \quad (16.2\text{-}8)$$

We still find the result (16.2-6), where $B(t) = [\exp(i\mathcal{L}t)]B$. We note that $\mathcal{L}$ is a *super*operator, acting on operators of the system. For the quantum case the state

[16] R. Kubo, Ref 1(b), Op. Cit., Eq. (2.27).

space of superoperators is the Liouville space $\mathcal{S}\otimes\bar{\mathcal{S}}$, where $\mathcal{S}$ is the Hilbert space spanned by the system Hamiltonian and $\bar{\mathcal{S}}$ is the dual space. All superoperators can be expressed by the form

$$S..=\sum\nolimits_{\alpha\beta}\left(A_\alpha\rightarrow..+..\leftarrow B_\beta\right), \tag{16.2-9}$$

where A_α and B_β are system operators, with A_α acting from the left and B_β acting from the right. Whereas ordinary operators have a binary matrix representation, superoperators have a tetradic representation, $S=\{S_{\alpha\beta|\alpha'\beta'}\}$; we refer to the literature for details. The properties of $\mathcal{L}$ in the Liouville space have been well studied by Fano.[17] Since $\mathcal{L}$ is Hermitean in Liouville space, the superoperator $\exp(i\mathcal{L}t)$ is unitary; thus, Kubo's "natural motion", as expressed by (16.2-7) and now written as $[\exp(i\mathcal{L}t)]B$, is a rotation in Liouville space. Properties of other superoperators will be met later.

We return to (16.2-6) and introduce the response function

$$\phi_{BA}(t)=(1/\hbar i)\mathrm{Tr}\{[A,B(t)]\rho_{eq}\}, \tag{16.2-10}$$

so that we have the convolution integral

$$\langle\Delta B(t)\rangle=\int_0^t dt'\phi_{BA}(t-t')F(t'). \tag{16.2-11}$$

Clearly, ϕ_{BA} is a memory function. With $\hat{F}(s)$, $b(s)$ and $\chi_{BA}(s)$ being the Laplace transforms of $F(t)$, $\langle\Delta B(t)\rangle$ and $\phi_{BA}(t)$, respectively, (16.2-11) yields

$$b(s)=\chi_{BA}(s)\hat{F}(s). \tag{16.2-12}$$

Thus, for the susceptibility [or *generalized susceptance*] χ_{BA} we have

$$\chi_{BA}(s)=\int_0^\infty dt\,e^{-st}\phi_{BA}(t)=(1/\hbar i)\int_0^\infty dt\,e^{-st}\,\mathrm{Tr}\{[A,B(t)]\rho_{eq}\}. \tag{16.2-13}$$

We note *that we did not do much physics*, since we said nothing about the processes which caused the response; we found, however, a general result in terms of a commutator of a correlation expression.

Often one seeks the response of a flux or current $d\mathcal{B}/dt$ represented by an operator $\dot{B}$. Then analogous to (16.2-11)

$$\langle\Delta\dot{B}(t)\rangle=\int_0^t dt'\phi_{\dot{B}A}(t-t')F(t'), \tag{16.2-14}$$

with the response function

$$\phi_{\dot{B}A}(t)=(1/\hbar i)\mathrm{Tr}\{[A,\dot{B}(t)]\rho_{eq}\}. \tag{16.2-15}$$

---

[17] U. Fano, Rev. Mod. Phys. **29**, 74 (1959); also in "Lectures on the Many-Body Problem", Vol. 2 (E.R. Caniello, Ed.), Acad. Press, NY 1964.

For this case it is more physical (though not necessary) to describe the response by a *generalized conductance* L_{BA}. Thus, making a Laplace transform of (16.2-14) we have

$$\dot{b}(s) = L_{BA}(s)\hat{F}(s), \tag{16.2-16}$$

with

$$L_{BA}(s) = \int_0^\infty dt\, e^{-st}\phi_{\dot{B}A}(t) = (1/\hbar i)\int_0^\infty dt\, e^{-st}\,\mathrm{Tr}\{[A,\dot{B}(t)]\rho_{eq}\}. \tag{16.2-17}$$

Now we have the following identities

$$\begin{aligned}\phi_{\dot{B}A}(t) &= (1/\hbar i)\mathrm{Tr}\{[A,\dot{B}(t)]\rho_{eq}\} \\ &= (1/\hbar i)\mathrm{Tr}\{[A(-t),\dot{B}]\rho_{eq}\} = -\frac{1}{\hbar i}\int_t^\infty dt'\frac{d}{dt'}\mathrm{Tr}\{[A(-t'),\dot{B}]\rho_{eq}\} \\ &= \frac{1}{\hbar i}\int_t^\infty dt'\mathrm{Tr}\{[\dot{A}(-t'),\dot{B}]\rho_{eq}\} = \frac{1}{\hbar i}\int_t^\infty dt'\mathrm{Tr}\{[\dot{A},\dot{B}(t')]\rho_{eq}\}.\end{aligned} \tag{16.2-18}$$

In the various transitions we used the following two properties:

(i) *Stationarity*. As for any two-point correlation, we have for two operators C and D

$$\mathrm{Tr}\{\rho_{eq}C(t_1)D(t_2)\} = \mathrm{Tr}\{\rho_{eq}C(t_1-t)D(t_2-t)\}; \tag{16.2-19}$$

The proof follows immediately from the definition (16.2-7) of the Heisenberg operators and from cyclic permutivity. In particular, we have

$$\mathrm{Tr}\{\rho_{eq}[A(0),\dot{B}(t)]\} = \mathrm{Tr}\{\rho_{eq}[A(-t),\dot{B}(0)]\}, \tag{16.2-20}$$

where we noted that $A(0)$ is the Schrödinger operator A.

(ii) *The mixing property*. It is assumed that

$$\lim_{|t_1-t_2|\to\infty} \mathrm{Tr}\{\rho_{eq}C(t_1)D(t_2)\} = \langle C\rangle\langle D\rangle; \tag{16.2-21}$$

in particular, this indicates that the limit of a commutator vanishes,

$$\lim_{|t|\to\infty} \mathrm{Tr}\{\rho_{eq}[A(-t),\dot{B}(0)]\} = 0. \tag{16.2-22}$$

The mixing property is much discussed by Kubo.[18] In its generality the mixing property has not been proven yet; a proof for harmonic crystals has been given by Lanford and Lebowitz.[19] We will come back to the mixing property later.

We substitute the last line of (16.2-18) into (16.2-17). We then obtain for the generalized conductance

---

[18] R. Kubo, J. Phys. Soc. Japan, Op. Cit., p. 577.

[19] O.E. Lanford III and J.L. Lebowitz, "Lecture Notes in Physics", Vol. 38, Springer 1974, p. 144.

$$L_{BA}(s) = \frac{1}{\hbar i}\int_0^\infty dt \int_t^\infty dt' e^{-st}\,\mathrm{Tr}\{[\dot{A}, \dot{B}(t')]\rho_{eq}\} = \frac{1}{\hbar i}\int_0^\infty dt' \int_0^{t'} dt\, e^{-st}\,\mathrm{Tr}\{[\dot{A}, \dot{B}(t')]\rho_{eq}\}$$

$$= -\int_0^\infty dt \frac{e^{-st}-1}{s\hbar i}\mathrm{Tr}\{[\dot{A}, \dot{B}(t')]\rho_{eq}\} = -\int_0^\infty dt \frac{e^{-st}-1}{s}\phi_{\dot{B}\dot{A}}(t)\,; \qquad (16.2\text{-}23)$$

we changed the order of integration in the second step. The diagram for the areas is indicated in Fig. 16-1. The above result applies in particular to electrical conduction. In that case $\mathbf{A} = \Sigma_i q_i(\mathbf{r}_i - \mathbf{r}_{i0})$, $\dot{\mathbf{A}} = \Sigma_i q_i \mathbf{v}_i = \dot{\mathbf{B}}$. The current density for a crystal of volume V_0 containing N electrons is $\mathbf{J} = N^{-1}\Sigma_i q_i \mathbf{v}_i(N/V_0) = \dot{\mathbf{A}}/V_0$. since L (being a tensor) is defined by $\mathbf{J}V_0 = \mathsf{L}_{AA}\cdot\mathbf{E}$, while also $\mathbf{J} = \sigma\cdot\mathbf{E}$, we see that $\sigma = \mathsf{L}_{AA}/V_0$. So, from (16.2-23), using Cartesian component notation, denoted by Greek subscripts,

$$\sigma_{\mu\nu}(s) = -V_0\int_0^\infty dt \frac{e^{-st}-1}{s\hbar i}\mathrm{Tr}\{[J_\nu, J_\mu(t)]\rho_{eq}\}. \qquad (16.2\text{-}24)$$

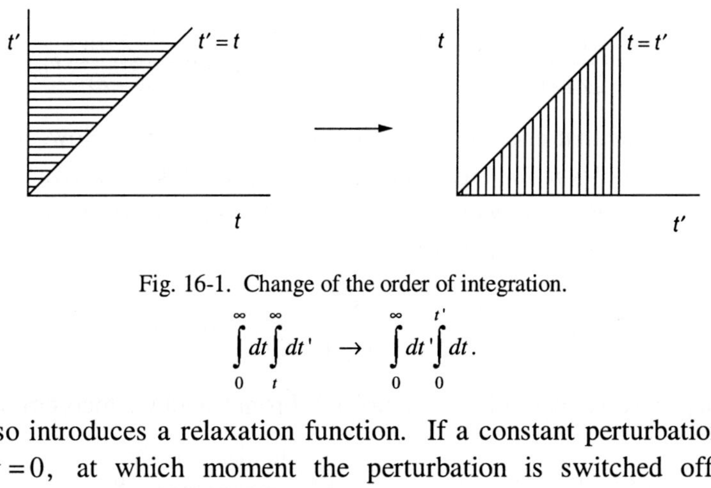

Fig. 16-1. Change of the order of integration.

$$\int_0^\infty dt \int_t^\infty dt' \;\rightarrow\; \int_0^\infty dt' \int_0^{t'} dt.$$

Kubo also introduces a relaxation function. If a constant perturbation acts from $t = -\infty$ to $t = 0$, at which moment the perturbation is switched off, we have $F(t) = F_0[1 - \Theta(t)]$; the response is given by

$$\langle \Delta B(t)\rangle = \int_{-\infty}^{t} dt' \phi_{BA}(t-t')F_0[1-\Theta(t')] = \int_{-\infty}^{0} dt' \phi_{BA}(t-t')F_0$$

$$= \int_t^\infty d\bar{t}\,\phi_{BA}(\bar{t})F_0 \equiv \psi_{BA}(t)\,F_0\,. \qquad (16.2\text{-}25)$$

The function ψ_{BA} is the relaxation function. We have the following connection with the response function

$$\psi_{BA}(t) = \int_t^\infty dt' \phi_{BA}(t') = \frac{1}{\hbar i}\int_t^\infty dt' \mathrm{Tr}\{[A, B(t')]\rho_{eq}\}, \qquad (16.2\text{-}26)$$

or also

$$\frac{d}{dt}\psi_{BA}(t) = -\frac{1}{\hbar i}\mathrm{Tr}\{[A, B(t)]\rho_{eq}\}. \qquad (16.2\text{-}27)$$

16.3 The Frequency Domain; Various Forms

16.3.1 *The Commutator Form*

For sinusoidal excitation we have $F(t) = \Theta(t)F_{0,\omega}e^{i\omega t}$. For $t \to \infty$ the response is also sinusoidal, i.e., $\langle \Delta B(t) \rangle = B_{0,\omega}e^{i\omega t}$, with

$$B_{0,\omega} = \chi(i\omega)F_{0,\omega}, \qquad (16.3\text{-}1)$$

where $\chi(i\omega)$ is the same function as $\chi(s)$. This result is, of course, well known from electrical network theory. For a direct proof we compute $b(s)$ for the above excitation. Thus with

$$\hat{F}(s) = \int_0^\infty dt\, e^{-st}\Theta(t)F_{0,\omega}e^{i\omega t} = F_{0,\omega}/(s - i\omega) \qquad (16.3\text{-}2)$$

we have $b(s) = \chi(s)F_{0,\omega}/(s - i\omega)$, giving for the response

$$\langle \Delta B(t) \rangle = \frac{1}{2\pi i}\int_{\gamma - i\infty}^{\gamma + i\infty} ds \frac{e^{st}\chi(s)}{s - i\omega}F_{0,\omega} = \frac{1}{2\pi i}\oint_C dw \frac{e^{iwt}\chi(iw)}{w - \omega}F_{0,\omega}, \qquad (16.3\text{-}3)$$

where C is a contour that encircles all poles counterclockwise. Now, *for a passive system $\chi(s)$ has no poles in the right half of the complex plane, or $\chi(iw)$ has no poles below the real axis of the complex frequency plane (causality principle).*[20] Thus, letting $w_k = \omega_k + i\mu_k$ denote the poles, we have

$$\chi(iw) = \sum_k \frac{g_k}{w - (\omega_k + i\mu_k)}, \quad \mu_k \geq 0. \qquad (16.3\text{-}4)$$

(For simplicity, we consider only single poles.) From Cauchy's theorem we find

$$\langle \Delta B(t) \rangle = e^{i\omega t}\left(\chi(i\omega) + \sum_k \frac{g_k e^{i(\omega_k - \omega)t} e^{-\mu_k t)}}{\omega_k - \omega + i\mu_k} \right) F_{0,\omega}. \qquad (16.3\text{-}5)$$

The summand damps out for $\mu_k > 0$. [If $\mu_k = 0$, we consider an excitation $\lim_{\varepsilon \to 0+} F_{0,\omega}\exp(i\omega t - \varepsilon t)$.] This completes the proof of (16.3-1).

The results of the s-plane of the previous section are now carried over in the frequency domain. The complex susceptance becomes the *Fourier-Laplace transform* (or one-sided Fourier transform) of the response function:

$$\chi_{BA}(i\omega) = \int_0^\infty dt\, e^{-i\omega t}\phi_{BA}(t) = (1/\hbar i)\int_0^\infty dt\, e^{-i\omega t}\,\mathrm{Tr}\{[A, B(t)]\rho_{eq}\}. \qquad (16.3\text{-}6)$$

---

[20] While the statement for the poles in the s-plane is binding, its consequence for the complex frequency plane depends on the choice of Fourier-Laplace transform. Many authors have $s \to -i\omega$, in which case there are no poles in the upper half plane; see e.g. Jackson, "Classical Electrodynamics" 3rd Ed., p.333ff.

Generally we write $\chi(i\omega) = \chi'(\omega) + i\chi''(\omega)$. From the causality principle one easily shows that the two parts are each other's Hilbert transforms, i.e., we have the Kramers–Kronig relations

$$\chi'(\omega) = -\frac{1}{\pi}\mathcal{P}\int_{-\infty}^{\infty} \frac{\chi''(\omega')}{\omega'-\omega}d\omega', \quad \chi''(\omega) = \frac{1}{\pi}\mathcal{P}\int_{-\infty}^{\infty} \frac{\chi'(\omega')}{\omega'-\omega}d\omega'; \tag{16.3-7}$$

here $\mathcal{P}$ denotes the Cauchy principal value.

We shall give the proof of (16.3-7) and discuss some other related formulae. Since $\chi(iw)$ is analytic in the lower half plane, let us take the function $\chi(iw)/(w-u)$ and a contour in the lower half plane consisting of a large semi-circle, the real axis and a small negative semi-circle of radius r about the point $w = u$, cf. Fig. 16-2(a). Since the chosen function is analytic within the contour and goes to zero on the large arc if its radius tends to infinity, we have from Cauchy's contour-integral theorem

$$\int_{-\infty}^{u-r} \frac{\chi(i\omega)}{\omega-u}d\omega + \int_{u+r}^{\infty} \frac{\chi(i\omega)}{\omega-u}d\omega + \int_{\pi}^{2\pi} \frac{\chi(re^{i\varphi})}{re^{i\varphi}} rie^{i\varphi} d\varphi = 0, \tag{16.3-8}$$

where the last contribution stems from the little semi-circle and gives $\pi i\chi(iu)$. The first two parts together define the Cauchy principal value when $r \to 0$. We thus obtain

$$\chi(iu) = -\frac{1}{\pi i}\mathcal{P}\int_{-\infty}^{\infty} \frac{\chi(i\omega)}{\omega-u}d\omega. \tag{16.3-9}$$

Splitting (16.3-9) in real and imaginary parts, Eqs. (16.3-7) follow. In a slightly different method we displace the real axis by an infinitesimal amount $i\eta$ into the upper plane, see Fig. 16-2(b). The contour integral now becomes by Cauchy's residue theorem

$$\oint_C \frac{\chi(iw)}{w-u}dw = \int_{-\infty}^{\infty} \frac{\chi(i\omega)}{\omega-u+i\eta}d\omega = -2\pi i\chi(iu), \tag{16.3-10}$$

Next, we use the frequently occurring limit (called by some the 'Dirac relation')

$$\lim_{\eta\to 0} \frac{1}{\omega-u\pm i\eta} = \mathcal{P}\frac{1}{\omega-u} \mp \pi i\,\delta(\omega-u). \tag{16.3-11}$$

Substituting into (16.3-10) we have

$$\mathcal{P}\int_{-\infty}^{\infty} \frac{\chi(i\omega)}{\omega-u}d\omega - \pi i\chi(iu) = -2\pi i\chi(iu), \tag{16.3-12}$$

which is equivalent to (16.3-9), leading therefore once more to the Kramers–Kronig relations. The proof of (16.3-11) is left to the problems at the end of this section.

We return to Kubo's developments. For the conductance we find likewise

$$L_{BA}(i\omega) = \int_0^\infty dt\, e^{-i\omega t} \phi_{\dot{B}A}(t) = (1/\hbar i)\int_0^\infty dt\, e^{-i\omega t}\, \mathrm{Tr}\{[A, \dot{B}(t)]\rho_{eq}\}, \qquad (16.3\text{-}13)$$

with $L(i\omega) = L'(\omega) + iL''(\omega)$, the two parts satisfying the Kramers–Kronig relations. Finally, the complex conductance is expressible as a transform of $\phi_{\dot{B}\dot{A}}(t)$. From (16.2-23) we obtain

$$L_{BA}(i\omega) = \int_0^\infty dt \frac{e^{-i\omega t} - 1}{\hbar\omega} \mathrm{Tr}\{[\dot{A}, \dot{B}(t')]\rho_{eq}\} = -\int_0^\infty dt \frac{e^{-i\omega t} - 1}{i\omega} \phi_{\dot{B}\dot{A}}(t). \qquad (16.3\text{-}14)$$

This is the standard form, see e.g., van Velsen.[21]

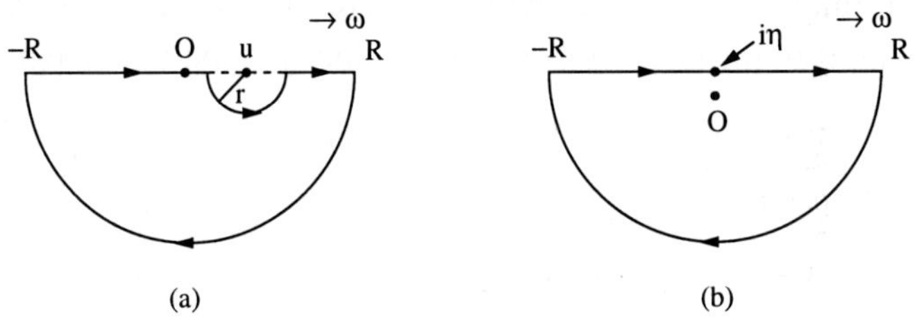

Fig. 16-2. Contours for the Kramers–Kronig relations.

16.3.2 *The Kubo Form and the Fujita Form*

The commutator expressions of the last subsection will be converted into correlation expressions involving real time as well as imaginary time. The response function (16.2-10) will be rewritten, employing invariance under cyclic permutation of the factors in the trace. We thus have

$$\phi_{BA}(t) = -(1/\hbar i)\mathrm{Tr}\{[A, \rho_{eq}]B(t)\}. \qquad (16.3\text{-}15)$$

Next we use *Kubo's identity*:

$$\frac{d}{d\tau}\left(e^{\tau\mathcal{H}} A e^{-\tau\mathcal{H}}\right) = e^{\tau\mathcal{H}}[\mathcal{H}, A]e^{-\tau\mathcal{H}}, \qquad (16.3\text{-}16)$$

as is verified by direct computation. This is integrated from 0 to β, yielding

$$e^{\beta\mathcal{H}} A e^{-\beta\mathcal{H}} - A = \int_0^\beta d\tau\, e^{\tau\mathcal{H}}[\mathcal{H}, A]e^{-\tau\mathcal{H}}; \qquad (16.3\text{-}17)$$

---

[21] J.W. van Velsen, "On linear response theory and area-preserving mappings", Ph. D Thesis, Univ. of Utrecht, 1976.

after pre-multiplication with $e^{-\beta\mathcal{H}}$ we have

$$[A, e^{-\beta\mathcal{H}}] = e^{-\beta\mathcal{H}} \int_0^\beta d\tau\, e^{\tau\mathcal{H}} [\mathcal{H}, A] e^{-\tau\mathcal{H}} . \tag{16.3-18}$$

We now recognise that the integrand at the rhs is the Heisenberg operator for imaginary time $t = -i\hbar\tau$ (this operator was also frequently used in equilibrium theory, cf. Section 12.9). Employing also the Heisenberg equation of motion, we have for this integrand $[\mathcal{H}, A(-i\hbar\tau)] = -i\hbar\, \dot{A}(-i\hbar\tau)$. So we arrive at *Kubo's Lemma*:

$$[A, e^{-\beta\mathcal{H}}] = -i\hbar \int_0^\beta e^{-\beta\mathcal{H}} d\tau\, \dot{A}(-i\hbar\tau) . \tag{16.3-19}$$

[Note that $\dot{A}(-i\hbar\tau)$ means $(d/dt)A(t)|_{t=-i\hbar\tau}$.] We divide this still by the canonical partition function $\mathcal{Z}$ and substitute into (16.3-15). This yields

$$\phi_{BA}(t) = \int_0^\beta d\tau \mathrm{Tr}\{\rho_{eq} \dot{A}(-i\hbar\tau) B(t)\} . \tag{16.3-20}$$

This leads us to the following Kubo-forms (or Kubo-Green forms) for the generalized susceptance and the generalized conductance:

$$\chi_{BA}(i\omega) = \int_0^\infty dt\, e^{-i\omega t} \int_0^\beta d\tau \mathrm{Tr}\{\rho_{eq} \dot{A}(-i\hbar\tau) B(t)\} , \tag{16.3-21}$$

$$L_{BA}(i\omega) = \int_0^\infty dt\, e^{-i\omega t} \int_0^\beta d\tau \mathrm{Tr}\{\rho_{eq} \dot{A}(-i\hbar\tau) \dot{B}(t)\} . \tag{16.3-22}$$

The Kubo expressions have the advantage that they are easily linked to the fluctuation-correlation functions. Let us write

$$A = \Delta A + \langle A \rangle, \qquad B = \Delta B + \langle B \rangle . \tag{16.3-23}$$

Since the averages are constants the time derivatives are equal, $\dot{A} = \Delta\dot{A}$, etc. Thus, in (16.3-22) we can use the fluctuation-correlation function instead of the regular correlation function. But this can also be done in(16.3-21) since $\mathrm{Tr}\{\rho_{eq} \Delta\dot{A}(-i\hbar\tau)\langle B\rangle\}$ $= \langle \Delta\dot{A}(-i\hbar\tau)\rangle \langle B\rangle = 0$. So, in both Kubo expressions we can replace the variables by their fluctuations. Let us now consider 'classical frequencies', by which we mean that $|i\hbar\tau| < |i\hbar\beta| \ll t \sim 1/\omega$, or $\hbar\omega \ll k_B T$. Since in a stationary process the correlations depend on $|t + i\hbar\tau|$, for classical frequencies the correlation function involves only $|t|$. We thus can write

$$\phi_{BA}(t) = \beta \mathrm{Tr}\{\rho_{eq} \Delta\dot{A} \Delta B(t)\} = \beta \langle \Delta\dot{A} \Delta B(t)\rangle \equiv \beta \Phi^{cl}_{B\dot{A}}(t) , \tag{16.3-24}$$

$$\phi_{\dot{B}A}(t) = \beta \mathrm{Tr}\{\rho_{eq} \Delta\dot{A} \Delta\dot{B}(t)\} = \beta \langle \Delta\dot{A} \Delta\dot{B}(t)\rangle \equiv \beta \Phi^{cl}_{\dot{B}\dot{A}}(t) , \tag{16.3-25}$$

$$\psi_{BA}(t) = \beta \mathrm{Tr}\{\rho_{eq} \Delta A \Delta B(t)\} = \beta \langle \Delta A \Delta B(t)\rangle \equiv \beta \Phi^{cl}_{BA}(t) , \tag{16.3-26}$$

where for the last relation we used (16.2-26), (16.3-24) and the mixing property. Eq. (16.3-26) is important in that it illustrates *Onsager's principle*: *the fluctuations decay on average in a similar way as the effect of an externally imposed disturbance.* Note, however, that this once regarded 'obvious' principle only holds for sufficiently low (i.e., classical) frequencies. The connection between the response and relaxation functions with the fluctuation-correlation function for quantum frequencies is considerably more complex and will be taken up in the next subsection.

The Kubo form can be further transformed into the Fujita form.[22] Let us introduce the fictitious Hamiltonian $\mathcal{H}' = \mathcal{H} - \alpha \dot{A}$. One easily shows

$$\frac{\partial}{\partial \alpha}\frac{1}{\mathcal{H}'-z} = \frac{1}{\mathcal{H}'-z}\dot{A}\frac{1}{\mathcal{H}'-z}, \tag{16.3-27}$$

where $1/(\mathcal{H}'-z)$ is the resolvent. By Cauchy's theorem

$$\frac{\partial}{\partial \alpha}e^{-\beta\mathcal{H}'} = -\frac{1}{2\pi i}\oint dz\, e^{-\beta z}\frac{\partial}{\partial \alpha}\frac{1}{\mathcal{H}'-z}. \tag{16.3-28}$$

Substituting (16.3-27) we have by convolution

$$\frac{\partial}{\partial \alpha}e^{-\beta\mathcal{H}'} = \int_0^\beta d\tau\, e^{-(\beta-\tau)\mathcal{H}'}\dot{A}\, e^{-\tau\mathcal{H}'} = \int_0^\beta d\tau\, e^{-\beta\mathcal{H}'}\dot{A}(-i\hbar\tau). \tag{16.3-29}$$

Hence, the Kubo forms can be transformed to

$$\chi_{BA}(i\omega) = \lim_{\alpha\to 0}\frac{\partial}{\partial \alpha}\int_0^\infty dt\, e^{-i\omega t}\,\mathrm{Tr}\{\rho_A' B(t)\}, \tag{16.3-30}$$

$$L_{BA}(i\omega) = \lim_{\alpha\to 0}\frac{\partial}{\partial \alpha}\int_0^\infty dt\, e^{-i\omega t}\,\mathrm{Tr}\{\rho_A' \dot{B}(t)\}, \tag{16.3-31}$$

where

$$\rho_A' = e^{-\beta\mathcal{H}'}/\mathrm{Tr}\, e^{-\beta\mathcal{H}'}. \tag{16.3-32}$$

16.3.3 *The Correlation form*

We proceed to obtain the connection of the response function with the fluctuation-correlation function for quantum frequencies.[23] The quantum mechanical fluctuation-correlation function is defined as the Hermitized product

$$\begin{aligned}\Phi_{BA}(t) &\equiv \langle \Delta\mathcal{B}(t)\Delta\mathcal{A}(0)\rangle = \tfrac{1}{2}\mathrm{Tr}\{\rho_{eq}[\Delta B(t),\Delta A]_+\}\\ &= \tfrac{1}{2}\big(\mathrm{Tr}\{\rho_{eq}\Delta B(t)\Delta A\} + \mathrm{Tr}\{\rho_{eq}\Delta A\Delta B(t)\}\big).\end{aligned} \tag{16.3-33}$$

---

[22] S. Fujita, Physica **26**,1161 (1969); also, "Introduction to Non-equilibrium Statistical Mechanics, R.E. Krieger Publ. Co, Malabar, FL, 1983.

[23] For a similar connection with the relaxation function, see LRT I, Ref. 15.

Whereas the response function is a commutator, the correlation function is an anticommutator. These functions can nevertheless be expressed into each other by:

$$\phi_{BA}(t)=\beta\int_{-\infty}^{\infty}dt'\Gamma(t-t')\Phi_{B\dot{A}}(t'), \tag{16.3-34a}$$

$$\phi_{\dot{B}A}(t)=\beta\int_{-\infty}^{\infty}dt'\Gamma(t-t')\Phi_{\dot{B}\dot{A}}(t'), \tag{16.3-34b}$$

where $\Gamma(t)$ is a kernel, even in *t*, to be obtained below. The connection is based on the following lemma (cf. Mazo, op. cit):

Lemma: The two-sided Fourier transforms of the expectations $\langle\Delta C(t)\Delta D\rangle$ and $\langle\Delta D\Delta C(t)\rangle$ for any two operators *C* and *D* are related by[24]

$$\int_{-\infty}^{\infty}dt\,e^{-i\omega t}\,\mathrm{Tr}\{\rho_{eq}\Delta C(t)\Delta D\}=e^{-\beta\hbar\omega}\int_{-\infty}^{\infty}dt\,e^{-i\omega t}\,\mathrm{Tr}\{\rho_{eq}\Delta D\Delta C(t)\}. \tag{16.3-35}$$

The proof goes as follows. Let us consider the rhs; it can be rewritten by analytic continuation in the complex plane. Thus, multiplying with the partition function,

$$\begin{aligned}\int_{-\infty}^{\infty}dt\,e^{-i\omega t}\,\mathrm{Tr}\{e^{-\beta\mathcal{H}}\Delta D\Delta C(t)\}&=\int_{-\infty}^{\infty}dt\,e^{-i\omega t}\,\mathrm{Tr}\{e^{-\beta\mathcal{H}}\Delta D\,e^{i\mathcal{H}t/\hbar}\Delta C e^{-i\mathcal{H}t/\hbar}\}\\ &=\int_{-\infty}^{\infty}dt\,e^{-i\omega t}\,\mathrm{Tr}\{e^{i\mathcal{H}t/\hbar}\Delta C e^{-i\mathcal{H}t/\hbar}e^{-\beta\mathcal{H}}\Delta D\}\\ &=e^{\beta\hbar\omega}\int_{-\infty}^{\infty}dt\,e^{-i\omega(t-i\hbar\beta)}\,\mathrm{Tr}\{e^{-\beta\mathcal{H}}e^{i\mathcal{H}(t-i\hbar\beta)/\hbar}\Delta C e^{-i\mathcal{H}(t-i\hbar\beta)/\hbar}\Delta D\}\\ &=e^{\beta\hbar\omega}\int_{-\infty-i\hbar\beta}^{\infty-i\hbar\beta}dz\,e^{-i\omega z}\,\mathrm{Tr}\{e^{-\beta\mathcal{H}}\Delta C(z)\Delta D\}.\end{aligned} \tag{16.3-36}$$

We assume analyticity on $0\le \mathrm{Im}\,z\le\hbar\beta$ and we make a contour integration along the real axis from $-R$ to $+R$, the line $z=x-i\hbar\beta$ and the lines $z=\pm R-iy$. Assuming the mixing property to be valid, the contributions along the latter vanish if $R\to\infty$ since $\langle\Delta C\rangle=\langle\Delta D\rangle=0$. The final integral of (16.3-36) can thus be replaced by the real axis integral $e^{\beta\hbar\omega}\int_{-\infty}^{\infty}dt...$ and (16.3-35) follows.

Two-sided Fourier transforms will henceforth be denoted by a caret. From (16.2-10) and stationarity we have $d\phi_{BA}/dt=-(1/\hbar i)\mathrm{Tr}\{\rho_{eq}[\Delta\dot{A},\Delta B(t)]\}$, which gives the Fourier equivalent

$$\hat{\phi}_{BA}(\omega)=-\frac{1}{\hbar\omega}\int_{-\infty}^{\infty}dt\,e^{-i\omega t}\left(\mathrm{Tr}\{\rho_{eq}\Delta B(t)\Delta\dot{A}\}-\mathrm{Tr}\{\rho_{eq}\Delta\dot{A}\Delta B(t)\}\right). \tag{16.3-37}$$

Also, from (16.3-33) we have, replacing *A* by $\dot{A}$

[24] For classical frequencies $e^{-\beta\hbar\omega}\to 1$ and we can with impunity commute the operators $\Delta C(t)$ and ΔD.

$$\hat{\Phi}_{B\dot{A}}(\omega) = \tfrac{1}{2}\int_{-\infty}^{\infty} dt\, e^{-i\omega t}\left(\mathrm{Tr}\{\rho_{eq}\Delta B(t)\Delta\dot{A}\} + \mathrm{Tr}\{\rho_{eq}\Delta\dot{A}\Delta B(t)\}\right). \tag{16.3-38}$$

With the lemma (16.3-35) we then obtain the connection

$$\hat{\phi}_{BA}(\omega) = [\mathcal{E}(\omega,T)]^{-1}\hat{\Phi}_{B\dot{A}}(\omega), \tag{16.3-39}$$

with $\mathcal{E}(\omega,T)/\beta$ being the 'quantum noise correction factor' compared to (16.3-24); the quantity $\mathcal{E}(\omega,T)$ is the mean energy of a quantum oscillator mode,

$$\mathcal{E}(\omega,T) = \frac{\hbar\omega}{2}\frac{1+e^{-\beta\hbar\omega}}{1-e^{-\beta\hbar\omega}} = \frac{\hbar\omega}{e^{\beta\hbar\omega}-1} + \frac{1}{2}\hbar\omega = \frac{\hbar\omega}{2}\coth\left(\frac{\beta\hbar\omega}{2}\right). \tag{16.3-40}$$

We note that the zero-point energy is included, a point that we further discuss below. By the convolution theorem for Fourier transforms (also often called the faltung theorem) the inverse relationships, (16.3-34a and b), then follow with

$$\Gamma(t) = \frac{1}{2\pi}\int_{-\infty}^{\infty} d\omega e^{i\omega t}\,\frac{\tanh(\beta\hbar\omega/2)}{\beta\hbar\omega/2} = \frac{2}{\beta\hbar\pi}\ln\coth\left(\frac{\pi}{2\beta\hbar}|t|\right). \tag{16.3-41}$$

[The proof, based on a simple contour integral and Mittag–Leffler's partial fraction theorem for meromorphic functions, is left to the problems.] We still note that for classical frequencies $\Gamma(t) \to \delta(t)$, so that (16.3-24) is confirmed.

The transport coefficients can now also be related to the fluctuation-correlation function. We employ (16.3-6) and (16.3-13), together with (16.3-34a and b) to obtain

$$\chi_{BA}(i\omega) = \beta\int_{0}^{\infty} dt\, e^{-i\omega t}\int_{-\infty}^{\infty} dt'\,\Gamma(t-t')\Phi_{B\dot{A}}(t'), \tag{16.3-42}$$

$$L_{BA}(i\omega) = \beta\int_{0}^{\infty} dt\, e^{-i\omega t}\int_{-\infty}^{\infty} dt'\,\Gamma(t-t')\Phi_{\dot{B}\dot{A}}(t'). \tag{16.3-43}$$

These results are fully equivalent to the Kubo expressions (16.33-21) and (16.3-22); in both cases we have a double integral, but in the present forms we stay with real-time correlations. Whatever type of form is more suitable depends on the application.

16.3.4 *The Fluctuation-Dissipation Theorem*

We shall go back to (16.3-39). By the Wiener–Khintchine theorem, the spectral density $\mathcal{S}_{BA}(\omega)$ of the cross-correlation covariance $\langle\Delta B\Delta A\rangle$ is just twice the Fourier transform $\hat{\Phi}_{BA}(\omega)$; the derivation will be given in the next section. First, let us consider $\Phi_{BA}(t) = \int_t^{\infty} dt'\langle\Delta B(t')\Delta\dot{A}\rangle$; Fourier analysis yields, employing (16.3-39),

$$\mathcal{S}_{BA}(\omega) = -(2/i\omega)\mathcal{E}(\omega,T)[\hat{\phi}_{BA}(\omega) - \hat{\phi}_{BA}(0)]. \tag{16.3-44}$$

The last part is connected to the dc susceptibility χ_{dc} and is of no concern. The correlation function, being real, satisfies the important *transposition property*:

$$\Phi_{BA}(t) = \Phi_{AB}(-t)\,, \tag{16.3-45}$$

as follows from the definition (16.3-33) and stationarity. Also, from the definition of the response function as a commutator, we note that $\phi_{BA}(t) = -\phi_{AB}(-t)$; moreover, since the commutator of Hermitean operators in anti-Hermitean, we conclude that $\phi_{BA}(t)$ *is real.*[25] Returning to (16.3-44), we have

$$\begin{aligned} \mathcal{S}_{BA}(\omega) &= -(2/i\omega)\mathcal{E}(\omega,T)\int_{-\infty}^{\infty} dt\, e^{-i\omega t}\phi_{BA}(t) \\ &= -(2/i\omega)\mathcal{E}(\omega,T)\left[\int_0^{\infty} dt\, e^{-i\omega t}\phi_{BA}(t) - \int_0^{\infty} dt\, e^{i\omega t}\phi_{AB}(t)\right]. \end{aligned} \tag{16.3-46}$$

Substituting the susceptance $\chi(i\omega)$, see (16.3-6), we find

$$\mathcal{S}_{BA}(\omega) = -(2/i\omega)\mathcal{E}(\omega,T)\left[\left(\chi'_{BA}(\omega) + i\chi''_{BA}(\omega)\right) - \left(\chi'_{AB}(\omega) - i\chi''_{AB}(\omega)\right)\right]. \tag{16.3-47}$$

We now introduce the symmetric and antisymmetric expressions

$$[\chi'_{BA}]^a \equiv \tfrac{1}{2}\left(\chi'_{BA} - \chi'_{AB}\right), \quad [\chi''_{BA}]^s \equiv \tfrac{1}{2}\left(\chi''_{BA} + \chi''_{AB}\right). \tag{16.3-48}$$

We then obtain

$$\mathcal{S}_{BA}(\omega) = -(4/\omega)\mathcal{E}(\omega,T)\left\{[\chi''_{BA}(\omega)]^s - i[\chi'_{BA}(\omega)]^a\right\}, \tag{\textbf{16.3-49}}$$

which is the fluctuation-dissipation theorem for susceptance-type processes.

For conductance-type processes, consider (16.3-39) for $B \to \dot{B}$. We thus have

$$\hat{\Phi}_{\dot{B}A}(\omega) = \mathcal{E}(\omega,T)\,\hat{\phi}_{\dot{B}A}(\omega)\,. \tag{16.3-50}$$

For the spectral density we obtain

$$\begin{aligned} \mathcal{S}_{\dot{B}A}(\omega) &= 2\mathcal{E}(\omega,T)\left[\int_0^{\infty} dt\, e^{-i\omega t}\phi_{\dot{B}A}(t) + \int_0^{\infty} dt\, e^{i\omega t}\phi_{A\dot{B}}(t)\right] \\ &= 2\mathcal{E}(\omega,T)\left[\left(L'_{BA}(\omega) + iL''_{BA}(\omega)\right) + \left(L'_{AB}(\omega) - iL''_{AB}(\omega)\right)\right], \end{aligned} \tag{16.3-51}$$

where we noted that now $\phi_{\dot{B}A}(t) = \phi_{A\dot{B}}(-t)$. Next we introduce

$$[L'_{BA}]^s \equiv \tfrac{1}{2}\left(L'_{BA} + L'_{AB}\right), \quad [L''_{BA}]^a \equiv \tfrac{1}{2}\left(L''_{BA} - L''_{AB}\right). \tag{16.3-52}$$

This leads to

$$\mathcal{S}_{\dot{B}A}(\omega) = 4\mathcal{E}(\omega,T)\left\{[L'_{BA}(\omega)]^s + i[L''_{BA}(\omega)]^a\right\}. \tag{\textbf{16.3-53}}$$

---

[25] This was noted by Kubo, although his proof is less lucid than ours above cf., Kubo, § 6, Theorem 4.

Finally, we consider the simplified case of a self-variance current-spectral density, $\mathcal{S}_{\dot{A}\dot{A}}(\omega)$. Then we have the basic result

$$\mathcal{S}_{\dot{A}\dot{A}}(\omega) = 4\mathcal{E}(\omega,T)\,\mathrm{Re}\,L_{AA}(i\omega)\,, \tag{16.3-54}$$

which is the generalized Nyquist result for the thermal noise of a passive device. To obtain Nyquist's original result, let $J = \dot{A}/V_0$, while also $\sigma = L_{AA}/V_0$. Hence,

$$\mathcal{S}_{JJ}(\omega) = \frac{1}{V_0^2}\mathcal{S}_{\dot{A}\dot{A}} = \frac{4}{V_0}\mathcal{E}(\omega,T)\,\mathrm{Re}\,\sigma(\omega)\,. \tag{16.3-55}$$

Consider a bar of cross section A and length ℓ, then $I = JA$ and $G = A\,\mathrm{Re}\,\sigma/\ell$, where $G = \mathrm{Re}\,Y(i\omega)$. This yields for the 'short-circuit current noise generator' (Norton generator) and its cousin, the 'open-circuit voltage generator' (Thévénin generator):

$$\mathcal{S}_I(\omega) = 4\mathcal{E}(\omega,T)\,\mathrm{Re}\,Y(i\omega), \quad \mathcal{S}_V(\omega) = 4\mathcal{E}(\omega,T)\,\mathrm{Re}\,Z(i\omega). \tag{16.3-56}$$

For low frequencies $\mathcal{E}(\omega,T) \to k_B T$. For the engineering-minded these generators – as equivalent linear signals in a bandwidth Δν – are pictured in Fig. 16-3(a) and (b).

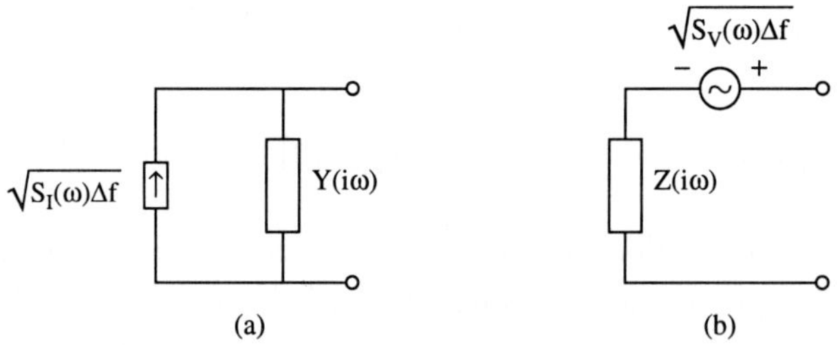

Fig. 16-3. Equivalent circuits for thermal noise, based on the Norton generator of a short-circuited sample (a) and the Thévénin generator of an open-circuited sample (b).

More intriguing is the fact that the present derivation, based on linear response theory, involves the full energy of an oscillator mode, *including zero-point energy.* This feature, absent in Nyquist's original derivation based on the energy trapped in a transmission line,[26] has long been a point of much debate.[27,28] From an experimental point of view, the problem arises if we couple an antenna to a matched electrical load. The antenna sees blackbody radiation, i.e., the energy density it receives is

---

[26] H. Nyquist, Phys. Rev. **32**, 110,1928.

[27] W.J. Kleen, in "Proc. Eighth Intern. Conf. on Noise in Physical Systems"(.A D'Amico and P. Mazetti, Eds.), North Holland Publ. Co., NY and Amsterdam 1985, p. 331.

[28] C.M. Van Vliet "On Planck's formula of Blackbody Radiation and the Quantum Correction Factor in Nyquist's Formula" in "Proc. Ninth Intern. Conf. on Noise in Physical Systems" (C.M. Van Vliet. Ed.), World Scient. Publ. Co., Singapore 1987, p. 191.

given by Planck's law for the mean energy density, Eq. (8.1-2). The radiation resistance of a Hertz dipole of length L is $R_r(\omega) = \omega^2 L^2 / 6\pi c^3$. The mean square electric field in vacuum is given by

$$\tfrac{1}{2}\langle H^2\rangle + \tfrac{1}{2}\langle E^2\rangle = \langle E^2\rangle \equiv \langle u(\omega)\rangle$$
$$= \frac{2\hbar\omega^3}{\pi c^3}\frac{1}{e^{\hbar\omega/k_BT}-1} = \frac{12R_r}{L^2}\frac{\hbar\omega}{e^{\hbar\omega/k_BT}-1}. \tag{16.3-57}$$

Noting that there are three possible directions of polarization, this yields

$$\mathcal{S}_V(\omega) = \tfrac{1}{3}\langle E^2\rangle L^2 = 4R_r\,\hbar\omega/[e^{\hbar\omega/k_BT}-1], \tag{16.3-58}$$

which is thermal noise minus the zero-point energy contribution. Heuristically, the discrepancy can be resolved by noting that the signal received by the antenna is more than the radiation density of the vacuum, in that we observe a truncated phasor of arbitrary phase,

$$\Delta V(t) = A\exp[i(\omega_0 t + \varphi)], \quad -\Delta t/2 \le t \le \Delta t/2. \tag{16.3-59}$$

The Fourier power of the signal in one side-band is

$$|F(i\omega)|^2 = A^2(\Delta t)^2\{\sin^2[\tfrac{1}{2}(\omega-\omega_0)\Delta t]/[\tfrac{1}{2}(\omega-\omega_0)\Delta t]^2\}, \tag{16.3-60}$$

which yields an uncertainty $(\Delta\omega)_{rms}(\Delta t)_{rms} = \pi$ or $(\Delta\mathcal{E})_{rms}(\Delta t)_{rms} = \pi\hbar$. Now $\Delta\varphi = \omega\,\Delta t = 2\pi$, which finally yields for the uncertainty of the antenna signal

$$(\Delta\mathcal{E})_{rms} = \tfrac{1}{2}\hbar\omega. \tag{16.3-61}$$

This energy must be added to the energy absorbed by the antenna from the thermal radiation field. Of course, one may consider this to be a contrived argument. It implies that zero-point energy basically constitutes zero-point energy *fluctuations*, a point generally perceived to be correct. However, the essential question is whether zero-point energy is 'extractable' or 'exchangeable'. While the conventional opinion holds that it is not, it is well-known that the Casimir force[29] between perfectly conducting electrical plates, as well as the van der Waals forces of non-polar substances, can be fully explained as caused by zero-point energy fluctuations.[30]

From an experimental point of view, direct measurements of thermal noise in the M-band (95 *Ghz*) and $1K$ by van der Ziel, Schmidt and the author at the University of Florida proved to be unsuccessful. However, heterodyned noise in Josephson junctions by Koch, Van Harlingen and Clarke at the University of California showed that the zero-point fluctuations were necessary in order to account for the data.[31]

[29] H.B.G. Casimir and D. Polder, Phys. Rev. **73**, 360 (1948); also, Versl. Koninkl. Akad. Wetensch. **51**, 793 (1948).

[30] For a recent discussion, see S.K. Lamoreaux, Physics Today, February 2007, p. 40.

[31] R.H. Koch, D.J. Van Harlingen and J. Clarke, Phys. Rev. Letts. **47**, 1216 (1981).

16.4 The Wiener–Khintchine Theorem

The connection between the spectral density of stochastic variables and the correlation function for the process was established in two classic papers by Norbert Wiener and A.I. Khintchine in 1930 and 1934, respectively.[32] We note hereby the definition of 'stochastic', as found with Khintchine: *A stochastic variable is a random variable with the time as a parameter.* Conform to our first introduction of such variables in Section 1.5, we shall denote them by $a_i(t)$ and their fluctuations by $\alpha_i(t) = a_i(t) - \langle a_i \rangle$. Generally, we are concerned with a number of such variables, which we shall represent by the column matrix $\mathbf{a}(t)$ or $\boldsymbol{\alpha}(t) = \mathbf{a}(t) - \langle \mathbf{a} \rangle$. Most treatments follow more or less Norbert Wiener's approach, in that they commence with Fourier analysis of the stochastic variables themselves. However, *likely*, few researchers (and textbook writers) have read Wiener's extremely complex paper; while 'almost all' functions have a Fourier transform,[33] the existence of the transform for a stochastic variable requires an extremely cumbersome mathematical analysis. We shall therefore first draw from Khintchine's very readable paper.

A stochastic variable $a(t)$ can only properly be described with a series of probability functions $W_n(a_n,t_n;a_{n-1},t_{n-1};...a_1,t_1)$, being the joint distribution that we observe a_n at time t_n, a_{n-1} at $t_{n-1},\ldots$, a_1 at t_1, with ultimately n going to infinity (this will be described in detail in Section 18.2). Yet, it can be shown that the correlation function, being a time-average or an appropriate ensemble average, is a well behaved function of the time increment between two observation points. For the stochastic matrix mentioned above, the (square) correlation matrix is defined by

$$\Phi(t) \equiv \langle \boldsymbol{\alpha}(t)\tilde{\boldsymbol{\alpha}}(0)\rangle \equiv \langle \boldsymbol{\alpha}(t)\boldsymbol{\alpha}(0)\rangle . \qquad (16.4\text{-}1)$$

Here, the tilde denotes the transpose, i.e., we multiply the column matrix with a row matrix; the last rhs is a short-cut notation with the tilde being implied. In any event, $\Phi_{ij}(t) = \langle \alpha_i(t)\alpha_j(0)\rangle$. In contrast with Section 1.5, in the context of LRT, we do not necessarily assume that the α's commute; thus the product is supposed to be Hermitized, as was done in the previous section. Further, we have the already stated transposition property,

$$\Phi(t) = \tilde{\Phi}(-t), \text{ or } \Phi_{ij}(t) = \Phi_{ji}(-t). \qquad (16.4\text{-}2)$$

It is easily shown that the correlation functions are well behaved and that a Fourier transform exists. Thus we write

$$\Phi(t) = \int_{-\infty}^{\infty} \hat{\Phi}(\omega) e^{i\omega t} d(\omega/2\pi), \qquad (16.4\text{-}3)$$

---

[32] N. Wiener, Acta Mathematica **55**, 117 (1930). A.I. Khintchine, Math. Ann. **109**, 604 (1934).

[33] Whereas extremely general sufficient conditions for the existence of a Fourier transform can be stated, no necessary conditions are known.

with Fourier inversion

$$\hat{\Phi}(\omega) = \int_{-\infty}^{\infty} \Phi(t) e^{-i\omega t} d\omega = \hat{\Phi}^*(-\omega). \tag{16.4-4}$$

With the above, the Fourier integral (16.4-3) can also be written as

$$\Phi(t) = \int_0^{\infty} \hat{\Phi}(\omega) e^{i\omega t} d\nu + \int_0^{\infty} \hat{\Phi}^*(\omega) e^{-i\omega t} d\nu, \tag{16.4-5}$$

where ν is the regular frequency $\omega/2\pi$. Khintchine establishes that both integrals are uniformly convergent for 'most' stochastic processes, so that we can take the limit $t \to 0$ under the integral; we thus obtain

$$\begin{aligned} \langle \boldsymbol{\alpha}\boldsymbol{\alpha} \rangle = \langle \Delta\mathbf{a}\Delta\mathbf{a} \rangle &= \int_0^{\infty} [\hat{\Phi}(\omega) + \hat{\Phi}^*(\omega)] d\nu \\ &= 2\int_0^{\infty} \mathrm{Re}\,\hat{\Phi}(\omega) d\nu = \int_0^{\infty} \mathrm{Re}\,\mathbb{S}(\omega) d\nu. \end{aligned} \tag{16.4-6}$$

We thus obtained a spectral resolution for the variances and covariances under discussion. Generally, we have $\mathbb{S}(\omega) = \mathbb{G}(\omega) + i\mathbb{H}(\omega)$, although, in our view, only the real spectra have physical significance. However, we will follow the usual, most obvious possibility, in that we set $\mathbb{S}(\omega) = 2\hat{\Phi}(\omega)$. With (16.4-4) we then arrive at the usual form of the Wiener-Khintchine theorem, announced in the previous section,

$$\mathbb{S}(\omega) = 2\int_{-\infty}^{\infty} \Phi(t) e^{-i\omega t} dt. \tag{16.4-7}$$

A few more remarks are in order. Later, in connection with the Onsager-Casimir relations, we shall distinguish between variables that are invariant under time reversal (α-variables) and those that change sign under time reversal (β-variables). When we have only either α-variables or β-variables, it follows from (16.4-2) that $\mathbb{S}$ is a Hermitean matrix. However, one must also reverse the magnetic field, i.e., $\mathbb{S}(\omega,\mathbf{H}) = \mathbb{S}^\dagger(\omega,-\mathbf{H})$. When dealing with a mixture of variables, the relations are a bit more complicated. We have $\mathcal{S}_{\alpha\beta}(\omega,\mathbf{H}) = -\mathcal{S}^*_{\beta\alpha}(\omega,-\mathbf{H})$. The real spectral part $\mathbb{G}(\omega)$ is a symmetrical matrix and follows similar rules as $\mathbb{S}(\omega)$. The imaginary part, however has opposite signs. For details we refer to de Groot and Mazur.[34]

Next, we consider Norbert Wiener's method, although we shall refrain from any mathematical justification. We start with the Fourier transform for a truncated set of stochastic variables (being zero outside $(-T/2 \le t \le T/2)$), writing

$$\hat{\boldsymbol{\alpha}}(\omega,T) = \int_{-T/2}^{T/2} \boldsymbol{\alpha}(t) e^{-i\omega t} dt = \hat{\boldsymbol{\alpha}}^*(-\omega,T). \tag{16.4-8}$$

The Fourier integral for the stochastic variable then reads

[34] S. R. de Groot and P. Mazur, Op. Cit., Chapter VIII.

$$\boldsymbol{\alpha}(t) = \operatorname*{Lim}_{T\to\infty} \int_{-T/2}^{T/2} \hat{\boldsymbol{\alpha}}(\omega,T) e^{i\omega t} d(\omega/2\pi), \tag{16.4-9}$$

where Lim means 'limit-in-probability'.[35] The spectral density is now *defined* as

$$\mathbb{S}(\omega) = \operatorname*{Lim}_{T\to\infty} \frac{2}{T} \hat{\boldsymbol{\alpha}}(\omega,T) \hat{\boldsymbol{\alpha}}^\dagger(\omega,T). \tag{16.4-10}$$

The physical meaning of $\mathbb{S}(\omega)$ is made clear from a generalized Parseval theorem,

$$\langle \boldsymbol{\alpha}(t)\boldsymbol{\alpha}(t)\rangle = \lim_{T\to\infty} \frac{1}{T} \int_{-T/2}^{T/2} [\boldsymbol{\alpha}(t)\tilde{\boldsymbol{\alpha}}(t)]dt = \operatorname*{Lim}_{T\to\infty} \frac{1}{T} \int_{-\infty}^{\infty} \hat{\boldsymbol{\alpha}}(\omega,T)\hat{\boldsymbol{\alpha}}^\dagger(\omega,T)\, d(\omega/2\pi)$$

$$= \operatorname*{Lim}_{T\to\infty} \frac{1}{T} \int_0^\infty [\hat{\boldsymbol{\alpha}}(\omega,T)\hat{\boldsymbol{\alpha}}^\dagger(\omega,T) + \hat{\boldsymbol{\alpha}}^*(\omega,T)\tilde{\hat{\boldsymbol{\alpha}}}(\omega,T)]d(\omega/2\pi) = \int_0^\infty \mathrm{Re}\,\mathbb{S}(\omega)\, d\nu. \tag{16.4-11}$$

Note that in the first line we assumed ergodicity. The far rhs reveals again the real part of the spectral density, $\mathbb{G}(\omega)$. Equation (16.4-10) also defines its imaginary part, although it plays no role in most ordinary stochastic processes. For an auto-variance, $\mathcal{S}_{ii}(\omega) \propto |\hat{\alpha}_i(\omega,T)|^2$ is always real. Further, from (16.4-10) and (16.4-8) we find

$$\mathbb{S}(\omega) = \operatorname*{Lim}_{T\to\infty} \frac{2}{T} \int_{-T/2}^{T/2} dt \int_{-T/2}^{T/2} dt' e^{-i\omega(t-t')} \boldsymbol{\alpha}(t)\tilde{\boldsymbol{\alpha}}(t'). \tag{16.4-12}$$

As new variables we take t' and $\vartheta = t - t'$. Then we get

$$\mathbb{S}(\omega) = \operatorname*{Lim}_{T\to\infty} \frac{2}{T} \int_{-T/2}^{T/2} dt' \int_{-\frac{1}{2}T-t'}^{\frac{1}{2}T-t'} d\vartheta e^{-i\omega\vartheta} \boldsymbol{\alpha}(t'+\vartheta)\tilde{\boldsymbol{\alpha}}(t')$$

$$= \int_{-\infty}^{\infty} d\vartheta e^{-i\omega\vartheta} \lim_{T\to\infty} \frac{2}{T} \int_{-T/2}^{T/2} dt' \boldsymbol{\alpha}(t'+\vartheta)\tilde{\boldsymbol{\alpha}}(t')$$

$$= 2\int_{-\infty}^{\infty} d\vartheta e^{-i\omega\vartheta} \langle \boldsymbol{\alpha}(t'+\vartheta)\tilde{\boldsymbol{\alpha}}(t')\rangle = 2\int_{-\infty}^{\infty} d\vartheta e^{-i\omega\vartheta} \boldsymbol{\Phi}(\vartheta) = 2\hat{\boldsymbol{\Phi}}(\omega), \tag{16.4-13}$$

as had to be proven; in the next to last transition we used once more ergodicity.

Basically, from a physics point of view, there is no need for the more elaborate procedure associated with Wiener's approach. The Fourier method offers new insights, however, if we consider the transmission of noise via a linear network. Since the spectral density is defined as a product of Fourier signals, it is now clear that the transmission of noise by a network with two-port transimpedance $Z_{BA}(i\omega)$ is

$$\mathcal{S}_{BB}(\omega) = Z_{BA}(i\omega) Z^*_{BA}(i\omega) \mathcal{S}_{AA}(\omega). \tag{16.4-14}$$

The Fourier method also helps to solve linear stochastic differential equations, such as the Langevin equation, to be discussed in Section 18.6.

---

[35] W.B. Davenport, Jr and W.L. Root "An introduction to the Theory of Random Signals and Noise" McGraw Hill, NY 1958, Chapter 6. They note that the limit sign in (16.4-9) indicates that the stochastic expression on the lhs converges *in probability* to the expression at the right; also, it is superfluous to add expectation brackets to $\hat{\boldsymbol{\alpha}}\hat{\boldsymbol{\alpha}}^\dagger$ in (16.4-10) as is often done – cf., A van der Ziel's various books on noise.

16.5 Density-Density Correlations and the Dynamic Structure Factor

16.5.1 *General Considerations*

The response formalism can also be employed for density fluctuations in quantum gases. Let $\rho(\mathbf{r}) = \Psi^\dagger(\mathbf{r})\Psi(\mathbf{r})$ be the quantum operator for the particle density, cf. Eq. (7.3-31) for a Bose gas and Section 7.6 for a Fermi gas; in the latter case a spin sum is assumed to be implied. The response Hamiltonian is now written as

$$\mathcal{H}_{total} = \mathcal{H} - \int d^3r\, \rho(\mathbf{r})\varsigma(\mathbf{r},t), \tag{16.5-1}$$

where the stimulus $\varsigma(\mathbf{r},t)\Theta(t)$ may be thought of as an externally invoked change in chemical or electrochemical potential; e.g., in a Fermi gas, a shift in quasi-Fermi level would change the particle density at all points in the system. The von Neumann equation is similar as in (16.2-2). For the density operator solution we have the equivalent Volterra equation, once iterated so as to obtain a linear response,

$$\rho_{dens}(t) = \rho_{can} + (i/\hbar)\int_0^t dt' e^{-i\mathcal{H}(t-t')/\hbar} \int d^3r' [\rho(\mathbf{r}',t'), \rho_{can}] e^{i\mathcal{H}(t-t')/\hbar} \varsigma(\mathbf{r}',t'). \tag{16.5-2}$$

For the density response we thus obtain – analogously to (16.2-6),

$$\begin{aligned}\langle\Delta\rho(\mathbf{r},t)\rangle &= (1/\hbar i)\mathrm{Tr}\left\{\int_0^t dt' \int d^3r' [\rho(\mathbf{r}',t'), \rho(\mathbf{r},t)]\,\rho_{can}\right\}\varsigma(\mathbf{r}',t') \\ &= (1/\hbar i)\,\langle\int_0^t dt' \int d^3r' [\rho(\mathbf{r}',t'), \rho(\mathbf{r},t)]\rangle \varsigma(\mathbf{r}',t'),\end{aligned} \tag{16.5-3}$$

where we used stationarity; the particle density operators are now in Heisenberg form. Let the momentum states $V_0^{-1/2} e^{i\mathbf{q}\cdot\mathbf{r}}$ serve as a normalized set of eigenfunctions; we then use the Fourier-series forms

$$\rho(\mathbf{r},t) = \frac{1}{V_0}\sum_{\mathbf{q}} \bar{\rho}(\mathbf{q},t) e^{i\mathbf{q}\cdot\mathbf{r}}, \quad \bar{\rho}(\mathbf{q},t) = \int d^3r\, \rho(\mathbf{r},t) e^{-i\mathbf{q}\cdot\mathbf{r}} = \bar{\rho}^\dagger(-\mathbf{q},t), \tag{16.5-4}$$

$$\varsigma(\mathbf{r},t) = \frac{1}{V_0}\sum_{\mathbf{q}} \bar{\varsigma}(\mathbf{q},t) e^{i\mathbf{q}\cdot\mathbf{r}}, \quad \bar{\varsigma}(\mathbf{q},t) = \int d^3r\, \varsigma(\mathbf{r},t) e^{-i\mathbf{q}\cdot\mathbf{r}}. \tag{16.5-5}$$

Also, for a translationally invariant system, the correlations depend only on $\mathbf{r} - \mathbf{r}'$. So,

$$\langle[\rho(\mathbf{r}',t'), \rho(\mathbf{r},t)]\rangle = \frac{1}{V_0^2}\sum_{\mathbf{q}} \langle[\bar{\rho}(\mathbf{q},t'), \bar{\rho}(-\mathbf{q},t)]\rangle e^{i\mathbf{q}\cdot(\mathbf{r}-\mathbf{r}')}. \tag{16.5-6}$$

For the spatial convolution this yields

$$\frac{1}{V_0^3}\sum_{\mathbf{q}}\sum_{\mathbf{q}'} \underbrace{\int d^3r' e^{i\mathbf{r}'\cdot(\mathbf{q}'-\mathbf{q})}}_{V_0\delta_{\mathbf{q},\mathbf{q}'}} \langle[\bar{\rho}(\mathbf{q},t'), \bar{\rho}(-\mathbf{q},t)]\rangle \bar{\varsigma}(\mathbf{q}',t') e^{i\mathbf{q}\cdot\mathbf{r}}$$

$$=\frac{1}{V_0^2}\sum_{\mathbf{q}}\langle[\bar{\rho}(\mathbf{q},t'),\bar{\rho}(-\mathbf{q},t)]\rangle\bar{\varsigma}(\mathbf{q},t')e^{i\mathbf{q}\cdot\mathbf{r}}. \tag{16.5-7}$$

Comparing with (16.5-3) we thus obtain

$$\langle\Delta\bar{\rho}(\mathbf{q},t)\rangle=(1/V_0\hbar i)\int_0^t dt'\langle[\bar{\rho}(\mathbf{q},t'),\bar{\rho}(-\mathbf{q},t)]\rangle\,\bar{\varsigma}(\mathbf{q},t'). \tag{16.5-8}$$

We introduce the response function

$$\phi(\mathbf{q},t-t')=(1/V_0\hbar i)\langle[\bar{\rho}(\mathbf{q},t'),\bar{\rho}(-\mathbf{q},t)]\rangle,\ \text{or} \tag{16.5-9a}$$

$$\phi(\mathbf{q},t)=(1/V_0\hbar i)\langle[\bar{\rho}(\mathbf{q},0),\bar{\rho}(-\mathbf{q},t)]\rangle. \tag{16.5-9b}$$

Next, we define Fourier–Laplace transforms similarly as in Section 16.2:

$$\chi(\mathbf{q},i\omega)=\int_0^\infty dt\,e^{-i\omega t}\phi(\mathbf{q},t)=(1/V_0\hbar i)\int_0^\infty dt\,e^{-i\omega t}\langle[\bar{\rho}(\mathbf{q},0),\bar{\rho}(-\mathbf{q},t)]\rangle \tag{16.5-10}$$

with similar integrals for the transforms $\hat{\varsigma}(\mathbf{q},i\omega)$ and $\langle\Delta\hat{\rho}(\mathbf{q},i\omega)\rangle$. Eq. (16.5-8) then yields with the Laplace-transform convolution theorem,

$$\langle\Delta\hat{\rho}(\mathbf{q},i\omega)\rangle=\chi(\mathbf{q},i\omega)\hat{\varsigma}(\mathbf{q},i\omega), \tag{16.5-11}$$

where $\chi(\mathbf{q},i\omega)$ as given by (16.5-10) is the complex susceptibility in **q**-space.

The susceptibility can usually not be measured in this form, but we shall relate it to the dynamical structure factor for quantum liquids, introduced previously. To that end we consider the retarded density-density correlation function in **r**- and **q**-space,

$$D(\mathbf{r},t;\mathbf{r}',t')\equiv\frac{1}{2}\langle[\Delta\rho(\mathbf{r},t),\Delta\rho(\mathbf{r}',t')]_+\rangle=\frac{1}{2V_0^2}\sum_{\mathbf{q}}\{\langle\Delta\bar{\rho}(\mathbf{q},t)\Delta\bar{\rho}(-\mathbf{q},t')\rangle e^{i\mathbf{q}\cdot\mathbf{R}}+hc\}$$

$$=\frac{1}{2V_0^2}\sum_{\mathbf{q}}\{\langle\Delta\bar{\rho}(\mathbf{q},t)\Delta\bar{\rho}(-\mathbf{q},t')\rangle e^{i\mathbf{q}\cdot\mathbf{R}}+\langle\Delta\bar{\rho}(\mathbf{q},t')\Delta\bar{\rho}(-\mathbf{q},t)\rangle e^{-i\mathbf{q}\cdot\mathbf{R}}\}$$

$$=\frac{1}{2V_0^2}\sum_{\mathbf{q}}\{\langle\Delta\bar{\rho}(\mathbf{q},t)\Delta\bar{\rho}(-\mathbf{q},t')\rangle+\langle\Delta\bar{\rho}(-\mathbf{q},t')\Delta\bar{\rho}(\mathbf{q},t)\rangle\}e^{i\mathbf{q}\cdot\mathbf{R}}$$

$$=\frac{1}{2V_0^2}\sum_{\mathbf{q}}\langle[\Delta\bar{\rho}(\mathbf{q},t)\,,\Delta\bar{\rho}(-\mathbf{q},t')]_+\rangle e^{i\mathbf{q}\cdot\mathbf{R}}\equiv\frac{1}{V_0^2}\sum_{\mathbf{q}}\Phi_{\rho\rho}(\mathbf{q},t-t')e^{i\mathbf{q}\cdot\mathbf{R}}. \tag{16.5-12}$$

The anticommutator was employed in order to Hermitize the operator product, while '*hc*' denotes the Hermitean conjugate. Clearly, we also have an anticommutator bracket in **q**-space. The inversion of (16.5-12) gives for the temporal correlation function in **q**-space,

$$\Phi_{\rho\rho}(\mathbf{q},t-t')=\tfrac{1}{2}\langle[\Delta\bar{\rho}(\mathbf{q},t),\Delta\bar{\rho}(-\mathbf{q},t')]_+\rangle=V_0\int d^3R\,e^{-i\mathbf{q}\cdot\mathbf{R}}\tfrac{1}{2}\langle[\Delta\rho(\mathbf{r},t),\Delta\rho(\mathbf{r}',t')]_+\rangle. \tag{16.5-13}$$

Since this is an autocorrelation function, its spectrum is real. The spectral density, being twice the two-sided Fourier transform therefore is

$$\mathcal{S}_{\rho\rho}(\mathbf{q},\omega) = \int_{-\infty}^{\infty} dt\, e^{-i\omega t} \{\langle \Delta\bar{\rho}(-\mathbf{q},0)\Delta\bar{\rho}(\mathbf{q},t)\rangle + \langle \Delta\bar{\rho}(\mathbf{q},t)\Delta\bar{\rho}(-\mathbf{q},0)\rangle\}. \quad (16.5\text{-}14)$$

The two parts in (16.5-14) are related by the previous lemma (16.3-5) [set $\Delta\bar{\rho}(\mathbf{q},t) = \Delta C(t)$ and $\Delta\bar{\rho}(-\mathbf{q},0) = \Delta D(0)$]. We can thus relate both parts in (16.5-14) by

$$\int_{-\infty}^{\infty} dt\, e^{-i\omega t} \langle \Delta\bar{\rho}(\mathbf{q},t)\Delta\bar{\rho}(-\mathbf{q},0)\rangle = e^{-\beta\hbar\omega} \int_{-\infty}^{\infty} dt\, e^{-i\omega t} \langle \Delta\bar{\rho}(-\mathbf{q},0)\Delta\bar{\rho}(\mathbf{q},t)\rangle. \quad (16.5\text{-}15)$$

Since this relationship involves a real exponential, we conclude that the separate parts are real and semi-positive definite.

We also introduce the *dynamical structure factor* by its usual definition [cf. (12.8-1)]:

$$S_d(\mathbf{q},\omega) = \int_{-\infty}^{\infty} dt\, e^{i\omega t} \langle \bar{\rho}(\mathbf{q},t)\bar{\rho}^{\dagger}(\mathbf{q},0)\rangle. \quad (16.5\text{-}16)$$

To connect with the previous, we shall add the deltas; this will only cause a spike $\delta(\omega)$ in the spectrum which we shall ignore. Also, we replace $t \to -t$, to obtain

$$S_d(\mathbf{q},\omega) = \int_{-\infty}^{\infty} dt\, e^{-i\omega t} \langle \Delta\bar{\rho}(\mathbf{q},0)\Delta\bar{\rho}(-\mathbf{q},t)\rangle$$

$$\overset{\mathbf{q}\to-\mathbf{q}}{=} \int_{-\infty}^{\infty} dt\, e^{-i\omega t} \langle \Delta\bar{\rho}(-\mathbf{q},0)\Delta\bar{\rho}(\mathbf{q},t)\rangle. \quad (16.5\text{-}16')$$

Hence, S_d is the first integral in (16.5-14). Moreover, employing (16.5-15), we find that S_d and $\mathcal{S}_{\rho\rho}$ are connected by

$$\mathcal{S}_{\rho\rho}(\mathbf{q},\omega) = S_d(\mathbf{q},\omega)\left[1 + e^{-\beta\hbar\omega}\right]. \quad (16.5\text{-}17)$$

It is interesting to note that for $T \to 0$, we have $\mathcal{S}_{\rho\rho}(\mathbf{q},\omega) \to S_d(\mathbf{q},\omega)$, so that the structure factor is the same as the spectral density. Also, the energy of an oscillator mode $\mathcal{E}(\omega,T)$ approaches $\frac{1}{2}\hbar\omega$, as noted from (16.3-40). Therefore, *at very low temperatures all quantum fluctuations, including the density fluctuations considered here, are driven by zero-point energy fluctuations.*

In the light of these statements we shall reconsider the low temperature excitations in interacting Bose and Fermi quantum liquids. Linear response theory and diagrammatic evaluation are presently the best tools to obtain approximate theoretical results.

16.5.2 *Another Form of the Fluctuation-Dissipation Theorem*

Although we are dealing with density-density fluctuations, the spectra mix the **q** and **−q** vectors, so that the expressions obtained look like cross-spectral densities. So, in order to employ our previous results from Section 16.3, let $\bar{\rho}(-\mathbf{q},t) \equiv A(t)$ and $\bar{\rho}(\mathbf{q},t) \equiv B(t)$. The desired connection between the spectral density and the two-sided Fourier transform of the response function is given in (16.3-44). We thus have at once,

$$\mathcal{S}_{\rho\rho}(\mathbf{q},\omega) = \int_{-\infty}^{\infty} dt\, e^{-i\omega t}\{\langle \Delta B(t)\Delta A(0)\rangle + \langle \Delta A(0)\Delta B(t)\rangle\} = rhs \text{ of } (16.5\text{-}14)$$

$$= -\frac{2}{i\omega}\frac{\hbar\omega}{2}\frac{1+e^{-\beta\hbar\omega}}{1-e^{-\beta\hbar\omega}}\hat{\phi}_{BA}(\omega) = i\hbar\frac{1+e^{-\beta\hbar\omega}}{1-e^{-\beta\hbar\omega}}\int_{-\infty}^{\infty} dt\, e^{-i\omega t}\frac{1}{\hbar i}\langle[\bar{\rho}(-\mathbf{q},0),\bar{\rho}(\mathbf{q},t)]\rangle$$

$$\stackrel{\mathbf{q}\to-\mathbf{q}}{=} i\hbar\frac{1+e^{-\beta\hbar\omega}}{1-e^{-\beta\hbar\omega}}\int_{-\infty}^{\infty} dt\, e^{-i\omega t}\frac{1}{\hbar i}\langle[\bar{\rho}(\mathbf{q},0),\bar{\rho}(-\mathbf{q},t)]\rangle. \qquad (16.5\text{-}18)$$

The expression at the far right-hand side entails the complex susceptibility, cf. (16.5-10). As usual, we split the integral in intervals $(-\infty,0)$ and $(0,\infty)$; this yields

$$\frac{1}{\hbar i}\int_{-\infty}^{\infty} dt\, e^{-i\omega t}\langle[\bar{\rho}(\mathbf{q},0),\bar{\rho}(-\mathbf{q},t)]\rangle = \frac{1}{\hbar i}\int_{0}^{\infty} dt\, e^{-i\omega t}\langle[\bar{\rho}(\mathbf{q},0),\bar{\rho}(-\mathbf{q},t)]\rangle$$

$$+\frac{1}{\hbar i}\int_{-\infty}^{0} dt\, e^{-i\omega t}\langle[\bar{\rho}(\mathbf{q},0),\bar{\rho}(-\mathbf{q},t)]\rangle = \frac{1}{\hbar i}\int_{0}^{\infty} dt\, e^{-i\omega t}\langle[\bar{\rho}(\mathbf{q},0),\bar{\rho}(-\mathbf{q},t)]\rangle + \frac{1}{\hbar i}$$

$$\times\int_{0}^{\infty} dt\, e^{i\omega t}\langle[\bar{\rho}(-\mathbf{q},t),\bar{\rho}(\mathbf{q},0)]\rangle = V_0[\chi(\mathbf{q},i\omega) - \chi(\mathbf{q},-i\omega)] = 2iV_0\,\mathrm{Im}\,\chi(\mathbf{q},i\omega). \quad (16.5\text{-}19)$$

Substituting into (16.5-18) we arrive at

$$\mathcal{S}_{\rho\rho}(\mathbf{q},\omega) = -2\hbar V_0\frac{1+e^{-\beta\hbar\omega}}{1-e^{-\beta\hbar\omega}}\mathrm{Im}\,\chi(\mathbf{q},i\omega). \qquad (16.5\text{-}20)$$

This is the new fluctuation-dissipation theorem. Usually, however, we connect with the dynamical structure factor, which can be experimentally obtained from inelastic neutron scattering, see subsection 12.8.1. With (16.5-17) we then find

$$S_d(\mathbf{q},\omega) = -[2\hbar V_0/(1-e^{-\beta\hbar\omega})]\,\mathrm{Im}\,\chi(\mathbf{q},i\omega). \qquad (16.5\text{-}21)$$

We still note that for most measurements done below $1K$ and with $\hbar\omega$ of order of milli-electronvolts, $\beta\hbar\omega \gg 1$. In the quantum limit we always have

$$\mathcal{S}_{\rho\rho}(\mathbf{q},\omega) \approx S_d(\mathbf{q},\omega) \approx -2\hbar V_0\,\mathrm{Im}\,\chi(\mathbf{q},i\omega). \qquad (16.5\text{-}22)$$

A measurement of S_d is the most direct way to obtain information on the interactions in a quantum liquid as already noted in Section 12.8.

16.5.3 *Thermodynamics and Sum-rules*

So far we have followed the theory as it was developed at its inception. In particular, with Kubo we have used the canonical density operator to describe the system. However, in the various examples of Chapter XII we have seen that the number of particles in a quantum fluid generally is not a constant of motion, thus necessitating the use of the grand-canonical ensemble. The basic operator is then the grand Hamiltonian, which we designate by $K = K^0 + \lambda\mathcal{V}$, with $\lambda \to 1$ when the interactions are fully 'turned on'. The 'modified Heisenberg operators' are now

$$A(t) = e^{iKt/\hbar} A e^{-iKt/\hbar}, \tag{16.5-23}$$

as was already amply indicated in Sections 12.9 ff. We shall, however, not explicitly employ the diagrammatic methods of those sections here; only our methodology will be in a similar vein. The previous results of the present chapter all remain valid in the modified Heisenberg picture, the substitution of K for $\mathcal{H}$ then being implied. The eigenstates of K will be denoted by $|\bar{\eta}_n\rangle$ and the eigenvalues by κ_n. The various parts of K will be expressed in second quantized form, whereby presently we shall employ the field operators $\Psi(\mathbf{r})$ and $\Psi^\dagger(\mathbf{r})$, rather than the creation operators $\hat{e}_k$ and $\hat{e}_k^\dagger$. In a Fermi liquid the spin indices must be added. The retarded density-density function is readily written down in terms of the field operators and the connection with the retarded Green's function is immanent but will not be pursued [see, however, subsection 16.6.4]. Rather, we shall indicate that the thermodynamic functions of interest follow easily from the dynamic and static structure factors as defined before.

The energy K^0 contains the kinetic and chemical energy, $\Sigma_i(\varepsilon_i - \varsigma)$, or in field form

$$K^0 = \int d^3r \left(\frac{\hbar^2}{2m} \nabla\Psi^\dagger(\mathbf{r}) \cdot \nabla\Psi(\mathbf{r}) - \varsigma\Psi^\dagger(\mathbf{r})\Psi(\mathbf{r}) \right), \tag{16.5-24}$$

see (7.3-35, (7.6-23). Likewise, for the two-body interaction energy $\mathcal{V} = \frac{1}{2}\Sigma_{ij} v(r_i - r_j)$ we have

$$\mathcal{V} = \tfrac{1}{2} \iint d^3r\, d^3r' \Psi^\dagger(\mathbf{r})\Psi^\dagger(\mathbf{r}') v(\mathbf{r} - \mathbf{r}')\Psi(\mathbf{r}')\Psi(\mathbf{r}), \tag{16.5-25}$$

where we employed the order of the field operators *for fermions* (being naturally also valid for bosons for which two ψ's commute rather than anticommute). The Gibbs function or grand potential can be found from integration of $\langle\lambda\mathcal{V}\rangle_{gcan}$, as was shown in Eq. (12.11-16):

$$\Omega(T,V,\varsigma) = \Omega_0(T,V,\varsigma) + \int_0^1 d\lambda\, \lambda^{-1} \langle\lambda\mathcal{V}\rangle_{gcan}, \tag{16.5-26}$$

where Ω_0 is the grand potential for a non-interacting system. We proceed to obtaining the connection of $\langle\lambda\mathcal{V}\rangle_{gcan}$ with the dynamical structure factor. In terms of the particle density operator $\rho(\mathbf{r})$ (16.5-25) reads

$$\langle \mathcal{V} \rangle = \tfrac{1}{2}\int\!\!\int d^3r\, d^3r'\, v(\mathbf{R})\rho(\mathbf{r})\rho(\mathbf{r}'), \tag{16.5-27}$$

where $\mathbf{R} = \mathbf{r} - \mathbf{r}'$ and where we moved $\psi(\mathbf{r})$ to the left with two transpositions and assumed that v is not velocity dependent. We write $v(\mathbf{R})$ in a Fourier series, $v(\mathbf{R}) = (1/V_0)\Sigma_{\mathbf{q}} e^{i\mathbf{q}\cdot\mathbf{R}} V_{\mathbf{q}}$ and we use the Fourier series – see (16.5-4),

$$\rho(\mathbf{r}) = (1/V_0)\sum\nolimits_{\mathbf{q}'} e^{i\mathbf{q}'\cdot\mathbf{r}}\bar{\rho}(\mathbf{q}'), \qquad \rho(\mathbf{r}') = (1/V_0)\sum\nolimits_{\mathbf{q}''} e^{i\mathbf{q}''\cdot\mathbf{r}'}\bar{\rho}(\mathbf{q}''). \tag{16.5-28}$$

This results in

$$\langle \mathcal{V} \rangle = \frac{1}{2V_0}\sum_{\mathbf{q}} V_{-\mathbf{q}} \langle \bar{\rho}(\mathbf{q})\bar{\rho}(-\mathbf{q})\rangle . \tag{16.5-29}$$

Now from (16.5-16)

$$\langle \bar{\rho}(\mathbf{q})\bar{\rho}(-\mathbf{q})\rangle = \lim_{t\to 0}\frac{1}{2\pi}\int_{-\infty}^{\infty} d\omega\, e^{-i\omega t} S_d(\mathbf{q},\omega) = \frac{1}{2\pi}\int_{-\infty}^{\infty} d\omega\, S_d(\mathbf{q},\omega). \tag{16.5-30}$$

Roughly speaking, this frequency integral equals the static structure factor $S_s(\mathbf{q})$, times the average particle number $\langle N\rangle$ in the system. Some refinements, however, are necessary since the static structure factor is related to the pair-correlation function rather than the two-particle distribution function, cf. Eq. (4.3-76) ff. We then obtain the result

$$\langle \mathcal{V} \rangle = \frac{1}{2}\langle n\rangle \left(\sum_{\mathbf{q}\neq 0} V_{-\mathbf{q}}[S_s(\mathbf{q}) - 1] + V_{\mathbf{q}=0}\langle N\rangle \right). \tag{16.5-31}$$

We proceed to obtain some sum rules regarding the oscillator strengths. They can be found in the nonequilibrium volume of the monographs by Kubo, Toda and Hashitsume.[36] Let us take the first Kramers–Kronig relation in (16.3-7) for $\omega = 0$. We then find

$$\int_{-\infty}^{\infty} \frac{d\omega}{\pi} \mathcal{P}\frac{\chi''(\omega)}{\omega} = -\chi'(0) = -\chi_{dc}\,. \tag{16.5-32}$$

This identity is always true, as examples confirm. This 'sum-rule' therefore gives no new information. However, by multiplying the integrand of the Kramers–Kronig's relation by ω, integrating by parts and employing Abel's limit theorem, one can show

$$\int_{-\infty}^{\infty} \frac{d\omega}{\pi}\chi''(\omega) = -\varphi(0)\,, \tag{16.5-33}$$

where φ is a slightly different form of the response function (defined by KTH).

[36] R. Kubo, M. Toda and N. Hashitsume, "Statistical Physics" II (Nonequilibrium Statistical Mechanics), Springer Verlag, Berlin and Heidelberg 1985, Section 3.7.

Generally, they prove:

$$\int_{-\infty}^{\infty} \frac{d\omega}{\pi} \omega^{2n+1} \chi''(\omega) = (-1)^{n+1} \varphi^{(2n+1)}(0)\,, \quad \int_{-\infty}^{\infty} \frac{d\omega}{\pi} \omega^{2n} \chi'(\omega) = (-1)^{n} \varphi^{(2n)}(0)\,. \tag{16.5-34}$$

[Similar rules for the generalized conductance are left to the problems.] Presently, we take $n = 0$ and aim to establish the *f*-sum-rule

$$\int_{-\infty}^{\infty} \frac{d\omega}{\pi} \omega \operatorname{Im} \chi(\mathbf{q}, i\omega) = -\dot{\varphi}(\mathbf{q}, 0) = \frac{q^2 \langle n \rangle}{m}\,. \tag{16.5-35}$$

We start with (16.5-21), which has

$$\operatorname{Im} \chi(\mathbf{q}, i\omega) = -S_d(\mathbf{q}, \omega)(1 - e^{-\beta\hbar\omega}) / 2\hbar V_0\,. \tag{16.5-36}$$

Next, we find a new result for S_d by evaluating the ensemble average in the grand-canonical ensemble, using the eigenstates of the grand Hamiltonian for the trace; so,

$$\begin{aligned} S_d(\mathbf{q}, \omega) &= \frac{1}{\mathcal{Z}} \int_{-\infty}^{\infty} dt\, e^{i\omega t} \operatorname{Tr}\{e^{-\beta K} [\bar{\rho}(\mathbf{q}, t) \bar{\rho}(-\mathbf{q})]\} \\ &= \frac{1}{\mathcal{Z}} \int_{-\infty}^{\infty} dt\, e^{i\omega t} \sum_{n,m} e^{-\beta \kappa_n} e^{i\kappa_n t/\hbar} \langle \bar{\eta}_n | \bar{\rho}(\mathbf{q}) | \bar{\eta}_m \rangle e^{-i\kappa_m t/\hbar} \langle \bar{\eta}_m | \bar{\rho}^\dagger(\mathbf{q}) | \bar{\eta}_n \rangle \\ &= \frac{1}{\mathcal{Z}} \sum_{n,m} e^{-\beta\kappa_n} \int_{-\infty}^{\infty} dt\, e^{it[\omega + (\kappa_n - \kappa_m)/\hbar]} |\langle \bar{\eta}_n | \bar{\rho}(\mathbf{q}) | \bar{\eta}_m \rangle|^2 \\ &= \frac{2\pi}{\mathcal{Z}} \sum_{n,m} e^{-\beta\kappa_n} |\langle \bar{\eta}_n | \bar{\rho}(\mathbf{q}) | \bar{\eta}_m \rangle|^2 \delta[\omega - (\kappa_m - \kappa_n)/\hbar]\,. \end{aligned} \tag{16.5-37}$$

From the last two results, the integral of (16.5-35) becomes

$$\begin{aligned} &\int_{-\infty}^{\infty} \frac{d\omega}{\pi} \omega \operatorname{Im} \chi(\mathbf{q}, i\omega) \\ &= -\frac{1}{\mathcal{Z}\hbar V_0} \sum_{n,m} e^{-\beta\kappa_n} |\langle \bar{\eta}_n | \bar{\rho}(\mathbf{q}) | \bar{\eta}_m \rangle|^2 \int_{-\infty}^{\infty} d\omega\, \omega [1 - e^{-\beta\hbar\omega}] \delta[\omega - (\kappa_m - \kappa_n)/\hbar] \\ &= \frac{1}{\mathcal{Z}\hbar^2 V_0} \sum_{n,m} |\langle \bar{\eta}_n | \bar{\rho}(\mathbf{q}) | \bar{\eta}_m \rangle|^2 (\kappa_n - \kappa_m) \left(e^{-\beta\kappa_n} - e^{-\beta\kappa_m} \right). \end{aligned} \tag{16.5-38}$$

Finally, let us consider the double commutator $[K, \bar{\rho}(\mathbf{q})], \bar{\rho}(\mathbf{q})]$. We will first find its value directly using the field-operator expressions given before. We saw already that $\bar{\rho}(\mathbf{q})$ commutes with $\mathcal{V}$, while in Chapter VII we noted that the ρ commutes with the number operator $\mathbf{N}$. So the commutator simplifies to $[\mathcal{H}^0, \bar{\rho}(\mathbf{q})], \bar{\rho}(\mathbf{q})]$, $\mathcal{H}^0$ representing kinetic energy. A straightforward computation gives $[K, \bar{\rho}(\mathbf{q})], \bar{\rho}(\mathbf{q})] = \hbar^2 q^2 N / m$. For a grand ensemble in which N fluctuates, we have the average

$$\langle[K,\bar{\rho}(\mathbf{q})],\bar{\rho}(\mathbf{q})]\rangle = \hbar^2 q^2\langle N\rangle/m. \tag{16.5-39}$$

Of course, this commutator-average can also obtained by finding the trace in the representation of the grand Hamiltonian eigenstates. We easily find

$$\mathrm{Tr}\{[K\bar{\rho}(\mathbf{q})-\bar{\rho}(\mathbf{q})K]\bar{\rho}(\mathbf{q})-\bar{\rho}(\mathbf{q})[K\bar{\rho}(\mathbf{q})-\bar{\rho}(\mathbf{q})K]\}$$
$$=\frac{1}{\mathcal{I}}\sum_{n.m}|\langle\bar{\eta}_n|\bar{\rho}(\mathbf{q})|\bar{\eta}_m\rangle|^2(\kappa_n-\kappa_m)\left(e^{-\beta\kappa_n}-e^{-\beta\kappa_m}\right). \tag{16.5-40}$$

This is the same as the expression in (16.5-38), except for the factor $(1/\hbar^2 V_0)$. Combining with (16.5-39), the rhs of (16.5-35) is confirmed.

16.6 A Return to Quantum Liquids

16.6.1 *Self-consistent Field Approximation*

The results presented so far are what many researchers consider '*exact*', i.e., within the framework of the linear response approach. The quest whether the so obtained results are intrinsically meaningful, will be taken up in Section 16.8. Here we only remark that in order to *apply* the results to actual systems, approximations that in one way or another consider the random aspects of the gas or quantum liquid must be made. In other words, the *linearization* becomes only meaningful if afterwards a *randomization* is applied (called by KTD[37] 'stochasticization').

The present section seeks to apply the density-response results to quantum fluids without knowing the precise two-body interactions in detail. The method about to be set forth will be called the '*self-consistent field method*'; basically, it takes a variety of forms, such as the Hartree–Fock method, the random phase-approximation (RPA), the Lindhard approximation and others, discussed previously in the diagrammatic formalism, cf. Sections 12.12 and 12.13. The essential assumption here is that a fluid constituent (pseudo-particle) does not *directly* feel the applied external perturbation but rather reacts to an effective field that consists of a mixture of the *applied field* and its locally induced *polarization field.* So, let the two-body potential $v(\mathbf{r}-\mathbf{r}')$ have a Fourier coefficient

$$V_{\mathbf{q}}=\int d^3R\, e^{-i\mathbf{q}\cdot\mathbf{R}}v(\mathbf{R}),\quad \mathbf{R}=\mathbf{r}-\mathbf{r}'. \tag{16.6-1}$$

The locally induced potential can then be written as the averaged spatial convolution

$$\varsigma_{ind}(\mathbf{r},t)=\int d^3r'v(\mathbf{r}-\mathbf{r}')\langle\Delta\rho(\mathbf{r}',t)\rangle, \tag{16.6-2}$$

or in **q**-space

[37] R. Kubo, M. Toda and N. Hashitsume, Op. Cit. p. 196.

$$\overline{\varsigma}_{ind}(\mathbf{q},t) = \int d^3r \int d^3r' v(\mathbf{r}-\mathbf{r}')\langle\Delta\rho(\mathbf{r}',t)\rangle e^{-i\mathbf{q}\cdot\mathbf{r}}$$
$$= \int d^3r \int d^3r' v(\mathbf{r}-\mathbf{r}')V_0^{-1}\sum_{\mathbf{q}'}\langle\Delta\overline{\rho}(\mathbf{q}',t)\rangle e^{i\mathbf{q}'\cdot\mathbf{r}'}e^{-i\mathbf{q}\cdot\mathbf{r}} = \langle\Delta\overline{\rho}(\mathbf{q},t)\rangle V_{\mathbf{q}} \,. \quad (16.6\text{-}3)$$

The effective potential is now

$$\overline{\varsigma}_{eff}(\mathbf{q},t) = \overline{\varsigma}(\mathbf{q},t) + \overline{\varsigma}_{ind}(\mathbf{q},t) \,. \quad (16.6\text{-}4)$$

For a sinusoidal externally applied potential we use Fourier-Laplace transforms. Then from (16.6-4) and (16.6-3)

$$\hat{\varsigma}_{eff}(\mathbf{q},i\omega) = \hat{\varsigma}(\mathbf{q},i\omega) + \hat{\varsigma}_{ind}(\mathbf{q},i\omega)$$
$$= \hat{\varsigma}(\mathbf{q},i\omega) + \langle\Delta\hat{\rho}(\mathbf{q},i\omega)\rangle V_{\mathbf{q}} \,. \quad (16.6\text{-}5)$$

Now the response $\langle\Delta\hat{\rho}(\mathbf{q},i\omega)\rangle$ is related to the effective field by the previously defined susceptibility for density fluctuations, i.e., $\langle\Delta\hat{\rho}(\mathbf{q},i\omega)\rangle = \chi_0(\mathbf{q},i\omega)\varsigma_{eff}(\mathbf{q},i\omega)$. Substituting into (16.6-5), we find that the effective potential is related to the applied potential by

$$\hat{\varsigma}_{eff}(\mathbf{q},i\omega)[1-\chi_0(\mathbf{q},i\omega)V_{\mathbf{q}}] = \varsigma(\mathbf{q},i\omega) \,. \quad (16.6\text{-}6)$$

As noticed previously, *every pseudo-particle is dressed with the interactions.* In terms of the applied potential we thus have the dressed susceptibility

$$\langle\Delta\hat{\rho}(\mathbf{q},i\omega)\rangle = \chi_{dr}(\mathbf{q},i\omega)\hat{\varsigma}(\mathbf{q},i\omega) \,. \quad (16.6\text{-}7)$$

Consequently,

$$\chi_{dr}(\mathbf{q},i\omega) = \frac{\chi_0(\mathbf{q},i\omega)}{1-V_{\mathbf{q}}\chi_0(\mathbf{q},i\omega)} \,. \quad (16.6\text{-}8)$$

This is exactly the form given (without proof) in Section 12.8.

To link the susceptibility to the dynamic structure factor we shall now go back to (16.5-10) and obtain some other, very useful, relationships; we write[38]

$$\chi(\mathbf{q},i\omega) = \frac{1}{V_0\hbar i}\int_{-\infty}^{\infty} dt\, e^{-i\omega t}\Theta(t)[\langle\overline{\rho}(\mathbf{q},0)\overline{\rho}(-\mathbf{q},t)\rangle - \langle\overline{\rho}(-\mathbf{q},t)\overline{\rho}(\mathbf{q},0)\rangle]$$

$$\stackrel{\mathbf{q}\to-\mathbf{q}}{=} \frac{1}{V_0\hbar i}\int_{-\infty}^{\infty} dt\, e^{-i\omega t}\,\Theta(t)\,[\langle\overline{\rho}(-\mathbf{q},0)\overline{\rho}(\mathbf{q},t)\rangle - \langle\overline{\rho}(\mathbf{q},t)\overline{\rho}(-\mathbf{q},0)\rangle] \,. \quad (16.6\text{-}9)$$

Employing the integral representation for the Heaviside function

$$\Theta(t-t') = -\lim_{\eta\to 0}\frac{1}{2\pi i}\int_{-\infty}^{\infty} d\omega' \frac{e^{-i\omega'(t-t')}}{\omega'+i\eta} \,, \quad (16.6\text{-}10)$$

---

[38] Cf. M. Plischke and B. Bergersen, "Equilibrium Statistical Physics" 3rd Ed., World Scient. Publ. Co Singapore 2006, Chapter 12. [Their $\chi(i\omega)$ corresponds with our – and Kubo's – $\chi(-i\omega)$, cf. footnote 20.]

we have

$$\mathcal{X}(\mathbf{q},i\omega)=\lim_{\eta\to 0}\frac{1}{2\pi\hbar V_0}$$

$$\times\int_{-\infty}^{\infty}dt\,e^{-i\omega t}\int_{-\infty}^{\infty}d\omega'\frac{e^{-i\omega' t}}{\omega'+i\eta}[\langle\overline{\rho}(-\mathbf{q},0)\overline{\rho}(\mathbf{q},t)\rangle-\langle\overline{\rho}(\mathbf{q},t)\overline{\rho}(-\mathbf{q},0)\rangle]$$

$$=\lim_{\eta\to 0}\frac{1}{2\pi\hbar V_0}\int_{-\infty}^{\infty}d\omega''\frac{S_d(\mathbf{q},\omega'')-S_d(\mathbf{q},-\omega'')}{\omega''-\omega+i\eta},\qquad\textbf{(16.6-11)}$$

where we interchanged the integrals and used the new variable $\omega''=\omega+\omega'$. A slightly different form obtains if we make the change $\omega''\to-\omega''$ in the second part,[39]

$$\mathcal{X}(\mathbf{q},i\omega)=\lim_{\eta\to 0}\frac{1}{2\pi\hbar V_0}\int_{-\infty}^{\infty}d\omega''S_d(\mathbf{q},\omega'')\left\{\frac{1}{\omega+\omega''-i\eta}-\frac{1}{\omega-\omega''-i\eta}\right\}.\qquad\textbf{(16.6-12)}$$

If $\omega\to\infty$, combining the two parts in the curly brackets shows that χ becomes real and has as magnitude

$$\mathcal{X}(\mathbf{q},i\omega)=-\frac{1}{2}\lim_{\omega\to\infty}\lim_{\eta\to 0}\frac{1}{\pi\hbar V_0}\int_{-\infty}^{\infty}\omega''d\omega''S_d(\mathbf{q},\omega'')/[(\omega-i\eta)^2-\omega''^2]$$

$$=\frac{1}{\pi}\lim_{\omega\to\infty}\int_{-\infty}^{\infty}\omega''d\omega''\,\mathrm{Im}\,\mathcal{X}(\mathbf{q},\omega'')/[(\omega-i\eta)^2-\omega''^2]\sim\frac{q^2\langle n\rangle}{m\omega^2},\qquad(16.6\text{-}13)$$

where we used the *f*-sum rule (16.3-35). Returning to the general form (16.6-12) and using the relation (16.3-11), we find for the separate real and imaginary parts,

$$\mathcal{X}'(\mathbf{q},\omega)=-\frac{1}{\pi\hbar V_0}\int_{-\infty}^{\infty}d\omega''S_d(\mathbf{q},\omega'')\mathcal{P}\frac{2\omega''}{(\omega^2-\omega''^2)},\qquad(16.6\text{-}14)$$

$$\mathcal{X}''(\mathbf{q},\omega)=\frac{1}{\hbar V_0}\{S_d(\mathbf{q},-\omega)-S_d(\mathbf{q},\omega)\}.\qquad(16.6\text{-}15)$$

We shall now consider two applications in more detail.

16.6.2 *Excitations in the Bose Liquid*

Let the Bose liquid be close to $T = 0$ and let at first the particles be non-interacting. For a pure state we then have

$$\overline{\rho}(\mathbf{q})=\int d^3r\Psi^\dagger(\mathbf{r})\Psi(\mathbf{r})e^{-i\mathbf{q}\cdot\mathbf{r}}=\sum_{\mathbf{k}}a^\dagger_{\mathbf{k}-\mathbf{q}}a_{\mathbf{k}},\qquad(16.6\text{-}16)$$

where we used the expansion of the field operators, Eqs (7.3-17). Now let the ground state be indicated by $|N,0,0....\rangle$; then

---

[39] D. Pines and P. Nozières, "The Theory of Quantum Liquids", Benjamin, NY 1966, Chapter 2.

$$\bar{\rho}(\mathbf{q})\,|\,N,0,0...\rangle = \sum\nolimits_{\mathbf{k}} a^{\dagger}_{\mathbf{k}-\mathbf{q}}\delta_{\mathbf{k},0}\sqrt{N}\,|\,N-1,0,0...\rangle = \sqrt{N}\,|\,N-1,...,1_{-\mathbf{q}},0,...\rangle. \quad (16.6\text{-}17)$$

So, $\bar{\rho}(\mathbf{q})$ yields an excitation of a single particle of momentum $-\mathbf{q}$ and energy $\varepsilon^0(\mathbf{q}) = \hbar^2 q^2/2m$. This leads to the expectation value $\langle\bar{\rho}(-\mathbf{q},0)\rho(\mathbf{q},t)\rangle\rangle = \langle N\rangle e^{-i\varepsilon^0(\mathbf{q})t/\hbar}$, where we averaged over N, going to a grand-canonical ensemble for $T > 0$. For the dynamical structure factor for the non-interacting fluid we now obtain with (16.5-16')

$$S_d^0(\mathbf{q},\omega) = \langle N\rangle \int_{-\infty}^{\infty} dt\, e^{-i\omega t} e^{-i\varepsilon^0(\mathbf{q})t/\hbar} = 2\pi\hbar\langle N\rangle\delta[\varepsilon^0(\mathbf{q}) + \hbar\omega]. \quad (16.6\text{-}18)$$

This will now be used for the susceptibility associated with the structure factor of (16.6-18). We thus obtain with (16.6-11):

$$\chi_0(\mathbf{q},i\omega) = \langle n\rangle \lim_{\eta\to 0}\int_{-\infty}^{\infty} d\omega''\frac{\delta[\varepsilon^0(\mathbf{q})+\hbar\omega''] - \delta[\varepsilon^0(\mathbf{q})-\hbar\omega'']}{\omega''-\omega+i\eta}. \quad (16.6\text{-}19)$$

Or also,

$$\chi_0[\mathbf{q},i(\omega - i0)] = \langle n\rangle\left\{\frac{-1}{\hbar\omega+\varepsilon^0(\mathbf{q})} + \frac{1}{\hbar\omega-\varepsilon^0(\mathbf{q})}\right\} = \langle n\rangle\,\frac{2\varepsilon^0(\mathbf{q})}{(\hbar\omega)^2 - [\varepsilon^0(\mathbf{q})]^2}. \quad (16.6\text{-}20)$$

Note the two poles on the real axis and the branch cut just above the real axis for positive ω. Let now the interactions be 'turned on'. The energy of the dressed particle becomes $\varepsilon^1(\mathbf{q})$ (similar as in the Hartree–Fock treatment, see subsection 12.13.1); hence we have the dressed susceptibility

$$\chi_{dr}[\mathbf{q},i(\omega - i0)] = \langle n\rangle\frac{\varepsilon^0(\mathbf{q})}{\varepsilon^1(\mathbf{q})}\left\{\frac{-1}{\hbar\omega+\varepsilon^1(\mathbf{q})} + \frac{1}{\hbar\omega-\varepsilon^1(\mathbf{q})}\right\} = \langle n\rangle\,\frac{2\varepsilon^0(\mathbf{q})}{(\hbar\omega)^2 - [\varepsilon^1(\mathbf{q})]^2}. \quad (16.6\text{-}21)$$

Using the general relationship (16.6-8) minor algebra gives the connection,

$$\hbar\omega_{\mathbf{q}} \equiv \varepsilon^1(\mathbf{q}) = \sqrt{[\varepsilon^0(\mathbf{q})]^2 + 2\langle n\rangle V_{\mathbf{q}}\varepsilon^0(\mathbf{q})}\,, \quad (16.6\text{-}22)$$

or approximately

$$\omega_{\mathbf{q}} \simeq \sqrt{2[\varepsilon^0(\mathbf{q})/\hbar^2]V_{\mathbf{q}}\langle n\rangle} = q\sqrt{\langle n\rangle V_{\mathbf{q}}/m} \equiv qu\,, \quad (16.6\text{-}23)$$

where $\hbar\omega_{\mathbf{q}}$ are the sound-wave excitations caused by the interactions; the result (16.6-26) is exactly the relationship found in Bogoliubov's theory for ^{4}He from the grand Hamiltonian, cf. Eq. (12.3-22). [Note that the Fourier coefficients $V_{\mathbf{q}}$ defined here differ by a factor V_0 (the volume) from those in Section 12.3.] For the dynamic structure factor of the interacting quantum liquid we find

$$S_d(\mathbf{q},\omega) = 2\pi\hbar\langle N\rangle[\varepsilon^0(\mathbf{q})/\varepsilon^1(\mathbf{q})]\,\Delta[\varepsilon^1(\mathbf{q}) + \hbar\omega]\,, \quad (16.6\text{-}24)$$

where Δ denotes a collision-broadened Lorentzian centred on $\varepsilon^1(\mathbf{q})$.

16.6.3 *Fermi Liquids*

In Fermi liquids we can have both density fluctuations and spin fluctuations, as we noted in subsection 12.8.1. From a linear response point of view, they are quite similar. For the density fluctuations we need to sum over the two nuclear spin orientations. Thus, consider the one-particle density operator, $\rho(\mathbf{r}) = \sum_\sigma \int d^3r \times \Psi_\sigma^\dagger(\mathbf{r})\Psi_\sigma(\mathbf{r})$. Expanding in momentum states, we have

$$\bar{\rho}(-\mathbf{q}) = V_0^{-1} \sum_{\mathbf{k},\mathbf{k}',\sigma} \underbrace{\int d^3r\, e^{i\mathbf{q}\cdot\mathbf{r}} e^{i(\mathbf{k}-\mathbf{k}')\cdot\mathbf{r}}}_{V_0\,\delta_{\mathbf{q},\mathbf{k}'-\mathbf{k}}} c_{\mathbf{k}',\sigma}^\dagger c_{\mathbf{k},\sigma} = \sum_{\mathbf{k},\sigma} c_{\mathbf{k}+\mathbf{q},\sigma}^\dagger c_{\mathbf{k},\sigma}. \quad (16.6\text{-}25)$$

For the non-interacting liquid we want to find the expectation value in the quasi-vacuum state $|\mathrm{O}\rangle$ for which all states are filled to the Fermi radius k_F. Therefore $c_{\mathbf{k},\sigma}$ must annihilate a state with $|\mathbf{k}| < k_F$, while $c_{\mathbf{k}+\mathbf{q},\sigma}^\dagger$ must create a state outside the Fermi sphere, $|\mathbf{k}+\mathbf{q}| > k_F$. The excitations thus are hole-electron pairs. The energy involved being $\Delta\mathscr{E} = \varepsilon_{\mathbf{k}+\mathbf{q}}^0 - \varepsilon_{\mathbf{k}}^0$, we obtain for the expectation value

$$\begin{aligned}\langle \bar{\rho}(\mathbf{q},0)\bar{\rho}(-\mathbf{q},t)\rangle &\equiv \langle \mathrm{O}|\bar{\rho}(\mathbf{q},0)\bar{\rho}(-\mathbf{q},t)|\mathrm{O}\rangle \\ &= 2\sum\nolimits_{\mathbf{k}} \Theta(k_F - |\mathbf{k}|)\Theta(|\mathbf{k}+\mathbf{q}| - k_F)\, e^{-i(\varepsilon_{\mathbf{k}+\mathbf{q}}^0 - \varepsilon_{\mathbf{k}}^0)t/\hbar}, \end{aligned} \quad (16.6\text{-}26)$$

the factor two stemming from the summation over the spin. From (16.5-16') we obtain for the dynamic structure factor of density fluctuations in the absence of interactions

$$S_d^0(\mathbf{q},\omega) = 4\pi\hbar \sum\nolimits_{\mathbf{k}} \Theta(k_F - |\mathbf{k}|)\Theta(|\mathbf{k}+\mathbf{q}| - k_F)\,\delta(\hbar\omega + \varepsilon_{\mathbf{k}+\mathbf{q}}^0 - \varepsilon_{\mathbf{k}}^0). \quad (16.6\text{-}27)$$

We now use (16.6-11) to obtain the susceptibility; thus we get

$$\chi_d^0(\mathbf{q},i\omega) = \lim_{\eta\to 0} \frac{2}{V_0} \sum_{|\mathbf{k}|<k_F,\,|\mathbf{k}+\mathbf{q}|>k_F} \left\{ \frac{1}{-\hbar\omega - (\hbar^2/2m)[(\mathbf{k}+\mathbf{q})^2 - k^2] + i\eta} - \frac{1}{-\hbar\omega + (\hbar^2/2m)[(\mathbf{k}+\mathbf{q})^2 - k^2] + i\eta} \right\}. \quad (16.6\text{-}28)$$

In the second fraction we make the substitution $\mathbf{k} \to -(\mathbf{k}+\mathbf{q})$; the term then is obviously the same as the first fraction, so that we end up with

$$\chi_d^0(\mathbf{q},i\omega) = \lim_{\eta\to 0} \frac{2}{V_0} \sum_{\mathbf{k}} \left\{ \frac{\Theta(k_F - |\mathbf{k}|) - \Theta(k_F - |\mathbf{k}+\mathbf{q}|)}{-\hbar\omega - (\hbar^2/2m)[(\mathbf{k}+\mathbf{q})^2 - k^2] + i\eta} \right\}, \quad (16.6\text{-}29)$$

where we used

$$\Theta(k_F - k)\Theta(p - k_F) - \Theta(k - k_F)\Theta(k_F - p) = \Theta(k_F - k) - \Theta(k_F - p). \quad (16.6\text{-}30)$$

The function given by the rhs of (16.6-29) is called the Lindhard function; it showed up in subsection 12.8.1. After we turn on the interactions, the final susceptibility will be the dressed susceptibility, which, in accordance with (16.6-8) now reads

$$\chi_d^{dr}(\mathbf{q},i\omega) = \frac{\chi_d^0(\mathbf{q},i\omega)}{1 - V_{\mathbf{q},d}\,\chi_d^0(\mathbf{q},i\omega)}, \tag{16.6-31}$$

where $V_{\mathbf{q},d}$ represents the Fourier series coefficient of the two-particle density interaction potential.

We shall evaluate the Lindhard function for collective density fluctuations, which occur for $|\omega| \gg qv_F$ (while also $|\omega| < v_F q + \hbar q^2/2m$), where v_F is the Fermi velocity. We return to (16.6-28) and make the change $\mathbf{k} \to \mathbf{k} - \mathbf{q}$ in the second term, to obtain

$$\chi_d^0(\mathbf{q},i\omega) = \lim_{\eta\to 0} \frac{2}{V_0} \Theta(k_F - k) \sum_{\mathbf{k}} \left\{ \frac{1}{-\hbar\omega - (\hbar^2/2m)[(\mathbf{k}+\mathbf{q})^2 - k^2] + i\eta} - \frac{1}{-\hbar\omega + (\hbar^2/2m)[k^2 - (\mathbf{k}-\mathbf{q})^2] + i\eta} \right\}. \tag{16.6-32}$$

We write $(\mathbf{k} \pm \mathbf{q})^2 - k^2 = q^2 \pm 2\mathbf{k}\cdot\mathbf{q}$ and the sums will be combined; also, since $v_F q/|\omega| < 1$, an expansion in this parameter will be made with ascending powers of $\mathbf{k}\cdot\mathbf{q}$. The imaginary part will not be explicitly determined, but the delta functions resulting from the limit in (16.6-32) are non-zero for the specified range of $|\omega|$ and contribute to $\mathrm{Im}\,\chi^0(\mathbf{q},\omega)$ and $S_d(\mathbf{q},\omega)$. For the real part the result is found to be

$$\chi_d^0(\mathbf{q},i\omega) = \frac{2q^2}{m\omega^2 V_0} \int \frac{V_0}{8\pi^3} d^3r\, \Theta(k_F - k) \left\{ 1 + 2\frac{\hbar\mathbf{k}\cdot\mathbf{q}}{m\omega} + 3\left(\frac{\hbar\mathbf{k}\cdot\mathbf{q}}{m\omega}\right)^2 + \ldots \right\}. \tag{16.6-33}$$

The angular integrations are easily carried out; for the k-integral we integrate the Heaviside function by parts and we use $\langle n \rangle = k_F^3/3\pi^2$, cf. (8.4-11). So we obtain,

$$\mathrm{Re}\,\chi^0(\mathbf{q},i\omega) = \frac{\langle n\rangle q^2}{m\omega^2} \left\{ 1 + \frac{3}{5}\left(\frac{v_F q}{\omega}\right)^2 + \ldots \right\}. \tag{16.6-34}$$

In what follows we shall only carry the main term. For the dressed susceptibility this yields

$$\chi_d^{dr}(\mathbf{q},i\omega) = \frac{\langle n\rangle q^2/m}{\omega^2 - V_{\mathbf{q}}\langle n\rangle q^2/m} \equiv \frac{\bar{u}^2 q^2 V_{\mathbf{q}}^{-1}}{\omega^2 - \bar{u}^2 q^2}, \tag{16.6-35}$$

which shows a resonance at $\omega = \bar{u}q$; here

$$\bar{u} = \sqrt{\langle n\rangle V_{\mathbf{q}}/m} \approx \sqrt{\langle n\rangle V_{\mathbf{q}\to 0}/m} \tag{16.6-36}$$

is the velocity of *zero sound.* The dynamical structure factor, linked to $\mathrm{Im}\chi$, has a broad resonance at the same place. Note the similarity with the bosonic excitations found in (16.6-19). For small q these results can be compared with the RPA; more on that below. Further, with the exact interaction potential being unknown, the aim is to express the results with the Landau parameters given in Section 12.7. So we postulate the connection $V_{\mathbf{q}\to 0} \approx \rho_F^{-1} F_0^s$, where ρ_F is the density of states at the Fermi surface for the dressed particles and F_0^s is the spin-symmetric Fermi liquid interaction parameter. Employing further that $\langle n\rangle = k_F^3/3\pi^2$ and $\rho_F = m^* k_F/\pi^2\hbar^2$, one easily finds that

$$\bar{u}^2 = u^2(m/m^*)F_0^s = \frac{v_F^2}{3}\frac{F_0^s}{1+F_1^s/3}, \tag{16.6-37}$$

where u is the velocity of first sound, cf. (12.7-28). Further, the numerator of (16.6-28) is just $\omega^2 P(\mathbf{q},\omega)$; here $P(\mathbf{q},\omega)$ is the polarization insertion of the RPA method.

We now turn to spin fluctuations, the subject of paramagnon theory. In that case the coupling involves particles of opposite spin, cf., Eq. (12.8-7). Since the bare mass energies are independent of the spin, the result for the non-interacting susceptibility, $\chi_s^0(\mathbf{q},i\omega)$, is the same as before, i.e., it is given by the Lindhard function. The interaction potential, however, is now spin-dependent. The dressed susceptibility for the paramagnon excitations is given as

$$\chi_s^{dr}(\mathbf{q},i\omega) = \frac{\chi_s^0(\mathbf{q},i\omega)}{1 - V_{\mathbf{q},s}^a\,\chi_d^0(\mathbf{q},i\omega)}, \tag{16.6-38}$$

where $-V_{\mathbf{q}\to 0}^a/V_0 = I > 0$ is the paramagnon energy. We shall evaluate $\chi_s^0(\mathbf{q},i\omega)$ in the RPA and long-wavelength limit. For spin excitations generally $|\omega| < qv_F$. We go back to the original form of the Lindhard function, (16.6-29). Separating into real and imaginary parts, we have

$$\begin{aligned}
\mathrm{Re}\,\chi_s^0(\mathbf{q},i\omega) &= \frac{2}{V_0}\sum_{\mathbf{k}}\left\{\frac{\Theta(k_F-|\mathbf{k}|)-\Theta(k_F-|\mathbf{k}+\mathbf{q}|)}{-\hbar\omega-(\hbar^2/2m)[(\mathbf{k}+\mathbf{q})^2-k^2]}\right. \\
&\overset{\omega\to 0}{=} -\frac{m}{2\hbar^2\pi^3}\mathcal{P}\int d^3k\left\{\frac{\Theta(k_F-|\mathbf{k}|)}{(\mathbf{k}+\mathbf{q})^2-k^2}-\frac{\Theta(k_F-|\mathbf{k}+\mathbf{q}|)}{(\mathbf{k}+\mathbf{q})^2-k^2}\right\} \\
&= -\frac{m}{\hbar^2\pi^3}\mathcal{P}\int d^3k\frac{\Theta(k_F-|\mathbf{k}|)}{q^2+2kq\cos\theta},
\end{aligned} \tag{16.6-39}$$

where in the second term we set $\mathbf{k}\to -\mathbf{k}-\mathbf{q}$ and we took $\mathbf{k}$ as polar axis. The integrations are straightforward. Write $d^3k = -k^2 dk\, d\varphi\, d(\cos\theta)$ to find

$$\mathrm{Re}\,\chi_s^0(\mathbf{q},i\omega) = -\frac{m}{q\hbar^2\pi^2}\mathcal{P}\int_0^{k_F} k\,dk\ln\frac{q+2k}{q-2k}$$

$$= -\frac{m}{q\hbar^2\pi^2}\mathcal{P}\int_{-k_F}^{k_F} k\,dk \ln|q+2k|\,. \tag{16.6-40}$$

A further integration by parts yields

$$\mathrm{Re}\,\chi_s^0(\mathbf{q},i\omega) = -\frac{mk_F}{\hbar^2\pi^2}\left(\frac{1}{2}+\frac{4k_F^2-q^2}{8qk_F}\ln\left|\frac{q+2k_F}{q-2k_F}\right|\right) \sim -\rho_F\,, \tag{16.6-41}$$

where ρ_F is again the energy-density of states at the Fermi surface.

The imaginary part requires some more unorthodox approximations. Gencrally it is given as

$$\mathrm{Im}\,\chi_s^0(\mathbf{q},i\omega) = -\frac{m}{2\pi\hbar^2}\int k^2 dk \int \sin\theta\, d\theta \{\Theta(k_F-k)-\Theta(k_F-|\mathbf{k}+\mathbf{q}|)\}$$
$$\times\delta\left\{\frac{-m\omega}{\hbar}-\frac{1}{2}\left[|\mathbf{k}+\mathbf{q}|^2-k^2\right]\right\}. \tag{16.6-42}$$

If we consider the static case, $\omega=0$, the delta function shows that both Heaviside functions cancel, so that there is no imaginary part. However, we look for the dynamic regime with structure factor $S_d^0(\mathbf{q},\omega) \propto -\mathrm{Im}\,\chi(\mathbf{q},i\omega)$. First, let us focus on the integral

$$\mathcal{I} = -\frac{m}{2\pi\hbar^2}\int k^2 dk \int \sin\theta\, d\theta\, \delta\left\{\frac{-m\omega}{\hbar}-\frac{1}{2}q^2-kq\cos\theta\right\}. \tag{16.6-43}$$

We chose as new variables

$$y = \frac{-m\omega}{\hbar}-\frac{q^2}{2}-kq\cos\theta,\quad K=k.$$

The Jacobian is found to be $\partial(y,K)/\partial(\theta,k) = kq\sin\theta$. This leaves

$$\mathcal{I} = \frac{m}{2\pi\hbar^2 q}\int K dK \int dy\, \delta(y+0)\,, \tag{16.6-44}$$

The limits on K are found from $\cos\theta = \pm 1$, $K \geq 0$ and setting $y = 0$ afterwards, similar as in Fig. 15-3. Hence, K lies in the interval $0 \leq K \leq (m|\omega|/\hbar q)+q/2 \equiv K_m$. The y-integration is performed, leaving the integral over dK.

Next, we deal with the Heaviside functions; by Taylor expansion

$$\Theta(k_F-k)-\Theta(k_F-|\mathbf{k}+\mathbf{q}|) = \Theta(k_F^2-k^2)-\Theta(k_F^2-|\mathbf{k}+\mathbf{q}|^2)$$
$$= \Theta(k_F^2-k^2)-\Theta(k_F^2-k^2-2m\omega/\hbar)$$
$$= \delta(k_F^2-k^2)(2m\omega/\hbar)\,. \tag{16.6-45}$$

This result supposes that the increment is small with respect to $k_F^2-k^2 = (k_F-k)\times(k_F+k) \simeq qk_F$, or $|\omega| < (\hbar k_F q/m) = qv_F$. Further note $\delta(k_F^2-k^2) = \delta(k-k_F)/2k_F$.

Equations (16.6-44) and (16.6-45) now leave for $\mathrm{Im}\chi(\mathbf{q},i\omega)$, providing also $K_m > k_F$,

$$\mathrm{Im}\,\chi_s^0(\mathbf{q},i\omega) = \frac{m^2\omega}{2\pi\hbar^3 q k_F}\int_0^{K_m} KdK\,\delta(K-k_F)$$

$$= \pi\frac{\rho_F}{2}\frac{m\omega}{q\hbar k_F}\Theta(qv_F - |\omega|). \tag{16.6-46}$$

Combining (16.6-41) and (16.6-46), we obtained

$$\chi_s^0(\mathbf{q},-i\omega) = -\rho_F\left(1 - \frac{i\pi}{2}\frac{\omega}{qv_F}\Theta(qv_F - |\omega|)\right), \tag{16.6-47}$$

which is the same as the expression for the polarization mentioned in Eq. (12.8-9).

The connection with Landau's Fermi-liquid parameters was already stated in subsection 12.8.1. The imaginary part of the dressed susceptibility was obtained in (12.8-10). Because of the repulsive nature of the paramagnon opposite spin-pair interaction, there are no resonances; just a broadened peak, as observed in the experimental neutron-scattering data.

16.6.4 *Real Time Green's Functions and the Diagrammatic Evaluation*

In equilibrium statistical mechanics we discussed quantum liquids from a general vantage point and sometimes, as for Fermi liquids, from a phenomenological point of view. A more rigorous diagrammatic presentation was provided in the second part of Chapter XII, to help the more advanced reader appreciate the abundant literature in this field. Basically, books like Pines and Nozières, Mahan and others, all use these methods. When it comes to non-equilibrium theory, one can likewise choose between a more intuitive and sometimes heuristic approach, and the more advanced coverage that relies on a diagrammatic evaluation. We have chosen here to present the former, based *in casu* on linear response theory. In Section 16.8 we will sound some warnings, but we submit that the method is profitable, although in need of *a posteriori* justification.

Yet, diagrammatic approaches are far more versatile, so we make some comments on these powerful methods, although they will not be included in this text for space-restrictions reasons. Whereas imaginary times are sometimes used, like in the Kubo forms (16.3-21) and (16.3-22), mostly we work in the real time domain. To obtain the retarded density-density correlation, we should start with the unperturbed real-time propagator or Green's function, $G_0^R(\mathbf{r},t;\mathbf{r}',t')$. It is defined by the expectation value

$$G_0^R(\mathbf{r},t;\mathbf{r}',t') = -i\langle\Phi_0|\tilde{\Psi}(\mathbf{r},t)\tilde{\Psi}^\dagger(\mathbf{r}',t')|\Phi_0\rangle, \quad t > t', \tag{16.6-48}$$

where $|\Phi_0\rangle$ denotes the normalized non-interacting ground state and the tilde-provided field operators are interaction operators, based on the unperturbed energies. The factor "$-i$" is provided in view of the properties of Fourier-transformed Green's functions, both in space and time; thus the gamut of possible forms is similar as in 'temperature Green's functions'. [Needless to say that Green's functions in mathematical physics, like those associated with the Laplace or Helmholtz operator are usually real.] Besides the retarded propagator there is an advanced propagator, denoted by $G_0^A(\mathbf{r},t;\mathbf{r}',t')$: replace $-i$ by i and set $t'>t$. We find little use for it; on the contrary, mostly one employs '*the*' Green's function,

$$G_0(\mathbf{r},t;\mathbf{r}',t') = -i\langle\Phi_0\,|\,\mathcal{T}[\tilde{\Psi}(\mathbf{r},t)\tilde{\Psi}^\dagger(\mathbf{r}',t')]\,|\,\Phi_0\rangle\,, \tag{16.6-49}$$

where $\mathcal{T}$ is the time-ordering operator which orders and permutes, using the proper parity for fermion transpositions.

The goal is as always to obtain the full Green's function, based on the Heisenberg picture,

$$G(\mathbf{r},t;\mathbf{r}',t') = -i\langle\Phi\,|\,\mathcal{T}[\Psi(\mathbf{r},t)\Psi^\dagger(\mathbf{r}',t')]\,|\,\Phi\rangle\,, \tag{16.6-50}$$

where $|\,\Phi\rangle$ is the new ground state. Generally the free propagators and the full Green's function are connected by perturbation expansions, which we discussed both for real and imaginary time in Section 12.9. The most useful tool is again the connection given by Dyson's equation (cf. Section 12.12), which requires the diagrammatic determination of the self-energy. The various schemes discussed for temperature Green's functions – Hartree, Hartree–Fock, the random phase approximation (ring-diagram summation) – are also useful in the present context.

For the problem of density-density correlations in Fermi liquids, discussed in the last subsection, the RPA is the most often employed scheme. Basically, the method is far more direct, since we can dispense with the external perturbation Hamiltonian. So we seek the effective or "dressed" interaction $\mathbf{V_q}$ in terms of the 'bare' interaction, $V_\mathbf{q}$. They are connected by the polarization insertion $\Pi(\mathbf{q})$, cf. (12.13-36). The polarization Π plays the role of the susceptibility $\chi^0(\mathbf{q},\omega)$. It was obtained in (12.13-39). For the real-time analogue, replace $i\nu_n$ by $-\hbar\omega$, so that

$$\Pi_{RPA}(\mathbf{q},i\omega) = 2\int d^3k\left(\frac{1}{8\pi^3}\right)\frac{n^0_{\mathbf{q+k}} - n^0_\mathbf{q}}{-\hbar\omega - (\varepsilon^0_{\mathbf{k+q}} - \varepsilon^0_\mathbf{k}) + i\eta}\,, \tag{16.6-51}$$

where the $+i\eta$ stems from the branch cut in the complex frequency plane[40]; note that the unperturbed Fermi-Dirac distributions $n^0_{\mathbf{q+k}}$ and $n^0_\mathbf{q}$ mimic the Heaviside functions of our present form (16.6-29). There is now full equivalence, which goes to show that for the case considered, the self-consistent field method yields exactly the RPA results. We will leave it to the reader to pursue this in greater detail.

---

[40] Cf. A.L. Fetter and J.D. Walecka, "Quantum Theory of Many-Particle Systems, McGraw-Hill, NY 1971, Fig. 7.1.

16.7 *Kubo-Theory Conductivity Computations*

We shall consider the electrical conductivity, for which the field $F(t) \to E_\nu(t)$, while we seek the response for the current flux $J_\mu(t)$, the subscripts denoting Cartesian components. Truly collective effects, like plasmon oscillations, were considered in principle previously–see also problem 16.9 and 16.10; for conductivity computations the ordinary Hartree or Hartree–Fock procedures suffice. In that sense then, the Hamiltonian is separable in a sum of one-electron Hamiltonians $h(p,q)$, which are themselves composed of a free particle contribution and an interaction potential associated with impurity scattering or electron-phonon interaction. Note that in these mid-twentieth century theories the interaction is considered on the one-particle plane, whereas in our "modified Kubo-theory", to be discussed in the second part of this chapter, we shall do the perturbational computations on the many-body level, which has two advantages: a) all manipulations involve *linear operators* for which the rules are well established; b) the results are general and need not be redone for the various processes considered in Refs. 6,7 and 9-11, such as 'randomly distributed scatterers', elastic and non-elastic phonon collisions, etc. Therefore, in a review of these older attempts, we only indicate the principles involved and we will not pursue the often intricate mathematics to obtain the final results.

First then, we carry out the reduction of the Kubo–Green formulas to the one-particle level, contemplated by Chester and Thellung[6] and Verboven[7]. To start with, we have

$$\sigma_{\mu\nu}(i\omega) = V_0 \int_0^\infty dt\, e^{-i\omega t} \int_0^\beta d\tau \operatorname{Tr}\{\rho_{gcan} J_\nu(-i\hbar\tau) J_\mu(t)\}. \qquad (16.7\text{-}1)$$

Considering classical frequencies only, we are at liberty to commute the operators; the tensor is clearly symmetrical in the absence of a magnetic field and for the low frequency conductivity we can write

$$\sigma_{\mu\nu} = \frac{1}{4} \lim_{T\to\infty} V_0 \int_{-T}^{T} dt \int_0^\beta d\tau \operatorname{Tr}\{\rho_{gcan}[J_\nu(0)J_\mu(t) + J_\mu(0)J_\nu(t)]\}. \qquad (16.7\text{-}2)$$

The integral over $d\tau$ can be carried out and simply gives a factor β. The reduction to one-particle operators is most easily carried out by writing the J's in second quantized form. With a complete set of Bloch states $\{|k\rangle\}$ where $k = (\mathbf{k},\sigma)$ includes the spin, we find

$$\sigma_{\mu\nu} = \frac{\beta}{4} \lim_{T\to\infty} V_0 \int_{-T}^{T} dt \sum_{k\ell} \operatorname{Tr}\left\{\rho_{gcan} \mathbf{n}_k \mathbf{n}_\ell \left[\langle k | j_\nu | k\rangle\langle \ell | j_\mu(t) | \ell\rangle + cc\right]\right\}, \qquad (16.7\text{-}3)$$

where j denotes the one-electron current density and $\mathbf{n}_k = c_k^\dagger c_k$, *etc.*; cc means the complex conjugate. The required grand-canonical average is already found in subsection 4.3.1; from the covariance (4.3-17) we have

$$\langle n_k n_\ell \rangle_{eq} = \langle n_k \rangle_{eq} [1 - \langle n_\ell \rangle_{eq}] \delta_{k\ell} + \langle n_k \rangle_{eq} \langle n_\ell \rangle_{eq} \,. \tag{16.7-4}$$

Carrying out the sum over k and ℓ for the last term, we find zero, since the equilibrium currents vanish. The first term admits the usual F–D result

$$\langle n_k \rangle_{eq} [1 - \langle n_\ell \rangle_{eq}] \delta_{k\ell} = -(1/\beta)(\partial f_k / \partial \varepsilon_k) \delta_{k\ell} \,. \tag{16.7-5}$$

So with little effort we have established the result employed by the cited authors

$$\sigma_{\mu\nu} = -\frac{1}{4} \lim_{T \to \infty} V_0 \int_{-T}^{T} dt \sum_{k\ell} \frac{\partial f_k}{\partial \varepsilon_k} \delta_{k\ell} \left[\langle k | j_\nu | \ell \rangle \langle \ell | j_\mu(t) | k \rangle + \langle k | j_\nu(t) | \ell \rangle \langle \ell | j_\mu | k \rangle \right] . \tag{16.7-6}$$

Of course, this can be rewritten as a trace:

$$\sigma_{\mu\nu} = -\frac{1}{4} \lim_{T \to \infty} V_0 \int_{-T}^{T} dt \, \mathrm{Tr} \left\{ \frac{\partial f}{\partial h_e} \left[j_\nu(0)\, j_\mu(t) + j_\nu(t)\, j_\mu(0) \right] \right\} . \tag{16.7-7}$$

Here h_e is the one-particle Hamiltonian of the free electron gas. Each electron sees, however, a larger Hamiltonian, viz.

$$h = h_e + \lambda v_{e-i} \,, \quad \text{or} \quad h = h_e + \lambda v_{e-ph} \,, \tag{16.7-8}$$

where the second part represents interactions with impurities or phonons. These interactions randomize the motion and give the observed dissipative behaviour. Physically this means that (16.7-7) is in need of re-interpretation: the operators $j_\nu(t)$ and $j_\mu(t)$ are no longer pure Heisenberg operators but reflect this random motion. The evaluation of these modified operators has been the subject of many earlier papers that tried to somehow recoup the Boltzmann-equation solutions. Two schemes have been in use since the mid-fifties:

(a) *The repeated random phase approximation* (*RRPA*);
(b) *The Van Hove* [*weak-coupling, long-time*] *limit* (*VHL*).

We briefly comment on both procedures. In the *RRPA* it is assumed that the density operator is at all times diagonal, i.e., the off-diagonal elements are erased. The method was employed by Peierls[41] and is implicit in Pauli's 1928 "*Festschrift*"[42]. This assumption is clearly at odds with the regular quantum procedure. For a pure state the density operator is always a projector; in an ensemble we can formally write $\rho(t) = \overline{|\Psi(t)\rangle\langle\Psi(t)|}$. An *initial* random phase assumption, rendering $\rho(t=0)$ to be diagonal, is quite acceptable; however, its behaviour for $t > 0$ is then determined by

[41] R.E. Peierls, Zeitschr. für Phys. **88**, 786 (1934); ibid. Helv. Phys. Acta **24**, 645 (1934). Also, R. Peierls, "The Quantum Theory of Solids", Clarendon Press, Oxford 1955.
[42] W. Pauli, "*Festschrift zum 60sten Geburtstage A. Sommerfelds*" Hirzel, Leipzig 1928, p. 30.

the Schrödinger equation or by the von Neumann equation. Notwithstanding, it may be argued that the random causes we deal with have the property that the memory is erased after *each* macroscopic time interval $\Delta t \simeq \tau_r$, where τ_r is the relaxation time, so that $\rho(t)$ is diagonal at all times – the underlying assumption for a mesoscopic Markov process (cf. Section 18.3). This is the position taken by van Kampen.[43]

The Van Hove limit entails a more complex procedure; it was first set forth as an alternative to the *RRPA* by Leon Van Hove in a 'tough' paper published in 1955.[44] The limit to be considered reads:

$$\lambda \to 0, \quad t \to \infty, \quad \lambda^2 t \simeq \lambda_0^2 \tau_r = \textit{finite}. \tag{16.7-9}$$

The middle member of the third statement is *ours*: of course, some finite weak coupling λ_0 must remain and the action-time involved never exceeds by much the macroscopically observed relaxation time τ_r. So, the Van Hove limit implies that we should distinguish two time regimes. Small times are those for which $\tau_t < t < \tau_r$, where τ_t is the transition time given by time-dependent perturbation theory; macroscopic time, on the other hand, relates to the regime $t \gtrsim \tau_r$. The detailed procedure involving the Van Hove limit will be indicated in the second part of this chapter. While the original procedure is extremely lengthy in that we must extract all terms $(\lambda^2 t)^n$, $n = 1, 2, \ldots$, from the complete solution of the von Neumann equation, much faster results can be obtained with the help of projection operators on the diagonal and nondiagonal subspaces of the Liouville space.

It is not useful to reminisce here on the further developments of Chester and Thellung's and Verboven's papers, since they are entirely based on Van Hove's original method. So, the following comments must suffice. The Van Hove limit alters the Heisenberg operators so as to become of the form

$$B_H(t) = e^{i\mathcal{L}t} B(0) \to e^{-\Lambda_d(t)} B(0) + B_{nd}(t) \,, \tag{16.7-10}$$

where Λ_d is the *master-superoperator*, to be introduced in the Section 16.9. It can be shown to be semi-positive definite, having real eigenvalues, of which the dominant one will be denoted by $1/\tau_r$. [The nondiagonal part, B_{nd}, will not be discussed here.] So, let us substitute this into (16.7-6). The time integral to be evaluated is then

$$\lim_{T\to\infty} \int_{-T}^{T} dt\, e^{-|t|/\tau_r} = 2\int_0^\infty dt\, e^{-t/\tau_r} = 2\tau_r \,. \tag{16.7-11}$$

This leads to the familiar Boltzmann-like result [cf. Eq. (15.2-7)]:

$$\sigma_{\mu\nu} = -V_0 \sum_{\mathbf{k},\sigma} (\partial f / \partial \varepsilon_{\mathbf{k}}) \tau(\varepsilon_{\mathbf{k}}) \big[\langle \mathbf{k} \,|\, j_\mu \,|\, \mathbf{k} \rangle \langle \mathbf{k} \,|\, j_\nu \,|\, \mathbf{k} \rangle \big]$$

---

[43] N.G. van Kampen, "Stochastic Processes in Physics and Chemistry", North-Holland Publ. Co., Amsterdam and NY 1981, § III.2-**c**; also, 3rd Ed., Elsevier's, Amsterdam and NY 2007, § III.2-**h**.

[44] L. Van Hove, Physica **21**, 517 (1955); ibid. Physica **23**, 441 (1957).

$$= -\frac{V_0^2}{4\pi^3}\int d^3k \frac{\partial f}{\partial \varepsilon_{\mathbf{k}}} \tau(\varepsilon_{\mathbf{k}})\Big[\langle \mathbf{k} \,|\, j_\mu \,|\, \mathbf{k}\rangle \langle \mathbf{k} \,|\, j_\nu \,|\, \mathbf{k}\rangle\Big]. \qquad (16.7\text{-}12)$$

Two questions are still pressing. First of all, can the present method improve over the Boltzmann-like result by taking higher-order interaction terms into account. The answer is affirmative. The weak coupling limit holds as long as $\hbar/\tau_r < \varepsilon_F$. Higher order terms are computed by Verboven. Likewise, Loss and Thellung have considered several results beyond the Van Hove limit.[45]

Secondly, various researchers have argued that *in final analysis* the Van Hove limit is equivalent to the repeated random phase approximation, since his master equation (to be derived later) is just the Pauli equation. In that sense then Van Hove's method would only be a *more gratifying* procedure in which the nondiagonal elements are not removed 'à bout portant' (by brute force). However, we shall show that there is a nondiagonal contribution to the master equation that is of *paramount* importance for a number of transport phenomena, including the renowned quantum Hall effect; the full 'reduced operators' [cf. the non-discussed term in (16.7-10)] and the full master equation are *not* obtainable with the *RRPA*.

16.8 Criticism of Linear Response Theory

16.8.1 *Van Kampen's Objections*

In an article of the Norwegian Physical Society already cited[8], honouring Prof. Wergeland's 60th birthday, van Kampen states his unease with the premises of linear response theory as follows: "Linear response theory does provide expressions for the phenomenological coefficients, but I assert that *it arrives at these expressions by a mathematical exercise rather than by describing the actual mechanism which is responsible for the response.*" Indeed, we fully agree that the physical content of Kubo's linear response theory *in its original form* is minimal.

Yet we have seen that LRT *has* been very successful in a number of areas. For the case of density fluctuations it has delivered solid results for the excitations in Bose liquids, correct results for collective density fluctuations in Fermi liquids and for plasma oscillations and cohesiveness of electron gases in metals, without requiring elaborate diagrammatic techniques. As to the phenomenological coefficients of simple systems, it *has* led to a many-body approach that issued into Boltzmann-like results but allowed to go beyond. The successes of LTR should not be discounted.

Van Kampen's main concern is the application of the linearity assumption with regard to the *microscopic* motion of response over prolonged times. He makes the following estimate. A particle undergoes by the field a shift of $d = \frac{1}{2}|\mathcal{q}|Et^2/m^*$,

[45] D. Loss, Physica A **139**, 505 and 526 (1986); D. Loss and A. Thellung, Physica A **144**, 17 (1987).

where q is the charge, E the electric field and m^* the effective mass. This shift should be small compared to the diameter d_0 of the scatterers. Taking for the latter 100 $Å$, setting $m^* = 0.1m$ and taking $t = 1$ s, one finds that the electric field should not exceed 10^{-18} V/cm! [not a printing error.]

Of course, the crucial point is *over which time period* the linearization must apply. The one second is clearly excessive. Macroscopic time needs to be only of order of a few collision times. If we can argue that the theory – after randomization is introduced, see below – exhibits a relaxation time τ_r of, say 10^{-10} s, fields of 100 V/cm would be permissible. In an attempt to lay the matter to rest, we have given a reformulation of the Kubo–Green formulas which leads to exact but field-dependent phenomenological coefficients. It has then been evaluated what fields are allowed, in order that the linear term suffices. The above estimates have been confirmed; for the interested reader, the study will be dealt with in Section 16.15.

16.8.2 *Our Criticism*

At present – some thirty years after we began our reflections on linear response theory – we are less disturbed by the founders' approach than in the past, in particular since they themselves have admitted (see Ref. 37) that the theory is *incomplete* until a *randomization* is applied, as is the case in all examples discussed in the previous sections. Hence, we content ourselves that both linearization and randomization do occur in all viable versions of the theory, albeit it that they are carried out in *reverse order* compared to other non-equilibrium approaches.

Yet, we wish to look somewhat closer at the original theory. The fluctuation-dissipation theorem is derived, but *a priori* it is not clear what fluctuates and *nowhere is the physics commensurate with dissipative processes introduced*; or so it seems. The secret is clearly that the system after all is not closed. The use of the Liouville or von Neumann equation under those conditions is not forbidden – note that Kubo's results can also be obtained from the Schrödinger equation – but what is significant is that the density operator *must* be either ρ_{can} or ρ_{gcan}. In other words, interactions with a thermal bath are presumed. These, then, should allow for *equilibrium fluctuations* in the chosen ensemble.

The linkage to a *dissipative* conductance or susceptance is, however, another matter. Neither Callen and Welton, nor Kubo show that their quantum admittances exhibit heat losses. In this respect, LRT is a 'hollow shell' *until* the randomization is actually imposed. As noted, Heisenberg operators only represent rotations in the Liouville space. So, we must, at some instant, introduce a partitioned Hamiltonian $\mathcal{H} = \mathcal{H}^0 + \lambda\mathcal{V}$, where $\lambda\mathcal{V}$ is a perturbation that randomizes 'the motion proper', contained in $\mathcal{H}^0$. This may be accomplished with the Van Hove limit or by other means. Yet, no irreversible behaviour will ever occur, unless a randomizing agent, however small, is present.

2. Reduced Operators and Convergent Forms

16.9 The Master Operator in Liouville Space

The Van Hove limit is quite different from the ordinary Born approximation in that it indicates that, when the interactions grow weaker, the time over which they work must increase. It should have been stated in *dimensionless* parameters. So we rephrase: let the Hamiltonian be $\mathcal{H} = \mathcal{H}^0 + \lambda\mathcal{V}$; it will then be assumed that

$$\lambda \to 0, \quad t/\tau_t \to \infty, \quad \lambda^2 t = \textit{finite}. \tag{16.9-1}$$

Here, τ_t is the time associated with a microscopic transition. As indicated already, from the limit process relaxation times will arise, the largest of which will be denoted by τ_r. This will give rise to two time regimes; 'small times'Δt and 'large times' t with

$$(i) \quad \tau_t \ll \Delta t \ll \tau_r, \quad (ii) \quad t \gg \tau_r. \tag{16.9-2}$$

We shall first be concerned with small times, for which we use the designation Δt. This interval of time, though 'microscopically large' is 'macroscopically small'.

Besides the Van Hove limit we consider as usual the thermodynamic or large system limit. This means that the quantum numbers for the ES $|\gamma\rangle$ of $\mathcal{H}^0$ are near continuous; at times we shall – with Van Hove – take the energy as one quantum number and write $|\gamma\rangle = |\mathcal{E}\alpha\rangle$. The perturbations will give transitions according to Fermi's golden rule,

$$W_{\gamma\gamma'} = (2\pi\lambda^2/\hbar)\,|\langle\gamma|\mathcal{V}|\gamma'\rangle|^2\,\delta(\mathcal{E}_\gamma - \mathcal{E}_{\gamma'}). \tag{16.9-3}$$

The more precise expressions for the transition probability indicate that energy is conserved only in energy cells of a width $\delta\mathcal{E}$[46]; intercell transitions do not occur. As to $\mathcal{V}$, we assume that it has no diagonal elements; if it does, then that part should be included in $\mathcal{H}^0$.[47]

Consider now a function space based on the set of variables $\{\gamma\}$ and let $f(\gamma)$ be an arbitrary function labelled by γ. We then introduce the "master operator" M in function space

$$M f(\gamma) \equiv -\sum_{\gamma''} [W_{\gamma''\gamma} f(\gamma'') - W_{\gamma\gamma''} f(\gamma)]\,. \tag{16.9-4}$$

Note that, at this moment, this is simply a definition; the master equation (ME) is as yet undisclosed. With each $f(\gamma)$ we will associate an operator $\hat{f}$ such that

---

[46] A. Messiah, "Quantum Mechanics II ", Op. Cit., Chapter XVII, § 4.

[47] The Hamiltonian $\mathcal{H}^0$ *is the largest Hamiltonian that can be diagonalized*; I owe this remark to Max Dresden at the "Vosbergen Conference", 1968.

$$\hat{f} = \sum_{\gamma} |\gamma\rangle f(\gamma) \langle\gamma| \quad \text{or} \quad \langle\gamma|\hat{f}|\gamma\rangle = f(\gamma). \tag{16.9-5}$$

The operators $\hat{f}$ are elements of a subspace of the Liouville space, referred to as the "diagonal Liouville space". Next we define the "master operator in diagonal Liouville space", Λ_d, by

$$\begin{aligned}\Lambda_d \hat{f} &\equiv \sum_{\gamma} |\gamma\rangle\langle\gamma| \, Mf(\gamma) \\ &= -\sum_{\gamma\gamma''} |\gamma\rangle\langle\gamma| \{W_{\gamma''\gamma} \langle\gamma''|\hat{f}|\gamma''\rangle - W_{\gamma\gamma''} \langle\gamma|\hat{f}|\gamma\rangle\}.\end{aligned} \tag{16.9-6}$$

Clearly, Λ_d is a superoperator, acting on ordinary operators, which are themselves 'elements' of the Liouville space. In this respect it is similar as the Liouville operator $\mathcal{L}$, defined previously, cf. (16.2-8). We recall its definition here,

$$\mathcal{L}K = \frac{1}{\hbar}[\mathcal{H}, K]. \tag{16.9-7}$$

Likewise, we define the interaction Liouvillian,

$$\mathcal{L}^0 K = \frac{1}{\hbar}[\mathcal{H}^0, K]. \tag{16.9-8}$$

Note that all super-operators have the general form (16.2-9).

We must say a bit more about the Liouville space. Besides the diagonal subspace, it is useful to introduce its complement; formally, we need the projection operators $\mathcal{P}$ and $1 - \mathcal{P}$. Every element of the Liouville space (i.e., ordinary operators) can be written as a sum of a diagonal and nondiagonal part, as we now show. From closure of the states $\{|\gamma\rangle\}$ we have always

$$\begin{aligned}K &= \sum_{\gamma\gamma'} |\gamma\rangle\langle\gamma|K|\gamma'\rangle\langle\gamma'| \\ &= \sum_{\gamma} |\gamma\rangle\langle\gamma|K|\gamma\rangle\langle\gamma| + \sum_{\gamma\neq\gamma'} |\gamma\rangle\langle\gamma|K|\gamma'\rangle\langle\gamma'| \\ &= \mathcal{P}K + (1-\mathcal{P})K \equiv K_d + K_{nd}.\end{aligned} \tag{16.9-9}$$

These equations define the projection on the diagonal Liouville space (the "diagonal part") and the projection on the compliment space ("nondiagonal part"). We also defined the projector $\mathcal{P}$ by

$$\mathcal{P} .. = \sum_{\gamma} \{|\gamma\rangle\langle\gamma| .. |\gamma\rangle\langle\gamma|\}. \tag{16.9-10}$$

Denoting also $\mathcal{P}$ and $1-\mathcal{P}$ by $\mathcal{P}$ and $\mathcal{Q}$, we note the properties of projector operators:

$$(i) \;\; \mathcal{P}^n = 1, \quad (ii) \;\; \mathcal{Q}^m = 1, \quad (iii) \;\; \mathcal{P}\mathcal{Q} = 0. \tag{16.9-11}$$

Finally, we come to the purpose of our dwelling on these superoperators. We claim the following main result: Let the Van Hove limit be denoted by $\lim_{\lambda,t}$; then

$$\lim_{\lambda,t} B_{H,d}(t) = \lim_{\lambda,t} e^{i\mathcal{L}t} B_{S,d} \equiv B_d^R(t) = e^{-\Lambda_d t} B_{S,d}. \tag{16.9-12}$$

For clarity we added here the subscripts H for Heisenberg and S for Schrödinger. The important point is that the diagonal part of any operator of the ordinary state space has no longer an 'oscillatory' behaviour; on the contrary, the *reduced operator*, i.e., after the van Hove limit, has a 'decaying' behaviour. Note, however, that Λ_d has an eigenvalue zero, so that the operator for large t approaches the Schrödinger operator[48]. The "master operator in Liouville space", Λ_d, was first introduced by the author in 1976 (cf. LRT I), in order to profit from linear algebra as developed for Hilbert spaces[49]; the equivalence with the function-space representation, *in casu* the Pauli master equation (ME), will be demonstrated later. The description in function space is far less transparent, however.

16.9.1 *Results for Small Times*

The Heisenberg operators in the real time domain are formally written as

$$B(t) = e^{i\mathcal{L}t} B = e^{i(\mathcal{H}^0 + \lambda\mathcal{V})t/\hbar} B e^{-i(\mathcal{H}^0 + \lambda\mathcal{V})t/\hbar}. \tag{16.9-13}$$

Since the two operators in the exponent do not commute, this form has no practical use. Rather, we need a perturbation series, based on $U(t,t_0)$ in terms of $U^0(t,t_0)$ $= \exp[-i\mathcal{H}^0(t-t_0)/\hbar]$. Generally we have, setting $t_0 = 0$,

$$U(t,0) = U^0(t,0) + \sum_{n=1}^{\infty} U^{(n)}(t,0), \tag{16.9-14}$$

where $U^{(n)}(t,0)$ is given as a time-ordered perturbation series, cf. Eq. (12.9-17) or Eq. (A.3-8) of Appendix A. Actually, we must compute the matrix elements of the reduced Heisenberg operators, so we repeat here Eq. (A.3-9), which reads, omitting the bar over the U since the perturbations $\lambda\mathcal{V}$ are time-independent:

$$\langle\gamma|U^{(n)}(t)|\gamma_0\rangle = \left(\frac{\lambda}{\hbar i}\right)^n \sum_{\gamma_1\cdots\gamma_{n-1}} \int_0^t dt_n \int_0^{t_n} dt_{n-1} \ldots \int_0^{t_2} dt_1$$
$$\times U^0(t-t_n)\langle\gamma|\mathcal{V}(t_n)|\gamma_{n-1}\rangle U^0(t_n - t_{n-1})\langle\gamma_{n-1}|\mathcal{V}(t_{n-1})|\gamma_{n-2}\rangle \ldots$$

---

[48] The operator Λ_d is semi-positive definite, like its equivalent in function space, M. See J.O. Vigfussen, "Time relaxation of the solutions of master equations for large systems", J. Statist. Phys. **27**, 339 (1982).

[49] Among the many excellent books on operator- and functional analysis, in particular for operators that are *non-bounded or that go beyond the state space proper*, we recommend F. Riesz and B. Sz.-Nagy, Leçons d'Analyse Fonctionnelle, 4^e^ Ed., Gauthiers–Villars, Paris, 1965. [The English translation, "Functional Analysis", Ungar 1960, Dover 1990, is based on the 2nd Ed. and does not contain the above.]

$$\times U^0(t_2 - t_1)\langle\gamma_1|\mathcal{V}(t_1)|\gamma_0\rangle U^0(t_1)\,. \tag{16.9-15}$$

To fully grasp the meaning of these results, please study the pictorial representation of the interactions in Fig. A-2.

For the present small time computations, two terms in the series expansion suffice, i.e., two-time interactions occur. Then we have, in particular

$$\begin{aligned} U^{(1)}(\Delta t) &= (\lambda / i\hbar)\int_0^{\Delta t} dt_1 e^{-i\mathcal{H}^0(\Delta t - t_1)/\hbar}\mathcal{V}e^{-i\mathcal{H}^0 t_1/\hbar}\,, \\ U^{(2)}(\Delta t) &= (\lambda / i\hbar)^2 \int_0^{\Delta t} dt_2 \int_0^{\Delta t} dt_1 e^{-i\mathcal{H}^0(\Delta t - t_2)/\hbar}\mathcal{V}e^{-i\mathcal{H}^0(t_2 - t_1)/\hbar}\mathcal{V}e^{-i\mathcal{H}^{(0)}t_1/\hbar}\,. \end{aligned} \tag{16.9-16}$$

We thus consider the matrix element

$$\langle\gamma|B_d(\Delta t)|\gamma'\rangle = \langle\gamma|U^\dagger(\Delta t)B_d U(\Delta t)|\gamma'\rangle. \tag{16.9-17}$$

Since we have an evolution operator on either side, in principle six terms are needed, although some terms can be combined or do not contribute. To start with, we have

$$\langle\gamma|U^{(0)\dagger}(\Delta t)B_d U^{(0)}(\Delta t)|\gamma'\rangle = e^{i(\mathcal{E}_\gamma - \mathcal{E}_{\gamma'})\Delta t/\hbar}\langle\gamma|B_d|\gamma'\rangle = \langle\gamma|B_d|\gamma\rangle\delta_{\gamma\gamma'}; \tag{16.9-18}$$

note that the diagonal operator B_d and $\mathcal{H}^0$ commute. Next, we consider

$$\begin{aligned} \langle\gamma|U^{(1)\dagger}(\Delta t)B_d U^{(1)}(\Delta t)|\gamma'\rangle &= \frac{\lambda^2}{\hbar^2}\int_0^{\Delta t} dt_1 \int_0^{\Delta t} dt_1' e^{i\mathcal{E}_\gamma t_1'/\hbar} \\ &\times\langle\gamma|\mathcal{V}\left\{e^{i\mathcal{H}^0(\Delta t - t_1')/\hbar}B_d e^{-i\mathcal{H}^0(\Delta t - t_1)/\hbar}\right\}\mathcal{V}|\gamma'\rangle e^{-i\mathcal{E}_{\gamma'}t_1/\hbar}\,. \end{aligned} \tag{16.9-19}$$

The diagonal operator in the curly brackets we denote by C. We now use Van Hove's functional rule, i.e., we write down a linear functional relationship in as yet unknown kernels X and Y,

$$\begin{aligned} \langle\gamma|\mathcal{V}C\mathcal{V}|\gamma'\rangle &= \langle\gamma|\mathcal{V}C\mathcal{V}|\gamma'\rangle_d + \langle\gamma|\mathcal{V}C\mathcal{V}|\gamma'\rangle_{nd} \\ &= \delta_{\gamma\gamma'}\sum\nolimits_{\gamma''}\langle\gamma''|C|\gamma''\rangle X(\gamma'',\gamma) + \sum\nolimits_{\gamma''}\langle\gamma''|C|\gamma''\rangle Y(\gamma'';\gamma,\gamma'). \end{aligned} \tag{16.9-20}$$

The reason for this splitting is that the diagonal and nondiagonal parts behave basically very differently: the diagonal terms will survive, whereas the nondiagonal terms represent interference effects which die out.[50] It is useful to identify the kernel X right away. Employing for C the particular diagonal operator $|\gamma^0\rangle\langle\gamma^0|$ we find

$$X(\gamma^0,\gamma) = |\langle\gamma^0|\mathcal{V}|\gamma\rangle|^2\,, \tag{16.9-21}$$

i.e., the kernel X is nothing but the matrix element squared as it occurs in the golden rule. For (16.9-19) we now obtain

---

[50] Van Hove employs the delta function $\delta(\gamma - \gamma')$ instead of the Kronecker delta $\delta_{\gamma\gamma'}$ and refers to the first term in (16.9-20) as the "diagonal singularity", to which he attributes the resulting irreversibility.

$$\langle\gamma|U^{(1)\dagger}(\Delta t)B_d U^{(1)}(\Delta t)|\gamma'\rangle$$
$$=\frac{\lambda^2}{\hbar^2}\delta_{\gamma\gamma'}\sum_{\gamma''}\int_0^{\Delta t}dt_1\int_0^{\Delta t}dt_1' e^{i(\mathcal{E}_\gamma-\mathcal{E}_{\gamma''})(t_1'-t_1)/\hbar}\langle\gamma''|B_d|\gamma''\rangle X(\gamma'',\gamma)$$
$$+\frac{\lambda^2}{\hbar^2}\sum_{\gamma''}\int_0^{\Delta t}dt_1\int_0^{\Delta t}dt_1' e^{i(\mathcal{E}_\gamma-\mathcal{E}_{\gamma''})t_1'/\hbar}e^{-i(\mathcal{E}_{\gamma'}-\mathcal{E}_{\gamma''})t_1/\hbar}\langle\gamma''|B_d|\gamma''\rangle Y(\gamma'';\gamma,\gamma'). \quad (16.9\text{-}22)$$

For the first line we have

$$\left|\int_0^{\Delta t}dt\, e^{i(\mathcal{E}_\gamma-\mathcal{E}_{\gamma''})t/\hbar}\right|^2=\frac{\sin^2[(\mathcal{E}_\gamma-\mathcal{E}_{\gamma''})\Delta t/2\hbar]}{[(\mathcal{E}_\gamma-\mathcal{E}_{\gamma''})/2\hbar]^2}\simeq 2\pi\hbar\Delta t\,\delta(\mathcal{E}_\gamma-\mathcal{E}_{\gamma''}), \quad (16.9\text{-}23)$$

providing $(\mathcal{E}_\gamma-\mathcal{E}_{\gamma''})\Delta t/\hbar \gg 1$. We will thus identify the two limits foreseen previously

$$\tau_t \approx \hbar/(\mathcal{E}_\gamma-\mathcal{E}_{\gamma''})=\hbar/\delta\mathcal{E}, \qquad \tau_r \sim \hbar\delta\mathcal{E}/(\lambda\mathcal{V})^2. \quad (16.9\text{-}24)$$

For the other part in (16.9-22) we note

$$\int_0^{\Delta t}dt_1'\, e^{i(\mathcal{E}_\gamma-\mathcal{E}_{\gamma''})t_1'/\hbar}\sim 2\pi\hbar\delta_-(\mathcal{E}_\gamma-\mathcal{E}_{\gamma''})$$
$$=\pi\hbar\delta(\mathcal{E}_\gamma-\mathcal{E}_{\gamma''})+\pi\hbar i\mathcal{P}/(\mathcal{E}_\gamma-\mathcal{E}_{\gamma''}), \quad (16.9\text{-}25)$$

with a similar result for the other integral (change $\gamma\to\gamma'$). Since $\gamma\neq\gamma'$, the delta functions multiply to zero and the principal parts are of no concern. The essential feature is that the first part of (12.9-22) goes with $\lambda^2\Delta t$, while the second part goes with λ^2; in the Van Hove limit only the first part (with the 'diagonal singularity') survives. The final result is, employing also (16.9-21),

$$\lim_{\lambda,t}\langle\gamma|U^{(1)\dagger}(\Delta t)B_d U^{(1)}(\Delta t)|\gamma'\rangle$$
$$=\frac{2\pi\lambda^2\Delta t}{\hbar}\delta_{\gamma\gamma'}\sum_{\gamma''}\langle\gamma''|B_d|\gamma''\rangle|\langle\gamma''|\mathcal{V}|\gamma\rangle|^2\,\delta(\mathcal{E}_{\gamma''}-\mathcal{E}_\gamma). \quad \underline{(16.9\text{-}26)}$$

Next, we have the combinations

$$\langle\gamma|U^{(0)\dagger}(\Delta t)B_d U^{(1)}(\Delta t)|\gamma'\rangle+\langle\gamma|U^{(1)\dagger}(\Delta t)B_d U^{(0)}(\Delta t)|\gamma'\rangle. \quad (16.9\text{-}27)$$

These are of order λ. Yet, we easily find the exact form, by inserting the closure relation $\Sigma|\gamma''\rangle\langle\gamma''|=1$ before the $\mathcal{V}$ of $U^{(1)}$. For $\gamma\neq\gamma'$ we have an interference term that averages to zero. For $\gamma=\gamma'$ the result is

$$4\lambda\sum_{\gamma''}\frac{\sin^2[(\mathcal{E}_\gamma-\mathcal{E}_{\gamma''})\Delta t/2\hbar]}{\mathcal{E}_\gamma-\mathcal{E}_{\gamma''}}\langle\gamma|B_d|\lambda''\rangle\langle\gamma''|\mathcal{V}|\gamma\rangle$$
$$\sim 2\lambda\sum_{\gamma''\neq\gamma}\langle\gamma|B_d|\lambda''\rangle\langle\gamma''|\mathcal{V}|\gamma\rangle/(\mathcal{E}_\gamma-\mathcal{E}_{\gamma''}); \quad (16.9\text{-}28)$$

there is no singularity and the terms can be dismissed. Finally, we have the contributions

$$\langle\gamma|U^{(0)\dagger}(\Delta t)B_dU^{(2)}(\Delta t)|\gamma'\rangle+\langle\gamma|U^{(2)\dagger}(\Delta t)B_dU^{(0)}(\Delta t)|\gamma'\rangle$$
$$=-\frac{\lambda^2}{\hbar^2}\sum_{\gamma'''}\int_0^{\Delta t}dt_2\int_0^{t_2}dt_1e^{i\mathcal{E}_\gamma\Delta t/\hbar}\langle\gamma|B_d|\gamma'''\rangle e^{-i\mathcal{E}_{\gamma'''}(\Delta t-t_2)/\hbar}$$
$$\times\langle\gamma'''|\mathcal{V}\{e^{-i\mathcal{H}^0(t_2-t_1)/\hbar}\}\mathcal{V}|\gamma'\rangle e^{-i\mathcal{E}_{\gamma'}t_1/\hbar}+hc. \tag{16.9-29}$$

Again we use Van Hove's functional rule, noting that there is no Y-part for this case. Hence,

$$above=-\frac{2\lambda^2}{\hbar^2}\delta_{\gamma\gamma'}\operatorname{Re}\sum\nolimits_{\gamma''}\langle\gamma|B_d|\gamma\rangle|\langle\gamma''|\mathcal{V}|\gamma\rangle|^2\int_0^{\Delta t}dt_2\int_0^{t_2}dt_1e^{i(\mathcal{E}_\gamma-\mathcal{E}_{\gamma''})(t_2-t_1)/\hbar}.$$

The time-ordered integral is

$$\mathcal{I}=\frac{\sin^2[(\mathcal{E}_\gamma-\mathcal{E}_{\gamma''})\Delta t/2\hbar]}{[(\mathcal{E}_\gamma-\mathcal{E}_{\gamma''})/2\hbar]^2}\simeq2\pi\hbar\Delta t\,\delta(\mathcal{E}_\gamma-\mathcal{E}_{\gamma''}). \tag{16.9-30}$$

This, then, yields

$$\lim_{\lambda,t}\langle\gamma|U^{(0)\dagger}(\Delta t)B_dU^{(2)}(\Delta t)|\gamma'\rangle+\langle\gamma|U^{(2)\dagger}(\Delta t)B_dU^{(0)}(\Delta t)|\gamma'\rangle$$
$$=-\frac{2\pi\lambda^2\Delta t}{\hbar}\sum\nolimits_{\gamma''}\langle\gamma|B_d|\gamma\rangle|\langle\gamma''|\mathcal{V}|\gamma\rangle|^2\delta(\mathcal{E}_\gamma-\mathcal{E}_{\gamma''}). \tag{16.9-31}$$

We now collect terms, cf. Eqs (6.9-18), (6.9-26) and (6.9-31). We established,

$$\lim_{\lambda,t}\langle\gamma|B_d(\Delta t)|\gamma'\rangle=\langle\gamma|B_d|\gamma'\rangle$$
$$+\frac{2\pi\lambda^2\Delta t}{\hbar}\delta_{\gamma\gamma'}\sum_{\gamma''}\{\langle\gamma''|B_d|\gamma''\rangle-\langle\gamma|B_d|\gamma\rangle\}|\langle\gamma''|V|\gamma\rangle|^2\,\delta(\mathcal{E}_\gamma-\mathcal{E}_{\gamma''}). \tag{16.9-32}$$

Finally, the Kronecker delta is rewritten, so that for any $F(\gamma)$

$$\delta_{\gamma\gamma'}F(\gamma)=\sum_{\bar\gamma}\delta_{\gamma\bar\gamma}\delta_{\bar\gamma\gamma'}F(\bar\gamma)=\sum_{\bar\gamma}\langle\gamma|\bar\gamma\rangle\langle\bar\gamma|\gamma'\rangle F(\bar\gamma)$$
$$=\langle\gamma|\left\{\sum\nolimits_{\bar\gamma}|\bar\gamma\rangle\langle\bar\gamma|F(\bar\gamma)\right\}|\gamma'\rangle. \tag{16.9-33}$$

[Sometimes, in higher education we write *simple things* in a complicated way!] This substitution for the Kronecker delta is entered in (16.9-32); we use (16.9-3) to obtain

$$\lim_{\lambda,t}\langle\gamma|B_d(\Delta t)|\gamma'\rangle=\langle\gamma|B_d|\gamma'\rangle$$
$$+\langle\gamma|\left\{\sum\nolimits_{\bar\gamma\gamma''}|\bar\gamma\rangle\langle\bar\gamma|[W_{\gamma''\bar\gamma}\Delta t\langle\gamma''|B_d|\gamma''\rangle-W_{\bar\gamma\gamma''}\Delta t\langle\bar\gamma|B_d|\bar\gamma\rangle]\right\}|\gamma'\rangle. \tag{16.9-34}$$

On the rhs of the last equation we recognise the master operator in Liouville space. Whence, the final result obtained reads

$$\lim_{\lambda,t}\langle\gamma|B_d(\Delta t)|\gamma'\rangle=\langle\gamma|B_d|\gamma'\rangle-\langle\gamma|\Lambda_d B_d\Delta t|\gamma'\rangle, \tag{16.9-35}$$

or also, given that $|\gamma\rangle$ and $|\gamma'\rangle$ are two arbitrary many-body states,

$$B_d^R(\Delta t)=B_d-\Lambda_d B_d\Delta t, \tag{16.9-36}$$

which is the small-time form of (16.9-12).

16.9.2 *Results for Large Times*

In the derivation for small times we assumed *a priori* that at $t=0$ the operator B was diagonal, i.e., $B(0)=B_S=B_d$. Such an 'initial random phase assumption' is warranted (although not necessary, as we shall see later). Now let us consider a time interval $[0,(n+1)\Delta t]$ and suppose that at the beginning of each new interval Δt a random phase assumption is made. We then find by repeated application of (16.9-36)

$$B_d^R(t+\Delta t)=B_d^R(t)-\Lambda_d B_d^R(t)\Delta t, \tag{16.9-37}$$

which can be rewritten as

$$\frac{B_d^R(t+\Delta t)-B_d^R(t)}{\Delta t}=-\Lambda_d B_d^R(t). \tag{16.9-38}$$

Since Δt is macroscopically small, we may interpret the lhs as dB_d^R/dt and integrate, to obtain

$$B_d^R(t)=e^{-\Lambda_d t}B_d. \tag{16.9-39}$$

The above "derivation" amounts to the *repeated random phase assumption* (*RRPA*), mentioned already in Section 16.7, but applied here to the operators $B(t)$, rather than to the density operator $\rho(t)$.[51] Of course, the *RRPA* is at variance with the Heisenberg equation of motion. So, we should try to find a solution for the reduced operators for all times, analogous to the procedure followed in the previous subsection. This has indeed been achieved by Van Hove using a direct procedure and by others employing alternate techniques.

Van Hove's procedure is daunting and extremely laborious. One must combine an infinite number of terms from the double evolution operator series and decide which ones are of order $(\lambda^2)^n$ and which leave parts of the form $(\lambda^2 t)^n$, where n is an integer, denumerably infinite! Besides in Van Hove's paper, the derivation is

---

[51]We always have the two forms for the expectation value of an operator

$$\langle B(t)\rangle=\mathrm{Tr}\{\rho B(t)\}=\mathrm{Tr}\{\rho(t)B\}. \tag{16.9-39'}$$

Thus, the RRPA applied to the Heisenberg operator $B(t)$ is equivalent to the RRPA applied to the density operator $\rho(t)$.

found in LRT I and in a slightly expanded form in a paper of ours in the Canadian J. of Physics.[52] One can also work in the Laplace s-domain; the transform of the evolution operator is called the resolvent. The straightforward analogue of Van Hove's procedure involves a 'two-resolvent' description.[53] We will, however, not discuss any of these procedures here and, fortunately, it is not necessary. Rather, we shall follow a projection-operator treatment, introduced by Zwanzig in 1960.[54] This will be a 'streamlined' procedure, after we develop a few theorems pertaining to the method. It will give us the reduced Heisenberg operators, when applied to the Heisenberg equation of motion. Also, it will convert the von Neumann equation into the Pauli–Van Hove master equation. In a discussion we had with Zwanzig at a conference in the seventies he called it "*a simple re-arrangement procedure!*" [55]

In *our* presentation of this method, we shall, however, go beyond the results of these authors and retain the nondiagonal parts, which are indispensable for a complete quantum-transport description. The so-obtained results are in no way equivalent to the *RRPA*.

We still note that all developments in Part E of this text deal with *microscopic*, many-body processes. In the next part F, pertaining to stochastic processes, the master equation will also show up. This *mesoscopic* ME is, however, coarse-grained and the diagonal master equation suffices.

16.10 Irreversible Transport Equations via Projector Operators

16.10.1 *Some Theorems*

We will need some theorems regarding the application of projector operators; also, we will have to be more specific with regard to the Dirac convention for contractions, since (unwittingly to many) the presence of a bilinear concomitant can invalidate the standard notation.

The Hamiltonian of departure has the perturbation form

$$\mathcal{H} = \mathcal{H}^0 + \lambda \mathcal{V}. \tag{16.10-1}$$

The state space pertaining to the sub-dynamics of $\mathcal{H}^0$ will be denoted by $\mathcal{S}$, with the eigenstates spanning these states being $\{|\gamma\rangle\}$; usually they will be given in occupation-number form, although we shall also formally employ a wave-mechanical

[52] K.M. Van Vliet, "On the derivation of the Pauli and van Hove master equations", Can. J. of Physics, 1345-1370 (1978).

[53] S. Fujita, " Introduction to Non-equilibrium Quantum Statistical Mechanics", Chapter V § 6.

[54] R. Zwanzig, "Statistical Mechanics of Irreversibility" in Lectures in Theoretical Physics, Vol. III, (W. E. Britten, B.W. Downs and J. Downs, Eds) Boulder Colorado 1960, Interscience, NY 1961, p. 116-141.

[55] R. Zwanzig, private communication.

realization in a function space, $\gamma(\{q_k\}) \equiv \langle\{q_k\}\,|\,\gamma\rangle$, $k = 1,2...3N$, as in standard Dirac notation. For γ we assume periodic boundary conditions over the volume $L^3 = V_0$ of the system, as customary in condensed-matter physics. We also assume that $\mathscr{S}$ is embedded in a larger space $\mathscr{S}'$ in which the scalar product is still defined as in $\mathscr{S}$, but without the restriction of periodic b.c.; this is necessary since many physical operators, i.c. functions of position, $P(\mathbf{q})$, carry the new state $P(\mathbf{q})|\,\gamma\rangle$ outside $\mathscr{S}$. So, let K be a general operator of $\mathscr{S}$. The decomposition of K into diagonal and nondiagonal parts was already carried out in (16.9-9). We repeat the final result here,

$$K = \mathscr{P}K + (1-\mathscr{P})K \equiv K_d + K_{nd}\,, \tag{16.10-2}$$

where the projection operator $\mathscr{P}$ acts in the Liouville space $\mathscr{S} \otimes \bar{\mathscr{S}}$, $\bar{\mathscr{S}}$ being the dual space; it has the form

$$\mathscr{P} = \sum_{\gamma} \{|\,\gamma\rangle\langle\gamma| \rightarrow \leftarrow |\gamma\rangle\langle\gamma|\}\,. \tag{16.10.3}$$

It projects on *diagonal* Liouville space and should not be confused with the identity operator of the Liouville space,

$$\mathscr{I} \equiv 1 = \sum_{\gamma,\gamma'} \{|\,\gamma\rangle\langle\gamma| \rightarrow \leftarrow |\gamma'\rangle\langle\gamma'|\}\,. \tag{16.10.4}$$

The last two equations together also define the operator $1-\mathscr{P} = \mathscr{Q}$, used to project on the compliment space. The Liouville super-operator $\mathscr{L}$, as well as the interaction Liouvillian $\mathscr{L}^0$ were already defined in (16.9-7) and (16.9-8); in particular we had

$$\mathscr{L}K = \frac{1}{\hbar}[\mathscr{H}, K] \quad \text{or} \quad \mathscr{L} = \frac{1}{\hbar}(\mathscr{H} \rightarrow \leftarrow I - I \rightarrow \leftarrow \mathscr{H})\,. \tag{16.10-5}$$

where I is the identity operator of $\mathscr{S}$. The following theorems are easily proven by series expansion:[56,57]

Theorem 1. *For the exponentiated Liouville operator we have the identity*

$$e^{\pm i\mathscr{L}t}K = e^{\pm i\mathscr{H}t/\hbar}Ke^{\mp i\mathscr{H}t/\hbar}\,. \tag{16.10-6}$$

Theorem 2. *We have the following identity involving diagonal parts*

$$e^{\sum_\gamma |\gamma\rangle\langle\gamma| f(\gamma)} = \sum_{\gamma} |\,\gamma\rangle\langle\gamma\,|\, e^{f(\gamma)}\,. \tag{16.10-7}$$

---

[56] C.M. Van Vliet, "Quantum Transport in Solids", in CRM Proceedings and Lecture Notes Vol. 11 (L. Vinet and Y. St.-Aubin, Eds.), The Am. Math. Soc., Providence, Rhode Island, 1997.

[57] C.M. Van Vliet and A. Barrios, "Quantum Electron Transport Beyond Linear Response," Physica **A** **315**, 493-536 (2002).

Theorem 3. *For nondiagonal parts it holds that*

$$e^{-it(1-\mathcal{P})\mathcal{L}}(1-\mathcal{P})K = (1-\mathcal{P})e^{-it\mathcal{L}(1-\mathcal{P})}K\,. \tag{16.10-8}$$

Theorem 4. *For* $\lambda \to 0$ *one has*

$$e^{-it(1-\mathcal{P})\mathcal{L}}K = [e^{-it\mathcal{L}^0} + \mathcal{O}(\lambda)]K, \quad \forall\ \mathcal{P}[K,\mathcal{H}^0]=0\,. \tag{16.10-9}$$

The proof is simple since

$$e^{-it(1-\mathcal{P})\mathcal{L}}K = e^{-it(1-\mathcal{P})\mathcal{L}^0}K + \mathcal{O}(\lambda) = \sum_{n=0}^{\infty}[(-it\mathcal{L}^0 + it\mathcal{P}\mathcal{L}^0)^n/n!]K + \mathcal{O}(\lambda); \tag{16.10-10}$$

but $\mathcal{P}\mathcal{L}^0 K = 0$ *if the condition of* (16.10-9) *is satisfied.*

At this moment it may surprise the reader that the *condition* $\mathcal{P}[K,\mathcal{H}^0]=0$ is *not* the same as $[\mathcal{P}K,\mathcal{H}^0]=0$. The problem here involves the *validity of the Dirac notation*, which assumes that there is no boundary term arising from Green's theorem, which in turn implies that no transformations occur that carry the states outside the space $\mathcal{S}$ determined by $\mathcal{H}^0$ and the boundary conditions. To appreciate the difficulty we shall revert to the standard algebraic notation for the scalar product in Hilbert space. We have if $|\varphi\rangle$ and $|\psi\rangle$ are arbitrary states,

$$\langle\varphi|\{K|\psi\rangle\} \equiv (K\psi,\varphi) \quad \text{and} \quad \{\langle\varphi|K\}|\psi\rangle \equiv (\psi,K^{\dagger}\varphi)\,, \tag{16.10-11}$$

where in the first operation K acts to the right in $\mathcal{S}$, while in the second operation K acts to the left, i.e., in the dual space $\overline{\mathcal{S}}$. Green's theorem reads

$$(K\psi,\varphi) - (\psi,K^{\dagger}\varphi) = Q_K[\psi,\varphi]\,, \tag{16.10-12}$$

where Q_K is the bilinear concomitant; it vanishes iff ψ satisfies the b.c. and φ the adjoint b.c. which, in view of the fact that K is self-adjoint – not just Hermitean – must be identical to those in $\mathcal{S}$. Thus when dealing with a transformation $K_1K_2\ldots K_n\gamma$, the bilinear concomitant $Q[K_1K_2\ldots K_n\gamma\,,\gamma] = 0$ iff $K_1K_2\ldots K_n\gamma$ satisfies the assumed periodic boundary conditions. We conclude that the standard Dirac projector algebra is valid *whenever the operators* K_1, K_2,... K_n *are invariant against translation over the dimensions of the system.* To avoid ambiguity we adopt the convention that in the Dirac bilinear form $\langle\varphi|K|\psi\rangle$ the operator K *shall act to the right*, unless otherwise indicated by curly brackets. So, the following rules apply. Let L_d be a diagonal operator; then we have

$$\begin{aligned}&\mathcal{P}(K_1...K_nL_d) = (\mathcal{P}K_1...K_n)L_d = L_d(\mathcal{P}K_1...K_n)\,,\\ &\mathcal{P}(L_dK_1...K_n) = L_d(\mathcal{P}K_1...K_n) + \textstyle\sum_\gamma|\gamma\rangle\langle\gamma|Q_{L_d}[K_1...K_n\gamma,\gamma]\,.\end{aligned} \tag{16.10-13}$$

To understand these rules, note that in the first of these equations we need for the superoperator $\mathcal{P}... = \sum_\gamma \{|\gamma\rangle\langle\gamma|...|\gamma\rangle\langle\gamma|\}$ the matrix element $\langle\gamma|K_1...K_n L_d|\gamma\rangle$, in which L_d works to the right; however, in the second equality of (16.10-13) L_d works to the left, so the bilinear concomitant must be added. For the commutator we obtain

Theorem 5. *The diagonal part of a commutator depends on the bilinear concomitant,*

$$\mathcal{P}[K_1...K_n, L_d] = -\sum_\gamma |\gamma\rangle\langle\gamma| Q_{L_d}[K_1...K_n\gamma, \gamma]. \qquad (16.10\text{-}14)$$

In particular for the commutator of the condition in (16.10-9) we have

$$\mathcal{P}[K, \mathcal{H}^0] = -\sum_\gamma |\gamma\rangle\langle\gamma| Q_{\mathcal{H}^0}[K\gamma, \gamma]. \qquad (16.10\text{-}15)$$

For the direct proof we employ the explicit form for $\mathcal{P}$ as given in (16.10-3); hence,

$$\begin{aligned}\mathcal{P}[K, \mathcal{H}^0] &= \sum_\gamma |\gamma\rangle\langle\gamma| \big(\langle\gamma|K\mathcal{H}^0|\gamma\rangle - \langle\gamma|\mathcal{H}^0 K|\gamma\rangle\big)\\ &= \sum_\gamma |\gamma\rangle\langle\gamma| \big(\langle\gamma|K\mathcal{H}^0|\gamma\rangle - \{\langle\gamma|\mathcal{H}^0\}K|\gamma\rangle - Q_{H^0}[K\gamma, \gamma]\big)\\ &= \sum_\gamma |\gamma\rangle\langle\gamma| \big(\langle\gamma|K|\gamma\rangle\mathcal{E}_\gamma - \mathcal{E}_\gamma\langle\gamma|K|\gamma\rangle - Q_{H^0}[K\gamma, \gamma]\big)\\ &= -\sum_\gamma |\gamma\rangle\langle\gamma| Q_{H^0}[K\gamma, \gamma]. \end{aligned} \qquad (16.10\text{-}16)$$

Therefore, the condition in (16.10-9) is fulfilled if and only if K *is an operator which is translationally invariant over the dimensions of the system.* In other cases one must either *first* evaluate the commutator and then find its diagonal part, or the bilinear concomitant must be obtained explicitly.

16.10.2 *Reduction of the Heisenberg Equation of Motion; Diagonal Part*

The quantum mechanical current $J_A \equiv dA/dt$ is governed by the Heisenberg equation of motion. So we have

$$\frac{dA}{dt} = \frac{\partial A}{\partial t} + \frac{1}{i\hbar}[A(t), \mathcal{H}] = \frac{\partial A}{\partial t} + i\mathcal{L}A(t), \qquad (16.10\text{-}17)$$

where $\mathcal{L}$ is the Liouville superoperator as before. In the interaction picture, $\mathcal{H} = \mathcal{H}^0 + \lambda\mathcal{V}$, the sub-Hamiltonian $\mathcal{H}^0$ may depend on applied fields, **E** and **B**, as we discuss in the next section in more detail. We assume that A does not depend on the time explicitly; then the Heisenberg equation reads

$$\frac{dA}{dt} - i\mathcal{L}A(t) = 0. \qquad (16.10\text{-}18)$$

We now write $A = A_d + A_{nd}$ and we apply projection operators $\mathcal{P}$ and $1-\mathcal{P}$ for the diagonal and nondiagonal Liouville space, respectively. Then (16.10-18) yields the following two equivalent equations:

$$\begin{aligned} &\frac{dA_d}{dt} - i\mathcal{P}\mathcal{L}A_d - i\mathcal{P}\mathcal{L}A_{nd} = 0, \\ &\frac{dA_{nd}}{dt} - i(1-\mathcal{P})\mathcal{L}A_d - i(1-\mathcal{P})\mathcal{L}A_{nd} = 0. \end{aligned} \tag{16.10-19}$$

For the term $\mathcal{P}\mathcal{L}A_d$ we have $\mathcal{P}\mathcal{L}^0 A_d + (\lambda/\hbar)\mathcal{P}[\mathcal{V}, A_d]$; $\mathcal{P}\mathcal{L}^0 A_d = 0$ since $\mathcal{L}^0$ (a commutator) destroys any diagonal part, while $\mathcal{V}$ and A_d often are both functions of position that commute (even if not, this part goes with λ) Thus, Eqs. (16.10-19) simplify to

$$\frac{dA_d}{dt} - i\mathcal{P}\mathcal{L}A_{nd} = 0, \quad \frac{dA_{nd}}{dt} - i\mathcal{L}A_d - i(1-\mathcal{P})\mathcal{L}A_{nd} = 0. \tag{16.10-20}$$

The formal solution of the last equation for $t > 0$ is

$$A_{nd}(t) = i\int_0^t dt' G^\dagger(t-t')\mathcal{L}A_d(t') + G^\dagger(t,0)A_{nd}, \tag{16.10-21}$$

where we employed the Green's operator

$$G(t,t') \equiv G(t-t') = \Theta(t-t')e^{-i(t-t')(1-\mathcal{P})\mathcal{L}}; \tag{16.10-22}$$

for $G^\dagger$ replace i by $-i$ since $\mathcal{L}$ and $\mathcal{P}$ are selfadjoint in Liouville space. This result is substituted into the first equation of (16.10-20), thereby yielding the integro-differential equation for the diagonal part,

$$\frac{dA_d}{dt} + \mathcal{P}\mathcal{L}\int_0^t dt' G^\dagger(t-t')\mathcal{L}A_d(t') - i\mathcal{P}\mathcal{L}G^\dagger(t,0)(1-\mathcal{P})A = 0. \tag{16.10-23}$$

This result is still exact, with the convolution integral containing the memory of all past times. Further, we note that the last term can be dismissed: employing Theorem 3, we can move the operator $1 - \mathcal{P}$ to the left through the Green's operator, where it is destroyed by the projector $\mathcal{P}$.

Proceeding now to the perturbation procedure, we use Theorem 4 of the preceding subsection, noting that $[A_d, \mathcal{H}^0] = 0$. For the Green's operator we then have

$$G(t-t') = \Theta(t-t')e^{-i(t-t')\mathcal{L}^0} + \mathcal{O}(\lambda) \to \Theta(t-t')G_0(t-t') + \mathcal{O}(\lambda) \tag{16.10-24}$$

where

$$G_0(t-t') = e^{-i(t-t')\mathcal{L}^0}. \tag{16.10-25}$$

In order to evaluate the convolution integral in (16.10-23), with $G^\dagger$ replaced by $G_0^\dagger$, we first note again that the $\mathcal{L}^0$ contributions vanish when acting on a diagonal operator. Next, the projection operator in front is written out in its explicit form given in (16.10-3). We thus have, employing also Theorem 1,

$$\mathcal{P}\mathcal{L}\int_0^t dt' e^{i(t-t')\mathcal{L}^0}\mathcal{L}A_d(t')$$

$$=\left(\frac{\lambda}{\hbar}\right)^2\int_0^t dt'\mathcal{P}[\mathcal{V},e^{i(t-t')\mathcal{L}^0}[\mathcal{V},A_d(t')]]$$

$$=\sum_\gamma|\gamma\rangle\langle\gamma|\left(\frac{\lambda}{\hbar}\right)^2\int_0^t dt'\langle\gamma|[\mathcal{V},e^{i\mathcal{H}^0(t-t')/\hbar}[\mathcal{V},A_d(t')]e^{-i\mathcal{H}^0(t-t')/\hbar}]|\gamma\rangle. \quad (16.10\text{-}26)$$

The matrix element under the integral is easily evaluated by inserting closure, $\Sigma_{\gamma''}|\gamma''\rangle\langle\gamma''|=1$. Hence,

$$\langle\gamma|[\mathcal{V},...]|\gamma\rangle=\sum_{\gamma''}\{\langle\gamma|\mathcal{V}|\gamma''\rangle e^{i\mathcal{E}_{\gamma''}(t-t')/\hbar}[\langle\gamma''|\mathcal{V}|\gamma\rangle A_{d\gamma}(t')$$
$$-A_{d\gamma''}(t')\langle\gamma''|\mathcal{V}|\gamma\rangle]e^{-i\mathcal{E}_\gamma(t-t')/\hbar}-e^{i\mathcal{E}_\gamma(t-t')/\hbar}[\langle\gamma|\mathcal{V}|\gamma''\rangle A_{d\gamma''}(t')$$
$$-A_{d\gamma}(t')\langle\gamma|\mathcal{V}|\gamma''\rangle]e^{-i\mathcal{E}_{\gamma''}(t-t')/\hbar}\langle\gamma''|\mathcal{V}|\gamma\rangle\}$$
$$=2\sum_{\gamma''}|\langle\gamma|\mathcal{V}|\gamma''\rangle|^2\cos[(\mathcal{E}_{\gamma''}-\mathcal{E}_\gamma)(t-t')/\hbar][A_{d\gamma}(t')-A_{d\gamma''}(t')]; \quad (16.10\text{-}27)$$

here, $A_{d\gamma}\equiv\langle\gamma|A_d|\gamma\rangle$. Denoting now the lhs of (16.10-26) by $-(dA_d/dt)_{mem.}$ we have

$$\left(\frac{dA_d}{dt}\right)_{mem.}=-\sum_{\gamma,\gamma''}|\gamma\rangle\langle\gamma|\left(\frac{2\lambda^2}{\hbar^2}\right)|\langle\gamma|\mathcal{V}|\gamma''\rangle|^2$$
$$\times\int_0^t dt'\cos[(\mathcal{E}_{\gamma''}-\mathcal{E}_\gamma)(t-t')/\hbar][A_{d\gamma}(t')-A_{d\gamma''}(t')]. \quad (16.10\text{-}28)$$

Employing the convolution theorem, this has the Laplace transform

$$\Phi(s)=-\sum_{\gamma,\gamma''}|\gamma\rangle\langle\gamma|\left(\frac{2\lambda^2}{\hbar^2}\right)|\langle\gamma|\mathcal{V}|\gamma''\rangle|^2$$
$$\times\frac{s}{(\mathcal{E}_{\gamma''}-\mathcal{E}_\gamma)^2/\hbar^2+s^2}[\bar{A}_{d\gamma}(s)-\bar{A}_{d\gamma''}(s)]. \quad (16.10\text{-}29)$$

The Van Hove limit $t\to\infty$ amounts to $s\to 0+$. With the well-known limits

$$\lim_{s\to 0+}\frac{1}{(\mathcal{E}_{\gamma''}-\mathcal{E}_\gamma)/\hbar\pm is}=\hbar\mathcal{P}\frac{1}{\mathcal{E}_{\gamma''}-\mathcal{E}_\gamma}\mp\pi i\hbar\delta(\mathcal{E}_{\gamma''}-\mathcal{E}_\gamma) \quad (16.10\text{-}30)$$

we find

$$\lim_{s\to 0+}\frac{s}{(\mathcal{E}_{\gamma''}-\mathcal{E}_\gamma)^2/\hbar^2+s^2}$$
$$=\lim_{s\to 0+}\frac{1}{2i}\left[\frac{1}{(\mathcal{E}_{\gamma''}-\mathcal{E}_\gamma)/\hbar-is}-\frac{1}{(\mathcal{E}_{\gamma''}-\mathcal{E}_\gamma)/\hbar+is}\right]=\pi\hbar\delta(\mathcal{E}_{\gamma''}-\mathcal{E}_\gamma). \quad (16.10\text{-}31)$$

The inverse transform of (16.10-29) thus becomes

$$\left(\frac{dA_d}{dt}\right)_{mem.} = -\Lambda_d A_d(t)\,, \tag{16.10-32}$$

where in a *natural fashion* the master (super)-operator emerged:

$$\Lambda_d A_d = -\sum_{\gamma,\gamma''} |\gamma\rangle\langle\gamma| [W_{\gamma''\gamma}\langle\gamma''|A|\gamma''\rangle - W_{\gamma\gamma''}\langle\gamma|A|\gamma\rangle]\,, \tag{16.10-33}$$

with $W_{\gamma\gamma''}$ being the transition probability of Fermi's golden rule

$$W_{\gamma\gamma''} = \left(\frac{2\pi\lambda^2}{\hbar}\right) |\langle\gamma|\mathcal{V}|\gamma''\rangle|^2\, \delta(\mathcal{E}_\gamma - \mathcal{E}_{\gamma''}) = W_{\gamma''\gamma}\,. \tag{16.10-34}$$

The equality $W_{\gamma\gamma''} = W_{\gamma''\gamma}$ expresses the property of *microscopic reversibility.*

Going back to Eq. (16.10-23) for the diagonal part, we have shown that the reduced diagonal Heisenberg equation after the Van Hove limit has the form

$$\frac{dA_d}{dt} + \Lambda_d A_d(t) = 0\,. \tag{16.10-35}$$

This form is said to be *Markovian*, since there is no memory, except for *one* preceding time, usually taken for the initial condition. However, *not* all memory has been erased; see the next subsection. The solution of (16.10-35) is just the earlier announced result (16.9-12).

16.10.3 *The Full Reduced Heisenberg Equation and the Current Operator*

We return to the result for the nondiagonal part, Eq. (16.10-21). Using the Green's operator (16.10-25) of the Van Hove limit and differentiating (16.10-21), we find

$$\frac{dA_{nd}}{dt} - i\mathcal{L}^0 A_{nd}(t) = 0\,. \tag{16.10-36}$$

We note that differentiation to the upper limit of the integral yields a contribution $iG_0^\dagger(0)\mathcal{L}A_d(t) = i(\lambda/\hbar)[\mathcal{V}, A_d(t)]$, which is either 0 or of order λ and thus dismissed. Next, we add to (16.10-35) the term $\Lambda_d A_{nd} = 0$ and to (16.10-36) the term $-i\mathcal{L}^0 A_d = 0$; then upon summing we obtain the *full reduced Heisenberg equation of motion,*

$$\frac{dA}{dt} + (\Lambda_d - i\mathcal{L}^0)A(t) = 0\,. \tag{16.10-37}$$

The solution is

$$A^R(t) \equiv \lim_{\lambda,t} A^H(t) = e^{(i\mathcal{L}^0 - \Lambda_d)t} A = e^{-\Lambda_d t} A_d + e^{i\mathcal{L}^0 t} A_{nd}\,. \tag{16.10-38}$$

Finally, let **J** be the electrical flux in a sample of volume V_0. The Heisenberg form is

$$\mathbf{J}^H = \frac{\mathcal{q}}{V_0}\frac{d}{dt}\sum_i(\mathbf{r}_i - \mathbf{r}_{i,0}), \tag{16.10-39}$$

where the sum is over all electrons of the conduction band. In the interaction picture $i\mathcal{L}^0\Sigma_i(\mathbf{r}_i - \mathbf{r}_{i,0})$ amounts to

$$i\mathcal{L}^0\sum_i(\mathbf{r}_i - \mathbf{r}_{i,0}) = \frac{1}{\hbar i}\sum_i[(\mathbf{r}_i - \mathbf{r}_{i,0}),\mathcal{H}^0] = \sum_i \mathbf{v}_i\,, \tag{16.10-40}$$

where **v** is the velocity. So, from (16.10-37) we have for the reduced flux (16.10-39),

$$\mathbf{J}^R = \lim_{\lambda,t}\mathbf{J}^H = \frac{\mathcal{q}}{V_0}\left\{-\Lambda_d\left[\sum_i(\mathbf{r}_i - \mathbf{r}_{i,0})\right] + \sum_i \mathbf{v}_i\right\}. \tag{\textbf{16.10-41}}$$

The two parts of (16.10-41) will be denoted as *collisional current* and *ponderomotive current*. The latter is simply the 'regular' current as usually conceived. However, the collisional current is a 'new contribution' that *solely* arises from the interaction description. Its existence appears first in magnetoresistance computations of Adams and Holstein[58] and Argyres and Laura Roth[59]; it was foreseen by Titeica[60]. The consequences for the "Boltzmann current", $\mathrm{Tr}(\rho\mathbf{J})$, will be examined in Chapter XVII.

The result (16.10-38) [with (16.10-41)] should be appreciated for what it truly represents: the *single* interaction $\lambda\mathcal{V}$, stipulated for an interval $\Delta t \to 0$, has given rise to large time behaviour involving the exponentiated master operator $\exp(-\Lambda t)$. In terms of a diagrammatic approach, it represents all n^{th} order repeat interactions! The Van Hove limit procedure has proven to be a powerful many-body evaluation technique.

We finish this subsection with the quantum-field form for the current in reduced-dimensional systems, such as quantum wires. Although both ponderomotive and collisional current are amenable to a quantum-field description[61], we limit ourselves here to the ponderomotive current. For it to be non-zero, the states must be extended (i.e., non-localised) in at least one dimension, represented by a crystal momentum $\mathbf{k} = \{k_i\}, i = 1...\nu,\ \nu \ge 1$. In the other dimensions quantum confinement will occur when the dimensions are less than the De Broglie wavelength, thus involving a set of quantum numbers α; the spin is denoted by σ. The full one-particle states are denoted by $|\zeta) = |\mathbf{k},\alpha,\sigma)$ with wave-mechanical form

$$u_\zeta(\mathbf{q}) \equiv u_{\mathbf{k}\alpha\sigma}(\mathbf{q}) = (\mathbf{q}\,|\,\mathbf{k},\alpha,\sigma)\,. \tag{16.10-42}$$

The quantum-field description is required whenever the system is inhomogeneous.

---

58 E.N. Adams and T.D. Holstein, J. Phys. Chem. Solids **10**, 254 (1959)

59 P.N. Argyres and L.M. Roth, J. Phys. Chem. Solids **12**, 89 (1959).

60 V.S. Titeica, Ann. Physik **22**, 129 (1935).

61 C.M. Van Vliet, Ch.G. van Weert and A.H. Marshak , Physica **134 A**, 249 (1985).

With $\Psi(\mathbf{q})$ and $\Psi^\dagger(\mathbf{q})$ being the usual field operators, we have for the current flux

$$\mathbf{J}(\mathbf{q}) = \int d^\nu q' \Psi^\dagger(\mathbf{q}')\mathbf{j}(\mathbf{q},\mathbf{q}')\Psi(\mathbf{q}'), \tag{16.10-43}$$

where $\mathbf{j}$ is the Hermitized one-particle operator

$$\mathbf{j}(\mathbf{q},\mathbf{q}') = \frac{q}{2m^*}[\mathbf{p}\,\delta(\mathbf{q}-\mathbf{q}') + \delta(\mathbf{q}-\mathbf{q}')\mathbf{p}] - \frac{q^2}{m^* c}\mathbf{A}(\mathbf{q})\delta(\mathbf{q}-\mathbf{q}'), \tag{16.10-44}$$

where $\mathbf{A}$ is the vector potential. With $\mathbf{p} = (\hbar/i)\nabla_\mathbf{q}$ we have

$$\mathbf{J}(\mathbf{q}) = \frac{q\hbar}{2m^* i}\{[-\nabla_\mathbf{q}\Psi^\dagger(\mathbf{q})]\Psi(\mathbf{q}) + \Psi^\dagger(\mathbf{q})\nabla_\mathbf{q}\Psi(\mathbf{q})\} - \frac{q^2}{m^* c}\Psi^\dagger(\mathbf{q})\mathbf{A}(\mathbf{q})\Psi(\mathbf{q}). \tag{16.10-45}$$

Alternately, by expanding the field operators in one-particle wave functions, we find

$$\mathbf{J}(\mathbf{q}) = \sum_{\zeta,\zeta'} c^\dagger_{\zeta'} c_\zeta \left\{ \frac{q\hbar}{2m^* i}[u^*_{\zeta'}(\mathbf{q})\nabla_\mathbf{q} u_\zeta(\mathbf{q}) - u_\zeta(\mathbf{q})\nabla_\mathbf{q} u^*_{\zeta'}(\mathbf{q})] - \frac{q^2}{m^* c} u^*_{\zeta'}(\mathbf{q}) u_\zeta(\mathbf{q})\mathbf{A}(\mathbf{q}) \right\}. \tag{16.10-46}$$

Representing the states by plane waves $e^{i\mathbf{p}\cdot\mathbf{q}/\hbar}$, we have $\nabla_\mathbf{q} u_\zeta(\mathbf{q}) = (i/\hbar)\mathbf{p}\, u_\zeta(\mathbf{q})$. The two parts in (16.10-46) can now be recombined, writing

$$\hbar\mathbf{k} \equiv \boldsymbol{\pi} = \mathbf{p} - (q/c)\mathbf{A}(\mathbf{q}), \tag{16.10-47}$$

where $\boldsymbol{\pi}$ is the kinetic momentum. Restoring the other quantum indices of the states, we arrive at

$$\langle \mathbf{J}(\mathbf{q})\rangle_t = \mathrm{Tr}[\mathbf{J}(\mathbf{q})\rho(t)] = \frac{q\hbar}{2\tilde{m}} \sum_{\alpha\alpha'\sigma\sigma'} \sum_{\mathbf{k}\mathbf{k}'} \langle c^\dagger_{\mathbf{k}'\alpha'\sigma'} c_{\mathbf{k}\alpha\sigma}\rangle_t\, (\mathbf{k}+\mathbf{k}')[u^*_{\mathbf{k}'\alpha'\sigma'}(\mathbf{q}) u_{\mathbf{k}\alpha\sigma}(\mathbf{q})]. \tag{16.10-48}$$

In this form it looks like the effect of the magnetic field has disappeared; however, that is not so. First, the confinement associated with Landau states may modify the effective mass, now indicated by $\tilde{m}$. Secondly, the density of states in the direction $\mathbf{k}$ must be convoluted with the delta-function density of the transverse direction(s), which depend on the magnetic induction $\mathbf{B}$; see the basic article by Kahn and Frederikse (Ref. 2 of Chapter XIV).

In the true realm of quantum transport, e.g., when involving a strong magnetic field that leads to closed Landau 'orbits' – so-called asymptotic states – Eq.(16-10-48) must be evaluated employing a Wigner-function description. Treatises on the many-body Wigner function $\rho(\{p\},\{q\},t)$, which is the Weyl transform of the density operator $\rho(t)$, can be found in various quantum-theory texts; e.g. de Groot[62]

---

[62] S.R. de Groot, "La Transformation de Weyl et la fonction de Wigner : une forme alternative de la mécanique quantique". Les Presses de l'Université de Montréal, 1974.

and Zachos et al.[63] The function can be written in the form found with Balescu[64]

$$\rho(\{p\},\{q\},t)=\frac{1}{\hbar^{3N}N!}\int d^{3N}v\, e^{(i/\hbar)\sum_i^N \mathbf{p}_i\cdot\mathbf{v}_i}$$
$$\times\mathrm{Tr}\left\{\rho(t):\prod_j \Psi^\dagger(\mathbf{q}_j+\tfrac{1}{2}\mathbf{v}_j)\Psi(\mathbf{q}_j-\tfrac{1}{2}\mathbf{v}_j):\right\}, \qquad (16.10\text{-}49)$$

where $:\Pi:$ denotes a normal ordered product. Using normalized plane waves, we obtain the second quantization form

$$\rho(\{p\},\{q\},t)=\left(\frac{V_0}{8\pi^3\hbar^3}\right)^N\frac{1}{N!}\int d^{3N}u\, e^{i\mathbf{q}\cdot\mathbf{u}}\,\mathrm{Tr}\left\{\rho(t):\prod_j c^\dagger_{\mathbf{k}_j-\frac{1}{2}\mathbf{u}_j}c_{\mathbf{k}_j+\frac{1}{2}\mathbf{u}_j}:\right\}. \qquad (16.10\text{-}50)$$

Although p and q do not commute, this is the equivalent of the classical phase-space distribution.

For the present problem with the total current flux **J** being the sum of one-particle operators **j**, it suffices to employ the one-particle Wigner function. Setting $N = 1$ and restricting ourselves to ν degrees of freedom, we have

$$\rho_1(\mathbf{k},\mathbf{q},t)=\frac{V_\nu}{(2\pi)^\nu\hbar^\nu}\int d^\nu u\, e^{i\mathbf{q}\cdot\mathbf{u}}\langle c^\dagger_{\mathbf{k}-\mathbf{u}/2}c_{\mathbf{k}+\mathbf{u}/2}\rangle_t\,, \qquad (16.10\text{-}51)$$

with inversion

$$\langle c^\dagger_{\mathbf{k}-\mathbf{u}/2}c_{\mathbf{k}+\mathbf{u}/2}\rangle_t=\frac{h^\nu}{V_\nu}\int d^\nu q\, e^{-i\mathbf{q}\cdot\mathbf{u}}\rho_1(\mathbf{k},\mathbf{q},t). \qquad (16.10\text{-}52)$$

Now in (16.10-48) we replace the summation variables in **k**-space by **K** and **K**', and we make the transformation

$$\mathbf{k}-\tfrac{1}{2}\mathbf{u}=\mathbf{K}', \qquad \mathbf{k}+\tfrac{1}{2}\mathbf{u}=\mathbf{K}. \qquad (16.10\text{-}53)$$

Averaging over the confined coordinates and over the spin gives factors $\delta_{\alpha\alpha'}\delta_{\sigma\sigma'}$. Also we write $\Sigma_\mathbf{u}\to[V_\nu/(2\pi)^\nu]\int d^\nu u$ and we note that

$$\int d^\nu u\, e^{i\mathbf{u}\cdot(\mathbf{q}-\mathbf{q}')}=(2\pi)^\nu\,\delta(\mathbf{q}-\mathbf{q}'). \qquad (16.10\text{-}54)$$

So we obtain

$$\langle\mathbf{J}(\mathbf{q})\rangle_t=\frac{q\hbar}{\tilde{m}V_0}\sum_{\alpha\sigma}\sum_{\mathbf{k}}\mathbf{k}[h^\nu\rho_{1,\alpha\sigma}(\mathbf{k},\mathbf{q},t). \qquad (16.10\text{-}55)$$

We note that h^ν is the volume of a microcell in μ-space [cf. (2.4-19)], $\Delta^\nu p\,\Delta^\nu q$. Since

[63] C.K. Zachos, D.B. Fairlie and Th.L. Curtright, "Quantum Mechanics in Phase Space", World Scientific Publ. Co, NJ and Singapore, 2005.
[64] R. Balescu, "Statistical Mechanics of Charged Particles", Interscience, NY 1963, p. 292ff.

ρ_1 is not necessarily positive definite, we shall replace the quantity $h^\nu\rho_1$ by the "*Wigner occupancy function*" f, obtained by coarse-graining over a microcell,

$$f_{\alpha\sigma}(\mathbf{k},\mathbf{q},t) = \int_{\omega(\mathbf{p},\mathbf{q})=h^\nu} d\varpi\rho_1(\overline{\mathbf{k}},\overline{\mathbf{q}},t)\,. \tag{16.10-56}$$

For the electrical flux this yields

$$\langle\mathbf{J}(\mathbf{q})\rangle_t = \frac{q\hbar}{\tilde{m}V_0}\sum_{\alpha\sigma}\frac{V_\nu}{(2\pi)^\nu}\int d^\nu k\mathbf{k}\, f_{\alpha\sigma}(\mathbf{k},\mathbf{q},t)\,. \tag{16.10-57}$$

While this result could have been anticipated, it has now been justified by the quantum-field and Wigner-function procedure. We still notice that both $\tilde{m}$ and Σ_α generally depend on the magnetic induction **B**, if the confinement is due to localised Landau states. Examples will be presented in the next chapter.

16.10.4 *Consequences for the Many-Body Response Formulae*

Employing the reduced operators, we shall now obtain new convergent response formulas for quantum transport; the results sought for here are *in many-body form.* Our derivations were first reported in LRT I, Part C, although we confined ourselves there to the diagonal contributions. Then, as in the original Kubo theory, exemplified by the approaches of Chester and Thellung and Verboven discussed in Section 16.7, one may look for one-body response formulae, i.e., we could consider the equivalent results on the one-particle level; this was indeed carried out in LRT III.[65] However, the results obtained there are fully equivalent to similar results based on the quantum Boltzmann equation (QBE), to be set forth in Chapter XVII. We shall therefore in this subsection restrict ourselves to many-body results which, although not directly amenable to applications, serve to show the convergence of the modified Kubo forms as well as the dissipative nature of the conductances and susceptances so obtained.

We reiterate the Kubo formulae for the generalized susceptances, conductances and the electrical conductivity tensor, adding the superscript 'R' to the operators,

$$\chi_{BA}(i\omega) = \int_0^\infty dt\, e^{-i\omega t}\int_0^\beta d\tau \mathrm{Tr}\{\rho_{eq}\dot{A}^R(-i\hbar\tau)B^R(t)\}\,, \tag{16.10-58}$$

$$L_{BA}(i\omega) = \int_0^\infty dt\, e^{-i\omega t}\int_0^\beta d\tau \mathrm{Tr}\{\rho_{eq}\dot{A}^R(-i\hbar\tau)\dot{B}^R(t)\}\,, \tag{16.10-59}$$

$$\sigma_{\mu\nu}(i\omega) = V_0\int_0^\infty dt\, e^{-i\omega t}\int_0^\beta d\tau \mathrm{Tr}\{\rho_{eq}J_\nu^R(-i\hbar\tau)J_\mu^R(t)\}\,. \tag{16.10-60}$$

---

[65] M. Charbonneau, K.M. Van Vliet and P. Vasilopoulos, "Linear Response Theory Revisited III: One-Body Response Formulas and Generalized Boltzmann Equations", J. Math. Phys. **23**, 318-336 (1982).

Suppose this is evaluated in the representation $\{|\gamma\rangle\}$. Then the trace in the first formula reads:

$$\mathrm{Tr}\{\rho_{eq}\dot{A}^R(-i\hbar\tau)B^R(t)\} = \sum_{\gamma,\gamma''} p_{eq}(\gamma)\langle\gamma|\dot{A}^R(-i\hbar\tau)|\gamma''\rangle\langle\gamma''|B^R(t)|\gamma\rangle. \quad (16.10\text{-}61)$$

Now if one of the two operators is diagonal, the other one must be necessarily be also and *vice versa.* It is therefore not useful to find a 'full' result, but the expressions suggest that we consider either the diagonal contribution or the nondiagonal contribution. Moreover, it must be borne in mind that for $t \to 0$ $B^R(0)$ equals the Schrödinger operator, while $\dot{A}^R(0)$ is *more* than the Schrödinger operator, since we must add the collisional current of the previous subsection.

(a) *Diagonal response results*

We first note that

$$\dot{A}_d^R(-i\hbar\tau) = e^{\tau\mathcal{H}^0}\dot{A}_d^R(0)e^{-\tau\mathcal{H}^0} = \dot{A}_d(0). \quad (16.10\text{-}62)$$

This yields for the diagonal susceptance

$$\chi_{BA}^d(i\omega) = \beta\int_0^\infty dte^{-i\omega t}\,\mathrm{Tr}\{\rho_{eq}\dot{A}_d^R(0)B_d^R(t)\} = \beta\int_0^\infty dte^{-i\omega t}\,\mathrm{Tr}\{\rho_{eq}(\dot{A}^R)_d e^{-\Lambda_d t}B_d\}$$

$$= \beta\mathrm{Tr}\left\{\rho_{eq}(\dot{A}^R)_d \frac{1}{i\omega+\Lambda_d}B_d\right\}, \quad (16.10\text{-}63)$$

and for the conductances

$$L_{BA}^d(i\omega) = \beta\mathrm{Tr}\left\{\rho_{eq}(\dot{A}^R)_d \frac{1}{i\omega+\Lambda_d}\dot{B}_d\right\}, \quad \sigma_{\mu\nu}^d(i\omega) = V_0\beta\mathrm{Tr}\left\{\rho_{eq}J_{\mu,d}\frac{1}{i\omega+\Lambda_d}J_{\nu,d}\right\}.$$

(10.6-64,65)

These expressions involve the resolvent of the master operator Λ_d.

For the fluctuation-dissipation theorem we obtain similar expressions. But, since we need the two-sided Fourier transform, the interval of integration must be split and we employ the transposition property. The results then involve the resolvents $(\pm i\omega + \Lambda_d)^{-1}$; the details are left to the reader. The expressions are convergent, since the operator Λ_d is positive semi-definite.

(b) *Nondiagonal response results*

The nondiagonal response contributes at all frequencies, including quantum frequencies. As illustration we consider only the result for $L_{BA}^{nd}(i\omega)$. We then have

$$L_{BA}^{nd}(i\omega) = \int_0^\infty dt\,e^{-i\omega t}\int_0^\beta d\tau\,\mathrm{Tr}\{\rho_{eq}\dot{A}_{nd}^R(-i\hbar\tau)\dot{B}_{nd}^R(t)\}. \quad (16.10\text{-}66)$$

For the nondiagonal parts, note that $A_{nd}^R(0) \equiv (A^R)_{nd} = A_{nd}$, the latter being the Schrödinger operator, since the collisional current is diagonal. In the representation $\{|\gamma\rangle\}$ we have

$$L_{BA}^{nd}(i\omega) = \sum_{\gamma,\bar{\gamma}} \int_0^\infty dt\, e^{-i\omega t} \int_0^\beta d\tau\, p_{eq}(\gamma) \langle \gamma | e^{\tau \mathcal{H}^0} \dot{A}_{nd}\, e^{-\tau \mathcal{H}^0} | \bar{\gamma} \rangle \langle \bar{\gamma} | e^{i\mathcal{H}^0 t/\hbar} \dot{B}_{nd} e^{-i\mathcal{H}^0 t/\hbar} | \gamma \rangle \}$$

$$= \sum_{\gamma,\bar{\gamma}} \int_0^\infty dt\, e^{-i\omega t} e^{-it(\mathcal{E}_\gamma - \mathcal{E}_{\bar{\gamma}})/\hbar} \int_0^\beta d\tau e^{\tau(\mathcal{E}_\gamma - \mathcal{E}_{\bar{\gamma}})} p_{eq}(\gamma) \langle \gamma | \dot{A}_{nd} | \bar{\gamma} \rangle \langle \bar{\gamma} | \dot{B}_{nd} | \gamma \rangle \} \,. \quad (16.10\text{-}67)$$

The τ-integration can be carried out. For the t-integration we add a convergence factor $\exp(-\eta t)$, $\eta > 0$. This yields

$$L_{BA}^{nd}(i\omega) = \lim_{\eta \to 0} \sum_{\gamma\bar{\gamma}} p_{eq}(\gamma) \langle \gamma | \dot{A}_{nd} | \bar{\gamma} \rangle \langle \bar{\gamma} | \dot{B}_{nd} | \gamma \rangle \frac{e^{\beta(\mathcal{E}_\gamma - \mathcal{E}_{\bar{\gamma}})} - 1}{\mathcal{E}_\gamma - \mathcal{E}_{\bar{\gamma}}} \frac{-i\hbar}{\mathcal{E}_\gamma - \mathcal{E}_{\bar{\gamma}} + \hbar\omega - i\hbar\eta}$$

$$= \sum_{\gamma\bar{\gamma}} p_{eq}(\gamma) \langle \gamma | \dot{A}_{nd} | \bar{\gamma} \rangle \langle \bar{\gamma} | \dot{B}_{nd} | \gamma \rangle \frac{1 - e^{-\beta(\mathcal{E}_{\bar{\gamma}} - \mathcal{E}_\gamma)}}{\mathcal{E}_{\bar{\gamma}} - \mathcal{E}_\gamma}$$

$$\times \left(i\hbar \mathcal{P} \frac{1}{\mathcal{E}_{\bar{\gamma}} - \mathcal{E}_\gamma - \hbar\omega} - \pi\hbar\delta(\mathcal{E}_{\bar{\gamma}} - \mathcal{E}_\gamma - \hbar\omega) \right). \quad (16.10\text{-}68)$$

We note once more that the equivalent one-body forms of all results in this subsection can readily be obtained by writing the operators $\dot{A}$, B and $\dot{B}$ in second quantization form. For the diagonal part we must connect Λ_d with the Boltzmann operator $\mathcal{B}$, which is a complex task; it will be carried out in Chapter XVII.
Note. A great many researchers, however, *equate* Λ_d with the Boltzmann operator directly, i.e., they employ the master equation on the one-body level; this is an over simplification, since the final Boltzmann operator must be quadratic in the distribution function f, while Λ_d, being by nature a linear operator, can for that type of application only give a collision operator that is linear in f.

16.11 The Pauli–Van Hove Master Equation

The projector-operator method, initiated by Zwanzig[53], which we have already employed extensively, was originally devised for obtaining the Pauli master equation of 1928 via the Van Hove limit. Before we go to the derivation, it is essential to understand the nature of the largest sub-Hamiltonian $\mathcal{H}^0$. Since we do *not*, *a priori*, look for results that are linear in the applied fields – in contrast to the premises of linear response theory – The Hamiltonian $\mathcal{H}^0$ will *contain* these fields. i.e.,

$$\mathcal{H}^0 = \mathcal{H}^0(system(\mathbf{E}, \mathbf{B})) \,. \quad (16.11\text{-}1)$$

For example, in dealing with otherwise free electrons in an E-M field, we have

$$\mathcal{H}^0 = \sum_{\text{particles}} \{[\mathbf{p}_i - (q/c)\mathbf{A}(\mathbf{q}_i)]^2 + q\Phi(\mathbf{q}_i)\}. \qquad (16.11\text{-}2)$$

As to the interaction $\lambda\mathcal{V}$, it may also be field-dependent, as exemplified by the Poole–Frenkel effect, but that changes little in the formalism.

With this in mind, let us apply the projectors $\mathcal{P}$ and $1 - \mathcal{P}$ to the von Neumann equation; the derivation goes essentially parallel to that of subsection 16.10.2:

$$\begin{aligned} &\frac{\partial \rho_d}{\partial t} + i\mathcal{P}\mathcal{L}\rho_d + i\mathcal{P}\mathcal{L}\rho_{nd} = 0, \\ &\frac{\partial \rho_{nd}}{\partial t} + i(1-\mathcal{P})\mathcal{L}\rho_d + i(1-\mathcal{P})\mathcal{L}\rho_{nd} = 0. \end{aligned} \qquad (16.11\text{-}3)$$

We note that we used partials for the time derivative, since the density operator is a Schrödinger operator. Now $\mathcal{P}\mathcal{L}\rho_d = \mathcal{P}\mathcal{L}^0\rho_d + (\lambda/\hbar)\mathcal{P}[V,\rho_d]$. The first term is obviously zero since $\mathcal{H}^0$ and ρ_d commute. As to the second term, with regard to our previous Theorem 5, we note that any two-body interaction potential is invariant with respect to translation over the volume of the system; so there is no bilinear concomitant and $\mathcal{P}$ can be brought inside the commutator, giving zero. The equations (16.11-3) can thus be rewritten as

$$\begin{aligned} &\frac{\partial \rho_d}{\partial t} + i\mathcal{P}\mathcal{L}\rho_{nd} = 0, \\ &\frac{\partial \rho_{nd}}{\partial t} + i\mathcal{L}\rho_d + i(1-\mathcal{P})\mathcal{L}\rho_{nd} = 0. \end{aligned} \qquad (16.11\text{-}4)$$

The solution of the second equation is for $t > 0$,

$$\rho_{nd}(t) = -i\int_0^t dt' G(t-t')\mathcal{L}\rho_d(t') + G(t,0)\rho_{nd}(0), \qquad (16.11\text{-}5)$$

where G is the Green's operator (16.10-22). Substituting into the first equation, we have the integro-differential equation,[66]

$$\frac{\partial \rho_d}{\partial t} + \mathcal{P}\mathcal{L}\int_0^t dt' G(t-t')\mathcal{L}\rho_d(t') + i\mathcal{P}\mathcal{L}G(t,0)(1-\mathcal{P})\rho(0) = 0. \qquad (16.11\text{-}6)$$

Using Theorem 3, the factor $(1 - \mathcal{P})$ of the last term can be moved to the left, where it is destroyed by the projector $\mathcal{P}$. So, as for the Heisenberg equation, *no* initial random phase assumption, requiring $\rho(0)$ to be diagonal, is imposed. The above two equations (16.11-5) and (16.11-6) are still exact.

---

[66] This is a 'generalized' (non-Markovian) master equation. Similar results have been given by the Brussels group of Prigogine, but require a special diagrammatic technique; see Ref 64 and I. Prigogine and P. Résibois, Physica **27**, 629 (1961).

Next, we approximate $G(t-t')$ by $G^0(t-t')=\Theta(t-t')\exp[-i(t-t')\mathcal{L}^0]$, cf. Eq. (16.10-25). The integro-differential equation (16.11-6) becomes

$$\frac{\partial\rho_d}{\partial t}+\mathcal{P}\mathcal{L}\int_0^t dt' e^{-i(t-t')\mathcal{L}^0}\mathcal{L}\rho_d(t')$$
$$=\frac{\partial\rho_d}{\partial t}+\frac{\lambda^2}{\hbar^2}\mathcal{P}[\mathcal{V},\int_0^t dt' e^{-i(t-t')\mathcal{L}^0}[\mathcal{V},\rho_d(t')]]=0\,, \qquad (16.11\text{-}7)$$

where we noted that the contributions $\mathcal{P}\mathcal{L}^0...\mathcal{L}\rho_d$, $\mathcal{P}\mathcal{L}...\mathcal{L}^0\rho_d$ and $\mathcal{P}\mathcal{L}^0...\mathcal{L}^0\rho_d$ yield zero. (For the second and third part, this is obvious, since ρ_d and $\mathcal{H}^0$ commute; for the first part, we have $(\lambda/\hbar)\mathcal{P}\mathcal{L}^0...[\mathcal{V},\rho_d]=(\lambda/\hbar)\mathcal{L}^0...[\mathcal{P}\mathcal{V},\rho_d]=0$, where we carried $\mathcal{P}$ through to the right and used that $\mathcal{V}$ is translationally invariant.)

The remaining derivation goes as for the Heisenberg equation. So, writing the diagonal projector in its explicit form (16.10-3), we obtain, employing Theorem 1,

$$\frac{\lambda^2}{\hbar^2}\mathcal{P}[\mathcal{V},\int_0^t dt' e^{-i(t-t')\mathcal{L}^0}[\mathcal{V},\rho_d(t')]]$$
$$=\frac{\lambda^2}{\hbar^2}\int_0^t dt'\sum_\gamma|\gamma\rangle\langle\gamma|[\mathcal{V},e^{-i(t-t')\mathcal{H}^0/\hbar}[\mathcal{V},\rho_d(t')]e^{i(t-t')\mathcal{H}^0/\hbar}]|\gamma\rangle\langle\gamma|. \qquad (16.11\text{-}8)$$

For the matrix element under the integral we have

$$\langle\gamma|[V,..[V,\rho_d]..]|\gamma\rangle$$
$$=\sum_{\gamma''}\{\langle\gamma|\mathcal{V}|\gamma''\rangle e^{-i\mathcal{E}_{\gamma''}(t-t')/\hbar}[\langle\gamma''|\mathcal{V}|\gamma\rangle p(\gamma,t')-p(\gamma'',t')\langle\gamma''|\mathcal{V}|\gamma\rangle]e^{i\mathcal{E}_\gamma(t-t')/\hbar}$$
$$-e^{-i\mathcal{E}_\gamma(t-t')/\hbar}[\langle\gamma|\mathcal{V}|\gamma''\rangle p(\gamma'',t')-p(\gamma,t')\langle\gamma|\mathcal{V}|\gamma''\rangle]e^{i\mathcal{E}_{\gamma''}(t-t')/\hbar}\langle\gamma''|V|\gamma\rangle$$
$$=2\sum_{\gamma''}|\langle\gamma|\mathcal{V}|\gamma''\rangle|^2\cos[(\mathcal{E}_{\gamma''}-\mathcal{E}_\gamma)(t-t')/\hbar][p(\gamma,t')-p(\gamma'',t')]\,. \qquad (16.11\text{-}9)$$

We shall denote the lhs of (16.11-8) by $-(\partial\rho_d/\partial t)_{mem.}$; then, taking the Laplace transform and using the convolution theorem, we have for the transform

$$\Psi(s)=-\sum_{\gamma,\gamma''}|\gamma\rangle\langle\gamma|\left(\frac{2\lambda^2}{\hbar^2}\right)|\langle\gamma|\mathcal{V}|\gamma''\rangle|^2\frac{s}{(\mathcal{E}_{\gamma''}-\mathcal{E}_\gamma)^2/\hbar^2+s^2}[P(\gamma,s)-P(\gamma'',s)]\,.$$
$$(16.11\text{-}10)$$

The van Hove limit $t\to\infty$ means $s\to 0+$. The limit was carried out in (16.10-31), so

$$\Psi(s)=-\sum_{\gamma,\gamma''}|\gamma\rangle\langle\gamma|\left(\frac{2\pi\lambda^2}{\hbar}\right)|\langle\gamma|\mathcal{V}|\gamma''\rangle|^2\delta(\mathcal{E}_{\gamma''}-\mathcal{E}_\gamma)[P(\gamma,s)-P(\gamma'',s)]\,. \quad (16.11\text{-}11)$$

The inverse transform yields the master operator with $(\partial\rho_d/\partial t)_{mem.}=-\Lambda_d\rho_d(t)$. The final result for (16.11-7) is the Pauli–Van Hove master equation

$$\frac{\partial \rho_d}{\partial t} + \Lambda_d \rho_d(t) = 0. \tag{16.11-12}$$

This is the compact form for the description in Liouville space. It is readily converted to the more usual form in the γ-function space for $\langle \gamma | \rho_d(t) | \gamma \rangle = p(\gamma,t)$:

$$\frac{\partial p(\gamma,t)}{\partial t} + M[p(\gamma,t)] = 0, \tag{16.11-13}$$

or

$$\frac{\partial p(\gamma,t)}{\partial t} = \sum_{\gamma''} \{ W_{\gamma''\gamma} p(\gamma'',t) - W_{\gamma\gamma''} p(\gamma,t) \}. \tag{16.11-14}$$

The end result here is a Markovian equation, in which the memory is erased, except for one instant, that may be used to provide an initial condition.

As we noted already, the ME is important for stochastic processes, to be discussed in Part F of this text. Here, however, we are interested in quantum transport quantities, such as the susceptibility and the conductivity of systems with extended states or confined .i.e., localised states. Some four years after his original derivation of the master equation (cf. Ref. 44), at the 1959 summer school at Les Houches Van Hove showed that the master equation (16.11-14) leads in a straightforward fashion to a quantum mechanical form of the Boltzmann equation.[67] The master operator M then yields the quantum analogue of the collision integral. However, no streaming terms appear on the scene! Since then, the author and collaborators have conclusively shown that this *échec* is solely due to the neglect of the nondiagonal part. Hence, the ME (16.11-12) is manifestly incomplete.

16.12 The Full Master Equation (FME)

Two methods may be pursued in order to obtain the streaming terms, missing in Van Hove's treatment. Firstly, we can add the result for the nondiagonal part, which has been already obtained, but which has been (unjustifiably) 'thrown away'. We differentiate the non-diagonal result (16.11-5), replacing the Green's operator G by the operator G^0 (differentiation to the upper limit t of the integral gives no contribution if we make an initial random phase assumption $\rho_{nd} = 0$). We thus find

$$\frac{\partial \rho_{nd}}{\partial t} + i\mathcal{L}^0 \rho_{nd}(t) = 0. \tag{16.12-1}$$

Clearly, we can change the second term on the left to $i\mathcal{L}^0 \rho(t)$ since $i\mathcal{L}^0 \rho_d(t) = 0$. Also, in (16.11-12) we can replace $\Lambda_d \rho_d(t)$ by $\Lambda_d \rho(t)$. Adding the thus modified

---

[67] L. Van Hove in "La théorie des gaz neutres et ionisés", École d'été en physique théorique, Les Houches 1959, C. De Witt and J.-F. Detoeuf, Eds., Hermann, Paris, 1960.

equations (16.12-1) and (16.11-12) we establish the Pauli–Van Hove–Van Vliet *full master equation* (FME):

$$\frac{\partial \rho}{\partial t} + (\Lambda_d + i\mathcal{L}^0)\rho(t) = 0. \qquad (16.12\text{-}2)$$

We note the symmetry with the reduced Heisenberg equation, (16.10-37). The reduced operators can be summarised by

$$\lim_{\lambda,t} e^{\pm i\mathcal{L}t} = e^{(-\Lambda_d \pm i\mathcal{L}^0)t}. \qquad (16.12\text{-}3)$$

In these results both Λ_d and (certainly) $\mathcal{L}^0$ may depend on an externally applied field. The challenge is now to extract the streaming terms from the field-dependent operators Λ_d and $i\mathcal{L}^0$. In its generality this has not been carried out. However, in the next chapter we shall show that for *extended states* the quantum Boltzmann equation for the Wigner distribution derived from (16.12-2) contains the usual two streaming terms. For *localised states* some problems remain, in particular for the stationary state, as will be discussed there.

Secondly, we can start with a different Hamiltonian in the Liouville equation, a program carried out in LRT II.[68] Suppose that in the spirit of linear response theory the field Hamiltonian is split off *ab initio*, i.e., we write

$$\mathcal{H}_{total} = \bar{\mathcal{H}}^0(system(\mathbf{A})) + \lambda\mathcal{V} - \sum\nolimits_i q_i(\mathbf{r}_i - \mathbf{r}_{i,0})\cdot\mathbf{E}(t), \qquad (16.12\text{-}4)$$

where the effect of the electric field has been taken out of $\mathcal{H}^0$; the latter may still depend on the vector potential as usual. We now seek a new result for $\partial\rho/\partial t$. More generally we write

$$\mathcal{H}_{total} = \bar{\mathcal{H}}^0 + \lambda\mathcal{V} - AF(t). \qquad (16.12\text{-}5)$$

This gives the von Neumann equation

$$\frac{\partial \rho}{\partial t} + i\bar{\mathcal{L}}^0\rho(t) + (i\lambda/\hbar)[\mathcal{V},\rho(t)] = (i/\hbar)[A,\rho(t)]F(t). \qquad (16.12\text{-}6)$$

Note that we used now an overhead bar for the interaction Liouvillian based on the Hamiltonian $\bar{\mathcal{H}}^0$ from which the field has been split off. The term on the right can only easily be evaluated, if we apply Kubo's linearization procedure and replace the density operator at the rhs by ρ_{eq}; the commutator $[A, \rho_{eq}]$ can then be obtained with Kubo's Lemma, (16.3-19). The next step is to apply the projection operators $\mathcal{P}$ and $1-\mathcal{P}$ to all terms on both sides of (16.12-6), retaining the diagonal as well as the nondiagonal parts. We shall do the computation at the end of this section; here we just state the result,

---

[68] K.M. Van Vliet, "Linear Response Theory Revisited II. The Master Equation Approach", J. Math. Phys. **20**, 2573-2595 (1979), Sections 3 and 4.

$$\frac{\partial \rho}{\partial t} + (\overline{\Lambda}_d + i\overline{\mathcal{L}}^0)\rho(t) = F(t)\rho_{eq}\int_0^\beta d\tau\, e^{\hbar\tau\overline{\mathcal{L}}^0}\dot{A}^R\,, \tag{16.12-7}$$

where the rhs contains *two* streaming terms, since $\dot{A}^R$ contains both ponderomotive and collisional current. Equation (16.12-4), which henceforth will be referred to as the '*inhomogeneous master equation*' (IME), has been very useful, since it gives a quantum Boltzmann equation (see next chapter) with Van Hove's collision operator, as well as all streaming terms. Main drawback: the transport coefficients so obtained only apply to the linear (or low-field) regime of a near-equilibrium system.

Derivation of the inhomogeneous master equation

Upon applying the projectors $\mathcal{P}$ and $1-\mathcal{P}$ to the von Neumann equation (16.12-6), we have the two equivalent equations

$$\begin{aligned}&\frac{\partial \rho_d}{\partial t} + i\mathcal{P}\overline{\mathcal{L}}\rho_d + i\mathcal{P}\overline{\mathcal{L}}\rho_{nd} = \frac{i}{\hbar}F(t)\mathcal{P}[A,\rho]\,,\\ &\frac{\partial \rho_{nd}}{\partial t} + i(1-\mathcal{P})\overline{\mathcal{L}}\rho_d + i(1-\mathcal{P})\overline{\mathcal{L}}\rho_{nd} = \frac{i}{\hbar}F(t)(1-\mathcal{P})[A,\rho]\,.\end{aligned} \tag{16.12-8}$$

With the same reductions as in Section 16.11, this simplifies to

$$\begin{aligned}&\frac{\partial \rho_d}{\partial t} + i\mathcal{P}\overline{\mathcal{L}}\rho_{nd} = \frac{i}{\hbar}F(t)\mathcal{P}[A,\rho]\,,\\ &\frac{\partial \rho_{nd}}{\partial t} + i\overline{\mathcal{L}}\rho_d + i(1-\mathcal{P})\overline{\mathcal{L}}\rho_{nd} = \frac{i}{\hbar}F(t)(1-\mathcal{P})[A,\rho]\,.\end{aligned} \tag{16.12-9}$$

Using the Green's operator $\overline{G}(t,t')$,analogous to (16.10-22), the equation for ρ_{nd} is formally solved,

$$\rho_{nd}(t) = \int_0^t dt'\,\overline{G}(t-t')\{-i\overline{\mathcal{L}}\rho_d(t') + (i/\hbar)F(t')(1-\mathcal{P})[A,\rho(t')]\} + \overline{G}(t,0)\rho_{nd}(0)\,. \tag{16.12-10}$$

When this is substituted in the first equation of (16.12-9) we obtain the integro-differential equation

$$\begin{aligned}&\frac{\partial \rho_d}{\partial t} + \mathcal{P}\overline{\mathcal{L}}\int_0^t dt'\,\overline{G}(t-t')\overline{\mathcal{L}}\rho_d(t')\\ &= \frac{i}{\hbar}F(t)\mathcal{P}[A,\rho_{eq}] + \frac{i}{\hbar}\mathcal{P}\overline{\mathcal{L}}\int_0^t dt'F(t')\overline{G}(t-t')(1-\mathcal{P})[A,\rho_{eq}]\,.\end{aligned} \tag{16.12-11}$$

Note that on the rhs we made the Kubo assumption of LRT, replacing $\rho(t)$ by ρ_{eq}. The lhs is entirely similar to (16.11-6); hence, application of the Van Hove limit yields $(\overline{\Lambda}_d + i\overline{\mathcal{L}}^0)\rho(t)$. On the rhs two streaming terms occur. For conductivity problems, $\dot{A}$ is related to the momentum, which is translationally invariant; we can therefore bring the projection operator $\mathcal{P}$ inside the commutator bracket. With Kubo's lemma, we find right away

$$\text{first str. term} = \beta F(t)\rho_{eq}(\dot{A})_d\,, \tag{16.12-12}$$

where $\rho_{eq} \to \exp(-\beta\overline{\mathcal{H}}^0)/\mathcal{Z}$. For the second streaming term one ends up with

$$(\lambda/\hbar^2)\int_0^t dt'F(t')\mathcal{P}[V, e^{-i\overline{\mathcal{L}}^0(t-t')}(1-\mathcal{P})\int_0^\beta d\tau\,\rho_{eq}[A(-i\hbar\tau),\overline{\mathcal{H}}]\,. \tag{16.12-13}$$

The important factor is the $(1-\mathcal{P})$ in front of the last commutator. We consider separately the '1' and the '$-\mathcal{P}$'. For the former, the commutator integral is approximately $\beta[A,\mathcal{H}^0]$, which is essentially unbounded. We will assume that it averages to zero, when acted upon by $\exp[-i\overline{\mathcal{L}}^0(t-t')]$. For the other part we have

$$-\mathcal{P}\int_0^\beta d\tau\,\rho_{eq}[A(-i\hbar\tau),\overline{\mathcal{H}}]=-\hbar i\int_0^\beta d\tau\,\rho_{eq}(\dot{A}(-i\hbar\tau))_d$$
$$=-\int_0^\beta d\tau\,\rho_{eq}[(A^I(-i\hbar\tau))_d,\overline{\mathcal{H}}]=\beta\lambda\rho_{eq}[V,A_d]. \qquad (16.12\text{-}14)$$

Note that we first evaluated the commutator with the Heisenberg equation and then took its diagonal part, after which we went back to the commutator form. Writing out the projector $\mathcal{P}$ in front of (16.12-13) explicitly, we arrive at

$$\text{second str. term}=-\left(\frac{\lambda}{\hbar}\right)^2\beta\rho_{eq}\sum_\gamma|\gamma\rangle\langle\gamma|\int_0^t dt'\,F(t')[V,e^{-i\overline{\mathcal{L}}^0(t-t')}[V,A_d]]|\gamma\rangle. \qquad (16.12\text{-}15)$$

The further evaluation proceeds with Laplace transformation, as in previous cases. One obtains $-\overline{\Lambda}_d A_d$. The diagonal result is thus found to be

$$\frac{\partial\rho_d}{\partial t}+\overline{\Lambda}_d\rho_d(t)=\beta F(t)\rho_{eq}[(\dot{A})_d-\overline{\Lambda}_d A_d]. \qquad (16.12\text{-}16)$$

Having now found the reduced diagonal result, we return to the equation for the nondiagonal part ρ_{nd} given in (16.12-10). This is differentiated, next $\rho_{nd}(t)$ is replaced by ρ_{eq} and the streaming term is found with Kubo's lemma. The nondiagonal equation then becomes

$$\frac{\partial\rho_{nd}}{\partial t}+i\overline{\mathcal{L}}^0\rho_{nd}=F(t)\rho_{eq}\int_0^\beta d\tau\,\dot{A}(-i\hbar\tau)_{nd}. \qquad (16.12\text{-}17)$$

Finally, the two parts can be combined, to yield

$$\frac{\partial\rho}{\partial t}+(\Lambda_d+i\overline{\mathcal{L}}^0)\rho=F(t)\rho_{eq}\int_0^\beta d\tau\,\dot{A}^R(-i\hbar\tau), \qquad (16.12\text{-}18)$$

which is (16.12-4). We note that the reduced operator (also written as J_A) contains the diagonal and nondiagonal ponderomotive current, as well as the collisional current.

16.13 Approach to Equilibrium

It was previously shown that the entropy production, as computed from the original von Neumann equation, is zero; hence there is no trend towards equilibrium; see Section 3.8. On the contrary, the diagonal part of $\rho(t)$, after the Van Hove limit, gives a non-zero result away from equilibrium, as we will now show. Quite generally from $S=-k_B\,\mathrm{Tr}(\rho\ln\rho)$ we have

$$\frac{dS_d}{dt}=-k_B\mathrm{Tr}\{[1+\ln\rho_d]\frac{\partial\rho_d}{\partial t}=-k_B\iint d\mathcal{E}d\mathcal{E}'\sum_{\alpha,\alpha'}[1+\ln p(\mathcal{E},\alpha)]W_{\gamma'\gamma}[p(\mathcal{E}',\alpha')-p(\mathcal{E},\alpha)]$$
$$=-(k_B/2)\iint d\mathcal{E}d\mathcal{E}'\sum_{\alpha,\alpha'}[\ln p(\mathcal{E},\alpha)-\ln p(\mathcal{E}',\alpha')]W_{\gamma'\gamma}[p(\mathcal{E}',\alpha')-p(\mathcal{E},\alpha)], \quad (16.13\text{-}1)$$

where we interchanged the primed and unprimed quantities and took half the sum. Next, with $W_{\gamma\gamma'} = (2\pi\lambda^2/\hbar)|\langle\mathcal{E}\alpha|\mathcal{V}|\mathcal{E}'\alpha'\rangle|^2\delta(\mathcal{E}-\mathcal{E}')$, we integrate out the delta function,

$$\frac{dS_d}{dt} = -\frac{k_B\pi\lambda^2}{\hbar} \times\int d\mathcal{E}\sum_{\alpha,\alpha'}|\langle\mathcal{E}\alpha|\mathcal{V}|\mathcal{E}\alpha'\rangle|^2\sum_{\alpha,\alpha'}[\ln p(\mathcal{E},\alpha)-\ln p(\mathcal{E},\alpha')]\,[p(\mathcal{E},\alpha')-p(\mathcal{E},\alpha)]\geq 0\,, \tag{16.13-2}$$

by Klein's lemma. The entropy production is therefore semi-positive definite, being zero only in equilibrium, for which

$$p(\mathcal{E},\alpha) = p(\mathcal{E},\alpha')\,. \tag{16.13-3}$$

Note that the remaining integration in (16.13-2) is over a range of energy cells. Within a cell, each state is equally probable, in accord with the canonical distribution.

The results of this section uphold the interpretation of the separation in diagonal and nondiagonal parts presented in this chapter. Only the diagonal parts give rise to dissipation. As to the nondiagonal contributions, we note that these are non-dissipative 'quantum-interference' terms.

16.14 The Onsager–Casimir Reciprocity Relations

The symmetry relations involving time reversal, coupled with magnetic field reversal, and the ensuing Onsager–Casimir relations for the transport coefficients, will now be derived in a general fashion. Whereas we could work with the full reduced variables for the problem, this is unnecessarily cumbersome. So, as in subsection 16.10.4, we shall consider the diagonal and nondiagonal correlation functions separately. The proof for the diagonal contributions is quite similar to the original proof given by Onsager. To this end we will need some stochastic concepts. The considerations for the nondiagonal part, however, involve 'quantum frequencies', for which *Onsager's Principle* [see Eq. (16.3-26) and subsequent discussion] does no longer hold; the proof required here is entirely similar to that of Kubo's original paper, with $\mathcal{L}$ replaced by $\mathcal{L}^0$, cf. Kubo, Op. Cit., § 6.

16.14.1 *The Diagonal Correlation Functions*

Starting with the diagonal contribution, since the density operator has been randomized by the transitions involving the interaction potential, we shall interpret the term $\langle\gamma|\rho(t)|\gamma\rangle = p(\gamma,t)$ as a stochastic variable. Let further $P(\gamma,t\,|\,\gamma_0)$ be the *conditional probability* that we find the state $|\gamma\rangle$ at time *t*, given that we had the state $|\gamma_0\rangle$ at time zero. Then from standard probability considerations,

$$p(\gamma,t)=\sum_{\gamma_0} P(\gamma,t|\gamma_0)p(\gamma_0,0)=\int\Delta\gamma_0\, P(\gamma,t\,|\,\gamma)p(\gamma_0)\,, \qquad (16.14\text{-}1)$$

where the density of states $Z(\gamma_0)$ is absorbed in $\Delta\gamma_0$. We substitute this in the diagonal part of the ME, as given by (16.11-13); this yields

$$\frac{\partial}{\partial t}\int\Delta\gamma\, P(\gamma,t\,|\,\gamma_0)p(\gamma_0)+M_\gamma\int\Delta\gamma_0\, P(\gamma,t\,|\,\gamma)p(\gamma_0)=0, \quad t>0. \qquad (16.14\text{-}2)$$

Since this holds for any $p(\gamma_0)$, we also have, noting the initial condition for P,

$$\frac{\partial P(\gamma,t\,|\,\gamma_0)}{\partial t}+M_\gamma P(\gamma,t\,|\,\gamma_0)=\delta(t)\delta(\gamma-\gamma_0)/Z(\gamma)\,. \qquad (16.14\text{-}3)$$

(This equation is sometimes also referred to as 'the master equation'.)

We now note that Eq. (16.14-3) is also the defining equation of the Green's function for any integro-differential equation of the form

$$\frac{\partial f(\gamma,t)}{\partial t}+M_\gamma f(\gamma,t)=Q(\gamma,t)\,, \qquad (16.14\text{-}4)$$

where $Q(\gamma,t)$ is an inhomogeneous term that may include the initial conditions. Its formal solution is

$$f(\gamma,t)=\int_0^{t+0}dt'\int\Delta\gamma' g(\gamma,t;\gamma',t')Q(\gamma',t')\,, \qquad (16.14\text{-}5)$$

where we assumed that there are no boundary terms from the bilinear concomitant. It is easily shown that the master operator M_γ is self-adjoint, so that the Green's function is symmetric with $t\to -t$. In fact, we shall set[69]

$$P(\gamma,t\,|\,\gamma_0,t_0)=P(\gamma,t-t_0\,|\,\gamma_0)=g(\gamma_0,t_0;\gamma,t)\,. \qquad (16.14\text{-}6)$$

---

[69] With Van Hove's original method, the conditional probability was explicitly computed in LRT I and related to the Green's function, cf. loc. cit., Eqs. (7.5) and (7.27). This result is also foreseeable, however, from the original Heisenberg forms. Using the usual full evolution operator $U(t,t_0)$,we have

$$\begin{aligned} p(\gamma,t)\equiv\langle\gamma\,|\,\rho(t)\,|\,\gamma\rangle &= \sum_{\gamma',\gamma_0}\langle\gamma\,|\,U(t,t_0)\,|\,\gamma'\rangle\langle\gamma'\,|\,\rho(0)\,|\,\gamma_0\rangle\langle\gamma_0\,|\,U^\dagger(t,t_0)\,|\,\gamma\rangle \\ &= \sum_{\gamma_0}|\langle\gamma\,|\,U(t,t_0)\,|\,\gamma_0\rangle|^2\,\langle\gamma_0\,|\,\rho(0)\,|\,\gamma_0\rangle\,, \end{aligned} \qquad (16.14\text{-}7)$$

$$\begin{aligned} \langle\gamma\,|\,B(t)\,|\,\gamma\rangle &= \sum_{\gamma',\gamma_0}\langle\gamma\,|\,U^\dagger(t,t_0)\,|\,\gamma'\rangle\langle\gamma'\,|\,B(0)\,|\,\gamma_0\rangle\langle\gamma_0\,|\,U(t,t_0)\,|\,\gamma\rangle \\ &= \sum_{\gamma_0}|\langle\gamma_0\,|\,U(t,t_0)\,|\,\gamma\rangle|^2\,\langle\gamma_0\,|\,B(0)\,|\,\gamma_0\rangle\,, \end{aligned} \qquad (16.14\text{-}8)$$

where in both cases we made an *initial* random phase assumption. From (16.14-7) we note that the conditional probability (before the Van Hove limit) is given by $P(\gamma,t\,|\,\gamma_0,t_0)=|\langle\gamma\,|\,U(t,t_0)\,|\,\gamma_0\rangle|^2$. Likewise, from (16.14-8) we read out the Green's function; note that the states γ and γ_0 are switched.

We now consider the reduced operator $B_d^R(t)$ and apply Theorem 2. This yields

$$B_d^R(t) = e^{-\int \Delta\gamma |\gamma\rangle\langle\gamma| M_\gamma t} \langle\gamma | B_d | \gamma\rangle = \int \Delta\gamma |\gamma\rangle\langle\gamma| e^{-M_\gamma t} \langle\gamma | B_d | \gamma\rangle, \quad (16.14\text{-}9)$$

or also upon differentiation

$$\frac{\partial}{\partial t}\langle\gamma | B_d^R(t) | \gamma\rangle + M_\gamma \langle\gamma | B_d^R(t) | \gamma\rangle = \delta(t)\langle\gamma | B_d^R | \gamma\rangle. \quad (16.14\text{-}10)$$

With (16.14-5) and (16.14-6) we thus obtain

$$\langle\gamma | B_d^R(t) | \gamma\rangle = \int \Delta\gamma' P(\gamma',t | \gamma)\langle\gamma' | B_d | \gamma'\rangle. \quad (16.14\text{-}11)$$

This yields the diagonal correlation function

$$\begin{aligned}\Phi_{BA}^d(t) &= \mathrm{Tr}\{\rho_{eq}\Delta B_d(t)\Delta A_d\} \\ &= \iint \Delta\gamma\Delta\gamma' p(\gamma)P(\gamma',t | \gamma)\langle\gamma' | \Delta B_d^R | \gamma'\rangle\langle\gamma | \Delta A_d^R | \gamma\rangle. \end{aligned} \quad (16.14\text{-}12)$$

The first two factors can be combined to the joint two-point probability by Bayes' theorem

$$W_2(\gamma',t;\gamma,0) = p(\gamma)P(\gamma',t | \gamma), \quad (16.14\text{-}13)$$

giving for the correlation function

$$\Phi_{BA}^d(t) = \iint \Delta\gamma\Delta\gamma' W_2(\gamma',t;\gamma,0)\langle\gamma' | \Delta B_d^R | \gamma'\rangle\langle\gamma | \Delta A_d^R | \gamma\rangle. \quad (16.14\text{-}14)$$

Similarly we have

$$\Phi_{\dot B\dot A}^d(t) = \iint \Delta\gamma\Delta\gamma' W_2(\gamma',t;\gamma,0)\langle\gamma' | \Delta\dot B_d^R | \gamma'\rangle\langle\gamma | \Delta\dot A_d^R | \gamma\rangle. \quad (16.14\text{-}15)$$

Turning to the diagonal susceptance $\chi_{BA}^d(i\omega)$, it was already given in (16.10-63), first line. A more appropriate form for our purpose is obtained by integrating $\dot A_d$ and differentiating the exponential; we then obtain,

$$\chi_{BA}^d(i\omega) - \chi_{BA,dc}^d = -i\omega\beta\int_0^\infty dt\, e^{-i\omega t} \iint \Delta\gamma\Delta\gamma' W_2(\gamma',t;\gamma,0)\langle\gamma' | \Delta B_d^R | \gamma'\rangle\langle\gamma | \Delta A_d^R | \gamma\rangle. \quad (16.14\text{-}16)$$

Likewise, for the generalized diagonal conductance we have

$$L_{BA}^d(i\omega) = \beta\int_0^\infty dt\, e^{-i\omega t} \iint \Delta\gamma\Delta\gamma' W_2(\gamma',t;\gamma,0)\langle\gamma' | \Delta\dot B_d^R | \gamma'\rangle\langle\gamma | \Delta\dot A_d^R | \gamma\rangle. \quad (16.14\text{-}17)$$

We refer to these formulae as the *Schrödinger form*, since in this form the operators are fixed, while the time dependence is vested in the probability function W_2.

Let us use again the short-cut notations $\Delta B_{d\gamma} = \langle\gamma | \Delta B_d^R | \gamma\rangle$ and similarly for $\Delta A_{d\gamma}$. Equation (16.14-14) now reads

$$\Phi^d_{BA}(t) = \iint \Delta\gamma\,\Delta\gamma' W_2(\gamma',t;\gamma,0)\,\Delta B^R_{d\gamma'}\,\Delta A^R_{d\gamma}\,. \tag{16.14-14'}$$

Also, upon interchanging the integration variables and using that

$$W_2(\gamma,t;\gamma',0) = W_2(\gamma,0;\gamma',-t) = W_2(\gamma',-t;\gamma,0) \tag{16-14-18}$$

by stationarity and pair invariance, we have

$$\Phi^d_{BA}(t) = \iint \Delta\gamma\,\Delta\gamma' W_2(\gamma',-t;\gamma,0)\,\Delta B^R_{d\gamma}\,\Delta A^R_{d\gamma'}\,. \tag{16.14-19}$$

Comparing with (16.14-14'), we established once more the *transposition property*

$$\Phi^d_{BA}(t,\mathbf{H}) = \Phi^d_{AB}(-t,\mathbf{H})\,, \tag{16.14-20}$$

where we added the magnetic field **H** for comparison with what follows.

Finally, consider a process based on the same Hamiltonian, but going backward in time. Nothing in the physics then changes, provided the magnetic field is also reversed, $\mathbf{H} \to -\mathbf{H}$, to ensure invariance of the Hamiltonian. We thus have

$$P^{\mathbf{H}}(\gamma',t\,|\,\gamma) = P^{-\mathbf{H}}(\gamma',-t\,|\,\gamma) \quad \text{or} \quad W_2^{\mathbf{H}}(\gamma',t;\gamma,0) = W_2^{-\mathbf{H}}(\gamma',-t;\gamma,0)\,. \tag{16.14-21}$$

The time reversal further alters the Schrödinger operators A and B according to $\varepsilon_A A$ and $\varepsilon_B B$, with $\varepsilon = \pm 1$, depending on the even (+1) or odd (–1) character of the variables. For the correlation function we thus have the *time-reversal property*

$$\Phi^d_{BA}(t,\mathbf{H}) = \varepsilon_A \varepsilon_B \Phi^d_{BA}(-t,-\mathbf{H})\,. \tag{16.14-22}$$

Combining with (16.14-20) we also obtained

$$\Phi^d_{BA}(t,\mathbf{H}) = \varepsilon_A \varepsilon_B \Phi^d_{AB}(t,-\mathbf{H})\,. \tag{16.14-23}$$

The main Onsager[70]–Casimir[71] relations for the generalized diagonal susceptance are

$$\chi^d_{BA}(i\omega,\mathbf{H}) = \varepsilon_A \varepsilon_B\, \chi^d_{AB}(i\omega,-\mathbf{H})\,. \tag{16.14-24}$$

For the generalized diagonal conductance we note that $\varepsilon_{\dot{A}}\varepsilon_{\dot{B}} = \varepsilon_A \varepsilon_B$. Hence we find

$$L^d_{BA}(i\omega,\mathbf{H}) = \varepsilon_A \varepsilon_B\, L^d_{AB}(i\omega,-\mathbf{H})\,. \tag{16.14-25}$$

16.14.2 *The Nondiagonal Correlation Functions*

The argument for the nondiagonal correlation function and the implied transport coefficients follow an entirely different argument, there being no recourse to stochastic properties. First consider the relaxation function for negative time,

---

[70] L. Onsager, Phys. Rev. **37**, 405 (1937); ibid. **38**, 265 (1938).

[71] H.B.G. Casimir, Rev. Mod. Phys. **17**, 343 (1945).

$$\begin{aligned}\psi_{BA}^{nd}(-t) &= \int_0^\beta d\lambda \mathrm{Tr}\{\rho_{eq} A_{nd}^R(-i\hbar\lambda) B_{nd}^R(-t)\} \\ &= \int_0^\beta d\lambda \mathrm{Tr}\{\rho_{eq} A_{nd}^R B_{nd}^R(-t+i\hbar\lambda)\} \\ &= \int_0^\beta d\lambda \mathrm{Tr}\{\rho_{eq} A_{nd}^R B_{nd}^R[-t+i\hbar(\beta-\lambda)]\} \\ &= \int_0^\beta d\lambda \mathrm{Tr}\{\rho_{eq} A_{nd}^R(t+i\hbar\lambda) B_{nd}^R(i\hbar\beta)\}. \end{aligned} \tag{16.14-26}$$

In the transition to the third line we replaced the integration variable λ by $\beta-\lambda$, while in the last transition we shifted the variables based on presumed stationarity. Next, we note that $B_{nd}^R(i\hbar\beta) = e^{-\beta\mathcal{H}^0} B_{nd}^R e^{\beta\mathcal{H}^0}$. Moving this all to the front of the trace, using cyclic invariance, we arrive at

$$\begin{aligned}\psi_{BA}^{nd}(-t,\mathbf{H}) &= \int_0^\beta d\lambda \mathrm{Tr}\{\rho_{eq} B_{nd}^R A_{nd}^R(t+i\hbar\lambda)\} \\ &= \int_0^\beta d\lambda \mathrm{Tr}\{\rho_{eq} B_{nd}^R(-i\hbar\lambda) A_{nd}^R(t)\} = \psi_{AB}^{nd}(t,\mathbf{H}). \end{aligned} \tag{16.14-27}$$

The *transposition property* has hereby again been obtained; note $\psi_{BA}^{nd} = \int_t^\infty dt' \phi_{BA}^{nd}(t')$.

The argument about time reversal and simultaneous inversion of the magnetic field is the same as before. It leaves the Hamiltonian $\mathcal{H}$ as well as the sub-Hamiltonian $\mathcal{H}^0$ invariant. Hence we have again the *time-reversal property*

$$\psi_{BA}^{nd}(t,\mathbf{H}) = \varepsilon_A \varepsilon_B \psi_{BA}^{nd}(-t,-\mathbf{H}). \tag{16.14-28}$$

Combining with the above, we established the result

$$\psi_{BA}^{nd}(t,\mathbf{H}) = \varepsilon_A \varepsilon_B \psi_{AB}^{nd}(t,-\mathbf{H}). \tag{16.14-29}$$

For the nondiagonal susceptance we need the Fourier-Laplace transform of $\phi_{B\dot{A}}^{nd}(t)$. We thus move the t-dependence to $\dot{A}_{nd}$ and integrate by parts, giving

$$\begin{aligned}\chi_{BA}^{nd}(i\omega) - \chi_{BA,dc}^{nd} &= -i\omega \int_0^\infty dt\, e^{-i\omega t} \int_0^\beta d\lambda \mathrm{Tr}\{\rho_{eq} A_{nd}^R(-i\hbar\lambda) B_{nd}^R(t)\} \\ &= -i\omega \int_0^\infty dt\, e^{-i\omega t} \psi_{BA}^{nd}(t). \end{aligned} \tag{16.14-30}$$

Similarly, for the nondiagonal generalized conductance we have

$$L_{BA}^{nd}(i\omega) = \int_0^\infty dt\, e^{-i\omega t} \int_0^\beta d\lambda \mathrm{Tr}\{\rho_{eq} \dot{A}_{nd}^R(-i\hbar\lambda) \dot{B}_{nd}^R(t)\} = \int_0^\infty dt\, e^{-i\omega t} \phi_{\dot{B}\dot{A}}^{nd}(t). \tag{16.14-31}$$

With (16.14-28) we arrive at the Onsager–Casimir relations

$$\chi_{BA}^{nd}(i\omega,\mathbf{H}) = \varepsilon_A \varepsilon_B \chi_{AB}^{nd}(i\omega,-\mathbf{H}), \tag{16.14-32}$$

$$L_{BA}^{nd}(i\omega,\mathbf{H}) = \varepsilon_A \varepsilon_B L_{AB}^{nd}(i\omega,-\mathbf{H}). \tag{16.14-33}$$

In addition, from the forms (16.14-16), (16-14-17), (16.14-29) and (16.14-30) we find for the ordinary susceptibility and conductivity tensors $\chi_{\mu\nu}(i\omega)$ and $\sigma_{\mu\nu}(i\omega)$:

$$\operatorname{Re}\chi_{\mu\nu}(\omega,\mathbf{H}) = \operatorname{Re}\chi_{\mu\nu}(-\omega,\mathbf{H}) = \operatorname{Re}\chi_{\nu\mu}(\omega,-\mathbf{H}), \tag{16.14-34}$$

$$\operatorname{Im}\chi_{\mu\nu}(\omega,\mathbf{H}) = -\operatorname{Im}\chi_{\mu\nu}(-\omega,\mathbf{H}) = \operatorname{Im}\chi_{\nu\mu}(\omega,-\mathbf{H}), \tag{16.14-35}$$

and also

$$\operatorname{Re}\sigma_{\mu\nu}(\omega,\mathbf{H}) = \operatorname{Re}\sigma_{\mu\nu}(-\omega,\mathbf{H}) = \operatorname{Re}\sigma_{\nu\mu}(\omega,-\mathbf{H}), \tag{16.14-36}$$

$$\operatorname{Im}\sigma_{\mu\nu}(\omega,\mathbf{H}) = -\operatorname{Im}\sigma_{\mu\nu}(-\omega,\mathbf{H}) = \operatorname{Im}\sigma_{\nu\mu}(\omega,-\mathbf{H}). \tag{16.14-37}$$

We can also form the symmetric and antisymmetric parts,

$$\chi_{\mu\nu}^{s}(i\omega) = \tfrac{1}{2}[\chi_{\mu\nu}(i\omega) + \chi_{\nu\mu}(i\omega)], \quad \chi_{\mu\nu}^{a}(i\omega) = \tfrac{1}{2}[\chi_{\mu\nu}(i\omega) - \chi_{\nu\mu}(i\omega)], \tag{16.14-38}$$

and similarly for $\sigma_{\mu\nu}(i\omega)$. Generally, we conclude that for the change $\omega \to -\omega$, the real part is even while the imaginary part is odd. For the reversal of the magnetic field **H**, the symmetric part is even and the antisymmetric part is odd.

16.14.3 *Some Lemmas*

We will finish this section by giving some useful lemmas – that could have been used in the previous derivations – and which are very helpful for the derivations in the next section and in particular for the Boltzmann formulas of the next chapter.[72]

Lemma 1. For any two operators C and D we have

$$\operatorname{Tr}(C\Lambda_d D) = \operatorname{Tr}(D\Lambda_d C). \tag{16.14-39}$$

From the definition of the master operator Λ_d we have

$$\begin{aligned}\operatorname{Tr}(C\Lambda_d D) &= -\textstyle\sum_{\gamma\gamma''}\langle\gamma|C|\gamma\rangle\{W_{\gamma''\gamma}\langle\gamma''|D|\gamma''\rangle - W_{\gamma\gamma''}\langle\gamma|D|\gamma\rangle\} \\ &= \frac{1}{2}\sum_{\gamma,\gamma''}\{\langle\gamma''|C|\gamma''\rangle - \langle\gamma|C|\gamma\rangle\}\{W_{\gamma''\gamma}\langle\gamma''|D|\gamma''\rangle - W_{\gamma\gamma''}\langle\gamma|D|\gamma\rangle\},\end{aligned} \tag{16.14-40}$$

where we interchanged the summation indices and took half the sum. Because of microscopic reversibility the two W's are equal, so that the W's can be distributed over the matrix elements involving the operator C. This completes the proof.[73]

---

[72] K.M. Van Vliet, LRT II, Op. Cit., Appendix C.

[73] The scalar product of any two operators (elements) of the Liouville space is defined as $\{A,B\} = \operatorname{Tr}AB^{\dagger}$. The lemma therefore also reads $\{C,\Lambda_d D^{\dagger}\} = \{\Lambda_d C, D^{\dagger}\}$, which is satisfied since Λ_d is self-adjoint.

Lemma 2. For any two operators C and D we have

$$\mathrm{Tr}(Ce^{-\Lambda_d t}D) = \mathrm{Tr}(De^{-\Lambda_d t}C). \tag{16.14-41}$$

The statement follows from series expansion of the exponential and repeated application of lemma 1.

Lemma 3. For any two operators C and D we have

$$\mathrm{Tr}(C\mathscr{L}^0 D) = -\mathrm{Tr}(D\mathscr{L}^0 C). \tag{16.14-42}$$

The proof follows from the commutator form for $\mathscr{L}^0$ and the cyclic invariance of the trace.[74]

$$\begin{aligned}\mathrm{Tr}(C\mathscr{L}^0 D) &= \hbar^{-1}\mathrm{Tr}\{C[\mathcal{H}^0, D]\} = \hbar^{-1}\mathrm{Tr}\{C\mathcal{H}^0 D - CD\mathcal{H}^0\} \\ &= \hbar^{-1}\mathrm{Tr}\{C\mathcal{H}^0 D - \mathcal{H}^0 CD\} = \hbar^{-1}\mathrm{Tr}\{[C,\mathcal{H}^0]D\} \\ &= -\hbar^{-1}\mathrm{Tr}\{D[\mathcal{H}^0, C]\} = -\mathrm{Tr}(D\mathscr{L}^0 C).\end{aligned} \tag{16.14-43}$$

Moreover, by series expansion and/or Theorem 1 [Eq.(16.10-6)] one easily shows the following additional lemmas:

Lemma 4.

$$\mathrm{Tr}(Ce^{-i\mathscr{L}^0 t}D) = \mathrm{Tr}(De^{i\mathscr{L}^0 t}C). \tag{16.14-44}$$

Lemma 5.

$$\mathrm{Tr}[Ce^{-(\Lambda_d + i\mathscr{L}^0)t}D] = \mathrm{Tr}[De^{-(\Lambda_d - i\mathscr{L}^0)t}C]. \tag{16.14-45}$$

16.15 An Exact Response Result: Cohen–Van Vliet *

In Section 16.8 we have already voiced some reservations regarding Kubo's linear response theory. How can a simple manipulation of the equation of motion for the non-equilibrium density operator lead to profound transport results? The culprit apparently is the linearization procedure. Moreover, nowhere do the appropriate time scales, basic to nonequilibrium transport equations, occur. For clarity we have listed the various time regimes pictorially in Fig. 16-4 below. In the absence of such time regimes, the linearization approximation must in principle be valid for all macroscopic measurement times – an impossibility, as shown by van Kampen[8].

However, the von Neumann equation – or its classical analogue, the Liouville equation – for the response Hamiltonian (16.2-1) can easily be solved *exactly*, generally giving a nonlinear connection between the response and the applied field,

---

[74] The lemma also means $\{C, \mathscr{L}^0 D^\dagger\} = \{\mathscr{L}^0 C, D^\dagger\}$, where we note that $(\mathscr{L}^0 D^\dagger)^\dagger = -\mathscr{L}^0 D$.

as was shown by Cohen[75] for a classical gas. He then demonstrated that for a dilute gas, with appropriate dynamics giving rise to the time regimes indicated above, Kubo's linearization can be justified. His argument was subsequently repeated for quantum systems by the author, with a quite analogous result.[76] These considerations then allow us to evaluate the extent of validity of Kubo's linearization procedure.

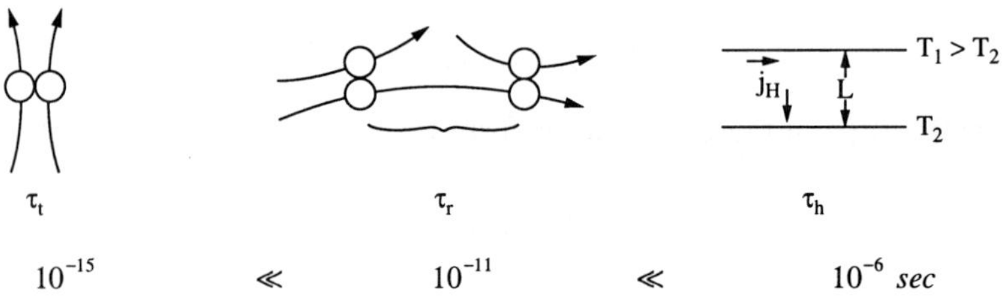

Fig. 16-4. The different time regimes: τ_t (transition time), τ_r (relaxation time), τ_h (hydrodynamic time).

The response Hamiltonian (16.2-1) will henceforth be denoted by $\mathcal{H}_F$ ('F' for 'full') and the corresponding Liouville operator by $\mathcal{L}_F$. The von Neumann equation now reads

$$\partial\rho/\partial t + i\mathcal{L}_F(t)\rho(t) = 0\,, \tag{16.15-1}$$

where

$$\mathcal{L}_F(t)K = (1/\hbar)[\mathcal{H}_F, K] = \mathcal{L}K - (1/\hbar)F(t)[A,K]\,. \tag{16.15-2}$$

With the field being switched on at $t = 0$, we have the formal solution

$$\rho(t) = \left\{\exp\left[-i\int_0^t dt'\mathcal{L}_F(t')\right]\right\}\rho_{eq}\,. \tag{16.15-3}$$

The exponential will be denoted by $W(t)$. A more appropriate solution is obtained by employing a perturbation procedure to all orders, similarly as formerly we did for the evolution operator $U(t)$. The full results are in our quoted paper[76].

Here we shall content ourselves with a far simpler exact result, obtained if we restrict ourselves to the dc conductance, $\omega \to 0$. So, let the field F be a constant; we then have the solution

$$\rho(t) = e^{-i\mathcal{L}_F t}\rho_{eq}\,, \tag{16.15-4}$$

where

---

[75] E.G.D. Cohen, "On two objections by van Kampen," Statistical mechanics meeting in honour of van Kampen's sixtieth birthday, University of Utrecht, June 1981.

[76] C.M. Van Vliet, J. of Statistical Physics, **53**, 49-60 (1988).

$$i\mathcal{L}_F K = (i\mathcal{L} + \mathcal{M})K, \quad \mathcal{M}K = (1/\hbar i) F[A, K]. \tag{16.15-5}$$

From (16.15-4) we obtain

$$\langle J_B(t)\rangle = \mathrm{Tr}\{\dot{B} e^{-i\mathcal{L}_F t} \rho_{eq}\}, \tag{16.15-6}$$

or also

$$\langle J_B(t)\rangle = \mathrm{Tr}\{\rho_{eq} e^{i\mathcal{L}_F t} \dot{B}\} = \langle e^{i\mathcal{L}_F t} J_B\rangle_{eq}. \tag{16.15-7}$$

Here a result analogous to Lemma 4 was employed,

$$\begin{aligned} \mathrm{Tr}\{C e^{-i\mathcal{L}_F t} D\} &= \mathrm{Tr}\{C e^{-i\mathcal{H}_F t/\hbar} D e^{i\mathcal{H}_F t/\hbar}\} \\ &= \mathrm{Tr}\{D e^{i\mathcal{H}_F t/\hbar} C e^{-i\mathcal{H}_F t/\hbar}\} = \mathrm{Tr}\{D e^{i\mathcal{L}_F t} C\}. \end{aligned} \tag{16.15-8}$$

Next, let us consider the resolvent (Laplace transform) of $e^{i\mathcal{L}_F t}$. Since $A^{-1} - B^{-1} = A^{-1}(B - A)B^{-1}$, we have the identity

$$\frac{1}{s - i\mathcal{L}_F} \equiv \frac{1}{s - i\mathcal{L} - \mathcal{M}} = \frac{1}{s - i\mathcal{L}} + \frac{1}{s - i\mathcal{L}} \mathcal{M} \frac{1}{s - i\mathcal{L} - \mathcal{M}}. \tag{16.15-9}$$

By inverse Laplace transformation, this yields the convolution

$$e^{i\mathcal{L}_F t} = e^{i\mathcal{L}t} + \int_0^t dt' e^{i\mathcal{L}(t-t')} \mathcal{M} e^{i\mathcal{L}_F t'}. \tag{16.15-10}$$

Substitution into (16.15-7) yields

$$\langle J_B(t)\rangle = \langle e^{i\mathcal{L}t} J_B\rangle_{eq} + \int_0^t dt' \langle e^{i\mathcal{L}(t-t')} \mathcal{M} e^{i\mathcal{L}_F t'} J_B\rangle_{eq}. \tag{16.15-11}$$

As in Kubo's theory, we take ρ_{eq} to be the canonical density operator $\rho_{eq} = e^{-\beta\mathcal{H}}/\mathcal{Q}$. Clearly $e^{i\mathcal{L}t}\rho_{eq} = \rho_{eq}$ and $\langle J_B\rangle_{eq} = 0$, so the first term in (16.15-11) vanishes, leaving

$$\langle J_B(t)\rangle = (F/\hbar i) \int_0^t dt' \langle e^{i\mathcal{L}(t-t')} [A, e^{i\mathcal{L}_F t'} J_B]\rangle_{eq}. \tag{16.15-12}$$

Furthermore,

$$\begin{aligned} \mathrm{Tr}\{\rho_{eq} e^{i\mathcal{L}(t-t')} [A, e^{i\mathcal{L}_F t'} J_B]\} &= \mathrm{Tr}\{[A, e^{i\mathcal{L}_F t'} J_B] e^{-i\mathcal{L}(t-t')} \rho_{eq}\} \\ &= \mathrm{Tr}\{[A, e^{i\mathcal{L}_F t'} J_B] e^{-\beta\mathcal{H}}\}/\mathcal{Q} = \mathrm{Tr}\{[e^{-\beta\mathcal{H}}, A] e^{i\mathcal{L}_F t'} J_B\}/\mathcal{Q}. \end{aligned} \tag{16.15-13}$$

Now by Kubo's identity (16.3-19) we have

$$[e^{-\beta\mathcal{H}}, A] = \hbar i \int_0^\beta d\tau\, e^{-\beta\mathcal{H}} J_A(-i\hbar\tau). \tag{16.15-14}$$

This and (16.15-13) into (16.15-12) yield

$$\langle J_B(t)\rangle = F \int_0^t dt' \int_0^\beta d\tau \langle J_A(-i\hbar\tau) e^{i\mathcal{L}_F t'} J_B\rangle_{eq}. \tag{16.15-15}$$

For $\hbar\omega \ll k_B T$, the argument of J_A can be taken to be zero. Letting further $t \to \infty$ and adding the thermodynamic limit, implied by the use of the canonical ensemble, we find for the dc conductivity

$$L_{BA} = \beta \lim_{t\to\infty} \lim_{\text{th}} \int_0^t dt' \langle J_A e^{i\mathcal{L}_F t'} J_B \rangle_{eq} \,. \tag{16.15-16}$$

This exact result replaces Kubo's linearized result (16.3-22), which with the present notation reads

$$L_{BA,\,Kubo} = \beta \lim_{t\to\infty} \lim_{\text{th}} \int_0^t dt' \langle J_A e^{i\mathcal{L}t'} J_B \rangle_{eq} \,. \tag{16.15-17}$$

For the electrical conductivity, in particular, we obtain for the conductivity tensor

$$\sigma_{\mu\nu} = \beta V_0 \lim_{t\to\infty} \lim_{\text{th}} \int_0^t dt' \langle J_\nu e^{i\mathcal{L}_F t'} J_\mu \rangle_{eq} \,. \tag{16.15-18}$$

The reduction from (16.15-16) to Kubo's form (16.15-17) cannot be justified, unless *new physical tenets* are introduced. We thus consider the weak coupling limit with Hamiltonian $\mathcal{H} = \mathcal{H}^0 + \lambda\mathcal{V}$. To appreciate the difference between the two results, we shall iterate the resolvent expression (16.15-9) to all orders. Substituting the original resolvent in the rhs, we obtain the second order result

$$\begin{aligned}\frac{1}{s - i\mathcal{L}_F} \equiv \frac{1}{s - i\mathcal{L} - \mathcal{M}} &= \frac{1}{s - i\mathcal{L}} + \frac{1}{s - i\mathcal{L}}\mathcal{M}\frac{1}{s - i\mathcal{L}} \\ &+ \frac{1}{s - i\mathcal{L}}\mathcal{M}\frac{1}{s - i\mathcal{L}}\mathcal{M}\frac{1}{s - i\mathcal{L} - \mathcal{M}},\end{aligned} \tag{16.15-19}$$

and so on. We thus can write $(s - i\mathcal{L}_F)^{-1}$ as an infinite sum, with the nth order term having n factors $\mathcal{M}$. Taking the inverse Laplace transform, we arrive at

$$W(t) \equiv e^{i\mathcal{L}_F t} = e^{i\mathcal{L}t} + \sum_{n=1}^{\infty} W^{(n)}(t)\,, \tag{16.15-20}$$

where the $W^{(n)}(t)$ are the time-ordered integrals

$$\begin{aligned}W^{(n)}(t) = \int_0^t dt_n \int_0^{t_n} dt_{n-1} \ldots \int_0^{t_2} dt_1 \\ \times e^{i\mathcal{L}(t-t_n)}\mathcal{M} e^{i\mathcal{L}(t_n - t_{n-1})}\mathcal{M} \ldots e^{i\mathcal{L}(t_2 - t_1)}\mathcal{M} e^{i\mathcal{L}t_1}\,.\end{aligned} \tag{16.15-21}$$

Since the operators $\mathcal{M}\ldots\mathcal{M}$ involve repeated commutators, we shall for simplicity consider the classical correspondence limit. For the case of electrical conduction, $F = qE_z$ and $A = \Sigma_i (z_i - z_{i,0})$. Replacing the commutator (16.15-5) by the Poisson bracket,

$$\mathcal{M}K \approx qE_z \{\textstyle\sum_i (z_i - z_{i,0}), K\} = \frac{qE_z}{m} \sum_i \frac{\partial K}{\partial v_{i,z}}\,. \tag{16.15-22}$$

The integrals (16.15-21) become

$$W^{(n)}(t) = \left(\frac{qE_z}{m}\right)^n \int_0^t dt_n \int_0^{t_n} dt_{n-1} \cdots \int_0^{t_2} dt_1$$

$$\times e^{i\mathcal{L}(t-t_n)} \sum_i (\partial/\partial v_z) e^{i\mathcal{L}(t_n-t_{n-1})} \cdots \sum_i (\partial/\partial v_z) e^{i\mathcal{L}(t_2-t_1)} \sum_i (\partial/\partial v_z) e^{i\mathcal{L}t_1} . \quad (16.15\text{-}23)$$

The Van Hove limit leads to the reduced operators as noted before,

$$K^R(t) = \lim_{\lambda,t} \lim_{\text{th}} e^{i\mathcal{L}t} K = e^{-(\Lambda_d - i\mathcal{L}^0)t} K . \quad (16.15\text{-}24)$$

This result is entered into all terms (16.15-23) of (16.15-20), after which the result is resummed. Hence, we established the final operator expression

$$\lim_{\lambda,t} \lim_{\text{th}} e^{\pm i\mathcal{L}_F t} = \exp\left[-(\Lambda_d \mp i\mathcal{L}^0)t - \frac{qE_z}{m} \sum_i \frac{\partial}{\partial v_{iz}} \right] . \quad (16.15\text{-}25)$$

In order to examine the consequences for the conductance (16.15-16), we go back to the original response result (16.15-6). The superoperator (16.15-25) is to act on $\rho_{eq} = \exp(-\beta\mathcal{H}^0)/\mathcal{Z}$, in which $\mathcal{H}^0$ contains a part $\frac{1}{2}\Sigma_i m v_{iz}^2$. So, $\partial/\partial v_{iz} \to m v_{iz}/k_B T$. Now let the reciprocal relaxation time τ_r^{-1} be the smallest eigenvalue of Λ_d. Then the field term in (16.15-25) is negligible if $1/\tau_r \gg qE_z \langle v_z \rangle / k_B T$. Or, if $\lambda \approx \langle v_z \rangle \tau_r$ is the mean free path, then we require

$$E_z \ll k_B T / q\lambda . \quad (16.15\text{-}26)$$

Let $\lambda = 1000\,\AA$ and $T = 300\ K$. Then this amounts to $E_z \ll 2.5 \times 10^3\ V/cm$. So, for reasonable electric fields, the replacement $\exp(\pm i\mathcal{L}_F t) \to \exp(\pm i\mathcal{L}t)$ is justified.

16.16 Problems to Chapter XVI

16.1 For a certain process the real part of the susceptibility is given by

$$\chi'(\omega) = A/(1 + \omega^2\tau^2) , \quad (1)$$

where τ is a relaxation time. Obtain the imaginary part $\chi''(\omega)$ from the Kramers–Kronig relations, as well as the full susceptibility $\chi(i\omega) = \chi'(\omega) + i\chi''(\omega)$. Also show that there are *no poles* in the upper half plane.

16.2 Repeat the above problem for given real part

$$\chi'(\omega) = \frac{A\tau_1\tau_2}{(1 + \omega^2\tau_1^2)(1 + \omega^2\tau_2^2)} . \quad (2)$$

16.3 Mazo's lemma [Eq. (16.3-35)] connects the two-sided Fourier transforms of the expectation values $\langle \Delta C(t)\Delta D\rangle$ and $\langle \Delta D\Delta C(t)\rangle$ by the relationship

$$\int_{-\infty}^{\infty} dt\, e^{-i\omega t}\, \mathrm{Tr}\{\rho_{eq}\Delta C(t)\Delta D\} = e^{-\beta\hbar\omega}\int_{-\infty}^{\infty} dt\, e^{-i\omega t}\, \mathrm{Tr}\{\rho_{eq}\Delta D\Delta C(t)\}. \tag{3}$$

Give a proof, based on evaluation of the trace in the representation $\{|\eta\rangle\}$, with $|\eta\rangle$ being an eigenstate of the Hamiltonian $\mathcal{H}$. [Evaluate the real integrals directly, without analytical continuation in the complex plane.]

16.4 (a) Show that the two-sided Fourier transform of the relaxation function, $\hat{\psi}_{BA}(t)$, and of the correlation function, $\hat{\Phi}_{BA}(t)$, are related by

$$\hat{\psi}_{BA}(\omega) = [\mathcal{E}(\omega,T)]^{-1}\hat{\Phi}_{BA}(\omega), \quad \text{with} \quad \mathcal{E}(\omega,T) = (\hbar\omega/2)\coth(\beta\hbar\omega/2). \tag{4}$$

(b) By convolution we then have (no proof required here):

$$\psi_{BA}(t) = \beta\int_{-\infty}^{\infty} dt'\Gamma(t-t')\Phi_{BA}(t'). \tag{5}$$

With contour integration, evaluate the kernel and show that

$$\Gamma(t,0) = \frac{1}{2\pi}\int_{-\infty}^{\infty} d\omega e^{i\omega t}\frac{\tanh(\beta\hbar\omega/2)}{\beta\hbar\omega/2} = \frac{2}{\beta\hbar\pi}\ln\frac{\pi}{2\beta\hbar}|t|. \tag{6}$$

Note: to sum the residues of this meromorphic function, use the series for $\ln[(1+x)/(1-x)]$.

(c) Show that for 'classical frequencies' $\hbar\omega \ll k_B T$, $\Gamma(t)\to\delta(t)$.

16.5 The delta-plus and delta-minus functions are defined by

$$\delta_+(\omega) = \tfrac{1}{2\pi}\int_0^{\infty} dt\, e^{-i\omega t}, \qquad \delta_-(\omega) = \tfrac{1}{2\pi}\int_0^{\infty} dt\, e^{+i\omega t}. \tag{7}$$

(a) Upon adding we have $\delta_+(\omega)+\delta_-(\omega) = \delta(\omega)$. Employing a convergence factor $\exp(\pm i\eta)$, show that $\delta_+(\omega)-\delta_-(\omega) = (-i/\pi)\mathcal{P}(1/\omega)$.

(b) Solve for both functions and establish that

$$\begin{aligned} \delta_+(\omega) &= \frac{1}{2\pi}\lim_{\eta\to 0}\frac{1}{\eta+i\omega} = -\frac{i}{2\pi}\mathcal{P}\frac{1}{\omega}+\frac{1}{2}\delta(\omega), \\ \delta_-(\omega) &= \frac{1}{2\pi}\lim_{\eta\to 0}\frac{1}{\eta-i\omega} = \frac{i}{2\pi}\mathcal{P}\frac{1}{\omega}+\frac{1}{2}\delta(\omega). \end{aligned} \tag{8}$$

(c) Also obtain the important often used relationships, cf. (16.3-11):

$$\lim_{\eta\to 0}\frac{1}{u\pm i\eta} = \mathcal{P}\frac{1}{u}\mp\pi i\delta(u). \tag{9}$$

16.6 Give a proof for the integral representation (16.6-10) of the Heaviside function.

16.7 Establish the following sum rule for the electrical conductivity of an electron gas with density n:

$$\frac{2}{\pi}\int_0^{\infty} \mathrm{Re}\,\sigma^s_{\mu\nu}(i\omega)\,d\omega = \frac{q^2}{nm}\delta_{\mu\nu}\,, \tag{10}$$

where $\sigma^s_{\mu\nu}(i\omega)$ is the symmetrical part of the conductivity tensor.

16.8 *Doob's theorem.* Correlation functions of a Gaussian Markovian equilibrium process have the form $\Sigma_i A_i \exp(-\alpha_i t)$, $\alpha_i > 0$.
(a) From the Wiener–Khintchine theorem show that the spectral density is a sum of Lorentzians, with high frequency asymptote $\mathcal{S} \propto \omega^{-2}$.
(b) Make a log-log sketch of the envelope spectrum, noting that the sum could be replaced by a weighted Stieltjes integral $\Sigma_i \to \int d\gamma(\omega)$.

16.9 Consider an electron gas with conductivity $\sigma(i\omega)=\sigma_0/(1+i\omega\tau)$, $\sigma_0 = ne^2\tau/m$. At optical frequencies, $\omega\tau \gg 1$, plasma oscillations may occur. Demonstrate this by the following *classical* arguments, based on the Drude–Sommerfeld model.
(a) From Maxwell's equations, recognising that the plasma is neutral (no net volume charge), compute $\nabla\times(\nabla\times\mathbf{E}) = -\nabla^2\mathbf{E}$, with $\mathbf{J}_\omega = \sigma(i\omega)\mathbf{E}_\omega$, where $\mathbf{J}_\omega$, etc., are phasors $\mathbf{J}_0 e^{i\omega t}$... Obtain the differential equation for the propagation of electromagnetic waves through the metal

$$\nabla^2\mathbf{E}_\omega + (\omega_p^2/c^2)\mathbf{E}_\omega = 0\,, \tag{11}$$

where ω_p is the plasma frequency, $\omega_p = \sqrt{ne^2/m}$. Note that alkali metals become transparent for EM waves in the ultraviolet.
(b) Consider a collective displacement of all electrons by an amount δx within a hypothetical rectangular volume ΔV. On the y-z surfaces there will be a bound charge $\sigma_b = np$, where p is the dipole moment $e\delta x$, giving rise to the field $\mathbf{E}$. Obtain the differential equation

$$m\,\delta\ddot{x} + ne^2\delta x = 0 \quad \text{or} \quad \delta\ddot{x} + \omega_p^2\delta x = 0\,, \tag{12}$$

which describes the plasma waves, with plasmons $\hbar\omega_p$ as excitations. [Note that *free* plasma waves, without a sustaining resonant EM field, will not occur, due to radiation damping $\propto \delta\dddot{x}$ and other causes.]
(c) Find the dielectric function

$$\varepsilon(\omega) = (E_\omega + P_\omega)/E_\omega = 1 - (\omega_p^2/\omega^2)\,; \tag{13}$$

Explain your results.

16.10 In linear response theory the Drude–Sommerfeld model leads to density fluctuations with a susceptibility $\chi_d^{dr}(i\omega)$ given by (16.6-35).
(a) Show that for Coulomb interactions, the Fourier coefficient $V_{\mathbf{q}} = e^2/q^2$.
(b) Prove that for $\omega \gg \hbar q^2/2m$ the bare susceptibility $\chi_d^0(\mathbf{q},i\omega)$ is real and given by $\chi_d^0(\mathbf{q},i\omega) \approx \langle n\rangle q^2/m\omega^2$; also show that in the self-consistent field approximation the dielectric function has the form

$$\varepsilon(\mathbf{q},i\omega) = 1-(e^2/q^2)\chi_d^0(i\omega) \approx 1-(\omega_p^2/\omega^2), \tag{14}$$

where ω_p is the plasma frequency given in the preceding problem.

16.11 In subsection 16.10.4 we computed separately the reduced diagonal and non-diagonal response results i.e., after the Van Hove limit. However, one can also obtain the full transport quantities, even if the results have only formal significance. In this vein show that the generalized conductance is given by

$$L_{BA}(i\omega) = \int_0^\beta d\tau \mathrm{Tr}\left\{ e^{\hbar\tau\mathcal{L}^0} J_A \frac{1}{\Lambda_d - i\mathcal{L}^0 + i\omega} J_B \right\}, \tag{15}$$

where J_A and J_B generally include both collisional and ponderomotive current. Also, establish this result from the solution of the master equation.

16.12 In Equation (16.10-16) we showed that the diagonal part of a commutator is not necessarily zero, depending on the bilinear concomitant.
(a) As a first example, consider the one-particle Hamiltonian for the harmonic oscillator with $U(q) = \frac{1}{2}\alpha q^2$. Show that the matrix element $\langle\gamma|\{[q,\mathcal{H}^0]|\gamma\rangle\}$ $=0$; next, show from Green's theorem that there is no bilinear concomitant for this case.
(b) Consider a particle in a 3D box, $\mathcal{H}^0 = p^2/2m + U_0$, with ES $\gamma(\mathbf{q}) = V_0^{-1/2}$ $\times e^{i\mathbf{k}_\gamma\cdot\mathbf{q}}$. Apply periodic b.c. From the commutator matrix elements show that $\langle\gamma|\{[\mathbf{q},\mathcal{H}^0]|\gamma\rangle\} = (\hbar^2 i/m)\mathbf{k}$. Next, from Green's theorem for the operator ∇^2, obtain the bilinear concomitant $Q_{\mathcal{H}^0}[\mathbf{q}\gamma,\gamma]$ and confirm (16.10-15).
Hints. Note that $\oint\!\!\oint \mathsf{I}\cdot d\mathbf{S} = \oint\!\!\oint d\mathbf{S} = 0$ because of periodic b.c. and observe that, with α,β denoting Cartesian coordinates, $\oint\!\!\oint q_\alpha dS_\beta = \delta_{\alpha\beta}V_0$ by Gauss' theorem.

Chapter XVII

The Quantum Boltzmann Equation and Some Applications of Modified Linear Response

1. THE QUANTUM BOLTZMANN EQUATION: SCOPE AND ESSENCE

17.1 From the Master Equation to the Quantum Boltzmann Equation

In the previous chapter we considered quantum transport on the many-body level. Linear response theory was shown to provide a very general many-body framework for the description of interacting gases or quantum liquids, as well as for transport properties of such systems in a near-equilibrium steady state. However, in order to justify the linear response description *and* to obtain meaningful results, a randomness assumption was introduced, either simultaneously or *a posteriori*. Moreover, in doing so, we found that in various instances the linearization assumption could be dropped altogether, with the more general results being equally well derivable from the full master equation (FME) for the density operator. Therefore, transport results based on the full master equation go *beyond linear response.* These results are easily expressed on the one-body level, since interactions between the constituents of the current fluxes are usually accounted for by other means, like e.g. a Hartree–Fock description.

In the classical phase-space description one commonly goes from the N-particle distribution to the one-particle distribution by integrating over $f(N-1)$ coordinates and momenta, where f is the number of degrees of freedom per molecule; the procedure is found in a great many books on the statistics of gases, see e.g. Delcroix.[1] In quantum theory the method is basically simpler. One evaluates

$$f_\zeta(t) \equiv \langle n_\zeta \rangle_t = \mathrm{Tr}[\rho(t)\mathbf{n}_\zeta], \quad \mathbf{n}_\zeta = c_\zeta^\dagger c_\zeta\,, \tag{17.1-1}$$

where the lhs is the average occupancy of the one-particle state $|\zeta\rangle$. The first computation of this kind was carried out by Van Hove.[2] As before, we use bold $\mathbf{n}_\zeta$ for

[1] J.L. Delcroix, "Physique des Plasmas", Vol. 1, Dunod, Paris, 1963.

[2] L. Van Hove in "La théorie des gaz neutres et ionisés", Chapter XVI Ref. 67, p. 159ff. We note that Van Hove stayed, however, within the **k**-space formalism, as do others who derive a so-called QBE (cf. Mahan, Op. Cit., Section 8.5, following a Wigner-type Green's function procedure in Kadanoff and Baym, "Quantum Stat. Mech.", Benjamin 1962). The QBE envisaged by us here is *far more general*, being applicable to transport involving *any type* of single particle states, be they extended or localised.

the occupation-number operator and italic n_ζ for its eigenvalue. For fermions we have the restriction $0 \le f_\zeta \le 1$. However, we will find that it is essential that the nondiagonal part of the master equation be included from the beginning in order to arrive at consistent results. Therefore, we shall start from the full master equation (16.12-2) and take the trace $\mathrm{Tr}[\rho(t) c^\dagger_{\zeta_0} c_\zeta]$, thus finding the average $\langle c^\dagger_{\zeta_0} c_\zeta \rangle_t$. The evaluation will be considerably simplified by employing Lemmas 1 and 3 of subsection 16.14.3 [Eqs. (16.14-39) and (16.14-42)], which allow us to switch the operators $\rho(t)$ and $c^\dagger_{\zeta_0} c_\zeta$, so that

$$\begin{aligned}\mathrm{Tr}[c^\dagger_{\zeta_0} c_\zeta (\Lambda_d + i\mathscr{L}^0)\rho(t)] &= \mathrm{Tr}[\rho(t)(\Lambda_d - i\mathscr{L}^0) c^\dagger_{\zeta_0} c_\zeta] \\ &= \mathrm{Tr}[\rho(t)\Lambda_d \mathbf{n}_{\zeta_0}]\delta_{\zeta_0\zeta} - i\mathrm{Tr}[\rho(t)\mathscr{L}^0 c^\dagger_{\zeta_0} c_\zeta]. \qquad (17.1\text{-}2)\end{aligned}$$

We are now prepared to proceed to the derivation of the general, nonlinear QBE.

Having said this, we will 'backtrack' a bit. The earlier stated challenge to extract the streaming terms out of the nonlinear QBE is a no mean task! For extended states a result for the Wigner function, fully equivalent to the semi-classical BTE with two streaming terms has been obtained, based on the Wannier Hamiltonian description. For localised states, both $|\zeta)$ and ε_ζ depend on the applied E-field and the streaming terms must be extracted with perturbation theory. This has so far been only partially successful. Basically, it is much simpler to employ the inhomogeneous master equation (IME) as given in (16.12-7). This will directly lead to a QBE *with* streaming terms, but valid only in the linear or low-field regime. Yet, results for nonlinear magneto-phonon effects have been duly obtained by using the generalized Calecki current flux, also presented in this chapter.

17.1.1 *The Quantum Boltzmann Equation for Binary Interactions*

Let us multiply the full master equation (16.12-2) by the operator $c^\dagger_{\zeta_0} c_\zeta$ and take the trace. Employing (17.1-2), the direct result is

$$\frac{\partial \langle c^\dagger_{\zeta_0} c_\zeta \rangle_t}{\partial t} + \mathrm{Tr}[\rho(t)\Lambda_d \mathbf{n}_{\zeta_0}]\delta_{\zeta_0\zeta} - i\mathrm{Tr}[\rho(t)\mathscr{L}^0 c^\dagger_{\zeta_0} c_\zeta] = 0. \qquad (17.1\text{-}3)$$

Our task is now to evaluate the two terms at the lhs, which is straightforward, although somewhat laborious. For definiteness, we assume that two-body interactions occur between the given system, assumed to be composed of fermions, and impurities, which we shall describe as bosons. The bases of one-particle states will be denoted by $\{|\zeta)\}$ and $\{|\eta)\}$, respectively. We need the second quantization form

$$\lambda\mathcal{V} = \sum_{\zeta\zeta'\eta\eta'} c^\dagger_{\zeta'} a^\dagger_{\eta'} (\zeta'\eta'|\lambda v|\zeta\eta) c_\zeta a_\eta . \qquad (17.1\text{-}4)$$

The many-body occupation-number states will be denoted by $|\gamma\rangle = |\{n_\zeta\},\{N_\eta\}\rangle$. We remind the reader of the raising and lowering rules for fermion and boson operators, discussed in Chapter VII:

$$\begin{aligned}
c_\zeta^\dagger\, |\{n_\zeta\},\{N_\eta\}\rangle &= (-1)^{\xi}\sqrt{1-n_\zeta}\,|n_1,n_2,...,1-n_\zeta,...,\{N_\eta\}\rangle,\\
c_\zeta\, |\{n_\zeta\},\{N_\eta\}\rangle &= (-1)^{\xi'}\sqrt{n_\zeta}\,|n_1,n_2,...,1-n_\zeta,...,\{N_\eta\}\rangle,\\
a_\eta^\dagger\, |\{n_\zeta\},\{N_\eta\}\rangle &= \sqrt{N_\eta+1}\,|\{n_\zeta\},N_1,N_2,...N_\eta+1,...\rangle,\\
a_\eta\, |\{n_\zeta\},\{N_\eta\}\rangle &= \sqrt{N_\eta}\,|\{n_\zeta\},N_1,N_2,...N_\eta-1,...\rangle.
\end{aligned} \tag{17.1-5}$$

To find $\Lambda_d \mathbf{n}_{\zeta_0}$ we need the matrix element $\langle\gamma|\lambda\mathcal{V}|\bar{\gamma}\rangle = \langle\{n_\zeta\},\{N_\eta\}|\,\lambda\mathcal{V}\,||\{\bar{n}_\zeta\},\{\bar{N}_\eta\}\rangle$. With the two-body interaction (17.1-4) we have explicitly

$$W_{\gamma\bar{\gamma}} = (2\pi\lambda^2/\hbar) \sum_{(\zeta'\zeta''\eta'\eta'')_{con.}} |\langle\{n_\zeta\},\{N_\eta\}|\, c_{\zeta''}^\dagger a_{\eta''}^\dagger c_{\zeta'} a_{\eta'}\, |\{\bar{n}_\zeta\},\{\bar{N}_\eta\}\rangle|^2$$
$$\times|(\zeta''\eta''|v|\zeta'\eta')|^2\, \delta(\mathcal{E}_{\bar{\gamma}} - \mathcal{E}_\gamma). \tag{17.1-6}$$

In principle we must sum over all sets of indices occurring in (17.1-4); however, the states $|\gamma\rangle$ and $|\bar{\gamma}\rangle$ are only *connected* for a restricted set of indices due to the raising and lowering rules. So, choosing a set of indices, for given $|\gamma\rangle$ we will denote a connected state by $|\bar{\gamma}\rangle_{\zeta''\zeta'\eta''\eta'}$. Non-zero matrix elements require that the action of the raising and lowering operators reproduce the original state $|\gamma\rangle$. So we see that $\bar{n}_\zeta$ will be unchanged for all $\zeta \neq \zeta', \zeta''$, while for $\zeta = \zeta'$ or ζ'', $\bar{n}_\zeta$ must be equal to 1 $- n_\zeta$. Summarising, with the use of the Kronecker delta, we have

$$\bar{n}_\zeta = n_\zeta(1-\delta_{\zeta\zeta'}-\delta_{\zeta\zeta''}) + (1-n_\zeta)(\delta_{\zeta\zeta'}+\delta_{\zeta\zeta''}) \text{ or } \bar{n}_\zeta - n_\zeta = (1-2n_\zeta)(\delta_{\zeta\zeta'}+\delta_{\zeta\zeta''}). \tag{17.1-7}$$

Likewise, the boson operators require that only η'' is raised and η' is lowered. Hence,

$$N_{\eta''} = \bar{N}_{\eta''}+1,\quad N_{\eta'} = \bar{N}_{\eta'}-1 \quad\text{or}\quad \bar{N}_\eta - N_\eta = \delta_{\eta\eta'} - \delta_{\eta\eta''}. \tag{17.1-8}$$

We thus obtain from (17.1-5) and (17.1-6) – (17.1-8)

$$\begin{aligned}
W_{\gamma\bar{\gamma}_{\zeta''\zeta'\eta''\eta'}} &= \frac{2\pi\lambda^2}{\hbar}(1-\bar{n}_{\zeta''})\bar{n}_{\zeta'}(1+\bar{N}_{\eta''})\bar{N}_{\eta'}\,|(\zeta''\eta''|v|\zeta'\eta')|^2\delta(\varepsilon_{\zeta''}-\varepsilon_{\zeta'}+E_{\eta''}-E_{\eta'})\\
&= \frac{2\pi\lambda^2}{\hbar}(1-n_{\zeta'})n_{\zeta''}(1+N_{\eta'})N_{\eta''}\,|(\zeta''\eta''|v|\zeta'\eta')|^2\delta(\varepsilon_{\zeta''}-\varepsilon_{\zeta'}+E_{\eta''}-E_{\eta'}),
\end{aligned} \tag{17.1-9}$$

where the delta function reflects the appropriate changes in energies. To further evaluate $\Lambda_d \mathbf{n}_{\zeta_0}$ we must multiply (17.1-9) with $-|\gamma\rangle\langle\gamma|(\bar{n}_{\zeta_0} - n_{\zeta_0})$, sum over all $|\gamma\rangle$, as well as over the set $(\zeta'\zeta''\eta'\eta'')$ which define the connected states $|\gamma\rangle_{\zeta'\zeta''\eta'\eta''}$. We will make the usual adiabatic assumption that the boson states remain in thermal equilibrium; hence, taking the trace over the boson bath states, we define the transition rates

$$w_{\zeta''\zeta'} = \sum_{\eta'\eta''} \langle N_{\eta''}(1+N_{\eta'})\rangle_{eq}\, Q(\zeta''\eta'';\zeta'\eta')$$
$$= \sum_{\eta'\eta''} \langle N_{\eta''}\rangle_{eq} \langle (1+N_{\eta'})\rangle\, Q(\zeta''\eta'';\zeta'\eta'), \qquad (17.1\text{-}10)$$

where

$$Q(\zeta''\eta'';\zeta'\eta') = \frac{2\pi\lambda^2}{\hbar} |(\zeta''\eta''|v|\zeta'\eta')|^2\, \delta(\varepsilon_{\zeta''} - \varepsilon_{\zeta'} + E_{\eta''} - E_{\eta'}). \quad (17.1\text{-}11)$$

From Bose–Einstein statistics and the delta function for the energies one easily corroborates

$$w_{\zeta''\zeta'} = w_{\zeta'\zeta''}\, e^{-\beta(\varepsilon_{\zeta'} - \varepsilon_{\zeta''})}. \qquad (17.1\text{-}12)$$

So we arrived at the boson-averaged expression $\langle \Lambda_d \mathbf{n}_{\zeta_0}\rangle_b = \sum_\gamma |\gamma\rangle\langle\gamma| \langle M n_{\zeta_0}\rangle_b$; left to evaluate

$$\langle M n_{\zeta_0}\rangle_b = -\sum_{\zeta'\zeta''} w_{\zeta''\zeta'}(1-n_{\zeta'})n_{\zeta''}(\bar{n}_{\zeta_0} - n_{\zeta_0})$$
$$\overset{(17.1\text{–}7)}{=} -\sum_{\zeta'\zeta''} w_{\zeta''\zeta'}(1-n_{\zeta'})n_{\zeta''}(1-2n_{\zeta_0})(\delta_{\zeta_0\zeta''} + \delta_{\zeta_0\zeta'})$$
$$= -\sum_{\zeta'} w_{\zeta_0\zeta'}(1-n_{\zeta'})n_{\zeta_0}(1-2n_{\zeta_0}) - \sum_{\zeta''} w_{\zeta''\zeta_0} n_{\zeta''}(1-n_{\zeta_0})(1-2n_{\zeta_0}). \quad (17.1\text{-}13)$$

Noting that $n_{\zeta_0}^2 = n_{\zeta_0}$, we have

$$n_{\zeta_0}(1-2n_{\zeta_0}) = -n_{\zeta_0}, \qquad (1-n_{\zeta_0})(1-2n_{\zeta_0}) = 1-n_{\zeta_0}, \qquad (17.1\text{-}14)$$

so that

$$\langle M n_{\zeta_0}\rangle_b = \sum_{\zeta'} [w_{\zeta_0\zeta'} n_{\zeta_0}(1-n_{\zeta'}) - w_{\zeta'\zeta_0} n_{\zeta'}(1-n_{\zeta_0})] \equiv \mathcal{B}_{\zeta_0} n_{\zeta_0}. \qquad (17.1\text{-}15)$$

Here $\mathcal{B}$ is the Boltzmann collision operator; note that, while on the many-body level Λ_d is linear, the Boltzmann collision operator is quadratic in the occupancies, since the Pauli exclusion principle is accounted for by the rules involving $c^\dagger$ and c. Hence, we established

$$\langle \Lambda_d \mathbf{n}_{\zeta_0}\rangle_b = \sum_{\{n_\zeta\}} |\{n_\zeta\}\rangle\langle\{n_\zeta\}| \mathcal{B}_{\zeta_0} n_{\zeta_0}. \qquad (17.1\text{-}16)$$

Lastly, we must average over the density operator of the fermion system. We find

$$\mathrm{Tr}[\rho(t)\Lambda_d \mathbf{n}_{\zeta_0}] = \sum_{\{n_\zeta\}} p(\{n_\zeta\},t) \langle \mathcal{B}_{\zeta_0} n_{\zeta_0}\rangle_b = \langle \mathcal{B}_{\zeta_0} n_{\zeta_0}\rangle_t \simeq \mathcal{B}_{\zeta_0}\langle n_{\zeta_0}\rangle_t; \quad (17.1\text{-}17)$$

in the last transition we used again the 'truncation rule'

$$\langle n_\zeta(1-n_{\zeta'})\rangle_t \simeq \langle n_\zeta\rangle_t (1-\langle n_{\zeta'}\rangle_t), \qquad (17.1\text{-}18)$$

which implies that there are no correlations, $\langle\Delta n_\zeta(t)\Delta n_{\zeta'}(t)\rangle = 0,\ \zeta\neq\zeta'$, a statement that is correct in an equilibrium grand-canonical ensemble. It corresponds to Boltzmann's assumption of 'molecular chaos'. The derivation given here is in essence equivalent to that of Van Hove[2], although carried out more consistently from the master equation in Liouville space.

We now turn to the last term in (17.1-3). Using the commutator form for $\mathcal{L}^0$, this term can be written as (we change the indices to ζ'',ζ'):

$$-\frac{i}{\hbar}\mathrm{Tr}\{\rho(t)[\mathcal{H}^0, c^\dagger_{\zeta''}c_{\zeta'}]\} =$$

$$=\frac{i}{\hbar}\sum_{\gamma\bar{\gamma}}\{\langle\gamma|\rho(t)|\bar{\gamma}\rangle\langle\bar{\gamma}|c^\dagger_{\zeta''}c_{\zeta'}\mathcal{H}^0|\gamma\rangle - \langle\gamma|\rho(t)|\bar{\gamma}\rangle\langle\bar{\gamma}|\,\mathcal{H}^0 c^\dagger_{\zeta''}c_{\zeta'}|\gamma\rangle\}$$

$$=\frac{i}{\hbar}\sum_{\gamma\bar{\gamma}}\langle\gamma|\rho(t)|\bar{\gamma}\rangle\langle\bar{\gamma}|c^\dagger_{\zeta''}c_{\zeta'}|\gamma\rangle(\mathcal{E}_\gamma - \mathcal{E}_{\bar{\gamma}}). \qquad (17.1\text{-}19)$$

Now with $|\gamma\rangle = |\{n_\zeta\},\{N_\eta\}\rangle$ the connecting states must be such that

$$\begin{aligned}&\bar{n}_{\zeta''} = 1 - n_{\zeta''}, \qquad \bar{n}_{\zeta'} = 1 - n_{\zeta'},\\ &\text{all other } \bar{n}_\zeta = n_\zeta, \quad \text{all } \bar{N}_\eta = N_\eta.\end{aligned} \qquad (17.1\text{-}20)$$

Since $n_{\zeta'}$ is lowered and $n_{\zeta''}$ is raised, we have $\mathcal{E}_{\bar{\gamma}} = \mathcal{E}_\gamma + \varepsilon_{\zeta''} - \varepsilon_{\zeta'}$. Further, in taking the trace in (17.1-19), the boson average sums over all bath states; the remaining average over the fermion density operator is once more denoted by $\langle..\rangle_t$. Whence,

$$-\frac{i}{\hbar}\mathrm{Tr}\{\rho(t)[\mathcal{H}^0, c^\dagger_{\zeta''}c_{\zeta'}]\} = \frac{i}{\hbar}\langle c^\dagger_{\zeta''}c_{\zeta'}\rangle_t(\varepsilon_{\zeta'} - \varepsilon_{\zeta''}). \qquad (17.1\text{-}21)$$

The resulting quantum Boltzmann equation becomes with $(\zeta'',\zeta')\to(\zeta,\zeta')$:

$$\frac{\partial\langle c^\dagger_\zeta c_{\zeta'}\rangle_t}{\partial t} + \mathcal{B}_\zeta\langle n_\zeta\rangle_t\delta_{\zeta\zeta'} + \frac{i}{\hbar}\langle c^\dagger_\zeta c_{\zeta'}\rangle_t(\varepsilon_{\zeta'} - \varepsilon_\zeta) = 0. \qquad (17.1\text{-}22)$$

While this is the final result, we believe that the sharp diagonal should not be taken as realistic, except in thermal equilibrium when the diagonal and nondiagonal parts are separately zero. Let $\Psi_{\zeta\zeta'}$ be an appropriate weighting function that couples to the density $\langle n_\zeta\rangle_t$. For some cases involving the steady state (or transient state) the coupling could represent the interaction with the electrodes, which serve as charge reservoirs, see Fig. 17-1. The relevant transport equation is now taken to be:

$$\sum_{\zeta'}\left[\frac{\partial\langle c^\dagger_\zeta c_{\zeta'}\rangle_t}{\partial t}\Psi_{\zeta\zeta'} + \mathcal{B}_\zeta\langle n_\zeta\rangle_t\Psi_{\zeta\zeta'} + \frac{i}{\hbar}\langle c^\dagger_\zeta c_{\zeta'}\rangle_t(\varepsilon_{\zeta'} - \varepsilon_\zeta)\Psi_{\zeta\zeta'}\right] = 0. \quad (17.1\text{-}23)$$

This will be further discussed in Section 17.2.

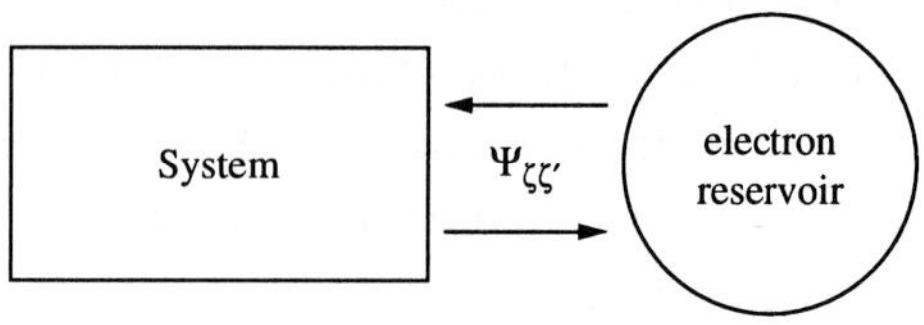

Fig. 17-1. Interaction with the electrodes electron reservoirs.

17.1.2 *The Quantum Boltzmann Equation for Electron-Phonon Interaction*

We shall briefly redo the argument for electron-phonon interaction. The interaction potential is now

$$\lambda V = (i) \sum_{\zeta'\zeta''\mathbf{q}'} \mathcal{F}(\mathbf{q}')[c^\dagger_{\zeta''} c_{\zeta'} a_{\mathbf{q}'} (\zeta''|e^{i\mathbf{q}'\cdot\mathbf{r}}|\zeta') \pm c^\dagger_{\zeta''} c_{\zeta'} a^\dagger_{\mathbf{q}'} (\zeta''|e^{-i\mathbf{q}'\cdot\mathbf{r}}|\zeta')], \quad (17.1\text{-}24)$$

where the inclusion of the factor (i) and the choice of $\pm$ depends on the type of phonons. For the transition rate we have instead of (17.1-6)

$$W_{\gamma\bar{\gamma}} = \frac{2\pi}{\hbar} \sum_{\zeta'\zeta''\mathbf{q}'} |\mathcal{F}(\mathbf{q}')|^2 \Big(|\langle\{n_\zeta\},\{N_\mathbf{q}\}| c^\dagger_{\zeta''} c_{\zeta'} a_{\mathbf{q}'} |\{\bar{n}_\zeta\},\{\bar{N}_\mathbf{q}\}\rangle|^2 |(\zeta''|e^{i\mathbf{q}'\cdot\mathbf{r}}|\zeta')|^2$$
$$\times \delta(\varepsilon_{\zeta'} - \varepsilon_{\zeta''} + \hbar\omega_{\mathbf{q}'}) + |\langle\{n_\zeta\},\{N_\mathbf{q}\}| c^\dagger_{\zeta''} c_{\zeta'} a^\dagger_{\mathbf{q}'} |\{\bar{n}_\zeta\},\{\bar{N}_\mathbf{q}\}\rangle|^2$$
$$\times |(\zeta''|e^{-i\mathbf{q}'\cdot\mathbf{r}}|\zeta')|^2\, \delta(\varepsilon_{\zeta'} - \varepsilon_{\zeta''} - \hbar\omega_{\mathbf{q}'}) \Big). \quad (17.1\text{-}25)$$

The diagrams for phonon absorption and emission were given in Fig. 8-14. To get connected states for the above matrix elements we need

$$\bar{n}_\zeta = n_\zeta, \quad \zeta \neq \zeta', \zeta'',$$
$$\bar{n}_{\zeta'} = 1 - n_{\zeta'}, \quad \bar{n}_{\zeta''} = 1 - n_{\zeta''}, \quad (17.1\text{-}26)$$
$$\bar{N}_\mathbf{q} = N_\mathbf{q} \pm \delta_{\mathbf{q},\mathbf{q}'},$$

with the upper sign for absorption and the lower sign for emission. Denoting the connected states as $|\bar{\gamma}\rangle_{\zeta'\zeta''\mathbf{q}'}$, with the above data we find

$$W^{abs}_{\gamma\bar{\gamma}_{\zeta'\zeta''\mathbf{q}'}} = Q(\zeta',\mathbf{q}' \to \zeta'')(1-n_{\zeta'}) n_{\zeta''} N_{\mathbf{q}'},$$
$$W^{em}_{\gamma\bar{\gamma}_{\zeta'\zeta''\mathbf{q}'}} = Q(\zeta' \to \zeta'',\mathbf{q}')(1-n_{\zeta'}) n_{\zeta''} (1+N_{\mathbf{q}'}), \quad (17.1\text{-}27)$$

where

$$\left.\begin{matrix} Q(\zeta',\mathbf{q}' \to \zeta'') \\ Q(\zeta' \to \zeta'',\mathbf{q}') \end{matrix}\right\} = \frac{2\pi}{\hbar} |\mathcal{F}(\mathbf{q}')|^2 |(\zeta''|e^{\pm i\mathbf{q}'\cdot\mathbf{r}}|\zeta')|^2\, \delta(\varepsilon_{\zeta'} - \varepsilon_{\zeta''} \pm \hbar\omega_{\mathbf{q}'}). \quad (17.1\text{-}28)$$

The fermionic transition rates are

$$w_{\zeta'\zeta''} = \sum_{\mathbf{q}} \{Q(\zeta',\mathbf{q} \to \zeta'')\langle N_{\mathbf{q}}\rangle_{eq} + Q(\zeta' \to \zeta'',\mathbf{q})[1+\langle N_{\mathbf{q}}\rangle_{eq}]\}, \quad (17.1\text{-}29)$$

where we assumed again that the phonons remain in equilibrium (no phonon drag). The relationship (17.1-12) still holds. Computing next the boson-averaged operator $\langle Mn_\zeta\rangle_b$, employing the above rates $w_{\zeta'\zeta''}$, one finds once more the Boltzmann operator (17.1-15). The form of the quantum Boltzmann equation is unchanged.

17.2 Discussion of the Equilibrium and Steady-State Distribution

Although the derivation of (17.1-22) was straightforward, a face-value acceptance of the results meets with considerable difficulty, as has been occasionally recognised in the literature.[3,4]

The transport through *extended states*, such as momentum states or Bloch states in condensed matter, was the intended subject of Van Hove's[2] original quantum Boltzmann equation for the time rate of change of the diagonal occupation $(\partial/\partial t)\langle n_\zeta\rangle_t$; since here $\zeta = \zeta'$, the streaming terms are altogether missing, certainly an odd feature for ponderomotive transport. Clearly, the nondiagonal contribution is an integral part of the QBE. The connective functional $\Psi_{\mathbf{kk'}}$ for this case turns out to be the Weyl transform, which ties the nondiagonal average $\langle c^\dagger_{\mathbf{k}-(1/2)\mathbf{u}}\, c_{\mathbf{k}+(1/2)\mathbf{u}}\rangle_t$ to the Wigner-occupancy function. The full semi-classical Boltzmann equation is then obtained, as will be shown in Section 17.3. In an equilibrium state, the collision terms and the streaming terms are separately zero, as was the case in Boltzmann's original approach.

More essential problems occur when we are dealing with *localised states*, as e.g. associated with a magnetic field (Landau states) or with quantum confinement in a 2D well in GaAlAs/GaAs hetero-junctions (2D electron-gas FETs). In equilibrium we conclude from (17.1-22) that the distribution is determined by

$$B_\zeta[f_{\zeta,eq}] = \sum_{\zeta'} [w_{\zeta\zeta'} f_{\zeta,eq}(1-f_{\zeta',eq}) - w_{\zeta'\zeta} f_{\zeta',eq}(1-f_{\zeta,eq})] = 0. \qquad (17.2\text{-}1)$$

The system can still be subject to an electric field, although there is no current flow; this situation is realised when the specimen is placed between the plates of a capacitor. The distribution function will be inhomogeneous, meaning that the centres of the localised 'orbits' will arrange themselves such that the negative charges congregate near the positive plate (although Coulomb repulsion will partially offset this displacement). The Gibbs' entropy $-k_B \Sigma_\zeta f_\zeta \ln f_\zeta$ has a maximum; from Klein's

[3] H.F. Budd, Phys. Rev. **175**, 241 (1968).

[4] D. Calecki, C. Lewiner and P. Nozières, J. Phys. **38**, 169 (1977).

lemma it then follows that each term of the summand is zero, so that there is detailed balance, as for transport through extended states.

Next, let us consider the steady state for the case that electrodes are attached. Current flow will now smooth the electron distribution. That means that the relevant solution $f_{\zeta,ss}$ *will only involve those quantum numbers* [a subset of $\{\zeta\}$] *that do not involve the localisation of the states*; explicit examples will be presented later. Therefore we *cannot* have

$$B_\zeta[f_{\zeta,ss}] = \sum_{\zeta'} [w_{\zeta\zeta'} f_{\zeta,ss}(1-f_{\zeta',ss}) - w_{\zeta'\zeta} f_{\zeta',ss}(1-f_{\zeta,ss})] = 0, \quad \text{[wrong]} \qquad (17.2\text{-}2)$$

although such a result is expected from (17.1-22) (the nondiagonal term giving zero) and claimed to be valid by many authors. From a practical point of view, we note that *if* (17.7-2) would hold, the collisional current, which is usually the only current for flow involving Landau states, or hopping between 'tail' states of impurity bands, etc., would be identically zero (!) – as elaborated in Section 17.4. To remedy the situation, we need again a summation over a weighting function $\Psi_{\zeta\bar{\zeta}}$ which depends on the applied field and removes the non-essential quantum numbers from the steady-state distribution.

More fundamentally, perhaps, in the presence of current flow, a *strict* one-electron distribution function may not exist. The dissipative effect of current flow entails an energy transfer to the electron gas, which, however slightly, heats up, resulting in an effective 'hot' electron temperature T_e. This quantity is a collective property of the electron gas. It is not our task here to develop such models. Without stringent proof, we believe, however, that this procedure will cause (17.2-2) to be replaced with a streaming-dependent form

$$\sum_{\zeta'} \Big\{ [w_{\zeta\zeta'} f_{\zeta,ss}(1-f_{\zeta',ss}) - w_{\zeta'\zeta} f_{\zeta',ss}(1-f_{\zeta,ss})] \\ + \mathcal{O}\Big(w_{\zeta\zeta'} f_{\zeta,ss}(1-f_{\zeta',ss})[q\mathbf{E}\cdot\Delta\mathbf{R}_{\zeta\zeta'}/\hbar\omega_0]^n \Big) \Big\} = 0, \qquad (17.2\text{-}3)$$

where $q\mathbf{E}\cdot\Delta\mathbf{r}_{\zeta\zeta'}$ is the energy transferred in a jump $\zeta \to \zeta'$, mediated by an optical phonon $\hbar\omega_0$; the exponent n is an appropriate power, depending on the model. One such a model was developed in the late seventies by Calecki et al. (Ref. 4); with a random walk picture they modified the lhs of (17.2-2) into a Fokker–Planck equation, whose solution led to a quasi-Maxwellian distribution with an electron temperature[5]

$$T_e = \frac{\hbar\omega_0}{2k_B} \coth(\hbar\omega_0/2k_B T)\Big[1 + (q\mathbf{E}\cdot\langle\Delta\mathbf{R}\rangle/\hbar\omega_0)^2\Big]. \qquad (17.2\text{-}4)$$

Note that for this model $n = 2$. For other models, see the references in Ref. 6 below.[6]

---

[5] Given in this form by Barker, J.R. Barker, Solid-State Electron. **21**, 197 (1978), Eq. (33).

[6] P. Vasilopoulos, M. Charbonneau and C.M. Van Vliet, Phys. Rev. **B 35**, 1334 (1987).

Lastly, we mention that even for small fields, i.e., in the context of LRT, the steady-state distribution does not satisfy Eq. (17.2-2), but obeys a form like (17.2-3), with the exponent n being one for this case; this will be shown in Section 17.7.

Altogether, while the QBE with slight coarse-graining is useful for linear and nonlinear flow through extended states, it needs additional models and assumptions for localised states. Fortunately, however, the form for the electron current flux, previously obtained from the reduced Heisenberg equation, can be employed *as is*; we will see that it contains all diagonal and nondiagonal contributions, both in the linear and nonlinear regime.

17.3. Extended States. Recovery of the BTE via the Wigner Formalism

Whenever we have extended states due to overlap of the atomic wave functions as in semiconductors and metals, regular band theory can be applied, with the charge carriers being largely free, except for an effective mass. Indeed, Bloch functions are in most transport treatises approximated by plane waves with $\mathbf{v_k} = \hbar^{-1}\nabla_{\mathbf{k}}\varepsilon(\mathbf{k}) \simeq \hbar\mathbf{k}/m^*$, m^* being the effective mass. The position matrix element for Bloch states is zero, so the current is fully ponderomotive. The transport equation was stated in (17.1-22) and (17.1-23). The collision integral of the semi-classical BTE is there, while the streaming terms come apparently from the nondiagonal quantum interference term. However, the nondiagonal part (ndp) does not have the right form to serve us here. We illustrate this for the case that the carriers move in a constant field in the x-direction. The potential energy is then a ramp $-qE_x x$ and the eigenstates are Airy functions. For a macroscopic sample, $L_x \gg$ de Broglie wavelength, the levels may be sufficiently dense for overlap to occur; yet, this is not the customary way to describe transport in condensed matter.

So, following Wannier and Slater, we employ the Wannier Hamiltonian, in which the band energy and potential energy due to external fields are superimposed,

$$h_W^0 = \varepsilon[-i\nabla - (q/\hbar c)\mathbf{A}(\mathbf{r})] + q\Phi(\mathbf{r}), \tag{17.3-1}$$

[see also (13.3-7)], where $\varepsilon(\mathbf{k})$ is the unperturbed band structure. Denoting the full system Hamiltonian by $\mathcal{H}^0$, we have the usual form

$$\mathcal{H}^0 = \sum_{i\text{ th particle}} h_{W,i}^0 = \sum_{\zeta_1\zeta_2} c_{\zeta_2}^\dagger c_{\zeta_1} (\zeta_2 | h_W^0 | \zeta_1). \tag{17.3-2}$$

Therefore,

$$\text{ndp} = -i\,\text{Tr}\{\rho(t)\mathcal{L}^0 c_{\zeta''}^\dagger c_{\zeta'}\} = -\frac{i}{\hbar}\sum_{\zeta_1\zeta_2} (\zeta_2 | h_W^0 | \zeta_1)\,\text{Tr}\{\rho(t)[c_{\zeta_2}^\dagger c_{\zeta_1}, c_{\zeta''}^\dagger c_{\zeta'}]\}. \tag{17.3-3}$$

For the commutator we have with the standard rules

$$\left[c_{\zeta_2}^\dagger c_{\zeta_1}, c_{\zeta''}^\dagger c_{\zeta'}\right] = c_{\zeta_2}^\dagger\left[c_{\zeta_1}, c_{\zeta''}^\dagger\right]c_{\zeta'} + c_{\zeta_2}^\dagger c_{\zeta''}^\dagger\left[c_{\zeta_1}, c_{\zeta'}\right]$$
$$+\left[c_{\zeta_2}^\dagger, c_{\zeta''}^\dagger\right]c_{\zeta'}c_{\zeta_1} + c_{\zeta''}^\dagger\left[c_{\zeta_2}^\dagger, c_{\zeta'}\right]c_{\zeta_1}. \tag{17.3-4}$$

We now write out the commutators and combine the four positive and negative terms to yield

$$c_{\zeta_2}^\dagger\left(c_{\zeta_1}c_{\zeta''}^\dagger + c_{\zeta''}^\dagger c_{\zeta_1}\right)c_{\zeta'} + \left(c_{\zeta_2}^\dagger c_{\zeta''}^\dagger + c_{\zeta''}^\dagger c_{\zeta_2}^\dagger\right)c_{\zeta'}c_{\zeta_1}$$
$$-c_{\zeta_2}^\dagger c_{\zeta''}^\dagger\left(c_{\zeta_1}c_{\zeta'} + c_{\zeta'}c_{\zeta_1}\right) - c_{\zeta''}^\dagger\left(c_{\zeta_2}^\dagger c_{\zeta'} + c_{\zeta'}c_{\zeta_2}^\dagger\right)c_{\zeta_1}. \tag{17.3-5}$$

This is simplified with the fermion-operator anti-commutation relationships

$$\left[c_\alpha, c_\beta^\dagger\right]_+ = c_\alpha c_\beta^\dagger + c_\beta^\dagger c_\alpha = \delta_{\alpha\beta},$$
$$\left[c_\alpha^\dagger, c_\beta^\dagger\right]_+ = c_\alpha^\dagger c_\beta^\dagger + c_\beta^\dagger c_\alpha^\dagger = 0,$$
$$\left[c_\alpha, c_\beta\right]_+ = c_\alpha c_\beta + c_\beta c_\alpha = 0. \tag{17.3-6}$$

We so obtain

$$\text{ndp} = (-i/\hbar)\sum_{\zeta_1\zeta_2}(\zeta_2|h_W^0|\zeta_1)\text{Tr}\left\{\rho(t)\left(c_{\zeta_2}^\dagger c_{\zeta'}\delta_{\zeta_1,\zeta''} - c_{\zeta''}^\dagger c_{\zeta_1}\delta_{\zeta_2,\zeta'}\right)\right\}$$
$$= (-i/\hbar)\sum_{\zeta}\left\{(\zeta|h_W^0|\zeta'')\langle c_\zeta^\dagger c_{\zeta'}\rangle_t - (\zeta'|h_W^0|\zeta)\langle c_{\zeta''}^\dagger c_\zeta\rangle_t\right\}. \tag{17.3-7}$$

For the states $|\zeta)$ we take plane waves, being the eigenstates of the band Hamiltonian in the effective mass approximation. For that part the matrix elements are diagonal. Thus we arrive at

$$\text{ndp} = (i/\hbar)\langle c_{\zeta''}^\dagger c_{\zeta'}\rangle_t(\varepsilon_{band,\zeta'} - \varepsilon_{band,\zeta''})$$
$$-(i/\hbar)\sum_{\zeta}\left[\langle c_\zeta^\dagger c_{\zeta'}\rangle_t(\zeta|q\Delta\Phi|\zeta'') - \langle c_{\zeta''}^\dagger c_\zeta\rangle_t(\zeta'|q\Delta\Phi|\zeta)\right], \tag{17.3-8}$$

where we used $\Delta\Phi$ since the extended states must be sufficiently close for transport to occur. Subsequently we write $q\Delta\Phi = q\nabla\Phi\cdot\Delta\mathbf{r} = -q\mathbf{E}(\mathbf{r})\cdot\Delta\mathbf{r}$. The QBE now takes the form

$$\frac{\partial}{\partial t}\langle c_{\zeta''}^\dagger c_{\zeta'}\rangle_t + \mathscr{B}_\zeta\langle n_\zeta\rangle_t\delta_{\zeta'\zeta''} + (i/\hbar)\langle c_{\zeta''}^\dagger c_{\zeta'}\rangle_t(\varepsilon_{band,\zeta'} - \varepsilon_{band,\zeta''})$$
$$+(i/\hbar)\sum_{\zeta}\left[\langle c_\zeta^\dagger c_{\zeta'}\rangle_t q\mathbf{E}\cdot(\zeta|\Delta\mathbf{r}|\zeta'') - \langle c_{\zeta''}^\dagger c_\zeta\rangle_t q\mathbf{E}\cdot(\zeta'|\Delta\mathbf{r}|\zeta)\right]. \tag{17.3-9}$$

Note that $\zeta = (\mathbf{k}, \sigma)$, where $\mathbf{k}$ is the crystal momentum and σ is the spin. The latter will be suppressed and taken into account in the density of states when necessary.

Eq. (17.3-9) will now be converted to a transport equation for the Wigner occupation function. Thus we set $\zeta'' \to \mathbf{k} - \frac{1}{2}\mathbf{u}$ and $\zeta' \to \mathbf{k} + \frac{1}{2}\mathbf{u}$; we multiply by

$(V_0/8\pi^3)e^{i\mathbf{u}\cdot\mathbf{r}}$ and we integrate over d^3u, to obtain (this is the envisaged $\Sigma_\zeta \Psi_{\zeta\zeta'}$):

$$\frac{V_0}{8\pi^3}\int d^3u e^{i\mathbf{u}\cdot\mathbf{r}}\left\{\frac{\partial}{\partial t}\langle c^\dagger_{\mathbf{k}-(1/2)\mathbf{u}} c_{\mathbf{k}+(1/2)\mathbf{u}}\rangle_t + \mathscr{R}_{\mathbf{k}}\langle n_{\mathbf{k}}\rangle_t \Delta^3 u\right.$$
$$+ (i/\hbar)\langle c^\dagger_{\mathbf{k}-(1/2)\mathbf{u}} c_{\mathbf{k}+(1/2)\mathbf{u}}\rangle_t (\varepsilon_{band,\mathbf{k}+(1/2)\mathbf{u}} - \varepsilon_{band,\mathbf{k}-(1/2)\mathbf{u}})$$
$$\left. +(i/\hbar)\sum_{\bar{\mathbf{k}}}[\langle c^\dagger_{\bar{\mathbf{k}}} c_{\mathbf{k}+(1/2)\mathbf{u}}\rangle_t \mathscr{q}\mathbf{E}\cdot(\bar{\mathbf{k}}\,|\,\Delta\mathbf{r}\,|\,\mathbf{k}-\tfrac{1}{2}\mathbf{u}) - \langle c^\dagger_{\mathbf{k}-(1/2)\mathbf{u}} c_{\bar{\mathbf{k}}}\rangle_t \mathscr{q}\mathbf{E}\cdot(\mathbf{k}+\tfrac{1}{2}\mathbf{u}\,|\,\Delta\mathbf{r}\,|\,\bar{\mathbf{k}})]\right\} = 0. \tag{17.3-10}$$

In this must be substituted the inverse transform for the one-particle Wigner distribution, cf. (16.10-52),

$$\langle c^\dagger_{\mathbf{k}-(1/2)\mathbf{u}} c_{\mathbf{k}+(1/2)\mathbf{u}}\rangle_t = (h^3/V_0)\int d^3\bar{r}\, e^{-i\mathbf{u}\cdot\bar{\mathbf{r}}}\rho_1(\mathbf{k},\bar{\mathbf{r}},t). \tag{17.3-11}$$

We now compute the four terms of (17.3-10). For the first term we have the simple result

$$\text{first term} = \frac{1}{8\pi^3}\int d^3\bar{r}\int d^3u\, e^{i\mathbf{u}\cdot(\mathbf{r}-\bar{\mathbf{r}})}[h^3\rho_1(\mathbf{k},\bar{\mathbf{r}},t)] = \frac{\partial f}{\partial t}, \tag{17.3-12}$$

where $f(\mathbf{k},\mathbf{r},t)$ is the (coarse grained) Wigner occupation function and where we used the well-known result

$$\int d^3u\, e^{i\mathbf{u}\cdot(\mathbf{r}-\bar{\mathbf{r}})} = 8\pi^3\delta(\mathbf{r}-\bar{\mathbf{r}}). \tag{17.3-13}$$

For the collision integral the result, when treating it as a sharp diagonal, would have been $\delta_{\mathbf{u},0}$, but in accord with our previous remarks, we broadened the diagonal to $\Delta^3 u$. The coarse-graining for this term must actually be carried out; it will be the subject of problem 17.4. The result is found to be

$$\text{second term} = \mathscr{R}_{\mathbf{k}} f(\mathbf{k},\mathbf{r},t)$$
$$= \sum_{\mathbf{k}'}\{w_{\mathbf{k}\mathbf{k}'} f(\mathbf{k},\mathbf{r},t)[1-f(\mathbf{k}',\mathbf{r},t)] - w_{\mathbf{k}'\mathbf{k}} f(\mathbf{k}',\mathbf{r},t)[1-f(\mathbf{k},\mathbf{r},t)]\}. \tag{17.3-14}$$

For the third term we need

$$\varepsilon_{band,\mathbf{k}+(1/2)\mathbf{u}} - \varepsilon_{band,\mathbf{k}-(1/2)\mathbf{u}} = \nabla_{\mathbf{k}}\varepsilon_{band}\cdot\mathbf{u} = \hbar\mathbf{v}_{\mathbf{k}}\cdot\mathbf{u}. \tag{17.3-15}$$

Differentiating (17.3-13) to $\bar{\mathbf{r}}$, we find

$$\int d^3u\,\mathbf{u}e^{i\mathbf{u}\cdot(\mathbf{r}-\bar{\mathbf{r}})} = 8i\pi^3\nabla_{\bar{\mathbf{r}}}\delta(\mathbf{r}-\bar{\mathbf{r}}). \tag{17.3-16}$$

Thus, we obtain

$$\text{third term} = \frac{i}{8\pi^3}\int d^3\bar{r}\int d^3u\,\mathbf{v}_{\mathbf{k}}\cdot\mathbf{u}\, e^{i\mathbf{u}\cdot(\mathbf{r}-\bar{\mathbf{r}})}[h^3\rho_1(\mathbf{k},\bar{\mathbf{r}},t)]$$
$$= -\mathbf{v}_{\mathbf{k}}\cdot\int d^3\bar{r}\,\nabla_{\bar{\mathbf{r}}}\delta(\mathbf{r}-\bar{\mathbf{r}})[h^3\rho_1(\mathbf{k},\bar{\mathbf{r}},t)] = \mathbf{v}_{\mathbf{k}}\cdot\nabla_{\mathbf{r}} f(\mathbf{k},\mathbf{r},t). \tag{17.3-17}$$

Finally, we will evaluate the fourth term. Needed are the matrix elements[7]

$$
\begin{aligned}
(\bar{\mathbf{k}}\,|\,\Delta\mathbf{r}\,|\,\mathbf{k}-\tfrac{1}{2}\mathbf{u}) &= \frac{1}{V_0}\int d^3r\,\mathbf{r}e^{i(\mathbf{k}-(1/2)\mathbf{u}-\bar{\mathbf{k}})\cdot\mathbf{r}} = \frac{8i\pi^3}{V_0}\nabla_{\bar{\mathbf{k}}}\delta(\mathbf{k}-\tfrac{1}{2}\mathbf{u}-\bar{\mathbf{k}}), \\
(\mathbf{k}+\tfrac{1}{2}\mathbf{u}\,|\,\Delta\mathbf{r}\,|\,\bar{\mathbf{k}}) &= \frac{1}{V_0}\int d^3r\,\mathbf{r}e^{i(\bar{\mathbf{k}}-(1/2)\mathbf{u}-\mathbf{k})\cdot\mathbf{r}} = -\frac{8i\pi^3}{V_0}\nabla_{\bar{\mathbf{k}}}\delta(\bar{\mathbf{k}}-\tfrac{1}{2}\mathbf{u}-\mathbf{k}),
\end{aligned}
\tag{17.3-18}
$$

(the integrals are analogous to (17.3-16)). Also we set $\sum_{\mathbf{k}} \to (V_0/8\pi^3)\int d^3k$. Then the fourth term of (17.3-10) takes the form

$$
\begin{aligned}
&(-1/\hbar)\int d^3\bar{k}\Big[\langle c_{\bar{\mathbf{k}}}^{\dagger}\, c_{\mathbf{k}+(1/2)\mathbf{u}}\rangle_t\, q\mathbf{E}\cdot\nabla_{\bar{\mathbf{k}}}\delta(\bar{\mathbf{k}}-\mathbf{k}+\tfrac{1}{2}\mathbf{u}) \\
&\quad+\langle c_{\mathbf{k}-(1/2)\mathbf{u}}^{\dagger}\, c_{\bar{\mathbf{k}}}\rangle_t\, q\mathbf{E}\cdot\nabla_{\bar{\mathbf{k}}}\delta(\bar{\mathbf{k}}-\mathbf{k}-\tfrac{1}{2}\mathbf{u})]\Big] \\
&=(q\mathbf{E}/\hbar)\cdot\Big[\langle(\nabla_{\mathbf{k}}c_{\mathbf{k}-(1/2)\mathbf{u}}^{\dagger})c_{\mathbf{k}+(1/2)\mathbf{u}}\rangle_t+\langle c_{\mathbf{k}-(1/2)\mathbf{u}}^{\dagger}\nabla_{\mathbf{k}}c_{\mathbf{k}+(1/2)\mathbf{u}}\rangle_t\Big] \\
&=(q\mathbf{E}/\hbar)\cdot\nabla_{\mathbf{k}}\langle c_{\mathbf{k}-(1/2)\mathbf{u}}^{\dagger}\, c_{\mathbf{k}+(1/2)\mathbf{u}}\rangle_t\,.
\end{aligned}
\tag{17.3-19}
$$

With (17.3-11) this yields for the fourth term

$$
\begin{aligned}
\text{fourth term} &= (q\mathbf{E}/\hbar)\cdot(1/8\pi^3)\int d^3\bar{r}\int d^3u\, e^{i\mathbf{u}\cdot(\mathbf{r}-\bar{\mathbf{r}})}[h^3\nabla_{\mathbf{k}}\rho_1(\mathbf{k},\bar{\mathbf{r}},t)] \\
&= (q\mathbf{E}/\hbar)\cdot\nabla_{\mathbf{k}}f(\mathbf{k},\mathbf{r},t)\,.
\end{aligned}
\tag{17.3-20}
$$

Note that both streaming terms are extracted from the nondiagonal part! Collecting, we arrive at the BTE for the Wigner occupation function in phase space

$$
\frac{\partial f(\mathbf{k},\mathbf{r},t)}{\partial t}+\mathbf{v}_{\mathbf{k}}\cdot\nabla_{\mathbf{r}}f(\mathbf{k},\mathbf{r},t)+\frac{q\mathbf{E}}{\hbar}\cdot\nabla_{\mathbf{k}}f(\mathbf{k},\mathbf{r},t)=-\mathcal{B}_{\mathbf{k}}f(\mathbf{k},\mathbf{r},t)\,. \tag{17.3-21}
$$

We have, therefore, given a definitive proof of Boltzmann's 1871 transport equation, deeply rooted in quantum mechanics, with irreversibility entering via the interaction picture and the Van Hove limit. The scheme followed is

$$
\begin{aligned}
&\text{von Neumann eq.}\xrightarrow[\text{picture}]{\text{interaction}}\text{master eq.}\xrightarrow[\text{quantization}]{\text{second}}\text{QBE} \\
&\xrightarrow[\text{transform}]{\text{Weyl}}\text{BTE for Wigner occupation function.}
\end{aligned}
\tag{17.3-22}
$$

The derivations of this section were meant to satisfy our curiosity as to whether modern quantum-field considerations could *allay* the fierce criticism that besieged Boltzmann in the 19th century. We believe that the answer is affirmative!

Lastly, we discuss the modifications when a magnetic field is present. One might believe that for 'small' B-fields this will justify the inclusion of the classical Lorentz

---

[7] We write $\Delta\mathbf{r} = \mathbf{r}-\mathbf{r}_0$. The integrals for $\mathbf{r}_0$ yield two imaginary terms that cancel.

force. This myth should be dispelled. Even when the 'orbits' are not closed due to collisions, plane waves and a three-dimensional k-space are out of the picture, thereby invalidating the use of the Weyl transform and the Wigner function. The states are now localised and any semi-classical description breaks down.[8] Actually, it was just with this in mind that we set out to derive the quantum Boltzmann equation and the associated quantum form for the current flux, described in the next section. If quantum confinement due to other causes occurs, the situation is exacerbated.

As to actual transport coefficient computations, the applications of this chapter (part 2) will abundantly show that even long established phenomena as the Hall effect need a quantum explanation. This is most strikingly clear with the 2D quantum Hall effect, discovered by von Klitzing (1980) at a time this theory was being developed.

17.4 Generalized Calecki Equation for the Nonlinear Current Flux

The *many-body form* for the current flux J_A associated with the flow of a variable A was derived in Section 16.10. In particular, the form for the electrical current flux was given in (16.10-41); we repeat it here

$$\mathbf{J}_e = \frac{\mathcal{q}}{V_0}\left\{-\Lambda_d\left[\sum_i(\mathbf{r}_i - \mathbf{r}_{i,eq})\right] + \sum_i \mathbf{v}_i\right\}. \tag{17.4-1}$$

The first part is the collisional current, which emerges from the interaction picture, while the second part is the usual ponderomotive current. We now write

$$\sum_i(\mathbf{r}_i - \mathbf{r}_{i,eq}) = \sum_{\zeta'\zeta''} c^\dagger_{\zeta''} c_{\zeta'}(\zeta''|(\mathbf{r}-\mathbf{r}_{eq})|\zeta'),$$

$$\sum_i \mathbf{v}_i = \sum_{\zeta'\zeta''} c^\dagger_{\zeta''} c_{\zeta'}(\zeta''|\mathbf{v}|\zeta'). \tag{17.4-2}$$

Substituting into (17.4-1), we obtain the *second quantization form*

$$\mathbf{J}_e = \frac{\mathcal{q}}{V_0}\sum_{\zeta'\zeta''}\left[-\Lambda_d \mathbf{n}_{\zeta'}(\zeta'|(\mathbf{r}-\mathbf{r}_{eq})|\zeta')\delta_{\zeta'\zeta''} + c^\dagger_{\zeta''} c_{\zeta'}(\zeta''|\mathbf{v}|\zeta')\right]. \tag{17.4-3}$$

This must now be averaged for a canonical or grand-canonical ensemble with density operator $\rho(t)$. For the first part we computed already the average in (17.1-17), or, for electron-phonon interaction in (17.1-25)ff. Thus, we obtain the *Boltzmann form*

$$\langle\mathbf{J}_e\rangle_t = \frac{\mathcal{q}}{V_0}\sum_{\zeta'\zeta''}\left[-\mathscr{B}_{\zeta'}\langle n_{\zeta'}\rangle_t(\zeta'|(\mathbf{r}-\mathbf{r}_{eq})|\zeta')\delta_{\zeta'\zeta''} + \langle c^\dagger_{\zeta''} c_{\zeta'}\rangle_t(\zeta''|\mathbf{v}|\zeta')\right]. \tag{17.4-4}$$

---

[8] Let the criterion be $\omega_c\tau \sim 0.1$, where $\omega_c = eB/m^*c$ is the cyclotron frequency and τ the collision time. Then for $m^* = 0.1m$, $\tau = 10^{-11}\,s$, we find that B should not exceed 100 *Gauss* (or 10^{-2} *Tesla*).

This expression will be referred to as the generalized Calecki equation for the current flux.[9] The ponderomotive current has both a diagonal and nondiagonal contribution; the latter is of paramount importance for the regular Hall effect and the quantum Hall effect, to be discussed in part 2.

We now study the collisional current in more detail. Writing out the Boltzmann operator, we have

$$\langle \mathbf{J}_{e,coll} \rangle_t = -\frac{q}{V_0} \sum_{\zeta\zeta'} [w_{\zeta\zeta'} \langle n_\zeta \rangle_t (1 - \langle n_{\zeta'} \rangle_t) - w_{\zeta'\zeta} \langle n_{\zeta'} \rangle_t (1 - \langle n_\zeta \rangle_t)] (\zeta | \mathbf{R} | \zeta),$$

$$= -\frac{q}{V_0} \sum_{\zeta} (\zeta | \mathbf{R} | \zeta) \sum_{\zeta'} [w_{\zeta\zeta'} \langle n_\zeta \rangle_t (1 - \langle n_{\zeta'} \rangle_t) - w_{\zeta'\zeta} \langle n_{\zeta'} \rangle_t (1 - \langle n_\zeta \rangle_t)]. \quad (17.4\text{-}5)$$

where we set $\mathbf{R} \equiv \mathbf{r} - \mathbf{r}_{eq}$. Now, *if the steady-state distribution satisfied* (17.2-2), *then* $\langle \mathbf{J}_{e,coll} \rangle_{ss}$ *would be identically zero*; this has escaped the attention of most authors, incl. Budd[3], Barker[5], and Vasilopoulos *et alii.*[6] So, once more, the result (17.2-2) has to be amended, as indicated in Section 17.2. Returning to (17.4-5), we interchange the summation indices and take half the sum of the so-obtained expressions. Denoting the steady state average by $\langle n_\zeta \rangle_{ss} = f_\zeta$ and using the notation $\langle \zeta | \mathbf{R} | \zeta \rangle = \mathbf{R}_\zeta$, we have

$$\langle \mathbf{J}_{e,coll} \rangle_{ss} = \frac{q}{2V_0} \sum_{\zeta\zeta'} [w_{\zeta\zeta'} f_\zeta (1 - f_{\zeta'}) - w_{\zeta'\zeta} f_{\zeta'} (1 - f_\zeta)] (\mathbf{R}_{\zeta'} - \mathbf{R}_\zeta), \quad (17.4\text{-}6)$$

which is the standard form. The last factor is also denoted by $\Delta\mathbf{R}_{\zeta\zeta'}$, being the 'orbit jump'. The 'jump current' was predicted by Titeica in 1935, as noted previously (XVI, Ref. 60); since these jumps are mostly mediated by optical phonons, we think that the denotation 'collisional current' is more appropriate. For small *E*-fields, (17.4-6) will lead to the 'collisional conductivity', to be shown in subsection 17.6.2.

17.5 The Linearized Quantum Boltzmann Equation*[10]

The derivation proceeds along the same lines as in Section 17.1. Our point of departure is now the inhomogeneous master equation (IME), Eq. (16.12-7). It is post-multiplied with $c^\dagger_{\zeta_1} c_{\zeta_2}$ and the trace is taken over a grand-canonical ensemble. For the lhs we find the same result as in (17.1-22). Although we will employ the same

---

[9] Calecki, Op. Cit., omits the ponderomotive current and has a linear Boltzmann operator without the proper factors (1 − *f*), stemming from the exclusion principle; this is true for most authors (like Budd, Barker), since they apply the von Neumann equation directly to the one-particle Hamiltonian. An exception is the older paper of Argyres and Roth, see Chapter XVI, Ref. 59.

[10] This equation is powerful for applications in that the streaming terms are explicitly there. However, its derivation is quite tedious. An alternate way to obtain the formulae for the various conductivities is shown in Sections 17.7 and 17.8.

symbol $\mathscr{B}_\zeta$ for the Boltzmann operator, it should be borne in mind that the field Hamiltonian has been split off, so that the many-body states $|\gamma\rangle$, as well as the one-particle states $|\zeta)$ and energies ε_ζ are different than for the previous case. As to the rhs, we must evaluate

$$\text{rhs} = F(t)\Big[\beta \text{Tr}\{\rho_{eq}[(-\overline{\Lambda}_d A_d + \dot{A}_d)]\mathbf{n}_{\zeta_1}\}\delta_{\zeta_1\zeta_2} + \int_0^\beta d\tau \text{Tr}\{(\rho_{eq} e^{\hbar\tau\mathscr{L}^0} \dot{A}_{nd}) c^\dagger_{\zeta_1} c_{\zeta_2}\}\Big]. \quad (17.5\text{-}1)$$

Herein we use the second quantization form for the variables $A = \Sigma_i a_i$, $\dot{A} = \Sigma_i \dot{a}_i$,

$$A_d = \sum_{\zeta'} \mathbf{n}_{\zeta'}(\zeta'|a|\zeta'),\ \dot{A}_d = \sum_{\zeta'} \mathbf{n}_{\zeta'}(\zeta'|\dot{a}|\zeta'),\ \dot{A}_{nd} = \sum_{\zeta'\neq\zeta'';\zeta''} c^\dagger_{\zeta'} c_{\zeta''}(\zeta'|\dot{a}|\zeta''). \quad (17.5\text{-}2)$$

With our previous results, the first term of (17.5-1) now involves

$$-\sum_{\zeta'} \langle n_{\zeta_1} \langle \overline{\Lambda}_d \mathbf{n}_{\zeta'} \rangle_b \rangle_{eq} (\zeta'|a|\zeta') = -\sum_{\zeta'} \langle n_{\zeta_1} \mathscr{B}_{\zeta} n_{\zeta'} \rangle_{eq} (\zeta'|a|\zeta'). \quad (17.5\text{-}3)$$

Writing out the Boltzmann operator, the term is evaluated as follows.[11]

$$-\sum_{\zeta'} \langle n_{\zeta_1} \mathscr{B}_{\zeta'} n_{\zeta'} \rangle_{eq} (\zeta'|a|\zeta')$$

$$= -\sum_{\zeta'\neq\zeta_1;\zeta''} (\zeta'|a|\zeta')[w_{\zeta'\zeta''} \langle n_{\zeta_1} n_{\zeta'} (1 - n_{\zeta''}) \rangle_{eq} - w_{\zeta''\zeta'} \langle n_{\zeta_1} n_{\zeta''} (1 - n_{\zeta'}) \rangle_{eq}] \quad (1)$$

$$-\sum_{\zeta''} (\zeta_1|a|\zeta_1)[w_{\zeta_1\zeta''} \langle n^2_{\zeta_1} (1 - n_{\zeta''}) \rangle_{eq} - w_{\zeta''\zeta_1} \langle n_{\zeta_1} n_{\zeta''} (1 - n_{\zeta_1}) \rangle_{eq}]. \quad (2)$$

Term (1) we split into two parts and the averages are simplified, using the truncation rules[12], so that

$$(1) = -\sum_{\zeta'\neq\zeta_1} (\zeta'|a|\zeta') \sum_{\zeta''\neq\zeta_1} [w_{\zeta'\zeta''} \langle n_{\zeta_1} \rangle_{eq} \langle n_{\zeta'} \rangle_{eq} (1 - \langle n_{\zeta''} \rangle_{eq})$$
$$- w_{\zeta''\zeta'} \langle n_{\zeta_1} \rangle_{eq} \langle n_{\zeta''} \rangle_{eq} (1 - \langle n_{\zeta'} \rangle_{eq})] \quad (1a)$$

$$-\sum_{\zeta'\neq\zeta_1} (\zeta'|a|\zeta')[w_{\zeta'\zeta_1} \langle n_{\zeta_1} (1 - n_{\zeta_1}) \rangle_{eq} \langle n_{\zeta'} \rangle_{eq}$$
$$- w_{\zeta_1\zeta'} \langle n^2_{\zeta_1} \rangle_{eq} (1 - \langle n_{\zeta'} \rangle_{eq})]. \quad (1b)$$

Detailed balance

$$w_{\zeta'\zeta''} \langle n_{\zeta'} \rangle_{eq} (1 - \langle n_{\zeta''} \rangle_{eq}) - w_{\zeta''\zeta'} \langle n_{\zeta''} \rangle_{eq} (1 - \langle n_{\zeta'} \rangle_{eq}) = 0, \quad (17.5\text{-}4)$$

shows that (1a) = 0. The restriction $\zeta' \neq \zeta_1$ can be omitted in (1b) since $w_{\zeta'\zeta_1} = 0$ for $\zeta' = \zeta_1$. Applying truncation to (2) and combining with (1b), we have

$$\sum_{\zeta'} [(\zeta_1|a|\zeta_1) - (\zeta'|a|\zeta')][w_{\zeta'\zeta_1} \langle n_{\zeta_1} (1 - n_{\zeta_1}) \rangle_{eq} \langle n_{\zeta'} \rangle_{eq} - w_{\zeta_1\zeta'} \langle n^2_{\zeta_1} \rangle_{eq} (1 - \langle n_{\zeta'} \rangle_{eq})]. \quad (3)$$

---

[11] LRT II, Eq. (8.43)ff. Our present rendering in this section is considerably simplified.

[12] In the grand-canonical ensemble we have $\langle n_{\zeta'} n_{\zeta''} \rangle_{eq} = \langle n_{\zeta'} \rangle_{eq} \langle n_{\zeta''} \rangle_{eq}\ \forall\, \zeta' \neq \zeta''$. However, for $\zeta' = \zeta''$ we cannot truncate, since $\langle n^2_\zeta \rangle_{eq} - \langle n_\zeta \rangle^2_{eq} = \langle \Delta n^2_\zeta \rangle_{eq} = \langle n_\zeta \rangle_{eq} (1 - \langle n_\zeta \rangle_{eq})$.

Now note that $n_{\zeta_1} = n_{\zeta_1}^2$; hence, in the second factor the first term is zero and the second term simplifies to $-w_{\zeta_1\zeta'}\langle n_{\zeta_1}\rangle_{eq}(1-\langle n_{\zeta'}\rangle_{eq})$. So, finally, we obtain for the first streaming term of (17.5-1)

$$1^{st}\text{ str. term} = -\beta F(t)\sum_{\zeta'}[(\zeta_1|a|\zeta_1)-(\zeta'|a|\zeta')][w_{\zeta_1\zeta'}\langle n_{\zeta_1}\rangle_{eq}(1-\langle n_{\zeta'}\rangle_{eq})]\delta_{\zeta_1\zeta_2}. \quad (17.5\text{-}5)$$

For the second term in (17.5.1) we have with (17.5-2),

$$\sum_{\zeta'}\langle n_{\zeta_1}n_{\zeta'}\rangle_{eq}(\zeta'|\dot{a}|\zeta') = \langle n_{\zeta_1}\rangle_{eq}\sum_{\zeta'\neq\zeta_1}\langle n_{\zeta'}\rangle_{eq}(\zeta'|\dot{a}|\zeta') + \langle n_{\zeta_1}^2\rangle_{eq}(\zeta_1|\dot{a}|\zeta_1). \quad (17.5\text{-}6)$$

Noting that the equilibrium current equals zero, we find that the first term on the rhs is $-\langle n_{\zeta_1}\rangle_{eq}^2(\zeta_1|\dot{a}|\zeta_1)$. So for both terms we get $\langle \Delta n_{\zeta_1}^2\rangle_{eq}(\zeta_1|\dot{a}|\zeta_1)$. Employing the grand-canonical variance (cf. footnote[12]), we arrive at

$$2^{nd}\text{ str. term}_{diag} = \beta F(t)\langle n_{\zeta_1}\rangle_{eq}(1-\langle n_{\zeta_1}\rangle_{eq})(\zeta_1|\dot{a}|\zeta_1)\delta_{\zeta_1\zeta_2}. \quad (17.5\text{-}7)$$

Rests the third term of (17.5-1). We are to compute

$$\mathrm{Tr}\{\rho_{eq}(e^{\hbar\tau\mathcal{L}^0}\dot{A}_{nd})c_{\zeta_1}^\dagger c_{\zeta_2}\} = \sum_{\zeta'\zeta''}\mathrm{Tr}\{\rho_{eq}e^{\tau\mathcal{H}^0}c_{\zeta'}^\dagger c_{\zeta''}e^{-\tau\mathcal{H}^0}c_{\zeta_1}^\dagger c_{\zeta_2}\}(\zeta'|\dot{a}|\zeta'')(1-\delta_{\zeta'\zeta''}). \quad (17.5\text{-}8)$$

Employing the states $\{|\gamma\rangle\}$ of $\mathcal{H}^0$, we have

$$\mathrm{Tr}\{\rho_{eq}e^{\tau\mathcal{H}^0}c_{\zeta'}^\dagger c_{\zeta''}e^{-\tau\mathcal{H}^0}c_{\zeta_1}^\dagger c_{\zeta_2}\} = \sum_{\gamma\bar{\gamma}}p_{eq}(\gamma)e^{-\tau(\mathcal{E}_{\bar{\gamma}}-\mathcal{E}_\gamma)}\langle\gamma|c_{\zeta'}^\dagger c_{\zeta''}|\bar{\gamma}\rangle\langle\bar{\gamma}|c_{\zeta_1}^\dagger c_{\zeta_2}|\gamma\rangle. \quad (17.5\text{-}9)$$

Let $|\gamma\rangle$ be fixed. The first matrix element shows that in the ket $|\bar{\gamma}\rangle$ the occupancy $\bar{n}_{\zeta''}$ is lowered and $\bar{n}_{\zeta'}$ is raised. From the second matrix element, however, in the bra $\langle\bar{\gamma}|$, $\bar{n}_{\zeta_1}$ is lowered and $\bar{n}_{\zeta_2}$ is raised. Since these states are each others dual, we must have that $\zeta'' = \zeta_1$ and $\zeta' = \zeta_2$, i.e., we include the Kronecker deltas $\delta_{\zeta''\zeta_1}\delta_{\zeta'\zeta_2}$ for the summations to be carried out in (17.5-8). Moreover, because of orthogonality, the new state $c_{\zeta'}^\dagger c_{\zeta''}|\bar{\gamma}\rangle$ must yield $|\gamma\rangle$, so that $\mathcal{E}_\gamma = \mathcal{E}_{\bar{\gamma}} + \varepsilon_{\zeta_2} - \varepsilon_{\zeta_1}$. Finally, from the rules (17.1-5) we note that the action of the raising and lowering operators produces the factors $[(-1)^{\xi_1+\xi_2}\sqrt{\bar{n}_{\zeta_1}(1-\bar{n}_{\zeta_2})}]^2 = (1-n_{\zeta_1})n_{\zeta_2}$. Hence, we obtain, doing also the integration over $d\tau$,

$$\int_0^\beta d\tau\,\mathrm{Tr}\{\rho_{eq}e^{\tau\mathcal{H}^0}c_{\zeta'}^\dagger c_{\zeta''}e^{-\tau\mathcal{H}^0}c_{\zeta_1}^\dagger c_{\zeta_2}\}$$

$$= \int_0^\beta d\tau\langle n_{\zeta_2}\rangle_{eq}(1-\langle n_{\zeta_1}\rangle_{eq})e^{-\tau(\varepsilon_{\zeta_1}-\varepsilon_{\zeta_2})} = \langle n_{\zeta_2}\rangle_{eq}(1-\langle n_{\zeta_1}\rangle_{eq})\frac{1-e^{-\beta(\varepsilon_{\zeta_1}-\varepsilon_{\zeta_2})}}{\varepsilon_{\zeta_1}-\varepsilon_{\zeta_2}}. \quad (17.5\text{-}10)$$

Carrying out the summation $\Sigma_{\zeta'\zeta''}(\text{above})\delta_{\zeta''\zeta_1}\delta_{\zeta'\zeta_2}$, in (17.5-8) we thus obtain for the nondiagonal part of the second streaming term as stated in (17.5-1):

$$2^{nd}\text{ str. term}_{nd} = F(t)\langle n_{\zeta_2}\rangle_{eq}(1-\langle n_{\zeta_1}\rangle_{eq})\frac{1-e^{-\beta(\varepsilon_{\zeta_1}-\varepsilon_{\zeta_2})}}{\varepsilon_{\zeta_1}-\varepsilon_{\zeta_2}}(\zeta_2|\dot{a}|\zeta_1)(1-\delta_{\zeta_1\zeta_2}). \quad (17.5\text{-}11)$$

This can be combined with the diagonal contribution, since

$$\lim_{\varepsilon_{\zeta_1}\to\varepsilon_{\zeta_2}}[1-e^{-\beta(\varepsilon_{\zeta_1}-\varepsilon_{\zeta_2})}]/(\varepsilon_{\zeta_1}-\varepsilon_{\zeta_2}) = \beta. \quad (17.5\text{-}12)$$

Upon adding both parts, the Kronecker deltas in (17.5-7) and (17.5-11) cancel out. Therefore, the final result is

$$2^{nd}\text{ str. term} = F(t)\langle n_{\zeta_2}\rangle_{eq}(1-\langle n_{\zeta_1}\rangle_{eq})\frac{1-e^{-\beta(\varepsilon_{\zeta_1}-\varepsilon_{\zeta_2})}}{\varepsilon_{\zeta_1}-\varepsilon_{\zeta_2}}(\zeta_2|\dot{a}|\zeta_1). \quad (17.5\text{-}13)$$

Collecting terms, we obtained the 'linearized QBE', having a collisional part and two streaming terms,

$$\begin{aligned}\frac{\partial\langle c^\dagger_{\zeta_1}c_{\zeta_2}\rangle_t}{\partial t} &+ \mathscr{B}_\zeta\langle n_{\zeta_1}\rangle_t\delta_{\zeta_1\zeta_2} + \frac{i}{\hbar}\langle c^\dagger_{\zeta_1}c_{\zeta_2}\rangle_t(\varepsilon_{\zeta_2}-\varepsilon_{\zeta_1})\\ &= \beta F(t)\sum_{\zeta'}[(\zeta'|a|\zeta')-(\zeta_1|a|\zeta_1)][w_{\zeta_1\zeta'}\langle n_{\zeta_1}\rangle_{eq}(1-\langle n_{\zeta'}\rangle_{eq})]\delta_{\zeta_1\zeta_2}\\ &+ F(t)\langle n_{\zeta_2}\rangle_{eq}(1-\langle n_{\zeta_1}\rangle_{eq})\frac{1-e^{-\beta(\varepsilon_{\zeta_1}-\varepsilon_{\zeta_2})}}{\varepsilon_{\zeta_1}-\varepsilon_{\zeta_2}}(\zeta_2|\dot{a}|\zeta_1). \quad (17.5\text{-}14)\end{aligned}$$

We note that the collision operator, as well as the energies in the above expression no longer depend on the field, since that was split off in the response Hamiltonian [properly speaking, the $\mathscr{B}$ and the epsilons should be denoted as $\overline{\mathscr{B}}$ and $\overline{\varepsilon}$]. Spatial dependence is contained in the wave mechanical form of the states $|\zeta)$; there is no streaming term like $\mathbf{v}\cdot\nabla_\mathbf{r}$, which is only defined for extended states. As to the two streaming terms on the rhs of (17.5-14), the ponderomotive term is as expected. The collisional streaming term is solely due to the fact that we work in the 'subdynamics' (Prigogine) of $\mathscr{H}^0$, associated with the interaction picture; in the full dynamics of $\mathscr{H}$, this term does not occur since all forces are on an equal footing. This, then, illumines anew the dialogue in Boltzmann's time. The streaming in his 1871 equation is fully ponderomotive, but the collision integral represents random dispersion forces of an underlying subdynamics, not accessible in classical theory.

17.6 Electrical Conductivities in the Linear Regime*

In this section we shall derive useful, closed-form expressions for the conductivities involving localised states. The conductivities due to ponderomotive current will be obtained in a straightforward way from the linearized QBE. The conductivity due to collisional current can be obtained either from the linearized QBE, or from the generalized Calecki equation.

17.6.1 *Ponderomotive Conductivities*

The ponderomotive current comprised in (17.4-4) involves the velocity matrix element $(\zeta|\mathbf{v}|\zeta')$, which (potentially) has both a diagonal ($\zeta=\zeta'$) and a nondiagonal ($\zeta\neq\zeta'$) component; the associated conductivities will be obtained separately.

We start with the diagonal part. In the linearized QBE we make the change of indices $(\zeta_1,\zeta_2)\to(\zeta,\zeta')$. For electrical problems, the response Hamiltonian is as usual $AF(t)\to q\mathbf{E}\cdot\Sigma_i(\mathbf{r}_i-\mathbf{r}_{i,eq})$. We take the field to be E_ν and look for the response of $\dot{a}_\mu=v_\mu$, where ν,μ refer to Cartesian components; also, it will be assumed that the position matrix element is zero. For $\zeta=\zeta'$ the QBE then reads

$$\partial\langle n_\zeta\rangle_t/\partial t+\mathcal{B}_\zeta\langle n_\zeta\rangle_t=\beta qE_\nu(t)\langle n_\zeta\rangle_{eq}(1-\langle n_\zeta\rangle_{eq})(\zeta|v_\nu|\zeta). \qquad (17.6\text{-}1)$$

We shall assume that $\mathcal{B}_\zeta$ is linear (as for impurity scattering and acoustical phonon interaction) or can be linearized. The solution is then

$$\langle n_\zeta\rangle_t=\langle n_\zeta\rangle_{eq}-q\int_0^t dt'e^{-\mathcal{B}_\zeta(t-t')}E_\nu(t')\frac{\partial\langle n_\zeta\rangle_{eq}}{\partial\varepsilon_\zeta}(\zeta|v_\nu|\zeta), \qquad (17.6\text{-}2)$$

where we used the well-known relation

$$\beta\langle n_\zeta\rangle_{eq}(1-\langle n_\zeta\rangle_{eq})=-\partial\langle n_\zeta\rangle_{eq}/\partial\varepsilon_\zeta. \qquad (17.6\text{-}3)$$

For the diagonal ponderomotive part of (17.4-4) we now find

$$\begin{aligned}\langle J^d_{e,\mu}\rangle_t&=\frac{q}{V_0}\sum_\zeta\langle n_\zeta\rangle_t(\zeta|v_\mu|\zeta)\\&=-\frac{q^2}{V_0}\int_0^t dt'\sum_\zeta e^{-\mathcal{B}_\zeta(t-t')}E_\nu(t')\frac{\partial\langle n_\zeta\rangle_{eq}}{\partial\varepsilon_\zeta}(\zeta|v_\nu|\zeta)(\zeta|v_\mu|\zeta).\end{aligned} \qquad (17.6\text{-}4)$$

This has the usual convolution form implying a response function

$$\langle J^d_{e,\mu}\rangle_t=\int_0^t dt'\phi_{\mu\nu}(t-t')E_\nu(t'). \qquad (17.6\text{-}5)$$

Thus, the electrical conductivity is the Fourier–Laplace transform of $\phi_{\mu\nu}(t)$. Assuming that a relaxation time exists, we have $\mathcal{B}_\zeta^{-1}\to\tau(\varepsilon_\zeta)$, so that

$$\sigma^d_{\mu\nu}(i\omega)=-\frac{q^2}{V_0}\int_0^\infty dt\sum_\zeta e^{-[i\omega+(1/\tau(\varepsilon_\zeta))]t}\frac{\partial\langle n_\zeta\rangle_{eq}}{\partial\varepsilon_\zeta}(\zeta|v_\nu|\zeta)(\zeta|v_\mu|\zeta)$$

$$= -\frac{q^2}{V_0}\sum_\zeta \frac{\partial\langle n_\zeta\rangle_{eq}}{\partial\varepsilon_\zeta}\frac{\tau(\varepsilon_\zeta)}{1+i\omega\tau(\varepsilon_\zeta)}(\zeta\,|\,v_\nu\,|\,\zeta)(\zeta\,|\,v_\mu\,|\,\zeta). \tag{17.6-6}$$

This expression bears resemblance to Verboven's conductivity of (16.7-12). However, it is the only result with this affinity; all other conductivities in this section are more complex.

We proceed to obtain the nondiagonal conductivity, which is of paramount importance for the Hall effect. For $\zeta' \neq \zeta$ the QBE yields

$$\frac{\partial\langle c_\zeta^\dagger c_{\zeta'}\rangle_t}{\partial t} + \frac{i}{\hbar}\langle c_\zeta^\dagger c_{\zeta'}\rangle_t(\varepsilon_{\zeta'} - \varepsilon_\zeta) = qE_\nu(t)\langle n_{\zeta'}\rangle_{eq}(1-\langle n_\zeta\rangle_{eq})\frac{1-e^{-\beta(\varepsilon_\zeta-\varepsilon_{\zeta'})}}{\varepsilon_\zeta-\varepsilon_{\zeta'}}(\zeta'\,|\,v_\nu\,|\,\zeta), \tag{17.6-7}$$

with solution

$$\langle c_\zeta^\dagger c_{\zeta'}\rangle_t = q\int_0^t dt' e^{-i(\varepsilon_{\zeta'}-\varepsilon_\zeta)(t-t')/\hbar}E_\nu(t')\langle n_{\zeta'}\rangle_{eq}(1-\langle n_\zeta\rangle_{eq})\frac{1-e^{-\beta(\varepsilon_\zeta-\varepsilon_{\zeta'})}}{\varepsilon_\zeta-\varepsilon_{\zeta'}}(\zeta'\,|\,v_\nu\,|\,\zeta). \tag{17.6-8}$$

For the nondiagonal ponderomotive current flux we have from (17.4-4):

$$\langle J_{e,\mu}^{nd}\rangle_t = \frac{q}{V_0}\sum_{\zeta\zeta'}\langle c_\zeta^\dagger c_{\zeta'}\rangle_t(\zeta\,|\,v_\mu\,|\,\zeta'). \tag{17.6-9}$$

With the result for $\langle c_\zeta^\dagger c_{\zeta'}\rangle_t$ from Eq. (17.6-8), this again defines a response function; from its Fourier–Laplace transform we obtain

$$\sigma_{\mu\nu}^{nd}(i\omega) = -\lim_{\eta\to 0+}\frac{q^2\hbar i}{V_0}\sum_{\zeta\zeta',spin}{}' \frac{1}{(\varepsilon_{\zeta'}-\varepsilon_\zeta)+\hbar\omega - i\hbar\eta}$$
$$\times\langle n_{\zeta'}\rangle_{eq}(1-\langle n_\zeta\rangle_{eq})\frac{1-e^{-\beta(\varepsilon_\zeta-\varepsilon_{\zeta'})}}{\varepsilon_\zeta-\varepsilon_{\zeta'}}(\zeta'\,|\,v_\nu\,|\,\zeta)(\zeta\,|\,v_\mu\,|\,\zeta'). \tag{17.6-10}$$

The prime on the sum means $\zeta' \neq \zeta$; we also added the sum over the spin index. The limit can be carried out with the Dirac relation (16.3-11). However, we shall be mainly interested in the dc conductivity; so we find

$$\sigma_{\mu\nu}^{nd}(0) = \frac{q^2\hbar i}{V_0}\sum_{\zeta\zeta',spin}{}' \langle n_{\zeta'}\rangle_{eq}(1-\langle n_\zeta\rangle_{eq})\frac{1-e^{-\beta(\varepsilon_\zeta-\varepsilon_{\zeta'})}}{(\varepsilon_\zeta-\varepsilon_{\zeta'})^2}(\zeta'\,|\,v_\nu\,|\,\zeta)(\zeta\,|\,v_\mu\,|\,\zeta'). \tag{17.6-11}$$

It is to be noted that this contribution does not depend on the collision operator.

17.6.2 *The Argyres–Roth Formula for the Collisional Conductivity*

The collisional conductivity formula is only straightforward for dc conditions. From the generalized Calecki equation (17.4-4) we have

$$\langle J_{\mu,coll}\rangle_t = -(q/V_0)\sum_\zeta \mathcal{B}_\zeta \langle n_\zeta\rangle_t (\zeta\,|\,r_\mu - r_{\mu,eq}\,|\,\zeta)\,; \tag{17.6-12}$$

here $\mathcal{B}_\zeta$ is the nonlinear Boltzmann operator. In the stationary regime the QBE gives

$$\mathcal{B}_\zeta \langle n_\zeta\rangle_{t\to\infty} = \beta q E_\nu \sum_{\zeta'} [(\zeta'\,|\,R_\nu\,|\,\zeta') - (\zeta\,|\,R_\nu\,|\,\zeta)]\, w_{\zeta\zeta'} \langle n_\zeta\rangle_{eq}(1-\langle n_{\zeta'}\rangle_{eq})\,, \tag{17.6-13}$$

where $\mathbf{R} \equiv \mathbf{r} - \mathbf{r}_{eq}$ and $\mathbf{E} \to E_\nu$. Substitution of (17.6-13) into (17.6-12) yields

$$\begin{aligned}\langle J_{\mu,coll}\rangle_{ss} &= -\frac{q^2\beta}{V_0} E_\nu \sum_{\zeta\zeta',spin} w_{\zeta\zeta'} \langle n_\zeta\rangle_{eq}(1-\langle n_{\zeta'}\rangle_{eq})[(\zeta'\,|\,R_\nu\,|\,\zeta') - (\zeta\,|\,R_\nu\,|\,\zeta)] \\ &\times (\zeta\,|\,R_\mu\,|\,\zeta) = \frac{q^2\beta}{2V_0} E_\nu \sum_{\zeta\zeta',spin} w_{\zeta\zeta'} \langle n_\zeta\rangle_{eq}(1-\langle n_{\zeta'}\rangle_{eq}) \\ &\times [(\zeta'\,|\,R_\nu\,|\,\zeta') - (\zeta\,|\,R_\nu\,|\,\zeta)][(\zeta'\,|\,R_\mu\,|\,\zeta') - (\zeta\,|\,R_\mu\,|\,\zeta)]\,,\end{aligned} \tag{17.6-14}$$

where we interchanged the summation indices and took half the sum of the obtained expressions; also, we explicited the sum over the spin for one of the variables, noting that spin is generally conserved in the transitions $w_{\zeta\zeta'}$. From (17.6-14) we obtain

$$\begin{aligned}\sigma_{\mu\nu,coll}(0) &= \frac{q^2\beta}{V_0} \sum_{\zeta\zeta'} w_{\zeta\zeta'} \langle n_\zeta\rangle_{eq}(1-\langle n_{\zeta'}\rangle_{eq}) \\ &\times [(\zeta'\,|\,R_\nu\,|\,\zeta') - (\zeta\,|\,R_\nu\,|\,\zeta)][(\zeta'\,|\,R_\mu\,|\,\zeta') - (\zeta\,|\,R_\mu\,|\,\zeta)]\,.\end{aligned} \tag{17.6-15}$$

This type of result was first obtained for the specific case of magnetic conductivity phenomena involving Landau states by Argyres and Roth, employing a perturbation treatment of the von Neumann equation for the density matrix, common in the late fifties; cf. Chapter XVI Ref. 59. Similar calculations in the same era were carried out by Adams and Holstein, cf. Chapter XVI Ref. 58, but they limited themselves to elastic collisions, in which case one may make the change

$$w_{\zeta\zeta'} \langle n_\zeta\rangle_{eq}(1-\langle n_{\zeta'}\rangle_{eq}) \to w_{\zeta\zeta'} \langle n_\zeta\rangle_{eq}(1-\langle n_\zeta\rangle_{eq})\,. \tag{17.6-16}$$

So, in the present subsection, we have recaptured their results, moreover showing that they are valid for *any type* of confinement leading to localised states. We still give the more simple result for longitudinal phenomena

$$\sigma_{xx,coll}(0) = \frac{q^2\beta}{V_0} \sum_{\zeta\zeta'} w_{\zeta\zeta'} \langle n_\zeta\rangle_{eq}(1-\langle n_{\zeta'}\rangle_{eq})\,[(\zeta'\,|\,R_x\,|\,\zeta') - (\zeta\,|\,R_x\,|\,\zeta)]^2\,. \tag{17.6-17}$$

Magnetic and other applications will be given in part 2 of this chapter.

17.7 Localised states: A Direct Perturbation Treatment

The collisional conductivity can readily be obtained from standard, first-order stationary perturbation theory, applied to the generalized Calecki equation for the current flux, (17.4-6), repeated below[13]

$$\langle \mathbf{J}_{coll} \rangle_{ss} = \frac{q}{2V_0} \sum_{\zeta\zeta',spin} [w_{\zeta\zeta'} f_\zeta (1-f_{\zeta'}) - w_{\zeta'\zeta} f_{\zeta'} (1-f_\zeta)](\mathbf{R}_{\zeta'} - \mathbf{R}_\zeta). \quad (17.7\text{-}1)$$

The last factor is also denoted by $\Delta\mathbf{R}_{\zeta'\zeta}$, being the jump. For the grand-Hamiltonian in the interaction picture pertaining to the steady state we now write

$$\mathcal{H}^0_{grand} = \mathcal{H}^0 - \mu \sum_\zeta \mathbf{n}_\zeta = \mathcal{H}^0 - \mu^0 \sum_\zeta \mathbf{n}_\zeta - \Delta\mu \sum_\zeta \mathbf{n}_\zeta . \quad (17.7\text{-}2)$$

Here μ is the quasi-Fermi level, while μ^0 is the equilibrium Fermi level. For the one-particle state $|\zeta)$ the perturbation is $\mathbf{n}_\zeta \Delta\mu$ with $\Delta\mu = q\Delta\Phi = q(\nabla\Phi \cdot \Delta\mathbf{r}) = -qE_\nu R_\nu$, the field being along $\hat{r}_\nu$. Due to the Legendre transform of the grand-canonical ensemble, the corresponding change in the one-particle Hamiltonian is $+qE_\nu R_{\nu,op}$.

So, for small electric fields the energies are changed in accordance with first-order perturbation theory

$$\varepsilon_\zeta = \varepsilon^0_\zeta + qE_\nu (\zeta^0 | R_{\nu,op} | \zeta^0) = \varepsilon^0_\zeta + qE_\nu R_{\zeta\nu} . \quad (17.7.3)$$

The near-equilibrium distribution is taken to have the quasi Fermi–Dirac form

$$f_\zeta = \frac{1}{e^{\beta[\varepsilon^0_\zeta - (\varepsilon^0_F - qE_\nu R_{\zeta\nu})]} + 1} \simeq f^0_\zeta + \left.\frac{df_\zeta}{dE_\nu}\right|_0 E_\nu = f^0_\zeta - \frac{df^0_\zeta}{d\varepsilon^0_F} qE_\nu R_{\zeta\nu} , \quad (17.7\text{-}4)$$

where we used first-order Taylor expansion. Likewise,

$$1 - f_0 \simeq 1 - f^0_\zeta + \frac{df^0_\zeta}{d\varepsilon^0_F} qE_\nu R_{\zeta\nu} . \quad (17.7\text{-}5)$$

We now compute the current flux $J_{coll,\mu}$. Knowing that the equilibrium current vanishes, we have up to first order in E_ν,

$$J_{coll,\mu} = \sum_{\zeta\zeta',spin} \left\{ w^0_{\zeta\zeta'} \left[f^0_\zeta \frac{df^0_{\zeta'}}{d\varepsilon^0_F} qE_\nu R_{\zeta'\nu} - (1-f^0_{\zeta'}) \frac{df^0_\zeta}{d\varepsilon^0_F} qE_\nu R_{\zeta\nu} \right] \right.$$

$$\left. - w^0_{\zeta'\zeta} \left[f^0_{\zeta'} \frac{df^0_\zeta}{d\varepsilon^0_F} qE_\nu R_{\zeta\nu} - (1-f^0_\zeta) \frac{df^0_{\zeta'}}{d\varepsilon^0_F} qE_\nu R_{\zeta'\nu} \right] \right\} \left(R_{\zeta'\mu} - R_{\zeta\mu} \right). \quad (17.7\text{-}6)$$

---

[13] For this section, see C.M. Van Vliet and A. Barrios, Physica **A 315**, 493 (2002), Section 8. [There are sign errors in Eqs. (8.5) and (8.9).]

Further, we need

$$\frac{df_\zeta^0}{d\varepsilon_F^0} = \beta f_\zeta^0 (1 - f_\zeta^0) . \tag{17.7-7}$$

This is substituted in (17.7-6), to yield

$$J_{coll,\mu} = \beta \sum_{\zeta\zeta',spin} \Big\{ w_{\zeta\zeta'}^0 \Big[f_\zeta^0 f_{\zeta'}^0 (1 - f_{\zeta'}^0) \mathcal{q} E_\nu R_{\zeta'\nu} - (1 - f_{\zeta'}^0) f_\zeta^0 (1 - f_\zeta^0) \mathcal{q} E_\nu R_{\zeta\nu} \Big]$$
$$- w_{\zeta'\zeta}^0 \Big[f_{\zeta'}^0 f_\zeta^0 (1 - f_\zeta^0) \mathcal{q} E_\nu R_{\zeta\nu} - (1 - f_\zeta^0) f_{\zeta'}^0 (1 - f_{\zeta'}^0) \mathcal{q} E_\nu R_{\zeta'\nu} \Big] \Big\} \big(R_{\zeta'\mu} - R_{\zeta\mu} \big) . \tag{17.7-8}$$

We use the property of detailed balance to eliminate the terms with $w_{\zeta'\zeta}^0$, which causes all terms that are cubic in f to cancel. Next, summing over spin, we obtain with minor algebra

$$J_{coll,\mu} = \frac{\beta \mathcal{q}^2 E_\nu}{V_0} \sum_{\zeta\zeta'} w_{\zeta\zeta'}^0 f_\zeta^0 (1 - f_{\zeta'}^0)(R_{\zeta'\nu} - R_{\zeta\nu})(R_{\zeta'\mu} - R_{\zeta\mu}) , \tag{17.7-9}$$

which implies a linear conductivity

$$\sigma_{\mu\nu,coll} = \frac{\beta \mathcal{q}^2}{V_0} \sum_{\zeta\zeta'} w_{\zeta\zeta'}^0 f_\zeta^0 (1 - f_{\zeta'}^0)(R_{\zeta'\nu} - R_{\zeta\nu})(R_{\zeta'\mu} - R_{\zeta\mu}) . \tag{17.7-10}$$

In particular, for the longitudinal conductivity we have, letting $\mu = \nu = x$,

$$\sigma_{xx,coll} = \frac{\beta \mathcal{q}^2}{V_0} \sum_{\zeta\zeta'} w_{\zeta\zeta'}^0 f_\zeta^0 (1 - f_{\zeta'}^0)(R_{\zeta'x} - R_{\zeta x})^2 . \tag{17.7-10'}$$

This is again the renowned result obtained for the magnetoresistance by Argyres and Roth. Clearly, complete agreement with Eqs. (17.6-15) and (17.6-17), based on the linearized quantum Boltzmann equation, is established.

Finally, it is instructive to employ the same technique to compute the collision sum $\mathcal{B}_\zeta f_\zeta$. We thus find

$$\mathcal{B}_\zeta f_\zeta = \sum_{\zeta'} \big[w_{\zeta\zeta'} f_\zeta (1 - f_{\zeta'}) - w_{\zeta'\zeta} f_{\zeta'} (1 - f_\zeta) \big]$$
$$= \sum_{\zeta'} \Bigg\{ w_{\zeta\zeta'}^0 \Bigg[f_\zeta^0 \frac{df_{\zeta'}^0}{d\varepsilon_F^0} \mathcal{q} E_\nu R_{\zeta'\nu} - (1 - f_{\zeta'}^0) \frac{df_\zeta^0}{d\varepsilon_F^0} \mathcal{q} E_\nu R_{\zeta\nu} \Bigg]$$
$$- w_{\zeta'\zeta}^0 \Bigg[f_{\zeta'}^0 \frac{df_\zeta^0}{d\varepsilon_F^0} \mathcal{q} E_\nu R_{\zeta\nu} - (1 - f_\zeta^0) \frac{df_{\zeta'}^0}{d\varepsilon_F^0} \mathcal{q} E_\nu R_{\zeta'\nu} \Bigg] \Bigg\}$$

$$= \beta \sum_{\zeta'} \Big\{ w^0_{\zeta\zeta'} \Big[f^0_\zeta f^0_{\zeta'} (1 - f^0_{\zeta'}) q E_\nu R_{\zeta'\nu} - (1 - f^0_{\zeta'}) f^0_\zeta (1 - f^0_\zeta) q E_\nu R_{\zeta\nu} \Big]$$

$$- w^0_{\zeta'\zeta} \Big[f^0_{\zeta'} f^0_\zeta (1 - f^0_\zeta) q E_\nu R_{\zeta\nu} - (1 - f^0_\zeta) f^0_{\zeta'} (1 - f^0_{\zeta'}) q E_\nu R_{\zeta'\nu} \Big] \Big\}. \quad (17.7\text{-}11)$$

Again, we use detailed balance to express the second part in terms of $w^0_{\zeta\zeta'}$. This results in the following expression for the steady-state distribution

$$\sum_{\zeta'} \Big\{ \Big[w_{\zeta\zeta'} f_{\zeta,ss} (1 - f_{\zeta',ss}) - w_{\zeta'\zeta} f_{\zeta',ss} (1 - f_{\zeta,ss}) \Big]$$

$$- \beta w^0_{\zeta\zeta'} f^0_\zeta (1 - f^0_{\zeta'}) q \mathbf{E} \cdot (\mathbf{R}_{\zeta'} - \mathbf{R}_\zeta) \Big\} = 0. \qquad (17.7\text{-}12)$$

This confirms once more that the steady-state distribution does not satisfy (17.2-2), but rather follows a model that yields the streaming-dependent form (17.2-3); for the above case the exponent n equals one.

Lastly, we note that (17.7-12) is identical with Eq. (17.6-13) for a field along $\hat{r}_\nu$. Clearly, then, if the current is only collisional current, the quasi Fermi distribution of departure is the solution of the QBE of Section 17.6.

17.8 Diagonal and Nondiagonal Conductivities from Modified LRT

In Chapter XVI expressions were given for the diagonal and nondiagonal susceptances and conductances associated with the reduced operators. The diagonal expressions are in Eqs. (16.10-63) – (16.10-65). Here we shall only consider the electrical conductivity, the other ones going similarly. The many-body result for the diagonal ponderomotive conductivity can be written as, cf. (10.6-65),

$$\sigma^d_{\mu\nu}(i\omega) = \frac{\beta q^2}{V_0} \mathrm{Tr} \left\{ \rho_{eq} \hat{v}_{\mu,d} \frac{1}{i\omega + \Lambda_d} \hat{v}_{\nu,d} \right\}. \qquad (17.8\text{-}1)$$

where the velocities $\hat{v}$ are many-body operators and where $(i\omega + \Lambda_d)^{-1}$ is the resolvent.[14] Going back to the Laplace transform, we have

$$(i\omega + \Lambda_d)^{-1} = \mathscr{L}_{i\omega}[\exp(-\Lambda_d t)]. \qquad (17.8\text{-}2)$$

From the definition of Λ_d and Theorem 2 [Eq. (16.10-7)] we have

$$e^{-\Lambda_d t} = \sum_\gamma |\gamma\rangle\langle\gamma| e^{-M_\gamma t}, \qquad (17.8\text{-}3)$$

where M_γ is the master operator in function space. If we take an average over the

---

[14] Note that $(i\omega + \Lambda_d)^{-1}$ is a Green's operator in the extended sense; in its spectral resolution the eigenvalue zero is to be omitted.

boson bath states, we have by (17.1-15)

$$\left\langle e^{-\Lambda_d t}\right\rangle_b = \sum_\gamma |\gamma\rangle\langle\gamma| \sum_{k=0}^{\infty} \langle(-M_\gamma t)^k\rangle_b / k!$$
$$= \sum_{\{n_\zeta\}} |\{n_\zeta\}\rangle\langle\{n_\zeta\}| \sum_{k=0}^{\infty} (-\mathscr{B}_\zeta t)^k / k! = \sum_{\{n_\zeta\}} |\{n_\zeta\}\rangle\langle\{n_\zeta\}|\, e^{-\mathscr{B}_\zeta t}, \quad (17.8\text{-}4)$$

which is formally correct, but of no value unless the operator on the rhs can be computed. Although M is a linear operator, $\mathscr{B}_\zeta$ is not. Therefore, we will limit ourselves to the case that $\mathscr{B}$ is linear or linearized and allows the definition of a relaxation time, i.e., $\mathscr{B} \to [1/\tau(\varepsilon)]$. Substituting these results in (17.8-2) yields

$$\left\langle \frac{1}{i\omega + \Lambda_d} \right\rangle_b = \sum_{\{n_\zeta\}} |\{n_\zeta\}\rangle\langle\{n_\zeta\}| \frac{1}{i\omega + \tau^{-1}(\varepsilon_\zeta)}. \quad (17.8\text{-}5)$$

Returning to (17.8-1), we find with $\sum_{\{n_\zeta\}}$ included in the *grand-canonical* trace,

$$\sigma_{\mu\nu}^d(i\omega) = \frac{\beta q^2}{V_0} \mathrm{Tr}\left\{ \rho_{eq,el}\, |\{n_\zeta\}\rangle\langle\{n_\zeta\}|\, \hat{v}_{\mu,d} \frac{\tau(\varepsilon_\zeta)}{1 + i\omega\tau(\varepsilon_\zeta)} \hat{v}_{\nu,d} \right\}. \quad (17.8\text{-}6)$$

The diagonal many-particle velocity operators are as usual given by

$$\hat{v}_{\mu,d} = \sum_{\zeta'} \mathbf{n}_{\zeta'} (\zeta'|v_\mu|\zeta'), \quad \hat{v}_{\nu,d} = \sum_{\zeta''} \mathbf{n}_{\zeta''} (\zeta''|v_\nu|\zeta''). \quad (17.8\text{-}7)$$

In evaluating the trace, we need the expression

$$\sum_{\zeta'\zeta''} \langle n_{\zeta'} n_{\zeta''}\rangle_{eq} (\zeta'|v_\mu|\zeta')(\zeta''|v_\nu|\zeta'')$$
$$= \sum_{\zeta'\neq\zeta'',\zeta''} \langle n_{\zeta'}\rangle_{eq} \langle n_{\zeta''}\rangle_{eq} (\zeta'|v_\mu|\zeta')(\zeta''|v_\nu|\zeta'')$$
$$+ \sum_{\zeta'} \langle n_{\zeta'}^2\rangle_{eq} (\zeta'|v_\mu|\zeta')(\zeta'|v_\nu|\zeta')$$
$$= \sum_{\zeta'} \langle n_{\zeta'}\rangle_{eq} (\zeta'|v_\mu|\zeta') \sum_{\zeta''} \langle n_{\zeta''}\rangle_{eq} (\zeta''|v_\nu|\zeta'')$$
$$+ \sum_{\zeta'} [\langle n_{\zeta'}^2\rangle_{eq} - \langle n_\zeta\rangle_{eq}^2] (\zeta'|v_\mu|\zeta')(\zeta'|v_\nu|\zeta'). \quad (17.8\text{-}8)$$

The next to last line of (17.8-8) is zero, since the equilibrium current vanishes. The last line involves the variance,

$$\langle n_{\zeta'}^2\rangle_{eq} - \langle n_\zeta\rangle_{eq}^2 = \langle \Delta n_\zeta^2\rangle_{eq} = \langle n_\zeta\rangle_{eq}(1 - \langle n_\zeta\rangle_{eq}) = -\beta^{-1} \frac{\partial \langle n_\zeta\rangle_{eq}}{\partial \varepsilon_\zeta}. \quad (17.8\text{-}9)$$

The final result now becomes

$$\sigma_{\mu\nu}^d(i\omega) = -\frac{q^2}{V_0} \sum_\zeta \frac{\partial \langle n_\zeta\rangle_{eq}}{\partial \varepsilon_\zeta} \frac{\tau(\varepsilon_\zeta)}{1 + i\omega\tau(\varepsilon_\zeta)} (\zeta|v_\mu|\zeta)(\zeta|v_\nu|\zeta). \quad (17.8\text{-}10)$$

This result is exactly what we obtained earlier from the linearized QBE, cf. (17.6-6).

Next, we consider the nondiagonal ponderomotive conductivity. Its many-body form was implied by (16.10-68). So we have

$$\sigma_{\mu\nu}^{nd}(i\omega) = \frac{\mathcal{q}^2 \hbar i}{V_0} \lim_{\eta\to 0+} \sum_{\gamma\bar{\gamma}} p_{eq}(\gamma) \langle \gamma | \hat{v}_\nu^{nd} | \bar{\gamma} \rangle \langle \bar{\gamma} | \hat{v}_\mu^{nd} | \gamma \rangle \frac{1-e^{-\beta(\mathcal{E}_{\bar{\gamma}} - \mathcal{E}_\gamma)}}{\mathcal{E}_{\bar{\gamma}} - \mathcal{E}_\gamma} \frac{1}{\mathcal{E}_{\bar{\gamma}} - \mathcal{E}_\gamma - \hbar\omega + i\hbar\eta} . \tag{17.8-11}$$

Employing the second quantization forms

$$\begin{aligned} \hat{v}_\mu^{nd} &= \sum_{\zeta_1\zeta_2}{}' c_{\zeta_1}^\dagger c_{\zeta_2} (\zeta_1 | v_\nu | \zeta_2) , \\ \hat{v}_\mu^{nd} &= \sum_{\zeta_3\zeta_4}{}' c_{\zeta_3}^\dagger c_{\zeta_4} (\zeta_3 | v_\mu | \zeta_4) , \end{aligned} \tag{17.8-12}$$

we must evaluate

$$\sum_{\zeta_1\zeta_2}{}' \sum_{\zeta_3\zeta_4}{}' \sum_{\gamma\bar{\gamma}} p_{eq}(\gamma) \langle \gamma | c_{\zeta_1}^\dagger c_{\zeta_2} | \bar{\gamma} \rangle \langle \bar{\gamma} | c_{\zeta_3}^\dagger c_{\zeta_4} | \gamma \rangle . \tag{17.8-13}$$

Let at first the state $|\gamma\rangle$ be fixed. Eq. (17.8-13) indicates that in the ket $|\bar{\gamma}\rangle$ $\bar{n}_{\zeta_2}$ is lowered and $\bar{n}_{\zeta_1}$ is raised. Simultaneously, in the bra $\langle\bar{\gamma}|$ $\bar{n}_{\zeta_3}$ is lowered while $\bar{n}_{\zeta_4}$ is raised. In order that these states are *bona fide* duals, we must have $\zeta_2 = \zeta_3$ and $\zeta_1 = \zeta_4$. The action of these operators entails the factors

$$[(-1)^\xi (-1)^{\xi'} \sqrt{\bar{n}_{\zeta_2}(1-\bar{n}_{\zeta_1})}]^2 = (1-n_{\zeta_2}) n_{\zeta_1} . \tag{17.8-14}$$

For the matrix element to be nonzero, we require that the states are connected, such that

$$\mathcal{E}_\gamma = \mathcal{E}_{\bar{\gamma}} + \varepsilon_{\zeta_1} - \varepsilon_{\zeta_2} . \tag{17.8-15}$$

Upon substitution of these results in (17.8-11), we obtain

$$\begin{aligned} \sigma_{\mu\nu}^{nd}(i\omega) = \frac{\mathcal{q}^2 \hbar i}{V_0} \lim_{\eta\to 0+} \sum_{\zeta_1\zeta_2, spin}{}' \langle n_{\zeta_1} \rangle_{eq} (1 - \langle n_{\zeta_2} \rangle_{eq}) \frac{1-e^{-\beta(\varepsilon_{\zeta_2} - \varepsilon_{\zeta_1})}}{\varepsilon_{\zeta_2} - \varepsilon_{\zeta_1}} \\ \times \frac{1}{\varepsilon_{\zeta_2} - \varepsilon_{\zeta_1} - \hbar\omega + i\hbar\eta} (\zeta_1 | v_\nu | \zeta_2)(\zeta_2 | v_\mu | \zeta_1) . \end{aligned} \tag{17.8-16}$$

This final result is, remarkably, in full agreement with the result (17.6-10) obtained earlier via the linearized quantum Boltzmann equation. Clearly, the various approaches presented here form a truly consistent nonequilibrium formalism. Computations involving the susceptibility $\chi(i\omega)$ are entirely analogous and are left to the reader.

In the next part of this chapter the theory will be applied to a number of transport phenomena in which localised states constitute the prevalent description.

2. Some Applications of Modified Linear Response Theory

17.9 Landau States: 3D Applications

17.9.1 *The Ordinary Hall Effect*

According to the classical picture, when charge carriers are subject to a magnetic field they experience the Lorentz force, which causes them to precess about the field. The Lorentz force involves no dissipation, since $(\mathbf{v}\times\mathbf{B})\cdot\mathbf{v} = -\mathbf{B}\cdot(\mathbf{v}\times\mathbf{v}) = 0$. In the usual arrangement of crossed electric and magnetic fields (Fig. 17-2), one takes **B** along the z-axis and **E** (so also **J**) along the x-axis; the deflection in the y-direction gives the Hall voltage V_{Hall}, the connection being [see Table 15-2][15]

$$(V_{Hall}/d) = E_y = -\rho_{xy} J_{e,x} \equiv R_H J_{e,x} B\,, \tag{17.9-1}$$

where d is the sample width in the y-direction and R_H is the Hall coefficient. We will obtain the ponderomotive conductivity σ_{yx}, which for high B-fields is related to ρ_{yx} by

$$\rho_{xy} = (\sigma)^{-1}_{xy} = -\sigma_{xy}/(\sigma_{xx}\sigma_{yy} - \sigma_{xy}\sigma_{yx}) \sim 1/\sigma_{yx}\,. \tag{17.9-2}$$

From these two equations, we note that

$$R_{H,\,\text{high}\,B} = -1/(\sigma_{yx}B)\,. \tag{17.9-3}$$

Since the Hall effect is non-dissipative, we expect that it is given by the nondiagonal part of the conductivity; this will indeed be borne out by the computation below, since the velocity matrix element has no diagonal components – which fact alone already proves that classical considerations are bound to fail.

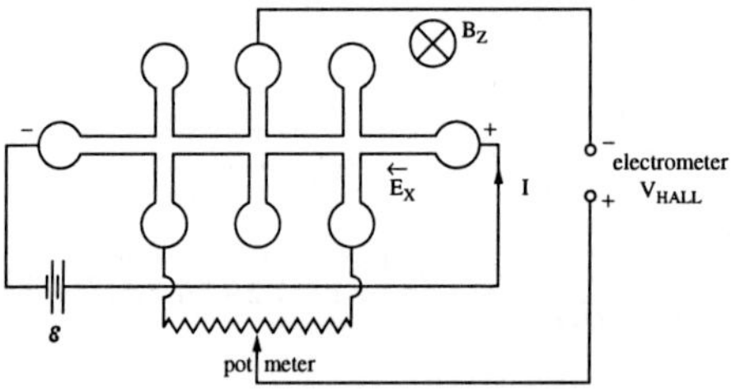

Fig. 17-2. Set-up for the Hall effect.

Proceeding to the quantum computation, the Hamiltonian of departure is

$$h^0 = [\mathbf{p} - (q/c)\mathbf{A}]^2/2m^* + U(z)\,, \qquad \mathbf{A} = (0, Bx, 0)\,, \tag{17.9-4}$$

---

[15] Although one might expect that in (17.9-1) the resistivity ρ_{yx} would appear, it should be appreciated that the orientation of the axes changes when we invert a relation $x\to y$ into a relation $y\to x$ (in chemistry, 'cis'→'trans'). Whereas the Onsager–Casimir relations state that ρ is a symmetrical tensor if we also let **B**→ –**B**, we note that for *fixed* **B** ρ is antisymmetrical, which entails $\rho_{yx}(\mathbf{B}) = -\rho_{xy}(\mathbf{B})$.

where we employed the usual Landau gauge[16]. For the present problem, the particles are free in the z-direction [$U(z) = 0$], with periodic boundary conditions applied over the dimension L_z; the width will henceforth be denoted by $d = L_y$. Writing out the scalar product in (17.9-4), we have the more detailed Hamiltonian

$$h^0 = \frac{p_x^2}{2m^*} + \frac{(p_y \mp m^*\omega_c x)^2}{2m^*} + \frac{p_z^2}{2m_{\|}}, \quad \mathbf{p} = \frac{\hbar}{i}\nabla . \tag{17.9-5}$$

Here $\omega_c = |q| B/m^*c = eB/m^*c$ is the cyclotron frequency,[17] the upper sign is for holes and the lower sign for electrons. The eigenfunctions of the Schrödinger equation are

$$\zeta_{N,k_y,k_z}(x,y,z) = (L_y L_z)^{-\frac{1}{2}} \phi_N(x - x_0) e^{ik_y y} e^{ik_z z} , \tag{17.9-6}$$

where the ϕ's are harmonic oscillator eigenfunctions, $x_0 = \pm \ell^2 k_y$ and $\ell = \sqrt{\hbar / m^* \omega_c}$ is the magnetic length [or smallest 'orbit radius']; the quantization is given by

$$N = 0,1,2,..., \quad k_y = 2\pi n_y / L_y \text{ with } n_y = 0, \pm 1,..., \quad k_z = 2\pi n_z / L_z \text{ with } n_z = 0, \pm 1,... . \tag{17.9-7}$$

The eigenvalues are degenerate with the expected form

$$\varepsilon_{N,k_z} = (N + \tfrac{1}{2})\hbar\omega_c + \hbar^2 k_z^2 / 2m_{\|} . \tag{17.9-8}$$

The matrix elements for the velocities are based on the matrices for p_x and x of the harmonic oscillator, as found in standard quantum mechanics texts[18]; the connection with the elements for v_x and v_y is indicated by Kahn and Frederikse[18a]. We have

$$\begin{aligned} (\zeta | v_x | \zeta') &= \pm i (\hbar\omega_c / 2m^*)^{1/2} [(N+1)^{1/2} \delta_{N',N+1} - N^{1/2} \delta_{N',N-1}] \delta_{kk'} , \\ (\zeta | v_y | \zeta') &= (\hbar\omega_c / 2m^*)^{1/2} [(N+1)^{1/2} \delta_{N',N+1} + N^{1/2} \delta_{N',N-1}] \delta_{kk'} , \end{aligned} \tag{17.9-9}$$

where $\delta_{kk'} = \delta_{k_y,k_y'} \delta_{k_z,k_z'}$. We note that there are only entries adjacent to the diagonal. There is therefore no diagonal ponderomotive current, as in the classical treatment! From (17.9-9) we find the required nondiagonal expression:

$$\begin{aligned} (\zeta | v_x | \zeta')(\zeta' | v_y | \zeta) &= \pm i (\hbar\omega_c / 2m^*) \\ &\times [(N+1)^{1/2} (N')^{1/2} \delta_{N',N+1} - (N)^{1/2} (N'+1)^{1/2} \delta_{N',N-1}] \delta_{kk'} . \end{aligned} \tag{17.9-10}$$

---

[16] Attempts to use the 'symmetrical' or Dingle gauge $\mathbf{A} = (-\frac{1}{2}By, \frac{1}{2}Bx, 0)$ have not been successful for most magnetic phenomena, largely because of integrals involving the associated Laguerre polynomials.

[17] Vectorially, $\boldsymbol{\omega}_c = -q\mathbf{B}/m^*c$, giving a counter-clockwise rotation in the x-y plane for electrons ($q = -e$) and a clockwise rotation for holes ($q = e$).

[18] H.A. Kramers, "Quantentheorie des Elektrons und der Strahlung", Akad. Verlag Leipzig, 1938 p. 432.

[18a] We have the following correspondence: $v_x = p_x/m^*$, $v_y = \hbar k_y / m^* \rightarrow (x - x_0)\omega_c$. For details, cf. A.H. Kahn and H.P.R. Frederikse, Solid State Physics (F. Seitz and D. Turnbull, Eds.) **9**, 257–293 (1959).

We also have for the allowed transitions

$$\varepsilon_{\zeta'} - \varepsilon_\zeta = \pm\hbar\omega_c \quad \text{for} \quad N' = N \pm 1. \tag{17.9-11}$$

With these data we obtain for the nondiagonal conductivity (for $\omega = 0$) from (17.8-16) for $\sigma_{yx}^{nd}(0) = \sigma_{yx}$ (omitting '*nd*' since this is the full contribution):

$$\sigma_{yx} = -\frac{qc}{2BV_0}\sum_{k,\sigma}\sum_{N=0,1,2...}(N+1)\Big\{\langle n_N\rangle_{eq}(1-\langle n_{N+1}\rangle_{eq})(1-e^{-\beta\hbar\omega_c})$$
$$-\langle n_{N+1}\rangle_{eq}(1-\langle n_N\rangle_{eq})(1-e^{\beta\hbar\omega_c})\Big\}. \tag{17.9-12}$$

(Note that we changed $N \to N+1$ in the summation for the second part of (17.9-10) and we suppressed the index 'k'; the sum over spin is denoted by 'σ'.) For ease in notation we will set $\langle n_N\rangle_{eq} \equiv f_N$. The result (17.9-12) is now split up as follows

$$\sigma_{yx} = -\frac{qc}{2BV_0}\sum_{k,\sigma}\sum_N (N+1)$$
$$\times\Big\{\Big[f_N(1-f_{N+1}) + f_{N+1}(1-f_N)e^{\beta\hbar\omega_c}\Big]$$
$$-\Big[f_N(1-f_{N+1})e^{-\beta\hbar\omega_c} + f_{N+1}(1-f_N)\Big]\Big\}. \tag{17.9-13}$$

It is easily proven that the two parts inside each square bracket are equal; for nondegenerate materials like most semiconductors, this was shown in a 1988 article on applications of the QBE.[19] The proof for degenerate materials like metals goes similarly. We have with $\varepsilon_F - \hbar^2k_z^2/2m_\parallel = \varepsilon_F'$

$$f_{N+1}(1-f_N)e^{\beta\hbar\omega_c} = \frac{1}{e^{\beta(N+3/2)\hbar\omega_c-\varepsilon_F'}+1}\,\frac{e^{\beta(N+1/2)\hbar\omega_c-\varepsilon_F'}}{e^{\beta(N+1/2)\hbar\omega_c-\varepsilon_F'}+1}\,e^{\beta\hbar\omega_c}$$
$$= \frac{1}{e^{\beta(N+1/2)\hbar\omega_c-\varepsilon_F'}+1}\,\frac{e^{\beta(N+3/2)\hbar\omega_c-\varepsilon_F'}}{e^{\beta(N+3/2)\hbar\omega_c-\varepsilon_F'}+1} = f_N(1-f_{N+1}). \tag{17.9-14}$$

The proof for the two terms inside the second square bracket goes the same. This gives

$$\sigma_{yx} = -\frac{qc}{BV_0}\sum_{k,\sigma}\sum_N (N+1)\big[f_N(1-f_{N+1}) - f_{N+1}(1-f_N)\big]$$
$$= -\frac{qc}{BV_0}\sum_{k,\sigma}\sum_N (N+1)(f_N - f_{N+1}). \tag{17.9-15}$$

Changing $N+1 \to N$ in the last term yields

---

[19] C.M. Van Vliet and P. Vasilopoulos, J. Phys. Chem. Solids, **49**, 639 (1988).

$$\sigma_{yx} = -\frac{qc}{BV_0}\sum_{k,\sigma}\sum_{N}\langle n_N\rangle_{eq} = -\frac{qc}{BV_0}\langle n_{total}\rangle = -\frac{qc\bar{\rho}_0}{B}, \tag{17.9-16}$$

where $\bar{\rho}_0$ is the equilibrium carrier density (n_0 or p_0). From (17.9-3) we find that the Hall coefficient for large B-fields is given by

$$R_{Hall} = \frac{1}{qc\bar{\rho}_0} \rightarrow \begin{cases} -1/ecn_0 & \text{(electrons)}, \\ +1/ecp_0 & \text{(holes)}. \end{cases} \tag{17.9-17}$$

This is in full agreement with the classical result, see e.g. Kittel.[20]

Although we included the effective mass in the Hamiltonian of departure, it dropped out of the final result. This is gratifying, since for most nondegenerate semiconductors the approximation of the band structure by appropriate transverse and longitudinal masses is adequate, providing **B** is along one of the symmetry axes. For simple metals obeying the Sommerfeld – Drude model like the alkali metals, m^* should be taken as the free electron mass m. Experimental values are in good agreement with the known electron density. For other metals, like cobalt, the sign of the Hall coefficient reverses, which indicates that the particular band structure and Fermi surface play a significant role. Clearly, the simple Landau states of this section are not applicable to solids with intricate energy-band structures.

17.9.2 *Transverse Magnetoresistance*

It was noted in subsection 15.4.2 that the classical single band effective mass description yields no magnetoresistance. This failure is easily explained with our present approach: there is no ponderomotive conductivity for a transverse field $\mathbf{B} = B\hat{z}$, i.e., $\sigma_{xx,pond} = \sigma_{yy,pond} = 0$. For the diagonal conductivity this is at once clear, since the velocity has no diagonal matrix elements as we saw in the preceding subsection. For the nondiagonal contribution it will be shown now. For σ_{xx}^{nd} we need

$$(\zeta|v_x|\zeta')(\zeta'|v_x|\zeta)$$
$$= (\hbar\omega_c/2m^*)[(N+1)^{1/2}(N')^{1/2}\delta_{N',N+1} + N^{1/2}(N'+1)^{1/2}\delta_{N',N-1}]\delta_{kk'}. \tag{17.9-18}$$

With the rule (17.9-11) one finds for σ_{xx}^{nd}

$$\sigma_{yx} = \frac{qc}{2BV_0}\sum_{k,\sigma}\sum_{N=0,1,2...}(N+1)\Big\{f_N(1-f_{N+1})(1-e^{-\beta\hbar\omega_c})$$
$$+ f_{N+1}(1-f_N)(1-e^{\beta\hbar\omega_c})\Big\}, \tag{17.9-19}$$

---

[20] Kittel gives his result in electromagnetic units, the result being the same as ours; cf., C. Kittel, "Introduction to Solid State Physics", 3rd Ed., Wiley & Sons, NY, third corrected printing 1968, p. 243.

which differs from (17.9-12) in that both terms in the curly brackets are positive. Next, we split the result to obtain

$$\sigma_{xx}^{nd} = const \times \Big\{ \Big[f_N(1-f_{N+1}) - f_{N+1}(1-f_N)e^{\beta\hbar\omega_c} \Big] - \Big[f_N(1-f_{N+1})e^{-\beta\hbar\omega_c} - f_{N+1}(1-f_N) \Big] \Big\} = 0, \qquad (17.9\text{-}20)$$

since – as shown before – the two terms in each square bracket are equal.

Consequently, the magnetoresistance stems from the 'orbit-jumps' that give the collisional conductivity (17.7-10'). We now need the position matrix element

$$(\zeta|x|\zeta') = x_0\delta_{N,N'}\delta_{kk'} \pm i(\hbar/2m^*\omega_c)^{1/2}[(N+1)^{1/2}\delta_{N',N+1} - N^{1/2}\delta_{N',N-1}]\delta_{kk'}; \qquad (17.9\text{-}21)$$

obviously, only the first term is needed for the collisional conductivity. So we obtain

$$\sigma_{xx} = \frac{\beta q^2}{V_0}\sum_{\zeta\zeta'} w_{\zeta\zeta'} f_\zeta(1-f_{\zeta'})(x_{0\zeta'} - x_{0\zeta})^2 . \qquad (17.9\text{-}22)$$

Herein we substitute

$$x_{0,N'k_y'} - x_{0,Nk_y} = \pm(\hbar/m^*\omega_c)(k_y' - k_y). \qquad (17.9\text{-}23)$$

The jumps involve phonon absorption or emission; so, the $w_{\zeta\zeta'}$ must be evaluated for whatever kind of phonons is involved, acoustical, optical, etc. In all cases momentum is conserved, so that $k_y' - k_y = \pm q_y$ and $\sigma_{xx} \propto q_y^2$. To symmetrize the results of the Landau gauge, we take half the sum of σ_{xx} and σ_{yy}. We thus obtain

$$\sigma_{xx} = \frac{\beta q^2\hbar^2}{2V_0 m^{*2}\omega_c^2}\sum_{\zeta\zeta'} w_{\zeta\zeta'} f_\zeta(1-f_{\zeta'})q_\perp^2 . \qquad (17.9\text{-}24)$$

To be noted is that the 'overall' magnetoresistance will be proportional to B^2, as in the Sondheimer–Wilson model considered before. However, the detailed behaviour will be oscillatory with period $\propto (1/B)$. The full computation is laborious[21] and will not be done here, particularly since the similar 2D effect is of greater current interest; it will be worked out in subsection 17.10.2. Part of the oscillatory behaviour stems from the density of states which is needed to carry out the sums. Since we have different quantization in the z-direction and the perpendicular direction, we have by convolution

$$Z(\varepsilon) = \int\int Z_1(\varepsilon_z)Z_2(\varepsilon_\perp)\delta(\varepsilon - \varepsilon_z - \varepsilon_\perp)d\varepsilon_z d\varepsilon_\perp = \int Z_1(\varepsilon - \varepsilon_\perp)Z_2(\varepsilon_\perp)d\varepsilon_\perp. \qquad (17.9\text{-}25)$$

---

[21] P. Vasilopoulos and C.M. Van Vliet, "Linear Response Theory Revisited IV. Applications", J. Math Phys. **25**, 1391-1403 and 3358 (err.), (1984).

Now $Z_1 = (L_z / \pi\hbar)(2m_{\|}/\varepsilon_z)^{½}$, $Z_\perp = \Sigma_N \delta[\varepsilon_\perp - (N + ½)\hbar\omega_c](L_x L_y / 2\pi\ell^2)$, so that

$$Z(\varepsilon) = \frac{V_0 \omega_c (2m_{\|})^{1/2} m^*}{\pi^2 \hbar^2} \sum_{N=0}^{N_{\max}} [\varepsilon - (N + ½)\hbar\omega_c]^{-½}, \qquad (17.9\text{-}26)$$

where the spin summation has been included.

In Fig. 17-3 we give a sketch of the so-obtained density of states.[22] In Fig. 17-4 a plot is given of the computed magnetoconductance vs. $1/B$ for polar optical phonon interaction, after Vasilopoulos.[23]

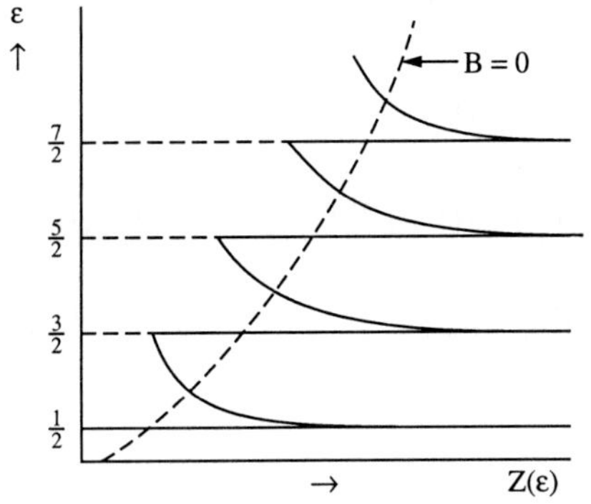

Fig. 17-3. Density of states vs. $\varepsilon / \hbar\omega_c$.[22]

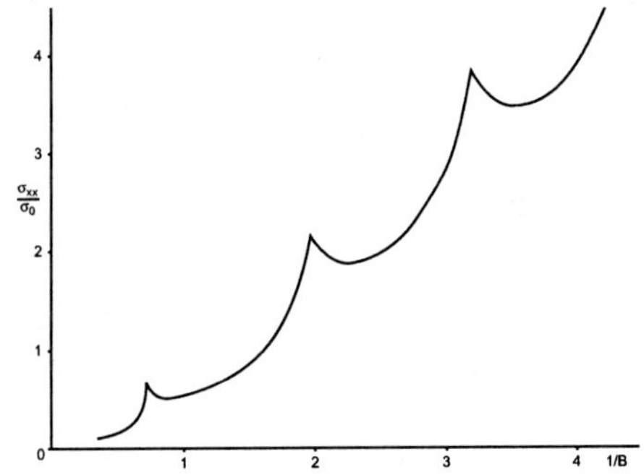

Fig. 17-4. Magnetoconductance (σ_{xx}/σ_0) vs. $1/B$.[23]

17.10 Landau States: 2D and 1D Applications

17.10.1 *The Quantum Hall Effect*

In two dimensions the Hall effect is quantized, i.e., the conductance is an integer number of the unit conductance, e^2/h. The effect was discovered by von Klitzing, Dorda and Pepper in 1980[24]; a Nobel prize was awarded in 1985. The present theory is 'tailored' to give their result, including the effect of dissipation.[25]

The Hamiltonian is the same as in (17.9-4), except that there is no energy $U(z)$. The states are also as before, whereby we omit the $\exp(ik_z z)$ while the quantization is now

$$\varepsilon_N = (N + ½)\hbar\omega_c. \qquad (17.10\text{-}1)$$

Entirely analogous to (17.9-12) we thus obtain, restricting ourselves to electron current,

[22] After Kahn and Frederikse, Solid State Physics (F. Seitz and D. Turnbull, Eds.) **9**, 257-293 (1959).
[23] P. Vasilopoulos, Ph. D. Thesis, Université de Montréal 1982.
[24] K. von Klitzing, G. Dorda and M. Pepper, Phys. Rev. Lett. **45**, 494 (1980).
[25] P. Vasilopoulos, Phys. Rev. **B 32**, 771 (1985); ibid. and C.M. Van Vliet, Phys. Rev. **B34** 1057 (1986).

$$\sigma_{yx} = \frac{ec}{2BA_0} \sum_{k_y} \sum_{N=0,1,2...} (N+1)\Big\{\langle n_N\rangle_{eq}(1-\langle n_{N+1}\rangle_{eq})(1-e^{-\beta\hbar\omega_c})$$

$$-\langle n_{N+1}\rangle_{eq}(1-\langle n_N\rangle_{eq})(1-e^{\beta\hbar\omega_c})\Big\}, \qquad (17.10\text{-}2)$$

where A_0 is the area $L_x L_y$. For the summation over k_y we write

$$\sum_{k_y} \to \frac{L_y}{2\pi}\int_{-L_x/2\ell^2}^{L_x/2\ell^2} dk_y = \frac{A_0}{2\pi\ell^2}. \qquad (17.10\text{-}3)$$

This number is the degeneracy of each Landau energy ε_N. For σ_{yx} we now find

$$\sigma_{yx} = \frac{e^2}{2h} \sum_{N=0,1,2...} (N+1)\Big\{\langle n_N\rangle_{eq}(1-\langle n_{N+1}\rangle_{eq})(1-e^{-\beta\hbar\omega_c})$$

$$-\langle n_{N+1}\rangle_{eq}(1-\langle n_N\rangle_{eq})(1-e^{\beta\hbar\omega_c})\Big\}. \qquad (17.10\text{-}4)$$

Note that here $h = 2\pi\hbar$ is Planck's constant. Setting $\langle n_N\rangle_{eq} = f_N$, we easily show that the two parts are equal:

$$f_N(1-f_{N+1})(1-e^{-\beta\hbar\omega_c})$$

$$= \frac{e^{\beta[(N+\frac{3}{2})\hbar\omega_c-\varepsilon_F]} - e^{\beta[(N+\frac{1}{2})\hbar\omega_c-\varepsilon_F]}}{\{e^{\beta[(N+\frac{1}{2})\hbar\omega_c-\varepsilon_F]}+1\}\{e^{\beta[(N+\frac{3}{2})\hbar\omega_c-\varepsilon_F]}+1\}} = -f_{N+1}(1-f_N)(1-e^{\beta\hbar\omega_c}). \quad (17.10\text{-}5)$$

So the final result is

$$\sigma_{yx} = \frac{e^2}{h} \sum_{N=0,1,2...} (N+1) f_N(1-f_{N+1})(1-e^{-\beta\hbar\omega_c}). \qquad (17.10\text{-}6)$$

We shall consider near zero temperatures; the last factor then becomes unity and the Fermi level is pinned in the gap between the broadened Landau levels, Fig. 17-5,

$$\varepsilon_N < \varepsilon_F < \varepsilon_{N+1}, \qquad (17.10\text{-}7)$$

while for the conductivity we have

$$\sigma_{yx} = \sum\nolimits_{N'} (N'+1) f_{N'}(1-f_{N'+1})(e^2/h). \qquad (17.10\text{-}8)$$

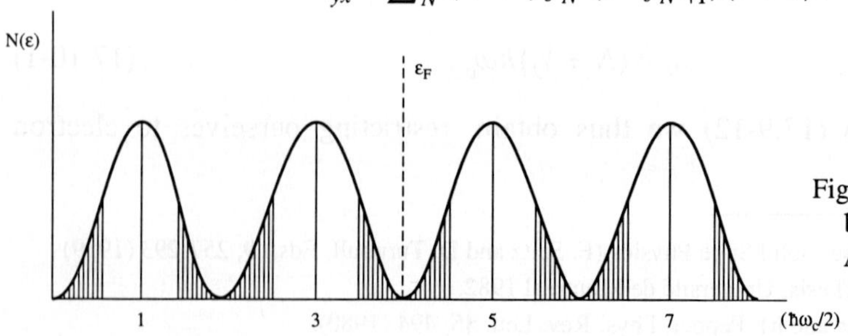

Fig. 17-5. Density of states of broadened Landau levels. After Vasilopoulos [25].

Let us first consider all $N' < N$ so that N' is at most $N-1$, or $N'+1$ is at most N; then

$$\varepsilon_{N'+1} \le \varepsilon_N < \varepsilon_F \rightarrow f_{N'+1} = 1, \tag{17.10-9}$$

showing that these terms do not contribute. Alternatively, for the terms with $N' > N$ we have that N' is at least $N+1$ so that

$$\varepsilon_{N'} \ge \varepsilon_{N+1} > \varepsilon_F \rightarrow f_{N'} = 0; \tag{17.10-10}$$

hence these terms do not contribute. Rests that $N' = N$, so that $f_{N'}=1$ and $f_{N'+1}=0$. This yields for σ_{yx}

$$\lim_{T\to 0} \sigma_{yx} = (N+1)\,(e^2/h), \quad N = 0,1,2... \tag{17.10-11}$$

So we have a *plateau* as long as (17.10-7) holds; when with increasing B the level ε_N empties out we move to a new, lower plateau. This is the integer quantum-Hall effect.

For finite temperatures we can expand in terms of $\delta = e^{-\beta\hbar\omega_c}$. Let also $0 < b = (\varepsilon_F - \varepsilon_N)/\hbar\omega_c < 1$ where ε_F now lies in a region of localised states. We then obtain

$$\sigma_{yx} = (N+1)(e^2/h)[1 - \delta^b - \delta^{1-b} + \delta^{2b} - \delta^{2(1-b)} + ...]. \tag{17.10-12}$$

Comparison with experimental plateau values[26,27] gives agreement to within at least the fifth decimal place! The accuracy becomes poorer, however, for low magnetic fields, typically due to impurity scattering affecting the Landau states. Also, the dissipation term $[\rho_{yx} = -\sigma_{yx}/(\sigma_{xx}^2 + \sigma_{yx}^2)]$ plays a role at lower B fields.[28] Some experimental data are presented in Fig. 17-6 (cf. Prange and Girvin).[29]

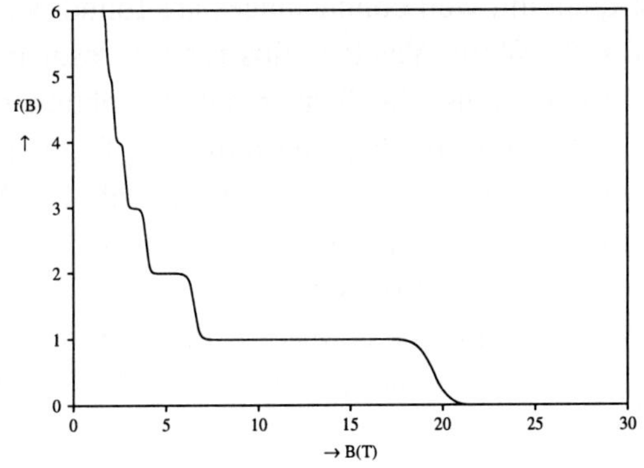

Fig. 17-6. Filling function $f(B)$ of Landau levels from measured QHE vs. B (in *Tesla*).

---

[26] D.C. Tsui, H.L. Stormer and A.C. Gossard, Phys. Rev. Lett. **48**, 1559 (1982); D.C. Tsui and A.C. Gossard, Appl. Phys. Lett. **38**, 550 (1981).

[27] T. Englert, Proc. Conf. on "Applications of High Magnetic Fields in Semiconductors", (Grenoble), Springer-Verlag, Heidelberg 1982.

[28] P. Vasilopoulos and C.M. Van Vliet, Phys. Rev. **B 34**, 1057 (1986).

[29] "The Quantum Hall Effect", R.E. Prange and S.M. Girvin, Eds., Springer, NY and Berlin, 1987.

17.10.2 *Magnetophonon Resonances*

(a) *The Linear Case, Small E-fields*

The electron gas in quantum wells is twodimensional. Such a well may be realised in MODFETs (also called HEMTs) at the interface of $Ga_xAl_{1-x}As/GaAs$ junctions or other III-V compound devices. Although some wells are nearly square, the well for most junctions is triangular. Here we shall, however, assume that we deal with a square well of dimension L_z. The B-field is taken to be $\mathbf{B} = B\hat{z}$, while $\mathbf{E} = E\hat{x}$. The Landau gauge is as in the previous sections $\mathbf{A} = (0, Bx, 0)$.

This gauge also must serve for confinement in quantum wires, which are basically onedimensional structures, since the well area is $L_y \times L_x$, with $\ell < L_y \ll L_x$. The first condition, requiring a very small nanometre size magnetic length, is realizable only for very high magnetic fields; the second condition is what defines a (quasi)-1D structure. [We note that some treatments[30] employ the gauge $\mathbf{A} = (-By, 0, 0)$ in order to combine the Landau Hamiltonian with a parabolic confinement potential $\frac{1}{2}m^*\varpi^2 y^2$; however, this does not yield collisional current in the x-direction.]

Magnetophonon resonances have been observed in GaAs 2D wells, as well as in quantum wires. In most cases the resonances are associated with polar optical phonon absorption or emission, with the resonances typically occurring for $P\omega_c = \omega_0$, where ω_0 is the longitudinal polar optical frequency and P is an integer, indicating a period proportional to $1/B$. Some recent experimental results for quantum wires[31] will be shown later. Our problem here is to quantitatively describe these phenomena. Early computations of quantum well conductances are found with Chaubey and Van Vliet[32] and Vasilopoulos.[33] While Vasilopoulos reports resonances for all types of phonons, incl. acoustical phonons, the former authors obtain resonances only for optical or polar optical phonons, as observed hitherto. Computations for quantum wires using the appropriate gauge were carried out in ref. 13, Appendix B. In the present section we shall first deal with optical phonons and next with polar optical phonons, leaving some details to the problems; we mainly follow Refs. 13 and 33.

The conductivity to be found is the collisional conductivity σ_{xx}, as given in Eqs. (17.7-10') or (17.6-17). For the Hamiltonian (17.9-4) with $U(z)$ representing a square well, the eigenfunctions are

$$\zeta_{N,k_y,n}(x,y,z) = (2/L_y L_z)^{1/2} \phi_N(x - x_0) e^{ik_y y} \sin(k_z z), \qquad (17.10\text{-}13)$$

with $N = 0,1,\ldots$ $k_y = 2\pi n_y / L_y$, $n_y = 0, \pm 1, \ldots$ $k_z = n\pi / L_z$, $n = 1, 2, \ldots$ Further, as before

[30] N. Mori, H. Momose and C. Hamaguchi, Phys. Rev. **B 45**, 4536 (1992).
[31] C. Brink, D. Schneider, G. Ploner, G. Strasser and E. Gornik, Physica **E 12**, 446 (2002).
[32] M.P. Chaubey and C.M. Van Vliet, Phys. Rev. **B 33**, 5617 (1986).
[33] P. Vasilopoulos, Phys. Rev. **B 33**, 8587 (1986).

$x_0 = \pm k_y \ell^2$, where $\ell = \sqrt{\hbar / m^* \omega_c}$ is the magnetic length, with the upper sign being for holes and the lower sign for electrons. The eigenvalues are

$$\varepsilon_{N,n} = (N + \tfrac{1}{2})\hbar\omega_c + n^2 \varepsilon_0, \quad \varepsilon_0 = \pi^2 \hbar^2 / 2m_{\|} L_z^2; \tag{17.10-14}$$

here N is the Landau state quantum number and n is the subband index. For quantum wells L_y and L_x are comparable, whereas for quantum wires we have $L_y \equiv d \ll L_x$, but as long as $\ell < d$, the *formal* results should be the same. The x-matrix element was given before, see Eqs. (17.9-21) and (17.9-23). Rests to obtain the matrix elements that go into $w_{\zeta\zeta'}$ due to electron-phonon interaction. We have

$$|(\zeta' | e^{\pm i\mathbf{q}\cdot\mathbf{r}} | \zeta)|^2 = |F_{n,n'}(\pm q_z)|^2 |J_{N,N'}(u)|^2 \, \delta_{k_y', k_y \pm q_y}, \tag{17.10-15}$$

where $u = \frac{1}{2}\ell^2 q_\perp^2$. Further, as shown by Enck et al.[34]

$$\begin{aligned} |J_{N,N'}(u)|^2 &= (N'!/N!) e^{-u} u^{N-N'} [\mathscr{L}_{N'}^{N-N'}(u)]^2, \quad N' \le N, \ N - N' \le N', \\ |J_{N,N'}(u)|^2 &= (N!/N'!) e^{-u} u^{N'-N} [\mathscr{L}_{N}^{N'-N}(u)]^2, \quad N \le N', \ N' - N \le N, \end{aligned} \tag{17.10-16}$$

where $\mathscr{L}_N^M(u)$ is an associated Laguerre polynomial. For $F_{n,n'}$ one finds with Parseval's theorem

$$\int_{-\infty}^{\infty} dq_z \, |F_{n,n'}(\pm q_z)|^2 = \frac{\pi}{L_z}(2 + \delta_{n,n'}). \tag{17.10-17}$$

Also, for the summation over $\mathbf{q}$ we write,

$$\sum\nolimits_{\mathbf{q}} \to \frac{A_0 L_z}{4\pi^2} \int dq_z \int q_\perp dq_\perp, \text{ noting } \int q_\perp dq_\perp = \frac{2}{\ell^2} \int du \,. \tag{17.10-18}$$

We are now ready to write down the transition rates; we obtain from (17.1-28), (17.1-29) and the above results

$$\begin{aligned} w_{Nk_y n, N'k_y' n'} &= \frac{A_0 L_z}{2\pi\hbar\ell^2} \int dq_z \int du \, |\mathscr{F}(\mathbf{q})|^2 \, |J_{NN'}(u)|^2 \\ &\times \Big\{ \langle N_\mathbf{q} \rangle_{eq} |F_{nn'}(q_z)|^2 \, \delta[(N - N')\hbar\omega_c + (n^2 - n'^2)\varepsilon_0 + \hbar\omega_\mathbf{q}] \delta_{k_y', k_y + q_y} \\ &+ (1 + \langle N_\mathbf{q} \rangle_{eq}) |F_{nn'}(-q_z)|^2 \, \delta[(N - N')\hbar\omega_c + (n^2 - n'^2)\varepsilon_0 - \hbar\omega_\mathbf{q}] \delta_{k_y', k_y - q_y} \Big\}. \end{aligned} \tag{17.10-19}$$

We also need the 'jump distance',

$$(R^0_{N'k_y'n',x} - R^0_{Nk_y n,x})^2 = (k_y' - k_y)^2 \ell^4 = q_y^2 \ell^4 , \tag{17.10-20}$$

where we used the y-momentum conservation as expressed by the Kronecker deltas.

---

[34] R.C. Enck, A.S. Saleh and H.Y. Fan, Phys. Rev. **182**, 790 (1969).

This expression needs to be symmetrized since we used the Landau gauge. Hence, we set $q_y^2 \to \frac{1}{2}(q_x^2 + q_y^2) = \frac{1}{2}q_\perp^2 = u/\ell^2$, so that

$$(R^0_{N'k_y'n',x} - R^0_{Nk_yn,x})^2 = (k_y' - k_y)^2 \ell^4 = u\ell^2 . \tag{17.10-20'}$$

For the summation over k_y we have the degeneracy result $\Sigma_{k_y} = A_0/2\pi\ell^2$. Finally, we set $N' - N = M$, $M = 0,1,..$ in the absorption term and $N - N' = M$, $M = 0,1...$ in the emission term, while the equilibrium phonon distribution is denoted by $\mathcal{N}_0$. For the collisional conductivity we now have quite generally

$$\sigma_{xx} = \frac{\beta q^2}{\hbar} \frac{A_0}{2\pi\ell^2} \sum_{N,N',n,n'} f^0_{Nn}(1 - f^0_{N'n'}) \int u\,du \int dq_z \,|\mathcal{F}(\mathbf{q})|^2 |J_{NN'}(u)|^2 |F_{nn}(\pm q_z)|^2$$

$$\times \Big\{ \mathcal{N}_0 |F_{nn'}(q_z)|^2 \,\delta[-M\hbar\omega_c + (n^2 - n'^2)\varepsilon_0 + \hbar\omega_\mathbf{q}]$$

$$+ (1 + \mathcal{N}_0) |F_{nn'}(-q_z)|^2 \delta[M\hbar\omega_c + (n^2 - n'^2)\varepsilon_0 - \hbar\omega_\mathbf{q}] \Big\}. \tag{17.10-21}$$

This is the final result. We now consider some specific types of phonons.

Optical phonons

We assume that $\hbar\omega_\mathbf{q} = \hbar\omega_0$ in the entire Brillouin zone, ω_0 being the longitudinal optical frequency. The coupling is given by [see (8.6-20)]

$$|\mathcal{F}(\mathbf{q})|^2 = \frac{\hbar D_0^2}{2A_0 L_z \rho\omega_0} \equiv \frac{D'}{A_0 L_z}, \tag{17.10-22}$$

where D' is a constant. The integral over dq_z was given before, cf. (17.10-17), while for the integration over du we will need the integral [see (A.4) of Ref. 21]

$$\int du\, e^{-u} u^{M+1} [\mathcal{L}_N^M(u)]^2 = \frac{(N+M)!}{N!}(2N + M + 1) . \tag{17.10-23}$$

Neglecting inter-subband transitions, the integral (17.10-17) yields $3\pi/L_z$. Whence,

$$\sigma_{xx}(\text{opt}) = \frac{3\beta q^2 D'}{2\pi\hbar L_z^2 \ell^2} \sum_{NMn} \Big\{ f^0_{Nn}(1 - f^0_{N+M,n}) \mathcal{N}_0 (2N + M + 1)$$

$$+ f^0_{Nn}(1 - f^0_{N-M,n})(1 + \mathcal{N}_0)(2N - M + 1) \Big\}\, \delta(M\hbar\omega_c - \hbar\omega_0) . \tag{17.10-24}$$

We note that for gallium arsenide ($m^* = 0.07\ m$) with L_z = 100 Å, the subband separation $\varepsilon_0 \approx 50\, meV$, while $\hbar\omega_c \approx 1.7B\, meV$ with B in *Tesla*. Therefore, even with a very strong magnetic induction of 10 *T*, inter-subband transitions are unlikely. Further, we will assume that only the lowest subband is occupied and that at sufficiently high temperatures, for which optical phonons are important, we can use nondegenerate statistics, so that $1-f^0_{N\pm M,1}=1$ and $f^0_N = e^{-\beta(\varepsilon_{N,1}-\varepsilon_F)}$. Then (17.10-24) reduces to

$$\sigma_{xx}(\text{opt}) = \frac{3\beta q^2 D'}{2\pi\hbar L_z^2 \ell^2} \times \sum_{N,M} e^{-\beta(\varepsilon_{N,1}-\varepsilon_F)}[(2\mathcal{N}_0+1)(2N+1)-M]\,\delta(M\hbar\omega_c - \hbar\omega_0). \quad (17.10\text{-}25)$$

The delta singularities show resonances at $M\omega_c = \omega_0$, $M = 1,2,\ldots$[35]. Of course, these extreme singularities do not occur due to collision broadening. So we replace the delta functions by Lorentzians, noting that we have always

$$\delta(\varepsilon - \hbar\omega_0) = \lim_{\Gamma\to 0+}\frac{1}{\pi}\text{Im}\frac{1}{\varepsilon - \hbar\omega_0 - i\Gamma}. \quad (17.10\text{-}26)$$

However, the limit will *not* be taken; Γ determines the line width $\hbar/\tau$, τ being the collision time. Hence, we have[35a]

$$\delta(M\hbar\omega_c - \hbar\omega_0) \to \frac{1}{\pi}\frac{\Gamma}{(M\hbar\omega_c - \hbar\omega_0)^2 + \Gamma^2}. \quad (17.10\text{-}27)$$

We now employ Poisson's summation formula in the form

$$\sum_{M=0}^{\infty} f(M+\tfrac{1}{2}) = \int_0^{\infty} f(x)dx + 2\sum_{s=1}^{\infty}(-1)^s \int_0^{\infty} f(x)\cos(2\pi s x)\,dx. \quad (17.10\text{-}28)$$

The integrals are easily obtained with contour integration. We find (note $\Gamma \ll \hbar\omega_0$):

$$\sum_{M=0}^{\infty}\delta(M - \omega_0/\omega_c) \to \frac{1}{\pi}\sum_{M=0}^{\infty}\frac{\Gamma/\hbar\omega_c}{[M-(\omega_0/\omega_c)]^2 + (\Gamma/\hbar\omega_c)^2}$$
$$= 1 + 2\sum_{s=1}^{\infty} e^{-2\pi s(\Gamma/\hbar\omega_c)}\cos[2\pi s(\omega_0/\omega_c)]. \quad (17.10\text{-}29)$$

For large N we can perform the summation over N as well by writing $\Sigma_N N e^{-\alpha N} = -(\partial/\partial\alpha)\Sigma_N e^{-\alpha N}$ and summing a geometric series. So we arrive at

$$\sigma_{xx}(\text{opt}) = \frac{3\beta q^2 D'}{2\pi\hbar^2\omega_c L_z^2\ell^2} e^{\beta(\varepsilon_F - \varepsilon_0)}\left[(2\mathcal{N}_0+1) - \frac{\omega_0}{2\omega_c}\right]\frac{\coth(\beta\hbar\omega_c/2)}{\sinh(\beta\hbar\omega_c/2)}$$
$$\times\left\{1 + 2\sum_{s=1}^{\infty} e^{-2\pi s(\Gamma/\hbar\omega_c)}\cos[2\pi s(\omega_0/\omega_c)]\right\}. \quad (17.10\text{-}30)$$

---

[35] $M = 0$ is now excluded since there are now no subband transitions to absorb or emit a phonon; however, it does not affect the delta function and we shall carry it along for Poisson's summation.

[35a] The expression on the rhs of (17.10-27) is also called the Breit–Wigner formula (especially in nuclear physics); the width Γ is related to the life time τ of the level by $\Gamma\tau = \hbar$.

Polar optical phonons

The computation for polar optical phonons goes quite similarly. We have now a different coupling function, which will result is some modified integrals. The coupling is given by Fröhlich's formula [(8.6-29)]:

$$|\mathcal{F}(\mathbf{q})|^2 = \frac{e^2\hbar\omega_0}{2A_0L_z\kappa_0}\left(\frac{1}{\kappa_\infty}-\frac{1}{\kappa_s}\right)\frac{1}{q^2} = \frac{\mathscr{C}}{A_0L_z(q_\perp^2+q_z^2)} \approx \frac{\mathscr{C}}{A_0L_zq_\perp^2}. \quad (17.10\text{-}31)$$

Here κ_∞ and κ_s are the dynamic and static dielectric constants. As in most treatments, we made the approximation shown in the far rhs of (17.10-18), approximating q^2 by $q_\perp^2$. Clearly, this is not quite correct, but inclusion of q_z leads to extremely laborious formulas[36] that would obscure the basic result to be obtained.

With $q_\perp^2 = 2u/\ell^2$ we note that now the factor u in the first line of (17.10-21) will cancel out, so that we need the integral[37]

$$\int_0^\infty du\, e^{-u}u^M [\mathscr{L}_N^M(u)]^2 = (N+M)!/N!, \quad M>0. \quad (17.10\text{-}32)$$

Noting from (17.10-16) that we still must multiply by $(N'!/N!)$ [absorption] or $(N!/N'!)$ [emission], the integrations always yield 1. Thus, the analogue of (17.10-24) becomes

$$\sigma_{xx}(\text{polar opt}) = \frac{3\beta q^2\mathscr{C}}{4\pi\hbar L_z^2}\sum_{NMn}\Big\{f_{Nn}^0(1-f_{N+M,n}^0)\mathscr{N}_0 + f_{Nn}^0(1-f_{N-M,n}^0)(1+\mathscr{N}_0)\Big\}\,\delta(M\hbar\omega_c - \hbar\omega_0). \quad (17.10\text{-}33)$$

Restricting ourselves again to nondegenerate statistics and occupation of the first subband only, the following result is found

$$\sigma_{xx}(\text{polar opt}) = \frac{3\beta q^2\mathscr{C}}{4\pi\hbar L_z^2}\sum_{N,M} e^{-\beta(\varepsilon_{N,1}-\varepsilon_F)}(2\mathscr{N}_0+1)\,\delta(M\hbar\omega_c-\hbar\omega_0). \quad (17.10\text{-}34)$$

Replacing the delta functions by Lorentzians and performing the summations with Poison's formula and a geometrical progression as before, we readily obtain

$$\sigma_{xx}(\text{polar opt}) = \frac{3\beta q^2\mathscr{C}}{8\pi\hbar^2\omega_cL_z^2}e^{\beta(\varepsilon_F-\varepsilon_0)}(2\mathscr{N}_0+1) \times \frac{1}{\sinh(\beta\hbar\omega_c/2)}\left\{1+2\sum_{s=1}^\infty e^{-2\pi s(\Gamma/\hbar\omega_c)}\cos[2\pi s(\omega_0/\omega_c)]\right\}. \quad (17.10\text{-}35)$$

---

[36] M.A. Stroscio and M. Dutta, "Phonons in Nanostructures", Cambridge Univ. Press, 2001, p. 140, 146.
[37] I.S. Gradshteyn and I.M. Ryzhik, "Table of Integrals...", Op. Cit., Eq. 7.414-3.

Experimental results for a GaAs n^+nn^+ 'sandwich' by Eaves et al.[38] are shown in Fig. 17-7 for fields up to 20 *Tesla*. The lowest occupied Landau level has $N = 2$ for $B = 11.3$ *Tesla*. For the ground state $N = 0$ unavailable magnetic fields would be required.

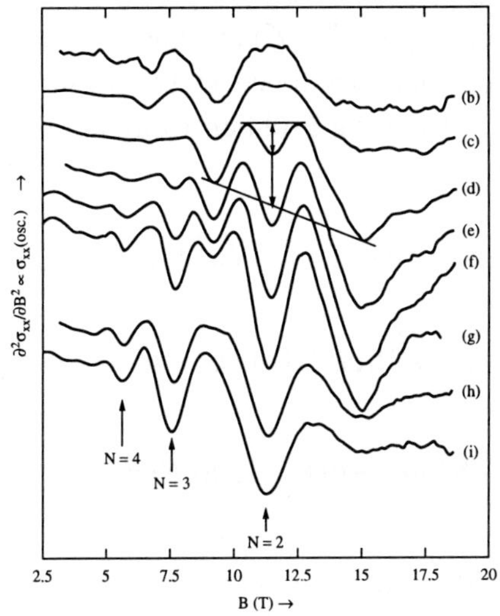

Fig. 17-7. Transverse magnetoconductivity vs. B after Eaves et al.[38] [With permission.]

(b) *The Nonlinear Case*

We shall follow the developments in Ref. 13; for an alternate approach, see Chaubey.[39] For the general nonlinear case the electric field must be incorporated in the Hamiltonian. So we have,

$$\begin{aligned} h^0 &= [\mathbf{p} - (q/c)\mathbf{A}]^2 / 2m^* - qEx + U(z) \\ &= \frac{p_x^2}{2m^*} + \frac{(p_y \mp m^* \omega_c x)^2}{2m^*} + \frac{p_z^2}{2m_{\parallel}} - qEx + U(z). \end{aligned} \qquad (17.10\text{-}36)$$

The eigenfunctions are given by Kahn and Frederikse[18,22]; they still have the form (17.10-13), but now $x_0 = \pm k_y \ell^2 + qE\ell^2 / \hbar\omega_c$ while the eigenvalues are

$$\varepsilon_{N,k_y,n} = \left(N + \frac{1}{2}\right)\hbar\omega_c + n^2\varepsilon_0 - \hbar k_y c\left(\frac{E}{B}\right) + \frac{1}{2}m^* c^2 \left(\frac{E}{B}\right)^2 . \qquad (17.10\text{-}37)$$

We note that the degeneracy over k_y has been lifted.

The collisional current was given in (17.4-6). The two terms add since the indices may be interchanged. Hence,

---

[38] L. Eaves, P.S.S. Guimaraes, J.C. Portal, T.P. Pearsall and G. Hill, Phys. Rev. Lett. **53**, 608 (1984).

[39] M.P. Chaubey, Il Nuovo Cimento, **14**, 379 (1992).

$$J_{e,x}^{ss} = \frac{q}{V_0} \sum_{NN'k_yk_y'nn'} w_{Nk_yn,N'k_y'n'} f_{Nn}^{ss}(1 - f_{N'n'}^{ss})(x_{0,k_y'} - x_{0,k_y}). \qquad (17.10\text{-}38)$$

This expression is now linear in the orbit jump $\Delta x_{0,k_y}$. Further, quite astonishingly, the phonon matrix elements $(\zeta'| e^{\pm i\mathbf{q}\cdot\mathbf{r}} |\zeta)$ are not affected by the E-field. We shall only consider polar optical phonons since most measurements are done on GaAs layers or quantum wires. Using the Fröhlich formula (17.10-31) into the scattering rates (17.10-19) we obtain

$$J_{e,x}^{ss} = \frac{2\pi q \mathscr{C}}{\hbar V_0^2} \sum_{\mathbf{q}} \sum_{NN'k_yk_y'nn'} |J_{NN'}(u)|^2 (\Delta x_{0,k_y}/q_\perp^2) f_{Nn}^{ss}(1 - f_{N'n'}^{ss}) \Big\{ \langle N_{\mathbf{q}}\rangle_{eq} |F_{nn'}(q_z)|^2$$
$$\times \delta[(N-N')\hbar\omega_c - \hbar(k_y - k_y')(cE/B) + (n^2 - n'^2)\varepsilon_0 + \hbar\omega_{\mathbf{q}}]\delta_{k_y',k_y+q_y} + (1 + \langle N_{\mathbf{q}}\rangle_{eq})$$
$$\times |F_{nn'}(-q_z)|^2 \delta[(N-N')\hbar\omega_c - \hbar(k_y - k_y')(cE/B) + (n^2 - n'^2)\varepsilon_0 - \hbar\omega_{\mathbf{q}}]\delta_{k_y',k_y-q_y} \Big\}. \qquad (17.10\text{-}39)$$

The reader will have noticed that we did not subscript the steady-state distribution with k_y since we argued in Section 17.2 that the distribution should be homogeneous and not depend on the orbit centres x_{0,k_y}. Thus we retain the energies (17.10-37), but we replace the momentum $\hbar k_y$ by a mean value $\langle \hbar k_y \rangle = m^* v_d$, where $v_d = (cE/B)$ is the drift velocity associated with the Lorentz force (compensated by the Hall field qE_H). So, for a nondegenerate gas we write

$$f_{Nn}^{ss} = \exp\Big\{-\beta\Big[(N + \tfrac{1}{2})\hbar\omega_c + n^2\varepsilon_0 - \tfrac{1}{2}m^*(cE/B)^2 - \varepsilon_F\Big]\Big\}. \qquad (17.10\text{-}40)$$

The kinetic energy term is tantamount to electron heating; in fact,

$$(k_BT)^{-1}\Big[\varepsilon_{N,n} - \tfrac{1}{2}m^*(cE/B)^2 - \varepsilon_F\Big] \sim (k_BT_e)^{-1}\Big[\varepsilon_{N,n} - \varepsilon_F\Big], \qquad (17.10\text{-}41)$$

with

$$T_e \approx T + m^*(cE/B)^2/2k_B, \qquad (17.10\text{-}42)$$

which is reminiscent of the Druyvesteyn distribution for plasma heating, cf. (15.8-12). Strict momentum conservation has been observed in $|J_{NN'}(u)|^2$ but will be ignored in the energy delta functions. Assuming a mean orbit jump $\overline{\Delta x}$, we set $k_y' - k_y = \mp\overline{\Delta x}/\ell^2$. Lastly, to render the problem tractable, we write

$$(\Delta x_{0,k_y}/q_\perp^2) \approx (\Delta x_{0,k_y}^2/q_\perp^2)/\overline{\Delta x} = (q_y^2/q_\perp^2)\ell^4/\overline{\Delta x} \to \ell^4/2\overline{\Delta x}, \qquad (17.10\text{-}43)$$

where the last step involves symmetrization for the Landau gauge.

Returning to (17.10-39), the $\mathbf{q}$-summation can now be carried out entirely similarly to the linear case. Thus, assuming again that only the lowest subband is occupied and that no inter-subband transitions occur, we arrive at

$$J_{e,x}^{ss} = \frac{3q\mathscr{E}}{4\pi\hbar L_z^2\overline{\Delta x}} \sum_{N,M} e^{-\beta_e[(N+\frac{1}{2})\hbar\omega_c+\varepsilon_0-\varepsilon_F]} (2\mathscr{N}_0+1)\delta\left[M\hbar\omega_c \mp \frac{\hbar\overline{\Delta x}}{\ell^2}\left(\frac{cE}{B}\right) - \hbar\omega_0\right], \tag{17.10-44}$$

where $\beta_e = 1/k_B T_e$. Now $B\ell^2 = \pm\hbar c/q$. We introduce $\varpi = \omega_0 + qE\overline{\Delta x}/\hbar$, so that the delta functions become $\delta(M\hbar\omega_c - \hbar\varpi)$. Further, we replace these functions by zero-shift Lorentzians as before. Summing the series over M with Poisson's formula and that over N as a geometrical progression, we finally obtain

$$J_{e,x}^{ss} = \frac{3q\mathscr{E}}{8\pi\hbar^2\omega_c L_z^2\overline{\Delta x}} e^{\beta_e(\varepsilon_F-\varepsilon_0)}(2\mathscr{N}_0+1)$$
$$\times \frac{1}{\sinh(\beta_e\hbar\omega_c/2)}\left\{1+2\sum_{s=1}^{\infty} e^{-2\pi s\Gamma/\hbar\omega_c}\cos[2\pi s(\varpi/\omega_c)]\right\}. \tag{17.10-45}$$

While this *seems* quite similar to the linear case, resonances now occur for

$$P\omega_c = \varpi = \omega_0 + qE\overline{\Delta x}/\hbar\,, \tag{17.10-46}$$

where P is an integer. Clearly, the resonances are displaced by the electric field. When the displacement term equals a half odd integer times $\omega_c/2$, maxima convert to minima and *vice versa.* Such effects have been observed by Eaves et al.[38] in GaAs quantum wells and more recently by Brink et al.[31] in quantum wires, cf. Fig. 17-8.

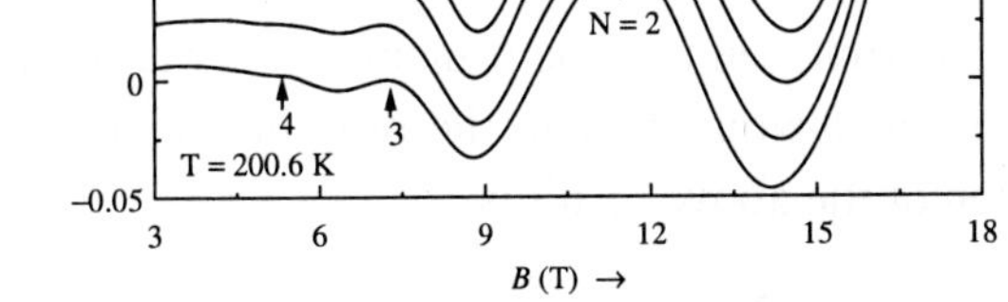

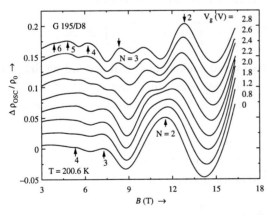

Fig. 17-8. Effect of electric field on the current maxima.[31] [With permission.]

There are many more applications that will not be discussed because of space limitations. When the 2D electron gas has a periodic lateral modulation, both collisional and ponderomotive (called by them 'diffusive') conduction occur. Also, the thermal conductance shows oscillatory behaviour; cf. the papers by Peeters and Vasilopoulos.[39a,39b]

[39a] P. Vasilopoulos and F.M. Peeters, Phys. Rev. Lett. **63**, 2120 (1989).
[39b] F.M. Peeters and P. Vasilopoulos, Phys. Rev. **B46**, 4667 (1992).

17.11 Slightly Disordered Metals. The Aharonov–Bohm Effect

In 1959 Y. Aharonov and David Bohm wrote a basic paper about the role of the scalar and vector potentials in quantum theory. Whereas in classical electrodynamics $\Phi(\mathbf{r},t)$ and $\mathbf{A}(\mathbf{r},t)$ are mathematical tools to derive the fields, in quantum theory these potentials assume an identity of their own: they are part and parcel of the canonical formalism required to describe the behaviour of charged particles in an electromagnetic field. Thus, as indicated in their paper, systems *not* subject to any field and not experiencing the basic Lorentz force, are affected by the potentials when present in space and time.[40] As first possibility they consider a particle in a Faraday cage whose outside carries a potential $\Phi(t)$, which gives rise to a term $-e\Phi(t)$ in the Hamiltonian of an electron inside the cage. The resulting wave function will be

$$\psi(\mathbf{r},t)=\psi_0(\mathbf{r})e^{iS(t)/\hbar}, \quad S(t)=\int_{-\infty}^{t} e\Phi(t')dt'. \tag{17.11-1}$$

Of course, for this case the field has no physical effect, the change in the wave function being just a change in phase.

The situation changes, however, if an electron beam in free space (region I) is allowed to have its path split by some means, e.g., a grating with slits, and afterwards enters two separate spaces (regions II) involving two hollow metal cylinders subject to time-varying potentials as before, being then united to interfere (region III); the situation is pictured in Fig. 17-9(a). The final wave function is now

$$\psi(\mathbf{r},t)=\psi_{10}e^{iS_1(t)/\hbar}+\psi_{20}e^{iS_2(t)/\hbar}, \quad S_1=e\int^{t}\Phi_1\,dt', \quad S_2=e\int^{t}\Phi_2\,dt'. \tag{17.11-2}$$

It is clear that the interference of the two wave packets at F will depend on the phase difference $(S_1-S_2)/\hbar=(e/\hbar)\oint\Phi\,dt$. The electrons 'feel' the effect of the scalar potential, although they are not subject to any field. This *quantum phenomenon* has become known as the 'Aharonov–Bohm effect'.

From relativistic considerations it is easily seen that the principle of covariance of the description demands that a similar conclusion apply to the presence of a vector potential. The relativistic generalization of the above result is

$$\Delta\varphi=\frac{e}{\hbar}\oint\left(\Phi\,dt-\frac{\mathbf{A}}{c}\cdot d\mathbf{r}\right), \tag{17.11-3}$$

where the integral is over a closed path in space-time d^4x. For a vector potential to be present, while the particles are not subject to a magnetic field, it suffices to install a small solenoid which fully encloses the magnetic induction $\mathbf{B}$ in the region II, prior to interference of the partial beams; the set-up is pictured in Fig. 17-9(b).

---

[40] Y. Aharonov and D. Bohm, Phys. Rev. **115**, 485 (1959).

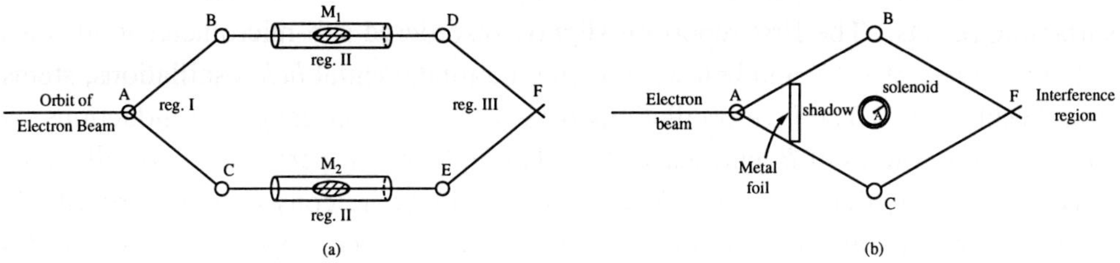

Fig. 17-9. Schematic experiment to demonstrate interference with time-dependent scalar potential or with a vector potential; A,B,C,D,E: suitable devices to separate and divert the beam; F: point of interference. (a): Scalar potential applied to cylinders M_1 and M_2; (b): Solenoid placed in the region II to create a vector potential. After Aharonov and Bohm [40]. [With permission.]

Although the above experiment has been carried out in the sixties, we are presently interested in the Aharonov–Bohm effect in slightly disordered metals. In geometries involving hollow cylindrical wires or annuli (rings) it has been predicted as well as observed that the magnetoresistance shows oscillatory behaviour as a function of the magnetic flux ϕ which, ideally, is confined to the inside region of the cylinder or the hole of the ring. From Stokes' theorem we have

$$\phi \equiv \iint \mathbf{B} \cdot d\mathbf{S} = \oint \mathbf{A} \cdot d\mathbf{l}. \tag{17.11-4}$$

Let us at first consider the case that we have a ring with two contacts on opposite ends, to which a voltage is applied; in this situation the **E**- and **B**-fields are crossed, with the **E**-field being tangential, i.e., $\mathbf{E} = E_\theta \hat{\theta}$. Clearly, **A** is also tangential, with $A_\theta = \phi / 2\pi R$, R being the radius and $\delta \ll R$ being the thickness of the ring. The wave picture presented previously is applicable if the electrons in the metal form a coherent wave, as is the case in a *mesoscopic* metallic conductor, for which the phase-breaking diffusion length $L_\varphi = \sqrt{D\tau_\varphi}$ is large (or at most comparable) with respect to the circumference $2\pi R$. Here τ_φ is, roughly speaking, the relaxation time associated with inelastic collisions; in the absence of these the normal resistance is due to (near) elastic collisions with impurities of the slightly disordered metal. Starting at one contact, the electrons can be thought of as interfering at the opposite contact, whereby each partial wave yields a phase difference $\pm(e/\hbar c)\int \mathbf{A} \cdot d\mathbf{l}$; they constructively interfere for

$$\frac{e}{\hbar c}\oint \mathbf{A} \cdot d\mathbf{l} = n2\pi, \quad \text{or} \quad \frac{e}{hc}\phi = n. \tag{17.11-5}$$

We shall denote the elementary *flux quantum* by $\phi_0 = hc/e$; it is twice as big as that in Josephson junctions in superconductors, given that we here have single electrons instead of Cooper pairs. According to the simple picture presented, the flux should be quantized in units ϕ_0 and the magnetoconductance will oscillate with that period.

Yet, experiments performed on metal rings in the mid-eighties have shown some surprising results. The first report on slightly disordered sub-micrometer gold rings at temperatures of ~ $1K$ and below, showing the fundamental *hc/e* oscillations, stems from Webb et al.[41] The *hc/e* oscillations persist to very high magnetic inductions (B ~1 T). At low values of B, curiously, they also find the half period *hc/2e* oscillations, predicted by early theories (see below). Their findings are consistent with results by Chandrasekhar et al. on aluminium and thin-film silver rings: the fundamental oscillations of period *hc/e* are only seen at high fields; the temperature dependences for the two types are also different.[42] These authors also offer an elementary picture for the half period oscillations. With reference to Fig. 17-10 (a) and (b), it is assumed that in the latter case the constructive interference takes place at the injecting contact, with the electrons being back-scattered at the other contact; the path length is

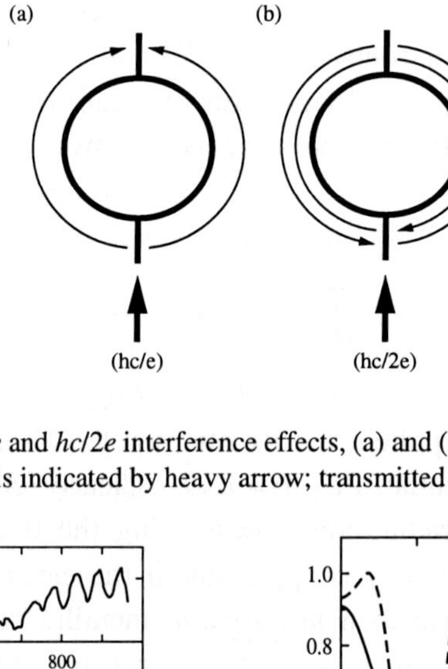

Fig. 17-10. Paths for *hc/e* and *hc/2e* interference effects, (a) and (b), respectively. Input contact for electrons is indicated by heavy arrow; transmitted wave by light arrows.

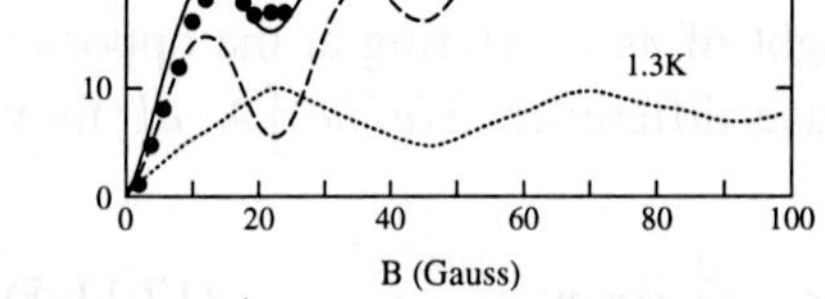

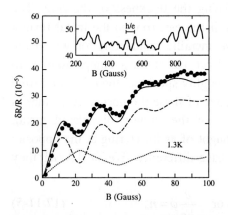

Fig.17-11. Magnetoresistance of a silver ring. Dots: experimental data. Dashed line: Al'shuler's *hc/2e* contribution. Dotted line: assumed *hc/e* term[42].

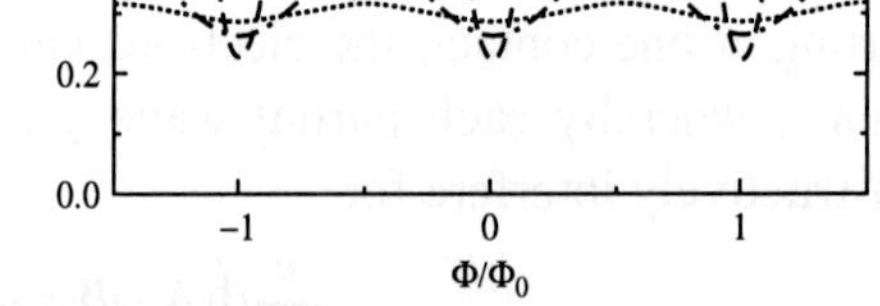

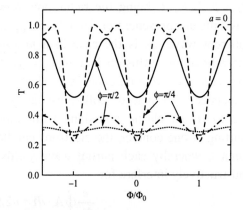

Fig.17-12.Transmission T for L–B model with one scatterer in one arm and none in other for different phases φ, defined by $t_1 = t_2 = \exp(i\varphi)$. After[49b].

[41] R.A. Webb, S. Washburn, C.P. Umbach and R.B. Laibowitz, Phys. Rev. Lett. **54**, 2696 (1985).
[42] V. Chandrasekhar, M.J. Rooks, S. Wind and D.E. Prober, Phys. Rev. Lett. **55**, 1610 (1985).

therefore twice that computed before. Some of their measurements for the relative magnetoresistance [note that $|\Delta G| = \Delta R / R^2$] are shown in Fig. 17-11.

We will now discuss the earlier results on cylinders, for which we can have either a longitudinal or a transverse electrical field and we shall dwell on some quite different theoretical aspects. In the transverse field configuration we basically have a number of rings in parallel and no new behaviour is expected. For the longitudinal arrangement one employs quantum wires of nanometre dimensions. Measurements have been reported by a variety of groups, notably in Belgium (Katholieke Universiteit Leuven[43]) and in Russia (Institute of Solid State physics, Moscow[44]). At liquid He temperatures and below oscillations with a period of 10 – 25 G are pronounced. From the known geometry it was concluded that a period of $hc/2e$ applies. Gijs et al. claim quantitative agreement with the theoretical expressions derived by Al'tshuler et al., discussed below.

This brings us back to the theoretical aspects of the Aharonov–Bohm effect in solids. Three types of theories have been advanced, as will now be discussed.

17.11.1 *Landauer–Büttiker (L–B) Models*

These authors emphasize the coherence of the wave function for electrons in a mesoscopic solid.[45] In a seminal paper in 1970 Landauer[46] introduced the model in which the current flow is likened to the quantum mechanical transmission of electrons through a barrier, the transmission and reflection coefficients being denoted by T and R, respectively. The theory was extended to the many-channel case by Büttiker et al. and by Fisher and Lee as well as Benamira[47]; the application to the A–B effect is found in later articles.[48,49,49a] Landauer's formula for the conductance reads

$$G = 2(e^2/h)T/R = 2(e^2/h)T/(1-T). \qquad (17\text{-}11\text{-}6)$$

We note that e^2/h is the quantum conductance unit also found in the quantum Hall

[43] M. Gijs, C. Van Haesendonck and Y. Bruynseraede, Phys. Rev. Lett. **52**, 2069 (1984); ibid. Physica **B 127**, 450 (1984).

[44] D. Yu. Sharvin and Yu. V. Sharvin, Pis'ma Zh. Eksp. Teor. Fiz. **34**, 285 (1981) [JETP Lett. **34**, 272].

[45] We note that mesoscopic solids *per se* do not exist; rather, a solid shows 'mesoscopic behaviour' if a coherence region occurs for which $\lambda \ll L \ll L_\varphi$, where L is the characteristic dimension, λ is the elastic scattering mean free path and L_φ is the inelastic phase-breaking diffusion length.

[46] R. Landauer, Phil. Mag. **21**, 863 (1970); also IBM J. Res. Dev. **1**, 223 (1957).

[47] D.S. Fisher and P.A. Lee, Phys. Rev. **B23**, 6851 (1981). Their result does not have the analogue of the denominator $1-T$; the full result is in F. Benamira, Ph. D. Thesis, Université de Montréal, 1996.

[48] M. Büttiker, Y. Imry and R. Landauer, Phys. Lett. **96A**, 365 (1983); also, M. Büttiker, Y. Imry, R. Landauer and S. Pinhas, Phys. Rev. **B31**, 6207 (1985).

[49] Y. Gefen , Y. Imry and M. Ya. Azbel, Phys. Rev. Lett. **52**, 129 (1984); M. Büttiker, Y. Imry and M.Ya. Azbel, Phys. Rev. **A30**, 1982 (1984); D. Stone and Y. Imry, Phys. Rev. Lett. **56**, 189 (1986).

[49a] Y. Imri, "Introduction to Mesoscopic Physics", Oxford University Press, Oxford and NY, 1997.

effect; in *SI* units $e^2/h = 3.87\times10^{-5}\ \Omega^{-1}$. Gefen, Imry and Azbel [49] have applied Landauer's formula to rings, which they consider to be two-channel 'scatterers'. They computed the transmission and reflection amplitudes r_i and t_i, $i = 1,2$. The overall transmission amplitude yields $T = tt^*$; they find

$$T = 4\frac{\alpha + \beta\cos 2\pi(\phi/\phi_0) + \beta'\sin 2\pi(\phi/\phi_0)}{\gamma + \delta\cos 2\pi(\phi/\phi_0) + \varepsilon\cos 4\pi(\phi/\phi_0)}. \tag{17.11-7}$$

Herein $\phi_0 = hc/e$ and $\alpha..\varepsilon$ are coefficients. The result clearly shows the expected periodicity with resonances for $\phi = n\phi_0$, $n = \pm1, \pm2, ...$ [The ε-term has period ½ ϕ_0.]

A more general theory of the Landauer type has been presented by Stone and Imry[49]. The formula of departure is the Fisher and Lee result[47]:

$$G = 2(e^2/h)\int d\varepsilon(-df/d\varepsilon)\,\mathrm{tr}[t(\varepsilon)t^*(\varepsilon)] \equiv 2(e^2/h)g(\varepsilon, B), \tag{17.11-8}$$

where $f(\varepsilon)$ is the Fermi–Dirac distribution. To obtain the transmission-coefficient matrix they consider a 2D tight-binding nearest neighbour model, subject to a perpendicular magnetic field. Employing further a recurrent Green's function procedure, they compute the trace of the matrix. According to these authors, the occurrence of the half period $\phi_0/2$ is due to ensemble averaging, which is simulated by the energy averaging in (17-11-8).

Very recent results employing the *S*-matrix to obtain the *R*- and *T* coefficients in rings with asymmetric injection have been published by Vasilopoulos et al.[49b] Some of their findings for the transmission *T* with one scatterer in one arm and none in the other one are shown in Fig. 17-12 [placed beside Fig. 17-11 above]. The main period is ϕ_0, with a trace of the half period $\phi_0/2$ still discernable.

17.11.2 *Diagrammatic Methods*

While the Landauer–Büttiker method uses the wave description of the originators of the A–B effect, their model, however sophisticated, is basically *ad hoc,* since the 'obstacles' responsible for the transmission and reflection are not specified. Depending on the temperature and other variables, there may not be a coherent wave[45]. So it seems to us that one should also consider the quantum mechanical description as for other magnetoconductance phenomena, putting in the relevant interactions in the perturbation potential $\mathcal{V}$. The problem then requires appropriate many-body techniques and a diagrammatic evaluation. This will not be carried out here, but it is well known that the main contribution to the magneto-conductivity stems from maximally crossed diagrams (connecting electron lines with impurity sites); see the paper of Langer and Neal[50] and a number of communications by

---

[49b] P. Vasilopoulos, O. Kálmán, F.M. Peeters and M.G. Benedict, Phys. Rev. **B75**, 035304 (2007).

[50] J.S. Langer and T. Neal, Phys. Rev. Lett. **16**, 984 (1966); also, Mahan, Op. Cit. p. 509ff.

Al'tshuler, Aronov and Spivak.[51] Their result has the form:

$$\frac{\Delta\sigma}{\sigma} = -\frac{2}{\pi\rho(\varepsilon_F)}\sum_{\omega,\mathbf{q}} C(\omega,\mathbf{q}), \tag{17.11-9}$$

where $\rho(\varepsilon_F)$ is the density of states at the Fermi level and the C's are the diagrams mentioned above. For a cylinder of height ℓ and radius R with longitudinal fields $E \parallel B$ this leads to the magnetoconductance

$$\Delta G = \frac{e^2}{\pi h}\frac{2\pi R}{\ell}\left[\left(\frac{1}{2}+\gamma\right) Z_\phi\left(L_\varphi(B)\right) - \frac{3}{2}\left(\tilde{L}_\varphi(B)\right)\right], \tag{17.11-10}$$

where

$$Z_\phi(L_\varphi) = 2\ln\frac{L_\varphi(B)}{L_\varphi(0)} + 4\sum_{n=1}^{\infty} K_0(2\pi Rn / L_\varphi)\cos 4\pi n(\phi/\phi_0). \tag{17.11-11}$$

The normalizing flux here is as before ($\phi_0 = hc/e$), so that oscillations occur with the half period $\phi_0/2$, as observed in most experiments with the longitudinal cylindrical configuration. Further, K_0 is the modified Bessel function of order zero, whose asymptotic expansion indicates that the amplitude decreases exponentially if the diffusion length $L_\varphi \ll 2\pi R$. The quantities L_φ and $\tilde{L}_\varphi$ are field-dependent diffusion lengths defined in Ref. 51(3). Clearly, these results are far more detailed than those of the previously described models.

17.11.3 *Modified Linear Response Results*

Finally we shall discuss the applicability of our earlier obtained linear response formulae for the magneto-conductivity. We must start with the one-particle Hamiltonian h_0 in the absence of interactions. For a cylindrical geometry we have[52]

$$h_0 = -\frac{\hbar^2}{2mR^2}\left(\frac{\partial}{\partial\theta} - i\frac{\phi}{\phi_0}\right)^2 - \frac{\hbar^2}{2m}\frac{\partial^2}{\partial z^2}, \quad \phi_0 = \frac{hc}{e}. \tag{17.11-12}$$

Note that we use θ for the azymuthal angle. The eigenfunctions and eigenvalues are

$$\zeta_{n,k_z}(\mathbf{r}) = \frac{1}{\sqrt{2\pi R\ell}} e^{i(n\theta + k_z z)}, \quad \varepsilon_{n,k_z} = \frac{\hbar^2}{2mR^2}(n - \phi/\phi_0)^2 + \frac{\hbar^2 k_z^2}{2m}. \tag{17.11-13,14}$$

---

[51] B.L. Al'tshuler, A.G. Aronov and B.Z. Spivak, Pis'ma Zh. Eksp. Teor. Fiz. **33**, 101 (1981) [JETP Lett. **33**, 94 (1981)]; ibid. Pis'ma Zh. Eksp. Teor. Fiz. **33**, 515 (1981) [JETP Lett. **33**, 499 (1981)]; also, ibid. & others, Pis'ma Zh Eksp. Teor. Fiz. **35**, 476 (1982) [JETP Lett. **35**, 588 (1982)].

[52] Most articles give the Hamiltonian for positive charge, even though they apply the results to electrons in metals. We will conform to this usage...which goes back to *A–B*. For a *correct* treatment, replace the normalized flux ϕ everywhere by $-\phi$ [so, $+i\phi/\phi_0$ in (17.11-13)]. The conductivities will be unaltered, see problem 17.6 for details.

For n we can restrict ourselves to positive integers, $n = 1, 2, \ldots$ while for k_z we have the usual quantization $k_z = 2\pi n_z / \ell$, $n_z = 0, \pm 1, \pm 2, \ldots$. For rings we take $\ell \to 1$ and $k_z = 0$. Henceforth we will set $\bar{\phi} = \phi / \phi_0$. An outline of the linear response method applied to the A–B effect in disordered metals is found in two articles cited below.[53]

First we shall consider the transverse configuration, in which the electrical field is along the cylinder surface (E_θ) while the magnetic induction is along the cylinder axis (B_z). For the conductivity the connection with Cartesian coordinates gives

$$\sigma_{\theta\theta} = (\sigma_{xx} + \sigma_{yy}) - \sin\theta(\sigma_{yx} + \sigma_{xy}) = 2\sigma_{xx}, \tag{17.11-15}$$

where the latter rhs follows from the symmetry relations discussed previously. For the velocity matrix elements one finds

$$(\zeta' | v_\mu | \zeta) = \frac{n\hbar}{2imR}(\delta_{n,n'+1} \mp \delta_{n,n'-1})\delta_{k_z,k_z'}, \tag{17.11-16}$$

where the – sign corresponds to $\mu = x$ and the + sign to $\mu = y$. For a ring, omit the delta of the z-direction. Clearly, there are no diagonal contributions. As to the nondiagonal contributions, the cross-components σ_{yx}^{nd} and σ_{xy}^{nd} are of no concern, cf. (17.11-15), while the like-components σ_{xx}^{nd} and σ_{yy}^{nd} based on (17.8-16) for $\omega \to 0$ are found to vanish.[54] Rests therefore the collisional conductivities σ_{xx}^{coll} and σ_{yy}^{coll}. From (17.7-10') we have

$$\sigma_{xx}^{coll} = \frac{\beta e^2}{2\pi R\ell} \sum_{n,n',k_z,k_z'} w_{nk_z,n'k_z'} f_{nk_z}(1 - f_{n'k_z'})(x_{nk_z} - x_{n'k_z'})^2, \tag{17.11-17}$$

where x_{nk_z} and $x_{n'k_z'}$ are diagonal matrix elements of the states $|nk_z)$ and $|n'k_z')$, respectively. If θ is allowed to run from $-\pi$ to $+\pi$, these matrix elements are zero. However, when the mean free path is less than $2\pi R$ – at the onset we have said that $\lambda \ll 2\pi R$ – then for these matrix elements we must limit ourselves to $-\theta_0 \le \theta \le \theta_0$ and the matrix elements are finite. Moreover, on average we will have $\langle x_{nk_z} - x_{n'k_z'} \rangle = \lambda(\varepsilon)$. The result is therefore

$$\sigma_{\theta\theta} = \sigma_{\theta\theta}^{coll} = \frac{\beta e^2}{\pi R\ell\delta} \sum_{n,n',k_z,k_z'} w_{nk_z,n'k_z'} f_{nk_z}(1 - f_{n'k_z'})\lambda^2(\varepsilon_{nk_z}). \tag{17.11-18}$$

We now compute the transition rate w, assuming that the interaction with the impurities involves a screened Coulomb potential, $v(R) = (e^2/4\pi R)e^{-\kappa R}$. Resolving into a Fourier series we have $V_{\mathbf{q}} = (1/V_0)[e^2/(q^2 + \kappa^2)]$ where $\mathbf{q} = (q_\theta, q_z)$. For the

---

[53] P. Vasilopoulos and C.M. Van Vliet, Solid State Comm. **61**, 121 (1987) (impurity scattering); ibid. Phys. Rev. **B34**, 5217 (1986) (acoustical phonon scattering). Various details have been revised.

[54] Of course, one should not expect a contribution from these terms, since they are independent of the particular scattering mechanisms envisaged.

matrix element we obtain

$$(n'k_z'\,|\,v(R)\,|\,nk_z)$$
$$=\frac{1}{2\pi R\ell}\sum_{\mathbf{q}}V_{\mathbf{q}}\int_{-\pi}^{\pi}\int_{0}^{\ell}Rd\theta dz\,e^{i\theta(n'-n+q_\theta R)}e^{iz(k_z'-k_z+q_z)}=V_{(n-n')/R,k_z-k_z'}\,. \quad (17.11\text{-}19)$$

The transition probability is found to be

$$w_{nk_z,n'k_z'}=\frac{N_I}{\hbar R\delta\ell}\left(\frac{e^2}{R^{-2}(n-n')^2+(k_z-k_z')^2+\kappa^2}\right)^2\delta(\varepsilon_{nk_z}-\varepsilon_{n'k_z'}),\ (17.11\text{-}20)$$

where N_I is the number of impurities per unit volume; note that the thickness δ of the cylinder must be finite. Observing that we have elastic collisions, the factor $(1-f_{n'k_z'})$ can be replaced by $(1-f_{nk_z})$;[55] further we set as usual $f_{nk_z}(1-f_{nk_z})$ $=\beta^{-1}df/d\varepsilon_F$. This yields for $\sigma_{\theta\theta}$

$$\sigma_{\theta\theta}=\frac{e^2}{\pi\hbar}\frac{N_I\lambda^2(\varepsilon_F)}{R^2\ell^2\delta^2}\sum_{n,n',k_z,k_z'}\frac{df_{nk_z}}{d\varepsilon_F}\left(\frac{e^2}{R^{-2}(n-n')^2+(k_z-k_z')^2+\kappa^2}\right)^2\delta(\varepsilon_{nk_z}-\varepsilon_{n'k_z'});$$
$$(17.11\text{-}21)$$

we took the mean free path out of the sum since the scattering will occur only near the Fermi energy. [One may verify that the factors after $e^2/\pi\hbar$ have the dimension of inverse length so that the conductance is $\sim e^2/\pi\hbar$.]

The factor of paramount importance here is the *energy conservation* as expressed by the delta function. Writing it out, one readily finds

$$\delta(\varepsilon_{nk_z}-\varepsilon_{n'k_z'})=(2mR^2/\hbar^2)\delta[(n-n')(n+n'-2\overline{\phi})+R^2(k_z^2-k_z'^2)].\quad (17.11\text{-}22)$$

Consider the case that $k_z=\pm k_z'$. This is always fulfilled for a ring since then $k_z=k_z'=0$. For the cylinder this condition renders the eigenvalues degenerate. Generally, the assumption of 'coherent backscattering', $\mathbf{k}+\mathbf{k}'=0$, has been assumed to apply. In our representation the components of the canonical momentum are p_z, $p_\theta-(q/c)A_\theta$ or p_z, $p_\theta-\overline{\phi}/R$, with eigenvalues $\hbar k_z$ and $(\hbar/R)(n-\overline{\phi})$. If in a transition coherent backscattering occurs, it entails $k_z+k_z'=0$ and $n+n'-2\overline{\phi}=0$. As evident from (17.11-22) this causes a resonance of the conductance. Clearly then, resonances will occur for $2\overline{\phi}=m$, where m is an integer. This indicates *a prima vista* disagreement with the experimental observations for rings which show both $\phi_0/2$ and ϕ_0 periodicities.[56]

Now it could be argued that perhaps there is a fundamental flaw in the quantum

---

[55] I.e., we can employ the Adams–Holstein result instead of the Argyres –Roth formula.

[56] There is no dependence as yet on the phase-breaking length L_φ since we have not incorporated inelastic collisions in our approach; these collisions can be incorporated in the damping constant Γ.

approach – as contrasted with the wave approach – since a *full* cylindrical geometry is presumed. Thus, the results from this method would apply if the collecting contact were placed at 2π radians of the ring, i.e., near the injecting contact, thereby mimicking the wave-picture situation of Fig. 17(b). However, as we noted, the mean free path for elastic scattering is many times smaller than the circumference of the cylinder, which makes the above argument mute. So, in order to obtain the observed ϕ_0 periodicity it should be shown that a suppression of odd or even harmonics occurs.

With this in mind we will proceed with the full evaluation of (17.11-22). While the integrations over dk_z and dk_z' can be carried out, we shall restrict ourselves to rings. The delta functions will be summed with Poisson's summation formula. Let generally we have $f(n'+1/2) \rightarrow f(x)$. Then $\delta[(n-n')(n+n'-2\bar{\phi})]$ causes $f(x)$ to have two roots, $x_1 = n + \frac{1}{2}$ and $x_2 = 2\bar{\phi} - n + \frac{1}{2}$. Since $n \neq n'$ only the second root will contribute. With the property

$$\delta[g(x)] = \sum_j \frac{\delta(x - x_j)}{|g'(x_j)|}, \tag{17.11-23}$$

where x_j are the roots of $g(x) = 0$, we can proceed. The remaining delta functions are replaced by Lorentzians with zero shift and width Γ. The following result is easily verified:

$$\sigma_{\theta\theta} = \frac{e^2}{\pi\hbar} \frac{2mN_I\lambda^2(\varepsilon_F)}{\hbar^2\ell^2\delta^2} \sum_n \frac{df_n}{d\varepsilon_F} \left(\frac{e^2}{4R^{-2}(n-\bar{\phi})^2 + \kappa^2} \right)^2 \frac{1}{|n-\bar{\phi}|}$$
$$\times \left\{ 1 + 2\sum_{s=1}^{\infty} e^{-2\pi sC} \cos(4\pi s\bar{\phi}) \right\}. \tag{17.11-24}$$

Here $C = 2\pi mR^2\Gamma/\hbar^2$ is the damping constant (see [56]). The result clearly shows the periodicity $\phi = p(hc/2e)$, $p =$ integer.

As to the remaining sum, it can generally not be performed, except at near-zero temperature. In that event we can write $df_n/d\varepsilon_F \approx \delta(\varepsilon_n - \varepsilon_F)$. With (17.11-23) we have

$$\delta(\varepsilon_n - \varepsilon_F) = \frac{mR^2}{\hbar^2} \frac{1}{|n-\bar{\phi}|} \{\delta[n-(\bar{\phi}+\eta)] + \delta[n-(\bar{\phi}-\eta)]\}, \tag{17.11-25}$$

where $\eta = \sqrt{2mR^2\varepsilon_F/\hbar^2}$. Although there is no *physical* collision broadening, we will replace the deltas formally by Lorentzians of width Γ' in order to avoid the singularities. The summation can now be done for each series of delta functions with Poisson's summation formula and later on combined. The straightforward result is,

$$\sigma_{\theta\theta} = \frac{e^2}{\pi\hbar} \frac{4mN_I\lambda^2(\varepsilon_F)}{\hbar^2\ell^2\delta^2\varepsilon_F} \left(\frac{e^2R^2}{4\eta^2 + R^2\kappa^2} \right)^2$$

$$\times\left\{1+2\sum_{s=1}^{\infty}e^{-2\pi sC}\cos(4\pi s\bar{\phi})\right\}\left\{1+2\sum_{m=1}^{\infty}e^{-2\pi mC'}\cos(2\pi m\bar{\phi})\cos(2\pi m\eta)\right\}. \quad (17.11\text{-}26)$$

This result shows that at very low temperatures we have both $\phi_0/2$ and ϕ_0 oscillations with different amplitudes. Suppose that η is close to an integer, say $|\delta| < ¼$ away; then for $\phi = p\phi_0/2$ and $\phi = p\phi_0$ (or $\bar{\phi} = p/2$, $\bar{\phi} = p$) the peak values are, respectively,[57]

$$\begin{aligned} A_1 &= A\coth(\pi C)\left\{1+2\sum\nolimits_m(-1)^{mp}e^{-2\pi mC'}\cos 2\pi m\delta\right\} \xrightarrow[p\text{ odd}]{C'\to 0} 0\,, \\ A_2 &= A\coth(\pi C)\left\{1+2\sum\nolimits_m e^{-2\pi C'}\cos 2\pi m\delta\right\} \sim A\coth(\pi C)/\pi C'. \end{aligned} \quad (17\text{-}11\text{-}27)$$

Hence, the odd harmonics are wiped out! With the damping factor C' finite, the results will be less dramatic, but clearly the odd harmonics of the $\phi_0/2$ cycle will be greatly reduced. [If η is close to a half integer, the even harmonics will disappear.] Encouragingly, this behaviour is in agreement with the observations, cf. Fig. 17-11, and with recent computations shown in Fig. 17-12. The above was valid at very low temperatures. At higher temperatures the Fermi distribution gets smeared out and the ϕ_0 oscillations will no longer show up.

Finally, we briefly consider the longitudinal field arrangement, with both **E** and **B** along the axis of the cylinder, $\hat{z}$. The Hamiltonian, eigenfunctions and eigenvalues are as before. The (diagonal) conductivity is as in (17.8-10), which for $\omega = 0$ with velocity matrix elements $(n'k_z'|v_z|nk_z) = (\hbar k_z/m)\delta_{nn'}\delta_{k_zk_z'}$ yields

$$\sigma_{zz} = \frac{e^2}{2\pi R\ell\delta}\sum_{n,k_z}\frac{\hbar^2k_z^2}{m^2}\frac{df(\varepsilon_{nk_z})}{d\varepsilon_F}\tau(p)\,, \quad (17.11\text{-}28)$$

where we noted $df/d\varepsilon = -df/d\varepsilon_F$. For this case it is easier to compute the resistivity; noting that in the tensor for σ the *zz*-component is not related to the planar components, we find

$$\rho_{zz} = \frac{2\pi R\ell\delta m^2}{e^2\hbar^2}\left\{\sum_{n,k_z}k_z^2\frac{df(\varepsilon_{nk_z})}{d\varepsilon_F}\left[\frac{1}{\tau(p)}\right]^{-1}\right\}^{-1}. \quad (17.11\text{-}29)$$

For τ we need the momentum-loss relaxation time. For the momentum loss we take

$$\frac{(\Delta p)_{rms}}{|p|} = \frac{[(k_z-k_z')^2+R^{-2}(n-n')^2]^{½}}{k_F} \quad (17.11\text{-}30)$$

so that

$$\frac{1}{\tau(p_{n,k_z})} = \sum_{n'k_z'}w_{nk_z,n'k_z'}\left\{1-[(k_z-k_z')^2+R^{-2}(n-n')^2]^{½}/k_F\right\}. \quad (17.11\text{-}31)$$

---

[57] The alternating series $\mathrm{Re}\,\Sigma_m(-1)^m e^{-2\pi(C'-i\delta)m}$ is convergent by Leibniz' criterion and for $C'\to 0$ by Abel's limit-theorem.

To simplify matters we shall make the common assumption of coherent backscattering with respect to k_z at the Fermi surface; then $k_z + k_z' = 0$ and $|\mathbf{k}| = k_F$, so that with the previously obtained scattering rates w [see (17.11-20)] we now have

$$\frac{1}{\tau(p_{n,k_z})} = \sum_{n'} w_{nk_z,n'-k_z} \left\{1 - R^{-1}|n-n'|/k_F\right\} = \frac{N_I}{\hbar R\ell\delta}$$

$$\times \sum_{n'} \left(\frac{e^2R^2}{(n-n')^2 + R^2\kappa^2} \right)^2 \left(1 - \frac{|n-n'|}{Rk_F}\right) \delta\left[\frac{\hbar^2}{2mR^2}(n-n')(n+n'-2\bar{\phi}) \right]. \quad (17.11\text{-}32)$$

The evaluation of this sum proceeds similarly as in the transverse case. With Γ being the width due to collision broadening of the delta functions, we obtain

$$\frac{1}{\tau(p_{n,k_z})} = \frac{N_I}{\hbar R\ell\delta}\frac{2mR^2}{\hbar^2}\left(\frac{e^2R^2}{4(n-\bar{\phi})^2 + R^2\kappa^2}\right)^2 \left(1 - \frac{2|n-\bar{\phi}|}{Rk_F}\right)\left\{1 + \sum_{s=1}^{\infty} e^{-2\pi C}\cos(4\pi s\bar{\phi})\right\}\alpha$$

$$(17.11\text{-}33)$$

where $\alpha = 1/2|n-\bar{\phi}|$. This is to be substituted into (17.11-29). We then arrive at

$$\rho_{zz} = \frac{2\pi R^2 m^3 N_I}{e^2\hbar^5}\left\{1 + \sum_{s=1}^{\infty} e^{-2\pi C}\cos(4\pi s\bar{\phi})\right\}$$

$$\times \left\{ \sum_{n,k_z} k_z^2 \frac{df(\varepsilon_{nk_z})}{d\varepsilon_F} \left[\left(\frac{e^2R^2}{4(n-\bar{\phi})^2 + R^2\kappa^2} \right)^2 \left(1 - \frac{2|n-\bar{\phi}|}{Rk_F}\right) \frac{1}{|n-\bar{\phi}|} \right]^{-1} \right\}^{-1}. \quad (17.11\text{-}34)$$

The remaining sums can only be analytically performed for very low temperatures. Then $df/d\varepsilon_F = \delta(\varepsilon_{nk_z} - \varepsilon_F)$. The sum over k_z is replaced by an integral $(\ell/2\pi)\int dk_z$, after which the sum over n is carried out with Poisson's summation formula; in the slowly-varying factors above that contain $n-\bar{\phi}$ we substitute $Rp_{F\theta}\hbar^{-1} \equiv R\bar{k}_{F\theta}$. We finally obtain

$$\rho_{zz} = \frac{4\pi m^2R^3N_I}{e^2\hbar^3\ell}\left(\frac{e^2}{4(\bar{k}_{F\theta})^2 + \kappa^2}\right)^2\left\{1 + \sum_{s=1}^{\infty} e^{-2\pi C}\cos(4\pi s\bar{\phi})\right\}\frac{1}{|R\bar{k}_{F\theta}|}$$

$$\times\left(1 - \frac{2|\bar{k}_{F\theta}|}{k_F}\right)\left\{1 + 2\sum_{m=1}^{\infty}\cos(2\pi m\bar{\phi})J_1(2\pi m\eta)/2\pi m\eta\right\}^{-1}. \quad (17.11\text{-}35)$$

From the asymptotic formula for the first-order Bessel function the last bracket gives

$$\{1+\ldots\} = 1 + 2\sqrt{\frac{2}{\pi}}\sum_{m=1}^{\infty}\cos(2\pi m\bar{\phi})\cos(2\pi m\eta - 3\pi/4)/(2\pi m\eta)^{3/2}. \quad (17.11\text{-}36)$$

With $\eta \sim 10^2$ we note that the $m = 1$ term is less than 0.1% of the constant term (1); so the last summand does not contribute. Hence, *only* oscillations with period $\phi_0/2$ will be noticeable, in agreement with experimental results for longitudinal cylinders.

3. THE MASTER HIERARCHY

17.12 Kinetic Equations for Quantum Systems with Binary Interactions

In classical systems kinetic equations for n-particle distribution functions ($n = 2,3,...$) can be derived from the Liouville equation; these equations go beyond the Boltzmann equation – which, properly speaking, contains the two-body distribution function if the 'closure assumption' of molecular chaos in not made. The hierarchy, of which we met the equilibrium version in Section 4.8, is known as the BBGKY hierarchy.[58] We could have presented it in Chapter XIII, but dispensed with it since, like the classical Boltzmann equation, the results do not apply to quantum systems. On the contrary, we should start from the master equation and obtain a set of equations in which: (a) the particles are not 'tagged' but indistinguishable from the start; (b) no **k**-space is presumed. In this section we shall proceed to obtain the hierarchy; in particular, the second-order result is of importance for stochastic processes, to be described in the next Part. We follow two earlier articles.[59]

We have already noted at several occasions that Van Hove was the first one to derive the Boltzmann equation from the master equation[2]. Unfortunately, there are no streaming terms in his result, since he employed the original ME as presented in Section 16.11. The full master hierarchy has been derived in the same vein by Ernst and Cohen – also without any connection to external fields.[60] So, in order to obtain kinetic equations that incorporate applied fields, at least in the approximation of LRT, we will present here a hierarchy based on the inhomogeneous ME of Section 16.12. The first-order kinetic equation is the linearized quantum Boltzmann equation of Section 17.5. The present hierarchy, useful for stochastic phenomena, will only employ the diagonal ME since the nondiagonal part of the linearized QBE has no relevance for mesoscopic fluctuations of coarse-grained variables, cf. Section 1.5. In the cited articles Vasilopoulos and the author investigated the kinetic equations both for a system of fermions and a system of bosons. In each case we assumed there were binary interactions with a 'bath' of other bosons, which remained in thermal equilibrium (adiabatic approximation). The details of the interactions need not be known; interactions with phonons or other fermions would lead to similar results.

17.12.1 *Fermion Moment Equations and Fokker–Planck Moments*

If in Eq. (16.12-7) we omit the nondiagonal Liouville term and take the diagonal matrix element in the representation $\{|\gamma\rangle\}$, being a basis of eigenstates pertaining to

[58] H. Grad in "Encyclop. of Physics" Vol. 12, (S. Flügge, Ed.), Springer, NY 1958; J.L. Delcroix, Ref. 1.
[59] C.M. Van Vliet and P. Vasilopoulos, Can. J. Phys. **75**, 401 (1997); P. Vasilopoulos and C.M. Van Vliet, Can. J. Phys, **61**, 102 (1983).
[60] M.H. Ernst and E.G.D. Cohen, J. of Statistical Physics **25**, 153(1981).

the unperturbed many-body Hamiltonian $\mathcal{H}^0$, we then have the probabilistic equation similar to Van Hove's ME (16.11-14) but with streaming terms

$$\frac{\partial p(\gamma,t)}{\partial t}=\sum_{\bar\gamma}[W_{\bar\gamma\gamma}p(\bar\gamma,t)-W_{\gamma\bar\gamma}p(\gamma,t)]+\beta F(t)p_{eq}(\gamma)\langle\gamma|(\dot A)_d|\gamma\rangle. \qquad (17.12\text{-}1)$$

As in the previous sections, we incorporated a response Hamiltonian $-AF(t)$ for coupling with an electrical field $A=\Sigma_i(\mathbf{r}_i-\mathbf{r}_{i0})$, $F(t)=q\mathbf{E}(t)\cdot$, while more generally we write $A=\Sigma_i a_i$, where a_i are one-particle operators. The derivative operator $(\dot A)_d$ is more complex; it contains a collisional part and a ponderomotive part, cf. (16.10-44) and preceding formulas: $(\dot A)_d=-\Lambda_d A_d+\dot A_d$. To obtain the moment equations, we shall use the generating function according to Laplace for the discrete variables $\{n_\zeta\}$; we multiply both sides of (17.12-1) by $\exp(-s^\zeta n_\zeta)\equiv\exp(-\sum_\zeta s_\zeta n_\zeta)$ and sum over the states $|\gamma\rangle=|\{n_\zeta\},\{|\hat N_\eta\}\rangle$, ($\{\hat N_\eta\}$ denoting the boson occupations):

$$\frac{\partial}{\partial t}\sum_\gamma e^{-s^\zeta n_\zeta}p(\gamma,t)$$

$$=\sum_{\gamma\bar\gamma}e^{-s^\zeta n_\zeta}[W_{\bar\gamma\gamma}p(\bar\gamma,t)-W_{\gamma\bar\gamma}p(\gamma,t)]+\beta F(t)\sum_\gamma e^{-s^\zeta n_\zeta}p_{eq}(\gamma)\langle\gamma|(\dot A)_d|\gamma\rangle$$

$$=\sum_{\gamma\bar\gamma}(e^{-s^\zeta\bar n_\zeta}-e^{-s^\zeta n_\zeta})\,p(\gamma,t)W_{\gamma\bar\gamma}+\beta F(t)\sum_\gamma e^{-s^\zeta n_\zeta}p_{eq}(\gamma)\langle\gamma|(\dot A)_d|\gamma\rangle, \qquad (17.12\text{-}2)$$

where we interchanged γ and $\bar\gamma$ in the first term of the double sum. Further we write

$$e^{-s^\zeta\bar n_\zeta}-e^{-s^\zeta n_\zeta}=e^{-s^\zeta n_\zeta}(e^{-s^\zeta(\bar n_\zeta-n_\zeta)}-1)$$

$$=e^{-s^\zeta n_\zeta}\sum_{k=1}^{\infty}\frac{(-1)^k}{k!}\sum_{\zeta_1\dots\zeta_k}s_{\zeta_1}\dots s_{\zeta_k}(\bar n_{\zeta_1}-n_{\zeta_1})\dots(\bar n_{\zeta_k}-n_{\zeta_k}), \qquad (17.12\text{-}3)$$

where we used $[\Sigma_\zeta s_\zeta(\bar n_\zeta-n_\zeta)]^k=\Sigma_{\zeta_1\dots\zeta_k}s_{\zeta_1}\dots s_{\zeta_k}(\bar n_{\zeta_1}-n_{\zeta_1})\dots(\bar n_{\zeta_k}-n_{\zeta_k})$. However, from the definition of generalized Fokker–Planck moments we have[61]

$$F_k(\{n_\zeta\})=\sum_{\bar\gamma}(\bar n_{\zeta_1}-n_{\zeta_1})\dots(\bar n_{\zeta_k}-n_{\zeta_k})W_{\gamma\bar\gamma}\,. \qquad (17.12\text{-}4)$$

Employing these moments and writing $\langle X\rangle_t=\sum_\gamma X\,p(\gamma,t)$ we find for (17.12-2):

$$\frac{\partial\langle e^{-s^\zeta n_\zeta}\rangle_t}{\partial t}=\sum_{k=1}^{\infty}\langle e^{-s^\zeta n_\zeta}\frac{(-1)^k}{k!}\sum_{\zeta_1\dots\zeta_k}s_{\zeta_1}\dots s_{\zeta_k}F_k(\{n_\zeta\})\rangle_t+\langle e^{-s^\zeta n_\zeta}\beta F(t)\langle\gamma|(\dot A)_d|\gamma\rangle\rangle_{eq}\,.$$
$$(17.12\text{-}5)$$

The k-th order equation is now easily found from (17.12-5); suffice it to take the k-th

[61] K.M. Van Vliet, J. Math. Phys. **12**, 1981 (1971). The Fokker–Planck equation will be introduced in Chapter XVIII; in this section we just use the definition of the moments.

order derivative of both sides of (17.12-5) with respect to $s_{\zeta_1}...s_{\zeta_k}$ and set afterwards all s_ζ equal to zero. Only terms up to order k need to be considered since the terms of order larger than k do not contribute. Some examples follow.

(i) *Boltzmann equation*

Taking the derivative of (17.12-5) with respect to $s_{\zeta_1} = s_{\zeta'}$ we obtain

$$\frac{\partial\langle -n_{\zeta'} e^{-s^\zeta n_\zeta}\rangle_t}{\partial t} = \langle -n_{\zeta'} e^{-s^\zeta n_\zeta}(-1)\sum_\zeta s_{\zeta'} F_1(n_{\zeta'})\rangle_t + \langle e^{-s^\zeta n_\zeta}(-1)F_1(n_{\zeta'})\rangle_t + \sum_{k=2}^{\infty}...$$

$$+\langle -n_{\zeta'} e^{-s^\zeta n_\zeta}\beta F(t)\langle\gamma|(\dot{A})_d|\gamma\rangle\rangle_{eq}\,. \tag{17.12-6}$$

Setting now $s_{\zeta'}\to 0,\ s_{\zeta_k}\to 0,\ k\geq 2,$ we note that the first term on the rhs is zero, together with all terms for $k\geq 2$. The result is

$$\frac{\partial\langle n_{\zeta'}\rangle_t}{\partial t} = \langle F_1(n_{\zeta'})\rangle_t + \beta F(t)\langle n_{\zeta'}\langle\gamma|(\dot{A})_d|\gamma\rangle\rangle_{eq}\,. \tag{17.12-7}$$

Now writing $p(\gamma,t) = p(\{n_\zeta\})P_{eq}(\{\hat{N}_\eta\})$ the boson average for the first term is

$$\langle F_1(n_{\zeta'})\rangle_b = -\langle (n_{\zeta'} - \bar{n}_{\zeta'})W_{\gamma\bar{\gamma}}\rangle_b = -\langle M n_{\zeta'}\rangle_b = -\mathscr{B}_{\zeta'} n_{\zeta'}\,, \tag{17.12-8}$$

where we used the master operator M in function space and the earlier proven connection with the Boltzmann operator (17.1-15). We thus arrive at

$$\frac{\partial\langle n_{\zeta_1}\rangle_t}{\partial t} + \langle\mathscr{B}_{\zeta_1}(n_{\zeta_1})\rangle_t = -\beta F(t)\sum_\zeta\langle n_{\zeta_1}\mathscr{B}_\zeta n_\zeta\rangle_{eq}a_\zeta + \beta F(t)\sum_\zeta\langle n_{\zeta_1}n_\zeta\rangle_{eq}\dot{a}_\zeta\,, \tag{17.12-9}$$

where we set $(\zeta|a|\zeta) = a_\zeta$ and $(\zeta|\dot{a}|\zeta) = \dot{a}_\zeta$. This is the previously obtained linearized quantum Boltzmann equation with two streaming terms; the averages on the rhs have been evaluated in Section 17.5, cf. Eqs. (17.5-5) and (17.5-7). In this evaluation we have used the truncation rule of footnote 12, which is exact for the equilibrium averages in the grand-canonical ensemble. However, the Boltzmann term (second term on the lhs) involves a two particle non-equilibrium average, so that the first moment equation is coupled to the next one as in most hierarchies; the usual truncation assumption is $\langle\mathscr{B}_{\zeta_1}(n_{\zeta_1})\rangle_t \approx \mathscr{B}_{\zeta_1}\langle(n_{\zeta_1})\rangle_t$.

(ii) *Equation for* $\langle n_{\zeta_1}n_{\zeta_2}\rangle_t$

We take the second-order derivative $\partial^2/\partial s_{\zeta'}\partial s_{\zeta''}$ of both sides of (17.2-5) and let afterwards all s_ζ go to zero; first, we assume that $\zeta'\neq\zeta''$. The result is

$$\frac{\partial\langle n_{\zeta'}n_{\zeta''}\rangle_t}{\partial t} = \langle n_{\zeta'}F_1(n_{\zeta''})\rangle_t + \langle n_{\zeta''}F_1(n_{\zeta'})\rangle_t + \langle F_2(n_{\zeta'},n_{\zeta''})\rangle_t$$

$$-\beta F(t)\sum_{\zeta}\langle n_{\zeta'}n_{\zeta''}B_{\zeta}n_{\zeta}\rangle_{eq}a_{\zeta}+\beta F(t)\sum_{\zeta}\langle n_{\zeta'}n_{\zeta''}n_{\zeta}\rangle_{eq}\dot{a}_{\zeta}\,. \qquad (17.12\text{-}10)$$

This kinetic equation contains both the first-order and second-order Fokker–Planck moments. Before we further evaluate the result, we must investigate the case of confluence of the two occupancies. Thus, let $\zeta' = \zeta''$. Differentiating twice to $s_{\zeta'}$ one easily finds

$$\frac{\partial\langle n_{\zeta'}^2\rangle_t}{\partial t}=2\langle n_{\zeta'}F_1(n_{\zeta'})\rangle_t+\langle F_2(n_{\zeta'},n_{\zeta'})\rangle_t-\beta F(t)\sum_{\zeta}\{\langle n_{\zeta'}^2\mathscr{B}_{\zeta}n_{\zeta}\rangle_{eq}a_{\zeta}-\langle n_{\zeta'}^2 n_{\zeta}\rangle_{eq}\dot{a}_{\zeta}\}. \qquad (17.12\text{-}11)$$

Since for fermions $n_{\zeta'}^2=n_{\zeta'}$, this result should reduce to the linearized QBE of (17.12-9) above. For the streaming terms this is at once clear; for the first two terms, we need the full second-order Fokker–Planck moment, to be considered now.

The evaluation goes as follows (Ref. 59$^{(1)}$). From its definition (17.12-4) and from (17.1-9), (17.1-10), (17.1-11), and in particular (17.1-7) for $\overline{n}_{\zeta}-n_{\zeta}$ we find:

$$\begin{aligned}\langle F_2(n_{\zeta_1},n_{\zeta_2})\rangle_b &= \sum_{\zeta'\zeta''} w_{\zeta'\zeta''}n_{\zeta'}(1-n_{\zeta''})(1-2n_{\zeta_1})(1-2n_{\zeta_2})(\delta_{\zeta_1\zeta'}+\delta_{\zeta_1\zeta''})(\delta_{\zeta_2\zeta'}+\delta_{\zeta_2\zeta''})\\ &=\sum_{\zeta'\zeta''} w_{\zeta'\zeta''}n_{\zeta'}(1-n_{\zeta''})(1-2n_{\zeta_1})(1-2n_{\zeta_2})\\ &\quad\times[(\delta_{\zeta_1\zeta'}\delta_{\zeta_2\zeta'}+\delta_{\zeta_1\zeta''}\delta_{\zeta_2\zeta''})+(\delta_{\zeta_1\zeta'}\delta_{\zeta_2\zeta''}+\delta_{\zeta_1\zeta''}\delta_{\zeta_2\zeta'})]\,. \end{aligned}\qquad (17.12\text{-}12)$$

Focussing on the two terms in the last parentheses of the brackets [ ], we see that the cross terms $\zeta_1\neq\zeta_2$ require that $\zeta_1=\zeta'$ and $\zeta_2=\zeta''$ or $\zeta_1=\zeta''$ and $\zeta_2=\zeta'$; in addition, the first parentheses in [ ] allow for $\zeta'=\zeta_1=\zeta_2$ and $\zeta''=\zeta_1=\zeta_2$. Clearly, this involves the Kronecker delta $\delta_{\zeta_1}\delta_{\zeta_2}$, thus yielding the diagonal contribution. Using further $n_{\zeta}(1-2n_{\zeta})=-n_{\zeta}$, the total result is found to be

$$\begin{aligned}\langle F_2(n_{\zeta_1},n_{\zeta_2})\rangle_b &= -[w_{\zeta_1\zeta_2}n_{\zeta_1}(1-n_{\zeta_2})+w_{\zeta_2\zeta_1}n_{\zeta_2}(1-n_{\zeta_1})]\\ &\quad+\delta_{\zeta_1\zeta_2}\sum_{\zeta}[w_{\zeta_1\zeta}n_{\zeta_1}(1-n_{\zeta})+w_{\zeta\zeta_1}n_{\zeta}(1-n_{\zeta_1})]\,.\end{aligned}\qquad (17.12\text{-}13)$$

In the next chapter we shall see that this microscopic result is in accord with the F–P moments for such macroscopic Markov processes as Brownian motion and generation-recombination noise. We still note that many results in the literature are either erroneous[62] or incomplete [i.e., are missing the degeneracy factors $(1-n_{\zeta})$][63].

We return to the confluence equation (17.12-11). With the first two F–P moments given by (17.12-8) and (17.12-13) [diagonal part] we find

$$2\langle n_{\zeta_1}F_1(n_{\zeta_1})\rangle_t+\langle F_2(n_{\zeta_1},n_{\zeta_1})\rangle_t=-2\sum_{\zeta}[w_{\zeta_1\zeta}\langle n_{\zeta_1}^2(1-n_{\zeta})\rangle_t-w_{\zeta\zeta_1}\langle n_{\zeta}\underbrace{n_{\zeta_1}(1-n_{\zeta_1})}_{0}\rangle]$$

---

[62] J.-P. Nougier and J.C. Vassière, Phys. Rev. **B 37**, 8882 (1988).

[63] S.V. Gantsevich, V.L. Gurevich and R. Katilius, Nuovo Cimento Rivista, **2**, 1 (1979).

$$+\sum_{\zeta}[w_{\zeta_1\zeta}\langle n_{\zeta_1}(1-n_{\zeta})\rangle_t + w_{\zeta\zeta_1}\langle n_{\zeta}(1-n_{\zeta_1})\rangle_t]$$

$$=-\sum_{\zeta}[w_{\zeta_1\zeta}\langle n_{\zeta_1}(1-n_{\zeta})\rangle_t - w_{\zeta\zeta_1}\langle n_{\zeta}(1-n_{\zeta_1})\rangle_t] = -\langle \mathscr{B}_{\zeta_1} n_{\zeta_1}\rangle_t \simeq -\mathscr{B}\langle n_{\zeta_1}\rangle_t \,. \quad (17.12\text{-}14)$$

So the confluence equation (17.11-11) is just the first moment QBE, as expected.

The computation of the streaming terms is a bit tedious, although the structure is easily deduced, following the previous developments for the linearized QBE. For the first streaming term we have, changing ζ' to ζ_1 and ζ'' to ζ_2,

$$\sum_{\zeta}\langle n_{\zeta_1}n_{\zeta_2}\mathscr{B}_{\zeta}n_{\zeta}\rangle_{eq}a_{\zeta} = \sum_{\zeta}a_{\zeta}\langle n_{\zeta_1}n_{\zeta_2}\sum_{\zeta'}[w_{\zeta\zeta'}(1-n_{\zeta'})n_{\zeta} - w_{\zeta'\zeta}(1-n_{\zeta})n_{\zeta'}]\rangle_{eq}$$

$$= \sum_{\zeta}\sum_{\zeta'}{}' a_{\zeta}\langle n_{\zeta_1}n_{\zeta_2}[w_{\zeta\zeta'}(1-n_{\zeta'})n_{\zeta} - w_{\zeta'\zeta}(1-n_{\zeta})n_{\zeta'}]\rangle_{eq}$$

$$+\sum_{\zeta}a_{\zeta}\langle n_{\zeta_1}n_{\zeta_2}[w_{\zeta\zeta_1}(1-n_{\zeta_1})n_{\zeta} - w_{\zeta_1\zeta}(1-n_{\zeta})n_{\zeta_1}]\rangle_{eq}$$

$$+\sum_{\zeta}a_{\zeta}\langle n_{\zeta_1}n_{\zeta_2}[w_{\zeta\zeta_2}(1-n_{\zeta_2})n_{\zeta} - w_{\zeta_2\zeta}(1-n_{\zeta})n_{\zeta_2}]\rangle_{eq} \,, \quad (17.12\text{-}15)$$

where the prime on the sum means $\zeta' \neq \zeta_1 \neq \zeta_2$. The double sum of the first line is rewritten as

$$\sum_{\zeta}\sum_{\zeta'}{}' a_{\zeta}\langle n_{\zeta_1}n_{\zeta_2}[w_{\zeta\zeta'}(1-n_{\zeta'})n_{\zeta} - w_{\zeta'\zeta}(1-n_{\zeta})n_{\zeta'}]\rangle_{eq}$$

$$= \sum_{\zeta}{}'\sum_{\zeta'}{}' a_{\zeta}\langle n_{\zeta_1}n_{\zeta_2}[w_{\zeta\zeta'}(1-n_{\zeta'})n_{\zeta} - w_{\zeta'\zeta}(1-n_{\zeta})n_{\zeta'}]\rangle_{eq}$$

$$+\sum_{\zeta'}{}' a_{\zeta_1}\langle n_{\zeta_1}n_{\zeta_2}[w_{\zeta_1\zeta'}(1-n_{\zeta'})n_{\zeta_1} - w_{\zeta'\zeta_1}(1-n_{\zeta_1})n_{\zeta'}]\rangle_{eq}$$

$$+\sum_{\zeta'}{}' a_{\zeta_2}\langle n_{\zeta_1}n_{\zeta_2}[w_{\zeta_2\zeta'}(1-n_{\zeta'})n_{\zeta_2} - w_{\zeta'\zeta_2}(1-n_{\zeta_2})n_{\zeta'}]\rangle_{eq} \,. \quad (17.12\text{-}16)$$

With $\langle n_{\zeta_1}n_{\zeta_2}n_{\zeta}n_{\zeta'}\rangle_{eq} = \langle n_{\zeta_1}n_{\zeta_2}\rangle_{eq}\langle n_{\zeta}n_{\zeta'}\rangle_{eq}$ we see that the first line is zero because of detailed balance. As to the sums over ζ' the second part of the second line and of the third line are zero since $n_{\zeta}(1-n_{\zeta}) = 0$ for whatever ζ. Likewise, for the sums over ζ of the last two lines of (17.12-15) the first parts are zero. We are therefore left with

$$\sum_{\zeta}\langle n_{\zeta_1}n_{\zeta_2}\mathscr{B}_{\zeta}n_{\zeta}\rangle_{eq}a_{\zeta} = \sum_{\zeta'}{}'\langle n_{\zeta_1}n_{\zeta_2}(1-n_{\zeta'})\rangle_{eq}(a_{\zeta_1}w_{\zeta_1\zeta'} + a_{\zeta_2}w_{\zeta_2\zeta'})$$

$$-\sum_{\zeta}a_{\zeta}\langle n_{\zeta_1}n_{\zeta_2}(1-n_{\zeta})\rangle_{eq}(w_{\zeta_1\zeta} + w_{\zeta_2\zeta}) \,. \quad (17.12\text{-}17)$$

The prime on the first sum can be removed since the added terms yield zero. Also, the triple correlations can be split up for the same reason. So, the final result for the first streaming term is

$$-\beta F(t)\sum_{\zeta}\langle n_{\zeta_1}n_{\zeta_2}\mathscr{B}_{\zeta}n_{\zeta}\rangle_{eq}a_{\zeta} = -\beta F(t)\sum_{\zeta}\langle n_{\zeta_1}n_{\zeta_2}\rangle_{eq}(1-\langle n_{\zeta}\rangle)_{eq}[(a_{\zeta_1}-a_{\zeta})w_{\zeta_1\zeta} + (a_{\zeta_2}-a_{\zeta})w_{\zeta_2\zeta}] \,.$$

$$(17.12\text{-}18)$$

Finally, we need the last term of (17.12-10), the second streaming term. We have

$$\sum_{\zeta}\dot{a}_{\zeta}\langle n_{\zeta_1}n_{\zeta_2}n_{\zeta}\rangle_{eq} = \sum_{\zeta}{}'\dot{a}_{\zeta}\langle n_{\zeta_1}n_{\zeta_2}\rangle_{eq}\langle n_{\zeta}\rangle_{eq} + \dot{a}_{\zeta_1}\langle n_{\zeta_1}n_{\zeta_2}\rangle_{eq} + \dot{a}_{\zeta_2}\langle n_{\zeta_1}n_{\zeta_2}\rangle_{eq} \,. \quad (17.12\text{-}19)$$

We now use the fact that the equilibrium current is zero; hence

$$\sum_{\zeta}\langle n_\zeta\rangle_{eq}\dot{a}_\zeta=\sum_{\zeta}{}'\langle n_\zeta\rangle_{eq}\dot{a}_\zeta+\langle n_{\zeta_1}\rangle_{eq}\dot{a}_{\zeta_1}+\langle n_{\zeta_2}\rangle_{eq}\dot{a}_{\zeta_2}=0\,. \tag{17.12-20}$$

Multiplying with $\langle n_{\zeta_1}n_{\zeta_2}\rangle_{eq}$ and combining the last two equations, we arrive at

$$\beta F(t)\sum_{\zeta}\dot{a}_\zeta\langle n_{\zeta_1}n_{\zeta_2}n_\zeta\rangle_{eq}=\beta F(t)\langle n_{\zeta_1}n_{\zeta_2}\rangle_{eq}\{(1-\langle n_{\zeta_1}\rangle_{eq})\dot{a}_{\zeta_1}+(1-\langle n_{\zeta_2}\rangle_{eq})\dot{a}_{\zeta_2}\}\,. \tag{17.12-21}$$

The second moment equation is now known by entering the first-order and diagonal second-order Fokker–Planck moment and the results (17.12-18) and (17.12-21) into Eq (17.12-10).

The full second-order moment equation reads:

$$\frac{\partial\langle n_{\zeta_1}n_{\zeta_2}\rangle_t}{\partial t}+\langle n_{\zeta_1}\mathscr{B}_{\zeta_2}n_{\zeta_2}\rangle_t+\langle n_{\zeta_2}\mathscr{B}_{\zeta_1}n_{\zeta 1}\rangle_t+w_{\zeta_1\zeta_2}\langle n_{\zeta_1}(1-n_{\zeta_2})\rangle_t+w_{\zeta_2\zeta_1}\langle n_{\zeta_2}(1-n_{\zeta_1})\rangle_t$$
$$=-\beta F(t)\sum_{\zeta}\langle n_{\zeta_1}n_{\zeta_2}\rangle_{eq}\langle(1-n_\zeta)\rangle_{eq}[(a_{\zeta_1}-a_\zeta)w_{\zeta_1\zeta}+(a_{\zeta_2}-a_\zeta)w_{\zeta_2\zeta}]$$
$$+\beta F(t)\langle n_{\zeta_1}n_{\zeta_2}\rangle_{eq}[(1-\langle n_{\zeta_1}\rangle_{eq})\dot{a}_{\zeta_1}+(1-\langle n_{\zeta_2}\rangle_{eq})\dot{a}_{\zeta_2}]\,, \tag{17.12-22}$$

where $\mathscr{B}_\zeta$ is the usual Boltzmann operator. We note again that the two Boltzmann terms on the lhs involve triple particle correlations; the truncation is easily obtained.

(iii) *Equations for* $\langle n_{\zeta_1}...n_{\zeta_k}\rangle_t$, $k\geq 3$

These results are readily obtained since the *essential* F–P moments of the third-order and higher are zero, as we now will show. The basic form is

$$\langle F_3(n_{\zeta_1},n_{\zeta_2},n_{\zeta_3})\rangle_b=\sum_{\zeta'\zeta''}w_{\zeta'\zeta''}n_{\zeta'}(1-n_{\zeta''})(1-2n_{\zeta_1})(1-2n_{\zeta_2})(1-2n_{\zeta_3})$$
$$\times(\delta_{\zeta_1\zeta'}+\delta_{\zeta_1\zeta''})(\delta_{\zeta_2\zeta'}+\delta_{\zeta_2\zeta''})(\delta_{\zeta_3\zeta'}+\delta_{\zeta_3\zeta''})\,. \tag{17.12-23}$$

This can be simplified by considering all triple Kronecker delta products – eight terms – on the rhs of (17.12-23). It is directly evident that *no moment exists* for $\zeta_1\neq\zeta_2\neq\zeta_3$. However, there are doubly confluent and triply confluent contributions. A straightforward evaluation yields

$$\langle F_3(n_{\zeta_1},n_{\zeta_2},n_{\zeta_3})\rangle_b=\delta_{\zeta_1\zeta_2}\delta_{\zeta_1\zeta_3}\mathscr{B}_{\zeta_1}n_{\zeta_1}+\sum_{\text{cyclic}}\delta_{\zeta_1\zeta_2}[w_{\zeta_1\zeta_3}n_{\zeta_1}(1-n_{\zeta_3})-w_{\zeta_3\zeta_1}n_{\zeta_3}(1-n_{\zeta_1})]\,, \tag{17.12-24}$$

where we sum over the cyclic permutations of the indices 1,2,3. The essential moment equation ($\zeta_1\neq\zeta_2\neq\zeta_3$) will be contained in the general result below. As to the confluent contributions, from the first term in (17.12-23) one sees at once that the equation for $\partial\langle n_{\zeta_1}^3\rangle_t/\partial t$ is the Boltzmann equation. For the doubly confluent terms, one recovers the second-order moment equation. Thus, while the confluent contributions *are* important for the full F–P moment, only the essential part, *being zero*, is of interest in the third-order moment equation. Needless to say that for the higher order F–P moments, *the essential parts are likewise zero*. Thus, in all equations we have only terms involving the first and second-order F–P moments. The structure of the streaming terms is likewise easily surmised from the foregoing. Altogether, the following result is obtained:

$$\frac{\partial\langle n_{\zeta_1}...n_{\zeta_N}\rangle_t}{\partial t}+\sum_{j=1}^{N}\left\langle\prod_{j\neq i}^{N-1}n_{\zeta_j}\mathscr{B}_{\zeta_i}n_{\zeta_i}\right\rangle_t-\sum_{i=1}^{N-1}\sum_{j>i}^{N}\left\langle\prod_{k\neq i,k\neq j}^{N-2}n_{\zeta_k}F_2(n_{\zeta_i},n_{\zeta_j})\right\rangle_t$$

$$=-\beta F(t)\sum_{\zeta}\langle n_{\zeta_1}...n_{\zeta_N}\mathscr{B}_{\zeta}n_{\zeta}\rangle_{eq}a_{\zeta}+\beta F(t)\sum_{\zeta}\langle n_{\zeta_1}...n_{\zeta_N}n_{\zeta}\rangle_{eq}\dot{a}_{\zeta}. \quad (17.12\text{-}25)$$

For N = 1,2 the previous results are readily verified. The streaming terms can be further explicated along the lines indicated before.

17.12.2 *Boson-Moment Equations*

We consider bosons with occupation-number states $\{|\eta\rangle\}$ interacting with boson bath states $\{|\hat{\eta}\rangle\}$, the latter remaining in thermal equilibrium. The ensemble probability is now denoted by $p(\gamma,t)=p(\{N_\eta\},t)P_{eq}(\{\hat{N}\})$ and an average over the bath states is again denoted by $\langle..\rangle_b$. The procedure with the generating function is entirely analogous. However, for the F–P moments we need the boson transition rule (17.7-8) in lieu of the fermion rule (17.1-7). The Boltzmann operator is found to be

$$\overline{\mathscr{B}}_\eta N_\eta=\sum\nolimits_{\eta'}[w_{\eta\eta'}N_\eta(1+N_{\eta'})-w_{\eta'\eta}N_{\eta'}(1+N_\eta)]. \quad (17.12\text{-}26)$$

The first two moment equations for $\eta_1\neq\eta_2$ are found to be

$$\frac{\partial\langle N_{\eta_1}\rangle_t}{\partial t}+\overline{\mathscr{B}}_{\eta_1}\langle N_{\eta_1}\rangle_t=-\beta F(t)\sum_{\eta}\langle N_{\eta_1}\overline{\mathscr{B}}_\eta N_\eta\rangle_{eq}a_\eta+\beta F(t)\sum_{\eta}\langle N_{\eta_1}N_\eta\rangle_{eq}\dot{a}_\eta, \quad (17.12\text{-}27)$$

$$\frac{\partial\langle N_{\eta_1}N_{\eta_2}\rangle_t}{\partial t}+\langle N_{\eta_1}\overline{\mathscr{B}}_{\eta_2}N_{\eta_2}\rangle_t+\langle N_{\eta_2}\overline{\mathscr{B}}_{\eta_1}N_{\eta_1}\rangle_t+\langle w_{\eta_1\eta_2}N_{\eta_1}(N_{\eta_2}+1)+w_{\eta_2\eta_1}N_{\eta_2}(N_{\eta_1}+1)\rangle_t$$

$$=-\beta F(t)\sum_{\eta}\langle N_{\eta_1}N_{\eta_2}\overline{\mathscr{B}}_\eta N_\eta\rangle_{eq}a_\eta+\beta F(t)\sum_{\eta}\langle N_{\eta_1}N_{\eta_2}N_\eta\rangle_{eq}\dot{a}_\eta. \quad (17.12\text{-}28)$$

The N-th order moment equation for distinct states is likewise entirely analogous to (17.12-25). As to the F–P moments, there are no essential (i.e., non-confluent) contributions for $k\geq 3$. However, for confluent states, there is no reduction to lower order since the exclusion principle does not apply. In Ref. 59[(2)] the fully confluent equations for $\partial\langle N_\eta^N\rangle_t/\partial t$ were investigated. There are confluent contributions to all F–P moments, $k=1...N$. However, for all odd k they are identical to F_1, i.e., the Boltzmann term, while for even k they are equal to the diagonal part of F_2, fourth term to the left in (17.12-28); this is due to the rule $[\delta_{\eta,\eta'}-\delta_{\eta\eta''}]^k=\delta_{\eta,\eta'}+(-1)^k\delta_{\eta\eta''}$.

17.13 Problems to Chapter XVII

17.1 Assume that the fermions of Section 17.1 have binary interactions with atomic nuclei that are fermions and act as a thermal bath, remaining in equilibrium. Let the test-fermions have states $\{|\zeta)\}$ and the bath-fermions have states $\{|\alpha)\}$. Obtain the general quantum Boltzmann equation and show that the same result as in (17.1-23) is found. Examine all steps in the derivation carefully!

17.2 Consider a systems of fermions with plane-wave states $\{|\mathbf{k})\}$ interacting with bosonic nuclei having momentum states $\{|\mathbf{K})\}$. Let $V_\mathbf{q}$ be the Fourier transform of the two-body interaction potential $\mathcal{V}(\mathbf{r}-\mathbf{R})$. Derive the diagonal QBE. NB. This limited goal was Van Hove's objective in Ref. 2.

17.3 Carry out the truncation for the two-particle correlation function of Eq. (17.12-22), i.e., resolve the triple particle averages $\langle n_{\zeta_i}\mathcal{B}_{\zeta_j}n_{\zeta_j}\rangle_t$ at the lhs, assuming the usual quadratic form for the Boltzmann operator $\mathcal{B}$.

17.4 Consider the collision integral $(V_0/8\pi^3)\int d^3u e^{i\mathbf{u}\cdot\mathbf{r}}\mathcal{B}_\mathbf{k}\langle n_\mathbf{k}\rangle_t\Delta^3u$ of Eq. (17.3-10). Take first the linear terms of $\mathcal{B}_\mathbf{k}$ and replace $\langle n_\mathbf{k}\rangle_t$ by $\langle c^\dagger_{\mathbf{k}-\mathbf{u}/2}c_{\mathbf{k}+\mathbf{u}/2}\rangle_t$, where $\mathbf{u}$ is a small extension $\mathbf{u}\in\Delta^3u=\Pi_{i=x,y,z}\Delta u_i$; substitute (17.3-11). Show that:

$$\int d^3u e^{i\mathbf{u}\cdot(\mathbf{r}-\bar{\mathbf{r}})}\delta_{\mathbf{u},0}\to\int_{\Delta^3u}d^3u e^{i\mathbf{u}\cdot(\mathbf{r}-\bar{\mathbf{r}})}=\prod_{xyz}\frac{\sin[\Delta u_x(\mathbf{r}_x-\bar{\mathbf{r}}_x)/2]}{(\mathbf{r}_x-\bar{\mathbf{r}}_x)/2}. \tag{1}$$

The rhs has its maximum value Δu_x for $\bar{\mathbf{r}}_x=\mathbf{r}_x$. The x-direction width is (*show*):

$$\frac{1}{\Delta u_x}\int_{-\infty}^{\infty}dr_x\frac{\sin[\Delta u_x(\mathbf{r}_x-\bar{\mathbf{r}}_x)/2]}{(\mathbf{r}_x-\bar{\mathbf{r}}_x)/2}=\frac{2\pi}{\Delta u_x}. \tag{2}$$

We can thus replace the rhs of (1) by a function that has the value Δ^3u in a rectangular box of volume $8\pi^3/\Delta^3u$ centred on $\mathbf{r}$ and zero elsewhere. Now integrate (1) over $d^3\bar{r}$ and obtain the linear terms of (17.3-14), observing (16.10-56) and noting that a microcell has a volume $8\pi^3\hbar^3$. Repeat the procedure for the quadratic terms.

17.5 For the magnetophonon resonances perform the Poisson series for (17.10-29) and for (17.10-34), using contour integration for the integrals. Also, perform the sum over N and obtain the full result (17.10-35) for polar optical phonons.

17.6 For the Aharonov–Bohm effect involving *electrons,* use the canonical momentum $\mathbf{p}+(e/c)\mathbf{A}$ and obtain (17.11-13) with $-i\bar{\phi}\to i\bar{\phi}$. Obtain the EF and EV $[\varepsilon\propto(n+\bar{\phi})^2]$ and re-establish (17.11-26) for a ring in a transverse field.

PART F

STOCHASTIC PHENOMENA

Chapter XVIII

Brownian Motion and the Mesoscopic Master Equation

18.1 Introduction to Fluctuations and Stochastic Phenomena

In the last part of this text we shall try to set forth the main methods pertaining to the description of stochastic processes. This is not a small task, since the field of *noise and fluctuations* – the more common name for this endeavour – is vast and could easily encompass a book of its own. Also, the diversity of the methods and the great variety of applications in physics, chemistry, biology and engineering would require a large number of separate chapters, which would render this portion of the present text unwieldy and poorly conceived. Consequently, we decided to restrict the number of divisions and to lump together quite different phenomena when the tools involved are mathematically similar. Thus "Brownian motion" will be understood in the sense of the earlier quoted thesis of Mrs. de Haas–Lorentz [Chapter XVI, Ref.5], which carries as title: "*...und einige anverwandte Erscheinungen"– and some related phenomena.* While in her time the *Brownian motion-like* phenomena were few (heat fluctuations were cited in Chapter V), presently the list is tedious and long; we mention temperature fluctuations in bolometers, carrier-density fluctuations in solids (also partially discussed in Chapter V), noise in lasers and masers, photon fluctuations, etc. Some of these phenomena will be dealt with in this chapter, while others will fall by the wayside... Our main emphasis will nevertheless be on Brownian motion proper, in particular with regard to velocity fluctuations and diffusion, since the older physics literature on stochastic phenomena involves these particular processes; a summary of the most important papers in this area was compiled long ago in a Dover reprint collection, comprising the papers of Uhlenbeck and Ornstein, Ming Chen Wang and Uhlenbeck, Chandrasekhar, Kac, Doob and Rice.[1] No serious student in this field should do without this valuable publication.

[1] "Noise and Stochastic Processes", N. Wax, Ed., Dover Publ. NY 1954. Contributions of interest are: (a): "On the Theory of Brownian Motion I", G.E. Uhlenbeck and L.S. Ornstein, Phys. Rev. **36**, 823-841 (1930); (b): "On the Theory of Brownian Motion II", Ming Chen Wang and G.E. Uhlenbeck, Rev. Mod. Phys. **17**, 323-342 (1945); (c): S. Chandrasekhar,"Stochastic Problems in Physics and Astronomy", Rev. Mod. Phys. **15**, 1-89 (1943); (d): M. Kac, "Random Walk and the Theory of Brownian Motion", Am. Math. Monthly **54**, 369-391 (1947); (e): S.O. Rice, Mathematical Analysis of Random Noise", Bell Syst. Techn. J. **23**, 1-114 (1944); also ibid. **24**, 115-162 (1945); (f): J.L. Doob, Ann. Math. **43**, 351 (1942).

In the description of stochastic processes it is expedient to make a number of useful basic distinctions. Three types of divisions stand out, as we now indicate.

(a) *Microscopic fluctuations vs. mesoscopic fluctuations*

Most of the fluctuations encountered in the previous chapters (III, IV, V, VI, IX, XI, XII and XVI) involved the deviations of microscopic system operators from their mean values; this description is too 'fine-grained' for the purpose of this chapter. The stochastic variables of present concern are coarse-grained over a volume of phase space or a range of quantum states; this was discussed in Section 1.5. While most authors will refer to these variables as 'macroscopic', we believe that this is stretching this concept too far. To observe the irregular motion of colloidal suspended particles, predicted by kinetic theory and now called Brownian motion, the British Botanist Robert Brown used a powerful microscope in his 1827 observations. For modern Brownian motion-like phenomena, like e.g. pressure fluctuations of a microphone membrane, one employs a transducer and some 120 dB amplification prior to a fast Fourier transform wave analyzer; similarly for the measurement of Johnson (Nyquist) noise, bolometer fluctuations, photon noise, etc. All this indicates that the variables to be measured are not quite macroscopic; we thus referred to them as *mesoscopic* [noting that this name has nothing to do with the modern usage of this term in so-called mesoscopic conductors, cf. Section 17.11]. Generally, we denote them by $a(t)$ or, when there are a number of variables, $a_1 ... a_s$, by the vector $\mathbf{a}(t)$; the fluctuations will be denoted by $\boldsymbol{\alpha}(t) = \mathbf{a}(t) - \langle \mathbf{a} \rangle$.[2] Since these variables are coarse-grained, we assumed in Section 1.5 that they commute.

(b) *Thermal equilibrium vs. steady-state*

Fluctuations about the equilibrium state are considerably less complex than fluctuations involving a steady state. Of course, for the former the entire apparatus of equilibrium statistical ensembles is available, while for steady-state fluctuations we should start from non-equilibrium kinetic equations or specific stochastic methods must be developed. As it will turn out, the main and far-reaching advantage of the thermal equilibrium state is the property of detailed balance; in the steady state, only total balancing of flow rates is possible.

(c) *Finite vs. infinite-dimensional*

Stochastic processes that depend on a continuous parameter, such as $a(\mathbf{r},t)$ are infinite-dimensional. Special methods have been developed and will be discussed.

[2] In Section 16.4 we arranged the variables $a_1(t) ... a_s(t)$ in a column matrix. Presently, it is slightly more convenient to employ tensor analysis; so we assume that there are labelled axes in some space R^s.

1. THE MESOSCOPIC MASTER EQUATION AND THE MOMENT EQUATIONS

18.2 Probabilistic Description of Ornstein and Burger

We remind the reader of Khintchine's definition of a stochastic process, being *a random variable (or set of variables) with the time as a parameter,* $a_i(t)$, $i = 1...s$, cf. Section 16.4. Obviously, the 'function' here indicated is not a mathematical function in the usual sense. More appropriately, the stochastic variable, or set of variables comprised in the vector **a**(t), are described by $1\ldots n$-point distribution functions, being the (joint) probabilities, with $t_n > t_{n-1} > ... > t_1$, such that[3]

$$W_1(\mathbf{a}_1,t_1) = \text{ probability of finding } \mathbf{a}_1 \text{ at } t_1;$$

$$W_2(\mathbf{a}_2,t_2;\mathbf{a}_1,t_1) = \text{probability of finding } \mathbf{a}_2 \text{ at } t_2 \text{ and } \mathbf{a}_1 \text{ at } t_1;$$

.

.

$$W_n(\mathbf{a}_n,t_n;\mathbf{a}_{n-1},t_{n-1};...;\mathbf{a}_1,t_1) = \text{prob. of finding } \mathbf{a}_n \text{ at } t_n, \mathbf{a}_{n-1} \text{ at } t_{n-1} \text{ .... and } \mathbf{a}_1 \text{ at } t_1 . \qquad (18.2\text{-}1)$$

The probabilities $W_1, W_2,...,W_n...$ describe the stochastic process in successively more detail. In the limit $n \to \infty$ the process is fully described by the *probability functional* $W = W[\mathbf{a}(t)]$. Ornstein and Burger require the following conditions for a stationary or steady-state process (*a fortiori,* an equilibrium process):

(*a*) $W_1(\mathbf{a}_1,t_1) = W_1(\mathbf{a}_1)$;

(*b*) $W_n \geq 0$ and W_n is symmetric in the pairs $(\mathbf{a}_i,t_i)$ and $(\mathbf{a}_j,t_j)$, $i,j \leq n$;

(*c*) $$W_k(\mathbf{a}_k,t_k;...;\mathbf{a}_1,t_1) = \sum_{\mathbf{a}_n}\cdots\sum_{\mathbf{a}_{k+1}} W_n(\mathbf{a}_n,t_n;\mathbf{a}_{n-1},t_{n-1};...;\mathbf{a}_1,t_1). \qquad (18.2\text{-}2)$$

We can now classify stochastic processes as to the order n (possibly ∞), required to describe the process.

18.2.1 *Purely Random Processes*

By definition, for a *purely random* process all information is contained in $W_1(\mathbf{a}_1,t_1) \equiv W(\mathbf{a}_1)$ (assuming the process to be stationary). For this case we have

$$W_2(\mathbf{a}_2,t_2;\mathbf{a}_1,t_1) = W(\mathbf{a}_2)W(\mathbf{a}_1), \quad W_n(\mathbf{a}_n,t_n;\mathbf{a}_{n-1},t_{n-1};...;\mathbf{a}_1,t_1) = \prod_{i=1}^{n} W(\mathbf{a}_i). \quad (18.2\text{-}3)$$

For the moment functions (tensors) we have by definition

[3] L.S. Ornstein and H.C. Burger, Versl. Koninkl. Acad. v. Wetenschappen, Amsterdam **27**, 1146 (1919); ibid. **28**, 183 (1919).

$$\mathsf{M}_k(t_k,t_{k-1},...t_1) \equiv \langle \mathbf{a}(t_k)\mathbf{a}(t_{k-1})....\mathbf{a}(t_1)\rangle. \tag{18.2-4}$$

For the present case this yields

$$\mathsf{M}_k(t_k,t_{k-1},...t_1) = \prod_{i=1}^{k} \langle \mathbf{a}(t_i)\rangle = \langle \mathbf{a}\rangle^k. \tag{18.2-5}$$

The fluctuation moments are defined as in (18.2-4), but with Δ' s in front of the **a**' s:

$$\mathsf{F}_k(t_k,t_{k-1},...t_1) = \prod_{i=1}^{k} \langle \Delta\mathbf{a}(t_i)\rangle = 0. \tag{18.2-6}$$

In order to see what happens at the confluence of two times we consider Bayes' rule, valid for any probabilistic process:

$$W_2(\mathbf{a}_2,t_2;\mathbf{a}_1,t_1) = P(\mathbf{a}_2,t_2\,|\,\mathbf{a}_1,t_1)W_1(\mathbf{a}_1,t_1). \tag{\textbf{18.2-7}}$$

Here $P(\mathbf{a}_2,t_2\,|\,\mathbf{a}_1,t_1)$ is the conditional probability that we have $\mathbf{a}_2$ at t_2 *given* that we have $\mathbf{a}_1$ at t_1. For the present case we find, comparing (18.2-7) with (18.2-3),

$$P(\mathbf{a}_2,t_2\,|\,\mathbf{a}_1,t_1) = W(\mathbf{a}_2), \quad t_2 \neq t_1, \quad \text{or}, \tag{18.2-8a}$$

$$P(\mathbf{a}_2,\tau\,|\,\mathbf{a}_1,0) = W(\mathbf{a}_2), \quad \tau \neq 0. \tag{18.2-8b}$$

There is clearly no memory, so we set

$$P(\mathbf{a}_2,\tau\,|\,\mathbf{a}_1,0) = CW(\mathbf{a}_2)\delta(\tau), \tag{18.2-9}$$

where C is a constant. Considering further a single variable process we have for 'the' correlation function (= second order fluctuation-correlation function):

$$\langle \Delta a(\tau)\Delta a(0)\rangle = \sum\nolimits_{a_2}\sum\nolimits_{a_1} \Delta a_2 \Delta a_1\, P(a_2,\tau\,|\,a_1,0)W(a_1) = C\langle \Delta a^2\rangle \delta(\tau). \tag{18.2-10}$$

From the Wiener-Khintchine theorem (16.4-7) this yields for the spectral density

$$\mathcal{S}(\omega) = 2C\langle \Delta a^2\rangle \int_{-\infty}^{\infty} e^{-i\omega\tau}\delta(\tau)\,d\tau = 2C\langle \Delta a^2\rangle = const. \tag{18.2-11}$$

So the spectrum of a purely random process is *white*, i.e., independent of the frequency. However, strictly speaking, the variance does not exist, unless there is some cut-off at a frequency ω_h, thus rendering C finite.

18.2.2 *Markovian Random Processes*

For a Markovian random process all information is contained in $W_2(\mathbf{a}_2,t_2;\mathbf{a}_1,t_1)$. However, it *suffices* to stipulate the conditional probability $P(\mathbf{a}_2,\tau\,|\,\mathbf{a}_1,0)$. This is simply seen from Bayes' rule (18.2-7) and the 'memory-loss rule'[4]

$$\lim_{\tau\to\infty} P(\mathbf{a}_2,\tau\,|\,\mathbf{a}_1,0) = W_1(\mathbf{a}_2). \tag{18.2-12}$$

---

[4] As for the mixing property (16.2-21), this limit may not exist; in that case we must specify $W_1(\mathbf{a}_2)$.

So, if P is known, W_1 is known and thus W_2. More succinctly, let us write

$$W_n(\mathbf{a}_n,t_n;\mathbf{a}_{n-1},t_{n-1};...;\mathbf{a}_1,t_1)=P(\mathbf{a}_n,t_n\,|\,\mathbf{a}_{n-1},t_{n-1};...;\mathbf{a}_1,t_1)$$
$$\times W_{n-1}(\mathbf{a}_{n-1},t_{n-1};...;\mathbf{a}_1,t_1)\,. \tag{18.2-13}$$

A Markov process is a process that has no memory of the past beyond one given time-point, i.e.,

$$P(\mathbf{a}_n,t_n\,|\,\mathbf{a}_{n-1},t_{n-1};...;\mathbf{a}_1,t_1)=P(\mathbf{a}_n,t_n\,|\,\mathbf{a}_{n-1},t_{n-1})\,. \tag{18.2-14}$$

In other words: a Markov process is "as deterministic as feasible"; given the value of the stochastic vector **a** at some time-point, say t_{n-1}, we can predict its probable value for a later time $t_n > t_{n-1}$, without knowing its history prior to that time-point, $t < t_{n-1}$.

It is now easy to see that all W_n are determined by the conditional probability P. Substituting (18.2-14) into (18.2-13) and repeating the process we obtain

$$W_n(\mathbf{a}_n,t_n;\mathbf{a}_{n-1},t_{n-1};...\mathbf{a}_1,t_1)=P(\mathbf{a}_n,t_n\,|\,\mathbf{a}_{n-1},t_{n-1})W_{n-1}(\mathbf{a}_{n-1},t_{n-1};...\mathbf{a}_1,t_1)$$
$$=P(\mathbf{a}_n,t_n\,|\,\mathbf{a}_{n-1},t_{n-1})P(\mathbf{a}_{n-1},t_{n-1}\,|\,\mathbf{a}_{n-2},t_{n-2})...P(\mathbf{a}_2,t_2\,|\,\mathbf{a}_1,t_1)W_1(\mathbf{a}_1,t_1)\,, \tag{18.2-15}$$

where the single time distribution W_1 is given by (18.2-12); hence, all W_n are expressible in P.

While we assumed the existence of the long time limit (18.2-12), the short time limit follows from a combination of Bayes' rule and $W_1(\mathbf{a}_1,\tau)=\Sigma_{\mathbf{a}'}W_2(\mathbf{a}_2,\tau;\mathbf{a}',0)$. This leads to

$$W_1(\mathbf{a}_1,\tau)=\sum_{\mathbf{a}'}P(\mathbf{a},\tau\,|\,\mathbf{a}',0)W_1(\mathbf{a}',0)\,, \tag{18.2-16}$$

which is solved by

$$\lim_{\tau\to 0+}P(\mathbf{a},\tau\,|\,\mathbf{a}',0)=\delta_{\mathbf{a},\mathbf{a}'}\,. \tag{18.2-17}$$

We finally note that P is normalized. Summing (18.2-7) over both $\mathbf{a}_2$ and $\mathbf{a}_1$, noticing that all W_n are normalized, we obtain

$$\sum\nolimits_{\mathbf{a}_2}P(\mathbf{a}_2,t_2\,|\,\mathbf{a}_1,t_1)=1\,. \tag{18.2-18}$$

In the above we have tacitly assumed that the **a**'s are discrete variables, a statement that often is true. For the following developments we will revert, however, to the case that the **a**'s are continuous variables; the W_n then become probability density functions (pdf's). With respect to the last two equations, in (18.2-17) we now need a delta function and in (18.2-18) we have a normalization integral. Let us consider the three-point pdf W_3, writing

$$W_3(\mathbf{a}_3,t_3;\mathbf{a}_2,t_2;\mathbf{a}_1,t_1)=P(\mathbf{a}_3,t_3\,|\,\mathbf{a}_2,t_2)P(\mathbf{a}_2,t_2\,|\,\mathbf{a}_1,t_1)W_1(\mathbf{a}_1,t_1)\,. \tag{18.2-19}$$

This is integrated over $d\mathbf{a}_2$, to yield

$$W_2(\mathbf{a}_3,t_3;\mathbf{a}_1,t_1)=\int d\mathbf{a}_2P(\mathbf{a}_3,t_3\,|\,\mathbf{a}_2,t_2)P(\mathbf{a}_2,t_2\,|\,\mathbf{a}_1,t_1)W_1(\mathbf{a}_1,t_1)\,. \tag{18.2-20}$$

Next, both sides are divided by $W_1(\mathbf{a}_1,t_1)$. We thus arrive at

$$P(\mathbf{a}_3,t_3\,|\,\mathbf{a}_1,t_1) = \int d\mathbf{a}_2 P(\mathbf{a}_3,t_3\,|\,\mathbf{a}_2,t_2)\,P(\mathbf{a}_2,t_2\,|\,\mathbf{a}_1,t_1)\ . \qquad \textbf{(18.2-21)}$$

This integral equation, valid only for Markov processes, is the renown Smoluchowski equation, also called the *Chapman–Kolmogoroff equation* in the mathematical (and more recent physics) literature. The solution will be pursued in the next section.

18.3 Derivation of the Mesoscopic Master Equation

Since stochastic variables are generally coarse-grained we need only go back to the diagonal master equation of Pauli and Van Hove, Section 16.11. The diagonal ME in function space was given in (16.11-14) and (16.14-3). We thus have for $t > 0$:

$$\frac{\partial P(\gamma,t\,|\,\gamma',0)}{\partial t} = \sum_{\gamma''} [W_{\gamma''\gamma} P(\gamma'',t\,|\,\gamma',0) - W_{\gamma\gamma''} P(\gamma,t\,|\,\gamma',0)]\,. \qquad (18.3\text{-}1)$$

This is the microscopic ME for the conditional probability of encountering the system state $|\gamma\rangle$ at time t, given we have the state $|\gamma'\rangle$ at time 0. For a specification of the a_i within $(a_i + da_i, a_i)$ we must multiply with the density of states $\chi(a_i)$, cf. Section 2.4. Although the treatment there is given for a microcanonical ensemble, we note that the transition probability $W_{\gamma\gamma''}$ contains a delta function for the energies; so the equation (18.3-1) holds just as well for each energy cell $\mathcal{E}_{\gamma J}$ with leeway $d\mathcal{E}_{\gamma J}$ in the canonical ensemble. The mesoscopic probabilities are

$$\begin{aligned} P(\mathbf{a},t\,|\,\mathbf{a}',0) &= P(\gamma,t\,|\,\gamma',0)\,\chi(\mathcal{E}_{\gamma J},\mathbf{a})\,, \\ P(\mathbf{a}'',t\,|\,\mathbf{a}',0) &= P(\gamma'',t\,|\,\gamma',0)\,\chi(\mathcal{E}_{\gamma'' J},\mathbf{a}'')\,. \end{aligned} \qquad (18.3\text{-}2)$$

where we assumed that a given γ' engenders a given $\mathbf{a}'$ (the converse not being true). Multiplying each term in (18.3-1) with the density of states $\chi(\mathcal{E}_{\gamma J},\mathbf{a}) = \chi(\mathcal{E}_{\gamma'' J},\mathbf{a}) \equiv \chi(\mathbf{a})$, we arrive at

$$\frac{\partial P(\mathbf{a},t\,|\,\mathbf{a}',0)}{\partial t} = \sum_{\gamma''} \left[W_{\gamma''\gamma} \frac{\chi(\mathbf{a})}{\chi(\mathbf{a}'')} P(\mathbf{a}'',t\,|\,\mathbf{a}',0) - W_{\gamma\gamma''} P(\mathbf{a},t\,|\,\mathbf{a}',0) \right], \qquad (18.3\text{-}3)$$

a result that is clearly in an inconsistent form. So, applying coarse-graining to the sum over the microstates, we have

$$\begin{aligned} \sum_{\gamma''} W_{\gamma\gamma''} &\to \int d\mathbf{a}''\,\chi(\mathbf{a}'') W_{\gamma\gamma''} \equiv \int d\mathbf{a}''\, Q_{\mathbf{a}\mathbf{a}''}\,, \\ \sum_{\gamma''} W_{\gamma''\gamma}[\chi(\mathbf{a})/\chi(\mathbf{a}'')] &\to \int d\mathbf{a}''\,\chi(\mathbf{a}) W_{\gamma''\gamma} \equiv \int d\mathbf{a}''\, Q_{\mathbf{a}''\mathbf{a}}\,. \end{aligned} \qquad (18.3\text{-}4)$$

Note again that the state of departure, γ or γ'', gives rise to $\mathbf{a}$ or $\mathbf{a}''$, respectively; the density of states goes into the final state. We now established

$$\frac{\partial P(\mathbf{a},t\,|\,\mathbf{a}',0)}{\partial t} = \int d\mathbf{a}''[P(\mathbf{a}'',t\,|\,\mathbf{a}',0)\,Q_{\mathbf{a}''\mathbf{a}} - P(\mathbf{a},t\,|\,\mathbf{a}',0)\,Q_{\mathbf{a}\mathbf{a}''}], \qquad \mathbf{(18.3\text{-}5)}$$

which is the mesoscopic master equation; its Markovian nature is evident from the above form. Yet, it has been duly obtained from the microscopic von Neumann equation, the point of departure in Section 16.11. As indicated there, no random phase assumption has been made, although one could suppose without loss of generality that the density operator at $t = 0$ is diagonal.

We now turn to the 'standard derivation' of the stochastic literature, whereby the mesoscopic master equation is derived with a few simple steps from the Chapman–Kolmogoroff equation. Equation (18.2-21) is applied for a composite time interval that consists of an infinitesimal part Δt and a finite range. Let us make the changes:

$$\mathbf{a}_3,t_3 \to \mathbf{a},t''+\Delta t;\quad \mathbf{a}_2,t_2 \to \mathbf{a}'',t'';\quad \mathbf{a}_1,t_1 \to \mathbf{a}',0.$$

Moreover, the '0' behind the conditional bar will be omitted, as is usually done. We then have

$$\begin{aligned} P(\mathbf{a},t''+\Delta t\,|\,\mathbf{a}') &= \int d\mathbf{a}'' P(\mathbf{a},t''+\Delta t\,|\,\mathbf{a}'',t'')\,P(\mathbf{a}'',t''\,|\,\mathbf{a}') \\ &= \int d\mathbf{a}'' P(\mathbf{a},\Delta t\,|\,\mathbf{a}'')\,P(\mathbf{a}'',t''\,|\,\mathbf{a}'), \end{aligned} \qquad (18.3\text{-}6)$$

where we applied stationarity in the first P, shifting the time axis by t''. We subtract $P(\mathbf{a}'',t''\,|\,\mathbf{a}')$ from both sides and divide by Δt, taking the limit $\Delta t \to 0$, to obtain

$$\frac{\partial P(\mathbf{a},t''\,|\,\mathbf{a}')}{\partial t''} = \lim_{\Delta t\to 0}\frac{1}{\Delta t}\left[\int d\mathbf{a}''\,P(\mathbf{a}'',t''\,|\,\mathbf{a}')P(\mathbf{a},\Delta t\,|\,\mathbf{a}'') - P(\mathbf{a},t''\,|\,\mathbf{a}')\right]. \qquad (18.3\text{-}7)$$

We now replace t'' by t and we notice that $\int d\mathbf{a}''\,P(\mathbf{a}'',\Delta t\,|\,\mathbf{a}) = 1$ can be inserted in the last term. Thus we find

$$\frac{\partial P(\mathbf{a},t\,|\,\mathbf{a}')}{\partial t} = \lim_{\Delta t\to 0}\frac{1}{\Delta t}\int d\mathbf{a}''[P(\mathbf{a}'',t\,|\,\mathbf{a}')P(\mathbf{a},\Delta t\,|\,\mathbf{a}'') - P(\mathbf{a},t\,|\,\mathbf{a}')P(\mathbf{a}'',\Delta t\,|\,\mathbf{a})]. \qquad (18.3\text{-}8)$$

Next we assume the "stosszahlansatz" (scattering assumption):

$$P(\mathbf{a},\Delta t\,|\,\mathbf{a}'') = Q_{\mathbf{a}''\mathbf{a}}\Delta t + \delta(\mathbf{a}-\mathbf{a}'')\left[1 - \int d\bar{\mathbf{a}}\,Q_{\mathbf{a}''\bar{\mathbf{a}}}\Delta t\right]. \qquad (18.3\text{-}9)$$

Here $Q_{\mathbf{a}''\mathbf{a}}$ is the transition probability per unit time for a change $\mathbf{a}'' \to \mathbf{a}$, which must be known from the underlying physics of the process. The delta part is often (unjustifiably) omitted; it represents the probability that no transition takes place in Δt. For a discrete variable process, replace (18.3-9) by

$$P(\mathbf{a},\Delta t\,|\,\mathbf{a}'') = Q_{\mathbf{a}''\mathbf{a}}\Delta t + \delta_{\mathbf{a},\mathbf{a}''}\left[1 - {\textstyle\sum'_{\bar{\mathbf{a}}}}\,Q_{\mathbf{a}''\bar{\mathbf{a}}}\Delta t\right], \qquad (18.3\text{-}9')$$

where the prime on the sum sign indicates $\bar{\mathbf{a}} \neq \mathbf{a}''$. Substituting (18.3-9) or (18.3-9') into (18.3-8), one recovers the ME in the form (18.3-5).

We now discuss the essence of this derivation. Clearly, the underlying substratum of the Chapman–Kolmogoroff equation is a microscopic integral equation of the form

$$P(\gamma,t\,|\,\gamma',t') = \sum_{\gamma''} P(\gamma,t\,|\,\gamma'',t'')P(\gamma'',t''\,|\,\gamma',t'). \tag{18.3-10}$$

In Section 16.14, footnote [69], we showed that the conditional probability is uniquely related to the matrix elements of the evolution operator if at the origin of the interval an initial random phase assumption is made. We shall repeat these ideas here. From evolution on the interval $t' \to t$ we have

$$\begin{aligned} p(\gamma,t) &= \langle\gamma\,|\,U(t-t')\rho(t')U^\dagger(t-t')\,|\,\gamma\rangle \\ &= \sum_{\gamma'\gamma''} \langle\gamma\,|U(t-t')\,|\,\gamma'\rangle\langle\gamma'\,|\,\rho(t')\,|\,\gamma''\rangle\langle\gamma''\,|U^\dagger(t-t')\,|\,\gamma\rangle. \end{aligned} \tag{18.3-11}$$

Making an *initial random phase assumption* at t' so that $\rho(t')$ is diagonal, we have

$$\begin{aligned} p(\gamma,t) &= \sum_{\gamma'\gamma''} \langle\gamma\,|U(t-t')\,|\,\gamma'\rangle p(\gamma',t')\delta_{\gamma'\gamma''}\langle\gamma''\,|U^\dagger(t-t')\,|\,\gamma\rangle \\ &= \sum_{\gamma'} |\langle\gamma\,|U(t-t')\,|\,\gamma'\rangle|^2\; p(\gamma',t'). \end{aligned} \tag{18.3-12}$$

But we also have

$$p(\gamma,t) = \sum_{\gamma'} W_2(\gamma,t;\gamma',t') \overset{Bayes'r.}{=} \sum_{\gamma'} P(\gamma,t\,|\,\gamma',t')p(\gamma',t'), \tag{18.3-13}$$

so that

$$P(\gamma,t\,|\,\gamma',t') = |\langle\gamma\,|U(t-t')\,|\,\gamma'\rangle|^2. \tag{18.3-14}$$

Now let t'' be an intermediate time $t > t'' > t'$ for which the state is γ''. Making another random phase assumption at t'' we have $P(\gamma,t\,|\,\gamma'',t'') = |\langle\gamma\,|U(t-t'')\,|\,\gamma''\rangle|^2$; it follows that

$$\begin{aligned} p(\gamma'',t'') &= \sum_{\gamma'} |\langle\gamma''\,|U(t''-t')\,|\,\gamma'\rangle|^2\; p(\gamma',t'), \\ p(\gamma,t) &= \sum_{\gamma''} |\langle\gamma\,|U(t-t'')\,|\,\gamma''\rangle|^2\; p(\gamma'',t''). \end{aligned} \tag{18.3-15}$$

Combining we find,

$$p(\gamma,t) = \sum_{\gamma'\gamma''} |\langle\gamma\,|U(t-t'')\,|\,\gamma''\rangle|^2 |\langle\gamma''\,|U(t''-t')\,|\,\gamma'\rangle|^2\; p(\gamma',t'). \tag{18.3-16}$$

Now let us work this out; we have

$$p(\gamma,t) = \sum_{\gamma'\gamma''} |\langle\gamma\,|U(t-t'')\,|\,\gamma''\rangle\langle\gamma''\,|U(t''-t')\,|\,\gamma'\rangle|^2\; p(\gamma',t')$$

$$= \sum_{\gamma'\gamma''} |\langle\gamma|\gamma''\rangle e^{i\mathcal{E}_{\gamma''}(t-t')/\hbar}\langle\gamma''|\gamma'\rangle|^2\ p(\gamma',t') = \sum_{\gamma'\gamma''} |\,\delta_{\gamma\gamma''}\,\delta_{\gamma'\gamma''}\,e^{i\mathcal{E}_{\gamma''}(t-t')/\hbar}\,|^2\ p(\gamma',t')$$

$$= \sum_{\gamma'} |\,\delta_{\gamma\gamma'}\,e^{i\mathcal{E}_{\gamma'}(t-t')/\hbar}\,|^2\ p(\gamma',t') = \sum_{\gamma'} |\langle\gamma|U(t-t')|\gamma'\rangle|^2\ p(\gamma',t')\ . \qquad (18.3\text{-}17)$$

We now compare (18.3-16) and (18.3-17), rewriting the$|\langle\ |U|\ \rangle|^2$'s in terms of the P's. This yields

$$\sum_{\gamma'\gamma''} P(\gamma,t\,|\,\gamma'',t'')\,P(\gamma'',t''\,|\,\gamma',t')\,p(\gamma',t') = \sum_{\gamma'} P(\gamma,t\,|\,\gamma',t')\,p(\gamma',t')\,. \qquad (18.3\text{-}18)$$

Since this must be an identity for all $p(\gamma',t')$, the integral equation (18.3-10) follows.

We went at some length to establish this and to show the connection of the standard derivation with the implied assumptions for the non-equilibrium density operator. Two random phase assumptions had to be made to arrive at the microscopic Smoluchowski equation. In the derivation of the master equation from the Smoluchowski equation the intermediate point t'' was shifted to Δt from an endpoint, so that the assumption must be remade for every interval Δt. This situation, then, is similar to what we met in subsection 16.9.2: the standard derivation involves a *repeated random phase assumption* (RRPA), which, in essence, is at variance with the von Neumann equation, being a first-order differential equation that allows for a one-time initial condition. It goes without saying that *we feel* that the derivation via the Pauli–Van Hove microscopic master equation is far better founded!

18.4 The Kramers–Moyal Expansion and the Fokker–Planck Equation

Let us consider the generating function according to Laplace (cf. the developments in Section 17.12). If

$$\Psi(\mathbf{s},t) = \int d\mathbf{a}\, e^{-\mathbf{s}\cdot\mathbf{a}} P(\mathbf{a},t\,|\,\mathbf{a}')\,, \qquad (18.4\text{-}1)$$

we have from (18.3-5)

$$\frac{\partial\Psi(\mathbf{s},t)}{\partial t} = \iint d\mathbf{a}\,d\mathbf{a}''\left(e^{-\mathbf{s}\cdot\mathbf{a}} - e^{-\mathbf{s}\cdot\mathbf{a}''}\right)P(\mathbf{a}'',t\,|\,\mathbf{a}')\,Q_{\mathbf{a}''\mathbf{a}}\,, \qquad (18.4\text{-}2)$$

where we interchanged the integration variables in one part of the master equation. We further write

$$e^{-\mathbf{s}\cdot\mathbf{a}} - e^{-\mathbf{s}\cdot\mathbf{a}''} = \left[e^{-\mathbf{s}\cdot(\mathbf{a}-\mathbf{a}'')} - 1\right]e^{-\mathbf{s}\cdot\mathbf{a}''} = \sum_{n=1}^{\infty}\frac{(-1)^n}{n!}\,\mathbf{s}^n : (\mathbf{a}-\mathbf{a}'')^n e^{-\mathbf{s}\cdot\mathbf{a}''}\,, \qquad (18.4\text{-}3)$$

where **:** means summing over all corresponding indices, so that a scalar results. E.g.,

$$\mathbf{s}^3 : \mathbf{a}^3 = \sum\nolimits_{ijk} s_i s_j s_k a_i a_j a_k = \left(\sum\nolimits_i s_i a_i\right)^3\,,\ \text{etc.}$$

We now define the n-th order Fokker–Planck moment as the tensor of rank n :

$$\mathsf{F}_n(\mathbf{a}'') = \int d\mathbf{a}\,(\mathbf{a}-\mathbf{a}'')^n Q_{\mathbf{a}''\mathbf{a}}$$
$$= \lim_{\Delta t\to 0}\frac{1}{\Delta t}\int d\mathbf{a}\,(\mathbf{a}-\mathbf{a}'')^n P(\mathbf{a},\Delta t\,|\,\mathbf{a}'') = \lim_{\Delta t\to 0}\frac{1}{\Delta t}\langle[\mathbf{a}(\Delta t)-\mathbf{a}'']^n\rangle_{\mathbf{a}''}. \quad (18.4\text{-}4)$$

(The sub **a**" means a conditional average for an ensemble with **a**" fixed.) The F–P moments give the rate of change for $\Delta t \to 0$. Substitution of (18.4-4) and (18.4-3) into (18.4-2) gives

$$\frac{\partial \Psi(\mathbf{s},t)}{\partial t} = \int d\mathbf{a}''e^{-\mathbf{s}\cdot\mathbf{a}''}\sum_{n=1}^{\infty}\frac{(-1)^n}{n!}\mathbf{s}^n : \mathsf{F}_n(\mathbf{a}'')P(\mathbf{a}'',t\,|\,\mathbf{a}'). \quad (18.4\text{-}5)$$

We replace the integration variable **a**" by **a** and transform back; we then arrive at

$$\frac{\partial}{\partial t}P(\mathbf{a},t\,|\,\mathbf{a}') = \sum_{n=1}^{\infty}\frac{(-1)^n}{n!}\left(\frac{\partial}{\partial\mathbf{a}}\right)^n : \mathsf{F}_n(\mathbf{a})P(\mathbf{a},t\,|\,\mathbf{a}'). \quad (18.4\text{-}6)$$

This is the Kramers–Moyal expansion of the master equation.[4a]

All F–P moments can ultimately be deduced from the corresponding microscopic moments which were evaluated in Section 17.12. We noted there that there are no non-confluent moments above the second order. As to the confluent contributions, often they are of no concern or of vanishing magnitude. The above expansion then reduces to the Fokker–Planck equation, in which only first and second-order moments are retained. The Fokker–Planck equation reads

$$\frac{\partial P(\mathbf{a},t\,|\,\mathbf{a}')}{\partial t} = -\sum_i\frac{\partial}{\partial a_i}[A_i(\mathbf{a})P] + \frac{1}{2}\sum_{ij}\frac{\partial^2}{\partial a_i\partial a_j}[B_{ij}(\mathbf{a})P], \quad (18.4\text{-}7)$$

where **A** and B are the first and second-order Fokker–Planck moments, respectively.

18.5 The Phenomenological Equations and the Fluctuation-Relaxation Theorem

Although we shall furnish some simple examples later in this section of solutions of the master equation, as a rule the equation is not solved, but the pertinent information is obtained from the moment equations. To that end we go back to the Laplace form (18.4-5). We shall write out the first few terms on the rhs, obtaining

$$\frac{\partial}{\partial t}\left\langle e^{-\mathbf{s}\cdot\mathbf{a}(t)}\right\rangle_{\mathbf{a}'} = \left\langle e^{-\mathbf{s}\cdot\mathbf{a}(t)}[-\mathbf{s}\cdot\mathbf{A}(\mathbf{a}) + \tfrac{1}{2}\mathbf{s}\cdot\mathsf{B}(\mathbf{a})\cdot\mathbf{s} + \tfrac{1}{6}\mathsf{F}_3(\mathbf{a}):\mathbf{sss}+...\right\rangle_{\mathbf{a}'}. \quad (18.5\text{-}1)$$

To obtain the moment equations, we differentiate repeatedly to **s** [i.e., we apply the derivatives $(\partial/\partial s_i)(\partial/\partial s_j)$....]. This yields for the first two equations (to verify the

[4a] H.A. Kramers, Physica **7**, 284 (1940); J.E. Moyal, J. Roy. Statist. Soc. **B11**, 150 (1949).

order of the factors, it is advised to write out in component form):

$$\frac{\partial}{\partial t}\left\langle e^{-\mathbf{s}\cdot\mathbf{a}(t)}\mathbf{a}\right\rangle_{\mathbf{a}'} = \left\langle e^{-\mathbf{s}\cdot\mathbf{a}(t)}\mathbf{a}[-\mathbf{s}\cdot\mathbf{A}(\mathbf{a}) + \tfrac{1}{2}\mathbf{s}\cdot\mathsf{B}(\mathbf{a})\cdot\mathbf{s} + \tfrac{1}{6}\mathsf{F}_3(\mathbf{a}):\mathbf{sss} + \ldots\right\rangle_{\mathbf{a}'}$$
$$-\left\langle e^{-\mathbf{s}\cdot\mathbf{a}(t)}[-\mathbf{A}(\mathbf{a}) + \tfrac{1}{2}\mathsf{B}(\mathbf{a})\cdot\mathbf{s} + \tfrac{1}{2}\mathbf{s}\cdot\mathsf{B}(\mathbf{a}) + \left(\tfrac{1}{2}\mathsf{F}_3(\mathbf{a}):\mathbf{ss}\right)_{sym} + \ldots\right\rangle_{\mathbf{a}'}, \quad (18.5\text{-}2)$$

$$\frac{\partial}{\partial t}\left\langle e^{-\mathbf{s}\cdot\mathbf{a}(t)}\mathbf{aa}\right\rangle_{\mathbf{a}'} = \left\langle e^{-\mathbf{s}\cdot\mathbf{a}(t)}\mathbf{aa}[-\mathbf{s}\cdot\mathbf{A}(\mathbf{a}) + \tfrac{1}{2}\mathsf{B}(\mathbf{a})\cdot\mathbf{s} + \tfrac{1}{2}\mathbf{s}\cdot\mathsf{B}(\mathbf{a}) + \ldots]\right\rangle_{\mathbf{a}'}$$
$$+\left\langle e^{-\mathbf{s}\cdot\mathbf{a}(t)}\mathbf{a}[\mathbf{A}(\mathbf{a}) - \tfrac{1}{2}\mathsf{B}(\mathbf{a})\cdot\mathbf{s} - \tfrac{1}{2}\mathbf{s}\cdot\mathsf{B}(\mathbf{a}) + \ldots]\right\rangle_{\mathbf{a}'}$$
$$+\left\langle e^{-\mathbf{s}\cdot\mathbf{a}(t)}[\mathbf{A}(\mathbf{a}) - \tfrac{1}{2}\mathsf{B}(\mathbf{a})\cdot\mathbf{s} - \tfrac{1}{2}\mathbf{s}\cdot\mathsf{B}(\mathbf{a}) + \ldots]\mathbf{a}\right\rangle_{\mathbf{a}'}$$
$$+\left\langle e^{-\mathbf{s}\cdot\mathbf{a}(t)}[\mathsf{B}(\mathbf{a}) + \ldots]\right\rangle_{\mathbf{a}'}. \quad (18.5\text{-}3)$$

Next, we let $\mathbf{s} \to 0$. We thus obtain

$$\frac{\partial}{\partial t}\langle \mathbf{a}(t)\rangle_{\mathbf{a}'} = \langle \mathbf{A}[\mathbf{a}(t)]\rangle_{\mathbf{a}'}, \quad (18.5\text{-}4)$$

$$\frac{\partial}{\partial t}\langle \mathbf{a}(t)\mathbf{a}(t)\rangle_{\mathbf{a}'} = \langle \mathbf{a}(t)\mathbf{A}[\mathbf{a}(t)]\rangle_{\mathbf{a}'} + \langle \mathbf{A}[\mathbf{a}(t)]\mathbf{a}(t)\rangle_{\mathbf{a}'} + \langle \mathsf{B}[\mathbf{a}(t)]\rangle_{\mathbf{a}'}. \quad (18.5\text{-}5)$$

The first (set of) equation(s) is (are) the *phenomenological equation*(*s*), albeit it in a somewhat obscure form. These state the average behaviour or *response* conditional to a disturbance characterized by the initial state $\mathbf{a}'(0)$. Let us expand about the equilibrium or steady-state average $\langle \mathbf{a}\rangle \equiv \mathbf{a}_0$; so we write

$$\mathbf{A}[\mathbf{a}(t)] = \mathbf{A}(\mathbf{a}_0) - \mathsf{M}\cdot\Delta\mathbf{a}(t) + \mathcal{O}(\Delta a^2), \quad (18.5\text{-}6)$$

where M is the phenomenological relaxation tensor (or matrix with the $\cdot$ on the rhs omitted). Specifically we have

$$\mathsf{M} = -(\partial \mathbf{A}[\mathbf{a}]/\partial \mathbf{a})_{\mathbf{a}=\mathbf{a}_0}, \quad \text{or} \quad M_{ij} = -\partial A_i[\mathbf{a}_0]/\partial a_j^0. \quad (18.5\text{-}7)$$

Considering the limit of (18.5-4) for $t \to \infty$, the lhs yields zero and the rhs tends towards the no-memory average $\mathbf{A}(\mathbf{a}_0)$; this quantity is therefore also equal to zero. Consequently, the linear(ized) phenomenological equations become – reverting further to matrix notation:

$$\frac{\partial}{\partial t}\langle \Delta\mathbf{a}(t)\rangle_{\mathbf{a}'} = -\mathsf{M}\langle \Delta\mathbf{a}(t)\rangle_{\mathbf{a}'}. \quad \mathbf{(18.5\text{-}8)}$$

These equations embody Onsager's principle, which states that fluctuations in a 'conditional ensemble' for which the values $\mathbf{a}'(0)$ are fixed, decay according to the phenomenological laws.[4b] It can be shown that for thermal equilibrium systems M is positive definite, by which we mean that $(\mathsf{M}\beta,\beta) = \Sigma_{ij}\, \mathsf{M}_{ij}\beta_i\beta_j > 0$ for arbitrary β.

[4b] Quantum frequencies are *a priori* excluded due to omission of the nondiagonal parts of the operators.

Next we consider the stationary second moment result (18.5-5). For $t \to \infty$ we easily obtain, using once more (18.5-6) and minor algebra

$$\langle \Delta\mathbf{a}\Delta\mathbf{a}\rangle \tilde{\mathsf{M}} + \mathsf{M}\langle \Delta\mathbf{a}\Delta\mathbf{a}\rangle = \mathsf{B}(\mathbf{a}_0)\,. \qquad \textbf{(18.5-9)}$$

Here $\tilde{\mathsf{M}}$ denotes the transpose matrix, while the covariance matrix $\langle \Delta\mathbf{a}\Delta\mathbf{a}\rangle$, strictly speaking, should be written as $\langle \Delta\mathbf{a}\Delta\tilde{\mathbf{a}}\rangle$, cf. Section 16.4. The above relationship determines the covariances uniquely. It was first obtained for a system of coupled random harmonic oscillators by Ming Chen Wang and Uhlenbeck (Ref. 1(b), Section 11) and later on from the Fokker–Planck equation by Van Vliet and Blok for multivariate carrier-density fluctuations in semiconductors and photoconductors.[5,6]. Since for Brownian motion the matrix B is related to the diffusivity, some have referred to (18.5-9) as the 'generalized Einstein relation' (Melvin Lax[7]). We have often referred to it as the generalized g-r theorem (see later examples). In this text we shall refer to this relationship as the *fluctuation-relaxation theorem.*[8] [A similar theorem has been found in transport noise involving continuous (infinite dimensional) stochastic processes, called the Λ-theorem, cf. Section 19.6.] While for equilibrium processes the variances and covariances can be found by many other means, for fluctuations about the steady state the theorem (18.5-9) affords the only direct computation; examples will be met. A general solution of (18.5-9) is easily given; one finds[7],

$$\langle \Delta\mathbf{a}\Delta\mathbf{a}\rangle = \int_0^\infty dt\, e^{-\mathsf{M}t}\mathsf{B}\, e^{-\tilde{\mathsf{M}}t}\,. \qquad (18.5\text{-}10)$$

To verify this solution, insert it into (18.5-9) and integrate one term by parts. For practical purposes, (18.5-10) is not useful; solving the matrix relations with Cramer's rule is the appropriate way.

In thermal equilibrium a great simplification is possible since both terms on the lhs are equal! So, we have the direct result

$$\langle \Delta\mathbf{a}\Delta\mathbf{a}\rangle = \tfrac{1}{2}\mathsf{M}^{-1}\mathsf{B}(\mathbf{a}_0)\,. \qquad (18.5\text{-}11)$$

Two proofs will be presented. Lax[7] employs the fact that the correlation function is time reversible if all variables are either odd or even under time and magnetic field reversal, cf. (16.14-22). Then, using the correct matrix form for the covariance matrix $\langle \Delta\mathbf{a}\Delta\mathbf{a}\rangle \to \langle \Delta\mathbf{a}\Delta\tilde{\mathbf{a}}\rangle$,

$$\langle \boldsymbol{\alpha}(t)\tilde{\boldsymbol{\alpha}}(0)\rangle \overset{\text{time rev.}}{=} \langle \boldsymbol{\alpha}(-t)\tilde{\boldsymbol{\alpha}}(0)\rangle \overset{\text{station.}}{=} \langle \boldsymbol{\alpha}(0)\tilde{\boldsymbol{\alpha}}(t)\rangle\,, \qquad (18.5\text{-}12)$$

---

[5] K.M. Van Vliet and J. Blok, "Electronic noise in semiconductors", Physica XXII, 231-242 (1956).

[6] Ibid. "Electronic noise in photoconducting insulators", Physica XXII, 525-540 (1956).

[7] M. Lax, "Fluctuations from the Non-equilibrium Steady State", Rev. Mod. Phys. **32**, 25-64 (1960).

[8] C.M. Van Vliet, "Noise out of Equilibrium", Proc. 7th Int. Conf. on Noise in Physical Systems (M. Savelli, G. LeCoy and J.-P. Nougier, Eds.) North Holland, Amsterdam/NY 1983, p.7-14.

where $\boldsymbol{\alpha} = \Delta \mathbf{a}$. The solution of the phenomenological equations (18.5-8) yields

$$\langle \boldsymbol{\alpha}(0)\tilde{\boldsymbol{\alpha}}(t)\rangle = \sum_{\boldsymbol{\alpha},\boldsymbol{\alpha}'} \boldsymbol{\alpha}'\tilde{\boldsymbol{\alpha}} W_2(\boldsymbol{\alpha},t;\boldsymbol{\alpha}',0) = \sum_{\boldsymbol{\alpha}'} \boldsymbol{\alpha}' W_1(\boldsymbol{\alpha}') \sum_{\boldsymbol{\alpha}} \tilde{\boldsymbol{\alpha}} P(\boldsymbol{\alpha},t\,|\,\boldsymbol{\alpha}')$$

$$= \langle \boldsymbol{\alpha}' \langle \tilde{\boldsymbol{\alpha}}(t)\rangle_{\boldsymbol{\alpha}'}\rangle = \langle \boldsymbol{\alpha}'\tilde{\boldsymbol{\alpha}}'\rangle e^{-\tilde{\mathsf{M}}t} = \langle \boldsymbol{\alpha}\tilde{\boldsymbol{\alpha}}\rangle e^{-\tilde{\mathsf{M}}t}\,, \tag{18.5-13a}$$

$$\langle \boldsymbol{\alpha}(t)\tilde{\boldsymbol{\alpha}}(0)\rangle = e^{-\mathsf{M}t}\langle \boldsymbol{\alpha}'\tilde{\boldsymbol{\alpha}}'\rangle = e^{-\mathsf{M}t}\langle \boldsymbol{\alpha}\tilde{\boldsymbol{\alpha}}\rangle\,. \tag{18.5-13b}$$

We note that the correlation average was computed by first taking an average over a conditional ensemble and then over an unrestricted ensemble, using Bayes' rule; this is often the preferred method. By (18.5-12) these expressions are equal. Now replace t by Δt and expand the exponentials. Then we get

$$\mathsf{M}\Delta t\langle \boldsymbol{\alpha}\tilde{\boldsymbol{\alpha}}\rangle = \langle \boldsymbol{\alpha}\tilde{\boldsymbol{\alpha}}\rangle \tilde{\mathsf{M}}\Delta t\,. \tag{18.5-14}$$

Now divide by Δt. [We cannot afterwards let Δt go to zero, since no physical process is Markovian for very small times; the bounds on Δt are actually the same as for the Van Hove limit, cf. subsection 16.9.1. Moreover, the exponential decay[9] as in Eqs. (18.5-13) does not hold near $t = 0$; the correlation function *does not* have a cusp at $t = 0$ but is rounded and continuous.]

A better proof relies on the connection with non-equilibrium thermodynamics; see Van Vliet and Fassett.[10] Let the α ' s be considered as extensive thermodynamic variables and let the corresponding entropic forces be Q_i. There will be macroscopic linear regression equations associated with the generalized forces; generally we will have $(Q_i)_{\rm total} = \sum_j R_{ij}\dot{\alpha}_j$. We may assume that the system is perturbed by a driving force $V_i(t)$ enacted via its reservoirs and an internal restoring force $-\sum_j s_{ij}^0 \alpha_j$ in the system. Here s_{ij}^0 is the matrix of second-order derivatives, $\mathsf{s}^0 = (\partial^2 S/\partial \mathbf{a}\partial \mathbf{a})_{\mathbf{a}_0}$. We find

$$\mathbf{V}(t) = \mathsf{s}^0 \boldsymbol{\alpha}(t) + \mathsf{R}\dot{\boldsymbol{\alpha}}\,. \tag{18.5-15}$$

If the perturbation $\mathbf{V}(t)$ is withdrawn, the system returns to equilibrium with relaxation matrix $\mathsf{M} = \mathsf{R}^{-1}\mathsf{s}^0$. Also, from the Einstein distribution (3.7-7) we have $\langle \boldsymbol{\alpha}\tilde{\boldsymbol{\alpha}}\rangle = k_B(\mathsf{s}^0)^{-1}$. Accordingly, in thermal equilibrium we have

$$\langle \boldsymbol{\alpha}\tilde{\boldsymbol{\alpha}}\rangle \tilde{\mathsf{M}} = k_B(\mathsf{s}^0)^{-1}\tilde{\mathsf{s}}^0\tilde{\mathsf{R}}^{-1} = k_B\tilde{\mathsf{R}}^{-1}\,, \tag{18.5-16}$$

where we used that by its definition s^0 is symmetrical. Moreover, from the Onsager relations we know that the conductance R^{-1} is a symmetrical matrix. Hence, we can transpose the lhs, obtaining the other term of (18.5-9); so both parts are equal, QED.

There are other properties that follow from the symmetry of (18.5-14). It is easily shown that it causes the spectra $\mathcal{S}_{\alpha\alpha}(\omega)$ to be real.

---

[9] The decay of the separate variables a_i is a sum of exponentials, determined by the eigenvalues of M.

[10] K.M. Van Vliet and J.R. Fassett, "Fluctuations due to Electronic Transitions and Transport in Solids" in "Fluctuation Phenomena in Solids" (R.E. Burgess, Ed.), Academic Press, NY 1965, pp. 267-354.

Alternative treatment

Because of the importance of the moment equations, we shall give an alternative, more direct treatment.[10] We will assume that the variables are discrete and represented by column matrices; the changes when the variables are continuous are self-evident. The first two Fokker–Planck moments are defined by

$$\mathbf{A}(\mathbf{a}'') = \lim_{\Delta t\to 0}\frac{1}{\Delta t}\sum_{\mathbf{a}}(\mathbf{a}-\mathbf{a}'')P(\mathbf{a},t\,|\,\mathbf{a}'',0) = \sum_{\mathbf{a}}{}'(\mathbf{a}-\mathbf{a}'')Q_{\mathbf{a}''\mathbf{a}}\,, \tag{18.5-17}$$

$$\mathsf{B}(\mathbf{a}'') = \lim_{\Delta t\to 0}\frac{1}{\Delta t}\sum_{\mathbf{a}}(\mathbf{a}-\mathbf{a}'')(\tilde{\mathbf{a}}-\tilde{\mathbf{a}}'')P(\mathbf{a},t\,|\,\mathbf{a}'',0) = \sum_{\mathbf{a}}{}'(\mathbf{a}-\mathbf{a}'')(\tilde{\mathbf{a}}-\tilde{\mathbf{a}}'')Q_{\mathbf{a}''\mathbf{a}}\,. \tag{18.5-18}$$

We now multiply the master equation by $\mathbf{a}$ and sum over all $\mathbf{a}$, denoting the average in a conditional ensemble by $\langle \mathbf{a}(t)\,|\,\mathbf{a}',0\rangle = \langle \mathbf{a}\rangle_{\mathbf{a}'}$; note that we suppressed the t-dependence in the latter short-cut notation. We obtain

$$\frac{\partial\langle \mathbf{a}\rangle_{\mathbf{a}'}}{\partial t} = \sum_{\mathbf{a}''}\sum_{\mathbf{a}}(\mathbf{a}-\mathbf{a}'')P(\mathbf{a}'',t\,|\,\mathbf{a}',0)\,Q_{\mathbf{a}''\mathbf{a}}\,, \tag{18.5-19}$$

where we interchanged the summation variables in one part. With the definition (18.5-15) this reads

$$\frac{\partial\langle \mathbf{a}\rangle_{\mathbf{a}'}}{\partial t} = \langle \mathbf{A}(\mathbf{a})\rangle_{\mathbf{a}'}\,. \tag{18.5-20}$$

The $\mathbf{A}$ is now expanded about the equilibrium value $\langle \mathbf{a}\rangle = \lim_{t\to\infty}\langle \mathbf{a}\rangle_{\mathbf{a}'} \equiv \mathbf{a}_0$. Hence,

$$\mathbf{A}(\mathbf{a}) = \mathbf{A}(\mathbf{a}_0) - \mathsf{M}\Delta\mathbf{a} + \mathcal{O}(\Delta a^2)\,. \tag{18.5-21}$$

This is substituted into (18.5-20); taking then the limit for $t\to\infty$, the lhs goes to zero and $\lim_{t\to\infty}\langle \Delta\mathbf{a}\rangle_{\mathbf{a}'} = 0$. Consequently,

$$\langle \mathbf{A}(\mathbf{a})\rangle \simeq \mathbf{A}(\mathbf{a}_0) = 0\,, \tag{18.5-22}$$

while (18.5-20) yields the linear(ized) phenomenological equations

$$\frac{\partial\langle \Delta\mathbf{a}\rangle_{\mathbf{a}'}}{\partial t} = -\mathsf{M}\langle \Delta(\mathbf{a})\rangle_{\mathbf{a}'}\,. \tag{18.5-23}$$

Next, we multiply the master equation by $\mathbf{a}\tilde{\mathbf{a}}$. This gives

$$\frac{\partial\langle \mathbf{a}\tilde{\mathbf{a}}\rangle_{\mathbf{a}'}}{\partial t} = \sum_{\mathbf{a}''}\sum_{\mathbf{a}}(\mathbf{a}\tilde{\mathbf{a}}-\mathbf{a}''\tilde{\mathbf{a}}'')\,P(\mathbf{a}'',t\,|\,\mathbf{a}',0)\,Q_{\mathbf{a}''\mathbf{a}}\,. \tag{18.5-24}$$

We replace the first matrix by

$$\begin{aligned}\mathbf{a}\tilde{\mathbf{a}} &= (\mathbf{a}-\mathbf{a}''+\mathbf{a}'')(\tilde{\mathbf{a}}-\tilde{\mathbf{a}}''+\tilde{\mathbf{a}}'')\\ &= (\mathbf{a}-\mathbf{a}'')(\tilde{\mathbf{a}}-\tilde{\mathbf{a}}'') + \mathbf{a}''\tilde{\mathbf{a}}'' + \mathbf{a}''(\tilde{\mathbf{a}}-\tilde{\mathbf{a}}'') + (\mathbf{a}-\mathbf{a}'')\tilde{\mathbf{a}}''\,.\end{aligned} \tag{18.5-25}$$

With the definition of both F–P moments (18.5-24) then yields

$$\frac{\partial\langle \mathbf{a}\tilde{\mathbf{a}}\rangle_{\mathbf{a}'}}{\partial t} = \langle \mathsf{B}(\mathbf{a})\rangle_{\mathbf{a}'} + \langle \mathbf{a}\tilde{\mathbf{A}}(\mathbf{a})\rangle_{\mathbf{a}'} + \langle \mathbf{A}(\mathbf{a})\tilde{\mathbf{a}}\rangle_{\mathbf{a}'}\,. \tag{18.5-26}$$

This is a complicated equation for the conditional second order moments; fortunately, we are only interested in the behaviour for $t \to \infty$. We thus obtain the connection for the non-conditional averages

$$\langle \mathbf{a}\tilde{\mathbf{A}}(\mathbf{a})\rangle + \langle \mathbf{A}(\mathbf{a})\tilde{\mathbf{a}}\rangle = -\langle \mathsf{B}(\mathbf{a})\rangle . \tag{18.5-27}$$

From (18.5-22) we also have

$$\langle \mathbf{a}_0\tilde{\mathbf{A}}(\mathbf{a})\rangle + \langle \mathbf{A}(\mathbf{a})\tilde{\mathbf{a}}_0\rangle = 0 . \tag{18.5-28}$$

Subtracting the last two equations and observing (18.5-21), we arrive at

$$\langle \boldsymbol{\alpha}\tilde{\boldsymbol{\alpha}}\rangle \tilde{\mathsf{M}} + \mathsf{M}\langle \boldsymbol{\alpha}\tilde{\boldsymbol{\alpha}}\rangle = \mathsf{B}(\mathbf{a}_0) + \mathcal{O}(\alpha^3) . \tag{18.5-29}$$

The reader will notice that we have given due attention to using the appropriate transposes in all these expressions. In further applications, however, we will usually set $\langle \boldsymbol{\alpha}\tilde{\boldsymbol{\alpha}}\rangle \to \langle \boldsymbol{\alpha}\boldsymbol{\alpha}\rangle$.

18.5.1 *One-Variable Master Equation; Birth-Death Rate Processes*

Let $a = n$ be a discrete stochastic variable, representing a population, a number of particles, etc. The master equation for this case is written as

$$\frac{\partial P(n,t\,|\,m)}{\partial t} = \sum_k{}' [P(k,t\,|\,m)Q_{kn} - P(n,t\,|\,m)Q_{nk}] . \tag{18.5-30}$$

The Q's are the transition rates associated with changes in $\Delta t \to 0$. Clearly, no more than one transition will take place – or no transitions at all. So in Q_{kn} and Q_{nk} we have $k = n \pm 1$, the possibility $k = n$ being excluded from the sum in the ME by the prime on Σ. Rewriting, we find

$$\frac{\partial P(n,t\,|\,m)}{\partial t} = P(n-1,t\,|\,m)Q_{n-1,n} + P(n+1,t\,|\,m)Q_{n+1,n} - P(n,t\,|\,m)[Q_{n,n+1} + Q_{n,n-1}] . \tag{18.5-31}$$

Generally we shall write $g(n)$ for the *gain* (or generation) rate and $r(n)$ for the *retrieval* (or recombination) rate. So we have

$$\begin{aligned} Q_{n-1,n} &= g(n-1), \qquad & Q_{n+1,n} &= r(n+1), \\ Q_{n,n+1} &= g(n), \qquad & Q_{n,n-1} &= r(n) . \end{aligned} \tag{18.5-32}$$

This yields

$$\frac{\partial P(n,t\,|\,m)}{\partial t} = P(n-1,t\,|\,m)g(n-1) + P(n+1,t\,|\,m)r(n+1) - P(n,t\,|\,m)[g(n)+r(n)] . \tag{18.5-33}$$

For the stationary state $\partial P/\partial t = 0$ we have

$$W(n-1)g(n-1) + W(n+1)r(n+1) = W(n)[g(n)+r(n)] . \tag{18.5-34}$$

This functional equation has the solution (Burgess[11]):

[11] R.E. Burgess, Physica XX, 1007 (1954).

$$W(n)=W(0)\prod_{\nu=0}^{n-1} g(\nu)\Big/\prod_{\nu=1}^{n} r(\nu)\,. \tag{18.5-35}$$

Example (*a*)

Let there be a constant gain rate, $g(n)=\alpha$, and a loss rate with a mean lifetime τ, so that $r(n)=n/\tau$. Equation (18.5-35) now yields

$$W(n)=W(0)(\alpha\tau)^n/n!\,,\qquad \alpha\tau=\langle n\rangle\,, \tag{18.5-36}$$

where the last equality follows from the balance equation

$$\langle g(n)\rangle=\langle r(n)\rangle\,. \tag{18.5-37}$$

From normalization we find $W(0)=e^{-\langle n\rangle}$, indicating that the final result is a Poisson distribution,

$$W(n)=e^{-\langle n\rangle}\langle n\rangle^n/n!\,. \tag{18.5-38}$$

Example (*b*)

Let the population be governed by a random telegraph signal (RTS). As a model we will focus on the exchange of electrons between the conduction band and a set of impurity levels. Then, $g(n)=(N-n)/\tau_1$ and $r(n)=n/\tau_2$, where N is the number of available carriers and τ_1 is the average time spent in the centres and τ_2 in the conduction band. If a current is supplied, its microscopic form $i(t)$ will have the appearance of a random telegraph signal, see Fig. 18-1. Such noise patterns are frequently observed and are known as *'burst noise'*.

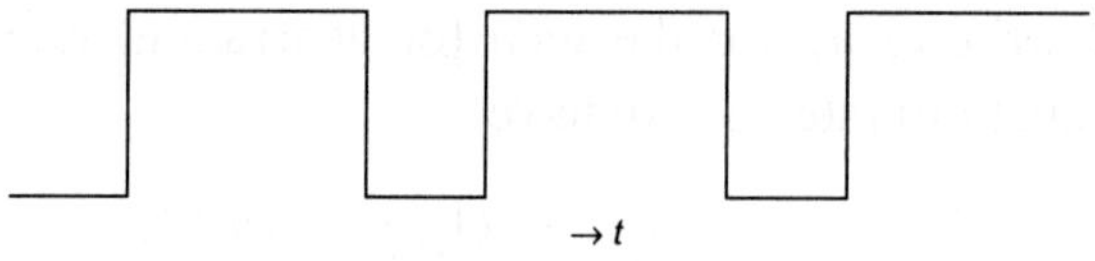

Fig. 18-1. RTS signal, as observed in burst noise.

Using the balance equation (18.5-37), we obtain for the mean occupancies

$$\langle n\rangle=\frac{\tau_2}{\tau_1+\tau_2}N=\lambda N\,,\qquad \langle n_t\rangle\equiv N-\langle n\rangle=\frac{\tau_1}{\tau_1+\tau_2}N=(1-\lambda)N\,. \tag{18.5-39}$$

For the distribution (18.5-35) we obtain

$$W(n)=W(0)\frac{N(N-1)\ldots(N-n+1)}{n!}\left(\frac{\tau_2}{\tau_1}\right)^n$$

$$= \left[W(0) \left(\frac{\tau_1 + \tau_2}{\tau_1} \right)^N \right] \frac{N!}{(N-n)!n!} \left(\frac{\tau_2}{\tau_1 + \tau_2} \right)^n \left(\frac{\tau_1}{\tau_1 + \tau_2} \right)^{N-n} . \quad (18.5\text{-}40)$$

This is the binomial distribution, as expected. From normalization it follows that the factor in the big square brackets must be equal to unity. The variance is given by

$$\langle \Delta n^2 \rangle = \langle \Delta n_t^2 \rangle = N \frac{\tau_1 \tau_2}{(\tau_1 + \tau_2)^2} = N\lambda(1-\lambda) . \quad (18.5\text{-}41)$$

While this type of distribution may apply to burst noise associated with defects in a space-charge region, it *does not* apply to electronic noise due to transitions in charge-neutral solids, notwithstanding that it was devised for that purpose by Machlup in a paper that appeared in 1954.[12] This was indicated by Burgess in the paper already cited (Ref.10), presented at the second "International Conference on the Physics of Semiconductors" held in Amsterdam and attended by this author, also in 1954. The variance is actually smaller than predicted by the binomial distribution due to the exclusion principle [see also 5-4-21)]. The variance from Burgess' approach is obtained as follows. Consider the logarithmic form of (18.5-35). Then,

$$\ln W(n) = \ln W(0) + \sum_{\nu=0}^{n-1} \ln g(\nu) - \sum_{\nu=1}^{n} \ln r(\nu)$$
$$\sim \ln W(0) + n \ln g(n) - n \ln r(n) . \quad (18.5\text{-}42)$$

This distribution peaks for $g(n_0) = r(n_0)$, as one finds by considering

$$(d/dn)[n \ln r(n) - n \ln g(n)] = \ln r(n) - \ln g(n) + [nr'(n)/r(n)] - [ng'(n)/g(n)] .$$

Generally each rate will follow a power law, $r(n) = cn^\gamma \rightarrow nr'(n)/r(n) = \gamma \ll \ln r(n)$. So the terms in square brackets will be neglected and the steady-state condition follows. Expanding up to second order, we have

$$\ln W(n) = \ln W(0) - \tfrac{1}{2}(n - n_0)^2 [r'(n_0) - g'(n_0)] / g(n_0) , \quad (18.5\text{-}43)$$

which is a normal distribution with variance

$$\langle \Delta n^2 \rangle = \frac{g(n_0)}{r'(n_0) - g'(n_0)} , \quad (18.5\text{-}44)$$

which is Burgess' *g-r- theorem.* For the case under study, the generation rate is mono-molecular, $g(n) = \alpha(N - n)$ and the recombination rate is bimolecular, $r(n) = \beta n^2$. One easily obtains

$$\langle \Delta n^2 \rangle = \lambda(1-\lambda)N/(2-\lambda), \quad \lambda = n_0/N . \quad (18.5\text{-}45)$$

---

[12] S. Machlup, J. Appl. Phys. **25**, 341 (1954).

Since $\lambda < 1$, the variance is reduced, as expected from the restrictions of the exclusion principle. We leave it to the reader to also obtain this from the Fermi statistics via the methods developed in Section 5.4.

We now consider the Fokker–Planck moments. We have

$$\begin{aligned} A(n) &= \sum_k (k-n)Q_{nk} = Q_{n,n+1} - Q_{n,n-1} = g(n) - r(n), \\ B(n) &= \sum_k (k-n)^2 Q_{nk} = Q_{n,n+1} + Q_{n,n-1} = g(n) + r(n). \end{aligned} \tag{18.5-46}$$

Expanding in a Taylor series

$$\begin{aligned} g(n) &= g(n_0) + g'(n_0)\Delta n + \tfrac{1}{2} g''(n_0)(\Delta n^2) + \ldots \\ r(n) &= r(n_0) + r'(n_0)\Delta n + \tfrac{1}{2} r''(n_0)(\Delta n^2) + \ldots \end{aligned} \tag{18.5-47}$$

and noting that $g(n_0) = r(n_0) + \mathcal{O}(\Delta n^2)$ by virtue of (18.5-37) and the averaged values of (18.5-47), we find the linearized phenomenological equation

$$A(n) = -\Delta n/\tau + \mathcal{O}(\Delta n^2), \quad 1/\tau = [r'(n_0) - g'(n_0)]. \tag{18.5-48}$$

Clearly, $1/\tau$ is the eigenvalue of M, i.e. $M - 1/\tau = 0$. For the second moment we have

$$B(n_0) = g(n_0) + r(n_0) = 2g(n_0) + \mathcal{O}(\Delta n^2). \tag{18.5-49}$$

In this result $2g(n_0) \gg \mathcal{O}(\Delta n^2)$, so that we can neglect the extra terms. Hence, from the fluctuation-relaxation theorem we find

$$2\langle \Delta n^2 \rangle [r'(n_0) - g'(n_0)] = 2g(n_0), \tag{18.5-50}$$

in accord with Burgess' result (18.5-44).

To obtain the correlation function it is not necessary to go back to the time-dependent second moment equation; instead, it suffices to use the phenomenological, first moment equation. We have

$$\begin{aligned} &\langle \Delta n(t) \rangle_{n_0} = \Delta n_0 \, e^{-t/\tau}, \text{ or} \\ &\langle \Delta n(t) \Delta n(0) \rangle = \langle \langle \Delta n(t) \rangle_{n_0} \Delta n_0 \rangle = \langle \Delta n_0^2 \rangle e^{-t/\tau} = \langle \Delta n^2 \rangle e^{-t/\tau}. \end{aligned} \tag{18.5-51}$$

By the Wiener–Khintchine theorem we have for the spectrum the Lorentzian

$$\mathcal{S}_n(\omega) = 4\langle \Delta n^2 \rangle \tau/(1+\omega^2\tau^2) = 4g(n_0)\,\tau^2/(1+\omega^2\tau^2), \tag{18.5-52}$$

where we used the result for the variance as well as (18.5-48) for τ.

18.5.2 *Multivariate Gain-Loss Processes*

Multivariate gain-loss processes are best exemplified by electronic transitions in solids. Let there be s 'levels', representing the nondegenerate conduction band, the valence band and impurity levels, donor sites, traps, etc. The transitions between them are characterized by rate functions p_{ij}. Usually there is a charge-neutrality constraint, so that only $s-1$ variables n_i are independent. The process is pictured in Fig. 18-2.

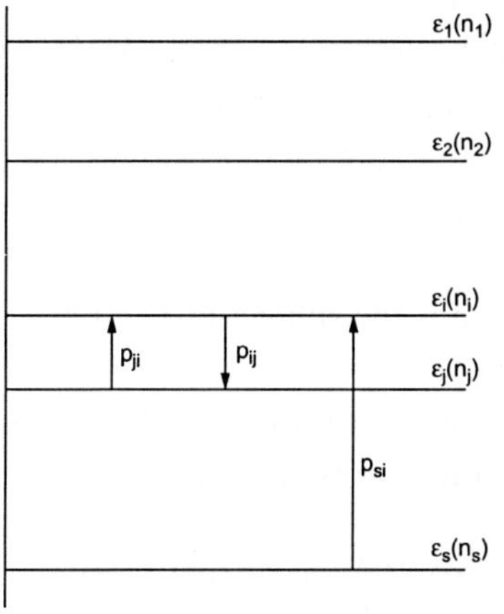

Fig. 18-2. Electronic transitions in solids.

Let us now compute the Fokker–Planck moments, cf. Refs. 5 and 10. We have,

$$\lim_{\Delta t\to 0}(1/\Delta t)P(\mathbf{a},\Delta t\,|\,\mathbf{a}',0)=Q_{\mathbf{a}'\mathbf{a}}=0 \quad \text{if} \quad |n_i-n_i'|\geq 2, \tag{18.5-53}$$

$$Q_{\mathbf{aa}'}=p_{ij}(n_1,\ldots,n_s) \quad (i,j=1\ldots s)$$
$$\text{if } \mathbf{a}=\{n_1,n_2,\ldots,n_i,n_j,\ldots.n_s\} \text{ and } \mathbf{a}'=\{n_1,n_2,\ldots,n_i-1,n_j+1,\ldots.n_s\}. \tag{18.5-54}$$

For symmetry purposes we have added the dependent variable $n_s=n_s(n_1,\ldots,n_{s-1})$. Then, from (18.5-17) and (18.5-18) we obtain [note $\mathbf{a}''\to\mathbf{a}$, $\mathbf{a}\to\mathbf{a}'$] :

$$A_i(\mathbf{a})=\sum_{j=1}^{s}{}'[p_{ji}(n_1\ldots n_s)-p_{ij}(n_1\ldots n_s)], \tag{18.5-55}$$

$$\begin{aligned} B_{ii}(\mathbf{a})&=\sum_{j=1}^{s}{}'[p_{ji}(n_1\ldots n_s)+p_{ij}(n_1\ldots n_s)]=2\sum_{j=1}^{s}{}'p_{ji}(n_1\ldots n_s), \\ B_{ij}(\mathbf{a})&=B_{ji}(\mathbf{a})=-p_{ij}(n_1\ldots n_s)-p_{ji}(n_1\ldots n_s). \end{aligned} \tag{18.5-56}$$

The phenomenological equations are the well-known rate equations for $a_i=n_i$:

$$\frac{\partial\langle n_i\rangle_{\mathbf{a}_0}}{\partial t}=\langle A_i(\mathbf{a})\rangle_{\mathbf{a}_0}=\sum_{j=1}^{s}{}'[\langle p_{ji}(n_1\ldots n_s)\rangle_{\mathbf{a}_0}-\langle p_{ij}(n_1\ldots n_s)\rangle_{\mathbf{a}_0}]. \tag{18.5-57}$$

In general these are nonlinear equations, the rates being usually mono-molecular or bimolecular in the numbers or densities. In linearized form they read,

$$\frac{\partial\langle\Delta n_i\rangle_{\mathbf{a}_0}}{\partial t}=-\sum_{k=1}^{s-1}M_{ik}\langle\Delta n_k\rangle_{\mathbf{a}_0}, \tag{18.5-58}$$

with M determined by

$$M_{ik} = \sum_{j=1}^{s}{}' \left[\frac{\partial p_{ij}}{\partial n_k} - \frac{\partial p_{ji}}{\partial n_k} \right]_{\{n_\ell\}=\{n_\ell^0\}} \doteq \sum_{j=1}^{s}{}' \left[\frac{\partial p_{ij}^0}{\partial n_k} - \frac{\partial p_{ji}^0}{\partial n_k} \right], \tag{18.5-59}$$

with the last rhs being a shortcut notation. From the fluctuation-relaxation theorem the variances and covariances can now be found. Then, from the linearized regression equations the correlation functions are obtained and via the Wiener–Khintchine theorem the spectra follow, being a sum of Lorentzians. While this is in principle straightforward, the details are often cumbersome and more so outside equilibrium when neither detailed balance nor the simplification embodied in (18.5-14) applies; the general treatment of *g-r* noise is the subject of Section 18.14.

Before we leave this subsection, we shall, however, give some *formal* results. We obtained matrix-form solutions for the correlation functions in (18-5-13). So, with the transposition property, we find for the spectral density matrix

$$\begin{aligned}\mathbb{S}_{\boldsymbol{\alpha\alpha}}(\omega) &= 2\int_0^\infty dt e^{-i\omega t} \langle \boldsymbol{\alpha}(t)\boldsymbol{\alpha}(0)\rangle + 2\int_0^\infty dt e^{i\omega t} \langle \boldsymbol{\alpha}(0)\boldsymbol{\alpha}(t)\rangle \\ &= 2(\mathsf{M}+i\omega\mathsf{I})^{-1}\langle\boldsymbol{\alpha\alpha}\rangle + 2\langle\boldsymbol{\alpha\alpha}\rangle(\tilde{\mathsf{M}}-i\omega\mathsf{I})^{-1}.\end{aligned} \tag{18.5-60}$$

Because of the symmetry relation discussed earlier, in thermal equilibrium both parts are equal and we have

$$\mathbb{S}_{\boldsymbol{\alpha\alpha}}(\omega) = 4\,\mathrm{Re}(\mathsf{M}+i\omega\mathsf{I})^{-1}\langle\boldsymbol{\alpha\alpha}\rangle. \quad \text{(thermal equil.)} \tag{18.5-61}$$

But a general explicit form, valid in the steady state, can also be obtained. By pre-multiplying (18.5-60) with $(\mathsf{M}+i\omega\mathsf{I})$ and post-multiplying with $(\tilde{\mathsf{M}}-i\omega\mathsf{I})$ we find

$$\begin{aligned}(\mathsf{M}+i\omega\mathsf{I})\mathcal{S}_{\boldsymbol{\alpha\alpha}}(\omega)(\tilde{\mathsf{M}}-i\omega\mathsf{I}) &= 2[\langle\boldsymbol{\alpha\alpha}\rangle(\tilde{\mathsf{M}}-i\omega\mathsf{I}) + (\mathsf{M}+i\omega\mathsf{I})\langle\boldsymbol{\alpha\alpha}\rangle] \\ &= 2[\langle\boldsymbol{\alpha\alpha}\rangle\tilde{\mathsf{M}} + \mathsf{M}\langle\boldsymbol{\alpha\alpha}\rangle] = 2\mathsf{B},\end{aligned} \tag{18.5-62}$$

where we used the full fluctuation-relaxation theorem. From the inverse procedure we then establish

$$\mathbb{S}_{\boldsymbol{\alpha\alpha}}(\omega) = 2(\mathsf{M}+i\omega\mathsf{I})^{-1}\mathsf{B}(\tilde{\mathsf{M}}-i\omega\mathsf{I})^{-1}. \quad \text{(general)} \tag{18.5-63}$$

18.5.3 *Electronic Fluctuations out of Equilibrium*

Since the vast majority of stochastic processes in physics occur in equilibrium, for which one has always recourse to regular equilibrium ensemble theory, we will give two examples of fluctuations out of equilibrium in this and the next subsection, for which the discussed theory is indispensable. We shall also demonstrate by the present example that non-equilibrium fluctuations lead to considerably more complex results than equilibrium expressions.

Early experimental results[13] indicated that the relative variance of the optically induced free electron population in the conduction band of photoconductors such as CdS or CdSe, $\langle \Delta n^2 \rangle / n_0$ is often larger than unity, with magnitudes as high as 10-100, a fact which *a prima vista* seems to be at variance with Fermi statistics.

So, in 1964[14] we examined the 'simple' problem of optical pumping in photoconductive insulators. The model considered involves a three level system, in which incident absorbed light causes transitions from the valence band to the conduction band (2.4 eV or 5240 Å for CdS), with subsequent recombination via impurity centres with holes in the valence band. The centres, being hole traps, are the usual activators used in these materials, such as Ag^+ centres or cat-ion vacancies. If these hole traps are 'deep', we can neglect any return transitions to the conduction band as well as thermal electronic transitions from the valence band to the activator centres. The model is pictured in Fig. 18-3(a). [In an actual photoconductor there are also electron traps that communicate mainly with the conduction band, as shown in Fig. 18-3(b); however, such a model cannot be handled fully analytically and requires Monte Carlo or other computational methods, hence it will not be discussed here.] The advantage of the three-level model proposed here is that the various capture cross sections or ensuing rate constants can be related through the steady-state balance equations, so that they do not occur in the final result for $\langle \Delta n^2 \rangle / n_0$.

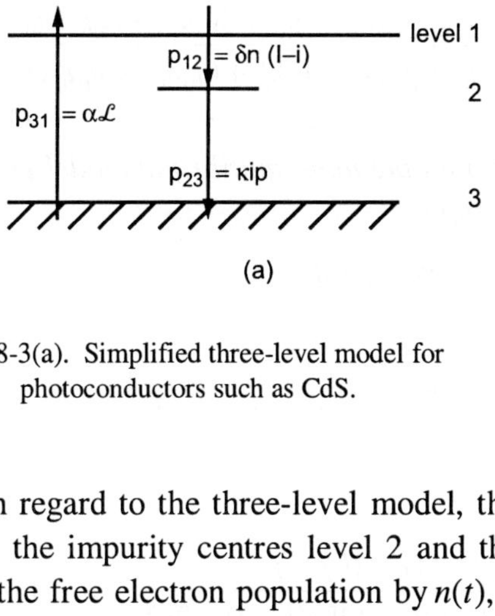

Fig. 18-3(a). Simplified three-level model for photoconductors such as CdS.

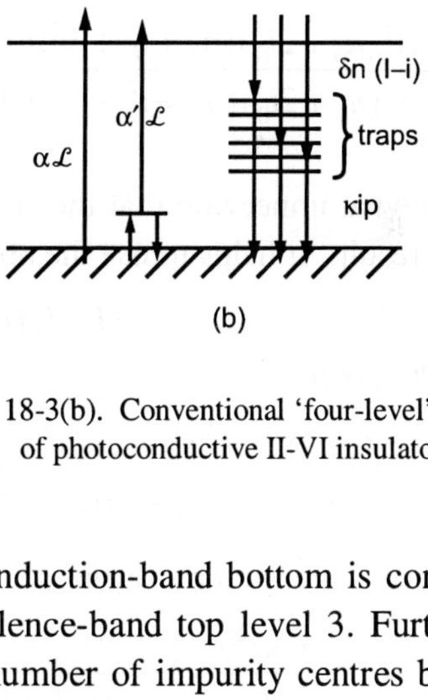

Fig. 18-3(b). Conventional 'four-level' model of photoconductive II-VI insulators.

With regard to the three-level model, the conduction-band bottom is considered level 1, the impurity centres level 2 and the valence-band top level 3. Further, we denote the free electron population by $n(t)$, the number of impurity centres by I and the number of electrons in these centres by $i(t)$, while the number of holes in the valence band is given as $p(t)$. From charge neutrality it follows that $n + i = p$, so that this is a bivariate stochastic process. With $\mathscr{L}$ being the light intensity, the rate equations are:

[13] K.M. Van Vliet, J. Blok, C. Ris and J. Steketee, Physica XXII, 723-740 (1956).

[14](a) K.M. Van Vliet, Phys Rev. **133A**, A1182-A1187 (1964); (b) also, Phys. Rev. **138 AB**, AB3 (1965).

$$dn/dt = \alpha\mathcal{L} - \delta n(I-i), \qquad di/dt = \delta n(I-i) - \kappa i(n+i). \tag{18.5-64}$$

The steady-state conditions give the relations

$$\alpha\mathcal{L} = \delta n_0(I-i_0) = \kappa i_0(n_0+i_0). \tag{18.5-65}$$

Linearizing the rate equations, we obtain the elements of the relaxation matrix; hence

$$\mathsf{M} = \begin{pmatrix} \delta(I-i_0) & -\delta n_0 \\ -\delta I + \delta i_0 + \kappa i_0 & \delta n_0 + \kappa n_0 + 2\kappa i_0 \end{pmatrix}. \tag{18.5-66}$$

For the matrix of the second-order Fokker–Planck moments we note that there are only three rates with $p_{13}^0 = p_{21}^0 = p_{32}^0 = 0$. This yields $B_{11} = 2p_{12}^0$, $B_{22} = 2p_{12}^0$, $B_{12} = B_{21} = -p_{12}^0$, or

$$\mathsf{B} = \begin{pmatrix} 2\delta n_0(I-i_0) & -\delta n_0(I-i_0) \\ -\delta n_0(I-i_0) & 2\delta n_0(I-i_0) \end{pmatrix}. \tag{18.5-67}$$

From (18.5-65) the ratio of δ/κ is expressed in the numbers I, i_0 and n_0; this is used in the above matrices, which are then substituted in the full fluctuation-relaxation theorem, (18.5-9). Solving the three coupled equations for $\langle \Delta n^2 \rangle$, one finds[14(b)]:

$$\langle \Delta n^2 \rangle / n_0 = \frac{(I-i_0)^2(n_0+2i_0)(n_0^2+i_0^2+3n_0i_0) + (I-i_0)n_0i_0(n_0^2+3i_0^2+4i_0n_0) + n_0i_0^2(n_0+i_0)^2}{(I-i_0)^2(n_0+2i_0)(n_0^2+i_0^2+3n_0i_0) + (I-i_0)n_0i_0(2n_0^2+3i_0^2+6i_0n_0) + n_0^2i_0^2(n_0+i_0)}. \tag{18.5-68}$$

It is somewhat unnerving that the *simplest possible non-equilibrium model* gives this complex result. It helps to use the abbreviations

$$(I-i_0)/i_0 = k, \quad n_0/i_0 = q. \tag{18.5-69}$$

The result is then

$$\frac{\langle \Delta n^2 \rangle}{n_0} = \frac{k^2(q+2)(q^2+3q+1) + kq(q^2+4q+3) + q(q+1)^2}{k^2(q+2)(q^2+3q+1) + kq(2q^2+6q+3) + q^2(q+1)}, \tag{18.5-70}$$

or, in a form suggested by Burgess[15]

$$\frac{\langle \Delta n^2 \rangle}{n_0} = 1 + \frac{-q^3k + q^2(1-2k) + q}{[q^2(k+1) + q(3k+2) + k][q(k+1) + 2k]}. \tag{18.5-70'}$$

First, we briefly discuss the photoconductivity as a function of $\mathcal{L}$. For very low light intensities, $n_0 \ll i_0$, $i_0 \ll I$ or $k \gg 1$, $q \ll 1$. This gives immediately

$$n_0 = \alpha\mathcal{L}/\delta I, \tag{18.5-71}$$

---

[15] R.E. Burgess, Private communication.

indicating a linear photoconductance curve. For high light intensity, on the contrary, we have from the second relationship in (18.5-61)

$$i_0 = \frac{-n_0(\kappa+\delta)}{2\kappa} + \frac{1}{2\kappa}[(\kappa+\delta)^2 + 4\delta\kappa n_0 I]^{1/2}. \tag{18.5-72}$$

If $\delta n_0 \gg 4\kappa I$, the square root can be expanded, yielding $i_0/I \simeq \delta/(\kappa+\delta)$. Substitution of this into the first relation of (18.5-61) gives

$$n_0 = [\alpha\mathcal{L}/\delta I](\kappa+\delta)/\kappa]. \tag{18.5-73}$$

Again, the photoconductance-light relationship is linear, but the proportionality constant is larger than in the low light case. Accordingly, the photoconductance-light relationship is 'superlinear' and exhibits the well-known *S*-form. Computations on a double logarithmic scale are given in Fig. 18-4 below.

We now return to the variance. For low light, $k \gg 1$, $q \ll 1$, the relative variance is close to 1. The same is easily shown to be true for very high light intensities, when the centres are filled ($k \ll 1$) and the free carriers are abundant ($q \gg 1$). A super-Poissonian region occurs for

$$1 + q(1-2k) - q^2 k > 0. \tag{18.5-74}$$

A *sufficient* condition for a *large* super-Poissonian variance is that $kq \ll 1$, which occurs in part of the superlinear photoconductance range. A plot of the relative variance as a function of the light intensity $\mathcal{L}$ is shown in Fig. 18-6 and a plot versus the free carrier density n_0 is given in Fig. 18-5. Values close to 10^2 are obtained. Surely, these steady-state results are quite distinct from equilibrium behaviour!

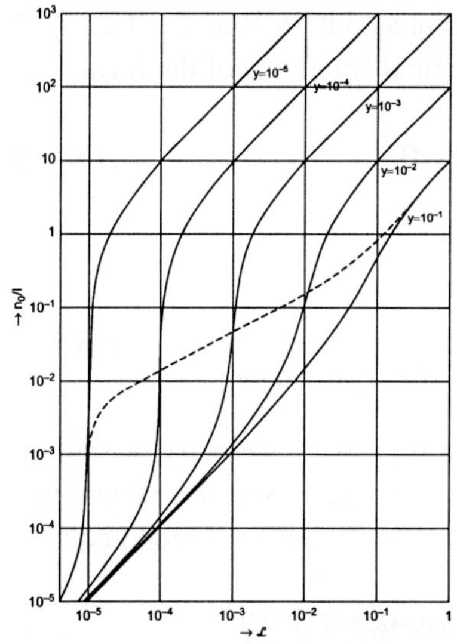

Fig. 18-4. Relative photoconductance vs. normalized light intensity for various values of $y = \kappa/\delta$.

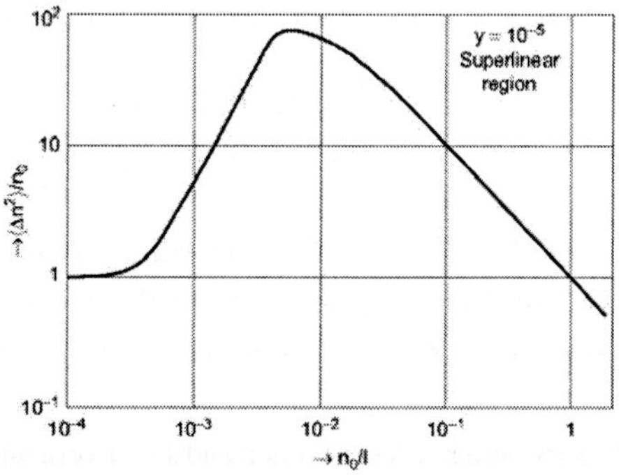

Fig. 18-5. Relative variance vs. normalized free carrier density n_0.

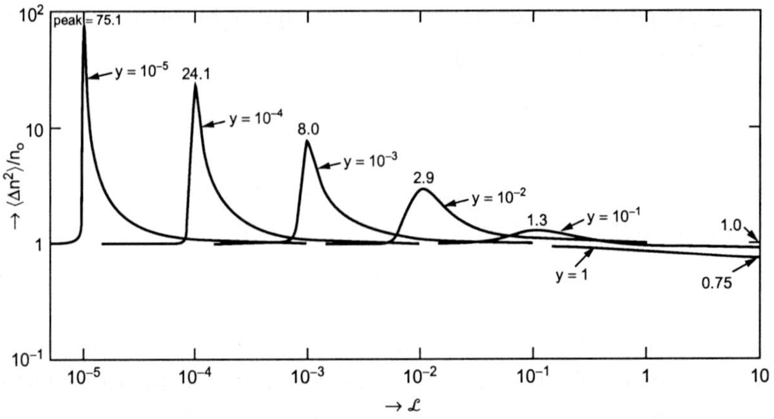

Fig.18-6. Relative variance $\langle \Delta n^2 \rangle / n_0$ vs. normalized light intensity.[Control data 1604 computer results.]

18.5.4 *Fluctuations About the Hydrodynamic Steady-State; Brillouin Scattering*

As a second non-equilibrium example we consider Brillouin scattering in a fluid with a fluctuating stress-field in a small uniform temperature gradient, first treated by Kirkpatrick et al.[16] and later by Tremblay et al.[17] These papers predict that some small but qualitatively unexpected effects in statically stressed fluids should be measurable by light scattering. In fact, in the presence of a small temperature gradient the two Brillouin peaks will be unequal in height. Tremblay et al. derive their result with coupled Langevin equations, such as considered in the next section. However, the results are also obtainable from the (full) fluctuation-relaxation theorem and the concomitant spectra, Eq. (18.5-63), as we have pointed out in Ref. 8. Generally, hydrodynamic modes are governed by phenomenological equations of the form

$$\frac{\partial \mathbf{A}(\mathbf{q},t)}{\partial t} + \mathsf{M}(\mathbf{q}) \cdot \mathbf{A}(\mathbf{q},t) = 0. \tag{18.5-75}$$

The particular forms pertaining to light scattering are[18]

$$\begin{aligned} &\partial P(\mathbf{q},t)/\partial t = -i\rho u^2\, \mathbf{q}\cdot\mathbf{v}\,, \\ &\partial(\mathbf{q}\cdot\mathbf{v})/\partial t = -i(q^2/\rho)P(\mathbf{q},t) - [(\vartheta + \tfrac{4}{3}\eta)/\rho]q^2\mathbf{q}\cdot\mathbf{v}\,, \end{aligned} \tag{18.5-76}$$

where $P(\mathbf{q},t)$ is the Fourier transformed local pressure, u is the velocity of sound, ρ is the fluid density, ϑ is the bulk viscosity and η is the shear viscosity. There is no second-order F–P moment associated with the first equation, but the second equation

[16] T. Kirkpatrick, E.G.D. Cohen and J.R. Dorfman, Phys. Rev. Lett. **42**, 862 (1972).

[17] A.-M.S. Tremblay, E.D. Siggia and M.R. Arai, Phys. Lett.**76A**, 57 (1980).

[18] L.E. Reichl, "A Modern Course in Statistical Physics", 2nd Ed., Wiley-VCH, 2004, 10.C and S10.F; also, L.D. Landau and E.M. Lifshitz, Fluid Mechanics", Pergamon Press, Oxford, 1959.

has the F–P kernel

$$B_{22}(\mathbf{q},\mathbf{q}') = (1/\rho^2)\mathbf{qq} : \mathsf{S}(\mathbf{q},\mathbf{q}') : \mathbf{q}'\mathbf{q}' , \tag{18.5-77}$$

where S represents the fluctuating stress field with

$$S_{ijkl}(\mathbf{q},\mathbf{q}') = k_B T(\mathbf{q}-\mathbf{q}')[\eta(\delta_{ik}\delta_{jl}+\delta_{il}\delta_{jk}) + (\vartheta - \tfrac{2}{3}\eta)\delta_{ij}\delta_{kl}] . \tag{18.5-78}$$

(In terms of Langevin sources, 2S is the tetradic spectral density of the fluctuating dyadic stress tensor.) The spectrum of the pressure fluctuations follows from (18.5-63) , now amended to read

$$\mathbb{S}_{\alpha\alpha}(\mathbf{q},\mathbf{q}';\omega) = 2[\mathsf{M}(\mathbf{q}) + i\omega\mathsf{I}]^{-1}\mathsf{B}(\mathbf{q},\mathbf{q}')[\tilde{\mathsf{M}}(\mathbf{q}') - i\omega\mathsf{I}]^{-1} , \tag{18.5-79}$$

and from the M-tensor implied by (18.5-76) plus the B-tensor (18.5-77). Leaving the algebra aside, the result is

$$\mathcal{S}_{PP}(\mathbf{q},\mathbf{q}';\omega) = 2\frac{u^4 k_B T(\mathbf{q}-\mathbf{q}')[2\eta(q\cdot q')^2 + (\vartheta - \tfrac{2}{3}\eta)q^2 q'^2]}{(\omega^2 - u^2q^2 - i\omega D_\ell q^2)(\omega^2 - u^2 q'^2 + i\omega D_\ell q'^2)} , \tag{18.5-80}$$

where $D_\ell = (\vartheta + \tfrac{4}{3}\eta)/\rho$. Due to the temperature gradient the two Brillouin peaks have a different magnitude; the $\delta\mathcal{S}_{PP}$ is easily obtained from (18.5-80), cf. Ref. 16.

18.6 The Langevin Equation

18.6.1 *General Procedure*

The Langevin equation was introduced in 1908[19] to provide an alternative treatment to Brownian motion, as treated previously by Einstein[20] and by Smoluchowski.[21] The new perspective was to treat the randomness of the velocity of a Brownian particle by adding a stochastic 'source term' to the usual dynamic equation; i.c., Langevin postulated

$$du(t)/dt = -\beta u + \xi(t) , \tag{18.6-1}$$

where $u(t)$ is the velocity of the particle, β is the friction coefficient as given by

$$\beta = 6\pi a\eta / m , \tag{18.6-2}$$

(Stokes' law) where a is the particle radius, m its mass, η is the viscosity of the

---

[19] P. Langevin, "Sur la théorie du mouvement Brownien", Comptes Rendus Acad. des Sciences, Paris, **146**, 530 (1908).

[20] A. Einstein, " Über die von der molekularkinetischen Theorie der Wärme geforderte Bewegung von in ruhenden Flüssigkeiten suspendierten Teilchen", Ann. der Physik **17**, 549 (1905); also, "Zur Theorie der Brownschen Bewegung", **19**, 371 (1906).

[21] M. von Smoluchowski, "Drie Vortrage über Diffusion, Brownsche Bewegung und Koagulation von Kolloidteilchen", Physik. Zeitschr. **17**, 557, 585 (1916); also, Ann. der Physik **21**, 756 (1906).

surrounding fluid and $\xi(t)$ is a purely random variable (subsection 18.2.1).

More generally, we deal with the phenomenological equations as discussed previously, Eq. (18.5-4). The regression to equilibrium (or the steady state) is a consequence of the elastic, ponderomotive forces acting on the system; also, there are random, dispersion forces working in the system, that cause the system to depart from average behaviour and exhibit fluctuations. Therefore, if we omit the averaging brackets in (18.5-4), the instantaneous behaviour is expressed by the Langevin equation, being a *stochastic differential equation*

$$\frac{d\mathbf{a}(t)}{dt} = \mathbf{A}[\mathbf{a}(t)] + \xi(t). \tag{18.6-3}$$

Linearizing we have,

$$\frac{d\Delta\mathbf{a}(t)}{dt} = -\mathsf{M}\Delta\mathbf{a}(t) + \xi(t). \tag{18.6-4}$$

There are those who hold that the Langevin source term should only be applied to the linear equations (18.6-4). Moreover, linear*ization* has its pitfalls, in that the phenomenological equation, being the average of (18.6-3), is not identical with the equation for the behaviour of the average $\langle\mathbf{a}(t)\rangle$; the literature on this is abundant.[22] However, applications in this chapter will always be concerned with the form (18.6-4) and the form (18.6-3) can be considered as symbolic if so desired.

Since averaging of (18.6-4) over an ensemble with given initial conditions must result in (18.5-8), the first requirement for the Langevin source term is that the conditional average $\langle\xi(t)\rangle_{\text{cond}} = 0$ for all $t > 0$. Moreover, this average can not depend on the value specified at t = 0, the source being totally independent of the response. Put otherwise, the source $\xi(t)$ is not correlated with any given quantity at an earlier time. Therefore, the two necessary and sufficient requirements are

$$\begin{aligned} &(i) \quad \langle\xi(t)\rangle = 0 \quad \text{(av. over any } T \text{ with } 0 < T < \infty\text{)}, \\ &(ii) \quad \langle\xi(t)\xi(t')\rangle = \Xi\,\delta(t-t'). \end{aligned} \tag{18.6-5}$$

The latter condition implies that the spectra are white,

$$\mathbb{S}_\xi = 2\int_{-\infty}^{\infty} \Xi\,\delta(t-t')e^{-i\omega(t-t')}dt' = 2\Xi. \tag{18.6-6}$$

We note the important connection with the second-order Fokker–Planck moments

$$\mathbb{S}_\xi = 2\Xi = 2\mathsf{B}(\mathbf{a}_0). \tag{18.6-7}$$

The proof proceeds as follows. We integrate (18.6-4) over a small time Δt, thus obtaining

---

[22] N.G. van Kampen, "Fluctuations in Nonlinear Systems", in "Fluctuation Phenomena in Solids" (R.E. Burgess, Ed.), Acad. Press NY 1965, pp. 139-177; see also, C.Th.J. Alkemade, N.G. van Kampen and D.K.C. MacDonald, Proc. Royal Soc. **A271**, 449 (1963).

$$\Delta\mathbf{a}(t) - \Delta\mathbf{a}(0) = \mathbf{a}(t) - \mathbf{a}(0) = -\mathsf{M}\Delta t + \int_0^{\Delta t} \xi(t)dt. \qquad (18.6\text{-}8)$$

When this divided by Δt and averaged over a conditional ensemble with $\mathbf{a}(0) = \mathbf{a}'$, we regain the phenomenological equations (18.5-8), using the definition of the first-order F–P moment, (18.5-17). Next, we multiply (18.6-8) by its transpose and we average; this yields

$$\langle(\mathbf{a}(\Delta t) - \mathbf{a}')(\tilde{\mathbf{a}}(\Delta t) - \tilde{\mathbf{a}}')\rangle_{\mathbf{a}'} = \int_0^{\Delta t}\int_0^{\Delta t} \langle\xi(t')\tilde{\xi}(t'')\rangle dt'dt'' + \mathcal{O}(\Delta t^2). \qquad (18.6\text{-}9)$$

We make the change of variables $t' - t'' = u,\ t'' = v$. We then find

$$\langle(\mathbf{a}(\Delta t) - \mathbf{a}')(\tilde{\mathbf{a}}(\Delta t) - \tilde{\mathbf{a}}')\rangle_{\mathbf{a}'} = \int_0^{\Delta t} dv \int_{-v}^{\Delta t - v} du \langle\xi(u)\tilde{\xi}(0)\rangle + \mathcal{O}(\Delta t^2). \qquad (18.6\text{-}10)$$

Here we omitted the conditional value on the source correlation function, in accord with the property (ii). However, since the rhs applies to an unrestricted ensemble, the same should be the case for the lhs; so we set $\mathbf{a}' \to \mathbf{a}_0$, where $\mathbf{a}_0 \equiv \langle\mathbf{a}\rangle$.[23] The area of integration is indicated in Fig. 18-7. Now we note that the exact area of the parallelogram is of no concern, since we only need to integrate over an infinitesimal strip about $u = 0$, because of the delta-type correlation. We thus obtain

$$\langle(\mathbf{a}(\Delta t) - \mathbf{a}_0)(\tilde{\mathbf{a}}(\Delta t) - \tilde{\mathbf{a}}_0)\rangle = \Xi\,\Delta t + \mathcal{O}(\Delta t^2). \qquad (18.6\text{-}11)$$

Dividing by Δt, letting $\Delta t \to 0$ and employing the definition of the second-order F–P moment in (18.5-18) – see also (18.4-4) – we arrive at the contemplated result

$$\mathsf{B}(\mathbf{a}_0) = \Xi. \qquad (18.6\text{-}12)$$

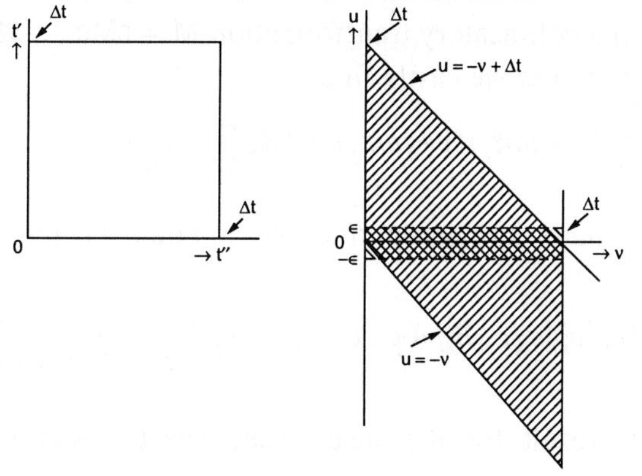

Fig. 18-7. Change of areas of integration for Eqs. (18.6-9) and (18.6-10).

[23] The inconsistency met here is removed by averaging both sides over $W_1(\mathbf{a}')$; this has no effect on the rhs, since $\Sigma_{\mathbf{a}'} W_1(\mathbf{a}') = 1$, but will change the lhs of (18.6-10) to the unrestricted average of (18.6-11). The result (18.6-12) is exact within the assumptions of the Langevin approach, as shown below.

The purpose of the Langevin method is to provide an *alternate method* to that of the master equation or the Fokker–Planck equation. In the next sections on Brownian motion, we will indeed obtain the correlation function directly from the Langevin equation. The only second moment results going into the computation are the spectral matrices (or kernels) of the source functions. In many engineering applications the spectra of the stochastic variables are most easily obtained from Fourier analysis (Rice: 'harmonic analysis') of the Langevin equation(s). Thus, writing the truncated variables $\boldsymbol{\xi}(t)$ and $\mathbf{a}(t)$ as a Fourier integral over an interval $(-T/2 \le t \le T/2)$, we have the connection

$$(\mathsf{M}+i\omega\mathsf{I})\Delta\hat{\mathbf{a}}(\omega,T)=\hat{\boldsymbol{\xi}}(\omega,T)\,, \qquad (18.6\text{-}13a)$$

where, as in Section 16.4 the transforms are denoted by a caret. We also consider the Hermitean conjugate

$$\hat{\boldsymbol{\xi}}^\dagger(\omega,T)=\Delta\hat{\mathbf{a}}^\dagger(\omega,T)(\tilde{\mathsf{M}}-i\omega\mathsf{I})\,. \qquad (18.6\text{-}13b)$$

Eq. (18.6-13a) is post-multiplied with (18.6-13b); employing (16.4-10) we obtain[24]

$$(\mathsf{M}+i\omega\mathsf{I})\mathbb{S}_{\mathbf{a}}(\omega)(\tilde{\mathsf{M}}-i\omega\mathsf{I})=\mathbb{S}_{\xi}\,. \qquad (18.6\text{-}14)$$

Pre-multiplying with $(\mathsf{M}+i\omega\mathsf{I})^{-1}$ and post-multiplying with $(\tilde{\mathsf{M}}-i\omega\mathsf{I})^{-1}$, we then find

$$\mathbb{S}_{\mathbf{a}}(\omega)=(\mathsf{M}+i\omega\mathsf{I})^{-1}\mathbb{S}_{\xi}(\tilde{\mathsf{M}}-i\omega\mathsf{I})^{-1}\,. \qquad (18.6\text{-}15)$$

Observing (18.6-7), this is just our earlier obtained general result (18.5-63); we note that we did *not* use the fluctuation-relaxation theorem.

Obviously, the result (18.5-63) as well as (18.6-15) are not in a suitable form for practical calculations. Thus, let us have new variables $\boldsymbol{\beta}=\mathsf{c}\boldsymbol{\alpha}$ and let c be a matrix that diagonalizes M in a collineatory transformation $\mathsf{M}'=\mathsf{c}\mathsf{M}\mathsf{c}^{-1}$, $M'_{ij}=(1/\tau_i)\delta_{ij}$. For the spectra of the β-variables we easily find

$$G'_{ij}(\omega)=2\,\mathrm{Re}\,B'_{ij}\tau_i\tau_j/(1+i\omega\tau_i)(1-i\omega\tau_j)=2\,\mathrm{Re}\sum_{kl}c_{ik}c_{jl}B_{kl}\tau_i\tau_j/(1+i\omega\tau_i)(1-i\omega\tau_j)\,.$$

Expanding the denominator into partial fractions and transforming back, one obtains

$$\mathbb{G}_{mn}(\omega)=2\sum_{ijkl}[c^{-1}_{mi}c^{-1}_{nj}c_{ik}c_{jl}B_{kl}+c^{-1}_{mj}c^{-1}_{ni}c_{jk}c_{il}B_{kl}]\left(\frac{\tau_i\tau_j}{\tau_i+\tau_j}\right)\left(\frac{\tau_i^2}{1+\omega^2\tau_i^2}\right). \qquad (18.6\text{-}16)$$

This rather complex result for the steady-state spectra was first established by

---

[24] Obviously, there is something amiss, since the Langevin equation has been given for $t > 0$ only. We can remedy this by using the *adjunct operator* $-(\partial/\partial t)+M$ for negative times. The main requirement for the spectrum is that for negative frequencies – not accessible by measurement – we have that $\mathbb{S}^*(-\omega)=\mathbb{S}(\omega)$, which is satisfied by (18.6-15). Alternatively, one can use truncated Fourier series on an interval $0<t\le T$. The spectral density then is given by $\mathbb{S}(\omega_n)=\lim_{T\to\infty}[T(\langle a_n^2\rangle+\langle b_n^2\rangle]$.

Ikhsanov and Uritskii.[25(a)] For thermal equilibrium conditions this can be greatly simplified. From (18.6-61) and (18.5-11) we find

$$\mathcal{G}_{mn}(\omega) = 2\sum_{ik} c_{mi}^{-1} c_{ik} B_{kn}\, \tau_i^2 / (1+\omega^2\tau_i^2)\,. \tag{18.6-17}$$

The fact that this expression is (seemingly only) not symmetrical in the indices *m,n* has been discussed elsewhere.[25(b)] These expressions are not too useful. If we have no more than two variables to contend with an expansion in partial fractions gives the required results forthwith; see Section 18.14.

18.6.2 *The Sources of Gain-Loss Processes and of G-R Noise*

Omitting the conditional averaging in (18.5-57), we have the stochastic equations

$$\frac{dn_i}{dt} = \sum_{j=1}^{s}{}' [p_{ji}(\mathbf{n}) - p_{ij}(\mathbf{n})] + \gamma_i(t)\,, \tag{18.6-18}$$

where γ_i are Langevin sources. The spectra are white and given by $\mathbb{S}_\gamma = 2\mathrm{B}(\mathbf{n}_0)$. Here $\mathbb{S}_\gamma$ is a square matrix consisting of all self- and cross-spectra. From (18.5-65),

$$\mathcal{S}_{\gamma_i,\gamma_i} = 2B_{ii}(\mathbf{n}_0) = 4\sum_{k=1}^{s} p_{ik}^0(n_1 ... n_s)\,, \tag{18.6-19}$$

$$\mathcal{S}_{\gamma_i,\gamma_j} = 2B_{ij}(\mathbf{n}_0) = -2p_{ji}^0(n_1 ... n_s) - 2p_{ij}^0(n_1 ... n_s)\,. \tag{18.6-20}$$

Whereas in the above we used lower case symbols n_i for occupations, describing *number* fluctuations (leaving capitals like *I* and *N* for available states), it should be borne in mind that as a rule lower case symbols denote densities. So, let $n_i(\mathbf{r},t)$ denote carrier densities; the corresponding sources are now $\gamma_i(\mathbf{r},t)$. Transitions leave the position **r** unchanged. Since carrier-density fluctuations are always singular for the confluence of two points, we must introduce a delta function to describe them properly. So, as an extension of (18.6-20) we have for the source-density spectra

$$\begin{aligned} \mathcal{S}_{\gamma_i,\gamma_i}(\mathbf{r},\mathbf{r}') &= 4\delta(\mathbf{r}-\mathbf{r}')\sum_{k=1}^{s} p_{ik}^0[n_0(\mathbf{r})]\,, \\ \mathcal{S}_{\gamma_i,\gamma_j}(\mathbf{r},\mathbf{r}') &= -2\delta(\mathbf{r}-\mathbf{r}')\{p_{ji}^0[n_0(\mathbf{r})] + p_{ij}^0[n_0(\mathbf{r})]\}\,. \end{aligned} \tag{18.6-21}$$

The factors behind the delta functions are expressed as $K_{ii}(\mathbf{r})$ and $K_{ij}(\mathbf{r})$. These are called source densities, having the dimension of $cm^3 s^{-1}$.

The equations (18.6-20) *suggest* that the rates p_{ij} have simple shot noise, i.e., each rate has as spectral density $2p_{ij}^0$;[26] the sources $\xi_i(t)$ that govern the rates to and fro for the level '*i*' have self-spectra with *double shot noise*, $\Sigma_j 4p_{ij}^0$, whereas the cross

---

[25] (a) R.N. Ikhsanov and Z.I. Uritskii, Fiz. Tverd. Tela **5**, 247 (1963); Sov. Phys. Solid State **5**, 247 (1963). (b) K.M. Van Vliet, Phys. Lett. **8**, 22 (1964).

[26] Shot noise for single and multiple variables will be treated in detail in subsection 18.10-1.

spectra have shot noise with a minus sign since the densities of levels i and j are negatively correlated due to the charge constraint. However, the transitions are either mediated by phonons or photons (exempting impact ionization and certain Auger processes), so that the rates must have *boson noise*, which is super-Poissonian, cf. (5.3-27) (bosons 'bunch'). This problem has been the subject of several papers.[27]

To describe the essence of the problem, let us assume that a two-level solid with populations n_2 (upper level) and n_1 (lower level) is in equilibrium with a blackbody radiation field and that both generation and recombination involve radiative transitions. Let the line shape be rectangular, of width $d\nu$. Denoting by u_ν the radiant energy density of frequency ν, we have the rate equation

$$dn_2/dt = B_{12}u_\nu d\nu n_1 - B_{21}u_\nu d\nu n_2 - A_{21}n_2 \,, \qquad (18.6\text{-}22)$$

where the terms on the rhs denote generation due to absorption, recombination associated with stimulated emission and recombination associated with spontaneous emission, respectively. The Einstein B's are related by the boson factor (5.3-28)[28]

$$B_{21}/B_{12} = (n_1/n_2)[B/(1+B)], \quad B = 1/(e^{h\nu/k_BT} - 1)\,. \qquad (18.6\text{-}23)$$

Denoting the Langevin sources by $\xi(B_{12})$, $\xi(B_{21})$ and $\xi(A_{21})$ in an obvious notation, we have

$$\gamma(t) = \xi(B_{12}) - \xi(B_{21}) - \xi(A_{21}) \equiv \xi_L - \xi_{sp}\,. \qquad (18.6\text{-}24)$$

The spectral density of the total source $\gamma(t)$ will be

$$\mathbb{S}_\gamma = 2\{G_0(B_{12}) + R_0(B_{21}) + R_0(A_{21})\}(1+B) - 2\mathbb{S}[\xi(B_{12}),\xi(B_{21})]\,, \quad (18.6\text{-}25)$$

the last term representing the cross-correlation spectrum of the generation rate and the stimulated emission rate; ξ_L and ξ_{sp} are clearly independent. Since the total rates balance, the first three contributions yield the spectral density $4G_0(1+B)$. But from (18.6-23) we also find

$$R_0(B_{21})/G_0(B_{12}) = R_0(B_{21})/[R_0(B_{21}) + R_0(A_{21})] = B/(1+B)\,. \qquad (18.6\text{-}26)$$

This ratio gives the fraction of the recombination rate that is correlated to the generation rate. The correlation therefore contributes to the Langevin-source spectrum $-4G_0(1+B)[B/(1+B)] = -4G_0B$. The final spectrum thus shows 'double shot noise', $\mathbb{S}_\gamma = 4G_0$.

So, the boson statistics of the photon gas can only be reconciled with the fermion statistics of the electron gas because of the occurrence of stimulated emission. Many more quite cumbersome details are found in Ref. 27.

---

[27] K.M. Van Vliet and R.J.J. Zijlstra, Physica **89A**, 353 (1977); ibid. and N.G. van Kampen in "Proc. 5th Intern. Conf. on Noise in Phys. Systems", D. Wolf, Ed., Springer, Berlin 1977, p. 309.

[28] A. Einstein, Physik. Zeitschr. **18**, 121 (1917).

2. Brownian Motion Proper. Velocity Fluctuations and Diffusion

18.7 Diffusion and Random Walk

18.7.1 *Einstein's Result*

Random walk has already been considered from a simple mathematical point of view in subsection 1.6.1. Now we want to connect this with diffusion in Brownian motion, i.e., we consider the classical problem of the random displacements of a colloidally suspended particle in a viscous fluid. While this problem is exhaustively studied in the cited paper by Chandrasekhar, cf. Ref. 1(c), we shall at first pursue a simple approach. (see also problem 18.10). The assumptions to be made are:

(i) When the particle is at a certain location, it has no memory of the past, i.e., the next move is independent of the previous ones and depends only on the last position;
(ii) The probability for a displacement Δ in a given direction is equal to that in the opposite direction;
(iii) The *a priori* probability for the particle to be after the i-th displacement in an interval $(\mathbf{r}, \mathbf{r}+d\mathbf{r})$ is spherically symmetrical, i.e., depends only on $|\mathbf{r}-\mathbf{r}_i|$.

We seek to know the conditional probability $P(\mathbf{r},t\,|\,\mathbf{r}_0)$. Because of (ii) the first-order Fokker–Planck moments are zero; further, to satisfy (iii) we write for the second-order Fokker–Planck moments $\mathsf{B}=2D_0\mathsf{I}$, where I is the unit tensor (or matrix). Hence, $B_{11}=B_{22}=B_{33}=2D_0$, $B_{ij}=0, i\neq j$. The Fokker–Planck equation (18.4-7) now becomes

$$\frac{\partial P(\mathbf{r},t\,|\,\mathbf{r}_0)}{\partial t}=D_0\nabla^2 P(\mathbf{r},t\,|\,\mathbf{r}_0)\,, \tag{18.7-1}$$

which is the ordinary diffusion equation. To solve it, we take the Fourier transform, i.e. we represent P by its characteristic function

$$\Phi(\mathbf{k},t)=\int d^3r e^{i\mathbf{k}\cdot\mathbf{r}}P(\mathbf{r},t\,|\,\mathbf{r}_0)\,. \tag{18.7-2}$$

The transformed diffusion equation is

$$d\Phi/dt=-D_0k^2\Phi(\mathbf{k},t)\,. \tag{18.7-3}$$

The initial condition is [see (18.2-17)]

$$\Phi(\mathbf{k},0)=\int d^3r\, e^{i\mathbf{k}\cdot\mathbf{r}}\delta(\mathbf{r}-\mathbf{r}_0)=e^{i\mathbf{k}\cdot\mathbf{r}_0}\,. \tag{18.7-4}$$

The solution of (18.7-3), observing (18.7-4), is

$$\Phi(\mathbf{k},t)=e^{i\mathbf{k}\cdot\mathbf{r}_0-D_0k^2t}=\prod_{xyz}e^{ikx_0-D_0k_x^2t}\,. \tag{18.7-5}$$

This is the characteristic function of a normal distribution with the centre at $\mathbf{r}_0$; the inverse we found previously, cf. Eqs. (1.7-9) and (1.7-12). So, by that procedure,

$$P(\mathbf{r},t\,|\,\mathbf{r}_0)=\left(\frac{1}{2\sqrt{\pi D_0 t}}\right)^3 e^{-|\mathbf{r}-\mathbf{r}_0|^2/4D_0 t}\,. \tag{18.7-6}$$

For the mean square displacement we obtain

$$\begin{aligned}\langle(x-x_0)^2\rangle &= \left(\frac{1}{2\sqrt{\pi D_0 t}}\right)^3 \int (x-x_0)^2 e^{-|\mathbf{r}-\mathbf{r}_0|^2/4D_0 t} d^3 r \\ &= \frac{1}{2\sqrt{\pi D_0 t}} \int (x-x_0)^2 e^{-(x-x_0)^2/4D_0 t} dx = 2D_0 t\,,\end{aligned} \tag{18.7-7}$$

which is Einstein's result[28]. Because of assumption (iii), this holds for the displacement in any direction. However, from molecular dynamics we see at once that this is only an asymptotic solution, valid for large t. For small t the mean square displacement must become proportional to t^2. So, what went wrong? Basically, as shown by Uhlenbeck and Ornstein, the problem of random walk is *not* Markovian: the displacement of a Brownian particle is the 'projection' of a bivariate Markovian process, the variables being $\mathbf{r}$ and $\mathbf{v}$.

18.7.2 *Langevin Approach of Uhlenbeck and Ornstein*

At first they consider one-dimensional motion with random velocity $u(t)$, satisfying the Langevin equation (18.6-1). The formal solution is

$$u(t)=u_0 e^{-\beta t}+e^{-\beta t}\int_0^t e^{\beta t'}\xi(t')dt'. \tag{18.7-8}$$

This is squared and averaged over a conditional ensemble with velocity u_0 at $t = 0$. This yields

$$\langle[u(t)]^2\rangle_{u_0}=u_0^2 e^{-2\beta t}+e^{-2\beta t}\int_0^t\int_0^t e^{\beta(t'+t'')}\langle\xi(t')\xi(t'')\rangle dt'dt''. \tag{18.7-9}$$

We make the change of variables $t'-t''=w,\ t''=v$. We then have

$$\langle[u(t)]^2\rangle_{u_0}=u_0^2 e^{-2\beta t}+e^{-2\beta t}\int_0^t dv\int_{-v}^{t-v} dw e^{\beta(w+2v)}\langle\xi(w)\xi(0)\rangle\,; \tag{18.7-10}$$

note that the change of variables is similar as for Eq. (18.6-9), treated previously with the area of integration being as in Fig. 18-7. Again, the integration over w involves an infinitesimal strip near $w = 0$, so that the limits on w can be taken to be $(-\infty,\infty)$. With the $\delta(w)$ correlation for the sources, we obtain

$$\langle[u(t)]^2\rangle_{u_0}=u_0^2 e^{-2\beta t}+\Xi(1-e^{-2\beta t})/2\beta\,. \tag{18.7-11}$$

This result shows the approach to equilibrium. For $t \to \infty$ we must have the Maxwell distribution, with the lhs being $k_B T/m$. We thus obtain for the source strength:

$$\Xi = 2\beta k_B T/m. \tag{18.7-12}$$

Next, Uhlenbeck and Ornstein proceed to obtain the higher order moments by the same method, see their Appendix, Note I. For these moments they find that they are related to the variance $\langle [\Delta u(t)]^2 \rangle_{u_0}$ by the relationship (1.7-14), thus confirming that the distribution is Gaussian (or normal), with a mean value $\langle u(t) \rangle_{u_0} = u_0 e^{-\beta t}$. The variance thus becomes,

$$\sigma^2 = \langle [u(t)]^2 \rangle_{u_0} - [\langle u(t) \rangle_{u_0}]^2 = \frac{k_B T}{m}\left(1 - e^{-2\beta t}\right). \tag{18.7-13}$$

For the distribution we now find

$$P_u(u,t\,|\,u_0) = \left[\frac{m}{2\pi k_B T(1-e^{-2\beta t})}\right]^{1/2} \exp\left[-\frac{m}{2k_B T}\frac{(u-u_0 e^{-\beta t})^2}{1-e^{-2\beta t}}\right]. \tag{18.7-14}$$

This rather complex result is, however, not the full story, since we are ultimately interested in the displacement $x(t)$. The Langevin equation becomes second order,

$$\frac{d^2 x}{dt^2} = -\beta \frac{dx}{dt} + \xi(t). \tag{18.7-15}$$

This equation is as such not useful, since the molecular dynamics dictate that only the bivariate process $\{u(t), x(t)\}$ can be Markovian. Thus, more appropriately, we have two coupled Langevin equations

$$\begin{aligned} du/dt &= -\beta u + \xi(t), \\ dx/dt &= u(t) \quad . \end{aligned} \tag{18.7-16}$$

While these equations can be solved by a similar procedure as carried out above, it is more expedient to solve the Fokker–Planck equation, discussed below.

18.7.3 *Fokker–Planck Solution for the Bivariate Process* $\{\mathbf{v}(t), \mathbf{r}(t)\}$

The first-order F–P moments can be read from (18.7-16) and the only non-zero second-order F–P moment follows from (18.6-12) and (18.7-12); hence, we have

$$\begin{aligned} A_1 &= -\beta u, \qquad A_2 = u \\ B_{11} &= 2\beta k_B T/m, \quad B_{12} = B_{21} = B_{22} = 0. \end{aligned} \tag{18.7-17}$$

The Fokker–Planck equation for $P(x,u,t\,|\,x_0,u_0)$ thus reads

$$\frac{\partial P}{\partial t} = \beta \frac{\partial}{\partial u}(uP) - \frac{\partial}{\partial x}(uP) + \frac{\beta k_B T}{m}\frac{\partial^2 P}{\partial u^2}. \tag{18.7-18}$$

First we should verify that this yields for the Markov probability for u alone, denoted by P_u, the solution (18.7-14), as obtained from the Langevin-equation method. Thus, omitting the term $-\partial(uP_u)/\partial x$ and making the substitutions $\tau = \beta t$ and $w = (m/k_BT)^{1/2}u$, we have the partial differential equation

$$\frac{\partial P_u}{\partial \tau} = P_u + w\frac{\partial P_u}{\partial w} + \frac{\partial^2 P_u}{\partial w^2}. \tag{18.7-19}$$

Solving this by separation of variables, with initial condition $P(u,0\,|\,u_0) = \delta(u-u_0)$ and boundary condition $\lim_{u\to\pm\infty} P_u = 0$, we find

$$P(w,\tau) = \sum_{n=0}^{\infty} A_n e^{-n\tau} D_n(w) e^{-w^2/4}, \tag{18.7-20}$$

where $D_n(w)$ is the Weber function of n-th order.[29] With the initial condition, the coefficients A_n are given by

$$A_n = \frac{1}{n!\sqrt{2\pi}}\int_{-\infty}^{\infty} D_n(\eta) f(\eta) e^{-\eta^2/4} d\eta, \quad f(\eta) \to \delta(\eta - w_0), \tag{18.7-21}$$

giving for the solution

$$P(w,\tau) = \frac{1}{(2\pi)^{1/2}}\int_{-\infty}^{\infty} d\eta f(\eta) e^{(-\eta^2-w^2)/4} \sum_{n=0}^{\infty} \frac{D_n(w)D_n(\eta)}{n!} e^{-n\tau}. \tag{18.7-22}$$

Uhlenbeck and Ornstein (their Note II) now proceed with the evaluation of the sum

$$M(\tau) = \sum_n D_n(w) D_n(\eta) e^{-n\tau}/n! \tag{18.7-23}$$

for which they use the recurrent relation of the Weber functions

$$D_{n+1}(z) = zD_n(z) - nD_{n-1}(z). \tag{18.7-24}$$

With a few steps this yields the differential equation

$$(1-e^{-2\tau})^2 \frac{dM}{d\tau} = M\left\{-w\eta e^{-\tau} + (w^2+\eta^2-1)e^{-2\tau} - w\eta e^{-3\tau} + e^{-4\tau}\right\}. \tag{18.7-25}$$

This first order ODE is readily integrated to yield, using the stated bc,

$$M = \frac{e^{(w^2+\eta^2)/4}}{(1-e^{-2\tau})^{1/2}} \exp\left\{-\frac{w^2+\eta^2-2w\eta e^{-\tau}}{2(1-e^{-2\tau})}\right\}. \tag{18.7-26}$$

This is substituted into (18.7-22), the integration over $d\eta$ is carried out in virtue of the delta function, to obtain

$$P(w,\tau;w_0) = \frac{1}{[2\pi(1-e^{-2\tau})]^{1/2}} \exp\left\{-\frac{(w-w_0e^{-\tau})^2}{2(1-e^{-2\tau})}\right\}. \tag{18.7-27}$$

---

[29] Cf., R.T. Whittaker and G.N. Watson, "A Course of Modern Analysis", 4th Ed., Cambridge Univ. Press, 1958, p. 347.

When this result is rewritten into the original variables, it just yields the Langevin solution (18.7-14). Although various researchers[30] have expressed reservations with respect to the Langevin method – this author included – one cannot deny that it provides rapid answers for many problems.[31]

We return to the more general F–P equation (18.7-18), but will consider the three-dimensional case and add a force term $K \equiv F/m$ to the first equation of (18.7-16); we thus deal with the 3D F–P equation $[P = P(\mathbf{r},\mathbf{v},t\,|\,\mathbf{r}_0,\mathbf{v}_0)]$

$$\frac{\partial P}{\partial t} + \nabla_{\mathbf{r}}\cdot(\mathbf{v}P) + \boldsymbol{K}\cdot\nabla_{\mathbf{v}}P = \beta\nabla_{\mathbf{v}}\cdot(\mathbf{v}P) + \frac{\beta k_B T}{m}\nabla_{\mathbf{v}}^2 P. \qquad (18.7\text{-}28)$$

The general solution is obtained by solving for the characteristic integrals of the first-order partial differential equation that results when the Laplacian term is omitted; we refer to Chandrasekhar, Ref. 1(c), for details [see his result (286)]. In the absence of an external field – free Brownian particle – the solution for the displacements alone, obtained by integrating over all velocities and expressed as $P_r(\mathbf{r},t\,|\,\mathbf{r}_0,\mathbf{v}_0)$ is found to be – in accord with another result by Uhlenbeck and Ornstein,

$$P(\mathbf{r},t\,|\,\mathbf{r}_0,\mathbf{v}_0) = \left\{\frac{m\beta^2}{2\pi k_B T[2\beta t - 3 + 4e^{-\beta t} - e^{-2\beta t}]}\right\}^{3/2} \exp\left\{-\frac{m\beta^2\,|\mathbf{r}-\mathbf{r}_0-\mathbf{v}_0(1-e^{-\beta t})/\beta|^2}{2k_B T[2\beta t - 3 + 4e^{-\beta t} - e^{-2\beta t}]}\right\}. \qquad (18.7\text{-}29)$$

This result is quite different from the normal distribution (18.7-6) associated with diffusion! To show that the latter is the asymptotic form of (18.7-29), we follow a device due to Chandrasekhar. Restoring the external force, we write (18.7-28) as follows – verified by direct algebra –

$$\frac{\partial P}{\partial t} = \beta\left[\nabla_{\mathbf{v}} - \frac{1}{\beta}\nabla_{\mathbf{r}}\right]\cdot\left[\mathbf{v}P + \frac{k_B T}{m}\nabla_{\mathbf{v}}P - \frac{\boldsymbol{K}}{\beta}P + \frac{k_B T}{m\beta}\nabla_{\mathbf{r}}P\right] + \nabla_{\mathbf{r}}\cdot\left[\frac{k_B T}{m\beta}\nabla_{\mathbf{r}}P - \frac{\boldsymbol{K}}{\beta}P\right]. \qquad (18.7\text{-}30)$$

This is now integrated over the line $\mathbf{r} + \mathbf{v}/\beta = \mathbf{r}' = const.$ (a straight line in the 6D μ-space) with $\mathbf{v}$ going from $-\infty$ to ∞. We note that for any $f(\mathbf{r},\mathbf{v})$ we have

$$\begin{aligned}\int \beta\left[\nabla_{\mathbf{v}} - \beta^{-1}\nabla_{\mathbf{r}}\right] f(\mathbf{r},\mathbf{v})\cdot d\ell &= \int\left[\nabla_{\mathbf{v}} - \beta^{-1}\nabla_{\mathbf{r}}\right] f[\mathbf{r}(\mathbf{v}),\mathbf{v}]\cdot d\mathbf{v} \\ &= \int\left[\nabla_{\mathbf{v}} f\cdot d\mathbf{v} - \frac{1}{\beta}(\nabla_{\mathbf{r}} f)\cdot\frac{d\mathbf{v}}{d\mathbf{r}}d\mathbf{r}\right] = \int\left[\frac{\partial f}{\partial \mathbf{v}}\cdot d\mathbf{v} - \frac{\partial f}{\partial \mathbf{r}}\cdot d\mathbf{r}\right] \\ &= \int_\ell df = f[\mathbf{r}(\mathbf{v}),\mathbf{v}]\Big|_{-\infty}^{\infty} = 0. \end{aligned} \qquad (18.7\text{-}31)$$

Thus, applying this to (18.7-30), we obtain

---

[30] F. Zernike in "Handbuch der Physik" (H. Geiger and K. Scheel, Eds.), First Ed., Springer 1928, p.483.

[31] The real simplification is of course in that U–O knew beforehand that the process is Gaussian.

$$\frac{\partial}{\partial t}\int P\left(\mathbf{r}'-\beta^{-1}\mathbf{v},\mathbf{v},t\,|\,\mathbf{r}_0,\mathbf{v}_0\right)d\mathbf{v}$$
$$=\frac{k_BT}{m\beta}\nabla_{\mathbf{r}}\cdot\int\left\{\nabla_{\mathbf{r}}P\left(\mathbf{r}'-\beta^{-1}\mathbf{v},\mathbf{v},t\,|\,\mathbf{r}_0,\mathbf{v}_0\right)-\frac{m\boldsymbol{K}}{k_BT}P\left(\mathbf{r}'-\beta^{-1}\mathbf{v},\mathbf{v},t\,|\,\mathbf{r}_0,\mathbf{v}_0\right)\right\}d\mathbf{v}. \tag{18.7-32}$$

Considering displacements $\gg|\mathbf{v}|/\beta$, we arrive at the differential equation

$$\frac{\partial}{\partial t}P(\mathbf{r}',t\,|\,\mathbf{r}_0,\mathbf{v}_0)=\frac{k_BT}{m\beta}\nabla_{\mathbf{r}}\cdot\left\{\nabla_{\mathbf{r}}P(\mathbf{r}',t\,|\,\mathbf{r}_0,\mathbf{v}_0)+\frac{m}{k_BT}(\nabla_{\mathbf{r}}\mathcal{V})P(\mathbf{r}',t\,|\,\mathbf{r}_0,\mathbf{v}_0)\right\}, \tag{18.7-33}$$

where we set $\boldsymbol{K}=-\nabla_{\mathbf{r}}\mathcal{V}/m$. This is the *Smoluchowski–Kramers* equation. In the absence of an external potential we recover the diffusion equation with

$$D_0=k_BT/m\beta, \tag{18.7-34}$$

which is the Einstein relation. [If the particles were charged, the Einstein relation would read $\mathcal{q}D_0=k_BT\mathcal{q}\tau/m=k_BT\mu$, μ being the mobility.]

Finally, some spectral notions follow from a consideration of the average of $v(t)$ over a small but macroscopic time θ. We then have the new variable ('short-time average')

$$v_\theta(t)\equiv\frac{1}{\theta}\int_t^{t+\theta}v(t')\,dt'=\Delta x_\theta(t)/\theta. \tag{18.7-35}$$

With Einstein's result $\langle[\Delta x_\theta(t)]^2\rangle=2D_0\theta$, we have $\langle[v_\theta(t)]^2\rangle=2D_0/\theta$. To obtain the velocity-fluctuation spectrum for frequencies small compared to $\beta=\tau^{-1}$, we employ Milatz' theorem which connects the spectral density with the variance of the short-time average, cf. Section 18.10. We then find

$$\mathcal{S}_v(\omega\to 0)=2\theta\langle[v_\theta(t)]^2\rangle=4D_0. \tag{18.7-36}$$

Sometimes this relationship is taken as the definition of the dc diffusion constant.

All relations and results of this section pertain to classical, coarse-grained stochastic variables. However, the *microscopic* velocity is a quantum mechanical operator that deserves a more correct treatment. So, in the Section 18.8 we shall be concerned with velocity fluctuations in solids, rather than in colloidal suspensions and more profound relationships will be obtained.

18.7.4 *Harmonically Bound Brownian Particle*

A full treatment of the Brownian harmonic oscillator again requires the solution of a bivariate F–P equation for $P(y,u,t\,|\,y_0,u_0)$. However, we shall only consider the Smoluchowski–Kramers equation for $P_y(y,t\,|\,y_0,u_0)$; needless to say that as such this is not a Markov probability, since the full problem involves both y and u. The stochastic differential equation for this process is

$$d^2y/dt^2 + \beta dy/dt + \omega^2 y = \xi(t). \tag{18.7-37}$$

For its direct solution we refer to Ming Chen Wang and Uhlenbeck, Ref 1(b). At present, we consider (18.7-33), rewritten for the appropriate variables. With the potential energy being $\mathcal{V}(y) = \frac{1}{2}m\omega^2 y^2$, the Smoluchowski–Kramers equation reads

$$\frac{\partial P_y(y,t\,|\,y_0,u_0)}{\partial t} = \frac{\omega^2}{\beta}\frac{\partial(yP_y)}{\partial y} + D_0\frac{\partial^2 P_y}{\partial y^2}. \tag{18.7-38}$$

Writing $w = (\omega^2/D_0\beta)^{1/2}y$, $\tau = (\omega^2/\beta)t$, we re-obtain equation (18.7-19); the problem is entirely analogous to the solution for P_u that we have already pursued. The result is therefore given by

$$P_y(y,t\,|\,y_0,u_0) = \left(\frac{\omega^2}{2\pi\beta D_0[1-e^{-2(\omega^2/\beta)t}]}\right)\exp\left\{-\frac{\omega^2}{2\beta D_0}\frac{[y-y_0e^{-(\omega^2/\beta)t}]}{[1-e^{-(\omega^2/\beta)t}]}\right\}. \tag{18.7-39}$$

The variance is clearly

$$\langle[\Delta y(t)]^2_{y_0}\rangle = \langle[y(t)]^2_{y_0}\rangle - \langle[y(t)]_{y_0}\rangle^2 = \frac{k_BT}{m\omega^2}\left(1-e^{-2(\omega^2/\beta)t}\right). \tag{18.7-40}$$

For $t \to \infty$ we find the classical average energy $m\omega^2\langle\Delta y^2\rangle = k_BT$. However, to obtain specific results for the periodic, critically damped and overdamped case, the full bivariate problem should be considered; we refer to the cited literature.

18.8 Velocity Fluctuations and Diffusion in Condensed Matter

We seek to find a microscopic expression for the diffusivity $D(i\omega)$ which for low frequencies, or for that matter, dc conditions, agrees with (18.7-36), while the Einstein relationship – amended to take care of degeneracy – will be valid for all frequencies. The problem was already considered by Kubo (Chapter XVI, Ref. 1,§ 9) and later by Van Vliet and van der Ziel.[32] At issue here is the placement of the quantum correction factor

$$p(\omega) = \frac{\hbar\omega/k_BT}{e^{\hbar\omega/k_BT}-1} + \frac{1}{2}\frac{\hbar\omega}{k_BT} = \frac{\hbar\omega/2k_BT}{\tanh(\hbar\omega/2k_BT)}. \tag{18.8-1}$$

At high ('quantum') frequencies, this factor is dominated by the zero-point energy, cf. Fig. 18-8, taken from Ref. 32(b); we shall, however, follow the later treatment of Ref. 32(a) and 32(c), which preserves the validity of the Einstein relation for quantum frequencies.

---

[32] (a) K.M. Van Vliet and A. van der Ziel, Physica **99A**, 337 (1979); (b) ibid. Solid State Electr. **20**, 931 (1977); (c) C.M. Van Vliet, Physica **133A**, 35 (1985).

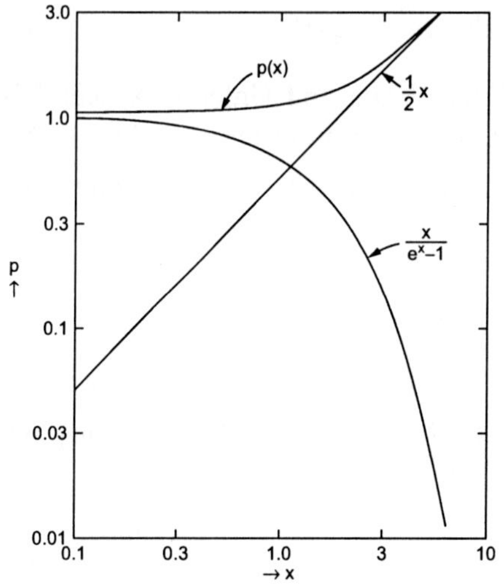

Fig. 18-8. The quantum correction factor $p(\omega)$ as a function of normalized frequency $x = \hbar\omega/k_BT$.

It is expedient to define the complex diffusivity $D(i\omega)$ as the Fourier–Laplace transform of the Kubo form for the velocity autocorrelation function,

$$D(i\omega) = \beta^{-1}\int_0^\infty dt\, e^{-i\omega t}\int_0^\beta d\tau\langle\Delta v_x(-i\hbar\tau)\Delta v_x(t)\rangle\,. \tag{18.8-2}$$

For classical frequencies with $\omega \ll k_BT/\hbar = 1/\hbar\beta$ we have $t \sim \omega^{-1} \gg \hbar\beta > \hbar\tau$, the correlation function reduces to

$$\langle\Delta v_x(-i\hbar\tau)\Delta v_x(t)\rangle = \langle\Delta v_x(0)\Delta v_x(t+i\hbar\tau)\rangle \to \langle\Delta v_x(0)\Delta v_x(t)\rangle\,, \tag{18.8-3}$$

so that

$$D(i\omega_{\rm class}) = \int_0^\infty dt e^{-i\omega t}\langle\Delta v_x(0)\Delta v_x(t)\rangle\,. \tag{18.8-4}$$

For $\omega \to 0$ this gives *a fortiori*

$$D_0 = \frac{1}{2}\int_{-\infty}^{\infty}\langle\Delta v_x(t)\Delta v_x(0)\rangle dt\,. \tag{18.8-5}$$

Note that for classical frequencies the two variables can be switched with impunity. To demonstrate the equivalence with D_0 as given in (18.7-36), we note that

$$\langle\Delta x^2\rangle_t = \int_0^t\int_0^t du\, dw\langle\Delta v_x(u)\Delta v_x(w)\rangle = \int_0^t dw\int_{-w}^{t-w} d\eta\langle\Delta v_x(\eta)\Delta v_x(0)\rangle$$

$$= \int_0^t dw\int_{-\infty}^{\infty} d\eta\langle\Delta v_x(\eta)\Delta v_x(0)\rangle = 2D_0 t\,, \tag{18.8-6}$$

(where we set $u - w = \eta$) in accord with Einstein's result.

Kramers–Kronig relations

The response function involving velocity fluctuations has the Kubo form

$$\phi_v(t) = \beta^{-1}\int_0^\beta d\tau\langle\Delta v_x(-i\hbar\tau)\Delta v_x(t)\rangle. \tag{18.8-7}$$

The diffusivity is the Fourier–Laplace transform of $\phi_v(t)$ so, in accord with (18.8-2),

$$D(i\omega) = \int_0^\infty dt\, e^{-i\omega t}\phi_v(t). \tag{18.8-8}$$

We also introduce the Fourier transform

$$F(i\omega) = \int_{-\infty}^\infty dt\, e^{-i\omega t}\phi_v(t). \tag{18.8-9}$$

It is easily shown that $F(i\omega)$ is real. In fact we find with (16.3-39) and (16.3-40)

$$F(i\omega) = 2\frac{\tanh(\beta\hbar\omega/2)}{\beta\hbar\omega/2}\int_0^\infty dt\cos\omega t\,\Phi_v(t), \tag{18.8-10}$$

where $\Phi_v(t)$ is the symmetrized velocity correlation function. We thus set $F(i\omega) = \hat{F}(\omega)$. Substituting the inverse of (18.8-9) into (18.8-8), we have

$$D(i\omega) = \frac{1}{2\pi}\int_0^\infty dt\int_{-\infty}^\infty d\omega'\hat{F}(\omega')e^{-i(\omega-\omega')}$$
$$= \int_{-\infty}^\infty d\omega'\hat{F}(\omega')\delta_+(\omega-\omega') = \tfrac{1}{2}\hat{F}(\omega) + \tfrac{1}{2\pi}i\mathcal{P}\int_{-\infty}^\infty d\omega'\hat{F}(\omega')/(\omega'-\omega). \tag{18.8-11}$$

The two terms are the real part $D'(\omega)$ and the imaginary part $D''(\omega)$, respectively. Clearly, they are each others Hilbert transforms; thus, our diffusivity as defined in (18.8-2), satisfies the Kramers–Kronig relations.

Also, from (18.8-11) we see that $\hat{F}(\omega) = 2D'(\omega)$. So, with the spectral density of the velocity fluctuations being given by the integral on the rhs of (18.8-10), we have

$$\mathcal{S}_v(\omega) = 4\int_0^\infty dt\cos\omega t\,\Phi_v(t) = 4D'(\omega)\,p(\omega). \tag{18.8-12}$$

Einstein relation

Next, we indicate that a generalized Einstein relation applies. From (16.3-22) we have for the conductivity the Kubo form

$$\sigma(i\omega) = V_0\int_0^\infty dt\, e^{-i\omega t}\int_0^\beta d\tau\langle\Delta J_e(-i\hbar\tau)\Delta J_e(t)\rangle, \tag{18.8-13}$$

where J_e is the electrical current density, $J_e = q\Sigma_k^N v_k/V_0$. If the carriers are independent, as in a Boltzmann gas, the velocity correlation functions and the noise spectra add, so that

$$\sigma(i\omega) \doteq q^2(\langle N\rangle/V_0)\int_0^\infty dt\, e^{-i\omega t}\int_0^\beta d\tau\langle\Delta v_x(-i\hbar\tau)\Delta v_x(t)\rangle\,. \tag{18.8-14}$$

For a degenerate electron gas this must be corrected by the factor $\langle \Delta N^2 \rangle / \langle N \rangle$. With $\sigma = q\mu n$, we find for the mobility

$$\mu(i\omega) = q[\langle \Delta N^2 \rangle / \langle N \rangle]^{-1} \int_0^\infty dt\, e^{-i\omega t} \int_0^\beta d\tau \langle \Delta v_x(-i\hbar\tau) \Delta v_x(t) \rangle . \quad (18.8\text{-}15)$$

Comparing with (18.8-2), we have the connection

$$qD(i\omega) = k_B T \mu(i\omega) [\langle \Delta N^2 \rangle / \langle N \rangle]^{-1} . \quad (18.8\text{-}16)$$

For the relative variance in a grand-canonical ensemble we have (4.2-23). Hence, we arrived at the Einstein relation

$$qD(i\omega) = \mu(i\omega) \left(\frac{\partial \varepsilon_F}{\partial \ln n} \right)_T . \quad (18.8\text{-}17)$$

We note that this is entirely the same as the classical result (15.2-18)! However, in the present formulas of departure the velocity fluctuations $\Delta v_x(-i\hbar\tau)$ and $\Delta v_x(t)$ are quantum mechanical, non-commuting operators.

The diffusion source

Finally, we consider the Langevin source associated with diffusion current,[32(c)] employing a simple corpuscular argument. If the carriers are electrons, we have

$$\mathbf{J}(\mathbf{r},t) = -en\mathsf{\mu} \cdot \nabla \Phi + e\mathsf{D} \cdot \nabla n + \boldsymbol{\eta}(\mathbf{r},t) . \quad (18.8\text{-}18)$$

To obtain the spectrum of $\boldsymbol{\eta}(\mathbf{r},t)$ we consider the fluctuations in a small volume δV, with $\delta \mathbf{J} = -\Sigma_k^N e \Delta \mathbf{v}_k / \delta V$. With the number of carriers being $N = n(\mathbf{r}) \delta V$, and taking into account the correlation factor $\langle \Delta N^2 \rangle / \langle N \rangle = k_B T \,\partial \ln \langle n(\mathbf{r}) \rangle / \partial \varepsilon_F$, we find for $\delta V \to 0$:

$$\mathbb{S}_\eta(\mathbf{r},\mathbf{r}',i\omega) = e^2 \langle n(\mathbf{r}) \rangle \frac{\partial \ln \langle n(\mathbf{r}) \rangle}{\partial (\beta \varepsilon_F)} \delta(\mathbf{r}-\mathbf{r}') \mathbb{S}_\mathbf{v}(\mathbf{r},i\omega) . \quad (18.8\text{-}19)$$

As an extension of (18.8-12) we now find, using the transposition property $\Phi_\mathbf{v}(-t) = \tilde{\Phi}_\mathbf{v}(t)$,

$$\mathbb{S}_\mathbf{v}(\mathbf{r},i\omega) = 2\int_{-\infty}^\infty \Phi_\mathbf{v}(t) e^{-i\omega t} dt = 2[\mathsf{D}(\mathbf{r},i\omega) + \mathsf{D}^\dagger(\mathbf{r},i\omega)]\, p(\omega) . \quad (18.8\text{-}20)$$

Defining symmetric and antisymmetric tensors as in (16.3-48) we have for $\mathbb{S}_\eta$:

$$\mathbb{S}_\eta(\mathbf{r},\mathbf{r}',i\omega) = 4e^2 \langle n(\mathbf{r}) \rangle \frac{\partial \ln \langle n(\mathbf{r}) \rangle}{\partial (\beta \varepsilon_F)} \delta(\mathbf{r}-\mathbf{r}') \{ [\mathsf{D}'(\mathbf{r},\omega)]^s + i[\mathsf{D}''(\mathbf{r},\omega)]^a \} p(\omega) . \quad (18.8\text{-}21)$$

Employing the Einstein relation (18.8-17), we also have the Nyquist form

$$\mathbb{S}_\eta(\mathbf{r},\mathbf{r}',i\omega) = 4k_B T e \langle n(\mathbf{r}) \rangle \delta(\mathbf{r}-\mathbf{r}') \{ [\mathsf{\mu}'(\mathbf{r},\omega)]^s + i[\mathsf{\mu}''(\mathbf{r},\omega)]^a \} p(\omega) . \quad (18.8\text{-}22)$$

Since $\sigma = en\mu$, this is in full accord with (16.3-53); for $\omega\tau_{\text{coll}} \ll 1$ $\mathbb{S}_\eta$ is white.

3. SPECTRAL ANALYSIS

18.9 Overview. Wiener–Khintchine Theorem

In this part of Chapter XVIII we shall consider a number of methods and associated theorems, designed for obtaining the spectral density of stochastic variables. While the Wiener–Khintchine theorem still occupies a primary place, the theorem is not applicable in many instances when the spectrum exists, but fails to converge for either low or high frequencies. This is most evident when the spectrum is rigorously white, as for a purely random process. Furthermore, there are various forms of 'pathological noise', the most well-known phenomenon being the ubiquitous $1/f$ or $1/\omega$ noise, being 'scale-invariant' and divergent at either end.

For noise that is white or 'mostly white', we shall introduce MacDonald s' and Milatz' theorems (1948, 1941), extended by us to many-variable processes and set forth in Section 18.10. Instead of the correlation function, the variance of the 'short-time average' must be known. These theorems have been proven to be indispensable for shot noise-like phenomena and for wave-interaction noise, to be described in Chapter XX. Also, there is the 'method of elementary events', applicable when a stochastic variable is a sum of random microscopic occurrences, for which the wave-form is known. The method is described in detail in Rice's papers, already cited [Ref. 1(e)]. The basic theorem, due to Carson (1931), goes back to Campbell's 1909 paper and – when dealing with wave-trains – even to the Rayleigh–Schuster analysis of 1894. We will present the basic ideas in Section 18.11.

In contrast to these quite old and established methods, there is the more recent 'Allan-variance theorem' (1966), in which a random variable is sampled in adjacent time-intervals. The inversion of this theorem was given by us in 1982, using the theory of Mellin transforms and Fredholm integral equations. The theorem is extremely useful for counting statistics and for 'highly pathological noise', with strongly divergent spectra. It will be discussed in Section 18.12.

This does not by any means exhaust the applicable analyses for a description of random processes! Other methods based on the compounding theorem, the variance theorem and other methods, will be described and applied in Chapter XIX.

Altogether, we have made a choice of analyses that have been powerful and applicable to a great many physical random processes, that might be helpful for the study of stochastic phenomena. Although tempting, we have refrained from delving much into 'quite mathematical' literature; so, books like, "Feller", "Stratonovich", and others, however useful, have only sparingly been employed. The interested reader might find more on these approaches in van Kampen's thorough exposé.[33]

---

[33] N.G. van Kampen, "Stochastic Processes in Physics and Chemistry", first edition North Holland, Amsterdam and NY,1981; 3rd Ed. Elsevier, Amsterdam and NY, 2007.

Before we leave this section, we restate here the Wiener–Khintchine theorem. Let us have a number of stochastic variables, which are arranged in a column matrix $\mathbf{a}(t)$ and let the correlation function be defined as $\Phi_{\alpha\alpha}(t) = \langle \alpha(t)\tilde{\alpha}(0)\rangle \equiv \langle \alpha(t)\alpha(0)\rangle$, where $\alpha = \Delta\mathbf{a}$. The spectral density matrix and the correlation matrix are each others Fourier transforms, with the pre-factors chosen such that

$$\mathbb{S}_{\alpha\alpha}(\omega) = 2\int_{-\infty}^{\infty} e^{-i\omega t}\Phi_{\alpha\alpha}(t)\,dt\,, \tag{18.9-1a}$$

$$\Phi_{\alpha\alpha}(t) = (1/4\pi)\int_{-\infty}^{\infty} e^{i\omega t}\,\mathbb{S}_{\alpha\alpha}(\omega)\,d\omega. \tag{18.9-1b}$$

Generally we will write $\mathbb{S}_{\alpha\alpha} = \mathbb{G}_{\alpha\alpha} + i\mathbb{H}_{\alpha\alpha}$. For the real parts of the spectra we then have the 'real Wiener–Khintchine' transform pair

$$\mathbb{G}_{\alpha\alpha}(\omega) = 2\int_0^{\infty} \cos\omega t\,[\Phi_{\alpha\alpha}(t) + \tilde{\Phi}_{\alpha\alpha}(t)]\,dt\,, \tag{18.9-2a}$$

$$\Phi_{\alpha\alpha}(t) + \tilde{\Phi}_{\alpha\alpha}(t) = 2\int_0^{\infty} \cos\omega t\;\mathbb{S}_{\alpha\alpha}(\omega)\,d(\omega/2\pi). \tag{18.9-2b}$$

18.10 The Short-Time Average. MacDonald's Theorem and Milatz's Theorem

The average of a fluctuating quantity [vector or column matrix] $\alpha(t)$ over a 'short' period θ is itself a new stochastic variable[34] denoted by $\alpha_\theta(t)$ and defined as

$$\alpha_\theta(t) = (1/\theta)\int_t^{t+\theta} \alpha(t')\,dt'. \tag{18.10-1}$$

The covariance matrix for the short-time averages is

$$\Psi_{\alpha\alpha}(\theta) = \frac{1}{\theta^2}\int_0^{\theta}\int_0^{\theta} \langle \alpha(\xi)\tilde{\alpha}(\eta)\rangle\, d\xi\, d\eta\,, \tag{18.10-2}$$

where we assumed the process to be stationary.[35] We shall now make the change of variables $\xi - \eta = v$, $\eta = u$. The area to be integrated over is indicated in Fig. 18-9, which is similar to Fig. 18-7, but we do not now have a delta-function correlation. The area will be split up into the triangles I and II of the figure. The contributions are

$$I_I = \int_0^{\theta} du \int_0^{-u+\theta} dv \langle \alpha(u+v)\tilde{\alpha}(u)\rangle = \int_0^{\theta} d\bar{u}\int_0^{\bar{u}} dv \langle \alpha(\theta - \bar{u} + v)\tilde{\alpha}(\theta - \bar{u})\rangle$$
$$= \int_0^{\theta} d\bar{u}\int_0^{\bar{u}} dv \langle \alpha(v)\tilde{\alpha}(0)\rangle\,, \tag{18.10-3}$$

---

[34] By 'short' we mean an interval that is large compared to the collision time of the process, but small compared to the macroscopic period T used for averaging in the ergodic theorem, $\tau \ll \theta \ll T$.

[35] D.K.C. MacDonald, Rep. Progress Physics **12**, 56 (1948); K.M. Van Vliet, Physica **86A**, 130 (1977).

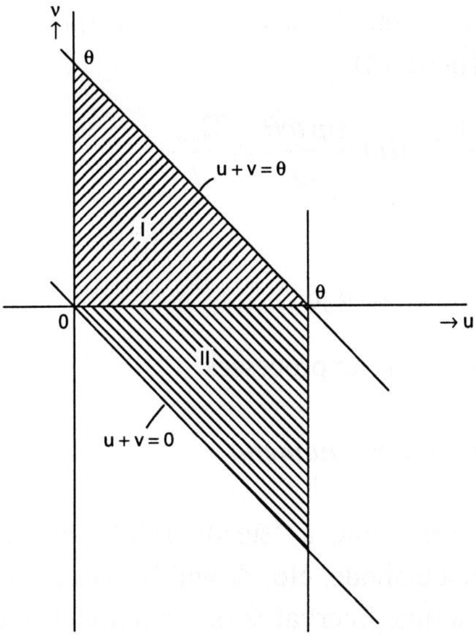

Fig. 18-9. New area of integration.

where we changed the integration variable $\bar{u} = \theta - u$ and applied stationarity. Likewise

$$I_{II} = \int_0^\theta du \int_{-u}^0 dv \langle \boldsymbol{\alpha}(u+v)\tilde{\boldsymbol{\alpha}}(u)\rangle = \int_0^\theta du \int_0^u d\bar{v} \langle \boldsymbol{\alpha}(u)\tilde{\boldsymbol{\alpha}}(u+\bar{v})\rangle$$

$$= \int_0^\theta du \int_0^u d\bar{v} \langle \boldsymbol{\alpha}(0)\tilde{\boldsymbol{\alpha}}(\bar{v})\rangle, \tag{18.10-4}$$

where we set $\bar{v} = -v$ and applied stationarity. Upon adding we obtain

$$\Psi_{\alpha\alpha}(\theta) = \frac{1}{\theta^2} \int_0^\theta du \int_0^u dv [\boldsymbol{\Phi}_{\boldsymbol{\alpha\alpha}}(v) + \tilde{\boldsymbol{\Phi}}_{\boldsymbol{\alpha\alpha}}(v)], \tag{18.10-5}$$

$$\frac{\partial}{\partial\theta}[\theta^2 \Psi_{\boldsymbol{\alpha\alpha}}(\theta)] = \int_0^\theta dv [\boldsymbol{\Phi}_{\boldsymbol{\alpha\alpha}}(v) + \tilde{\boldsymbol{\Phi}}_{\boldsymbol{\alpha\alpha}}(v)]. \tag{18.10-6}$$

We now apply the real Wiener–Khintchine theorem (18.9-2b). Then, changing the order of integration, we obtain

$$\frac{\partial}{\partial\theta}[\theta^2 \Psi_{\boldsymbol{\alpha\alpha}}(\theta)] = \frac{1}{\pi} \int_0^\infty d\omega \sin\omega\theta \, \frac{\mathbb{G}_{\boldsymbol{\alpha\alpha}}(\omega)}{\omega}, \tag{18.10-7}$$

with inversion

$$\mathbb{G}_{\boldsymbol{\alpha\alpha}}(\omega) = 2\omega \int_0^\infty d\theta \sin\omega\theta \, \frac{\partial}{\partial\theta}[\theta^2 \Psi_{\boldsymbol{\alpha\alpha}}(\theta)]. \tag{18.10-8}$$

this transform pair constitutes MacDonald's theorem. In the multivariate case, it does not provide for the imaginary part $\mathbb{H}_{\boldsymbol{\alpha\alpha}}(\omega)$. In thermal equilibrium all spectra are real and $\mathbb{H}_{\boldsymbol{\alpha\alpha}}(\omega) \equiv 0$; in the steady state in the absence of a magnetic field $\mathbb{H}_{\boldsymbol{\alpha\alpha}}(\omega) = 0$, providing all α's have the same time-reversal symmetry, cf. de Groot and Mazur, Chapter XIII, Ref. 17.

We now come to Milatz's theorem.[36] For a purely random process the spectrum is white, i.e., $\mathbb{G}_{\alpha\alpha}$ is constant. Then from (18-10-13)

$$\frac{\partial}{\partial\theta}[\theta^2 \Psi_{\alpha\alpha}(\theta)] = \mathbb{G}_{\alpha\alpha} \frac{1}{\pi}\int_0^\infty d\omega \, \frac{\sin\omega\theta}{\omega} = \frac{\mathbb{G}_{\alpha\alpha}}{2}, \tag{18.10-9}$$

or, integrating with respect to θ,

$$\mathbb{G}_{\alpha\alpha} = 2\theta\Psi_{\alpha\alpha}(\theta) = 2\theta\langle \boldsymbol{\alpha}_\theta \tilde{\boldsymbol{\alpha}}_\theta \rangle. \tag{18.10-10}$$

This is Milatz's theorem, generalized to multivariate processes.

18.10.1 *Application to Shot Noise and Similar Phenomena*

Let us consider a stationary *rate process*, like a 'steady rain', the arrival of electrons on the anode after emission from a cathode, etc. It will be assumed that the number of rain drops hitting the ground in a time interval θ, or the number of electron arrivals in θ, is governed by a Poisson distribution,

$$W(n_\theta) = e^{-\langle n_\theta \rangle} \langle n_\theta \rangle^{n_\theta} / n_\theta! \tag{18.10-11}$$

Hence, as usual, we have $\langle \Delta n_\theta^2 \rangle = \langle n_\theta \rangle$. Now also, if g_0 is the mean rate for the process, $\langle n_\theta \rangle = g_0\theta$. We thus find

$$\langle \Delta n_\theta^2 \rangle = g_0\theta. \tag{18.10-12}$$

The stochastic variable of primary interest is the arrival rate $m(t)$, whereby

$$n_\theta(t) = \int_t^{t+\theta} m(t')dt'. \tag{18.10.13}$$

[note that $\langle m(t) \rangle = g_0$]. The short-time average of $m(t)$ is by definition, cf. (18.10-1),

$$m_\theta(t) = (1/\theta)\int_t^{t+\theta} m(t')dt' = n_\theta(t)/\theta. \tag{18.10-14}$$

From Milatz's theorem we find

$$\mathbb{S}_m = 2\theta\langle \Delta m_\theta^2 \rangle = (2/\theta)\langle \Delta n_\theta^2 \rangle \overset{(18.10\text{-}18)}{=} 2g_0. \tag{18.10-15}$$

The spectrum of a Poissonian rate process is twice the mean rate. This is the basic result for shot noise. If the particles carry a charge e, the mean current is $I_0 = eg_0$. With the instantaneous current being $i(t) = em(t)$, we thus have

$$\mathbb{S}_i = e^2\mathbb{S}_m = 2e^2 g_0 = 2eI_0, \tag{18.10-16}$$

which is Walter Schottky's basic 1918 result; we note hereby that spectral analysis was not known at that time, so that he actually computed the noise as integrated over

[36] J.M.W. Milatz, Nederl. Tijdschrift voor Natuurk. **8**, 19 (1941).

the response of an LCR circuit.[37]

For a multivariate example, we return to the problem of the spectra of g-r noise, cf. subsection 18.6.2. Let $n_{ij\theta}$ be the number of particles travelling from level i to level j in time θ. For a given level 'i', particles are coming and going to all other levels $j = 1...i-1, i+1,...,s-1$. The distribution will be assumed to be multinomial Poisson, i.e.,

$$W(n_{i1\theta}, n_{i2\theta}, ..., n_{is\theta}) = \frac{\langle n_{i1\theta}\rangle^{n_{i1\theta}} ... \langle n_{is\theta}\rangle^{n_{is\theta}}}{n_{i1\theta}! ... n_{is\theta}!} e^{-\sum_k^{s}{}'\langle n_{ik\theta}\rangle} . \qquad (18.10\text{-}17)$$

From the many-variable generating function one easily proves that all numbers $n_{ij\theta}$ for different j are uncorrelated. Now, if $\gamma_i(t)$ represents the Langevin source for level i, then

$$\gamma_{i\theta} = \sum_{k=1}^{s}{}'[\langle n_{ki\theta}\rangle - \langle n_{ik\theta}\rangle]/\theta . \qquad (18.10\text{-}18)$$

For the variances and covariances we find

$$\langle \Delta\gamma_{i\theta}^{2}\rangle = \sum_{k=1}^{s}{}'[\langle \Delta n_{ki\theta}^{2}\rangle + \langle \Delta n_{ik\theta}^{2}\rangle]/\theta^2 , \qquad (18.10\text{-}19)$$

$$\langle \Delta\gamma_{i\theta}\Delta\gamma_{j\theta}\rangle = \sum_{k=1}^{s}{}'\sum_{\ell=1}^{s}{}'\langle(\Delta n_{ki\theta} - \Delta n_{ik\theta})\rangle\langle(\Delta n_{\ell i\theta} - \Delta n_{i\ell\theta})\rangle/\theta^2$$

$$= [\langle \Delta n_{ji\theta}\Delta n_{ij\theta}\rangle - \langle \Delta n_{ji\theta}\Delta n_{ji\theta}\rangle - \langle \Delta n_{ij\theta}\Delta n_{ij\theta}\rangle + \langle \Delta n_{ij\theta}\Delta n_{ji\theta}\rangle]/\theta^2$$

$$= -[\langle \Delta n_{ji\theta}^{2}\rangle + \langle \Delta n_{ij\theta}^{2}\rangle]/\theta^2 . \qquad (18.10\text{-}20)$$

The remaining variances in the above results are equal to their means, which in turn are equal to the stationary rates times θ; hence,

$$\langle \Delta\gamma_{i\theta}^{2}\rangle = \sum_{k=1}^{s}{}'[p_{ik}^{0} + p_{ki}^{0}]/\theta , \qquad (18.10\text{-}21)$$

$$\langle \Delta\gamma_{i\theta}\Delta\gamma_{j\theta}\rangle = -[p_{ij}^{0} + p_{ji}^{0}]/\theta . \qquad (18.10\text{-}22)$$

For the self-spectra of the sources we find 'double shot noise', while the cross spectra show anti-correlation:

$$\mathcal{S}_{\gamma_i} = 2\sum_{k=1}^{s}{}'[p_{ik}^{0} + p_{ki}^{0}], \quad \mathcal{S}_{\gamma_i,\gamma_j} = -2[p_{ij}^{0} + p_{ji}^{0}] \overset{\text{therm.eq.}}{=} -4p_{ij}^{0} . \qquad (18.10\text{-}23)$$

These confirm the results of subsection 18.6.2. However, as shown there, in essence the rates are *not* Poissonian; there is correlation due to stimulated emission.

---

[37] W. Schottky, Ann. der Physik, **57**, 541 (1918). The integral proved troublesome and was re-evaluated by his father, F. Schottky; for the final result, see W. Schottky, Ann. der Physik, **68**, 157 (1922).

18.10.2 *Modulated Emission Noise and Wave-Interaction Noise*

Often an emission process is not Poissonian, but allows for a description with a 'compound Poisson distribution'; in such a case the average rate of emission is itself subject to fluctuations. The oldest example is perhaps that of flicker noise in diodes, suggested by Schottky in 1926: he assumed that there were emission centres in the cathode that varied in number due to some chemical process (equilibrium reaction) in the oxide layer, which thereby modulated the emission rate.[38] While that may not be of interest to physicists today, a similar but more fundamental process was proposed by Mandel to account for wave-interaction noise in quantum optics;[39] the same type of compound Poisson distribution is used for its description.

So, to start with, let us consider the emission from N emission centres that emit particles with an average rate λ. We seek the stationary distribution $W(M,\theta)$ for the emission of M particles in the time interval $(t, t+\theta)$.[40] Noticing that the instantaneous rate of emission in dt is $\lambda N(t)dt$, we write

$$W(M,\theta)=\sum_{\{N(t)\}}\frac{\left[\lambda\int_t^{t+\theta}N(t')dt'\right]^M}{M!}\exp\left\{-\lambda\int_t^{t+\theta}N(t')dt'\right\}P[N(t)], \quad (18.10\text{-}24)$$

where $P[N(t)]$ is the functional distribution for $N(t)$. Using an overhead bar for $W(M,\theta)$ and angular brackets for averages involving $P[N(t)]$, we find

$$\overline{M_\theta}=\sum_M W(M,\theta)=\left\langle\lambda\int_t^{t+\theta}N(t')dt'\right\rangle=\lambda\theta\langle N\rangle, \quad (18.10\text{-}25)$$

$$\overline{M_\theta(M_\theta-1)}=\sum_M M(M-1)W(M,\theta)=\lambda^2\int_t^{t+\theta}\int_t^{t+\theta}\langle N(t_1)N(t_2)\rangle dt_1dt_2\,. \quad (18.10\text{-}26)$$

For the variance this yields

$$\begin{aligned}\overline{\Delta M_\theta^2}&=\overline{M_\theta(M_\theta-1)}-\overline{M_\theta}\,(\overline{M_\theta}-1)\\&=\lambda\theta\langle N\rangle+\lambda^2\int_0^\theta\int_0^\theta\langle\Delta N(t_1)\Delta N(t_2)\rangle dt_1dt_2\,.\end{aligned} \quad (18.10\text{-}27)$$

For the emission centres we assume the correlation function

$$\langle\Delta N(t_1)\Delta N(t_2)\rangle=\langle\Delta N^2\rangle e^{-|t_1-t_2|/\tau}\,. \quad (18.10\text{-}28)$$

To compute the integral, we make the change of variables $t_1-t_2=u,\ t_2=w$; the region of integration is once more as in Fig. 18-9. Because of the absolute sign in

---

[38] W. Schottky, Phys. Rev. **28** II, 74 (1926).

[39] L. Mandel, Proc. Phys. Soc. London **81**, 1104 (1963).

[40] K.M. Van Vliet et alii, "Superstatistical Emission Noise", Physica **108A**, 511 (1981), Sections 1-3. [Section 4, dealing with so-called 'quantum-1/*f* noise' in emission, is presently thought to be incorrect.]

(18-10-32) we must separately integrate over regions I and II. One easily obtains

$$\langle \Delta M_\theta^2 \rangle = \lambda\theta\langle N\rangle + 2\lambda^2\tau\langle \Delta N^2\rangle \int_0^\theta dw(1-e^{-w/\tau}). \qquad (18.10\text{-}29)$$

Now let the instantaneous emission rate be $m(t)$. Then, for its short-time average $m_\theta(t) = (1/\theta)\int_t^{t+\theta} m)t')dt' = M_\theta/\theta$ we have $\langle \Delta m_\theta^2\rangle = \langle \Delta M_\theta^2\rangle/\theta^2$ or from (18.10-33):

$$\frac{\partial}{\partial\theta}[\theta^2\langle \Delta m_\theta^2\rangle] = \lambda\langle N\rangle + 2\lambda^2\tau\langle \Delta N^2\rangle(1-e^{-\theta/\tau}). \qquad (18.10\text{-}30)$$

When this is put into MacDonald's theorem (18.10-8) we find

$$\mathcal{S}_m(\omega) = 2\omega\lambda\langle N\rangle\int_0^\infty d\theta \sin\omega\theta + 4\omega\lambda^2\tau\langle \Delta N^2\rangle\int_0^\infty d\theta \sin\omega\theta(1-e^{-\theta/\tau}). \qquad (18.10\text{-}31)$$

For the integrals in this expression we note

$$\int_0^\infty d\theta\, e^{\pm i\omega\theta} = 2\pi\,\delta_\mp(\omega) = \pi\,\delta(\omega) \pm i\mathcal{P}\frac{1}{\omega}, \quad \text{or} \quad \int_0^\infty d\theta\, \sin\omega\theta = \mathcal{P}\frac{1}{\omega}, \qquad (18.10\text{-}32)$$

where $\mathcal{P}$ denotes the Cauchy principal value. So, the first part yields $2\lambda\langle N\rangle$. For the second integral we find by integration by parts $1/\omega(1+\omega^2\tau^2)$. We thus obtain

$$\mathcal{S}_m(\omega) = 2\lambda\langle N\rangle + 4\lambda^2\tau\langle \Delta N^2\rangle/(1+\omega^2\tau^2). \qquad (18.10\text{-}33)$$

The first term represents shot noise and the Lorentzian represents modulation noise. These effects are – as usually assumed without proof – additive, although the same particles partake in the 'regular' and the modulation process. When multiplied by e^2, the first term is the Schottky term $2eI_0$, while the second term gives the flicker noise.

More fundamental is the wave-interaction noise. Photon noise spectra consist of two parts, a shot noise part and a frequency-dependent part which goes to zero for frequencies above the reciprocal coherence time. The need of compound statistics for this case was indicated by Mandel, as stated above. Let the light intensity be represented by its Sudarshan–Glauber transform $\mathcal{I}(t)$ ("*P*-representation"), which we shall discuss in Chapter XX; for now it suffices that it is linearly related to the classical intensity $I(t)$. The mean emission count in an interval $(t, t+\mathcal{T})$ is modulated by the fluctuating light intensity; thus, for the distribution of having M emissions in this interval we expect:

$$W(M,\mathcal{T}) = \int_0^\infty \frac{(\gamma\mathcal{U})^M}{M!} e^{-\gamma\mathcal{U}} W(\mathcal{U})d\mathcal{U}, \qquad (18.10\text{-}34)$$

where $\mathcal{U} = \int_t^{t+\mathcal{T}} \mathcal{I}(t')dt'$ and γ is a constant; one recognises the correspondence with (18.10-24).[41] The spectrum of the photon rate is found from MacDonald's theorem. For the count variance in $(t, t+\mathcal{T})$ we obtain, analogous to the previous case

[41] Properly speaking, this is a functional integration over the space of functions $\mathcal{I}(t)$ comprised in $\mathcal{U}$.

$$\frac{\partial}{\partial \mathcal{T}}\langle \mathcal{T}^2 \Delta m_{\mathcal{T}}^2\rangle = \gamma\langle \mathcal{I}\rangle + 2\gamma^2 \int_0^{\mathcal{T}} dv \langle \Delta\mathcal{I}(v)\Delta\mathcal{I}(0)\rangle . \qquad (18.10\text{-}35)$$

Let now $\overline{m}$ be the mean photon rate an let $\hat{\mathcal{S}}_{\Delta\mathcal{I}}(\omega)$ be the normalized spectrum $\mathcal{S}_{\Delta\mathcal{I}}(\omega)/\langle\mathcal{I}\rangle^2$. With $\gamma\langle\mathcal{I}\rangle = \overline{m}$ and employing MacDonald's theorem, we find

$$\mathcal{S}_m(\omega) = 2\omega\overline{m}\int_0^\infty d\mathcal{T}\sin\omega\mathcal{T} + 4\omega\overline{m}^2\langle\mathcal{I}\rangle^{-2}\int_0^\infty d\mathcal{T}\sin\omega\mathcal{T}\int_0^{\mathcal{T}} dv\langle\Delta\mathcal{I}(v)\Delta\mathcal{I}(0)\rangle. \quad (18.10\text{-}36)$$

The first integral yields $2\overline{m}$ by (18.10-32). The second integral will be integrated by parts:

$$2^{\text{nd}}\ \text{term} = 4\overline{m}^2\langle\mathcal{I}\rangle^{-2}\Big[(-\cos\omega\mathcal{T})\int_0^{\mathcal{T}} dv\langle\Delta\mathcal{I}(v)\Delta\mathcal{I}(0)\rangle\Big]_0^\infty$$

$$+4\overline{m}^2\langle\mathcal{I}\rangle^{-2}\int_0^\infty d\mathcal{T}\cos\omega\mathcal{T}\,\langle\Delta\mathcal{I}(\mathcal{T})\Delta\mathcal{I}(0)\rangle . \qquad (18.10\text{-}37)$$

The first term will be dismissed since $\cos\omega\mathcal{T}$ averages to zero for $\mathcal{T}\to\infty$. The other term is $\overline{m}^2\hat{\mathcal{S}}_{\Delta\mathcal{I}}(\omega)$ by the Wiener–Khintchine theorem. The total result becomes

$$\mathcal{S}_m(\omega) = 2\overline{m}[1+\tfrac{1}{2}\overline{m}\hat{\mathcal{S}}_{\Delta\mathcal{I}}(\omega)] . \qquad (18.10\text{-}38)$$

The last part is the wave-interaction noise. For a Lorentzian signal, the line profile is

$$S_{E^+}(\nu) = \frac{2\Gamma}{4\pi^2(\nu-\nu_0)^2+\Gamma^2} , \qquad (18.10\text{-}39)$$

where the width $\Gamma = 1/\tau_c$ is the reciprocal coherence time. The resulting intensity-fluctuations are found from convolution, cf. Section 20.2,

$$\hat{\mathcal{S}}_{\Delta\mathcal{I}}(\omega) = 2\int_0^\infty d\nu\,\hat{S}_{E^+}(\nu)\hat{S}_{E^+}(\nu+f) \simeq \frac{2\Gamma}{\pi^2 f^2+\Gamma^2} = 4\frac{(\tau_c/2)}{1+\omega^2(\tau_c/2)^2} , \quad (18.10\text{-}40)$$

where $\omega = 2\pi f$. In this result we assumed that the light is measured within a coherence cone Ω_c, i.e., the light is unresolved. In the opposite case the result must be *multiplied* by Ω_c/Ω. The familiar boson form is obtained by considering a time $\mathcal{T}$ in which the photons are received. The number of boson cells associated with a measurement in $(\mathcal{T},\Omega)$ is $Z = (\Omega/\Omega_c)(\mathcal{T}/\tau_c)$. The boson factor is the relative occupancy $B = \langle M_{\mathcal{T}}\rangle/Z = \overline{m}\mathcal{T}/Z$. From this and (18.10-40) we thus obtain the 'corpuscular form' of the wave-interaction noise:

$$\mathcal{S}_m(\omega) = 2\overline{m}\left[1+\frac{B(\nu_0)}{1+\omega^2(\tau_c/2)^2}\right] . \qquad (18.10\text{-}41)$$

We finally mention that for blackbody radiation the line width is usually set by the monochromator. The spectrum has the form

$$\mathcal{S}_m(\omega) = 2\overline{m}[1+B(\nu_0)](1-f/\Delta\nu) . \qquad (18.10\text{-}42)$$

18.11 Method of Elementary Events. Campbell's Theorem and Carson's Theorem

In this Section we shall turn to the "method of elementary events". Campbell's theorem has to do with the superposition of elementary responses.[42] It was formulated in 1909 and goes back to the Rayleigh–Schuster theorem of 1894 for the superposition of wave trains in optics.[43] A survey of these older methods is found with Ballantine.[44] In this text we follow the formulation of Fowler and of Rice.[45]

Let $f(t)$ be the response of a linear instrument due to a particle registered by the detector, arriving there at $t = 0$. Let $\bar{N}$ be the rate of arrivals per second and let the arrivals be independent of each other. The detector signal will be the sampling variable $y(t)$, being the sum of all arrivals in an interval T. Labelling the particles by $k = 1, 2, \ldots$ we will generally have

$$y(t) = \sum_{k=1}^{\infty} f(t - t_k), \tag{18.11-1}$$

where the kth electron arrives at t_k and where the sum is assumed to converge. The theorem now states that the average and variance of the stochastic variable $y(t)$ are given by

$$\langle y(t) \rangle = \bar{N} \int_{-\infty}^{\infty} f(t)\,dt, \tag{18.11-2}$$

$$\langle [\Delta y(t)]^2 \rangle = \bar{N} \int_{-\infty}^{\infty} [f(t)]^2\,dt. \tag{18.11-3}$$

Proof of Campbell's Theorem

The particles under consideration are 'classical particles', that show neither bunching nor anti-bunching. Thus, the probability that the arrival time t_k of particle 'k' lies within the interval $(t_k, t_k + dt_k)$ is dt_k/T. Let us now first consider all intervals T for which *exactly* K particles are registered by the detector. We then find, averaging over all K arrival times

$$\bar{y}_K = \int_0^T \frac{dt_1}{T} \int_0^T \frac{dt_2}{T} \ldots \int_0^T \frac{dt_K}{T} \sum_{k=1}^{K} f(t - t_k). \tag{18.11-4}$$

Clearly, all integrals, except the one over dt_k can be carried out, yielding unity. Remains,

[42] A. Campbell, Proc. Cambridge Phil. Soc. **15**, 117-136; ibid. 310-328 (1909).

[43] Lord Rayleigh, Phil. Mag. **27**, 466 (1889); A. Schuster, Phil. Mag. **37**, 509 (1894). See also next ref.

[44] S. Ballantine, Proc. IRE **18**, 1377 (1930).

[45] R.H. Fowler, "Statistical Mechanics", 2nd Ed., Cambridge Univ. Press, Cambridge 1936, § 20.71; S.O. Rice, Ref. 1(e), Part I, pp. 13-29.

$$\overline{y}_K = \sum_{k=1}^{K} \int_0^T \frac{dt_k}{T} f(t - t_k) = \frac{1}{T} \sum_{k=1}^{K} \int_{t-T}^{t} dt' f(t'), \tag{18.11-5}$$

where we set $t' = t - t_k$; note that the upper limit on t_k now becomes the lower limit on t' and *vice versa.* However, we can formally take the limits to be $\pm\infty$, if t' is large compared to the individual response times. Finally, replacing $t' \to t$, we arrive at

$$\overline{y}_K = \frac{K}{T} \int_{-\infty}^{\infty} f(t)\, dt. \tag{18.11-6}$$

Next, we allow K to be variable and we assume Poisson statistics for the number of arrivals in T to be K. Hence,

$$\langle y \rangle = \sum_{K=0}^{\infty} \overline{y}_K W(K) = \sum_{K=0}^{\infty} K \frac{(\overline{N}T)^K}{K!} e^{-\overline{N}T} \frac{1}{T} \int_{-\infty}^{\infty} f(t)\, dt = \overline{N} \int_{-\infty}^{\infty} f(t)\, dt. \tag{18.11-7}$$

This proves the first part of the theorem.

We proceed to obtain the second moment. First we consider all intervals with exactly K arrivals; for these we have

$$\overline{y_K^2} = \int_0^T \frac{dt_1}{T} \int_0^T \frac{dt_2}{T} \ldots \int_0^T \frac{dt_K}{T} \sum_{k=1}^{K} f(t - t_k) \sum_{\ell=1}^{K} f(t - t_\ell). \tag{18.11-8}$$

If $k = \ell$, all integrations can be carried out, yielding unity, except the one over dt_k; we make the change in variable $t' = t - t_k$ and note that there are K such terms. If, on the other hand, $k \neq \ell$, then all integrations except those over dt_k and dt_ℓ can be carried out; we set $t' = t - t_k$ and $t'' = t - t_\ell$; there are $K(K-1)$ such terms. Remains therefore,

$$\begin{aligned}
\overline{y_K^2} &= K \int_0^T \frac{dt_k}{T} [f(t - t_k)]^2 + K(K-1) \int_0^T \frac{dt_k}{T} f(t - t_k) \int_0^T \frac{dt_\ell}{T} f(t - t_\ell) \\
&= K \int_{t-T}^{t} \frac{dt'}{T} [f(t')]^2 + K(K-1) \int_{t-T}^{t} \frac{dt'}{T} f(t') \int_{t-T}^{t} \frac{dt''}{T} f(t'') \\
&= K \int_{-\infty}^{\infty} \frac{dt}{T} [f(t)]^2 + K(K-1) \left[\int_{-\infty}^{\infty} \frac{dt}{T} f(t) \right]^2 .
\end{aligned} \tag{18.11-9}$$

Next, we must abandon the fixed value of K and average the number of arrivals over a Poisson distribution. We shall not again carry this out explicitly, but appeal to Eq. (1.6-40). Thus, with $\langle K \rangle = \overline{N}T$ and $\langle K(K-1) \rangle = \langle K \rangle^2 = \overline{N}^2 T^2$, we obtain

$$\langle \Delta y^2 \rangle = \langle y^2 \rangle - \langle y \rangle^2 = \langle K \rangle \int_{-\infty}^{\infty} \frac{dt}{T} [f(t)]^2 = \overline{N} \int_{-\infty}^{\infty} dt [f(t)]^2 , \tag{18.11-10}$$

which completes the proof.

Extended Campbell Theorem

Rice also considers the case that $y(t)$ is a more involved stochastic variable:

$$y(t) = \sum_{k=1}^{\infty} a_k f(t - t_k), \qquad (18.11\text{-}11)$$

where the a_k are independent random variables, having all the same distribution. We now just repeat the essential elements of the previous proof. The following results are then easily confirmed:

$$\langle y(t) \rangle = \bar{N} \langle a \rangle \int_{-\infty}^{\infty} f(t)\, dt, \qquad (18.11\text{-}12)$$

$$\langle [\Delta y(t)]^2 \rangle = \bar{N} \langle a^2 \rangle \int_{-\infty}^{\infty} [f(t)]^2\, dt. \qquad (18.11\text{-}13)$$

In fact, Rice shows that similar expressions hold for all cumulants. Denoting the latter by c_n, the following can be proven:

$$c_n(y) = \bar{N} \langle a^n \rangle \int_{-\infty}^{\infty} [f(t)]^n\, dt. \qquad (18.11\text{-}14)$$

The proof requires a far more complete treatment than we have presented so far. Due to space limitations we shall refrain from providing the details here.

Carson's Theorem

We must now connect with spectral analysis. Let $F(\omega)$ be the Fourier transform of the elementary event $f(t)$, i.e.

$$F(\omega) = \int_{-\infty}^{\infty} f(t) e^{-i\omega t}\, dt, \quad \omega = 2\pi f. \qquad (18.11\text{-}15)$$

Since $f(t)$ is real, we have $F^*(\omega) = F(-\omega)$. From Parseval's theorem we have

$$\int_{-\infty}^{\infty} [f(t)]^2 dt = \int_{-\infty}^{\infty} |F(\omega)|^2 d(\omega / 2\pi) = 2\int_0^{\infty} |F(\omega)|^2 df. \qquad (18.11\text{-}16)$$

We used here the regular frequency, rather than the radial frequency, since it is to be borne in mind that the spectral density is defined as the 'variance per unit bandwidth' with the latter given in Hertz. Combining with Campbell's theorem, we note that the integrand of the last rhs is the spectral density.[46] We arrive at Carson's theorem,[47]

$$\mathcal{S}_y(\omega) = 2\bar{N} |F(\omega)|^2. \qquad (18.11\text{-}17)$$

---

[46] While this may not seem to be a mathematical necessity, physically it must be so, since (18.11-16) remains valid if the signal is filtered with a narrow-band filter; note that the spectrum is positive definite.

[47] J.R. Carson, Bell Syst. Techn. J. **10**, 374 (1931).

A word of caution is still in order. The functions $f(t)$ which figure in Campbell's theorem must truly denote' elementary' or *microscopic* random events. In this respect one is advised to read the detailed exposition and justification given in the original papers! As a negative example, we mention that several researchers in the sixties and seventies looked for events that had a Fourier transform of shape $1/\sqrt{f}$ in order to explain 1/*f* noise. They came up with one-dimensional diffusion. However, diffusion is a *collective* phenomenon; it cannot serve as an elementary event.[48]

Example

Consider electrons moving in a vacuum between two electrodes spaced at distance *L*. Then by Ramo's theorem the current induced in an external circuit is

$$\begin{aligned} &i(t) = e\bar{v}/L, \quad t_0 \le t \le t_0 + \tau_d, \\ &i(t) = 0 \ \text{elsewhere}. \end{aligned} \tag{18.11-18}$$

Here τ_d is the drift time. The situation is pictured in Fig. 18-10.

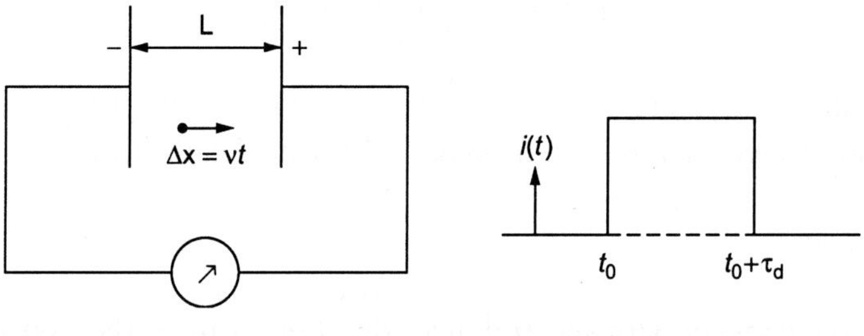

Fig. 18-10(a). Electron motion between two electrodes. Fig. 18-10(b). Current pulse in external circuit.

For the Fourier transform we find

$$F(\omega) = 2e^{-i\omega(t_0+\tau_d/2)} \sin(\tfrac{1}{2}\omega\tau_d)\, e\bar{v}/\omega L. \tag{18.11-19}$$

Thus, for the spectrum with $\bar{N}$ pulses per second, with $\tau_d = L/\bar{v}$,

$$\mathcal{S}_{\Delta i}(\omega) = 2\bar{N}e^2 \frac{\sin^2(\omega\tau_d/2)}{(\omega\tau_d/2)^2} = 2eI_0 \frac{\sin^2(\omega\tau_d/2)}{(\omega\tau_d/2)^2}. \tag{18.11-20}$$

This is shot noise with a high frequency turnover due to transit-time effects. In fact, this is probably the most *physical* derivation of shot noise. If the pulses are replaced by delta functions, $f(t-t_k) = \delta(t-t_k)$, $i(t) = ef(t)$, the Fourier transform is e^2 and the spectrum is white for all frequencies, $\mathcal{S}_{\Delta i} = 2eI_0$.

[48] See C.M. Van Vliet, "Random walk and 1/f Noise", Physica **A303**, 421 (2002) and references therein.

18.12 The Allan- Variance Theorem

The previous spectral theorems have been helpful for white noise and Lorentzians, or a mixture thereof; for many spectra, however, the theorems are not useful because of divergence of the integrals involved. So we now discuss another theorem stemming from the mid-sixties, which seems to do the trick. Following Allen and Barnes [49], who considered rate fluctuations in atomic clocks, in particular their 'flicker floor', we introduce the *two-sample variance* or *Allen variance*, which we will denote by $\hat{\sigma}_y^2(\mathcal{T})$ and define as follows. For the stochastic variable $y(t)$ let $y_{\mathcal{T}}^{(1)}$ be the average of y over an interval $(t, t+\mathcal{T})$ and let $y_{\mathcal{T}}^{(2)}$ be the average over the adjacent interval $(t+\mathcal{T}, t+2\mathcal{T})$. Or,

$$y_{\mathcal{T}}^{(1)} = (1/\mathcal{T})\int_t^{t+\mathcal{T}} dt'\, y(t'), \qquad y_{\mathcal{T}}^{(2)} = (1/\mathcal{T})\int_{t+\mathcal{T}}^{t+2\mathcal{T}} dt'\, y(t'). \qquad (18.12\text{-}1)$$

The Allen variance is then defined by [50,51]

$$\hat{\sigma}_y^2(\mathcal{T}) = \frac{1}{2}\langle [y_{\mathcal{T}}^{(1)} - y_{\mathcal{T}}^{(2)}]^2 \rangle. \qquad (18.12\text{-}2)$$

The theorem now connects the Allen variance with the spectral density $\mathbb{S}_y(\omega)$ of the stochastic variable $y(t)$ by

$$\hat{\sigma}_y^2(\mathcal{T}) = \frac{4}{\pi}\int_0^\infty \frac{d\omega}{\omega^2 \mathcal{T}^2}\, \mathbb{S}_y(\omega) \sin^4 \frac{\omega \mathcal{T}}{2}. \qquad \mathbf{(18.12\text{-}3)}$$

Because of the $\sin^4(\omega\mathcal{T}/2)$ term it has excellent convergence properties. The theorem is easily derived, as we shall now show. As any transform pair, it also has an inversion, obtained by the author[52]; that result will be stated later.

Let $q(t)$ be the 'cumulative sampling' of $y(t)$, i.e.

$$q(t) = \int_{-\infty}^t y(t')\, dt'. \qquad (18.12\text{-}4)$$

Further, let $A_q(\vartheta)$ be the autocorrelation function of $q(t)$. Then by the Wiener–Khintchine theorem we have, noting that $\mathbb{S}_q(\omega) = \mathbb{S}_y(\omega)/\omega^2$,

---

[49] (a) D.W. Allen, Proc. IEEE **54**, 221 (1966); (b) J.A. Barnes, Proc. IEEE **54**, 207 (1966). We note that this paper precedes that of Allen, but the concept used is more restrictive.

[50] We define here the Allen variance far more general than done usually for its primary application to counting statistics, to be dealt with at the end of this section.

[51] The concept of two-sample variance has been little known to physicists until the eighties. However, it has been implied in several studies on counting statistics; we mention here the work of Alkemade et al., who speak of "paired readings"; cf. C.Th.J. Alkemade, W. Snelleman, G.D. Boutilier, B.D. Pollard, J.D. Winefordner, T.L. Chester and N. Omeneto, Spectrochimica Acta **33B**, 383 (1978), Section II, C.2 ex.2.

[52] C.M. Van Vliet, " General Solution Methods for Fredholm Integral Equations with Kernel $K(z, z_0) = g(zz_0)$ on the interval (0,∞)", Annales des Sciences Mathématiques du Québec **6**, 197-214 (1982).

$$A_q(\vartheta) \equiv \langle q(t)q(t+\vartheta)\rangle = \int_0^\infty \mathcal{S}_q(\omega)\cos\omega\vartheta \frac{d\omega}{2\pi} = \lim_{\varepsilon\to 0+}\int_0^\infty \frac{\mathcal{S}_y(\omega)}{\omega^2}\cos\omega\vartheta\frac{d\omega}{2\pi}. \quad (18.12\text{-}5)$$

The limit may not exist, so it will not be carried out at this moment. From the definition of Allen variance we have

$$\hat{\sigma}_y^2(\mathcal{T}) = \frac{1}{2\mathcal{T}^2}\langle [q(\mathcal{T}+t) - q(t) - q(2\mathcal{T}+t) + q(\mathcal{T}+t)]^2\rangle$$
$$= \frac{1}{2\mathcal{T}^2}[2A_q(2\mathcal{T}) - 8A_q(\mathcal{T}) + 6A_q(0)], \quad (18.12\text{-}6)$$

where we assumed the process to be stationary, so that $\langle q(2\mathcal{T}+t)q(\mathcal{T}+t)\rangle = A_q(\mathcal{T})$, etc. Substituting (18.12-5) we have

$$\hat{\sigma}_y^2(\mathcal{T}) = \frac{1}{4\pi}\lim_{\varepsilon\to 0+}\int_\varepsilon^\infty \frac{d\omega}{\omega^2\mathcal{T}^2}\mathcal{S}_y(\omega)[2\cos 2\omega\mathcal{T} - 8\cos\omega\mathcal{T} + 6]. \quad (18.12\text{-}7)$$

This expression is meaningful if the limit for the *total* rhs exists. We now notice

$$2\cos 2\omega\mathcal{T} - 8\cos\omega\mathcal{T} + 6 = -2(1-\cos 2\omega\mathcal{T}) + 8(1-\cos\omega\mathcal{T})$$
$$= -16\sin^2(\omega\mathcal{T}/2)\cos^2(\omega\mathcal{T}/2) + 16\sin^2(\omega\mathcal{T}/2) = 16\sin^4(\omega\mathcal{T}/2).$$

Substituting into (18.12-7), the statement (18.12-4) follows, QED.

As a first example we try the theorem for the divergent $1/f$ noise spectrum. So let $\mathcal{S}_y(\omega) = C/|f|$. Then we find[49(b)],

$$\hat{\sigma}_y^2(\mathcal{T}) = 8C\int_0^\infty \frac{1}{\omega^2\mathcal{T}^2}\sin^4\left(\frac{\omega\mathcal{T}}{2}\right)\frac{d\omega}{\omega} = 2C\ln 2. \quad (18.12\text{-}8)$$

We shall also introduce the associated 'sampling variable' $z_{\mathcal{T}} = \int_0^{\mathcal{T}} dt'\, y(t')$, with Allen variance $\hat{\sigma}_z^2(\mathcal{T})$; clearly, $\hat{\sigma}_z^2(\mathcal{T}) = \mathcal{T}^2\hat{\sigma}_y^2(\mathcal{T})$. For $1/f$ noise this gives

$$\hat{\sigma}_z^2(\mathcal{T}) = 2\mathcal{T}^2 C\ln 2. \quad (18.12\text{-}8')$$

Since the transforms (18.12-3) are not always easily performable, we also will give another form. Let $F(s)$ be the Laplace transform of $\hat{\sigma}_z^2(\mathcal{T})$,

$$F(s) = \mathcal{L}[\hat{\sigma}_z^2(\mathcal{T})] = \int_0^\infty d\mathcal{T}\, e^{-s\mathcal{T}}\hat{\sigma}_z^2(\mathcal{T}). \quad (18.12\text{-}9)$$

The transform of $\sin^4(\omega\mathcal{T}/2)$ is easily found by rewriting the argument in its cosine form; we thus obtain

$$F(s) = \frac{6}{\pi}\int_0^\infty d\omega \frac{\omega^2}{s(s^2+4\omega^2)(s^2+\omega^2)}\mathcal{S}_y(\omega). \quad (18.12\text{-}10)$$

For spectra that are even and analytic this can be gainfully written as

$$F(s) = \frac{3}{\pi}\oint_C d\omega \frac{\omega^2}{s(s^2+4\omega^2)(s^2+\omega^2)} \mathcal{S}_y(\omega), \tag{18.12-11}$$

where C is a ccw contour consisting of the real axis and a semicircle in the upper plane, with its radius going to ∞. For $1/f$ noise (18.12-11) is not useful, since $1/|\omega|$ is not analytic in the plane. However, using partial fraction expansion, one easily verifies $F(s) = 4C(\ln 2)s^{-3}$, which upon inversion yields (18.12-8').

For white noise, $\mathcal{S}_y(\omega) = 2A$, (18.12-11) has only poles at $\omega = is$ and $\omega = is/2$, so that $F(s) = A/s^2$, or $\hat{\sigma}_z^2(\mathcal{T}) = A\mathcal{T}$. Finally, we consider Lorentzian flicker noise of the form $\mathcal{S}_y(\omega) = 4\alpha B/(\alpha^2+\omega^2)$. There is now an extra pole at $\omega = i\alpha$. From the residue theorem one finds

$$F(s) = 2B\left[\frac{4\alpha}{s^2(s^2-4\alpha^2)} - \frac{2\alpha}{s^2(s^2-\alpha^2)} - \frac{6\alpha^2}{s(s^2-4\alpha^2)(s^2-\alpha^2)}\right]. \tag{18.12-12}$$

For the inverse transform one obtains

$$\hat{\sigma}_z^2(\mathcal{T}) = (B/\alpha^2)[4e^{-\alpha\mathcal{T}} - e^{-2\alpha\mathcal{T}} + 2\alpha\mathcal{T} - 3]. \tag{18.12-13}$$

The various results are summarised in Table 18-1 below.

Table 18-1. Spectral density, Laplace transformed result and Allen variance for various $\mathcal{S}_y(\omega)$.

	$\mathcal{S}_y(\omega)$	$F(s)$	$\hat{\sigma}_z^2$
Poissonian shot noise	$2A$	A/s^2	$A\mathcal{T}$
general shot noise	$2A\hat{\sigma}_y^2(\mathcal{T})\mathcal{T}$	$\hat{\sigma}_y^2(\mathcal{T})\mathcal{T}/s^2$	$\hat{\sigma}_z^2(\mathcal{T}) = \kappa\langle y\rangle\mathcal{T}$
$1/f$ noise	$2\pi C/\lvert\omega\rvert$	$4C(\ln 2)/s^3$	$2C\mathcal{T}^2 \ln 2$
Lorentzian flicker noise	$\dfrac{4\alpha B}{(\alpha^2+\omega^2)}$	$2B\left[\dfrac{4\alpha}{s^2(s^2-4\alpha^2)} - \dfrac{2\alpha}{s^2(s^2-\alpha^2)} - \dfrac{6\alpha^2}{s(s^2-4\alpha^2)(s^2-\alpha^2)}\right]$	$\dfrac{B}{\alpha^2}[4e^{-\alpha\mathcal{T}} - e^{-2\alpha\mathcal{T}} + 2\alpha\mathcal{T} - 3]$
'Pathological noise'	$L/\lvert\omega\rvert^{\lambda-1}$ $0<\lambda<4;\ \lambda\neq 2$	$\dfrac{L(1-2^{\lambda-2})s^{-\lambda-1}}{\sin(\pi\lambda/2)}$	$\dfrac{L\mathcal{T}^{\lambda}(1-2^{\lambda-2})}{\sin(\pi\lambda/2)\Gamma(\lambda+1)}$

We still note that for white noise the Allen variance is equal to the regular variance. For 'pathological noise' we performed the computations from the inverse theorem.

18.12.1 *Inversion of the Allen Variance Theorem*

It is generally believed that a constant relative Allen variance $\hat{\sigma}_z^2/\langle z\rangle^2$ implies the presence of $1/f$ noise; or, more generally, one expects that a given Allen variance determines uniquely the shape of the spectrum. This will now be shown, by inverting the equation (18.12-3), which constitutes a Fredholm integral equation of the first kind. To this purpose we restate (18.12.10) in the form

$$F(s) = \frac{6}{\pi}\int_0^\infty \frac{d\omega}{\omega} \frac{1}{(s/\omega)[(s/\omega)^2+4][(s/\omega)^2+1]}\, \chi(\omega), \qquad (18.12\text{-}14)$$

where $\chi(\omega) = \mathcal{S}_y(\omega)/\omega^2$. We will take the Mellin transform of this equation and follow a method discussed by Morse and Feshbach.[53] The transform has to be taken piecewise, since the full transform does not usually exist for the functions $F(s)$ encountered in noise problems. We have seen in all cases (cf. Table 18.1) that for

$$\begin{aligned} s \to 0, \quad & F(s) = \mathcal{O}(s^{-\sigma_0}), \quad \sigma_0 > 0, \\ s \to \infty, \quad & F(s) = \mathcal{O}(s^{-\tau_0}), \quad \tau_0 > 0. \end{aligned} \qquad (18.12\text{-}15)$$

For the various noise processes the values of σ_0 and τ_0 are given in Table 18-2. Consequently, let[54]

$$\begin{aligned} \Phi_-(p) &= \int_0^1 F(s)s^{p-1}ds \qquad [\text{exists for } \operatorname{Re} p > \sigma_0], \\ \Phi_+(p) &= \int_1^\infty F(s)s^{p-1}ds \qquad [\text{exists for } \operatorname{Re} p < \tau_0]. \end{aligned} \qquad (18.12\text{-}16)$$

Since for most cases $\tau_0 = \sigma_0$, there is no region in the complex plane where both partial transforms exist together. Only the Lorentzian behaves better; it has a complete Mellin transform $\Phi(p)$. Other cases may occur where $\tau_0 < \sigma_0$.[55] We now recall the Mellin transform convolution theorem,

$$\mathcal{M}\left[\int_0^\infty v(s/\omega)g(\omega)\, d\omega/\omega\right] = V(p)G(p), \qquad (18.12\text{-}17)$$

where $\mathcal{M}$ denotes the Mellin transform; capitals refer to the corresponding transformed functions. Noting that (18.12-14) has the form of the [ ] at the lhs, we find for the transformed equation $V(p)X(p)$ with for the transformed kernel

[53] P.M. Morse and H. Feshbach, "Methods of Theoretical Physics" Vol. I, McGraw-Hill, NY 1953, Section 8.5.

[54] Usually, for the existence one considers the Lebesgue integral $\int_0^1 |F(s)|^2 s^{2\sigma-1}ds$; if this is finite for $\sigma > \sigma_0$, the transform Φ_- exists for $\operatorname{Re} p \geq \sigma$. Likewise for Φ_+.

[55] The Mellin transform presents a dimensional anomaly, since s has the dimension of sec^{-1}. This can be circumvented by defining appropriate dimensionless quantities, cf. subsection 1.7.3.

Table 18-2. Behaviour for the Mellin transform.

Noise	σ_0	τ_0	Inversion applies	Type of transform
white	2	2	yes	partial
$1/f$	3	3	yes	partial
Lorentzian	2	4	yes	full
$1/f^3$	5	5	no	—

$$V(p) = \frac{6}{\pi}\int_0^\infty ds \frac{s^{p-2}}{(s^2+4)(s^2+1)} = \frac{1-2^{p-3}}{\sin\frac{1}{2}(p-1)\pi}, \quad 1 < \mathrm{Re}\, p < 5. \tag{18.12-18}$$

We thus obtain

$$\Phi_-(p) + \Phi_+(p) = \frac{1-2^{p-3}}{\sin\frac{1}{2}(p-1)\pi}[X_-(p) + X_+(p)]. \tag{18.12-19}$$

In order that there is a region in which both $V(p)$ and $\Phi_-(p)$ or $\Phi_+(p)$ exist, it is necessary that $\tau_0 > 1$ and $\sigma_0 < 5$. The situation is depicted in Fig. 18-11.

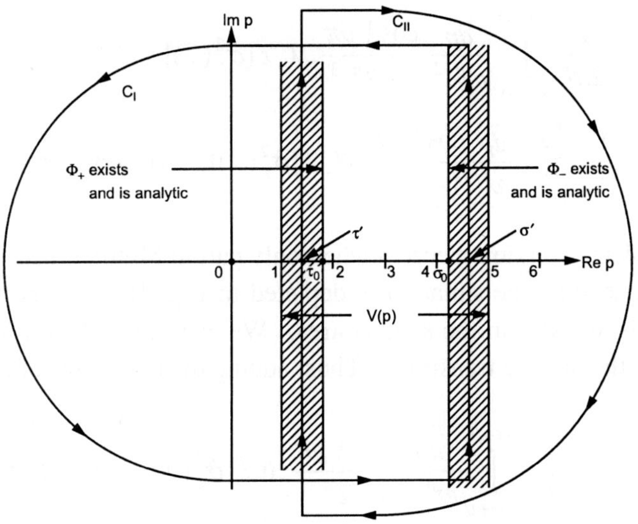

Fig. 18-11. Shaded area is domain of analyticity for the transformed equation (18.12-19). We pictured the case $\tau_0 < \sigma_0$. If $\tau_0 = \sigma_0$, the shaded areas touch but still have no region in common.

From (18.12-19) we deduce

$$\Phi_-(p) - X_-(p)\frac{1-2^{p-3}}{\sin\frac{1}{2}(p-1)\pi} = -\Phi_+(p) + X_+(p)\frac{1-2^{p-3}}{\sin\frac{1}{2}(p-1)\pi}. \quad (18.12\text{-}20)$$

This equality which initially is valid in the shaded area only can be made to hold in the entire plane except at singularities by analytic continuation. Besides this, the integrals over each of the members, going along a line $\mathrm{Re}\, p = \sigma'$ for the lhs and $\mathrm{Re}\, p = \tau'$ for the rhs, are equal due to the inverted transform equality. Using a theorem of Morse and Feshbach (Op. Cit. p. 463) we conclude that each member equals a function $R(p)$ which is analytic in the entire area $1 < \mathrm{Re}\, p < 5$. We now find by transforming back:

$$\begin{aligned}\chi(\omega) &= (\mathcal{M}_-)^{-1} X_-(p) + (\mathcal{M}_+)^{-1} X_+(p) \\ &= -\frac{1}{2\pi i}\int_{-i\infty+\sigma'}^{i\infty+\sigma'} \frac{dp}{\omega^p}\Phi_-(p)\frac{\cos p\pi/2}{1-2^{p-3}} - \frac{1}{2\pi i}\int_{-i\infty+\tau'}^{i\infty+\tau'} \frac{dp}{\omega^p}\Phi_+(p)\frac{\cos p\pi/2}{1-2^{p-3}} \\ &+ \oint_C \frac{dp}{\omega^p} R(p)\frac{\cos p\pi/2}{1-2^{p-3}}, \end{aligned} \quad (18.12\text{-}21)$$

where C is a ccw contour made up of the lines $\mathrm{Re}\, p = \sigma'$ and $\mathrm{Re}\, p = \tau'$, indicated in Fig. 18-11. Since $R(p)$ is analytic and the factor following $R(p)$ is regular (at $p = 3$, both numerator and denominator are zero, but their ratio is finite) the contour integral is zero; so the solution of the inhomogeneous equation is unique. The solution (18.12-21) becomes, reverting to $\mathcal{S}_y(\omega) = \chi(\omega)\omega^2$:

$$\begin{aligned}\mathcal{S}_y(\omega) &= -\frac{1}{2\pi i}\int_{-i\infty+\sigma'}^{i\infty+\sigma'} \frac{dp}{\omega^{p-2}}\frac{\cos\frac{1}{2}p\pi}{1-2^{p-3}}\mathcal{M}_-\mathcal{L}[\hat{\sigma}_z^2(\mathcal{T})] \\ &- \frac{1}{2\pi i}\int_{-i\infty+\tau'}^{i\infty+\tau'} \frac{dp}{\omega^{p}}\frac{\cos\frac{1}{2}p\pi}{1-2^{p-3}}\mathcal{M}_+\mathcal{L}[\hat{\sigma}_z^2(\mathcal{T})], \quad \sigma' > \sigma_0, \ \tau' < \tau_0. \end{aligned} \quad \mathbf{(18.12\text{-}22)}$$

This is the complete inversion theorem when only partial Mellin transforms exist.

For the case $\sigma_0 < \tau_0$ the situation is depicted in Fig. 18-12. The full transform now exists and is analytic in the shaded area. We can now select a line $\mathrm{Re}\, p = \beta$, $\sigma_0 < \beta < \tau_0$, for the inverse transform. Thus, adding the two terms of (18.12-22), we obtain

$$\mathcal{S}_y(\omega) = -\frac{1}{2\pi i}\int_{-i\infty+\beta}^{i\infty+\beta} \frac{dp}{\omega^{p-2}}\frac{\cos\frac{1}{2}p\pi}{1-2^{p-3}}\mathcal{M}\mathcal{L}[\hat{\sigma}_z^2(\mathcal{T})], \quad \sigma_0 < \beta < \tau_0. \quad (18.12\text{-}23)$$

Usually the Mellin and Laplace transforms are interchangeable. This allows the simple expression:

$$\mathbb{S}_y(\omega) = -\frac{1}{2\pi i}\int_{-i\infty+\beta}^{i\infty+\beta} \frac{dp}{\omega^{p-2}} \frac{\cos\frac{1}{2}p\pi}{1-2^{p-3}} \Gamma(p) \int_0^\infty \frac{d\mathcal{T}}{\mathcal{T}^p} \hat{\sigma}_z^2(\mathcal{T}), \qquad \textbf{(18.12-24)}$$

where β is in the domain of analyticity of $\Gamma(p)\int_0^\infty d\mathcal{T}\,\mathcal{T}^{-p}\hat{\sigma}_z^2(\mathcal{T})$.

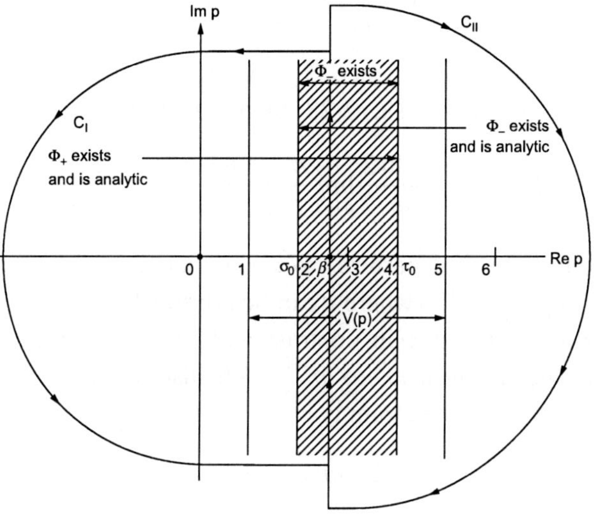

Fig. 18-12. Case $\sigma_0 < \tau_0$. Shaded area is the domain of existence for the full Mellin transform $\Phi(p)$. The values of τ_0 and σ_0 are those for Lorentzian noise.

We will show the application of (18.12-22) to an Allen variance of the form $K\mathcal{T}^\lambda$, $0<\lambda<4$. Then $F(s) = K\Gamma(\lambda+1)s^{-\lambda-1}$ and

$$\Phi_-(p) = \frac{K}{p-\lambda-1}\Gamma(\lambda+1), \quad \mathrm{Re}\, p > \sigma_0 = 1+\lambda, \qquad (18.12\text{-}25)$$

$$\Phi_+(p) = -\frac{K}{p-\lambda-1}\Gamma(\lambda+1), \quad \mathrm{Re}\, p < \tau_0 = 1+\lambda. \qquad (18.12\text{-}26)$$

These Φ's are substituted into (18.12-22). We consider first '$\omega < 1$'. In view of our previous footnote[55], this means that we have $\omega < \omega_c$. Since ω_c can be chosen arbitrarily large, the Φ_- solution alone should cover the entire spectrum! We now close the contour in Fig. 18.11 with a large semicircle in the left-hand plane; on this semicircle the integrand goes sufficiently fast to zero. Moreover, since $\Phi_+(p)$ as well as $\cos\frac{1}{2}p\pi/(1-2^{p-3})$ are analytic for $p<\tau$, the contour integral for this part vanishes. We are thus left with

$$\mathbb{S}_y(\omega) = -\frac{1}{2\pi i}\oint_{C_I} \frac{dp}{\omega^{p-2}} \frac{\cos\frac{1}{2}p\pi}{1-2^{p-3}} \frac{K\Gamma(\lambda+1)}{p-\lambda-1}. \qquad (18.12\text{-}27)$$

We note that there is only one pole, $p = \lambda+1$. Hence,

$$\mathcal{S}_y(\omega) = -\frac{K}{\omega^{\lambda-1}} \lim_{p\to 1+\lambda} \frac{\cos\frac{1}{2}p\pi}{1-2^{p-3}}\Gamma(\lambda+1)\,. \tag{18.12-28}$$

For $\lambda \neq 2$ the limit is straightforward; then

$$\mathcal{S}_y(\omega) = \frac{K}{\omega^{\lambda-1}} \frac{\sin\frac{1}{2}\lambda\pi}{1-2^{\lambda-2}}\Gamma(\lambda+1)\,. \tag{18.12-29}$$

For white noise, $K = \langle y \rangle$, $\lambda = 1$, hence $\mathcal{S}_y(\omega) = 2\langle y \rangle$, as expected. For 1/*f* noise $K = 2C\ln 2$ and $\lambda = 2$. Then from (18.12-28) by de l'Hôpital's rule,

$$\mathcal{S}_y(\omega) = \frac{4C\ln 2}{\omega} \lim_{p\to 3} \frac{\pi}{2} \frac{\sin\frac{1}{2}p\pi}{\ln 2 e^{(p-3)\ln 2}} = \frac{2\pi C}{\omega}\,, \tag{18.12-30}$$

thus confirming our point of departure. We note that since λ runs from zero to four, spectra from $\omega^{1-\varepsilon}$ up to $\omega^{-3+\varepsilon}$ (where ε can be arbitrarily small) can be handled with the present transform pair; this in contrast with the Wiener–Khintchine theorem, which allows only spectra from ω^0 to $\omega^{-1+\varepsilon}$. We also notice from (18.12-29) that, if *we know in advance* that the spectrum is

$$\mathcal{S}_y(\omega) = L/\omega^{\lambda-1}\,, \tag{18.12-31}$$

then the Allen variance is $(\lambda \neq 2)$

$$\hat{\sigma}_z^2(\mathcal{T}) = L\mathcal{T}^\lambda[(1-2^{\lambda-2})/\sin\tfrac{1}{2}\lambda\pi\,\Gamma(\lambda+1)]\,. \tag{18.12-32}$$

Finally, we remark that we can obtain the same result from Φ_+. We then take '$\omega > 1$', i.e., $\omega > \omega_c$, where ω_c can be taken to be arbitrarily small. The contour is then closed with a large clockwise arc in the right-hand plane, C_{II}. The contour integral over Φ_- now vanishes; so, the same result is obtained.

As an instructive exercise we mention the case of Lorentzian noise, for which a full Mellin transform must be employed. We leave this as a problem at the end of this chapter.

18.12.2 *Counting Experiments and Non-Adjacent Sampling*

The theory of the Allen variance transform has been mainly developed in view of counting procedures and associated noise. Thus, the variable $y(t)$ is often the counting rate $m(t)$ and the associated sampling variable $z(t)$ is the interval count $M(t)$, defined accordingly, $M_{\mathcal{T}}(t) = \int_t^{t+\mathcal{T}} m(t')dt'$. The Allen variance theorem then reads

$$\hat{\sigma}_M^2(\mathcal{T}) = \frac{4}{\pi}\int_0^\infty \frac{d\omega}{\omega^2}\mathcal{S}_m(\omega)\sin^4\frac{\omega\mathcal{T}}{2}\,. \tag{18.12-33}$$

Experimental determination of the Allen variance $\hat{\sigma}_M^2(\mathcal{T})$ is in principle straightforward. In particular, one measures the relative Allen variance by dividing by the average count squared. Assuming the presence of white noise, $1/f$ noise and Lorentzian noise, we then have

$$R(\mathcal{T}) \equiv \frac{\langle (M_{\mathcal{T}}^{(1)} - M_{\mathcal{T}}^{(2)})^2 \rangle}{2\langle M_{\mathcal{T}} \rangle^2}$$

$$= \frac{\kappa}{m_0 \mathcal{T}} + 2\frac{C \ln 2}{m_0^2} + \frac{B}{m_0^2 \mathcal{T}^2 \alpha^2}\left[4e^{-\alpha\mathcal{T}} - e^{-2\alpha\mathcal{T}} + 2\alpha\mathcal{T} - 3\right], \quad (18.12\text{-}34)$$

where m_0 is the mean rate $\langle m \rangle$. In all flicker noise and 1/f noise theories the extra noise is proportional to m_0^2; thus we write $B = B_0 m_0^2$, $C = C_0 m_0^2$. For $\mathcal{T} \gg \alpha^{-1}$ we get

$$R(\mathcal{T}) \approx \frac{\kappa}{m_0 \mathcal{T}} + 2C_0 \ln 2 + \frac{2B_0}{\alpha \mathcal{T}}, \quad (18.12\text{-}35)$$

whereas for $\mathcal{T} \ll \alpha^{-1}$ we have

$$R(\mathcal{T}) \approx \frac{\kappa}{m_0 \mathcal{T}} + 2C_0 \ln 2 + \frac{2B_0}{3}\alpha\mathcal{T}. \quad (18.12\text{-}36)$$

(The latter situation is realised for very slow flicker noise.) In a double logarithmic plot of $R(\mathcal{T})$ *vs.* $1/\mathcal{T}$ the noise will decrease with a slope of –1 until the 'flicker floor', $2C_0 \ln 2$, is reached. If very slow Lorentzians are present, then for very large $\mathcal{T}$ the relative Allen variance may increase again, as expressed by (18.12-36).

We shall still comment on an extension of the Allen variance theorem when strict adjacent sampling is not possible due to the dead time of the registering instrument. Let the latter be denoted by τ. We define a modified Allen variance $\bar{\sigma}_M^2(\mathcal{T},\tau)$ by

$$\bar{\sigma}_M^2(\mathcal{T},\tau) = \tfrac{1}{2}\langle [M_{\mathcal{T}}(\mathcal{T}+\tau+t) - M_{\mathcal{T}}(t)]^2 \rangle. \quad (18.12\text{-}37)$$

Analogous to (18.12-6) we find

$$\bar{\sigma}_M^2(\mathcal{T},\tau) = A_q(2\mathcal{T}+\tau) - 2A_q(\mathcal{T}+\tau) - 2A_q(\mathcal{T}) + A_q(\tau) + 2A_q(0). \quad (18.12\text{-}38)$$

Using the W–K theorem, this gives

$$\bar{\sigma}_M^2(\mathcal{T},\tau) = \int_0^\infty (d\omega/2\pi)\mathcal{S}_m(\omega)\omega^{-2}[\cos\omega(2\mathcal{T}+\tau) - 2\cos\omega(\mathcal{T}+\tau) - 2\cos\omega\mathcal{T} + \cos\omega\tau + 2]. \quad (18.12\text{-}39)$$

With some trigonometry this yields

$$\bar{\sigma}_M^2(\mathcal{T},\tau) = \frac{4}{\pi}\int_0^\infty \frac{d\omega}{\omega^2}\mathcal{S}_m(\omega)\sin^2\frac{\omega\mathcal{T}}{2}\sin^2\frac{\omega(\mathcal{T}+\tau)}{2}, \quad (18.12\text{-}40)$$

which is a straightforward extension of the Allen variance theorem (18.12-3). The Allen variance of $1/f$ noise is found to be

$$\bar{\sigma}_M^2(\mathcal{T},\tau) = \tfrac{1}{2}C\mathcal{T}^2[(2+r)^2\ln(2+r) - 2(1+r)^2\ln(1+r) + r^2\ln r] \approx 2C\mathcal{T}^2(1+r)\ln 2, \quad (18.12\text{-}41)$$

where $r = \tau/\mathcal{T}$ and where the latter approximation holds if $r < 0.2$. This is the correction for (18.12-8').

18.13 On the Origin of 1/f-like Noise

The quest for the origin of *1/f*-like noise, first observed by Johnson in oxide-cathode vacuum tubes and by Williams and Thatcher in metallic resistors[56], while later on studied extensively by Bernamont,[57] has been arduous, long and bizarre. Around the middle of the twentieth century dozens of far-fetched theories saw the light, which are not worthy to mention – let alone describe – here; we recall, however, the valiant attempt to summarise many developments by the late J.J. Brophy.[58] Most likely, most *1/f* noise, notwithstanding the many decades over which it has been observed, is due to a summation of Lorentzians. After this had been suggested by many authors, the theme was picked up by Dutta and Horn.[59] Basically, one needs a distribution of life times, based on a distribution of activation energies. So, if $W(\mathcal{E})d\mathcal{E} = C\, d\mathcal{E}$ and $\tau = \tau_0 \exp(\beta\mathcal{E})$, one finds for uncorrelated contributions,

$$\begin{aligned}\mathcal{S}(\omega) &= 4\int_{\tau_1}^{\tau_2} d\tau W(\tau)\,\frac{\tau}{1+\omega^2\tau^2} = 4\int_{\tau_1}^{\tau_2} d\tau W(\mathcal{E})\frac{d\mathcal{E}}{d\tau}\frac{\tau}{1+\omega^2\tau^2}\\ &= 4C\int_{\tau_1}^{\tau_2} d\tau\,\frac{k_BT}{\tau}\,\frac{\tau}{1+\omega^2\tau^2} = 4C\,k_BT\frac{1}{\omega}\int_{\omega\tau_1}^{\omega\tau_2} d(\omega\tau)\,\frac{1}{1+\omega^2\tau^2}\\ &= 4C\,k_BT\frac{1}{\omega}[\arctan(\omega\tau_2) - \arctan(\omega\tau_1)] \approx 2\pi C\,k_BT\frac{1}{\omega},\quad \frac{1}{\tau_2} < \omega < \frac{1}{\tau_1}. \qquad (18.13\text{-}1)\end{aligned}$$

If the distribution over the activation energies is not uniform but skewed, a slope different from unity can be obtained; slopes between 0.8 and 1.2 are not uncommon. In Fig. 18.13 we exhibit a spectrum measured by us long ago;[60] *the slope is not unity.* In fact, we ascertained $\mathcal{S}_I(\omega) \propto 1/f^{\,1.055\pm0.015}$.

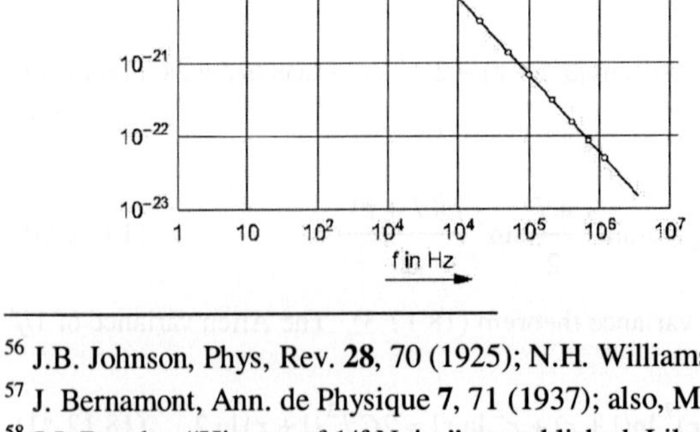

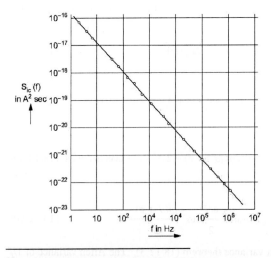

Fig. 18.13. Current noise of a carbon resistor over the frequency range 1 Hz to 1.2 MHz.

---

[56] J.B. Johnson, Phys, Rev. **28**, 70 (1925); N.H. Williams and E.W. Thatcher, Phys. Rev. **40**, 121 (1932).

[57] J. Bernamont, Ann. de Physique **7**, 71 (1937); also, M.M. Surdin, J. Phys. Radium, **10**, 188 (1939).

[58] J.J. Brophy, "History of 1/f Noise", unpublished, Library of the University of Utah.

[59] D. Dutta and P.M. Horn, Rev. Mod. Phys. **53**, 497 (1981).

[60] K.M. Van Vliet, C.J. van Leeuwen, J. Blok and C. Ris, Physica XX, 481-496 (1954).

18.14 The Spectra of Generation-Recombination Noise

Finally, in this chapter we return to the presently well understood and established results for generation-recombination (g-r) noise. As indicated earlier, the spectra can always be analyzed as a sum of Lorentzians. In many good pure single crystals there may be as many as five or six of these partial spectra, although at a given temperature usually no more than two Lorentzians are discernable. To compute the noise, some investigators have developed computer programs that do the analysis of the result (18.6-16) or (18.6-17). The main point of the analysis is to deduct the time constants τ_i, which usually depend exponentially on the energy of the pertinent levels below the conduction band, or above the valence band, as well as the magnitude of the Lorentzians, which also reveal details about the transitions involved; see the next subsection. First, however, we will do some analytical work and derive the spectra for frequently occurring two-Lorentzian spectra.

18.14.1 *Three-Level Systems*

For a semiconductor in thermal equilibrium we have from (18-5-61) and (18.5-11)

$$\mathcal{S}_{\mathbf{n}}(\omega) = 2\mathrm{Re}[\mathsf{M} + i\omega\mathsf{I}]^{-1}\mathsf{M}^{-1}\mathsf{B}\,, \tag{18.14-1}$$

where M is the phenomenological relaxation matrix and B the matrix of the second-order Fokker–Plank moments with I being the unit matrix. Let us now alter this as follows, to read

$$\begin{aligned}\mathcal{S}_{\mathbf{n}}(\omega) &= 2\omega^{-2}\,\mathrm{Re}[-i\omega(\mathsf{M}+i\omega\mathsf{I})^{-1}(\mathsf{M}+i\omega\mathsf{I}-\mathsf{M})\mathsf{M}^{-1}\mathsf{B} \\ &= 2\omega^{-2}\,\mathrm{Re}[-i\omega\mathsf{M}^{-1}\mathsf{B}+i\omega(\mathsf{M}+i\omega\mathsf{I})^{-1}\mathsf{B}] = 2\omega^{-2}\,\mathrm{Re}[(\mathsf{I}+\mathsf{M}/i\omega)^{-1}\mathsf{B}]\,,\end{aligned} \tag{18.14-2}$$

where we omitted a purely imaginary term. To invert the above matrix we simply write it out for a bivariate process:

$$\omega^2[\mathsf{I}+\mathsf{M}/i\omega] = \begin{pmatrix} \omega^2 - i\omega M_{11} & -i\omega M_{12} \\ -i\omega M_{21} & \omega^2 - i\omega M_{22} \end{pmatrix}, \tag{18.14-3a}$$

with reciprocal

$$\begin{pmatrix} \omega^2 - i\omega M_{22} & i\omega M_{12} \\ i\omega M_{21} & \omega^2 - i\omega M_{11} \end{pmatrix} \frac{1}{\omega^4 - i\omega^3(M_{11}+M_{22}) - \omega^2\Delta}; \tag{18.14.3b}$$

here $\Delta = M_{11}M_{22} - M_{12}M_{21}$. The eigenvalues satisfy

$$\begin{vmatrix} M_{11} - 1/\tau & M_{12} \\ M_{21} & M_{22} - 1/\tau \end{vmatrix} = 0\,, \quad \text{or} \quad \frac{1}{\tau^2} - \frac{1}{\tau}(M_{11}+M_{22}) + \Delta = 0\,. \tag{18.14-4}$$

We notice that the roots of the last factor in (18.14-3) are $\omega_{1,2} = i/\tau_{1,2}$. The following partial fraction expansion is easily corroborated:

$$\frac{1}{\omega^2 - i\omega(M_{11}+M_{22}) - \Delta} = \frac{\tau_1\tau_2}{\tau_2-\tau_1}\left[\frac{\tau_1}{1+i\omega\tau_1} - \frac{\tau_2}{1+i\omega\tau_2}\right]. \tag{18.14-5}$$

So we find

$$\omega^{-2}\,\mathrm{Re}[(\mathsf{I}+\mathsf{M}/i\omega)^{-1}] = \frac{\tau_1\tau_2}{\tau_2-\tau_1}$$
$$\times\left\{\frac{\tau_1}{1+\omega^2\tau_1^2}\begin{pmatrix}1-\tau_1M_{22} & \tau_1M_{12}\\ \tau_1M_{21} & 1-\tau_1M_{11}\end{pmatrix} - \frac{\tau_2}{1+\omega^2\tau_2^2}\begin{pmatrix}1-\tau_2M_{22} & \tau_2M_{12}\\ \tau_2M_{21} & 1-\tau_2M_{11}\end{pmatrix}\right\}. \tag{18.14-6}$$

This is to be multiplied by 2B in order to obtain the spectral densities; we arrive at

$$\mathcal{S}_{11}(\omega) = 2\sum_{1,2}\phi(\omega;\tau_1,\tau_2)\left[\left((1/\tau_1)-M_{22}\right)B_{11} + M_{12}B_{12}\right],$$
$$\mathcal{S}_{22}(\omega) = 2\sum_{1,2}\phi(\omega;\tau_1,\tau_2)\left[\left((1/\tau_1)-M_{11}\right)B_{22} + M_{21}B_{12}\right],$$
$$\mathcal{S}_{12}(\omega) = 2\sum_{1,2}\phi(\omega;\tau_1,\tau_2)\left[\left((1/\tau_1)-M_{22}\right)B_{12} + M_{12}B_{22}\right],$$
$$= \mathcal{S}_{21}(\omega) = 2\sum_{1,2}\phi(\omega;\tau_1,\tau_2)\left[\left((1/\tau_1)-M_{11}\right)B_{12} + M_{21}B_{11}\right]. \tag{18.14-7}$$

where we used the abbreviation

$$\phi(\omega;\tau_1,\tau_2) = \frac{\tau_1\tau_2}{\tau_2-\tau_1}\frac{\tau_1^2}{1+\omega^2\tau_1^2}; \tag{18.14-8}$$

here $\sum_{12}$ means that a similar term has to be added with τ_1 and τ_2 interchanged. The symmetry of $\mathcal{S}_{12}$ requires that the following condition be satisfied:

$$B_{12}(M_{22}-M_{11}) + M_{21}B_{11} - M_{12}B_{22} = 0. \tag{18.14-9}$$

This is a consequence of the principle of microscopic reversibility.[61]

We now use the B-values obtained before, cf. (18.5-56), where for thermal equilibrium conditions, $p_{ij}^0 = p_{ji}^0$. Letting levels 1 and 2 refer to the independent variables of the system, with level 3 having $\Delta n_3 = -\Delta n_1 - \Delta n_2$, we find for the spectra

$$\mathcal{S}_{11}(\omega) = 4\sum_{1,2}\phi(\omega;\tau_1,\tau_2)\left[p_{12}^0\left((1/\tau_1)-M_{22}-M_{12}\right) + p_{13}^0\left((1/\tau_1)-M_{22}\right)\right],$$
$$\mathcal{S}_{22}(\omega) = 4\sum_{1,2}\phi(\omega;\tau_1,\tau_2)\left[p_{12}^0\left((1/\tau_1)-M_{11}-M_{21}\right) + p_{23}^0\left((1/\tau_1)-M_{11}\right)\right],$$

---

[61] The corresponding bivariate steady-state result is more complicated (and *more symmetrical*); one could proceed in a similar manner.

$$\mathcal{S}_{12}(\omega) = 4\sum_{1,2}\phi(\omega;\tau_1,\tau_2)\left[p_{12}^0\left(-(1/\tau_1)+M_{22}+M_{12}\right)+p_{23}^0 M_{12}\right]$$
$$= \mathcal{S}_{21}(\omega) = 4\sum_{1,2}\phi(\omega;\tau_1,\tau_2)\left[p_{12}^0\left(-(1/\tau_1)+M_{11}+M_{21}\right)+p_{13}^0 M_{21}\right]. \quad (18.14\text{-}10)$$

These expressions are not as general as one might think. Firstly, we have excluded Auger processes, in which one transition may trigger another one. Secondly, we have excluded *multiple* electron processes, such as occur for transitions to impurity levels that can accommodate two or more electrons, like e.g., double acceptors cf. Fig. 5-4. This is not particularly worrisome, but the *A*'s and *B*'s assume a more elaborate form; one brief example will follow below.

First a note on the variances and covariances. Let level 1 refer to a non-degenerate conduction band (occupancy variable *n*), level 2 refer to a non-degenerate valence band (occupancy variable –*p*) and level 3 represent *I* impurity levels of which the occupancy is denoted by *i*. From the fluctuation-relaxation theorem in its equilibrium form (18.5-11), or from the Boltzmann–Einstein distribution (3.7-5) one easily obtains, denoting equilibrium values by the subscript '0':[62,63]

$$\langle \Delta n^2 \rangle = A\left[\frac{1}{p_0}+\frac{1}{i_0}+\frac{1}{I-i_0}\right], \quad \langle \Delta p^2 \rangle = A\left[\frac{1}{n_0}+\frac{1}{i_0}+\frac{1}{I-i_0}\right], \quad (18.14\text{-}11a)$$

$$\langle \Delta n \Delta p \rangle = A\left[\frac{1}{i_0}+\frac{1}{I-i_0}\right], \quad A = \left[\frac{1}{n_0 p_0}+\left(\frac{1}{i_0}+\frac{1}{I-i_0}\right)\left(\frac{1}{n_0}+\frac{1}{p_0}\right)\right]^{-1}. \quad (18.14\text{-}11b)$$

Generally, we have no need of these results, since the spectra have been given directly in terms of the transition rates and their derivatives (necessary for the components of *M*). However, since we shall make approximations, the integrated spectra *might* not yield the (co)-variances as given above.

As a first example we consider the case of a nondegenerate near-intrinsic semiconductor with recombination centres, using the Shockley–Read model. With the level designations as above, we have the following mass-action laws, Fig. 18-14:

$$\begin{aligned} &p_{12} = \alpha, && p_{21} = \beta np, \\ &p_{13} = \delta n(I-i), && p_{31} = \lambda i \equiv \delta n_1 i, \\ &p_{32} = \kappa ip && p_{23} = \mu(I-i) \equiv \kappa p_1(I-i). \end{aligned} \quad (18.14\text{-}12)$$

Some rates appear as monomolecular, since we assumed the number of available places in the conduction band and valence band to be very large so that their occupancies do not affect certain transitions. Apart from the entry for p_{12}, we made, however, all rates *seemingly* bimolecular by introducing the Shockley–Read

---

[62] R.E. Burgess, Proc. Phys. Soc. (London) **B68**, 661 (1955).

[63] K.M. Van Vliet, Physica XXIII, 248 (1957).

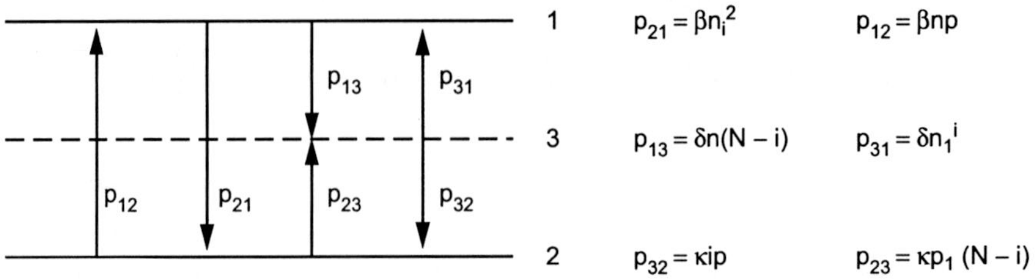

Fig. 18-14. Bivariate g-r process; two nondegenerate bands and recombination centres.

quantities n_1 and p_1; their meaning will be made clear below. The rate equations are

$$\begin{aligned} dn/dt &= \alpha + \delta n_1 i - \delta n(I-i) - \beta np, \\ dp/dt &= \alpha + \kappa p_1(I-i) - \kappa ip - \beta np, \\ di/dt &= \delta n(I-i) - \delta n_1 i + \kappa p_1(I-i) - \kappa ip. \end{aligned} \qquad (18.14\text{-}13)$$

[Note that all transitions are given as *electron*-transition rates; the appropriate variable for the valence band is the electron occupancy or $-p$.] The third rate equation is superfluous, for the system satisfies the constraint $(d/dt)(n+i-p)=0$, or $n+i-p=const.$ Differentiating to n and $-p$ one obtains the phenomenological relaxation matrix

$$\mathsf{M} = \begin{pmatrix} \delta(n_0+n_1+I-i_0)+\beta p_0 & \delta(n_0+n_1)-\beta n_0 \\ \kappa(p_0+p_1)-\beta p_0 & \kappa(p_0+p_1+i_0)+\beta n_0 \end{pmatrix}. \qquad (18.14\text{-}14)$$

We now comment on the Shockley–Read quantities, n_1 and p_1. From detailed balance $(p_{13}^0 = p_{31}^0,\ p_{32}^0 = p_{23}^0)$ we have

$$\begin{aligned} n_1/n_0 &= (I/i_0) - 1 = g\exp[(\varepsilon_3 - \varepsilon_F)/k_BT], \\ p_1/p_0 &= [I/(I-i_0)] - 1 = g^{-1}\exp[-(\varepsilon_3 - \varepsilon_F)/k_BT], \end{aligned} \qquad (18.14\text{-}15)$$

where we used (5.6-8); g is the spin-degeneracy of the centres. Now setting the spin degeneracy equal to unity[64], one notes that n_1 *would be* the number of electrons in the conduction band if the Fermi level were to coincide with the trap level; *mutatis mutandis for* p_1. These numbers (or densities) are therefore also called 'would-be' densities. It is customary to also define the lifetime of the free electrons prior to capture and similarly for the holes; hence,

---

[64] This factor was apparently not known at the time to Shockley and Read; cf. W. Shockley and W.T. Read, Phys. Rev. **87**, 835 (1952).

$$\tau_{n0} = 1/\sigma_n \langle v_n \rangle I = 1/\delta I, \qquad \tau_{p0} = 1/\sigma_p \langle v_p \rangle I = 1/\kappa I, \tag{18.14-16}$$

where the σ's are the capture cross sections. Omitting the direct recombination β, we rewrite the phenomenological relaxation matrix as

$$\mathsf{M} = \begin{pmatrix} (I\tau_{n0})^{-1}(n_0 + n_1 + I - i_0) & (I\tau_{n0})^{-1}(n_0 + n_1) \\ (I\tau_{p0})^{-1}(p_0 + p_1) & (I\tau_{p0})^{-1}(p_0 + p_1 + i_0) \end{pmatrix}. \tag{18.14-17}$$

Now Shockley and Read assumed that either $n_0 \gg I - i_0$ or $p_0 \gg i_0$. Then, for the eigenvalues

$$1/\tau_{1,2} = \tfrac{1}{2}(M_{11} + M_{22})\{1 \pm [1 - 4\Delta/(M_{11} + M_{22})^2]^{\frac{1}{2}}\} \tag{18.14-18}$$

we can expand the square root; this yields

$$\tau_1 \simeq \frac{M_{11} + M_{22}}{M_{11}M_{22} - M_{12}M_{21}}, \qquad \tau_2 \simeq \frac{1}{M_{11} + M_{22}}. \tag{18.14-19}$$

For the determinant we find from (18.14-17) and detailed balance

$$M_{11}M_{22} - M_{12}M_{21} \simeq (\tau_{n0}\tau_{p0}I)^{-1}(n_0 + p_0), \tag{18.14-20}$$

resulting in the relaxation times[80,65]

$$\tau_1 \equiv \tau_{SR} \simeq \frac{\tau_{n0}(p_0 + p_1) + \tau_{p0}(n_0 + n_1)}{n_0 + p_0}, \qquad \tau_2 \simeq \frac{\tau_{n0}\tau_{p0}I}{\tau_{SR}(n_0 + p_0)}. \tag{18.14-21}$$

The main lifetime is the Shockley–Read lifetime; the second lifetime $\tau_2 \ll \tau_{SR}$ will give rise to a Lorentzian at much higher frequencies, ~ 1MHz or higher and is usually not observed. The spectra can now be evaluated. One obtains with some algebra the symmetrical result

$$\left.\begin{matrix} \mathcal{S}_{nn}(\omega) \\ \mathcal{S}_{pp}(\omega) \end{matrix}\right\} \approx 4\frac{n_0 p_0}{n_0 + p_0}\frac{\tau_{SR}}{1 + \omega^2\tau_{SR}^2} + 4\frac{n_0 p_0}{n_0 + p_0}\left(\frac{\tau_{p0}(n_0 + n_1)}{\tau_{n0}(p_0 + p_1)}\right)^{\pm 1}\frac{\tau_2}{\tau_{SR}}\frac{\tau_2}{1 + \omega^2\tau_2^2}, \tag{18.14-22}$$

(upper sign *n*, lower sign *p*). The first Lorentzian is as for near-intrinsic g-r noise, except that τ is replaced by τ_{SR}; the second term is *not* that much smaller at the place where it occurs, the plateau ratio for $\omega > 1/\tau_2$ being $[\tau_{n0}(p_0 + p_1)/\tau_{p0}(n_0 + n_1)]$. The main problem is that it may drown in the thermal noise. A measurement of the resulting normalized current noise $\mathcal{S}_i/\langle i \rangle^2$ for n-type germanium single crystals is given in Fig. 18-15(a). The inferred Shockley–Read lifetime is 1.85 μsec.[66]

In the above computations a nondegenerate conduction band and valence band were assumed. The restriction can easily be removed. We must then integrate over

---

[65] F.W. Rose and D.J. Sandiford, Proc. Phys. Soc. (London) **B68**, 894 (1955).

[66] J.E. Hill and K.M. Van Vliet, J. Appl. Phys. **29**, 177 (1958).

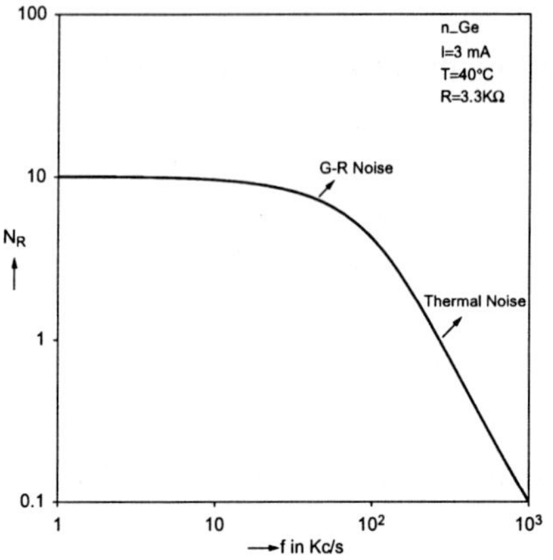

Fig. 18-15(a). Noise in near intrinsic n-type Ge. After Ref. 66.

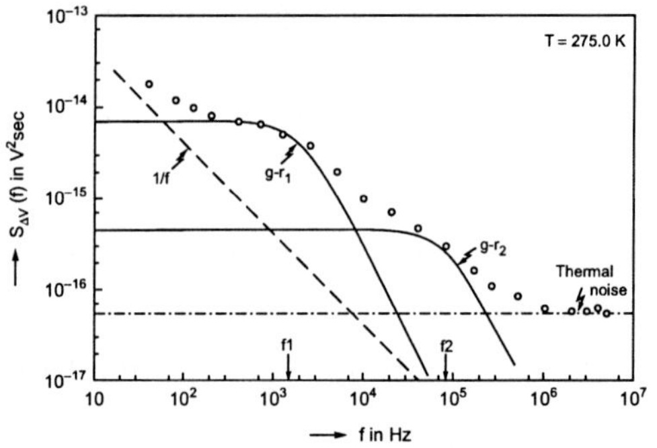

Fig. 18-15(b). G-R noise in p-type Si. After G. Bosman, Ref. 70.

the states in each band. The treatment is most easily done with the thermodynamic method, in which the relaxation matrix is replaced by $\mathsf{M} = \mathsf{R}^{-1}\mathsf{s}^0$, cf. Eqs. (18-5-5) and (18.5-16); in this method the mass-action laws for the transition rates are *not employed.* The space is lacking here, so we refer the interested reader to Ref. 9, Section IV-D. With the degeneracy factors ξ_n and ξ_p, being a ratio of Fermi integrals, see (5.4-28), we then find the following modified results:

$$\tau'_{SR} \simeq \xi_n \xi_p [\tau_{n0}(p_0 + p_1) + \tau_{p0}(n_0 + n_1)]/(\xi_n n_0 + \xi_p p_0), \qquad (18.14\text{-}21')$$

$$\mathcal{S}_{nn}(\omega) \approx 4[\xi_n \xi_p (n_0 p_0)/(\xi_n n_0 + \xi_p p_0)]\tau'_{SR}/(1 + \omega^2 \tau_{SR}^2). \qquad (18.14\text{-}22')$$

Half a dozen other examples involving three-level systems have been given in the literature, cf. Klaassen et al.[67] One more interesting example involving double acceptors will be given here. The band diagram is in Fig. 18-16. The particular merit of this example is that almost no approximations need to be

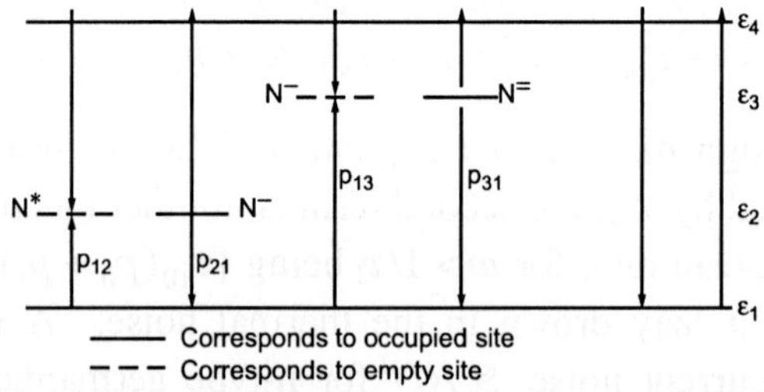

Fig. 18-16. Transitions in a double acceptor system.

made. The acceptors give rise to neutral sites (N^*), singly ionized sites (N^-) and doubly ionized sites ($N^=$). At low temperatures ($\leq 150\,K$) transitions to the conduction band are near absent; we then have

---

[67] F.M. Klaassen, K.M. Van Vliet and J. Blok, Physica **26**, 605 (1960).

a three-level problem. In contrast to Klaassen et al, we consider the variables $\mathbf{a} = \{a_1 a_2 a_3\}$ with $-p = a_1$, $N^- = a_2$, $N^= = a_3$; here $-p$ and N^- are taken as independent variables. We then have for the relevant transitions

$$p_{12} = \delta_2 p_2 (N - N^- - N^=), \qquad p_{21} = \delta_2 N^- p, \qquad p_{13} = \delta_1 p_1 N^-, \qquad p_{31} = \delta_1 N^= p. \tag{18.14-23}$$

Here p_1 and p_2 are 'would-be' quantities; also, note that $p_{23} = p_{32} = 0$. The Fokker–Planck A's now are slightly different, since a doubly ionized centre appears when an electron from a singly ionized centre recombines with a hole from the valance band and *vice versa.* Denoting by A^* the result for our previous one-electron transitions, we have

$$A_1 = p_{31} + p_{21} - p_{13} - p_{12}, \qquad A_2 = A_2^* - A_3^* = (p_{12} - p_{21}) - (p_{13} - p_{31}). \tag{18.14-24}$$

Differentiating to $-p$ and N^-, we find for the M-matrix

$$\mathsf{M} = \begin{pmatrix} \delta_1 (N_0^= + \frac{1}{2} p_0) + \delta_2 (N_0^- + \frac{1}{2} p_2) & \delta_1 (p_1 + \frac{1}{2} p_o) - \delta_2 (p_0 + \frac{1}{2} p_2) \\ \delta_1 (N_0^= + \frac{1}{2} p_0) - \delta_2 (N_0^- + \frac{1}{2} p_2) & \delta_1 (p_1 + \frac{1}{2} p_0) + \delta_2 (p_0 + \frac{1}{2} p_2) \end{pmatrix}. \tag{18.14-25}$$

At low temperatures, when the Fermi level is in the vicinity of the lower levels we expect

$$p_2 \sim p_0 < N_0^-, \qquad p_1 \ll p_0, \qquad N_0^= \ll p_0.$$

With these *reasonable* approximations, the M-matrix can be simplified to

$$\mathsf{M} = \begin{pmatrix} \frac{1}{2} \delta_1 p_0 + \delta_2 (N_0^- + \frac{1}{2} p_2) & \delta_1 p_1 - \delta_2 (p_0 + \frac{1}{2} p_2) \\ \frac{1}{2} \delta_1 p_0 - \delta_2 (N_0^- + \frac{1}{2} p_2) & \delta_1 p_1 + \delta_2 (p_0 + \frac{1}{2} p_2) \end{pmatrix}. \tag{18.14-26}$$

The eigenvalues can now be obtained without further approximations since the discriminant of the secular equation is a complete square; so we obtain

$$1/\tau_1 = \delta_1 p_0, \qquad 1/\tau_2 = \delta_2 (N_0^- + p_0 + p_2). \tag{18.14-27}$$

The second-order F–P moments are also somewhat different than for single electron transitions. One easily corroborates

$$B_{11} = 2(\overset{0}{p}_{12} + \overset{0}{p}_{13}), \quad B_{12} = -2\overset{0}{p}_{12} + 2\overset{0}{p}_{13} = B_{21}, \quad B_{22} = 2(\overset{0}{p}_{12} + \overset{0}{p}_{13}). \tag{18.14-28}$$

Hence, writing the spectra of the hole fluctuations as $\mathcal{S}_{pp} = \Sigma_{1,2} K_1 /(1 + \omega^2 \tau_1^2)$, we obtain,

$$K_1 = 4\tau_1 N_0^= \left[\frac{\delta_2 (2p_0 + p_2) - \delta_1 p_0}{\delta_2 (N_0^- + p_0 + p_2) - \delta_1 p_0} \right], \qquad K_2 = 4\tau_2 p_0 \left[\frac{N_0^-}{N_0^- + p_0 + p_2} \right]. \tag{18.14-29}$$

These results were first obtained by Klaassen et al., who chose $-p$ and $N^=$ as independent variables.[68] It is easily seen that at low temperatures the plateau value K_2 is much larger than K_1. So, the Lorentzian with time constant τ_2 will be observed. If compensating centres n_c are present, the centres are partially filled at absolute zero and $N_0^- \simeq n_c \gg p_0 + p_2$. The time constant then is independent of temperature

---

[68] These variables do not have anomalous F–P moments, i.e., the moments are as for single electron states; however, the approximations to obtain the relaxation times now involve an expansion for the square root. Cf. F.M. Klaassen, J. Blok and H.C. Booy, Physica **27**, 48 (1961).

and the noise has a range with a simple Lorentzian component of the form (18.5-52), p_0 replacing n_0. At higher temperatures the time constant τ_2 becomes exponentially dependent on temperature and, ultimately, the Lorentzian for the time constant τ_1 will dominate the observed noise. Detailed graphs are given in the cited papers. Lax[69] has argued that the cross section $\sigma_p^{=}$ will be much larger than σ_p^{-} on account of the charge on the centres. This means that $\delta_1 >> \delta_2$ and $\tau_1 << \tau_2$. It is not likely that both modes will be observed simultaneously in some temperature range.

18.14.2 *General Structure of Multi-Level G-R Noise*

Finally, we shall turn to some more general ideas regarding the structure of the computed and observed generation-recombination noise spectra and *what we can learn from this type of study.* There is a trade-off between accuracy and versatility. In many solids there are as many as four to seven impurity levels that will contribute to g-r noise in a given temperature range, with usually two or three Lorentzians being visible. Bosman and Zijlstra used a computer program to plot the results from (18.5-61).[70] Some typical data of theirs are shown in the preceding Fig. 18-15(b). Here we shall present a more simple picture that is quite useful in most cases, if one does not mind an error in magnitude of up to a factor two (often of no concern when the data are plotted on a log-log scale).[71] Let us consider a solid with transitions between $s-1$ sets of impurity centres (traps for short) that interact with the conduction band but not with each other; the results will also apply to hole traps that interact with the valence band, *mutatis mutandis.* The population of the conduction band is denoted by $n \equiv n_s$ and that of the traps by n_i, $i = 1 ... s-1$; the trap occupancies are taken as the independent variables, with rates to and fro $p_{si} = \delta n (N_i - n_i)$, $p_{is} = \delta n_{SRi} n_i$, respectively where n_{SRi} are Shockley – Read 'would-be' quantities. From detailed balance and F–D statistics for the equilibrium occupancies n_{i0} we have as before,

$$n_{i0} / N_i = 1/[g_i e^{(\varepsilon_i - \varepsilon_F)/k_B T} + 1], \tag{18.14-30}$$

$$n_{SRi} = g_i n_0 e^{(\varepsilon_i - \varepsilon_F)/k_B T} = n_0 e^{(\varepsilon_i + k_B T \ln g_i - \varepsilon_F)/k_B T}, \tag{18.14-31}$$

showing once more that n_{SRi} is the number of free carriers that *would be* in the conduction band if the Fermi level coincided with $\varepsilon_i + k_B T \ln g_i$ (or roughly, when the Fermi level passes through the trap level). Since for the present case $p_{ij} = 0$ unless $i = s$ or $j = s$, the first-order and second-order F–P moments are simple:

$$A_i = p_{si}(n_1 ... n_s) - p_{is}(n_1 ... n_s), \quad i = 1 ... s-1, \tag{18.14-32}$$

---

[69] M. Lax, J. Phys. Chem. Solids **8**, 66 (1959); ibid. Phys. Rev. **119**, 1502 (1960).

[70] G. Bosman and R.J.J. Zijlstra, Solid-State Electron. **25**, 273 (1982).

[71] A recent paper with quite general but approximate results, as contrasted with previous exact studies[10], is found in C.M. Van Vliet, "Electronic noise due to multiple trap levels in homogeneous solids and space-charge layers", J. Appl. Phys. **93**, 6068 (2003).

$$B_{ii} = 2p_{si}(n_1 \ldots n_s) \approx 2p_{si}(n_{10} \ldots n_{s0}), \quad i = 1 \ldots s-1,$$
$$B_{ij} = 0 \,. \tag{18.14-33}$$

Next, we will find the phenomenological relaxation matrix. For the differentiation we note that $\partial/\partial n_j$ involves terms $(\partial/\partial n)dn/dn_j = -(\partial/\partial n)$. We also set $\delta_i = \sigma_i\langle v_n\rangle$, where σ_i is the capture cross section. This, then, yields

$$M_{ii} = \sigma_i\langle v_n\rangle(n_{i1} + n_0 + N_i - n_{i0}),$$
$$M_{ij} = \sigma_i\langle v_n\rangle(N_i - n_{i0}) \,. \tag{18.14-34}$$

If the traps are not too numerous – henceforth assumed – we can neglect the terms $N_i - n_{i0}$, thus rendering M approximately diagonal. The eigenvalues are now

$$1/\tau_i = \sigma_i\langle v_n\rangle(n_{i1} + n_0) \simeq \sigma_i\langle v_n\rangle g_i N_c \exp[-(\varepsilon_c - \varepsilon_i)/k_B T]; \tag{18.14-35}$$

we assumed temperatures $T > T_{\min}$ for which $n_{SRi} = n_0$ and we used (18.14-31) together with $n_0 = N_c \exp[(\varepsilon_F - \varepsilon_c)/k_B T]$, $N_c = 2(2\pi m^* k_B T/h^2)^{3/2}$ being the statistical weight of the conduction band. Since $\langle v_n\rangle \propto T^{1/2}$, we note that an *Arrhenius plot* of $\log(\tau T^2)$ vs. 1000/T gives a straight line whose slope gives the energy of the impurity levels with respect to the conduction band edge, $\varepsilon_c - \varepsilon_i$.

We proceed to find the plateau values. From Eqs. (18.5-11), (18.14-33) and the above results for M, we see that the (co)variance matrix is approximately diagonal. The variances are found to be

$$\langle \Delta n_i^2 \rangle = \frac{n_0(N_i - n_{i0})}{n_{SRi} + n_0 + N_i - n_{i0}} \approx \frac{n_0(N_i - n_{i0})}{n_{SRi} + n_0} = N_i \frac{1 - n_{i0}/N_i}{1 + n_{SRi}/n_0} \,. \tag{18.14-36}$$

Substituting from (18.14-30) and (18.14-31), this gives

$$\langle \Delta n_i^2 \rangle = N_i \frac{g_i e^{(\varepsilon_i - \varepsilon_F)/k_B T}}{[1 + g_i e^{(\varepsilon_i - \varepsilon_F)/k_B T}]^2} \to N_i f_i (1 - f_i) \,. \tag{18.14-37}$$

The expression obtained is the grand-canonical result, rather than the canonical result; this is clearly caused by the approximations we made. The corrected result is found in Section 5.4, Eq. (5.4-24); the error is around 30% and will at present be ignored. Equation (18.14-37) is rewritten as

$$\langle \Delta n_i^2 \rangle = \frac{N_i}{[\exp(\frac{1}{2}\gamma_i) + \exp(-\frac{1}{2}\gamma_i)]^2} = \tfrac{1}{4} N_i \,\mathrm{sech}^2(\tfrac{1}{2}\gamma_i) \,, \tag{18.14-38}$$

where

$$\gamma_i(\varepsilon_F) = (\varepsilon_i - \varepsilon_F)/k_B T + \ln g_i \,. \tag{18.14-39}$$

To obtain the spectra, it is expedient to not employ (18.14-1), but go back to (18.5-61). Since both M and the covariance matrix $\langle \alpha\tilde{\alpha}\rangle = \langle \Delta n \Delta n\rangle$ are near-diagonal, we find immediately

$$\mathcal{S}_{nn}(\omega) = \sum_{i=1}^{s-1} \mathcal{S}_{n_i n_i}(\omega) = \sum_{i=1}^{s-1} 4N_i\tau_i \exp[\pm\gamma_i(\varepsilon_F)]/(1+\omega^2\tau_i^2), \qquad (18.14\text{-}40)$$

where the – sign holds for high T and the + sign for low T. The resulting fluctuations

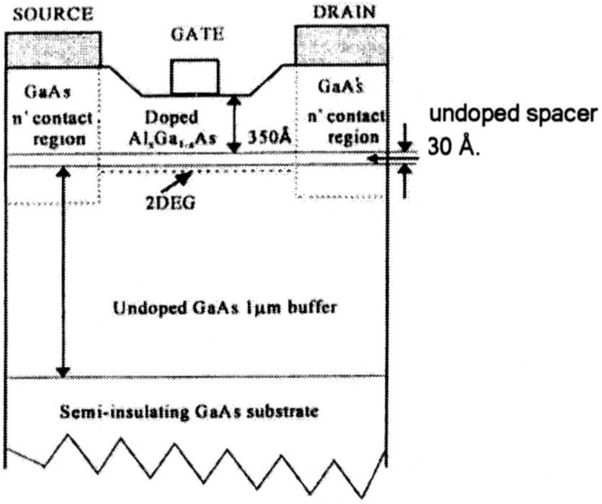

Fig. 18-17. Cross section view of 2D- electron gas FET (TEGFET) of $Al_xGa_{1-x}As/GaAs$.

in the current are obtained by multiplying by $(I_0/n_0)^2$. We eliminate ε_F using $\exp(\varepsilon_F - \varepsilon_c) = n_0/N_c$, so that

$$\exp[\pm\gamma_i(\varepsilon_F)] = (g_i)^{\pm1}(N_c/n_0)^{\pm1}\exp[\pm(\varepsilon_i - \varepsilon_c)/k_BT]; \qquad (18.14\text{-}41)$$

we then arrive at

$$[\mathcal{S}_i/I_0^2]_{\text{plateau},i} = \begin{cases} 4\tau_i(N_i/n_0N_cg_i)\exp[(\varepsilon_c-\varepsilon_i)/k_BT], & \text{(high } T) \\ 4\tau_i(N_iN_cg_i/n_0^3)\exp[(\varepsilon_i-\varepsilon_c)/k_BT], & \text{(low } T). \end{cases} \qquad (18.14\text{-}42)$$

Although τ_i and n_0 are temperature-dependent, a semi-logarithmic plot of the plateau values vs. 1000/T will result is 'inverted triangles'. The maximum occurs for T_{0i}, at which temperature $\gamma_i(\varepsilon_F) = 0$. This allows us to once more obtain $\varepsilon_c - \varepsilon_i$ if g_i is known, or, alternately, to obtain the value of the spin/valley degeneracy factor g_i. Plots of this nature have been made by Van Rheenen et al.[72] and by Bosman[70].

Here we shall report on recent measurements by Chen et al. for g-r noise in $Al_xGa_{1-x}As/GaAs$ HEMT's (high-electron mobility transistors), also called TEGFET's (two-dimensional electron-gas field-effect transistors)[73]; similar measurements have been made on blue laser GaN structures by Duran et al.[74] A cross-section view for this submicron hetero-structure device is shown in Fig. 18-17. The spectra were measured with a HP 3589A FFT/analogue wave analyzer between

---

[72] A.D. Van Rheenen, G. Bosman and R.J.J. Zijlstra, Solid-State Electron. **25**, 30 (1987).

[73] Yuping Chen, G.L. Larkins Jr. and H. Morkoç, IEEE Trans. El. Dev. **ED-47**, 2045 (2000).

[74] R.S. Duran, G.L. Larkins Jr., C.M. Van Vliet and H. Morkoç, J. Appl. Phys. **93**, 5337 (2003).

10 Hz and 10 MHz in the temperature range 78-295 *K*. Two or three Lorentzians were simultaneously observed. A computer program was used to find the least-square fit. An Arrhenius plot for $\log(\tau T^2)$ vs. $1000/T$ is given in Fig. 18-18(a) and a semi-log plot of the plateau values in shown in Fig. 18-18(b). Four activation energies are deduced, the values being given in the caption. Excellent agreement (within 5%) with the values from the maxima of the inverted triangles was obtained.

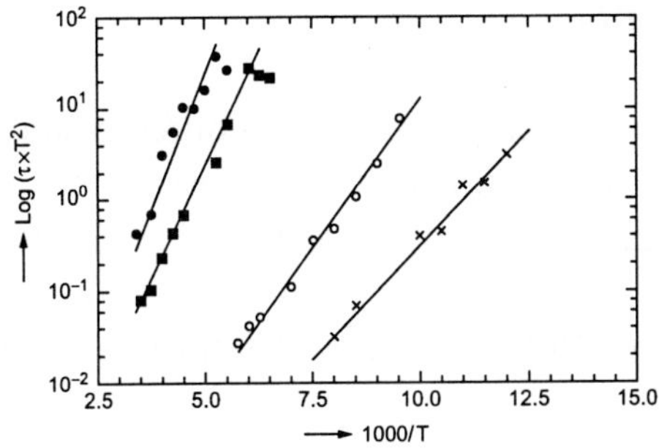

Fig. 18-18(a). Arrhenius plots for non-gated GaAs device. Symbols •: 240 meV, ∗: 200 meV, ∘: 130 meV, ×: 100 meV.

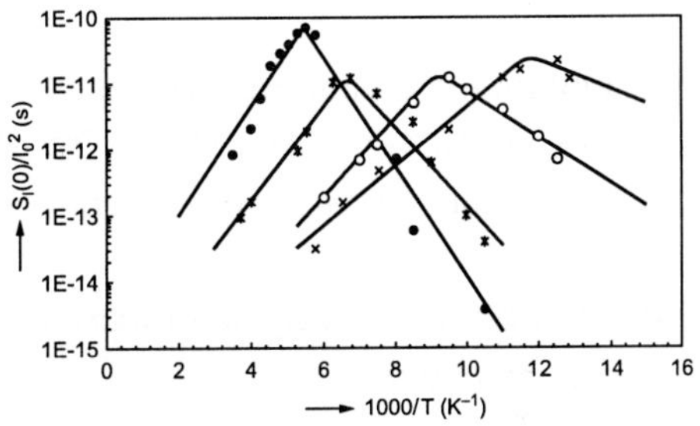

Fig. 18-18(b). Normalized plateau values for the same spectra vs. 1000/*T*.

Finally, a note on the assumption that the number of traps is small with respect to the free majority carriers. *If* the contrary is true, the traps will not act 'in series', but in parallel; then the reciprocal life times add and a single Lorentzian is observed.

Altogether, from a theoretical point of view, generation-recombination noise is an example *par excellence* of a multivariate Markov process; whereas Brownian motion represents fluctuations due to *intra-band* scattering, g-r noise involves fluctuations stemming from *interband* transitions. From an experimental point of view, g-r noise analysis is an eminent tool for impurity level spectroscopy. The principal problem is the quest for *pure* samples, so that 1/*f* noise is low. The high frequency limit is usually set by 'drowning' in the thermal noise, so that measurements seldom extend beyond ~ 30 *MHz*.

18.15 Problems to Chapter XVIII [75]

18.1 With a repeated random phase assumption (RRPA), the microscopic Chapman–Kolmogoroff equation was derived in Eq. (18.3-10). Assuming a density of states $\mathcal{X}(\mathbf{a})$, obtain from this the mesoscopic Chapman–Kolmogoroff equation.

18.2 Employing the property $\mathsf{M}\langle\boldsymbol{\alpha}\tilde{\boldsymbol{\alpha}}\rangle = \langle\boldsymbol{\alpha}\tilde{\boldsymbol{\alpha}}\rangle\tilde{\mathsf{M}}$, show that the thermal equilibrium spectra $\mathbb{S}_{\boldsymbol{\alpha\alpha}}(\omega) = 2(\mathsf{M} + i\omega\mathsf{I})^{-1}\langle\boldsymbol{\alpha\alpha}\rangle + 2\langle\boldsymbol{\alpha\alpha}\rangle(\tilde{\mathsf{M}} - i\omega\mathsf{I})^{-1}$ are real.

18.3 Obtain the conditional average $(\partial/\partial t)\langle e^{-\mathbf{s}\cdot\mathbf{a}(t)}\mathbf{aaa}\rangle_{\mathbf{a}'}$ by three-fold differentiation to $\mathbf{s}$ of (18.4-5). Analyze the result and let $t \to \infty$; obtain an expression for the third-order fluctuation moments $\langle\Delta\mathbf{a}\Delta\mathbf{a}\Delta\mathbf{a}\rangle$ in terms of M and the FP moments.

18.4 Multiply the steady-state master equation (19.5-34) with an arbitrary function $f(n)$, to obtain

$$\langle[f(n+1) - f(n)]g(n)\rangle = \langle[f(n) - f(n-1)]r(n)\rangle. \qquad (1)$$

Choose $f(n) = n$ and $f(n) = n^2$, respectively. Now expand $g(n)$ and $r(n)$ in a Taylor series about the average value n_0. Obtain Burgess'g-r theorem (18.5-44).

18.5 Brillouin scattering from a viscous fluid in a temperature gradient: Accept Eqs. (18.5-76) as given. Obtain the relaxation matrix M and the FP moment matrix B

$$M = \begin{pmatrix} 0 & iu^2\rho \\ i(q^2/\rho & D_\ell q^2/\rho \end{pmatrix}, \quad B = \begin{pmatrix} 0 & 0 \\ 0 & \rho^{-2}\mathbf{qq}:\mathsf{S}:\mathbf{q'q'} \end{pmatrix}. \qquad (2)$$

Given a temperature gradient $T(\mathbf{r})$, obtain the result for the spectral density of the pressure fluctuations in $\mathbf{q}$-space of the text, Eq. (18.5-80). Also, find an expression for the difference of the two peaks, $\Delta\mathcal{S}_{PP}(\mathbf{q},\mathbf{q}',\omega)$.

18.6 Assume that you have obtained the solution for the displacement $\mathbf{r}$ in an ensemble with given velocity $\mathbf{v}_0$ and position $\mathbf{r}_0$, as given by Uhlenbeck and Ornstein in (18.7-29). Obtain the mean square displacement $\langle s^2(t)\rangle_{\mathbf{v}_0} = \langle|\mathbf{r}(t) - \mathbf{r}_0|^2\rangle_{\mathbf{v}_0}$ and let $t \gg 1/\beta$; confirm Einstein's result. Also, for small t make a Taylor expansion up to second order and show that $\langle s^2(t)\rangle_{\mathbf{v}_0} \propto t^2$.

18.7 A Wiener–Lévy process[76] $y(t)$ is defined by its probabilities P and W_1, given as

[75] The bold printed problems are *essential* extensions of this chapter.

[76] N. Wiener, J. Math. and Phys. **2**, 131 (1923); P. Lévy, "Théorie de l'addition des variables aléatoires", Gauthier–Villars, Paris, 1937. See also N.G. van Kampen, Op. Cit. Chapter IV, Section 2; ibid. 3rd Ed. Chapter IV, Section 2.

$$P(y,t\,|\,y_0,t_0) = [1/2\pi\eta(t-t_0)]^{\frac{1}{2}} e^{-(y-y_0)^2/2\eta(t-t_0)}, \quad W_1(y_0) = \delta(y_0). \quad (3)$$

(a) Show that the above conditional probability satisfies the Chapman–Kolmogoroff equation [hint: employ the characteristic function];
(b) Obtain the (non-stationary) probability $W_1(y,t)$, using (18.2-16);
(c) Employing the characteristic function show that $\lim_{t\to 0+} P(y,t\,|y_0) = \delta(y-y_0)$.

18.8 A general Ornstein–Uhlenbeck process is defined by the stationary conditional probability

$$P(y,t\,|\,y_0) = \frac{1}{[2\pi(1-e^{-2\alpha\tau})]^{\frac{1}{2}}} \exp\left[-\frac{(y-y_0e^{-\alpha t})^2}{2(1-e^{-2\alpha\tau})}\right]. \quad (4)$$

(a) Show that the limit (18.2-12) exists and find $W_1(y)$ [a normal distribution];
(b) Show that it satisfies the Chapman–Kolmogoroff equation;
(c) Find $\langle y(t)\rangle_{y_0}{=}y_0$, the stationary correlation function $\phi_y(t)$ and $\mathcal{S}_y(\omega)$, being a Lorentzian.
(d) Prove Doob's theorem which states that the O–U process is the only one-variable Markovian process that is stationary and normal; cf. Refs. 1(f) and 33.

18.9 Suppose a plot of the relative Allen variance reveals the probability for Lorentzian noise, i.e., you fit the data by Eq. (18.12-13). Perform the Mellin transform (18.12-24) to obtain the spectrum. Hint. Note the reflection property $\Gamma(p)\Gamma(1-p) = \pi/\sin\pi p$. Sum over the residues of the meromorphic function $1/\sin(p\pi/2)$.

18.10 A random walk on an infinite cubic lattice of sides a is described by a Markov chain with $\mathbf{r} = \mathbf{n}a$, $t = s\tau$, where $\mathbf{n} = (n_x, n_y, n_z)$ is the number of steps of length a and duration τ. The usual result $W(\mathbf{r},t) = \int d^3r' P(\mathbf{r},t\,|\,\mathbf{r}',t')W(\mathbf{r}',t')$ then reads

$$W\big(\mathbf{n}a,(s+1)\tau\big) = \sum_{m=(-\infty,-\infty,-\infty)}^{(\infty,\infty,\infty)} P\big(\mathbf{n}a,(s+1)\tau\,|\,\mathbf{m}a,s\tau\big)W(\mathbf{m}a,s\tau). \quad (5)$$

Given are the *a priori* probabilities

$$P\big(\mathbf{n}a,(s+1)\tau\,|\,\mathbf{m}a,s\tau\big) = \tfrac{1}{6}\sum_{xyz}[\delta_{n_x,m_x+1} + \delta_{n_x,m_x-1}]. \quad (6)$$

Now substitute into (5), subtract $W(\mathbf{n}a,s\tau)$ and divide by τ. Expand the rhs in a Taylor series about $\mathbf{r}{=}\mathbf{n}a$; let $a,\tau\to 0$ with $D_0 = a^2/6\tau$ being the finite diffusion coefficient, in accord with Einstein's result$[\langle|\Delta\mathbf{r}|^2\rangle \equiv a^2 = 6D_0\tau]$. Obtain the diffusion equation (18.7-1).

18.11 Prove the generalization of Campbell's theorem, involving the n-th order cumulant, stated by Rice, Eq. (18.11-14).

18.12 Find the moments $\langle y(t_1)y(t_2)y(t_3)y(t_4)\rangle$ for the Wiener–Lévy process (3), using the rules for a normal distribution.

18.13 If the variables a_i to describe a stochastic Markov process are *incomplete*, the stochastic vector $\mathbf{a}(t)$, describing the 'projection' of a Markov process, may still satisfy a Langevin equation, but the spectrum of the source-function $\xi(t)$ is no longer white, i.e., it has *coloured noise.* Using Fourier analysis (or the Fourier–Laplace transform, see footnote[24]), obtain the spectral densities $\mathbb{S}_{\mathbf{a}}(\omega)$ in terms of $\mathbb{S}_{\xi}(\omega)$. Give sufficient conditions for the correlation function $\Phi_{\mathbf{a}}(\vartheta)$ to decay for $\vartheta \to \infty$.

18.14 Non-Markovian stochastic processes can often be described with the *generalized* Langevin equation with memory function $\gamma(t)$ [77,78,79]

$$du/dt = \int_0^t dt'\gamma(t-t')u(t') + \xi(t). \tag{7}$$

(a) Find the spectrum $\mathbb{S}_u(\omega)$ in terms of $\mathbb{S}_\xi(\omega)$, employing Fourier–Laplace transforms;

(b) Assuming that $\xi(t)$ is still Gaussian and zero-centred, show that the existence of a stationary equilibrium state requires that

$$\langle \xi(t)\xi(t') = (k_B T/m)\,\gamma(t-t'). \tag{8}$$

Note that this replaces the ordinary relationship (18.7-12). For further details, cf. Kubo et al.[80]

(c) Also establish the correlations

$$\langle u(t)u(t+\vartheta)\rangle = \frac{k_B T}{2\pi m}\int_{-\infty-i\varepsilon}^{\infty-i\varepsilon} \frac{e^{i\omega\vartheta}}{i\omega + \hat{\gamma}(i\omega)}\,d\omega\,, \tag{9}$$

where the integration path is just below the real axis. Assuming further that $\lim_{|\omega|\to\infty} \hat{\gamma}(i\omega) = \text{finite}$, show by contour integration that

$$\lim_{\vartheta\to 0+} \langle u(t)u(t+\vartheta)\rangle \equiv \langle u^2\rangle = k_B T/m\,. \tag{10}$$

---

[77] H. Mori, Progress Theor. Physics **33**, 423 (1965).

[78] Katja Lindenberg and B.J. West, "The Nonequilibrium Statistical Mechanics of Open and Closed Systems", Wiley-VCH, NY and Weinheim 1990, Chapter 4.

[79] R. Kubo, J. Math. Phys. **4**, 174 (1963).

[80] R. Kubo, M. Toda and N. Hahitsume, "Statistical Physics II" Springer Verlag, Berlin and Heidelberg, 1985, p. 31ff.

Chapter XIX

Branching Processes and Continuous Stochastic Phenomena

1. THE COMPOUNDING THEOREM AND APPLICATIONS

19.1 The Compounding Theorem, Variance Theorem and Addition Theorem

Quite often one encounters the problem of finding the distribution for a particle stream J_m, which results from a primary stream J_n that undergoes a random process involving reduction, multiplication or branching. Explicitly, let the 'output particles', sampled over a period $(t, t+\theta)$ be $m(t)$, the random process being $\{X_k\}$ and the 'input particles' in the same period be $n(t)$. Assuming that the distribution of the $n(t)$ is known, together with the *a priori* probabilities for the process $\{X_k\}$, we seek to obtain the distribution for the particles $m(t)$.[1,2] The answer lies in the *compounding theorem* as found in Feller's book,[3] which reads as follows. Let $\chi(z)$ be the Fowler generating function for the distribution $W(n)$, $\phi(z)$ that for the process $p(X)$ and $\Phi(z)$ that for the distribution $W(m)$; then we have the connection

$$\Phi(z) = \chi[\phi(z)]. \tag{19.1-1}$$

Obviously, this is a very powerful theorem, which is the cornerstone for the description of all reduction, multiplication and branching processes.

To prove it, we can either employ an *elementary event* description, or a *collective description*; both methods have their merits. Starting with the former, we have

$$m(t) = \sum_{k=1}^{n(t)} X_k . \tag{19.1-2}$$

Now for independent X_k we have for the generating function:

$$\Phi(z) = \sum_n W(n) \sum_m z^m \sum_{\{X_k\}} \prod_{k=1}^{n} P(X_k) \delta_{\Sigma_k X_k, m} = \sum_n W(n) \sum_{\{X_k\}} z^{\Sigma_k X_k} \prod_{k=1}^{n} P(X_k)$$

---

[1] As usual in physics, the words probability and distribution will be used interchangeably.

[2] Our notation is different from that in subsection 18.12.2: we use $m(t)$ for what there was called $M_\theta(t)$ and $J_m(t)$ for what there was called the rate $m(t)$. M is reserved for the multiplication.

[3] W. Feller, " An introduction to Probability Theory and its Applications", Vol. I, Wiley and Sons, NY 1950, Chapter XII.

$$= \sum_n W(n) \sum_{\{X_k\}} \prod_{k=1}^n z^{X_k} P(x_k) = \sum_n W(n) \prod_{k=1}^n \sum_{X_k} z^{X_k} P(X_k)$$

$$= \sum_n W(n)[\phi(z)]^n = \chi[\phi(z)]. \qquad \text{QED}$$

For the collective point of view, we start with $W(n)$ as resulting from Bayes' rule

$$W(m) = \sum_n P(m|n) W(n), \tag{19.1-3}$$

or,

$$\Phi(z) = \sum_n W(n) \sum_m z^m P(m|n). \tag{19.1-4}$$

The last sum is the generating function for fixed n; as shown in (1.7-39) this is simply

$$\sum_m z^m P(m|n) = \sum_m z^m P\left(m = \sum_{k=1}^n X_k \right) = \prod_{k=1}^n \langle z^{X_k} \rangle = [\phi(z)]^n. \tag{19.1-5}$$

Substituting into (19.1-4), the same result is obtained.

Consequences

The compound distribution follows from, cf. (1.7-42),

$$W(m) = \frac{1}{m!} \left(\frac{\partial^m \Phi}{\partial z^m} \right)_{z=0}, \tag{19.1-6}$$

while the factorial moments are given by

$$\langle m(m-1)...(m-k+1) \rangle = (\partial^k \Phi / \partial z^k)_{z=1}. \tag{19.1-7}$$

In particular, for the mean value we have

$$\langle m \rangle = \Phi'(1) = \left(\frac{\partial \chi}{\partial \phi} \right)_{\phi=\phi(1)} \left(\frac{\partial \phi}{\partial z} \right)_{z=1} = \left(\frac{\partial \chi}{\partial \phi} \right)_{\phi=1} \left(\frac{\partial \phi}{\partial z} \right)_{z=1}, \tag{19.1-8}$$

since by definition of generating function $\phi(1) = 1$. This yields $\langle m \rangle = \langle n \rangle \langle X \rangle$. Likewise, for the second factorial moment

$$\langle m(m-1) \rangle = \left(\frac{\partial^2 \chi}{\partial \phi^2} \right)_{\phi=1} \left(\frac{\partial \phi}{\partial z} \right)_{z=1}^2 + \left(\frac{\partial \chi}{\partial \phi} \right)_{\phi=1} \left(\frac{\partial^2 \phi}{\partial z^2} \right)_{z=1}$$

$$= \langle X \rangle^2 \langle n(n-1) \rangle + \langle n \rangle \langle X(X-1) \rangle, \tag{19.1-9}$$

from which we obtain Burgess' *variance theorem*[4]

[4] R.E. Burgess, Disc. Faraday Soc. **28**, 151 (1959).

$$\begin{aligned}\langle \Delta m^2 \rangle &= \langle m(m-1) \rangle - \langle m \rangle (\langle m \rangle - 1) \\ &= \langle X \rangle^2 \langle \Delta n^2 \rangle + \langle n \rangle \langle \Delta X^2 \rangle .\end{aligned} \tag{19.1-10}$$

The spectrum of the rate J_m is now also readily obtained. For the short time average:

$$J_{m\theta} = (1/\theta) \int_t^{t+\theta} J_m(t')dt' = m/\theta . \tag{19.1-11}$$

So,

$$\langle \Delta J_{m\theta}^2 \rangle = \langle \Delta m^2 \rangle / \theta^2 . \tag{19.1-12}$$

If the spectra for both rates are white, we can apply Milatz's theorem,

$$\mathbb{S}_{J_m} = 2\theta \langle \Delta J_{m\theta}^2 \rangle = (2/\theta) \langle \Delta m^2 \rangle , \tag{19.1-13}$$

and likewise

$$\mathbb{S}_{J_n} = 2\theta \langle \Delta J_{n\theta}^2 \rangle = (2/\theta) \langle \Delta n^2 \rangle . \tag{19.1-14}$$

Thus, multiplying both sides of (19.1-10) by $(2/\theta)$, we find

$$\mathbb{S}_{J_m} = \langle X \rangle^2 \mathbb{S}_{J_n} + 2 \langle J_n \rangle \langle \Delta X^2 \rangle , \tag{19.1-15}$$

which is the variance theorem for the spectra. In particular, if $I = eJ$ is an electrical current, we find

$$\mathbb{S}_{I_m} = \langle X \rangle^2 \mathbb{S}_{I_n} + 2e \langle I_n \rangle \langle \Delta X^2 \rangle . \tag{19.1-16}$$

The first term is reduced or multiplied shot noise; the second term is extra noise stemming from the coupling process. Because of the form of departure (19.1-2), the process $\{X\}$ does not affect the spectral shape. More generally, we may have a modulated emission process with $\{X\} = \{X(t)\}$, see subsection 18.10.2; then the spectral density of the output may be non-white, even if that of the input rate is white, cf. (18.10-37). Needless to say that the compounding theorem is 'too static' to handle those processes. While our previous method is more general, many processes can be handled with the compounding theorem; moreover, branching processes are already very complex in the present description and would be prohibitively so if the method of modulated emission processes were employed.

One other theorem is still needed, called the *addition theorem*. Let

$$X = y_0 + y_1 + \ldots + y_N . \tag{19.1-17}$$

If the generating functions of the y_k are $g_k(z)$, then the generating function of X is, with the y's being independent variables,

$$\phi(z) = \sum_X z^{\Sigma_k y_k} P(y_1) \ldots P(y_N) = \sum_{\{y_k\}} \prod_{k=1}^{N} z^{y_k} P(y_k) = \prod_{k=1}^{N} g_k(z) . \tag{\textbf{19.1-18}}$$

Note that if $y_0 = 1$ (being its only probability with $P = 1$), then $g_0 = z$.

19.2 Bernoulli and Geometric Compounding

In this section we shall consider several straightforward applications of the compounding theorem and connect them with statistical mechanical problems.

Reduction Process Involving Bernoulli Trials

(i) Suppose $W(n)$ is a *Poisson process* and X is a Bernoulli trial. Such a process often occurs. For instance, a cathode may emit electrons in a Poisson stream, but only a fraction η reaches the anode, with $1-\eta$ going elsewhere. The generating function for a Bernoulli trial is

$$\phi(z)=\sum\nolimits_X z^X P(X)=\eta z+1-\eta\,. \tag{19.2-1}$$

Therefore

$$\Phi(z)=\mathcal{X}[\eta(z-1)+1]\,. \tag{19.2-2}$$

Since for a Poisson process

$$\mathcal{X}(z)=\exp[-\langle n\rangle(1-z)]\,, \tag{19.2-3}$$

we find

$$\begin{aligned}\Phi(z)&=\exp\{-\langle n\rangle[1-\mathcal{X}(z)]\}\\&=\exp\{-\langle n\rangle[1-(\eta(z-1)+1)]\}\\&=\exp[-\langle m\rangle(1-z)]\,.\end{aligned} \tag{19.2-4}$$

We conclude: *a Poisson process is invariant against Bernoulli compounding.*

(ii) The situation is different if the input process is *not* Poissonian. E.g., let us have a photon stream that falls on a partial absorber. Describing the latter again as a Bernoulli trial, we have, using this time the variance theorem (19.1-12):

$$\begin{aligned}\langle\Delta m^2\rangle&=\eta^2\langle n\rangle(1+B)+\langle n\rangle\eta(1-\eta)\\&=\eta\langle n\rangle(1+\eta B)=\langle m\rangle(1+\eta B)\,.\end{aligned} \tag{19.2-5}$$

Thus, the boson factor *degrades*, $B\rightarrow\eta B$.

(iii) *Collective point of view.* We use the result from Bayes' rule (19.1-5). The conditional probability $P(m\,|\,n)$ is the binomial distribution [see problem 1.4] :

$$P(m\,|\,n)=\frac{n!}{(n-m)!m!}\eta^m(1-\eta)^m\,. \tag{19.2-6}$$

Indeed, this gives the same result, for now

$$W(m) = \sum_n \frac{n!}{(n-m)!m!} \eta^m (1-\eta)^m W(n), \tag{19.2-7}$$

so that

$$\Phi(z) = \sum_m \sum_n \frac{n!}{(n-m)!m!} z^m \eta^m (1-\eta)^m W(n)$$
$$= \sum_n (z\eta + 1 - \eta)^n W(n) = \mathcal{X}[\eta(z-1)+1], \tag{19.2-8}$$

in accord with (19.2-2).

Multiplication Process Involving Geometric Variables

The process is indicated in Fig. 19-1. An incoming particle produces offspring; the *a priori* probability for each production is assumed to be constant regardless of previous productions. The total offspring of one particle forms the event X_k; the totality of all offsprings forms the output m. This may pertain to the process of secondary emission and photomultipliers; we shall however, not discuss these applications here.[5]

(i) *Independent events point of view.* The probability for each event is given by the geometric distribution ($\beta < 1$) :

$$P(X=0) = C, \quad P(X=1) = C\beta,$$
$$P(X=2) = C\beta^2, \ldots, \quad P(X=\ell) = C\beta^\ell . \tag{19.2-9}$$

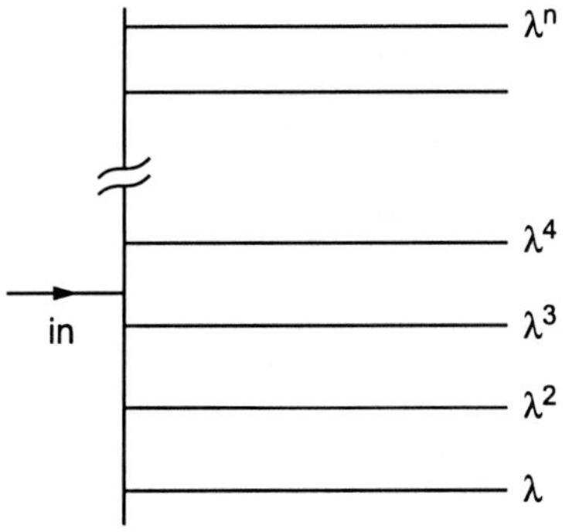

Fig. 19-1. Geometric multiplication process.

[5] K.M. Van Vliet and L.M. Rucker, "Noise associated with reduction, multiplication and branching processes", Physica **95A**, 117 (1979).

The normalization constant is found to be $C = 1-\beta$. So, also

$$P(X=\ell)=(1-\beta)\beta^{\ell}\,. \tag{19.2-9'}$$

From this we find for the cumulative distribution

$$P(X \le n-1)=1-\beta^{n}\,,\quad P(X \ge n)=\beta^{n}\,. \tag{19.2-10}$$

Thus the geometric distribution reflects that the probability for producing at least one particle is β, for producing at least two particles is β^2, etc. The generating function is found to be

$$\begin{aligned}\phi(z) &= \sum_X z^X P(X) = \sum_{\ell=1}^{\infty} z^{\ell}\beta^{\ell}(1-\beta)\\ &= (1-\beta)/(1-\beta z)\,.\end{aligned} \tag{19.2-11}$$

For the mean multiplication per particle we obtain

$$M \equiv \langle X\rangle = \phi'(1) = \beta/(1-\beta)\,. \tag{19.2-12}$$

Expressing β as $\beta = M/(1+M)$, we have the alternate form for the geometric distribution and its generating function

$$P(X_{\ell}) = \frac{M^{\ell}}{(M+1)^{\ell+1}}\,,\qquad \phi(z)=\frac{1}{M(1-z)+1}\,. \tag{19.2-13}$$

The compound distribution for m is now

$$\Phi(z)=\chi[(M(1-z)+1)^{-1}]\,. \tag{19.2-14}$$

Differentiating, we have

$$\begin{aligned}\Phi'(z) &= \chi'(\phi)\frac{M}{[M(1-z)+1]^2}\,,\\ \Phi''(z) &= \chi''(\phi)\frac{M^2}{[M(1-z)+1]^4}+\chi'(\phi)\frac{2M^2}{[M(1-z)+1]^3}\,.\end{aligned} \tag{19.2-15}$$

From this one finds $\langle m\rangle = M\langle n\rangle$, as expected and

$$\langle \Delta m^2\rangle = M^2\langle \Delta n^2\rangle + M(M+1)\langle n\rangle\,, \tag{19.2-16}$$

with corresponding results for the white noise spectra.

(ii) *Collective point of view.* This adds some more clout to this distribution and establishes a link with statistical mechanics proper. To obtain $P(m\,|\,n)$ we need a distribution that gives the number of output particles, with no limit the store of m, i.e., the number of offspring produced. This problem is the same as for the occupancy of boson states studied before [cf. subsection 2.5.2(b), Eq. (2.5-29)]. Hence, we have

$$P(m\,|\,n) = \frac{(m+n-1)!}{m!(n-1)!}\beta^m (1-\beta)^n . \tag{19.2-17}$$

In the statistical literature this distribution is also known as the *negative binomial* distribution since with[6] the reflection property $(z-1)! = \pi/\sin(\pi z)(-z)!$, the result is

$$P(m\,|\,n) = (-1)^m \frac{(-n)!}{m!(-n-m)!}\beta^m (1-\beta)^n . \tag{19.2-17'}$$

The generating function for this distribution is

$$G(z) = \sum_{m=0}^{\infty} z^m P(m\,|\,n) = \left[1 + \frac{\langle m \rangle}{n}(1-z)\right]^{-n} . \tag{19.2-18}$$

Here $\langle m\rangle/\langle n\rangle \equiv \langle m\,|\,n\rangle/\langle n\rangle = M$. The factor in $[1+M(1-z)]^{-1}$ is clearly $\phi(z)$. Thus, the generating function for the output particles $\Phi(z)$ is again (19.2-14); in all examples the elementary events method and the collective method are complementary, the latter having some extra information, but leading to the same results.

One generation of offspring

In the simplest branching process, which may occur in solids at the onset of avalanching, one considers the incoming particle (zeroth generation) and one generation of offspring. The output is now the offspring plus the original particle, these being usually indistinguishable in energy or otherwise. Let p_1 be the membership of the first generation and let the chance for offspring production be a Bernoulli trial with *a priori* probability λ; $\omega(z) = \lambda(z+1)+1$. Now the random event is $X = 1 + p_1$. Hence,

$$\phi(z) = \sum_X z^X P(X) = \sum_{p_1} z^{1+p_1} P(p_1) . \tag{19.2-19}$$

It is left to the reader to find the generating function for the output, $\Phi(z)$. The overall output variance is found to be

$$\langle \Delta m^2 \rangle = M^2 \langle \Delta n^2 \rangle + (M-1)(2-M)\langle n \rangle , \tag{19.2-20}$$

where $M = \langle X \rangle = 1 + \lambda$. The process is sub-Poissonian since $(M-1)(2-M) < M$. The effect has been observed in avalanching of carriers in reverse biased p-n junctions and JFETs by Rucker.[7]

---

[6] The factorials in these formulae are to be interpreted as Gamma functions.

[7] L.M. Rucker, "An Investigation of Some Properties of Carriers in Field-effect Devices", Ph. D. Thesis, University of Florida (1977).

2. METHOD OF RECURRENT GENERATING FUNCTIONS

19.3 Preamble

Avalanche multiplication occurs in a variety of semiconductor structures. Most known are reverse-biased p-n junctions in which the process is usually started by impact ionization; often the process involves only one type of carriers, say electrons, whereby the other type of carriers (holes) are carried off at an electrode. The collected carriers, consisting of the inciting carriers plus all generations of offspring, are collected in the output current that flows in the external circuit. The device is characterized by its mean multiplication M and by the magnitude of the fluctuation spectrum $\mathbb{S}_{I_a}$ of the final current. Secondly, there are photo-avalanche diodes, employed for the detection of weak optical signals. Obviously, these are only useful if the multiplication factor is high and the output noise is low. Almost always the avalanche process involves two types of carriers. The incident light creates a hole-electron pair, but they need not both be free. So, the avalanche process is initiated by either an electron or a hole, depending on the design of the avalanche region and wavelength of the light. The photo-liberated carrier (zeroth generation) than produces hole-electron pairs (first generation) which in turn ionize new pairs (second generation) and so on. Statisticians call this process branching, while the physicist speaks of an avalanche process. An accurate description is of paramount importance both for the statistician and the physicist or engineer. We will make some initial observations in order to classify the possible processes.

First of all, processes involving the ionization by one type of carrier only, causing singly-incited branching, are far simpler than processes in which both types of carriers can initiate ionization. In the former case the 'tree' is maintained by a single type of ionization process, identical for all generations if there is no dependence on carrier energy and the carriers ionize with a fixed *a priori* probability. This is the case that will be dealt with in Section 19.4. If, on the other hand, both carriers can ionize, usually with different *a priori* probabilities, we have doubly-incited branching with the tree being maintained by two types of offspring; the description for that case is immeasurably more complicated. It will be discussed in Section 19.5.

While physicists have given theories for avalanche multiplication and noise since the mid-sixties[8], we noted at the time that the methods employed were (a) often deficient from a statistical point of view; (b) do generally not apply to present-day smaller structures. All theories cited assume that the ionization trajectory has infinite length with no limit to the number of possible ionizations per transit. So, encouraged

---

[8] A.S. Tager, Sov. Phys.–Solid State, **6**, 1919-1925 (1965); R.J. McIntyre, IEEE Trans. El. Dev. **ED-13**, 164-168 (1965); ibid. **ED-19**, 703-713 (1973); S.D. Personick, Bell Syst. Tech. J. **50**, 167-190, (1971).

by new experimental results and partial theories by Rucker[7] and Lukaszek[9] we decided to provide a novel, discrete theory for both one-carrier and two-carrier avalanche processes. Simultaneously, we believed it to be imperative to connect more closely with the mathematical description of statistical branching processes – rather than to give a physical *ad hoc* theory – since branching processes are often embedded in stochastic phenomena. Whereas for singly-incited branching processes suitable existing techniques involving multiple-compounded generating functions are applicable (covered in Ref. 5), for more complex branching a new scheme was devised, called the *method of recurrent generating functions.*[10,11]

As a point of reassurance, it was established that the older *continuum* theories keep their validity as 'asymptotic' theories: in the limit that the number of ionizations per transit becomes infinite, those results are recovered from the new *discrete-process* approach. Therefore, this more general treatment will be presented here.

19.4 Singly-Incited Branching Processes. One-Carrier Avalanche

The multiplication process is pictured in Fig. 19-2. We consider N possible ionizations per transit through the avalanche region, labelling the first ionization by N, the next one by $N-1$, etc., up to the last one, labelled as 1. This labelling is chosen so that the designation indicates the number of ionizations which is possible from the stated position onward. The various generations are indicated as the zeroth,

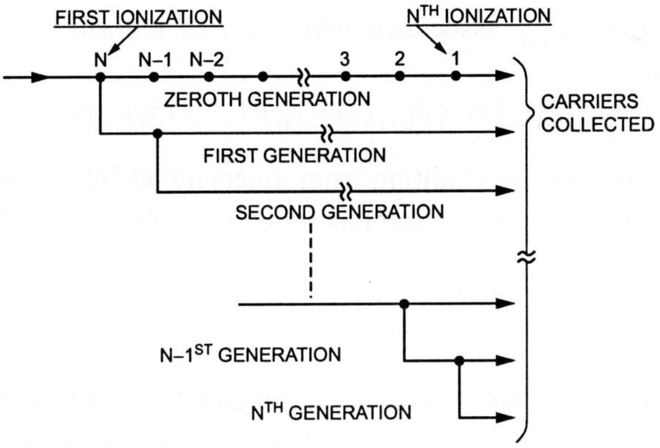

Fig. 19-2. Impact ionization process involving N possible ionizations per primary carrier transit.

---

[9] W.A. Lukaszek, "Conduction Mechanisms in Low Breakdown Voltage Silicon P-N$^+$ Junctions", Ph.D. Thesis University of Florida (1974).

[10] K.M. Van Vliet and L.M. Rucker, "Theory of Carrier Multiplication and Noise in Avalanche Devices–Part I: One-Carrier Processes", IEEE Trans. El. Dev. **ED-26**, 746-751 (1979); Also, Ibid. and A. Friedmann, "–Part II: Two-Carrier Processes", IEEE Trans. El. Dev. **ED-26**, 752-764 (1979).

[11] Since the literature on branching processes is vast, we cannot be certain that similar algorithms have not been reported elsewhere.

the first...the N-th generation. We note that the number of generations is finite, since all carriers are successively carried off at the output electrode.[12] Denoting by p_i the various generations, we have for the tree X of the total branching process:

$$X = 1 + p_1 + p_2 + ... + p_N. \qquad (19.4\text{-}1)$$

The generating function for X is denoted as $\phi_N(z)$. Since the various generations are correlated, a calculation of ϕ_N based on the decomposition (19.4-1) is cumbersome; we therefore do not present this method.

We now describe the method of recurrent generating functions. Let us consider the subtrees arising from the first, second, ..., Nth ionization event of the incoming particle; these subtrees are denoted by y_N, y_{N-1},..., y_1, respectively. Clearly, this allows for another decomposition

$$X = X_N = 1 + y_N + y_{N-1} + ... + y_1. \qquad (19.4\text{-}2)$$

Hereby it is observed that the subtree y_N is identical to the tree X, except that (i) only $N-1$ ionizations can occur and that (ii) this subtree is 'fed' by a Bernoulli trial with *a priori* chance for ionization λ_1. Thus, if g_N denotes the generating function of y_N, then by the compounding theorem

$$g_N(z) = \omega[\phi_{N-1}(z)], \qquad (19.4\text{-}3)$$

where ω is the generating function for this Bernoulli trial $\omega(z) = \lambda z + 1 - \lambda$. Hence,

$$g_N(z) = \lambda_1 \phi_{N-1}(z) - \lambda_1 + 1. \qquad (19.4\text{-}4)$$

Similarly, the subtree y_{N-i+1} associated with the i th ionization event leads to the generating function

$$g_{N-i+1}(z) = \lambda_i\, \phi_{N-i}(z) - \lambda_i + 1, \qquad i = 1...N. \qquad (19.4\text{-}5)$$

Since X is built up by subtrees resulting from independent Bernoulli trials, it follows from (19.4-2), (19.4-5) and the addition theorem (19.1-18) that

$$\begin{aligned}\phi_N(z) &= z g_N(z) g_{N-1}(z) ... g_1(z)\\ &= z[\lambda_1 \phi_{N-1}(z) - \lambda_1 + 1][\lambda_2 \phi_{N-2}(z) - \lambda_2 + 1]...[\lambda_N \phi_0(z) - \lambda_N + 1]. \qquad (19.4\text{-}6)\end{aligned}$$

This is the basic recurrent relation sought for. It can, however, be greatly simplified by writing down the same result for $N \to N-1$. Call this result (19.4-6'). Now divide (19.4-6) by (19.4-6'), to obtain

$$\phi_N(z) = \phi_{N-1}(z)[\lambda_1 \phi_{N-1}(z) - \lambda_1 + 1]. \qquad \mathbf{(19.4\text{-}7)}$$

---

[12] In the stylized process of Fig. 19-2, to be discussed in more detail below, we spaced the ionizations evenly, the rational being that a carrier coming from the left looses all its energy upon ionizing, after which it needs the same mean free path to ionize a second time, and so on. So, the number of generations is equal to the number of possible ionizations per transit. However, this is not essential for the mathematical framework of the theory; later below we consider position-dependent mean free paths.

Likewise, writing (19.4-6) down for $N \to N-i+1$ and for $N \to N-i$ we find, upon dividing the two results,

$$\phi_{N-i+1}(z) = \phi_{N-i}(z)[\lambda_i \phi_{N-i}(z) - \lambda_i + 1], \quad i = 1 \ldots N. \qquad \textbf{(19.4-7')}$$

Eqs. (19.4-7) and (19.4-7') are the final forms of the recurrent relation for the generating function. As boundary condition we look for the tree of the *N*th ionization. If it takes place, the resulting particle cannot ionize again but is collected at the electrode. This event, then, exists with certainty 1, indicating that

$$\phi_0(z) = z. \qquad (19.4\text{-}8)$$

Multiplication for constant ionization probability

First, let us consider the case that all λ_i are equal, say λ. Differentiating (19.4-7), we obtain

$$\phi_N'(z) = \phi_{N-1}'(z)[\lambda\phi_{N-1}(z) - \lambda + 1] + \phi_{N-1}(z)\lambda\phi_{N-1}'(z), \qquad (19.4\text{-}9)$$

from which, noting that

$$\phi_N(1) = \phi_{N-1}(1) = \ldots = \phi_0(1) = 1 \qquad (19.4\text{-}10)$$

we find

$$\phi_N'(1) = (1+\lambda)\phi_{N-1}'(1). \qquad (19.4\text{-}11)$$

From the bc (19.4-8), we have also $\phi_0'(1) = 1$. We thus obtain

$$M = \phi_N'(1) = (1+\lambda)^N. \qquad (19.4\text{-}12)$$

It is also possible to solve for $\phi_N(z)$ itself, but we shall have no need for this. As to the asymptotic result, see the more general form below.

Multiplication for position-dependent ionization probability

Usually $\lambda = \lambda(E)$ depends on position, since the accelerating electric field depends on position. Repeating the previous arguments, one readily obtains the more general result

$$M = \prod_{i=1}^{N}(1+\lambda_i), \qquad (19.4\text{-}13)$$

where λ_i is the *a priori* probability at the site of the i th possible ionization. Now let l_i be the path over which the particle must accelerate after the $(i-1)$ th ionization before the i th ionization can occur. We will write $\lambda_i = \alpha_i l_i$, where α_i is the ionization probability per unit length at site i. We also introduce an average mean free path for ionization in the total width w, $l = w/N$. We have the identity

$$\lambda_i = \alpha_i (l_i/l)(w/N). \qquad (19.4\text{-}14)$$

In this relationship, the factor (l_i/l) is considered to be bounded. For the asymptotic form of large N we then have

$$\ln M = \sum_{i=1}^{N} \ln\left[1+\frac{\alpha_i l_i}{l}\frac{w}{N}\right] = \frac{1}{N}\sum_{i=1}^{N} \ln\left[1+\frac{\alpha_i l_i}{l}\frac{w}{N}\right]^N$$
$$\sim \frac{1}{N}\sum_{i=1}^{N} \ln\exp\left(\frac{\alpha_i l_i}{l} w\right) = \sum_{i=1}^{N} \alpha_i l_i \,. \qquad (19.4\text{-}15)$$

Going over to a Riemann integral, we obtain

$$M(0,x) \sim \exp\left[\int_0^x \alpha(x')dx'\right], \qquad M(0,w) \sim \exp\left[\int_0^w \alpha(x)dx\right]. \qquad (19.4\text{-}16)$$

This is the result of the continuum theories cited in Ref. 8.

The variance

For simplicity we consider here only the case that all λ_i are equal. Differentiating (19.4-9) once more, we have

$$\phi_N''(z) = \phi_{N-1}''(z)[\lambda\phi_{N-1}(z) - \lambda + 1] + 2[\phi_{N-1}'(z)]^2\lambda + \phi_{N-1}(z)\phi_{N-1}''(z)\lambda\,, \quad (19.4\text{-}17)$$

or with (19.4-10) and (19.4-12)

$$\phi_N''(1) = \phi_{N-1}''(1)(1+\lambda) + 2\lambda(1+\lambda)^{2N-2}\,, \qquad (19.4\text{-}18)$$

with the bc $\phi_0''(1) = 0$. This recurrent relation is easily solved. One obtains

$$\phi_N''(1) = 2\lambda(1+\lambda)^{N-1}[1+(1+\lambda)+...+(1+\lambda)^{N-1}]$$
$$= 2(1+\lambda)^{N-1}[(1+\lambda)^N - 1]. \qquad (19.4\text{-}19)$$

This yields for the variance

$$\operatorname{var} X = \phi_N''(1) - [\phi_N'(1)]^2 + [\phi_N'(1)]$$
$$= 2(1+\lambda)^{N-1}[(1+\lambda)^N - 1] - (1+\lambda)^{2N} + (1+\lambda)^N$$
$$= (1-\lambda)(1+\lambda)^{N-1}[(1+\lambda)^N - 1] = \frac{1-\lambda}{1+\lambda} M(M-1). \qquad (19.4\text{-}20)$$

Substituting for λ in terms of M, this can also be written as

$$\operatorname{var} X = (2 - M^{1/N})M^{1-1/N}(M-1). \qquad (19.4\text{-}21)$$

For $N = 1$, applicable at the onset of ionization, this yields $\operatorname{var} X = (2-M)(M-1)$, in accord with results of Refs. 7 and 9. For large N, one recovers a result by Tager[8],

$$\operatorname{var} X \sim M(M-1). \qquad (19.4\text{-}22)$$

From (19.1-16) one sees that in the asymptotic case the multiplied primary noise and the noise from the avalanche process contribute about equally.

19.5 Doubly-Incited Branching Processes. Two-carrier Avalanche

In this section it will be assumed that both carriers of the electron-hole pair, liberated by photon absorption or otherwise, can ionize, with *a priori* probability λ and μ, respectively. We thus have a doubly-incited branching process. If one of these, say μ, goes to zero, the treatment becomes simplified and the results of the previous section should be recovered. Also, if $\lambda = \mu$, a much simpler situation prevails which, for the continuum case was treated by Tager, Op. Cit.; however, the fluctuations in the branching process var$X$ become of order M^3 and the process degrades rapidly with increasing gain; this situation is to be avoided. As in the previous section we will allow for position dependence of the ionization process in the computation of the mean multiplication or gain M; for the noise we shall, however, assume that mean values of the ionization probabilities can be assigned.

The physical picture in sketched in Fig. 19-3. The range (width) of the avalanche region is denoted by w, with the negative electrode being at the left and the positive electrode being at the right. We assume that the magnitude of the field $|E|$, if not constant, increases with x, i.e., toward the right. Electrons are propelled to the positive electrode and may ionize when the energy gained from the field $e|E| \sim \varepsilon_G$, where ε_G is the bandgap. The reason that $\lambda_i < 1$, where λ_i is the local ionization probability, is that energy can be lost in other collisions, e.g., with phonons; however, we assume that at the end of the mean free path l_i all energy is lost and the momentum is randomized, with the electron embarking on its way to the next possible ionization and so on. For the holes a similar description applies, but they move to the positive electrode at the left. At the electrodes all arriving carriers are carried off and contribute to the output current of the avalanche diode. While – naturally – this picture is oversimplified, we will see that the results are astonishingly complicated. More refined descriptions could be the subject of new doctoral studies!

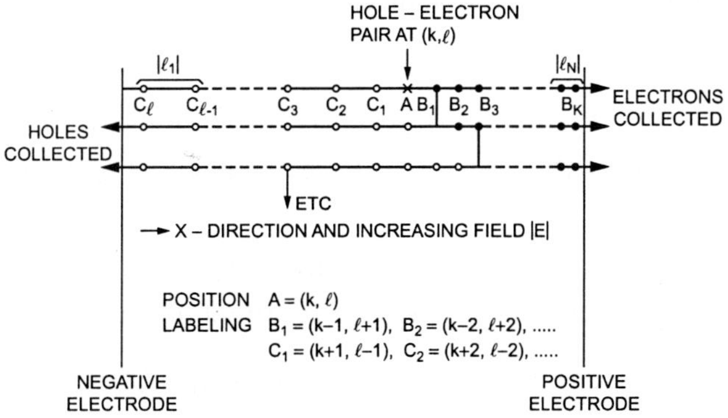

Fig. 19-3. Ionization by electrons (•) and by holes (○). The incoming hole-electron pair is at position A. The electron travels to the right, possibly causing ionizations at B_1, B_2, etc. The hole travels to the left, possibly causing ionizations at C_1, C_2, etc. The first generation for an event at B_1 is also shown.

We assume that the photo-induced pair production region is to the *left* of the avalanche region; the avalanche process is then started by an electron, the primary hole being carried off. If the converse is true, the results change, *mutatis mutandis.* In order to connect with the continuum theories at a later stage, we also introduce the ionization coefficients per unit length:

$$\alpha(x_i) \equiv \alpha_i = \lambda_i / l_i, \qquad \beta(x_i) \equiv \beta_i = \mu_i / l_i. \tag{19.5-1}$$

A mean value for the *a priori* probabilities is often useful:

$$\lambda = (1/N)\sum\nolimits_N \lambda_i, \qquad \mu = (1/N)\sum\nolimits_N \mu_i. \tag{19.5-2}$$

Each created hole-electron pair is endowed with two indices, (k,l). Here k is the number of possible ionizations that the electron can cause while moving to the right, while l is the number of possible hole-caused ionizations while it travels to the left; further, if N is the total number of ionizations possible in w, then $k+l=N$. We also note that the index l labels appropriately positions along the x-axis with increasing x.

We shall now proceed with the statistical description. Let us consider an incoming hole-electron pair at position $\mathrm{A}=(k,l)$ in Fig. 19-3, being the zeroth generation. Its tree will be denoted by X, its statistical distribution by $W(X)$, and its generating function by $\phi_{k,l}(z)$. We note, as in the previous section, that X can be decomposed into the original pair plus the subtrees of the first generation offspring, occurring at B_1, B_2, ... for electron-triggered events and the subtrees at C_1, C_2,... for hole events. Hence,

$$\begin{aligned} X_{k,l} &= 1 + y_{k-1,l+1} + y_{k-2,l+2} + \ldots + y_{0,N} \\ &\quad + \overline{y}_{k+1,l-1} + \overline{y}_{k+2,l-2} + \ldots + \overline{y}_{N,0}\,. \end{aligned} \tag{19.5-3}$$

Noting that these trees are subject to a Bernoulli trial for their realization, we have the following result for the associated generating functions, based on the compounding theorem and the addition theorem,

$$\begin{aligned} &\phi_{k,l}(z) = \\ &\quad z[\lambda_{l+1}\phi_{k-1,l+1}(z) - \lambda_{l+1} + 1][\lambda_{l+2}\phi_{k-2,l+2}(z) - \lambda_{l+2} + 1] \times \ldots \times [\lambda_N \phi_{0,N}(z) - \lambda_N + 1] \\ &\quad \times [\mu_{l-1}\phi_{k+1,l-1}(z) - \mu_{l-1} + 1][\mu_{l-2}\phi_{k+2,l-2}(z) - \mu_{l-2} + 1] \times \ldots \times [\mu_0 \phi_{N,0}(z) - \mu_0 + 1]\,. \end{aligned} \tag{19.5-4}$$

We also write down the shifted relationship for $(k,l) \to (k+1, l-1)$ and divide the new relationship by the above; the result is the recurrent relation

$$\phi_{k,l}(z)[\lambda_l\, \phi_{k,l}(z) - \lambda_l + 1] = \phi_{k+1,l-1}(z)[\mu_{l-1}\, \phi_{k+1,l-1}(z) - \mu_{l-1} + 1]\,. \tag{19.5-5}$$

When $k=N$, $l=0$ the hole-electron pair comes in at the left, with the hole being carried off and the electron starting its journey; thus $\phi_{N,0}$ *is the result sought for*; it gives the number of electrons collectable at right when one electron comes in at left.

Now we need a boundary condition. To that purpose, we apply (19.5-4) for $k=0, N=l$:

$$\phi_{0,N}(z) = z[\mu_{N-1}\phi_{1,N-1}(z)-\mu_{N-1}+1][\mu_{N-2}\phi_{2,N-2}(z)-\mu_{N-2}+1]\times\ldots\times[\mu_0\phi_{N,0}(z)-\mu_0+1]. \tag{19.5-6}$$

This is a self-consistent bc, since it contains the unknown. Physically, it provides a feedback between the two sides of the avalanche region.; the adjustment of the carriers resulting from this feedback takes care of the convergence of the procedure.

The multiplication factor M_N

We differentiate the recurrent relation (19.5-5); this gives

$$\phi_{k,l}'(z)[\lambda_l\,\phi_{k,l}(z)-\lambda_l+1]+\lambda_l\,\phi_{k,l}(z)\phi_{k,l}'(z)$$
$$=\phi_{k+1,l-1}'(z)[\mu_{l-1}\,\phi_{k+1,l-1}(z)-\mu_{l-1}+1]+\mu_{l-1}\,\phi_{k+1,l-1}(z)\phi_{k+1,l-1}'(z). \tag{19.5-7}$$

This is evaluated for $z=1$ whereby we write $\phi_{k,l}'(1)\equiv\phi_{k,l}'$ and note $\phi_{k,l}(1)=1$, giving

$$\phi_{k,l}'(1+\lambda_l)=\phi_{k+1,l-1}'(1+\mu_{l-1}). \tag{19.5-8}$$

So, starting with $k=0, l=N$ we have

$$\phi_{1,N-1}'=\phi_{0,N}'\frac{1+\lambda_N}{1+\mu_{N-1}}. \tag{19.5-9}$$

Repeating, we arrive at

$$\phi_{k,l}'=\phi_{0,N}'\frac{1+\lambda_N}{1+\mu_{N-1}}\frac{1+\lambda_{N-1}}{1+\mu_{N-2}}\times\cdots\times\frac{1+\lambda_{l+1}}{1+\mu_l}, \tag{19.5-9'}$$

$$\phi_{N,0}'=\phi_{0,N}'\frac{1+\lambda_N}{1+\mu_{N-1}}\frac{1+\lambda_{N-1}}{1+\mu_{N-2}}\times\cdots\times\frac{1+\lambda_1}{1+\mu_0}. \tag{19.5-10}$$

Next, we differentiate the closure condition (19.5-6):

$$\phi_{0,N}'(z)=[\mu_{N-1}\phi_{1,N-1}-\mu_{N-1}+1][\mu_{N-2}\phi_{2,N-2}-\mu_{N-2}+1]\times\ldots\times[\mu_0\phi_{N,0}-\mu_0+1]$$
$$+z\mu_{N-1}\phi_{1,N-1}'(z)[\mu_{N-2}\phi_{2,N-2}-\mu_{N-2}+1]\times\ldots\times[\mu_0\phi_{N,0}-\mu_0+1]$$
$$+z[\mu_{N-1}\phi_{1,N-1}(z)-\mu_{N-1}+1]\mu_{N-2}\phi_{2,N-2}'\times\ldots\times[\mu_0\phi_{N,0}-\mu_0+1]$$
$$\vdots$$
$$+z[\mu_{N-1}\phi_{1,N-1}(z)-\mu_{N-1}+1][\mu_{N-2}\phi_{2,N-2}-\mu_{N-2}+1]\times\ldots\times\mu_0\phi_{N,0}'(z). \tag{19.5-11}$$

Herein we set $z=1$; this yields

$$\phi_{0,N}'=1+\mu_{N-1}\phi_{1,N-1}'+\mu_{N-2}\phi_{2,N-2}'+\ldots+\mu_0\phi_{N,0}'. \tag{19.5-12}$$

Substituting in here (19.5-9'), we obtain

$$\phi'_{0,N} = R_N^{-1} \text{ with} \tag{19.5-13}$$

$$R_N = 1 - \left[\mu_{N-1}\frac{1+\lambda_N}{1+\mu_{N-1}} + \mu_{N-2}\frac{1+\lambda_N}{1+\mu_{N-1}}\frac{1+\lambda_{N-1}}{1+\mu_{N-2}} + \ldots + \mu_0 \frac{1+\lambda_N}{1+\mu_{N-1}}\frac{1+\lambda_{N-1}}{1+\mu_{N-2}} \times \ldots \times \frac{1+\lambda_1}{1+\mu_0}\right]. \tag{19.5-14}$$

From (19.5-10) we now find the desired result

$$M_N = \phi'_{N,0}\frac{1+\lambda_N}{1+\mu_{N-1}}\frac{1+\lambda_{N-1}}{1+\mu_{N-2}} \times \ldots \times \frac{1+\lambda_1}{1+\mu_0} R_N^{-1}. \tag{19.5-15}$$

We shall show that this gives the known asymptotic result of the continuum theory. With $\lambda_i = \alpha_i l_i = \alpha(x_i)\Delta x_i$ and $\mu_i = \beta_i l_i = \beta(x_i)\Delta x_i$ and $N \to \infty$ (or $\Delta x_i \to 0$) we have

$$\frac{1+\lambda_i}{1+\mu_{i-1}} = \exp\ln\left[\frac{1+\alpha(x_i)\Delta x_i}{1+\beta(x_i - \Delta x_i)\Delta x_i}\right] \sim \exp\{[\alpha(x_i) - \beta(x_i)]\Delta x_i\}, \tag{19.5-16}$$

$$\frac{1+\lambda_N}{1+\mu_{N-1}}\frac{1+\lambda_{N-1}}{1+\mu_{N-2}} \times \ldots \times \frac{1+\lambda_{l+1}}{1+\mu_l} \sim \prod_{i=l+1}^{N} \exp\{[\alpha(x_i) - \beta(x_i)]\Delta x_i\}$$
$$= \exp\{\sum_{i=l+1}^{N} [\alpha(x_i) - \beta(x_i)]\Delta x_i\}. \tag{19.5-17}$$

Going over to Riemann integrals, we arrive at

$$R_\infty = 1 - \int_0^w dx\, \beta(x) \exp\left\{\int_x^w dx'[\alpha(x') - \beta(x')]\right\}, \tag{19.5-18}$$

and

$$M_\infty = \frac{\exp\left\{\int_0^w [\alpha(x) - \beta(x)]\,dx\right\}}{1 - \int_0^w dx\, \beta(x) \exp\left\{\int_x^w [\alpha(x') - \beta(x')]\,dx'\right\}}, \tag{19.5-19}$$

in accord with McIntyre[8(b)] and Stilman and Wolfe.[13] These rather complex result are hereby fully confirmed. However, the present results are much more complete, as we discuss below. For now, we will still simplify the results for the case that a mean λ and μ can be assigned. One easily finds with

$$Q = (1+\lambda)/(1+\mu), \tag{19.5-20}$$

[13] G.E. Stilman and C.M. Wolfe, "Avalanche Photodiodes", in Semiconductors and Semimetals, **12** (R.K. Willardson and A.C. Beer, Eds.), Acad. Press, NY 1977, Chapter 5.

$$R_N = 1 - \mu \sum_{j=1}^{N} Q^j = 1 - \mu Q \frac{1 - Q^N}{1 - Q}, \tag{19.5-21}$$

$$M_N = \frac{(\lambda - \mu) Q^N}{\lambda(1+\mu) - \mu(1+\lambda) Q^N}. \tag{19.5-22}$$

For convergence it is necessary that $M > 1$. Therefore not all N are possible to yield a certain M. For example, taking N = 10, λ = 0.1 and μ = 0.05 yields M = 4.6. For later reference we also give the following expressions

$$\phi'_{0,N} = \frac{\lambda - \mu}{\lambda(1+\mu) - \mu(1+\lambda) Q^N} = \frac{M}{Q^N}, \tag{19.5-23}$$

while (19.5-22) shows that Q is expressible in M by

$$Q^N = \frac{\lambda(1+\mu) M_N}{(\lambda - \mu) + \mu(1+\lambda) M_N}. \tag{19.5-24}$$

The asymptotic results are also much simplified if the probabilities for ionization are constants. We write $k = \mu/\lambda = \beta/\alpha$. Further, let $\int_0^w \alpha(x)dx = \delta$. Then we have

$$\lim_{N\to\infty} Q^{-N} = \lim_{N\to\infty} \left(\frac{1+\mu}{1+\lambda} \right)^N = \lim_{N\to\infty} \left(\frac{1 + k\delta/N}{1 + \delta/N} \right)^N = e^{-(1-k)\delta}. \tag{19.5-25}$$

Then from (19.5-22) we obtain, as found with McIntyre,[14]

$$M_\infty \approx \frac{1-k}{e^{-(1-k)\delta} - k}, \tag{19.5-26}$$

The variance var$X$ *for any* N *and mean values for* λ_i *and* μ_i.

First we differentiate once more (19.5-7) and evaluate in the point z = 1; using also (19.5-9') for the case of mean probabilities λ and μ and writing $\phi''_{k,l}(1) \equiv \phi''_{k,l}$, we obtain the recurrent relation

$$\phi''_{k+1,l-1} = Q \phi''_{k,l} + \frac{2(\lambda - \mu)(1 - \lambda\mu)}{(1+\mu)^3} Q^{2k} \phi'^2_{0,N}, \tag{19.5-27}$$

with Q still given by (19.5-20). Starting with $k = 0$, $l = N$, we easily generate the solution of (19.5-27), the result being

$$\phi''_{k,l} = Q^k \phi''_{0,N} + \frac{2(\lambda - \mu)(1 - \lambda\mu)}{(1+\mu)^3} \phi'^2_{0,N} (1 + Q + \ldots + Q^{k-1}) Q^{k-1}. \tag{19.5-28}$$

---

[14] R.J. McIntyre, IEEE Trans. El. Dev. **E-D 19**, 703 (1972), Eq. (13).

Evaluating the sum and employing (19.5-23), this yields

$$\phi''_{k,l} = Q^k \phi''_{0,N} + \frac{2(1-\lambda\mu)}{(1+\mu)^2}\frac{M^2}{Q^{2N}} Q^{k-1}(Q^k - 1) . \qquad (19.5\text{-}29)$$

In particular, for $k = N,\ l = 0$

$$\phi''_{N,0} = Q^N \phi''_{0,N} + \frac{2(1-\lambda\mu)}{(1+\mu)^2}\frac{M^2(Q^N - 1)}{Q^{N+1}} . \qquad (19.5\text{-}30)$$

Next, we differentiate once more the result from the closure condition, (19.5-11), the result, evaluated in $z = 1$, is

$$\phi''_{0,N} = 2\mu(\phi'_{1,N-1} + \phi'_{2,N-2} + ... + \phi'_{N,0}) + \mu(\phi''_{1,N-1} + \phi''_{2,N-2} + ... + \phi''_{N,0})$$
$$+\mu^2(\phi'_{1,N-1} + \phi'_{2,N-2} + ... + \phi'_{N,0})^2 - \mu^2(\phi'^2_{1,N-1} + \phi'^2_{2,N-2} + ... + \phi'^2_{N,0}) . \qquad (19.5\text{-}31)$$

Now for the ϕ ' we have $\phi'_{k,l} = MQ^{k-N}$; for the ϕ'' we substitute (19.5-29). For the various sums we use

$$Q + Q^2 + ... + Q^N = \frac{1+\lambda}{\lambda-\mu}(Q^N - 1),$$
$$Q^2 + Q^4 + ... + Q^{2N} = \frac{(1+\lambda)^2}{(\lambda-\mu)(2+\lambda+\mu)}(Q^{2N} - 1) . \qquad (19.5\text{-}32)$$

The rhs of (19.5-31) then becomes an expression in the only unknown $\phi''_{0,N}$; we thus have a self-consistent result, with explicit solution

$$\phi''_{0,N} = \frac{M}{Q^N}\Bigg\{\frac{2\mu(1+\lambda)}{\lambda-\mu}\frac{(Q^N-1)M}{Q^N} + \frac{\mu^2(1+\lambda)^2}{(\lambda-\mu)^2}\frac{(Q^N-1)^2M^2}{Q^{2N}}$$
$$-\left[\frac{\mu^2(1+\lambda)^2}{(\lambda-\mu)(2+\lambda+\mu)} - \frac{2\mu(1-\lambda\mu)(1+\lambda)}{(1+\mu)(\lambda-\mu)(2+\lambda+\mu)}\right]\frac{(Q^{2N}-1)M^2}{Q^{2N}}$$
$$-\frac{2\mu(1-\lambda\mu)}{(1+\mu)(\lambda-\mu)}\frac{(Q^N-1)M^2}{Q^{2N}}\Bigg\} . \qquad (19.5\text{-}33)$$

This, finally, is substituted into (19.5-30), which connects the generating function's second derivatives for the two ends. So, for the variance $\operatorname{var} X_N = \phi''_{N,0} - M_N^2 + M_N$:

$$\operatorname{var} X_N = M_N\Bigg\{\frac{2\mu(1+\lambda)}{\lambda-\mu)}\frac{(Q^N-1)M_N}{Q^N} + \frac{\mu^2(1+\lambda)^2}{(\lambda-\mu)^2}\frac{(Q^N-1)^2M_N^2}{Q^{2N}}$$
$$+\left[\frac{2\mu(1-\lambda\mu)(1+\lambda)}{(1+\mu)(\lambda-\mu)(2+\lambda+\mu)} - \frac{\mu^2(1+\lambda)^2}{(\lambda-\mu)(2+\lambda+\mu)}\right]\frac{(Q^{2N}-1)M_N^2}{Q^{2N}}$$
$$-\frac{2\mu(1-\lambda\mu)}{(1+\mu)(\lambda-\mu)}\frac{(Q^N-1)M_N^2}{Q^{2N}} + \frac{2(1-\lambda\mu)}{(1+\mu)^2}\frac{(Q^N-1)}{Q^{N+1}} - M_N + 1\Bigg\} . \qquad (19.5\text{-}34)$$

Herein Q^N must still be expressed in M_N via (19.5-24). Leaving the terms in the same order, we find with some algebra, setting also $\mu = k\lambda$,

$$\operatorname{var} X_N = M_N(M_N-1)\left\{\frac{2k(1+\lambda)}{1+k\lambda}+\frac{k^2(1+\lambda)^2}{(1+k\lambda)^2}(M_N-1)\right.$$
$$+\left[\frac{2k(1-k\lambda^2)(1+\lambda)}{(1+k\lambda)(2+\lambda+k\lambda)}-\frac{k^2\lambda(1+\lambda)^2}{2+\lambda+k\lambda}\right]\left[\frac{1-k+M_N(1+k+2k\lambda)}{(1+k\lambda)^2}\right]$$
$$\left.-\frac{2k(1-k\lambda^2)}{(1+k\lambda)^3}[1-k+k(1+\lambda)M_N]+\frac{2(1-k\lambda^2)(1-k)}{(1+k\lambda)^2(1+\lambda)}-1\right\}. \quad (19.5\text{-}35)$$

This is the complete result. With much algebra this can be further simplified to

$$\operatorname{var} X_N = \frac{M_N(M_N-1)(1-k)}{2+\lambda+k\lambda}\left\{-\lambda+2\frac{1-k\lambda^2}{1+k\lambda}\left[M_N k\frac{1+\lambda}{1-k}+\frac{1}{1+\lambda}\right]\right\}. \quad \mathbf{(19.5\text{-}36)}$$

This is the final result! We still note that a similar expression can be found for the situation that holes are injected as primary particles at $x = w$ and the noise is computed due to holes flowing out of the negative electrode at $x = 0$. Then, in (19.5-36), replace λ by μ and k by $1/k$.

Various checks are easily accomplished. First, let us take $k = 0$; we then have one type of carriers. This yields indeed (19.4-20), derived in the previous section. Secondly, we can obtain the result for $k = 1$ with the generation-sum method; this is carried out in Ref. 10, II, Section III. The result there obtained

$$\operatorname{var} X_N = M_N(M_N-1)(1-\lambda) = M_N(M_N-1)\left(M_N-\frac{M_N-1}{N}\right), \quad (19.5\text{-}37)$$

is fully confirmed by the present, general result. Lastly let us consider the asymptotic case, $N \to \infty$; then $\lambda = \delta/N \to 0$. One finds directly

$$\operatorname{var} X_\infty = \frac{M(M-1)(1-k)}{2}\left\{\frac{2Mk}{1-k}+2\right\} = M(M-1)+kM(M-1)^2, \quad (19.5\text{-}38)$$

which is McIntyre's result.

Discussion of results and comparison with experimental data

The main result for the variance is in (19.5-36). The multiplication M_N is in Eq. (19.5-22). Using the parameter k, it reads

$$M_N = \frac{(1+\lambda)^N(1-k)}{(1+k\lambda)^{N+1}-k(1+\lambda)^{N+1}}. \quad (19.5\text{-}39)$$

The cubic term M^3 in var X_N really degrades the noise so it is important to have $k < 1$.

For silicon the accepted value is $k = 0.028$.[15] The expressions for var X_N and M_N are to be used in conjunction. For given N and desired M_N, it is possible to obtain $\lambda = \lambda(M_N;N)$. These data are used to compute curves for var $X_N(M_N;N)$. The results so obtained, var X_N vs. M_N, are plotted in Figs. 19-4(a) and 19-4(b). As seen there, for given N the curves reach a maximum and then terminate; this means that the required λ attains the value unity. For higher gains M more ionizations per transit occur. Generally, it was assumed that the transition $N \to N+1$ is realised when $\lambda = 0.5$. Thus, in an actual photo-avalanche diode, one expects that var X_N will for low M (low diode current) follow the curve for $N = 1$; then when λ reaches a value ~ 0.5, the characteristic switches over to the range $N=2$, with two possible ionizations per transit; and so on. The overall characteristic will thus be composed of parts of all the curves for finite N till at high N it joins the McIntyre asymptotic curve for var X_∞.

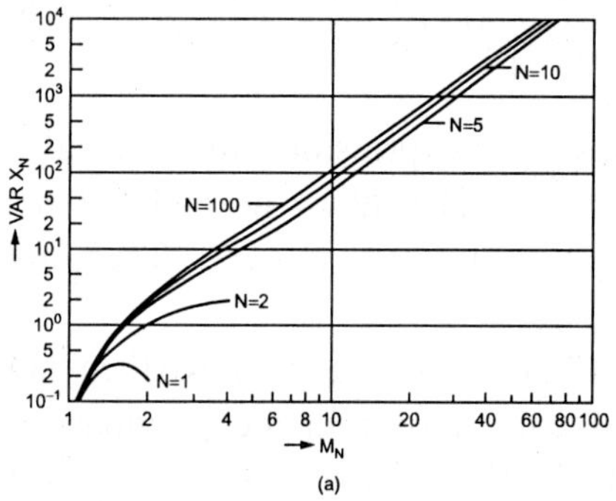

Fig. 19-4(a). var X_N vs. M_N for various N.

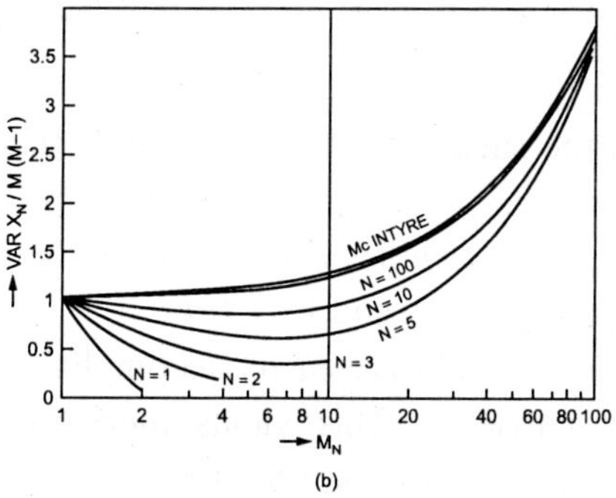

Fig. 19-4(b). var $X_N/M(M-1)$ vs. M_N.

Two conclusions are evident. (i) The noise from the discrete theory is always lower than the McIntyre curve; (ii) Breakpoints occur at the switch-over $N \to N+1$. This theoretical behaviour is pictured in Fig. 19-5 and experimental data in Fig. 19-6.

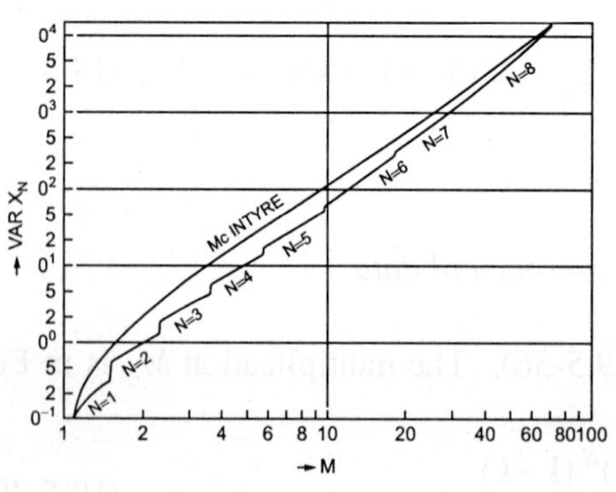

Fig. 19-5. Overall characteristic of var X vs. M.

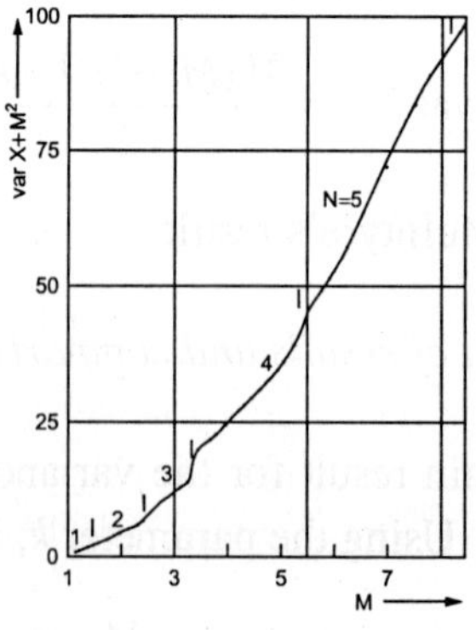

Fig. 19-6. Experimental data of J.Gong[15a] I_{photo}= 96nA.

[15] J. Conradi, IEEE Trans. El. Dev. **ED 19**, 713 (1972). [15a] J. Gong et al., Phys. Stat. Sol. **a 63**, 445 (1981).

3. Transport Fluctuations

19.6 On the Two Green's Function Procedures for Transport Noise

In the next few sections we shall be concerned with stochastic variables that depend on a set of continuous parameters **x**, so that the process is denoted by $a(\mathbf{x},t)$, or if more variables play a role, by $\mathbf{a}(\mathbf{x},t)$. Most often **x** is the spatial variable **r**, but it can also represent the phase-space variables **k,r** or apply to other situations. Basically we now have an infinite-dimensional stochastic process. The correlation 'function' is a kernel or matrix-kernel $\Phi_{ij}(\mathbf{x},\mathbf{x}',\vartheta) \equiv \langle a_i(\mathbf{x},t)\, a_j(\mathbf{x}',t+\vartheta)\rangle$. It can presumably be linked to the covariance 'function' $\Gamma_{ij}(\mathbf{x},\mathbf{x}') \equiv \langle \alpha_i(\mathbf{x})\alpha_j(\mathbf{x}')\rangle$, $\alpha_i = \Delta a_i$, if the regression solutions of the phenomenological equations for the process are known. With the Wiener–Khintchine theorem we can then obtain the spectral densities $\mathcal{S}_{ij}(\mathbf{x},\mathbf{x}',\omega)$. This method will be called 'the correlation method'. Alternately, we may presume that Langevin equations can be constructed for this process and that the white spectra of the sources are known. Then, upon Fourier transforming the source terms and the variables of the process, one can once more obtain the spectra $\mathcal{S}_{ij}(\mathbf{x},\mathbf{x}',\omega)$. This method will be denoted as the 'response method'.

Before we go into details, let us describe the oldest method to treat fluctuations in continuous systems, based on 'probability after effects, introduced by Smoluchowski. Let V_s be a subvolume of a large expanse V, on the surface of which appropriate Dirichlet or Neumann boundary conditions are specified. The number of particles in V_s fluctuates due to crossings over the fictitious boundary with the remainder of V, being governed by a Poisson distribution,

$$W(N) = e^{-N_0} N_0^N / N!, \quad N_0 \equiv \langle N \rangle. \tag{19.6-1}$$

Let now $R(\delta)$ be the probability that a particle is removed out of V_s in a time δ. Then, for the conditional average of particles at $t + \delta$, given we had $N(t)$ at time t, we can write

$$\langle N(t+\delta)\rangle_{N(t)} = N(t)[1 - R(\delta)] + N_0 R(\delta), \text{ or} \tag{19.6-2}$$

$$\langle \Delta N(t+\delta)\rangle_{N(t)} = \Delta N(t)[1 - R(\delta)]. \tag{19.6-3}$$

Next, averaging over all values of $N(t)$, we find for the correlation function,

$$\langle \Delta N(t+\delta)\Delta N(t)\rangle = \langle [\Delta N(t)]^2\rangle[1 - R(\delta)] = N_0 P_{\text{after}}(\delta), \tag{19.6-4}$$

where P_{after} is the 'after-effect probability' that the particle stays in V_s during $(t, t+\delta)$, cf. Chandrasekhar, Ref. 1(c), Chapter III; clearly we have

$$P_{\text{after}}(\delta) = \int_{V_s}\int_{V_s} P(\mathbf{r}, t+\delta \,|\, \mathbf{r}', t)\, d^3r\, d^3r'. \tag{19.6-5}$$

The conditional probability in this equation we computed before, see (18.7-29) and the diffusion approximation for times larger than the reciprocal friction constant, (18.7-6). However, $P(\mathbf{r},t+\delta|\mathbf{r}',t)$ [16] is *not* the Markov probability for the process, which is the functional $\mathcal{P}[n(\mathbf{r},t+\delta)\,|\,n(\mathbf{r}',t)]$. Yet, $P(\mathbf{r},t+\delta|\mathbf{r}',t)$ can be interpreted as the spatial-temporal Green's function for the process. Below we shall show that it suffices, however, to find the Laplace transformed Green's function and that the time-domain Green's function and the often formidable integrals arising from the Wiener–Khintchine theorem can be avoided altogether.

19.6.1 *The Correlation Method and Uniqueness*

We shall start with transport fluctuations involving a single species of particles, following a heuristic procedure as given by Fassett.[17] Let an expanse V be divided in K small volumes Δv_i, $i = 1\ldots K$. Let $\Delta N_i(t)$ be the fluctuating part of the number of particles in Δv_i at time t. Then, for the correlation,

$$\langle \Delta N_i(0)\Delta N_j(t)\rangle = \sum_{N_i}\sum_{M_j} \Delta N_i \Delta M_j W(N_i,M_j)\,, \tag{19.6-6}$$

where $W(N_i,M_j)$ is the joint probability that $N_i(0)=N_i$ and $N_j(t)=M_j$. This is rewritten as follows:

$$\begin{aligned}\langle \Delta N_i(0)\Delta N_j(t)\rangle &= \sum_{N_1}\sum_{N_2}\cdots\sum_{N_K}\sum_{M_j} \Delta N_i \Delta M_j W(N_1,N_2,\ldots,N_K,M_j)\\ &= \sum_{N_1}\sum_{N_2}\cdots\sum_{N_K} \Delta N_i\, W(N_1,N_2,\ldots,N_K)\sum_{M_j}\Delta M_j P(M_j\,|\,N_1,\ldots,N_K)\\ &= \sum_{N_1}\sum_{N_2}\cdots\sum_{N_K} \Delta N_i\, W(N_1,N_2,\ldots,N_K)\langle \Delta N_j(t)\rangle_{N_1\ldots N_K}\,,\end{aligned} \tag{19.6-7}$$

where $\langle \Delta N_j(t)\rangle_{N_1\ldots N_K}$ is the average fluctuation in Δv_i at time t, following from the linear(ized) macroscopic transport equation, due to a specified disturbance in each volume Δv_k centred on $\mathbf{r}_k$ at $t = 0$. Thus, by definition of Green's function, we have

$$\langle \Delta N_j(t)\rangle_{N_1\ldots N_K} = \sum_{k=1}^{K} g(\mathbf{r}_j,t;\mathbf{r}_k)\Delta v_k \Delta N_k\,. \tag{19.6-8}$$

For (19.6-6) we now have

$$\begin{aligned}\langle \Delta N_i(0)\Delta N_j(t)\rangle &= \sum_{k=1}^{K} \Delta v_k\, g(\mathbf{r}_j,t;\mathbf{r}_k)\sum_{N_1}\cdots\sum_{N_K} \Delta N_i \Delta N_k W(N_1\ldots N_K)\\ &= \sum_{k=1}^{K} \Delta v_k\, g(\mathbf{r}_j,t;\mathbf{r}_k)\langle \Delta N_i(0)\Delta N_k(0)\rangle\,.\end{aligned} \tag{19.6-9}$$

---

[16] Note that we left in the variable t, since the Wiener–Lévy process is not stationary, cf. problem 18.7.

[17] J.R. Fassett, Ph. D. Thesis, University of Minnesota, 1962.

Letting all Δv_i go to zero, we obtain for the density-density correlation function the Riemann integral

$$\Phi(\mathbf{r}',\mathbf{r},t) \equiv \langle \Delta n(\mathbf{r}',0)\Delta n(\mathbf{r},t)\rangle = \int_V d^3r'' g(\mathbf{r},t;\mathbf{r}'')\langle \Delta n(\mathbf{r}',0)\Delta n(\mathbf{r}'',0)\rangle . \quad (19.6\text{-}10)$$

To find the spectra, we split the time integral, employing the transposition property $\Phi(\mathbf{r}',\mathbf{r},-t) = \Phi(\mathbf{r},\mathbf{r}',t)$. We thus obtain

$$\mathcal{S}(\mathbf{r}',\mathbf{r},\omega) = 2\int_V d^3r'' \Big\{ \langle \Delta n(\mathbf{r}')\Delta n(\mathbf{r}'')\rangle \int_0^\infty e^{-i\omega t} g(\mathbf{r},t;\mathbf{r}'')$$

$$+ \langle \Delta n(\mathbf{r})\Delta n(\mathbf{r}'')\rangle \int_0^\infty e^{i\omega t} g(\mathbf{r}',t;\mathbf{r}'') \Big\} . \quad (19.6\text{-}11)$$

Let us now introduce the Laplace-transformed Green's function

$$G(\mathbf{r},\mathbf{r}'',s) = \int_0^\infty e^{-st} g(\mathbf{r},t;\mathbf{r}'')\,dt . \quad (19.6\text{-}12)$$

We also introduce the covariance function $\Gamma(\mathbf{r},\mathbf{r}') \equiv \langle \Delta n(\mathbf{r})\Delta n(\mathbf{r}')\rangle$. This, then, yields with $s \to i\omega + 0$,

$$\mathcal{S}(\mathbf{r}',\mathbf{r},\omega) = 2\int_V d^3r'' \{ \Gamma(\mathbf{r}',\mathbf{r}'')G(\mathbf{r},\mathbf{r}'',i\omega) + \Gamma(\mathbf{r},\mathbf{r}'')G(\mathbf{r}',\mathbf{r}'',-i\omega)\} . \quad (19.6\text{-}13)$$

This is the *correlation form.* Although we followed a simple argument, the result is basically correct, but possibly, incomplete. Also, we need a differential equation for the covariance function.

We will now go into more detail.[18] Let the transport operator be $\mathcal{L} = (\partial/\partial t) + \Lambda_{\mathbf{r}}$, being first order in the time for a single variable Markov process.[19] The time domain Green's function is defined as usual by

$$\mathcal{L} g(\mathbf{r},t;\mathbf{r}') = \partial g(\mathbf{r},t;\mathbf{r}')/\partial t + \Lambda_{\mathbf{r}} g(\mathbf{r},t;\mathbf{r}') = \delta(\mathbf{r}-\mathbf{r}')\delta(t) . \quad (19.6\text{-}14)$$

Here Λ is a spatial operator, like e.g. the diffusion operator, $-D\nabla^2$. We will also need the adjoint Green's function, which satisfies

$$\tilde{\mathcal{L}} \tilde{g}(\mathbf{r},t;\mathbf{r}') = -\partial \tilde{g}(\mathbf{r},t;\mathbf{r}')/\partial t + \tilde{\Lambda}_{\mathbf{r}} \tilde{g}(\mathbf{r},t;\mathbf{r}') = \delta(\mathbf{r}-\mathbf{r}')\delta(t) . \quad (19.6\text{-}15)$$

Upon Fourier–Laplace transforming, the corresponding results are, respectively,

$$L_{\mathbf{r}} G(\mathbf{r},\mathbf{r}',i\omega) \equiv i\omega G(\mathbf{r},\mathbf{r}',i\omega) + \Lambda_{\mathbf{r}} G(\mathbf{r},\mathbf{r}',i\omega) = \delta(\mathbf{r}-\mathbf{r}') , \quad (19.6\text{-}16)$$

$$\tilde{L}_{\mathbf{r}} \tilde{G}(\mathbf{r},\mathbf{r}',i\omega) \equiv -i\omega \tilde{G}(\mathbf{r},\mathbf{r}',i\omega) + \tilde{\Lambda}_{\mathbf{r}} \tilde{G}(\mathbf{r},\mathbf{r}',i\omega) = \delta(\mathbf{r}-\mathbf{r}') . \quad (19.6\text{-}17)$$

---

[18] (a) M. Lax and P. Mengert, J. Phys. Chem. Solids **14**, 248 (1960); (b) K.M. Van Vliet, Physica **99**, 345 (1979); (c) Ibid. and H. Mehta, Phys. Stat. Solidi (**b**) **106**, 11 (1981).

[19] If it is of higher order, then the variables can always be arranged in a column matrix, so that $\mathcal{L}\,\mathbf{a}(\mathbf{r},t) = (\partial/\partial t)\mathbf{a}(\mathbf{r},t) + \Lambda_{\mathbf{r}}\mathbf{a}(\mathbf{r},t)$ is first order in the time. The generalization to matrix equations is in most instances self-evident; however, for important results, the details will be stated explicitly.

The adjointness of these operators leads to Green's theorem

$$(L\phi,\psi)-(\phi,\tilde{L}\psi)=(\Lambda\phi,\psi)-(\phi,\tilde{\Lambda}\psi)=C[\phi,\psi^*], \tag{19.6-18}$$

where C is the bilinear concomitant. We give an example that we shall need later. Let Λ represent drift and diffusion; then

$$L=i\omega+\Lambda=i\omega+\nabla\cdot\mu\mathbf{E}-D\nabla^2, \tag{19.6-19}$$

$$\tilde{L}=-i\omega+\tilde{\Lambda}=-i\omega-\mu\mathbf{E}\cdot\nabla-D\nabla^2; \tag{19.6-20}$$

Green's theorem now reads

$$\begin{aligned}(L\phi,\psi)&-(\phi,\tilde{L}\psi)\\ &=\int_V d^3r\left\{\psi^*\left(i\omega\phi+\nabla\cdot\mu\mathbf{E}\phi-D\nabla^2\phi\right)-\phi\left(i\omega\psi^*-\mu\mathbf{E}\cdot\nabla\psi^*-D\nabla^2\psi^*\right)\right\}\\ &=\oint\!\!\oint_S d\mathbf{S}\cdot\left[\mu\mathbf{E}\phi\psi^*-\left(\psi^*D\nabla\phi-\phi D\nabla\psi^*\right)\right]=C[\phi,\psi^*].\end{aligned} \tag{19.6-21}$$

The bilinear concomitant can be made to vanish if appropriate bc are applied. E.g., we can have the mixed Dirichlet–Neumann direct bc $(\mu\mathbf{E}-D\nabla)\phi=0$ and the Neumann adjoint bc $\nabla\psi=0$. For non-homogeneous bc, the bilinear concomitant will contribute. In the case of particle diffusion with zero boundary current, the direct and adjoint bc for C to vanish of are $\nabla\langle n(\mathbf{r},t)|_{\text{surf}}\rangle=0$. This is valid in homogeneous systems, but not necessarily so in inhomogeneous structures, cf. Section 19.8.

With the aid of the adjoint Green's function we can now solve any inhomogeneous transport equation of the kind

$$\mathscr{L}a(\mathbf{r},t)=\upsilon(\mathbf{r},t)\quad\text{or}\quad L\hat{a}(\mathbf{r},i\omega)=\hat{\upsilon}(\mathbf{r},i\omega). \tag{19.6-22}$$

Following the standard procedure of Morse and Feshbach (Chapter 7), we multiply (19.6-22) by $G^*(\mathbf{r},\mathbf{r}',i\omega)=G(\mathbf{r},\mathbf{r}',-i\omega)$ and the complex conjugate relation corresponding to (19.6-17) by $a(\mathbf{r},t)$, after which we subtract and integrate over all d^3r; this yields the result

$$\begin{aligned}\int d^3r&\left\{L\hat{a}(\mathbf{r},i\omega)\tilde{G}^*(\mathbf{r},\mathbf{r}',i\omega)-\hat{a}(\mathbf{r},i\omega)\tilde{L}\tilde{G}^*(\mathbf{r},\mathbf{r}',i\omega)\right\}\\ &=\int d^3r\,\tilde{G}^*(\mathbf{r},\mathbf{r}',i\omega)\hat{\upsilon}(\mathbf{r},i\omega)-\hat{a}(\mathbf{r}',i\omega).\end{aligned} \tag{19.6-23}$$

We now interchange $\mathbf{r}$ and $\mathbf{r}'$ and we use Green's theorem (19.6-18); we then obtain

$$\hat{a}(\mathbf{r},i\omega)=\int d^3r'\tilde{G}^*(\mathbf{r}',\mathbf{r},i\omega)\hat{\upsilon}(\mathbf{r}',i\omega)-C[\hat{a}(\mathbf{r}',i\omega),\tilde{G}^*(\mathbf{r}',\mathbf{r},i\omega)]_{\mathbf{r}'\in S}. \tag{19.6-24}$$

Using reciprocity, $\tilde{G}(\mathbf{r}',\mathbf{r},-i\omega)=G(\mathbf{r},\mathbf{r}',i\omega)$, this finally gives

$$\hat{a}(\mathbf{r},i\omega)=\int d^3r'G(\mathbf{r},\mathbf{r}',i\omega)\hat{\upsilon}(\mathbf{r}',i\omega)-C[\hat{a}(\mathbf{r}_0,i\omega),G(\mathbf{r},\mathbf{r}_0,i\omega)], \tag{19.6-25}$$

where $\mathbf{r}_0$ is a surface coordinate. We note that the adjoint Green's function must *exist*, but needs not to be known; only the direct Green's function occurs in the final result. As to the surface term: $\tilde{G}$ satisfies the adjoint bc; however, $\hat{a}(\mathbf{r},i\omega)$ may not satisfy the direct bc, in which case a boundary contribution remains. Also, there may be surface Langevin sources (see next subsection), giving rise to stochastic boundary conditions. Their contribution can only be given in the response form.

We return to our original elementary result, (19.6-13). Assuming the absence of stochastic sources at the surface of V, the result still may have to be amended with a surface contribution, to wit,

$$\mathcal{S}(\mathbf{r}',\mathbf{r},\omega) = 2\int_V d^3r''\{G(\mathbf{r},\mathbf{r}'',i\omega)\Gamma(\mathbf{r}',\mathbf{r}'') + G(\mathbf{r}',\mathbf{r}'',-i\omega)\Gamma(\mathbf{r},\mathbf{r}'')\}$$
$$+2\int_V d^3r''G(\mathbf{r},\mathbf{r}'',i\omega)C[\Gamma(\mathbf{r}'',\mathbf{r}_0),G(\mathbf{r}',\mathbf{r}_0,-i\omega)] + hcj\,; \qquad (19.6\text{-}26)$$

here hcj indicates that a similar part with $\mathbf{r}\leftrightarrow\mathbf{r}'$ and $i\omega\rightarrow -i\omega$ is to be added. The first line of (19.6-26) is called the *volume contribution* and denoted as $\mathcal{S}^{\text{vol}}(\mathbf{r},\mathbf{r}',i\omega)$. For the case of coupled variables, $\mathbf{a}(\mathbf{r},t)$, $\mathbb{S}$, G and Γ are matrices.

The Lambda-Theorem

The transport operator Λ is the analogue of the phenomenological relaxation matrix M. Although we will not formally introduce the second-order F–P kernel, it is found to be related to the white spectral density of the Langevin source spectrum by the relationship $B(\mathbf{r},\mathbf{r}') = \frac{1}{2}\mathcal{S}_\xi(\mathbf{r},\mathbf{r}')$. The analogue of the fluctuation-relaxation theorem (18.5-9) is therefore the *lambda theorem*

$$\Lambda_\mathbf{r}\Gamma(\mathbf{r},\mathbf{r}') + \Lambda_{\mathbf{r}'}\Gamma(\mathbf{r},\mathbf{r}') = \tfrac{1}{2}\mathcal{S}_\xi(\mathbf{r},\mathbf{r}'). \qquad \textbf{(19.6-27)}$$

Note that this is a theorem in the six-dimensional space of $\mathbf{r},\mathbf{r}'$. For multivariate processes the following theorem holds:

$$\overrightarrow{\Lambda_\mathbf{r}}\Gamma(\mathbf{r},\mathbf{r}') + \Gamma(\mathbf{r},\mathbf{r}')\overleftarrow{\Lambda}{}^{\text{tr}}_{\mathbf{r}'} = \tfrac{1}{2}\mathbb{S}_\xi(\mathbf{r},\mathbf{r}'). \qquad \textbf{(19.6-28)}$$

In the second term the operator works from the right; super 'tr' means the transpose matrix. The detailed proof is in our 1971 J. Math. Phys. paper.[20(I)]

We shall now show that the correlation form (19.6-26) is unique; i.e., for Γ one can take any *particular solution* of the lambda theorem, since contributions $\Psi(\mathbf{r},\mathbf{r}')$ that satisfy the homogeneous equation $\Lambda_\mathbf{r}\Psi(\mathbf{r},\mathbf{r}') + \Lambda_{\mathbf{r}'}\Psi(\mathbf{r},\mathbf{r}') = 0$ do not contribute.

Proof. Let $\Lambda_\mathbf{r}\varphi(\mathbf{r}) = 0$; then $\Psi(\mathbf{r},\mathbf{r}') = p\,\varphi(\mathbf{r})\varphi(\mathbf{r}')$ is a solution of the homogeneous equation, p being a constant. We now multiply the complex conjugate of (19.6-17) by $\varphi(\mathbf{r})$ and we multiply $\Lambda_\mathbf{r}\varphi(\mathbf{r}) = 0$ by $\tilde{G}(r,r',-i\omega)$, subtract and integrate over d^3r; employing Green's theorem and interchanging $\mathbf{r}$ and $\mathbf{r}'$ we find

[20] K.M. Van Vliet, "Markov Approach to Density Fluctuations Due to Transport and Scattering". I. Mathematical Formalism. J. Math. Phys. **12**, 1981-1998 (1971); II Applications. Ibid. 1998-2012 (1971).

$$\varphi(\mathbf{r}) = -C[\varphi(\mathbf{r}_0), \tilde{G}(\mathbf{r}_0, \mathbf{r}, -i\omega)] + i\omega \int d^3r'' \varphi(\mathbf{r}'') \tilde{G}(\mathbf{r}'', \mathbf{r}, -i\omega). \tag{19.6-29}$$

This is multiplied by $p\,\varphi(\mathbf{r}'')$ and we use reciprocity; the result is

$$\Psi(\mathbf{r},\mathbf{r}'') = -C[\Psi(\mathbf{r}_0,\mathbf{r}''), G(\mathbf{r},\mathbf{r}_0,i\omega)] + i\omega \int d^3r''' \Psi(\mathbf{r}''',\mathbf{r}'') G(\mathbf{r},\mathbf{r}''',i\omega). \tag{19.6-30}$$

Likewise, upon changing $\mathbf{r} \to \mathbf{r}'$, $i\omega \to -i\omega$,

$$\Psi(\mathbf{r}',\mathbf{r}'') = -C[\Psi(\mathbf{r}_0,\mathbf{r}''), G(\mathbf{r}',\mathbf{r}_0,-i\omega)] - i\omega \int d^3r''' \Psi(\mathbf{r}''',\mathbf{r}'') G(\mathbf{r}',\mathbf{r}''',-i\omega). \tag{19.6-31}$$

Next, we multiply (19.6-30) by $2G(\mathbf{r}',\mathbf{r}'',-i\omega)$ and (19.6-31) by $2G(\mathbf{r},\mathbf{r}'',i\omega)$ and we add and integrate over d^3r''; this yields

$$\begin{aligned}
&2\int d^3r'' \{G(\mathbf{r},\mathbf{r}'',i\omega)\Psi(\mathbf{r}',\mathbf{r}'') + G(\mathbf{r}',\mathbf{r}'',-i\omega)\Psi(\mathbf{r},\mathbf{r}'')\} + 2G(\mathbf{r},\mathbf{r}'',i\omega) \\
&\quad \times C[\Psi(\mathbf{r}_0,\mathbf{r}''), G(\mathbf{r}',\mathbf{r}_0,-i\omega)] + 2G(\mathbf{r}',\mathbf{r}'',-i\omega) C[\Psi(\mathbf{r}_0,\mathbf{r}''), G(\mathbf{r},\mathbf{r}_0,i\omega)] \\
&\quad = 2i\omega \int\int d^3r'' d^3r''' \Psi(\mathbf{r}''',\mathbf{r}'') G(\mathbf{r},\mathbf{r}''',i\omega) G(\mathbf{r}',\mathbf{r}'',-i\omega) \\
&\quad -2i\omega \int\int d^3r'' d^3r''' \Psi(\mathbf{r}''',\mathbf{r}'') G(\mathbf{r}',\mathbf{r}''',-i\omega) G(\mathbf{r},\mathbf{r}'',i\omega) = 0,
\end{aligned} \tag{19.6-32}$$

by interchanging the integration variables in the last line. The first line is the effect of the solution of the homogeneous lambda theorem to (19.6-26); it does not contribute, so that any particular solution of the lambda theorem will do. The choice is usually obvious for a given transport operator, as we shall see in the applications.

Biorthogonal Expansion of the Green's Function

The biorthogonal series expansion for any non-self-adjoint Green's function can be found in Morse and Feshbach, Loc. Cit. Let $\varphi_k(\mathbf{r})$ be the eigenfunctions of the operator Λ (or, for multivariate processes, let $\boldsymbol{\varphi}_k(\mathbf{r})$ be the eigenvectors – column matrices – of the square matrix operator Λ); let further $\psi_k(\mathbf{r})$ be the eigenfunctions (or $\boldsymbol{\psi}_k(\mathbf{r})$ be the eigenvectors) of $\tilde{\Lambda}$ (or of $\tilde{\Lambda}$); we thus have

$$\Lambda\varphi_k(\mathbf{r}) = \lambda_k \varphi_k(\mathbf{r}), \qquad \tilde{\Lambda}\psi_k(\mathbf{r}) = \lambda_k^* \psi_k(\mathbf{r}). \tag{19.6-33}$$

These sets of eigenfunctions (vectors) are biorthogonal, so we have

$$(\varphi_k, \psi_l) = \int d^3r\, \varphi_k(\mathbf{r}) \psi_l^*(\mathbf{r}) = \delta_{kl}\,; \tag{19.6-34}$$

for the case that the eigenvectors are column matrices, the scalar product is defined as

$$(\boldsymbol{\varphi}_k, \boldsymbol{\psi}_l) = \int d^3r\, \tilde{\boldsymbol{\psi}}_l^*(\mathbf{r}) \boldsymbol{\varphi}_k(\mathbf{r}) = \delta_{kl}\,. \tag{19.6-34'}$$

Expanding G in the EF $\varphi_k(\mathbf{r})$,

$$G(\mathbf{r},\mathbf{r}',i\omega) = \sum\nolimits_k A_k(\mathbf{r}')\varphi_k(\mathbf{r}), \tag{19.6-35}$$

(19.6-16) yields

$$\sum_k (\lambda_k + i\omega) A_k(\mathbf{r}')\varphi_k(\mathbf{r}) = \delta(\mathbf{r}-\mathbf{r}') . \qquad (19.6\text{-}36)$$

Now, multiplying with $\psi^*_{k'}(\mathbf{r}')$ and integrating over d^3r, we find for A_k with (19.6-34) : $A_{k'}(\mathbf{r}') = \psi_{k'}(\mathbf{r}')/(\lambda_{k'} + i\omega)$. We thus obtained the biorthogonal series

$$G(\mathbf{r},\mathbf{r}',i\omega) = \sum_k \frac{\varphi_k(\mathbf{r})\psi^*_k(\mathbf{r}')}{\lambda_k + i\omega} . \qquad (19.6\text{-}37)$$

Sometimes we deal with particle fluctuations and are interested in the fluctuations in the total number $\mathcal{N}$ in a subvolume V_s of V. Assuming that the bilinear concomitant vanishes, we obtain from (19.6-26):

$$\mathcal{S}_{\mathcal{N}}(\omega) = 4\,\mathrm{Re}\frac{\langle \Delta\mathcal{N}^2\rangle}{V_s}\sum_k \frac{\int_{V_s}\int_{V_s} d^3r d^3r'\,\varphi_k(\mathbf{r})\psi^*_k(\mathbf{r}')}{\lambda_k + i\omega} . \qquad (19.6\text{-}38)$$

In particular, let us consider diffusion fluctuations in an infinite expanse, $\mathcal{N}$ being the particle number in the finite ν-dimensional volume V_s. With the EF being plane waves and the diffusion operator having EV Dk^2, we find for the spectrum

$$\mathcal{S}_{\mathcal{N}}(\omega) = 4\frac{\langle \Delta\mathcal{N}^2\rangle}{(2\pi)^\nu V_s}\int \frac{d^\nu k\, Dk^2}{D^2k^4 + \omega^2}\left|\int_{V_s} d^\nu r e^{i\mathbf{k}\cdot\mathbf{r}}\right|^2 . \qquad (19.6\text{-}39)$$

This is Richardson's formula.[21] We leave it to the problems to show that the high frequency asymptote in any dimension[22] is $\omega^{-3/2}$.

19.6.2 *The Response Form or Langevin Form*

The solution via the Langevin equation is in many respects far more direct. However, it yields a result that is quadratic in the Green's function, which complicates matters. Yet, if no surface noise sources are present, the result is easily converted to the previously found correlation form.

The Langevin equation reads

$$\mathcal{L}a(\mathbf{r},t) = [(\partial/\partial t) + \Lambda]a(\mathbf{r},t) = \xi(\mathbf{r},t) . \qquad (19.6\text{-}40)$$

Now, as indicated already by the first-order time derivative, the above relationship holds for $t > 0$; for negative times, we need the *adjunct operator* $-(\partial/\partial t) + \Lambda$ with adjunct Green's function $G^*(\mathbf{r},\mathbf{r}',i\omega)$.[23] The Fourier transforms of the variables $a(\mathbf{r},t)$ and $\xi(\mathbf{r},t)$ will as usual be denoted with a caret; so we find for $t > 0$,

---

[21] J.M. Richardson, Bell Syst. Techn. J. **29**, 117 (1950).

[22] Richardson – and others[18(a)] – obtained this result based on the Markov probability (18.7-6). Since this formula was shown to be incorrect for *very small* times, in reality the 'final' asymptote is ω^{-2}.

[23] This problem was discussed in footnote 24 of Chapter XVIII. As indicated there, one can also avoid this problem by employing truncated Fourier series on the positive time interval $0 < t \le T$.

$$L\hat{a}(\mathbf{r},i\omega) = [i\omega + \Lambda]\hat{a}(\mathbf{r},i\omega) = \hat{\xi}(\mathbf{r},i\omega)\,, \tag{19.6-41}$$

with the complex conjugate relationship for $t < 0$. This equation is of the form (19.6-22), second equality; the solution is therefore as in (19.6-25):

$$\hat{a}(\mathbf{r},i\omega) = \int d^3r' G(\mathbf{r},\mathbf{r}',i\omega)\hat{\xi}(\mathbf{r}',i\omega) - C[\hat{a}(\mathbf{r}_0,i\omega),\tilde{G}^*(\mathbf{r}_0,\mathbf{r},i\omega)]\,. \tag{19.6-42}$$

In some cases we have stochastic boundary conditions; i.e., we have

$$m\hat{a}(\mathbf{r}_0,i\omega) = \hat{\zeta}(\mathbf{r}_0,i\omega)\,, \tag{19.6-43}$$

where m generally is a linear surface operator of the mixed Dirichlet–Neumann type while $\hat{\zeta}(\mathbf{r}_0,i\omega)$ is a surface source. This may occur when there is surface diffusion or surface recombination. If so, the effect of this source, contained in the bilinear concomitant of the source and the surface Green's function must be taken along. Here we will only compute the effect of the volume Langevin source, i.e. we set $\hat{\zeta}(\mathbf{r}_0,i\omega) = 0$. It is then always possible to stipulate adjoint bc for the adjoint Green's function, so that the bilinear concomitant vanishes. With a spectral representation according to (16.4-10) for both the source and the response, we thus obtain

$$\mathcal{S}_a(\mathbf{r},\mathbf{r}',\omega) = \int\!\!\int d^3r_1 d^3r_2 G(\mathbf{r},\mathbf{r}_1,i\omega) G(\mathbf{r}',\mathbf{r}_2,-i\omega)\mathcal{S}_\xi(\mathbf{r}_1,\mathbf{r}_2)\,. \tag{19.6-44}$$

For the case that we have a set of random variables represented by $\mathbf{a}(\mathbf{r},t)$ we must be more careful with the order of the matrix Green's functions. The equivalent of (19.6-44) then reads

$$\mathbb{S}_\mathbf{a}(\mathbf{r},\mathbf{r}',\omega) = \int\!\!\int d^3r_1 d^3r_2 \mathbb{G}(\mathbf{r},\mathbf{r}_1,i\omega)\mathbb{S}_\xi(\mathbf{r}_1,\mathbf{r}_2)\mathbb{G}^\dagger(\mathbf{r}',\mathbf{r}_2,i\omega)\,. \tag{19.6-44'}$$

We note that this still satisfies the requirement that for negative frequencies we have $\mathbb{S}^*_\mathbf{a}(\mathbf{r},\mathbf{r}',-\omega) = \mathbb{S}^\dagger_\mathbf{a}(\mathbf{r},\mathbf{r}',\omega) = \mathbb{S}_\mathbf{a}(\mathbf{r},\mathbf{r}',\omega)$. Continuing with (19.6-44), we substitute for the source spectrum from the lambda theorem, (19.6-27), obtaining

$$\begin{aligned}\mathcal{S}_a(\mathbf{r},\mathbf{r}',\omega) &= 2\int d^3r_1 G(\mathbf{r},\mathbf{r}_1,i\omega)\int d^3r_2 G(\mathbf{r}',\mathbf{r}_2,-i\omega)(-i\omega + \Lambda_{\mathbf{r}_2})\Gamma(\mathbf{r}_1,\mathbf{r}_2)\\ &\quad + 2\int d^3r_2 G(\mathbf{r}',\mathbf{r}_2,-i\omega)\int d^3r_1 G(\mathbf{r},\mathbf{r}_1,i\omega)(i\omega + \Lambda_{\mathbf{r}_1})\Gamma(\mathbf{r}_1,\mathbf{r}_2)\,;\end{aligned} \tag{19.6-45}$$

note that the extra terms in $\pm i\omega$ cancel. Using now Green's theorem, we make the operators to work on the Green's function instead of on the covariance function. So,

$$\begin{aligned}&\mathcal{S}_a(\mathbf{r},\mathbf{r}',\omega)\\ &= 2\int d^3r_1 G(\mathbf{r},\mathbf{r}_1,i\omega)\Big\{\int d^3r_2 \Gamma(\mathbf{r}_1,\mathbf{r}_2)\tilde{L}_{\mathbf{r}_2} G(\mathbf{r}',\mathbf{r}_2,-i\omega) + C[\Gamma(\mathbf{r}_1,\mathbf{r}_0), G(\mathbf{r}',\mathbf{r}_0,-i\omega)]\Big\}\\ &+ 2\int d^3r_2 G(\mathbf{r}',\mathbf{r}_2,-i\omega)\Big\{\int d^3r_1 \Gamma(\mathbf{r}_1,\mathbf{r}_2)\tilde{L}^*_{\mathbf{r}_1} G(\mathbf{r},\mathbf{r}_1,i\omega) + C[\Gamma(\mathbf{r}_2,\mathbf{r}_0), G(\mathbf{r},\mathbf{r}_0,i\omega)]\Big\}\,.\end{aligned} \tag{19.6-46}$$

Next, we note that $\tilde{L}^*_{\mathbf{r}_1} G(\mathbf{r},\mathbf{r}_1,i\omega) = \tilde{L}^*_{\mathbf{r}_1} \tilde{G}(\mathbf{r}_1,\mathbf{r},-i\omega) = \delta(\mathbf{r}-\mathbf{r}_1)$ and similarly for $\tilde{L}_{\mathbf{r}_2} G(\mathbf{r}',\mathbf{r}_2,-i\omega)$. Performing the integrations over the delta functions, we arrive at

$$\mathcal{S}_a(\mathbf{r},\mathbf{r}',\omega) = 2\int_V d^3r'' G(\mathbf{r},\mathbf{r}'',i\omega)\{\Gamma(\mathbf{r}',\mathbf{r}'') + C[\Gamma(\mathbf{r}'',\mathbf{r}_0), G(\mathbf{r}',\mathbf{r}_0,-i\omega)]\}$$
$$+2\int_V d^3r'' G(\mathbf{r}',\mathbf{r}'',-i\omega)\{\Gamma(\mathbf{r},\mathbf{r}'') + C[\Gamma(\mathbf{r}'',\mathbf{r}_0), G(\mathbf{r},\mathbf{r}_0,i\omega)]\}. \quad (19.6\text{-}47)$$

This is just the correlation form (19.6-26), stated previously. The two methods are therefore entirely equivalent. Obviously, in most cases the correlation form is easier to use, since it is linear, rather than quadratic in the Green's function.

Once more we come back to the particular problem that we seek to find the fluctuations in the number of particles in a subvolume V_s. For this case we need to integrate (19.6-47) over both coordinates $\mathbf{r},\mathbf{r}' \in V_s$. Between V_s and the rest of V we have *fictitious* boundary conditions. So, what happens on the boundary of V is usually irrelevant. On dimensional grounds the density-density covariance function must have a delta-function part. Let us suppose that this is the only part, i.e., $\Gamma(\mathbf{r},\mathbf{r}'') = A\delta(\mathbf{r}-\mathbf{r}'')$. Keeping $\mathbf{r}''$ as an internal coordinate slightly away from the surface, the covariance function $\Gamma(\mathbf{r}'',\mathbf{r}_0) = 0$. There is therefore no contribution from the bilinear concomitant for this case. We now easily obtain

$$\mathcal{S}_{\mathcal{N}}(\omega) = 4[\langle\Delta\mathcal{N}^2\rangle / V_s]\mathrm{Re}\int_{V_s}\int_{V_s} d^3r\, d^3r' G(\mathbf{r},\mathbf{r}',i\omega). \quad (19.6\text{-}48)$$

19.7 Applications: Rayleigh Diffusion and Ambipolar Sweep-out

We shall start with the simple macroscopic example in which deviations from the steady-state density $\langle n(\mathbf{r})\rangle$ of a one-component gas are governed by diffusion; that is, we consider diffusion of the *fluctuations* $\Delta n(\mathbf{r},t)$, following the original model of Lord Rayleigh.[24] It involves scattering in position space over distances $\mathbf{a}_i$ and as such is well applicable to ionic diffusion. To apply it to electronic or heat diffusion, one must stretch the imagination and compare $|\mathbf{a}|$ to the mean free path. Let $Q(\mathbf{r},\mathbf{r}')d^3r d^3r'$ be the rate of scattering from $\mathbf{r}$ (within d^3r) to $\mathbf{r}'$ (within d^3r'). Also, let f_i be the probability for scattering per second over a lattice distance $\pm\mathbf{a}_i$ (we assume equal probabilities for forward or backward scattering). Then we have

$$Q(\mathbf{r},\mathbf{r}') = n(\mathbf{r})\sum_i f_i[\delta(\mathbf{r}-\mathbf{r}'-\mathbf{a}_i) + \delta(\mathbf{r}-\mathbf{r}'+\mathbf{a}_i)]. \quad (19.7\text{-}1)$$

For the first and second-order F–P moments we find, employing their basic definition,[25]

$$A_{\mathbf{r}} = \sum_i f_i[n(\mathbf{r}+\mathbf{a}_i) + n(\mathbf{r}-\mathbf{a}_i) - 2n(\mathbf{r})], \quad (19.7\text{-}2)$$

[24] Lord Rayleigh, Phil. Mag. 47, 246 1899.

[25] See also (17.12-13), omitting the factors $(1 - n)$ stemming from the exclusion principle.

$$B_{\mathbf{rr}'} = -\sum_i f_i\{[n(\mathbf{r})+n(\mathbf{r}')][\delta(\mathbf{r}-\mathbf{r}'-\mathbf{a}_i)+\delta(\mathbf{r}-\mathbf{r}'+\mathbf{a}_i) \\ -\delta(\mathbf{r}-\mathbf{r}')[n(\mathbf{r}+\mathbf{a}_i)+n(\mathbf{r}-\mathbf{a}_i)+2n(\mathbf{r})]\}. \tag{19.7-3}$$

The macroscopic diffusion equation is recovered for lengths $\gg|\mathbf{a}_i|$, for by expansion of the delta functions to second order we have

$$A_{\mathbf{r}} = \sum_i f_i \nabla\nabla n(\mathbf{r}) : \mathbf{a}_i\mathbf{a}_i = \mathsf{D} : \nabla\nabla n(\mathbf{r}), \tag{19.7-4}$$

where we defined the diffusion tensor by $\mathsf{D} = \sum_i f_i\mathbf{a}_i\mathbf{a}_i$; notice the dimension of cm^2/s. Considering further the isotropic case that $\mathsf{D} = D\mathsf{I}$, we have for the phenomenological equation

$$\partial[\langle\Delta n(\mathbf{r},t)\rangle_{\text{cond}}/\partial t] - D\nabla^2\langle\Delta n(\mathbf{r},t)\rangle_{\text{cond}} = 0, \tag{19.7-5}$$

which is of course the regular diffusion equation. For the second-order F–P moment we find likewise,

$$B_{\mathbf{rr}'} = 2\mathsf{D} : \nabla\nabla'[n(\mathbf{r})\delta(\mathbf{r}-\mathbf{r}')] \doteq 2D\nabla\cdot\nabla'[n(\mathbf{r})\delta(\mathbf{r}-\mathbf{r}')], \tag{19.7-6}$$

where the last rhs pertains to the isotropic case; further, $\nabla \equiv \nabla_{\mathbf{r}}$ and $\nabla' \equiv \nabla_{\mathbf{r}'}$. As could have been expected, this is also equal to one-half times the source-spectrum $\mathcal{S}_\xi(\mathbf{r},\mathbf{r}')$ associated with diffusive transport; from Eqs. (18.8-18) *ff* we gather that $\xi(\mathbf{r},t) = \nabla\boldsymbol{\eta}(\mathbf{r},t)$, so that for a steady (possibly non-equilibrium) state

$$\mathcal{S}_\xi(\mathbf{r},\mathbf{r}') = \nabla\nabla' : \mathbb{S}_\eta(\mathbf{r},\mathbf{r}') = 4\mathsf{D} : \nabla\nabla'\langle n(\mathbf{r})\rangle\delta(\mathbf{r}-\mathbf{r}'). \tag{19.7-7}$$

Adding a possible drift term to (19.7-5), the Λ-theorem for this problem reads

$$[\mathbf{v}_d\cdot\nabla + \mathbf{v}_d\cdot\nabla' - \mathsf{D}:\nabla\nabla - \mathsf{D}:\nabla'\nabla']\,\Gamma(\mathbf{r},\mathbf{r}') = 2\mathsf{D}:\nabla\nabla'\langle n(\mathbf{r})\rangle\delta(\mathbf{r}-\mathbf{r}'). \tag{19.7-8}$$

By inspection we see that a solution with the required symmetry is

$$\Gamma(\mathbf{r},\mathbf{r}') = \langle n(\mathbf{r})\rangle\delta(\mathbf{r}-\mathbf{r}'), \tag{19.7-9}$$

provided

$$[\mathbf{v}_d\cdot\nabla - \mathsf{D}:\nabla\nabla]\langle n(r)\rangle = \nabla\cdot\langle\mathbf{J}(\mathbf{r})\rangle = 0. \tag{19.7-10}$$

This simply means that the current should be solenoidal, i.e., displacement current is negligible. *A fortiori*, this is always true in an equilibrium state with $\langle\mathbf{J}\rangle = 0$.

Let us now look at one-dimensional diffusion. The transport equation is

$$[\partial\langle\Delta n(x,t)\rangle_{\text{cond}}/\partial t] - D[\partial^2\langle\Delta n(x,t)\rangle_{\text{cond}}/\partial x^2] = 0, \tag{19.7-11}$$

with Fourier–Laplace transformed Green's function

$$i\omega G(x,x',i\omega) - D[\partial^2 G(x,x't)/\partial x^2] = \delta(x-x'). \tag{19.7-12}$$

The Green's function can be found from standard procedures, e.g. from the bounded homogeneous solutions for $x < x'$ and $x > x'$, requiring continuity at $x = x'$ and the jump condition[26]

$$\left[\frac{\partial G}{\partial x}\right]_{x'-0}^{x'+0} = -\frac{1}{D}. \qquad (19.7\text{-}13)$$

The solution is found to be

$$G(x,x',i\omega) = (i/2u_0 D)\exp(iu_0\,|x-x'|)\,, \qquad (19.7\text{-}14)$$

where $u_0 = \sqrt{(\omega/D)}\,e^{3i\pi/4}$. For the fluctuations of the particles in a strip (0, L) this gives[27]

$$\mathcal{S}_{\Delta\mathcal{N}}(\omega) = [\langle\mathcal{N}\rangle L^2/D\theta^3]\left\{1-e^{-\theta}[\cos\theta+\sin\theta]\right\},\quad \theta = L\sqrt{\omega/2D}\,. \qquad (19.7\text{-}15)$$

For low frequencies this goes as $\omega^{-1/2}$, while for high frequencies $\mathcal{S} \propto \omega^{-3/2}$.

For 3D diffusion the Green's function $g(\mathbf{r},t;\mathbf{r}')$ is the Gaussian diffusion distribution. The time Fourier–Laplace transformed Green's function satisfies the Helmholtz equation

$$(\nabla^2 + k^2)G(\mathbf{r},\mathbf{r}',k) = D^{-1}\,\delta(\mathbf{r}-\mathbf{r}')\,, \qquad (19.7\text{-}16)$$

where $k = \sqrt{(\omega/D)}\,e^{3i\pi/4}$, $\mathbf{r}$ and $\mathbf{r}'$ are coordinates from the origin of a spherical coordinate system. This yields the well-known result

$$G(\mathbf{r},\mathbf{r}',k) = e^{ik|\mathbf{r}-\mathbf{r}'|}/4\pi D\,|\mathbf{r}-\mathbf{r}'|\;. \qquad (19.7\text{-}17)$$

Assuming that we have a small sphere of radius R_0 embedded in an infinite expanse with fictitious bc on the surface of the sphere, we find from (19.6-48) after a somewhat tedious integration

$$\mathcal{S}_{\Delta\mathcal{N}}(\omega) = [12\langle\mathcal{N}\rangle R_0^2 D\vartheta^5]$$
$$\times\{\vartheta^2 - 2 + e^{-\vartheta}[\vartheta^2(\cos\vartheta+\sin\vartheta) + 4\vartheta\cos\vartheta + 2(\cos\vartheta-\sin\vartheta)]\}\,, \qquad (19.7\text{-}18)$$

where $\vartheta = 2R_0\sqrt{(\omega/2D)}$. For low frequencies the spectrum is a constant and the high frequency asymptote is as before.

Lastly, we consider ambipolar sweep-out. In a single charge-type solid a disturbance in the number of carriers in a small strip $\delta\ell$ stays 'put', even though individual carriers move under the influence of an applied field. The reason is that, when carriers move out of the strip, other ones move in 'instantaneously' (i.e. within the dielectric relaxation time) to preserve neutrality. The situation for an electron disturbance in an n-type semiconductor is pictured in Fig. 19-7(a).

---

[26] Generally, for the non-self adjoint 1D differential equation $\alpha_2(x)G_x'' + \alpha_1(x)G_x' + \alpha_0(x)G(x,\xi) = \delta(x-\xi)$ the discontinuity must satisfy $G_x'|_{\xi-0}^{\xi+0} = 1/\alpha_2(\xi)$. The Green's function can be constructed as $G(x,\xi) = u(x)v(\xi)/W(\xi)$, $x \le \xi$, with the two variables interchanged for $x \ge \xi$; W is the Wronskian.

[27] G.G. MacFarlane, Proc. Phys. Soc. **B63**, 807 (1950); R.E. Burgess, Proc. Phys. Soc. **B66**, 334 (1953).

A different situation prevails, however, if minority carriers are introduced in an n-type two-band semiconductor. Thus, let us generate a hole-electron pair excess by shining light of frequency $h\nu \geq \overline{\varepsilon}_G$ (optical bandgap) in a strip $\delta\ell$, see Fig. 19-7(b). The electrons will still move to the left (positive electrode) and the holes to the right (negative electrode). The electrons are, however, immediately replaced by others flowing in; as to the holes, their displacement must also be compensated to preserve charge neutrality, but, lacking an abundance of holes outside the strip, their motion is met by a lesser influx of electrons. In other words: the electron density adjusts itself at any spot, such that the hole flow to the right is accompanied by a similar excess of electrons, until the holes with their electron cloud are swept out at the negative electrode. Basically, the motion is *ambipolar.*

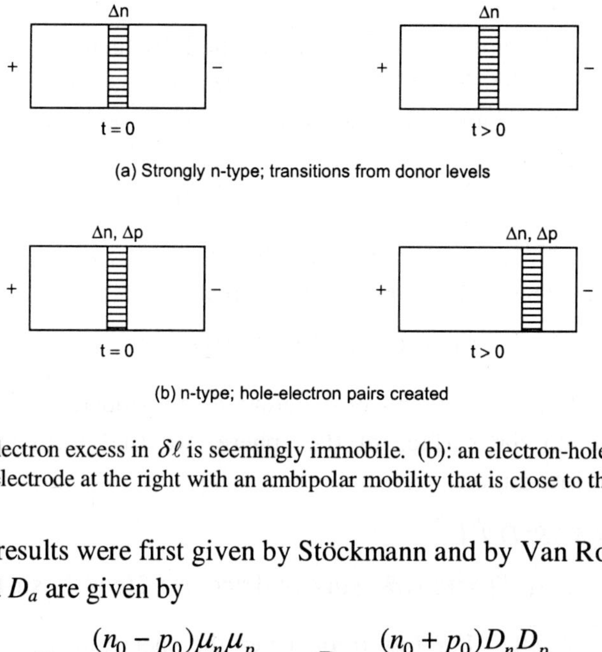

Fig. 19-7. (a): an electron excess in $\delta\ell$ is seemingly immobile. (b): an electron-hole pair excess moves to the negative electrode at the right with an ambipolar mobility that is close to the hole mobility.

Quantitative results were first given by Stöckmann and by Van Roosbroeck.[28] The quantities μ_a and D_a are given by

$$\mu_a = \frac{(n_0 - p_0)\mu_n\mu_p}{\mu_n n_0 + \mu_p p_0}, \qquad D_a = \frac{(n_0 + p_0)D_n D_p}{D_n n_0 + D_p p_0}. \tag{19.7-19}$$

Note that for $n_0 \gg p_0$, $\mu_a \to \mu_p$ and $D_a \to D_p$. Also, for $n_0 = p_0$, $\mu_a = 0$. Ambipolar sweep-out (neglecting diffusion) accompanied by generation-recombination gives rise to the following Langevin transport equation

$$\frac{\partial \Delta p}{\partial t} + q\mu_a E \frac{\partial \Delta p}{\partial x} + \frac{\Delta p(x,t)}{\tau} = \gamma(x,t) + \frac{\partial \eta(x,t)}{\partial x}, \tag{19.7-20}$$

with the Langevin source stemming both from g-r processes and from diffusion. Defining the corresponding Green's function $g(x,t;x')$, we find for the Fourier–

[28] (a) F. Stöckmann, Z. für Physik, **147**, 544 (1957); (b) W. Van Roosbroeck, Phys. Rev. **91**, 282 (1953).

Laplace transform $G(x,x',i\omega)$:

$$G(x,x',i\omega) = (1/\mu_a E)\Theta(x-x')e^{-[(i\omega+1/\tau)/\mu_a E](x-x')}. \tag{19.7-21}$$

The Heaviside function indicates that the observation coordinate x is larger than the source coordinate x', i.e. the disturbance moves to the right only, in accordance with the qualitative argument given before. For the number of holes in a region of length L (generally realised between two probes of a four-contact specimen), one finds

$$\mathcal{S}_{\Delta\mathcal{P}}(\omega) = 4\langle\Delta\mathcal{P}^2\rangle\tau_a W(\theta,\phi), \quad \text{with} \tag{19.7-22}$$

$$W(\theta,\phi) = \frac{\theta^2/\phi^2}{1+\theta^2}\left\{\frac{\phi}{\theta} - \frac{1-\theta^2}{1+\theta^2} + \frac{\exp(-\phi/\theta)}{1+\theta^2}[(1-\theta^2)\cos\phi - 2\theta\sin\phi]\right\}, \tag{19.7-23}$$

with $\theta = \omega\tau$, $\phi = \omega\tau_a$, where $\tau_a = L/|\mu_a|E$ is the sweep-out time. If $\tau \gg \tau_a$, the spectrum simplifies considerably and takes the familiar form (converted to the current-fluctuation spectrum):

$$\mathcal{S}_{\Delta I}(\omega) = 2qI\left[\frac{b+2+1/b}{|n_0/p_0 - p_0/n_0|}\right]\left[\frac{\sin\frac{1}{2}\omega\tau_a}{\frac{1}{2}\omega\tau_a}\right]^2. \tag{19.7-24}$$

Here $b = \mu_n/\mu_p$. The result for the spectrum (19.7-24) is pictured in Fig. 19-8(a). The sharp resonances will not be observed as such, however, due to the finite bandwidth. So, in Fig. 19-8(b) we give the result integrated over a 10% bandwidth; also, a plot of (19.7-23) for $\tau = \tau_a$ is shown. Experimentally, ambipolar sweep-out in p-type Ge has been observed by Hill.[29] The measurements were in good agreement with the known mobilities. Most convincing was the fact that sweep-out was entirely absent in intrinsic Ge; for that case a pure Lorentzian due to g-r noise was observed.

Altogether, transport noise is presently well understood. Multivariate processes are of relevance for the physics of space-charge limited conduction in insulators.

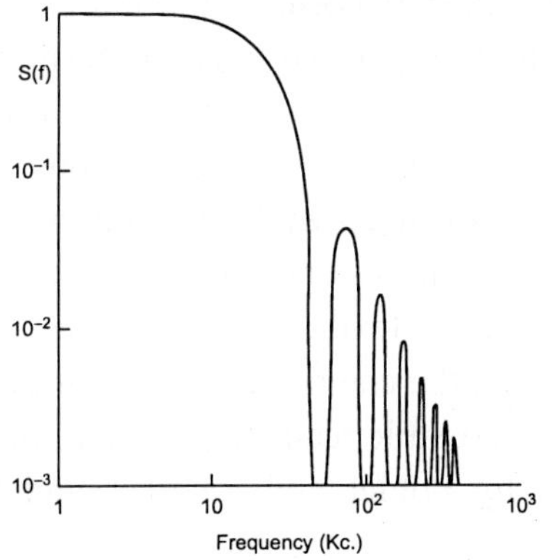

Fig. 19-8(a). Ambipolar sweep-out spectrum, $\tau \gg \tau_a$.

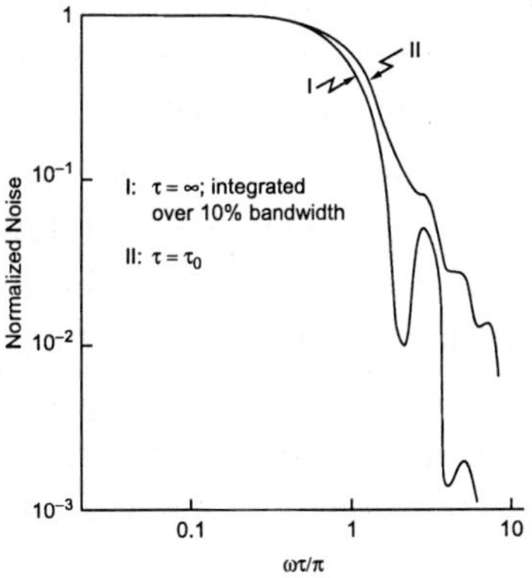

Fig. 19-8(b). Spectra for 10% bandwidth.[29]

[29] J.E. Hill and K.M. Van Vliet, Physica XXIV, 709-720 (1958).

19.8 Inhomogeneous Systems, Effect of Boundary Terms, Examples

Carrier transport in p-n junctions and bipolar transistors has been a vexing problem from its inception by Shockley, Brattain and Bardeen. At low frequencies a simple corpuscular argument will give the noise. Let us consider a p^+-n junction. At forward bias, holes from the p^+ region are injected into the n-region. The dc characteristic for low injection levels is given by the Boltzmann-type formula

$$I = I_0\left(e^{eV/k_BT} - 1\right), \tag{19.8-1}$$

where $\mathcal{q} = e$ is the charge of the holes. While physicists may (and should) marvel at the fact that this result holds for applied voltages of several tenths of volts, i.e., for ~ 10 kT_B/e at room temperature, for the electronics engineer this is a simple fact of life... The two terms in (19.8-1) are due to the injected hole current, $\exp(eV/k_BT)$ and the reverse current due to minority carriers generated in the n-region. When crossing the junction barrier, these carriers independently give shot noise. Hence,

$$\begin{aligned} \mathcal{S}_I &= 2eI_0e^{eV/k_BT} + 2eI_0 \\ &= 2e(I + I_0) + 2eI_0 = 2eI + 4eI_0 . \end{aligned} \tag{19.8-2}$$

The low frequency conductance is given by

$$G_0 = dI/dV = [e/k_BT]e^{eV/k_BT} = (e/k_BT)(I + I_0). \tag{19.8-2'}$$

Combining the last two formulas, one has

$$\mathcal{S}_I = 4k_BTG_0 - 2eI . \tag{19.8-3}$$

As is clear from the formula of departure, (19.8-2), this is always positive, while for I = 0 we have just thermal noise. For high frequencies, however, this corpuscular argument is mute and we must solve for the transport noise of the holes in the n-region. At the boundary of the junction and the n-region the injection formula with the Boltzmann factor still applies; at the ohmic contact on the other side (or at the collector junction in a transistor) the holes recombine. The first solutions for this behaviour were given by Petritz, Solow, North and van der Ziel using a transmission-line analogue description[30]; they idealised the structure as one-dimensional, a picture still sometimes found in introductory device physics textbooks. However, modern vhf transistors have a 3D-structure associated with the chemical alloying or diffusion of impurities in the substrate material; we give a possible realistic picture in Fig.19-9. A Green's function solution was later given by Polder and Baelde[31] and by the

[30] R.L. Petritz, Proc. IRE **40**, 1440 (1952); also, Phys. Rev. **91**, 204 and 231 (1953); M. Solow, Ph.D. Thesis, Catholic Univ. of America (1957); D.O. North (RCA), unpublished; A. van der Ziel, Proc. IRE **46**, 1019 (1958).

[31] D. Polder and A. Baelde, Solid-State Electr. **6**, 103 (1963); A. Baelde, Ph. D. Thesis, Delft (1964).

author. Below we describe the essential results as obtained by us.[32]

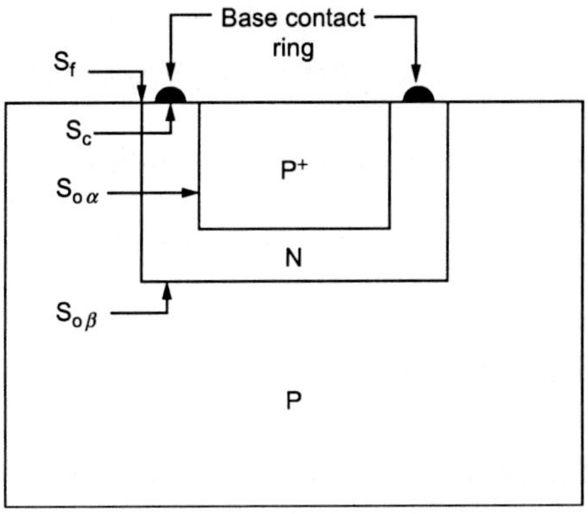

Fig. 19-9. Schematic geometry of a vhf transistor.

Although in modern transistors recombination in the bulk is virtually absent, we will take it along by adding a term $(1/\tau)$ to the operators in Eqs. (19.6-19) and (19.6-20); Green's theorem (19.6-21) is unaffected by this change. As noted in the figure, there are three types of boundary surfaces. First, there is the injecting surface, denoted by S_0 (or $S_{0\alpha}$ in transistors). Next, there is the ohmic contact or base-ring contact (transistors), denoted as S_c; finally, there is the surface barrier with the substrate or the collector junction, denoted as S_b or $S_{0\beta}$ (transistors). At the ohmic contact there is usually infinite surface recombination, so that there $\Delta p = 0$. Also, at the barrier with the substrate or the collector junction all carriers experience a sink, so that there also $\Delta p = 0$. With similar adjoint bc at these surfaces, the bilinear concomitant vanishes for these surfaces.[33] Rest the injecting contact, which can be short-circuited or open-circuited. We begin with the former; since for the injected density at S_0 we have $p(\mathbf{r}_0,t) = C\exp[eV(t)/k_BT]$ it follows that $\Delta p(\mathbf{r}_0,t) = \langle p(\mathbf{r}_0)\rangle \times e\Delta V(t)/k_BT$, so that $\Delta\hat{p}(r_0,i\omega) = 0$ for $\Delta\hat{V}(i\omega) = 0$; these are the direct bc. The Green's function will be denoted by $G^S(\mathbf{r},\mathbf{r}',i\omega)$. It satisfies the direct bc for the surface observation-point coordinate and its adjoint will satisfy the adjoint bc:

$$\tilde{G}^S(\mathbf{r}_0,\mathbf{r},i\omega) = 0, \text{ or } \quad G^S(\mathbf{r},\mathbf{r}_0,-i\omega) = 0; \tag{19.8-4}$$

note that G^S is also zero for a surface source-point coordinate. The surface terms in (19.6-26) from the bilinear concomitant at the injection surface do not contribute.

Let us now write down the basic stochastic transport equation for the holes injected into the n-region:

---

[32] K.M. Van Vliet, Solid-State Electr. **15**, 1033-1053 (1972).

[33] If here is finite surface recombination, there will be noise from the surface Langevin sources; it has been shown, however, in Ref. 32 that these contributions have no ultimate effect on the junction noise.

$$\frac{\partial \Delta p(\mathbf{r},t)}{\partial t} = -\frac{\Delta p(\mathbf{r},t)}{\tau} - \frac{1}{e}\nabla \cdot \Delta \mathbf{J}_p(\mathbf{r},t) + \gamma_p(\mathbf{r},t), \tag{19.8-5}$$

with

$$\Delta \mathbf{J}_p(\mathbf{r},t) = e\mu_p \Delta p(\mathbf{r},t)\mathbf{E}(\mathbf{r}) - eD_p \nabla \Delta p(\mathbf{r},t) - e\boldsymbol{\eta}(\mathbf{r},t). \tag{19.8-6}$$

The subscript p will henceforth be dropped. Substituting (19.8-6) into (19.8-5) gives

$$\mathscr{L} = \partial/\partial t + 1/\tau + \nabla \cdot \mu \mathbf{E} - D\nabla^2 , \tag{19.8-7}$$

with the Fourier–Laplace transform and its adjoint as in (19.6-19) and (19.6-20), the term $1/\tau$ being added. The total Langevin source is $\xi(\mathbf{r},t) = \gamma(\mathbf{r},t) + \nabla \cdot \boldsymbol{\eta}(\mathbf{r},t)$. The g-r source is *not* $4g(p_s)$ – a matter of contention in the early days[30] – since there is no detailed balance; rather the recombination is balanced by minority-carrier generation *and* transport. So, from the steady-state form of (19.8-5) and assumed shot noise for the recombination rate p_s/τ (where $p_s \equiv \langle p \rangle$) and the generation rate p_0/τ (where p_0 is the thermal hole density in the n-region), we have

$$\mathbb{S}_\xi = \left\{[4p_s(\mathbf{r})/\tau] - 2D\nabla^2 p_s(\mathbf{r}) + 2\nabla \cdot [\mu \mathbf{E}(\mathbf{r}) p_s(\mathbf{r})]\right\}\delta(\mathbf{r}-\mathbf{r}')$$
$$+4D\nabla \cdot \nabla'[p_s(r)\delta(\mathbf{r}-\mathbf{r}')]. \tag{19.8-8}$$

This is substituted into the lambda theorem

$$\left[(2/\tau) + \nabla \cdot \mu \mathbf{E}(\mathbf{r}) - D\nabla^2 + \nabla' \cdot \mu \mathbf{E}(\mathbf{r}') - D\nabla'^2\right]\Gamma(\mathbf{r},\mathbf{r}') = \tfrac{1}{2}\mathbb{S}_\xi . \tag{19.8-9}$$

observing the rules

$$\begin{aligned} &\nabla \delta(\mathbf{r}-\mathbf{r}') = -\nabla' \delta(\mathbf{r}-\mathbf{r}'), \\ &\varphi(\mathbf{r}')\nabla \delta(\mathbf{r}-\mathbf{r}') = \varphi(\mathbf{r})\nabla \delta(\mathbf{r}-\mathbf{r}') + [\nabla \varphi(\mathbf{r})]\delta(\mathbf{r}-\mathbf{r}'), \end{aligned} \tag{19.8-10}$$

(for any φ) one finds – amazingly – that (19.8-9) still admits the simple solution

$$\Gamma(\mathbf{r},\mathbf{r}') = p_s(\mathbf{r})\delta(\mathbf{r}-\mathbf{r}'). \tag{19.8-11}$$

The noise spectra are found from (19.9-26), first line. Substitution of (19.8-11) gives

$$\mathbb{S}_{\Delta p}(\mathbf{r},\mathbf{r}',\omega) = 2G^S(\mathbf{r},\mathbf{r}',i\omega)p_s(\mathbf{r}') + 2G^S(\mathbf{r}',\mathbf{r},-i\omega)p_s(\mathbf{r}). \tag{19.8-12}$$

We return to the current density, (19.8-6). We shall split it as follows:

$$\Delta \mathbf{J}(\mathbf{r},t) = \Delta \mathbf{J}^{\text{o}}(\mathbf{r},t) - e\boldsymbol{\eta}(\mathbf{r},t) = e(\mu \mathbf{E} - D\nabla)\Delta p(\mathbf{r},t) - e\boldsymbol{\eta}(\mathbf{r}_0,t). \tag{19.8-13}$$

The spectrum for $\Delta \mathbf{J}^{\text{o}}$ becomes, employing the last two equations,

$$\begin{aligned} \mathbb{S}_{\Delta \mathbf{J}^{\text{o}}}(\mathbf{r},\mathbf{r}',\omega) &= 2e^2(\mu \mathbf{E} - D\nabla)(\mu \mathbf{E}' - D\nabla')\mathbb{S}_{\Delta p}(\mathbf{r},\mathbf{r}',\omega) \\ &= 2e^2(\mu \mathbf{E} - D\nabla)(\mu \mathbf{E}' - D\nabla')[G^S(\mathbf{r},\mathbf{r}',i\omega)p_s(\mathbf{r}') + G^S(\mathbf{r}',\mathbf{r},-i\omega)p_s(\mathbf{r})] \\ &= -2e^2 p_s(\mathbf{r})(\mu \mathbf{E} - D\nabla)D\nabla' G^S(\mathbf{r},\mathbf{r}',i\omega) + hcj \end{aligned}$$

$$+2e^2[(\mu\mathbf{E}'-D\nabla')p_s(\mathbf{r}')](\mu\mathbf{E}-D\nabla)G^S(\mathbf{r},\mathbf{r}',i\omega)+hcj\,. \qquad (19.8\text{-}14)$$

The above is obviously not yet the complete result, since we need the spectrum of $\Delta\mathbf{J}(\mathbf{r},t)=\Delta\mathbf{J}^{\circ}(\mathbf{r},t)-e\boldsymbol{\eta}(\mathbf{r},t)$. The spectrum of the source is a spike term that can be dismissed, see below. For the cross correlation one readily obtains, using Fourier analysis of the transport equation [see also Ref. 20, I, Eq. (6.20)],

$$\mathbb{S}_{\Delta p,\boldsymbol{\eta}}(\mathbf{r},\mathbf{r}',\omega)=4D\int d^3r''G^S(\mathbf{r},\mathbf{r}'',i\omega)\nabla''[p_s(\mathbf{r}')\delta(\mathbf{r}''-\mathbf{r}')]$$
$$=-4Dp_s(\mathbf{r}')\nabla'G^S(\mathbf{r},\mathbf{r}',i\omega)\,. \qquad (19.8\text{-}15)$$

(The surface integral in this 'integration by parts' vanishes.) This, then, yields the cross-correlation spectrum

$$\mathbb{S}_{\Delta\mathbf{J}^{\circ},-e\boldsymbol{\eta}}(\mathbf{r},\mathbf{r}',i\omega)=4e^2(\mu\mathbf{E}-D\nabla)[Dp_s(\mathbf{r}')\nabla'G^S(\mathbf{r},\mathbf{r}',i\omega)]$$
$$=4e^2p_s(\mathbf{r}')(\mu\mathbf{E}-D\nabla)D\nabla'G^S(\mathbf{r},\mathbf{r}',i\omega)\,. \qquad (19.8\text{-}16)$$

Likewise, we need the spectrum for $\mathbb{S}_{-e\boldsymbol{\eta},\Delta\mathbf{J}^{\circ}}(\mathbf{r},\mathbf{r}',i\omega)$, which is simply the Hermitean conjugate. When these spectra are added to (19.8-14), they *change the sign* of the first part. We therefore find for the spectrum of the full $\Delta\mathbf{J}$:

$$\mathbb{S}_{\Delta\mathbf{J}}(\mathbf{r},\mathbf{r}',\omega)=2e^2p_s(\mathbf{r})(\mu\mathbf{E}-D\nabla)D\nabla'G^S(\mathbf{r},\mathbf{r}',i\omega)+hcj$$
$$+2e^2[(\mu\mathbf{E}'-D\nabla')p_s(\mathbf{r}')](\mu\mathbf{E}-D\nabla)G^S(\mathbf{r},\mathbf{r}',i\omega)+hcj\,. \qquad (19.8\text{-}17)$$

This result is needed at the injecting surface, the beginning of the n-region. To avoid meaningless results, we let $\mathbf{r}'\to\mathbf{r}_0$ and $\mathbf{r}\to\mathbf{r}_0^+$, where $\mathbf{r}_0^+$ is taken to be slightly inside the surface. Also, we need $\Delta i(\mathbf{r}_0,t)=-\int\Delta\mathbf{J}(\mathbf{r}_0,t)\cdot d\boldsymbol{\sigma}_0$, whereby we note that $d\boldsymbol{\sigma}_0$ is along the outside normal, whereas the hole current flows *into* the n-region. Finally, with the dc current given as

$$I_0=-\int d\boldsymbol{\sigma}\cdot(\mu\mathbf{E}_0-D\nabla_0)p_s(\mathbf{r}_0)\,, \qquad (19.8\text{-}18)$$

we obtain for $\mathcal{S}_{\Delta i^{\circ}}(\omega)$:

$$\mathcal{S}_{\Delta i}(\omega)=2e^2p_s(\mathbf{r}_0)\int\int d\boldsymbol{\sigma}_0^+d\boldsymbol{\sigma}_0:(\mu\mathbf{E}_0^+-D\nabla_0^+)D\nabla_0)G^S(\mathbf{r}_0^+,\mathbf{r}_0,i\omega)+hcj$$
$$-2eI_0\left\{\int d\boldsymbol{\sigma}_0^+\cdot(\mu\mathbf{E}_0^+-D\nabla_0^+)G^S(r_0^+,r_0,i\omega)+\int d\boldsymbol{\sigma}_0\cdot(\mu\mathbf{E}_0-D\nabla_0)G^S(\mathbf{r}_0,\mathbf{r}_0^+,-i\omega)\right\}. \qquad (19.8\text{-}19)$$

The G^S in the third term is zero, because of the adjoint (source-point) bc. The last integral is equal to unity, as is found by integrating over a 'box' between $\mathbf{r}_0$ and $\mathbf{r}_0^+$, employing Gauss' theorem and (19.6-16). So, the last line of (19.8-19) yields $-2eI_0$. As to the source term's self spectrum, the integration over the surface is performed similarly,

$$\mathbb{S}_{\Delta i\text{ due to }e\boldsymbol{\eta}}=e^2\int\int d\boldsymbol{\sigma}_0^+d\boldsymbol{\sigma}_0:\mathbb{S}_{\boldsymbol{\eta}}(\mathbf{r}_0,\mathbf{r}_0^+)=0. \qquad (19.8\text{-}20)$$

Rests to associate the first line with the admittance. Let us apply a small ac voltage to the junction, $V_1(t) = V_{1,\omega} e^{i\omega t}$, resulting in a hole signal $p_1(\mathbf{r}_0,t) = (e/k_BT) \times p_s(\mathbf{r}_0)V_1(t)$. Accordingly, in the Fourier–Laplace domain, we have $\hat{p}_1(\mathbf{r}_0,-i\omega) \doteq (e/k_BT)p_s(\mathbf{r}_0)V_{1,\omega}$. This is a boundary term. We thus need the full solution (19.6-25) wherein the bilinear concomitant is given by (19.6-21).[34] So we find

$$\hat{p}_1(\mathbf{r},-i\omega) = -(eV_{1,\omega}/k_BT)$$
$$\times\int d\boldsymbol{\sigma}_0 \cdot \left\{[(\mu\mathbf{E}_0 - D\nabla_0)p_s(\mathbf{r}_0)]G^S(\mathbf{r},\mathbf{r}_0,-i\omega) + p_s(\mathbf{r}_0)D\nabla_0 G^S(\mathbf{r},\mathbf{r}_0,-i\omega)\right\}. \quad (19.8\text{-}21)$$

The first term is zero because of the adjoint bc for the Green's function. For the ac current density at **r** we now have

$$\mathbf{J}_1(\mathbf{r},-i\omega) = -(e^2V_{1,\omega}p_{s0}/k_BT)(\mu\mathbf{E}_\mathbf{r} - D\nabla)\int d\boldsymbol{\sigma}_0 \cdot D\nabla_0 G^S(\mathbf{r},\mathbf{r}_0,-i\omega), \quad (19.8\text{-}22)$$

where we took the $p(\mathbf{r}_0)$ out of the integral since the junction is an equipotential surface. To obtain the ac current $I_{1,\omega}$ we must let $\mathbf{r} \to \mathbf{r}_0^+$ and integrate over $-d\boldsymbol{\sigma}_0^+$. Hence, we obtain

$$Y(-i\omega) = (e^2 p_{s0}/k_BT)\int\int d\boldsymbol{\sigma}_0^+ d\boldsymbol{\sigma}_0 : (\mu\mathbf{E}_0^+ - D\nabla_0^+)D\nabla_0 G^S(\mathbf{r}_0^+,\mathbf{r}_0,-i\omega). \quad (19.8\text{-}23)$$

Finally, comparing (19.8-19) with (19.8-23), we arrive at the form for junction noise

$$\mathcal{S}_{\Delta i}(\omega) = 4k_BT\,\mathrm{Re}\,Y(i\omega) - 2eI_0\,. \quad (19.8\text{-}24)$$

This result, first derived with a one-dimensional transmission line analogue, has hereby been shown to be valid for any realistic 3D geometry. We also note that this is the correct high frequency extension of (19.8-4). It remains baffling, however, that the present complete *collective argument for minority-carrier transport* gives the same result (for low frequencies) as the simple *corpuscular picture of crossings* over the junction. Perhaps, the shot noise argument, as for the sources of generation-recombination noise discussed in 18.6.2, is simply a 'fluke'. We hereby still note that both results must be modified for high injection due to correlations.

Lastly, we briefly consider the alternative that the junction at the injection surface is open-circuited. Thus, in each point of the junction surface we assume as direct bc $\Delta\hat{\mathbf{J}}^\circ(\mathbf{r}_0,i\omega) = (\mu\mathbf{E}_0 - D\nabla_0)\Delta\hat{p}(\mathbf{r}_0,i\omega) = 0$. The open-circuited Green's function will be denoted by $G^O(\mathbf{r},\mathbf{r}',i\omega)$. It will be assumed to satisfy the adjoint bc, which now means that

[34] The result (19.6-25) is sometimes written as

$$\hat{a}(\mathbf{r},i\omega) = \int d^3r' G(\mathbf{r},\mathbf{r}',i\omega)\hat{\eta}(\mathbf{r}',i\omega) + \oint d\sigma_0 H(\mathbf{r},\mathbf{r}_0,i\omega)\hat{\zeta}(\mathbf{r}_0,i\omega)\,, \quad (19.8\text{-}21')$$

where H is called the surface Green's function. For the present case $H(\mathbf{r},\mathbf{r}_0,i\omega) = -\hat{\mathbf{v}}\cdot D\nabla G(\mathbf{r},\mathbf{r}_0,i\omega)$.

$$\nabla_0\tilde{G}^O(\mathbf{r}_0,\mathbf{r},i\omega)=0, \text{ or } \quad \nabla_0 G^O(\mathbf{r},\mathbf{r}_0,-i\omega)=0. \tag{19.8-25}$$

The volume spectrum, first line of (19.6-26), is presently not the full result, but there are a number of contributions stemming from surface integrals. We shall list them separately (the super 'O' on the Green's function will be added later).

(i) The transformation of the response form via the lambda theorem to the correlation form has left the boundary terms comprised in the second line of (19.6-26). Writing out the bilinear concomitant and using the covariance kernel (19.8-11), we have the two contributions

$$\begin{aligned}
&\mathcal{S}_{\Delta p}^{\text{bound},1}(\mathbf{r},\mathbf{r}',i\omega)\\
&=2\int d^3r''\int d\boldsymbol{\sigma}\cdot G(\mathbf{r},\mathbf{r}'',i\omega)G(\mathbf{r}',\mathbf{r}_0,-i\omega)(\mu\mathbf{E}_0-D\nabla_0)[p_s(\mathbf{r}_0)\delta(\mathbf{r}''-\mathbf{r}_0)]+hcj,\\
&\mathcal{S}_{\Delta p}^{\text{bound},2}(\mathbf{r},\mathbf{r}',i\omega)\\
&=2\int d^3r''\int d\boldsymbol{\sigma}\cdot G(\mathbf{r},\mathbf{r}'',i\omega)[D\nabla_0 G(\mathbf{r}',\mathbf{r}_0,-i\omega)]\,p_s(\mathbf{r}_0)\delta(\mathbf{r}''-\mathbf{r}_0)+hcj. \qquad (19.8\text{-}26)
\end{aligned}$$

(ii) Since at the injecting surface $\Delta\mathbf{J}(\mathbf{r}_0,t)=0$, we have the stochastic surface source $(\mu\mathbf{E}_0-D\nabla_0)\Delta p(\mathbf{r}_0,t)=\boldsymbol{\eta}(\mathbf{r}_0,t)$. In the response expressions we now have a surface contribution, given by the bilinear concomitant in (19.6-42). Noting the adjoint bc (19.8-25) this leads to the 'spike' spectrum

$$\begin{aligned}
\mathcal{S}_{\Delta p}^{\text{bound},3}(\mathbf{r},\mathbf{r}',i\omega)&=\int d\boldsymbol{\sigma}\int d\boldsymbol{\sigma}':G(\mathbf{r},\mathbf{r}_0,i\omega)G(\mathbf{r}',\mathbf{r}_0',-i\omega)\mathbb{S}_{\eta}(\mathbf{r}_0,\mathbf{r}_0')\\
&=4\int d\sigma\int d\sigma' G(\mathbf{r},\mathbf{r}_0,i\omega)G(\mathbf{r}',\mathbf{r}_0',-i\omega)Dp_s(\mathbf{r}_0)\delta(\mathbf{r}_0-\mathbf{r}_0'). \quad (19.8\text{-}27)
\end{aligned}$$

(iii) Finally, the two response terms of (19.6-42) leave a cross spectral density. One easily finds

$$\begin{aligned}
&\mathcal{S}_{\Delta p}^{\text{bound},4}(\mathbf{r},\mathbf{r}',i\omega)\\
&=-4\int d^3r''\int d\boldsymbol{\sigma}\cdot G(\mathbf{r},\mathbf{r}'',i\omega)G(\mathbf{r}',\mathbf{r}_0,-i\omega)\,D\nabla''[p_s(\mathbf{r}'')\delta(\mathbf{r}''-\mathbf{r}_0)]+hcj. \quad (19.8\text{-}28)
\end{aligned}$$

These terms are added and simplified. Some algebra gives for the boundary noise[35]

$$\begin{aligned}
\mathcal{S}_{\Delta p}^{\text{bound}}(\mathbf{r},\mathbf{r}',i\omega)&=2\int d\boldsymbol{\sigma}\cdot G(\mathbf{r},\mathbf{r}_0,i\omega)G(\mathbf{r}',\mathbf{r}_0,-i\omega)(\mu\mathbf{E}_0-D\nabla_0)p_s(\mathbf{r}_0)\\
&+2\int d\boldsymbol{\sigma}\cdot p_s(\mathbf{r}_0)D\nabla_0[G(\mathbf{r},\mathbf{r}_0,i\omega)G(\mathbf{r}',\mathbf{r}_0,-i\omega)]. \qquad (19.8\text{-}29)
\end{aligned}$$

For the open-circuited junction the last integral is zero. [We also note that for the short-circuited case, *both* final boundary terms are zero.]

Proceeding with the open-circuited case, the volume contribution remains as in (19.8-12), with $G^S \to G^O$. This is evaluated at the surface where the voltage noise is related to the hole-density spectrum by $\mathcal{S}_{\Delta V}(\omega) = [k_BT/ep_s(\mathbf{r}_0)]^2\mathcal{S}_{\Delta p}(\mathbf{r}_0^+,\mathbf{r}_0'^+,\omega)$, in an obvious notation. One thus obtains

---

[35] Cf., Ref. 32, Appendix B.2.

$$\mathcal{S}_{\Delta V}(\mathbf{r}_0^+, \mathbf{r}_0'^+, \omega) = 2[k_B^2 T^2 / e^2 p_{s0}]\{G^O(\mathbf{r}_0^+, \mathbf{r}_0'^+, i\omega) + G^O(\mathbf{r}_0'^+, \mathbf{r}_0^+, -i\omega)$$
$$+ \int d\boldsymbol{\sigma}_0 \cdot G^O(\mathbf{r}_0^+, \mathbf{r}_0, i\omega) G^O(\mathbf{r}_0'^+, \mathbf{r}_0, -i\omega)(\mu \mathbf{E}_0 - D\nabla_0) p_s(\mathbf{r}_0)\}. \quad (19.8\text{-}30)$$

The computation of the impedance goes along the same lines as previously; we obtain[36]

$$Z(-i\omega) = [k_B T / e^2 p_{s0}] G^O(\mathbf{r}_0^+, \mathbf{r}_0, -i\omega). \quad (19.8\text{-}31)$$

this finally gives

$$\mathcal{S}_{\Delta V}(\omega) = 4k_B T \,\mathrm{Re}\, Z(i\omega) - 2eI_0 Z(i\omega) Z^*(i\omega), \quad (19.8\text{-}32)$$

in full harmony with (19.8-24).

In a similar fashion the correlation spectrum between two junctions can be found. Thus, for the current-current correlation at the junction surfaces S_α and S_β:

$$\mathcal{S}_{\Delta i_\alpha, \Delta i_\beta}(\omega) = 2k_B T[Y_{\alpha\beta}(i\omega) + Y_{\beta\alpha}^*(i\omega)] - \delta_{\alpha\beta} 2eI_\alpha. \quad (19.8\text{-}33)$$

This is an important result for bipolar transistors. A host of other details will be left to the researcher in device physics.

19.9 Problems to Chapter XIX

19.1 Show the validity of the binomial theorem in the form

$$\sum_{m=0}^{\infty} \frac{(-n)!}{m!(-n-m)!}(-q)^m = (1-q)^{-n}. \quad (1)$$

Using this, from (19.2-17') obtain the generating function for $P(m|n)$ as given by (19.2-18). Also, compound with the incoming particle distribution; assuming this to be Poissonian, obtain the full generating function for $W(m)$.

19.2 In a branching process with one generation of offspring, let $\omega(z)$ be the generating function for the distribution of offspring (assumed to be a Bernoulli trial). Show that the generating function $\phi(z)$ for the number of particles X is

$$\phi(z) = z\omega(z) = \lambda z(z-1) + z. \quad (2)$$

Find $\langle X \rangle$ and $\mathrm{var}\, X$ and confirm (19.2-20).

19.3 Repeat the above problem for just *two* generations of offspring.

---

[36] The impedance here is the same for any surface position. In a refinement of this model one may consider an injection distribution, $\gamma(\mathbf{r}_0)$. In particular for transistors this is necessary to account for 'emitter crowding'.

19.4 Consider an infinite branching process, with each particle producing at most one descendant. The nth and (n + 1)th generation are connected by $\omega_{n+1}(z) = \omega_n[\omega(z)]$. Show that n-fold compounding leads to

$$\omega_n(z) = \lambda^n(z-1)+1. \tag{3}$$

With p_n being the membership of the nth generation and $X = \sum_{n=0}^{\infty} p_n$, obtain $M = \langle X \rangle$ and the variance

$$\langle \Delta X^2 \rangle = \sum_{n=0}^{\infty} \langle \Delta p_n^2 \rangle + 2 \sum_{n,n';n>n'}^{\infty} \langle \Delta p_n \Delta p_{n'} \rangle. \tag{4}$$

Show that $\langle \Delta X^2 \rangle = [1/(1-\lambda)][\lambda/(1-\lambda)] = M(M-1)$, in agreement with (19.4-22), obtained with the method of recurrent generating functions.

19.5 Use Richardson's formula, (19.6-39), to show that in any dimension the high frequency asymptote for a diffusion process is $\sim\omega^{-3/2}$. Example: In one dimension

$$\left|\int_0^L e^{ikx}dx\right|^2 = \frac{4\sin^2 kL/2}{k^2} \sim \frac{2}{k^2}, \quad \int_{-\infty}^{\infty} \frac{dk\, Dk^2}{(D^2k^4+\omega^2)k^2} = \frac{\sqrt{D}}{\omega^{3/2}} \int_{-\infty}^{\infty} \frac{dy}{y^4+1}. \tag{5}$$

Repeat this procedure for two (Bessel-function integral) and three dimensions.

19.6 Let N particles be subject to a 1D diffusion process in $-L \le x \le L$, with perfectly reflecting boundaries at $\pm L$. The Fourier–Laplace transformed Green's function satisfies $(d^2G/dx^2) + k^2G(x,x',k) = -(1/D)\delta(x-x')$, subject to zero derivative at $\pm L$, where $k = \sqrt{-i\omega/D}$. Show that G is given by (see[26]):

$$G(x,x') = \begin{cases} -\cos k(L+x')\cos k(L-x)/kD\sin 2kL, & x \ge x', \\ -\cos k(L+x)\cos k(L-x')/kD\sin 2kL, & x \le x'. \end{cases} \tag{6}$$

Find the spectrum for the Poisson-distributed particles inside $-d \le x \le d$ and obtain the low frequency and asymptotic frequency limits.

19.7 Obtain the result (19.8-29) for the complete boundary spectral density of the particles. N.B. First show that for any function $\psi: \int d^3r\psi(\mathbf{r})\delta(\mathbf{r}-\mathbf{r}_0) = \frac{1}{2}\psi(\mathbf{r}_0)$, where $\mathbf{r}_0$ is a surface coordinate [set $\delta(\mathbf{R}) = -(1/4\pi)\nabla^2(1/R)$ and use Gauss' theorem].

19.8 Realistically, diffusion plays a role in ambipolar sweep-out, thereby broadening the resonances. Add the diffusion term, let $\tau \to \infty$ and solve for the time-dependent Green's function $g(\mathbf{r},t;\mathbf{r},t')$. Using Chandrasekhar's after-effect probability, evaluate the spectral density of the carrier fluctuations in a subvolume of the sample and compare the results with (19.7-24) of the text.

Chapter XX

Stochastic Optical Signals and Photon Fluctuations

20.1 Introductory Remarks

Optical fields are related to their sources by Maxwell's equations. While these equations in their classical form can be thought of as being deterministic, the sources inherently are not. Also, even if they were, the transmission through any medium other than vacuum may introduce randomness. So, optical signals are in essence to be described as *stochastic fields*. The extent of this randomness will be quantitatively stated in this chapter by means of coherence functions (or tensors) of various orders. Two descriptions will be introduced. Firstly, we will follow a standard classical method developed decades ago, known as *analytic signals*. Secondly, we will use a Heisenberg description of the fluctuating fields in a way consistent with the developments in Chapter XVI and XVII of this text. However, we shall employ *coherent states*, being the eigenstates of the annihilation operators of the photon modes, rather than occupation-number states, as we have done in most of this text. By doing so, a pseudo-classical description of the stochastic quantum fields known as the *Sudarshan–Glauber transform or P-representation* will emerge which puts the description of stochastic optical fields in an extremely useful form, both for its theoretical formulation and for comparison with experimental data. The *P*-representation is a remarkable union between classical and quantum formalisms.

Our primary goal, however, is of a physical nature and concerns the quest of how to measure the statistical properties of optical signals with appropriate detection techniques. These may involve some of the oldest detection methods using photodiodes and photomultipliers or the more modern photo-avalanche diodes described in the preceding chapter. We note that so far it has mostly been assumed that the primary flux has shot noise, giving in the output of the detector rise to multiplied shot noise; to this the noise of the detection process is simply added on. If the light is of a thermal nature, then there is a boson factor B, which degrades upon partial absorption [cf. (19.2-5)] and not much of the boson nature of the light is observed in the output process. In photoconductors, the situation is even bleaker. While we have shown that no trace of the photon field remains in the output of a *thermal* detector due to stimulated emission, which reduces the final noise of the Langevin sources to just double shot noise (subsection 18.6.2), we have fully ignored that an effect of the incoming radiation field *should show up* in the output of a photoconductor operated in a non-equilibrium steady state. This means that, strictly speaking, the carrier-density fluctuations in a photoconducting solid illuminated by an optical signal can not be treated as an independent Markov process, separate from the statistical properties of the incident radiation-field. Criticism against the use of the generalized *g-r* theorem, or what we have called the fluctuation-relaxation theorem [cf. (18.5-9)], has been voiced by various researchers, among others by Cook, Blok and van Kampen.[1] This goes not to say that this state of affairs was not in principle known, cf. R.C. Jones (1953), Van Vliet (1959) and Alkemade (1959).[2] However, we have so far tacitly assumed that the effects of the radiation field are negligibly small.

The method used by Cook et al. is based on an expansion in the inverse volume of the part of the photoconductor that is illuminated by the light, following a similar procedure as developed by van Kampen in Ref. 22 of Chapter XVIII. A much more detailed procedure is found with Ubbink in a later publication.[3] Be it noted that the correlation matrix of the output fluctuations does not change in form; however, the covariance matrix of the carrier populations responsible for the magnitudes of the spectra is changed with respect to the 'collective process' results of the *g-r* theorem.

For those who study photo-emissive or photoconducting devices it is perhaps gratifying that the effect of the light fluctuations on the output noise is usually very small. However, for those interested in the statistical properties of light it would be a sad story if photo-detectors – whether consisting of standard optical instruments or of electronic apparatus – would not by a study of their output statistics yield detailed information on the optical radiation field. And of course, they do. However, the usual spectral noise measurements, referred to as self-beating spectroscopy in the optical literature, give information only on the intensity-fluctuation correlations, i.e. on the measure of second-order coherence. Thus, other methods, like photon counting techniques are required to obtain information that is far more detailed.

With the advent of the laser and other non-thermal sources a great new impetus was given to the study of stochastic photon fields and extensive new experimental research methods have been developed, some of which will be described here. As to the literature pertaining to the present chapter, several pages could be filled. We therefore mention only a few sources. For the measurement of the statistical properties of optical fields, the reader may want to consult the extensive article by Arecchi and Degiorgio, while for the theory of photon counting we refer to C.L.

---

[1] J.G. Cook, J. Blok and N.G. van Kampen, Physica **35**, 241-257 (1967).

[2] R.C. Jones, "Advances in Electronics", Vol. V, Acad. Pres, NY 1953; K.M. Van Vliet, Proc. IRE **46**, 1004 (1958); C.Th.J. Alkemade, Physica **25**, 1145-1158 (1955).

[3] J.T. Ubbink, Physica **53**, 253-278 (1971); similar results have been obtained by Zijlstra with a Langevin-equation procedure, R.J.J. Zijlstra, unpublished.

Metha.[4,5] For analytic signals, see Beran and Parrent[6] and Wolf.[7] Earlier work by the author is found in a survey article[8]; also, we have benefited from course notes.[9]

20.2 Analytic Signals and Coherence

At this moment we will not specify whether we use the vector potential $\mathbf{A}(\mathbf{r},t)$ or the electrical field $\mathbf{E}(\mathbf{r},t)$ to characterize the field; rather we shall denote the stochastic signal by $\mathbf{V}(\mathbf{r},t)$, which can be either. Moreover, we will at first assume that the signal is linearly polarized, so that we can deal with the scalar function $V(\mathbf{r},t)$. Let this signal be represented by a Fourier series on $(-T/2,T/2)$, split into a positive frequency part, $\exp(-i\omega_n t)$, and negative frequency part, $\exp(i\omega_n t)$, whereby $\omega_n = 2\pi n/T$ with $n=(0),1,2,...,\infty$. Hence, we write

$$V(\mathbf{r},t) = V^+(\mathbf{r},t) + V^-(\mathbf{r},t) = \sum_n a_n(\mathbf{r})e^{-i\omega_n t} + \sum_n a_n^*(\mathbf{r})e^{i\omega_n t}, \qquad (20.2\text{-}1)$$

with inversion $a_n(\mathbf{r}) = (1/T)\int_{-T/2}^{T/2} dt' e^{i\omega_n t'} V(\mathbf{r},t')$. This yields for V^+:

$$V^+(\mathbf{r},t) = \frac{1}{T}\int_{-T/2}^{T/2} dt' V(\mathbf{r},t')\sum_n e^{-i\omega_n(t-t')}. \qquad (20.2\text{-}2)$$

Writing now $\Delta\omega = 2\pi/T$ and letting $T\to\infty$, we change $\sum_n$ to an integral and note that

$$\frac{1}{2\pi}\int_0^\infty d\omega e^{-i\omega(t-t')} = \delta_+(t-t') = \frac{1}{2}\left[\delta(t-t') - \frac{i}{\pi}\mathcal{P}\frac{1}{t-t'}\right], \qquad (20.2\text{-}3)$$

cf. problem 16.5, Eq. (8). Substituting (20.2-3) into (20.2-2), we find

$$V^+(\mathbf{r},t) = \int_{-\infty}^{\infty} dt' V(\mathbf{r},t')\delta_+(t-t') = \frac{1}{2}[V(\mathbf{r},t) + i\mathcal{H}[V(\mathbf{r},t)]], \qquad (20.2\text{-}4)$$

where $\mathcal{H}$ denotes the Hilbert transform. A similar result can be obtained for $V^-(\mathbf{r},t) = V^{+*}(\mathbf{r},t)$. We call $V^+(\mathbf{r},t)$ the *analytic signal associated with* $V(\mathbf{r},t)$; it is a complex signal, obtained from $V(\mathbf{r},t)$ by adding as imaginary part its Hilbert transform and taking one half of this result. Two advantages of this description are at once clear. First, the analytic signal has only positive Fourier frequencies. Secondly,

---

[4] F.T. Arecchi and V. Degiorgio in "Laser Handbook I" , North Holland, Amsterdam 1972, p. 191-264.
[5] C.L. Mehta, "Theory of Photo-electron Counting", in Progress in Optics VIII (E. Wolf, Ed.), North Holland, Amsterdam 1970, p. 373-440.
[6] M.J. Beran and G.B. Parrent, Jr., "Theory of Partial Coherence", Prentice Hall, N.J. 1964.
[7] E. Wolf, "Introd. to the Theory of Coherence and Polarization of Light", Cambridge U. Press, 2007.
[8] K.M. Van Vliet, "Photon Fluctuations and Their Interactions with Solids", Physica **83 B**, 52-69 (1976).
[9] C.Th.J. Alkemade and F.C. van Rijswijk, "Stralingsfluctuaties Seminarium Quantum Optica, Chapter III, Sterrewacht, University of Utrecht, 1973-1974). [In Dutch.]

since it is a complex signal, we can employ Maxwell's equations in their form for complex phasors, as is customarily done.

Some basic properties of analytic signals will now be discussed. Henceforth, the signal will be denoted by $E^+(t) = \frac{1}{2}[u(t) + iv(t)]$, with the spatial variable being suppressed and where $u(t) = E(t)$ and $v(t) = \mathcal{H}[E(t)]$. Firstly we note that for the convolution of two real functions u_1 and u_2 we have

$$\int_{-\infty}^{\infty} dt\, u_1(t) u_2(t+\theta) = \int_{-\infty}^{\infty} dt\, v_1(t) v_2(t+\theta)\,, \tag{20.2-5}$$

where the v's are the Hilbert transforms. Hints for the proof will be stated in Problem 20.1. This property has a direct bearing on the correlation function of two analytic signals. With $\Phi_{u_1u_2}(\theta) = \lim_{T\to\infty}(1/T)\int_{-T/2}^{T/2} dt\, u_1(t)u_2(t+\theta)$, we find

$$\langle u_1(t) u_2(t+\theta)\rangle = \langle v_1(t) v_2(t+\theta)\rangle\,. \tag{20.2-6}$$

Also we have, noting that $-u_2$ is the Hilbert transform of v_2,

$$\langle u_1(t) v_2(t+\theta)\rangle = -\langle v_1(t) u_2(t+\theta)\rangle\,. \tag{20.2-7}$$

In particular, for $\theta = 0$,

$$\langle u_1(t) v_2(t)\rangle = -\langle v_1(t) u_2(t)\rangle\,. \tag{20.2-8}$$

If now u_1 and v_2 refer to the *same* analytic signal, it follows that

$$\langle u(t)v(t)\rangle = 4\langle \mathrm{Re}\, E^+(t)\ \mathrm{Im}\, E^+(t)\rangle = 0\,. \tag{20.2-9}$$

Thus, the two parts have zero-covariance and – going back to (20.2-5) – we see that $u(t)$ and $v(t)$ are orthogonal to each other.

Secondly, for the correlation of two analytic signals $E_1^+(t)$ and $E_2^+(t)$ we find

$$\widehat{\Phi}_{12}(\theta) \equiv \langle E_1^-(t) E_2^+(t+\theta)\rangle = \tfrac{1}{2}[\langle u_1(t)u_2(t+\theta)\rangle + i\langle u_1(t)v_2(t+\theta)\rangle]\,, \tag{20.2-10}$$

where we remember that $E^-(t) = E^{+*}(t)$; be it further noted that the real and imaginary parts are again each other's Hilbert transforms. Also we observe that the transposition property $\widehat{\Phi}_{12}(-\theta) = \widehat{\Phi}_{21}^*(\theta)$ is still valid for analytic signals, as it was for the original fields. Most important, since the correlation function is itself an analytic signal, we find that its Fourier transform has only positive frequencies. Let us now denote by $\mathcal{S}_{12}(\omega)$ the *usual* cross-spectral density of the original fields $E_1(t)$ and $E_2(t)$ and by $S_{12}(\omega)$ the Fourier transform of $\widehat{\Phi}_{12}(\theta)$; the Wiener–Khintchine theorem gives $\mathcal{S}_{12}(\omega)$ as twice the Fourier transform of $\langle E_1(t)E_2(t+\theta)\rangle$ but, due to the factor ½ on the rhs of (20.2-10), $S_{12}(\omega)$ is the spectral density of the analytic signals associated with the fields without the factor of two. Moreover,

$$S_{12}(\omega) = \begin{cases} \mathcal{S}_{12}(\omega), & \omega > 0, \\ 0, & \omega < 0. \end{cases} \tag{20.2-11}$$

Clearly, these rules greatly simplify the spectral properties of optical fields; *a fortiori*, Eq. (20.2-11) applies to the self-spectral density of a field.

Next, we discuss the line shape of an optical signal and the wave-interaction noise spectrum, alluded to already in Chapter XVIII, subsection 18.10.2. The intensity will be denoted by $I(t) = E^-(t)E^+(t)$. Clearly we also have

$$\langle I\rangle = \tfrac{1}{4}\langle(u-iv)(u+iv)\rangle = \tfrac{1}{4}[\langle u^2\rangle + \langle v^2\rangle] = \tfrac{1}{2}\langle E^2\rangle. \qquad (20.2\text{-}12)$$

With $S_{E^+}(\omega)$ being the Fourier transform of $\widehat{\Phi}(\theta)$, and its inverse given by

$$\widehat{\Phi}(\theta) = (1/2\pi)\int_0^\infty S_{E^+}(\omega)e^{i\omega\theta}d\omega = \int_0^\infty S_{E^+}(\omega)e^{i\omega\theta}d\nu\,, \qquad (20.2\text{-}13)$$

we see that for $\theta = 0$ we have

$$\langle I\rangle = \int_0^\infty S_{E^+}(\nu)d\nu\,. \qquad (20.2\text{-}14)$$

Thus, $S_{E^+}(\nu)$ is also the optical line shape.

We now turn to the intensity fluctuations. For the correlation function we obtain

$$\Phi_I(\theta) = \langle E^-(t)E^+(t)E^-(t+\theta)E^+(t+\theta)\rangle = \tfrac{1}{16}[\langle u^2(t)u^2(t+\theta)\rangle$$
$$+\langle v^2(t)v^2(t+\theta)\rangle + \langle u^2(t)v^2(t+\theta)\rangle + \langle v^2(t)u^2(t+\theta)\rangle]\,. \qquad (20.2\text{-}15)$$

Further, let us assume that the quantities of interest have a normal distribution; for the four variables involved, we can then employ the property (2) of Problem 1.12. Thus, noticing (20.2-6) and (20.2-7), we arrive at

$$\Phi_I(\theta) = \tfrac{1}{4}[\langle u^2\rangle^2 + \langle u(t)u(t+\theta)\rangle^2 + \langle u(t)v(t+\theta)\rangle^2]\,. \qquad (20.2\text{-}16)$$

With (20.2-12) $[\langle I\rangle = \tfrac{1}{2}\langle u^2\rangle]$, we find for the fluctuation-correlation function

$$\Psi_I(\theta) \equiv \langle \Delta I(t)\Delta I(t+\theta)\rangle = \Phi_I(\theta) - \langle I\rangle^2$$
$$= \tfrac{1}{4}[\langle u(t)u(t+\theta)\rangle^2 + \langle u(t)v(t+\theta)\rangle^2]\,. \qquad (20.2\text{-}17)$$

Employing also (20.2-10), this finally yields

$$\Psi_I(\theta) = \widehat{\Phi}_{E^+}(\theta)\widehat{\Phi}^*_{E^+}(\theta)\,. \qquad (20.2\text{-}18)$$

The spectral density of the intensity fluctuations follows from the convolution theorem for Fourier transforms[10] (by the W–K th. we need *twice* the transform for $\mathcal{S}_{\Delta I}$):

$$\mathcal{S}_{\Delta I}(f) = 2\int_{-\infty}^\infty S^*_{E^+}(\xi)S_{E^+}(f-\xi)d\xi\,. \qquad (20.2\text{-}19)$$

Lastly, noting $S^*_{E^+}(\xi) = S_{E^+}(-\xi)$ and setting $\xi \to -\nu$, we proved the result (18.10-40),

$$\mathcal{S}_{\Delta I}(f) = 2\int_0^\infty S_{E^+}(\nu)S_{E^+}(\nu+f)d\nu\,. \qquad (20.2\text{-}20)$$

[10] Cf. I. Sneddon, "Fourier Transforms", McGraw Hill, NY 1952, p. 24.

Coherence Tensors

Presently the position dependence will be restored. Moreover, we will use the 'coordinate' x to denote both spatial and temporal dependence, i.e. $x = (\mathbf{r}, t)$. With the signal still being represented by $E^{\pm}(\mathbf{r}, t)$ the Nth order correlation tensor is defined as

$$\mathsf{G}^{(N)}(\bar{x}_1, \bar{x}_2, \ldots, \bar{x}_N \,\|\, x_N, x_{N-1}, \ldots, x_1) = \langle \mathbf{E}^-(\bar{x}_1)\mathbf{E}^-(\bar{x}_2)\ldots\mathbf{E}^-(\bar{x}_N)\mathbf{E}^+(x_N)\mathbf{E}^+(x_{N-1})\ldots\mathbf{E}^+(x_1)\rangle. \qquad (20.2\text{-}21)$$

Note that we will adhere to the ordering

$$\bar{t}_1 < \bar{t}_2 < \ldots < \bar{t}_N, \quad t_N > t_{N-1} > \ldots > t_1, \qquad \text{(time ordering)}$$

$$\mathbf{E}^- \text{ precedes } \mathbf{E}^+. \qquad \text{(normal ordering)}$$

In particular, we will need the pair-correlation or *coherence tensor*, derived from $\mathsf{G}^{(N)}$ if $\bar{x}_i = x_i$ for all i; its elements are the *coherence functions*,

$$\mathsf{G}^{(N)}(x_1, x_2, \ldots, x_N \,\|\, x_N, x_{N-1}, \ldots, x_1) = \langle \mathbf{E}^-(x_1)\mathbf{E}^-(x_2)\ldots\mathbf{E}^-(x_N)\mathbf{E}^+(x_N)\mathbf{E}^+(x_{N-1})\ldots\mathbf{E}^+(x_1)\rangle. \qquad (20.2\text{-}22)$$

Let further $(\text{tr})^N$ denote the operation of contracting all pairs $\mathbf{E}^-(x_i)\mathbf{E}^+(x_i)$ of the $2N$th order tensor $\mathsf{G}^{(N)}$; we then have the basic identity relating $\mathsf{G}^{(N)}$ to intensity correlations

$$(\text{tr})^N \mathsf{G}^{(N)}(x_1, x_2, \ldots, x_N \,\|\, x_N, x_{N-1}, \ldots, x_1) = \langle I(x_N)I(x_{N-1})\ldots I(x_1)\rangle. \qquad (20.2\text{-}23)$$

For polarized light $\mathbf{E}$ is represented by the scalar E. The operation $(\text{tr})^N$ can then be omitted; in that event we simply write $(\text{tr})^N\mathsf{G}^{(N)} = G^{(N)}$. *The coherence functions are the main source of statistical information about the field.*

To connect with the previous, the following special cases are noted regarding $\mathsf{G}^{(1)}$ and $\mathsf{G}^{(2)}$. If $\bar{x} = (\mathbf{r}', t+\theta)$ and $x = (\mathbf{r}, t)$, we have the Fourier-transform pair:

$$\mathsf{G}^{(1)}(\mathbf{r}', t+\theta; \mathbf{r}, t) = \langle \mathbf{E}^-(\mathbf{r}', t+\theta)\mathbf{E}^+(\mathbf{r}, t)\rangle = \int_0^\infty \mathsf{S}_{E^+}(\mathbf{r}', \mathbf{r}, \omega) e^{i\omega\theta} d\omega / 2\pi, \quad (20.2\text{-}24)$$

and

$$\mathsf{S}_{E^+}(\mathbf{r}', \mathbf{r}, \omega) = \int_{-\infty}^{\infty} e^{-i\omega\theta} \mathsf{G}^{(1)}(\mathbf{r}', t+\theta \,\|\, \mathbf{r}, t)\, d\theta. \qquad (20.2\text{-}25)$$

In particular, for $\mathbf{r}' = \mathbf{r}$ and $\theta = 0$ we have $\langle I(\mathbf{r}, t)\rangle = \int_0^\infty S_{E^+}(\mathbf{r}, \omega)\, d\nu$, showing once more that $S_{E^+}(\mathbf{r}, \omega)$ is the line shape, cf. (20.2-14). Further, the Fourier transform of $(\text{tr})^2\mathsf{G}^{(2)} - \langle I(\mathbf{r}')\rangle\langle I(\mathbf{r})\rangle$ gives the intensity-fluctuation noise as in (20.2-19).

First-order (ordinary) optical coherence occurs if $\mathsf{G}^{(1)}$ is factorizable as $\mathbf{E}^*(\bar{x})\mathbf{E}(x)$ where $\mathbf{E}(x)$ is a non-stochastic positive frequency solution of Maxwell's equations; it is easily shown (see below) that this implies maximum fringe contrast in interference experiments. Likewise, *higher order coherence* means that $\mathsf{G}^{(N)}$ is factorizable, $N \geq 2$; higher-order coherence curtails intensity fluctuations. *Complete coherence* (all N) means that the intensity is fully fixed, i.e. $W(I) = \delta(I - I_0)$.

Let us conduct an interference experiment similar as in Young's classical set-up. We let a radiation field fall on an opaque screen with two tiny pinholes, positioned at $\mathbf{r}_1$ and $\mathbf{r}_2$, with a counter being placed behind the screen at $\mathbf{r}$ in the far field, such that the distances to the pinholes are s_1 and s_2, respectively. The corresponding times of the wavelets in the pinholes are then $t_1 = t - (s_1/c)$ and $t_2 = t - (s_2/c)$. Also, we set $x_1 = (\mathbf{r}_1, t_1)$ and $x_2 = (\mathbf{r}_2, t_2)$. The intensity detected by the counter

$$I(\mathbf{r},t) = \langle E^-(\mathbf{r},t)E^+(\mathbf{r},t)\rangle = G^{(1)}(\mathbf{r},t \,\|\, \mathbf{r},t) \tag{20.2-26}$$

is proportional to that of the wave front at the screen, i.e.

$$\begin{aligned}&\langle [E^-(x_1) + E^-(x_2)][E^+(x_1) + E^+(x_2)]\rangle \\ &\qquad = G^{(1)}(x_1 \,\|\, x_1) + G^{(1)}(x_2 \,\|\, x_2) + 2\,\mathrm{Re}\, G^{(1)}(x_1 \,\|\, x_2)\,.\end{aligned} \tag{20.2-27}$$

The first two terms on the rhs represent the intensities that would be measured due to one hole if the other one were closed; the last term therefore contains the interference contribution. Writing now $G^{(1)} = |G^{(1)}|\,e^{i\varphi}$, the last term becomes $2|G^{(1)}(x_1 \,\|\, x_2)|\cos\varphi$. The oscillations of $\cos\varphi$ correspond to the intensity fringes that are observed when the counter moves along the screen. The degree of contrast is determined by the magnitude of the modulus $|G^{(1)}(x_1 \,\|\, x_2)|$. The latter has, however, an upper bound, that is given by Schwartz inequality.[11] Since every ensemble average can be conceived as the mean of a sequence of n observations, Schwartz inequality yields for this sequence

$$|G^{(1)}(x_1 \,\|\, x_2)|^2 \le G^{(1)}(x_1 \,\|\, x_1)\, G^{(1)}(x_2 \,\|\, x_2)\,. \tag{20.2-28}$$

Maximum contrast is therefore obtained if the equal sign holds. This, in turn requires that the correlation function can be factorized as

$$G^{(1)}(x_1 \,\|\, x_2) = E^*(x_1)E(x_2), \tag{20.2-29}$$

where $E(x_1)$ and $E(x_2)$ are positive frequency solutions of Maxwell's equations.

The converse statement is also true. If $G^{(1)}$ factorizes as in (20.2-29), the equality form of (20.2-28) applies and we have maximum fringe contrast.[12] The property of factorability thus implies the condition of coherence, as conventionally understood. In the present context, since only $G^{(1)}$ is involved, we speak of first-order coherence. Likewise, higher-order coherence occurs if (and only if) we have the factorization

$$G^{(N)}(\bar{x}_1, \bar{x}_2, \ldots \bar{x}_N \,\|\, x_N, x_{N-1}, \ldots, x_1) = E^*(\bar{x}_1)\ldots E^*(\bar{x}_N)\, E(x_N)\ldots E(x_1)\,. \tag{20.2-30}$$

Some final remarks on the classical solution are still in order. Suppose that the field is expanded in vector solutions of the Helmholtz equation $(\nabla^2 + q^2)\mathbf{u}_\mathbf{q} = 0$, subject to the transversality condition div $\mathbf{u}_\mathbf{q} = 0$. Suitable solutions are $\mathbf{u}_\mathbf{q} = \hat{\mathbf{e}}_{\mathbf{q}\lambda} V_0^{-1/2} e^{i\mathbf{q}\cdot\mathbf{r}}$ where $\lambda = 1, 2$ denotes the polarization index, so that

$$\mathbf{E}(\mathbf{r},t) = \Sigma_{n\mathbf{q}\lambda}[c_{n\mathbf{q}\lambda} u_{\mathbf{q}\lambda}(\mathbf{r}) e^{-i\omega_n t} + c^*_{n\mathbf{q}\lambda} u^*_{\mathbf{q}\lambda}(\mathbf{r}) e^{i\omega_n t}]\,. \tag{20.2-31}$$

We now define $C(\mathbf{q},\lambda,t) = \Sigma_n c_{n\mathbf{q}\lambda} e^{-i(\omega_n - \omega_\mathbf{q})t}$ with $\omega_\mathbf{q} = c|\mathbf{q}|$ and $\mathbf{q} \in$ source modes. The classical analytic signal (wave) is then

$$\mathbf{E}^+(\mathbf{r},t) = \Sigma_{\mathbf{q}\lambda} C(\mathbf{q},\lambda,t)\mathbf{u}_{\mathbf{q}\lambda}(\mathbf{r})\, \exp(-i\omega_\mathbf{q} t)\,. \tag{20.2-32}$$

For the stochastic process $C(\mathbf{q},\lambda,t)$ we have amplitude and phase modulation, or stochastic mode-broadening caused by the coupling to the sources. However, as stressed by Arecchi and Degiorgio,[4] we can also study the field divorced from the sources.

---

[11] R.J. Glauber, Phys. Rev. **130**, 2529 (1963).

[12] U.M. Titulaer and R.J. Glauber, Phys. Rev. **140**, B676 (1965).

20.3 The Quantum Field

In the quantum description the field Hamiltonian

$$\mathcal{H} = \tfrac{1}{2}\int dv[E^2 + (\text{curl}\, A)^2] \tag{20.3-1}$$

must be quantized. The solution to this problem was stated in Eqs. (7.4-27). Since we have chosen the field **E** as the representative variable, we repeat the result here:

$$\mathbf{E}(\mathbf{r},t) = \sum_{\mathbf{q},\lambda} (\hbar\omega_{\mathbf{q}}/2V_0)^{1/2}\left[i a_{\mathbf{q}\lambda}\mathbf{u}_{\mathbf{q}\lambda}(\mathbf{r})e^{-i\omega_{\mathbf{q}}t} + (-i)a^{\dagger}_{\mathbf{q}\lambda}\mathbf{u}^{*}_{\mathbf{q}\lambda}(\mathbf{r})e^{i\omega_{\mathbf{q}}t}\right], \tag{20.3-2}$$

where the $\mathbf{u}_{\mathbf{q}\lambda}$ are again the vector solutions of the Helmholtz equation $(\nabla^2 + q^2)\mathbf{u}_{\mathbf{q}} = 0$, subject to the transversality condition $\text{div}\,\mathbf{u}_{\mathbf{q}} = 0$. The **q**'s are determined by the periodic bc applied to the volume V_0 of space under consideration. As before, $\lambda = 1,2$ indicates the polarization; we will also set $q = (\mathbf{q},\lambda)$. The two parts of (20.3-2) will be denoted by $\mathbf{E}^{\pm}(\mathbf{r},t)$; we note that they are each other's Hermitean conjugates. The positive frequency operator is the analogue of the analytic signal; hence

$$\mathbf{E}^{+}(\mathbf{r},t) = \sum_{\mathbf{q},\lambda} i(\hbar\omega_{\mathbf{q}}/2V_0)^{1/2} a_{\mathbf{q}\lambda}\mathbf{u}_{\mathbf{q}\lambda}(\mathbf{r})e^{-i\omega_{\mathbf{q}}t}. \tag{20.3-3}$$

Whereas throughout this text we have used occupation number states to describe the quantum system, this is *not* a good idea for optical fields. In order to ensure that a classical wave with amplitude and phase emerges in the correspondence limit, the N-representation is wanting; basically there is an uncertainty relation between photon number and phase [see also the discussion in the last paragraphs of Section 16.3.4]. Thus, if the phase is to be determinate within a certain spread, then the number of photons must also be determinate within a certain spread. *It is the merit of the founders of quantum optics to have recognised that the natural representation for the field operator* $\mathbf{E}^{+}(\mathbf{r},t)$ *is based on the eigenstates of the annihilation operators* $\{a_q\}$ *occurring in* (20.3-3). These states are called "*coherent states*". Moreover, the choice of employing the ES of the $\{a_q\}$, rather than those of the $\{a_q^{\dagger}\}$, has a logical (though not stringent) physical basis. In the photo-electric effect – which is the basis of photon counting with all photo-emissive detectors – a photon is absorbed and an electron is set free. The reverse process involving stimulated emission is far rarer.

Proceeding, let us denote by $|\alpha_q\rangle$ the ES of a_q and by α_q the corresponding EV; also, let $|\{\alpha_q\}\rangle$ be the tensor product state for all $\{a_q\}$; we thus have

$$a_q|\{\alpha_q\}\rangle = \alpha_q|\{\alpha_q\}\rangle. \tag{20.3-4}$$

Since the operators a_q are non-Hermitean, the EV are complex and not observable in a measurement. In fact, there is an uncertainly relation $\langle(\Delta\text{Re}\,\alpha)(\Delta\,\text{Im}\,\alpha)\rangle \geq 1$. A corollary is that we cannot prepare the system in an exact coherent state, except in the correspondence limit of high quantum numbers, when $[\mathbf{E}^+,\mathbf{E}^-]\sim 0$.

Employing the coherent states,[13] the field is now obtained in the "*P*-representation", or (more correctly) by its Sudarshan–Glauber transform:

$$\mathbf{E}^+(\mathbf{r},t)\,|\{\alpha_q\}\rangle=\sum\nolimits_q i(\hbar/2\omega_q V_0)^{\frac{1}{2}}\alpha_q\mathbf{u}_q(\mathbf{r})e^{-i\omega_q t}\,|\{\alpha_q\}\rangle\equiv\mathcal{E}^+(\mathbf{r},t)\,|\{\alpha_q\}\rangle. \quad (20.3\text{-}5)$$

Accordingly, the 'pseudo-classical field' takes the form

$$\mathcal{E}^+(\mathbf{r},t)=\sum\nolimits_q i(\hbar/2\omega_q V_0)^{\frac{1}{2}}\alpha_q\mathbf{u}_q(\mathbf{r})e^{-i\omega_q t}. \quad (20.3\text{-}6)$$

Comparing with (20.2-32), we have that $C(q,\lambda,t)\leftrightarrow i(\hbar\omega_q/2V_0)^{1/2}\alpha_q$, where in the *P*-representation the randomness is inherent in the store of values for α_q. Further, Eq. (20.3-6) indicates that there is a one-to-one correspondence between the positive frequency solutions of Maxwell's equations $\mathcal{E}^+(\mathbf{r},t)$ and the coherent states of the field $|\{\alpha_q\}\rangle$. However, one should not conclude that the classical picture is vindicated by the present approach; *a physical wave picture arises only in the correspondence limit.*

Coherent states can always be constructed by a superposition of occupation-number states. We have Klauder's expansion[14]

$$|\{\alpha_q\}\rangle=\sum_{\{n_q\}}\prod_q\frac{\alpha_q^{\,n_q}}{(n_q!)^{1/2}}e^{-\frac{1}{2}\sum_{q'}|\alpha_{q'}|^2}\,|\{n_q\}\rangle. \quad (20.3\text{-}7)$$

Thus, the probability for finding n_q photons in a mode q is given by

$$P(n_q)=|\langle n_q\,|\,\alpha_q\rangle|^2=(|\alpha_q|^{2n_q}/n_q!)e^{-|\alpha_q|^2}, \quad (20.3\text{-}8)$$

which is a Poisson distribution with mean $\langle n_q\rangle=|\alpha_q|^2$. Coherent states are normalized but *not* orthogonal.[14a] The classical limit is only reached if $|\alpha_q|^2\gg 1$. Further, coherent states have an extremely simple form of time dependence. Let us consider a unitary transformation based on the single mode-Hamiltonian $\mathcal{H}_q=a_q^\dagger a_q\hbar\omega_q$. Then,

$$\begin{aligned}|\alpha_q,t\rangle&=e^{-i\mathcal{H}_q t/\hbar}\,|\alpha_q\rangle=e^{-ia_q^\dagger a_q\omega_q t}e^{-\frac{1}{2}|\alpha_q|^2}\sum_{n_q}\frac{\alpha_q^{\,n_q}}{(n_q!)^{1/2}}|n_q\rangle\\&=e^{-\frac{1}{2}|\alpha_q|^2}\sum_{n_q}\frac{[\alpha_q e^{-i\omega_q}]^{n_q}}{(n_q!)^{1/2}}|n_q\rangle=|\alpha_q e^{-i\omega t}\rangle=|\alpha_q\rangle e^{-i\omega t}. \quad (20.3\text{-}9)\end{aligned}$$

The state remains coherent at all times. The projections of the EV $\alpha_q(t)$ on the real and imaginary axes exhibit simple harmonic oscillator motion.

---

[13] R.J. Glauber in "Fundamental Problems in Statistical Mechanics" Vol. II, E.G.D. Cohen Ed., North Holland, Amsterdam 1968, pp. 140-187.

[14] J.R. Klauder, Ann. Phys. **11**, 123 (1960). [14a] From (20.3-7) we see $\langle\alpha\,|\,\beta\rangle=e^{\alpha^*\beta-\frac{1}{2}|\alpha|^2-\frac{1}{2}|\beta|^2}$. (20.3-8').

20.4 The Pseudo-classical Field

20.4.1 *Sudarshan–Glauber Transform of the Statistical Density Operator*

The eigenstates of the annihilation operator a_q do not form an orthonormal basis. The test of usefulness for these states lies – as in standard quantum mechanics – in the possibility of a *decomposition of unity.* Indeed, from (20.3-7) one easily shows:

$$(1/\pi)\int|\alpha_q\rangle\langle\alpha_q\rangle d^2\alpha_q = 1, \qquad (20.4\text{-}1)$$

the integration being over the complex α_q-plane, $d^2\alpha_q = d(\mathrm{Re}\,\alpha_q)d(\mathrm{Im}\,\alpha_q)$. In view of this, Sudarshan[15] and Glauber[16] proposed the following form for the density operator ρ

$$\rho = \int\ldots\int|\{\alpha_q\}\rangle P(\{\alpha_q\})\langle\{\alpha_q\}|\prod_q d^2\alpha_q, \qquad (20.4\text{-}2)$$

the integration being over all complex α-planes. Assuming that such a $P(\{\alpha_q\})$ exists, we have for any operator A:

$$\begin{aligned}\langle A\rangle = \mathrm{Tr}(\rho A) &= \int\ldots\int\prod_q d^2\alpha_q P(\{\alpha_q\})\mathrm{Tr}[\,|\{\alpha_q\}\rangle\langle\{\alpha_q\}|A] \\ &= \int\ldots\int\prod_q d^2\alpha_q P(\{\alpha_q\})\langle\{\alpha_q\}|\,A\,|\{\alpha_q\}\rangle. \qquad (20.4\text{-}3)\end{aligned}$$

[We applied the corollary (1.2-10).] Let us now associate a pseudo-classical variable $\mathcal{A}(\{\alpha_q\})$ with the operator A, defined by the last bracket $\langle\ \rangle$ in (20.4-3); we then have

$$\langle A\rangle = \int\ldots\int\prod_q d^2\alpha_q P(\{\alpha_q\})\langle\{\alpha_q\}|\,A\,|\{\alpha_q\}\rangle = \overline{\mathcal{A}}, \qquad (20.4\text{-}4)$$

where the overhead bar will denote a complex α-space (or 'phase space') average. Note that both P and $\mathcal{A}$ are *functions*, so the Sudarshan–Glauber transform allows us to use classical calculus instead of operator algebra; herein lies its principal value.

The function $P(\{\alpha_q\})$ is a quasi-probability, i.e. only a probability function in appearance. It is not necessarily positive definite, but may become so if we coarse-grain over areas large compared to $\Delta^{(2)}\alpha_q$. Like the Wigner function, P is rather to be seen as a transform of ρ than a 'representation'. In particular, $P(\{\alpha_q\})$ is *not* a diagonal representation of ρ in the basis $|\{\alpha_q\}\rangle$, as is sometimes read in the literature. On the contrary, $\rho_{\alpha\alpha}$ and $P(\alpha)$ are Gauss transforms,

$$\langle\alpha|\rho|\alpha\rangle = \int e^{-|\alpha-\beta|^2}P(\beta)d^2\beta. \qquad (20.4\text{-}5)$$

---

[15] E.C.G. Sudarshan, Phys. Rev. Lett. **10**, 277 (1963).
[16] R.J. Glauber, Phys. Rev. **131**, 2766 (1963).

20.4.2 *Pseudo-classical Form of the Coherence Tensors*

For the coherence tensors, Eqs. (20.2-21)ff, the Sudarshan–Glauber transforms are

$$\begin{aligned}
&\mathsf{G}^{(N)}(\bar{x}_1,\bar{x}_2,\dots,\bar{x}_N \,\|\, x_N, x_{N-1},\dots,x_1) \\
&= \mathrm{Tr}\{\rho \mathbf{E}^-(\bar{x}_1)\mathbf{E}^-(\bar{x}_2)\dots\mathbf{E}^-(\bar{x}_N)\mathbf{E}^+(x_N)\mathbf{E}^+(x_{N-1})\dots\mathbf{E}^+(x_1)\} \\
&= \int\dots\int\prod_q d^2\alpha_q P(\{\alpha_q\}) \\
&\times \mathrm{Tr}[|\{\alpha_q\rangle\langle\{\alpha_q|\mathbf{E}^-(\bar{x}_1)\mathbf{E}^-(\bar{x}_2)\dots\mathbf{E}^-(\bar{x}_N)\mathbf{E}^+(x_N)\mathbf{E}^+(x_{N-1})\dots\mathbf{E}^+(x_1)] \\
&= \int\dots\int\prod_q d^2\alpha_q P(\{\alpha_q\}) \\
&\times \langle\{\alpha_q|\mathbf{E}^-(\bar{x}_1)\mathbf{E}^-(\bar{x}_2)\dots\mathbf{E}^-(\bar{x}_N)\mathbf{E}^+(x_N)\mathbf{E}^+(x_{N-1})\dots\mathbf{E}^+(x_1)\,|\{\alpha_q\rangle. \qquad (20.4\text{-}6)
\end{aligned}$$

We now employ the pseudo-classical field, given as

$$\mathbf{E}^+(x)\,|\{\alpha_q\}\rangle = \mathcal{E}(x,\{\alpha_q\})\,|\{\alpha_q\}\rangle, \qquad \mathbf{E}^-(x)\,|\{\alpha_q\}\rangle = \mathcal{E}^*(x,\{\alpha_q\})\,|\{\alpha_q\}\rangle. \qquad (20.4\text{-}7)$$

The correspondence requires that $\mathcal{E}$ contains positive frequencies only. It is to be noted that the $\mathcal{E}$'s depend on the relevant values $\{\alpha_q\}$; there is a one-to-one correspondence between the positive frequency solutions of the Maxwell equations $\mathcal{E}(\mathbf{r},t)$ and the coherent states of the field $|\{\alpha_q\}\rangle$. Thus, as stressed by Glauber, there are just as many coherent states as there are classical (i.e. non-operator) solutions to the Maxwell equations.

Returning to the coherence tensors, the use of the pseudo-classical field vectors yields

$$\begin{aligned}
&\mathsf{G}^{(N)}(\bar{x}_1,\bar{x}_2,\dots,\bar{x}_N \,\|\, x_N, x_{N-1},\dots,x_1) \\
&= \int\dots\int\prod_q d^2\alpha_q P(\{\alpha_q\})\big(\mathcal{E}^*(\bar{x}_1)\mathcal{E}^*(\bar{x}_2)\dots\mathcal{E}^*(\bar{x}_N)\mathcal{E}(x_N)\mathcal{E}(x_{N-1})\dots\mathcal{E}(x_1)\big) \\
&= \overline{\mathcal{E}^*(\bar{x}_1)\mathcal{E}^*(\bar{x}_2)\dots\mathcal{E}^*(\bar{x}_N)\mathcal{E}(x_N)\mathcal{E}(x_{N-1})\dots\mathcal{E}(x_1)}. \qquad (20.4\text{-}8)
\end{aligned}$$

In the last expression we have a function average of the pseudo-classical field vectors. For the pair-coherence functions we have likewise

$$\begin{aligned}
&(\mathrm{tr})^N \mathsf{G}^{(N)}(x_1,x_2,\dots,x_N \,\|\, x_N, x_{N-1},\dots,x_1) \\
&= (\mathrm{tr})^N\, \mathrm{Tr}\{\rho \mathbf{E}^-(x_1)\mathbf{E}^-(x_2)\dots\mathbf{E}^-(x_N)\mathbf{E}^+(x_N)\mathbf{E}^+(x_{N-1})\dots\mathbf{E}^+(x_1)\} \\
&= \mathrm{Tr}\{\rho : I(x_N) I(x_{N-1})\dots I(x_1):\}, \qquad (20.4\text{-}9)
\end{aligned}$$

in which $:O:$ means normal ordering of O. Upon introducing the pseudo-classical field intensity $\mathcal{I} = \mathcal{E}^* \cdot \mathcal{E}$, we arrive at

$$(\mathrm{tr})^N \mathsf{G}^{(N)}(x_1,x_2,\dots,x_N \,\|\, x_N, x_{N-1},\dots,x_1) = \overline{\mathcal{I}(x_N)\mathcal{I}(x_{N-1})\dots\mathcal{I}(x_1)}, \qquad (20.4\text{-}10)$$

the order of the factors now being immaterial. Needless to say that $\langle I \rangle = \overline{\mathcal{I}}$.

For the intensity-fluctuation correlation function we find

$$\Psi_I(\mathbf{r},\mathbf{r}',\theta) = \langle \Delta\mathcal{I}(\mathbf{r},t)\Delta\mathcal{I}(\mathbf{r}',t+\theta)\rangle$$
$$= (\mathrm{tr})^2 \mathrm{G}^{(2)}(\mathbf{r},0 \,\|\, \mathbf{r}',\theta) - \overline{\mathcal{I}(\mathbf{r})}\;\overline{\mathcal{I}(\mathbf{r}')}. \tag{20.4-11}$$

The Fourier transform, multiplied by 2, yields as usual the spectral density of the intensity fluctuations; *we note that the intensity-fluctuation spectrum is a second-order coherence effect.*[17] The relationship (20.2-20) remains valid, with the classical analytic field now being replaced by $\mathcal{E}$. We thus arrive at

$$\mathcal{S}_{\Delta I}(\mathbf{r},\mathbf{r}',f) = 2\int_0^\infty S^*_{\mathcal{E}^*}(\mathbf{r},\nu)\, S_{\mathcal{E}}(\mathbf{r}',\nu+f)\, d\nu. \tag{20.4-12}$$

Because of the correspondence between the coherent states and the pseudo-classical fields, the probability distributions for the field at multiple times are solely determined by the coherent state distribution, $P(\{\alpha_q\}) \to W_n(\mathcal{E}_n,t_n;\mathcal{E}_{n-1},t_{n-1};\ldots\mathcal{E}_1,t_1)$. These distributions, in turn, give rise to distribution functions derived from the field, like $W(\mathcal{I})$ for the intensity distribution and $W(\mathcal{U})$ for the time-integrated intensity distribution. Examples will be given in the next section. A flow chart for the various distributions is given below. The sign $\xrightarrow{\sim}$ indicates the correspondence limit of large $|\alpha_q|$.

<u>Flow Chart</u>

				quantum field $\mathbf{E} = \mathbf{E}^+ + \mathbf{E}^-$ $\mathbf{E}^+(\mathbf{r},t)\,|\{\alpha_q\}\rangle =$ $\mathcal{E}(\mathbf{r},t\{\alpha_q\}|\,|\{\alpha_q\}\rangle$
				$\downarrow$
density operator ρ	$\leftrightarrow$	coherent state distribution $P(\{\alpha_q\})$	$\leftrightarrow$	pseudo-classical field sequence of pdf 's $W_n(\mathcal{E}_n,t_n;\ldots;\mathcal{E}_1,t_1)$
$\sim\downarrow$		$\sim\downarrow$		$\sim\downarrow$
photon distribution $P(\{n_q\})$	$\leftrightarrow$	classical mode distribution $P(\{C_q\})$	$\leftrightarrow$	classical field sequence of p.d.f.'s $W_n(\mathbf{E}_n^+,t_n;\ldots;\mathbf{E}_1^+,t_1)$

Further:

$W(\mathcal{E},t) \to W(\mathcal{I}) \xrightarrow{\sim} W(I)$, etc.

$W_n(\mathcal{I}_n,t_n;\mathcal{I}_{n-1},t_{n-1};\ldots;\mathcal{I}_1,t_1) \to W(\mathcal{U})$, etc.

---

[17] In classical optical literature (see e.g. Wolf, Ref.7) another nomenclature is often followed. Our tensor $\mathrm{G}^{(N)}$, having $2N$ field quantities, is then associated with $2N$th order coherence. 'Ordinary coherence' is then synonymous with second order coherence, while intensity fluctuation noise is a fourth order effect.

20.5 Examples for Thermal and Non-thermal Radiation Fields

Black-body radiation and other chaotic fields

The canonical partition function for thermally emitted photons was given in Eq. (8.1-8). Whence, the density operator is given by

$$\rho = \left(\prod\nolimits_q [1-e^{-\beta\hbar\omega_q}]\right) e^{-\beta\sum_{q'}\hbar\omega_{q'} a^\dagger_{q'} a_{q'}}, \quad \beta = 1/k_B T. \tag{20.5-1}$$

For the distribution of photons over all modes $q = (\mathbf{q}, \lambda)$ we thus have

$$P(\{n_q\}) = \langle \{n_q\} | \rho | \{n_q\} \rangle = \prod_q \left([1-e^{-\beta\hbar\omega_q}]\, e^{-\beta\hbar\omega_q n_q}\right). \tag{20.5-2}$$

The mean occupancy of a mode is the B–E distribution, $\langle n_q \rangle = [\exp(\beta\hbar\omega_q) - 1]^{-1}$. With this the distribution (20.5-2) can be rewritten as

$$P(\{n_q\}) = \prod_q \frac{1}{1+\langle n_q\rangle}\left[\frac{\langle n_q\rangle}{1+\langle n_q\rangle}\right]^{n_q}; \tag{20.5-3}$$

this is a geometric distribution, met also in Section 19.2. For the covariances of the mode-occupancy fluctuations, one obtains Einstein's result, cf. (4.3-12)

$$\langle \Delta n_q \Delta n_{q'} \rangle = \langle n_q \rangle [1 + \langle n_q \rangle] \delta_{qq'}. \tag{20.5-4}$$

Because of the density of modes it is useful to do some coarse-graining. Let $N_\kappa = \Sigma_{q=1}^{Z_\kappa} n_q$. The distribution of the N_κ's is most easily found from the (Fowler) generating function, the result being the boson distribution

$$P(\{N_\kappa\}) = \prod_\kappa \frac{(N_\kappa + Z_\kappa - 1)!}{N_\kappa!(Z_\kappa - 1)!}\left(\frac{1}{1+B_\kappa}\right)^{Z_\kappa}\left(\frac{B_\kappa}{1+B_\kappa}\right)^{N_\kappa}, \tag{20.5-5}$$

where B_κ is the boson factor $B_\kappa = \langle N_\kappa \rangle / Z_\kappa = 1/(e^{\beta\hbar\omega_q} - 1)$, $q \in \kappa$. The boson distribution is also called the negative binomial distribution, cf. (19.2-17'). For the variance one finds the expected super-Poissonian result

$$\langle \Delta N_\kappa^2 \rangle = \langle N_\kappa \rangle [1 + \langle N_\kappa \rangle / Z_\kappa] = \langle N_\kappa \rangle [1 + B_\kappa]. \tag{20.5-6}$$

The Sudarshan–Glauber function for one mode $P(\alpha)$ [we omit the subscript q] can in principle be obtained from $P(n)$ by inversion of the Gauss transform (20.4-5). This requires the introduction of the characteristic function or Fourier transform in the complex plane, which will be carried out below. The following result is found

$$P(\alpha) = (1/\pi\langle n\rangle) e^{-|\alpha|^2/\langle n\rangle}. \tag{20.5-7}$$

Since the modes are independent, we also have

$$P(\{\alpha_q\}) = \prod_q \frac{1}{\pi\langle n_q\rangle} \exp\{-\sum_q [\,|\alpha_q|^2/\langle n_q\rangle]\}. \tag{20.5-8}$$

This will be referred to as a Gaussian field. Let us assume that we deal with polarized light. Then, from the P-distribution we can readily obtain the probability density function for the pseudo-classical field, $W(\mathscr{E})$. We recall that

$$\begin{aligned}\mathscr{E}(\mathbf{r},t) &= \langle\{\alpha_q\}|\,E^+(\mathbf{r},t)\,|\{\alpha_q\}\rangle \\ &= \sum_q i(\hbar\omega_q/2V_0)^{1/2}\alpha_q u_q(\mathbf{r})e^{-i\omega_q t} \equiv c_q(\mathbf{r})e^{-i\omega_q t}\alpha_q\,. \end{aligned} \tag{20.5-9}$$

The time dependence is of no concern, since it can be absorbed in the phase of c_q. Formally, the pdf for $\mathscr{E}$ is now given by ($\mathscr{E}_q = c_q\alpha_q$):

$$W(\mathscr{E}) = \int\!\ldots\!\int \prod d^2\alpha_q \prod \{(1/\pi\langle n_q\rangle|c_q|^2)e^{-\mathscr{E}_q^*\mathscr{E}_q/|c_q|^2\langle n_q\rangle}\}\delta^{(2)}(\mathscr{E} - \Sigma_q\mathscr{E}_q)\,. \tag{20.5-10}$$

The evaluation can be done with the characteristic function (Problem 20.3); however, one can surmise that the result will again be a Gaussian,

$$W(\mathscr{E}) = (1/\pi\langle I\rangle)e^{-\mathscr{E}_q^*\mathscr{E}_q/\langle I\rangle}\,. \tag{20.5-11}$$

The intensity distribution is immediate. The differential area in the complex plane has the form $d(\mathrm{Re}\mathscr{E})d(\mathrm{Im}\mathscr{E}) = |\mathscr{E}|\,d|\mathscr{E}|\,d\phi = \frac{1}{2}d\mathscr{I}d\phi$. Integrating over the phase, we find the negative exponential distribution

$$W(\mathscr{I}) = (1/\langle I\rangle)e^{-\mathscr{I}/\langle I\rangle}\,. \tag{20.5-12}$$

For non-polarized radiation the distribution is more complex, cf. Mandel.[18]

For many other sources which do not have the blackbody spectrum but are still thermal in origin (e.g. a tungsten ribbon lamp), or are chaotic for purely statistical reasons, the results (20.5-8) through (20.5-12) remain valid, as can be shown by a simple entropy argument; this is left to the problems to establish.

Digression. For the details we must pause somewhat longer on the meaning and implications of the P-distribution.[19] We start by inserting the decomposition of unity ('closure relation') (20.4-1) into the density operator from both sides, giving

$$\rho = (1/\pi^2)\iint |\alpha\rangle\langle\alpha|\rho|\beta\rangle\langle\beta|\,d^2\alpha\, d^2\beta\,. \tag{20.5-13}$$

In order to obtain the weight function $\langle\alpha|\rho|\beta\rangle$ we use closure of the n-states so that

[18] L. Mandel, Proc. Phys. Soc. **81**, 1104 (1963).

[19] R.J. Glauber, Ref. 13, Op. Cit., Sections 7 and 8.

$$\rho = \sum_{n,m} |n\rangle \rho_{nm} \langle m| = \sum_{n,m} \rho_{nm} (n!m!)^{-1/2} a^{\dagger n} |0\rangle\langle 0| a^m . \tag{20.5-14}$$

This form suggests that the following complex-valued function might be useful

$$R(\alpha^*, \beta) \equiv \sum_{n,m} \rho_{nm} (n!m!)^{-1/2} \alpha^{*n} \beta^m . \tag{20.5-15}$$

In particular we have

$$R(\alpha^*, \alpha) = \sum_{n,m} \rho_{nm} (n!m!)^{-1/2} \alpha^{*n} \alpha^m . \tag{20.5-15'}$$

The weighting function now takes the form

$$\begin{aligned} \langle \alpha | \rho | \beta \rangle &= R(\alpha^*, \beta) \langle \alpha | 0 \rangle \langle 0 | \beta \rangle \\ &= R(\alpha^*, \beta) \exp[-\tfrac{1}{2} |\alpha|^2 - \tfrac{1}{2} |\beta|^2], \end{aligned} \tag{20.5-16}$$

where we used (20.3-8') for the scalar products in the first line. Since the density operator is a bounded operator, meaning that

$$\mathrm{Tr}\{\rho^2\} = \sum_{n,m} |\rho_{nm}|^2 \leq 1, \tag{20.5-17}$$

the series (20.5-15) for the bilinear function R converges for all finite values of α^* and β, being therefore an *entire function* in both of its arguments.[20] Inserting (20.5-16) into (20.5-13), we obtain

$$\rho = (1/\pi^2) \int\int |\alpha\rangle\langle\beta| R(\alpha^*, \beta) e^{-\frac{1}{2}|\alpha|^2 - \frac{1}{2}|\beta|^2} d^2\alpha \, d^2\beta . \tag{20.5-18}$$

Next let us consider the normally ordered and antinormally ordered characteristic functions of ρ defined, respectively, by

$$\Phi_N(u) = \mathrm{Tr}\{\rho e^{ua^\dagger} e^{-u^* a}\} = \langle e^{ua^\dagger} e^{-u^* a} \rangle , \tag{20.5-19}$$

$$\Phi_A(u) = \mathrm{Tr}\{\rho e^{-u^* a} e^{ua^\dagger}\} = \langle e^{-u^* a} e^{ua^\dagger} \rangle . \tag{20.5-19'}$$

Because of the relationship $e^A e^B = e^{A+B+\frac{1}{2}[A,B]}$, valid for any two operators that commute with their commutator, cf. footnote 12 of Chapter XI, we have the connection

$$\Phi_N(u) = e^{|u|^2} \Phi_A(u) . \tag{20.5-20}$$

Although we will not use it, the non-ordered characteristic function is defined as $\langle e^{ua^\dagger - u^* a} \rangle$. Since the exponent in the P-representation is purely imaginary, the characteristic function is referred to as the 2D Fourier transform in the complex plane.

---

[20] An entire function is analytic everywhere in the complex plane, except for a possible pole at ∞. Dyadic expansions of the type considered here are useful for other types of operators, including unbounded ones, while having the same kind of analyticity properties as R.

Employing the cyclic property of the trace and inserting the decomposition of unity for the coherent states, the antinormally ordered characteristic function can be evaluated as follows

$$\begin{aligned}\Phi_A(u) &= (1/\pi)\int d^2\alpha \mathrm{Tr}\{|\alpha\rangle\langle\alpha| e^{ua^\dagger}\rho e^{-u^*a}\} \\ &= (1/\pi)\int d^2\alpha \langle\alpha| e^{ua^\dagger}\rho e^{-u^*a}|\alpha\rangle \\ &= (1/\pi)\int d^2\alpha\, e^{u\alpha^*-u^*\alpha}\langle\alpha|\rho|\alpha\rangle. \end{aligned} \tag{20.5-21}$$

While the function $\langle\alpha|\rho|\alpha\rangle$ is positive definite, it is another quasi-probability since the coherent states are not orthogonal; the quantity $\langle\alpha|\rho|\alpha\rangle$ cannot be inferred from a direct physical measurement. The inversion of the Fourier integral may be obtained from the Fourier property of the complex 2D delta function,

$$(1/\pi^2)\int d^2\alpha\, e^{u(\alpha^*-\beta^*)-u^*(\alpha-\beta)} = \delta^{(2)}(\alpha-\beta). \tag{20.5-22}$$

This leads to the inverse transform

$$\langle\alpha|\rho|\alpha\rangle = (1/\pi)\int d^2u\, e^{u^*\alpha-u\alpha^*}\Phi_A(u)\,. \tag{20.5-23}$$

Another connection is needed. Writing down (20.5-16) for the case $\alpha=\beta$, we have

$$\langle\alpha|\rho|\alpha\rangle = R(\alpha^*,\alpha)e^{-|\alpha|^2}\,. \tag{20.5-24}$$

For the normally ordered characteristic function the transform pair is more direct. assuming that the P-representation exists, the far rhs of (20.5-19) can be evaluated as

$$\begin{aligned}\Phi_N(u) &= \int d^2\alpha \langle\alpha| e^{ua^\dagger-u^*a}|\alpha\rangle P(\alpha) \\ &= \int d^2\alpha\, e^{u\alpha^*-u^*\alpha}P(\alpha)\end{aligned} \tag{20.5-25}$$

with inversion

$$P(\alpha) = (1/\pi^2)\int d^2u\, e^{u^*\alpha-u\alpha^*}\Phi_N(u)\,. \tag{20.5-26}$$

Finally, we can return to the case of blackbody radiation. The density operator [Eqs. (20.5-1) - (20.5-5)] is rewritten as

$$\rho = \prod_q \frac{1}{1+\langle n_q\rangle}\left[\frac{\langle n_q\rangle}{1+\langle n_q\rangle}\right]^{a_q^\dagger a_q} = \prod_q \frac{1}{1+\langle n_q\rangle}\sum_{\ell_q}\left[\frac{\langle n_q\rangle}{1+\langle n_q\rangle}\right]^{\ell_q}|\ell_q\rangle\langle\ell_q|\,, \tag{20.5-27}$$

where in the final rhs we made a spectral decomposition in terms of the projectors $|\ell_q\rangle\langle\ell_q|$. Restricting ourselves to a single mode, we obtain for $R(\alpha^*,\beta)$

$$R(\alpha^*,\beta) = \frac{1}{1+\langle n\rangle}\exp\left(\frac{\langle n\rangle}{1+\langle n\rangle}\alpha^*\beta\right). \tag{20.5-28}$$

Hence, from (20.5-24) the function $\langle\alpha|\rho|\alpha\rangle$ is given by

$$\langle\alpha|\rho|\alpha\rangle = \frac{1}{1+\langle n\rangle}\exp\left(-\frac{|\alpha|^2}{1+\langle n\rangle}\right). \tag{20.5-29}$$

The Fourier transform is found to be $\Phi_A(u) = e^{-(1+\langle n\rangle)|u|^2}$. Accordingly, we also have

$$\Phi_N(u) = e^{-\langle n\rangle|u|^2}. \tag{20.5-30}$$

The inversion is now straightforward, yielding the earlier stated result:

$$P(\alpha) = (1/\pi\langle n\rangle)e^{-|\alpha|^2/\langle n\rangle}. \tag{20.5-31}$$

Spectral density of the intensity fluctuations

Noise measurements are usually made with photo-emissive devices, either based on photocells and photomultipliers, or on solid-state detectors such as cooled InSb diodes or avalanche diodes. The spectrum based on compound Poisson statistics and (20.4-12) was already computed in subsection 18.10.2. We assume that the light is resolved and that the interval $\mathcal{T}$ on which the counting variance is based is larger than the coherence time τ_c ('slow detection'). The number of boson cells in such a measurement is $Z = (\Omega/\Omega_c)(\mathcal{T}/\tau_c)$. The boson factor is then $B = \langle M_{\mathcal{T}}\rangle/Z = \langle m\rangle\mathcal{T}/Z$. For a rectangular line set by a monochromator with slit width $\Delta\nu$ centred on ν_0 the spectrum of the incident quanta per second m was found to be [(18.10-42)]

$$\mathcal{S}_m(f) = 2\langle m\rangle[1+B(\nu_0)](1-f/\Delta\nu). \tag{20.5-32}$$

For all practical arrangements $f \ll \Delta\nu$. Let the quantum efficiency for detection be η so that the instantaneous photocurrent is $i(t) = e\eta m(t)$. The boson factor degrades due to binomial compounding as noted in Section 19.2. Whence, the noise at the detector side is

$$\mathcal{S}_i(f) = 2e\langle i\rangle[1+\eta B(\nu_0)]. \tag{20.5-33}$$

The excess noise ratio with respect to Poissonian shot noise is

$$\delta \equiv [\mathcal{S}_i(f) - 2e\langle i\rangle]/2e\langle i\rangle = \eta B(\nu_0). \tag{20.5-34}$$

It is readily seen that the boson factor – in the absence of transmission losses – is the same as the boson factor of the source with radiation temperature T_s. For a detector with surface area S subtending a solid angle Ω the detected radiation is coherent within a solid angle $\Omega_c = \lambda_0^2/S$. The wave-interaction noise on the detector side involves the stream

$$B\,\Delta\nu = \frac{\langle m\rangle \mathcal{I}}{Z}\Delta\nu = \frac{\langle m\rangle \mathcal{I}}{(\Omega/\Omega_c)(\mathcal{I}/\tau_c)}\Delta\nu = \frac{\langle m\rangle \lambda_0^2}{\Omega S}, \tag{20.5-35}$$

where we used $\Delta\nu\,\tau_c = 1$, valid for a rectangular slit. As is well known, in optical imaging the product ΩS is conserved; hence, at the source $B\,\Delta\nu = \langle m\rangle\lambda_0^2/\Omega_s S_s$. We note that $\langle m\rangle h\nu\, d\nu/\Omega_s S_s$ is the radiation energy coming off one cm^2 of source surface per second per steradian in $(\nu, \nu+d\nu)$; this is the quantity denoted by $K(\nu, T_s)d\nu$ in subsection 8.1.2. It has the same Planck distribution as the mode occupancy, cf. Eqs. (8.1-11) and (8.1-15). Thus, the variance has the same super-Poissonian boson factor as considered in this section, Eq. (20.5-6), with $T \to T_s$.

The measurement of the excess noise ratio (20.5-34) is technically usually done with a cross-correlator, to eliminate most of the extraneous noise, following the well-known set-up of Hanbury–Twiss and Brown.[21] Results by Kattke and van der Ziel[22] are shown in Fig. 20-1. Similar results by Alkemade et al.[23] are given in Fig. 20-2. We note that no adjustment was made for radiation-transmission losses by Kattke and van der Ziel; this may explain that the measured deltas are slightly smaller than the theoretical values based on B–E statistics. Alkemade et al. adjusted the results, employing an effective efficiency $\eta' = \zeta\eta$ with $\zeta < 1$.

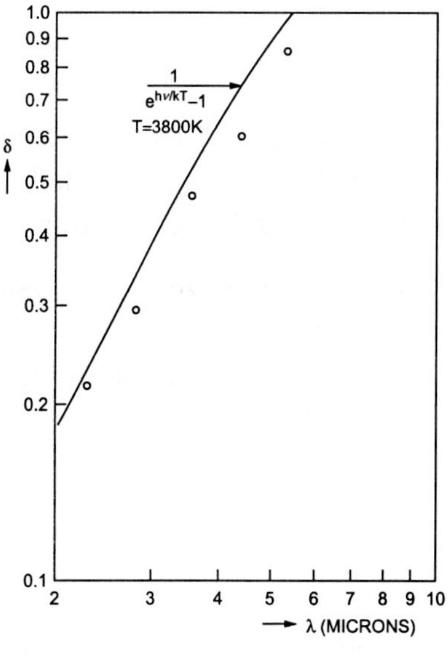

Fig. 20-1. Fractional excess noise measured with a cross-correlator vs. λ.[22] [With permission.]

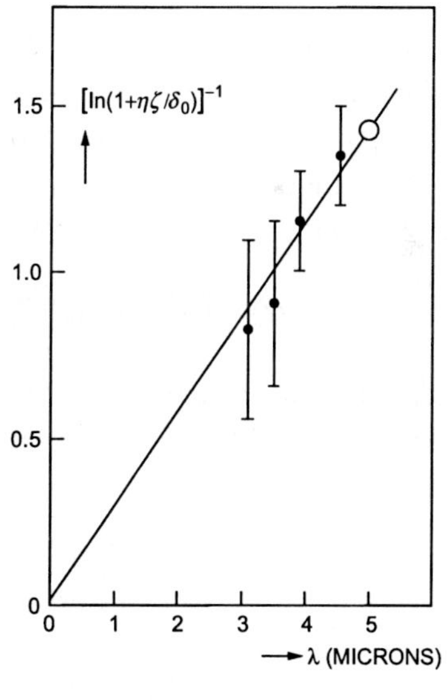

Fig. 20-2. Single-beam measurement of excess noise vs. λ with cooled InSb.[23] [With permission.]

[21] R. Hanbury Brown and R.Q. Twiss, Proc. Roy. Soc. **A 242**, 300 (1957); ibid. **A 243**, 291 (1958).

[22] G.W. Kattke and A. van der Ziel, "Verification of Einstein's formula for fluctuations in thermal radiation", Physica **49**, 461 (1970).

[23] C.Th.J. Alkemade, P.T. Bolwijn and J.H.C. van der Veer, "Single-beam measurement of Bose–Einstein fluctuations in a natural Gaussian radiation field", Phys. Lett. **22**, 70 (1966).

Lasers

To obtain the statistics of laser radiation we cannot rely on thermal equilibrium ensembles. On the contrary, the standard non-equilibrium methods for stochastic phenomena set forth in Chapter XVIII must be employed, such as the master equation (Agarwal[24]), Fokker–Planck equation (Risken[25]) and others. It is entirely outside our scope to describe these applications of the general stochastic literature in any detail. We therefore confine ourselves to a summary of a few models.

(a) In the *academic model* of an ideal laser it is assumed that there is a single mode with coherence to all orders. The probability of the *P*-representation is that of a pure state, $P(\alpha_q) = \delta^{(2)}(\alpha_q - \beta)$. For the normally ordered characteristic function we find

$$\begin{aligned}\Phi_N(u) &= \langle e^{ua_q^\dagger} e^{-u^* a_q} \rangle \\ &= \int d^2\alpha_q \delta^{(2)}(\alpha_q - \beta) \langle \alpha_q | e^{ua_q^\dagger} e^{-u^* a_q} | \alpha_q \rangle = e^{u\beta^* - u^* \beta} . \end{aligned} \tag{20.5-36}$$

This yields for the density operator quasi-probability $\langle \alpha | \rho | \alpha \rangle = e^{-|\alpha - \beta|^2}$. The pdf for the intensity is a delta function as is directly found from $P(\alpha_q)$,

$$W(\mathcal{I}) = \delta(\mathcal{I} - \mathcal{I}_0) . \tag{20.5-37}$$

(b) The *random phase model* is more realistic with $P(\alpha_q) = (1/2\pi\alpha_0)\delta(|\alpha_q| - |\beta|)$. The intensity distribution is the same as in (20.5-37).

(c) In the *signal plus noise model* the laser field is described by a superposition, $\alpha = \alpha_s + \alpha_n$, of a coherent mode $[P(\alpha) = \delta^{(2)}(\alpha - \alpha_0)]$ and a Gaussian mode $[P(\alpha)$ given by (20.5-7)$]$. The total state quasi-probability is found by convolution,

$$P(\alpha) = \int d^2\alpha' P_s(\alpha') P_n(\alpha - \alpha') = (1/\pi\sigma^2) e^{-|\alpha - \alpha_0|^2/\sigma^2} . \tag{20.5-38}$$

(d) Risken (Op. Cit.) has given a treatment of the *van der Pol oscillator model*, based on the nonlinear Fokker–Planck equation. The result is the following intensity distribution ($\mathcal{I} \propto |\alpha|^2$):

$$W(\mathcal{I}) = \frac{2}{\sqrt{\pi}\langle I \rangle} \frac{w + \phi(w)}{1 + \text{erf } w} \exp\left\{ -\left[\frac{(\mathcal{I} - \langle I \rangle) w + \mathcal{I}\phi(w)}{\langle I \rangle} \right]^2 \right\}, \tag{20.5-39}$$

where

$$\phi(w) = (1/\sqrt{\pi})\, e^{-w^2} / [1 + \text{erf } w] , \tag{20.5-40}$$

with w having large negative values below threshold and large positive values well above threshold. For $w \gg 1$ $W(\mathcal{I})$ approaches a Gaussian with width $\sigma^2 = \langle I \rangle^2 / 2w^2$. For $w \to \infty$ the distribution becomes a delta function as for an ideal laser.

---

[24] G.S. Agarwal in "Progress in Optics" Vol. XI, E. Wolf, Ed., North Holland, Amsterdam 1973, p. 3.

[25] H. Risken in "Progress in Optics" Vol. VIII, E. Wolf, Ed., North Holland, Amsterdam 1970, p. 241; also, Fortschritte der Physik **16**, 261 (1968).

20.6 Photon Counting. Theory and Some Experimental Results

A set of discrete events which occur at stochastic time points on the time axis comprise a *stochastic point process*;[26] such a process is characterized by a sequence of pdf's on the time scale, $f_1(t_1)$, $f_2(t_1,t_2)$, ... , $f_k(t_1,t_2,...,t_k)$, where $f_k(t_1,t_2,...,t_k)$ $\times dt_1...dt_k$ is the probability of having an absorption in (t_1,t_1+dt_1), another one in (t_2,t_2+dt_2), ..., and one in (t_k,t_k+dt_k). The arrival and subsequent absorption of photons at the detector surface can be seen as a physical example of a stochastic point process, with the absorptions taking place at stochastically positioned time points. Whereas for the case of shot noise it was assumed that the arrivals were independent of each other (see Campbell's theorem in Section 18.11), presently *no* such assumption will be made since generally photons tend to 'bunch'. We assume the process to be stationary, i.e., the time for all points $\{t_i\}$ can be shifted by a fixed time interval τ. Intuitively, for a stationary process one expects the following connection with the average field intensity,

$$f_1(t) = \gamma\langle I(t)\rangle = \gamma\overline{\mathcal{I}(t)}\,, \tag{20.6-1}$$

where γ represents the sensitivity of the detector. Although this result is correct, it is not as trivial as it may appear. A detailed quantum mechanical computation, using first order perturbation theory, has been given by Glauber.[27] More precisely, he obtains for the probability of absorption of one photon in a time interval (t_0,t):

$$P[1,(t_0,t)] = \frac{1}{2\pi}\int_0^\infty \mathsf{R}(\omega)d\omega \int_{t_0}^{t}\int_{t_0}^{t} dt'dt'' e^{-i\omega(t''-t')} : \mathsf{G}^{(1)}(\mathbf{r},t''\,||\,\mathbf{r},t')\,. \tag{20.6-2}$$

Here $\mathsf{R}(\omega)$ is the frequency response of the detector for light of given polarization vector, while $\mathsf{G}^{(1)}$ is the first order coherence tensor. For broad band detectors with $\mathsf{R}(\omega) = \gamma\mathsf{I}$, ($\mathsf{I}$ being the unit tensor) one obtains for a stationary process:

$$P[1,(t_0,t)] = \gamma\int_{t_0}^{t} dt'\overline{\mathcal{I}(t')} = (\gamma\mathcal{T})\,\overline{\mathcal{I}(t)}\,, \tag{20.6-3}$$

($\mathcal{T} = t - t_0$) from which (20.6-1) follows since $f_1 = dP(1,\mathcal{T})/d\mathcal{T}$. In like manner the probability for the absorption of k photons in specified time intervals can be obtained. For broad band detectors the result is

$$f_k(t_1,t_2,...,t_k) = \gamma^k(\mathrm{tr})^k\mathsf{G}^k(t_1,t_2,...,t_k\,||\,t_k,...,t_1) = \gamma^k\,\overline{\mathcal{I}(t_k)...\mathcal{I}(t_1)}\,; \tag{20.6-4}$$

this shows complete correspondence with the classical result of Mandel.

---

[26] R.L. Stratonovich, "Topics in the Theory of Random Noise", Vol. I, Gordon and Breach, NY and London 1963, Chapter 6.

[27] R.J. Glauber in "Quantum Optics and Electronics", Les Houches 1964, C. De Witt et al. Eds., Gordon and Breach, NY and London 1965, p. 63-185.

Counting statistics

The relationship between the intensity moments and the counting probabilities is best pursued by using appropriate transforms[9,26]. We will choose an interval $(t,t+\mathcal{T})$ having N stochastic time points $\{t_k\}$; the generating functional is defined by (cf. Chapter I footnote 26)

$$\chi_{\mathcal{T}}[\lambda(t)] = \langle \prod_{k=1}^{N} (1-\lambda(t_k)) \rangle . \tag{20.6-5}$$

For the case that $N = 1$, the rhs of (20.6-5) is clearly $1-\int_t^{t+\mathcal{T}} f_1(t_1)\lambda(t_1)dt_1$. In a similar manner we have for any arbitrary stochastic N within $(0,\infty)$ the series

$$\chi_{\mathcal{T}}[\lambda(t)] = 1 + \sum_{k=1}^{\infty} \frac{(-1)^k}{k!} \int_t^{t+\mathcal{T}} \cdots \int_t^{t+\mathcal{T}} dt_1 ... dt_k f_k(t_1,...,t_k)\lambda(t_1)...\lambda(t_k) . \tag{20.6-6}$$

Let now the transform variable be time-independent, i.e. $\lambda(t) \to \lambda$. We then have the generating function

$$\chi_{\mathcal{T}}(\lambda) = \langle \prod_{k=1}^{N} (1-\lambda) \rangle = \sum_{N=0}^{\infty} (1-\lambda)^N P(N,\mathcal{T}) . \tag{20.6-7}$$

Note that f_k is now replaced by the probability of having N events in $\mathcal{T}$, $P(k,\mathcal{T})$, since the positioning of the time points is immaterial. The inversion is direct:

$$P(N,\mathcal{T}) = \frac{(-1)^N}{N!} \frac{d^N}{d\lambda^N} \chi_{\mathcal{T}}(\lambda) \bigg|_{\lambda=1} . \tag{20.6-8}$$

Next, let us consider the Laplace generating functional for the integrated light intensity, $\mathcal{U}(t) = \int_t^{t+\mathcal{T}} \mathcal{I}(t')dt'$. We have

$$\begin{aligned} \Psi[\lambda(t)] &= \langle \exp\{-\int_t^{t+\mathcal{T}} dt' \mathcal{I}(t')\lambda(t')\} \rangle \\ &= 1 + \sum_{k=1}^{\infty} \frac{(-1)^k}{k!} \int_t^{t+\mathcal{T}} \cdots \int_t^{t+\mathcal{T}} dt_1 ... dt_k \overline{\mathcal{I}(t_1)...\mathcal{I}(t_k)}\, \lambda(t_1)...\lambda(t_k) . \end{aligned} \tag{20.6-9}$$

Appealing now to the connection of the point process pdf's with the intensity correlations as established in (20.6-4) and comparing (20.6-9) with (20.6-6), we note the identity

$$\chi_{\mathcal{T}}[\lambda(t)] = \Psi[\gamma\lambda(t)] . \tag{20.6-10}$$

Finally, letting again $\lambda(t) \to \lambda$, we have *a fortiori*

$$\chi_{\mathcal{T}}(\lambda) = \langle e^{-\lambda\gamma\mathcal{U}} \rangle = \int_0^{\infty} e^{-\lambda\gamma\mathcal{U}} W(\mathcal{U}) d\mathcal{U} . \tag{20.6-11}$$

From (20.6-8) the counting distribution follows; we obtain:

$$P(N,\mathcal{T}) = \int_0^\infty \frac{(\gamma\mathcal{U})^N}{N!} e^{-\gamma\mathcal{U}} W(\mathcal{U})\, d\mathcal{U}\,, \tag{20.6-12}$$

being the Sudarshan–Glauber analogue of Mandel's compound Poisson distribution, introduced *ad hoc* in (18.10-34). From (20.6-10) we can easily obtain the factorial moments of the counting distribution. We find

$$\langle N_{\mathcal{T}}(N_{\mathcal{T}}-1)\ldots(N_{\mathcal{T}}-k+1)\rangle = \gamma^k \int_0^{\mathcal{T}}\ldots\int_0^{\mathcal{T}} \overline{\mathcal{I}(t_1)\ldots\mathcal{I}(t_k)}\, dt_1\ldots dt_k\,. \tag{20.6-13}$$

For the variance (20.6-13) yields

$$\langle \Delta N_{\mathcal{T}}^2\rangle = \gamma\mathcal{T}\bar{\mathcal{I}} + \gamma^2 \int_0^{\mathcal{T}}\int_0^{\mathcal{T}} \overline{\Delta\mathcal{I}(t_1)\Delta\mathcal{I}(t_2)}\, dt_1 dt_2\,. \tag{20.6-14}$$

We will also need the probability function for an empty interval $(0,\mathcal{T})$. From (20.6-12) we note

$$P(0,\mathcal{T}) = \int_0^\infty e^{-\gamma\mathcal{U}} W(\mathcal{U})\, d\mathcal{U} = \Psi_{\mathcal{U}}(\gamma)\,, \tag{20.6-15}$$

where $\Psi_{\mathcal{U}}(\gamma)$ is the Laplace generating function for the integrated light intensity. We also introduce the cumulant distribution by setting

$$\ln \Psi_{\mathcal{U}}(\gamma) = \sum_{n=1}^\infty \frac{(-\gamma)^n}{n!} c_n\,, \tag{20.6-16}$$

where c_n are the cumulants of $\mathcal{U}$. Hence we obtain

$$P(0,\mathcal{T}) = \exp\left(-\gamma\mathcal{T}\bar{\mathcal{I}} + \sum_{n=2}^\infty \frac{(-\gamma)^n}{n!} \int_0^{\mathcal{T}}\ldots\int_0^{\mathcal{T}} \overline{\Delta\mathcal{I}(t_1)\ldots\Delta\mathcal{I}(t_n)}\, dt_1\ldots dt_n \right). \tag{20.6-17}$$

For 'fast detection processes' (i.e. counting times $\mathcal{T}$ much less than the coherence time τ_c) factors $\mathcal{T}^n$ can be joined with γ^n; the intensity correlations (cumulants) then follow from a measurement of $P(0,\mathcal{T})$:

$$\overline{\Delta\mathcal{I}(t_1)\ldots\Delta\mathcal{I}(t_n)} = \gamma^{-n}(-1)^n \left.\frac{d^n \ln P(0,\mathcal{T})}{d\mathcal{T}^n}\right|_{\mathcal{T}=0}\,. \tag{20.6-18}$$

Lastly we introduce the 2nd order coherence quantity $\Upsilon_2(\mathcal{T}) \equiv \langle N_{\mathcal{T}}(N_{\mathcal{T}}-1)\rangle - \langle N_{\mathcal{T}}\rangle^2$. One may verify that

$$\Upsilon_2(\mathcal{T}) = \gamma^2 \langle I\rangle^2 \int_0^{\mathcal{T}} (\mathcal{T}-\tau)\hat{R}(\tau)\, d\tau\,, \tag{20.6-19}$$

where $\hat{R}(\tau)$ is the normalized fluctuation-correlation function $\overline{\Delta\mathcal{I}(t_1)\Delta\mathcal{I}(t_2)}\,/\,\langle I\rangle^2$.

Interval statistics

Information on the intensity fluctuations is also obtainable from the distribution of intervals between an arbitrary fixed time point and the first incoming pulse, $w(\mathcal{T})$, or between two successive pulses, $w_c(\mathcal{T})$. The following relations apply:[28]

$$w(\mathcal{T}) = -(d/d\mathcal{T})P(0,\mathcal{T}), \tag{20.6-20}$$

$$w_c(\mathcal{T}) = (\mathcal{T}/\langle n\rangle)(d^2/d\mathcal{T}^2)P(0,\mathcal{T}). \tag{20.6-21}$$

Illustrations

The counting distribution is only simple if $\mathcal{U} \approx \mathcal{T}\mathcal{I}$, valid for $\mathcal{T} \ll \tau_c$, which is experimentally achievable with fast gating circuits. For thermal radiation with $W(\mathcal{I})$ given by (20.5-12) the counting distribution $P(N,\mathcal{T})$ is found to be geometric. For radiation of an ideal laser, for which the intensity distribution is a delta function, the counting distribution is pure Poisson. For a laser with the signal plus noise model having the Gaussian intensity distribution (20.5-38) $[\mathcal{I} \propto |\alpha|^2]$ one easily shows[9]

$$P(N,\mathcal{T}) = \frac{\langle N_{\mathcal{T}n}\rangle^N}{1+\langle N_{\mathcal{T}n}\rangle^{N+1}} \exp\left[-\frac{\langle N_{\mathcal{T}s}\rangle}{1+\langle N_{\mathcal{T}n}\rangle}\right] \mathcal{L}_N\left[-\frac{\langle N_{\mathcal{T}s}\rangle}{\langle N_{\mathcal{T}n}\rangle(1+\langle N_{\mathcal{T}n}\rangle)}\right], \tag{20.6-22}$$

where the subs 'n' and 's' refer to the two parts of the signal plus noise model and $\mathcal{L}_N$ denotes the Laguerre polynomial of order N.

The three distributions, as measured by Arecchi[29] are reproduced in Fig. 20-3.

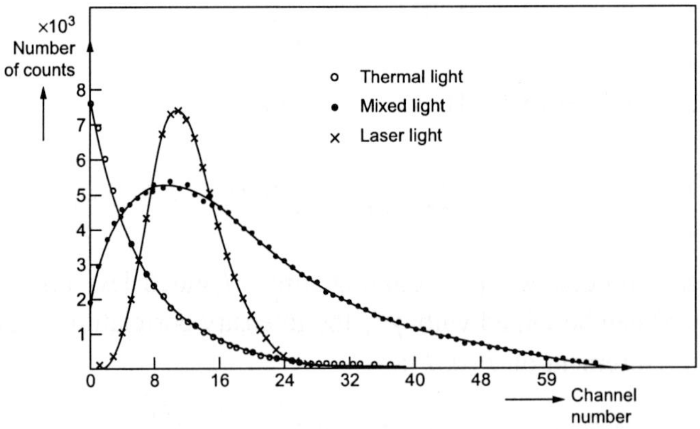

Fig. 20-3. Photocount distribution for various kinds of light after Arecchi, Ref. 30. [With permission.]

Next, we discuss an experiment by Meltzer and Mandel.[30] The reduced second factorial moment given in (20.6-19) was measured for a He-Ne laser far below threshold with variable $\mathcal{T}$. The result is pictured in Fig. 20-4, whereby we note that the limiting behaviour for $\mathcal{T} \gg \tau_c$ is linear in $\mathcal{T}$: $\Upsilon_2 \approx \gamma^2\langle I\rangle^2 \mathcal{T}\int_{-\infty}^{\infty} \hat{R}(\tau)d\tau = \gamma^2\langle I\rangle^2 \mathcal{T}\tau_c$.

---

[28] R.J. Glauber, Ref. 13, Op. Cit., p. 185-186.

[29] F.T. Arecchi, Phys. Rev. Lett. **15**, 912 (1965).

[30] D. Meltzer and L. Mandel, IEEE J. Quantum Electr. **QE-6**, 661 (1970).

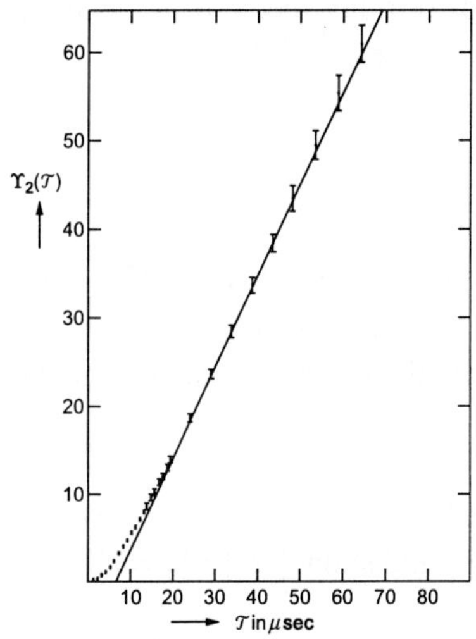

Fig. 20-4. The reduced second factorial moment Υ_2 vs. $\mathcal{T}$, pertaining to a He-Ne laser far below threshold; after Ref. 30. [With permission.]

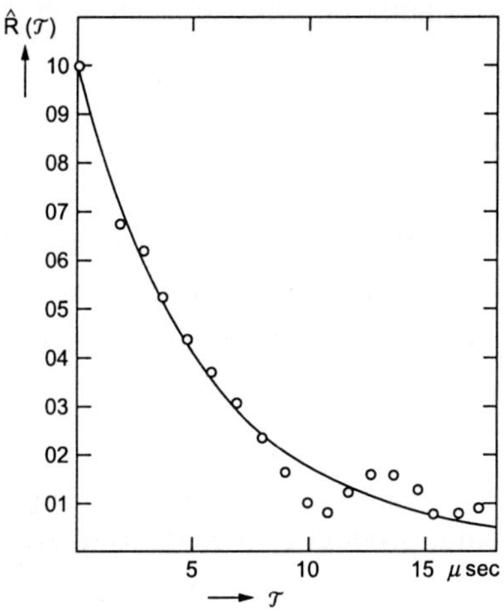

Fig. 20-5. Normalized intensity fluctuation correlation function $\hat{R}(\mathcal{T})$, derived from Fig. 20-4; after Ref. 30. [With permission.]

From differentiation they obtained the normalized intensity-fluctuation correlation function, $\hat{R}(\mathcal{T})$; it is shown in Fig. 20-5.

Finally, in Fig. 20-6 we give the interval distribution $w(\mathcal{T})$ for cathodo-luminescence from a YVO_4-Eu^{3+} phosphor, as measured by van Rijswijk and Zijlstra.[31]

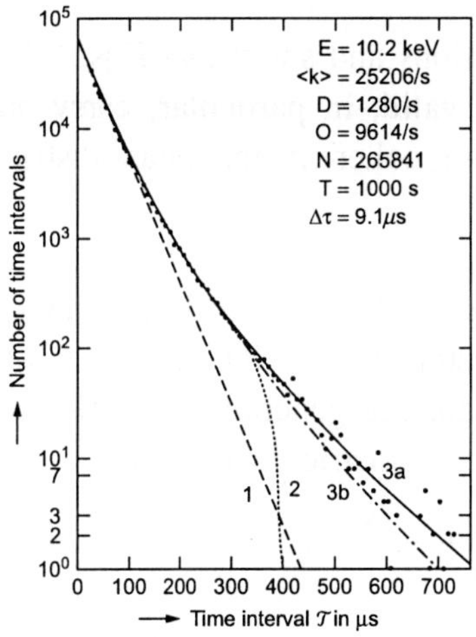

Fig. 20-6. Photon interval-time distribution for cathode-luminescence from a YVO_4-Eu^{3+} phosphor after Ref. 31. Curves (1) and (2): first and second-order cumulant expansion. Curves 3a and b are based on special models. [With permission.]

Photon noise measurements are most easily performed for fields with long coherence times. As such, they are especially important for the determination of narrow line widths. Investigations of the statistical properties of radiation (stemming

[31] F.C. van Rijswijk and R.J.J. Zijlstra, Phys. Lett. **51A**, 271 (1975).

from lasers or thermal sources) scattered by fluids or other media have been reported by various authors; referral to the literature must suffice.[32,33,34,35]

20.7 Problems to Chapter XX

20.1 Prove the convolution property (20.2-5). *Hint*. On the lhs of (20.2-5) replace $u_2(t+\theta)$ by its Hilbert transform $-v_2(t+\theta)$ and interchange the order of integration of the resulting double integral; then change the time-integration variables.

20.2 Give a proof for Klauder's expansion (20.3-7). Also prove the result for the scalar product of the coherent states (20.3-8') and show $|\langle\alpha|\beta\rangle|^2 = e^{-|\alpha-\beta|^2}$.

20.3 *Given* the quasi-distribution (20.5-8), fill in the details to obtain the field distribution (20.5-10); employ the characteristic function in the complex plane,

$$\Phi(u) \equiv \langle e^{u\mathscr{E}^* - u^*\mathscr{E}} \rangle = \int d^2\mathscr{E} e^{u\mathscr{E}^* - u^*\mathscr{E}} W(\mathscr{E}) . \tag{1}$$

20.4 The most common state is the chaotic state that maximizes the entropy function $S = -\mathrm{Tr}\{\rho \ln \rho\}$, subject to $\mathrm{Tr}\,\rho = 1$ and to fixed $\langle n \rangle = \mathrm{Tr}\{\rho a^\dagger a\}$. Use two Lagrangian multipliers and show that

$$\rho = \exp[-\lambda_1 - \lambda_2 a^\dagger a]. \tag{2}$$

Obtain λ_1 and λ_2 from the auxiliary conditions and show that Eqs. (20.5-27) [for one mode] through (20.5-31) are still valid; in particular, carry out the various transforms to obtain the corresponding coherent state quasi-distribution $P(\alpha)$ and show that it is a Gaussian.

20.5 Consider the case that a single mode of the field is excited in a chaotic state with the quasi-probability (20.5-31) and intensity distribution (20.5-12); let further $\mathcal{U} = \mathcal{T}\mathcal{I}$. Employing (20.6-11), obtain the generating function $\chi_{\mathcal{T}}(\lambda)$ and show that the counting distribution is the geometric distribution

$$P(N,\mathcal{T}) = [\langle m \rangle \mathcal{T}]^N / [1 + \langle m \rangle \mathcal{T}]^{N+1} , \tag{3}$$

with $\langle m \rangle = \gamma \bar{\mathcal{I}}$ being the mean counting rate.

---

[32] J.B. Lastovka and G.B. Benedek, Phys. Rev. Lett. **17**, 1039 (1966).

[33] B. Crosignani, P. Di Porto and M. Bertolotti, "Statistical Properties of Scattered Light", Acad. Press, NY and London 1975, in particular Chapters V and VI.

[34] E. Jakeman and R.J.A. Tough, Adv. in Physics **37**, 471-529 (1988).

[35] M. Mujat, A. Dogariu and G.S. Agarwal, Optics Letters **29**, 1539 (2004).

Appendix A

The Schrödinger, Heisenberg and Interaction Picture

A.1 Schrödinger Form

The most well-known description of quantum mechanics employs the Schrödinger form. Observables $\mathcal{A}_i$ are represented by operators A_i that act upon states $|\phi_k\rangle$, which are elements in a state space $\mathcal{S}$. Ideally, this space is a Hilbert space spanned by a basis of a complete set of commuting operators $\{B_i\}$. Each operator gives rise to a particular set of transformations of $\mathcal{S}$

$$A_i\,|\phi\rangle = |\psi\rangle. \tag{A.1-1}$$

The elements may be chosen to be spatial functions $\phi_k(\mathbf{r})$ – sometimes denoted as $\langle \mathbf{r}\,|\,\phi_k\rangle$ – i.e., the Hilbert space is the function space $\mathcal{L}^2(R_3)$; yet (A.1-1) expresses no dynamics. In the Schrödinger picture the operators generally do not depend on time, but the time dependence, necessary to describe dynamics, is vested in the dynamical state $|\Psi_S(t)\rangle$ or, in wave-mechanical form, in $\Psi_S(\mathbf{r},t)$. The operators must satisfy the quantum conditions; as shown in Dirac's book[1]

$$[A,B] = \hbar i\{A_i, B_i\}, \tag{A.1-2}$$

where the square bracket is the commutator bracket and the curly bracket denotes the Poisson bracket,

$$\{A,B\} = \sum\nolimits_k \left(\frac{\partial A}{\partial q_k}\frac{\partial B}{\partial p_k} - \frac{\partial A}{\partial p_k}\frac{\partial B}{\partial q_k} \right). \tag{A.1-3}$$

The proportionality $[A,B] \propto \{A,B\}$ is easily proven from the algebra pertaining to both brackets. Further, the commutator of Hermitean operators is itself anti-Hermitean and the commutator must vanish in the classical limit; the proportionality factor $\hbar i$ is thus self-evident. As a special quantum condition for coordinates and momenta one finds from the Poisson bracket, $[p_i, q_j] = (\hbar/i)\delta_{i,j}$, which is satisfied by the usual choice

$$p_i = (\hbar/i)(\partial/\partial q_i), \qquad q_i = q_i \times. \tag{A.1-4}$$

---

[1] P.A.M. Dirac, "The Principles of Quantum Mechanics", 4th Ed., Oxford Univ. Press 1958.

The above choice comprises the wave mechanical Schrödinger q-form. Mostly, however, in this text we do not commit ourselves to a $\mathscr{L}^2$ state-space and we use the abstract Dirac form, with the Schrödinger equation being written as

$$\mathcal{H}|\Psi_S(t)\rangle = \hbar i\partial|\Psi_S(t)\rangle/\partial t. \tag{A.1-5}$$

The Hamiltonian is derived from the classical Hamiltonian, $\mathcal{H}(\{p_i(t)\},\{q_i(t)\},t)$. Explicit time dependence is usually absent in quantum mechanics, except in time-dependent perturbation theory. The potential energy then contains a contribution $\lambda\mathcal{V}(t)$, but this is still a Schrödinger operator, most other operators being time-independent.

The Schrödinger form of quantum mechanics has many similarities with the Hamilton-Jacobi form of classical mechanics in which a canonical transformation has been made to remove the time dependence of the dynamical variables p_i and q_i, so that the generating function, S, becomes time-dependent and solves the dynamical problem as contained in the Hamilton-Jacobi equation. The dynamical state $\Psi(\mathbf{r},t)$ can be related to the Hamilton-Jacobi function $S(\mathbf{r}, t)$ by

$$\Psi_S(\mathbf{r},t) = A(\mathbf{r},t)\exp[iS(\mathbf{r},t)/\hbar], \tag{A.1-6}$$

which is used in the BKW approximation and other occasions, see Section 12.1. If the Hamiltonian does not depend on time, (A.1-5) has the *formal* solution

$$|\Psi_S(t)\rangle = e^{-i(t-t_0)\mathcal{H}/\hbar}|\Psi_S(t_0)\rangle \equiv U(t-t_0)|\Psi_S(t_0)\rangle. \tag{A.1-7}$$

In case there is explicit time dependence in $\mathcal{H}$ this is modified to read

$$|\Psi_S(t)\rangle = e^{-i\int_{t_0}^{t} dt'\mathcal{H}(t')/\hbar}|\Psi_S(t_0)\rangle \equiv \bar{U}(t-t_0)|\Psi_S(t_0)\rangle. \tag{A.1-8}$$

In both cases the evolution operator satisfies

$$\hbar i\frac{dF(t,t_0)}{dt} = \mathcal{H}F(t,t_0), \qquad F = U \text{ or } \bar{U}. \tag{A.1-9}$$

The expectation value of a Schrödinger operator A_S at a time t is given by

$$\langle A_S(t)\rangle = \langle\Psi_S(t)|A_S|\Psi_S(t)\rangle. \tag{A.1-10}$$

We note that the time dependence stems from the Schrödinger state $|\Psi_S(t)\rangle$.

A.2 The Heisenberg Picture, Unitary transformation and Connection with Classical Mechanics

To obtain the Heisenberg picture a unitary transformation is made, such that the dynamical state which represents the system is time-independent while the operators become dependent on time – as in the ordinary classical treatment of a dynamical system. The time dependence will be established through the usual unitary transformation

$$A_H(t) = U^\dagger(t-t_0)A_S U(t-t_0)\,, \tag{A.2-1}$$

with U being the unitary matrix defined in (A.1-7) or (A.1-8). By differentiation we obtain the Heisenberg equation of motion

$$\frac{dA_H}{dt} = \frac{1}{\hbar i}[A_H(t),\mathcal{H}]. \tag{A.2-2}$$

Or, if A is an operator with explicit time dependence, like $\lambda V(t)$ in time-dependent perturbation theory, we have

$$\frac{dA_H}{dt} = \frac{\partial A_H}{\partial t} + \frac{1}{\hbar i}[A_H(t),\mathcal{H}]. \tag{A.2-3}$$

The dynamical state Ψ plays a lesser role in the Heisenberg picture, since it must undergo the same unitary transformation, i.e.,

$$|\Psi_H\rangle = U^\dagger(t-t_0)|\Psi_S(t)\rangle = |\Psi_S(t_0)\rangle\,, \tag{A.2-4}$$

where we substituted (A.1-7). The Heisenberg dynamical state $|\Psi_H\rangle$ is a (largely insignificant) constant. For a time-dependent Hamiltonian, this remains true, but in (A.2-4) we need $\overline{U}^\dagger(t-t_0)$ and we substitute (A.1-8).

We note that the dynamical state has no direct physical meaning in *any picture*, except in that it gives the expectation value of observables. We easily verify that the two pictures are consistent, for

$$\begin{aligned}\langle A_H(t)\rangle &\equiv \langle \Psi_H | A_H(t) | \Psi_H\rangle \\ &= \langle \Psi_S(t)\underbrace{U(t-t_0)U^\dagger(t-t_0)}_{I}| A_S | \underbrace{U(t-t_0)U^\dagger(t-t_0)}_{I}\Psi_S(t)\rangle \\ &= \langle \Psi_S(t)| A_S | \Psi_S(t)\rangle = \langle A_S(t)\rangle \equiv \langle A(t)\rangle\,.\end{aligned} \tag{A.2-5}$$

From what we have done, it is clear that the Heisenberg description corresponds with Hamilton's equations in classical mechanics, while the Schrödinger picture corresponds with the Hamilton-Jacobi approach, as already noted above. This is most clearly borne out by Ehrenfest's theorem for the derivative of expectation values:

$$\frac{d}{dt}\langle A(t)\rangle = \left\langle \frac{\partial A_H}{\partial t}\right\rangle + \frac{1}{\hbar i}\langle [A_H(t),\mathcal{H}]\rangle \to \frac{\partial A}{\partial t} + \{A,\mathcal{H}\}\,. \tag{A.2-6}$$

For Ehrenfest's form, see Messiah.[2] For the classical derivative of $A[p_i(t),q_i(t),t]$,

[2] A. Messiah, "Quantum Mechanics", North-Holland Publ. Co, Amsterdam, 1961, p. 210 and 216.

see Goldstein;[3] Hamilton's equations $\dot{q}_i = \partial\mathcal{H}/\partial p_i$ and $\dot{p}_i = -\partial\mathcal{H}/\partial q_i$ are implied. Note that (A.2-6) is confirmed by (A.1-2). In Fig. A-1 we give a flow-diagram of the various schemes.

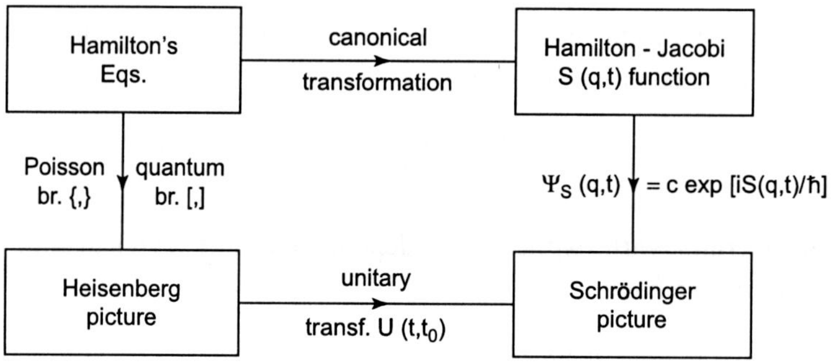

Fig. A-1. Connections of classical and quantum pictures.

A.3 Interaction Picture

The two forms for the evolution operator, (A.1-7) and (A.1-8), contain the exponentiated Hamiltonian. While those forms are formally correct, they are not useful, unless the Hamiltonian is diagonal – which is seldom the case. Therefore, let us write

$$\mathcal{H} = \mathcal{H}^0 + \lambda\mathcal{V}(t)\,, \qquad \text{(A.3-1)}$$

where $\mathcal{H}^0$ is "the largest Hamiltonian that can be diagonalized"[4] and $\lambda\mathcal{V}$ is a perturbation, which may depend on t. The differential equation for $U(t)$ [or $\overline{U}(t)$] (A.1-9) will be solved by iteration. We will do this for $U(t)$, but everything holds nearly verbatim for $\overline{U}(t)$. Hence, consider

$$\frac{dU(t,t_0)}{dt} = \frac{1}{\hbar i}(\mathcal{H}^0 + \lambda V)U(t,t_0), \qquad \text{(A.3-2)}$$

with the zero-order operator satisfying

$$\frac{dU^0(t,t_0)}{dt} = \frac{1}{\hbar i}U^0(t,t_0)\,. \qquad \text{(A.3-3)}$$

---

[3] H. Goldstein, "Classical Mechanics", 2nd Ed., Addison Wesley, N.Y., 1981, p. 405.

[4] M. Dresden, Private Communication (Vosbergen Conference, The Netherlands, 1967).

The solution is

$$U^0(t,t_0) = \exp[-i\mathcal{H}^0(t-t_0)/\hbar]\,, \tag{A.3-4}$$

as expected. Next the interaction operators U_I and $\mathcal{V}_I$ are introduced by

$$U_I(t,t_0) = U^{0\dagger}(t,t_0)U(t,t_0), \qquad \mathcal{V}_I(t) = U^{0\dagger}(t,t_0)\mathcal{V}U^0(t,t_0)\,. \tag{A.3-5}$$

From (A.3-2) – (A.3-3) one easily obtains

$$\frac{dU_I}{dt} = \frac{1}{\hbar i}\lambda\mathcal{V}_I(t)U_I(t,t_0)\,. \tag{A.3-6}$$

This differential equation is solved by a time-ordered perturbation series $U_I(t,t_0) = 1 + \Sigma_{n=1}^{\infty} U_I^{(n)}(t,t_0)$, where

$$U_I^{(n)}(t,t_0) = \left(\frac{\lambda}{\hbar i}\right)^n \int_{t_0}^{t} dt_n \int_{t_0}^{t_n} dt_{n-1} \ldots \int_{t_0}^{t_2} dt_1 \mathcal{V}_I(t_n)\mathcal{V}_I(t_{n-1}) \ldots \mathcal{V}_I(t_1)\,. \tag{A.3-7}$$

This, in turn, yields the perturbation series $U(t,t_0) = 1 + \Sigma_{n=1}^{\infty} U^{(n)}(t,t_0)$ with

$$U^{(n)}(t,t_0)$$
$$= \left(\frac{\lambda}{\hbar i}\right)^n \int_{t_0}^{t} dt_n \int_{t_0}^{t_n} dt_{n-1} \ldots \int_{t_0}^{t_2} dt_1 U^0(t,t_n)\mathcal{V}(t_n)U^0(t_n,t_{n-1})\mathcal{V}(t_{n-1}) \ldots U^0(t_2,t_1)\mathcal{V}(t_1)U^0(t_1,t_0). \tag{A.3-8}$$

The solution is more illuminating if we compute the matrix elements between two many-body states $|\gamma\rangle$ and $|\gamma_0\rangle$. Using closure, $\Sigma_\gamma |\gamma\rangle\langle\gamma| = 1$, we can insert intermediate states in (A.3-8); we then have for the general case that the Hamiltonian has explicit time dependence with $\langle\gamma|\bar{U}(t,t_0)|\gamma_0\rangle = \Sigma_n \langle\gamma|\bar{U}^{(n)}(t,t_0)|\gamma_0\rangle$,

$$\langle\gamma|\bar{U}^{(n)}(t,t_0)|\gamma_0\rangle = \left(\frac{\lambda}{\hbar i}\right)^n \sum_{\gamma_1\ldots\gamma_{n-1}} \int_{t_0}^{t} dt_n \int_{t_0}^{t_n} dt_{n-1} \ldots \int_{t_0}^{t_2} dt_1$$
$$\times \bar{U}^0(t-t_n)\langle\gamma|\mathcal{V}(t_n)|\gamma_{n-1}\rangle\bar{U}^0(t_n-t_{n-1})\langle\gamma_{n-1}|\mathcal{V}(t_{n-1})|\gamma_{n-2}\rangle \ldots$$
$$\times \bar{U}^0(t_2-t_1)\langle\gamma_1|\mathcal{V}(t_1)|\gamma_0\rangle\bar{U}^0(t_1-t_0)\,. \tag{A.3-9}$$

This result is summarized in the diagram of Fig. A-2. The states are represented by lines, going from t to the first vertex, then to the next vertex, …, and finally to t_0. At each vertex t_i a perturbation $\lambda\mathcal{V}(t_i)$ enters the system; between vertices the motion is guided by $\bar{U}^0(t_i - t_{i-1})$. In statistical treatments we often also need the evolution operator for imaginary times $-i\hbar\tau$, where τ is a real parameter; this is discussed in Section 12.9*ff.*

The interaction picture employs 'intermediate' time-dependent operators, formed by the transformation

$$A_I(t) = U^{0\dagger}(t-t_0) A_S U^0(t-t_0) . \tag{A.3-10}$$

They satisfy the Heisenberg equation based on $\mathcal{H}^0$,

$$\frac{dA_I}{dt} = \frac{1}{\hbar i}[A_I(t), \mathcal{H}^0] . \tag{A.3-11}$$

The Heisenberg operators can easily be expressed in the interaction operators. This is most useful for the creation and annihilation operators. One may verify:

$$a_{k,H}^\dagger(t) = U_{k,I}^\dagger(t) a_{k,I}^\dagger U_{k,I}(t), \qquad a_{k,H}(t) = U_{k,I}^\dagger(t) a_{k,I} U_{k,I}(t). \tag{A.3-12}$$

This is just a summary of results. Detailed connections are encountered in many parts of this text, especially when dealing with interacting systems, *in casu*, Parts C and E.

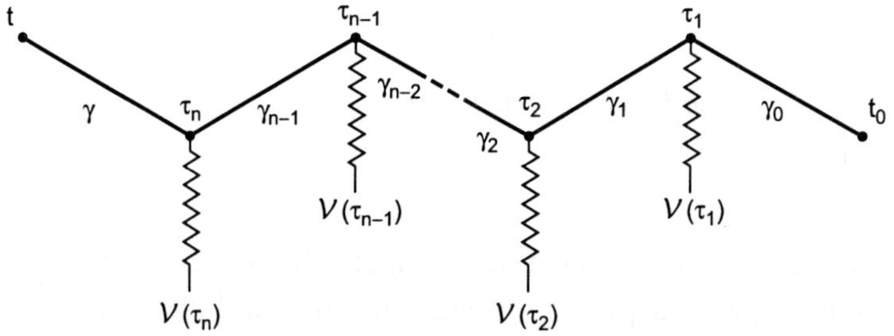

Fig. A-2. Diagram for the n-th order matrix elements in time-dependent perturbation theory.

Appendix B

Spin and Statistics

In Sections 7.1 and 7.7 we stated that systems of particles only occur in two kinds according to the symmetrization postulate, viz., with creation and annihilation operators satisfying commutator-bracket rules or anticommutator-bracket rules. The latter brackets were absent in early quantum theory, until they first showed up in commutation relations for the Pauli spin matrices of an electron. Whereas a relationship between spin and statistics may be surmised from mathematical analogies, the real connection lies quite deep. First of all, as for all symmetries of a fundamental nature, we must go to a relativistic framework. Secondly, we must abandon one-particle theories for the well-known reason of non-definiteness of mass and energy, and employ a field theory with field operators $\phi_k(x)$ and $\phi_k^\dagger(x)$, with x being a four-vector $x = (t,\mathbf{r})$, that either locally commute or anticommute. The connection between spin and statistics is then expressed by the vanishing of the fields if the 'wrong' type of commutator is used in a theory with a given type of spin, which is invariant against the restricted inhomogeneous Lorentz group, also called Poincaré group. The first proof of this nature was given by Pauli in 1940. It is repeated in more detail in the book by Akhieser and Beretstetskii.[1] The proof is rather lengthy and only valid for free particle fields. Moreover, non-parallel approaches are necessary for bosons and fermions. A completely general proof had to await the more refined axiomatic development of quantum field theory, as undertaken by Wightman, Jost, Ruelle and others; in addition, important theorems concerning the analytic continuation of functions from the real Lorentz group to the complex group have removed the necessity of *ad hoc* assumptions.[2] The argumentation of the method is closely related to that of the Lüders-Pauli theorem regarding PCT (parity, charge, time) invariance, as shown by Jost. The first independent proofs of the spin-statistics connection along the above lines are found with Lüders and Zumino[3] and with Burgoyne.[4] This completely general proof is also found in the well-known book by

---

[1] A.I. Akhieser and V.B. Berestetskii, "Quantum Electrodynamics", Interscience, NY 1965; W. Pauli, Phys. Rev. **58**, 716 (1940).

[2] A.S. Wightman, Phys Rev. **101**, 860 (1956); D. Hall and A.S. Wightman, Kgl Danske Videnskab Selskab Mat.-fys. Medd. **31**, no. 5, 1-41 (1957); R. Jost, Helv. Phys. Acta **30**, 409 (1957).

[3] G. Lüders and B. Zumino, Phys. Rev. **110**, 1450 (1958).

[4] N. Burgoyne, Nuovo Cimento **8**, 6007 (1958).

Streater and Wightman.[5]

In what follows we shall first present some generalities about fields, then we give the proof for scalar fields for particles with zero-spin, which is straightforward, if we take some mathematical theorems for granted; after that we shall consider fields for any spin.

B.1 Generalities on Fields

We recall that the general inhomogeneous Lorentz group of transformations[6]

$$x' = \Lambda x + a, \qquad x = (t, \mathbf{r}) = (x_0, x_1, x_2, x_3, x_4) \tag{B.1-1}$$

has four subsets which cannot be connected to each other by a family of continuous transformations $\Lambda(t)$, as distinguished by the possibilities $\det \Lambda = \pm 1$ and $\operatorname{sgn} \Lambda_0^0 = \pm 1$. The most important group is the *restricted* Lorentz group (or proper orthochronous Lorentz group) of (real) matrices Λ_ν^μ containing only proper rotations of four-vectors ($\det \Lambda = 1$) and with no time reversal ($\operatorname{sgn} \Lambda_0^0 = 1$). We further use the denotations *orthochronous* Lorentz group for transformations with $\det \Lambda = \pm 1$ and $\operatorname{sgn} \Lambda_0^0 = 1$, *proper* Lorentz group for transformations with $\det \Lambda = 1$ and $\operatorname{sgn} \Lambda_0^0$ = ±1 and the name *general* Lorentz group if all four options are allowed. The behaviour of these groups can be visualized with respect to the light cone defined by[7]

$$x^\mu x_\mu = x^2 = 0, \quad \text{or} \quad t^2 - x_1^2 - x_2^2 - x_3^2 = 0. \tag{B.1-2}$$

The light cone divides the four-dimensional space into two areas, the inside of the cone with $x^2 > 0$ (time-like vectors) and the outside of the cone with $x^2 < 0$ (space-like vectors. The 4-space is further divided into the V_+ space $(x_0 > 0)$ and the V_- space $(x_0 < 0)$. Every Lorentz transformation leaves the inside and the outside invariant, i.e., transforms these regions into themselves. The proper Lorentz transformations have as further restrictions that no spatial reflections are allowed, whereas the restricted group preserves in addition the identity of V_+ and V_-. The inhomogeneous Lorentz transformations also allow for a displacement of the origin by a vector a.

Next, we consider the field operators $\phi_k(x)$. In order that the theory is invariant operators must transform according to a linear representation (Darstellung) of the

---

[5] R.F. Streater and A.S. Wightman, "PCT, Spin and Statistics, and all that", Benjamin, NY and Amsterdam 1964.

[6] We set c = 1, $\hbar$ = 1.

[7] $x^\mu x_\mu = g_{\mu\nu} x^\mu x^\nu$, where $g_{\mu\nu}$ is the metric tensor with $g_{00} = 1$, $g_{11} = g_{22} = g_{33} = -1$, $g_{\mu\nu} = 0$, $\mu \neq \nu$. Further, we use the summation convention to sum over repeated indices.

inhomogeneous Lorentz group.[8] We are only concerned with finite dimensional decomposable representations and we shall therefore *refer to a field when the components* $\phi_k(x)$, $k = 1,2,\ldots p$, *transform according to some irreducible representation of the restricted Lorentz group* (or sometimes a more general group). For fixed x these operators therefore form a set of elements in a p-dimensional representation space and they can be given as a column matrix $\{\phi_1\,\phi_2\ldots\phi_p\}$. If in this space $S(\Lambda)$ is a matrix representation group, homomorphic to Λ, then the transformation law is[9]

$$\phi_j'(x) = \sum_k S_{jk}(\Lambda^{-1})\phi_k(\Lambda x + a), \tag{B.1-3}$$

where j and k each describe p indices. For a scalar field $S = 1$ and we simply have

$$\phi'(x) = \phi(\Lambda x + a), \tag{B.1-4}$$

i.e., the field is invariant under the transformation (A.1-1).[10]

The field operators ϕ_k and their adjoins $\phi_k^\dagger$ are defined on a state space $\mathscr{S}$ with state vectors Ψ. In particular, we denote the vacuum state by Ψ_0. According to Wightman, this space is a Hilbert space which is separable and which has a definite metric, though others (Heisenberg and Dürr) take a wider point of view. In this state space we associate with each inhomogeneous restricted Lorentz transformation a unitary operator $U(\Lambda, a)$ affecting the transformations

$$\Psi' = U(\Lambda, a)\Psi \tag{B.1-5a}$$

$$\phi_k'(x) = U(\Lambda, a)\phi_k U^\dagger(\Lambda, a). \tag{B.1-5b}$$

The operator U is in essence defined by (B.1-3) and by (B.1-5b). The field equations, such as the Dirac equation, must be such that they are invariant under the indicated transformations. Naturally the scalar product is conserved, as are the field expectation values.[11] The proof is trivial:

$$\begin{aligned}&\left(\Psi, \phi_a'(x_1)\chi_b'(x_2)\ldots\Phi'\right) = \left(U\Psi, U\phi_a(x_1)U^\dagger U\chi_b(x_2)U^\dagger U\ldots U\Phi\right)\\ &= \left(U\Psi, U\phi_a(x_1)\chi_b(x_2)\ldots\Phi\right) = \left(U^\dagger U\Psi, \phi_a(x_1)\ldots\Phi\right) = \left(\Psi, \phi_a(x_1)\ldots\Phi\right).\end{aligned} \tag{B.1-6}$$

---

[8] The word "representation" in group theory has nothing to do with representation in quantum mechanics, where it denotes the choice of basis in the state space. The German word 'Darstellung' more properly indicates that a group is "represented" (*depicted*) by a linearly related group, being homomorphic to the original one.

[9] Details on the representation $S(\Lambda)$ are considered in subsection A.3.

[10] The above description of the field is not entirely correct, since we actually need 'smeared out' functionals $\phi_k[f] = \int \phi_k(x) f(x) d^4x$, defined on a compact support. Thus point divergences requiring delta functions or other generalized functions are avoided.

[11] We use the (,) notation rather than the Dirac convention, as is common in the mathematical literature.

In particular, we shall be interested in vacuum expectation values. The vacuum state is invariant under any transformation (except for phase); so $U\Psi_0 = \Psi_0$. The transformation of vacuum expectation values is therefore determined by the irreducible transformation matrices of the fields, i.e., if $S^{(\phi)}, S^{(\chi)}, \ldots$ are the relevant matrices in the representation spaces, we have

$$\begin{aligned}&\left(\Psi_0, \phi_a'(x_1)\chi_b'(x_2)\ldots\Psi_0\right)\\&= \sum_{a'b'..} S_{aa'}^{(\phi)}(\Lambda^{-1})S_{bb'}^{(\chi)}(\Lambda^{-1})\ldots\left(\Psi_0, \phi_{a'}(\Lambda x_1 + a)\chi_{b'}(\Lambda x_2 + a)\ldots\Psi_0\right).\end{aligned} \tag{B.1-7a}$$

We may also write this in a different form by operating on both sides with $\Pi S(\Lambda)$. Using the invariance of the vacuum expectation values as stated in (B-6), we obtain the following identity under matrix transformations representing the restricted Lorentz group

$$\begin{aligned}&\sum_{a'b'..} S_{aa'}^{\phi}(\Lambda)S_{bb'}^{\chi}(\Lambda)\ldots\left(\Psi_0, \phi_{a'}(x_1)\chi_{b'}(x_2)\ldots\Psi_0\right)\\&= \left(\Psi_0, \phi_a(\Lambda x_1 + a)\,\chi_b(\Lambda x_2 + a)\ldots\Psi_0\right).\end{aligned} \tag{B.1-7b}$$

These vacuum expectation values are properly defined for the smeared fields, i.e., the functionals $\phi_k[f]$. Because of the invariance against translations $(\Lambda = \mathrm{I},\ a \neq 0)$, Eqs. (B.1-7a,b) show that these quantities are only distributions in $x_j - x_{j+1}$. In particular, for the two-point quantities of one field,

$$\left(\Psi_0, \phi_k(x)\phi_\ell^\dagger(y)\Psi_0\right) = F_{k\ell}(x-y) \equiv F_{k\ell}(\xi), \tag{B.1-8a}$$

$$\left(\Psi_0, \phi_k^\dagger(x)\phi_\ell(y)\Psi_0\right) = G_{k\ell}(x-y) \equiv G_{k\ell}(\xi). \tag{B.1-8b}$$

We now consider the commutators. The symmetrization postulate in quantum-field theory reads: *For a given field, either the commutator brackets or the anticommutators brackets are zero for space-like distances* (so-called local commutivity), *i.e., one or the other holds:*

$$[\phi_k(x), \phi_\ell(y)]_\pm = 0, \text{ all } k,\ell \quad \text{for } (x-y)^2 < 0. \tag{B.1-9a}$$

Clearly, this also entails for the adjoint operators in dual space

$$[\phi_k^\dagger(x), \phi_\ell^\dagger(y)]_\pm = 0, \text{ all } k,\ell \quad \text{for } (x-y)^2 < 0. \tag{B.1-9b}$$

Perhaps, more surprising, from (B.1-9a) it also follows that

$$[\phi_k(x), \phi_\ell^\dagger(y)]_\pm = 0, \text{ all } k,\ell \quad \text{for } (x-y)^2 < 0, \tag{B.1-10}$$

where we notice that the singular delta function is removed when these results are conceived as functionals on a support, like stated above. The proof is direct.

Consider the vacuum state norm for two test functions f and g:

$$\|\phi_\ell[g]\phi_k[f]\Psi_0\|^2=\left(\Psi_0,\phi_k^\dagger[f]\phi_\ell^\dagger[g]\phi_\ell[g]\phi_k[f]\Psi_0\right)\geq 0. \qquad \text{(B.1-11)}$$

Let us assume that (B.1-10) is not true, i.e., according to the symmetrization postulate the converse holds $[\phi_k,\phi_\ell^\dagger]_\mp=0$; or, using this and (B.1-9a)

$$\phi_k^\dagger\phi_\ell^\dagger\phi_\ell\phi_k=\mp\phi_\ell^\dagger\phi_k^\dagger\phi_\ell\phi_k=\mp\phi_\ell^\dagger(\phi_k^\dagger\phi_\ell)\phi_k=-\phi_\ell^\dagger\phi_\ell\phi_k^\dagger\phi_k\,. \qquad \text{(B.1-12)}$$

If the support of g runs to infinity, the cluster decomposition theorem gives[12]

$$-\left(\Psi_0,\phi_\ell^\dagger\phi_\ell\,\phi_k^\dagger\phi_k\Psi_0\right)=-\left(\Psi_0,\phi_\ell^\dagger\phi_\ell\,\Psi_0\right)\left(\Psi_0,\phi_k^\dagger\phi_k\Psi_0\right)=-\|\phi_\ell\Psi_0\|^2\|\phi_k\Psi_0\|^2, \qquad \text{(B.1-13)}$$

in contradiction with (B.1-11).

B.2 Statistics for a Scalar Spin-zero Field

We consider a scalar, i.e., spin-zero field, which does not identically vanish. The spin-statistics theorem then reads that the field commutators $[\phi(x),\phi(y)]_-$ and $[\phi(x),\phi^\dagger(y)]_-$ are zero for space-like distances; the statistics is then Bose–Einstein.

In order to prove this theorem, let us assume the *wrong* connection, i.e.,

$$[\phi(x),\phi^\dagger(y)]_+=0, \quad (x-y)^2<0. \qquad \text{(B.2-1)}$$

We take the vacuum expectation value for this commutator. Using the definitions (B.1-8) we obtain

$$F(\xi)+G(-\xi)=0, \quad \varsigma^2=(x-y)^2<0. \qquad \text{(B.2-2)}$$

We have already seen that F and G transform according to an irreducible representation of the restricted (real) Lorentz group, $\mathcal{L}_r(R)$, and, in fact, are invariant. We need some further mathematical theorems.

Theorem 1. The above distributions are boundary values of analytic functions in a space of four complex variables, C^4:

$$G(\xi)=\lim_{\eta\to 0}G(\xi-i\eta)=\lim_{\operatorname{Im} z\to 0}G(z)\,, \qquad \text{(B.2-3)}$$

with a similar result for F. The domain of analyticity is $T(\text{all }\xi,\eta\in V_+)$. We also consider the domain T', obtained by applying all complex Lorentz transformations $\Lambda_c z=z'$ to the vectors z of T. We then have (Hall and Wightman):

[12] Streater and Wightman, Op. Cit., theorem 3-4, p.111. The theorem is obvious when put in the Dirac notation: $-\langle\Psi_0\,|\,\phi_k^\dagger\phi_k\,\phi_\ell^\dagger\phi_\ell\,|\,\Psi_0\rangle=-\langle\Psi_0\,|\,\phi_k^\dagger\phi_k\,|\,\Psi_0\rangle\langle\Psi_0\,|\phi_\ell^\dagger\phi_\ell\,|\,\Psi_0\rangle=-\|\phi_\ell\psi_0\|^2\|\phi_k\psi_0\|^2\leq 0.$

Theorem 2. The functions $F(z)$ and $G(z)$ have a unique analytical continuation in T' and transform according to irreducible transformations of the proper complex Lorentz group $\mathcal{L}_+(C)$. The proofs, based on properties of the Laplace transformed quantities, will not be given here.

Although T contains no real points, the domain T' does: a connection z (complex) $\in T \to x'(\text{real}) \in T'$ with $x' = \Lambda_c z$, is easily realized. Further, all real points of T' are space-like, shown as follows. Since Λ_c preserves the norm, $x'^2 = z^2 = \xi^2 - \eta^2 - 2i\xi^\mu\eta_\mu$. In order that x' be real, we have $\xi^\mu\eta_\mu = 0$, so that ξ, being orthogonal to the time-like vector η, is space-like. Consequently, $\xi^2 < 0$ and $\eta^2 > 0$, hence $x'^2 < 0$, as was to be proven.[13] The relationship (B.2-2) holds therefore for all real points of T', and by analytic continuation, everywhere. Whence,

$$F(z') + G(-z') = 0, \quad z' \in T'. \tag{B.2-4}$$

We now consider the complex Lorentz transformation

$$\Lambda_c(t) = \begin{pmatrix} \cosh it & 0 & 0 & \sinh it \\ 0 & \cos t & -\sin t & 0 \\ 0 & \sin t & \cos t & 0 \\ \sinh it & 0 & 0 & \cosh it \end{pmatrix}. \tag{B.2-5}$$

We have

$$\det \Lambda_c = 1, \quad 0 \le t \le \pi; \quad \Lambda_c(0) = \mathsf{I}, \quad \Lambda_c(\pi) = -\mathsf{I}. \tag{B.2-6}$$

Under this family of transformations F and G, being vacuum expectation values, remain invariant. Ergo,

$$\Lambda_c(0)F(z') + \Lambda_c(\pi)G(-z') = F(z') + G(z') = 0, \quad z' \in T'. \tag{B.2-7}$$

Finally, we let the imaginary part of z' go to zero, without crossing the boundaries of the open domain T'[14]; we approach thereby all real vectors ξ of the space R^4:

$$F(\xi) + G(\xi) = 0, \quad \xi \in R^4. \tag{B.2-8}$$

We now readily arrive at a contradiction. Consider the functionals

$$\phi[f] = \int d^4x\, f(x)\phi(x), \quad \phi^\dagger[f] = \int d^4x\, f^*(x)\phi^\dagger(x),$$
$$\phi[g] = \int d^4x\, f(-x)\phi(x) = \int d^4x\, f(x)\phi(-x),$$

---

[13] Stated differently, for ordinary complex w with $(z')^2 = w$, there is a branch cut along the positive real axis (cf. Burgoyne, Op. Cit.), but the negative real axis is included in the domain of analyticity.

[14] Or, without crossing the branch cut $\mathrm{Re}(z'^2) \ge 0$ in the w-plane; since $F + G = 0$ everywhere, there is no jump, however, so the result is unique, even along the branch cut.

$$\phi^\dagger[g]=\int d^4x f^*(-x)\phi^\dagger(x)=\int d^4x f^*(x)\phi^\dagger(-x)\,, \tag{B.2-9}$$

where we set $f(-x)=g(x)$. We then have

$$\begin{aligned}\|\phi^\dagger[f]\Psi_0\|^2+\|\phi[g]\Psi_0\|^2&=\left(\Psi_0,\phi[f]\phi^\dagger[f]\Psi_0\right)+\left(\Psi_0,\phi^\dagger[g]\phi[g]\Psi_0\right)\\&=\iint d^4x d^4y\, f(x)f^*(y)\left\{\left(\Psi_0,\phi(x)\phi^\dagger(y)\Psi_0\right)+\left(\Psi_0,\phi^\dagger(-y)\phi(-x)\Psi_0\right)\right\}\\&=\iint d^4x d^4y\, f(x)f^*(y)\{F(\xi)+G(\xi)\}=0\,.\end{aligned} \tag{B.2-10}$$

Since this is true for all test functions f and g, we arrive at

$$\phi^\dagger[f]\Psi_0=\phi[g]\Psi_0=0\,. \tag{B.2-11}$$

The vacuum state cannot be annihilated by a field, however. Whence the field is identically zero, contrary to our point of departure.

B.3 The Connection for a Field of General Spin

The proof is quite similar as for scalar fields. However, although the vacuum expectation values $F_{k\ell}(\xi)$ and $G_{k\ell}(\xi)$ are invariant under the restricted Lorentz group, they are not so under the complex Lorentz group. In order to find the transformation law for any spin, we first deal briefly with the usual representations of the Lorentz group.[15]

Homomorphisms for the real restricted Lorentz group

The restricted Lorentz group $\mathcal{L}_r(R)$ has a two-fold representation (Darstellung) by the group of complex two by two matrices of determinant one, denoted as SC_2. Consider a space of two dimensions with basic vectors u_1 and u_2. A general bi-vector of this space is denoted by $\{a_\alpha\}$, i.e. $a=a_1u_1+a_2u_2$. The transformation matrices affecting a change of basis are denoted as $\tilde{A}_{\alpha\beta}$, $\det\tilde{A}=\det A=1$, where the tilde denotes the transpose. Under this transformation

$$a_\alpha{}'=\sum\nolimits_\beta A_{\alpha\beta}a_\beta\,, \tag{B.3-1}$$

where $\mathsf{A}\in SC_2$. We also consider the dual space, with basis vectors $\overset{*}{u_1}\equiv\dot{u}_1$, $\overset{*}{u_2}\equiv\dot{u}_2$. The dot is used to avoid too many superscripts. The transformation is by A^* :

[15] For those familiar with German, the best readable short account still is, in our opinion, B.L. van der Waerden, "Die Gruppentheoretische Methode in der Quantenmechanik", Springer, Berlin 1932. Two other books, both translated from the Russian, are: C.Ya Lyubarskii, "The Applications of Group Theory in Physics", Pergamon Press, London 1960; I.M. Gel'fand, R.A. Minlos and Z.Ya. Shapiro, "Representations of the Rotation and Lorentz Group and their Applications", Macmillan, NY 1963.

$$\dot{a}_\alpha' = \sum_\beta A^*_{\alpha\beta}\dot{a}_\beta \,. \tag{B.3-2}$$

Representations of this group in other vector spaces are obtained by forming basis vectors of the homogeneous powers of $u_1, u_2, \dot{u}_1$ and $\dot{u}_2$, such as $u_1^{v-k}u_2^k\,\dot{u}_1^{w-\ell}\dot{u}_2^\ell$. The linear combinations $\{c(vwk\ell)\}$ form a space of bipolynomials of dimension (i.e., number of basis vectors) $(v+1)(w+1)$. It is customary to set $v = 2j$ and $w = 2k$, for reasons which will be clear below. A change in basis in the original space will affect the 'coordinates' of the bipolynomials $\{c(vwk\ell)\}$, so that we have a set of representations of SC_2 in these spaces. The customary notation for these representations is $\tau(jk)$. The transformation laws will be given shortly.

We first consider the bilinear forms

$$C = c_{1\dot{1}}u_1\dot{u}_1 + c_{1\dot{2}}u_1\dot{u}_2 + c_{2\dot{1}}u_2\dot{u}_1 + c_{2\dot{2}}u_2\dot{u}_2. \tag{B.3-3}$$

In coordinate form $\mathsf{C} = \{c_{\alpha\dot{\beta}}\}$. Transforming the basis vectors with $\tilde{A}$ and $\tilde{A}^*$ we find that the c's transform as

$$c'_{\alpha\beta} = \sum_{\rho\dot{\sigma}} A_{\alpha\rho}A^*_{\dot{\beta}\dot{\sigma}}c_{\rho\dot{\sigma}} \text{ or } \mathsf{C}' = \mathsf{A}\mathsf{C}\mathsf{A}^\dagger. \tag{B.3-4}$$

It may be noted that the transformation leaves the determinant invariant. The connection with the Lorentz transformation is made by setting

$$\mathsf{C} = \begin{pmatrix} x^0 + x^3 & x^1 - ix^2 \\ x^1 + ix^2 & x^0 - x^3 \end{pmatrix}. \tag{B.3-5}$$

The matrix C is Hermitean; (B.3-4) is a unitary transformation of C. Moreover, $\det\mathsf{C} = x^\mu x_\mu$. Thus the bilinear form with coordinates $\{c_{\alpha\beta}\}$ is the representation $\tau(\frac{1}{2}\frac{1}{2})$ of the restricted real Lorentz group $\mathscr{L}_r(R)$. It contains the identity $x' = x$, if $\mathsf{A} = \pm\mathbf{1}$. It is left to the reader to show that the correspondence $\mathsf{A}(\Lambda)$ is twofold; if A is a transformation representing $x' = \Lambda x$, then $-\mathsf{A}$ also represents $x' = \Lambda x$ and there are no other matrices with this property. Given the matrix C, we can recover the four-vector x, via the recipe

$$x^\mu = \tfrac{1}{2}\mathrm{Tr}(\mathsf{C}\sigma^\mu), \tag{B.3-6}$$

where σ^μ are the Pauli spin matrices and the unit matrix,

$$\sigma^0 = \mathbf{1} = \begin{pmatrix} 1 & 0 \\ 0 & 1 \end{pmatrix},\quad \sigma^1 = \begin{pmatrix} 0 & 1 \\ 1 & 0 \end{pmatrix},\quad \sigma^2 = \begin{pmatrix} 0 & -i \\ i & 0 \end{pmatrix},\quad \sigma^3 = \begin{pmatrix} 1 & 0 \\ 0 & -1 \end{pmatrix}. \tag{B.3-7}$$

Altogether, we have obtained a large number of representations of the restricted Lorentz group, viz. all $\tau(jk)$. We note that $\tau(00)$ is the identity representation, $\tau(0\frac{1}{2})$ and $\tau(\frac{1}{2}0)$ are twofold representations by SC_2 and $\tau(\frac{1}{2}\frac{1}{2})$ is the bilinear representation we just discussed. The $\tau(jk)$ for other jk are more general representations, which are either single or twofold.

In order to show that the above representations with $j,k=0,\frac{1}{2},1,\frac{3}{2},\ldots$ are irreducible, one considers the Lie algebra for the infinitesimal transformations. We are then led to the same commutation rules as for two angular momenta $\mathbf{J}$ and $\mathbf{J}'$. Abstract spaces isomorphic with the spaces of the bipolynomials result from this picture.[16] The quantum numbers jk occur naturally in this approach. The normalized basis vectors of $\tau(jk)$ can be realised as

$$e_{mn}^{(jk)}=\frac{u_1^{j+m}u_2^{j-m}\dot{u}_1^{k+n}\dot{u}_2^{k-n}}{\{(j+m)!(j-m)!(k+n)!(k-n)!\}^{1/2}}, \tag{B.3-8}$$

with $m=-j,-j+1,\ldots j$ and $n=-k,-k+1,\ldots k$, the dimension (still) being $(2j+1)(2k+1)$. One can of course also arrange the basic vectors as $e_\ell^{(jk)}$, with $\ell=0,1,\ldots(2j+1)(2k+1)$. For the fields we have

$$\phi(x)\to\{\phi_{mn}\}\quad\text{where }\phi(x)=\sum_{mn}^{(2j+1)(2k+1)}\phi_{mn}e_{mn}^{(jk)} \tag{B.3-9}$$

or

$$\phi(x)\to\{\phi_\ell\}\quad\text{where }\phi(x)=\sum_{\ell}^{(2j+1)(2k+1)}\phi_\ell e_\ell^{(jk)}. \tag{B.3-10}$$

Spinor form

The representation (in quantum mechanical sense) of the field by the projections on the basis vectors as in the last two equations is not practical, since the transformation law is not obvious. Rather, we represent any element Ξ in the space of $\tau(jk)$ as a multi-component tensor

$$\Xi\to\{\xi_{\alpha_1\alpha_2\ldots\alpha_{2j}\dot{\beta}_1\dot{\beta}_2\ldots\dot{\beta}_{2k}}\}, \tag{B.3-11}$$

symmetric in the $2j$ undotted indices and symmetric in the $2k$ dotted indices, with each index α and β taking the values 1 and 2. From the transformations of the u_α and $\dot{u}_\beta$ vectors we find the transformation law

$$\xi'_{\alpha_1\alpha_2\ldots\alpha_{2j}\dot{\beta}_1\dot{\beta}_2\ldots\dot{\beta}_{2k}}=\sum_{(\rho)(\sigma)}A_{\alpha_1\rho_1}\ldots A_{\alpha_{2j}\rho_{2j}}$$

$$\times A^*_{\dot{\beta}_1\dot{\sigma}_1}\ldots A^*_{\dot{\beta}_{2k}\dot{\sigma}_{2k}}\xi_{\rho_1\rho_2\ldots\rho_{2j}\dot{\sigma}_1\dot{\sigma}_2\ldots\dot{\sigma}_{2k}}. \tag{B.3-12}$$

For instance, in $\tau(\frac{1}{2}0)$ we have the spinor $\xi_\alpha=a_\alpha$, in $\tau(0\frac{1}{2})$ we have $\xi_{\dot{\alpha}}=a_{\dot{\alpha}}$, in $\tau(\frac{1}{2}\frac{1}{2})$ we have $\xi_{\alpha\dot{\beta}}=c_{\alpha\dot{\beta}}$, corresponding to four-vectors, in $\tau(1,\frac{1}{2})$ we have the

---

[16] Van der Waerden, Op. Cit., pp. 83-87.

spinor $\xi_{\alpha\beta\dot{\gamma}}$, etc. Naturally, all spinors can be rewritten as a single column $\{\xi_\ell\}$. The transformation (B.3-12) then fixes the symbolic matrix $S_{\ell\ell'}(\Lambda)$ of subsection B.1.

For the field operators we have a corresponding spinor form $\phi_{\alpha\ldots\alpha\dot{\beta}\ldots\dot{\beta}}$, if the field transforms according to a representation of $\mathscr{L}_r(R)$. The shorter single column form $\{\phi_\ell\}$ was already discussed before.

The above discussion has not included improper Lorentz transformations so far. To include reflections, it turns out that it is sufficient to consider u_α and $u_{\dot{\alpha}}$ ($\alpha = 1,2$) as basic vectors in non-corresponding spaces. Thus the number of independent elements in a given representation is doubled. There is a linear closure of $2(2j+1)(2k+1)$ vectors which transform invariantly with respect to the orthochronous Lorentz group. Generally, the group is denoted by the tensor sum $\tau(jk) \oplus \tau(kj)$, $j \neq k$.[17]

The Dirac field is invariant under the orthochronous group, a well as against time reversal and charge conjugation. Its field is a spinor in $\tau(\frac{1}{2}0) \oplus \tau(0\frac{1}{2})$. It can be arranged as the symplectic column matrix

$$\phi(x) = \begin{pmatrix} \phi_\alpha(x) \\ \phi_{\dot{\alpha}}(x) \end{pmatrix}. \tag{B.3-12'}$$

The Dirac equation in this form is seen to be Lorentz invariant at first glance, in contrast to the more usual four component form.

Complex Lorentz group

For the proper complex Lorentz group we only need the analogue of the transformation law (B.3-12). The group $\mathscr{L}_+(C)$ can be associated with *pairs* of matrices of rank two and determinant value one, $SC_2 \otimes SC_2$. The correspondence is $(\mathsf{A},\mathsf{B}) \to \Lambda(\mathsf{A},\mathsf{B})$, while $\Lambda(\mathsf{A},\mathsf{B}) \to (\pm\mathsf{A},\pm\mathsf{B})$. In particular, the transformation $\Lambda = -1$ is represented by $(\mathsf{A},\mathsf{B}) = (-\mathbf{1},\mathbf{1})$ or $(\mathbf{1},-\mathbf{1})$. The transformation rule for the spinor $\xi_{\alpha\beta} = c_{\alpha\beta}$ now reads

$$\mathsf{C}' = \mathsf{A}\mathsf{C}\tilde{\mathsf{B}}, \tag{B.3-13}$$

while a general spinor transforms under a complex Lorentz transformation $\Lambda(\mathsf{A},\mathsf{B})$ as

$$\xi'_{\alpha_1\alpha_2\ldots\alpha_{2j}\dot{\beta}_1\dot{\beta}_2\ldots\dot{\beta}_{2k}} = \sum_{(\rho)(\sigma)} A_{\alpha_1\rho_1}\cdots A_{\alpha_{2j}\rho_{2j}}$$

$$\times B_{\dot{\beta}_1\dot{\sigma}_1}\cdots B_{\dot{\beta}_{2k}\dot{\sigma}_{2k}}\,\xi_{\rho_1\rho_2\ldots\rho_{2j}\dot{\sigma}_1\dot{\sigma}_2\ldots\dot{\sigma}_{2k}}. \tag{B.3-14}$$

If a spinor is written in column form $\{\xi_\ell\}$, then (B.3-14) defines a matrix $S_{\ell\ell'}(\mathsf{A},\mathsf{B})$, *which contains as many A matrices as there are undotted indices and as many B matrices as there are dotted indices.*

The analogue of the transformation law (B.1-7b) for the vacuum expectation value

---

[17] Lyubarskii, Op. Cit., p.303ff.

of the fields under a complex Lorentz transformation is straightforward. We have

$$\sum_{a'b'..} S^{\phi}_{aa'}[\Lambda(\mathsf{A},\mathsf{B})]S^{\chi}_{bb'}[\Lambda(\mathsf{A},\mathsf{B})]...\left(\Psi_0,\phi_{a'}(x_1)\chi_{b'}(x_2)...\Psi_0\right)$$
$$=\left(\Psi_0,\phi_a(\Lambda(\mathsf{A},\mathsf{B})x_1+a)\chi_b(\Lambda(\mathsf{A},\mathsf{B})x_2+a)...\Psi_0\right). \quad \text{(B.3-15)}$$

We consider especially the transformation for $\Lambda=-1$, i.e., $\Lambda(\pm\mathbf{1},\mp\mathbf{1})$. The vacuum expectation value changes then by $(-1)^{2K}$ or $(-1)^{2J}$ for upper and lower signs, respectively, where $2K$ is the number of undotted indices and $2J$ the total number of dotted indices of $\phi_a\,\phi_b....$ In particular, for the twofold expectation values $F_{k\ell}$ and $G_{k\ell}$ of a field $\phi\in\tau(jk)$ [as defined in Eqs. (B.1-8a,b)], we note that they change by a factor $(-1)^{2(j+k)}$ under both forms of $\Lambda=-1$. *They conserve sign if* $j+k$ *is integral and they change sign if* $j+k$ *is half integral.*

Field-spin connection

The connection of spin with the form for the field is elaborated in standard texts. One finds that the angular momentum density of the field comes into two parts, an orbital part associated with $\mathbf{r}\times\nabla\phi$ and a spin part, associated with the number of plane waves that are possible in the absence of interactions. If under a restricted Lorentz transformation the field transforms irreducibly under $\tau(jk)$, it turns out that the spin is always integral if $j+k$ is integral and half integral if $j+k$ is half integral. Thus, the transformation law for $F_{k\ell}$ and $G_{k\ell}$ is unique for a field of given spin.

We now state the spin-statistics theorem for a field of general spin: *A non-vanishing spinor field satisfies the commutation rule*

$$[\phi_k(x),\phi^\dagger_\ell(y)]_-=0, \quad (x-y)^2<0, \quad \text{(B.3-16a)}$$

for fields with integral spin, and

$$[\phi_k(x),\phi^\dagger_\ell(y)]_+=0, \quad (x-y)^2<0, \quad \text{(B.3-16b)}$$

for fields with half integral spin.

For the proof, we assume the wrong connection, i.e.,

$$[\phi_k(x),\phi^\dagger_\ell(y)]_\pm=0, \quad (x-y)^2<0, \quad \text{(B.3-17)}$$

where we make the convention throughout this proof: upper sign ⇔ integral spin and lower sign ⇔ half integral spin. Taking the vacuum expectation value, we have

$$F_{k\ell}(\xi)\pm G_{\ell k}(-\xi)=0, \quad \xi^2=(x-y)^2<0. \quad \text{(B.3-18)}$$

These distributions are again limits of holomorphic functions, which can be analytically continued in an open domain T' of complex Lorentz transformed vectors z'. Carrying out the complex transformation $\Lambda=-1$, we find with $G_{\ell k}(-z')$

$\rightarrow \pm G_{\ell k}(z')$:

$$F_{k\ell}(z') + G_{\ell k}(z') = 0, \quad z' \in T', \tag{B.3-19}$$

or, also at real points

$$F_{k\ell}(\xi) + G_{\ell k}(\xi) = 0, \quad \xi \in R^4. \tag{B.3-20}$$

This relation has uniquely the + sign for all spinors. Now we define the functionals

$$\begin{aligned}
\phi[f] &= \int d^4x\, f^\ell(x)\phi_\ell(x), \\
\phi^\dagger[f] &= \int d^4x\, f^{\ell*}(x)\phi_\ell^\dagger(x), \\
\phi[g] &= \int d^4x\, f^\ell(-x)\phi_\ell(x) = \int d^4x\, f^\ell(x)\phi_\ell(-x), \\
\phi^\dagger[g] &= \int d^4x\, f^{\ell*}(-x)\phi_\ell^\dagger(x) = \int d^4x\, f^{\ell*}(x)\phi_\ell^\dagger(-x),
\end{aligned} \tag{B.3-21}$$

where the $f^\ell(x)$ belong to a space of test functions invariant under $\tau(jk)$. As in the scalar case we find

$$\begin{aligned}
&\|\phi^\dagger[f]\Psi_0\|^2 + \|\phi[g]\Psi_0\|^2 \\
&\quad = \iint d^4x d^4y\, f^k(x) f^{\ell*}(y)\{F_{k\ell}(\xi) + G_{\ell k}(\xi)\} = 0.
\end{aligned} \tag{B.3-22}$$

This entails that the field is identically zero. Thus, the impossibility of the wrong connection has been established.

Herewith we have completed the proof for one of the most fundamental consequences of symmetry in physics. The spin-statistics theorem leads to the dichotomy of statistical mechanics into the separate domains of Bose–Einstein and Fermi–Dirac systems.

Author Index *

---

* References to publications are marked by italicised numbers.

Subject Index